SYMBOL	DESCRIPTION	PAGE
a_i	Action i in a decision analysis	1048
A^c	Complement of event A	163
A, B	Events	150
$A \cup B$	Union of events A and B	160
$A \cap B$	Intersection of events A and B	160
α (alpha)	Probability of rejecting H_0 if in fact H_0 is true (Type I error)	357
b	Number of blocks in a randomized block design	879
β (beta)	Probability of accepting H_0 if in fact H_0 is false (Type II error)	359
β_0	y-intercept in regression models	489
$\hat{\beta}_0$	Least squares estimator of β_0	494
β_1	Slope of straight-line regression model	489
$\hat{\beta}_1$	Least squares estimator of β_1	494
β_i	Coefficient of independent variable x_i in a multiple regression model	556
$\hat{\beta}_i$	Least squares estimator of β_i	558
C_t	Cyclical effect in a time series	796
CS	Cost of sampling	1110
χ^2 (chi-square)	Probability distribution of various test statistics	967
df	Degrees of freedom for the t, χ^2, and F distributions	387
$E(n_{ij})$	Expected count in cell (i, j) of a contingency table	1014
$\hat{E}(n_{ij})$	Estimated expected count in cell (i, j) of a contingency table	1014
$E(x)$	Expected value of the random variable x	210
$EOL(a_i)$	Expected opportunity loss for action i in a decision analysis	1100
ENGS	Expected net gain of sampling in a decision analysis	1110
$EP(a_i)$	Expected payoff for action i in a decision analysis	1058
EPNS	Expected payoff for no sampling in a decision analysis	1108
EPS	Expected payoff of sampling in a decision analysis	1108
$EU(a_i)$	Expected utility for action i in a decision analysis	1070
EVPI	Expected value of perfect information in a decision analysis	1101
EVSI	Expected value of sample information	1109
ε (epsilon)	Random error component in regression models	489
f_i	Frequency for category i in a qualitative data set	19
$f(x)$	Probability density function for a continuous random variable x	260
F	Test statistic used to compare two variances, to compare p population means, and to test several terms in a multiple regression	441
F_r	Test statistic for the Friedman nonparametric analysis of variance	975
F_α	Value of F distribution with area α to its right	442
H	Test statistic for the Kruskal–Wallis nonparametric analysis of variance	968
H_a	Alternative (or research) hypothesis	360
H_0	Null hypothesis	360

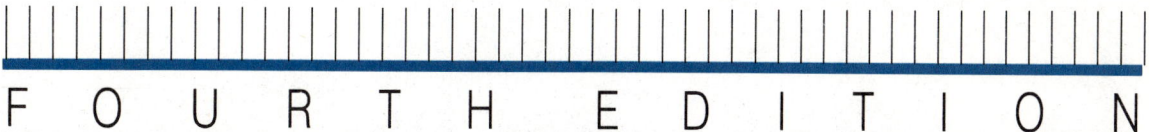

FOURTH EDITION

STATISTICS FOR BUSINESS AND ECONOMICS

FOURTH EDITION

STATISTICS FOR BUSINESS AND ECONOMICS

JAMES T. McCLAVE
College of Business Administration
University of Florida

P. GEORGE BENSON
Curtis L. Carlson School of Management
University of Minnesota

DELLEN PUBLISHING COMPANY
San Francisco

COLLIER MACMILLAN PUBLISHERS
London

divisions of Macmillan, Inc.

On the cover: "Skyhook Boca Raton #30" was executed by Los Angeles artist Peter Shire. The work measures $21 \times 20 \times 8\frac{1}{2}$ inches. One of the foremost ceramic sculptors in the United States, Shire is a key member of the Memphis Group. This particular piece of work is a new medium for Shire; the piece is painted wood and wire. Shire's work may be seen in the collections of the Los Angeles County Museum and the Los Angeles Museum of Contemporary Art. He is represented by Saxon Lee Gallery in Los Angeles, California.

© Copyright 1988 by Dellen Publishing Company, a division of Macmillan, Inc.

Printed in the United States of America

All rights reserved. No part of this book may be reproduced or transmitted in any form or by any means, electronic or mechanical, including photocopying, recording, or any information storage and retrieval system, without permission in writing from the Publisher.

Permissions: Dellen Publishing Company
400 Pacific Avenue
San Francisco, California 94133

Orders: Dellen Publishing Company
c/o Macmillan Publishing Company
Front and Brown Streets
Riverside, New Jersey 08075

Collier Macmillan Canada, Inc.

LIBRARY OF CONGRESS CATALOGING-IN-PUBLICATION DATA

McClave, James T.
  Statistics for business and economics.

  Includes index.
  1. Commercial statistics   2. Economics—Statistical methods.   3. Statistics.   I. Benson, P. George, 1946–   . II. Title.
HF1017.M36   1988   519.5   87–20195

Printing: 4 5 6 7 8   Year: 9 0 1 2

ISBN 0-02-379020-2

C O N T E N T S

The fourth edition of *Statistics for Business and Economics* finds the application of statistics to business and economic problems in the midst of a number of significant changes. The availability of microcomputers with a wide range of statistical software has dramatically affected the amount of time we spend teaching computational formulas. The use of statistical quality control in American industry is perhaps the fastest-growing application of statistics in any area. The same corporate structure which nearly 40 years ago forced W. Edwards Deming to take his revolutionary ideas to Japan has welcomed him back as a conquering hero. Deming and his disciples are in constant demand in both academic and corporate circles, and the impact of their philosophy on the statistical education of business students is and will be profound. Academic professional organizations are paying increased attention to the content of business statistics courses and to the philosophy and manner in which they are taught.

In light of these and other developments, we have decided that the time has come to acknowledge that the text we wrote 10 years ago is not the same one we would write today. For example, 10 years ago the calculator had taken the "arithmetic" out of analysis of variance, but not the formulas. The computer gives us an additional option: We can use precious time teaching the interpretation of ANOVA results rather than cookbook formulas that produce those results. This fourth edition exercises that option (and many others like it).

Before detailing the specifics of the fourth edition, we stress that the underlying philosophy of the text is unchanged, despite the many new features and additional options. Our original intent was to write a text *for the students* that stressed *inference making*. If anything, the rapid evolution of "business statistics" is making that easier to accomplish.

1. **Using the Computer.** A new feature has been added at the end of most chapters to encourage the use of computers in the analysis of real data. A demographic data base, consisting of 1,000 observations on 15 variables, has been described in Appendix C and is available on diskette from the publisher. Each "Using the Computer" section provides one or more computer exercises that utilize the data in Appendix C and enhance the new material covered in the chapter.

2. **Quality Control.** Numerous case studies, examples, and exercises that focus on statistical quality control applications have been added throughout the text. Since the application of statistical ideas to quality management and quality control has become so pervasive, we decided that the subject needed to be stressed throughout the text, as the relevant statistical concepts are covered, rather than concentrated in a few pages or chapter devoted to the subject.

   Examples include a quality control perspective on the explosion of the space shuttle *Challenger* (Case Study 5.3); quality control problems associated with mass production (Example 6.13); quality comparisons of rental cars (Case Study 7.1); introduction to quality control charts (Case Studies 8.3 and 8.4); quality control problems in the telecommunications industry following deregulation

(Example 8.12); comparison of manual and automated quality inspection procedures (Exercise 9.41); analysis of the American Society for Quality Control survey of upper-level executives (Exercise 9.101); and the use of regression analysis in the quality control process (Exercise 10.21)

3. **Exploratory Data Analysis.** The computer permits the analyst to look at the data from every angle in order to extract all the information they have to offer. The use of microcomputer packages to generate stem and leaf displays, box plots, and the more traditional histograms is introduced in Chapters 2 and 3, and utilized throughout the text. We are careful to delineate the difference between descriptive and inferential statistics, and the student is constantly reminded of the necessity to consider the reliability of any statistical procedure used.

4. **Computers in Regression and Residual Analysis.** Ten years ago we set out to include the most complete treatment of regression analysis on the market, because we believed that modeling and forecasting were two of the most important applications of statistics in business. Our view has not changed, and the computer has enabled aspects of modeling to be realized that were only theoretical constructs when we wrote the original text. In keeping with our goal of a thorough, modern treatment of regression analysis, the fourth edition introduces computer printouts in the simple linear regression chapter, adds significantly to the number of examples and exercises that use the computer for regression analyses, and incorporates residual analyses into the model-building process. Although our first three editions made substantial use of the computer in the treatment of regression modeling, this edition adds significantly to that base. Microcomputer printouts are included, more exercises show printouts instead of requiring the student to perform the calculations, and more exercises with real data requiring the student to use the computer have been added. In addition, both Chapters 11 and 12 contain new sections showing how to use the computer to perform residual analyses, and, more importantly, how to integrate them into the model-building process. The residual analyses lead naturally to a more complete discussion of transformations, including significantly more space devoted to the important multiplicative model.

5. **Tests of Hypotheses.** Much current debate centers on how and what, if anything, to teach about testing hypotheses. Our view is that a modern text must present the subject, because much research and application in business and economics continues to make use of statistical testing procedures. In this fourth edition, we have expanded the explanatory material about testing, including a new (optional) section on the computation of Type II error probabilities (β) and on the power of tests. We have also added exercises that require the student to confront the issue of which inferential procedure—testing or confidence intervals—provides more information about a parameter. Thus, the fourth edition offers an option as to the depth of treatment the instructor selects for testing hypotheses, while at the same time encouraging the student to contrast and to evaluate the statistical procedures employed.

6. **Normal Distribution and Sampling Distributions.** The normal distribution remains the cornerstone of applied statistics. Although robustness rightfully receives increased attention, the Central Limit Theorem ensures the continued

importance of the normal distribution. We have added more explanatory examples on the use of the normal distribution in Chapter 6, and we have replaced the more difficult two-sample material in the sampling distribution chapter (Chapter 7) with an expanded treatment of the one-sample case, including more computer simulations of sampling distributions and an introduction to the sampling distribution of the sample proportion for a binomial random variable.

7. **Analysis of Variance.** The fourth edition adopts a computerized rather than a "cookbook" approach to ANOVA. All calculation formulas are relegated to Appendix D, and the emphasis is changed to the understanding and interpretation of designed experiments. The terminology of designed experiments is defined and utilized throughout the chapter, with constant reinforcement by example. Computer printouts are presented for each type of analysis covered, with space that was formerly occupied by tedious calculations now devoted to interpretation of the statistical results produced by the software. The Bonferroni multiple comparisons procedure is introduced early and is utilized during each analysis, where it would naturally occur, rather than in a separate section at the end of the chapter. Since virtually every ANOVA model can be written as a regression model (certainly every one in this text, with a single random component), we develop the relationship between regression and ANOVA models in an optional section.

8. **Determination of Sample Size.** Part of the quality control aspect of the statistical revolution in business applications is increased attention to sampling efforts necessary to achieve specified statistical objectives. In accordance with this, we have included separate sections for determining the sample size for estimating a population mean and a binomial probability. Sample size formulas in terms of both the bound on the error of estimation and the total width of the confidence interval are provided.

9. **Normal Approximation to Binomial Probabilities.** The rationale behind and the conditions under which the normal distribution can be used for approximating binomial probabilities are significantly expanded in Chapter 6.

10. **Real Applications of Statistics in Business and Economics.** From the start we have attempted to include many examples and exercises that draw on the extensive (and growing) literature containing real applications of statistics in business and economics. The fourth edition adds to and updates the many examples, exercises, and case studies that represent applications of statistics to real business problems. It should be noted, however, that constructed exercises and examples often better serve our pedagogic objectives, and we therefore continue to make use of them where appropriate.

In spite of the fact that we consider the fourth edition a major revision, the flexibility of past editions is maintained. Sections that are not prerequisite to succeeding sections and chapters are marked "(Optional)." For example, an instructor who wishes to devote significant time to exploratory data analysis might cover all topics in Chapters 2 and 3. In contrast, an instructor who wishes to move rapidly into inferential procedures might omit optional sections, devoting only several lectures to these chapters.

We have maintained the features of this text that we believe make it unique among introductory statistics texts for business courses. These features, which assist the student in achieving an overview of statistics and an understanding of its relevance in the solution of business problems, are as follows:

1. **Case Studies.**   (See the list of case studies on page xxi.) Many important concepts are emphasized by the inclusion of case studies, which consist of brief summaries of actual business applications of the concepts and are often drawn directly from the business literature. These case studies allow the student to see business applications of important statistical concepts immediately after the introduction of the concepts. The case studies also help to answer by example the often asked questions, "Why should I study statistics? Of what relevance is statistics to business?" Finally, the case studies constantly remind the student that each concept is related to the dominant theme—statistical inference.

2. **Where We've Been . . . Where We're Going . . .**   The first page of each chapter is a "unification" page. Our purpose is to allow the student to see how the chapter fits into the scheme of statistical inference. First, we briefly show how the material presented in previous chapters helps us to achieve our goal (Where We've Been). Then, we indicate what the next chapter (or chapters) contributes to the overall objective (Where We're Going). This feature allows us to point out that we are constructing the foundation block by block, with each chapter an important component in the structure of statistical inference. Furthermore, this feature provides a series of brief résumés of the material covered as well as glimpses of future topics.

3. **Many Examples and Exercises.**   We believe that most students learn by doing. The text contains many worked examples to demonstrate how to solve various types of problems. We then provide the student with a large number (more than 1,300) of exercises. The answers for most are included at the end of the text. The exercises are of two types:

   a. **Learning the Mechanics.**   These exercises are intended to be straightforward applications of the new concepts. They are introduced in a few words and are unhampered by a barrage of background information designed to make them "practical," but which often detracts from instructional objectives. Thus, with a minimum of labor, the student can recheck his or her ability to comprehend a concept or a definition.

   b. **Applying the Concepts.**   The mechanical exercises described above are followed by realistic exercises that allow the student to see applications of statistics to the solution of problems encountered in business and economics. Once the mechanics are mastered, these exercises develop the student's skills at comprehending realistic problems that describe situations to which the techniques may be applied.

4. **On Your Own . . .**   The chapters end with an exercise entitled "On Your Own . . . ." The intent of this exercise is to give the student some hands-on experience with a business application of the statistical concepts introduced in the chapter. In most cases, the student is required to collect, analyze, and interpret data relating to some business phenomenon.

5. **A Simple, Clear Style.** We have tried to achieve a simple and clear writing style. Subjects that are tangential to our objective have been avoided, even though some may be of academic interest to those well-versed in statistics. We have not taken an encyclopedic approach in the presentation of material.

6. **An Extensive Coverage of Multiple Regression Analysis and Model Building.** This topic represents one of the most useful statistical tools for the solution of business problems. Although an entire text could be devoted to regression modeling, we feel that we have presented a coverage that is understandable, usable, and much more comprehensive than the presentations in other introductory business statistics texts. We devote three chapters to discussing the major types of inferences that can be derived from a regression analysis, showing how these results appear in computer printouts and, most important, selecting multiple regression models to be used in an analysis. Thus, the instructor has the choice of a one-chapter coverage of simple regression, a two-chapter treatment of simple and multiple regression, or a complete three-chapter coverage of simple regression, multiple regression, and model building. The following two chapters on index numbers and time series analysis are closely tied to the three chapters on multiple regression analysis because they present an introduction to forecasting based on time-dependent data. This extensive coverage of such useful statistical tools will provide added evidence to the student of the relevance of statistics to the solution of business problems.

7. **Footnotes and Appendix A.** Although the text is designed for students with a noncalculus background, footnotes explain the role of calculus in various derivations. Footnotes are also used to inform the student about some of the theory underlying certain results. Appendix A presents some useful counting rules for the instructor who wishes to place greater emphasis on probability. Consequently, we think the footnotes and Appendix A provide an opportunity for flexibility in the mathematical and theoretical level at which the material is presented.

8. **Decision Analysis.** We have included a two-chapter treatment of decision analysis. In Chapter 18 the classic decision problem is presented. In addition to the standard expected payoff criterion using prior information, several other decision-making criteria are presented. This includes a rather complete introduction to the role of utility functions in decision analysis. In Chapter 19 Bayes' Rule is used both to compute the expected value of sample information before it is purchased and to revise the prior probabilities after sample information is obtained. Throughout both chapters we stress (by setting off in boxes) a step-by-step approach for all calculations. This allows the student to devote time to understanding the concepts and philosophies of decison analysis and to avoid being caught up in the necessary tedious calculations that accompany these analyses.

9. **Supplementary Material.** A solutions manual, a study guide, a Minitab supplement, an integrated companion software system, a computer-generated test system, and a 1,000-observation demographic data base are available.

   a. **Solutions Manual** (by Nancy Shafer). The solutions manual presents the solutions to most odd-numbered exercises in the text. Many points are clarified and expanded to provide maximum insight into and benefit from each exercise.

b. Study Guide (by Susan L. Reiland).   For each chapter, the study guide includes (1) a brief summary that highlights the concepts and terms introduced in the textbook; (2) section-by-section examples with detailed solutions; and (3) exercises (with answers provided at the end of the study guide) that allow the student to check mastery of the material in each section.

c. Minitab Supplement (by David D. Krueger and Ruth K. Meyer).   The Minitab computer supplement was developed to be used with Minitab Release 5.1, a general-purpose statistical computing system. The supplement, which was written especially for the student with no previous experience with computers, provides step-by-step descriptions of how to use Minitab effectively as an aid in data analysis. Each chapter begins with a list of new commands introduced in the chapter. Brief examples are then given to explain new commands, followed by examples from the text illustrating the new and previously learned commands. Where appropriate, simulation examples are included. Exercises, many of which are drawn from the text, conclude each chapter.

   A special feature of the supplement is a chapter describing a survey sampling project. The objectives of the project are to illustrate the evaluation of a questionnaire, provide a review of statistical techniques, and illustrate the use of Minitab for questionnaire evaluation.

d. DellenStat (by Michael Conlon).   DellenStat is an integrated statistics package consisting of a workbook and an IBM PC floppy diskette with software and example sets of data. The system contains a file creation and management facility, a statistics facility, and a presentation facility. The software is menu-driven and has an extensive help facility. It is completely compatible with the text.

   The DellenStat workbook describes the operation of the software and uses examples from the text. After an introductory chapter for new computer users, the remaining chapters follow the outline of the text. Additional chapters show how to create new sets of data. Technical appendices cover material for advanced users and programmers.

   DellenStat runs on any IBM PC or close compatible with at least 256K of memory and at least one floppy disk drive.

e. DellenTest.   This unique computer-generated random test system is available to instructors without cost. Utilizing an IBM PC computer and a number of commonly used dot-matrix printers, the system will generate an almost unlimited number of quizzes, chapter tests, final examinations, and drill exercises. At the same time, the system produces an answer key and student worksheet with an answer column that exactly matches the column on the answer key.

f. Data Base.   A demographic data set was assembled based on a systematic random sample of 1,000 U.S. zip codes. Demographic data for each zip code area selected were supplied by CACI, an international demographic and market information firm. Fifteen demographic measurements (including population, number of households, median age, median household income, variables related to the cost of housing, educational levels, the work force, and purchasing potential indexes based on the Bureau of the Census Consumer Expenditure Surveys) are presented for each zip code area.

Some of the data are referenced in the "Using the Computer" sections. The objectives are to enable the student to analyze real data in a relatively large sample using the computer, and to gain experience using the statistical techniques and concepts on real data.

ACKNOWLEDGMENTS

As with the first three editions, we owe thanks to the many people who assisted in reviewing and preparing this edition. Their names are listed below. We particularly acknowledge the editorial assistance of Susan L. Reiland, the outstanding administrative support of Jane Oas Benson, and the typing and assistance of Brenda Dobson. Without these three, we never could have completed this work.

Gordon J. Alexander
University of Minnesota

Larry M. Austin
Texas Tech University

Clarence Bayne
Concordia University

Carl Bedell
Philadelphia College of Textiles and Science

David M. Bergman
University of Minnesota

William H. Beyer
University of Akron

Atul Bhatia
University of Minnesota

Jim Branscome
University of Texas at Arlington

Francis J. Brewerton
Middle Tennessee State University

Daniel Brick
College of St. Thomas

Robert W. Brobst
University of Texas at Arlington

Michael S. Broida
Miami University of Ohio

Edward Carlstein
University of North Carolina at Chapel Hill

John M. Charnes
University of Minnesota

Chih-Hsu Cheng
University of Minnesota

Larry Claypool
Oklahoma State University

Edward R. Clayton
Virginia Polytechnic Institute and State University

Ronald L. Coccari
Cleveland State University

Ken Constantine
University of New Hampshire

Jim Daly
California State Polytechnic Institute

Dileep Dhavale
University of Northern Iowa

Carol Eger
Stanford University

Robert Elrod
Georgia State University

Douglas A. Elvers
University of North Carolina at Chapel Hill

Susan Flach
General Mills, Inc.

Alan E. Gelfand
University of Connecticut

Joseph Glaz
University of Connecticut at Storrs

Michael E. Hanna
University of Texas at Arlington

Don Holbert
Oklahoma State University

James Holstein
University of Missouri

Warren M. Holt
Southeastern Massachusetts University

Steve Hora
Texas Tech University

Iris B. Ibrahim
Clemson University

Marius Janson
University of Missouri

Ross H. Johnson
Madison College

Timothy J. Killeen
University of Connecticut

David D. Krueger
St. Cloud State University

Richard W. Kulp
Wright-Patterson AFB, Air Force Institute of
Technology

Martin Labbe
State University of New York College at New Paltz

James Lackritz
California State University at San Diego

Philip Levine
William Patterson College

Eddie M. Lewis
University of Southern Mississippi

Pi-Erh Lin
Florida State University

Ruth K. Meyer
St. Cloud State University

Paul I. Nelson
Kansas State University

Paula M. Oas
General Furniture Leasing

William M. Partian
Fordham College

Vijay Pisharody
University of Minnesota

P. V. Rao
University of Florida

Jan Saraph
University of Minnesota

Craig W. Slinkman
University of Texas at Arlington

Donald N. Steinnes
University of Minnesota at Duluth

Virgil F. Stone
Texas A and I University

Chipei Tseng
Northern Illinois University

Pankaj Vaish
Arthur Andersen & Company

Charles F. Warnock
Colorado State University

William J. Weida
United States Air Force Academy

T. J. Wharton
University of New Hampshire

Kathleen M. Whitcomb
University of Minnesota

Edna White
Texas A & M University.

Steve Wickstrom
University of Minnesota

James Willis
Louisiana State University

Douglas A. Wolfe
Ohio State University

Dilek Yeldan
University of Minnesota

Fike Zahroon
Moorhead State University

C A S E   S T U D I E S

FOURTH EDITION

STATISTICS FOR BUSINESS AND ECONOMICS

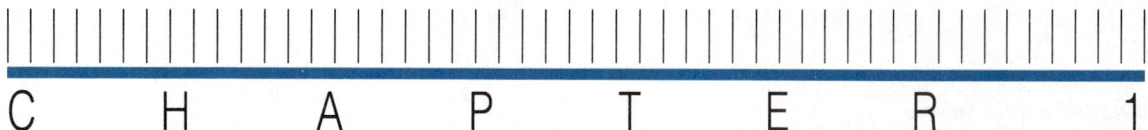

C H A P T E R   1

WHERE WE'RE GOING...

Statistics? Is it a field of study, a group of numbers that summarize some business operation, or, as the title of a popular book (Tanur et al., 1978) suggests, "a guide to the unknown"? We attempt to answer this question in Chapter 1. Throughout the remainder of the text, we will show you how statistics can be used to aid in making business decisions.

CONTENTS

WHAT IS STATISTICS?

STATISTICS: WHAT IS IT?

What does statistics mean to you? Does it bring to mind batting averages, the Dow Jones Average, unemployment figures, numerical distortions of facts (lying with statistics!), or simply a college requirement you have to complete? We hope to convince you that statistics is a meaningful, useful science with a broad, almost limitless scope of application to business and economic problems. We also want to show that statistics lie only when they are misapplied. Finally, our objective is to paint a unified picture of statistics to leave you with the impression that your time was well spent studying a subject that will prove useful to you in many ways.

Statistics means "numerical descriptions" to most people. The Dow Jones Average, monthly unemployment figures, and the fraction of women executives in a particular industry are all statistical descriptions of large sets of data collected on some phenomenon. Most often, the purpose of calculating these numbers goes beyond the description of the particular set of data. Frequently, the data are regarded as a sample selected from some larger set of data. For example, a sampling of unpaid accounts for a large merchandiser would allow you to calculate an estimate of the average value of unpaid accounts. This estimate could be used as an audit check on the total value of all unpaid accounts held by the merchandiser. So, the applications of statistics to business can be divided into two broad areas: (1) describing large masses of data and (2) making inferences (estimates, decisions, predictions, etc.) about some set of data based on sampling. Let us examine some case studies that illustrate applications of statistics in business and government.

CASE STUDY 1.1
THE CONSUMER PRICE INDEX

A data set of interest to virtually all Americans is the set of prices charged for goods and services in the U.S. economy. The general upward movement in this set of prices is referred to as *inflation*; the general downward movement is referred to as *deflation*. In order to *estimate* the change in prices over time, the Bureau of Labor Statistics (BLS) of the U.S. Department of Labor developed the Consumer Price Index (CPI). Each month, the BLS collects price data about a specific collection of goods and services (called a *market basket*) from 85 urban areas around the country. Statistical procedures are used to compute the CPI from this sample price data and other information about consumers' spending habits. By comparing the level of the CPI at different points in time, it is possible to *estimate* the rate of inflation (or deflation) over particular time intervals and to compare the purchasing power of a dollar at different points in time.

One major use of the CPI as an index of inflation is as an indicator of the success or failure of government economic policies. A second use of the CPI is to escalate income payments. Millions of workers have *escalator clauses* in their collective bargaining contracts; these clauses call for increases in wage rates based on increases in the CPI. In addition, the incomes of Social Security beneficiaries and retired military and federal civil service employees are tied to the CPI. It has been estimated that a 1% increase in the CPI can trigger an increase of over $1 billion in income payments. Thus, it can be said that the very livelihoods of millions of Americans depend on the behavior of a statistical estimator, the CPI (U.S. Department of Labor, 1978). [*Note:* We will discuss the Consumer Price Index in greater detail in Chapter 13.]

CASE STUDY 1.2

TASTE-PREFERENCE
SCORES FOR BEER

Two sets of data of interest to a firm's marketing department are (1) the set of taste-preference scores given by consumers to its product and to competitors' products when all brands are clearly labeled and (2) the taste-preference scores given by the same set of consumers when all brand labels have been removed and the consumer's only means of product identification is taste. With such information, the marketing department should be able to determine whether taste preference arose because of perceived physical differences in the products or as a resut of the consumer's image of the brand (brand image is, of course, largely a result of a firm's marketing efforts). Such a determination should help the firm develop marketing strategies for its product.

A study using these two types of data was conducted by Ralph Allison and Kenneth Uhl (1965) in an effort to determine whether beer drinkers could distinguish among major brands of unlabeled beer. A sample of 326 beer drinkers was randomly selected from the set of beer drinkers identified as males who drank beer at least three times a week. During the first week of the study, each of the 326 participants was given a six-pack of unlabeled beer containing three major brands and was asked to taste-rate each beer on a scale from 1 (poor) to 10 (excellent). During the second week, the same set of drinkers was given a six-pack containing six major brands. This time, however, each bottle carried its usual label. Again, the drinkers were asked to taste-rate each beer from 1 to 10. From a statistical analysis of the two sets of data yielded by the study, Allison and Uhl concluded that the 326 beer drinkers studied could not distinguish among brands by taste on an overall basis. This result enabled them to infer statistically that such was also the case for beer drinkers in general. Their results also indicated that brand labels and their associations did significantly influence the tasters' evaluations. These findings suggest that physical differences in the products have less to do with their success or failure in the marketplace than the image of the brand in the consumers' minds. As to the benefits of such a study, Allison and Uhl note, "to extent that product images, and their changes, are believed to be a result of advertising . . . the ability of firms' advertising programs to influence product images can be more thoroughly examined."

CASE STUDY 1.3

MONITORING THE
UNEMPLOYMENT RATE

The employment status (employed or unemployed) of each individual in the U.S. work force is a set of data that is of interest to economists, businesspeople, and sociologists. These data provide information on the social and economic health of our society. In order to obtain information about the employment status of the work force, the U.S. Bureau of the Census conducts what is known as the *Current Population Survey*. Each month approximately 1,500 interviewers visit about 59,000 of the 79.1 million households in the United States and question the occupants over 14 years of age about their employment status. Their responses enable the Bureau of the Census to *estimate* the percentage of people in the labor force who are unemployed (the *unemployment rate*). Thus, a *statistical estimator* serves as a monthly indicator of the nation's economic welfare.

Perhaps you are wondering how a reliable estimate of this percentage can be obtained from a sample that includes only about .1% of the households in the

United States. The answer lies in the method used to select the sample of house-holds. The method was designed to enable the Bureau of the Census to control the precision of its estimate while obtaining a sample that is representative of the set of all households in the country. That reliable estimates of nationwide char-acteristics can be obtained from relatively small sample sizes is an illustration of the power of statistics (U.S. Department of Commerce, 1978). [*Note:* We will discuss sampling methods in detail in Chapter 20.]

CASE STUDY 1.4

AUDITING PARTS AND EQUIPMENT FOR AIRLINE MAINTENANCE

The United Airlines Maintenance Base in San Francisco is responsible for the maintenance and overhaul of all United Airlines aircraft. Its storeroom receives, stores, and distributes all the parts needed for maintenance of the aircraft. To control the stock of spare parts and to determine the value of parts on hand, *inventory counts* of the number of each item in stock are taken. It is the responsibility of the Auditing Division of United Airlines to verify the accuracy of the inventory counts. Rather than verifying the accuracy of the counts by recounting all of the inventory item groups, the accountants sample a small number of these groups and recount them. If they find a large number of discrepancies between the original counts and their sample counts, they infer that many of the rest of the item counts (those not sampled and recounted) are also in error. They conclude that the original inventory counts are unacceptable and must be recounted. On the other hand, if they find only a small number of discrepancies, they infer that most of the item counts not rechecked are accurate and conclude that the original inventory counts are satisfactory.

Before this inferential statistical procedure was implemented, the Auditing Divi-sion verified inventory counts by recounting all the items in stock. The inferential procedure enables the accountants to maintain the quality of their verifications with a substantial reduction in work-hours (Hunz, 1956).

CASE STUDY 1.5

THE DECENNIAL CENSUS OF THE UNITED STATES

The following description is quoted from the U.S. Bureau of the Census, *Statistical Abstract of the United States: 1981:*

> The U.S. Constitution provides for a census of the population every 10 years, primarily to establish a basis for apportionment of members of the House of Representatives among the states. For over a century after the first census in 1790, the census organization was a temporary one, created only for each decennial census. In 1902, the Bureau of the Census was established as a permanent federal agency, responsible for enumerating the population and also for compiling statistics on other subjects.
>
> The census of the population is a complete count. That is, an attempt is made to account for every person, for each person's residence, and for other characteristics (sex, age, family relationships, etc.). Since the 1940 census, however, some data have been obtained from representative samples of the population rather than a complete count. In the 1980 census, two sampling rates were employed. For most of the country, one in every six households (about 17%) received the long form or sample questionnaire; in areas estimated to have fewer than 2,500 inhabitants, every other household (50%) received the sample questionnaire to enhance the reliability of sample data in small areas. Exact agreement is not expected between sample data and the complete census count.

Census statistics regarding total numbers of people in various age groups are examples of numerical descriptions that require no statistical inference, since they are (purportedly) complete counts. However, income data collected by the Bureau of the Census from "representative samples" might be used to make inferences about the incomes of *all* persons. You will learn that the reliability of statistical inferences is dependent on the sampling procedure, characteristics of the data, and the methodology employed to make the inferences.

Why study statistics in a business program? The quantification of business research and business operations (quality control, statistical auditing, forecasting, etc.) has been truly astounding over the past several decades. Econometric modeling, market surveys, and the creation of indexes such as the Consumer Price Index all represent relatively recent attempts to quantify economic behavior. It is extremely important that today's business graduate understand the methods and language of statistics, since the alternative is to be swamped by a flood of numbers that are more confusing than enlightening to the untutored mind. The business student should develop a discerning sense of rational thought that will distill the information contained in these numbers so it can be used to make intelligent decisions, inferences, and generalizations. We believe that the study of statistics is essential to the ability to operate effectively in the modern business environment.

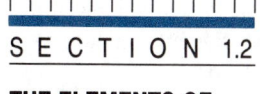

S E C T I O N 1.2

THE ELEMENTS OF STATISTICS

Although applications of statistics abound in almost every area of human endeavor, there are certain elements common to all statistical problems. The foundation of every statistical problem is a **population**:

DEFINITION 1.1

The **population** is a set of data that characterizes some phenomenon (in our situation, some business phenomenon).

Our definition of *population* is broader than the usual one. We are not referring only to a group of people. For example, the employment status of every person in the U.S. labor force is a population, as is the weekly profit figure for the entire time (past and future) a firm is in business. Other examples of populations are the number of errors on each page in an accountant's ledger and the daily Dow Jones Average, past and future. Thus, we think of a population as being a large—perhaps infinitely large—collection of measurements.

The second element of a statistical problem is the **sample**:

DEFINITION 1.2

A **sample** is a subset of data selected from the population.

The sample is a subset (part of) the population.* The collections of daily Dow Jones Averages for the past 5 years, the monthly unemployment figures for the past 18 months, the weekly sales of a firm over the past year, and the number of errors per page on 10 pages of a 100-page ledger are all samples of the respective populations.

The usefulness of the sample is clarified by considering the third element of a statistical problem—the **inference**:

DEFINITION 1.3

A **statistical inference** is a decision, estimate, prediction, or generalization about the population based on information contained in a sample.

That is, *we use the information in the smaller set of measurements (the sample) to make decisions, predictions, or generalizations about the large or whole set of measurements (the population)*. For example, we might use the number of accounting errors in a 10-page sample of a ledger to estimate the number of errors on all 100 pages of the ledger. Or we could use the past 18 months' unemployment figures to predict the next month's unemployment rate. We might try to infer this year's total sales from last year's weekly sales figures. Finally, we could predict the Dow Jones Average a year from now based on the sample of daily Dow Jones Averages over the past 5 years. In each case, we are using the information in a sample to make inferences about the corresponding population.

The preceding definitions identify three of the four elements of a statistical problem. The fourth, and perhaps the most important, is the topic of Section 1.3.

EXAMPLE 1.1

Cola wars is the popular media term for the intense competition between the marketing campaigns of Coca-Cola and Pepsi. The campaigns have featured movie and television stars, rock videos, athlete endorsements, and claims of consumer preference based on taste tests. Suppose 1,000 consumers are asked to state their preferences in a "blind" taste test, i.e., a taste test in which the brand names are disguised.

a. Describe the population.
b. Describe the sample.
c. Describe the inference of interest.

---

*In everyday usage, the word *sample* implies a collection of objects—e.g., a sample of 1,200 people from a city, or a sample of 10 transistors from a day's production. When selecting people or objects from some group, we will sometimes use this terminology; that is, we will speak of a sample of objects rather than the collection of measurements made on the objects. Whether we are speaking of a collection of measurements (our definition) or the collection of objects on which the measurements are made (everyday usage) will be clear from the context of the discussion.

SOLUTION

a. The population of interest is defined as the set of *data* that characterizes some phenomenon. Although it is tempting to describe the population corresponding to the taste test as "the collection of all cola consumers," the description is technically incorrect because it describes a set of *people*, not *data*. The population is more precisely described as the collection of *preferences* of all cola consumers. That is, the population consists of a collection of measurements or observations (in this case, the cola preferences) made on a group of people, *not* the people themselves.*

b. The sample must be a subset of the population. In this case the cola preferences of the 1,000 consumers who take the taste test represent a sample. Note that, like the population, the sample consists of data (cola preferences), not people.

c. The inference of interest is to *generalize* the cola preferences of the 1,000 sampled consumers to the population of all consumer preferences. Specifically, the sample results will be used to *estimate* the percentage of all cola consumers who prefer each brand. ∎

| | | | | | | | | | | | | |

S E C T I O N 1.3

STATISTICS: WITCHCRAFT OR SCIENCE?

We have identified the *primary objective of statistics*: *making inferences about a population based on information contained in a sample*. However, inference-making constitutes only part of our story. The only way we could be completely certain that an inference about a population is correct would be to include the entire population in our sample. Since it may be too costly, time-consuming, or even impossible to do this, we sample only a portion of the observations in the population. This introduces an element of uncertainty associated with inferences based on this partial information about the population. Consequently, we will want to measure and report the **reliability** of each inference made; this is the fourth element of a statistical problem.

The measure of reliability that accompanies an inference separates the science of statistics from the art of fortune-telling. A palm reader, like a statistician, may examine a sample (your hand) and make inferences about the population (your life). However, no measure of reliability can be attached to the reader's inferences. On the other hand, we always assess the reliability of our statistical inferences. For example, if we use a sample of previous profit figures to predict a firm's future profits, we will give a **bound** on our **prediction error**. This bound is simply a number that the error of our prediction is not likely to exceed. Thus, the uncertainty of our prediction is measured by the size of the bound on the prediction errors. The reliability of our statistical inferences will be discussed throughout this text. For now, we simply want you to realize that an inference is incomplete without a measure of its reliability.

---

*We have already alluded to the fact that the terms *population* and *sample* are often used more generally to refer to objects rather than data. For these introductory examples and exercises, we will encourage the more precise usage to help build a solid foundation.

We conclude with a summary of the elements of a statistical problem and several examples to illustrate the elements.

> **FOUR ELEMENTS COMMON TO ALL STATISTICAL PROBLEMS**
>
> 1. The population of interest, with a procedure for sampling the population
> 2. The sample and analysis of the information in the sample
> 3. The inferences about the population, based on information contained in the sample
> 4. A measure of reliability for the inference

EXAMPLE 1.2

Refer to Example 1.1, in which 1,000 consumers stated their cola preferences in a taste test. Describe how the reliability of an inference concerning the preferences of all cola consumers could be measured.

SOLUTION

When the preferences of 1,000 consumers are used to estimate the preferences of all consumers, the estimates will not be exact. For example, if the taste test shows that 56% of the 1,000 consumers prefer Pepsi, it does not follow (nor is it likely) that exactly 56% of *all* cola consumers prefer Pepsi. Nevertheless, we may be able to use sound statistical reasoning to assure that the sampling procedure used will generate estimates that are almost certainly within a specified limit of the true percentage of all consumers who prefer Pepsi. For example, such reasoning might assure us that the estimate of the preference for Pepsi is almost certainly within 5% of the actual population preference. The implication is that the actual preference for Pepsi is between 51% [i.e., (56 − 5)%] and 61% [i.e., (56 + 5)%]. This interval represents a measure of reliability for the inference.   ■

EXAMPLE 1.3

Research is conducted to compare the starting salaries of male and female MBAs (Masters of Business Administration) who enter the job market after graduation. The starting salaries of 200 recent MBA graduates (100 males and 100 females) are obtained.

a. Describe the population for this research.
b. Describe the sample.
c. Describe the inference of interest.
d. Discuss how the reliability of the inference could be measured.

SOLUTION

a. We first recall that the population consists of data (not MBA graduates), and then recognize that the data of interest in this research are the starting salaries of MBA graduates. Because the objective of this research is to compare these data for males and females, two populations are pertinent. The first consists of starting salaries for all female MBAs, the second for all male MBAs.

b. Two samples were selected, one from each population. The first sample consists of the starting salaries of the 100 female MBAs, and the second of the starting salaries of the 100 male MBAs.

c. One inference of interest is the comparison of the average* starting salaries for all male MBAs with that of all female MBAs.

d. The reliability of the inference must address the issue of the precision with which the difference between the average salaries of the two populations is estimated. For example, suppose the two samples indicate that, on average, females start at $3,000 per year less than males. To assert that this implies that the average starting salary of all male MBAs exceeds that of all female MBAs is premature; we must first know the precision associated with the difference between the sample averages. For example, suppose that sound statistical reasoning is used to show that the sample estimate of the difference is almost certainly within $1,500 of the true difference between the population averages. This implies that the true difference is (almost certainly) between $1,500 [i.e., $(3,000 - 1,500)$] and $4,500 [i.e., $(3,000 + 1,500)$], which supports the inference that recent male MBA graduates' starting salaries, on average, exceed those of females by between $1,500 and $4,500. This interval is therefore a measure of reliability for the inference.

■

| | | | | | | | | | | | | |

S E C T I O N   1.4

THE ROLE OF STATISTICS IN MANAGERIAL DECISION-MAKING

Managers frequently rely on input from statistical analyses to help them make decisions. The role statistics can play in managerial decision-making is indicated in the flow diagram in Figure 1.1 (page 10). Every managerial decision-making problem begins with a real-world problem. This problem is then formulated in managerial terms and framed as a managerial question. The next sequence of steps (proceeding counterclockwise around the flow diagram) identifies the role that statistics can play in this process. The managerial question is translated into a statistical question, the sample data are collected and analyzed, and the statistical question is answered. The next step in the process is using the answer to the statistical question to reach an answer to the managerial question. The answer to the managerial question will suggest a reformulation of the original managerial problem, suggest a new managerial question, or lead to the solution of the managerial problem.

One of the most difficult steps in the decision-making process—one that requires a cooperative effort among managers and statisticians—is the translation of the managerial question into statistical terms (that is, into a question about a population of data). This question must be formulated so that, when answered, it will provide the key to the answer to the managerial question. Thus, as in the game of chess, you must formulate the statistical question with the end result, the solution to the managerial question, in mind.

---

*Although we will not make the notion of "average" precise until Chapter 3, "typical" or "middle" can be substituted here without confusion.

FIGURE 1.1
Flow Diagram Showing
the Role of Statistics in
Managerial Decision-
Making.
Source: Chervany,
Benson, and Iyer (1980).

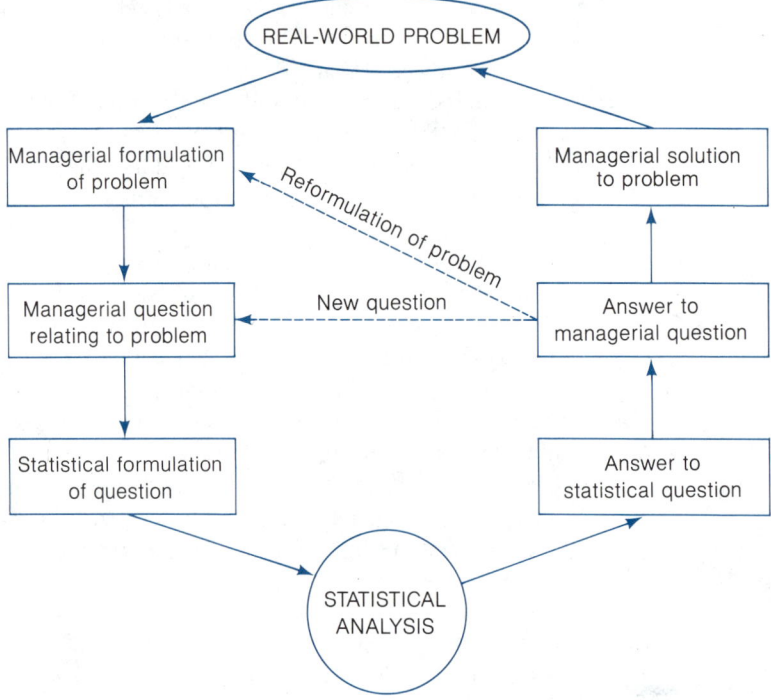

EXERCISES 1.1–1.11

APPLYING THE CONCEPTS

1.1 Manufacturers of consumer goods rely on the information provided by consumer preference surveys to guide both the design and the marketing of new products. In the winter
of 1986, Onan Corporation, a manufacturer of built-in generators for recreational vehicles
(RVs), was considering developing and marketing a portable generator. Such a product
could potentially be marketed to both RV owners and RV manufacturers. To determine
RV owners' preferences with respect to the features of a portable generator (e.g., size,
manual or electric start, and so forth), 3,000 questionnaires were mailed to RV owners
in the continental United States. One thousand fifty-two (1,052) responses were received
by Onan.*

a. Identify the population, the sample, and the inferences of interest to Onan.

b. In Chapters 7 and 8, we will see that the reliability of an inference is related to the
size of the sample used. In addition to sample size, what other factors might affect
the reliability of inferences based on the responses to a mailed questionnaire?

1.2 A problem frequently faced by the management-training specialist is the lack of management enthusiasm for in-plant training programs. Robert J. House (1962) studied the
effects of certain changes in the management-training program of a large engineering
firm on the enthusiasm for and acceptance of the program by management trainees. In
particular, he made the program more challenging and less permissive by increasing the

*Information by personal communication with Thomas J. Roess, Manager, Market Analysis, Onan
Corporation, Fridley, Minnesota.

requirements that had to be met for graduation and using a more authoritative teaching style. In order to investigate the effectiveness of these changes, he gave the revised program to 51 trainees and compared their rate of absence with that of a class of 49 trainees who had taken the course under the old teaching policies. He found that the number of absences per person in his trainee group was significantly lower than that for the group trained under the old policies. He concluded that the revised teaching policies should increase class attendance in future courses as well. Thus, instead of speculating informally about the effectiveness of the revised teaching policies, House used statistical analysis to formally assess their impact. Identify the populations studied by House, the two samples, and the inference made by House.

1.3 Suppose you work for a major public opinion pollster and you wish to estimate the proportion of adult citizens who think the president is doing a good job in handling the nation's economy. Clearly define the population you wish to sample.

1.4 An insurance company would like to determine the proportion of all medical doctors who have been involved in one or more malpractice suits. The company selects 500 doctors at random and determines the number in the sample who have ever been involved in a malpractice suit. Identify the population of interest to the insurance company. Describe the sample and identify the type of inference the insurance company wishes to make.

1.5 *Corporate merger* is a means through which one firm (the bidder) acquires control of the assets of another firm (the target). Carol E. Eger (1982) identified a total of 497 mergers between firms listed on the New York Stock Exchange that resulted in the delisting of the acquired (target) firm's stock during the period 1958–1980. She sampled 38 of these mergers and evaluated the effects of the merger on the value of the holdings of the bidder firm's bondholders. In particular, she wanted to learn whether the value of the holdings increased or decreased as a result of the merger. Identify the population studied, the sample used, and the types of inferences we might wish to make about the population.

1.6 *Job-sharing* is an innovative employment alternative that originated in Sweden and is becoming very popular in the United States. Firms that offer job-sharing plans allow two or more persons to work part-time, sharing one full-time job on different days. For example, two job-sharers might alternate work weeks, with one working while the other is off. Job-sharers never work at the same time and may not even know each other. Job-sharing is particularly attractive to working mothers and to people who frequently lose their jobs due to fluctuations in the economy ("Your Job in the 1980's," 1980). In order to evaluate employers' satisfaction with job-sharing plans, a government agency contacted 100 firms that offer job-sharing. Each firm's director of personnel was asked whether the firm was satisfied with the productivity of workers with shared jobs. Describe the population from which the sample was selected and the type of inference the government agency wishes to make.

1.7 The checking of all accounts payable invoices for errors is a costly and time-consuming procedure. A method for effectively and economically checking for errors involves selecting a portion of the invoices and using the percentage of examined invoices that contain errors to estimate the percentage of all invoices that are in error. Identify the population, sample, and type of statistical inference to be made for this problem.

1.8 To compute their yearly income, trading stamp companies must determine their liability for unredeemed stamps. This requires an estimate of the fraction of all stamps issued that have not been redeemed. Davidson and colleagues (1967) have developed a method for estimating this fraction by studying the time lapse between the issue and redemption of a small percentage of the stamps in circulation. Identify the population, sample, and type of statistical inference to be made for this problem.

1.9 In the mid-1960's, engineers at the Whirlpool Corporation noted that the problem of household garbage disposal was one that few households had satisfactorily solved. Before undertaking a costly research project to attempt to devise a product to solve the problem, Whirlpool conducted a survey of households to determine what proportion of households were in fact concerned about the problem of garbage disposal. It was determined that the proportion was large enough that Whirlpool would probably be able to successfully market whatever product evolved from its research efforts; thus, they proceeded with a research study. The product that resulted was the portable trash compactor (McGuire, 1973). Identify the population of interest to Whirlpool. Describe the sample and the nature of the inference made by Whirlpool.

The next two exercises are designed to examine your current thinking on the subject of reliability. We will formally address this subject in subsequent chapters.

1.10 Refer to Case Study 1.3. Suppose you were asked to assess the reliability of the monthly estimate of the percentage of workers in the labor force who are unemployed. What information do you think would help you in making your assessment?

1.11 Refer to Exercise 1.8. Suppose you were an accountant for a trading stamp company and were responsible for preparing the firm's end-of-year income statement. You would be interested in knowing the firm's liability for unredeemed stamps. Suppose the firm's statistician estimated the fraction of stamps unredeemed out of all those sold during the year to be $\frac{1}{12}$. Suppose also that the statistician failed to give you an indication of the reliability of this estimate (a very serious oversight for a statistician). Since the reliability of your income statement would depend on the reliability of the statistician's estimate, you would be interested in knowing the reliability of the estimate. What information would help you measure the reliability of the statistician's estimate?

ON YOUR OWN . . .

If you could start your own business right now, what kind would it be? Identify a set of business data that would be of interest to you and your firm. Is the data set you identified a sample or population? How could you use this data set to help your business operate more efficiently?

REFERENCES

Allison, R. I. and Uhl, K. P. "Influence of beer brand identification on taste perception." *Journal of Marketing Research*, Aug. 1965, 36–39.

Careers in Statistics. Washington, D.C.: American Statistical Association and the Institute of Mathematical Statistics, 1974.

Chervany, N. L., Benson, P. G., and Iyer, R. K. "The planning stage in statistical reasoning." *The American Statistician*, Nov. 1980, 222–226.

Davidson, H. J., Neter, J., and Petras, A. S. "Estimating the liability for unredeemed stamps." *Journal of Accounting Research*, 1967, 5, 186–207.

Eger, C. E. "Corporate mergers: An analytical analysis of the role of risky debt." Unpublished Ph.D. dissertation. University of Minnesota, 1982.

House, R. J. "An experiment in the use of management training standards." *Journal of the Academy of Management*, 1962, 5.

Hunz, E. "Application of statistical sampling to inventory audits." *The Internal Auditor*, 1956, 13, 38.

McGuire, E. P. *Evaluating New Product Proposals.* New York: National Industrial Conference Board, 1973, p. 42.

Tanur, J. M., Mosteller, F., Kruskal, W. H., Link, R. F., Pieters, R. S., and Rising, G. R. *Statistics: A Guide to the Unknown.* San Francisco: Holden-Day, 1978.

U.S. Bureau of the Census. *Statistical Abstract of the United States: 1981.* 102nd ed. Washington, D.C.: U.S. Government Printing Office, 1981.

U.S. Department of Commerce. *An Error Profile: Employment as Measured by the Current Population Survey.* Statistical Policy Working Paper 3. Washington, D.C.: U.S. Government Printing Office, 1978.

U.S. Department of Labor. *The Consumer Price Index: Concepts and Content over the Years.* Bureau of Labor Statistics, Report 517. Washington, D.C.: U.S. Government Printing Office, May 1978.

Willis, R. E. and Chervany, N. L. *Statistical Analysis and Modeling for Management Decision-making.* Belmont, Calif.: Wadsworth, 1974, Chapter 1.

"Your job in the 1980's." *Consumer's Digest*, Nov.–Dec. 1980, 32–36.

WHERE WE'VE BEEN . . .

By examining typical examples of the use of statistics in business, we listed four elements that are common to every business statistical problem: a population, a sample, an inference, and a measure of the reliability of the inference. The last two elements identify the goal of statistics—using sample data to make an inference (a decision, estimate, or prediction) about a population.

WHERE WE'RE GOING . . .

Before we make an inference, we must be able to describe a data set. Graphical methods that provide a compact description of a data set are the topic of this chapter.

CONTENTS

GRAPHICAL DESCRIPTIONS OF DATA

Before we can use the information in a sample to make inferences about a population, we must be able to extract the relevant information from the sample. That is, we need methods to summarize and describe the sample measurements. For example, if we look at last year's sales for 100 randomly selected companies, we are unlikely to extract much information by looking at the set of 100 sales figures. We would get a clearer picture of the data by calculating the average sales for all 100 companies, by determining the highest and lowest company sales, by drawing a graph showing the average monthly sales over the 12-month period, or in general, by using some technique that will extract and summarize relevant information from the data and, at the same time, allow us to obtain a clearer understanding of the sample.

In this chapter we first define two different types of business data and then present some graphical methods for describing data of each type. You will see that graphical methods for describing data are intuitively appealing descriptive techniques that can be used to describe either a sample or a population. However, as we will begin to demonstrate in Chapter 3, numerical methods for describing data are the keys that unlock the door to population inference-making.

As you will subsequently see, some of the descriptive measures discussed in this chapter are primarily of interest for describing large data sets. If statistical inference is your goal, only Sections 2.1, 2.4, and 2.5 are essential. For this reason, Sections 2.2, 2.3, 2.6, and 2.7 are marked "optional."

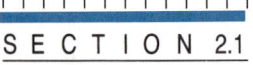

S E C T I O N 2.1

TYPES OF BUSINESS DATA

Although the number of business phenomena that can be measured is almost limitless, business data can generally be classified as one of two types: **quantitative** or **qualitative.**

DEFINITION 2.1

Quantitative data are observations that are measured on a numerical scale.

The most common type of business data is quantitative data, since many business phenomena are measured on numerical scales. Examples of quantitative business data are:

1. The daily Dow Jones Industrial Average
2. The monthly unemployment percentage
3. Last year's sales for selected firms
4. The number of women executives in an industry

The measurements in these examples are all numerical.

All data that are not quantitative are qualitative (or categorical).

DEFINITION 2.2

If each measurement in a data set falls into one and only one of a set of categories, the data set is called **qualitative** (or **categorical**).

Qualitative data are observations that are nonnumerical. Examples of qualitative business data are:

1. The political party affiliations of 50 randomly selected business executives—each executive would have one and only one political party affiliation
2. The brand of gasoline last purchased by 74 randomly selected automobile owners—again, each measurement would fall into one and only one category
3. The state in which each of 32 randomly selected firms in the United States has its highest yearly sales

Notice that each of the examples has nonnumerical, or qualitative, measurements.

As you would expect, the method used for summarizing the information in a sample of measurements depends on the type of business data being collected. We devote the remainder of this chapter to the presentation of graphical methods for describing quantitative and qualitative data sets.

EXERCISES 2.1–2.5

APPLYING THE CONCEPTS

2.1 A food products company is considering marketing a new snack food. To see how consumers react to the product, the company conducted a taste test using 100 randomly selected shoppers at a suburban shopping mall. The shoppers were asked to taste the snack food and then fill out a short questionnaire that requested the following information:
a. What is your age?
b. Are you the person who typically does the food shopping for your household?
c. How many people are in your family?
d. How would you rate the taste of the snack food on a scale of 1 to 10, where 1 is least tasty?
e. Would you purchase this snack food if it were available on the market?
f. If you answered yes to question e, how often would you purchase it?

Each of these questions generates a data set of interest to the company. Classify the data in each data set as either quantitative or qualitative. Justify your classification.

2.2 Classify the following examples of data as either qualitative or quantitative. Justify your classification.
a. Ten college freshmen were asked to indicate the brand of jeans they prefer.
b. Fifteen television cable companies were asked how many hours of sports programming they carry in a typical week.
c. Fifty executives were asked what percentage of their workday is spent in meetings.
d. The number of long-distance phone calls made from each of 100 public telephone booths on a particular day was recorded.

2.3 Classify the following examples of data as either qualitative or quantitative:
a. The brand of calculator purchased by 20 business statistics students
b. The list price of calculators purchased by 20 business statistics students
c. The number of automobiles purchased during the past 5 years by the heads of 50 randomly selected households
d. The month indicated by each of 41 randomly selected business firms as the month during which it had the highest sales
e. The depth of tread remaining on each of 137 randomly selected automobile tires after 20,000 miles of wear

2.4 Classify the following examples of data as either qualitative or quantitative:
  a. The brand of stereo speaker for which each of 25 college students indicated a preference
  b. The loss (in dollars) incurred in each of the last 5 years by a department store as a result of shoplifting
  c. The color of interior house paint (other than white) that each of the five largest manufacturers of paint says generates the most sales revenue for the firm

2.5 Classify the following examples of data as either qualitative or quantitative:
  a. The number of corporate mergers during each of the last 15 years
  b. The change in the Consumer Price Index during each of the last 6 months
  c. The length of time before each of 30 dry-cell batteries goes dead
  d. The American automobile manufacturer that each of 25 service station mechanics indicated as producing the most reliable cars

SECTION 2.2

GRAPHICAL METHODS FOR DESCRIBING QUALITATIVE DATA: THE BAR CHART (OPTIONAL)

As we noted in Section 2.1, a qualitative observation falls into one and only one of a group of categories. For example, suppose a women's clothing store located in the downtown area of a large city wants to open a branch in the suburbs. To obtain some information about the geographic distribution of its present customers, the store manager conducts a survey in which each customer is asked to identify her place of residence with regard to the city's four quadrants: northwest (NW), northeast (NE), southwest (SW), or southeast (SE). Out-of-town customers are excluded from the survey.

The results of the survey—the responses of $n = 30$ randomly selected resident customers—might appear as in Table 2.1. (Note that the symbol n is used here and throughout the text to represent the sample size—i.e., the number of measurements in a sample.) You can see that each of the 30 measurements falls in one and only one of the four possible categories representing the four quadrants of the city.

TABLE 2.1
Customer Residence
Survey: $n = 30$

CUSTOMER	RESIDENCE	CUSTOMER	RESIDENCE	CUSTOMER	RESIDENCE
1	NW	11	NW	21	NE
2	SE	12	SE	22	NW
3	SE	13	SW	23	SW
4	NW	14	NW	24	SE
5	SW	15	SW	25	SW
6	NW	16	NE	26	NW
7	NE	17	NE	27	NW
8	SW	18	NW	28	SE
9	NW	19	NW	29	NE
10	SE	20	SW	30	SW

A natural and useful technique for summarizing qualitative data is to tabulate the **frequency** or **relative frequency** of each category.

DEFINITION 2.3

The **frequency** for a category is the total number of measurements that fall in the category. The frequency for a particular category, say category i, will be denoted by the symbol f_i.

DEFINITION 2.4

The **relative frequency** for a category is the frequency of that category divided by the total number of measurements; that is, the relative frequency for category i is

$$\text{Relative frequency} = \frac{f_i}{n}$$

where

n = Total number of measurements in the sample

f_i = Frequency for the ith category

The frequency for a category is the total number of measurements in that category, whereas the relative frequency for a category is the **proportion** of measurements in the category. Table 2.2 shows the frequency and relative frequency for the customer residences listed in Table 2.1. Note that the sum of the frequencies should always equal the total number of measurements in the sample and the sum of the relative frequencies should always equal 1 (except for rounding errors), as in Table 2.2.

TABLE 2.2
Frequencies and Relative Frequencies for Customer Residence Survey

CATEGORY	FREQUENCY	RELATIVE FREQUENCY
NE	5	$\frac{5}{30}$ = .167
NW	11	$\frac{11}{30}$ = .367
SE	6	$\frac{6}{30}$ = .200
SW	8	$\frac{8}{30}$ = .267
Total	30	1.000

A common means of graphically presenting the frequencies or relative frequencies for qualitative data is the **bar chart**. For this type of chart the frequencies (or relative frequencies) are represented by bars of equal width—one bar for each category. The height of the bar for a given category is proportional to the category frequency (or relative frequency). Usually the bars are placed in a vertical position with the base of the bar on the horizontal axis of the graph. The order of the bars on the horizontal axis is unimportant. Both a frequency bar chart and a relative frequency bar chart for the customer residence example are shown in Figure 2.1.

FIGURE 2.1

Bar Charts for Customer
Residence Example

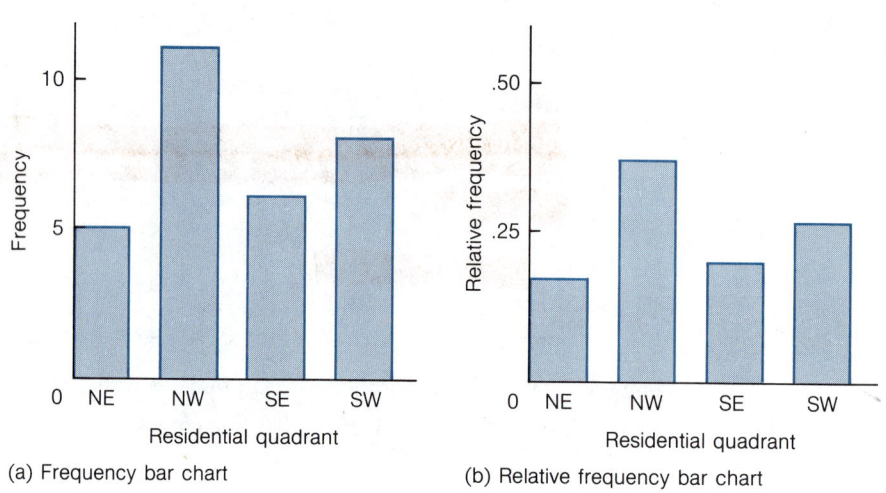

(a) Frequency bar chart   (b) Relative frequency bar chart

CASE STUDY 2.1

HOW AMERICANS
KEEP FIT

A recent article in the *Minneapolis Star and Tribune* (Winegar, 1983) described the fitness craze in the United States and the big business it has become:

> Two decades ago, fewer than one in four adult Americans exercised regularly; today around 72 million adults (approximately one out of three) claim to exercise on a regular basis, according to the President's Council on Physical Fitness and Sports.
>
> . . . more than 500 major U.S. corporations have created on-site fitness programs for their employees and thousands of smaller companies are cooperating with local fitness resources such as YWCA's or park and recreation offices to gain employee access to the exercise facilities.
>
> The fitness fad has created mini-booms in the fashion industry's exercise/dance-wear market as well as in the development and marketing of weight-training equipment and machines such as Nautilus, Universal, and Olympic.
>
> [The fitness boom has generated] an estimated $30 billion annually in sales of sports and health equipment, memberships, books, records, company fitness programs, health foods and diet aids.

The author of the article used the *bar chart* shown in Figure 2.2 to describe the athletic activities of Americans.

FIGURE 2.2

How Americans Keep Fit
Source: Winegar (1983).

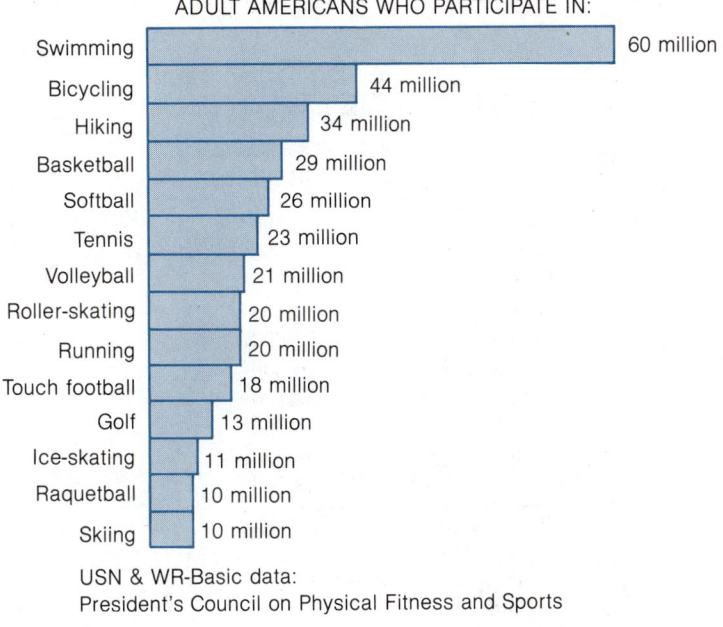

ADULT AMERICANS WHO PARTICIPATE IN:

Swimming — 60 million
Bicycling — 44 million
Hiking — 34 million
Basketball — 29 million
Softball — 26 million
Tennis — 23 million
Volleyball — 21 million
Roller-skating — 20 million
Running — 20 million
Touch football — 18 million
Golf — 13 million
Ice-skating — 11 million
Raquetball — 10 million
Skiing — 10 million

USN & WR-Basic data:
President's Council on Physical Fitness and Sports

EXERCISES 2.6–2.13

LEARNING THE MECHANICS

2.6 The popularity of automated bank tellers has been steadily increasing since their inception in the early 1970's. By the end of 1982, there were 35,721 automatic teller machines in the United States. Although they are popular as a means for withdrawing cash, bank officials have become concerned about the public's general reluctance to deposit funds through the machines. The table illustrates this problem. Construct a relative frequency bar chart for the data.

TYPE OF TRANSACTION	NUMBER OF TRANSACTIONS DURING 1982 (Billions)
Withdrawal	2.35
Deposit	.59
Transfer or other transaction	.16

Source: *Minneapolis Star and Tribune,* Mar. 12, 1984, p. 3M.

2.7 The Internal Revenue Service (IRS) is collecting more financial information than ever before about individual taxpayers. In addition to the tax returns that we all file, the IRS receives notifications of payments made to individuals by employers, banks, stock brokerage firms, and so forth. From this information the IRS has determined where the "well-to-do" reside. Construct a relative frequency bar chart for the data provided in the table.

STATE	NUMBER OF PEOPLE WITH GROSS ASSETS[a] GREATER THAN $500,000
California	301,500
Texas	204,800
Florida	151,800
New York	110,100
Illinois	108,000
Pennsylvania	86,800
Ohio	52,500
New Jersey	51,300
Iowa	50,800
Michigan	48,100

[a]Excludes mortgages and other debts.
Source: *U.S. News & World Report*, Mar. 18 1985, p.53. Copyright 1985, U.S. News & World Report.

2.8 Consider the accompanying bar chart, which shows 1985 cigarette sales (billions of cigarettes) by company.

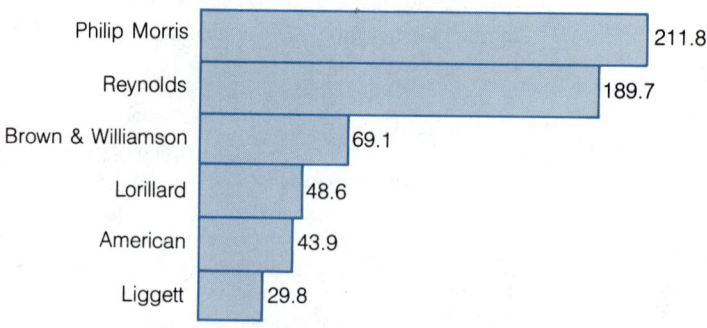

Source: *Business Week*, Dec. 23, 1985, p. 43.

a. In general, what is being described by the bar chart?
b. Which company sold the most cigarettes in 1985? Approximately how many cigarettes did the company sell?
c. Convert the bar chart to a relative frequency bar chart. Describe any problems you encounter in making the conversion.

2.9 The accompanying table lists the number of new housing units authorized (i.e., number of building permits granted) during 1983 for a sample of seven states. Construct a relative frequency bar chart for the data.

STATE	NUMBER OF HOUSING UNITS AUTHORIZED (Thousands)
Alaska	17.3
California	171.9
Florida	189.0
Kansas	14.2
New York	37.9
Ohio	26.6
Texas	276.2

Source: U.S. Bureau of the Census, *Statistical Abstract of the United States: 1985*, p. 726.

APPLYING THE CONCEPTS

2.10 A *bond* is a promissory note issued by a business or government unit in exchange for a specified amount of money. Typically, the note promises to repay the bondholder (*lender*) the face value of the bond (*par value*) at some future date (*maturity date*) plus a specified number of dollars in interest each year. Since the early 1900's, bonds have been given quality ratings that indicate the likelihood that the issuing firm will default on promises to bondholders. One of the major rating agencies is Moody's Investors Service. Moody's rating categories are shown here:

HIGH QUALITY	INVESTMENT GRADE	SUBSTANDARD	SPECULATIVE
Aaa, Aa	A, Baa	Ba, B	Caa, Ca, C

The ratings run from triple-A bonds (which are extremely safe) to C bonds (which have a high probability of default) (Brigham, 1982). The table on page 24 lists the maturity dates and the Moody rating for a sample of bonds listed in *Moody's Bond Record*. Construct a bar chart to describe the bond ratings of this sample of bonds.

BOND ISSUER	MATURITY DATE	RATING
Beatrice Foods	1994	A
Caterpillar Tractor	1992	A
Continental Group	1990	Baa
Diamond Shamrock	1994	A
Digicon, Inc.	1993	Caa
Eastern Airlines	2002	B
Eaton	1992	A
Essex Chemical	1998	B
First Bank System	1992	Aa
Ford Motor Co.	1993	A
General Electric	2004	Aaa
General Signal	1999	Aa
Johns–Manville	2004	Ca
Litton Industries, Inc.	2005	Baa
Marathon Oil	2006	Baa
Northern States Power	1990	Aaa
Norton Simon	1998	Baa
Philip Morris	1993	A
Pillsbury	1991	A
Procter & Gamble Co.	2002	Aaa
Quaker State Oil	1995	Baa
Ralston Purina Co.	1996	A
Sears & Roebuck Co.	1994	Aa
Singer	1999	Ba
Sun Oil	1990	Aa
Xerox	1999	A

Source: *Moody's Bond Record*, Moody's Investors Service, Vol. 53, No. 2, Feb. 1986.

2.11 During the 1970's the prices of single-family houses in the United States soared to record levels. The price of a typical new single-family home jumped from $23,400 in 1970 to $64,600 in 1980, and by 1983 had reached $75,300 (U.S. Bureau of the Census, 1985). No longer able to afford a house, many people opted for condominium living. As the demand for condominiums increased, owners of apartments found it profitable to convert their apartments to condominiums. For her analysis of the condominium market in the seven-county Minneapolis–St. Paul metropolitan area, Mary L. Bochnak (1982) constructed the accompanying table to describe the stock of condominiums in Dakota and Ramsey counties as of December 31, 1980.

a. Construct a frequency bar chart for the number of apartments converted to condominiums in the cities of Ramsey county.
b. Construct a relative frequency bar chart for the total number of condominiums in the cities of Ramsey county. Do the same for the cities of Dakota county.
c. According to your graphs of part **b,** which city in each county had the largest share of the county's stock of condominiums?

COUNTY	CITY	NUMBER OF APARTMENTS CONVERTED TO CONDOMINIUMS	NUMBER OF CONDOMINIUMS CONSTRUCTED
Dakota	Burnsville	409	135
	Eagan	8	128
	Farmington	0	36
	Inver Grove Heights	0	84
	Lilydale	0	139
	Mendota Heights	0	200
	West St. Paul	66	8
Ramsey	Little Canada	511	101
	Maplewood	0	252
	Mounds View	385	0
	New Brighton	54	0
	Roseville	767	30
	St. Anthony	0	148
	St. Paul	832	443
	Shoreview	192	8

Source: Bochnak (1982).

2.12 By the early 1980's the 16-bit microprocessor had replaced the 8-bit system as the "brains" of state-of-the-art personal computers. In the mid-1980's the next generation of microprocessors was introduced, the 32-bit processor. The 32-bit processor moved the technology of personal computers ever closer to the much larger and more expensive mainframe computers. The table contains a forecast of the demand for 32-bit processors in 1990.

APPLICATIONS	1990 DEMAND (Number of Units)
Office automation	3,950,000
Computer-aided design	375,000
Robots and factory systems	187,000
All other uses	188,000
Total	4,700,000

Data: Dataquest, Inc.
Source: *Business Week*, Oct. 28, 1985, p. 34.

a. Construct a frequency bar chart for these data.
b. Construct a relative frequency bar chart for these data.
c. The bar charts you constructed differ in terms of the information each conveys about a particular category (e.g., office automation). Explain.

2.13 Are we running out of oil? The data in the table at the top of page 26 describe the new barrels of oil added to the reserves of the four largest U.S. oil companies in 1982 and the barrels of oil withdrawn from these reserves in the same year.

COMPANY	BARRELS ADDED (Millions)	BARRELS WITHDRAWN (Millions)
Exxon	124	270
Texaco	55	127
Socal	57	121
Mobil	34	103

Source: *U.S. News & World Report*. Jan. 23, 1984, p. 59. Copyright 1984, U.S. News & World Report.

a. Construct two relative frequency bar charts for the data, one for barrels added and one for barrels withdrawn.
b. Combine the bar charts you constructed in part **a** by plotting the eight relative frequencies on the same bar chart. You can do this by drawing two bars side by side for each company listed on the horizontal axis of your chart. Such a chart facilitates comparison of the two data sets.
c. In 1982, were these four oil companies able to find oil as quickly as they used it? Explain your answer making reference to your bar charts.

SECTION 2.3

GRAPHICAL METHODS FOR DESCRIBING QUALITATIVE DATA: THE PIE CHART (OPTIONAL)

A second method of describing qualitative data sets—the pie chart—is often used in newspaper and magazine articles to depict budgets and other economic information. A complete circle (the pie) represents the total number of measurements. This is partitioned into a number of slices, with one slice for each category. The size of a slice is proportional to the relative frequency of a particular category. For example, since a complete circle spans 360°, if the relative frequency for a category is .30, the slice assigned to that category is 30% of 360 or (.30)(360) = 108°. See Figure 2.3.

FIGURE 2.3
The Portion of a Pie Chart Corresponding to a Relative Frequency of .30

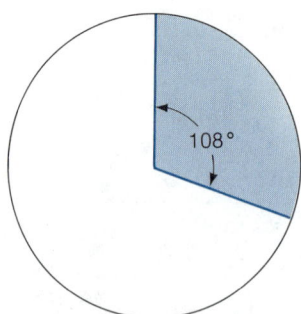

Figure 2.4 shows a pie chart for the customer residence data of Section 2.2. Notice that the sizes of the slices are proportional to the relative frequencies assigned to the four categories. A compass and calculator are needed if the pie chart is to be precisely drawn, which makes it somewhat inconvenient to construct. However, even if we only approximate the size of the wedges, the pie chart provides a useful picture of a qualitative data set.

FIGURE 2.4

Pie Chart for Customer
Residence Survey

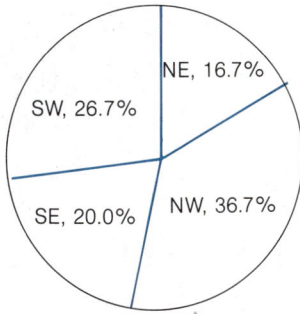

CASE STUDY 2.2

*STATISTICAL ABSTRACT
OF THE UNITED STATES*

Each year the Bureau of the Census (U.S. Department of Commerce) publishes a book entitled *Statistical Abstract of the United States* (hereafter referred to as the *Statistical Abstract*). This book, published yearly since 1878, contains a "summary of statistics on the social, political, and economic organization of the United States. It is designed to serve as a convenient volume for statistical reference and as a guide to other statistical publications and [data] sources" (*Statistical Abstract*, 1981, p. v).

The vast majority of the data included in the *Statistical Abstract* are reported using summary tables such as Table 2.3 (page 28). However, beginning with the 1981 edition, the *Statistical Abstract* began making extensive use of graphical descriptions of data in a section called "Recent Trends." Graphics employed include bar charts, pie charts, and time series graphs (we will discuss the time series graphs in Section 2.7 and again in Chapter 13). For example, the pie charts in Figure 2.5 describe the sources of energy used to produce electricity in the United States in 1970 and 1980. In addition, the effects of the oil crisis and our resulting desire to reduce our dependence on foreign oil are reflected in the decrease in the proportion of total electricity production provided by oil and the increase in the proportion provided by coal. Notice, however, that while oil's proportion has decreased since 1970, the total number of kilowatt-hours of electricity produced from oil has increased since 1970.

FIGURE 2.5

Electric Energy
Production, by Source
of Energy

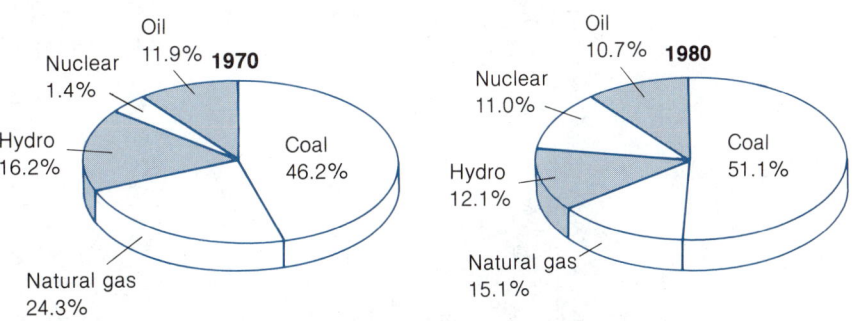

TABLE 2.3 Money in Circulation, By Denomination: 1960 to 1980 (in millions of dollars, as of December 31)

DENOMINATION	1960	1965	1970	1973	1974	1975	1976	1977	1978	1979	1980
Total[a]	32,869	42,056	57,093	72,497	79,743	86,547	93,717	103,811	114,645	125,600	137,244
Coin and small currency	23,521	29,842	39,639	48,288	51,606	54,865	57,645	62,543	66,693	70,693	73,893
Coin	2,427	4,027	6,281	7,759	8,332	8,959	9,483	10,071	10,739	11,658	12,419
$1[b]	1,533	1,908	2,310	2,639	2,720	2,809	2,858	3,038	3,194	3,308	3,499
$2	88	127	136	135	135	135	637	650	661	671	677
$5	2,246	2,618	3,161	3,614	3,718	3,841	3,905	4,190	4,393	4,549	4,635
$10	6,691	7,794	9,170	10,226	10,503	10,777	10,775	11,361	11,661	11,894	11,924
$20	10,536	13,369	18,581	23,915	26,197	28,344	29,987	33,233	36,045	38,613	40,739
Large currency	9,348	12,214	17,454	24,210	28,137	31,681	36,072	41,269	47,952	54,907	63,352
$50	2,815	3,540	4,896	6,514	7,444	8,157	9,026	10,079	11,279	12,585	13,731
$100	5,954	8,135	12,084	17,288	20,298	23,139	26,668	30,818	36,306	41,960	49,264
$500	249	245	215	185	179	175	172	169	167	164	163
$1,000	316	288	252	216	209	204	200	197	194	192	189
$5,000	3	3	3	2	2	2	2	2	2	2	2
$10,000	10	4	4	4	4	4	4	4	4	4	3

[a]Outside Treasury and Federal Reserve banks.
[b]Paper currency only; $1 silver coins reported under coin.
Source: 1960–1973, Board of Governors of the Federal Reserve System, *Federal Reserve Bulletin*, monthly; thereafter, U.S. Department of the Treasury, *Monthly Statement of United States Currency and Coin*, Form 1028.

| | | | | | | | | | | | |

EXERCISES 2.14–2.23

LEARNING THE MECHANICS

2.14 The disk-memory market for personal computers is booming. Demand continues to outpace whatever manufacturers can supply. Even though more than 80 companies build disk memories, the data in the table indicate that the market is dominated by a handful of firms. Construct a pie chart to describe the data.

FLOPPY DISK DRIVES 1983 TOTAL SALES: 4.7 MILLION UNITS	
Company	Market Share
Tandon	53%
Shugart	18%
Alps Electric	16%
Others	13%

Source: *Business Week*, Feb. 6, 1984, p. 69.

2.15 Business is booming for the National Basketball Association (NBA). Gross revenues grew from $108 million for the 1980–1981 season to $192 million for the 1984–1985 season.

Pro Basketball's 1984–1985 Revenue Sources

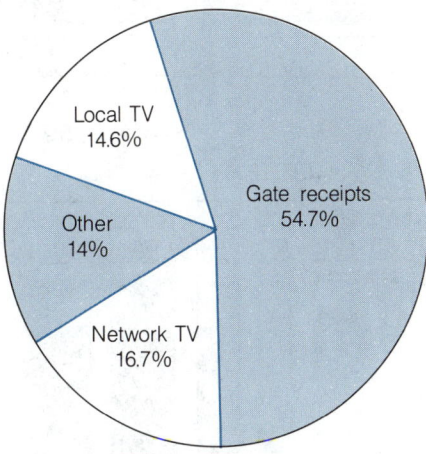

Data: National Basketball Association.
Source: *Business Week*, Oct. 28, 1985, p. 74.

a. Describe the information being conveyed by the pie chart.
b. What proportion of the NBA's gross revenues for the 1984–1985 season were derived from network television?
c. How much money did the NBA receive from gate receipts?

APPLYING THE CONCEPTS

2.16 When describing monetary data sets, newspapers and magazines sometimes use a *dollar chart* rather than a pie chart. The dollar charts shown here accompanied a series of articles in the *Minneapolis Star and Tribune* devoted to describing and analyzing President Reagan's proposed federal budget for 1984.

a. Convert the two dollar charts to pie charts.

b. Social insurance tax totals $242.9 billion and is less than the outlay for income security (Social Security, Medicare, etc.), which totals $282.4 billion. Is the proportion of total receipts due to the social insurance tax also less than the proportion of outlays due to income security? Explain.

Dollar Charts to Describe the 1984 Federal Budget. Source: "The 1984 Federal Budget" (1983).

2.17 From 1950 to 1980 the number of people living in suburbs in the United States grew by 66.3 million while the number living in central cities grew by only 18 million. As a result of its population growth, suburbia has become a center for jobs, shopping, and entertainment. The business ventures that once existed only in cities, now thrive in the suburbs. The table describes the growth of suburbia.

	U.S. POPULATION (MILLIONS)		
	1960	1970	1980
Suburbs	54.9	75.6	101.5
Cities	58.0	63.8	67.9
Other	66.4	63.8	57.1

Source: *U.S. News & World Report*, Mar. 12, 1984, p. 59. Copyright 1984, U.S. News & World Report.

a. For each year listed in the table, construct a pie chart to describe the residences of the U.S. population.

b. Using the information displayed in your pie charts, describe the relative shift in the population between cities and suburbs over the period 1960–1980.

2.18 Although Wendy's and Burger King have increased their market share more than McDonald's in recent years, McDonald's still dominates the fast-food hamburger market.

1983 Market Share (Percentage Point Change in Market Share Since 1981).
Source: *Business Week*, Jan. 30, 1984, p. 46.

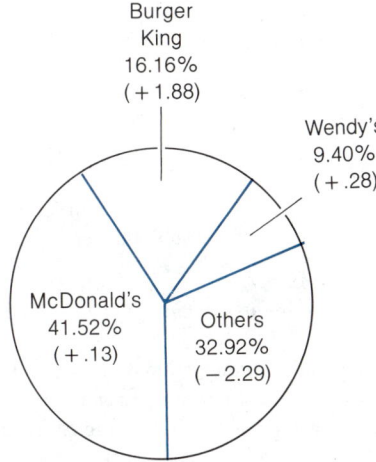

a. The market shares portrayed in the pie chart were based on 1983 sales. McDonald's 1983 sales were $8.16 billion. Find Burger King's and Wendy's 1983 sales.

b. Construct a pie chart that displays the market shares for the 1981 fast-food hamburger market.

2.19 In recent years, big-name chain stores such as J. C. Penney, Sears, Montgomery Ward, Target, and others have entered the $13-billion-a-year jewelry market. In fact, J. C. Penney has become the fourth-largest retail jewelry merchant in the United States, behind Zale's, Gordon Jewelry, and Best Products. Relying on heavy advertising and deep discounting, these retailers have brought mass-merchandising techniques to the jewelry business. As a result, Americans are changing the way they shop for jewelry, as illustrated by the accompanying chart.

Where Consumers Buy Jewelry.
Source: *Business Week,*
Feb. 6, 1984, p. 56.

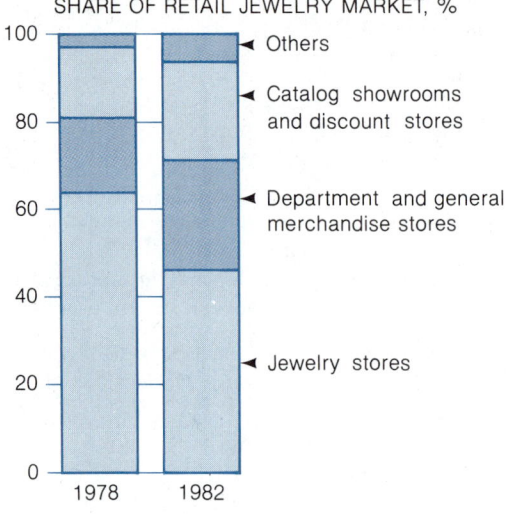

SHARE OF RETAIL JEWELRY MARKET, %

◄ Others

◄ Catalog showrooms and discount stores

◄ Department and general merchandise stores

◄ Jewelry stores

1978   1982

Data: Intergold Corp. N.W. Ayer, *Accent*

a. Construct a pie chart to describe the retail jewelry market in 1978.
b. Repeat part **a** for the 1982 jewelry market.

2.20 Refer to Exercise 2.19.
a. Construct relative frequency bar charts to characterize the two sets of market-share data.
b. With which graphical presentation do you find it easiest to compare the two market-share data sets: the graph given in Exercise 2.19, the pie charts you constructed in Exercise 2.19, or the bar charts constructed in part **a** of this exercise? Explain your reasoning.

2.21 Consider the pie chart shown on page 33.
a. What is the pie chart attempting to portray?
b. What does the pie chart tell you about the relationship between the cost of the ingredients in a box of cereal and the cost of the cereal's packaging?

The Breakfast Cereal
Dollar: Where It Goes

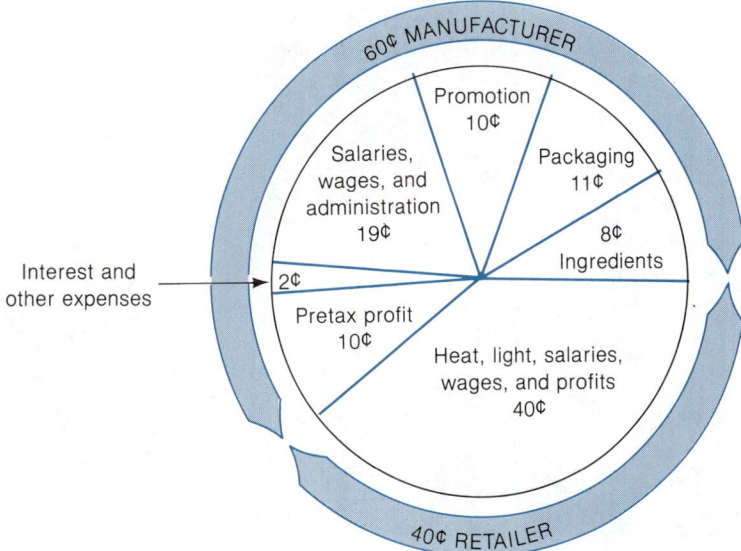

c. According to the pie chart, how much do the ingredients in a $1.75 box of cereal cost the manufacturer?

2.22 In 1985, American women spent over $8 billion for cosmetics and fragrances. The table describes the sales of these products by outlet.

OUTLET	SALES
	(Billions of Dollars)
Food stores	1.13
Discount stores	2.43
Chain drug stores	2.27
Independent drug stores	.73
Department stores	2.92

Data: Compact Disc Group of America.
Source: *Minneapolis Star and Tribune*, Jan. 26, 1986, p. 3F.

a. What percentage of the cosmetics and fragrances market was captured by drug stores in 1985?
b. Construct a pie chart that depicts market share by outlet for the cosmetics and fragrances industry.

2.23 The accompanying pie charts reveal that some dramatic changes occurred in the U.S. economy between 1950 and 1983.

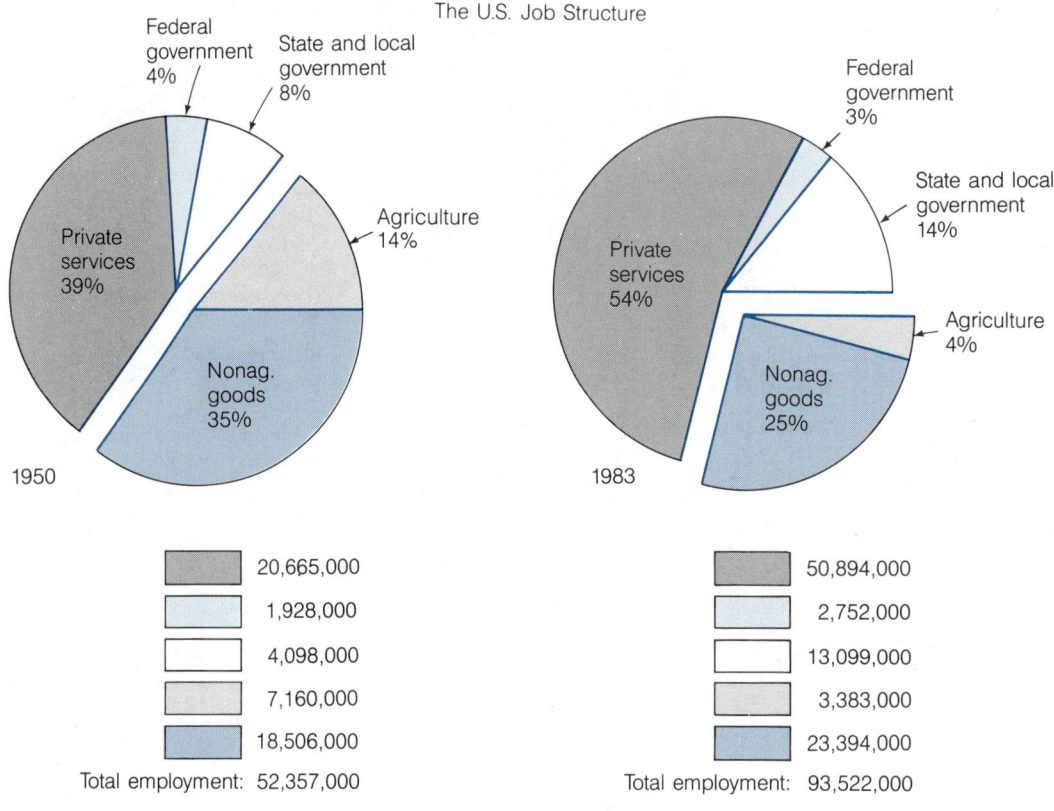

The U.S. Job Structure

Federal government 4%
State and local government 8%
Private services 39%
Agriculture 14%
Nonag. goods 35%

1950

	20,665,000
	1,928,000
	4,098,000
	7,160,000
	18,506,000

Total employment: 52,357,000

Federal government 3%
State and local government 14%
Private services 54%
Agriculture 4%
Nonag. goods 25%

1983

	50,894,000
	2,752,000
	13,099,000
	3,383,000
	23,394,000

Total employment: 93,522,000

Data: U.S. Bureau of Labor Statistics.
Source: *National Forum*, Phi Kappa Phi Journal, Spring 1984, p. 16.

    a. Describe the changes revealed by the pie charts.
    b. What proportion of the jobs in the United States were nonagricultural in 1950? In 1983?
    c. How many people worked for either the federal or state and local government in 1950? In 1983?

S E C T I O N 2.4

GRAPHICAL METHODS FOR DESCRIBING QUANTITATIVE DATA: STEM AND LEAF DISPLAYS

Quantitative data sets are those that consist of numerical measurements. Thus, a quantitative sample is simply a list of numerical values that result from observations taken on some variable x. Most business data are quantitative, so methods for summarizing quantitative data are especially important.

For example, suppose a financial analyst is interested in the amount of resources spent by computer hardware and software companies on research and development (R&D). She samples 50 of these high-technology firms, and calculates the amount each spent last year on R&D as a percentage of their total revenues. The results are given in Table 2.4. As numerical measurements, these percentages represent quantitative data. The analyst's initial objective is to describe these data.

TABLE 2.4 Percentage of Revenues Spent on Research and Development

COMPANY	PERCENTAGE	COMPANY	PERCENTAGE	COMPANY	PERCENTAGE	COMPANY	PERCENTAGE
1	13.5	14	9.5	27	8.2	39	6.5
2	8.4	15	8.1	28	6.9	40	7.5
3	10.5	16	13.5	29	7.2	41	7.1
4	9.0	17	9.9	30	8.2	42	13.2
5	9.2	18	6.9	31	9.6	43	7.7
6	9.7	19	7.5	32	7.2	44	5.9
7	6.6	20	11.1	33	8.8	45	5.2
8	10.6	21	8.2	34	11.3	46	5.6
9	10.1	22	8.0	35	8.5	47	11.7
10	7.1	23	7.7	36	9.4	48	6.0
11	8.0	24	7.4	37	10.5	49	7.8
12	7.9	25	6.5	38	6.9	50	6.5
13	6.8	26	9.5				

A useful graphical description of quantitative data is the **stem and leaf display**. To construct such a display, it is necessary to partition each measurement into two components: a **stem** and a **leaf**. In this example the *stem* is the part of each measurement to the left of the decimal place—the units and tens digits; the **leaf** is the part of each measurement to the right of the decimal place—the tenths digit. The stems and leaves for the R&D percentages 7.4, 10.5, and 13.2 are shown here:

STEM	LEAF		STEM	LEAF		STEM	LEAF
7	4		10	5		13	2

The stem and leaf display for all 50 R&D percentages is shown in Figure 2.6. Note that the leaves corresponding to each stem are arranged in ascending order, and a key is included with the display to specify the units of the leaf (and, by implication, the units of the stem).

FIGURE 2.6
Stem and Leaf Display for Fifty Computer Companies' Research and Development Percentages

STEM	LEAF
5	2 6 9
6	0 5 5 5 6 8 9 9 9
7	1 1 2 2 4 5 5 7 7 8 9
8	0 0 1 2 2 2 4 5 8
9	0 2 4 5 5 6 7 9
10	1 5 5 6
11	1 3 7
12	
13	2 5 5

Key: Leaf units are tenths.

Note that although the stem 12 has no leaves (meaning that none of the 50 observations fell in the range from 12.0 to 12.9), we include the 12 stem in the display so that this fact is visually obvious. Note also that the decimal point is not included in the display. When there is no confusion caused by its omission, we can usually obtain a less cluttered graphical description without it.

Several descriptive facts about these data are easily seen in the stem and leaf display. Most of the sampled computer companies (37 of 50) spent between 6.0% and 9.9% of their revenues on R&D, and 11 of them spent between 7.0% and 7.9%. Relative to the rest of the sampled companies, three spent a high percentage of revenues on R&D—in excess of 13%.

The selection of the stem and leaf that best display a set of data is not always clear-cut, and you may have to try several to obtain the best graphical description.

HOW TO CONSTRUCT A STEM AND LEAF DISPLAY

1. Define the stem and leaf that you will use. Choose the units for the stem so that the number of stems in the display is between 5 and 20.
2. Write the stems in a column arranged with the smallest stem at the top and the largest stem at the bottom. Include all stems in the range of the data, even if there are some stems with no corresponding leaves.
3. If the leaves consist of more than one digit, drop the digits after the first. You may round the numbers to be more precise, but this is not necessary for the graphical description to be useful.
4. Record the leaf for each measurement in the row corresponding to its stem. Order the leaves corresponding to each stem to obtain a more informative display. Omit decimals, and include a key that defines the units of the leaf.

EXAMPLE 2.1

A manufacturer of industrial wheels suspects that profitable orders are being lost because of the long time the firm takes to develop price quotes for potential customers. To investigate this possibility, 50 requests for price quotes were randomly selected from the set of all quotes made last year, and the processing time was determined for each quote. The processing times are displayed in Table 2.5, and each quote was classified according to whether the order was "lost" or not. Construct a stem and leaf display to describe these data. Shade each leaf that corresponds to a lost order, and interpret the display.

SOLUTION

If we use "days" (units and tens digits) as the stem, the tenths and hundredths digits to the right of the decimal would be the leaf. However, the hundredths digit will be dropped to make the display more visually effective. Thus, the first processing time in Table 2.5, 2.36 days, is partitioned as follows:

STEM	LEAF
2	3

TABLE 2.5

Price Quote Processing Times (Days)

REQUEST NUMBER	PROCESSING TIME	LOST?	REQUEST NUMBER	PROCESSING TIME	LOST?
1	2.36	No	26	3.34	No
2	5.73	No	27	6.00	No
3	6.60	No	28	5.92	No
4	10.05	Yes	29	7.28	Yes
5	5.13	No	30	1.25	No
6	1.88	No	31	4.01	No
7	2.52	No	32	7.59	No
8	2.00	No	33	13.42	Yes
9	4.69	No	34	3.24	No
10	1.91	No	35	3.37	No
11	6.75	Yes	36	14.06	Yes
12	3.92	No	37	5.10	No
13	3.46	No	38	6.44	No
14	2.64	No	39	7.76	No
15	3.63	No	40	4.40	No
16	3.44	No	41	5.48	No
17	9.49	Yes	42	7.51	No
18	4.90	No	43	6.18	No
19	7.45	No	44	8.22	Yes
20	20.23	Yes	45	4.37	No
21	3.91	No	46	2.93	No
22	1.70	No	47	9.95	Yes
23	16.29	Yes	48	4.46	No
24	5.52	No	49	14.32	Yes
25	1.44	No	50	9.01	No

The entire stem and leaf display is shown in Figure 2.7 (page 38). This display uses 20 stems, a rather large number for only 50 measurements. However, inspection of Figure 2.7 reveals that only 14 stems have corresponding leaves. The presence of one or more extreme observations, such as the one at 20.2 and the cluster from 13.4 to 16.2, stretches out the display, and the result is more stems than we would normally use for a more compact set of data. Extreme observations that are detached from the remainder of the data are called **outliers**, and they usually receive special attention in statistical analyses. Although outliers may represent legitimate measurements, they are more often "mistakes": incorrectly recorded observations, miscoded input into a computer, or measurements from a different population than the population from which the rest of the sample was selected. Stem and leaf displays are useful for identifying outliers.

When the leaves in Figure 2.7 are shaded to identify lost orders, it is apparent that the processing time and the success of the quote are related. Most of the lost orders correspond to long processing times, and one plausible explanation is that they correspond to a different population—one associated with lost orders. Management's objective will be to reduce substantially the size of this population by establishing processing time limits.

FIGURE 2.7
Stem and Leaf Display for
Processing Time Data

STEM	LEAF
1	2 4 7 8 9
2	0 3 5 6 9
3	2 3 3 4 4 6 9 9
4	0 3 4 4 6 9
5	1 1 4 5 7 9
6	0 1 4 6 7
7	2 4 5 5 7
8	2
9	0 4 9
10	0
11	
12	
13	4
14	0 3
15	
16	2
17	
18	
19	
20	2

EXAMPLE 2.2

Refer to Example 2.1 and the data on price quote processing time given in Table 2.5. Use a statistical computer software package to create a stem and leaf display for the price quote times.

SOLUTION

Most statistical software packages have stem and leaf options available. We used SPSS/PC® on a microcomputer to generate the stem and leaf display shown in Figure 2.8. The stem is defined as the number of days (as in Figure 2.7), but the leaf is now the tenths digit *rounded* from the hundredths digit. Probably the most significant difference between the stem and leaf displays in Figures 2.7 and 2.8 is that the SPSS/PC display uses only the even-numbered days as stems, so that odd-numbered days are effectively combined with the even. Thus, the stem of "2" is used to represent both 2- and 3-day times. You can ascertain the difference between measurements corresponding to the two days by noticing that the leaves are arranged in ascending order. Thus, the first five leaves on the 2 stem, 04569, correspond to observations of 2 days, while the last eight leaves, 23445699, correspond to observations of 3 days. In many cases, the measurements can be distinguished in this manner. However, when only a few leaves correspond to a stem, it may not be possible to determine the exact value(s) of the measurement(s). For example, we cannot tell, without looking at the original data, whether the single measurement corresponding to the 20 stem is 20.2 or 21.2.

FIGURE 2.8
SPSS/PC Stem and Leaf
Printout for Processing
Time Data

```
Stem-and-leaf display for variable .. TIME

 0 . 34799
 2 . 0456923445699
 4 . 044579115579
 6 . 0246834568
 8 . 2059
10 . 1
12 . 4
14 . 13
16 . 3
18 .
20 . 2
```

The SPSS/PC display in Figure 2.8 clearly indicates that several measurements are detached from the main body of data. We have already seen in Example 2.1 that most of these observations correspond to "lost" business.   ■

Stem and leaf displays possess a number of advantages over other forms of graphical description. First and foremost, each of the original measurements is usually "visible" in the display. The stem and leaf display arranges the data in ascending order, which enables us to label particular classes of observations, as we did with the "lost orders" in Example 2.1. Finally, for data sets that are not too large, the construction of stem and leaf displays is relatively simple.

Stem and leaf displays for large data sets can be generated by computer. Alternatively, a **relative frequency histogram** can be used to obtain a graphical description of large data sets by grouping the measurements before creating the display. This method is discussed in Section 2.5.

| | | | | | | | | | | | |

EXERCISES 2.24–2.32

LEARNING THE MECHANICS

[*Note: Starred (*) exercises require the use of a computer.*]

2.24 Construct a stem and leaf display for the following measurements:

2.6	3.3	2.4	1.1	.8	3.5	3.9	1.6	2.8	2.6
3.4	4.1	2.0	1.7	2.9	1.9	2.9	2.5	4.5	5.0

2.25 The SAS System was used to generate the following stem and leaf display:

```
Stem Leaf                      #
   5 1                         1
   4 457                       3
   3 00036                     5
   2 1134599                   7
   1 2248                      4
   0 012                       3
   ----+----+----+----+
Multiply Stem.Leaf by 10**+01
```

Note that SAS arranges the stems in descending order. Also, the instruction to "Multiply Stem.Leaf by 10∗∗+.01" indicates that each number should be multiplied by 10. For example, the top number in the display, 5.1, represents an observation of 10(5.1) = 51.

a. How many observations were in the original data set?

b. In the first row of the stem and leaf display, identify the stem, the leaves, and the numbers in the original data set represented by this stem and its leaves.

2.26 Construct a stem and leaf display for the following measurements:

.02   .08   .14   .32   .27   .08   .01   .11   .20   .06
.01   .03   .12   .22   .42   .33   .18   .09   .02   .07

2.27 Construct a stem and leaf display for the following measurements using the leftmost two digits of each number as the stem:

1,050   1,530   2,111   1,786   1,633   1,819   1,899
1,763   1,312   1,400   1,219   1,101   1,375   1,701
1,301   1,256   1,492   1,616   1,907   1,777   1,339
1,781   1,662   1,831   1,790   1,788   1,769   1,344

APPLYING THE CONCEPTS

2.28 Production processes may be classified as *make-to-stock processes* or *make-to-order processes*. Make-to-stock processes are designed to produce a standardized product that can be sold to customers from the firm's inventory. Make-to-order processes are designed to produce products according to customer specifications. The McDonald's and Burger King fast-food chains are classic examples of these two types of processes. McDonald's produces and stocks standardized hamburgers; Burger King—whose slogan is "Have It Your Way"—makes hamburgers according to the ingredients specified by the customer (Schroeder, 1985). In general, performance of make-to-order processes is measured by delivery time—the time from receipt of an order until the product is delivered to the customer. The following data set is a sample of delivery times (in days) for a particular make-to-order firm last year. The delivery times marked by an asterisk are associated with customers who subsequently placed additional orders with the firm.

50*   64*   56*   43*   64*
82*   65*   49*   32*   63*
44*   71   54*   51*   102
49*   73*   50*   39*   86
33*   95   59*   51*   68

a. Construct a stem and leaf display for the data.

b. Circle the individual leaves of your stem and leaf display that are associated with customers who did not place a subsequent order.

c. Concerned that they are losing potential repeat customers due to long delivery times, management would like to establish a guideline for the maximum tolerable delivery time. Using your stem and leaf display of part **b**, suggest a guideline. Explain your reasoning.

2.29 When two firms announce plans to merge, it frequently happens that within a few weeks one firm or the other becomes dissatisfied with the consequences of merging and the merger is canceled. Dodd (1980) reported that of 151 merger announcements that he identified, 80 were canceled. Thus, at the time a proposed merger is announced, there exists a great amount of uncertainty concerning whether the merger will take place. This uncertainty may persist for a considerable period of time and it may be many months after the announcement that the merger actually occurs. In her study of 38 mergers that were consummated, Eger (1982) reported the number of *trading days* (days the New York Stock Exchange is open for business) between the merger announcement (defined as the first mention of the potential merger in the *Wall Street Journal*) and the effective date of the merger. These data are listed below:

74	45	55	74	64	97	65	82	92	116
140	62	92	78	45	93	94	57	123	128
92	73	173	116	35	124	64	84	255	277
123	80	143	112	76	214	64	86		

a. Construct a stem and leaf display for these data.

b. Summarize the information reflected in your stem and leaf display concerning the number of trading days between announcement and the effective merger date for this sample of mergers.

*c. Use a statistical software package to generate a stem and leaf display for these data.

2.30 In a manufacturing plant a *work center* is a specific production facility that consists of one or more people and/or machines and is treated as one unit for the purposes of capacity requirements planning and job scheduling. If jobs arrive at a particular work center at a faster rate than they depart, the work center impedes the overall production process and is referred to as a *bottleneck* (Fogarty and Hoffmann, 1983). The data in the table were collected by an operations manager for use in investigating a potential bottleneck work center.

NUMBER OF ITEMS ARRIVING AT WORK CENTER PER HOUR			NUMBER OF ITEMS DEPARTING WORK CENTER PER HOUR		
155	115	156	156	109	127
150	159	163	148	135	119
172	143	159	140	127	115
166	148	175	122	99	106
151	161	138	171	123	135
148	129	135	125	107	152
140	152	139	111	137	161

a. Construct a stem and leaf display for the two sets of measurements.

b. Do your stem and leaf displays suggest that the work center may be a bottleneck? Explain.

2.31 Typically, the more attractive a corporate common stock is to an investor, the higher the stock's price–earnings ratio. For example, if investors expect the stock's future earnings per share to increase, the price of the stock will be bid up and a high price–earnings ratio will result. Thus, the level of a stock's price–earnings ratio is a function of both the current financial performance of the firm and an investor's expectation of future performance (Spiro, 1982). The table contains the 1986 price–earnings (P/E) ratios for samples of firms from the electronics industry and the auto parts industry.

AUTO PARTS		ELECTRONICS	
Firm	P/E Ratio	Firm	P/E Ratio
Lear Siegler	9	AMP Inc.	22
Genuine Parts	15	Raytheon	12
Federal–Mogul	12	Intel	85
PPG Industries	13	Avnet	24
Borg–Warner	13	Perkin Elmer	14
Hoover Universal	12	TRW Inc.	15
Libbey–Owens–Ford	8	Motorola	39
Dana	8	Hewlett–Packard	20
Champion Spark Plug	15	Honeywell	12
Dayco	10	American District	27
Sheller–Globe	7	Corning Glass Works	27
Arvin Industries	10	EG&G	19
Allen Group	13	Varian Associates	14
Eaton	9	M/A-Com, Inc.	14
Cummins Engine	10	Harris Corp.	19
Barnes Group	11	Texas Instruments	10
Echlin	14	Intl Tel & Tel	12
Johnson Controls	9	North American Philips	11
Rockwell Int.	9	GTE	8
Snap-on-Tools	14	Tektronix	16

Source: Stock reports (OTC, NYSE, American), Standard and Poor's 1986.

a. Construct a stem and leaf display for each of these data sets.
b. What do your stem and leaf displays suggest about the level of the P/E ratios of firms in the electronics industry as compared to firms in the auto parts industry? Explain.

2.32 The accompanying table contains the top salary offer received by each of a sample of 30 MBA students who graduated from the University of Minnesota in 1985 and had 1 year or less of work experience prior to entering the MBA program. Salaries offered to students who majored in management information systems (MIS) are indicated with an asterisk; offers made to students who majored in marketing are indicated with two asterisks.

SALARY OFFERS TO 1985 MBA GRADUATES				
$31,200*	$27,000	$32,810	$35,800*	$29,999
27,100	23,500	28,100**	26,900	30,590
28,000	24,000**	27,800	25,690	29,200
26,350	30,400	26,725**	30,850	26,666**
24,000	38,500	31,000	33,000	29,000**
26,500**	26,350	32,500*	27,000*	33,333*

Source: Placement Office, School of Management, University of Minnesota.

a. Construct a stem and leaf display for the data.

b. Circle the individual leaves of your stem and leaf display that are associated with MIS graduates. Draw boxes around the individual leaves associated with marketing graduates. What does the pattern of circles and boxes suggest about the relative magnitudes of the starting salaries in MIS and marketing? Explain.

*c. Use a statistical software package to generate a stem and leaf display for these data.

SECTION 2.5

GRAPHICAL METHODS FOR DESCRIBING QUANTITATIVE DATA: HISTOGRAMS

Relative Frequency Histogram like Sideways stem/Leaf
—Intervals instead of Numbers

The computer companies' percentages of revenues spent on research and development (R&D), first given in Table 2.4, are graphically described in Figure 2.9 using a **relative frequency histogram**. The histogram is similar to what would be obtained if the stem and leaf display in Figure 2.6 were rotated counterclockwise 90°. The major difference is that the stems have been converted to **measurement classes**—intervals along the horizontal axis. Measurement classes differ from stems in the respect that stems are actual digits of the measurements, whereas **measurement classes are intervals into which measurements fall**. The class intervals are constructed so that they are contiguous and of equal width, and the endpoints of the classes are selected so that no measurement falls on the boundary between two classes. The leaves of the stem and leaf display are converted to rectangles with the height of a given rectangle equal to the proportion of measurements falling in the corresponding measurement class.

The information provided by the relative frequency histogram is similar to that conveyed by the stem and leaf display. Although the values of the individual measurements cannot be seen in the histogram, we are able to see the proportion of measurements in a particular class by reading the height of its rectangle on the vertical axis. Also, we can more easily control the number of classes in the histogram, which gives us more control over the quality of the graphical description. The histogram in Figure 2.9 makes it clear that all measurements fall between 5.15% and 13.95%, and that the bulk of the companies spend between 6.25% and 9.55% of their revenues on R&D.

FIGURE 2.9
Relative Frequency Histogram for the 50 Computer Companies' R&D Percentages

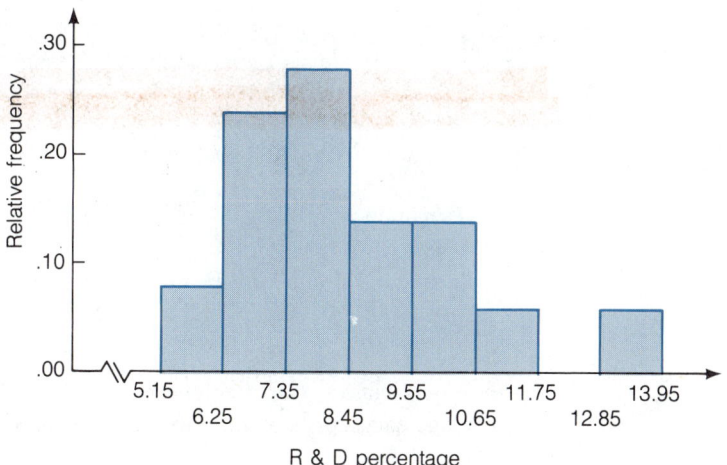

To construct the relative frequency histogram for the R&D percentages, we first choose the **class interval width** and then define the measurement classes. We usually choose between 5 and 20 measurement classes for histograms, the same general rule we followed for choosing the number of stems in a stem and leaf display. Through trial and error, we selected eight measurement classes for the R&D percentages. Once the number of classes is determined, the class width is calculated by

$$\frac{\text{Largest measurement} - \text{Smallest measurement}}{\text{Number of intervals}} = \frac{13.5 - 5.2}{8} = 1.04$$

We round upward to be certain of including all observations:

Class width = 1.1

Next, we determine the lower boundary of the first class by selecting the value 5.15, just .05 below the smallest measurement, 5.2. We use one additional decimal place for the boundaries so that no measurement falls on a class boundary. We then add the class width 1.1 to this boundary to find the upper boundary of the first class, and the lower boundary of the second class, $5.15 + 1.1 = 6.25$. The boundary between the second and third classes is obtained by adding 1.1 again, with the result 7.35. This process is repeated a total of eight times, to generate eight intervals, with the last boundary at 13.95.

The next step is to calculate the **class frequencies** and the **class relative frequencies**. These quantities are defined as follows:

DEFINITION 2.5

The **class frequency** for a given class is equal to the total number of measurements that fall in that class. The class frequency for class i is denoted by the symbol f_i.

DEFINITION 2.6

The **class relative frequency** for a given class is equal to the *proportion* of the total number of measurements that fall in that class. The formula for calculating the relative frequency for class i is

$$\text{Relative frequency for class } i = \frac{f_i}{n}$$

where n is the total number of measurements.

The measurement classes, class frequencies, and class relative frequencies for the R&D percentage data are shown in Table 2.6.

TABLE 2.6

Measurement Classes, Frequencies, and Relative Frequencies for the R&D Percentage Data

CLASS	MEASUREMENT CLASS	CLASS FREQUENCY	CLASS RELATIVE FREQUENCY
1	5.15– 6.25	4	$\frac{4}{50}$ = .08
2	6.25– 7.35	12	$\frac{12}{50}$ = .24
3	7.35– 8.45	14	$\frac{14}{50}$ = .28
4	8.45– 9.55	7	$\frac{7}{50}$ = .14
5	9.55–10.65	7	$\frac{7}{50}$ = .14
6	10.65–11.75	3	$\frac{3}{50}$ = .06
7	11.75–12.85	0	$\frac{0}{50}$ = .00
8	12.85–13.95	3	$\frac{3}{50}$ = .06
			1.00

The final step in the construction of a histogram is to plot the measurement classes on a horizontal axis and the frequency (or relative frequency) of each class on a vertical axis. The frequency (or relative frequency) is plotted as a rectangle with a base width equal to that of the measurement class and a height equal to the frequency (or relative frequency). The result is the relative frequency histogram shown in Figure 2.9.

The steps for constructing histograms for quantitative data sets are summarized in the box.

Histogram is a bar chart using intervals instead of numbers.

HOW TO CONSTRUCT A HISTOGRAM

1. Arrange the data in increasing order, from the smallest to the largest measurement.
2. Divide the interval from the smallest to the largest measurement into between 5 and 20 equal subintervals, making sure that:
   a. Each measurement falls into one and only one measurement class.
   b. No measurement falls on a measurement class boundary.
   Use a small number of measurement classes if you have a small amount of data; use a larger number of classes for a larger amount of data.
3. Compute the frequency (or relative frequency) of measurements in each measurement class.
4. Using a vertical axis of about three-fourths the length of the horizontal axis, plot each frequency (or relative frequency) as a rectangle over the corresponding measurement class.

By looking at a histogram (say, the relative frequency histogram in Figure 2.9), you can see two important facts. First, note the total area under the histogram, and then note the proportion of the total area that falls over a particular interval of the horizontal axis. You will see that the proportion of the total area that falls above an interval is equal to the relative frequency of the measurements that fall in the interval.* For example, the relative frequency for the class interval 5.15– 6.25 is .08. Consequently, the rectangle above that interval contains 8% of the total area under the histogram.

Second, you can imagine the appearance of the relative frequency histogram for a very large set of data (say, a population). As the number of measurements in a data set is increased, you can obtain a better description of the data by decreasing the width of the class intervals. When the class intervals become small enough, a relative frequency histogram will (for all practical purposes) appear as a smooth curve (see Figure 2.10).

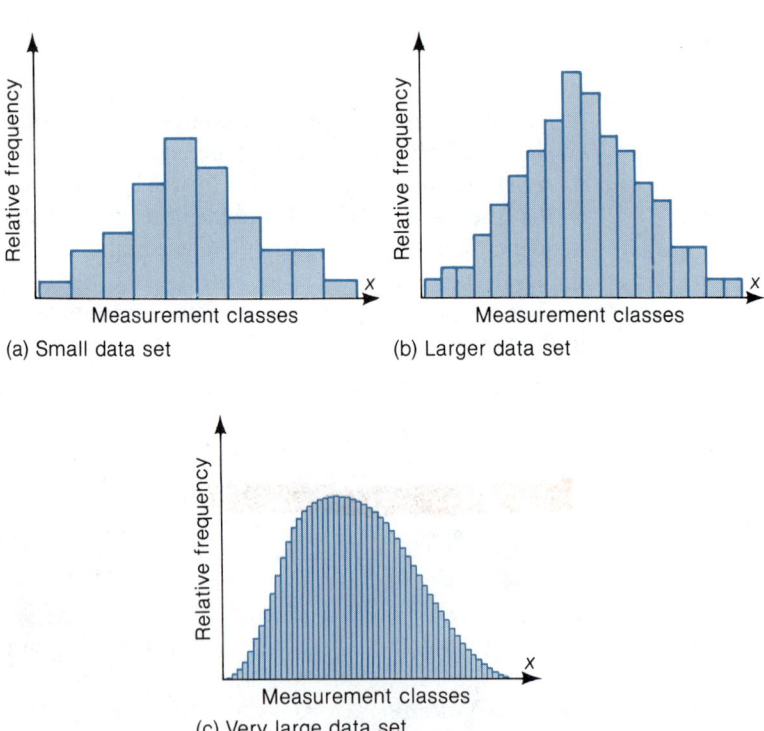

(a) Small data set

(b) Larger data set

(c) Very large data set

*Some histograms are constructed with all class intervals of equal width except the first and last, which are open-ended. The proportionality between area and relative frequency will not hold for such histograms. We will restrict our attention to histograms that have equal-sized class intervals, because later we will want to establish a correspondence between relative frequency histograms and probability distributions.

EXAMPLE 2.3

Refer to Examples 2.1 and 2.2, in which 50 processing times for price quotes were analyzed by a manufacturer of industrial wheels. The data are given in Table 2.5 (page 37). Construct a frequency histogram for the data.

SOLUTION

There are 50 measurements, and we choose to use 10 classes. We emphasize that the final selection of the number of classes is based on trial and error, and the exact number of classes is, therefore, judgmental. Using 10 classes, we calculate

$$\text{Class width} = \frac{\text{Largest measurement} - \text{Smallest measurement}}{\text{Number of classes}}$$

$$= \frac{20.23 - 1.25}{10} = \frac{18.98}{10} = 1.898, \quad \text{or } 1.9$$

Since the times are recorded to the nearest hundredth, we subtract .005 from the smallest value, 1.25, and establish the first class boundary at 1.245. The class width of 1.9 is then repeatedly added to this lower boundary to obtain each of the subsequent class boundaries. The measurement classes and class frequencies are shown in Table 2.7. Note that we do not need to calculate the relative frequencies, because we are creating a *frequency* histogram. Nevertheless, we include the relative frequencies in Table 2.7 for illustration.

TABLE 2.7

Measurement Classes and Frequencies for the Processing Time Data of Table 2.5

CLASS	MEASUREMENT CLASS	CLASS FREQUENCY	CLASS RELATIVE FREQUENCY	
1	1.245– 3.145	10	$\frac{10}{50} =$	.20
2	3.145– 5.045	14	$\frac{14}{50} =$	.28
3	5.045– 6.945	11	$\frac{11}{50} =$	.22
4	6.945– 8.845	6	$\frac{6}{50} =$	.12
5	8.845–10.745	4	$\frac{4}{50} =$	.08
6	10.745–12.645	0	$\frac{0}{50} =$	.00
7	12.645–14.545	3	$\frac{3}{50} =$	.06
8	14.545–16.445	1	$\frac{1}{50} =$	.02
9	16.445–18.345	0	$\frac{0}{50} =$	.00
10	18.345–20.245	1	$\frac{1}{50} =$	.02
				1.00

We now plot the frequencies to obtain a frequency histogram for the processing time data, as shown in Figure 2.11. Note that most of the measurements are bunched at the lower end of the display, while only a few measurements exceed 12.645 (all of them "lost orders"; see Table 2.5). We have shaded that portion of the frequency histogram corresponding to "lost orders."

FIGURE 2.11

Frequency Histogram for the Quote Processing Time Data

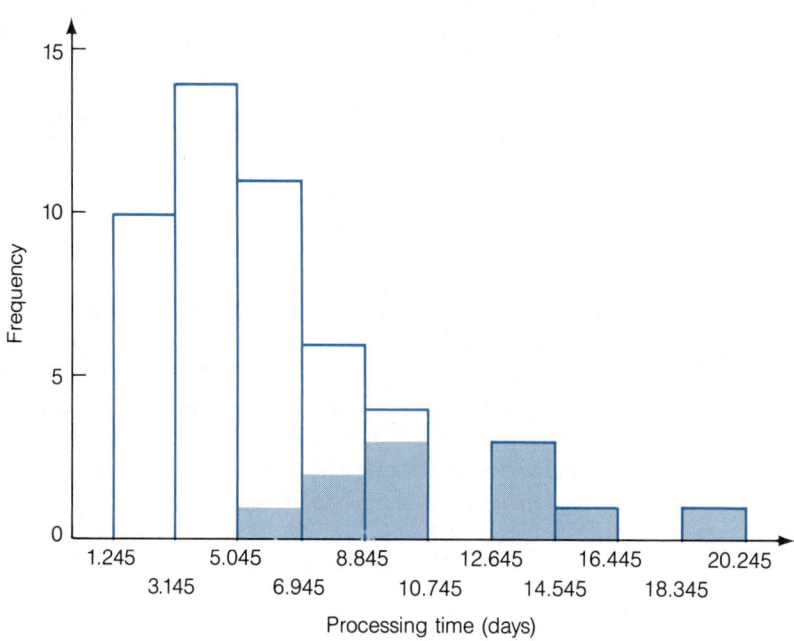

EXAMPLE 2.4

Refer to Example 2.3. Use a statistical software package to create a relative frequency histogram for the price quote processing time data.

SOLUTION

We used the SAS System on a microcomputer to generate the relative frequency histogram in Figure 2.12. Note that six measurement classes were formed by the SAS program, illustrating the subjective aspect of this selection. Note, too, that the SAS program labels the vertical axis "Percentage" rather than "Relative frequency." Perhaps the most significant difference between Figures 2.11 and 2.12 is that the measurement classes formed by the SAS program are identified by their midpoints rather than their endpoints. Thus, the first interval has a midpoint of 3, the second has a midpoint of 6, and so forth. The corresponding class intervals are therefore 1.5 to 4.5, 4.5 to 7.5, and so forth.

Compare the computer-generated stem and leaf display (Figure 2.8) to the computer-generated histogram for these data (Figure 2.12). Which do you prefer as a graphical description of the data?

FIGURE 2.12 SAS Relative Frequency Histogram for the Processing Time Data of Table 2.5

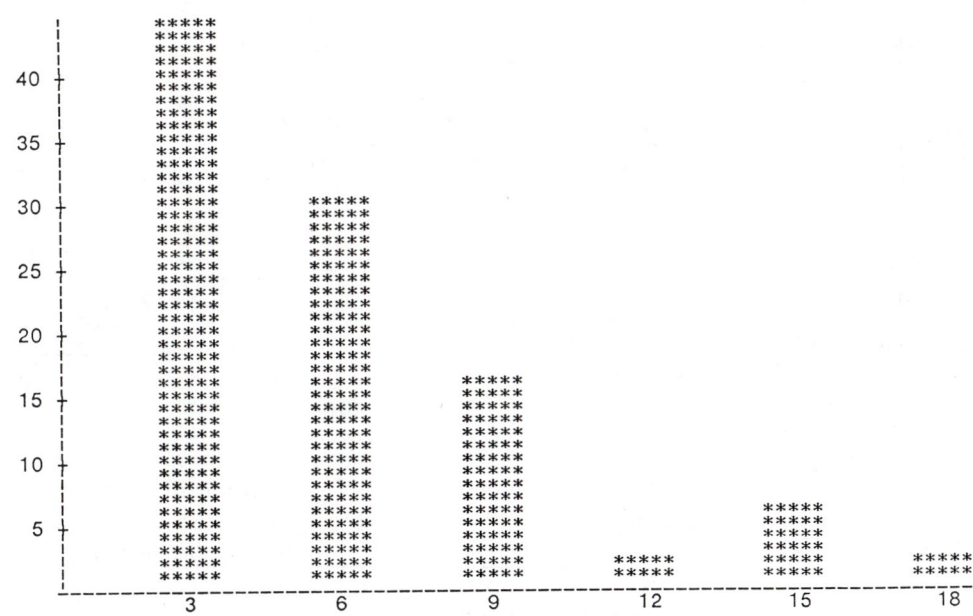

CASE STUDY 2.3

APPRAISING THE MARKET VALUE OF AN ASSET

The *market value* of an asset is the price negotiated by a willing buyer and a willing seller of the asset, each acting rationally in his or her own self-interest. The *book value* of an asset is the value of the asset as shown in its owner's accounting records. Generally speaking, it is the amount the owner paid for the asset, less any depreciation expense (Davidson, Stickney, and Weil, 1979).

Robert R. Sterling and Raymond Radosevich (1969) examined the hypothesis that accountants generally agree on the book value of a depreciable asset, but do not agree on its current market value. A questionnaire was prepared in which the installment purchase of a depreciable asset was described and the respondent was asked to determine the market value of the asset. The questionnaire also contained a series of questions relating to the book value of the asset. These questions enabled Sterling and Radosevich to calculate a book value for the asset for each of the respondents. The questionnaire was mailed to 500 randomly selected Certified Public Accountants (CPAs) in the United States; 114 and 99 usable book value and market value responses, respectively, were returned.

The frequency distributions of book values and market values obtained from the returned questionnaires appear in Figure 2.13 (page 50). In both histograms, the intervals from $150 to $200 and $600 to $650 include all responses less than $200 and greater than $600, respectively. The histograms suggest disagreement among the CPAs as to both the book value and the market value of the asset. Thus, Sterling and Radosevich rejected the hypothesis that accountants tend to agree on book values and to disagree on market values.

FIGURE 2.13   Frequency Histograms for Book and Market Values as Assessed by CPAs

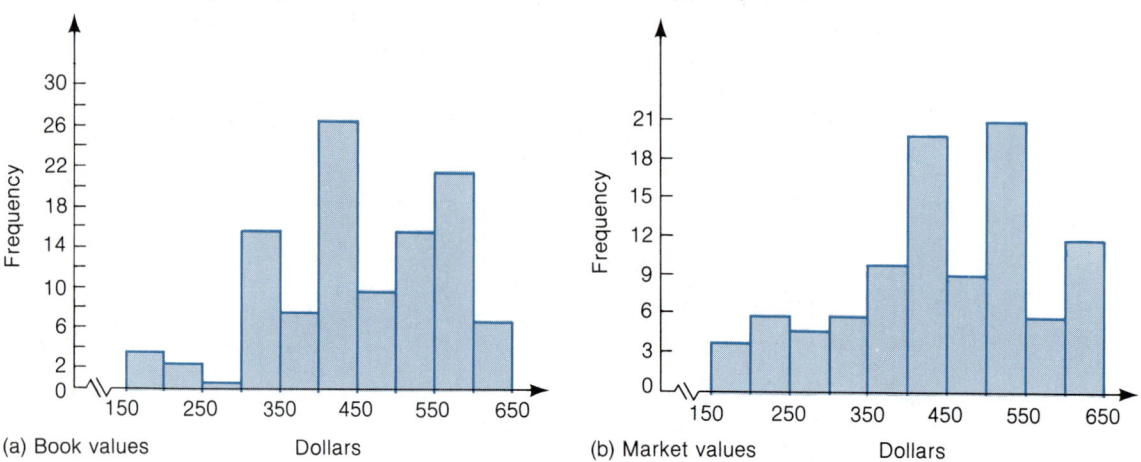

(a) Book values   Dollars

(b) Market values   Dollars

Note that decisions based on a visual comparison of histograms (among other graphical descriptions) are risky because they are subject to an unknown probability of error. For example, we might wonder whether disagreement among the CPAs really exists or whether the difference we see in the histogram is due to random variation that would be present from sample to sample. We will begin to answer questions of this type in Chapter 7.

EXERCISES 2.33–2.43

LEARNING THE MECHANICS

[*Note: Starred exercises (*) require the use of a computer.*]

2.33 One of the three basic operating units in the gas industry is the producer who explores and drills for the gas. Recently, producers have experienced sharp cost increases. To offset these increases, the producers charge higher prices at the wellhead when drilling. The following is a sample of wellhead prices (in cents per thousand cubic feet) for 25 gas producers last year:

47	53	58	52	55
56	57	54	49	57
62	51	54	59	61
60	62	53	57	63
58	59	60	58	56

a. Construct a relative frequency histogram for the data. Use six class intervals, each with a width of 3¢. Begin the first interval at 46.5.

b. What proportion of wellhead prices exceed 55¢? Now examine the total area under the relative frequency histogram. What proportion of this area lies to the right of 55¢?

c. Explain the correspondence between areas under the relative frequency histogram and relative frequencies.

2.34 The statistical package Minitab was used to generate the following histogram:

MIDDLE OF INTERVAL	NUMBER OF OBSERVATIONS	
20	1	*
22	3	***
24	2	**
26	3	***
28	4	****
30	7	*******
32	11	***********
34	6	******
36	2	**
38	3	***
40	3	***
42	2	**
44	1	*
46	1	*

a. Is this a frequency histogram or relative frequency histogram? Explain.
b. How many measurement classes were used in the construction of this histogram?
c. How many measurements are there in the data set described by this histogram?

2.35 The annual incomes for 30 randomly selected secretaries are recorded below (in thousands of dollars):

9.8   10.1   13.2   15.4   10.9   13.6   11.2   9.7   12.6   10.3
11.0   17.6   8.9   9.4   8.9   10.6   10.4   10.9   10.5   10.2
11.8   8.1   12.1   12.0   13.2   12.2   12.4   14.5   10.5   9.7

Construct a relative frequency histogram for the data. Use a class interval width equal to 1.2, and use 8.05 as the lower boundary of the first class. The intervals will then be 8.05–9.25, 9.25–10.45, etc.

2.36 A large company is interested in determining the length of service of its employees. Twenty-five employees are randomly chosen, and the length of service (in years) is recorded for each. The data are as follows:

3.1	1.8	6.4	10.2	11.2
15.6	11.6	6.8	1.5	2.9
3.4	7.2	.5	7.7	8.4
.7	3.9	8.2	8.0	5.5
10.3	12.1	3.9	.9	4.3

Construct a relative frequency histogram for the data.

2.37 Twenty-four economists were asked to project the percentage change in the Consumer Price Index between now (September) and January 1 next year. The following are their projections:

+2%	−5%	+7%	+4%	+4%	+0%	+1%	+3%
−1%	−1%	−2%	−2%	+4%	−1%	+5%	+6%
+6%	+5%	+2%	+6%	+8%	+12%	+3%	−4%

a. Construct a relative frequency histogram for the data.
b. How might you summarize these 24 predictions without using a graph or a table?

APPLYING THE CONCEPTS

2.38 Considering the climate, is it economically feasible to start an orange grove in northern Florida? If the temperature falls below 32°F, oil-burning smudge pots must be lit to keep the orange trees from freezing. Suppose a prospective grower decides that a grove would be economically feasible if the pots have to be lit an average of 15 days or less each year. The grower selects 20 years since 1900 at random and obtains the total number of days per year that the temperature fell below 32°F:

20	15	13	25	12	13	18	14	6	13
9	14	16	10	28	16	17	12	11	15

a. Construct a relative frequency histogram for the data.

b. Based on the sample data, estimate the proportion of years in which the pots have to be lit 15 days or less. [*Note:* We will show you how to evaluate the reliability of this estimate in Chapter 8.]

2.39 Several different types of bonds are issued by businesses, two of which are described here. A *mortgage bond* is a promissory note in which the issuing company pledges certain real assets as security for the note in exchange for a specified amount of money. A *debenture* is an unsecured promissory note (Brigham, 1982). The accompanying table contains the "asking prices" (the price at which a bond is offered for sale) on December 31, 1985, for a sample of 30 publicly traded mortgage bonds issued by utility companies.

UTILITY COMPANY	ASKING PRICE	UTILITY COMPANY	ASKING PRICE
Metropolitan Edison Co.	$ 94	Duquesne Lt Co.	$ $68\frac{3}{8}$
Pennsylvania Pwr & Lt Co.	$86\frac{1}{2}$	Arkansas Pwr & Lt Co.	$78\frac{7}{8}$
Southwestern Elec Pwr Co.	$74\frac{1}{2}$	Washington Gas Lt Co.	$98\frac{7}{8}$
Wisconsin Pwr & Lt Co.	$88\frac{7}{8}$	Vermont Yankee Nuclear	$81\frac{1}{2}$
Central Ill. Pub Svc Co.	$91\frac{3}{4}$	Pacific Gas & Elec Co.	$64\frac{1}{8}$
Alabama Gas Corp.	90	Union Elec Co.	$97\frac{3}{4}$
Tampa Elec Co.	$97\frac{5}{8}$	Baltimore Gas & Elec Co.	$76\frac{7}{8}$
St. Joseph Lt & Pwr Co.	$88\frac{3}{8}$	Rochester Gas & Elec Co.	$94\frac{1}{8}$
Iowa Elec Lt & Pwr Co.	$75\frac{7}{8}$	California Elec Pwr Co.	$97\frac{1}{4}$
Dallas Pwr & Lt Co.	100	Eastern Edison Co.	119
Fitchburg Gas & Elec Lt	$85\frac{1}{4}$	Ohio Edison Co.	$89\frac{7}{8}$
Natural Gas Pipeline Co.	$98\frac{1}{2}$	Madison Gas & Elec Co.	90
Gas Svc Co.	$96\frac{1}{2}$	Lacledge Gas Co.	$87\frac{3}{4}$
Kansas City Pwr & Lt Co.	$85\frac{1}{4}$	Hackensack Wtr Co.	$96\frac{3}{8}$
Idaho Power Co.	$96\frac{1}{4}$	Otter Tail Power Co.	$90\frac{5}{8}$

Source: *Bank & Quotation Record* (a publication of the *Commercial & Financial Chronicle*), Vol. 59, Number 1, Jan. 1986, pp. 135–140.

a. Construct a relative frequency histogram for these data.
b. What proportion of the bonds in the sample had an asking price higher than $85? Shade in the area under your histogram of part a that corresponds to this proportion.

2.40 In order to better understand the interactions that take place between salespeople and customers, Ronald P. Willett and Allan L. Pennington (1966) monitored the interactions of appliance salespeople and customers on the floor of a large department store. Part of their research involved observing the length of time customers and salespeople interacted prior to the close of the sale or the departure of the customer. The data below, adapted from the article, are the lengths of time (in minutes) from the first customer–salesperson contact to the close of the sale or the customer's departure for 132 customers who completed their appliance purchase either at the time they were observed or within the following 2 weeks. Instances where a purchase was made by the customer at the time he or she was observed are denoted with an asterisk.

1.0*	33.3	37.0	40.1	6.0*	1.7	4.5*	3.0	5.1*
7.4*	.7	15.0*	9.7	27.2*	10.9	18.7*	13.3*	30.0*
41.3*	44.4*	15.0	12.3	7.0*	16.2	7.4*	17.6*	14.9
15.1*	32.2	1.9*	5.4*	8.4	8.1*	15.5	14.0*	40.0
6.1	28.7	38.1	30.5	7.6	7.9	4.1*	25.4*	12.2*
22.3	7.8*	29.2	30.5	34.6*	7.0*	10.3	10.0*	21.9
.4*	25.6*	3.3	26.4*	39.2*	42.3	1.1	10.1*	41.6
17.4*	14.9	20.1*	9.0*	27.7	42.1	48.6	11.1	31.8
11.0*	25.1*	12.0*	16.9*	26.0	27.7	47.6*	35.1	15.0*
35.4	49.1*	30.0*	14.2	19.2*	39.9	23.0	43.1	38.2
12.8*	10.9*	13.0	20.6*	7.7	50.1	8.0*	24.8	35.0*
8.1*	118.4	8.9*	77.1	60.2*	105.2*	11.0	15.9	20.1*
3.2	8.8	18.0*	.8	7.9*	12.5	69.1	11.1	30.0
12.4	1.5	14.2*	17.7*	.9	13.5	8.4	81.0	10.5
26.2	18.4	6.0	15.9	66.1*	98.2			

a. Construct a relative frequency histogram for each of the following data sets:
   (1) The complete set of 132 times
   (2) The set of times associated with customers who made appliance purchases at the time they were being observed
   (3) The set of times associated with customers who made the appliance purchases at a later date
b. Describe any differences you detect between the histograms of parts a(2) and a(3).
c. Suggest possible explanations for the differences you noted in part b.
*d. Use a statistical software package to generate the histogram of part a(1).

2.41 The table at the top of page 54 contains the populations (in thousands) of 30 "sunbelt" cities in 1960 and 1980.

a. Construct two frequency histograms, one for the 1960 data set and one for the 1980 data set. (Use the same width for the measurement classes of the two histograms.)

CITY	POPULATION		CITY	POPULATION	
	1960	1980		1960	1980
Albuquerque, NM	201	332	Las Vegas, NV	64	165
Anaheim, CA	104	219	Lubbock, TX	129	174
Arlington, TX	45	160	Miami, FL	292	347
Atlanta, GA	487	425	New Orleans, LA	628	558
Austin, TX	187	345	Oklahoma City, OK	324	403
Charlotte, NC	202	314	Orlando, FL	88	128
Dallas, TX	680	904	Phoenix, AZ	439	790
Fort Lauderdale, FL	84	153	St. Petersburg, FL	181	239
Fort Worth, TX	356	385	San Antonio, TX	588	786
Fresno, CA	134	218	San Diego, CA	573	876
Honolulu, HI	294	365	San Francisco, CA	740	679
Houston, TX	938	1,595	San Jose, CA	204	629
Huntington Beach, CA	11	171	Tampa, FL	275	272
Huntsville, AL	72	143	Tucson, AZ	213	331
Jacksonville, FL	201	541	Tulsa, OK	262	361

Source: U.S. Bureau of the Census, *Statistical Abstract of the United States: 1981* (pp. 21–23), *1985* (pp. 23–25).

b. Compare your histograms and describe what they reveal about the change in the populations of sunbelt cities between 1960 and 1980.

c. A better way to describe the population differences is to calculate the change in population for each city and then construct a frequency histogram of these changes. Do this and describe the information reflected in your histogram. Interpret the results.

*d. Use a statistical software package to generate the histograms of parts **a** and **c**.

2.42 The ability to fill a customer's order on time depends to a great extent on being able to estimate how long it will take to produce the product in question. In most production processes, the time required to complete a particular task will be shorter each time the task is undertaken. Furthermore, it has been observed that in most cases the task time will decrease at a decreasing rate the more times the task is undertaken. Thus, in order to estimate how long it will take to produce a particular product, a manufacturer may want to study the relationship between production time per unit and the number of units that have been produced. The line or curve characterizing this relationship is called a *learning curve* (Chase and Aquilano, 1977). Twenty-five employees, all of whom were performing the same production task for the tenth time, were observed. Each person's task-completion time (in minutes) was recorded. The same 25 employees were observed again the thirtieth time they performed the same task and the fiftieth time they performed the task. The resulting completion times are shown in the table.

a. Construct frequency histograms for each of the three data sets.

b. Compare the histograms. Does it appear that the relationship between task completion time and the number of times the task is performed is in agreement with the observations noted above about production processes in general? Explain.

*c. Use a statistical software package to generate the three histograms of part **a**.

TENTH PERFORMANCE		THIRTIETH PERFORMANCE		FIFTIETH PERFORMANCE	
15	19	16	11	10	8
21	20	10	10	5	10
30	22	12	13	7	8
17	20	9	12	9	7
18	19	7	8	8	8
22	18	11	20	11	6
33	17	8	7	12	5
41	16	9	6	9	6
10	20	5	9	7	4
14	22	15	10	6	15
18	19	10	10	8	7
25	24	11	11	14	20
23		9		9	

2.43 Construct relative frequency histograms for each of the two data sets in Exercise 2.30. Compare them with each other and with the corresponding stem and leaf displays of Exercise 2.30. Do they convey the same information about the data sets? Explain.

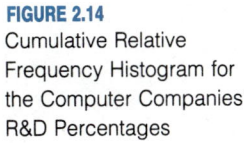

S E C T I O N 2.6

CUMULATIVE RELATIVE FREQUENCY DISTRIBUTIONS (OPTIONAL)

A **cumulative relative frequency distribution** is derived from the relative frequency histogram and provides another graphical description of quantitative data. The cumulative relative frequency distribution for the fifty computer companies' percentages of revenues spent on research and development (R&D) is shown in Figure 2.14.

FIGURE 2.14
Cumulative Relative Frequency Histogram for the Computer Companies' R&D Percentages

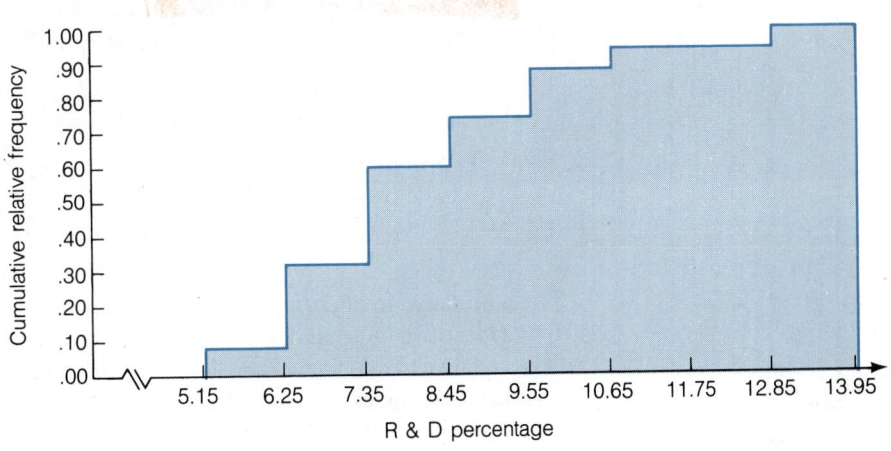

The cumulative relative frequency histogram is constructed in the same manner as the relative frequency histogram, except that the height of the rectangle constructed over each measurement class is the *cumulative* proportion of measurements less than or equal to the upper boundary of that class. Also, the right side of each rectangle (except the last) is eliminated to give the histogram the appearance of a stairway.

To compute the cumulative frequencies, proceed as follows. Suppose the classes are numbered 1, 2, 3, . . . from the smallest to the largest measurement class. Then the cumulative frequency for class 3 would equal the sum of the class frequencies corresponding to classes 1, 2, and 3:

$$(\text{Cumulative frequency for class 3}) = f_1 + f_2 + f_3$$

Then the cumulative relative frequency is calculated by converting the cumulative frequency into a proportion:

$$(\text{Cumulative relative frequency for class 3}) = \frac{f_1 + f_2 + f_3}{n}$$

where n is the number of measurements in the sample.

DEFINITION 2.7

The **class cumulative frequency** for a given class, say class i, is equal to the sum of the class frequencies up to and including the frequency for class i, i.e.,

$$(\text{Class cumulative frequency for class } i) = f_1 + f_2 + \cdots + f_i$$

DEFINITION 2.8

The **class cumulative relative frequency** for a given class, say class i, is equal to the class cumulative frequency divided by the total number n of measurements, i.e.,

$$(\text{Class cumulative relative frequency for class } i)$$
$$= \frac{\text{Class cumulative frequency}}{n}$$

The cumulative frequencies and cumulative relative frequencies can be calculated by adding two more columns to the table used to calculate the class frequencies and class relative frequencies. For example, Table 2.6, which was used to calculate the relative frequencies for the R&D percentage data, is reproduced with the two new columns (shaded) in Table 2.8.

TABLE 2.8

Cumulative Frequencies and Cumulative Relative Frequencies for the R&D Percentage Data

CLASS	MEASUREMENT CLASS	CLASS FREQUENCY	CLASS CUMULATIVE FREQUENCY	CLASS RELATIVE FREQUENCY	CLASS CUMULATIVE RELATIVE FREQUENCY
1	5.15– 6.25	4	4	$\frac{4}{50}$ = .08	$\frac{4}{50}$ = .08
2	6.25– 7.35	12	16	$\frac{12}{50}$ = .24	$\frac{16}{50}$ = .32
3	7.35– 8.45	14	30	$\frac{14}{50}$ = .28	$\frac{30}{50}$ = .60
4	8.45– 9.55	7	37	$\frac{7}{50}$ = .14	$\frac{37}{50}$ = .74
5	9.55–10.65	7	44	$\frac{7}{50}$ = .14	$\frac{44}{50}$ = .88
6	10.65–11.75	3	47	$\frac{3}{50}$ = .06	$\frac{47}{50}$ = .94
7	11.75–12.85	0	47	$\frac{0}{50}$ = .00	$\frac{47}{50}$ = .94
8	12.85–13.95	3	50	$\frac{3}{50}$ = .06	$\frac{50}{50}$ = 1.00
				1.00	

You can see that the cumulative relative frequency for a class is calculated by summing the frequencies of all classes up to and including that class and dividing by the total number of measurements, $n = 50$ in the current example. Note that the cumulative relative frequency can never decrease as you move to the right across the distribution, so that the graph has the appearance of a rising stairway. This type of display is called a **step function**, and the last step always has height equal to 1.0. The rapid rise of the stairs in the first few measurement classes in Figure 2.14 indicates that most of the measurements fall in those classes, and the very gradual rise at the right of the graph, corresponding to the larger measurement classes, indicates that only a few measurements fall in those classes.

The general guidelines to follow when constructing a cumulative relative frequency histogram are given in the box.

HOW TO CONSTRUCT A CUMULATIVE RELATIVE FREQUENCY DISTRIBUTION

1. Add two columns to the table used to calculate the class relative frequencies: a column for the class cumulative frequency and one for the class cumulative relative frequency.
2. Calculate the cumulative frequency and the cumulative relative frequency for each class.
3. Construct a graph by plotting the class cumulative relative frequency as a rectangle over the corresponding measurement class. Eliminate the right side of each rectangle, except the last, to give the histogram the appearance of a stairway, or step function. The height of the last step should equal 1.0.

| | | | | | | | | | | |

EXERCISES 2.44–2.51

LEARNING THE MECHANICS

2.44 Use the relative frequency distribution you prepared for Exercise 2.33 to construct a cumulative relative frequency distribution.

2.45 Consider the following data set:

2.3	2.5	4.2	5.6	2.8	3.3	6.7	3.4	4.7	2.9
1.6	3.0	2.0	1.8	1.4	2.1	1.0	3.1	2.8	2.8
1.8	2.1	5.2	4.3	6.0	7.1	5.2	4.9	3.5	1.6
1.0	2.2	3.6	4.5	5.1	6.3	1.9	2.7	7.3	8.3

a. Construct a frequency distribution (in table form) for these data.
b. Add a class cumulative frequency column to your table of part **a**.
c. Add a class cumulative relative frequency column to your table of part **b**.
d. Generally speaking, what question does a class frequency answer about a particular measurement class? A class cumulative frequency? A class cumulative relative frequency?

2.46 Consider the following distribution:

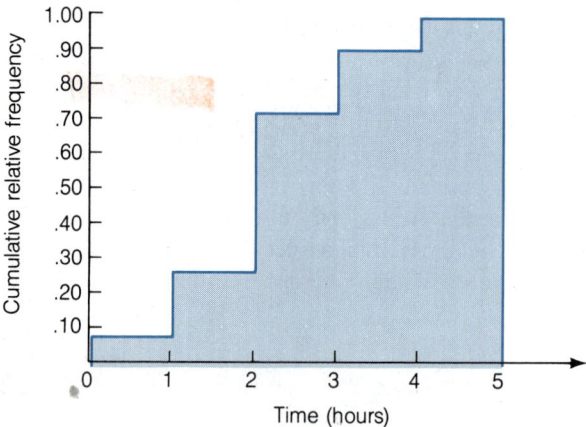

a. What type of distribution is this?
b. How many measurement classes were used in the construction of this distribution?
c. Is it possible to determine how many measurements are contained in the data set described by this distribution? Explain.
d. Explain why the value on the vertical axis associated with the highest measurement class is always 1.0 for this type of distribution.
e. Convert the distribution shown in the figure to a relative frequency histogram.

2.47 Use the relative frequency distribution you prepared for Exercise 2.35 to construct a cumulative relative frequency distribution.

2.48 Use the relative frequency distribution you prepared for Exercise 2.36 to construct a cumulative relative frequency distribution.

APPLYING THE CONCEPTS

2.49 Refer to the cumulative relative frequency distribution from Exercise 2.47.
  a. Find the proportion of the 30 annual incomes that are less than $14,050.
  b. Find the proportion that exceed $14,050.

2.50 Each year *Forbes* magazine publishes a list of the "200 Best Small Companies in America." (*Forbes* defines small companies as those with sales under $300 million.) The companies are ranked in terms of their 5-year average return on equity. Thirty-five of these firms were sampled and their sales (in millions of dollars) for the previous 12 months were recorded below:

11.2	50.9	3.8	195.1	49.5	17.3	31.9
18.8	26.6	200.0	103.0	83.8	108.2	48.7
219.4	124.4	19.4	19.6	18.5	66.6	143.3
37.4	36.0	83.4	135.2	62.9	283.8	14.2
102.9	52.4	67.1	180.1	19.9	139.1	35.3

  a. Construct a frequency distribution (in table form) for these data.
  b. Add class cumulative frequency and cumulative relative frequency columns to your table of part **a**.
  c. Construct a cumulative relative frequency histogram for these data.
  d. Use your graph from part **c** to determine what proportion of the firms in the sample had sales less than or equal to $80 million. Greater than $100 million.
  e. Use the sample data to obtain answers to the questions of part **d**.
  f. Compare your answers to parts **d** and **e**. Do they differ? Why or why not?

2.51 As part of a study of property values in Minneapolis, the Robinson Appraisal Company sampled 79 apartment buildings and determined, among other things, each building's size, age, and condition as of January 1982. The following data are the ages (in years) of the buildings:

82	70	69	19	71	70	13	70	70	23
62	53	32	65	66	55	79	15	21	69
59	18	21	66	72	19	65	57	24	71
51	50	82	13	13	82	82	54	64	36
18	19	21	22	18	21	67	22	14	10
67	50	58	71	63	74	73	82	3	56
72	64	82	79	69	23	76	67	57	82
56	82	18	82	75	70	82	55	82	

  a. Construct a cumulative relative frequency histogram for these data.
  b. Use your graph of part **a** to determine what percentage of the apartment buildings in the sample are more than 50 years old. Less than 28 years old.
  c. Use your graph of part **a** to determine the age that approximately 50% of the buildings exceed.
  d. Without counting the measurements in the data set, determine the number of apartments that are less than 75 years old.

**DISTORTING THE
TRUTH WITH
PICTURES (OPTIONAL)**

While it may be true in telling a story that a picture is worth a thousand words, it is also true that pictures can be used to convey a colored and distorted message to the viewer. So the old adage "Let the buyer (reader) beware" applies. Examine relative frequency histograms and, in general, all graphical descriptions with care.

We will mention a few of the pitfalls to watch for when analyzing a chart or graph. But first we should mention the **time series graph**, which is often the object of distortion. This type of graph records the behavior of some business variable over time, with the business variable plotted on the vertical axis and the time plotted on the horizontal axis. Examples of business variables commonly graphed as time series abound: economic indexes, profit, sales, supply, demand, etc. We will treat the subject of time series more completely in Chapters 13 and 14. For now, we will simply use some time series graphs to demonstrate several ways pictures may be distorted.

One common way to change the impression conveyed by a graph is to change the scale on the vertical axis, the horizontal axis, or both. For example, if you want to show that the change in firm A's market share over time is moderate, you could pack in a large number of units per inch on the vertical axis. That is, make the distance between successive units on the vertical scale small, as shown in Figure 2.15. You can see that the change in the firm's market share over time appears to be minimal.

FIGURE 2.15
Firm A's Market Share
from 1979 to 1984—
Packed Vertical Axis

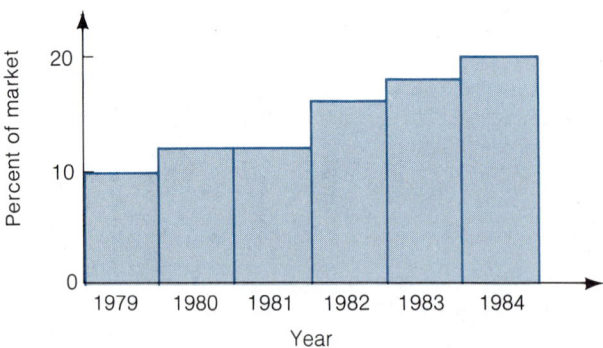

To make the changes in firm A's market share appear large, you could increase the distance between successive units on the vertical axis. That is, you stretch the vertical axis by graphing only a few units per inch, as shown in Figure 2.16. The telltale sign of stretching is a long vertical axis, but this is often hidden by starting the vertical axis at some point above 0, as shown in Figure 2.17(a). Or, the same effect can be achieved by using a broken line for the vertical axis, as shown in Figure 2.17(b).

FIGURE 2.16
Firm A's Market Share
from 1979 to 1984—
Stretched Vertical Axis

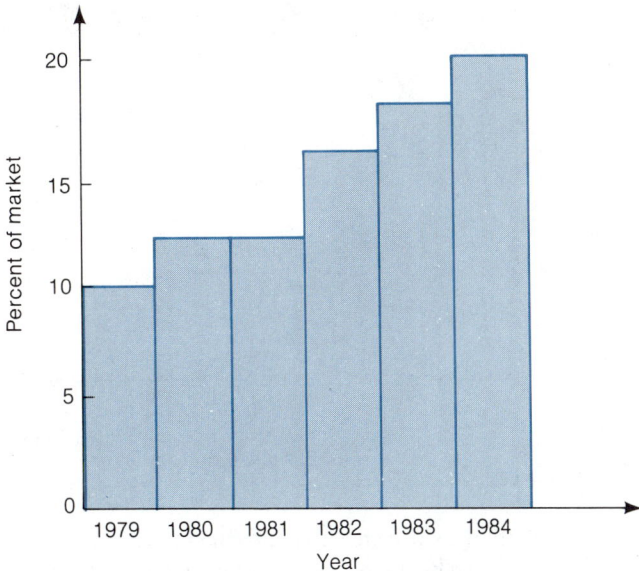

FIGURE 2.17
Daily Stock Sales on the
New York Stock Exchange
from May to July 1979

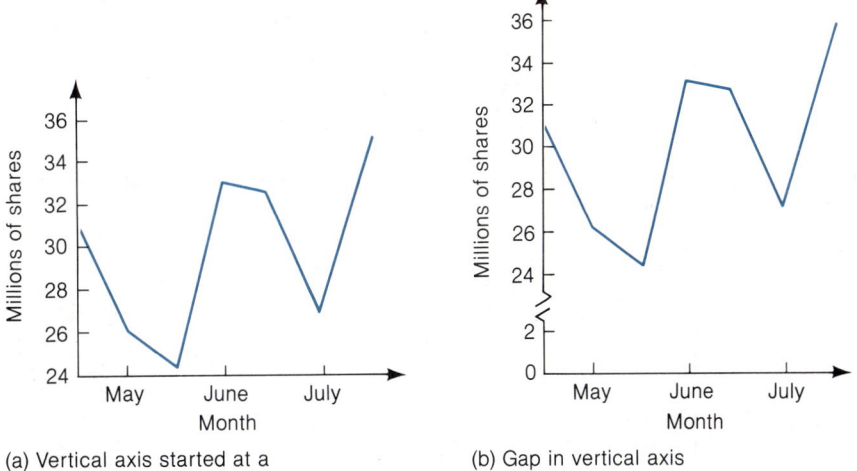

(a) Vertical axis started at a
point greater than 0

(b) Gap in vertical axis

Stretching the horizontal axis (increasing the distance between successive units) may also lead you to incorrect conclusions. For example, Figure 2.18(a) on page 62 depicts rental income in the United States from the first quarter of 1978 to the first quarter of 1980. If you increase the length of the horizontal axis, as in Figure 2.18(b), the change in the rental income over time seems to be less pronounced.

FIGURE 2.18   Rental Income from the First Quarter of 1978 to the First Quarter of 1980

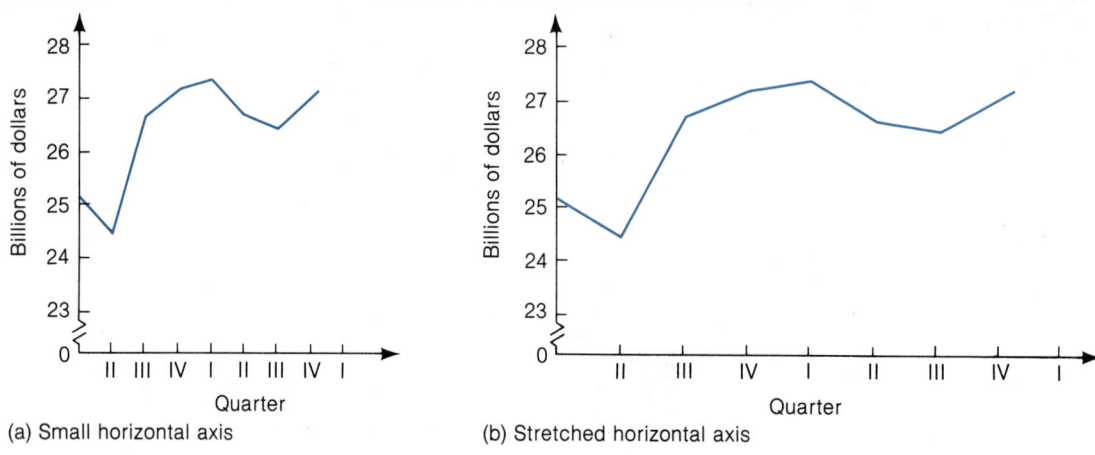

(a) Small horizontal axis

(b) Stretched horizontal axis

The changes in categories indicated by a bar chart can also be emphasized or deemphasized by stretching or shrinking the vertical axis. Another method of achieving visual distortion with bar charts is by making the width of the bars proportional to their height. For example, look at the bar chart in Figure 2.19(a), which depicts the percentage of a year's total automobile sales attributable to each of the four major manufacturers. Now suppose we make the width as well as the height grow as the market share grows. This is shown in Figure 2.19(b). The reader may tend to equate the *area* of the bars with the relative market share of each manufacturer. In fact, the true relative market share is proportional only to the height of the bars.

FIGURE 2.19   Relative Share of the Automobile Market for Each of Four Major Manufacturers

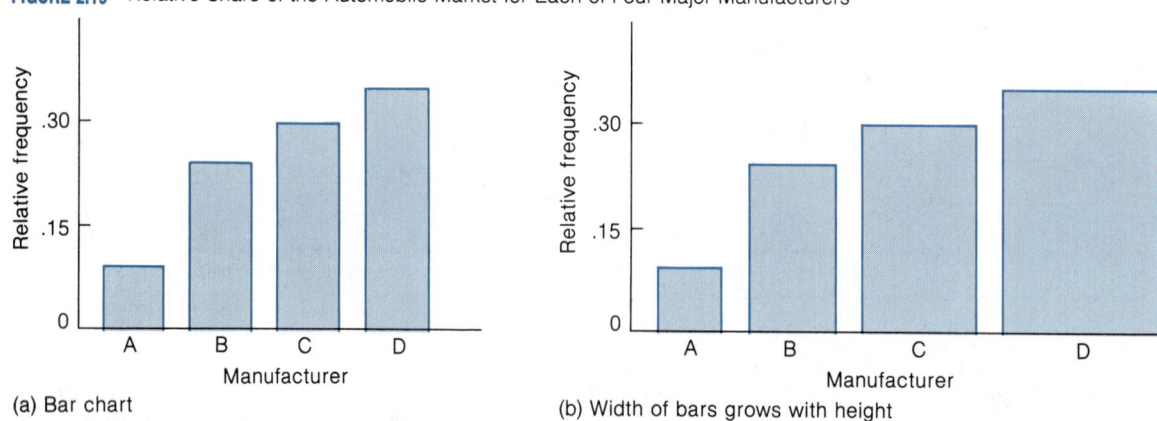

(a) Bar chart

(b) Width of bars grows with height

Sometimes, as noted by Selazny (1975), we do not need to manipulate the graph to distort the impression it creates. Modifying the verbal description that accompanies the graph can change the interpretation that will be made by the viewer. Figure 2.20 provides a good illustration of this ploy.

FIGURE 2.20
Changing the Verbal
Description to Change a
Viewer's Interpretation.
Source: Reprinted
by permission of the
publisher, from "Grappling
with Graphics," by Gene
Selazny, *Management
Review*, Oct. 1975, p. 7.
© 1975 American
Management Association,
New York. All rights
reserved.

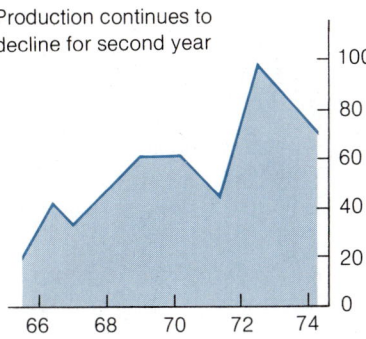

Production continues to
decline for second year

For our production, we need not even
change the chart, so we can't be
accused of fudging the data. Here
we'll simply change the title so that
for the Senate subcommittee, we'll
indicate that we're not doing as well
as in the past. . . .

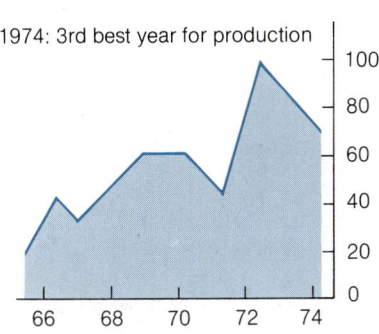

1974: 3rd best year for production

Whereas for the general public, we'll
tell them that we're still in the prime
years.

We have presented only a few of the ways that graphs can be used to convey misleading pictures of business phenomena. However, the lesson is clear. Examine all graphical descriptions of data with care. In particular, check the axes and the size of the units on each axis. Ignore visual changes and concentrate on the actual numerical changes indicated by the graph or chart.

SUMMARY

Business data can be classified as one of two types: **qualitative** or **quantitative**. In a qualitative data set each observation falls into one of a set of categories, whereas in a quantitative data set each observation is measured on a numerical scale.

Since we want to use sample data to make inferences about the population from which it is drawn, it is important for us to be able to describe the data. Graphical methods are important and useful tools for describing both types of data. The **bar chart** and **pie chart** are useful graphical methods for describing qualitative data. **Stem and leaf displays** and **relative frequency histograms** are graphical techniques used to describe quantitative data sets.

Our ultimate goal is to use the sample to make inferences about the population. We must be wary of using graphical techniques to accomplish this goal, since they do not lend themselves to a measure of reliability for an inference. Therefore, we need to develop numerical measures to describe a data set. That is the purpose of the next chapter.

[*Note: Starred (*) exercises require the use of a computer. Double-starred exercises (**) refer to optional sections in this chapter.*]

2.52 According to estimates of the International Data Corporation, 85,000 word processing keyboards were sold by U.S. manufacturers in 1979. The table shows each manufacturer's market share. Construct a relative frequency bar chart for the data.

COMPANY	MARKET SHARE	COMPANY	MARKET SHARE
Artec	1%	Wang	16%
Wordstream	1%	NBI	4%
Four-Phase	2%	Vydec	4%
A.B. Dick	2%	Lexitron	4%
Digital Equipment	3%	Olivetti	4%
CPT	4%	3M	3%
Xerox	4%	Burroughs	3%
AM International	6%	Datapoint	2%
IBM	8%	Micom	1%
Lanier/AES	13%	Others	15%

Source: *Fortune*, Sept. 22, 1980, p. 56.

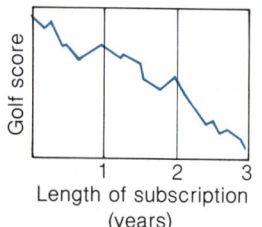

Golf score / Length of subscription (years)

**2.53 A graph similar to the one shown here appeared in a recent advertisement for a well-known golf magazine. One person might interpret the graph's message as being the longer you subscribe to the magazine, the better golfer you should become. Another person might interpret it as indicating that if you subscribe for 3 years, your game should improve dramatically.
a. Explain why the graph can be interpreted in more than one way.
b. How could the graph be altered to rectify the current distortion?

2.54 Classify the following examples of data as either qualitative or quantitative.
a. The style of music preferred by 30 randomly selected radio listeners
b. The length of time it takes each of 15 telephone installers to hook up a wall telephone
c. The population of each of 10 randomly selected cities in the United States

2.55 A questionnaire sent to the chief executives of the country's 500 largest industrial corporations and commercial banking companies by *Fortune* magazine revealed the information given in the table concerning their educational backgrounds. Construct a relative frequency bar chart for the data.

LEVEL OF EDUCATION	PERCENT
High school or less	4.5
Attended college	9.3
College graduate	27.9
Postgraduate study	18.6
Master's degree	24.2
Doctorate	15.5

Source: "A Group Profile of the *Fortune* 500 Chief Executive," May 1976.

2.56 Companies that locate and remove natural gas from the ground are known as *producers* of gas. Companies that buy the natural gas from the producers and transport it via pipelines are known as *suppliers* of gas. Companies that buy gas from the suppliers and sell it to homeowners are known as *distributors* of gas. Northern States Power (NSP), a large midwestern distributor, recently included a pamphlet with its mailing of monthly gas bills to explain how NSP spends the money collected from its customers. The pamphlet contained the information provided in the accompanying table.

HOW YOUR NATURAL GAS DOLLAR IS SPENT	AMOUNT
Labor, operation, maintenance, and depreciation	10.6 cents
Cost of capital	5.1
Taxes	4.8
Payments of gas suppliers	79.5

Source: Northern States Power Corporation.

a. Construct a relative frequency bar chart for these data.
b. Explain why it is not possible to construct a frequency bar chart for these data.
c. What do these data reveal about the relationship between the price NSP pays for gas and the price NSP's customers pay for gas?
d. A customer's gas bill was $80 last month. Use your bar chart to determine how much of the $80 went to the government in the form of taxes.

2.57 Construct a pie chart for the data in Exercise 2.55.

2.58 If it is not examined carefully, the graphical description of U.S. peanut production shown here can be misleading.

Source: *Gainesville Sun*, Sept. 11, 1976.

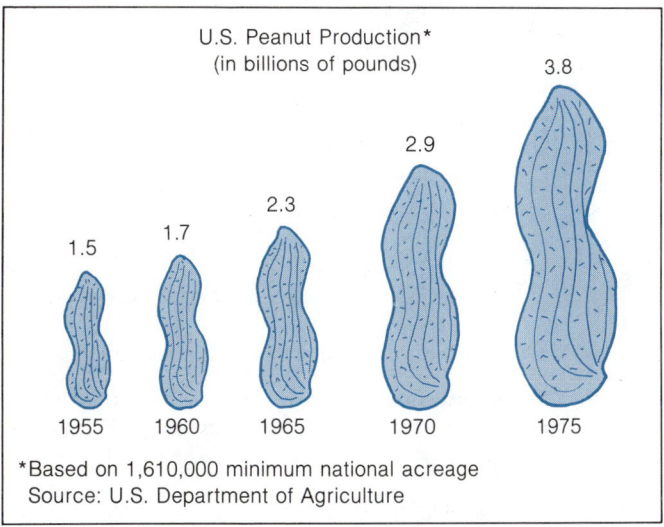

U.S. Peanut Production*
(in billions of pounds)

3.8
2.9
2.3
1.7
1.5

1955   1960   1965   1970   1975

*Based on 1,610,000 minimum national acreage
Source: U.S. Department of Agriculture

a. Explain why the graph may mislead some readers.
b. Construct an undistorted graph of U.S. peanut production for the given years.

2.59 The following is a list of the lengths (in inches) of 30 randomly selected golf tees produced by a machine designed to produce tees 1.5 inches long:

1.49	1.47	1.52	1.50	1.51	1.54	1.55	1.52	1.48	1.49
1.50	1.51	1.51	1.50	1.53	1.54	1.52	1.55	1.51	1.50
1.50	1.49	1.51	1.57	1.51	1.53	1.47	1.50	1.51	1.49

Construct a relative frequency histogram for the data.

**2.60 Refer to Exercise 2.59. Construct a cumulative frequency distribution and a cumulative relative frequency distribution for the data on the tee lengths.

**2.61 During the last decade, turkey processors began marketing a large variety of turkey products, including turkey franks, turkey sausage, turkey ham, turkey roasts, turkey rolls, and even ground turkey. The accompanying pie charts describe turkey consumption in the United States in 1960 and 1984.

Turkey Consumption by Quarter

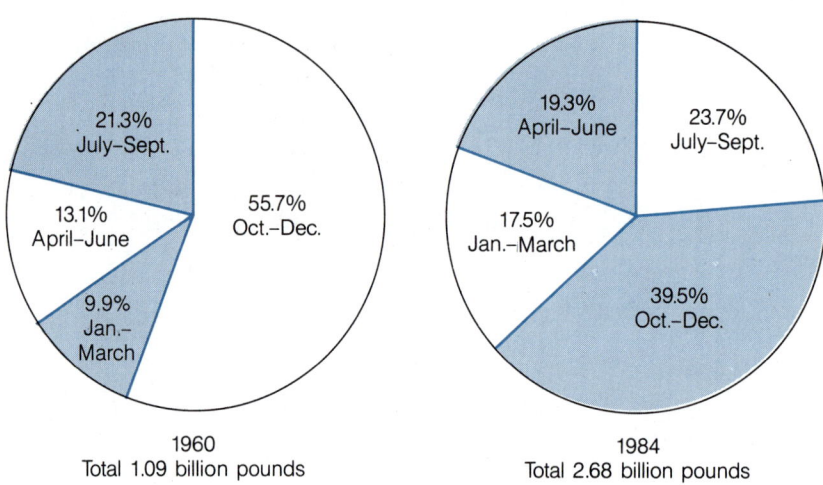

1960
Total 1.09 billion pounds

1984
Total 2.68 billion pounds

Source: U.S. Department of Agriculture.

    a. Describe the change in seasonal turkey consumption that occurred between 1960 and 1984.

    b. Why do you think this shift in consumption occurred?

**2.62 In experimenting with a new technique for imprinting paper napkins with designs, names, etc., a paper products company discovered that four different results were possible:

(A) Imprint successful

(B) Imprint smeared

(C) Imprint off-center to the left

(D) Imprint off-center to the right

To test the reliability of the technique, the company imprinted 1,000 napkins and obtained the results shown in the graph.

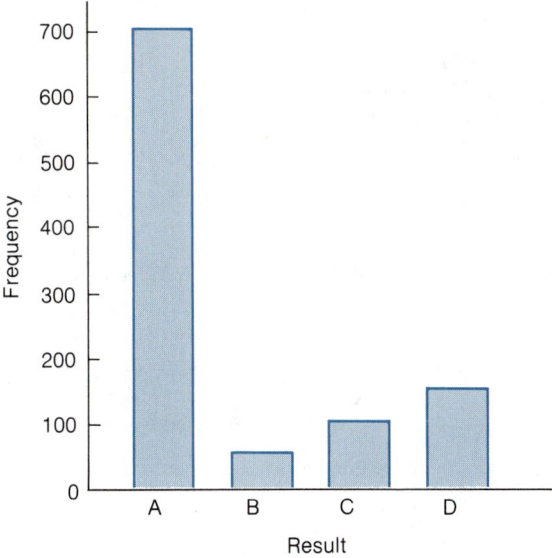

a. What type of graphical tool is the figure?
b. What information does the graph convey to you?
c. From the information provided by the graph, how might you numerically describe the reliability of the imprinting technique?

2.63 On a recent Friday, six of the major luxury hotels in Atlanta reported that the following number of rooms were occupied:

HOTEL	NUMBER OF ROOMS OCCUPIED
Atlanta Hilton	801
Fairmont	389
Hyatt Regency	699
Marriott	542
Omni International	521
Peachtree Plaza	1,002

a. Construct a relative frequency bar chart for the data.
b. Is it possible to determine which hotel had the greatest percentage of its rooms filled on that Friday? Explain.
c. What other information about the hotels would make the data more meaningful?

2.64 Businesses planning international expansion should be aware of the impending surge in the population of the world. It took until 1800 for the world's population to reach 1 billion. But between 1984 and 2000, the population is expected to grow from 4.6 billion to 6.1 billion. That's a jump of 1.5 billion in just 16 years. The table at the top of page 68 describes the expected increase in more detail.

LOCATION	POPULATION (MILLIONS)	
	1984	2000
Africa	513	851
Asia	2,730	3,564
Europe	489	511
Latin America	390	564
North America	259	302
Oceania	24	29
USSR	272	309
Total	4,677	6,130

Source: *U.S. News & World Report*, Jan. 9, 1984,
p. 53. Copyright 1984, U.S. News & World Report.

a. Construct two relative frequency bar charts, one for the 1984 data and one for the 2000 data.

b. Use the information on your bar charts to describe the pattern of shifts in population concentrations that are expected between 1984 and 2000.

2.65 Consider the graphical information portrayed here.

Regional Share of Single-Family Housing Starts in the United States

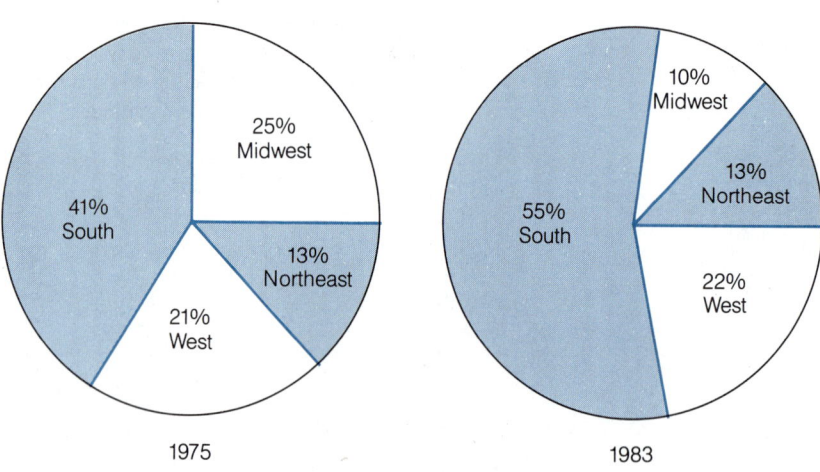

1975   1983

Source: U.S. Bureau of the Census, *Statistical Abstract of the United States: 1985.*

a. What type of graphical tool is shown here?

b. What do these figures describe?

c. Suppose you were given the *number* of housing starts for each region in 1975 and 1983. Explain how you would construct the given figures using such information.

d. Describe the shift in the pattern of housing starts revealed by the two figures.

2.66 Use one of the graphical methods presented in this chapter to describe the data set given in the table.

RENT (In Dollars)	PERCENTAGE OF RENTERS IN RENT CLASS
Less than 80	2%
80–99	2%
100–149	6%
150–199	8%
200–299	27%
300 and over	55%

Source: U.S. Bureau of the Census, *Statistical Abstract of the United States: 1985.*

**2.67 One measure of the value of the equity of a firm is the total value of the stock issued by the firm. This total is determined by multiplying the number of shares of stock issued by the current market value of a share of stock. In a recent survey commissioned by *Business Week* and conducted by Louis Harris & Associates, top executives of more than 600 U.S. corporations were asked the following:

Question: Do you feel that the current price of your company's stock is an accurate indicator of the real value of your company? If not, does the stock price undervalue the company or overvalue it?

RESPONSE	PERCENTAGE OF SAMPLE
Assigns real value	32%
Undervalues	60%
Overvalues	2%
Not sure	6%

Source: *Business Week*, Feb. 20, 1984, p. 14.

a. Construct a pie chart for the data.
b. Assume that the Harris organization received responses from 700 executives. Construct a frequency bar chart for the given data.

2.68 A manufacturer of industrial wheels is losing many profitable orders because of the long time it takes the firm's marketing, engineering, and accounting departments to develop price quotes for potential customers. To remedy this problem the firm's management would like to set guidelines for the length of time each department should spend developing price quotes. To help develop these guidelines, 50 requests for price quotes were randomly selected from the set of all price quotes made last year; the processing time was determined for each price quote for each department. These times are displayed in the table on page 70. The price quotes are also classified by whether they were "lost" (i.e., whether or not the customer placed an order after receiving the price quote).

Price Quote Processing Times (in Days)

REQUEST NUMBER	MARKETING	ENGINEERING	ACCOUNTING	LOST?	REQUEST NUMBER	MARKETING	ENGINEERING	ACCOUNTING	LOST?
1	7.0	6.2	.1	No	26	.6	2.2	.5	No
2	.4	5.2	.1	No	27	6.0	1.8	.2	No
3	2.4	4.6	.6	No	28	5.8	.6	.5	No
4	6.2	13.0	.8	Yes	29	7.8	7.2	2.2	Yes
5	4.7	.9	.5	No	30	3.2	6.9	.1	No
6	1.3	.4	.1	No	31	11.0	1.7	3.3	No
7	7.3	6.1	.1	No	32	6.2	1.3	2.0	No
8	5.6	3.6	3.8	No	33	6.9	6.0	10.5	Yes
9	5.5	9.6	.5	No	34	5.4	.4	8.4	No
10	5.3	4.8	.8	No	35	6.0	7.9	.4	No
11	6.0	2.6	.1	No	36	4.0	1.8	18.2	Yes
12	2.6	11.3	1.0	No	37	4.5	1.3	.3	No
13	2.0	.6	.8	No	38	2.2	4.8	.4	No
14	.4	12.2	1.0	No	39	3.5	7.2	7.0	Yes
15	8.7	2.2	3.7	No	40	.1	.9	14.4	No
16	4.7	9.6	.1	No	41	2.9	7.7	5.8	No
17	6.9	12.3	.2	Yes	42	5.4	3.8	.3	No
18	.2	4.2	.3	No	43	6.7	1.3	.1	No
19	5.5	3.5	.4	No	44	2.0	6.3	9.9	Yes
20	2.9	5.3	22.0	No	45	.1	12.0	3.2	No
21	5.9	7.3	1.7	No	46	6.4	1.3	6.2	No
22	6.2	4.4	.1	No	47	4.0	2.4	13.5	Yes
23	4.1	2.1	30.0	Yes	48	10.0	5.3	.1	No
24	5.8	.6	.1	No	49	8.0	14.4	1.9	Yes
25	5.0	3.1	2.3	No	50	7.0	10.0	2.0	No

a. Use a graphical technique to describe each of the following:
   (1) The processing time of the marketing department
   (2) The processing time of the engineering department
   (3) The processing time of the accounting department
   (4) The total processing time for individual price quotes
b. Using your results from part a, develop "maximum processing time" guidelines for each department that, if followed, will help the firm reduce the number of lost orders.
*c. Use a statistical software package to generate a stem and leaf display for the processing time of the engineering department.

**2.69 Referring to Exercise 2.68, construct a cumulative frequency distribution for the total processing time data set. It is said that a frequency distribution answers the question: How many measurements fell in class i? What question does your cumulative frequency distribution answer about the total processing time data set?

**2.70 In order to estimate the demand for a product, it may be important to identify the age groups to whom it will appeal, and the sizes of these groups. The table describes the resident population (in thousands) of the United States by age in 1960 and 1980.

AGE GROUP	1960	1980
Under 5 years	20,321	16,348
5–9	18,692	16,700
10–14	16,773	18,242
15–19	13,219	21,168
20–24	10,801	21,319
25–29	10,869	19,521
30–34	11,949	17,561
35–39	12,481	13,965
40–44	11,600	11,669
45–49	10,879	11,090
50–54	9,606	11,710
55–59	8,430	11,615
60–64	7,142	10,088
65 and over	16,560	25,550

Source: U.S. Bureau of the Census, *Statistical Abstract of the United States: 1981* (p. 26), *1985* (p. 28).

a. On the same graph (i.e., using the same axes), construct a cumulative relative frequency distribution for the 1960 data and another for the 1980 data.

b. Use your graph from part **a** to determine what changes occurred in the age distribution between 1960 and 1980.

c. From the end of World War II until about 1965, births in the United States soared to record levels. People born during this period are said to belong to the "baby-boom generation." In 1960, approximately what proportion of the population was comprised of baby-boomers? In 1980?

**2.71 The sequence of pie charts on page 72 portrays the evolution of the structure of the top 500 firms in the United States and the top 200 firms in the United Kingdom over the period 1950–1980. Describe the trends that are revealed by these pie charts.

Pie Charts for Exercise 2.71

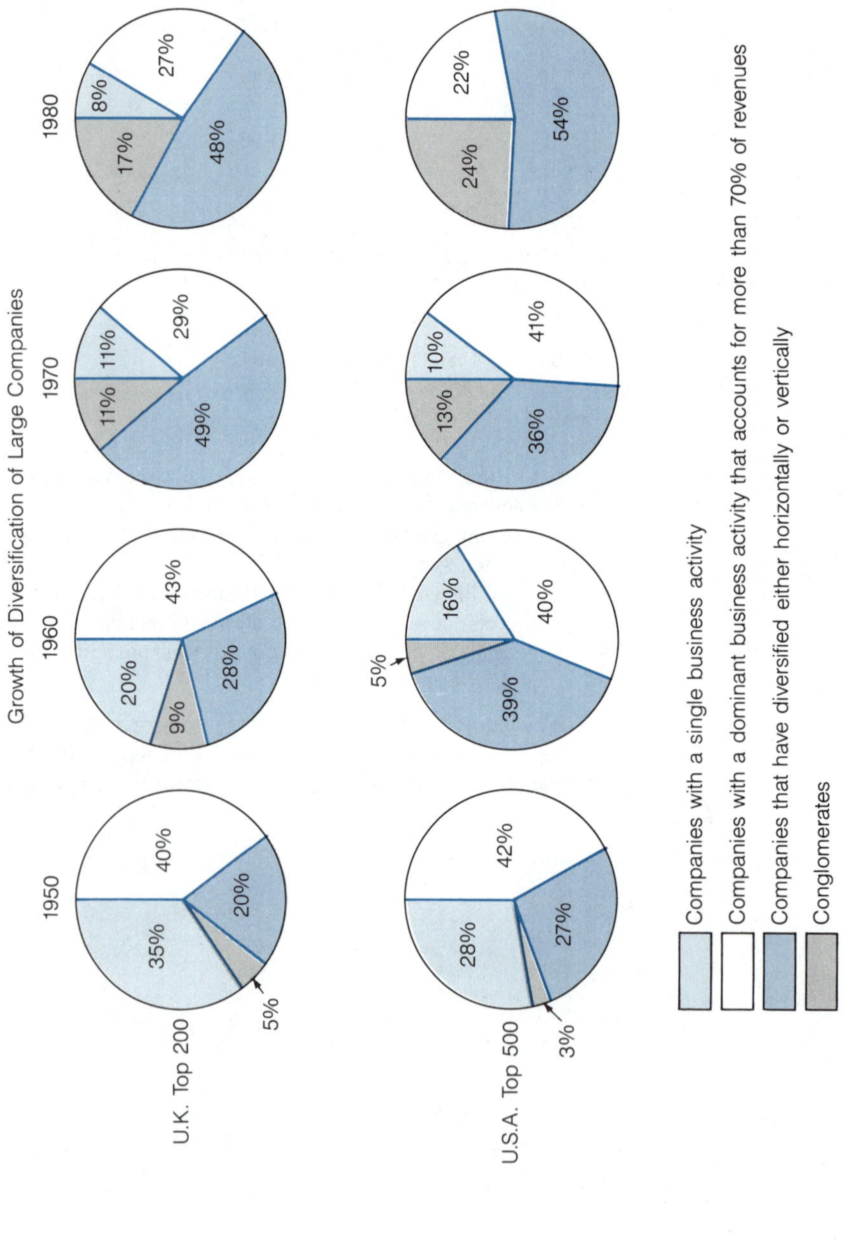

Growth of Diversification of Large Companies

U.K. Top 200

1950: 40%, 20%, 35%, 5%

1960: 43%, 28%, 20%, 9%

1970: 29%, 49%, 11%, 11%

1980: 27%, 48%, 8%, 17%

U.S.A. Top 500

1950: 42%, 27%, 28%, 3%

1960: 40%, 39%, 16%, 5%

1970: 41%, 36%, 10%, 13%

1980: 22%, 54%, 24%

☐ Companies with a single business activity

☐ Companies with a dominant business activity that accounts for more than 70% of revenues

☐ Companies that have diversified either horizontally or vertically

☐ Conglomerates

Source: Adapted from *Long Range Planning.* Vol 19, No. 1, 1986, pp. 52–60. Reprinted with permission. Copyright 1986, Pergamon Press, Ltd.

ON YOUR OWN . . .

Utilizing the data sources listed here, sources suggested by your instructor, or your own resourcefulness, find two real business-oriented data sets: one quantitative and one qualitative. Describe both data sets graphically using one or more of the graphical techniques presented in this chapter. These data sets and your graphs will be referred to in "On Your Own" sections in later chapters, so choose data sets of interest to you and be sure to keep copies of the data sets and your graphs.

SUGGESTED SECONDARY DATA SOURCES*

Board of Governors of the Federal Reserve System. *Federal Reserve Bulletin* (monthly).
Business Week (magazine).
Dun & Bradstreet's *Million Dollar Directory* (yearly).
Dun & Bradstreet's *Middle Market Directory* (yearly).
Forbes (magazine).
Fortune (magazine).
Standard & Poor's Trade and Securities Statistics (monthly supplements).
Standard & Poor's Corporation, *Industry Surveys.*
Target Group Index. Axiom Press (yearly).
U.S. Bureau of the Census. *Census of Manufacturers.*
U.S. Bureau of the Census. *County Business Patterns.*
U.S. Bureau of the Census. *Statistical Abstract of the United States* (yearly).
U.S. Department of Commerce, Office of Business Economics. *Business Statistics.*
U.S. Department of Commerce, Office of Business Economics. *Survey of Current Business* (monthly).
U.S. Department of Labor. *Monthly Labor Review.*
U.S. Department of Labor, Bureau of Labor Statistics. *Employment and Earnings.*
U.S. Department of Labor, Bureau of Labor Statistics. *National Survey of Professional Administrative, Technical, and Clerical Pay.*
Wall Street Journal (daily).
Your state's statistical abstract.

GUIDES TO FINDING SECONDARY DATA

Business Periodical Index.
Coman. *Sources of Business Information.*
Encyclopedia of Business Information Sources. Detroit: Gale Research Co.
Funk and Scott Index of Corporations and Industries.
Funk and Scott Index International: Industries, Countries, Companies.
Guide to Special Issues and Indexes of Periodicals.
Houser and Leonard. *Government Statistics for Business Use.*
New York Times Index.
Statistics Sources, edited by Paul Wasserman. Detroit: Gale Research Co.
U.S. Bureau of the Census. *Directory of Non-federal Statistics for States and Local Areas.*
Wall Street Journal Index.
Your local Chamber of Commerce.

*Primary data are data you (or someone in your organization) collect for the study at hand. *Secondary data* are data collected by someone outside your organization for purposes other than the study at hand.

USING THE COMPUTER . . .

We have supplied a set of data in Appendix C (also available on diskette from the publisher) that will be used as a source for the "Using the Computer" exercises at the end of most chapters. A complete description of the data can be found in Appendix C. Briefly, the data set includes such variables as population size, number of households, average household size, average income, percentage of college graduates, percentage of women in the work force, and purchasing-potential indexes for groceries, sporting goods, and home improvements for each of 1,000 U.S. zip codes.

a. Consider the percentage of women in the work force. Use a statistical software package to generate a stem and leaf display and a relative frequency histogram for these 1,000 percentages. Compare the graphical descriptions you obtain, and discuss what each reveals about the distribution of the percentage of women in the work force in the sample of 1,000 zip codes.

b. Repeat part **a** for one of the census regions, perhaps the one in which you currently reside. Compare the regional distribution to the national distribution you obtained in part **a**.

c. Finally, repeat part **a** for the zip codes corresponding to one state, perhaps the one in which you currently reside. Compare the state distribution to the national and regional distributions of parts **a** and **b**.

REFERENCES

Bochnak, M. L. "An analysis of the residential condominium conversion market." Unpublished Ph.D. dissertation. School of Management, University of Minnesota, 1982.

Brigham, E. F. *Financial Management Theory and Practice*, 3rd ed. Chicago: Dryden Press, 1982. Chapter 13.

Chase, R. B. and Aquilano, N. J. *Production and Operations Management*, rev. ed. Homewood, Ill.: Richard D. Irwin, 1977.

Chou, Ya-lun. *Statistical Analysis*, 2nd ed. New York: Holt, Rinehart, and Winston, 1975. Chapter 2.

Davidson, S., Stickney, C. P., and Weil, R. L. *Financial Accounting*, 2nd ed. Chicago: Dryden Press, 1979.

Dodd, P. "Merger proposals, management discretion, and stockholder wealth," *Journal of Financial Economics*, 1980, 8, 105–137.

Eger, C. E. "Corporate mergers: An empirical analysis of the role of risky debt." Unpublished Ph.D. dissertation. School of Management, University of Minnesota, 1982.

Fogarty, D. W. and Hoffmann, T. R. *Production and Inventory Management*. Cincinnati, Ohio: South-Western, 1983.

Huff, D. *How to Lie with Statistics*. New York: Norton, 1954.

Neter, J., Wasserman, W., and Whitmore, G. A. *Applied Statistics*, 2nd ed. Boston: Allyn & Bacon, 1982. Chapter 3.

Schroeder, R. G. *Operations Management*, 2nd ed. New York: McGraw-Hill, 1985. Chapter 5.

Selazny, G. "Grappling with graphics." *Management Review*, Oct. 1975, 7.

Spiro, H. T. *Finance for the Nonfinancial Manager*. New York: Wiley, 1982. Chapter 17.

Sterling, R. R. and Radosevich, R. "A valuation experiment." *Journal of Accounting Research*, Spring 1969, 90–95.

"The 1984 federal budget." *Minneapolis Star and Tribune*, Feb. 1, 1983, 4a.

Willett, R. P. and Pennington, A. L. "Customer and salesman: The anatomy of choice and influence in a retail setting." *Science, Technology and Marketing*, Proceedings of the 1966 Fall Conference of the American Marketing Association, Raymond M. Haas (ed.), 1966, 598–616.

Winegar, K. "No longer a fad, fitness has grown into healthy obsession." *Minneapolis Star and Tribune*, Jan 5, 1983, 1c.

WHERE WE'VE BEEN . . .

As we noted in Chapter 1, the goal of this course is to teach you how to use sample data to make inferences about a population data set. The first step in arriving at this goal is to learn how to describe a data set. As you learned in Chapter 2, graphical methods are advantageous because they convey a rapid and easily understood description of data sets.

WHERE WE'RE GOING . . .

There is a major drawback to using a graphical descriptive method for making an inference about a population from which a sample was selected. Namely, it is difficult to provide a measure of the reliability of the inference. How similar will the graphical description of the sample data be to the corresponding figure for the population? To answer this question, statisticians use one or more numbers to create a mental image of a data set. These numbers, called *numerical descriptive measures*, are the topic of this chapter.

NUMERICAL DESCRIPTIVE MEASURES

CONTENTS

Several types of numerical descriptive measures have been developed to characterize and help us create a mental picture of the relative frequency distribution of a data set. The two most important of these are measures of central tendency and measures of variation. Measures of central tendency are numbers computed from the data set that help us locate the "center" of a relative frequency distribution. Similarly, measures of variation describe the spread or dispersion of a set of data and therefore of its relative frequency distribution.

Numerical descriptive measures have also been devised to measure the skewness of a distribution (the tendency for a relative frequency distribution to stretch out in one direction) and the kurtosis (peakedness) of a distribution, as well as other characteristics of a data set. These measures do not play an important role in the statistical methods discussed in this text; thus, they are omitted from the discussion.

Numerical descriptive measures can be calculated for any data set, either for a sample or for a population. If statistical inference is our goal, as is often the case, we will ultimately use sample numerical descriptive measures to make inferences about the corresponding measures for the population.

SECTION 3.1

THE MODE: A MEASURE OF CENTRAL TENDENCY

If you visualize a relative frequency distribution, one measure of central tendency that immediately comes to mind is the value of x that locates the peak of the distribution—that is, the value of x that occurs with the greatest frequency. This value of x is called the mode.

DEFINITION 3.1

The mode is the measurement that occurs with greatest frequency in the data set.

Because it emphasizes data concentration, the mode has applications in marketing as well as in the description of large data sets collected by state and federal agencies. For example, a retailer of men's clothing would be interested in the modal neck size and sleeve length of potential customers. The modal income class of the laborers in the United States is of interest to the Labor Department. Thus, the mode provides a useful measure of central tendency for many business applications.

Unless a quantitative data set is very large, its mode may not be very meaningful. For example, consider the percentage of revenues spent on research and development (R&D) by 50 companies. These data, first presented in Table 2.4, were analyzed in the previous chapter. A reexamination of the data reveals that three of the measurements are repeated three times: 6.5%, 6.9%, and 8.2%. Thus, there are three modes in the sample, and none is particularly useful as a measure of central tendency.

We can calculate a more meaningful measure by first constructing a relative frequency histogram for the data. The measurement class containing the most measurements is called the modal class, and the mode is taken to be the midpoint of this interval. For the R&D data, the modal class is the one corresponding to the interval 7.35 to 8.45, as shown by the shaded rectangle in Figure 3.1. The mode is taken to be the midpoint of this interval,* that is, $(7.35 + 8.45)/2 = 7.90$. This modal class (and the mode itself) identifies the area in which the data are most concentrated, and in that sense is a measure of central tendency. However, for most applications involving quantitative data, other measures of central tendency provide more descriptive information than the mode.

FIGURE 3.1

Relative Frequency Histogram for the Computer Companies' R&D Percentages: The Modal Class

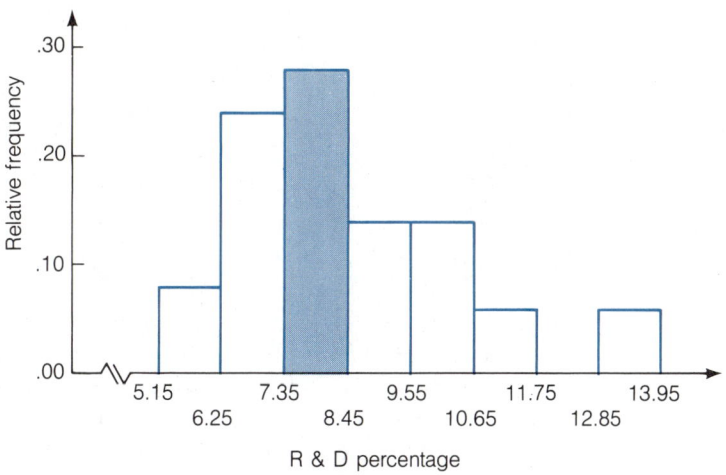

The most popular and best understood measure of central tendency for a quantitative data set is the **arithmetic mean** (or simply the **mean**):

| | | | | | | | | | | | |
S E C T I O N 3.2

THE ARITHMETIC MEAN: A MEASURE OF CENTRAL TENDENCY

DEFINITION 3.2

The **mean** of a set of quantitative data is equal to the sum of the measurements divided by the number of measurements contained in the data set.

Or, in nontechnical terms, the mean is the *average* value of the data set.

Before calculating the mean (or other numerical descriptive measures) of data sets, we present some shorthand notation that will simplify our calculation instructions. Remember that such notation is used for only one reason—to avoid having to repeat the same verbal descriptions over and over. If you mentally substitute the verbal definition of a symbol each time you read it, you will soon become accustomed to its use.

We will denote the measurements of a data set as follows:

$$x_1, x_2, x_3, \ldots, x_n$$

---

*There are several definitions for the mode of a relative frequency histogram. Our definition is one of the simplest, but it is adequate for an introductory discussion.

where x_1 is the first measurement in the data set, x_2 is the second measurement in the data set, x_3 is the third measurement in the data set, . . . , and x_n is the nth (and last) measurement in the data set. Thus, if we have five measurements in a set of data, we will write x_1, x_2, x_3, x_4, x_5 to represent the measurements. If the actual numbers are 5, 3, 8, 5, and 4, we have $x_1 = 5$, $x_2 = 3$, $x_3 = 8$, $x_4 = 5$, and $x_5 = 4$.

To calculate the mean of a set of measurements, we must sum them and divide by n, the number of measurements in the set. The sum of measurements x_1, x_2, $\cdots$, x_n is

$$x_1 + x_2 + \cdots + x_n$$

To shorten the notation, we will write this sum as

$$x_1 + x_2 + \cdots + x_n = \sum_{i=1}^{n} x_i$$

where Σ is the symbol for the summation. Verbally translate $\sum_{i=1}^{n} x_i$ as follows: "The sum of the measurements, whose typical member is x_i, beginning with the member x_1 and ending with the member x_n." The typical member will always appear following the Σ, the subscript of the first member of the summation will always appear below the Σ symbol, and the subscript of the last member of the summation will always appear above the Σ. For example,

$$\sum_{i=2}^{5} x_i = x_2 + x_3 + x_4 + x_5$$

where the typical element x_i follows the Σ symbol, the first element to appear in the sum is identified by the subscript 2 (shown below the Σ symbol), and the last element to appear in the sum is identified by the subscript 5 (shown above the Σ symbol).

Finally, we will denote the mean of a sample of measurements by $\bar{x}$ (read "x-bar"), and represent the formula for its calculation as follows:

$$\bar{x} = \frac{\sum_{i=1}^{n} x_i}{n}$$

EXAMPLE 3.1

Calculate the mean of the following five sample measurements: 5, 3, 8, 5, 6

SOLUTION

Using the definition of sample mean and the shorthand notation, we find

$$\bar{x} = \frac{\sum_{i=1}^{5} x_i}{5} = \frac{5 + 3 + 8 + 5 + 6}{5} = \frac{27}{5} = 5.4$$

∎

EXAMPLE 3.2

Refer to Table 3.1. Calculate the mean of the percentages of revenues spent by the 50 companies on research and development.

TABLE 3.1 Percentages of Revenues Spent on Research and Development

COMPANY	PERCENTAGE	COMPANY	PERCENTAGE	COMPANY	PERCENTAGE	COMPANY	PERCENTAGE
1	13.5	14	9.5	27	8.2	39	6.5
2	8.4	15	8.1	28	6.9	40	7.5
3	10.5	16	13.5	29	7.2	41	7.1
4	9.0	17	9.9	30	8.2	42	13.2
5	9.2	18	6.9	31	9.6	43	7.7
6	9.7	19	7.5	32	7.2	44	5.9
7	6.6	20	11.1	33	8.8	45	5.2
8	10.6	21	8.2	34	11.3	46	5.6
9	10.1	22	8.0	35	8.5	47	11.7
10	7.1	23	7.7	36	9.4	48	6.0
11	8.0	24	7.4	37	10.5	49	7.8
12	7.9	25	6.5	38	6.9	50	6.5
13	6.8	26	9.5				

SOLUTION

Using the data in Table 3.1, we have

$$\bar{x} = \frac{\sum_{i=1}^{50} x_i}{50} = \frac{13.5 + 8.4 + \cdots + 6.5}{50} = \frac{424.6}{50} = 8.49$$

The average expenditure on research and development for the 50 companies is 8.49% of revenues. Glancing at the relative frequency histogram for these data (Figure 3.1), we note that the mean falls in the middle of this data set, just above the modal class (7.35–8.45). ■

The sample mean will play an important role in accomplishing our objective of making inferences about populations based on sample information. For this reason, it is important to use a different symbol when we want to discuss the **mean of a population** of measurements—i.e., the mean of the entire set of measurements in which we are interested. We use the Greek letter μ (mu) for the population mean. We will adopt a general policy of using Greek letters to represent population numerical descriptive measures and Roman letters to represent corresponding descriptive measures for the sample.

> $\bar{x}$ = Sample mean μ = Population mean

The sample mean, $\bar{x}$, will often be used to estimate (make an inference about) the population mean, μ. For example, the population of percentages of revenues spent on research and development by *all* companies has a mean equal to some value, μ. Our sample of 50 percentages has a mean of $\bar{x} = 8.49$. If, as is usually the case, we did not have access to the population of measurements, we could use $\bar{x}$ as an estimator or approximator for μ. Then we would need to know something about the reliability of our inference. That is, we would need to know how accurately we might expect $\bar{x}$ to estimate μ. In Chapter 8, we will find that this accuracy depends on two factors:

1. *The size of the sample.* The larger the sample, the more accurate the estimate will tend to be.
2. *The variability or spread of the data.* All other factors remaining constant, the more variable the data, the less accurate the estimate.

In summary, the mean provides a valuable measure of the central tendency for a set of measurements. It is a very common tool in business and economic research, and therefore the mean will be the focus of much of our discussion of inferential statistics.

CASE STUDY 3.1

HOTELS: A RATIONAL METHOD FOR OVERBOOKING

The most outstanding characteristic of the general hotel reservation system is the option of the prospective guest, without penalty, to change or cancel his reservation or even to "no-show" (fail to arrive without notice). Overbooking (taking reservations in excess of the hotel capacity) is practiced widely throughout the industry as a compensating economic measure. This has motivated our research into the problem of determining policies for overbooking which are based on some set of rational criteria.

So said Marvin Rothstein (1974) in an article that appeared in *Decision Sciences*, a journal published by the Decision Sciences Institute. In this paper Rothstein introduces a method for scientifically determining hotel booking policies and applies it to the booking problems of the 133-room Sheraton Pocono Inn at Stroudsburg, Pennsylvania.

From the Sheraton Pocono Inn's records, the number of reservations, walk-ins (people without reservations who expect to be accommodated), cancellations, and no-shows were tabulated for each day during the period August 1–28, 1971. The inn's records for this period included approximately 3,100 guest histories. From the tabulated data, the mean or average number of room reservations per day for each of the 7 days of the week was computed, as shown in Table 3.2. In applying his booking policy decision method to the Sheraton's data, Rothstein used the means listed in Table 3.2 to help portray the inn's demand for rooms.

TABLE 3.2

Mean Number of Room Reservations, August 1–28, 1971, 133 Rooms

SUNDAY	MONDAY	TUESDAY	WEDNESDAY	THURSDAY	FRIDAY	SATURDAY
138	126	149	160	150	150	169

The mean number of Saturday reservations during the period August 1–28, 1971, is 169. This may be interpreted as an estimate of μ, the mean number of rooms demanded via reservations (walk-ins also contribute to the demand for rooms)

on a Saturday during 1971. If the reservation data for all Saturdays during 1971 had been tabulated, μ could have been computed. But, since only the August data are available, they were used to estimate μ. Can you think of some problems associated with using August's data to estimate the mean for the entire year?

CASE STUDY 3.2

MEASURING INVESTORS' REACTIONS TO A CORPORATE SELLOFF ANNOUNCEMENT: THE GENERAL ELECTRIC/ UTAH INTERNATIONAL CASE

In 1976, General Electric Company (GE) acquired Utah International Inc., an Australian mining company, in exchange for common stock valued at $2.17 billion. At the time, this was said to be the largest merger in U.S. history. In the Friday, January 28, 1983, edition of the *Wall Street Journal*, GE announced that it "agreed tentatively to sell most of its Utah International Inc. mining unit to an Australian natural resources company [Broken Hill Proprietary Co.] for $2.4 billion." When a parent firm sells a subsidiary, division, product line, or some other asset to another firm, the transaction is referred to as a *selloff* (or sometimes a *divestiture*). Payment is generally in the form of cash and/or marketable securities.

A question of interest to financial analysts with respect to GE's proposed selloff would be: How will investors interpret this information? As good news for GE? As bad news? Financial theory suggests that if investors perceive the announcement as good news, the stock's daily rate of return* will jump up on the day of the announcement or the following day, then subside to preannouncement level. Downward movement at the time of the announcement is an indication that the selloff is viewed as bad news. No movement suggests that the announcement contained little if any new information for investors.

Before examining GE's rates of return during the period in question, it is helpful to note that in studying stock prices (or rates of return) we make a distinction between (1) sampling from a population of prices and (2) sampling a consecutive series of prices. In the first case, we sample from a finite set of *existing* prices. For example, we might draw a sample of 20 closing prices from the last 100 closing prices of GE stock. This sample provides information about the population of 100 prices. In the second case, we think of prices as being generated or produced over time by a theoretical price-generating *process* in the same sense that manufactured goods are produced over time by a production *process*. Thus, when we observe a consecutive series of prices, we say we are sampling from the price-generating process. For example, we might decide to sample the next 20 closing prices of GE stock. This sample will provide information about the characteristics of the price-generating process.

One method that can be used to examine the effects of GE's announcement on GE's rates of return involves sampling GE's sequence of closing prices from, say, 20 days before the announcement to 20 days after the announcement. From these data, the daily rates of return over this period can be calculated. Then, GE's mean rate of return just prior to the announcement, $\bar{r}_1$, can be obtained, as well as the mean rate of return following the announcement, $\bar{r}_3$. These means can be compared

---

*If no dividends have been declared on the day in question, a stock's rate of return on day t is defined as $(P_t - P_{t-1})/P_{t-1}$, where P_t is the closing price of the stock on day t and P_{t-1} is its closing price the day before.

to the rate of return on the announcement day (and/or the mean rate of return for, say, the announcement day and the day following, $\bar{r}_2$) to examine the effect the announcement had on GE's rate of return. Table 3.3 shows both the rates of return during the period in question and the sample means described above. The fact that $\bar{r}_1$ and $\bar{r}_3$ are of approximately the same magnitude, while $\bar{r}_2$ is much larger, suggests that GE's selloff announcement was taken as good news by investors. For other more sophisticated approaches to using mean rates of return to study the effects of announcements such as GE's, see Brown and Warner (1980).

TABLE 3.3 Rates of Return on GE's Common Stock

DATE	RATE OF RETURN		DATE	RATE OF RETURN	
12/31/82	−.0078		1/28/83[a]	.0459	$\bar{r}_2 = .0424$
1/3/83	−.0329		1/31/83	.0389	
1/4/83	.0204		2/1/83	−.0254	
1/5/83	−.0067		2/2/83	−.0062	
1/6/83	.0255		2/3/83	−.0025	
1/7/83	.0131		2/4/83	.0050	
1/10/83	.0259		2/7/83	.0249	
1/11/83	−.0126		2/8/83	−.0121	
1/12/83	−.0038		2/9/83	.0074	
1/13/83	−.0077		2/10/83	.0110	
1/14/83	−.0052	$\bar{r}_1 = -.0017$	2/11/83	−.0084	
1/17/83	.0026		2/14/83	.0097	$\bar{r}_3 = .0026$
1/18/83	.0000		2/15/83	−.0169	
1/19/83	−.0091		2/16/83	−.0024	
1/20/83	−.0065		2/17/83	.0000	
1/21/83	−.0118		2/18/83	.0197	
1/24/83	−.0200		2/22/83	.0012	
1/25/83	−.0163		2/23/83	.0301	
1/26/83	−.0080		2/24/83	.0164	
1/27/83	.0270		2/25/83	−.0011	
			2/28/83	−.0014	

[a]Date selloff plans were announced in the *Wall Street Journal*.

THE MEDIAN: ANOTHER MEASURE OF CENTRAL TENDENCY

Another very important measure of central tendency is the **median** of a set of measurements:

DEFINITION 3.3

The **median** of a data set is the middle number when the measurements are arranged in ascending (or descending) order.

The median is of most value in describing large data sets. If the data set is characterized by a relative frequency histogram (see Figure 3.2), the median is the point on the x-axis such that half the area under the histogram lies above the median and half lies below. [*Note:* In Section 2.5, we observed that the relative frequency associated with a particular interval on the x-axis is proportional to the area under the histogram that lies above the interval.]

FIGURE 3.2

Location of the Median

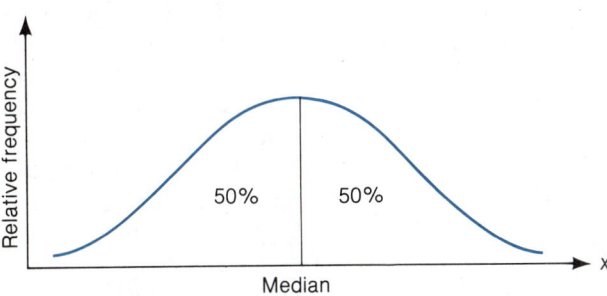

For a small, or even a large but finite, number of measurements, there may be many numbers that satisfy the property indicated in Figure 3.2. For this reason, we will arbitrarily calculate the median of a data set as follows:

CALCULATING A MEDIAN

Arrange the n measurements from the smallest to the largest.

1. If n is odd, the median is the middle number.
2. If n is even, the median is the mean (average) of the middle two numbers.

EXAMPLE 3.3

Consider the following sample of $n = 7$ measurements: 5, 7, 4, 5, 20, 6, 2

a. Calculate the median of this sample.
b. Eliminate the last measurement (the 2), and calculate the median of the remaining $n = 6$ measurements.

SOLUTION

a. The seven measurements in the sample are first ranked in ascending order:

   2, 4, 5, 5, 6, 7, 20

   Since the number of measurements is odd, the median is the middle measurement. Thus, the median of this sample is 5.

b. After removing the 2 from the set of measurements, we rank the sample measurements in ascending order as follows:

   4, 5, 5, 6, 7, 20

   Now the number of measurements is even, so we average the middle two measurements. The median is $(5 + 6)/2 = 5.5$. ∎

In certain situations, the median may be a better measure of central tendency than the mean. In particular, the median is less sensitive than the mean to extremely large or small measurements. To illustrate, note that all but one of the measurements in Example 3.3a center about $x = 5$. The single large measurement, $x = 20$, does not affect the value of the median, 5, but it shifts the mean, $\bar{x} = 7$, to the right of most of the measurements.

As another example, if you were interested in computing a measure of central tendency of the incomes of a company's employees, the mean might be misleading. If all blue- and white-collar employees' incomes are included in the data set, the high incomes of a few executives will influence the mean more than the median. Thus, the median will often provide a more accurate picture of the typical income for an employee. Similarly, the median yearly sales for a set of companies would locate the middle of the sales data. However, the very large yearly sales of a few companies would greatly influence the mean, making it deceptively large. That is, the mean could exceed a vast majority of the sample measurements, making it a misleading measure of central tendency.

For an example using more measurements, we have arranged the 50 R&D percentages in ascending order in Table 3.4. Since the number of measurements is even, the median equals the mean of the middle two numbers—that is, the mean of the 25th and 26th numbers in the ordered list:

$$\frac{8.0 + 8.1}{2} = 8.05$$

TABLE 3.4

Percentages of Revenues Spent on Research and Development in Ascending Order

PERCENTAGES			
5.2	7.1	8.2	9.9
5.6	7.2	8.2	10.1
5.9	7.2	8.2	10.5
6.0	7.4	8.4	10.5
6.5	7.5	8.5	10.6
6.5	7.5	8.8	11.1
6.5	7.7	9.0	11.3
6.6	7.7	9.2	11.7
6.8	7.8	9.4	13.2
6.9	7.9	9.5	13.5
6.9	8.0	9.5	13.5
6.9	8.0	9.6	
7.1	8.1	9.7	

Note that the median is smaller than the mean (8.49) for these data. This fact indicates that the data are **skewed** to the right—i.e., the measurements tend to be pulled toward the right tail of the distribution. This affects the mean more than the median, because the extreme values (large or small) are used explicitly in the calculation of the mean when all the measurements are summed (and divided by

COMPARING THE MEAN AND THE MEDIAN

1. If the median is less than the mean, the data set is skewed to the right:

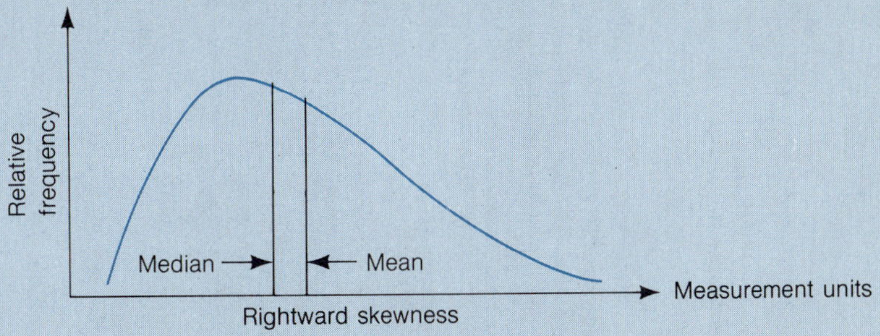

Relative frequency

Median ───► ◄─── Mean

Rightward skewness

Measurement units

2. The median will equal the mean when the data set is symmetric:

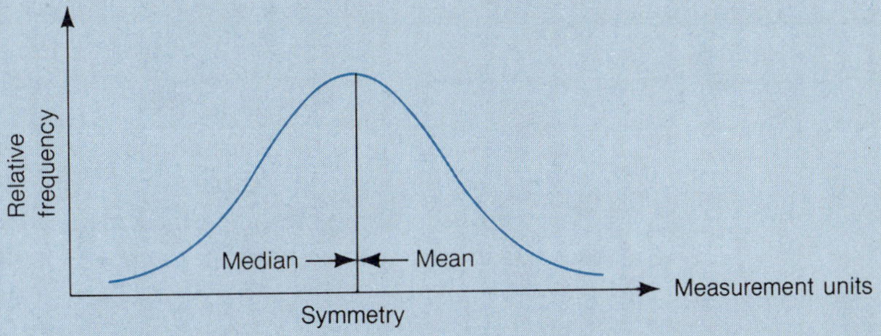

Relative frequency

Median ───► ◄─── Mean

Symmetry

Measurement units

3. If the median is greater than the mean, the data set is skewed to the left:

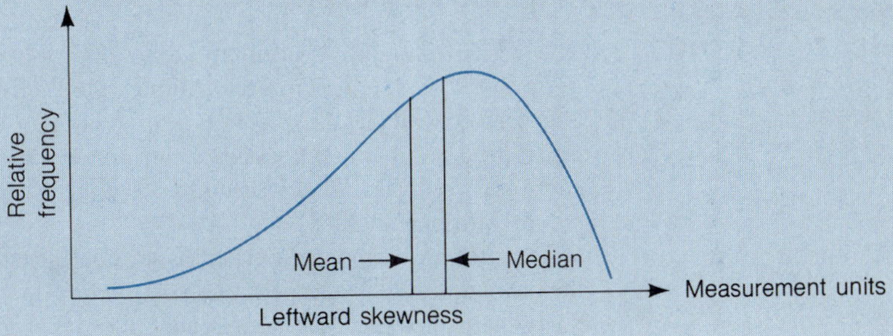

Relative frequency

Mean ───► ◄─── Median

Leftward skewness

Measurement units

the sample size). On the other hand, the median is not affected directly by extreme measurements, since the middle measurement(s) is (are) the only one(s) explicitly used to calculate the median. Consequently, if measurements are pulled toward one end of the distribution, the mean will shift toward that tail. The skewness of the R&D data set is evident in Figure 3.3 (page 86), where we show the median and mean of the R&D percentages.

FIGURE 3.3
Relative Frequency
Histogram for the R&D
Percentages: Mean and
Median

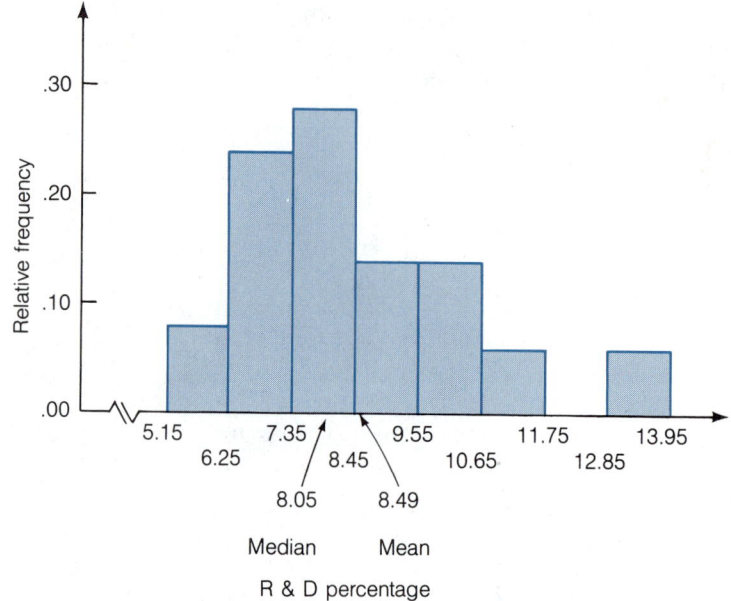

A comparison of the mean and median gives us a general method for detecting skewness in data sets, as shown in the preceding box.

CASE STUDY 3.3

THE DELPHI TECHNIQUE FOR OBTAINING A CONSENSUS OF OPINION

George T. Milkovich, Anthony J. Annoni, and Thomas A. Mahoney (1972) explain the delphi technique as follows:

The delphi technique, a set of procedures originally developed by the Rand Corporation in the late 1940's, is designed to obtain the most reliable consensus of opinion of a group of experts. Essentially, the delphi is a series of intensive interrogations of each individual expert (by a series of questionnaires) concerning some primary question interspersed with controlled feedback. The procedures are designed to avoid direct confrontation of the experts with one another.

The interaction among the experts is accomplished through an intermediary who gathers the data requests of the experts and summarizes them along with the experts' answers to the primary question. This mode of controlled interaction among the experts is a deliberate attempt to avoid the disadvantages associated with more conventional use of experts such as in round table discussions or direct confrontation of opposing views. The developers of the delphi argue the procedures are more conducive to independent thought and allow more gradual formulation to a considered opinion.

This article presents a study of the usefulness of the delphi procedure in projecting labor requirements in a low profit margin national retail firm. Seven company executives formed the panel of experts. Five questionnaires submitted at approximately 8-day intervals were used to interrogate the seven experts. On questionnaires 2–5, they were each asked the primary question: "How many buyers will the firm need 1 year from now?" Their individual responses along with the median response of the group for each questionnaire appear in Table 3.5.

QUESTIONNAIRE	EXPERTS							MEDIAN
	A	B	C	D	E	F	G	
2	55	35	33	35	55	33	32	35
3	45	35	41	35	41	34	32	35
4	45	38	41	35	41	34	34	38
5	45	38	41	35	45	34	34	38

TABLE 3.5
Projected Demand
for Buyers

Note that the median response increased from 35 on questionnaire 2 to 38 on questionnaires 4 and 5. This increase indicates an upward shift in the distribution of the experts' estimates as to the number of buyers the firm would need a year from now.

One conclusion of the study was that the delphi technique provided closer estimates of the actual number of buyers (37) needed by the firm 1 year later than did other more conventional estimating techniques.

EXERCISES 3.1–3.11

LEARNING THE MECHANICS

3.1 According to *Consumers' Digest* (Nov.–Dec. 1980), sugar is the leading food additive in the U.S. food supply. Sugar may be listed more than once on a product's ingredient list since it goes by different names depending on its source (e.g., sucrose, corn sweetener, fructose, and dextrose). Thus, when you read a product's label you may have to total up the sugar in the product to see how much sweetener it contains. The accompanying table gives a list of candy bars and the percentage of sugar they contain relative to their weight.

BRAND	PERCENTAGE OF SUGAR BY WEIGHT	BRAND	PERCENTAGE OF SUGAR BY WEIGHT
Baby Ruth	23.7	Power House	30.6
Butterfinger	29.5	Bit-O-Honey	23.5
Mr. Goodbar	34.2	Chunky	38.4
Milk Duds	36.0	Milk Chocolate Covered	
Mello Mint	79.6	Raisinettes	24.7
M & M Plain Chocolate		Oh Henry!	31.2
Candies	52.2	Borden Cracker Jack	14.7
Mars Chocolate Almond	36.4	Good & Plenty Licorice	28.2
Milky Way	26.8	Nestle's Crunch	43.5
Marathon	36.7	Planter Jumbo Block	
Snickers	28.0	Peanut Candy	21.5
3 Musketeers	36.1	Switzer Licorice	8.4
Junior Mints	45.3	Switzer Red Licorice	2.8
Pom Poms	29.5	Tootsie Pop Drops	54.1
Sugar Babies	41.0	Tootsie Roll	21.1
Sugar Daddy	22.0	Fancy Fruit Lifesavers	77.6
Almond Joy	20.0	Spear-O-Mint Lifesavers	67.6

Source: *National Confectioners Association Brand Name Guide to Sugar* (Nelson Hall Paperback).
Secondary Source: *Consumers' Digest*, Nov.–Dec. 1980, p. 11.

a. Calculate the mean percentage of sugar per bar for the candy bars listed.
b. Find the median for the data set.
c. What do the mean and median indicate about the skewness of the data set?
d. Construct a relative frequency histogram for the data set. Indicate the location of the mean, median, and modal class of the data set on your histogram.

3.2 During the 1980's the pharmaceutical industry has placed an increased emphasis on producing revolutionary new products. As a result, research and development (R&D) costs have increased and companies are taking a greater interest in R&D management. The table lists the 1984 R&D expenditures (in millions of dollars) of the world's largest pharmaceutical manufacturers.

COMPANY	R&D EXPENDITURES
Abbott	$110
American Home	90
Bayer	200
Boehringer Ingelheim	176
Bristol–Myers	162
Ciba–Geigy	230
Hoechst	274
Hoffman–LaRoche	363
Johnson & Johnson	187
Merck	290
Pfizer	159
Rhone–Poulenc	110
Sandoz	181
Schering–Plough	129
Smith Kline–Beckman	158
Squibb	114
Takeda	125
Upjohn	200
Warner–Lambert	162

Source: *Business Quarterly*, Fall 1985, p. 81.

a. Calculate the mean and median for this data set.
b. What do the mean and median indicate about the skewness of this data set?
c. Will the median of a data set always be equal to an actual value in the data set, as was the case in part a?

3.3 Thirty stocks were selected from the New York Stock Exchange Composite Transactions Table published in the March 5, 1986 edition of the *Wall Street Journal*. These stocks are listed in the table (using the *Wall Street Journal* abbreviations) along with their closing prices on March 4, 1986, and the change in each closing price from March 3 to March 4.

a. Construct a relative frequency histogram for the net changes in closing price for this sample of stocks.
b. Calculate the mean, median, and mode for the net change data and locate them on your histogram of part a.
c. According to your histogram, what is the modal class of the net change data set? Shade the appropriate rectangle of your histogram.

STOCK	CLOSING PRICE	NET CHANGE	STOCK	CLOSING PRICE	NET CHANGE
Alcoa	42	$-1\frac{1}{2}$	MCA	$51\frac{7}{8}$	$+\frac{7}{8}$
AT & T	$48\frac{7}{8}$	$-\frac{1}{8}$	Munsg	$19\frac{1}{4}$	$-\frac{3}{4}$
Avon	31	$-\frac{1}{4}$	Nwst	$52\frac{3}{4}$	$+\frac{1}{2}$
Bnk Am	$15\frac{1}{2}$	$-\frac{7}{8}$	PepsiCo	$78\frac{1}{4}$	$+\frac{5}{8}$
Boeing	53	$+\frac{1}{8}$	Polarid	68	$-\frac{5}{8}$
CBS	$136\frac{1}{4}$	$-5\frac{3}{4}$	QuaO	$110\frac{1}{2}$	$+1\frac{1}{2}$
Chryslr	$55\frac{3}{4}$	0	RepAir	$15\frac{1}{2}$	0
Citicrp	$54\frac{1}{8}$	$+\frac{3}{8}$	Rockwl	$38\frac{1}{8}$	$-\frac{7}{8}$
ColgPal	$35\frac{3}{8}$	$+\frac{5}{8}$	Seagrm	$51\frac{7}{8}$	$-1\frac{3}{8}$
DeltaAr	$41\frac{1}{2}$	$+\frac{1}{4}$	Sears	$44\frac{1}{2}$	$+\frac{1}{8}$
Exxon	$51\frac{7}{8}$	-1	Tndycft	15	$-\frac{1}{4}$
FedExp	$70\frac{7}{8}$	0	ToroCo	$21\frac{5}{8}$	$+\frac{1}{2}$
GnMills	$72\frac{7}{8}$	$+3\frac{5}{8}$	USSteel	$23\frac{5}{8}$	$+\frac{1}{8}$
Hershy	$56\frac{3}{8}$	$+\frac{3}{8}$	Upjohn	$149\frac{7}{8}$	$-1\frac{1}{8}$
Kroger	$26\frac{1}{4}$	$+\frac{1}{4}$	WinDix	$36\frac{5}{8}$	-2

Reprinted by permission of *Wall Street Journal,* © Dow Jones & Company, Inc. 1986. All rights reserved.

3.4 The table lists the mean age for each team in the National Basketball Association (NBA), along with the number of players on each team at the start of the 1982–1983 season. What was the population mean age at the start of the 1982–1983 season?

[*Hint:* $\sum_{i=1}^{n} x_i = n\bar{x}$]

TEAM	NUMBER OF PLAYERS	MEAN AGE	TEAM	NUMBER OF PLAYERS	MEAN AGE
1. Indiana	12	24.720	13. San Antonio	12	26.25
2. Portland	12	24.724	14. Los Angeles	13	26.42
3. Golden State	15	24.96	15. New York	14	26.45
4. Dallas	13	25.17	16. Cleveland	14	26.48
5. New Jersey	13	25.26	17. San Diego	13	26.71
6. Kansas City	12	25.27	18. Boston	12	26.94
7. Chicago	12	25.32	19. Seattle	11	27.09
8. Detroit	13	25.65	20. Atlanta	12	27.12
9. Philadelphia	12	25.96	21. Denver	11	27.59
10. Phoenix	13	26.10	22. Houston	12	29.21
11. Washington	13	26.10	23. Milwaukee	13	29.72
12. Utah	11	26.13			

Source: *Basketball Weekly,* Vol. 16, No. 5, Jan. 3, 1983, pp. 10–11.

APPLYING THE CONCEPTS

3.5 According to the American Automobile Association, the average price of self-service unleaded gasoline in the United States as of December 19, 1985, was 16 cents per gallon cheaper than the average price of full-service unleaded gas. The table lists the average price (in cents per gallon) of full-service unleaded gas in each of a sample of 20 states.

STATE	PRICE	STATE	PRICE
Alaska	142.0	Nevada	147.4
Arkansas	135.0	New Hampshire	131.2
Connecticut	134.6	New York	132.0
Delaware	131.2	North Dakota	139.0
Louisiana	142.6	Oklahoma	133.9
Maine	129.8	Oregon	127.0
Massachusetts	124.8	Pennsylvania	131.7
Michigan	141.6	Texas	137.5
Missouri	131.7	Wisconsin	139.0
Montana	135.9	Wyoming	136.9

Source: American Automobile Association *Fuel Gauge Report*, Dec. 19, 1985.

a. Calculate the mean, median, and mode of this data set.
b. Eliminate the highest price from the data set and repeat part **a**. What effect does dropping this measurement have on the measures of central tendency calculated in part **a**?
c. Arrange the 20 prices in order from lowest to highest. Next, eliminate the lowest two prices and the highest two prices from the data set and calculate the mean of the remaining prices. The result is called an **80% trimmed mean** since it is calculated using the central 80% of the values in the data set. An advantage of the trimmed mean is that it is not as sensitive as the arithmetic mean to extreme observations in the data set.

3.6 At the end of 1982, McDonald's had 7,300 restaurants and total sales for the year of $7.8 billion. Burger King had 3,400 restaurants and total sales of $2.4 billion. McDonald's opens new restaurants at the rate of 500 per year and Burger King at the rate of 200 per year (*Minneapolis Tribune*, July 3, 1983).
a. Calculate the average sales per restaurant for McDonald's in 1982 and compare it with the average sales per restaurant for Burger King in 1982.
b. On average, how many restaurants does McDonald's open per month? Per week? Per day?

3.7 In 1985, U.S. consumers redeemed 6.49 billion manufacturers' coupons worth a total of $2.24 billion (McCullough, 1986). Find the mean value per coupon.

3.8 The accompanying table contains the 1984 price per acre of farmland for a sample of states that includes nine eastern states and eleven western states (i.e., west of the Mississippi River).
a. Find the mean and median price per acre for the sample of 20 states.

STATE	PRICE PER ACRE	STATE	PRICE PER ACRE
Arizona	$ 265	Nebraska	$ 444
California	1,726	Nevada	229
Colorado	435	New Hampshire	1,419
Connecticut	3,208	New Jersey	3,525
Delaware	1,642	New Mexico	163
Florida	1,527	North Dakota	360
Kansas	466	Pennsylvania	1,510
Massachusetts	2,372	Rhode Island	3,335
Maryland	2,097	South Dakota	250
Montana	222	Wyoming	177

Data: U.S. Department of Agriculture.

b. Find and compare the mean prices per acre for the eastern states and the western states. Also, compare these means to the mean you found in part **a**. What do your comparisons reveal about the value of farmland in the United States?

c. As a measure of central tendency, the mean of a data set is frequently used to characterize the data set or to represent a typical measurement in the data set. Examine the data set and determine whether you would use the mean of these data to characterize a typical measurement. Explain.

3.9 In 1972, Kroger Corporation, the second largest supermarket chain in the United States, made a major strategic decision. Instead of continuing to hold prices high and trying to attract customers with weekend specials and heavy advertising, Kroger decided to emphasize price competition all week long. Its new strategy involved selling "brand-name groceries for less, on average, than its competitors were doing, and to advertise this fact strenuously" ("Keeping Up," 1979). For the situation described, explain what is meant by "on average" in the above quote.

3.10 Refer to Case Study 3.3. Practitioners of the delphi technique frequently use the median of the set of responses given by the panel of experts on the last questionnaire to describe "the opinion of the experts." Why do you suppose the median response is used rather than the mean response?

3.11 An *invoice* is a document that indicates the seller has completed his or her part of a sales contract. It states the names of the buyer and the seller and indicates what product or products have been shipped or delivered to the buyer, the date and method of shipment, and the prices, total sale figure, and terms of payment (Vatter, 1971).

A mail-order firm is interested in describing the size (in dollars) of the orders it received during the past fiscal year. To do so, the firm's accounting office randomly selected 100 invoices from the collection of all invoices written last year and recorded the "total sale" figure listed on each. These figures, rounded to the nearest dollar, are listed at the top of page 92.

a. Check the direction of skewness of the data set by comparing the median and mean.

b. Verify your answer to part **a** by constructing a frequency histogram of the data set. Locate the median and the mean of the data set on your histogram.

c. Suppose two more invoices were examined and their total sale figures, $20 and $250, were included in the data set. How would this affect the median and mean of the data set?

1	21	31	33	41	42	44	23	22	5
11	25	8	20	20	8	43	19	45	20
15	30	10	30	29	9	80	25	48	19
17	25	22	28	35	32	88	21	48	36
34	32	24	27	43	18	75	24	126	26
39	40	36	28	52	33	66	35	49	34
15	19	11	24	55	27	92	45	2	33
30	30	12	29	55	58	7	110	10	33
38	29	27	37	44	58	90	24	63	70
10	30	28	28	29	19	20	13	12	6

| | | | | | | | | | | | |

SECTION 3.4

THE RANGE: A MEASURE OF VARIABILITY

In the preceding sections, we presented some methods for measuring the central tendency of a quantitative data set. However, central tendency tells only part of the story. Our information is incomplete without a measure of the **variability** or **spread of the data set.** Note that in describing a data set, we refer to either the sample or the population. Ultimately (in Chapter 8) we will use the sample numerical descriptive measures (statistics) to make inferences about the corresponding descriptive measures for the population from which the sample was selected.

If you examine the two histograms in Figure 3.4, you will notice that both hypothetical data sets are symmetric, with equal modes, medians, and means. However, in data set 1 in Figure 3.4(a), the measurements occur with almost equal frequency in the measurement classes, while in data set 2 in Figure 3.4(b), most of the measurements are clustered about the center. For this reason, a measure of variability is needed, along with a measure of central tendency, to describe a data set.

FIGURE 3.4 Two Hypothetical Data Sets

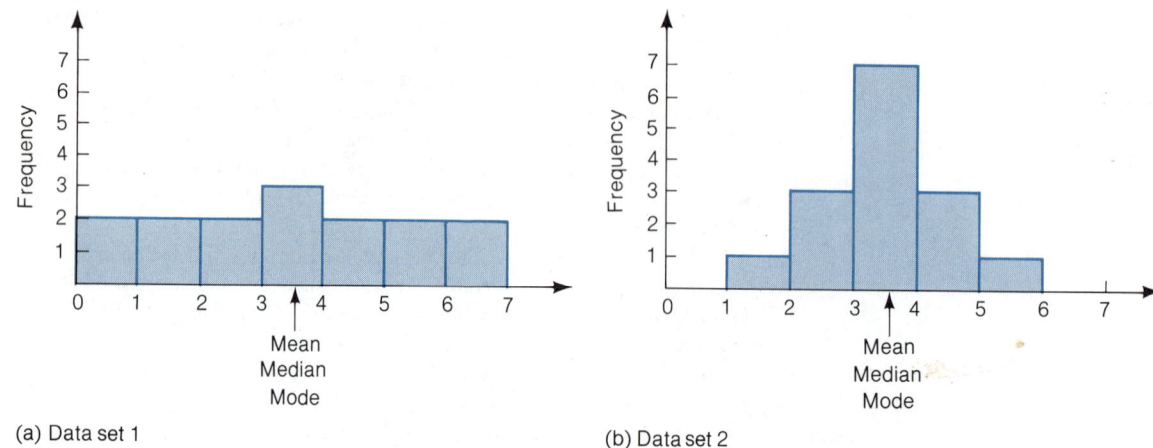

(a) Data set 1

(b) Data set 2

Perhaps the simplest measure of the variability of a quantitative data set is its **range**.

DEFINITION 3.4

The **range** of a data set is equal to the largest measurement minus the smallest measurement.

The range measures the spread of the data by measuring the distance between the smallest and largest measurements. For example, stock A may vary in price during a given year from $32 to $36, while stock B may vary from $10 to $58, as shown in Figure 3.5. The range in price of stock A is $36 − $32 = $4, while that for stock B is $58 − $10 = $48. A comparison of ranges tells us that the price of stock B was much more variable than the price of stock A.

FIGURE 3.5
Ranges of Stock Prices for Two Companies

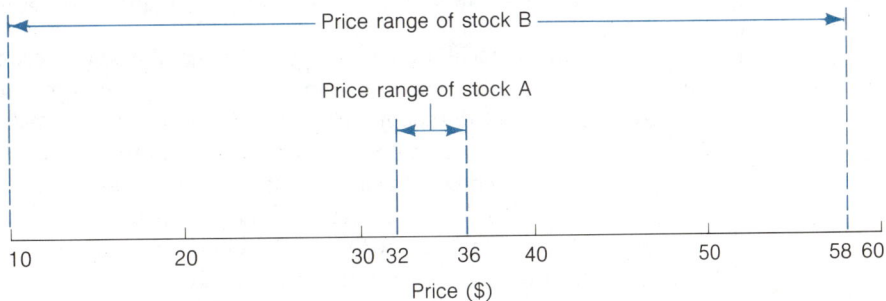

The range is not always a satisfactory measure of variability. For example, suppose we are comparing the profit margin (as a percentage of the total bid price) per construction job for 100 construction jobs for each of two cost estimators working for a large construction company. We find that the profit margins range from −10% (loss) to +40% (profit) for both cost estimators and therefore that the ranges for the two data sets, 40% − (−10%) = 50%, are equal. Because of this, we might be inclined to conclude that there is little or no difference in the performance of the two estimators.

But, suppose the histograms for the two sets of 100 profit margin measurements appear as shown in Figure 3.6 (page 94). Although the ranges are equal and all central tendency measures are the same for these two symmetric data sets, there is an obvious difference between the two sets of measurements. The difference is that estimator B's profit margins tend to be more stable—i.e., to pile up or to cluster about the center of the data set. In contrast, estimator A's profit margins are more spread out over the range, indicating a higher incidence of some high profit margins, but also a greater risk of losses. Thus, even though the ranges are equal, the profit margin record of estimator A is more variable than that of estimator B, indicating a distinct difference in their cost estimating characteristics. We therefore need to develop more informative numerical measures of variability than the range. In particular, we need a measure that takes into consideration the magnitude of all measurements, not just the largest and smallest.

FIGURE 3.6

Profit Margin Histograms
for Two Cost Estimators

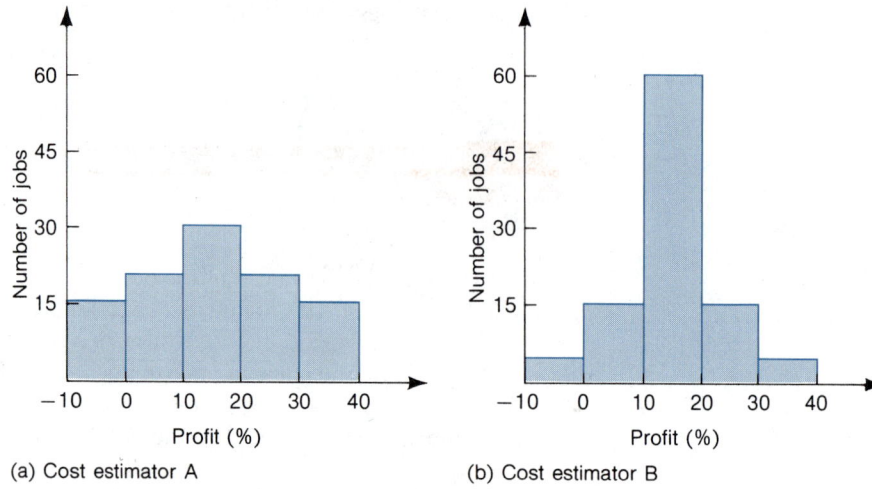

(a) Cost estimator A   (b) Cost estimator B

CASE STUDY 3.4

MORE ON THE DELPHI
TECHNIQUE

You will recall from Case Study 3.3 that the delphi technique is a set of procedures that may be used to obtain a consensus opinion from a group of experts through a series of questionnaires. Case Study 3.3 illustrated the use of the median as a measure of central tendency for the distribution of expert opinions elicited by the questionnaires. As a measure of variability of the data (i.e., the opinions), Milkovich et al. (1972) used the range. Table 3.5, showing the experts' opinions, is repeated here as Table 3.6, with the addition of the range of the distribution of opinions of each questionnaire in the right-hand column.

The range of 23 on questionnaire 2 indicates that at that time the experts' opinions were widely dispersed. The decrease in the range to 11 following questionnaire 4 indicates that as the experts received more information about the firm's needs and learned about one another's opinions, the variability in the distribution of their opinions decreased. Milkovich et al. noted that the decrease in the range was an indication that experts' opinions were converging.

TABLE 3.6

Projected Demand for
Buyers

QUESTIONNAIRE	EXPERTS							MEDIAN	RANGE
	A	B	C	D	E	F	G		
2	55	35	33	35	55	33	32	35	23
3	45	35	41	35	41	34	32	35	13
4	45	38	41	35	41	34	34	38	11
5	45	38	41	35	45	34	34	38	11

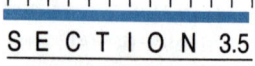

SECTION 3.5

**VARIANCE AND
STANDARD DEVIATION**

Recall that we represent the n measurements in a sample by the symbols x_1, x_2, . . . , x_n, and we represent their mean by $\bar{x}$. What would be the interpretation of $x_1 - \bar{x}$? It is the distance, or **deviation**, between the first sample measurement, x_1, and the sample mean, $\bar{x}$. If we were to calculate the distance for *every* measurement in the sample, we would create a set of distances from the mean:

$$x_1 - \bar{x}, \quad x_2 - \bar{x}, \quad x_3 - \bar{x}, \quad . . . , \quad x_n - \bar{x}$$

What information do these distances contain? If they tend to be large, the interpretation is that the data are spread out or highly variable. If the distances are mostly small, the data are clustered around the mean $\bar{x}$ and therefore do not exhibit much variability. As a simple example, consider the two samples in Table 3.7, which have five measurements (we have ordered the numbers for convenience). You will note that both samples have a mean of 3. However, a glance at the distances shows that sample 1 has greater variability—i.e., more large distances from $\bar{x}$—than sample 2, which is clustered around $\bar{x}$. You can see this clearly by looking at these distances in Figure 3.7. Thus, the distances provide information about the variability of the sample measurements.

TABLE 3.7

	SAMPLE 1	SAMPLE 2
Measurements	1, 2, 3, 4, 5	2, 3, 3, 3, 4
Mean	$\bar{x} = \dfrac{1 + 2 + 3 + 4 + 5}{5} = \dfrac{15}{5}$ $= 3$	$\bar{x} = \dfrac{2 + 3 + 3 + 3 + 4}{5} = \dfrac{15}{5}$ $= 3$
Distances from $\bar{x}$	$1-3,\ 2-3,\ 3-3,\ 4-3,\ 5-3,$ or $-2,\quad -1,\quad 0,\quad 1,\quad 2$	$2-3,\ 3-3,\ 3-3,\ 3-3,\ 4-3$ or $-1,\quad 0,\quad 0,\quad 0,\quad 1$

The next step is to condense the information on distances from $\bar{x}$ into a single numerical measure of variability. Simply averaging the distances from $\bar{x}$ will not help. For example, in samples 1 and 2 the negative and positive distances cancel, so that the average distance is 0. Since this is true for any data set—i.e., the sum of the deviations, $\sum_{i=1}^{n} (x_i - \bar{x})$, is always 0—we gain no information by averaging the distances from $\bar{x}$.

FIGURE 3.7
Distances from the Mean
for Two Data Sets

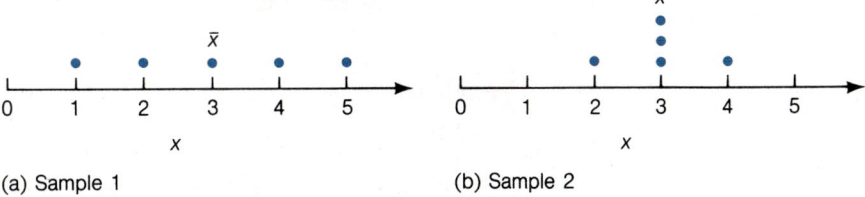

(a) Sample 1   (b) Sample 2

There are two methods for dealing with the fact that positive and negative distances from the mean cancel. The first is to treat all the distances as though they were positive, ignoring the sign of the negative distances. We will not pursue this line of thought because the resulting measure of variability (the mean of the absolute values of the distances) presents analytical difficulties beyond the scope of this text. A second method of eliminating the minus signs associated with the distances is to square them. The quantity we can calculate from the squared distances will provide a meaningful description of the variability of a data set.

To use the squared distances calculated from a data set, we first calculate the **sample variance**:

DEFINITION 3.5

The **sample variance** for a sample of n measurements is equal to the sum of the squared distances from the mean divided by $(n - 1)$. In symbols, using s^2 to represent the sample variance,

$$s^2 = \frac{\sum_{i=1}^{n} (x_i - \bar{x})^2}{n - 1}$$

Referring to the two samples in Table 3.7, you can calculate the variance for sample 1 as follows:

$$s^2 = \frac{(1 - 3)^2 + (2 - 3)^2 + (3 - 3)^2 + (4 - 3)^2 + (5 - 3)^2}{5 - 1}$$

$$= \frac{4 + 1 + 0 + 1 + 4}{4} = 2.5$$

The second step in finding a meaningful measure of data variability is to calculate the **standard deviation** of the data set.

DEFINITION 3.6

The **sample standard deviation**, s, is defined as the positive square root of the sample variance, s^2. Thus,

$$s = \sqrt{s^2} = \sqrt{\frac{\sum_{i=1}^{n} (x_i - \bar{x})^2}{n - 1}}$$

The **population variance**, denoted by the symbol σ^2 (sigma squared), is the average of the squared distances of the observations from the mean, μ, and σ (sigma) is the square root of this quantity. Since we never really compute σ^2 or σ from the population (the object of sampling is to avoid this costly procedure), we simply denote these two quantities by their respective symbols.

$$s^2 = \text{Sample variance}$$
$$\sigma^2 = \text{Population variance}$$
$$s = \text{Sample standard deviation}$$
$$\sigma = \text{Population standard deviation}$$

Notice that, in contrast to the variance, the standard deviation is expressed in the original units of measurement. For example, if the original measurements are in dollars, the standard deviation will be expressed in dollars. Second, you may wonder why we use the divisor $(n - 1)$ instead of n when calculating the sample variance. This is because using the divisor $(n - 1)$ yields a better* estimate of σ^2 than is obtained by dividing the sum of the squared distances by n. Since we will ultimately want to use sample statistics to make inferences about numerical descriptive measures of the corresponding population, $(n - 1)$ is preferred to n when defining the sample variance.

EXAMPLE 3.4

Calculate the standard deviation of the following sample: 2, 3, 3, 3, 4

SOLUTION

For this set of data, $\bar{x} = 3$. Then,

$$s = \sqrt{\frac{(2 - 3)^2 + (3 - 3)^2 + (3 - 3)^2 + (3 - 3)^2 + (4 - 3)^2}{5 - 1}}$$

$$= \sqrt{\frac{2}{4}} = \sqrt{.5} = .71 \qquad \blacksquare$$

Example 3.4 may have raised two thoughts in your mind. First, calculating s^2 and s can be very tedious if $\bar{x}$ is a number that contains a large number of significant figures or if there are a large number of measurements in a data set. Second, we have not explained how a sample standard deviation can be used to describe the variability of a data set. Fortunately, we have an easier method for calculating s^2 and s, and this method will be explained in Section 3.6. The interpretation of s will be the subject of Section 3.7.

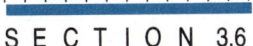

SECTION 3.6

CALCULATION FORMULAS FOR VARIANCE AND STANDARD DEVIATION

As the number of measurements in the sample becomes larger, the sample variance becomes more difficult to calculate. We must calculate the distance between each measurement and the mean, square it, sum the squared distances, and finally divide by $(n - 1)$. Fortunately, there is a shortcut formula for computing the sample variance:

SHORTCUT FORMULA FOR SAMPLE VARIANCE

$$s^2 = \frac{(\text{Sum of squares of sample measurements}) - \dfrac{(\text{Sum of sample measurements})^2}{n}}{n - 1}$$

$$= \frac{\displaystyle\sum_{i=1}^{n} x_i^2 - \dfrac{\left(\displaystyle\sum_{i=1}^{n} x_i\right)^2}{n}}{n - 1}$$

---

*Better here means that s^2 with a divisor of $(n - 1)$ is an **unbiased** estimator of σ^2. We will define and discuss unbiasedness of estimators in Chapter 7.

Note that the formula requires only the sum of the sample measurements, $\sum_{i=1}^{n} x_i$, and the sum of the squares of the sample measurements, $\sum_{i=1}^{n} x_i^2$. Be careful when you calculate these two sums. Rounding the values of x^2 that appear in $\sum_{i=1}^{n} x_i^2$ or rounding the quantity $\left(\sum_{i=1}^{n} x_i\right)^2 \Big/ n$ can lead to substantial errors in the calculation of s^2.

EXAMPLE 3.5

Use the shortcut formula to compute the variances of these two samples of five measurements each:

Sample 1: 1, 2, 3, 4, 5   Sample 2: 2, 3, 3, 3, 4

SOLUTION

We first work with sample 1. The two quantities needed are

$$\sum_{i=1}^{5} x_i = 1 + 2 + 3 + 4 + 5 = 15$$

and

$$\sum_{i=1}^{5} x_i^2 = 1^2 + 2^2 + 3^2 + 4^2 + 5^2 = 1 + 4 + 9 + 16 + 25 = 55$$

Then the sample variance for sample 1 is

$$s^2 = \frac{\sum_{i=1}^{5} x_i^2 - \frac{\left(\sum_{i=1}^{5} x_i\right)^2}{5}}{5 - 1} = \frac{55 - \frac{(15)^2}{5}}{4} = \frac{55 - 45}{4} = \frac{10}{4} = 2.5$$

Similarly, for sample 2 we get

$$\sum_{i=1}^{5} x_i = 2 + 3 + 3 + 3 + 4 = 15$$

and

$$\sum_{i=1}^{5} x_i^2 = 2^2 + 3^2 + 3^2 + 3^2 + 4^2 = 4 + 9 + 9 + 9 + 16 = 47$$

Then the variance for sample 2 is

$$s^2 = \frac{\sum_{i=1}^{5} x_i^2 - \frac{\left(\sum_{i=1}^{5} x_i\right)^2}{5}}{5 - 1} = \frac{47 - \frac{(15)^2}{5}}{4} = \frac{47 - 45}{4} = \frac{2}{4} = .5$$

Note that these results agree with our calculations in the previous section. ∎

EXAMPLE 3.6

The 50 companies' percentages of revenues spent on R&D are repeated here. Calculate the sample variance, s^2, and the standard deviation, s, for these measurements.

13.5	9.5	8.2	6.5	8.4	8.1	6.9	7.5	10.5	13.5
7.2	7.1	9.0	9.9	8.2	13.2	9.2	6.9	9.6	7.7
9.7	7.5	7.2	5.9	6.6	11.1	8.8	5.2	10.6	8.2
11.3	5.6	10.1	8.0	8.5	11.7	7.1	7.7	9.4	6.0
8.0	7.4	10.5	7.8	7.9	6.5	6.9	6.5	6.8	9.5

SOLUTION

The calculation of the sample variance, s^2, would be very tedious for this sample if we tried to use the formula

$$s^2 = \frac{\sum_{i=1}^{50} (x_i - \bar{x})^2}{50 - 1}$$

because it would be necessary to compute all 50 squared distances from the mean. However, for the shortcut formula we need compute only

$$\sum_{i=1}^{50} x_i = 13.5 + 8.4 + \cdots + 6.5 = 424.6$$

and

$$\sum_{i=1}^{50} x_i^2 = (13.5)^2 + (8.4)^2 + \cdots + (6.5)^2 = 3{,}797.92$$

Then

$$s^2 = \frac{\sum_{i=1}^{50} x_i^2 - \dfrac{\left(\sum_{i=1}^{50} x_i\right)^2}{50}}{50 - 1} = \frac{3{,}797.92 - \dfrac{(424.6)^2}{50}}{49}$$

$$= 3.9228$$

The standard deviation is

$$s = \sqrt{s^2} = \sqrt{3.9228} = 1.98$$

Notice that we retained all the decimal places in the calculation of the sum of squares of the measurements. This was done to reduce the rounding error in the calculations, even though the original data were accurate to only one decimal place. *

*The accuracy of the original data has nothing to do with the degree of accuracy used in computing s^2 and s. You should retain twice as many decimal places in s^2 as you want in s. For example, if you want to calculate s to the nearest hundredth, you should calculate s^2 to the nearest ten-thousandth.

EXAMPLE 3.7

Use a statistical software package to compute the median, mean, variance, and standard deviation of the R&D data given in Example 3.6.

SOLUTION

The SAS/PC printout is shown in Figure 3.8. The median, mean, variance, and standard deviation are shaded on the printout. Although the SAS procedure (PROC UNIVARIATE) generates many other descriptive statistics, some of which we will discuss later, we can easily pick out those of interest to us. Note that many decimal places are carried by the program, but when we round to the same number of decimal places used in the previous examples, we find

$$\text{Median} = 8.05$$
$$\bar{x} = 8.49$$
$$s^2 = 3.9228$$
$$s = 1.98$$

The answers are identical to those obtained by hand calculation.

FIGURE 3.8

SAS/PC Printout for Mean, Median, Variance, and Standard Deviation

```
                          Moments

N                      50  Sum Wgts              50
Mean                8.492  Sum                424.6
Std Dev          1.980604  Variance        3.922792
Skewness         .8546013  Kurtosis        .4192877
USS               3797.92  CSS             192.2168
CV               23.32317  Std Mean        .2800997
T:Mean=0         30.31778  Prob>¦T¦          0.0001
Sgn Rank            637.5  Prob>¦S¦          0.0001
Num ^= 0               50
W:Normal         .9328984  Prob<W            0.009


                   Quantiles(Def=5)

100% Max      13.5        99%        13.5
75% Q3         9.6        95%        13.2
50% Med       8.05        90%        11.2
25% Q1         7.1        10%         6.5
0% Min         5.2         5%         5.9
                           1%         5.2
```

EXERCISES 3.12–3.27

LEARNING THE MECHANICS

3.12 Given the following information about two data sets, compute $\bar{x}$, the sample variance, and the standard deviation for each:

a. $n = 25$, $\displaystyle\sum_{i=1}^{n} x_i^2 = 1{,}000$, $\displaystyle\sum_{i=1}^{n} x_i = 50$

b. $n = 80$, $\displaystyle\sum_{i=1}^{n} x_i^2 = 270$, $\displaystyle\sum_{i=1}^{n} x_i = 100$

3.13 For each of the following data sets compute $\sum_{i=1}^{n} x_i$, $\sum_{i=1}^{n} x_i^2$, and $\left(\sum_{i=1}^{n} x_i\right)^2$:

a. 5, 9, 6, 3, 7   **b.** 3, 1, 4, 3, 0, −2
c. 90, 12, 40, 15   **d.** −1, 4, 1, 0, 5
e. 1, 0, 0, 1, 0, 10

3.14 Compute $\bar{x}$, s^2, and s for each of the data sets in Exercise 3.13.

3.15 Compute $\bar{x}$, s^2, and s for each of the following data sets.
a. 4, 3, 6, 0, 5   **b.** 3, 7, 7, 1, 2, 6
c. 4, 1, 1, 5, 6, 3   **d.** 3, 0, 1, 5

3.16 Compute $\bar{x}$, s^2, and s for each of the following data sets:
a. 10, 1, 0, 0, 20   **b.** 5, 9, −1, 100

3.17 Compute $\bar{x}$, s^2, and s for each of the following data sets. If appropriate, specify the units in which your answer is expressed.
a. 3, 1, 10, 10, 4
b. 8 feet, 10 feet, 32 feet, 5 feet
c. −1, −4, −3, 1, −4, −4
d. $\frac{1}{5}$ ounce, $\frac{1}{5}$ ounce, $\frac{1}{5}$ ounce, $\frac{2}{5}$ ounce, $\frac{1}{5}$ ounce, $\frac{4}{5}$ ounce

3.18 The range, variance, and standard deviation provide information about the variation in a data set.
a. Describe the information each conveys.
b. Discuss the advantages and disadvantages of using each to measure the variability of a data set.

3.19 Using only integers between 0 and 10, construct two data sets with at least ten observations each that have the same mean but different variances. Construct dot diagrams for each of your data sets (see Figure 3.7), and mark the mean of each data set on its dot diagram.

3.20 Using only integers between 0 and 10, construct two data sets with at least ten observations each that have the same range but different variances. Construct a dot diagram for each of your data sets (see Figure 3.7).

3.21 Using only integers between 0 and 10, construct two data sets with at least ten observations each that have the same range but different means. Construct a dot diagram for each of your data sets (see Figure 3.7), and mark the mean of each data set on its dot diagram.

3.22 The ten states with the highest percentages of total land devoted to state parks are listed below:*

New York	9.40%	New Mexico	2.10%
Connecticut	6.09%	South Dakota	1.84%
New Jersey	5.25%	New Hampshire	1.82%
Massachusetts	4.66%	Rhode Island	1.77%
South Carolina	3.25%	Pennsylvania	1.02%

a. Calculate the range of this data set.
b. Calculate the variance and standard deviation.

*Source: "States' park land," USA Today, June 23, 1983. Copyright 1983, USA Today. Excerpted with permission.

c. Can the variance of a data set ever be smaller than the standard deviation? Explain.

d. Can the variance of a data set ever be negative? Explain.

3.23 Refer to Exercise 3.22. If the percentage of land in Pennsylvania devoted to state parks were .90% instead of 1.02%, would the variance of the data set increase or decrease? Why? If instead of 1.02% or .90%, Pennsylvania's percentage were 3.0%, how would the resulting variance of the data set compare with the original variance calculated in part **b** of Exercise 3.22? Explain.

3.24 The table lists the 1985 third-quarter profits (in millions of dollars) for a sample of seven airlines.

AIRLINE	PROFIT
Continental Airlines	22.3
Eastern Air Lines	22.3
NWA	39.0
Pan Am	21.1
Texas Air	36.7
Trans World Airlines	−13.5
Western Air Lines	15.1

Data: Standard & Poor's Compustat Services, Inc.
Source: *Business Week*, Nov. 18, 1985, p. 117.

a. Calculate the range, variance, and standard deviation of the data set.

b. Specify the units in which each of your answers to part **a** is expressed.

c. Suppose Trans World Airlines had a profit of $0 instead of a loss of $13.5 million. Would the range of the data set increase or decrease? Why? Would the standard deviation of the data set increase or decrease? Why?

APPLYING THE CONCEPTS

3.25 The Consumer Price Index (CPI) measures the price change of a constant market basket of goods and services. The Bureau of Labor Statistics publishes a national CPI (called the U.S. City Average Index) as well as separate indexes for each of 28 different cities in the United States. The national index and some of the city indexes are published monthly; the remainder of the city indexes are published bimonthly. The CPI is used in cost-of-living escalator clauses of many labor contracts to adjust wages for inflation (U.S. Department of Labor, 1978). For example, in the printing industry of Minneapolis and St. Paul, hourly wages are adjusted every 6 months (based on October and April values of the CPI) by 4¢ for every point change in the Minneapolis/St. Paul CPI.

The table lists the published values of the U.S. City Average Index and Minneapolis/St. Paul Index during 1984 and the first 11 months of 1985.

a. Calculate the mean values for the U.S. City Average Index and the Minneapolis/St. Paul Index.

b. Find the ranges of the U.S. City Average Index and the Minneapolis/St. Paul Index.

c. The standard deviation of the U.S. City Average Index over the 23 months described in the table is 6.40. Calculate the standard deviation for the Minneapolis/St. Paul Index over the time period described in the table.

MONTH	U.S. CITY AVERAGE INDEX	MINNEAPOLIS/ST. PAUL	MONTH	U.S. CITY AVERAGE INDEX	MINNEAPOLIS/ST. PAUL
January 1984	305.2	—	January 1985	316.1	—
February	306.6	319.6	February	317.4	330.4
March	307.3	—	March	318.8	—
April	308.8	322.0	April	320.1	333.6
May	309.7	—	May	321.3	—
June	310.7	324.1	June	322.3	336.7
July	311.7	—	July	322.8	—
August	313.0	324.8	August	323.5	338.8
September	314.5	—	September	324.5	—
October	315.3	328.0	October	325.5	340.6
November	315.3	—	November	326.6	—
December	315.5	327.9			

Source: U.S. Dept. of Labor, *CPI Detailed Report*, Mar. 1984–Jan. 1985.

d. Which index displays greater variation about its mean over the time period in question? Justify your response.

3.26 In order to set an appropriate price for a product, it is necessary to be able to estimate its cost of production. One element of the cost is based on the length of time it takes workers to produce the product. The most widely used technique for making such measurements is the *time study*. In a time study, the task to be studied is divided into measurable parts and each is timed with a stopwatch or filmed for later analysis. For each worker, this process is repeated many times for each subtask. Then the average and standard deviation of the time required to complete each subtask are computed for each worker. A worker's overall time to complete the task under study is then determined by adding his or her subtask-time averages (Chase and Aquilano, 1977). The data (in minutes) given in the table are the result of a time study of a production operation involving two subtasks.

REPETITION	WORKER A		WORKER B	
	Subtask 1	Subtask 2	Subtask 1	Subtask 2
1	30	2	31	7
2	28	4	30	2
3	31	3	32	6
4	38	3	30	5
5	25	2	29	4
6	29	4	30	1
7	30	3	31	4

a. Find the overall time it took each worker to complete the manufacturing operation under study.
b. For each worker, find the standard deviation of the seven times for subtask 1.
c. In the context of this problem, what are the standard deviations you computed in part **b** measuring?
d. Repeat part **b** for subtask 2.
e. If you could choose workers similar to A or workers similar to B to perform subtasks 1 and 2, which type would you assign to each subtask? Explain your decisions on the basis of your answers to parts **a–d**.

3.27 The accompanying table lists the yearly steel production (in thousands of metric tons) for 1976–1982 for the five leading steel-producing countries in Europe.

	1976	1977	1978	1979	1980	1981	1982
Germany	567	512	491	545	538	514	466
Italy	239	242	234	277	248	211	179
France	399	356	347	354	365	367	319
United Kingdom	489	475	441	395	373	308	345
Belgium	46	33	23	22	23	16	14

Source: *Industrial Statistics Yearbook*, 1982, p. 547.

a. Measure the variation of each country's yearly steel production using the range.
b. Repeat part **a** using the standard deviation as the measure of variation.
c. Rank the countries in terms of the variability of their production using the ranges from part **a**, and then rank them according to the standard deviations from part **b**.
d. Do your two sets of rankings agree? Will the range and standard deviation always yield rankings that agree? Explain.

SECTION 3.7

INTERPRETING THE STANDARD DEVIATION

As we have seen, if we are comparing the variability of two samples selected from a population, the sample with the larger standard deviation is the more variable of the two. Thus, we know how to interpret the standard deviation on a relative or comparative basis, but we have not explained how it provides a measure of variability for a single sample.

One way to interpret the standard deviation as a measure of variability of a data set would be to answer questions such as the following: How many measurements are within 1 standard deviation of the mean? How many measurements are within 2 standard deviations? For a specific data set, we can answer the questions by counting the number of measurements in each of the intervals. However, if we are interested in obtaining a general answer to these questions, the problem is more difficult.

In Table 3.8, we present two sets of guidelines to help answer the questions of how many measurements fall within 1, 2, and 3 standard deviations of the mean. The first set, which applies to any sample, is derived from a theorem proved by the Russian mathematician, Chebyshev. The second set, the Empirical Rule, is based on empirical evidence that has accumulated over time and applies to samples that possess mound-shaped frequency distributions—those that are approximately symmetric, with a clustering of measurements about the midpoint of the distribution (the mean, median, and mode should all be about the same) and that tail off as we move away from the center of the histogram. Thus, the histogram will have the appearance of a mound or bell, as shown in Figure 3.9. The percentages given for the various intervals (particularly the interval $\bar{x} - 2s$ to $\bar{x} + 2s$) in Table 3.8 provide remarkably good approximations even when the distribution of the data is slightly skewed or asymmetric.*

*It is our intention to imply that the Empirical Rule of Table 3.8 not only applies to normal distributions of data, but also applies very well to mound-shaped distributions of large data sets and to distributions that are moderately skewed.

TABLE 3.8

Aids to the Interpretation and Use of a Standard Deviation

1. A rule (from Chebyshev's theorem) that applies to any sample of measurements, regardless of the shape of the frequency distribution:
  a. It is possible that none of the measurements will fall within the interval $\bar{x} \pm s$ or $(\bar{x} - s, \bar{x} + s)$—i.e., within 1 standard deviation of the mean.
  b. At least $\frac{3}{4}$ of the measurements will fall within $(\bar{x} - 2s, \bar{x} + 2s)$—i.e., within 2 standard deviations of the mean.
  c. At least $\frac{8}{9}$ of the measurements will fall within $(\bar{x} - 3s, \bar{x} + 3s)$—i.e., within 3 standard deviations of the mean.
  d. Generally, at least $(1 - 1/k^2)$ of the measurements will fall within $(\bar{x} - ks, \bar{x} + ks)$ —i.e., within k standard deviations of the mean, where k is any number greater than 1.

2. A rule of thumb, called the Empirical Rule, that applies to samples with frequency distributions that are mound-shaped:
  a. Approximately 68% of the measurements will fall within the interval $\bar{x} \pm s$ or $(\bar{x} - s, \bar{x} + s)$—i.e., within 1 standard deviation of the mean.
  b. Approximately 95% of the measurements will fall within $(\bar{x} - 2s, \bar{x} + 2s)$—i.e., within 2 standard deviations of the mean.
  c. Essentially all the measurements will fall within $(\bar{x} - 3s, \bar{x} + 3s)$—i.e., within 3 standard deviations of the mean.

FIGURE 3.9

Histogram of a Mound-Shaped Sample

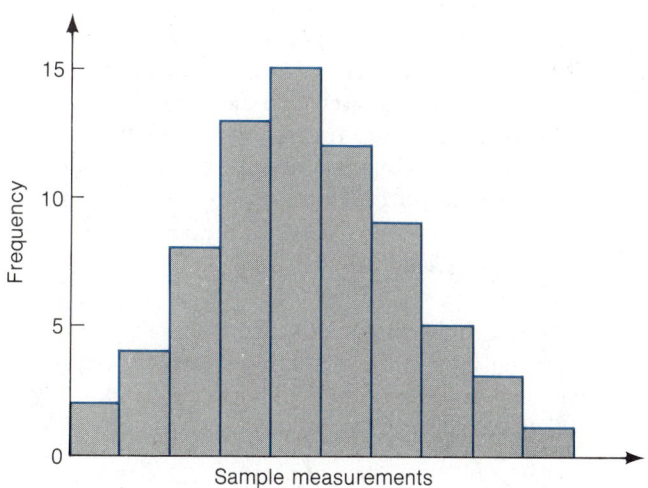

EXAMPLE 3.8

The 50 companies' percentages of revenues spent on R&D are repeated here:

13.5	9.5	8.2	6.5	8.4	8.1	6.9	7.5	10.5	13.5
7.2	7.1	9.0	9.9	8.2	13.2	9.2	6.9	9.6	7.7
9.7	7.5	7.2	5.9	6.6	11.1	8.8	5.2	10.6	8.2
11.3	5.6	10.1	8.0	8.5	11.7	7.1	7.7	9.4	6.0
8.0	7.4	10.5	7.8	7.9	6.5	6.9	6.5	6.8	9.5

We have previously shown that the mean and standard deviation of these data are 8.49 and 1.98, respectively. Calculate the fraction of these measurements that lie within the intervals $\bar{x} \pm s$, $\bar{x} \pm 2s$, and $\bar{x} \pm 3s$, and compare the results with those in Table 3.8.

SOLUTION

We first form the interval

$$(\bar{x} - s, \bar{x} + s) = (8.49 - 1.98, 8.49 + 1.98)$$
$$= (6.51, 10.47)$$

A check of the measurements reveals that 34 of the 50 measurements, or 68%, are within 1 standard deviation of the mean.

The interval

$$(\bar{x} - 2s, \bar{x} + 2s) = (8.49 - 3.96, 8.49 + 3.96)$$
$$= (4.53, 12.45)$$

contains 47 of the 50 measurements, or 94%.

The 3 standard deviation interval around $\bar{x}$,

$$(\bar{x} - 3s, \bar{x} + 3s) = (8.49 - 5.94, 8.49 + 5.94)$$
$$= (2.55, 14.43)$$

contains all the measurements.

In spite of the fact that the distribution of these data is skewed to the right (see Figure 3.3), the percentages within 1, 2, and 3 standard deviations (68%, 94%, and 100%) agree very well with the approximations of 68%, 95%, and 100% given by the Empirical Rule. You will find that unless the distribution is extremely skewed, the mound-shaped approximations will be reasonably accurate. Of course, no matter what the shape of the distribution, Chebyshev's theorem from Table 3.8 assures that at least 75% and at least 89% $\left(\frac{8}{9}\right)$ of the measurements will lie within 2 and 3 standard deviations, respectively, of the mean. ∎

EXAMPLE 3.9

Chebyshev's theorem and the Empirical Rule (Table 3.8) are useful as a check on the calculation of the standard deviation. For example, suppose we calculated the standard deviation of the R&D percentages to be 3.92. Are there any clues in the data that enable us to judge whether this number is reasonable?

SOLUTION

The range of the R&D percentages is $13.5 - 5.2 = 8.3$. From the Empirical Rule we know that most of the measurements (approximately 95% if the distribution is not extremely skewed) will be within 2 standard deviations of the mean. And from Chebyshev's theorem, regardless of the shape of the distribution, almost all of the measurements $\left(\text{at least } \frac{8}{9}\right)$ will fall within 3 standard deviations of the mean. Consequently, we would expect a range of between $4 (\pm 2)$ and $6 (\pm 3)$ standard deviations to cover the range of the measurements (see Figure 3.10). For the R&D data, this means that the standard deviation s should fall between

$$\frac{\text{Range}}{6} = \frac{8.3}{6} = 1.38 \quad \text{and} \quad \frac{\text{Range}}{4} = \frac{8.3}{4} = 2.08$$

FIGURE 3.10

The Relation Between the Range and the Standard Deviation

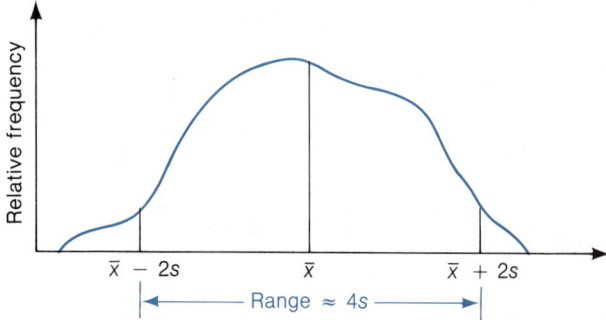

Thus, we would have reason to believe that a calculated standard deviation of 3.92 for these data is too large, since it well exceeds one-fourth the range, 2.08. A check of our work reveals that 3.92 is the variance s^2, not the standard deviation s (see Example 3.6). We "forgot" to take the square root, a common error. The correct value of 1.98 is between one-sixth and one-fourth the range. ∎

In examples and exercises we will sometimes use $s \approx \text{Range}/4$ to obtain a crude, and usually conservatively large, approximation for s. However, we stress that this is no substitute for calculating the exact value of s when possible.

Finally, and most important, we will use the concepts in Chebyshev's theorem and the Empirical Rule to build the foundation for statistical inference-making.

EXAMPLE 3.10

A manufacturer of automobile batteries claims that the average length of useful life for its grade A battery is 60 months. However, the guarantee on this brand is for only 36 months. Assume that the standard deviation of the lifelengths is known to be 10 months, and that the frequency distribution of the lifelengths is known to be mound-shaped.

a. Approximately what percentage of the manufacturer's grade A batteries will last more than 50 months, assuming the manufacturer's claim about the mean lifelength is true?

b. Approximately what percentage of the manufacturer's batteries will last less than 40 months, assuming that the manufacturer's claim about the mean lifelength is true?

c. Suppose that your grade A battery lasts 37 months. What would you infer about the manufacturer's claim that the mean lifelength is 60 months?

SOLUTION

Assuming that the distribution of lifelengths is approximately mound-shaped, with a mean of 60 months and a standard deviation of 10 months, it would appear as shown in Figure 3.11 (page 108). Note that we can take advantage of the fact that mound-shaped distributions are (approximately) symmetric about the mean, so that the percentages given by the Empirical Rule can be split equally between the halves of the distribution on each side of the mean. The approximations given in Figure 3.11 are more dependent on the assumption of a mound-shaped distribution than

FIGURE 3.11
Grade A Battery
Lifelength Distribution;
Manufacturer's Claim
Assumed True

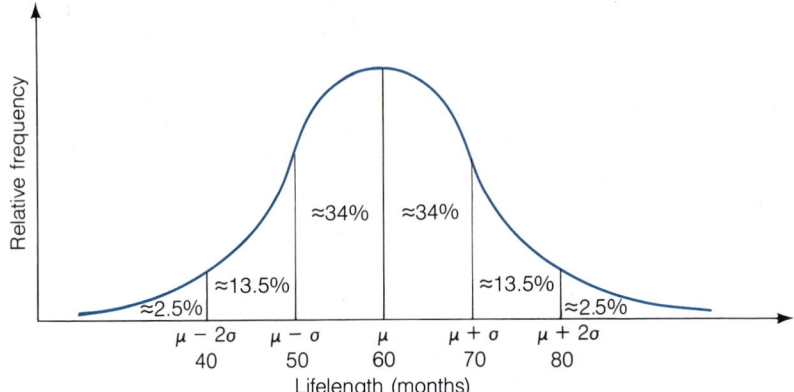

those given by the Empirical Rule (Table 3.8), because the approximations in Figure 3.11 depend on the (approximate) symmetry of the mound-shaped distribution. We saw in Example 3.8 that the Empirical Rule can yield good approximations even for skewed distributions. This will *not* be true of the approximations in Figure 3.11; the distribution must be mound-shaped and (approximately) symmetric.

For example, since approximately 68% of the measurements will fall within 1 standard deviation on both sides of the mean, the distribution's symmetry implies that approximately $\frac{1}{2}(68\%) = 34\%$ of the measurements will fall between the mean and 1 standard deviation on each side. This concept is demonstrated in Figure 3.11. Note that the 2.5% of the measurements beyond 2 standard deviations in each direction from the mean follows from the fact that, if approximately 95% of the measurements fall within 2 standard deviations, then about 5% fall outside 2 standard deviations. If the distribution is approximately symmetric, then about 2.5% fall beyond 2 standard deviations on each side of the mean.

a. Using Figure 3.11, it is easy to see that the percentage of batteries lasting more than 50 months is, approximately, 34% (between 50 and 60 months) plus 50% (greater than 60 months). Thus, approximately $(50 + 34)\% = 84\%$ of the batteries should last longer than 50 months.

b. The percentage of batteries that last less than 40 months can also be determined from Figure 3.11. Since 40 is 2 standard deviations below the claimed mean of 60 months, approximately 2.5% of the batteries should fail prior to 40 months, assuming the manufacturer's claim is true.

c. If you are so unfortunate that your grade A battery fails at 37 months, one of two inferences can be made. Either your battery was one of the approximately 2.5% that fail prior to 40 months, or the manufacturer's claim is not true. Because the chances are so small that a battery fails before 40 months if the claim is true, you would have good reason to have serious doubts about the manufacturer's claim. A mean smaller than 60 months and/or a standard deviation greater than 10 months would increase the likelihood of a battery's failure prior to 40 months. *

■

Example 3.10 is our initial demonstration of the statistical inference-making process. At this point you should realize that we will use sample information (in Example 3.10, your battery's failure at 37 months) to make inferences about the population of measurements (in Example 3.10, the manufacturer's claim about the lifelength population). We will build on this foundation as we proceed.

CASE STUDY 3.5

BECOMING MORE SENSITIVE TO CUSTOMER NEEDS

The degree of sensitization on the part of a firm to the needs and wants of its consumers is frequently an important factor in determining the firm's overall success. Namias (1964) presents a procedure for achieving such sensitivity. The procedure uses the rate of consumer complaints about a product to determine when and when not to conduct a search for specific causes of consumer complaints. For simplification, we will discuss Namias's paper as if the procedure described used the number of complaints per 10,000 units of a product sold to determine when and when not to conduct a search for specific causes of consumer complaints. The details of the procedure are discussed in Case Study 3.6.

Namias's procedure, given our simplification, makes use of the **distribution** of the number of consumer complaints received about a product per 10,000 units of the product sold. To visualize such a distribution, imagine that a company produces its product in lots of 10,000 units and keeps track of the number of complaints received about items in each lot. The company's complaint records will show a series of numbers, perhaps 100, 96, 145, 201, etc., each of which is the number of complaints received about a particular lot of 10,000 units. This series of numbers is a quantitative data set from which a relative frequency histogram can be drawn. The histogram constructed from this data set is a representation of the distribution of interest—i.e., the distribution of the number of consumer complaints received about a product per 10,000 units of the product sold. The variance and standard deviation of this distribution are measures of the variation in the number of consumer complaints received. Namias determined that this distribution was mound-shaped. Accordingly, it can be said that approximately 95% of the time the number of complaints about a product will be within 2 standard deviations of the mean number of complaints. It is upon this fact, as we shall see in Case Study 3.6, that Namias's procedure for determining when it would be worthwhile to conduct a search for specific causes of consumer complaints is founded.

If it could not have been determined that the distribution of the number of complaints was mound-shaped, it could have been said only that at least 75% of the time the number of complaints about a product will be within 2 standard deviations of the mean number of complaints.

*The assumption that the distribution is mound-shaped and symmetric may also be incorrect. However, if the distribution were skewed to the right, as lifelength distributions often tend to be, the percentage of measurements more than 2 standard deviations *below* the mean would be even less than 2.5%.

| | | | | | | | | | | | | |

EXERCISES 3.28–3.41

LEARNING THE MECHANICS

3.28 Given a data set with a largest value of 760 and a smallest value of 135, what would you estimate the standard deviation to be? Explain the logic behind the procedure you used to estimate the standard deviation.

3.29 As a result of government and consumer pressure, automobile manufacturers in the United States are deeply involved in research to improve their products' gasoline mileage. One manufacturer, hoping to achieve 40 miles per gallon on one of its compact models, measured the mileage obtained by 36 test versions of the model with the following results (rounded to the nearest mile for convenience):

43	45	42	42	40	40
40	40	42	41	40	37
38	41	41	38	40	41
39	41	39	43	41	40
35	39	42	37	41	44
40	37	36	39	38	40

a. If the manufacturer would be satisfied with a (population) mean of 40 miles per gallon, how would it react to the above test data?
b. Compute $\bar{x}$, s^2, and s for the data set.
c. Use the information in Table 3.8 to check your calculation of s in part b.
d. Construct a relative frequency histogram of the data set. Is the data set mound-shaped?
e. What percentage of the measurements would you expect to find within the intervals $\bar{x} \pm s$, $\bar{x} \pm 2s$, and $\bar{x} \pm 3s$?
f. Count the number of measurements that actually fall within the intervals of part e. Express each interval count as a percentage of the total number of measurements. Compare these results with your answers to part e.

3.30 A manufacturer of video cassette recorders is disturbed because retailers were complaining that they were not receiving shipments of recorders as fast as they had been promised. The manufacturer decided to run a check on the distribution network. Each of the 50 warehouses owned by the manufacturer throughout the country had been instructed to maintain at least 200 recorders in stock at all times so that a supply would always be readily available for retailers. The manufacturer checked the inventories of 20 of these warehouses and obtained the following numbers of video cassette recorders in stock:

40	10	44	142	14
301	175	0	38	202
220	32	400	78	16
99	0	176	5	86

a. What is the mean number of recorders in stock for the 20 warehouses checked?
b. Compute s^2 and s for the data set.

    c. Use the information in Table 3.8 to check your calculation of s in part **b**.

    d. According to Chebyshev's theorem, what percentage of the measurements would you expect to find outside the intervals $\bar{x} \pm s$, $\bar{x} \pm 2s$, and $\bar{x} \pm 3s$?

    e. Check the numbers of measurements that actually fall outside the intervals specified in part **d**. Express each count as a percentage of the total number of measurements. Compare these results with your answers to part **d**.

3.31 Twenty-five mergers were sampled from the population of mergers that occurred between firms (excluding railroads) listed on the New York Stock Exchange during the period 1958–1960. For each merger, the ratio of the target firm's sales for the preceding year to the bidder firm's sales was calculated. (See Exercise 1.5 for definitions of target and bidder firms.) The 25 sales ratios are listed below (Eger, 1982, p. 133):

.16	.14	.18	.15	.14
.05	.10	.05	.02	.30
.03	.08	.32	.34	.04
.06	.02	.14	.07	.10
.04	.09	.29	.79	.02

    a. What is the mean sales ratio, $\bar{x}$, for this sample of mergers? Also, find s^2 and s.

    b. According to Chebyshev's theorem, what percentage of the measurements would you expect to find in the intervals $\bar{x} \pm .75s$, $\bar{x} \pm 2.5s$, and $\bar{x} \pm 4s$?

    c. What percentage of measurements actually fall in the intervals of part **b**? Compare these results with the results of part **b**.

    d. What percentage of the mergers involved a target firm with larger sales than the bidder firm?

APPLYING THE CONCEPTS

3.32 Following World War II, births in the United States soared to record levels: from about 2.9 million in 1945, to 3.6 million in 1950, to over 4 million a year during the period 1955–1964, before declining in the mid-1960's (*Statistical Abstract of the United States: 1985*, p. 57). People born during this period are frequently referred to as belonging to the "baby-boom" generation. The identification of such demographic patterns is vitally important to both business and government. For example, government can use such information to help predict tax revenues; business can use it to predict spending patterns within the population (Curley, 1983). The table on page 112 lists the percentage change in the number of births between the first quarter of 1980 and the second quarter of 1983 for all 50 states and the District of Columbia.

    a. Would it be appropriate to use the Empirical Rule to describe this data set? Justify your answer.

    b. According to Chebyshev's theorem, what percentage of the measurements in the data set will fall more than 2 standard deviations from the mean of the data set?

    c. What percentage of measurements actually fall more than 2 standard deviations from the mean? Compare your answer to that of part **b**.

STATE	PERCENTAGE CHANGE IN NUMBER OF BIRTHS	STATE	PERCENTAGE CHANGE IN NUMBER OF BIRTHS
Alabama	1.7	Montana	3.8
Alaska	19.2	Nebraska	1.7
Arizona	9.0	Nevada	11.3
Arkansas	1.8	New Hampshire	4.1
California	6.4	New Jersey	1.4
Colorado	8.6	New Mexico	7.4
Connecticut	1.0	New York	.6
Delaware	1.9	North Carolina	3.4
District of Columbia	−2.4	North Dakota	4.3
Florida	9.6	Ohio	−.5
Georgia	4.9	Oklahoma	9.0
Hawaii	6.1	Oregon	1.1
Idaho	4.8	Pennsylvania	.3
Illinois	.5	Rhode Island	.9
Indiana	−.2	South Carolina	4.5
Iowa	−.3	South Dakota	1.3
Kansas	2.6	Tennessee	2.1
Kentucky	1.5	Texas	10.5
Louisiana	5.5	Utah	10.8
Maine	1.9	Vermont	2.7
Maryland	2.1	Virginia	3.8
Massachusetts	.5	Washington	4.1
Michigan	−2.1	West Virginia	.8
Minnesota	1.7	Wisconsin	1.0
Mississippi	2.6	Wyoming	9.5
Missouri	1.1		

Source: *Statistical Abstract of the United States: 1985*, p. 14.

3.33 How much does a new 1,600-square-foot home with $1\frac{1}{2}$ or 2 baths in a desirable neighborhood cost? To find out, Better Homes and Gardens® Real Estate Service conducted a nationwide survey of real estate agents. The table on page 113 contains a portion of the resulting sample data.

　a. In describing this data set, is it more appropriate to apply Chebyshev's theorem or the Empirical Rule? Justify your answer.

　b. Calculate the mean and standard deviation of this data set and interpret each in the context of the problem.

　c. What proportion of the sale prices in the sample would you expect to find within 2 standard deviations of the sample mean? More than 3 standard deviations from the mean? How many sale prices actually fall in these intervals?

　d. In terms of numbers of standard deviations, how far does the price of a home in Darien, Conn. lie from the mean price of the sample?

CITY	SALE PRICE	CITY	SALE PRICE
Albany, N.Y.	$ 72,500	Lake Tahoe, Nev.	$190,000
Allentown, Pa.	82,500	Little Rock, Ark.	85,000
Amarillo, Tex.	75,000	Manhattan, Kans.	85,000
Baltimore, Md.	100,000	Montgomery, Ala.	65,000
Baton Rouge, La.	100,000	Nashville, Tenn.	84,000
Biloxi, Miss.	78,500	Olympia, Wash.	75,000
Bowling Green, Ky.	67,400	Owensboro, Ky.	72,500
Cheyenne, Wyo.	82,000	Pittsburgh, Pa.	72,500
Cincinnati, Ohio	85,000	Pleasanton, Calif.	185,000
Columbus, Ga.	69,000	Raleigh, N.C.	94,000
Cranston, R.I.	98,500	Reno, Nev.	93,950
Darien, Conn.	290,000	Rockford, Ill.	76,000
Des Moines, Iowa	93,900	San Diego, Calif.	140,000
Fargo, N.D.	100,000	Sheboygan, Wis.	72,000
Greenville, S.C.	76,800	South Bend, Ind.	71,300
Honolulu, Hawaii	165,000	Springfield, Mass.	75,000
Jacksonville, Fla.	77,000	Trenton, N.J.	100,000
Joplin, Mo.	65,000	Vancouver, Wash.	68,000

Source: *Consumers Research*, Feb. 1985, p. 19. Courtesy Meredith Corporation, Des Moines, Iowa.

3.34 A company that bottles sparkling water has determined that it lost an average of 30.4 cases a week last year due to breakage in transit. The standard deviation of the number of cases lost per week was 3.8 cases. With only this information, what can you say about the number of weeks last year that the company lost more than 38 cases due to breakage in transit? Justify your answer.

3.35 A chemical company produces a substance composed of 98% cracked corn particles and 2% zinc phosphide for use in controlling rat populations in sugarcane fields. Production must be carefully controlled to maintain the zinc phosphide at 2% because too much zinc phosphide will damage the sugarcane and too little will be ineffective in controlling the rat population. Records from past production indicate that the distribution of the actual percentage of zinc phosphide present in the substance is approximately mound-shaped, with a mean of 2.0% and a standard deviation of .08%. If the production line is operating correctly and a batch is chosen at random from a day's production, what is the approximate probability that it will contain less than 1.84% zinc phosphide?

3.36 In 1978, when the airlines were deregulated, there was hope that more airline companies would enter the market. The table at the top of page 114 contains some recent annual revenue data for the new entrants.

   a. Find the sample mean and variance for revenues, profits, and passenger miles flown.

   b. Use the results of part **a** to sketch the distribution for each of the variables—revenue, profits, and passenger miles flown.

COMPANY	REVENUES ($ millions)	PROFITS ($ millions)	PASSENGER MILES (millions)
People Express	292	10.4	3,700
New York Air	130	4.5	657
Midway	104	−15.0	648
Muse Air	73	−2.0	651
Jet America	60	−8.0	610
American International	53	−11.8	295
Northeastern International	46	−.8	498
Air 1	20	−21.0	95
American West	18	−6.3	351

Source: *New York Times*, Mar. 11, 1984. Copyright © 1984 by The New York Times Company. Reprinted by permission.

3.37 For 50 randomly selected days, the number of vehicles that used a certain road was ascertained by a city engineer. The mean was 385; the standard deviation was 15. Suppose you are interested in the proportion of days that there were between 340 and 430 vehicles using the road. What does Chebyshev's theorem tell you about this proportion?

3.38 A boat dealer has determined that the frequency distribution of the number of outboard motor sales per month over the last 5 years is mound-shaped, with a sample mean of 30 and a sample variance of 4. Approximately what percentage of the recorded monthly sales figures of the past 5 years would be expected to be greater than 34? Less than 26? Greater than 36?

3.39 Solar energy is considered by many to be the energy of the future. A recent survey was taken to compare the cost of solar energy with the cost of gas or electric energy. Results of the survey revealed that the average monthly utility bill of a three-bedroom house using gas or electric energy was $125 and the standard deviation was $15.
  a. If nothing is known about the distribution of utility bills, what can you say about the fraction of all three-bedroom homes with gas or electric energy that have bills between $80 and $170?
  b. If it is reasonable to assume that the distribution of utility bills is mound-shaped, approximately what proportion of three-bedroom homes would have monthly bills less than $110?
  c. Suppose that three houses with solar energy units had the following utility bills: $78, $92, $87. Does this suggest that solar energy units might result in lower utility bills? Explain. [*Note:* We will present a statistical method for testing this conjecture in Chapter 8.]

3.40 When it is working properly, a machine that fills 25-pound bags of flour dispenses an average of 25 pounds per fill; the standard deviation of the amount of fill is .1 pound. To monitor the performance of the machine, an inspector weighs the contents of a bag coming off the machine's conveyor belt every half-hour during the day. If the contents of two consecutive bags fall more than 2 standard deviations from the mean (using the mean and standard deviation given above), the filling process is said to be out of control and the machine is shut down briefly for adjustments. The data given in the table are the weights measured by the inspector yesterday. Assume the machine is never shut down for more than 15 minutes at a time. At what times yesterday was the process shut down for adjustment? Justify your answer.

TIME	WEIGHT (pounds)	TIME	WEIGHT (pounds)
8:00 A.M.	25.10	12:30 P.M.	25.06
8:30	25.15	1:00	24.95
9:00	24.81	1:30	24.80
9:30	24.75	2:00	24.95
10:00	25.00	2:30	25.21
10:30	25.05	3:00	24.90
11:00	25.23	3:30	24.71
11:30	25.25	4:00	25.31
12:00	25.01	4:30	25.15
		5:00	25.20

3.41 A buyer for a lumber company must determine whether to buy a piece of land containing 5,000 pine trees. If 1,000 of the trees are at least 40 feet tall, the buyer will purchase the land; otherwise, he will not. The owner of the land reports that the distribution of the heights of the trees has a mean of 30 feet, and a standard deviation of 3 feet. Based on this information, what should the buyer decide?

| | | | | | | | | | | | | |

SECTION 3.8

CALCULATING A MEAN AND STANDARD DEVIATION FROM GROUPED DATA (OPTIONAL)

If your data have been grouped in classes of equal width and arranged in a frequency table, you can use the following formulas to calculate $\bar{x}$, s^2, and s:

FORMULAS FOR CALCULATING A MEAN AND STANDARD DEVIATION FROM GROUPED DATA

$$\bar{x} = \frac{\sum_{i=1}^{k} x_i f_i}{n}$$

$$s^2 = \frac{\sum_{i=1}^{k} (x_i - \bar{x})^2 f_i}{n - 1} \qquad s = \sqrt{s^2}$$

Shortcut formula: $$s^2 = \frac{\sum_{i=1}^{k} x_i^2 f_i - \frac{\left(\sum_{i=1}^{k} x_i f_i\right)^2}{n}}{n - 1}$$

where

x_i = Midpoint of the ith class

f_i = Frequency of the ith class

k = Number of classes

EXAMPLE 3.11

Compute the mean and standard deviation for the companies' R&D percentages based on the groupings used to construct the relative frequency histogram of Figure 3.3.

SOLUTION

The frequency table is repeated in Table 3.9, with a column showing the class midpoints added.

TABLE 3.9

Measurement Classes, Frequencies, and Class Midpoints for the R&D Percentage Data

CLASS	MEASUREMENT CLASS	CLASS MIDPOINT	CLASS FREQUENCY
1	5.15– 6.25	5.70	4
2	6.25– 7.35	6.80	12
3	7.35– 8.45	7.90	14
4	8.45– 9.55	9.00	7
5	9.55–10.65	10.10	7
6	10.65–11.75	11.20	3
7	11.75–12.85	12.30	0
8	12.85–13.95	13.40	3

Substituting the class midpoints and frequencies into the formula for the mean of grouped data, we obtain

$$\bar{x} = \frac{\sum_{i=1}^{k} x_i f_i}{n} = \frac{(5.70)(4) + (6.80)(12) + \cdots + (13.40)(3)}{50}$$

$$= \frac{422.5}{50} = 8.45$$

Next, using the shortcut formula for calculating the sample variance for grouped data, we obtain

$$s^2 = \frac{\sum_{i=1}^{k} x_i^2 f_i - \frac{\left(\sum_{i=1}^{k} x_i f_i\right)^2}{n}}{n-1}$$

$$= \frac{(5.70)^2(4) + (6.80)^2(12) + \cdots + (13.40)^2(3) - \frac{(422.5)^2}{50}}{50-1}$$

$$= \frac{3,754.65 - 3,570.125}{49} = 3.7658$$

$$s = \sqrt{3.7658} = 1.94$$

You will note that the sample mean and standard deviation for the grouped data, 8.45 and 1.94, agree well with the exact mean and standard deviation for the 50 measurements, $\bar{x} = 8.49$ and $s = 1.98$. As long as a reasonable number of class intervals is used, the approximations based on grouped data will usually be good.

■

SECTION   3.9

**MEASURES OF
RELATIVE STANDING**

As we have seen, numerical measures of central tendency and variability describe the general nature of a data set (either a sample or a population). We may also be interested in describing the relative location of a particular measurement within a data set. Descriptive measures of the relationship of a measurement to the rest of the data are called **measures of relative standing**.

One measure of the relative standing of a particular measurement is its **percentile ranking**:

DEFINITION 3.7

Let $x_1, x_2, \ldots, x_n$ be a set of n measurements arranged in increasing (or decreasing) order. The **pth percentile** is a number x such that $p\%$ of the measurements fall below the pth percentile and $(100 - p)\%$ fall above it.

For example, if oil company A reports that its yearly sales are in the 90th percentile of all companies in the industry, the implication is that 90% of all oil companies have yearly sales less than company A's, and only 10% have yearly sales exceeding company A's. This is demonstrated in Figure 3.12.

FIGURE 3.12

Relative Frequency
Distribution for Yearly
Sales of Oil Companies

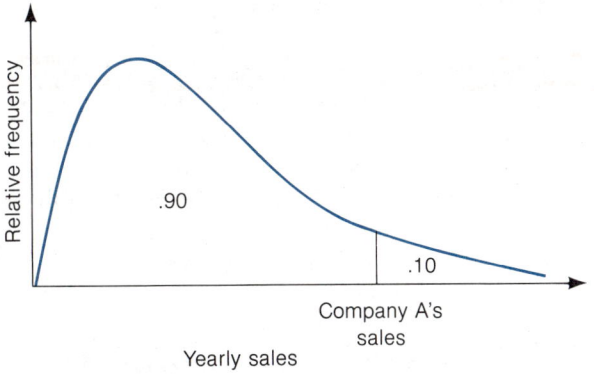

Another measure of relative standing in popular use is the **z-score**. As you can see in Definition 3.8, the z-score makes use of the mean and standard deviation of the data set in order to specify the location of a measurement:

DEFINITION 3.8

The **sample z-score** for a measurement x is

$$z = \frac{x - \bar{x}}{s}$$

The **population z-score** for a measurement x is

$$z = \frac{x - \mu}{\sigma}$$

Note that the z-score is calculated by subtracting $\bar{x}$ (or μ) from the measurement x and then dividing the result by s (or σ). The result, the z-score, represents the distance between a given measurement, x, and the mean, expressed in standard deviations.

EXAMPLE 3.12

Suppose 200 steelworkers are selected, and the annual income of each is determined. The mean and standard deviation are $\bar{x} = \$24{,}000$ and $s = \$2{,}000$. Suppose Joe Smith's annual income is $22,000. What is his sample z-score?

FIGURE 3.13
Annual Income of
Steelworkers

$18,000		$22,000	$24,000		$30,000
$\bar{x} - 3s$		Joe Smith's income	$\bar{x}$		$\bar{x} + 3s$

SOLUTION

Joe Smith's annual income lies below the mean income of the 200 steelworkers (see Figure 3.13). We compute

$$z = \frac{x - \bar{x}}{s} = \frac{\$22{,}000 - \$24{,}000}{\$2{,}000} = -1.0$$

which tells us that Joe Smith's annual income is 1.0 standard deviation *below* the sample mean, or, in short, his sample z-score is -1.0.   ∎

The numerical value of the z-score reflects the relative standing of the measurement. A large positive z-score implies that the measurement is larger than almost all other measurements, whereas a large negative z-score indicates that the measurement is smaller than almost every other measurement. If a z-score is 0 or near 0, the measurement is located near the middle of the sample or population.

We can be more specific if we know that the frequency distribution of the measurements is mound-shaped. In this case, the following interpretation of the z-scores can be given:

INTERPRETATION OF z-SCORES FOR MOUND-SHAPED DISTRIBUTIONS OF DATA

1. Approximately 68% of the measurements will have a z-score between -1 and 1.
2. Approximately 95% of the measurements will have a z-score between -2 and 2.
3. All or almost all the measurements will have a z-score between -3 and 3.

Note that this interpretation of z-scores is identical to that given in Table 3.8 for samples that exhibit mound-shaped frequency distributions. The statement that a measurement falls in the interval $(\mu - \sigma, \mu + \sigma)$ is identical to the statement that a measurement has a population z-score between -1 and 1, since all measurements between $(\mu - \sigma)$ and $(\mu + \sigma)$ are within 1 standard deviation of μ (see Figure 3.14).

FIGURE 3.14

Population z-Scores for a Mound-Shaped Distribution

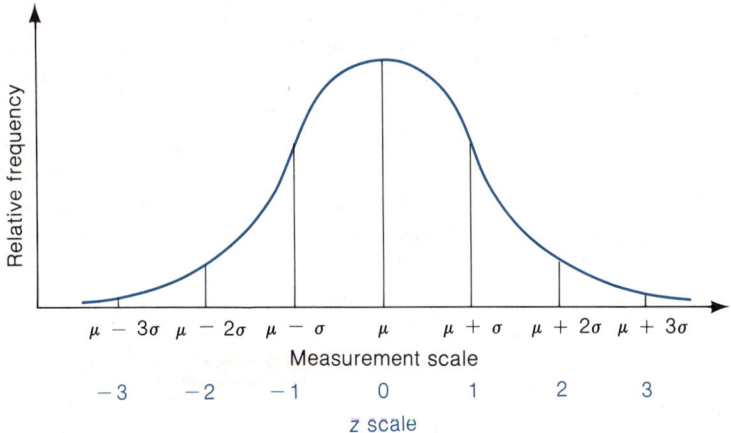

EXAMPLE 3.13

Suppose a female bank employee believes her salary is low as a result of sex discrimination. To try to substantiate her belief, she collects information on the salaries of her male counterparts in the banking business. She finds that their salaries have a mean of $34,000 and a standard deviation of $2,000. Her salary is $27,000. Does this information support her claim of sex discrimination?

SOLUTION

The analysis might proceed as follows. First, we calculate the z-score for the woman's salary with respect to those of her male counterparts. Thus,

$$z = \frac{\$27,000 - \$34,000}{\$2,000} = -3.5$$

The implication is that the woman's salary is 3.5 standard deviations *below* the mean of the male salary distribution. Furthermore, if a check of the male salary data shows that the frequency distribution is mound-shaped, we can infer that very few salaries in this distribution should have a z-score less than -3, as shown in Figure 3.15. Therefore, a z-score of -3.5 represents either a measurement from a distribution different from the male salary distribution or a very unusual (highly improbable) measurement for the male salary distribution.

FIGURE 3.15

Male Salary Distribution

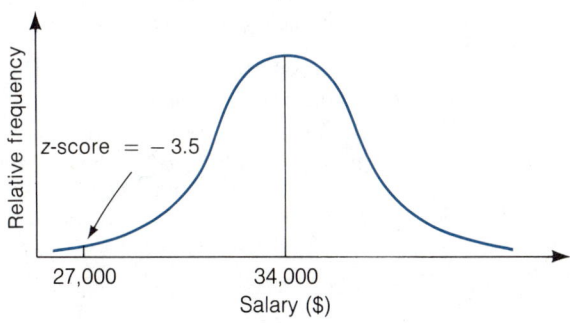

Well, which of the two situations do you think prevails? Do you think the woman's salary is simply an unusually low one in the distribution of salaries, or do you think her claim of salary discrimination is justified? Most people would probably conclude that her salary does not come from the male salary distribution. However, the careful investigator should require more information before inferring sex discrimination as the cause. We would want to know more about the data collection technique the woman used, and more about her competence at her job. Also, perhaps other factors such as the length of employment should be considered in the analysis. ■

The method of Example 3.13 exemplifies an approach to statistical inference that might be called the rare event approach. An experimenter hypothesizes a specific frequency distribution to describe a population of measurements. Then a sample of measurements is drawn from the population. If the experimenter finds it unlikely that the sample came from the hypothesized distribution, the hypothesis is concluded to be false. Thus, in Example 3.13 the woman believes her salary reflects sex discrimination. She hypothesizes that her salary should be just another measurement in the distribution of her male counterparts' salaries if no discrimination exists. However, it is so unlikely that the sample (in this case, her salary) came from the male frequency distribution that she rejects that hypothesis, concluding that the distribution from which her salary was drawn is different from the distribution for the men.

This rare event approach to inference-making is discussed further in later chapters. Proper application of the approach requires a knowledge of probability, the subject of our next chapter.

CASE STUDY 3.6

DECIDING WHEN TO RESPOND TO CONSUMER COMPLAINTS

Now we will finish the discussion, begun in Case Study 3.5, of a method proposed by Namias (1964) to determine when and when not to conduct a search for specific causes of consumer complaints.

The rate of consumer complaints about a product may change or vary as a result of merely chance or fate, or it may be due to some specific cause, such as a decline in the quality of the product. Concerning the former, Namias says:

> In any operation or production process, variability in the output or product will occur, and no two operational results may be expected to be exactly alike. Complete constancy of consumer rates of complaint is not possible, for the vagaries of fate and chance operate even within the most rigid framework of quality or operation control.*

Namias provides a decision rule with which to determine when the observed variation in the rate of consumer complaints is due to chance and when it is due to specific causes. If the observed rate is 2 standard deviations or less away from the mean rate of complaint, it is attributed to chance. If the observed rate is farther

*Reprinted from *Journal of Marketing Research*, published by the American Marketing Association. Namias, J. "A method to detect specific causes of consumer complaints," Aug. 1964, pp. 63–68.

than 2 standard deviations above the mean rate, it is attributed to a specific problem in the production or distribution of the product. The reasoning is that if there are no problems with the production and distribution of the product, 95% of the time the rate of complaints should be within 2 standard deviations of the mean rate. If the production and distribution processes were operating normally, it would be very unlikely for a rate higher than 2 standard deviations above the mean to occur. Instead, it is more likely that the high complaint rate is caused by abnormal operation of the production and/or distribution process; that is, something specific is wrong with the process.

Namias recommends searching for the cause (or causes) only if the observed variation in the rate of complaints is determined by the rule to be the result of a specific cause (or causes). The degree of variability due to chance must be tolerated. Namias says,

> As long as the results exhibit chance variability, the causes are common, and there is no need to attempt to improve the product by making specific changes. Indeed this may only create more variability, not less, and it may inject trouble where none existed, with waste of time and money. . . . On the other hand, time and money are again wasted through failure to recognize specific conditions when they arise. It is therefore economical to look for a specific cause when there is more variability than is expected on the basis of chance alone.

Namias collected data from the records of a beverage company for a 2-week period to demonstrate the effectiveness of the rule. Consumer complaints concerned chipped bottles that looked dangerous. For one of the firm's brands the mean complaint rate was determined to be 26.01 and the rate 2 standard deviations above the mean was determined to be 48.78 complaints per 10,000 bottles sold. The complaint rate observed during the 2 weeks under study was 93.12 complaints per 10,000 bottles sold. Since 93.12 is many more than 2 standard deviations above the mean rate, it was concluded that the high rate of complaints must have been caused by some specific problem in the production or distribution of the particular brand of beverage and that a search for the problem would probably be worthwhile. The problem was traced to rough handling of the bottled beverage in the warehouse by newly hired workers. As a result, a training program for new workers was instituted.

EXERCISES 3.42–3.53

LEARNING THE MECHANICS

3.42 What is the 50th percentile of a quantitative data set? What is another name for the 50th percentile?

3.43 In each of the following compute the z-score for the x value, and note whether your result is a sample z-score or a population z-score:
   a. $x = 31$, $s = 7$, $\bar{x} = 24$ **b.** $x = 95$, $s = 4$, $\bar{x} = 101$
   c. $x = 5$, $\mu = 2$, $\sigma = 1.7$ **d.** $\mu = 17$, $\sigma = 5$, $x = 14$

3.44 Consider the following data set:

0	5	6	3	1
2	3	2	1	8
1	7	1	5	4
4	3	1	6	0
5	0	20	5	2

a. Calculate $\bar{x}$ and s for the data set.

b. Suppose you drew an additional observation, and its value was 100. Calculate the z-score for this observation, and explain why you would or would not classify it as an unusually large measurement.

APPLYING THE CONCEPTS

3.45 In 1983 the United States imported merchandise valued at $258 billion and exported merchandise worth $200.5 billion. The difference between these two quantities (exports minus imports) is referred to as the *merchandise trade balance*. Since more goods were imported than exported in 1983, the merchandise trade balance was a *negative* $57.5 billion. The accompanying table lists the United States exports to and imports from a sample of ten countries in 1983 (in millions of dollars).

COUNTRY	EXPORTS	IMPORTS
Brazil	2,557	4,946
Egypt	2,813	303
France	5,961	6,025
Italy	3,908	5,455
Japan	21,894	41,183
Mexico	9,082	16,776
Panama	748	337
Soviet Union	2,003	347
Sweden	1,581	2,429
Turkey	783	320

Source: *Statistical Abstract of the United States: 1985*, pp. 815–819.

a. Calculate the U.S. merchandise trade balance with each of the ten countries. Express your answers in billions of dollars.

b. Use a z-score to identify the relative position of the U.S. trade balance with Japan within the data set you developed in part **a**. Do the same for the trade balance with the Soviet Union. Write a sentence or two that describes the relative positions of these two trade balances.

3.46 The accompanying table lists the unemployment rate in 1983 for a sample of nine countries.

a. Calculate the mean and standard deviation of this data set.

b. Calculate the z-scores of the unemployment rates of the United States, Australia, and Japan.

c. Describe the information conveyed by the sign (positive or negative) of the z-scores you calculated in part **b**.

COUNTRY	PERCENT UNEMPLOYED
Australia	10.0
Canada	11.9
France	8.8
Germany	7.3
Great Britain	13.4
Italy	5.2
Japan	2.7
Sweden	3.5
United States	9.6

Source: *Statistical Abstract of the United States: 1985*, p. 852.

3.47 In *Fortune* magazine's 1985 ranking of the 500 largest industrial corporations in the United States, Control Data Corporation ranked 71st in terms of 1984 sales. In 1981, it ranked 144th. Use percentiles to describe Control Data Corporation's position in each year's sales distribution.

3.48 A parking lot owner's accountant determined the owner's receipts for each of 100 randomly chosen days from the past year. The mean and standard deviation for the 100 days were $360 and $25, respectively. Yesterday's receipts amounted to $370.
  a. Find the sample z-score for yesterday's receipts.
  b. How many standard deviations away from the mean is the value of yesterday's receipts?
  c. Would you consider yesterday's receipts to be unusually high? Why or why not?

3.49 It is known that the frequency distribution of the number of video cassette recorders (VCRs) sold each week by a large department store in Atlanta is mound-shaped, with a mean of 35 and a variance of 9.
  a. Approximately what percentage of the measurements in the frequency distribution should fall between 32 and 38? Between 26 and 44?
  b. If the z-score for last week's sales was -1.33, how many VCRs did the store sell last week?
  c. If it is known that the number of VCRs sold each week by a rival department store has a mound-shaped frequency distribution with a mean of 35 and a standard deviation of 2, for which store is it more likely that more than 41 VCRs will be sold in a week? Why?

3.50 Suppose that 40 and 90 are two elements of a population data set and that their z-scores are -2 and 3, respectively. Using only this information, is it possible to determine the population's mean and standard deviation? If so, find them. If not, explain why it is not possible.

3.51 One of the ways the federal government raises money is through the sale of securities such as Treasury bonds, Treasury bills ("T-bills"), and U.S. savings bonds. Treasury bonds and bills are marketable (i.e., they can be traded in the securities market) long-term and short-term notes, respectively. U.S. savings bonds are nonmarketable notes; they can be purchased and redeemed only from the U.S. Treasury. On June 30, 1983, the interest rate on 3-month T-bills was 8.75%. Within the next week, the *Wall Street Journal* sampled 17 economists and asked them to forecast the interest rate of 3-month T-bills on September 30, 1983 (T-bills are offered for sale weekly by the government, and their interest rates typically vary with each offering). The forecasts obtained are listed in the table at the top of page 124.

ECONOMIST	INTEREST RATE FORECAST (%)	ECONOMIST	INTEREST RATE FORECAST (%)
Alan Greenspan	8.70	Robert Parry	8.50
Timothy Howard	8.75	John Paulus	9.50
Lacy H. Hunt	9.35	Norman Robertson	8.50
Edward Hyman	7.80	Francis Schott	8.50
David Jones	9.25	Stuart Schweitzer	9.00
Irwin Kellner	8.25	Allen Sinai	9.15
Alan Lerner	9.25	Thomas Thompson	9.25
Donald Maude	7.70	John Wilson	10.00
Anne Parker Mills	8.50		

Source: *Wall Street Journal*, July 5, 1983, p. 2. Reprinted by permission. © Dow Jones & Company, Inc. 1983. All rights reserved.

a. Calculate the *z*-scores of Alan Greenspan's forecast and John Wilson's forecast. What do the *z*-scores tell you about their forecasts relative to the forecasts of the other economists?

b. Write a sentence or two that summarizes the 17 forecasts. In your summary, use a measure of central tendency and a measure of variability.

3.52 The mean and standard deviation of the gross weekly income distribution of a local firm's 120 employees were determined to be $170 and $10, respectively.

a. Approximately what percentage of the employees would be expected to have incomes over $190 per week? Under $160 per week? Over $200 per week?

b. If you were employed by this firm and your weekly income was $185, what would your *z*-score be and how many standard deviations would your salary be away from the mean salary?

3.53 Refer to Exercise 3.52. Suppose it is known that the weekly gross income distribution is mound-shaped.

a. Approximately what percentage of the employees would be expected to have incomes over $190 per week? Under $160 per week? Over $200 per week?

b. If you and a friend both worked at the firm and your income was $160 per week and hers was $195 per week, how many standard deviations apart are your incomes?

c. If you randomly chose an employee of this firm, is it more likely that his or her gross income is over $190 per week or under $145 per week? Why?

SECTION 3.10

BOX PLOTS: GRAPHICAL DESCRIPTIONS BASED ON QUARTILES (OPTIONAL)

The **box plot**, a relatively recent introduction to the methodology of descriptive measures, makes use of the sample percentiles; it is based on the **quartiles** of a data set. Quartiles are values that partition the data set into four groups, each containing 25% of the measurements. The lower quartile Q_L is the 25th percentile, the middle quartile is the median M (the 50th percentile), and the upper quartile Q_U is the 75th percentile (see Figure 3.16).

DEFINITION 3.9

The **lower quartile** is the 25th percentile of a data set, the **middle quartile** is the median, and the **upper quartile** is the 75th percentile.

FIGURE 3.16

The Quartiles for a
Data Set

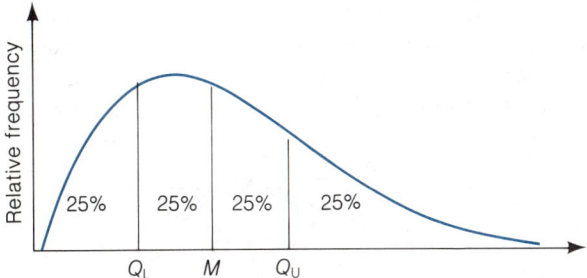

A box plot is based on the **interquartile range (IQR)**, the distance between the lower and upper quartiles:

$$IQR = Q_U - Q_L$$

DEFINITION 3.10

The **interquartile range** is the distance between the lower and upper quartiles:

$$IQR = Q_U - Q_L$$

The box plot for the 50 companies' percentages of revenues spent on R&D (Table 3.4) is given in Figure 3.17. The box plot shown there was generated by the Minitab statistical software package for personal computers.* Note that a rectangle (the **box**) is drawn, with the ends of the rectangle (the **hinges**, represented by the "I's" at the ends of the box drawn by the Minitab program) drawn at the quartiles Q_L and Q_U. By definition, then, the "middle" 50% of the observations—those between Q_L and Q_U—fall inside the box. For the R&D data, these quartiles appear to be at (approximately) 7.0 and 9.5. Thus,

$$IQR = 9.5 - 7.0 = 2.5 \quad \text{(approximately)}.$$

Note that the median is shown at about 8.0 by a "+" sign within the box.

FIGURE 3.17 Minitab Box Plot for R&D Percentages

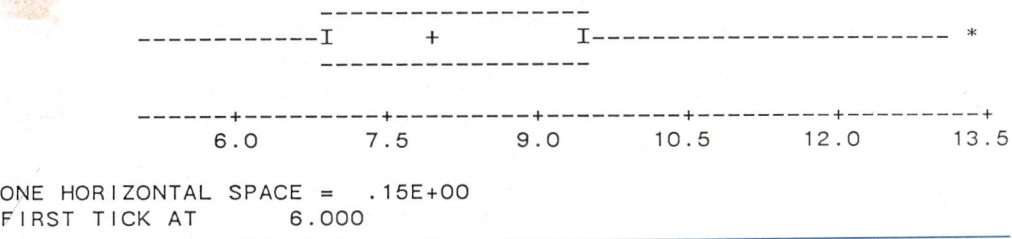

*Although box plots can be generated by hand, the amount of detail required makes them particularly well-suited for computer generation. We will use computer software to generate the box plots in this section. Numerical labels have been added to the horizontal axis of the Minitab printouts for ease of reference. Note that Minitab specifies the distance between the horizontal marks in scientific notation. The distance between horizontal tick marks in the Minitab box plot of Figure 3.17 is thus $.15 \times 10^0 = .15$.

The points that are 1.5(IQR) from the hinges are called the **inner fences** of the box plot. Emanating from each hinge of the box are dashed lines called the **whiskers**. The two whiskers will extend to the most extreme observation inside the inner fences. For example, the inner fence on the lower side of the R&D percentage plot is (approximately):

$$\text{Lower inner fence} = \text{Lower hinge} - 1.5(\text{IQR})$$
$$\approx 7.0 - 1.5(2.5)$$
$$= 7.0 - 3.75 = 3.25$$

The smallest measurement in the data set is 5.2, which is well inside this inner fence. Thus, the lower whisker extends to 5.2. On the upper end, the inner fence is at about $(9.5 + 3.75) = 13.25$. The largest measurement inside this fence is the third largest measurement, 13.2. Note that the longer upper whisker reveals the rightward skewness of the R&D distribution.

Values that are beyond the inner fences receive special attention, because they are extreme values that represent relatively rare occurrences. In fact, for mound-shaped distributions, fewer than 1% of the observations are expected to fall outside the inner fences. As discussed above, none of the R&D measurements fall outside the lower inner fence. However, the two measurements at 13.5 fall outside the upper inner fence at 13.25. These measurements are represented by asterisks (*), and they further emphasize the rightward skewness of the distribution. Note that the box plot does not reveal that there are *two* measurements at 13.5, since only a single symbol is used to represent both observations at that point.

Another pair of fences, the **outer fences**, are defined at a distance 3(IQR) from each end of the box. These fences are not shown unless one or more measurements fall beyond an outer fence. Such measurements are represented by zeros (0), and are very extreme measurements that require special attention and analysis. Less than one-hundredth of 1% (.01%, or .0001) of the measurements from mound-shaped distributions are expected to fall beyond the outer fences. Since no measurement in the R&D box plot (Figure 3.17) is represented by a 0, we know that none of the measurements fall outside the outer fences.

Generally, any measurements that fall beyond the inner fences, and certainly any that fall beyond the outer fences, are considered potential **outliers**. Outliers are extreme measurements that stand out from the rest of the sample, and which may be faulty—incorrectly recorded observations, members of a different population than the rest of the sample or, at the least, very unusual measurements from the same population. For example, the two R&D measurements at 13.5 may be considered outliers, because they exceed the inner fence ("*" representation in Figure 3.17). When we analyze these measurements, we find they are correctly recorded. However, it turns out that both represent R&D expenditures of relatively young and fast-growing companies. Thus, the outlier analysis may have revealed important factors that relate to the R&D expenditures of high-tech companies: their age and rate of growth. Outlier analysis often reveals useful information of this kind, and therefore plays an important role in the statistical inference-making process.

The elements (and nomenclature) of box plots are summarized in the box. Some aids to the interpretation of box plots are also given.

ELEMENTS OF A BOX PLOT

1. A rectangle (the **box**) is drawn with the ends (the **hinges**) drawn at the lower and upper quartiles (Q_L and Q_U). The median of the data is shown in the box, usually by a "+".
2. The points at distances $1.5(IQR)$ from each hinge mark the **inner fences** of the data set. Horizontal lines (the **whiskers**) are drawn from each hinge to the most extreme measurement inside the inner fence.
3. A second pair of fences, the **outer fences**, exist at a distance of 3 interquartile ranges, $3(IQR)$, from the hinges. One symbol (usually "*") is used to represent measurements falling between the inner and outer fences, and another (usually "0") is used to represent measurements beyond the outer fences. Thus, outer fences are invisible unless one or more measurements lie beyond them.
4. The symbols used to represent the median and the extreme data points (those beyond the fences) will vary depending on the software you use to construct the box plot. (You may use your own symbols if you are constructing a box plot by hand.) You should consult the program's documentation to determine exactly which symbols are used.

AIDS TO THE INTERPRETATION OF BOX PLOTS

1. Examine the length of the box. The IQR is a measure of the sample's variability, and is especially useful for the comparison of two samples (see Example 3.15).
2. Visually compare the lengths of the whiskers. If one is clearly longer, the distribution of the data is probably skewed in the direction of the longer whisker.
3. Analyze any measurements that lie beyond the fences. Fewer than 5% should fall beyond the inner fences, even for very skewed distributions. Measurements beyond the outer fences are probably **outliers**, with one of the following explanations:
   a. The measurement is incorrect. It may have been observed, recorded, or entered into the computer incorrectly.
   b. The measurement belongs to a population different from that from which the rest of the sample was drawn (see Example 3.15).
   c. The measurement may be correct and from the same population as the rest, but represents a rare event. Generally, we accept this explanation only after carefully ruling out all others.

EXAMPLE 3.14

In Example 2.3 we analyzed 50 processing times for the development of price quotes by a manufacturer of industrial wheels. The intent was to determine whether processing time was related to the success or failure in obtaining the business, and each quote that corresponds to "lost" business was so classified. The data are repeated in Table 3.10. Use a statistical software package to draw a box plot for these data.

TABLE 3.10

Price Quote Processing Times (Days)

REQUEST NUMBER	PROCESSING TIME	LOST?	REQUEST NUMBER	PROCESSING TIME	LOST?
1	2.36	No	26	3.34	No
2	5.73	No	27	6.00	No
3	6.60	No	28	5.92	No
4	10.05	Yes	29	7.28	Yes
5	5.13	No	30	1.25	No
6	1.88	No	31	4.01	No
7	2.52	No	32	7.59	No
8	2.00	No	33	13.42	Yes
9	4.69	No	34	3.24	No
10	1.91	No	35	3.37	No
11	6.75	Yes	36	14.06	Yes
12	3.92	No	37	5.10	No
13	3.46	No	38	6.44	No
14	2.64	No	39	7.76	No
15	3.63	No	40	4.40	No
16	3.44	No	41	5.48	No
17	9.49	Yes	42	7.51	No
18	4.90	No	43	6.18	No
19	7.45	No	44	8.22	Yes
20	20.23	Yes	45	4.37	No
21	3.91	No	46	2.93	No
22	1.70	No	47	9.95	Yes
23	16.29	Yes	48	4.46	No
24	5.52	No	49	14.32	Yes
25	1.44	No	50	9.01	No

SOLUTION

The SAS/PC box plot printout for these data is shown in Figure 3.18. Note that the SAS program draws the box plot vertically, and shows a scale for the processing times on the left. SAS uses a horizontal dashed line in the box to represent the median, and a plus (+) sign to represent the mean (SAS shows the mean in box plots, unlike many other statistical programs). Also, note that SAS uses the symbol "0" to represent measurements between the inner and outer fences, and "*" to represent observations beyond the outer fences.

Note that the upper whisker is longer than the lower whisker, and that the mean lies above the median; these characteristics reveal the rightward skewness of the data. However, the most important feature of the data is made very obvious by the box plot: there are at least two measurements between the inner and outer fences (in fact, there are three, but two are almost equal, and are represented by the same "0"), at least one beyond the outer fence, all on the upper end of the distribution.

FIGURE 3.18
SAS Box Plot for
Processing Time Data

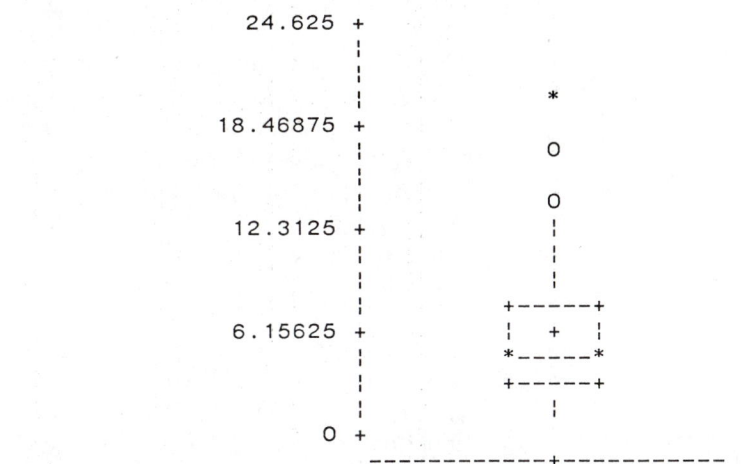

Variable=TIME

Thus, the distribution is extremely skewed to the right, and several measurements need special attention in our analysis. We offer an explanation for the outliers in the following example. ∎

EXAMPLE 3.15

The box plot for the 50 processing times (Figure 3.18) does not explicitly reveal the differences, if any, between the times corresponding to the success or failure to obtain the business. Box plots corresponding to the 39 "won" and 11 "lost" bids were generated using the SAS/PC program, and are shown in Figure 3.19. Interpret them.

FIGURE 3.19
Box Plots of Processing
Time Data: Won and
Lost Bids

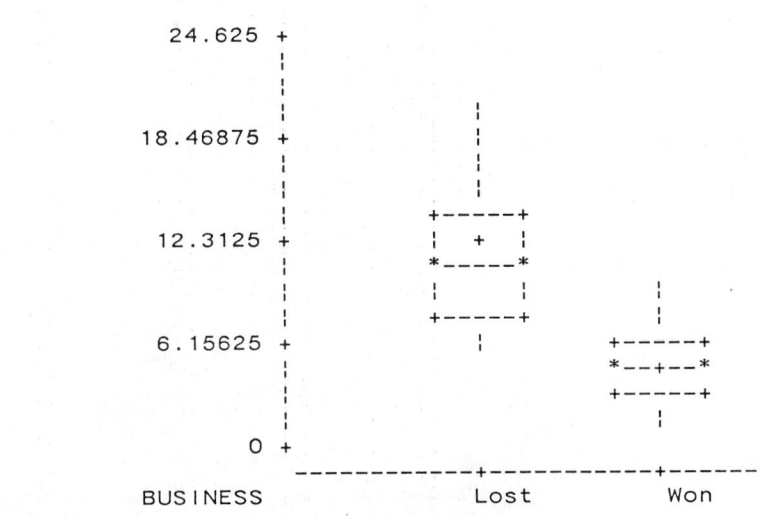

Variable=TIME

SOLUTION

The division of the data set into two parts, corresponding to won and lost bids, eliminates any observations that are beyond inner or outer fences. Furthermore, the skewness in the distributions has been reduced, as evidenced by the facts that the upper whiskers are only slightly longer than the lower, and that the means are closer to the medians than for the combined sample. The box plots also reveal that the processing times corresponding to the lost bids tend to exceed those of the won bids. A plausible explanation for the outliers in the combined box plot (Figure 3.18) is that they are from a different population than the bulk of the times. In other words, there are two populations represented by the sample of processing times—one corresponding to lost bids, and the other to won bids.

The box plots lend support to the conclusion that the price quote processing time and the success of acquiring the business are related. However, whether the visual differences between the box plots generalize to inferences about the populations corresponding to these two samples is a matter for inferential statistics, not graphical descriptions. We will discuss how to use samples to compare two populations using inferential statistics in Chapter 9.   ■

EXERCISES 3.54–3.65

LEARNING THE MECHANICS

[*Note: Starred (*) exercises require the use of a computer.*]

3.54 What is the 75th percentile of a data set?

3.55 Minitab was used to generate the following box plot:*

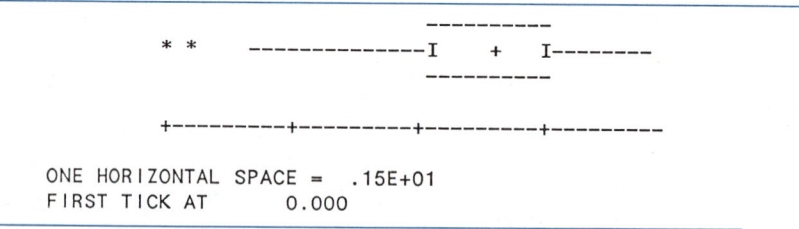

```
                                              ----------
        * *          ---------------I    +    I--------
                                              ----------


        +---------+---------+---------+---------+---------
```

```
    ONE HORIZONTAL SPACE =   .15E+01
    FIRST TICK AT        0.000
```

a. What is the median of the data set (approximately)?
b. What are the upper and lower quartiles of the data set (approximately)?
c. What is the interquartile range of the data set (approximately)?
d. Is the data set skewed to the left, skewed to the right, or symmetric?
e. What percentage of the measurements in the data set lie to the right of the median? To the left of the upper quartile?

3.56 Minitab was used to generate the accompanying box plots. Compare and contrast the frequency distributions of the two data sets. Your answer should include comparisons of the following characteristics: central tendency, variation, skewness, and outliers.

*Remember that Minitab uses scientific notation to express the distance between horizontal tick marks. This distance is .15 × 10¹ = 1.5 in the box plot for Exercise 3.55.

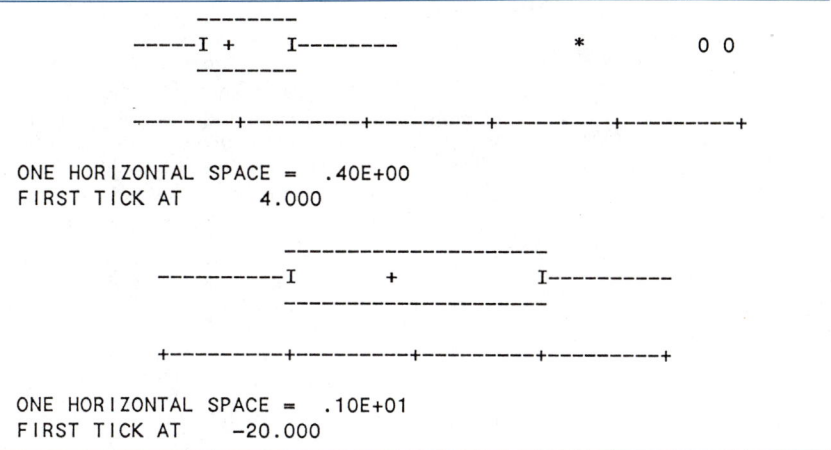

```
                          --------
                  -----I +    I--------            *        0 0
                          --------

              --------+---------+---------+---------+---------+
     ONE HORIZONTAL SPACE =  .40E+00
     FIRST TICK AT       4.000


                          ----------------------
              ----------I        +         I----------
                          ----------------------

              +---------+---------+---------+---------+
     ONE HORIZONTAL SPACE =  .10E+01
     FIRST TICK AT     -20.000
```

*3.57 Use a statistical software package to construct a box plot for the following set of sample measurements:

1.11	1.24	1.35	1.31	1.55	1.49
1.72	1.66	1.40	2.00	1.24	1.14
1.55	1.26	1.30	1.86	1.25	1.28
1.39	1.65	1.46	1.41	1.31	2.10
1.36	1.33	1.50	1.12	1.38	1.82

*3.58 Consider the following data set:

21	42	30	65	51	35	28	10	34	55
44	49	47	99	33	34	32	33	33	72

a. Use a statistical software package to construct a box plot for the data set.
b. Identify any outliers that may exist in the data set.

*3.59 Consider the following two sample data sets:

SAMPLE A			SAMPLE B		
121	171	158	171	152	170
173	184	163	168	169	171
157	85	145	190	183	185
165	172	196	140	173	206
170	159	172	172	174	169
161	187	100	199	151	180
142	166	171	167	170	188

a. Use a statistical software package to construct a box plot for each data set.
b. Using the information reflected in your box plots, describe the similarities and differences between the two data sets.
c. Identify any outliers that may exist in the two data sets.

APPLYING THE CONCEPTS

3.60 In 1985, a special issue of *Forbes* magazine ranked the nation's 500 largest companies by 1984 sales. American Brands fell at the upper quartile with sales of $5,333 million. Describe American Brands' location within the sales distribution.

3.61 The table contains the top salary offer (in thousands of dollars) received by each member of a sample of 50 undergraduate business majors (excluding accounting majors) who graduated from the School of Management at the University of Minnesota in 1985.

SALARY OFFERS TO 1985 GRADUATES									
20.1	19.8	17.6	17.7	17.3	18.1	13.4	13.7	20.9	15.2
17.9	21.5	23.6	26.5	21.1	15.0	15.9	20.9	19.2	16.8
18.1	20.0	32.3	28.0	16.1	18.5	22.8	15.7	15.4	17.6
16.9	19.8	24.8	16.0	20.7	12.6	18.2	26.0	16.3	16.0
15.9	17.7	5.8	17.5	52.0	24.1	20.9	18.2	14.8	21.9

Source: Placement Office, School of Management, University of Minnesota.

a. Find and interpret the z-score associated with the highest salary offer, the lowest salary offer, and the mean salary offer. Would you consider the highest offer to be unusually high? Why or why not?

*b. Use a statistical software package to construct a box plot for this data set. Which salary offers (if any) are potentially faulty observations? Explain.

*3.62 A firm's earnings per share (E/S) of common stock is a measure used by investors to monitor the financial performance of a firm. Thirty firms were sampled from *Fortune* magazine's 1985 listing of the 500 largest corporations in the United States, and their earnings per share are recorded in the table.

FIRM	E/S	FIRM	E/S
Illinois Tool Works	$ 2.40	Dow Jones	$ 2.01
Dayco	2.35	United Brands	3.43
Reynolds Metals	6.36	Washington Post	6.11
Scott Paper	3.83	Avon Products	2.16
Phelps Dodge	−11.27	Reichhold Chemicals	3.63
Westmoreland Coal	−6.01	American Cyanamid	4.41
Avery International	2.42	Todd Shipyards	4.84
Warner–Lambert	2.81	Asarco	−12.56
Burroughs	5.40	Snap-on Tools	2.93
General Electric	5.03	McCormick	4.40
Cooper Industries	2.13	Exxon	6.77
Lockheed	5.28	Georgia–Pacific	.97
Kellogg	3.35	Crown Cork & Seal	4.78
Midland–Ross	−3.03	DuPont (E.I.) de Nemours	5.93
Oxford Industries	2.17	United Merchants & Manufacturers	1.41

Source: *Fortune*, Apr. 29, 1985, pp. 266–284.

a. Use a statistical software package to construct a box plot for this data set. Identify any outliers that may exist in this data set.

b. For each outlier identified in part **a**, determine how many standard deviations it lies from the mean of the E/S data set.

*3.63 A manufacturer of minicomputer systems is interested in improving its customer support services. As a first step, its marketing department has been charged with the responsibility of summarizing the extent of customer problems in terms of system down time. The 40 most recent customers were surveyed to determine the amount of down time (in hours) they had experienced during the previous month. These data are listed in the table.

CUSTOMER NUMBER	DOWN TIME	CUSTOMER NUMBER	DOWN TIME	CUSTOMER NUMBER	DOWN TIME	CUSTOMER NUMBER	DOWN TIME
230	12	240	24	250	4	260	34
231	16	241	15	251	10	261	26
232	5	242	13	252	15	262	17
233	16	243	8	253	7	263	11
234	21	244	2	254	20	264	64
235	29	245	11	255	9	265	19
236	38	246	22	256	22	266	18
237	14	247	17	257	18	267	24
238	47	248	31	258	28	268	49
239	0	249	10	259	19	269	50

a. Use a statistical software package to construct a box plot for these data. Use the information reflected in the box plot to describe the frequency distribution of the data set. Your description should address central tendency, variation, and skewness.
b. Use your box plot to determine which customers are having unusually lengthy down times.
c. Find and interpret the z-scores associated with customers you identified in part **b**.

*3.64 A hydraulic metal stamping machine is set to produce disks that are between 15.75 and 16.25 centimeters in diameter. At the start of each work shift, 30 disks are sampled from the production process and measured by an inspector to be sure the machine is operating properly. The measurements (in centimeters) shown in the table were obtained at the start of the current work shift.

DISK DIAMETERS				
16.22	15.95	15.92	16.05	16.10
15.82	16.15	16.05	15.96	16.02
15.74	16.07	16.13	16.00	16.09
16.01	16.13	16.19	15.84	15.92
16.13	16.25	15.94	16.04	16.07
16.00	15.99	16.24	16.09	16.02

a. Use a statistical software package to construct a box plot for these data. What does your box plot reveal about the current operation of the stamping machine?
b. Does your box plot indicate that the set of 30 measurements contains any outliers? Explain.

3.65 The accompanying Minitab-generated box plots describe the U.S. Environmental Protection Agency's 1986 automobile mileage estimates for all models manufactured by Ford and Honda.

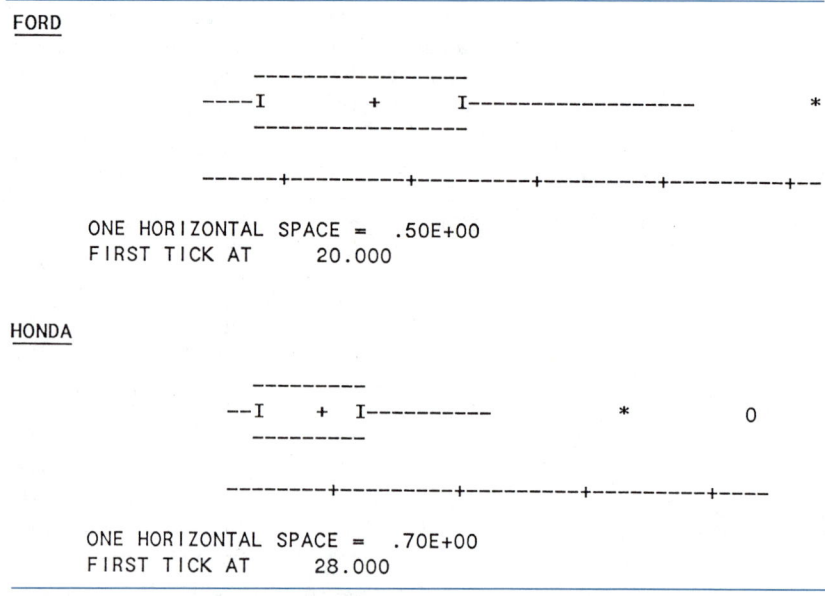

FORD

```
                            --------------------
                   ----I         +       I-------------------                       *
                            --------------------

                   ------+---------+---------+---------+---------+--

        ONE HORIZONTAL SPACE =  .50E+00
        FIRST TICK AT     20.000
```

HONDA

```
                       ----------
                  --I     + I-----------              *            O
                       ----------

                  --------+---------+---------+---------+----

        ONE HORIZONTAL SPACE =  .70E+00
        FIRST TICK AT     28.000
```

a. Which manufacturer has the higher median mileage estimate?
b. Which manufacturer's mileage estimates have the greater range?
c. Which manufacturer's mileage estimates have the greater interquartile range?
d. Which manufacturer has the model with the highest mileage estimate? Approximately what is that mileage?

SUMMARY

Numerical methods for describing quantitative data sets can be grouped as follows:

1. Measures of central tendency
2. Measures of variability

The mode, mean, and median of a data set are measures of central tendency. The **mode** is the most frequently observed member of the data set. The **mean** is the average of the data, obtained by summing the n measurements in the data set and dividing by n. The **median** is a number in the data set chosen so that half the measurements in the data set fall below the median and half fall above. The relationship between the mean and median provides information about the skewness of the frequency distribution. For making inferences about the population, the sample mean will usually be preferred to the other measures of central tendency.

The numerical description of a set of data requires more than a measure of central tendency. The **variability** of the data set must also be described. The **range, absolute deviation, variance,** and **standard deviation** all represent numerical measures of variability. Of these, the variance and standard deviation are used most commonly, especially when the ultimate objective is to make inferences about a population.

The mean and standard deviation may be used to make statements about the fraction of measurements in a given interval. For example, we know that at least 75% of the measurements in a data set will lie within 2 standard deviations of the mean. If the frequency distribution of the data set is mound-shaped, approximately 95% of the measurements will lie within 2 standard deviations of the mean.

Measures of relative standing provide still another dimension on which to describe a data set. The objective of these measures is to describe the location of a specific measurement relative to the rest of the data set. Percentiles, quartiles, and z-scores are important examples of measures of relative standing. Outliers in a data set may be detected through the construction of a box plot.

According to the rare event concept of statistical inference, if the chance that a particular sample came from a hypothetical population is very small, we can conclude either that the sample is extremely rare or that the hypothesized population is not the one from which the sample was drawn. The more unlikely it is that the sample came from the hypothesized population, the more strongly we favor the conclusion that the hypothesized population is not the true one. We need to be able to assess accurately the rarity of a sample, and this requires a knowledge of probability, the subject of Chapter 4.

| | | | | | | | | | | | |

SUPPLEMENTARY EXERCISES 3.66–3.96

[Note: Starred (*) exercises require the use of a computer. Double-starred (**) exercises refer to optional sections in this chapter.]

3.66 Discuss the conditions under which the median is preferred to the mean as a measure of central tendency.

3.67 Compute $\sum_{i=1}^{n} x_i^2$, $\sum_{i=1}^{n} x_i$, and $\left(\sum_{i=1}^{n} x_i\right)^2$ for each of the following data sets:

a. 11, 1, 2, 8, 7   **b.** 15, 15, 2, 6, 12
c. −1, 2, 0, −4, −8, 13   **d.** 100, 0, 0, 2

3.68 Compute s^2 and s for each of the data sets in Exercise 3.67.

3.69 Compute s^2 for data sets with the following characteristics:

a. $\sum_{i=1}^{n} x_i^2 = 246$, $\sum_{i=1}^{n} x_i = 63$, $n = 22$

b. $\sum_{i=1}^{n} x_i^2 = 666$, $\sum_{i=1}^{n} x_i = 106$, $n = 25$

c. $\sum_{i=1}^{n} x_i^2 = 76$, $\sum_{i=1}^{n} x_i = 11$, $n = 7$

3.70 For each of the following data sets, compute $\bar{x}$, s^2, and s:

a. 13, 1, 10, 3, 3   **b.** 13, 6, 6, 0
c. 1, 0, 1, 10, 11, 11, 15   **d.** 3, 3, 3, 3

3.71 Compute the range for each of the data sets in Exercise 3.70.

3.72 For each of the following data sets, compute $\bar{x}$, s^2, and s. If appropriate, specify the units in which your answers are expressed.
a. 4, 6, 6, 5, 6, 7
b. −$1, $4, −$3, $0, −$3, −$6
c. $\frac{3}{5}$ percent, $\frac{4}{5}$ percent, $\frac{2}{5}$ percent, $\frac{1}{5}$ percent, $\frac{1}{16}$ percent

3.73 Explain why we generally prefer the standard deviation to the range as a measure of variability for quantitative data.

*3.74 Compute the mean and variance of the following data sets:

a. CLASS	CLASS FREQUENCY		b. CLASS	CLASS FREQUENCY
1–5	2		.25– .50	0
6–10	5		.50– .75	5
11–15	12		.75–1.00	12
16–20	6		1.00–1.25	8
			1.25–1.50	6
			1.50–1.75	2

3.75 Under what circumstances might the standard deviation be preferred to the variance as a measure of the variation in a data set?

3.76 In reference to a measurement or observation, what is a measure of relative standing?

3.77 How does a z-score locate a measurement within a set of measurements? Explain.

3.78 In 1985, consumers spent an estimated average of $11,883 for a new car, compared with $5,414 in 1976 (*Minneapolis Star and Tribune*, Mar. 15, 1985, p. 2S). In order to be able to determine the 1985 average *exactly*, what information is needed?

3.79 The table lists the expenditures (in millions of dollars) on research and development (R&D) by industry group in 1980.

INDUSTRY GROUP	EXPENDITURES ON R&D
Food and kindred products	620
Paper and allied products	495
Chemicals and allied products	4,636
Petroleum refining	1,552
Rubber products	656
Stone, clay, and glass products	406
Primary metals	728
Fabricated metal products	550
Machinery	5,901
Electrical equipment	9,175
Motor vehicles and other transportation equipment	5,117
Aircraft and missiles	9,198
Professional and scientific instruments	3,029

Source: *Statistical Abstract of the United States: 1985*, p. 577.

    a. Compute the median of the R&D expenditures for this data set.
    b. What was the average amount spent per industry group on R&D in 1980?
    c. Explain why the mode cannot be effectively used as a measure of central tendency in the R&D expenditures data set.

3.80 A quality control inspector is interested in determining whether a metal lathe used to produce machine bearings is properly adjusted. He plans to do so by using 30 bearings selected randomly from the last 2,000 bearings produced by the machine to estimate the average diameter of bearings being produced. Define and explain the meanings of the following symbols *in the context of this problem:*
    a. μ **b.** $\bar{x}$ **c.** σ **d.** s **e.** $\mu + 2\sigma$

3.81 One hundred management trainees were given an examination in basic accounting. Their test scores were found to have a mean and variance of 75 and 36, respectively.
    a. Make a statement about the percentage of the test scores that would be expected to fall between 69 and 81.
    b. If a grade of 63 was required to pass the test, make a statement about the percentage of the trainees who would be expected to fail.

3.82 Redo Exercise 3.81 assuming that the distribution of test scores was determined to be mound-shaped.

*,**3.83 "In most industries, goods are shipped, the product works, and the customer pays. But in high-technology [industries], confusion abounds over when a sale is really a sale. Sometimes, the product is shipped, and a sale is recorded—but the product has bugs, and customers balk at paying" ("High-Tech Sales: Now You See Them, Now You Don't?", *Business Week*, Nov. 18, 1985, p. 106). The accompanying table lists a sample of high-technology companies and the number of days (on average) it takes each to collect payment for sales made.

COMPANY	AVERAGE NO. OF DAYS TO COLLECT ON SALE	COMPANY	AVERAGE NO. OF DAYS TO COLLECT ON SALE
CPT	155	Ungermann-Bass	131
Data Switch	143	Radiation Systems	184
C3	231	Porta Systems	141
Scan-Optics	144	Silicon General	140
Aydin	178	Network Systems	181
Computer Entry Systems	137	Masstor Systems	163
UTL	242	Intecom	131
Seagate Technology	143	Microdyne	154
DSC Communications	217	T-Bar	144
Computervision	132	Applied Data Research	126

Data: Standard & Poor's Compustat Services, Inc.

    a. Use a statistical software package to construct a box plot for these data.
    b. Based on the box plot, what can be said about the shape of the frequency distribution of the data set?
    c. Use the box plot to identify outliers in the data set. Which firm(s) have outlying values?

d. The standard deviation of the data set is 34.3. Find the z-score for the outlier(s) you identified in part **c**.

e. Considering the z-score you obtained in part **c**, is it likely that the average number of days to collect on a sale for a high-tech firm would be larger than 242? Explain.

3.84 The vice-president in charge of sales for the conglomerate you work for has asked you to evaluate the sale records of two of the firm's divisions. You note that the range of monthly sales for division A over the last 2 years is $50,000 and the range for division B is only $30,000. You compute each division's mean monthly sales for the same time period and discover that both divisions have a mean of $110,000. Assume that is all the information you have about the division's sales records. Would you be willing to say which of the divisions has a more consistent sales record? Why or why not?

3.85 Refer to Exercise 3.84.
a. Estimate the standard deviation of the monthly sales distribution of division B.
b. Based on your estimate in part **a**, would you say it is more likely that division B's sales next month will be over $120,000 or under $90,000?
c. Is it possible for division B's sales next month to be over $160,000? Explain.

3.86 Before purchasing stock in an electronics firm, the management of a mutual fund wants information concerning the price movements of the firm's stock during the past year. Thirty days of the past year were randomly selected and the closing price (to the nearest dollar) was recorded for each day:

$33	$20	$45	$41	$52	$36	$21	$33	$41	$36
49	26	51	28	32	35	32	28	50	35
48	29	50	28	33	39	30	30	28	29

The mutual fund's management has decided it should purchase the stock only if the mean closing price for last year is $41 or more.
a. Define the terms *mean*, *median*, and *mode* in the context of this problem.
b. Construct a relative frequency histogram for the data.
c. Compute the mean, median, and mode for the data set and locate them on the histogram.
d. Do you think the mutual fund should purchase the firm's stock? Explain. [*Note:* Simply looking at the sample and using your intuition could lead you to an erroneous conclusion. We will learn how to use the sample mean to make decisions about a population mean in Chapters 8 and 9.]

3.87 The manufacturers of an amazing new gadget claim in an advertisement for sales personnel that their first-year salespeople earn an average of $35,000. You call their main office and demand to know the standard deviation of first-year salespeople's incomes. The assistant personnel manager's secretary tells you the standard deviation is $9,000. What can be said about the fraction of the manufacturer's salespeople who make between $8,000 and $62,000 during their first year?

3.88 To help evaluate the impact of a proposed change in its credit policy, a department store wants to obtain a description of the ages of its accounts receivable. Accordingly, the store's internal auditor sampled 30 accounts and determined the age (number of days since each account had a balance of zero). The data are given in the table.

ACCOUNT NUMBER	AGE	ACCOUNT NUMBER	AGE	ACCOUNT NUMBER	AGE
133	8	2312	63	4480	28
398	10	2425	6	4618	0
502	22	2788	0	5021	38
553	12	3001	42	5118	9
992	0	3110	0	5226	18
1009	7	3333	80	5438	21
1351	25	3621	15	5506	3
1667	28	3981	0	5872	110
1993	15	4310	7	6095	54
2210	36	4322	45	6204	0

a. Find the z-scores for the ages of account numbers 5021 and 5872.

*,**b.** Use a statistical software package to construct a box plot of the data.

*,**c.** Locate the ages of account numbers 5021 and 5872 on the box plot. Do you consider both accounts to be outliers in terms of their ages? Explain.

3.89 Suppose you used the following formula as a measure of the variability (V) of a data set:

$$V = \frac{\sum\limits_{i=1}^{n} (x_i - \bar{x})}{n}$$

What information can be learned about the variability of a data set using this formula? Using the data in part **a** of Exercise 3.67, find V.

3.90 Many firms use on-the-job training to teach their employees computer programming. Suppose you work in the personnel department of a firm that just finished training a group of its employees to program and you have been requested to review the performance of one of the trainees on the final test that was given to all trainees. The mean and standard deviation of the test scores are 80 and 5, respectively, and the distribution of scores is mound-shaped.

a. The employee in question scored 65 on the final test. Compute the employee's z-score.

b. Approximately what percentage of the trainees will have z-scores equal to or less than the employee of part **a**?

c. If a trainee was arbitrarily selected from those who had taken the final test, is it more likely that he or she would score 90 or above, or 65 or below?

3.91 The 1984 advertising expenditures (in thousands of dollars) by media are described for the 16 top-selling brandies and cordials by the Minitab-generated box plots on page 140.

a. For which medium are expenditures the highest?

b. For which medium is the range of expenditures the largest? The smallest?

c. For which medium is the interquartile range the largest? The smallest?

d. Describe the shape of each data set's frequency distribution.

e. Do any of the media receive advertising expenditures from all 16 brands of brandies and cordials? Explain.

MAGAZINES

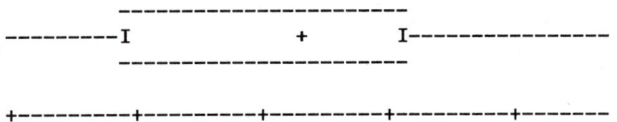

```
ONE HORIZONTAL SPACE = .10E+03
FIRST TICK AT        0.000
```

BILLBOARDS

```
ONE HORIZONTAL SPACE = .50E+02
FIRST TICK AT        0.000
```

NEWSPAPERS

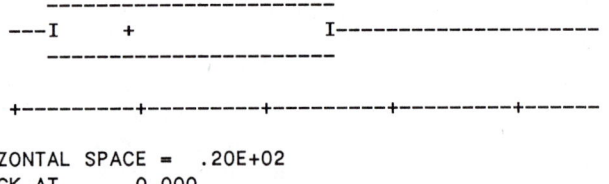

```
ONE HORIZONTAL SPACE = .20E+02
FIRST TICK AT        0.000
```

3.92 Chebyshev's theorem (presented in Table 3.8) states that at least $(1 - 1/k^2)$ of a set of measurements will lie within k standard deviations of the mean of the data set. Use Chebyshev's theorem to state the fraction of a set of measurements that will lie:

a. Within 2 standard deviations of the mean μ ($k = 2$).

b. Within 3 standard deviations of the mean.

c. Within 1.5 standard deviations of the mean.

d. More than 2.75 standard deviations from the mean.

**3.93 Calculate the mean and standard deviation of the following data sets:

a.

CLASS	CLASS FREQUENCY
10–19	15
20–29	12
30–39	8
40–49	5
50–59	2

b.

CLASS		CLASS FREQUENCY
$ −99 to $ −50		20
−49 to	0	55
1 to	50	102
51 to	100	63
101 to	150	18

**3.94 Consider the following data set:

1	31	43	49	52	55	59	66	72	89
12	31	45	50	52	55	59	67	76	92
20	33	46	50	53	56	60	68	76	96
24	39	47	51	53	57	62	69	82	103
28	40	48	52	53	58	63	71	84	109

a. Calculate the mean and standard deviation of this data set.
b. Group the data into classes and construct a tabular frequency distribution (see Table 3.9, for example) for this data set.
c. Using your frequency distribution of part **b** and the formulas of Section 3.8, calculate the mean and standard deviation of the data set.
d. Explain why your answers to part **a** and part **c** do not agree.

3.95 Seven workdays in June were selected, and the number of machine breakdowns per day in the woodworking shop of a furniture company was recorded. This procedure was repeated in August after the firm had replaced ten of its oldest machines. The data are listed here:

June: 8 3 0   0 10 4 9
August: 0 3 4 11   3 3 2

a. Which month exhibits more variability in the number of machines that break down per day as measured by the range? As measured by s^2? In this case, which measure, s^2 or the range, do you feel better represents the variability of the data sets? Explain.
b. Add 3 to each of the numbers of breakdowns per day in June and recompute s^2 for June. Compare your result with that obtained in part **a**. What is the effect on s^2 of adding a constant to each of the sample measurements?
c. Multiply each of the numbers of breakdowns per day in June by 3 and recompute s^2 for June. Compare your result with that obtained in part **a**. What is the effect on s^2 of multiplying each sample measurement by a constant?

3.96 Economic theory suggests that the vigor of competition in an industry is related to the number of firms in the industry. As a measure of competitiveness, however, the number of firms in an industry does not take into consideration the extent to which a few firms may dominate that industry. For example, in an industry with 100 firms, each may produce 1% of industry output, or three firms may dominate, producing 75% compared to the other 97 firms' 25%. The *market concentration ratio* is a measure of competitiveness that reflects such inequalities among firms in an industry. The market concentration ratio is usually defined as the percentage of total industry sales contributed by the largest few firms (usually three or four). A high concentration ratio indicates an industry dominated by a few firms. A low concentration ratio indicates an industry with much competition among many firms. The table on page 142 contains 1970 three-firm market concentration ratios for 12 industries in each of six different countries. Answer the following questions using the methods from Chapters 2 and 3 that you deem appropriate.

a. For each of the six countries, characterize the magnitude and variability of the sample of 12 market concentration ratios.
b. For each of the 12 industries, characterize the magnitude and variability of the sample of six market concentration ratios.

INDUSTRY	UNITED STATES	CANADA	UNITED KINGDOM	SWEDEN	FRANCE	WEST GERMANY
Brewing	39	89	47	70	63	17
Cigarettes	68	90	94	100	100	94
Fabric weaving	30	67	28	50	23	16
Paints	26	40	40	92	14	32
Petroleum refining	25	64	79	100	60	47
Shoes (except rubber)	17	18	17	37	13	20
Glass bottles	65	100	73	100	84	93
Cement	20	65	86	100	81	54
Ordinary steel	42	80	39	63	84	56
Antifriction bearings	43	89	82	100	80	90
Refrigerators	64	75	65	89	100	72
Storage batteries	54	73	75	100	94	82

Source: F. M. Scherer, Alan Beckenstein, Erich Kaufer, and R. D. Murphy, *The Economics of Multiplant Operation: An International Comparisons Study* (Cambridge, Mass.: Harvard University Press, 1975). Reprinted by permission.

c. Which nation has on average the most competition within its industries? The least competition?

d. Which are the three most competitive industries? The three least competitive industries?

ON YOUR OWN . . .

1. Find the variance, standard deviation, and range of the quantitative data set you found for the "On Your Own" section in Chapter 2.

2. Use Table 3.8 to describe the distribution of this data set.

3. Count the actual numbers of observations that fall within 1, 2, and 3 standard deviations of the mean of the data set and compare these counts with the description of the data set you developed above.

USING THE COMPUTER . . .

In Chapter 2 you used the computer to generate graphical descriptions of the percentage of women in the work force in 1986—nationally, for one census region, and for one state. Now compute the mean and standard deviation of the same data set over the 1,000 zip codes, and over the zip codes in the region you selected in Chapter 2. Use the computer to count the number of zip codes' percentages within the intervals $\bar{x} \pm s$, $\bar{x} \pm 2s$, and $\bar{x} \pm 3s$. Compare the results with those given by Chebyshev's theorem and the Empirical Rule (Table 3.8).

REFERENCES

Brown, S. J. and Warner, J. B. "Measuring security price performance," *Journal of Financial Economics*, Sept. 1980, *8*, 205–258.

Chase, R. B. and Aquilano, N. J. *Production and Operations Management*, rev. ed. Homewood, Ill.: Richard D. Irwin, 1977. Chapter 11.

Curley, J. "Some think a 'baby boom' spending spree could lead to strong economic recovery." *Wall Street Journal*, Jan. 24, 1983, 25.

Eger, C. E. "Corporate mergers: An empirical analysis of the role of risky debt." Unpublished Ph.D. dissertation, School of Management, University of Minnesota, 1982.

"The *Fortune* directory of the 500 largest U.S. industrial corporations." *Fortune*, Apr. 29, 1985, 266–284.

"Keeping up." *Fortune*, Aug. 13, 1979, 100.

McCullough, B. "Should you be a coupon clipper?" *Minneapolis Star and Tribune*, Mar. 17, 1986, 3C.

Mendenhall, W. *Introduction to Probability and Statistics*, 7th ed. Boston: Duxbury, 1987. Chapter 3.

Milkovich, G. T., Annoni, A. J., and Mahoney, T. A. "The use of the delphi procedures in manpower forecasting." *Management Science*, Dec. 1972, *19*, part 1, 381–388.

Namias, J. "A method to detect specific causes of consumer complaints." *Journal of Marketing Research*, Aug. 1964, 63–68.

Neter, J., Wasserman, W., and Whitmore, G. A. *Applied Statistics*. Boston: Allyn & Bacon, 1982. Chapter 3.

Postlewaite, S. "Salad bars sprout as fast-food battle turns from the burger." *Minneapolis Tribune*, July 3, 1983, 7D.

Rothstein, M. "Hotel overbooking as a Markovian sequential decision process." *Decision Sciences*, July 1974, *5*, 389–405.

U.S. Bureau of the Census. *Statistical Abstract of the United States: 1985*. Washington, D.C.: U.S. Government Printing Office, 1985.

U.S. Department of Labor. *The Consumer Price Index: Concepts and Content over the Years*. Bureau of Labor Statistics, Report 517, May 1978.

Vatter, W. J. *Accounting Measurements for Financial Reports*. Homewood, Ill.: Richard D. Irwin, 1971, p. 78.

WHERE WE'VE BEEN . . .

In Chapter 1, we identified inference from a sample to a population as the goal of statistics. In Chapters 2 and 3, we learned how to describe a set of measurements using graphical and numerical descriptive methods.

WHERE WE'RE GOING . . .

We now begin to consider the problem of making an inference. What permits us to make the inferential jump from sample to population and then to give a measure of reliability for the inference? As you will subsequently see, the answer is *probability*. This chapter is devoted to a study of probability—what it is and some of the basic concepts of the theory that surrounds it.

CONTENTS

PROBABILITY

You will recall that statistics is concerned with inferences about a population based on sample information. Understanding how this will be accomplished is easier if you understand the relationship between population and sample. This understanding is enhanced by reversing the statistical procedure of making inferences from sample to population. In this chapter we assume the population *known* and calculate the chances of obtaining various samples from the population. Thus, probability is the "reverse" of statistics: In probability we use the population information to infer the probable nature of the sample.

Probability plays an important role in decision-making. To illustrate, suppose you have an opportunity to invest in an oil exploration company. Past records show that for ten out of ten previous oil drillings (a sample of the company's experiences), all ten resulted in dry wells. What do you conclude? Do you think the chances are better than 50–50 that the company will hit a producing well? Should you invest in this company? We think your answer to these questions will be an emphatic no. If the company's exploratory prowess is sufficient to hit a producing well 50% of the time, a record of ten dry wells out of ten drilled is an event that is just too **improbable**. Do you agree?

As another illustration, suppose you are playing poker with what your opponents assure you is a well-shuffled deck of cards. In three consecutive five-card hands, the person on your right is dealt four aces. Based on this sample of three deals, do you think the cards are being adequately shuffled? Again, we think your answer will be no and that you will reach this conclusion because dealing three hands of four aces is just too **improbable**, assuming that the cards were properly shuffled.

Note that the decision concerning the potential success of the oil drilling company and the decision concerning the card shuffling were both based on probabilities—namely, the probabilities of certain sample results. Both situations were contrived so you could easily conclude that the probabilities of the sample results were small. Unfortunately, the probabilities of many observed sample results are not so easy to evaluate. For these cases, we will need the assistance of a theory of probability.

| | | | | | | | | | | | |

S E C T I O N 4.1

EVENTS, SAMPLE SPACES, AND PROBABILITY

Most sets of data that are of interest to the business community are generated by some **experiment**:

DEFINITION 4.1

An **experiment** is an act or process that leads to a single outcome that cannot be predicted with certainty.

Our definition of *experiment* is broader than that used in the physical sciences, where we might picture test tubes, microscopes, and other equipment. Examples of statistical experiments in business are recording whether a customer prefers one of two brands of coffee (say, brand A or brand B), measuring the change in the Dow Jones Average from one day to the next, recording the weekly sales of a business firm, and counting the number of errors on a page of an accountant's ledger.

A "single outcome" of an experiment is called a **simple event**.

DEFINITION 4.2

A **simple event** is an outcome of an experiment that cannot be decomposed into a simpler outcome.

In Table 4.1 we present three examples of experiments and their simple events. We begin with simple coin and dice examples because they are most likely to be familiar to you. Experiment **a** in Table 4.1 is to toss a coin and observe the up face. You will undoubtedly agree that the most basic possible outcomes of this experiment are Observe a head and Observe a tail. Experiment **b** in Table 4.1 is to toss a die and observe the up face. Notice that the simple events Observe a 1, Observe a 2, etc., cannot be decomposed into simpler outcomes. However, the outcome Observe an even number can be decomposed into Observe a 2, Observe a 4, and Observe a 6. Thus, Observe an even number is not a simple event. The reasoning is similar for experiment **c** in Table 4.1.

TABLE 4.1
Experiments and Their
Simple Events

---

a. Experiment: Toss a coin and observe the up face.
  Simple events: 1. Observe a head
             2. Observe a tail

b. Experiment: Toss a die and observe the up face.
  Simple events: 1. Observe a 1
             2. Observe a 2
             3. Observe a 3
             4. Observe a 4
             5. Observe a 5
             6. Observe a 6

c. Experiment: Toss two coins and observe the up faces.
  Simple events: 1. Observe H_1, H_2
             2. Observe H_1, T_2
             3. Observe T_1, H_2
             4. Observe T_1, T_2
  (where H_1 means "Head on coin 1," H_2 means "Head on coin 2," etc.)

Outcomes such as Observe an even number, which can be decomposed into simpler outcomes (Observe a 2, Observe a 4, Observe a 6), are called **events**.

DEFINITION 4.3

An **event** is a collection of one or more simple events.

Thus, in experiment **b** of Table 4.1 three examples of events are Observe an even number, Observe a number less than 4, and Observe a 6. Note that the last event,

Observe a 6, cannot be decomposed into a simpler outcome, and so is also a simple event. If the experiment is counting the number of errors on a page of an accountant's ledger, three examples of events are Observe no errors, Observe fewer than five errors, and Observe more than ten errors. Only the first, Observe no errors, is also a simple event. Our goal is to be able to calculate the probability that a particular event will occur when an experiment is performed.

The first step in achieving this goal is to note that simple events have an important property. If the experiment is conducted once, you will observe one and only one simple event. For example, if the experiment is to toss a coin and observe the up face, you cannot Observe a tail and Observe a head on the same toss. Or, if you toss a die and Observe a 2, you cannot Observe a 6 on the same toss. That is, for a single performance of an experiment, one and only one of the simple events will occur. To see that this property does not hold for all events, consider the event Observe an even number. It is possible to Observe an even number and Observe a 2 on the same toss of a die.

The collection of all the simple events of an experiment is called the **sample space**:

> **DEFINITION 4.4**
>
> The **sample space** of an experiment is the collection of all its simple events.

For example, there are six simple events associated with experiment **b** in Table 4.1. These six simple events comprise the sample space for the experiment. Similarly, for experiment **c** in Table 4.1, there are four simple events in the sample space.

A graphical method, called the **Venn diagram**, is useful for presenting the sample space and its simple events. The sample space is shown as a closed figure, labeled S. This figure contains a set of points, called **sample points**, with each point representing a simple event. Figure 4.1 shows the Venn diagram for each of the three experiments in Table 4.1. Note that the number of sample points in a sample space S is equal to the number of simple events associated with the respective experiment: two for experiment **a**, six for experiment **b**, and four for experiment **c**.

Now that we have defined the terms *simple event* and *sample space*, we are prepared to define the **probabilities of simple events**. The probability of a simple event is a number that measures the likelihood that the event will occur when the experiment is performed. This number is usually taken to be the relative frequency of the occurrence of a simple event in a very long series of repetitions of an experiment. Or, when this information is not available, we select the number based on experience. For example, if we are assigning probabilities to the two simple events in the coin-toss experiment (Observe a head and Observe a tail), we might reason that if we toss a balanced coin a very large number of times, the simple events Observe a head and Observe a tail will occur with the same relative frequency of .5. Thus, the probability of each simple event is .5.

FIGURE 4.1

Venn Diagrams for the Three Experiments from Table 4.1

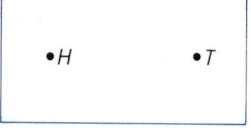

(a) Experiment: Observe the up face on a coin

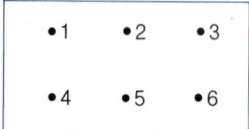

(b) Experiment: Observe the up face on a die

(c) Experiment: Observe the up faces on two coins

FIGURE 4.2

Experiment: Invest in a
Business Venture and
Observe Whether It
Succeeds or Fails

• S   • F

S

For some experiments we may assign probabilities to the simple events based on general information about the experiment. For example, if the experiment is to invest in a business venture and to observe whether it succeeds or fails, the sample space would appear as in Figure 4.2.

We are unlikely to be able to assign probabilities to the simple events of this experiment based on a long series of repetitions, since unique factors govern each performance of this kind of experiment. Instead, we may consider factors such as the personnel managing the venture, the general state of the economy at the time, the rate of success of similar ventures, and any other information deemed pertinent. If we finally decide that the venture has an 80% chance of succeeding, we assign a probability of .8 to the simple event Success. This probability can be interpreted as a measure of our degree of belief in the outcome of the business venture. Such subjective probabilities should be based on expert information and must be carefully assessed. Otherwise, we run the risk of being misled by uninformed and/or biased probability statements. That is, we may be misled on any decisions based on these probabilities or based on any calculations in which they appear.*

CASE STUDY 4.1

BLOOM COUNTY
PROBABILITIES

The issue of whether probability should be defined as the relative frequency in a long series of repetitions of an experiment or as a subjective measure of belief is one that has been debated for many years by probabilists, statisticians, and even philosophers. A considerably lighter side of this debate was illustrated in the accompanying Bloom County comic strip.

© 1986, Washington Post Writers Group, reprinted with permission.

No matter how you assign the probabilities to simple events, the probabilities assigned must obey two rules:

1. All simple event probabilities must lie between 0 and 1, inclusive.
2. The probabilities of all the simple events in the sample space must sum to 1.

*For a text that deals in detail with the subjective evaluation of probabilities, see Winkler (1972).

FIGURE 4.3

Die-Toss Experiment with
Event *A*: Observe an Even
Number

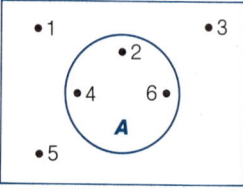

Recall that an event is a collection of one or more simple events. Let us examine this statement in greater detail. Consider the die-tossing experiment and the event Observe an even number. This event will occur if and only if one of the three simple events, Observe a 2, Observe a 4, or Observe a 6, occurs. Consequently, you can think of the event Observe an even number as the collection of the three simple events, Observe a 2, Observe a 4, and Observe a 6. This event, which we will denote by the symbol *A*, can be represented in a Venn diagram by a closed figure inside the sample space *S*. The closed figure *A* will contain the simple events that constitute event *A*, as shown in Figure 4.3.

How do you decide which simple events belong to the set associated with an event *A*? Test each simple event in the sample space *S*. If event *A* occurs when a particular simple event occurs, then that simple event is in the event *A*. For example, in the die-toss experiment, the event Observe an even number (event *A*) will occur if the simple event Observe a 2 occurs. By the same reasoning, the simple events Observe a 4 and Observe a 6 are in event *A*.

Now return to our original objective—finding the probability of any event. Consider the problem of finding the probability of observing an even number (event *A*) in the single toss of a die. You will recall that *A* will occur if one of the three simple events, toss a 2, 4, or 6, occurs. Since two or more simple events cannot occur at the same time, we can easily calculate the probability of event *A* by summing the probabilities of the three simple events. We would attach a probability equal to $\frac{1}{6}$ to each of the simple events (if the die is fair), so the probability of observing an even number (event *A*), denoted by the symbol *P(A)*, would be

$$P(A) = P(\text{Observe a 2}) + P(\text{Observe a 4}) + P(\text{Observe a 6})$$
$$= \tfrac{1}{6} + \tfrac{1}{6} + \tfrac{1}{6} = \tfrac{1}{2}$$

The previous example leads us to a general procedure for finding the probability of an event *A*:

> The probability of an event *A* is calculated by summing the probabilities of the simple events in *A*.

Thus, we can summarize the steps for calculating the probability of any event:

STEPS FOR CALCULATING PROBABILITIES OF EVENTS

1. Define the experiment.
2. List the simple events.
3. Assign probabilities to the simple events.
4. Determine the collection of simple events contained in the event of interest.
5. Sum the simple event probabilities to obtain the event probability.

EXAMPLE 4.1

Consider the experiment of tossing two coins and observing the up faces. Assume both coins are fair.

a. List the simple events and assign them reasonable probabilities.
b. Consider the events

> A: {Observe exactly one head}   B: {Observe at least one head}

and calculate $P(A)$ and $P(B)$.

SOLUTION

a. The simple events are

> H_1, H_2 H_1, T_2 T_1, H_2 T_1, T_2

where H_1 denotes Observe a head on coin 1, H_2 denotes Observe a head on coin 2, etc. If both coins are fair, we can again use the concept of relative frequency in a long series of experimental repetitions to conclude that each simple event should be assigned a probability of $\frac{1}{4}$.

FIGURE 4.4
Venn Diagram of the Two-Coin Toss

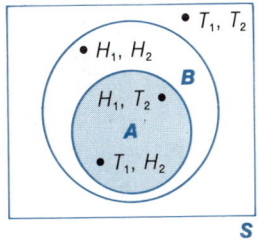

b. We use a Venn diagram to show the events A: {Observe exactly one head} and B: {Observe at least one head} (Figure 4.4). Using the collection of simple events in A and B, we may calculate the probabilities by adding the appropriate simple event probabilities:

$$P(A) = P(H_1, T_2) + P(T_1, H_2) = \tfrac{1}{4} + \tfrac{1}{4} = \tfrac{1}{2}$$
$$P(B) = P(H_1, H_2) + P(H_1, T_2) + P(T_1, H_2) = \tfrac{1}{4} + \tfrac{1}{4} + \tfrac{1}{4} = \tfrac{3}{4}$$

■

EXAMPLE 4.2

A retail computer store owner sells two basic types of microcomputers: IBM personal computers (IBM PCs) and IBM compatibles (PCs that run all or most of the same software as an IBM PC, but that are not manufactured by IBM). One problem facing the owner is deciding how many of each type of PC to stock. One important factor affecting the solution is the proportion of customers who purchase each type of PC. Show how this problem might be formulated in the framework of an experiment, with simple events and a sample space. Indicate how probabilities might be assigned to the simple events.

SOLUTION

Using the term *customer* to refer to a person who purchases one of the two types of PCs, we can define the experiment as the entrance of a customer and the observation of which type of PC is purchased. There are two simple events in the sample space corresponding to this experiment:

> Experiment: Observe the type of PC purchased by a customer.
>
> Simple events: 1. I: {The customer purchases an IBM PC}
>   2. C: {The customer purchases an IBM compatible}

The difference between this experiment and the coin-toss experiment becomes apparent when we attempt to assign probabilities to the two simple events. What probability should we assign to the simple event I? If you answer .5, you are assuming that the events I and C should occur with equal likelihood, just as the simple events Heads and Tails in the coin-toss experiment. The assignment of simple event probabilities for the PC purchase experiment is not so easy. Suppose a check of the store's records indicates that 80% of its customers purchase IBM PCs. Then it might be reasonable to approximate the probability of the simple event I as .8, and that of the simple event C as .2. The important points are that simple events are not always equally likely, and that the probabilities of simple events are not always easy to assign, particularly for experiments that represent real applications (as opposed to coin- and die-toss experiments). ∎

EXAMPLE 4.3

A poll of "computer-familiar" adults who do not own a home computer was conducted by *USA Today* (Sept. 25, 1985). Each adult was asked to identify which of ten electronic appliances was his or her highest priority purchase, if any. The results are summarized in Table 4.2.

TABLE 4.2
Electronic Appliances of Highest Priority to Purchase by Computer-Familiar Adults

ELECTRONIC APPLIANCE	PERCENT RESPONSE[a]
Home computer (HC)	24
Microwave oven (MO)	13
Compact disc player (CDP)	4
Phone-answering machine (PAM)	6
Car telephone (CT)	5
Programmable phone (PP)	3
Video cassette recorder (VCR)	17
Video camera (VC)	9
Big-screen TV (BSTV)	7
Movie camera (MC)	3
None (N)	9

[a]Response percentages in the *USA Today* article did not add to 100% due to rounding. We have added 1% to the two smallest responses to facilitate the solution to this example.

Source: "Buying a computer is in our budget," *USA Today*, Sept. 25, 1985. Copyright 1985, USA Today. Excerpted with permission.

a. Define the experiment that generated the data in Table 4.2 and list the simple events.
b. Assign probabilities to the simple events.
c. What is the probability that a telephonic appliance is of highest priority?
d. What is the probability that a video appliance is of highest priority?

SOLUTION

a. The experiment is the act of polling a "computer-familiar" adult. The simple events, the simplest outcomes of the experiment, are the 11 response categories listed in Table 4.2. They are shown in the Venn diagram in Figure 4.5.
b. In Example 4.1, the simple events were assigned equal probabilities. If we were to assign equal probabilities in this case, each of the response categories would be assigned a probability of $\frac{1}{11}$, or .09. However, you can see by examining Table 4.2 that equal probabilities are not reasonable in this case, because the response

FIGURE 4.5
Venn Diagram for
Electronic Appliance Poll

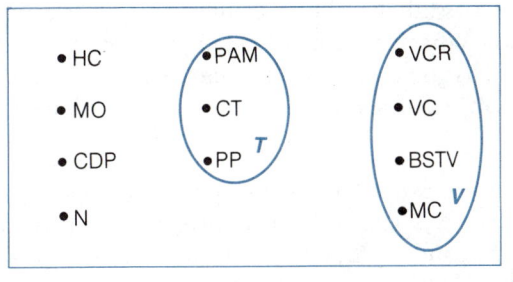

percentages were not the same in the eleven classifications. Instead, we assign a probability equal to the actual response percentage in each class, as shown in Table 4.3.

TABLE 4.3
Simple Event Probabilities
for Electronic Appliance
Poll

SIMPLE EVENT	PROBABILITY
HC	.24
MO	.13
CDP	.04
PAM	.06
CT	.05
PP	.03
VCR	.17
VC	.09
BSTV	.07
MC	.03
N	.09

c. The event T that a telephonic appliance is the highest priority is not a simple event, because it consists of more than one of the response classifications (the simple events). In fact, as shown in Figure 4.5, T consists of three simple events. The probability of T is defined to be the sum of the probabilities of the simple events in T:

$$P(T) = P(PAM) + P(CT) + P(PP)$$
$$= .06 + .05 + .03 = .14$$

d. The event V that a video appliance is identified as the highest priority purchase consists of four simple events, and the probability of V is the sum of the corresponding simple event probabilities:

$$P(V) = P(VCR) + P(VC) + P(BSTV) + P(MC)$$
$$= .17 + .09 + .07 + .03 = .36$$

∎

EXAMPLE 4.4

You have the capital to invest in two of four ventures, each of which requires approximately the same amount of investment capital. Unknown to you, two of the investments will eventually fail and two will be successful. You research the four ventures because you think that your research should increase your probability of a successful choice over a purely random selection, and you eventually decide on two. What is the lower limit of your probability of selecting the two best out of four? That is, if you used no information and selected two ventures at random, what is the probability that you would select the two successful ventures? At least one?

SOLUTION

Denote the two successful enterprises as S_1 and S_2 and the two failing enterprises as F_1 and F_2. The experiment involves a random selection of two out of the four ventures, and each possible pair of ventures represents a simple event. The six simple events that make up the sample space are

1. S_1, S_2 4. S_2, F_1
2. S_1, F_1 5. S_2, F_2
3. S_1, F_2 6. F_1, F_2

The next step is to assign probabilities to the simple events. If we assume that the choice of any one pair is as likely as any other, then the probability of each simple event is $\frac{1}{6}$. Now check to see which simple events result in the choice of two successful ventures. Only one such simple event exists—namely, S_1, S_2. Therefore, the probability of choosing two successful ventures out of the four is

$$P(S_1, S_2) = \tfrac{1}{6}$$

The event of selecting at least one of the two successful ventures includes all the simple events except F_1, F_2.

P(Select at least one success)

$$= P(S_1, S_2) + P(S_1, F_1) + P(S_1, F_2) + P(S_2, F_1) + P(S_2, F_2)$$

$$= \tfrac{1}{6} + \tfrac{1}{6} + \tfrac{1}{6} + \tfrac{1}{6} + \tfrac{1}{6} = \tfrac{5}{6}$$

Therefore, the worst that you could do in selecting two ventures out of four may not be too bad. With a random selection, the probability of selecting two successful ventures will be at least $\frac{1}{6}$ and the probability of selecting at least one successful venture out of two is at least $\frac{5}{6}$. ∎

The preceding examples have one thing in common: The number of simple events in each of the sample spaces was small; hence, the simple events were easy to identify and list. How can we manage this when the simple events run into the thousands or millions? For example, suppose you wish to select 5 business ventures from a group of 1,000. Then each different group of 5 ventures would represent a simple event. How can you determine the number of simple events associated with this experiment?

One method of determining the number of simple events for a complex experiment is to develop a counting system. Start by examining a simple version of the experiment. For example, see if you can develop a system for counting the number of ways to select 2 people from a total of 4 (this is exactly what was done in Example 4.4). If the ventures are represented by the symbols V_1, V_2, V_3, and V_4, the simple events could be listed in the following pattern:

$$V_1, V_2 \qquad V_2, V_3 \qquad V_3, V_4$$
$$V_1, V_3 \qquad V_2, V_4$$
$$V_1, V_4$$

Note the pattern and now try a more complex situation—say, sampling 3 ventures out of 5. List the simple events and observe the pattern. Finally, see if you can deduce the pattern for the general case. Perhaps you can program a computer to produce the matching and counting for the number of samples of 5 selected from a total of 1,000.

A second method of determining the number of simple events for an experiment is to use **combinatorial mathematics**. This branch of mathematics is concerned with developing counting rules for given situations. For example, there is a simple rule for finding the number of different samples of 5 ventures selected from 1,000. This rule is given by the formula

$$\binom{N}{n} = \frac{N!}{n!(N-n)!}$$

where N is the number of elements in the population; n is the number of elements in the sample; and the factorial symbol (!) means that, say,

$$n! = n(n-1)(n-2) \cdots \cdots 3 \cdot 2 \cdot 1$$

Thus, $5! = 5 \cdot 4 \cdot 3 \cdot 2 \cdot 1$. (The quantity $0!$ is defined to be equal to 1.)

EXAMPLE 4.5

Refer to Example 4.4, in which we selected two ventures from four in which to invest. Use the combinatorial counting rule to determine how many different selections can be made.

SOLUTION

For this example, $N = 4$, $n = 2$, and

$$\binom{4}{2} = \frac{4!}{2!2!} = \frac{4 \cdot 3 \cdot 2 \cdot 1}{(2 \cdot 1)(2 \cdot 1)} = 6$$

You can see that this agrees with the number of simple events obtained in Example 4.4.   ∎

EXAMPLE 4.6

Suppose you plan to invest equal amounts of money in each of 5 business ventures. If you have 20 ventures from which to make the selection, how many different samples of 5 ventures can be selected from the 20?

SOLUTION

For this example, $N = 20$ and $n = 5$. Then the number of different samples of 5 that can be selected from the 20 ventures is

$$\binom{20}{5} = \frac{20!}{5!(20-5)!} = \frac{20!}{5!15!}$$
$$= \frac{20 \cdot 19 \cdot 18 \cdots \cdots 3 \cdot 2 \cdot 1}{(5 \cdot 4 \cdot 3 \cdot 2 \cdot 1)(15 \cdot 14 \cdot 13 \cdots \cdots 3 \cdot 2 \cdot 1)} = 15,504$$ ∎

The symbol $\binom{N}{n}$, meaning the **number of combinations of N elements taken n at a time**, is just one of a large number of counting rules that have been developed by combinatorial mathematicians. This counting rule applies to situations in which the experiment calls for selecting n elements from a total of N elements, without replacing each element before the next is selected. If you are interested in learning other methods for counting simple events for various types of experiments, you will find a few of the basic counting rules in Appendix A. Others can be found in the references listed at the end of this chapter.

EXERCISES 4.1–4.17

LEARNING THE MECHANICS

4.1 What is the difference between a *simple event* and an *event*?

4.2 The diagram describes the sample space of a particular experiment and events A and B.

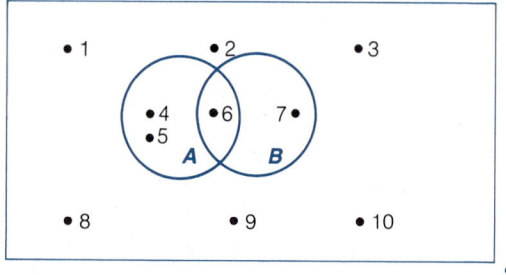

a. What is this type of diagram called?
b. Suppose the simple events are equally likely. Find $P(A)$ and $P(B)$.
c. Suppose $P(1) = P(2) = P(3) = P(4) = P(5) = \frac{1}{20}$ and $P(6) = P(7) = P(8) = P(9) = P(10) = \frac{3}{20}$. Find $P(A)$ and $P(B)$.

4.3 Consider the experiment of tossing a die and observing the up face.
a. Draw a Venn diagram for the experiment. On your diagram indicate the event Observe a number greater than 4. Call this event A. Also indicate the event Observe an even number. Call this event B.
b. We would all agree that the probability of observing a 3 on the toss of a "fair" die is $\frac{1}{6}$. Explain what it means for a die to be "fair" (or "balanced") and explain how knowing a die is fair leads us to conclude that $P(3) = \frac{1}{6}$.
c. For your Venn diagram of part **a**, assume the die is fair and find $P(A)$ and $P(B)$.
d. If you knew that a particular die were unfair (i.e., "loaded"), how would you determine the probability of observing a 3?

4.4 An experiment results in one of the following simple events: $E_1, E_2, E_3, E_4,$ and E_5.
a. Find $P(E_3)$ if $P(E_1) = .1$, $P(E_2) = .3$, $P(E_4) = .1$, and $P(E_5) = .1$.
b. Find $P(E_3)$ if $P(E_1) = P(E_3)$, $P(E_2) = .1$, $P(E_4) = .2$, and $P(E_5) = .2$.
c. Find $P(E_3)$ if $P(E_1) = P(E_2) = P(E_4) = P(E_5) = .1$.

4.5 Compute each of the following:
a. $\binom{10}{3}$ **b.** $\binom{6}{2}$ **c.** $\binom{8}{3}$ **d.** $\binom{5}{5}$ **e.** $\binom{4}{0}$

APPLYING THE CONCEPTS

4.6 The population of the amounts of 500 automobile loans made by a bank last year can be described as follows:

	AMOUNT OF LOAN ($)		
Under 1,000	1,000–3,999	4,000–5,999	6,000 or more
27	99	298	76

For the purpose of checking the accuracy of the bank's records, an auditor will randomly select one of these loans for inspection (i.e., each loan has an equal probability of being selected).
a. List the simple events in this experiment.
b. What is the probability that the loan selected will be for $6,000 or more?
c. What is the probability that the loan will be for less than $4,000?

4.7 A cigarette manufacturer has decided to market two new brands. An analysis of current market conditions and a review of the firm's past successes and failures with new brands have led the manufacturer to believe that the simple events and the probabilities of their occurrence in this marketing experiment are as listed in the table (S_1: Brand 1 succeeds, F_1: Brand 1 fails, etc.). Find the probability of each of the following events:

SIMPLE EVENTS	PROBABILITIES
S_1, S_2	.16
S_1, F_2	.24
F_1, S_2	.24
F_1, F_2	.36

    A: {Both new brands are successful in the first year}

    B: {At least one new brand is successful in the first year}

4.8 Of six cars produced at a particular factory between 8 and 10 A.M. last Monday morning, three are known to be "lemons." Three of the six cars were shipped to dealer A and the other three to dealer B. Just by chance, dealer A received all three lemons. What is the probability of this event occurring if, in fact, the three cars shipped to dealer A were selected at random from the six produced?

4.9 A buyer for a large metropolitan department store must choose two firms from the four available to supply the store's fall line of men's slacks. The buyer has not dealt with any of the four firms before and considers their products equally attractive. Unknown to the buyer, two of the four firms are having serious financial problems that may result in their not being able to deliver the slacks as soon as promised. The four firms are identified as G_1 and G_2 (firms in good financial condition) and P_1 and P_2 (firms in poor financial condition). Simple events identify the pair of firms selected. If the probability of the buyer selecting a particular pair from among the four is the same for each pair, the table gives the simple events and their probabilities for this buying experiment. Find the probability of each of the following events:

SIMPLE EVENTS	PROBABILITIES
G_1, G_2	$\frac{1}{6}$
G_1, P_1	$\frac{1}{6}$
G_1, P_2	$\frac{1}{6}$
G_2, P_1	$\frac{1}{6}$
G_2, P_2	$\frac{1}{6}$
P_1, P_2	$\frac{1}{6}$

    A: {Buyer selects two firms in good financial condition}

    B: {Buyer selects at least one firm in poor financial condition}

4.10 Simulate the experiment in Exercise 4.9 by marking four poker chips (or cards), one corresponding to each of the four firms. Mix the chips, randomly draw two, and record the results. Replace the chips. Now repeat the experiment a large number of times (at least 100).
a. Calculate the proportion of times event A occurs. How does this proportion compare with $P(A)$? Should the proportion equal $P(A)$? Explain.
b. Calculate the proportion of times event B occurs and compare that with $P(B)$.

4.11 The Value Line Survey, a service for common stock investors, provides its subscribers with up-to-date evaluations of the prospects and risks associated with the purchase of a large number of common stocks. Each stock is ranked 1 (highest) to 5 (lowest) according to Value Line's estimate of the stock's potential for price appreciation during the next 12 months and according to its safety as an investment.

Suppose you plan to purchase stock in three electrical utility companies from among seven that possess rankings of 2 for price appreciation. Unknown to you, two of the companies will experience serious difficulties with their nuclear facilities during the coming year. If you randomly select the three companies from among the seven, what is the probability that:

a. You select none of the companies with prospective nuclear difficulties?
b. You select one of the companies with prospective nuclear difficulties?
c. You select both of the companies with prospective nuclear difficulties?

4.12 Approximately 77 million Visa and 60 million Mastercard credit cards have been issued in the United States. Both are issued by thousands of banks, including Citicorp. Citicorp also issues competing cards of its own: Diners Club, Carte Blanche, and Choice are wholly owned by Citicorp. The accompanying table describes the population of credit cards issued by Citicorp.

CREDIT CARD	NUMBER ISSUED (in millions)
Visa and Mastercard	6.0
Diners Club	2.2
Carte Blanche	.3
Choice	1.0

Source: *Fortune*, Feb. 4, 1985, p. 21.

One Citicorp credit card customer is to be selected at random and the type of credit card will be recorded. (Assume that each customer has only one Citicorp-issued card.)

a. List the simple events in this experiment.
b. Find the probability of each simple event.
c. What is the probability that the customer selected uses one of Citicorp's own credit cards?

4.13 You are a lawyer for a client who has committed a felony, and there are seven judges who could hear your motion to set bail. Four judges are strict, and the other three are lenient. As you walk into the courtroom, judge A (a strict judge) is leaving to go home.

a. What is the probability of drawing a lenient judge for your client?
b. What is the probability of drawing a lenient judge for your client if the probability of getting judge B (a strict judge) is .3, the probability of getting judge C (a strict judge) is .4, and the probabilities of getting any of the other four judges are equally likely?
c. Suppose you know that judge D (a lenient judge) never follows judge A. What is the probability of drawing a lenient judge for your client if the probabilities in part **b** are valid?

4.14 Before placing a person in a highly skilled position, a company gives applicants a series of three examinations. The first is a physical examination, and each applicant is classified as satisfactory or unsatisfactory. The other two are verbal and quantitative examinations, and the scores are used to classify each applicant as high, medium, or low in each area. Thus, each individual will receive a health score, a verbal score, and a quantitative score.

a. List the different sets of classifications that can result from the battery of examinations.

b. If all applicants who take the examinations are equally qualified, and all the variation in test scores is due to random variation in the types of test questions, what is the probability that an applicant receives the lowest classification on all three examinations?

c. If an applicant scores in the highest category on at least two of the three examinations, the applicant will get a position. What is the probability that a randomly selected applicant will get a position?

4.15 In 1984, unemployment rates in the United States ranged from 4.3% in New Hampshire to 15% in West Virginia. The nationwide unemployment rate was 7.5%. The table lists the 1984 and 1983 unemployment rates for the 15 Atlantic coast states. Suppose one of these 15 states is to be selected and the direction and amount of change in its unemployment rate from 1983 to 1984 is to be observed. Assume that each state has an equal probability of being selected.

STATE	1984 (%)	1983 (%)	STATE	1984 (%)	1983 (%)
Connecticut	4.6	6.0	New Jersey	6.2	7.8
Delaware	6.2	8.1	New York	7.2	8.6
Florida	6.3	8.6	North Carolina	6.7	8.9
Georgia	6.0	7.5	Pennsylvania	9.1	11.8
Maine	6.1	9.0	Rhode Island	5.3	8.3
Maryland	5.4	6.9	South Carolina	7.1	10.0
Massachusetts	4.8	6.9	Virginia	5.0	6.1
New Hampshire	4.3	5.4			

Source: *Statistical Abstract of the United States: 1985, 1986*, p. 393.

a. What is the probability that Pennsylvania will be selected? Florida? Virginia?

b. What is the probability of selecting a state that had no change in its unemployment rate?

c. What is the probability of selecting a state whose unemployment rate increased? Decreased?

d. What is the probability of selecting a state whose unemployment rate increased 1% or more? Decreased 1% or more?

4.16 Corporation G, a manufacturer of razor blades, supplied a consumer with one pack of each of the three top name brands—G, S, and W—and asked him to use them and rank them in order of preference. The corporation was hoping the consumer would prefer its brand and rank it first, thereby giving them some material for an advertising campaign. If the consumer did not prefer one blade more than any other, but was still required to rank the blades, what is the probability that:

a. The consumer ranked brand G first? [*Hint:* One of the possible rankings would be GSW; i.e., G ranked 1, S ranked 2, and W ranked 3.]

b. The consumer ranked brand G last?

c. The consumer ranked brand G last and brand W second?

d. The consumer ranked brand W first, brand G second, and brand S third?

4.17 Probabilities are often expressed in terms of **odds**, especially in gambling settings. For example, handicappers for horse races express their belief about the probability of each horse winning a race in terms of odds. If the probability of event E is $P(E)$, then the **odds in favor of E** are $P(E)$ to $[1 - P(E)]$. Thus, if a handicapper assesses a probability of .25 that Snow Chief will win the Belmont Stakes, the odds in favor of Snow Chief are $\frac{25}{100}$ to $\frac{75}{100}$, or 1 to 3. It follows that the **odds against E** are $[1 - P(E)]$ to $P(E)$, or 3 to 1 against a win by Snow Chief. In general, if the odds in favor of event E are a to b, then $P(E) = a/(a + b)$.

a. A second handicapper assesses the probability of a win by Snow Chief to be $\frac{1}{3}$. According to the second handicapper, what are the odds in favor of a Snow Chief win?

b. A third handicapper assesses the odds in favor of Snow Chief to be 1 to 1. According to the third handicapper, what is the probability of a Snow Chief win?

c. A fourth handicapper assesses the odds against Snow Chief winning to be 3 to 2. Find this handicapper's assessment of the probability that Snow Chief will win.

S E C T I O N 4.2

COMPOUND EVENTS

An event can often be viewed as a composition of two or more other events. Such an event, called a **compound event**, can be formed (composed) in two ways:

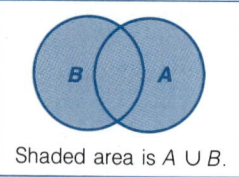

Shaded area is $A \cup B$.

DEFINITION 4.5

The **union** of two events A and B is the event that occurs if either A or B or both occur on a single performance of the experiment. We will denote the union of events A and B by the symbol $A \cup B$.

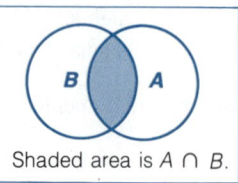

Shaded area is $A \cap B$.

DEFINITION 4.6

The **intersection** of two events A and B is the event that occurs if both A and B occur on a single performance of the experiment. We will write $A \cap B$ for the intersection of events A and B.

EXAMPLE 4.7

Consider the die-toss experiment. Define the following events:

A: {Toss an even number}

B: {Toss a number less than or equal to 3}

a. Describe $A \cup B$ for this experiment.

b. Describe $A \cap B$ for this experiment.

c. Calculate $P(A \cup B)$ and $P(A \cap B)$ assuming the die is fair.

SOLUTION

a. The union of A and B is the event that occurs if we observe an even number, a number less than or equal to 3, or both on a single throw of the die. Consequently, the simple events in the event $A \cup B$ are those for which A occurs,

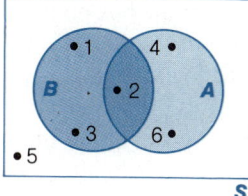

B occurs, or both A and B occur. Testing the simple events in the entire sample space, we find that the collection of simple events in the union of A and B is

$$A \cup B = \{1, 2, 3, 4, 6\}$$

b. The intersection of A and B is the event that occurs if we observe *both* an even number and a number less than or equal to 3 on a single throw of the die. Testing the simple events to see which imply the occurrence of *both* events A and B, we see that the intersection contains only one simple event:

$$A \cap B = \{2\}$$

In other words, the intersection of A and B is the simple event Observe a 2.

c. Recalling that the probability of an event is the sum of the probabilities of the simple events of which the event is composed, we have

$$P(A \cup B) = P(1) + P(2) + P(3) + P(4) + P(6)$$
$$= \tfrac{1}{6} + \tfrac{1}{6} + \tfrac{1}{6} + \tfrac{1}{6} + \tfrac{1}{6} = \tfrac{5}{6}$$

and

$$P(A \cap B) = P(2) = \tfrac{1}{6} \qquad \blacksquare$$

EXAMPLE 4.8

Many firms have undertaken direct marketing campaigns to promote their products. The campaigns typically involve mailing information to millions of households; the response rates are carefully monitored to determine the demographic characteristics of respondents. By studying tendencies to respond, the firm can better target future mailings to those segments of the population most likely to purchase the products.

Suppose a distributor of mail-order tools is analyzing the results of a recent mailing. The probability of response is believed to be related to income and age. The percentages of the total number of respondents to the mailing are given by income and age classification in Table 4.4.

TABLE 4.4
Percentage of Respondents in Age–Income Classes

		INCOME	
	< $25,000	$25,000–$50,000	> $50,000
< 30 yrs.	5%	12%	10%
AGE 30–50 yrs.	14%	22%	16%
> 50 yrs.	8%	10%	3%

Define the following events:

A: {A respondent's income is more than $50,000}

B: {A respondent's age is 30 years or more}

a. Find $P(A)$ and $P(B)$.
b. Find $P(A \cup B)$.
c. Find $P(A \cap B)$.

SOLUTION

Following the steps for calculating probabilities of events (given in the box on page 150), we first note that the objective is to characterize the income and age distribution of respondents to the mailing. To accomplish this, we define the experiment as selecting a respondent from the collection of all respondents, and observing which income and age class he or she occupies. The simple events are the nine different age–income classifications:

E_1: {<30 yrs., < $25,000}

E_2: {30–50 yrs., < $25,000}

$\vdots$ $\vdots$

E_9: {> 50 yrs., > $50,000}

Next, we assign probabilities to the simple events. If we blindly select one of the respondents, the probability that he or she will occupy a particular age–income classification is just the proportion, or relative frequency, of respondents in that classification. These proportions are given (as percentages) in Table 4.4. Thus,

$$P(E_1) = \text{Relative frequency of respondents in}$$
$$\text{age–income class } (< 30 \text{ yrs.}, < \$25,000)$$
$$= .05$$
$$P(E_2) = .14$$

and so forth. You may verify that the simple event probabilities add to 1.

a. To find $P(A)$ we first determine the collection of simple events contained in event A. Since A is defined as {> $50,000}, we see from Table 4.4 that A contains the three simple events represented by the last column of the table. In words, the event A consists of the income classification (> $50,000) and all three age classifications within that income class. The probability of A is the sum of the probabilities of the simple events in A:

$$P(A) = .10 + .16 + .03 = .29$$

Similarly, B consists of the six simple events in the second and third rows of Table 4.4:

$$P(B) = .14 + .22 + .16 + .08 + .10 + .03 = .73$$

b. The union of events A and B, $A \cup B$, consists of all simple events in *either* A or B or *both A and B*. That is, the union of A and B consists of all respondents whose income exceeds $50,000 *or* whose age is 30 or more. In Table 4.4 this is any simple event found in the third column *or* the last two rows. Thus,

$$P(A \cup B) = .10 + .14 + .22 + .16 + .08 + .10 + .03$$
$$= .83$$

c. The intersection of events A and B, $A \cap B$, consists of all simple events in *both A and B*. That is, the intersection of A and B consists of all respondents whose income exceeds $50,000 *and* whose age is 30 or more. In Table 4.4 this is any simple event found in the third column *and* the last two rows, i.e., the last two rows of the third column. Thus,

$$P(A \cap B) = .16 + .03 = .19$$ ∎

SECTION 4.3

COMPLEMENTARY EVENTS

FIGURE 4.6
Venn Diagram of
Complementary Events

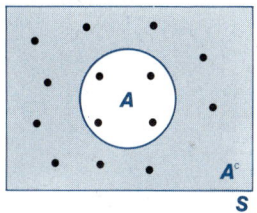

A very useful concept in the calculation of event probabilities is the notion of **complementary events**:

DEFINITION 4.7

The **complement** of any event A is the event that A does not occur. We will denote the complement of A by A^c.

Since an event A is a collection of simple events, the simple events included in A^c are just those that are not in A. Figure 4.6 demonstrates this. You will note from the figure that all simple events in S are included in either A or A^c, and that *no* simple event is in both A and A^c. This leads us to conclude that the probabilities of an event and its complement ***must sum to 1***:

The sum of the probabilities of complementary events equals 1; i.e.,

$$P(A) + P(A^c) = 1$$

In many probability problems, it will be easier to calculate the probability of the complement of the event of interest than the event itself. Then, since

$$P(A) + P(A^c) = 1$$

we can calculate $P(A)$ by using the relationship

$$P(A) = 1 - P(A^c)$$

EXAMPLE 4.9

Consider the experiment of tossing two fair coins. Calculate the probability of event A: {Observe at least one head} by using the complementary relationship.

SOLUTION

We know that the event A: {Observe at least one head} consists of the simple events

$$A = \{H_1, H_2;\quad H_1, T_2;\quad T_1, H_2\}$$

The complement of A is defined as the event that occurs when A does not occur. Therefore,

$$A^c = \{T_1, T_2\}$$

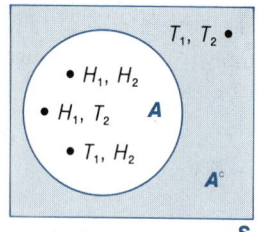

FIGURE 4.7
Complementary Events in
the Toss of Two Coins

This complementary relationship is shown in Figure 4.7. Assuming the coins are balanced,

$$P(A^c) = P(T_1, T_2) = \tfrac{1}{4}$$

and

$$P(A) = 1 - P(A^c) = 1 - \tfrac{1}{4} = \tfrac{3}{4}$$ ■

EXERCISES 4.18–4.33

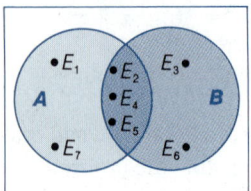

LEARNING THE MECHANICS

4.18 Consider the Venn diagram shown, where $P(E_1) = .13$, $P(E_2) = .05$, $P(E_3) = P(E_4) = .2$, $P(E_5) = .06$, $P(E_6) = .3$, and $P(E_7) = .06$. Find each of the following probabilities:
  a. $P(A^c)$ **b.** $P(B^c)$
  c. $P(A^c \cap B)$ **d.** $P(A \cup B)$
  e. $P(A \cap B)$ **f.** $P(A^c \cup B^c)$

4.19 Two dice are tossed and the following events are defined:

  A: {The total number of dots on the upper faces of the two dice is equal to 5}

  B: {At least one of the two dice has three dots on the upper face}

  a. Identify the simple events in each of the following events: A, B, A ∩ B, and A ∪ B.
  b. Find $P(A)$ and $P(B)$ by summing the probabilities of the appropriate simple events.
  c. Find $P(A \cap B)$ and $P(A \cup B)$ by summing the probabilities of the appropriate simple events.

4.20 Three coins are tossed and the following events are defined:

  A: {Observe at least one head}

  B: {Observe exactly two heads}

  C: {Observe exactly two tails}

  Assume that the coins are balanced. Calculate the following probabilities by summing the probabilities of the appropriate simple events:
  a. $P(A \cup B)$ **b.** $P(A \cap B)$ **c.** $P(A \cap C)$ **d.** $P(B \cap C)$

4.21 The accompanying table describes the adult population of a small suburb of a large southern city.

		INCOME		
		Under $20,000	$20,000–$50,000	Over $50,000
AGE	Under 25	950	1,000	50
	25–45	450	2,050	1,500
	Over 45	50	950	1,000

A marketing research firm plans to randomly select one adult from this suburb to evaluate a new food product. For this experiment the nine age–income categories are the simple events. Consider the following events:

A: {Person is under 25}

B: {Person is between 25 and 45}

C: {Person is over 45}

D: {Person has income under $20,000}

E: {Person has income of $20,000–$50,000}

F: {Person has income over $50,000}

Convert the frequencies in the table to relative frequencies and use them to calculate the following probabilities:

a. $P(B)$ **b.** $P(F)$ **c.** $P(C \cap F)$

d. $P(B \cup C)$ **e.** $P(A^c)$ **f.** $P(A^c \cap F)$

4.22 Refer to Exercise 4.21 and use the same event definitions to solve the following.

    a. Write the event that the person selected is under 25 with an income over $50,000 as an intersection of two of the events defined in Exercise 4.21.

    b. Write the event that a person of age 25 or more is selected as the union of two events. As the complement of an event.

APPLYING THE CONCEPTS

4.23 A state energy agency mailed questionnaires on energy conservation to 1,000 home-owners in the state capital. Five hundred questionnaires were returned. Suppose an experiment consists of randomly selecting one of the returned questionnaires. Consider the events:

A: {The home is constructed of brick}

B: {The home is more than 30 years old}

C: {The home is heated with oil}

Describe each of the following events in terms of unions, intersections, and complements (i.e., $A \cup B$, $A \cap B$, A^c, etc.):

a. The home is more than 30 years old and is heated with oil.

b. The home is not constructed of brick.

c. The home is heated with oil or is more than 30 years old.

d. The home is constructed of brick and is not heated with oil.

4.24 Identifying managerial prospects who are both talented and motivated is difficult. A personnel manager constructed the table shown here to define nine combinations of talent–motivation levels. The numbers in the table are the manager's estimates of the probabilities that a managerial prospect will be classified in the respective categories.

		TALENT		
		High	Medium	Low
	High	.05	.16	.05
MOTIVATION	Medium	.19	.32	.05
	Low	.11	.05	.02

Suppose the personnel manager has decided to hire a new manager. Define the following events:

A: {Prospect places in the high motivation category}

B: {Prospect places in the high talent category}

C: {Prospect rates medium or better in both categories}

D: {Prospect rates low in at least one of the categories}

E: {Prospect places high in both categories}

a. Does the sum of the probabilities in the table equal 1?

b. Find the probability of each event defined above.

c. Find $P(A \cup B)$, $P(A \cap B)$, and $P(A \cup C)$.

d. Find $P(A^c)$ and explain what this means from a practical point of view.

4.25 After completing an inventory of three warehouses, a golf club shaft manufacturer described its stock of 12,246 shafts with the percentages given in the table. Suppose a shaft is selected at random from the 12,246 currently in stock and the warehouse number and type of shaft are observed.

		TYPE OF SHAFT		
		Regular	Stiff	Extra Stiff
	1	19%	8%	3%
WAREHOUSE	2	14%	8%	2%
	3	28%	18%	0%

a. List all the simple events for this experiment.

b. What is the set of all simple events called?

c. Let C be the event that the shaft selected is from warehouse 3. Find $P(C)$ by summing the probabilities of the simple events in C.

d. Let F be the event that the shaft chosen is an extra stiff type. Find $P(F)$.

e. Let A be the event that the shaft selected is from warehouse 1. Find $P(A)$.

f. Let D be the event that the shaft selected is a regular type. Find $P(D)$.

g. Let E be the event that the shaft selected is a stiff type. Find $P(E)$.

4.26 Refer to Exercise 4.25. Define the characteristics of a golf club shaft portrayed by the following events, and then find the probability of each.

a. $A \cap F$ **b.** $C \cup E$ **c.** $C \cap D$ **d.** $A \cup F$ **e.** $A \cup D$

4.27 Refer to Exercise 4.6, in which a bank's loan records were being audited. Suppose each of the 500 automobile loans made by the bank last year is now classified according to two characteristics: amount of loan and length of loan. As before, an auditor is planning to choose one loan at random for inspection.

		AMOUNT OF LOAN ($)			
		Under 1,000	1,000–3,999	4,000–5,999	6,000 or more
	12	25	4	0	0
LENGTH OF	24	2	15	1	0
LOAN	36	0	27	92	2
(MONTHS)	42	0	53	93	50
	48	0	0	112	24

a. List the simple events in this experiment.
b. What is the probability that the loan selected will be for $6,000 or more? Does your answer agree with your answer to part **b** in Exercise 4.6?
c. What is the probability that the loan selected is a 3-year loan for more than $5,999?
d. What is the probability that the loan selected is a 3- or 4-year loan?
e. What is the probability that the loan selected is a 42-month loan for $1,000 or more?

4.28 The long-run success of a business depends on its ability to market products with superior characteristics that maximize consumer satisfaction and that give the firm a competitive advantage (Bagozzi, 1986). Ten new products have been developed by a food products firm. Market research has indicated that the ten products have the characteristics described by the Venn diagram shown here.

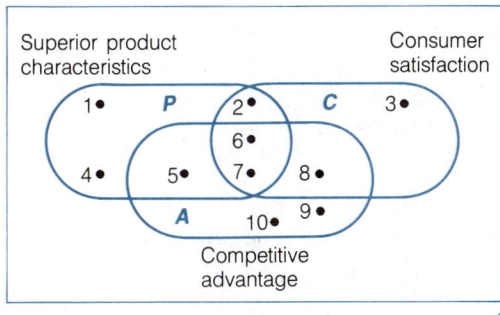

a. Write the event that a product possesses all the desired characteristics as an intersection of the events defined in the Venn diagram. Which products are contained in this intersection?
b. If one of the ten products were selected at random to be marketed, what is the probability that it would possess all the desired characteristics?
c. Write the event that the randomly selected product would give the firm a competitive advantage or would satisfy consumers as a union of the events defined in the Venn diagram. Find the probability of this union.
d. Write the event that the randomly selected product would possess superior product characteristics and satisfy consumers. Find the probability of this intersection.

4.29 A large research and development corporation is interested in providing an in-house continuing education program for its employees. It compiled the given table of percentages describing its 5,000 employees' current education level.

| | HIGHEST DEGREE OBTAINED | | | |
	High school diploma	Bachelor's	Master's	Ph.D.
Males	5%	20%	12%	11%
Females	18%	15%	14%	5%

Suppose an employee is selected at random from the firm's 5,000 employees and the following events are defined:

A: {Employee chosen is a male}

B: {Employee chosen is a female}

C: {Highest degree obtained by the chosen employee is Ph.D.}

D: {Highest degree obtained by the chosen employee is master's}

E: {Highest degree obtained by the chosen employee is bachelor's}

F: {Highest degree obtained by the chosen employee is high school diploma}

Describe the characteristics of an employee portrayed by the following events:
a. $A \cup C$ b. $B \cup F$ c. $A \cap D$ d. $E \cap B$

4.30 Refer to Exercise 4.29. Find the probabilities of the following events by summing the probabilities of the appropriate simple events:
a. A, B, C, D, E, F b. $A \cup B$ c. $B \cap C$
d. $A \cap F$ e. $A \cap B$ f. $C \cap D$

4.31 Whether purchases are made by cash or credit card is of concern to merchandisers because they must pay a certain percentage of the sale value to the credit agency. To better understand the relationship between type of purchase (credit or cash) and type of merchandise, a department store analyzed 10,000 sales and placed them in the categories shown in the table. Suppose a single sale is selected at random from the 10,000 and the following events are defined:

A: {Sale was paid by credit card}

B: {Merchandise purchased was women's wear}

C: {Merchandise purchased was men's wear}

D: {Merchandise purchased was sportswear}

	TYPE OF MERCHANDISE			
	Women's wear	Men's wear	Sportswear	Household
TYPE OF Cash	4%	7%	12%	7%
PURCHASE Credit card	37%	11%	4%	18%

Describe the characteristics of a sale implied by the following events, and find the probability of each.
a. $A \cup B$ b. $B \cup C$ c. $B \cap A$ d. $C \cap A$

4.32 Refer to Exercise 4.31. The following events are defined:

A: {Sale was paid by credit card}

B: {Merchandise purchased was women's wear}

a. Describe the events A^c and B^c.
b. Find $P(A^c)$.
c. Find $P(B^c)$.
d. Find the probability that the sale was in neither men's wear nor women's wear.
e. Find the probability that the sale was *not* a credit card purchase in the sportswear department.

4.33 The types of occupations of the 83,594,000 employed workers (age 25 years and older) in the United States in 1985 are described in the table, and their relative frequencies are listed. A worker is to be selected at random from this population and his or her occupation is to be determined. (Assume that each worker in the population has only one occupation.)

OCCUPATION	RELATIVE FREQUENCY	
Male worker	.555	
Managerial/professional		.167
Technical/sales/administrative		.114
Service		.047
Precision production		.121
Operators/fabricators		.106
Female workers	.445	
Managerial/professional		.120
Technical/sales/administrative		.198
Service		.074
Precision production		.011
Operators/fabricators		.042

Source: *Statistical Abstract of the United States: 1986*, p. 401.

a. What is the probability that the worker will be a male service worker?
b. What is the probability that the worker will be a manager or a professional?
c. What is the probability that the worker will be a female professional or a female operator/fabricator?
d. What is the probability that the worker will not be in a technical/sales/administrative occupation?

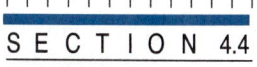

SECTION 4.4

CONDITIONAL PROBABILITY

FIGURE 4.8
Reduced Sample Space for the Die-Toss Experiment—Given That Event B Has Occurred

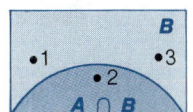

The probabilities we assign to the simple events of an experiment are measures of our belief that they will occur when the experiment is performed. When we assign these probabilities, we should make no assumptions other than those contained in or implied by the definition of the experiment. However, at times we will want to make assumptions other than those implied by the experimental description, and these extra assumptions may alter the probabilities we assign to the simple events of an experiment.

For example, we have shown that the probability of observing an even number (event A) on a toss of a fair die is $\frac{1}{2}$. However, suppose you are given the information that on a particular throw of the die the result was a number less than or equal to 3 (event B). Would you still believe that the probability of observing an even number on that throw of the die is equal to $\frac{1}{2}$? If you reason that making the assumption that B has occurred reduces the sample space from six simple events to three simple events (namely, those contained in event B), the reduced sample space is as shown in Figure 4.8.

Since the reduced sample space contains only three simple events (Observe a 1, Observe a 2, Observe a 3), each is assigned a new probability, called a **conditional probability**. Because the simple events for the die-toss experiment are equally likely, each of the three simple events in the reduced sample space is assigned a conditional probability of $\frac{1}{3}$. Since the only even number of the three numbers in the reduced sample space B is the number 2 and since the die is fair, we conclude that the probability that A occurs **given that B occurs** is one in three, or $\frac{1}{3}$. We will use the symbol $P(A \mid B)$ to represent the probability of event A given that event B occurs.

For the die-toss example,

$$P(A \mid B) = \tfrac{1}{3}$$

To get the probability of event A given that event B occurs, we proceed as follows: We divide the probability of the part of A that falls within the reduced sample space B—namely, $P(A \cap B)$—by the total probability of the reduced sample space—namely, $P(B)$. Thus, for the die-toss example with event A: {Observe an even number} and event B: {Observe a number less than or equal to 3}, we find

$$P(A \mid B) = \frac{P(A \cap B)}{P(B)} = \frac{P(2)}{P(1) + P(2) + P(3)} = \frac{\tfrac{1}{6}}{\tfrac{3}{6}} = \tfrac{1}{3}$$

This formula for $P(A \mid B)$ is true in general:

> To find the **conditional probability that event A occurs given that event B occurs**, divide the probability that *both* A and B occur by the probability that B occurs; that is,
>
> $$P(A \mid B) = \frac{P(A \cap B)}{P(B)}$$

This formula adjusts the probability of $A \cap B$ from its original value in the complete sample space S to a conditional probability in the reduced sample space B. If the simple events in the complete sample space are equally likely, then the formula will assign equal probabilities to the simple events in the reduced sample space, as in the die-toss experiment. If, on the other hand, the simple events have unequal probabilities, the formula will assign conditional probabilities proportional to the probabilities in the complete sample space.

EXAMPLE 4.10

Suppose you are interested in the probability of the sale of a large piece of earth-moving equipment. A single prospect is contacted. Let F be the event that the buyer has sufficient money (or credit) to buy the product and let F^c denote the complement of F (the event that the prospect does not have the financial capability to buy the product). Similarly, let B be the event that the buyer wishes to buy the product and let B^c be the complement of that event. Then the four simple events associated with the experiment are shown in Figure 4.9, and their probabilities are given in Table 4.5.

TABLE 4.5

Probabilities of Customer Desire to Buy and Ability to Finance

		DESIRE	
		To buy, B	Not to buy, B^c
ABLE TO	Yes, F	.2	.1
FINANCE	No, F^c	.4	.3

Find the probability that a single prospect will buy, given that the prospect is able to finance the purchase.

SOLUTION

Suppose you consider the large collection of prospects for the sale of your product and randomly select one person from this collection. What is the probability that the person selected will buy the product? In order to buy the product, the customer must be financially able and have the desire to buy, so this probability would correspond to the entry in Table 4.5 below B and next to F, or $P(B \cap F) = .2$. This is called the **unconditional probability** of the event $B \cap F$.

FIGURE 4.9

Sample Space for Contacting a Sales Prospect

In contrast, suppose you know that the prospect selected has the financial capability for purchasing the product. Now you are seeking the probability that the customer will buy given (the condition) that the customer has the financial ability to pay. This probability, the **conditional probability** of B given that F has occurred and denoted by the symbol $P(B \mid F)$, would be determined by considering only the simple events in the reduced sample space containing the simple events $B \cap F$ and $B^c \cap F$—i.e., simple events that imply the prospect is financially able to buy (this subspace is shaded in Figure 4.10). From our definition of conditional probability,

$$P(B \mid F) = \frac{P(B \cap F)}{P(F)}$$

FIGURE 4.10

Subspace (Shaded) Containing Sample Points Implying a Financially Able Prospect

where $P(F)$ is the sum of the probabilities of the two simple events corresponding to $B \cap F$ and $B^c \cap F$ (given in Table 4.5). Then

$$P(F) = P(B \cap F) + P(B^c \cap F) = .2 + .1 = .3$$

and the conditional probability that a prospect buys, given that the prospect is financially able, is

$$P(B \mid F) = \frac{P(B \cap F)}{P(F)} = \frac{.2}{.3} = .667$$

As we would expect, the probability that the prospect will buy, given that he or she is financially able, is higher than the unconditional probability of selecting a prospect who will buy. ∎

Note in Example 4.10 that the conditional probability formula assigns a probability to the event $(B \cap F)$ in the reduced sample space that is proportional to the probability of the event in the complete sample space. To see this, note that the two simple events in the reduced sample space, $(B \cap F)$ and $(B^c \cap F)$, have probabilities of .2 and .1, respectively, in the complete sample space S. The formula assigns conditional probabilities $\frac{2}{3}$ and $\frac{1}{3}$ (use the formula to check the second one) to these events in the reduced sample space F, so that the conditional probabilities retain the 2 to 1 proportionality of the original simple event probabilities.

EXAMPLE 4.11

The investigation of consumer product complaints by the Federal Trade Commission has generated much interest by manufacturers in the quality of their products. A manufacturer of an electromechanical kitchen aid conducted an analysis of a large number of consumer complaints and found that they fell into the six categories shown in Table 4.6 (page 172). If a consumer complaint is received, what is the probability that the cause of the complaint was product appearance, given that the complaint originated prior to the end of the guarantee period?

TABLE 4.6

Distribution of Product
Complaints

| | REASON FOR COMPLAINT | | |
	Electrical	Mechanical	Appearance
During guarantee period	18%	13%	32%
After guarantee period	12%	22%	3%

SOLUTION

Let A represent the event that the cause of a particular complaint was product appearance and let B represent the event that the complaint occurred during the guarantee period. Checking Table 4.6, you can see that $(18 + 13 + 32)\% = 63\%$ of the complaints occurred during the guarantee time. Hence, $P(B) = .63$. The percentage of complaints that were caused by appearance, A, *and* occurred during the guarantee period, B, is 32%. Therefore, $P(A \cap B) = .32$. Using these probability values, we can calculate the conditional probability $P(A \mid B)$ that the cause of a complaint is appearance given that the complaint occurred prior to the termination of the guarantee time:

$$P(A \mid B) = \frac{P(A \cap B)}{P(B)} = \frac{.32}{.63} = .51$$

Consequently, you can see that slightly more than half of the complaints that occurred during the guarantee period were due to scratches, dents, or other imperfections in the surface of the kitchen devices.   ∎

You will see in later chapters that conditional probability plays a key role in many business applications of statistics. For example, we may be interested in the probability that a particular stock gains 10% during the next year. We may estimate this probability (a statistical problem) by using information such as the past performance of the stock or the general state of the economy at present. However, our probability estimate may change drastically if we assume that the Gross National Product will increase by 10% in the next year. We would then be estimating the *conditional probability* that our stock gains 10% in the next year given that the GNP gains 10% in the same year. Thus, the probability of any event that is calculated or estimated based on an assumption that some other event occurs concurrently is a conditional probability.

CASE STUDY 4.2

PURCHASE PATTERNS AND THE CONDITIONAL PROBABILITY OF PURCHASING

In his doctoral dissertation Alfred A. Kuehn (1958) examined sequential purchase data to gain some insight into consumer brand switching. He analyzed the frozen orange juice purchases of approximately 600 Chicago families during 1950–1952. The data were collected by the *Chicago Tribune* Consumer Panel. Kuehn was interested in determining the influence of a consumer's last four orange juice purchases on the next purchase. Thus, sequences of five purchases were analyzed.

Table 4.7 contains a summary of the data collected for Snow Crop brand orange juice and part of Kuehn's analysis of the data. In the column labeled "Previous Purchase Pattern" an S stands for the purchase of Snow Crop by a consumer and an O stands for the purchase of a brand other than Snow Crop. Thus, for example, SSSO is used to represent the purchase of Snow Crop three times in a row followed

TABLE 4.7 Observed Approximate Probability of Purchasing Snow Crop, Given the Four Previous Brand Purchases

PREVIOUS PURCHASE PATTERN S = Snow Crop O = Other brand	SAMPLE SIZE	FREQUENCY	OBSERVED APPROXIMATE PROBABILITY OF PURCHASE	PREVIOUS PURCHASE PATTERN S = Snow Crop O = Other brand	SAMPLE SIZE	FREQUENCY	OBSERVED APPROXIMATE PROBABILITY OF PURCHASE
SSSS	1,047	844	.806	SOSO	163	66	.405
OSSS	277	191	.690	OSSO	181	75	.414
SOSS	206	137	.665	SSOO	256	78	.305
SSOS	222	132	.595	OOOS	500	165	.330
SSSO	296	144	.486	OOSO	404	77	.191
OOSS	248	137	.552	OSOO	433	56	.129
SOOS	138	78	.565	SOOO	557	86	.154
OSOS	149	74	.497	OOOO	8,442	405	.048

by the purchase of some other brand of frozen orange juice. The column labeled "Sample Size" lists the number of occurrences of the purchase sequences in the first column. The column labeled "Frequency" lists the number of times the associated purchase sequence in the first column led to the next purchase (i.e., the fifth purchase in the sequence) being Snow Crop.

The column labeled "Observed Approximate Probability of Purchase" contains the relative frequency with which each sequence of the first column led to the next purchase being Snow Crop. These relative frequencies, which give approximate probabilities, are computed for each sequence of the first column by dividing the frequency of the sequence by the sample size of the sequence. Notice that these approximate probabilities are really conditional probabilities. For the sequences of five purchases analyzed, each of the entries in the fourth column is the approximate probability that the next purchase is Snow Crop, given that the previous four purchases were as noted in the first column. For example, .806 is the approximate probability that the next purchase will be Snow Crop given that the previous four purchases were also Snow Crop.

An examination of the approximate probabilities in the fourth column indicates that both the most recent brand purchased and the number of times a brand is purchased have an effect on the next brand purchased. It appears that the influence on the next brand of orange juice purchased by the second most recent purchase is not so strong as the most recent purchase, but is stronger than the third most recent purchase. In general, it appears that the probability of a particular consumer purchasing Snow Crop the next time he or she buys orange juice is inversely related to the number of consecutive purchases of another brand he or she made since last purchasing Snow Crop and is directly proportional to the number of Snow Crop purchases among the four purchases.

Kuehn, of course, goes on to conduct a more formal statistical analysis of these data, which we will not pursue here. We simply want you to see that probability is a basic tool for making inferences about populations using sample data.

PROBABILITIES OF UNIONS AND INTERSECTIONS

FIGURE 4.11
Venn Diagram of Union

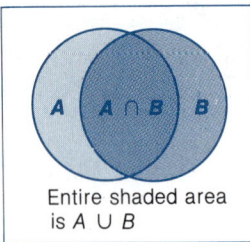

Entire shaded area
is $A \cup B$

Since unions and intersections of events are themselves events, we can always calculate their probabilities by adding the probabilities of the simple events that constitute them. However, if we know the probabilities of certain events related to the union or intersection, sometimes it is simpler to use special formulas to calculate their probabilities.

The union of two events will often contain many simple events, since the union occurs if either one or both of the events occur. By studying the Venn diagram in Figure 4.11, you can see that the probability of the union of two events A and B can be obtained by summing $P(A)$ and $P(B)$ and subtracting the probability corresponding to $A \cap B$. Therefore, the formula for calculating the probability of the union of two events is as given in the box.

ADDITIVE RULE OF PROBABILITY

The probability of the union of events A and B is the sum of the probabilities of events A and B minus the probability of the intersection of events A and B:

$$P(A \cup B) = P(A) + P(B) - P(A \cap B)$$

Note that we must subtract the probability of the intersection because, when we add the probabilities of A and B, the intersection probability is counted twice.

EXAMPLE 4.12

Consider the die-toss experiment. Define the events

A: {Observe an even number}

B: {Observe a number less than or equal to 3}

Assuming the die is fair, calculate the probability of the union of A and B by using the additive rule of probability.

SOLUTION

The formula for the probability of a union requires that we calculate the following:

$$P(A) = P(2) + P(4) + P(6) = \tfrac{1}{6} + \tfrac{1}{6} + \tfrac{1}{6} = \tfrac{3}{6}$$
$$P(B) = P(1) + P(2) + P(3) = \tfrac{1}{6} + \tfrac{1}{6} + \tfrac{1}{6} = \tfrac{3}{6}$$
$$P(A \cap B) = P(2) = \tfrac{1}{6}$$

Now we can calculate the probability of $A \cup B$:

$$P(A \cup B) = P(A) + P(B) - P(A \cap B) = \tfrac{3}{6} + \tfrac{3}{6} - \tfrac{1}{6} = \tfrac{5}{6} \qquad \blacksquare$$

If two events A and B do not intersect—i.e., when $A \cap B$ contains no simple events—we call the events A and B **mutually exclusive** events:

DEFINITION 4.8

Events A and B are **mutually exclusive** if $A \cap B$ contains no simple events.

FIGURE 4.12

Venn Diagram of Mutually
Exclusive Events

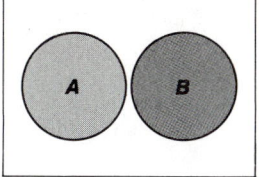

Figure 4.12 shows a Venn diagram of two mutually exclusive events. The events A and B have no simple events in common; i.e., A and B cannot occur simultaneously, and $P(A \cap B) = 0$. Thus, we have the following important relationship:

> If two events A and B are **mutually exclusive**, the probability of the union of A and B equals the sum of the probabilities of A and B:
>
> $$P(A \cup B) = P(A) + P(B)$$

Now we will develop a formula for calculating the probability of an intersection. Actually, we have already developed the formula in another context. You will recall that the formula for calculating the conditional probability of A given B is

$$P(A \mid B) = \frac{P(A \cap B)}{P(B)}$$

If we multiply both sides of this equation by $P(B)$, we get a formula for the probability of the intersection of events A and B:

> **MULTIPLICATIVE RULE OF PROBABILITY**
>
> $$P(A \cap B) = P(B)P(A \mid B) \quad \text{or, equivalently,} \quad P(A \cap B) = P(A)P(B \mid A)$$

The second expression in the box is obtained by multiplying both sides of the equation $P(B \mid A) = P(A \cap B)/P(A)$ by $P(A)$.

Before working an example, we emphasize that the intersection often contains only a few simple events, in which case the probability is easy to calculate by summing the appropriate simple event probabilities. However, the formula for calculating intersection probabilities plays a very important role in an area of statistics known as **Bayesian statistics**. (More complete discussions of Bayesian statistics are contained in Chapter 19 and in the references at the end of this chapter.)

EXAMPLE 4.13

Suppose an investment firm is interested in the following events:

   A: {Common stock in XYZ Corporation gains 10% next year}
   B: {Gross National Product gains 10% next year}

The firm has assigned the following probabilities on the basis of available information:

   $P(A \mid B) = .8 \qquad P(B) = .3$

That is, the investment company believes the probability is .8 that XYZ common stock will gain 10% in the next year *assuming that* the GNP gains 10% in the same time period. In addition, the company believes the probability is only .3 that the GNP will gain 10% in the next year. Use the formula for calculating the probability of an intersection to determine the probability that XYZ common stock *and* the GNP gain 10% in the next year.

SOLUTION   We want to calculate $P(A \cap B)$. The formula is

$$P(A \cap B) = P(B)P(A \mid B) = (.3)(.8) = .24$$

Thus, according to this investment firm, the probability is .24 that both XYZ common stock and the GNP will gain 10% in the next year.   ■

In the previous section we showed that the probability of an event A may be substantially altered by the assumption that the event B has occurred. However, this will not always be the case. In some instances the assumption that event B has occurred will not alter the probability of event A at all. When this is true, we call events A and B **independent**:

DEFINITION 4.9

Events A and B are **independent** if the assumption that B has occurred does not alter the probability that A occurs; i.e., events A and B are independent if

$$P(A \mid B) = P(A)$$

Equivalently, events A and B are **independent** if

$$P(B \mid A) = P(B)$$

Events that are not independent are said to be **dependent**.

EXAMPLE 4.14   Suppose that we decide to change the definition of event B in the die-toss experiment to {Observe a number less than or equal to 4} but we let event A remain an even number. Are events A and B independent (assuming a fair die)?

SOLUTION   The Venn diagram for this experiment is shown in Figure 4.13. We first calculate

FIGURE 4.13
Die-Toss Experiment,
Example 4.14

$$P(A) = \tfrac{1}{2}$$
$$P(B) = P(1) + P(2) + P(3) + P(4) = \tfrac{4}{6} = \tfrac{2}{3}$$
$$P(A \cap B) = P(2) + P(4) = \tfrac{2}{6} = \tfrac{1}{3}$$

Now assuming B has occurred, the conditional probability of A is

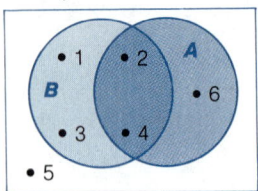

$$P(A \mid B) = \frac{P(A \cap B)}{P(B)} = \frac{\tfrac{1}{3}}{\tfrac{2}{3}} = \tfrac{1}{2} = P(A)$$

Thus, the assumption of the occurrence of event B does not alter the probability of observing an even number—it remains $\tfrac{1}{2}$. Therefore, the events A and B are independent. Note that if we calculate the conditional probability of B given A, our conclusion is the same:

$$P(B \mid A) = \frac{P(A \cap B)}{P(A)} = \frac{\frac{1}{3}}{\frac{1}{2}} = \frac{2}{3} = P(B)$$

∎

EXAMPLE 4.15

Refer to the consumer product complaint study in Example 4.11. The percentages of complaints of various types in the pre- and post-guarantee periods are shown in Table 4.6. Define the following events:

A: {Cause of complaint is product appearance}

B: {Complaint occurred during the guarantee term}

Are A and B independent events?

SOLUTION

Events A and B are independent if $P(A \mid B) = P(A)$. We calculated $P(A \mid B)$ in Example 4.11 to be .51, and from Table 4.6 we can see that

$$P(A) = .32 + .03 = .35$$

Therefore, $P(A \mid B)$ is not equal to $P(A)$, and A and B are not independent events.

∎

To gain an intuitive understanding of independence, think of situations in which the occurrence of one event does not alter the probability that a second event will occur. For example, suppose two small companies are being monitored by a financier for possible investment. If the businesses are of a different nature and they are otherwise unrelated, then the success or failure of one company may be *independent* of the success or failure of the other. That is, the event that company A fails may not alter the probability that company B will fail.

As a second example, consider an election poll in which 1,000 registered voters are asked to state their preference between two candidates. One objective of the pollsters is to select a sample of voters so that the responses will be independent. That is, the sample is selected so that the event that one polled voter prefers candidate A does not alter the probability that a second polled voter prefers candidate A.

We will make three final points about independence. The first is that the property of independence, unlike the mutually exclusive property, cannot be shown on or gleaned from a Venn diagram. In general, the only way to check for independence is by performing the calculations of the probabilities in the definition.

The second point concerns the relationship between the mutually exclusive and independence properties. Suppose that events A and B are mutually exclusive, as shown in Figure 4.12. Are these events independent or dependent? That is, does the assumption that B occurs alter the probability of the occurrence of A? It certainly does, because if we assume that B has occurred, it is impossible for A to have occurred simultaneously. Thus, mutually exclusive events are dependent events.

The third point is that the probability of the intersection of independent events is very easy to calculate. Referring to the formula for calculating the probability of an intersection, we find

$$P(A \cap B) = P(B)P(A \mid B)$$

Thus, since $P(A \mid B) = P(A)$ when A and B are independent, we have the following useful rule:

> **If events A and B are independent**, the probability of the intersection of A and B equals the product of the probabilities of A and B:
>
> $$P(A \cap B) = P(A)P(B)$$

In the die-toss experiment, we showed in Example 4.14 that the events A: {Observe an even number} and B: {Observe a number less than or equal to 4} are independent if the die is fair. Thus,

$$P(A \cap B) = P(A)P(B) = \left(\tfrac{1}{2}\right)\left(\tfrac{2}{3}\right) = \tfrac{1}{3}$$

This agrees with the result

$$P(A \cap B) = P(2) + P(4) = \tfrac{2}{6} = \tfrac{1}{3}$$

that we obtained in the example.

EXAMPLE 4.16

Almost every retail business has the problem of determining how much inventory to purchase. Insufficient inventory may result in lost business, and excess inventory will have a detrimental effect on profits. Suppose a retail computer store owner is planning to place an order for personal computers (PCs). She is trying to decide how many IBM PCs and how many IBM compatibles (personal computers that run all or most of the same software as the IBM PC, but that are not manufactured by IBM) to order. The owner's records indicate that 80% of the previous PC customers purchased IBM PCs, and 20% purchased compatibles.

a. What is the probability that the next two customers will purchase compatibles?
b. What is the probability that the next ten customers will purchase compatibles?

SOLUTION

a. Let C_1 represent the event that customer 1 will purchase a compatible, and C_2 the event that customer 2 will purchase a compatible. The event that *both* customers purchase compatibles is the intersection of these events, $C_1 \cap C_2$. From past records the store owner could reasonably conclude that the probability of C_1 is equal to .2 (based on the fact that 20% of past customers have purchased compatibles), and the same reasoning would apply to C_2. However, in order to compute the probability of the intersection $C_1 \cap C_2$, more information is needed. Either the records must be examined for the occurrence of consecutive purchases of compatibles, or some assumption must be made to enable the calculation of

$P(C_1 \cap C_2)$ from the multiplicative rule. It seems reasonable to make the assumption that the two events are independent, since the decision of the first customer is not likely to affect that of the second customer. Assuming independence, we have

$$P(C_1 \cap C_2) = P(C_1)P(C_2) = (.2)(.2) = .04$$

b. To see how to compute the probability that ten consecutive purchases will be of compatibles, first consider the event that three consecutive customers purchase compatibles. If C_3 represents the event that the third customer purchases a compatible, then we want to compute the probability of the intersection of $C_1 \cap C_2$ with C_3. Again assuming independence of the purchasing decisions, we have

$$P(C_1 \cap C_2 \cap C_3) = P(C_1 \cap C_2)P(C_3)$$
$$= (.2)^2(.2) = .008$$

Similar reasoning leads to the conclusion that the intersection of ten such events can be calculated as follows:

$$P(C_1 \cap C_2 \cap \cdots \cap C_{10}) = P(C_1)P(C_2) \cdots \cdots P(C_{10})$$
$$= (.2)^{10}$$
$$= .0000001024$$

Thus, the probability that ten consecutive customers purchase IBM compatibles is about 1 in 10 million, assuming that the probability of each customer's purchase of a compatible is .2, and that the purchase decisions are independent.   ■

EXERCISES 4.34–4.51

LEARNING THE MECHANICS

4.34 An experiment results in one of three mutually exclusive events, A, B, or C. It is known that $P(A) = .40$, $P(B) = .25$, and $P(C) = .35$. Find each of the following probabilities:
a. $P(A \cup B)$ **b.** $P(A \cap C)$ **c.** $P(A \mid B)$ **d.** $P(B \cup C)$
e. Are B and C independent events? Explain.

4.35 Consider the experiment depicted by the accompanying Venn diagram, with the sample space S containing five simple events.

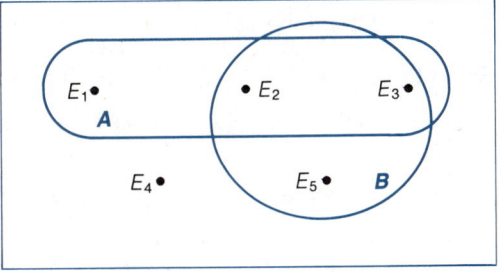

The simple events are assigned the following probabilities:

$$P(E_1) = .1 \quad P(E_2) = .1 \quad P(E_3) = .3 \quad P(E_4) = .4 \quad P(E_5) = .1$$

a. Calculate $P(A)$, $P(B)$, and $P(A \cap B)$.

b. Suppose we know that event A has occurred, so that the reduced sample space consists of the three simple events in A—namely, E_1, E_2, and E_3. Use the formula for conditional probability to adjust the probabilities of these three simple events for the knowledge that A has occurred. Verify that the conditional probabilities are in the same proportion to one another as the original simple event probabilities.

c. Calculate the conditional probability $P(B \mid A)$ in two ways: (1) Add the adjusted (conditional) probabilities of the simple events in the intersection $A \cap B$, since these represent the event that B occurs given that A has occurred. (2) Use the formula for conditional probability:

$$P(B \mid A) = \frac{P(A \cap B)}{P(A)}$$

Verify that the two methods yield the same result.

d. Are events A and B independent? Why or why not?

4.36 An experiment results in one of five simple events, with the following probabilities: $P(E_1) = .22$, $P(E_2) = .31$, $P(E_3) = .15$, $P(E_4) = .22$, and $P(E_5) = .10$. The following events have been defined:

$$A = \{E_1, E_3\} \quad B = \{E_2, E_3, E_4\} \quad C = \{E_1, E_5\}$$

Find each of the following probabilities:
a. $P(A)$ b. $P(B)$ c. $P(A \cap B)$
d. $P(A \mid B)$ e. $P(B \cap C)$ f. $P(C \mid B)$

4.37 Refer to Exercise 4.36. Which of the following pairs of events are independent? Explain.
a. A and B b. A and C c. B and C

4.38 Three coins are tossed and the following events are defined:

A: {Observe at least one head}

B: {Observe exactly two heads}

C: {Observe exactly two tails}

D: {Observe at most one head}

E: {Observe at least two tails}

Use the formulas of this section to calculate the following:
a. $P(A \cap B)$ b. $P(A \cup B)$ c. $P(C \cap A)$
d. $P(A \cup C)$ e. $P(A^c \cap C)$ f. $P(D \cap E)$

4.39 If $P(R) = \frac{1}{3}$, $P(S) = \frac{1}{3}$, and events R and S are mutually exclusive, find $P(R \mid S)$ and $P(S \mid R)$.

4.40 Two dice are tossed and the following events are defined:

A: {Sum of the numbers showing is an odd number}

B: {Sum of the numbers showing is 8, 9, 10, or 11}

Are events A and B independent? Why or why not?

4.41 Two hundred shoppers at a large suburban mall were asked two questions: (1) Did you see a television ad for the sale at department store X during the past 2 weeks? (2) Did you shop at department store X during the past 2 weeks? The responses to these questions are summarized in the table.

	Shopped at X	Did not shop at X
Saw ad	100	25
Did not see ad	25	50

One of the 200 shoppers questioned is to be chosen at random.

a. What is the probability that the person selected saw the ad?

b. What is the probability that the person selected saw the ad and shopped at store X?

c. Find the conditional probability that the person shopped at store X given that the person saw the ad.

d. What is the probability that the person selected shopped at store X?

e. Use your answers to parts **a, b,** and **d** to check the independence of the events Saw ad and Shopped at X.

f. Are the two events Did not see ad and Did not shop at X mutually exclusive? Explain.

APPLYING THE CONCEPTS

4.42 The table describes the 89.9 million U.S. federal tax returns filed with the Internal Revenue Service (IRS) in 1984 and the percentage of those returns that were audited by the IRS.

INCOME	NUMBER OF TAX FILERS (Millions)	PERCENTAGE AUDITED
Under $10,000	31.4	.34
$10,000–$24,999	30.7	.92
$25,000–$49,999	22.2	2.05
$50,000 or more	5.5	4.00

Source: *1984 Annual Report of Commissioner and Chief Counsel*, Internal Revenue Service, U.S. Department of Treasury (produced by the Office of Public Affairs), p. 60.

a. If a tax filer were to be randomly selected from this population of tax filers (i.e., each tax filer has an equal probability of being selected), what is the probability that the tax filer would have been audited?

b. If a tax filer were to be randomly selected from the population of tax filers described in the table, what is the probability that the tax filer had an income of $10,000–$24,999 in 1984 *and* was audited? What is the probability that the tax filer had an income of $50,000 or more in 1984 or was not audited?

c. Refer to part **b**. If it were known that the randomly selected tax filer had been audited, what is the probability that the person had an income of $50,000 or more? Under $10,000?

d. What is the probability that a tax filer with an income of $50,000 or more in 1984 would have been audited?

4.43 A particular automatic sprinkler system for high-rise apartment buildings, office buildings, and hotels has two different types of activation devices for each sprinkler head. One type has a reliability of .91 (i.e., the probability that it will activate the sprinkler when it should is .91). The other type, which operates independently of the first type, has a reliability of .87. Suppose a serious fire starts near a particular sprinkler head.
   a. What is the probability that the sprinkler head will be activated?
   b. What is the probability that the sprinkler head will not be activated?
   c. What is the probability that both activation devices will work properly?
   d. What is the probability that only the device with reliability .91 will work properly?

4.44 Which events in the following sets of events are mutually exclusive?
   a. The rate of inflation will increase next year.
       The rate of inflation will decrease next year.
       The rate of inflation will be the same next year as this year.
   b. Sales will increase by 10,000 units.
       Sales will increase by at least 5,000 units.
       Sales will decrease by 1,000 units.
   c. The consultant's fee will be at least $400 a day.
       The consultant's fee will be at least $600 a day.
       The consultant's fee will be at least $800 a day.

4.45 A soft drink bottler has two quality control inspectors independently check each case of soft drinks for chipped or cracked bottles before the cases leave the bottling plant. Having observed the work of the two trusted inspectors over several years, the bottler has determined that the probability of a defective case getting by the first inspector is .05 and the probability of a defective case getting by the second inspector is .10. What is the probability that a defective case gets by both inspectors?

4.46 Businesses that offer credit to their customers are inevitably faced with the task of collecting unpaid bills. Richard L. Peterson (1986) conducted a study of collection remedies used by creditors. As part of the study he asked samples of creditors in four states about how they deal with past-due bills. Their responses are tallied in the table. "Tough actions" included filing a legal action, turning the debt over to a third party such as an attorney or collection agency, garnishing wages, and repossessing secured property.

	Wisconsin	Illinois	Arkansas	Louisiana
Take tough action early	0	1	5	1
Take tough action late	37	23	22	21
Never take tough action	9	11	6	15

Suppose one of the creditors questioned is selected at random.
   a. What is the probability that the creditor is from Wisconsin or Louisiana?
   b. What is the probability that the creditor is not from Wisconsin or Louisiana?
   c. What is the probability that the creditor never takes tough action?
   d. What is the probability that the creditor is from Arkansas and never takes tough action?
   e. What is the probability that the creditor is from Arkansas, given that the creditor never takes tough action?

4.47 A fast-food restaurant chain with 700 outlets in the United States describes the geographic location of its restaurants with the given table of percentages. A restaurant is to be chosen at random from the 700 to test market a new style of chicken.

		REGION			
		NE	SE	SW	NW
POPULATION OF CITY	Under 10,000	5%	6%	3%	0%
	10,000–100,000	10%	15%	12%	5%
	Over 100,000	25%	4%	5%	10%

  a. Given the restaurant chosen is in a city with a population over 100,000, what is the probability that it is located in the Northeast?
  b. Given the restaurant chosen is in the Southeast, what is the probability that it is located in a city with a population under 10,000?
  c. If the restaurant selected is located in the Southwest, what is the probability that the city it is in has a population of 100,000 or less?
  d. If the restaurant selected is located in the Northwest, what is the probability that the city it is in has a population of 10,000 or more?

4.48 An article in *Business Week* (Sept. 12, 1983) reports on the problems that evolve from the failure to inform patients adequately of the proper application of prescription drugs and the precautions needed to avoid potential side effects. This failure results in numerous cases of serious illness and, in some cases, even death. One study revealed that 300,000 U.S. hospital admissions each year are caused by adverse reactions to prescription drugs. Another study concluded that 7% of all hospital admissions are related to drug-induced problems resulting from imprudent prescriptions. One method of increasing patients' awareness of the problem is for physicians to provide PMI (Patient Medication Instruction) sheets. However, the American Medical Association has found that only 20% of doctors who prescribe drugs frequently distribute such sheets to their patients. Assume that 20% of all patients receive the PMI sheet with their prescriptions and that 12% of all patients receive the PMI sheet and are hospitalized because of a drug-related problem. What is the probability that a person will be hospitalized for a drug-related problem given that the person has received the PMI sheet?

4.49 Even with strong advertising programs, new products are often unsuccessful. A company that produces a variety of household items found that only 18% of the new products it introduced over the last 10 years have become profitable. When two new products were introduced during the same year, only 5% of the time did both products become profitable. Suppose the company plans to introduce two new products, A and B, next year. If the percentages just cited define the probabilities of success, what is the probability that:
  a. Product A will become profitable?
  b. Product B will not become profitable?
  c. At least one of the two products will become profitable?
  d. Neither of the two products will become profitable?
  e. Either product A or product B (but not both) will become profitable?

4.50 Refer to Exercise 4.49.
  a. What is the probability that product A becomes profitable, given that product B is profitable?
  b. Given that at least one of the products will be profitable, what is the probability that the profitable product is A?

4.51 In order to understand the challenge that advertisers face in trying to persuade consumers to purchase their products, researchers have studied mental stages of a potential consumer from the point of initial exposure to an advertising message through the final decision on whether to buy the product. The currently popular *information-processing model of communication* includes five sequential stages: perception, comprehension, attitude formation, preference formation, and development of intention to buy or not (Bagozzi, 1986, p. 337). Assume that for a particular consumer, the probability of initial exposure to a particular ad is .9, and that the probability of *each* of the five mental stages occurring in such a way as to lead to the decision to purchase the product is .7. Assume also that the stages are statistically independent. What is the probability that the advertisement will lead the consumer to decide to purchase the product?

S E C T I O N 4.6

RANDOM SAMPLING

How a sample is selected from a population is of vital importance in statistical inference because the probability of an observed sample will be used to infer the characteristics of the sampled population. To illustrate, suppose you deal yourself 4 cards from a deck of 52 cards, and all 4 cards are aces. Do you conclude that your deck is an ordinary bridge deck, containing only 4 aces, or do you conclude that the deck is stacked with more than 4 aces? It depends on how the cards were drawn. If the 4 aces were always placed on the top of a standard bridge deck, drawing 4 aces would not be unusual—it would be certain. On the other hand, if the cards were thoroughly mixed, drawing 4 aces in a sample of 4 cards would be highly improbable. The point, of course, is that, in order to use the observed sample of 4 cards to make inferences about the population (the deck of 52 cards), you need to know how the sample was selected from the deck.

One of the simplest and most frequently used sampling procedures (implied in the previous examples and exercises) produces what is known as a **random sample**.

> **DEFINITION 4.10**
>
> If *n* elements are selected from a population in such a way that every possible combination of *n* elements in the population has an equal probability of being selected, the *n* elements are said to be a **random sample**.*

If a population is not too large and the elements can be marked on slips of paper or poker chips, you can physically mix the slips of paper or chips and remove *n* elements from the total. Then the elements that appear on the slips or chips selected would indicate the population elements to be included in the sample. Such a procedure would not guarantee a random sample because it is often difficult to achieve a thorough mix, but it provides a reasonably good approximation to random sampling.

Many samplers use a table of random numbers (see Table I in Appendix B). Random-number tables are constructed in such a way that every digit occurs with

*Strictly speaking, this is a **simple random sample**. There are many different types of random samples. The simple random sample is the most frequently employed.

(approximately) equal probability. To use a table of random numbers, we number the N elements in the population from 1 to N. Then we turn to Table I and haphazardly select a number in the table. Proceeding from this number across the row or down the column (either will do), remove and record n numbers from the table. Use only the necessary number of digits in each random number to identify the element to be included in the sample. If, in the course of recording the n numbers from the table, you obtain a number that has already been selected, simply discard the duplicate, and select a replacement at the end of the sequence. Thus, you may have to select more than n numbers to obtain a random sample of n unique numbers.

We illustrate this procedure with an example.

EXAMPLE 4.17

Suppose you wish to randomly sample 5 households (we will keep the number in the sample small to simplify our example) from a population of 100,000 households. Use Table I to select a random sample.

SOLUTION

First, number the households in the population from 1 to 100,000. Then, turn to a page of Table I, say, the first page. A reproduction of part of the first page of Table I is shown in Table 4.8. Now, commence with the random number that appears in the third row, second column. This number is 48360. Proceed down the second column to obtain the remaining four random numbers. The five selected random numbers are shaded in Table 4.8. Using the first five digits to represent the households from 1 to 99,999 and the number 00000 to represent household 100,000, you can see that the households numbered

48,360   93,093   39,975   6,907   72,905

should be included in your sample. ∎

TABLE 4.8

Reproduction of Part of Table I of Appendix B

COLUMN ROW	1	2	3	4	5	6
1	10480	15011	01536	02011	81647	91646
2	22368	46573	25595	85393	30995	89198
3	24130	48360	22527	97265	76393	64809
4	42167	93093	06243	61680	07856	16376
5	37570	39975	81837	16656	06121	91782
6	77921	06907	11008	42751	27756	53498
7	99562	72905	56420	69994	98872	31016
8	96301	91977	05463	07972	18876	20922
9	89579	14342	63661	10281	17453	18103
10	85475	36857	53342	53988	53060	59533
11	28918	69578	88231	33276	70997	79936
12	63553	40961	48235	03427	49626	69445
13	09429	93969	52636	92737	88974	33488
14	10365	61129	87529	85689	48237	52267
15	07119	97336	71048	08178	77233	13916

From 1948 through the early years of the Vietnam War, the Selective Service System drafted men into the military service by age—oldest first, starting with 25-year-olds. A network of local draft boards was used to implement the selection process. Then, on the evening of December 1, 1969, the Selective Service System conducted a lottery to determine the order of selection for 1970 in an attempt to overcome what many believed were inequities in the system. (Such lotteries had been used during World Wars I and II, but it had been 27 years since the last one.)

The objective of the lottery was to randomly order the induction sequence of men between the ages of 19 and 26. To do this, the 366 possible days in a year were written on slips of paper and placed in egg-shaped capsules that were stored in monthly lots. The monthly lots were placed one by one into a wooden box that was ". . . turned end over end several times to mix the numbers" ("Random or not? Judge studies lottery protest," *The National Observer*, Jan. 12, 1970, p. 2). The capsules were then dumped into a large glass bowl and drawn one by one to obtain the order of induction. All men born on the first day drawn would be inducted first; those born on the second day drawn would be drafted next, etc. Thus, the lottery assigned a rank to each of the 366 birthdays. The results of the lottery are shown in Table 4.9 (pages 188–189).

In order for a random sequence of numbers to be generated with this procedure, it is necessary for each (remaining) capsule in the bowl to have an equal probability of being selected on each draw. That is, by means of thorough mixing, each capsule must have an equal opportunity to come to rest precisely where the sampler's hand closes within the bowl. Although a mixing procedure with this property is almost impossible to achieve, the ideal can be closely approximated. Unfortunately, this was apparently not the case in the 1970 lottery. Even though the sequence of dates in Table 4.9 may appear to be random, there is ample statistical evidence to indicate a nonrandom selection of induction dates.*

To obtain an understanding of the problem with the 1970 lottery, we calculated the median rank for each month and plotted the medians in Figure 4.14(a). If the sequence of ranks were randomly generated, there should be no relationship between the size of the median ranks and the months of the year. The medians should vary randomly above and below a horizontal line with intercept 183.5 (the median of the integers 1 through 366). However, Figure 4.14(a) reveals a general downward trend in the medians. Men born later in the year were more likely to be drafted before men born early in the year. Furthermore, since not all men between the ages of 19 and 25 would be drafted in 1970, men born earlier in the year were more likely not to be drafted at all. Although it is possible to observe such a sequence of medians when random sampling is employed, it is highly unlikely. It is more likely that the capsules were not mixed thoroughly enough to give every capsule an equal chance of selection on each draw. The graph indicates that capsules tended to be drawn in monthly groups, and thus suggests that monthly lots of capsules in the wooden boxes were not mixed thoroughly enough after they were dumped into the large glass bowl (Williams, 1978).

*Formal statistical tests to detect nonrandomness are beyond the scope of this text.

FIGURE 4.14

Median Plots for Lottery
Results: 1970 and 1971

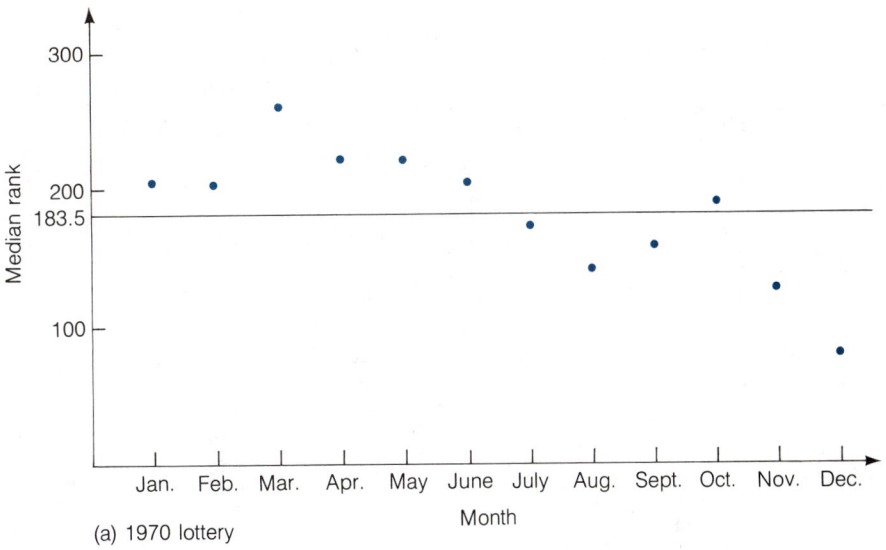

(a) 1970 lottery

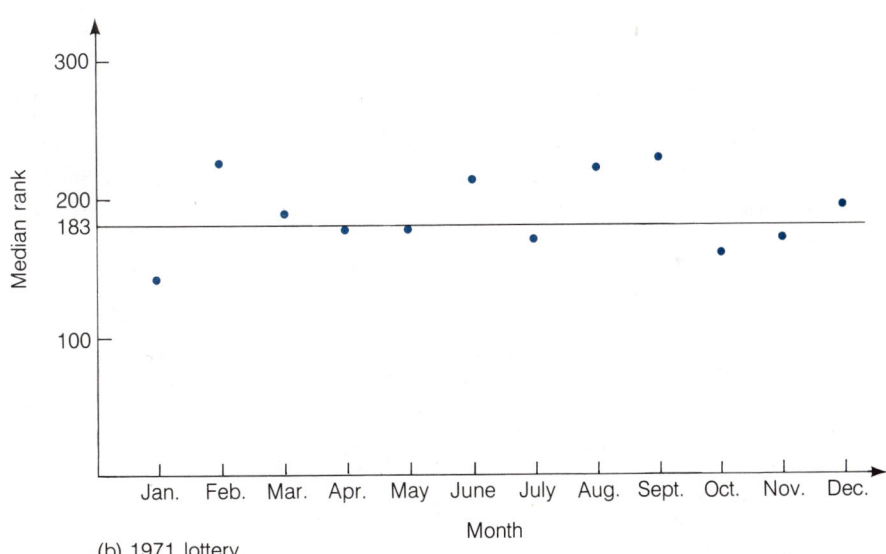

(b) 1971 lottery

The following year, the Selective Service System used more sophisticated mixing techniques to guard against the monthly clustering of selections. The median plot for the 1971 lottery is shown in Figure 4.14(b). [*Note:* The 1971 lottery involved only men born in 1951. Thus, only 365 birthdays were ranked and the sequence of monthly ranked medians should be compared to 183 instead of 183.5.] Note that no apparent trend remains; the new mixing technique appears to have been successful.

TABLE 4.9
1970 Draft Lottery Results

No.	Date	No.	Date	No.	Date	No.	Date	No.	Date	No.	Date
1	Sept. 14	46	Nov. 11	91	Feb. 7	136	Mar. 11	181	Feb. 8	226	May 29
2	April 24	47	Nov. 27	92	Jan. 26	137	June 25	182	Nov. 23	227	July 19
3	Dec. 30	48	Aug. 8	93	July 1	138	Oct. 13	183	May 20	228	June 2
4	Feb. 14	49	Sept. 3	94	Oct. 28	139	Mar. 6	184	Sept. 8	229	Oct. 29
5	Oct. 18	50	July 7	95	Dec. 24	140	Jan. 18	185	Nov. 20	230	Nov. 24
6	Sept. 6	51	Nov. 7	96	Dec. 16	141	Aug. 18	186	Jan. 21	231	April 14
7	Oct. 26	52	Jan. 25	97	Nov. 8	142	Aug. 12	187	July 20	232	Sept. 4
8	Sept. 7	53	Dec. 22	98	July 17	143	Nov. 17	188	July 5	233	Sept. 27
9	Nov. 22	54	Aug. 5	99	Nov. 29	144	Feb. 2	189	Feb. 17	234	Oct. 7
10	Dec. 6	55	May 16	100	Dec. 31	145	Aug. 4	190	July 18	235	Jan. 17
11	Aug. 31	56	Dec. 5	101	Jan. 5	146	Nov. 18	191	April 29	236	Feb. 24
12	Dec. 7	57	Feb. 23	102	Aug. 15	147	April 7	192	Oct. 20	237	Oct. 11
13	July 8	58	Jan. 19	103	May 30	148	April 16	193	July 31	238	Jan. 14
14	April 11	59	Jan. 24	104	June 19	149	Sept. 25	194	Jan. 9	239	Mar. 20
15	July 12	60	June 21	105	Dec. 8	150	Feb. 11	195	Sept. 24	240	Dec. 19
16	Dec. 29	61	Aug. 29	106	Aug. 9	151	Sept. 29	196	Oct. 24	241	Oct. 19
17	Jan. 15	62	April 21	107	Nov. 16	152	Feb. 13	197	May 9	242	Sept. 12
18	Sept. 26	63	Sept. 20	108	Mar. 1	153	July 22	198	Aug. 14	243	Oct. 21
19	Nov. 1	64	June 27	109	June 23	154	Aug. 17	199	Jan. 8	244	Oct. 3
20	June 4	65	May 10	110	June 6	155	May 6	200	Mar. 19	245	Aug. 26
21	Aug. 10	66	Nov. 12	111	Aug. 1	156	Nov. 21	201	Oct. 23	246	Sept. 18
22	June 26	67	July 25	112	May 17	157	Dec. 3	202	Oct. 4	247	June 22
23	July 24	68	Feb. 12	113	Sept. 15	158	Sept. 11	203	Nov. 19	248	July 11
24	Oct. 5	69	June 13	114	Aug. 6	159	Jan. 2	204	Sept. 21	249	June 1
25	Feb. 19	70	Dec. 21	115	July 3	160	Sept. 22	205	Feb. 27	250	May 21
26	Dec. 14	71	Sept. 10	116	Aug. 23	161	Sept. 2	206	June 10	251	Jan. 3
27	July 21	72	Oct. 12	117	Oct. 22	162	Dec. 23	207	Sept. 16	252	April 23
28	June 5	73	June 17	118	Jan. 23	163	Dec. 13	208	April 30	253	April 6
29	Mar. 2	74	April 27	119	Sept. 23	164	Jan. 30	209	June 30	254	Oct. 16
30	Mar. 31	75	May 19	120	July 16	165	Dec. 4	210	Feb. 4	255	Sept. 17
31	May 24	76	Nov. 6	121	Jan. 16	166	Mar. 16	211	Jan. 31	256	Mar. 23
32	April 1	77	Jan. 28	122	Mar. 7	167	Aug. 28	212	Feb. 16	257	Sept. 28
33	Mar. 17	78	Dec. 27	123	Dec. 28	168	Aug. 7	213	Mar. 8	258	Mar. 24
34	Nov. 2	79	Oct. 31	124	April 13	169	Mar. 15	214	Feb. 5	259	Mar. 13
35	May 7	80	Nov. 9	125	Oct. 2	170	Mar. 26	215	Jan. 4	260	April 17
36	Aug. 24	81	April 4	126	Nov. 13	171	Oct. 15	216	Feb. 10	261	Aug. 3
37	May 11	82	Sept. 5	127	Nov. 14	172	July 23	217	Mar. 30	262	April 28
38	Oct. 30	83	April 3	128	Dec. 18	173	Dec. 26	218	April 10	263	Sept. 9
39	Dec. 11	84	Dec. 25	129	Dec. 1	174	Nov. 30	219	April 9	264	Oct. 27
40	May 3	85	June 7	130	May 15	175	Sept. 13	220	Oct. 10	265	Mar. 22
41	Dec. 10	86	Feb. 1	131	Nov. 15	176	Oct. 25	221	Jan. 12	266	Nov. 4
42	July 13	87	Oct. 6	132	Nov. 25	177	Sept. 19	222	Jan. 28	267	Mar. 3
43	Dec. 9	88	July 28	133	May 12	178	May 14	223	Mar. 28	268	Mar. 27
44	Aug. 16	89	Feb. 15	134	June 11	179	Feb. 25	224	Jan. 6	269	April 5
45	Aug. 2	90	April 18	135	Dec. 20	180	June 15	225	Sept. 1	270	July 29

continued

TABLE 4.9 Continued

271 April 2	287 July 30	303 July 26	319 May 23	335 June 9	351 April 25
272 June 12	288 Oct. 17	304 Dec. 17	320 Dec. 15	336 April 19	352 Aug. 27
273 April 15	289 July 27	305 Jan. 1	321 May 8	337 Jan. 22	353 June 29
274 June 16	290 Feb. 22	306 Jan. 7	322 July 15	338 Feb. 9	354 Mar. 14
275 Mar. 4	291 Aug. 21	307 Aug. 13	323 Mar. 10	339 Aug. 22	355 Jan. 27
276 May 4	292 Feb. 18	308 May 28	324 Aug. 11	340 April 26	356 June 14
277 July 9	293 Mar. 5	309 Nov. 26	325 Jan. 10	341 June 18	357 May 26
278 May 18	294 Oct. 14	310 Nov. 5	326 May 22	342 Oct. 9	358 June 24
279 July 4	295 May 13	311 Aug. 19	327 July 6	344 Mar. 25	359 Oct. 1
280 Jan. 20	296 May 27	312 April 8	328 Dec. 2	344 Aug. 20	360 June 20
281 Nov. 28	297 Feb. 3	313 May 31	329 Jan. 11	345 April 20	361 May 25
282 Nov. 10	298 May 2	314 Dec. 12	330 May 1	346 April 12	362 Mar. 29
283 Oct. 8	299 Feb. 28	315 Sept. 30	331 July 14	347 Feb. 6	363 Feb. 21
284 July 10	300 Mar. 12	316 April 22	332 Mar. 18	348 Nov. 3	364 May 5
285 Feb. 29	301 June 3	317 Mar. 9	333 Aug. 30	349 Jan. 29	365 Feb. 26
286 Aug. 25	302 Feb. 20	318 Jan. 13	334 Mar. 21	350 July 2	366 June 8

This case study emphasizes the value of using a random-number table in the selection of a random sample. Most important, it points out the problems that may be encountered when attempting to acquire a random sample by a mechanical selection process.

EXERCISES 4.52–4.60

LEARNING THE MECHANICS

4.52 Suppose you wish to draw a sample of $n = 2$ elements from a population that contains $N = 10$ elements.

a. List all possible samples of $n = 2$ elements, and also use combinatorial mathematics (see Section 4.1) to count the number of different samples that can be selected. How many different samples of $n = 2$ elements can be selected?

b. If random sampling is employed, what is the probability of drawing a particular pair of elements in your sample?

4.53 Use Table I of Appendix B to select a random sample of size $n = 20$ from among the digits 0, 1, 2, . . . , 9. (Digits may repeat themselves.) Explain your procedure.

4.54 Suppose that a population contains $N = 200,000$ elements. Use Table I of Appendix B to select a random sample of $n = 10$ elements from the population. Explain how you selected your sample.

4.55 Consider the population of size $N = 12$ described in the table at the top of page 190.

a. If a random sample of size 1 is selected, what is the probability of selecting a person with a weekly income of $900? Of $1,500?

b. How many different samples of size 3 can be selected? [*Hint:* Use combinatorial mathematics (see Section 4.1).]

NAME	WEEKLY INCOME	NAME	WEEKLY INCOME
Norm	$2,000	Art	$1,900
Carl	1,500	John	1,800
Maureen	780	David	1,050
Paul	1,500	George	1,400
Rema	950	Chris	1,200
Shawn	950	Jerry	900

c. If a random sample of size $n = 3$ is selected, what is the probability of selecting Norm, Art, and John? Rema, Shawn, and David?

d. Use the random-number table (Table I of Appendix B) to select a random sample of size 3 from this population. Explain your procedure.

4.56 Consider the following population of savings accounts at a new bank:

ACCOUNT NUMBER	ACCOUNT BALANCE
0001	$ 1,000
0002	12,500
0003	850
0004	1,000
0005	3,450

Suppose you wish to draw a random sample of two accounts from this population.

a. List all possible different pairs of accounts that could be obtained.

b. What is the probability of selecting accounts 0001 and 0004?

c. What is the probability of selecting two accounts that each have a balance of $1,000? That each have a balance other than $1,000?

APPLYING THE CONCEPTS

4.57 To ascertain the effectiveness of their advertising campaigns, firms frequently conduct telephone interviews with consumers. Random samples of telephone numbers may be randomly or systematically selected from telephone directories, or a recent innovation called *random-digit dialing* may be employed. This approach involves using a random-number generator to mechanically create the sample of phone numbers to be called. An advantage of random-digit dialing is that it makes it possible to obtain a representative sample from the population of all households with telephones, whereas with telephone directory sampling, it is only possible to obtain a sample from the population of households that have *listed* telephone numbers (Glasser and Metzger, 1972).

a. Explain how the random-number table (Table I) could be used to generate a sample of seven-digit telephone numbers.

b. Use the procedure you described in part **a** to generate a sample of 10 seven-digit telephone numbers.

c. Use the procedure you described in part **a** to generate 5 seven-digit telephone numbers whose first three digits are 373.

4.58 When a company sells shares of stock to investors, the transaction is said to take place in the *primary market*. To enable investors to resell the stock when they wish, *secondary markets* called *stock exchanges* were created. Stock exchange transactions involve buyers and sellers exchanging cash for shares of stock, with none of the proceeds going to the companies that issued the shares (Greenleaf, Foster, and Prinsky, 1982). The results of the previous business day's transactions for stocks traded on the New York Stock Exchange (NYSE) and five regional exchanges—the Midwest, Pacific, Philadelphia, Boston, and Cincinnati stock exchanges—are summarized each business day in the NYSE–Composite Transactions table in the *Wall Street Journal*.

a. Examine the NYSE–Composite Transactions table in a recent issue of the *Wall Street Journal* and explain how to draw a random sample of stocks from the table.

b. Use the procedure you described in part **a** to draw a random sample of 20 stocks from a recent NYSE–Composite Transactions table. For each stock in the sample, list its name (i.e., the abbreviation given in the table), its sales volume, and its closing price.

4.59 Using the random-number table (Table I of Appendix B), draw a random sample of five nonbusiness telephone subscribers from page 1 of your local telephone directory.

a. List your sample.

b. Describe the procedure you used to obtain the sample.

c. Based on your experiences with one page of the telephone directory, what problems do you suppose would be encountered in drawing a random sample of size 1,000 from the entire directory?

4.60 From the late 1970's to the mid-1980's, the fastest growing and heaviest spending segment of U.S. consumers consisted of adults with household incomes of $50,000 or more. According to Census Bureau estimates, in 1979 there were 6,227,993 such affluent adults; in 1984 there were 23,188,100. The table shows that there has been a substantial increase in the number of young adults in this group.

AGE	ADULTS WITH HOUSEHOLD INCOME $50,000 OR MORE	
	1978	1984
18–39	43%	48%
40–64	50%	46%
65+	7%	6%

Source: *Marketing Communications*, Vol. 11, No. 2, Feb. 1986, p. 35.

a. If the year were 1984 and one adult was to be randomly selected from the population of affluent adults, what is the probability that a particular affluent adult would be selected? What is the probability that the affluent adult would be between 18 and 39 years of age?

b. Repeat part **a** for the year 1978, and compare the answers.

c. If two adults were to be randomly selected from the 1984 population of affluent consumers, what is the probability of selecting a particular pair of affluent adults?

SUMMARY

We have developed some of the basic tools of probability that enable us to assess the probabilities of various sample outcomes, given a specific population structure. Although many of the examples we presented were of no practical importance, they accomplished their purpose if you now understand the concepts and definitions necessary for a basic knowledge of probability.

In the next several chapters, we will present probability models that can be used to solve practical business problems. You will see that for most applications, we will need to make inferences about unknown aspects of these probability models; i.e., we will need to apply inferential statistics to the problem.

SUPPLEMENTARY EXERCISES 4.61–4.91

4.61 What are the two rules that probabilities assigned to simple events must obey?

4.62 Are mutually exclusive events also dependent events? Explain.

4.63 Given that $P(A \cap B) = .4$ and $P(A \mid B) = .8$, find $P(B)$.

4.64 Which of the following pairs of events are mutually exclusive? Justify your response.
a. The Dow Jones Industrial Average increases on Monday.
A large New York bank decreases its prime interest rate on Monday.
b. An IBM microcomputer is purchased.
An Apple microcomputer is purchased.
c. You reinvest all your dividend income for 1988 in a limited partnership.
You reinvest all your dividend income for 1988 in a money market fund.

4.65 A manufacturer of electronic digital watches claims that the probability of its watch running more than 1 minute slow or 1 minute fast after 1 year of use is .05. A consumer protection agency has purchased four of the manufacturer's watches with the intention of testing the claim.
a. Assuming that the manufacturer's claim is correct, what is the probability that all four of the watches are as accurate as claimed?
b. Assuming that the manufacturer's claim is correct, what is the probability that exactly two of the four watches fail to meet the claim?
c. Suppose that three of the four tested watches failed to meet the claim. What inference can be made about the manufacturer's claim? Explain.
d. Suppose that all four tested watches failed to meet the claim. Is it necessarily true that the manufacturer's claim is false? Explain.

4.66 The state legislature has appropriated $1 million to be distributed in the form of grants to individuals and organizations engaged in the research and development of alternative energy sources. You have been hired by the state's energy agency to assemble a panel of five energy experts whose task it will be to determine which individuals and organizations should receive the grant money. You have identified 11 equally qualified individuals who are willing to serve on the panel. How many different panels of five experts could be formed from these 11 individuals?

4.67 A research and development company surveyed all 200 of its employees over the age of 60 and obtained the information given in the table. One of these 200 employees is selected at random.

	UNDER 20 YEARS WITH COMPANY		OVER 20 YEARS WITH COMPANY	
	Technical staff	Nontechnical staff	Technical staff	Nontechnical staff
Plan to retire at age 65	31	5	45	12
Plan to retire at age 68	59	25	15	8

a. What is the probability that the person selected is on the technical staff?

b. If the person selected has over 20 years of service with the company, what is the probability that the person plans to retire at age 68?

c. If the person selected is on the technical staff, what is the probability that the person has been with the company less than 20 years?

d. What is the probability that the person selected has over 20 years with the company, is on the nontechnical staff, and plans to retire at age 65?

4.68 Refer to Exercise 4.67.

a. Consider the events A: {Plan to retire at age 68} and B: {On the technical staff}. Are events A and B independent? Explain.

b. Consider the event D: {Plan to retire at age 68 *and* on the technical staff}. Describe the complement of event D.

c. Consider the event E: {On the nontechnical staff}. Are events B and E mutually exclusive? Explain.

4.69 The set of securities (stocks, bonds, etc.) held by an individual or an organization is referred to as its *portfolio*. An investor wants to invest $2,000 in each of five different common stocks and has identified ten different stocks that she believes would be sound investments. How many different portfolios of five stocks could she form from among her list of ten stocks?

4.70 The performance of quality inspectors affects both the quality and the cost of outgoing products. A product that passes inspection is assumed to meet quality standards; a product that fails inspection may be reworked, scrapped, or reinspected. Quality engineers at Westinghouse Electric Corporation evaluated inspectors' performances in judging the quality of solder joints by comparing each inspector's classifications of a set of 153 joints with the consensus evaluation of a panel of experts (Meagher and Scazzero, 1985). Suppose the results for a particular inspector are as shown in the table.

		INSPECTOR'S JUDGMENT	
		Joint acceptable	Joint rejectable
COMMITTEE'S	Joint acceptable	101	10
JUDGMENT	Joint rejectable	23	19

One of the 153 solder joints is to be selected at random.

a. What is the probability that the inspector judges the joint to be acceptable? That the committee judges the joint to be acceptable?

b. What is the probability that both the inspector and the committee judge the joint to be acceptable? That neither judge the joint to be acceptable?

c. What is the probability that the inspector and the committee disagree? Agree?

4.71 An advertising agency was interested in whether a client's advertisement in the latest issue of a nutrition magazine was noticed by readers of the magazine. Accordingly, it asked 1,500 of the magazine's subscribers who said they had read the latest issue whether they had noticed the advertisement. The table of percentages describes their responses. Suppose one of the 1,500 readers is chosen at random.

		Noticed the ad	Did not notice the ad
	Under 30	25%	5%
AGE GROUP	30–50	20%	15%
	Over 50	10%	25%

a. List all the simple events for this experiment.
b. What is the set of simple events in part **a** called?
c. For each of the simple events in part **a**, find the probability that it will occur.

4.72 Refer to Exercise 4.71. The following events are defined:

A: {Reader questioned was under 30}

B: {Reader questioned was between 30 and 50}

C: {Reader questioned noticed the advertisement}

D: {Reader questioned did not notice the advertisement}

Describe a reader portrayed by each of the following events:
a. $A \cap C$ **b.** $B \cup D$ **c.** $A \cap B$

4.73 A local country club has a membership of 600 and operates facilities that include an 18-hole championship golf course and 12 tennis courts. Before deciding whether to accept new members, the club president would like to know how many members regularly use each facility. A survey of the membership indicates that 70% regularly use the golf course, 50% regularly use the tennis courts, and 5% use neither of these facilities regularly.
a. Construct a Venn diagram to describe the results of the survey.
b. If one club member is chosen at random, what is the probability that the member uses either the golf course or the tennis courts or both?
c. If one member is chosen at random, what is the probability that the member uses both the golf and the tennis facilities?
d. A member is chosen at random from among those known to use the tennis courts regularly. What is the probability that the member also uses the golf course regularly?

4.74 Insurance companies use *mortality tables* to help them determine how large a premium to charge a particular individual for a particular life insurance policy. The accompanying table shows the probability of survival to age 65 for persons of the specified ages.
a. For a person 20 years old, what is the probability that he or she will die before age 65?
b. Describe in words the trend indicated by the increasing probabilities in the second and fourth columns.

AGE	PROBABILITY OF SURVIVAL TO AGE 65	AGE	PROBABILITY OF SURVIVAL TO AGE 65
0	.72	40	.77
10	.74	45	.79
20	.74	50	.81
30	.75	55	.85
35	.76	60	.90

Source: U.S. Department of Health, Education, and Welfare, Public Health Service, National Center for Health Statistics, *United States Life Tables: 1969–71* (1973).

4.75 Explain why the following statement is or is not valid: If an individual is chosen at random from all U.S. citizens living in the 50 states, the probability that this individual lives in New Hampshire is $\frac{1}{50}$.

4.76 A manufacturer of 35-mm cameras knows that a shipment of 30 cameras sent to a large discount store contains six defective cameras. The manufacturer also knows that the store will choose two of the cameras at random, test them, and accept the shipment if neither is defective.

a. What is the probability that the first camera chosen by the store will be defective?

b. Given that the first camera chosen passed inspection, what is the probability that the second camera chosen will fail inspection?

c. What is the probability that the shipment will be accepted?

4.77 The accompanying figure is a schematic representation of a system comprised of three components. The system operates properly only if all three components operate properly. The three components are said to operate *in series*. The components could be mechanical or electrical; they could be work stations in an assembly process; or they could represent the functions of three different departments in an organization. The probability of failure for each component is listed in the table. Assume that the components operate independently of each other.

A System Comprised of Three
Components in Series

Input ⟶ (#1) ⟶ (#2) ⟶ (#3) ⟶ Output

COMPONENT	PROBABILITY OF FAILURE
#1	.12
#2	.09
#3	.11

a. Find the probability that the system operates properly.

b. What is the probability that at least one of the components will fail, and therefore that the system will fail?

4.78 The accompanying figure is a representation of a system comprised of two subsystems that are said to operate *in parallel*. Each subsystem has two components that operate in series (refer to Exercise 4.77). The system will operate properly as long as at least one of the subsystems functions properly. The probability of failure for each component in the system is .1. Assume that the components operate independently of each other.

A System Comprised of Two Parallel Subsystems

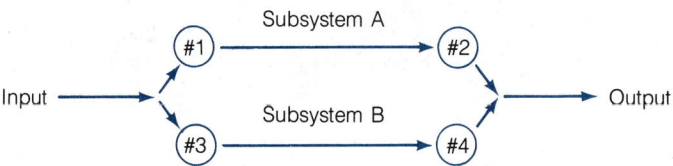

a. Find the probability that the system operates properly.
b. Find the probability that exactly one subsystem fails.
c. Find the probability that the system fails to operate properly.
d. How many parallel subsystems like the two shown here would be required to guarantee that the system would operate properly more than 99% of the time?

4.79 Your firm has decided to market two new products. The manager of the Marketing Department believes the probability of product A being accepted by the public and product B not being accepted is .3, of product B being accepted and product A not being accepted is .4, and of both products A and B being accepted is .2. Given these probabilities the manager has concluded that the probability of both products failing is .01. Do you agree with this conclusion? Explain.

4.80 Suppose only two daily newspapers are available in your town—a local paper and one from a nearby city—and that 1,000 people in town subscribe to a daily paper. Assume that 65% of the people in town who subscribe to a daily newspaper subscribe to the local paper and 40% of those who subscribe to a daily paper subscribe to the city paper.
a. Use a Venn diagram to describe the population of newspaper subscribers.
b. If one of the 1,000 subscribers is chosen at random, what is the probability that he or she subscribes to both newspapers?

4.81 Six people apply for two identical positions in a company. Four are minority applicants and the remainder are nonminority. Define the following events:

A: {Both persons selected for the positions are nonminority candidates}

B: {Both persons selected for the positions are minority candidates}

C: {At least one of the persons selected is a minority candidate}

If all the applicants are equally qualified and the choice is therefore a random selection of two applicants from the six available, find the following:
a. $P(A)$ b. $P(B)$ c. $P(C)$ d. $P(B \mid C)$
e. For the purpose of identification, assume that the minority candidates are numbered 1, 2, 3, and 4. Define the event D: {Minority candidate 1 is selected}. Find $P(D \mid C)$.

4.82 The table of percentages describing the location of a fast-food restaurant chain's franchises (Exercise 4.47) is reproduced here.

		REGION			
		NE	SE	SW	NW
POPULATION OF CITY	Under 10,000	5%	6%	3%	0%
	10,000–100,000	10%	15%	12%	5%
	Over 100,000	25%	4%	5%	10%

A restaurant is to be chosen at random from the 700 to test market a new style of chicken. The following events are defined:

A: {Chosen restaurant is in the Southeast}

B: {Chosen restaurant is in city of population under 10,000}

C: {Chosen restaurant is in city of population of 10,000 or over}

D: {Chosen restaurant is in the Northeast}

a. Find $P(A \mid B)$.   **b.** Find $P(B \mid A)$.

c. Find $P(A \mid C)$.   **d.** Find $P(C \mid D)$.

4.83 Refer to Exercise 4.82. Show that the following is true:

$$P(A \cap B) = P(A)P(B \mid A) = P(B)P(A \mid B) = .06$$

4.84 In conducting a market potential study for a manufacturer of portable generators, Miller Research Services of Minneapolis asked 829 owners of fifth-wheel and conventional travel trailers the following two questions: (1) Do you own or plan to buy a generator for your trailer? (2) What was the approximate cost of your trailer? The answers to the questions are summarized in the table.

	Own	Do not own but plan to buy	Do not own and do not plan to buy
Less than $6,000	1	5	14
$6,000–$11,999	52	65	160
$12,000–$17,999	73	78	175
$18,000–$23,999	29	36	62
$24,000–$29,999	21	18	21
$30,000 or more	3	10	6

Source: MBA Field Project Team Report, University of Minnesota, 1986.

One of the travel-trailer owners who does not currently own a generator is to be selected at random for further questioning. What is the probability that the person:

a. Will own a travel trailer worth $12,000–$17,999?

b. Will own a travel trailer worth more than $17,999?

c. Will own a trailer worth $12,000–$17,999 and be planning to buy a generator?

d. Will own a trailer worth $12,000–$17,999 or be planning to buy a generator?

4.85 Suppose there are 500 applicants for five equivalent positions at a factory and the company is able to narrow the field to 30 equally qualified applicants. Seven of the finalists are minority candidates. Assume that the five who are chosen are selected at random from this final group of 30.

a. What is the probability that none of the minority candidates is hired?

b. What is the probability that no more than one minority candidate is hired?

4.86 According to David Dreman (*Forbes*, Oct. 27, 1980, pp. 202–203), investment in *new issues* (the stock of newly formed companies) can be both suicidal and rewarding. Dreman based his comments on a Securities and Exchange Commission study of 500 new issues that went public during the 1961–1962 stock boom. The commission found that of the 500 companies, 43% went bankrupt, 25% were operating at a loss, and only 20% showed any profitability. Only 12 companies out of the 500 appeared to have outstanding prospects. Suppose that, back in 1961, you had invested $1,000 in each of three of these new issues and that, because the prospects of the stocks were unknown, for all practical purposes the three stocks were randomly selected from among the 500 new issues. What is the probability that:

a. All three of the new issues would be among the 12 with outstanding prospects?

b. All three of the new issues would be from among the 43% that went bankrupt?

c. At least one of the three would be among the 12 with outstanding prospects?

4.87 The probability that a microcomputer salesperson sells a computer to a prospective customer on the first visit to the customer is .4. If the salesperson fails to make the sale on the first visit, the probability that the sale will be made on the second visit is .65. The salesperson never visits a prospective customer more than twice. What is the probability that the salesperson will make a sale to a particular customer?

4.88 A credit counselor claims that the probability that at least two local firms go bankrupt next year is .15, and the probability that exactly two local firms go bankrupt is .20. Can this statement be true? Explain.

4.89 Use a Venn diagram to show that

$$P(A \cap B^c) = P(A) - P(A \cap B)$$

4.90 Use your intuitive understanding of independence to form an opinion about whether each of the following scenarios represents independent events.

a. The results of consecutive tosses of a coin

b. The opinions of randomly selected individuals in a pre-election poll

c. A major-league baseball player's results in two consecutive at-bats

d. The amount of gain or loss associated with investments in different stocks that are bought on the same day, and sold on the same day 1 month later

e. The amount of gain or loss associated with investments in different stocks that are bought and sold in different time periods, 5 years apart

f. The prices bid by two different development firms in response to a building construction proposal issued by a university.

4.91 A fair coin is flipped 20 times and 20 heads are observed. In such cases it is often said that a tail is due on the next flip. Is this statement true or false? Explain.

ON YOUR OWN . . .

Obtain a standard bridge deck of 52 cards and think of the cards as the 52 items your firm produces each day. Let the 4 aces and 4 kings in the deck represent defective items.

a. If one item is randomly sampled from a day's production, what is the probability of its being defective?

b. Shuffle the cards, draw one, and record whether it is a defective item. Then replace the card and repeat the process. After each draw, recalculate the proportion of the draws that have resulted in a defective item. Construct a graph with the proportion of defectives on the y-axis and the number of draws on the x-axis. Notice how the proportion defective stabilizes as the number of draws increases.

c. Draw a horizontal line on the graph in part **b** at a height equal to the probability you calculated in part **a**. Compare the calculated proportion of defectives to this probability. As the number of draws is increased, does the calculated proportion of defectives more closely approach the actual probability of drawing a defective?

USING THE COMPUTER . . .

Suppose a large bank is planning a national mailing to market a major credit card. However, the bank will first test its marketing materials by mailing to a single zip code.

a. If one of the 1,000 zip codes in Appendix C were to be selected randomly for a test mailing, what is the probability that the selected zone is one for which the average income exceeds $35,000?

b. Suppose the zip code were to be selected from the Northeast census region (among those in Appendix C). What is the probability that the selected zone is one for which the average income exceeds $35,000?

c. Are the events described in parts **a** and **b** independent? Why or why not? What are the practical implications of the independence, or non-independence, of the events?

REFERENCES

Bagozzi, R. P. *Principles of Marketing Management.* Chicago: SRA, 1986, p. 215.

"Drawing tonight will determine who is drafted," *New York Times,* Dec. 1, 1969, p. 1.

Feller, W. *An Introduction to Probability Theory and Its Applications,* 3rd ed. Vol. 1. New York: Wiley, 1968. Chapters 1, 4, and 5.

Glasser, G. J. and Metzger, G. D. "Random-digit dialing as a method of telephone sampling." *Journal of Marketing Research,* Feb. 1972, 9, 59–64.

Greenleaf, J., Foster, R., and Prinsky, R. "Understanding financial data in the *Wall Street Journal.*" *Wall Street Journal,* Special Education Edition, 1982, p. 19.

Kuehn, A. A. "An analysis of the dynamics of consumer behavior and its implications for marketing management." Unpublished doctoral dissertation, Graduate School of Industrial Administration, Carnegie Institute of Technology, 1958.

Meagher, J. J. and Scazzero, J. A. "Measuring inspector variability." *39th Annual Quality Congress Transactions*, American Society for Quality Control, May 1985, 75–81.

Parzen, E. *Modern Probability Theory and Its Applications*. New York: Wiley, 1960. Chapters 1 and 2.

Peterson, R. L. "Creditors' use of collection remedies." *Journal of Financial Research*, Vol. 9, No. 1, Spring 1986, 71–86.

"Random or not? Judge studies lottery protest," *The National Observer*, Jan. 12, 1970, p. 2.

Rosenblatt, J. R. and Filliben, J. J. "Randomization and the draft lottery." *Science*, *171*, 306–308.

Williams, B. *A Sampler on Sampling*. New York: Wiley, 1978. pp. 5–8.

Winkler, R. L. *An Introduction to Bayesian Inference and Decision*. New York: Holt, Rinehart and Winston, 1972. Chapter 2.

Winkler, R. L. and Hays, W. L. *Statistics: Probability, Inference, and Decision*, 2nd ed. New York: Holt, Rinehart and Winston, 1975. Chapters 1 and 2.

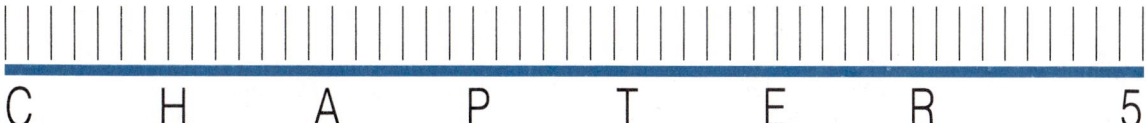
WHERE WE'VE BEEN . . .

By illustration, we indicated in Chapter 4 how probability would be used to make an inference about a population from information contained in a sample. We also noted that probability would be used to measure the reliability of the inference.

WHERE WE'RE GOING . . .

Most experimental events in Chapter 4 were described in words or denoted by capital letters. In real life, most sample observations are numerical—in other words, they are quantitative data. In this chapter, we will learn that business data are observed values of random variables. We will study several important random variables and will learn how to find the probabilities of specific numerical outcomes.

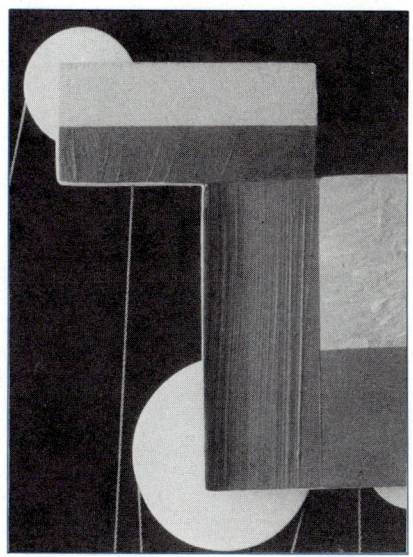

CONTENTS

DISCRETE RANDOM VARIABLES

You may have noticed that most of the examples of experiments given in Chapter 4 generated quantitative (numerical) data. This is frequently true; observations on many types of phenomena are numerical measurements. The Consumer Price Index, unemployment rate, number of sales made in a week, and yearly profit of a company are all examples of numerical measurements of some business phenomena. Thus, most experiments have simple events that correspond to values of some numerical variable.

DEFINITION 5.1

A **random variable** is a rule that assigns one (and only one) numerical value to each simple event of an experiment. *

The term *random variable* is more meaningful than the simpler term *variable* because the adjective *random* indicates that the experiment may result in one of the several possible values of the variable, according to the *random* outcome of the experiment. For example, if the experiment is to count the number of customers who use the drive-up window of a bank each day, the random variable (the number of customers) will vary from day to day, partly because of the random phenomena that influence whether customers use the drive-up window. Thus, the possible values of this random variable range from zero to the maximum number of customers the window could possibly serve in a day.

We define two different types of random variables, **discrete** and **continuous**, in Section 5.1. Then we spend the remainder of the chapter discussing specific types of discrete random variables and the aspects that make them important in business applications.

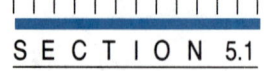

SECTION 5.1

TWO TYPES OF RANDOM VARIABLES

Recall that the simple event probabilities corresponding to an experiment must sum to 1. Assigning one unit of probability to the simple events in a sample space, and consequently to the values of a random variable, is not always as easy as the examples in Chapter 4 may lead you to believe. If the number of simple events is finite, the job is easy. If the number of simple events is infinite but you can list them in order (we call this **countable**), the task is still not too difficult. But if the simple events are numerical and correspond to the infinitely large number of points contained in an interval, the task is impossible. Why? Because you cannot assign a small portion of probability to each of the simple events in this infinitely large set—the sum of the probabilities will exceed 1 (in fact, the sum will be infinitely large). The consequences of this mathematical fact are important. We will have to use two different probability models, depending on whether the number of simple events in a sample space (or equivalently, the values that a random variable can assume) are countable or they correspond to the infinitely large number of points contained in one or more intervals.

*By *experiment*, we mean an experiment that yields random outcomes (as defined in Chapter 4).

DEFINITION 5.2

Random variables that can assume a *countable* number of values are called **discrete**.

DEFINITION 5.3

Random variables that can assume values corresponding to any of the points contained in one or more intervals are called **continuous**.

Examples of discrete random variables are:

1. The number of sales made by a salesperson in a given week: $x = 0, 1, 2, \ldots$.
2. The number of people in a sample of 500 who favor a particular product over all competitors: $x = 0, 1, 2, \ldots, 499, 500$.
3. The number of bids received in a bond offering: $x = 0, 1, 2, \ldots$.
4. The number of errors on a page of an accountant's ledger: $x = 0, 1, 2, \ldots$.
5. The number of customers waiting to be served in a restaurant at a particular time: $x = 0, 1, 2, \ldots$.

Note that each of the examples of discrete random variables begins with the words "the number of." This is very common because the discrete random variables most frequently observed are counts.

Examples of continuous random variables are:

1. The length of time between arrivals at a hospital clinic: $0 \leq x < \infty$ (infinity).
2. For a new apartment complex, the length of time from completion until a specified number of apartments are rented: $0 \leq x < \infty$.
3. The amount of carbonated beverage loaded into a 12-ounce can in a can filling operation: $0 \leq x \leq 12$.
4. The depth at which a successful oil drilling venture first strikes oil.
5. The weight of a food item bought in a supermarket.

In Section 5.2 we will discuss how to find the probability distribution for a discrete random variable. Then, several types of discrete random variables that play important roles in business decisions will be presented in subsequent sections. Probability distributions for some useful types of continuous random variables will be the subject of Chapter 6.

EXERCISES 5.1–5.10

APPLYING THE CONCEPTS

5.1 What is a random variable?

5.2 How do discrete and continuous random variables differ?

5.3 Which of the following describe continuous random variables, and which describe discrete random variables? Justify your answers.
   a. The number of houses sold by a real estate developer
   b. The amount of natural gas used per month for heating an apartment building
   c. The exact amount of milk in a quart container
   d. The number of accidents per week at a manufacturing plant

5.4 Which of the following describe continuous random variables, and which describe discrete random variables?
   a. The number of people per day who report for work at a manufacturing plant
   b. The number of errors found in an audit of a company's financial records
   c. The length of time a customer waits for service at a supermarket checkout counter
   d. The number of automobiles recalled by General Motors next year
   e. The actual number of ounces of cola drink in a 12-ounce bottle

5.5 Give two examples of a discrete business-oriented random variable. Do the same for a continuous random variable.

5.6 Give an example of a discrete random variable that would be of interest to a real estate salesperson.

5.7 Give an example of a continuous random variable that would be of interest to an economist.

5.8 Give an example of a discrete random variable that would be of interest to the manager of a hotel.

5.9 Give two examples of discrete random variables that would be of interest to the manager of a clothing store.

5.10 Give an example of a continuous random variable that would be of interest to a stockbroker.

| | | | | | | | | | | | |

S E C T I O N   5.2

PROBABILITY DISTRIBUTIONS FOR DISCRETE RANDOM VARIABLES

Since a random variable assigns a numerical value to each of the simple events associated with an experiment, a complete description of a random variable requires that we specify its **probability distribution**. Note that each simple event is assigned one and only one value of the random variable, and hence, the values of the random variable represent mutually exclusive events.

DEFINITION 5.4

The **probability distribution** of a discrete random variable is a graph, table, or formula that specifies the probability associated with each possible value the random variable can assume.

To illustrate, consider Example 5.1.

EXAMPLE 5.1

Recall the experiment of tossing two coins (Chapter 4), and let x be the number of heads observed. Find the probability distribution for the random variable x, assuming the two coins are fair.

SOLUTION

FIGURE 5.1

Venn Diagram for the
Two-Coin Toss Experiment

H_1, H_2	T_1, H_2
•	•
$x = 2$	$x = 1$
H_1, T_2	T_1, T_2
•	•
$x = 1$	$x = 0$

                                S

TABLE 5.1

Probability Distribution:
Tabular Form

x	0	1	2
$p(x)$	$\frac{1}{4}$	$\frac{1}{2}$	$\frac{1}{4}$

Recall from Chapter 4 that the sample space and simple events for this experiment are as shown in Figure 5.1, and the probability associated with each of the four simple events is $\frac{1}{4}$. The random variable x can assume values 0, 1, 2. Then, identifying the probabilities of the simple events associated with each of these values of x, we have

$$P(x = 0) = P(T_1, T_2) = \tfrac{1}{4}$$
$$P(x = 1) = P(T_1, H_2) + P(H_1, T_2) = \tfrac{1}{4} + \tfrac{1}{4} = \tfrac{1}{2}$$
$$P(x = 2) = P(H_1, H_2) = \tfrac{1}{4}$$

We will denote the probability of the random variable x by the symbol $p(x)$. Then, for this example, $p(0) = \frac{1}{4}$, $p(1) = \frac{1}{2}$, and $p(2) = \frac{1}{4}$. Table 5.1 shows the probability distribution of x in tabular form, and Figure 5.2 shows it in two alternative graphical forms. Figure 5.2(a) shows the probabilities concentrated at the points, $x = 0, 1$, and 2. The heights of the vertical line segments give the probabilities that correspond to each of these values of x. The probability distribution in Figure 5.2(b) is shown as a histogram with one class corresponding to each of the three values of x. Although the probabilities for discrete random variables are, in fact, concentrated at specific points, the probability histogram will be a convenient way of viewing the probability distribution for a discrete random variable when we attempt to approximate certain probabilities in Section 6.5.

FIGURE 5.2

Probability Distribution:
Graphical Forms

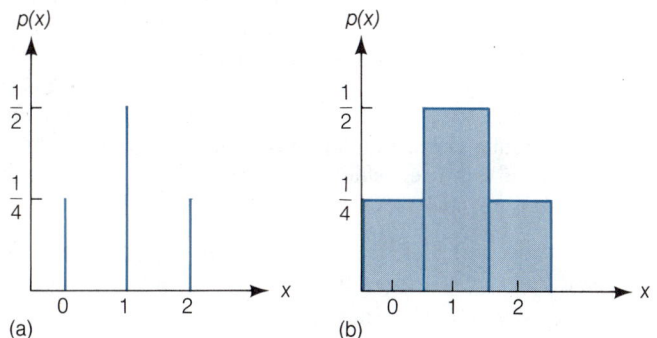

(a)   (b)

    We could also present the probability distribution for x as a formula, but this would unnecessarily complicate a very simple example. We will give the formulas for the probability distributions of some discrete random variables later in this chapter.

    Two requirements must be satisfied by all probability distributions for discrete random variables:

**REQUIREMENTS FOR THE PROBABILITY DISTRIBUTION
OF A DISCRETE RANDOM VARIABLE x**

1. $p(x) \geq 0$ for all values of x

2. $\displaystyle\sum_{\text{All } x} p(x) = 1$

Example 5.1 illustrates how the probability distribution for a discrete random variable can be derived, but for many practical examples of random variables, the task is much more difficult. Fortunately, experiments and associated discrete random variables with identical characteristics are found in many different areas of business. That is, although the data may be collected in an area of accounting, marketing, economics, or management, for all practical purposes the data may represent observed values of the same type of random variable. This fact simplifies the problem for the business statistician. All that must be done is to define the nature of these often repeated experiments, define the type of random variable, and give the probability distribution of the random variable.

In Sections 5.4–5.7 we will describe four important types of discrete random variables, give their probability distributions, and explain where and how they can be applied in business. (Mathematical derivations of the probability distributions will be omitted, but these details can be found in the references at the end of the chapter.)

But first, in Section 5.3, we will discuss some descriptive measures of these sometimes complex probability distributions. Since probability distributions are analogous to the relative frequency distributions of Chapter 2, it should be no surprise that the mean and standard deviation are useful descriptive measures.

CASE STUDY 5.1

ASSESSING THE EFFECTS OF THE DEADLY DUTCH ELM DISEASE

Since 1930 when the Dutch elm disease fungus was first discovered in the U.S. on the east coast, it has spread westward destroying elm trees all across the continent. It has been called ". . . the most destructive and widespread plague of trees of our time" (Webster, 1978). The fungus is spread primarily by English bark beetles which breed in diseased elm trees. Because of a lack of an effective chemical treatment, the major tactic in combating Dutch elm disease is a sanitation program which involves the removal of the diseased elm trees before they become a breeding ground for the fungus-carrying beetle (Chervany et al., 1980).

At the start of 1977, the city of Minneapolis had approximately 200,000 elms and had not lost more than 7,200 elms in any single year to Dutch elm disease. Since a major reforestation program had recently been inaugurated, this loss rate was not overly alarming. In 1977, however, Minneapolis lost 31,475 elms, approximately 16% of the existing elm population.

Prior to the 1978 growing season the Minneapolis Park and Recreation Board (Park Board), the organization responsible for managing the city's Dutch Elm Disease Sanitation Program, hired a team of management scientists to help in the design and operation of a sanitation program capable of dealing with the increased disease incidence. An integral part of the program proposed by the team involved forecasting the number of elms to be lost in the coming growing season. This information would enable the Park Board to plan staffing, evaluate potential bottlenecks in the elm removal operation, and inform the citizenry of the extent of the damage to the urban forest expected during the next year.

Because of the scarcity of historical data, the forecast was developed from the opinions of four disease experts. Each expert developed a discrete probability distribution to represent his beliefs regarding the number of elms that would be infected by the disease during the 1978 growing season. These four probability distributions were combined (by averaging using equal weights) to obtain a single discrete probability distribution reflecting the beliefs of all the experts regarding the number of elms to be infected. A slightly modified version of this probability distribution appears in the table.

NUMBER OF ELMS THAT WILL BE INFECTED IN 1978	
x	$p(x)$
5,000– 9,999	.02
10,000–14,999	.07
15,000–19,999	.16
20,000–24,999	.17
25,000–29,999	.12
30,000–34,999	.13
35,000–39,999	.11
40,000–44,999	.08
45,000–49,999	.07
50,000 or more	.07

This distribution can be used to make probabilistic forecasts of the number of trees that will be infected in 1978. It indicates that, according to the combined opinions of the disease experts, it is most likely that the number of trees to be infected will be between 20,000 and 24,999. Furthermore, probability statements like the following can be made about the number of trees to be infected, x:

$$P(15,000 \leq x \leq 29,999) = .45$$
$$P(15,000 \leq x \leq 39,999) = .69$$
$$P(10,000 \leq x \leq 49,999) = .91$$
$$P(x < 15,000) = .09$$
$$P(x \geq 40,000) = .22$$

It is interesting to note that the actual number of losses to Dutch elm disease in 1978 was 20,817 trees, which is within the modal interval of the forecast distribution (Chervany et al., 1980).

EXERCISES 5.11–5.19

LEARNING THE MECHANICS

5.11 Three coins are tossed. Let x equal the number of heads observed.
   a. Identify the simple events associated with this experiment and assign a value of x to each simple event, assuming the coins are fair.
   b. Calculate $p(x)$ for each value of x.
   c. Display the probability distribution of x in graphical form.

5.12 A die is tossed. Let x be the number of spots observed on the upturned face of the die.
   a. Find the probability distribution of x and display it in tabular form.
   b. Display the probability distribution of x in graphical form.

Explain why each of the following is or is not a valid probability distribution for a random variable x.

a.

x	0	1	2	3
p(x)	.1	.3	.3	.2

b.

x	−2	−1	0
p(x)	.25	.50	.25

c.

x	4	9	20
p(x)	−.3	.4	.3

d.

x	2	3	5	6
p(x)	.15	.15	.45	.35

5.14 The random variable x has the following discrete probability distribution:

x	10	11	12	13	14
p(x)	.1	.2	.5	.1	.1

Since the values that x can assume are mutually exclusive events, the event $\{x \le 12\}$ is the union of three mutually exclusive events: $\{x = 10\} \cup \{x = 11\} \cup \{x = 12\}$.
a. Find $P(x \le 12)$.
b. Find $P(x > 12)$.
c. Find $P(x \le 14)$.
d. Find $P(x = 14)$.
e. Find $P(x \le 11 \text{ or } x > 12)$.

5.15 The random variable x has the following discrete probability distribution:

x	0	1	2	3	4
p(x)	.10	.10	.25	.25	.30

a. Find $P(x \le 0)$.
b. Find $P(x < 0)$.
c. Find $P(x = 2)$.
d. Find $P(x \le 3)$.
e. Find $P(x > 0)$.
f. Find $P(2 \le x \le 4)$.

APPLYING THE CONCEPTS

5.16 A group of marketing managers were polled to assess the probabilities associated with the number, x, of sales that a company might expect per month for a new super-computer that was in the planning stages. These probabilities are shown here.

x	0	1	2	3	4	5	6	7	8
p(x)	.02	.08	.15	.19	.24	.17	.10	.04	.01

a. Display the probability distribution for x in graphical form.
b. Based on the marketing managers' probability distribution for x, what is the probability that the company will sell more than three computers per month? More than four?

5.17 Experience has shown that a builder of custom houses makes a profit on 95% of his contracts. Assume the event that the builder makes a profit on any one job is independent of whether he makes a profit on any other. The builder typically contracts to build three houses per month. Let x equal the number of houses per month that result in a profit.

a. Find $p(x)$.
b. Graph $p(x)$.
c. Find the probability that $x \geq 2$.

5.18 To entice potential real estate buyers to attend one of its sales presentations, Recreation Resorts of America mails advertising brochures which guarantee that all attendees will win one prize from the following groups of prizes:

PRIZE	VALUE OF PRIZE
Group 1: $5,000	$5,000.00
Group 2: Microwave oven	769.00
Group 3: 2-man inflatable raft	49.95
Group 4: 19″ color television	699.00
Group 5: Home computer	699.00
Group 6: Home video recorder	795.00
Group 7: Moped	500.00

The brochure contains a coded "winning number" that can be turned in at the sales presentation to determine which prize has been won. The fine print of the brochure explains that for each 100,000 brochures mailed, there is one $5,000 winner, one microwave oven winner, one television set winner, one home computer winner, one video recorder winner, one moped winner, and 99,994 inflatable raft winners. It also explains that the winning numbers have been "randomly assigned" to mailing labels.

a. Suppose 300,000 brochures were mailed and you received "winning number" 2823. Construct a probability distribution in graphical form that describes the potential value of your winning number.
b. What is the probability that you will not win an inflatable raft?
c. Does your answer to part **b** change if instead 1,000,000 brochures were mailed? Explain.

5.19 Suppose the product development manager for your firm plans to market two new products. She thinks it is possible that both will fail consumer market tests, it is more likely that one will pass, and it is even more likely that neither will fail the market tests. Let x represent the number of the two new products that pass market tests. Display in tabular form a possible representation of the product manager's probability distribution for x.

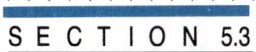

SECTION 5.3

EXPECTED VALUES OF DISCRETE RANDOM VARIABLES

If a discrete random variable, x, were observed a very large number of times and if the data generated were arranged in a relative frequency distribution, the relative frequency distribution would be indistinguishable from the probability distribution for the random variable. Thus, the probability distribution for a random variable can be viewed as a theoretical model for the relative frequency distribution of a population. To the extent that the two distributions are equivalent (and we will assume they are), the probability distribution for x possesses a mean μ and a variance σ^2 that are identical to the corresponding descriptive measures for the population.

The purpose of this section is to explain how you can find the mean value—or **expected value**, as it is called—for a random variable. We will illustrate the procedure with an example.

Examine the probability distribution for x (the number of heads observed in the toss of two fair coins) in Figure 5.3. Try to locate the mean of the distribution intuitively. We may reason as follows that the mean μ of this distribution is equal to 1: In a large number of experiments, $\frac{1}{4}$ should result in $x = 0$, $\frac{1}{2}$ in $x = 1$, and $\frac{1}{4}$ in $x = 2$ heads. Therefore, the average number of heads is

$$\mu = 0\left(\tfrac{1}{4}\right) + 1\left(\tfrac{1}{2}\right) + 2\left(\tfrac{1}{4}\right)$$
$$= 0 + \tfrac{1}{2} + \tfrac{1}{2} = 1$$

FIGURE 5.3

Probability Distribution for a Two-Coin Toss

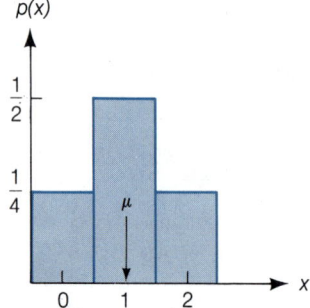

Note that to obtain the mean of the random variable x, we multiply each possible value of x by its probability $p(x)$, and then we sum this product over all possible values of x. The **mean of x** is also referred to as the **expected value of x**, denoted by $E(x)$:

DEFINITION 5.5

The **mean**, or **expected value**, of a discrete random variable x is

$$\mu = E(x) = \sum_{\text{All } x} xp(x)$$

The term *expected* is a mathematical term and should not be interpreted as it is typically used. Specifically, it is possible that a random variable might never equal its *expected* value. Rather, the expected value is the mean of the probability distribution, a measure of its central tendency.

EXAMPLE 5.2

Suppose you work for an insurance company and you sell a $10,000 whole-life insurance policy at an annual premium of $290. Actuarial tables show that the probability of death during the next year for a person of your customer's age, sex, health, etc., is .001. What is the expected gain (amount of money made by the company) for a policy of this type?

SOLUTION

The experiment is to observe whether the customer survives the upcoming year. The probabilities associated with the two simple events, Live and Die, are .999 and .001, respectively. The random variable you are interested in is the gain, x, which can assume the following values:

GAIN x	SIMPLE EVENT	PROBABILITY
$290	Customer lives	.999
$290 − $10,000	Customer dies	.001

If the customer lives, the company gains the $290 premium as profit. If the customer dies, the gain is negative because the company must pay $10,000, for a net "gain" of $(290 − 10,000)$. The expected gain is therefore

$$\mu = E(x) = \sum_{\text{All } x} x p(x)$$

$$= (290)(.999) + (290 - 10,000)(.001)$$

$$= 290(.999 + .001) - 10,000(.001)$$

$$= 290 - 10 = \$280$$

In other words, if the company were to sell a very large number of 1-year $10,000 policies to customers possessing the characteristics described above, it would (on the average) net $280 per sale in the next year.

This example illustrates that the expected value of a random variable x need not be a possible value of x. That is, the expected value is $280, but x will equal either $290 or $−$9,710 each time the experiment is performed (a policy is sold and a year elapses). The expected value is the mean, a measure of central tendency— but not necessarily a possible value of x. ■

We found in Chapter 3 that the mean, and other measures of central tendency, tell only part of the story about a set of data. The same is true about probability distributions. We need to measure variability as well. Since a probability distribution can be viewed as a representation of a population, we will use the population variance to measure its variability.

The **population variance**, σ^2, is defined as the average squared distance of x from the population mean, μ. Since x is a random variable, the squared distance, $(x - \mu)^2$, is also a random variable. Applying the same logic used to find the mean value of x, we find the mean value of $(x - \mu)^2$ by multiplying all possible values of $(x - \mu)^2$ by $p(x)$ and then summing over all possible x values. * This quantity,

$$E[(x - \mu)^2] = \sum_{\text{All } x} (x - \mu)^2 p(x)$$

---

*It can be shown that $E[(x - \mu)^2] = E(x^2) - \mu^2$, where $E(x^2) = \sum_{\text{All } x} x^2 p(x)$. Note the similarity between this expression and the shortcut formula $\sum_{i=1}^{n} (x_i - \bar{x})^2 = \sum_{i=1}^{n} x_i^2 - (\Sigma x_i)^2 / n$ given in Chapter 3.

is also called the **expected value of the squared distance from the mean**; i.e., $\sigma^2 = E[(x - \mu)^2]$. The **standard deviation** of x is defined as the square root of the variance.

DEFINITION 5.6

The **variance** of a discrete random variable x is

$$\sigma^2 = E[(x - \mu)^2] = \sum_{\text{All } x} (x - \mu)^2 p(x)$$

EXAMPLE 5.3

Suppose you invest a fixed sum of money in each of five business ventures. Assume you know that 70% of such ventures are successful, the outcomes of the ventures are independent of one another, and the probability distribution for the number, x, of successful ventures out of five is:

x	0	1	2	3	4	5
$p(x)$	.002	.029	.132	.309	.360	.168

a. Find $\mu = E(x)$.
b. Find $\sigma = \sqrt{E[(x - \mu)^2]}$.
c. Graph $p(x)$. Locate μ and the interval $\mu \pm 2\sigma$ on the graph. Explain how μ and σ can be used to describe $p(x)$.

SOLUTION

a. Applying the formula, we obtain

$$\mu = E(x) = \sum_{\text{All } x} xp(x)$$

$$= 0(.002) + 1(.029) + 2(.132) + 3(.309) + 4(.360) + 5(.168)$$

$$= 3.50$$

b. Now we calculate the variance of x:

$$\sigma^2 = E[(x - \mu)^2] = \sum_{\text{All } x} (x - \mu)^2 p(x)$$

$$= (0 - 3.5)^2(.002) + (1 - 3.5)^2(.029) + (2 - 3.5)^2(.132)$$
$$+ (3 - 3.5)^2(.309) + (4 - 3.5)^2(.360) + (5 - 3.5)^2(.168)$$

$$= 1.05$$

Thus, the standard deviation is

$$\sigma = \sqrt{\sigma^2} = \sqrt{1.05} = 1.02$$

c. The graph of $p(x)$ is shown in Figure 5.4. Note that the mean μ and the interval $\mu \pm 2\sigma$ are shown on the graph. We can use μ and σ to describe the probability distribution in the same way that we used $\bar{x}$ and s to describe a relative frequency

FIGURE 5.4

Graph of p(x) for
Example 5.3

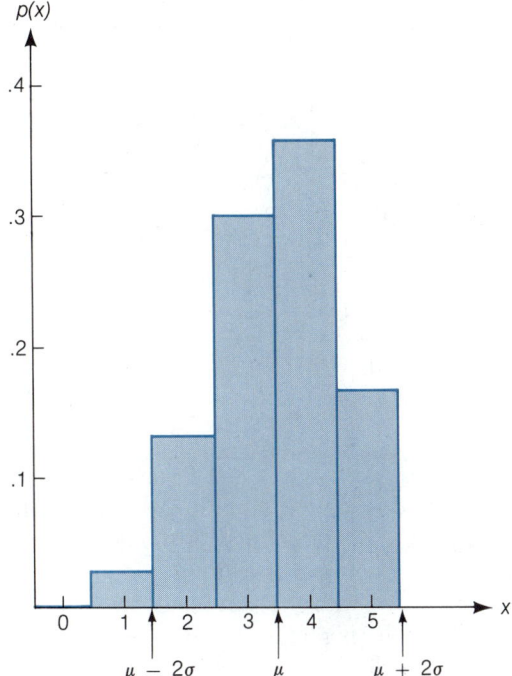

distribution in Chapter 3. Note in particular that $\mu = 3.5$ locates the probability distribution along the x-axis. If the investment is made in the five ventures, we expect to obtain a number x of successes near 3.5. Similarly, $\sigma = 1.02$ measures the spread of the probability distribution. Since this distribution is a theoretical relative frequency distribution that is moderately mound-shaped (see Figure 5.4), we expect (see Table 3.8) at least 75% and, more likely, near 95% of observed x values to fall in the interval $\mu \pm 2\sigma$—i.e., between 1.46 and 5.54. Compare this with the actual probability that x falls in the interval $\mu \pm 2\sigma$. From Figure 5.4 you can see that this probability includes the sum of $p(x)$ for all values of x except $p(0) = .002$ and $p(1) = .029$. Therefore, 96.9% of the probability distribution lies within 2 standard deviations of the mean. This percentage is consistent with Table 3.8. ∎

Investors—be they large corporations, banks, pension funds, mutual funds, or individuals—seldom hold a single financial asset; rather, they hold portfolios of financial assets. Thus, the investor should be less concerned with the rate of return achieved by, say, a particular stock in the portfolio (rate of return is defined in Case Study 3.2) than in the overall rate of return of the portfolio. Since the future rate of return of a portfolio is uncertain, a probability distribution can be used to characterize a portfolio's future rate of return. Two examples of such distributions are shown in Figure 5.5 (page 214). Alternatively, we might use just the mean and standard deviation of the probability distribution to characterize the future rate of

FIGURE 5.5
Rate of Return
Distributions

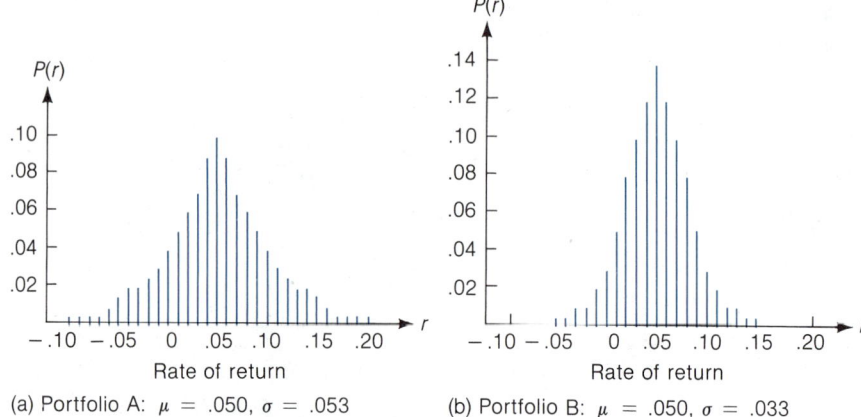

(a) Portfolio A: $\mu = .050$, $\sigma = .053$ (b) Portfolio B: $\mu = .050$, $\sigma = .033$

return. Notice that portfolios A and B have the same mean, but that A has the greater standard deviation. As a result, A has the higher probability of yielding a negative rate of return. Accordingly, it should not be surprising that the standard deviation of a portfolio's rate of return distribution is frequently used as a measure of the risk associated with the portfolio—the higher the standard deviation, the greater the risk of the portfolio (i.e., the greater the uncertainty of the portfolio's rate of return) and vice versa.

Typically, an investor can choose from among many different assets to form a portfolio. Or put another way, there are many different portfolios for the investor to choose among. But which portfolio should the investor select? This problem was addressed by Harry M. Markowitz (1952) in a classic article published in the *Journal of Finance*. Characterizing portfolios by the mean and standard deviation of their rates of return, Markowitz proposed a two-step procedure for choosing among portfolios. First, the set of all possible portfolios—the *feasible set*—must be reduced to an *efficient set* of portfolios. An efficient portfolio is one that provides the highest possible mean rate of return for any given degree of risk (i.e., any given standard deviation) or the lowest possible degree of risk for any given mean rate of return. Second, from the efficient set, the investor should choose the portfolio that best suits his or her needs.

The graph in Figure 5.6 shows the mean and standard deviation of the rate of return for various portfolios. Portfolios identified by a value of μ and σ that fall within the ellipse represent the feasible set of portfolios for a particular investor. The efficient set of portfolios is denoted by the boundary line *ABC* and is sometimes called the *efficient frontier*. Portfolios to the left of *ABC* are not obtainable because they fall outside the feasible set. Portfolios to the right of *ABC* are not efficient because there always exists a portfolio on the efficient frontier that could provide (1) a higher mean return for a given level of standard deviation of returns (compare points *B* and *D*), or (2) a lower standard deviation (lower degree of risk) for a given mean return (compare points *B* and *E*). For details on how to determine the efficient set of portfolios for an investor, see Elton and Gruber (1981) or Alexander and Francis (1986).

FIGURE 5.6
Efficient Set of Portfolios

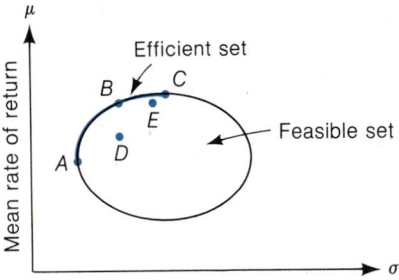

This case study illustrates that the probability distribution for the rate of return of a portfolio of financial assets provides a useful way to characterize the likelihood that the portfolio will yield an overall gain or loss. It also demonstrates that the process of selecting a portfolio of financial assets can be improved by comparing the means and standard deviations of the rates of return of alternative portfolios.

EXERCISES 5.20–5.34

LEARNING THE MECHANICS

5.20 Describe the differences in the meanings of the symbols μ, $\bar{x}$, and $E(x)$.

5.21 Consider the following probability distribution for the random variable x:

x	1	2	3	5	12
$p(x)$	.05	.15	.40	.30	.10

a. Find $E(x)$.
b. Graph $p(x)$ and locate $E(x)$ on the graph.
c. Interpret the value you obtained for $E(x)$.

5.22 Consider the following probability distribution for the random variable x:

x	−2	−1	0	1	2
$p(x)$	.05	.20	.20	.30	.25

a. Find μ.
b. Graph $p(x)$ and locate μ on the graph.
c. Interpret the value you obtained for μ.
d. In this case, can the random variable x ever assume the value μ? Explain.
e. In general, can a random variable ever assume a value equal to its expected value? Explain.

5.23 Consider the following probability distribution for the random variable x:

x	10	20	30	40	50	60
$p(x)$	.10	.25	.30	.20	.10	.05

a. Find μ, σ^2, and σ.
b. Graph $p(x)$.
c. Locate μ and the interval $\mu \pm 2\sigma$ on your graph. What is the probability that x will fall within the interval $\mu \pm 2\sigma$?

5.24 Consider the following probability distribution for the random variable x:

x	.1	.2	.3	.4	.5
$p(x)$	.05	.30	.35	.20	.10

a. Find μ, σ^2, and σ.
b. Graph $p(x)$.
c. Locate μ and the interval $\mu \pm \sigma$ on your graph. What is the probability that x will fall within the interval $\mu \pm \sigma$?
d. Locate the interval $\mu \pm 3\sigma$ on your graph. What is the probability that x falls within this interval?

APPLYING THE CONCEPTS

5.25 The economic risks taken by businesses can be classified as being either *pure risks* or *speculative risks*. A pure risk is faced when there is a chance of incurring an economic loss but no chance of gain. A speculative risk is faced when there is a chance of gain as well as a chance of loss. Risk is sometimes measured by computing the variance or standard deviation of the probability distribution that describes the potential gains or losses of the firm. This follows from the fact that the greater the variation in potential outcomes, the greater the uncertainty faced by the firm; the smaller the variation, the more predictable the firm's gains or losses (Williams and Heins, 1976). The two discrete probability distributions given in the table were developed from historical data. They describe the potential total physical damage losses next year to the fleets of delivery trucks of two different firms. Both firms have ten trucks, and both have the same expected loss next year.

FIRM A		FIRM B	
Loss next year	Probability	Loss next year	Probability
$ 0	.01	$ 0	.00
500	.01	200	.01
1,000	.01	700	.02
1,500	.02	1,200	.02
2,000	.35	1,700	.15
2,500	.30	2,200	.30
3,000	.25	2,700	.30
3,500	.02	3,200	.15
4,000	.01	3,700	.02
4,500	.01	4,200	.02
5,000	.01	4,700	.01

a. Verify that both firms have the same expected total physical damage loss.
b. Compute the standard deviation of both probability distributions, and determine which firm faces the greater risk of physical damage to its fleet next year.

c. Was part b concerned with measuring speculative risk or pure risk? Explain.

5.26 In 1965 the U.S. Weather Bureau (now called the National Weather Service) initiated a nationwide program in which precipitation probabilities were included in all public weather forecasts. This was the first time in any field of application that probabilities were issued on such a large scale. The program continues today. Precipitation forecasts indicate the likelihood of measurable precipitation ($\geq .01$ inch) at a specific point (the official rain gauge) during a given time period (Murphy and Winkler, 1984). Suppose that if a measurable amount of rain falls during the next 24 hours, a river will reach flood stage and a business will incur damages of $300,000. The National Weather Service has indicated that there is a 30% chance of rain during the next 24 hours.
a. Construct the probability distribution that describes the potential flood damages.
b. Find the firm's expected loss due to flood damage.

5.27 A stock market analyst believes that the probability of stock ABC increasing in price by the close of business tomorrow is .6, the probability of it decreasing in price is .2, and the probability of tomorrow's price remaining the same as today's is .2. Assume that when stock ABC's price changes, it does so by exactly $2.
a. Based on the analyst's assumptions, what is the expected change in price of ABC at the close of business tomorrow?
b. Can the change in ABC's price at the close of business tomorrow actually equal its expected value? Explain.

5.28 Banks and finance companies compete for personal loan customers. For purposes of product design and marketing, it is important for these financial institutions to understand the similarities and differences of the customers they attract. To further such an understanding, Robert W. Johnson and A. Charlene Sullivan (1981) sampled and questioned 488 bank borrowers and 87 finance company borrowers. One of the characteristics they investigated was the credit-worthiness of borrowers. They evaluated each borrower using a credit-scoring system of a large commercial bank. (Credit score is inversely related to credit risk.) The table summarizes the scores they obtained. The relative frequencies can be thought of as the approximate probabilities of a randomly selected borrower having the associated credit score.

CREDIT SCORE	RELATIVE FREQUENCY	
	Bank customers	Finance company customers
210	.109	.000
200	.117	.023
190	.109	.034
180	.113	.034
170	.219	.184
160	.102	.069
150	.102	.161
140	.074	.172
130	.035	.105
120	.020	.218

Source: Johnson, R. W. and Sullivan, A.C. "Segmentation of the consumer loan market." Reprinted from *Journal of Retail Banking*, Sept. 1981, Vol. III, No. 3, pp. 1–7. © 1981 Lafferty Publications, 3945 Holcomb Bridge Road, Suite 301, Norcross, Georgia 30092.

a. Find the expected credit score for bank borrowers. Find the expected credit score for finance company borrowers. Interpret both these values in the context of the problem.

b. Find the standard deviation for the credit scores of bank borrowers. For finance company borrowers.

c. Using the probability distributions in the table and your results from parts **a** and **b**, compare the credit-worthiness of bank borrowers and finance company borrowers.

5.29 Some managers believe frequent job changes will lead to more rapid advancement than if they remained at the same company. Other managers rarely leave their employers, changing no more than once every 7 years. An article on why and when managers move provided the accompanying data on the average time between moves for 1,191 managers. Assume the average time between moves, x, for each of the 12 time intervals is the midpoint of the interval (except for our estimated value of x for the last interval). These values are shown in the second column of the table. The percentages shown in the third column can be viewed as the approximate probabilities of the respective values of x.

AVERAGE TIME BETWEEN MOVES (Years)	x	MANAGERS WHO MOVE AT THIS RATE (%)
1–2	1.5	5.5
2–3	2.5	21.1
3–4	3.5	22.6
4–5	4.5	14.7
5–6	5.5	10.0
6–7	6.5	7.9
7–8	7.5	4.6
8–9	8.5	3.0
9–10	9.5	2.8
10–11	10.5	2.7
11–12	11.5	2.2
12 and over	15.5	2.9

Source: Veige, J. F. "Do managers on the move get anywhere?" *Harvard Business Review,* Mar.–Apr. 1981. Copyright © 1981 by the President and Fellows of Harvard College; all rights reserved.

a. Find the mean length of time managers stay on the job.

b. Find σ.

c. Find the approximate probability that the length of time a manager stays on the job falls within the interval $\mu \pm 2\sigma$.

5.30 To project the inventory required for a particular type of microwave oven, an appliance dealer analyzed the weekly number of sales over a long period of time. The appropriate probability distribution of the number, x, of sales per week is shown in the table.

x	0	1	2	3	4	5	6	7
$p(x)$	.01	.07	.18	.34	.24	.12	.03	.01

a. Find the expected number of sales per week.
b. Find σ^2 and σ.
c. Graph $p(x)$, and locate the interval $\mu \pm 2\sigma$ on the graph.
d. Find the probability that x falls in the interval $\mu \pm 2\sigma$.

PROFIT CONTRIBUTION	
x	$p(x)$
−$5,000[a]	.2
$10,000	.5
$30,000	.3

[a]A negative profit is a loss.

5.31 A company's marketing and accounting departments have determined that if the company markets its newly developed line of party favors, the probability distribution shown in the table will describe the contribution of the new line to the firm's profit during the next 6 months. The company has decided it should market the new line of party favors if the expected contribution to profit for the next 6 months is over $10,000. Based on the probability distribution, should the company market the new line?

5.32 Suppose you own a company that bonds financial managers. Based on past experience, you assess the probability that you will have to forfeit any particular bond to be .001. How much should you charge for a $1 million bond in order to break even on all such bonds?

5.33 A rock concert producer has scheduled an outdoor concert for Saturday, May 24. If it does not rain, the producer expects to make $20,000 profit from the concert. If it does rain, the producer will be forced to cancel the concert and will lose $12,000 (rock star's fee, advertising costs, stadium rental, administrative costs, etc.). The producer has learned from the National Weather Service that the probability of rain on May 24 is .4.
a. Find the producer's expected profit from the concert.
b. For a fee of $1,000, an insurance company has offered to insure the producer against all losses resulting from a rained-out concert. If the producer buys the insurance, what is her expected profit from the concert?
c. Assuming the National Weather Service's forecast is accurate, do you believe the insurance company has charged too much or too little for the policy? Explain.

5.34 One of the primary responsibilities of the personnel department of a firm is to maintain accurate and up-to-date professional and personal profiles of current employees and potential employees. Such information is utilized to match people to project assignments within the firm (Misshauk, 1979). A company is interested in hiring a person with an MBA degree and at least 2 years experience in a marketing department of a computer products firm. The company's personnel department has determined that it will cost the company $1,000 per job candidate to collect the required background information and to interview the candidate. As a result, it was decided to interview a maximum of three candidates. That is, the company will hire the first qualified person it finds, but will interview no more than three candidates. The company has received job applications from four persons who appear to be qualified but, unknown to the company, only one actually possesses the required background. Candidates to be interviewed will be randomly selected from the pool of four applicants.
a. Construct the probability distribution for the total cost to the firm of the interviewing strategy.
b. What is the probability that the firm's interviewing strategy will result in none of the four applicants being hired?
c. Calculate the mean of the probability distribution you constructed in part **a**.
d. What is the expected total cost of the interviewing strategy?

SECTION 5.4

**THE BINOMIAL
RANDOM VARIABLE**

A common source of business data is an opinion or preference survey. Many of these surveys result in dichotomous responses—i.e, responses that admit one of two possible alternatives, such as Yes–No. The number of Yes responses (or No responses) will usually have a **binomial probability distribution**. For example, suppose a random sample of consumers is selected from the totality of potential consumers of a particular product. The number of consumers in the sample who prefer the product to its competition is a random variable that has a binomial probability distribution.

All experiments that have the characteristics of the coin-tossing experiments of Chapter 4 and the preceding sections of this chapter yield **binomial random variables**. Imagine an experiment that is equivalent to tossing a coin n times. You are interested in observing the number of heads, x, in the n tosses. For such an experiment, x is a binomial random variable. In general, to decide whether a discrete random variable has a binomial probability distribution, check it against the characteristics listed in the box.

CHARACTERISTICS OF A BINOMIAL RANDOM VARIABLE

1. The experiment consists of n identical trials.
2. There are only two possible outcomes on each trial. We denote one outcome by S (for Success) and the other by F (for Failure).
3. The probability of S remains the same from trial to trial. This probability is denoted by p, and the probability of F is denoted by q. Note that $p + q = 1$.
4. The trials are independent.
5. The binomial random variable x is the number of S's in n trials.

EXAMPLE 5.4

For each of the following examples, decide whether x is a binomial random variable:

a. You randomly select three bonds out of a possible ten for an investment portfolio. Unknown to you, eight of the ten will maintain their present value, and the other two will lose value due to a change in their ratings. Let x be the number of the three bonds you select that lose value.

b. Before marketing a new product on a large scale, many companies conduct a consumer preference survey to determine whether the product is likely to be successful. Suppose a company develops a new diet soda and then conducts a taste-preference survey with 100 randomly chosen consumers stating their preference among the new soda and the two leading sellers. Let x be the number of the 100 who choose the new brand over the two others.

c. Some surveys are conducted using a method of sampling other than simple random sampling (defined in Chapter 4). For example, suppose a television cable company is trying to decide whether to establish a branch in a particular city. The company plans to conduct a survey to determine the fraction of households in the city that would use the cable television service. The sampling method is to choose a city block at random and then to survey every household on that block. This sampling technique is called **cluster sampling** and is discussed in Chapter 20. Suppose ten blocks are sampled in this manner, producing a total of 124 household responses. Let x be the number of the 124 households that would use the cable television service.

SOLUTION

a. In checking the binomial characteristics, a problem arises with independence (characteristic 4 in the box). Suppose the first bond you picked was one of the two that will lose value. This reduces the chance that the second bond you pick will lose value, since now only one of the nine remaining bonds are in that category. Thus, the choices you make are dependent, and therefore x, the number of the three bonds you select that lose value, is *not* a binomial random variable.

b. Surveys that produce dichotomous responses and use random sampling techniques are classic examples of binomial experiments. In our example, each randomly selected consumer either states a preference for the new diet soda or does not. The sample of 100 consumers is a very small proportion of the totality of potential consumers, so the response of one would be, for all practical purposes, independent of another. Thus, x is a binomial random variable.

c. This example is a survey with dichotomous responses (Yes or No to the cable service), but the sampling method is not simple random sampling. Again, the binomial characteristic of independent trials would probably not be satisfied. The responses of households within a particular block would almost surely be dependent, since households within a block tend to be similar with respect to income, race, and general interests. Thus, the binomial model would not be satisfactory for x if the cluster sampling technique was used. ∎

To see how to compute probabilities for binomial random variables, consider the following example.

EXAMPLE 5.5

A retail computer store sells IBM personal computers (PCs) and compatibles (personal computers that run all or most of the software that runs on IBM PCs, but that are not manufactured by IBM). Assume that 80% of the PCs that the store sells are IBM PCs, and 20% are compatibles.

a. Use the steps given in Chapter 4 (box on page 150) to find the probability that all of the next four PC purchases are compatibles.
b. Find the probability that three of the next four PC purchases are compatibles.
c. Let x represent the number of the next four PC purchases that are compatibles. Explain why x is a binomial random variable.
d. Use the answers to parts **a** and **b** to derive a formula for $p(x)$, the probability distribution of the binomial random variable, x.

SOLUTION

a. 1. The first step is to define the experiment. Here we are interested in observing the type of PC purchased by each of the next four (buying) customers: IBM (I) or compatible (C).

2. Next, we list the simple events associated with the experiment. Each simple event consists of the purchase decisions made by the four customers. For example, *IIII* represents the simple event that all four purchase IBM PCs, while *CIII* represents the simple event that customer 1 purchases a compatible, while customers 2, 3, and 4 purchase IBM PCs. The 16 simple events are listed in Table 5.2.

TABLE 5.2
Simple Events for
PC Experiment of
Example 5.5

IIII	*CIII*	*CCII*	*ICCC*	*CCCC*
	ICII	*CICI*	*CICC*	
	IICI	*CIIC*	*CCIC*	
	IIIC	*ICCI*	*CCCI*	
		ICIC		
		IICC		

3. We now assign probabilities to the simple events. Note that each simple event can be viewed as the intersection of four customers' decisions and, assuming the decisions are made independently, the probability of each simple event can be obtained as follows:

$$P(IIII) = P[(\text{customer 1 chooses IBM}) \cap (\text{customer 2 chooses IBM})$$
$$\cap (\text{customer 3 chooses IBM}) \cap (\text{customer 4 chooses IBM})]$$
$$= P(\text{customer 1 chooses IBM}) \times P(\text{customer 2 chooses IBM})$$
$$\times P(\text{customer 3 chooses IBM}) \times P(\text{customer 4 chooses IBM})$$
$$= (.8)(.8)(.8)(.8) = (.8)^4$$
$$= .4096$$

All other simple event probabilities are calculated using similar reasoning. For example,

$$P(CIII) = (.2)(.8)(.8)(.8) = (.2)(.8)^3$$
$$= .1024$$

You can check that this reasoning results in simple event probabilities that add to 1 over the 16 simple events in the sample space.

4. Finally, we add the appropriate simple event probabilities to obtain the desired event probability. The event of interest is that all four customers purchase compatibles. In Table 5.2 we find only one simple event, *CCCC*, contained in this event. All other simple events imply that at least one IBM is purchased. Thus,

$$P(\text{All four purchase compatibles}) = P(CCCC)$$
$$= (.2)^4 = .0016$$

That is, the probability is only 16 in 10,000 that all four customers purchase compatibles.

b. The event that three of the next four buyers purchase compatibles consists of the four simple events in the fourth column of Table 5.2: *ICCC, CICC, CCIC,* and *CCCI*. To obtain the event probability we add the simple event probabilities:

P(3 of next 4 customers purchase compatibles)

$= P(ICCC) + P(CICC) + P(CCIC) + P(CCCI)$

$= (.2)^3(.8) + (.2)^3(.8) + (.2)^3(.8) + (.2)^3(.8)$

$= 4(.2)^3(.8) = .0256$

Note that each of the four simple event probabilities is the same, because each simple event consists of three *C*s and one *I*; the order does not affect the probability because the customers' decisions are (assumed) independent.

c. We can characterize the experiment as consisting of four identical trials—the four customers' purchase decisions. There are two possible outcomes to each trial, *I* or *C*, and the probability of *C*, $p = .2$, is the same for each trial. Finally, we are assuming that each customer's purchase decision is independent of all others, so that the four trials are independent. Then it follows that *x*, the number of the next four purchases that are compatibles, is a binomial random variable.

d. The event probabilities in parts **a** and **b** provide insight into the formula for the probability distribution $p(x)$. First, consider the event that three purchases are compatibles (part **b**). We found that

$P(x = 3) = 4(.2)^3(.8)$

Recall that this probability is the sum of four simple event probabilities, each equal to $(.2)^3(.8)$. This line of reasoning leads to the following formula:

$P(x = 3) =$ (Number of simple events for which $x = 3$)

$\times (.2)^{\text{Number of compatibles purchased}}$

$\times (.8)^{\text{Number of IBMs purchased}}$

$= 4(.2)^3(.8)^1$

In general, we can use combinatorial mathematics to count the number of simple events. For example,

Number of simple events for which $x = 3$

$=$ Number of different ways of selecting 3 of the 4 trials for *C* purchases

$= \binom{4}{3} = \frac{4!}{3!(4-3)!} = \frac{4 \cdot 3 \cdot 2 \cdot 1}{(3 \cdot 2 \cdot 1) \cdot 1} = 4$

The formula that works for any value of *x* can be deduced as follows:

$P(x = 3) = \binom{4}{3}(.2)^3(.8)^1$

$= \binom{4}{x}(.2)^x(.8)^{4-x}$

The component $\binom{4}{x}$ counts the number of simple events with x compatibles, and the component $(.2)^x(.8)^{4-x}$ is the probability associated with each simple event having x compatibles.

For the general binomial experiment, with n trials and probability of Success p on each trial, the probability of x Successes is

$$p(x) = \binom{n}{x} \qquad p^x(1 - p)^{n-x}$$

$$\uparrow \qquad\qquad\qquad \uparrow$$

\# simple events   Probability of x S's
with x S's   and $(n - x)$ F's in
any simple event   ■

In theory, you could always resort to first principles to calculate binomial probabilities: list the simple events and sum their probabilities. However, as the number of trials (n) increases, the number of simple events grows very rapidly (the number of simple events is 2^n). Thus, we prefer the formula for calculating binomial probabilities, since its use avoids listing simple events.

The binomial probability distribution is summarized in the box.

THE BINOMIAL PROBABILITY DISTRIBUTION

$$p(x) = \binom{n}{x}p^x q^{n-x} \qquad (x = 0, 1, 2, \ldots, n)$$

where

p = Probability of a success on a single trial
q = $1 - p$
n = Number of trials
x = Number of successes in n trials

$$\binom{n}{x} = \frac{n!}{x!(n - x)!}$$

As noted in Chapter 4, the symbol 5! means $5 \cdot 4 \cdot 3 \cdot 2 \cdot 1 = 120$. Similarly, $n! = n(n - 1)(n - 2) \cdots \cdots 3 \cdot 2 \cdot 1$, and remember, $0! = 1$.

The mean, variance, and standard deviation for the binomial random variable x are shown in the box.

MEAN, VARIANCE, AND STANDARD DEVIATION FOR A BINOMIAL RANDOM VARIABLE

Mean: $\mu = np$
Variance: $\sigma^2 = npq$
Standard deviation: $\sigma = \sqrt{npq}$

As we demonstrated in Chapter 3, the mean and standard deviation provide measures of the central tendency and variability, respectively, of a distribution. Thus, we can use μ and σ to obtain a rough visualization of the probability distribution for x when the calculation of the probabilities is too tedious. To illustrate the use of the binomial probability distribution, consider Example 5.6.

EXAMPLE 5.6

A machine that produces stampings for automobile engines is malfunctioning and producing 10% defectives. The defective and nondefective stampings proceed from the machine in a random manner. If the next five stampings are tested, find the probability that three of them are defective.

SOLUTION

Let x equal the number of defectives in $n = 5$ trials. Then x is a binomial random variable with p, the probability that a single stamping will be defective, equal to .1, and $q = 1 - p = 1 - .1 = .9$. The probability distribution for x is given by the expression

$$p(x) = \binom{n}{x} p^x q^{n-x} = \binom{5}{x}(.1)^x(.9)^{5-x}$$

$$= \frac{5!}{x!(5-x)!}(.1)^x(.9)^{5-x} \qquad (x = 0, 1, 2, 3, 4, 5)$$

To find the probability of observing $x = 3$ defectives in a sample of $n = 5$, substitute $x = 3$ into the formula for $p(x)$ to obtain

$$p(3) = \frac{5!}{3!(5-3)!}(.1)^3(.9)^{5-3} = \frac{5!}{3!2!}(.1)^3(.9)^2$$

$$= \frac{5 \cdot 4 \cdot 3 \cdot 2 \cdot 1}{(3 \cdot 2 \cdot 1)(2 \cdot 1)}(.1)^3(.9)^2 = 10(.1)^3(.9)^2$$

$$= .0081$$

Note that the binomial formula tells us that there are 10 simple events having 3 defectives (check this by listing them), each with probability $(.1)^3(.9)^2$. ∎

EXAMPLE 5.7

Refer to Example 5.6 and find the values of $p(0)$, $p(1)$, $p(2)$, $p(4)$, and $p(5)$. Graph $p(x)$. Calculate the mean μ and standard deviation σ. Locate μ and the interval $\mu - 2\sigma$ to $\mu + 2\sigma$ on the graph. If the experiment were to be repeated many times, what proportion of the x observations would fall within the interval $\mu - 2\sigma$ to $\mu + 2\sigma$?

SOLUTION

Again, $n = 5$, $p = .1$, and $q = .9$. Then, substituting into the formula for $p(x)$:

$$p(0) = \frac{5!}{0!(5-0)!}(.1)^0(.9)^{5-0} = \frac{5 \cdot 4 \cdot 3 \cdot 2 \cdot 1}{(1)(5 \cdot 4 \cdot 3 \cdot 2 \cdot 1)}(1)(.9)^5$$

$$= .59049$$

$$p(1) = \frac{5!}{1!(5-1)!}(.1)^1(.9)^{5-1} = 5(.1)(.9)^4$$

$$= .32805$$

$$p(2) = \frac{5!}{2!(5-2)!}(.1)^2(.9)^{5-2} = (10)(.1)^2(.9)^3$$
$$= .07290$$

$$p(4) = \frac{5!}{4!(5-4)!}(.1)^4(.9)^{5-4} = 5(.1)^4(.9)$$
$$= .00045$$

$$p(5) = \frac{5!}{5!(5-5)!}(.1)^5(.9)^{5-5} = (.1)^5$$
$$= .00001$$

The graph of $p(x)$ is shown as a probability histogram in Figure 5.7 [$p(3)$ is taken from Example 5.6 to be .0081].

FIGURE 5.7
The Binomial Distribution:
$n = 5, p = .1$

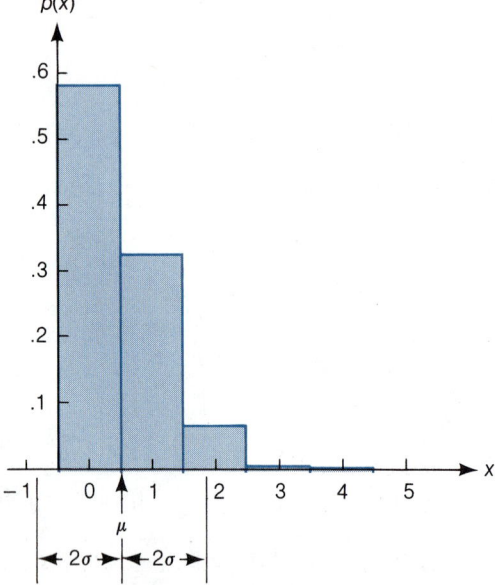

To calculate the values of μ and σ, substitute $n = 5$ and $p = .1$ into the following formulas:

$$\mu = np = (5)(.1) = .5 \qquad \sigma = \sqrt{npq} = \sqrt{(5)(.1)(.9)} = \sqrt{.45} = .67$$

To find the interval $\mu - 2\sigma$ to $\mu + 2\sigma$, we calculate

$$\mu - 2\sigma = .5 - 2(.67) = -.84 \qquad \mu + 2\sigma = .5 + 2(.67) = 1.84$$

If the experiment were to be repeated a large number of times, what proportion of the x observations would fall within the interval $\mu - 2\sigma$ to $\mu + 2\sigma$? You can see from Figure 5.7 that all observations equal to 0 or 1 will fall within the interval. The probabilities corresponding to these values are .5905 and .3280, respectively.

Consequently, you would expect .5905 + .3280 = .9185, or approximately 91.9%, of the observations to fall within the interval $\mu - 2\sigma$ to $\mu + 2\sigma$. This again emphasizes that for most probability distributions, observations rarely fall more than 2 standard deviations from μ.

■

CASE STUDY 5.3

THE SPACE SHUTTLE
CHALLENGER:
CATASTROPHE IN SPACE

On January 28, 1986, at 11:39.13 A.M., while travelling at Mach 1.92 at an altitude of 46,000 feet, the space shuttle *Challenger* was totally enveloped in an explosive burn that destroyed the shuttle and resulted in the deaths of all seven astronauts aboard. What happened? What was the cause of this catastrophe? This was the twenty-fifth shuttle mission. Each of the preceding 24 missions had been successful.

The report of the Presidential Commission assigned to investigate the accident concluded that the explosion was caused by the failure of the O-ring seal in the joint between the two lower segments of the right solid rocket booster. The seal is supposed to prevent superhot gases from leaking through the joint during the propellant burn of the booster rocket. The failure of the seal permitted a jet of white-hot gases to escape and to ignite the liquid fuel of the external fuel tank. The fuel tank fireburst destroyed the *Challenger*.

What were the chances of this event occurring? In a 1985 report, the National Aeronautics and Space Administration (NASA) claimed that the probability of such a failure was about $\frac{1}{60,000}$, or about once in every 60,000 flights. But a 1983 risk assessment study conducted for the Air Force assessed the probability of a shuttle catastrophe due to booster rocket "burn-through" to be $\frac{1}{35}$, or about once in every 35 missions.

If it is assumed that (1) p, the probability of shuttle catastrophe due to booster failure, remains the same from mission to mission, and (2) the performance of the booster rockets on one mission is independent of the performance of the boosters on other missions, then the number, x, of shuttle catastrophes due to booster failure in n missions can be treated as a binomial random variable. Accordingly, the probability that no disasters would have occurred during 25 missions is

$$P(x = 0) = \binom{25}{0} p^0 (1 - p)^{25-0} = \frac{25!}{0!25!} p^0 (1 - p)^{25}$$

$$= (1 - p)^{25}$$

If we use NASA's probability of shuttle catastrophe ($p = \frac{1}{60,000} = .0000167$), the probability of no catastrophes in 25 missions is approximately .9996. If we use the probability of catastrophe from the study prepared for the Air Force ($p = \frac{1}{35} = .02857$), the probability of no catastrophes in 25 missions is approximately .4845. Or, if we consider the complementary event that at least one catastrophe occurs in 25 missions, the chances are .0004, or about 4 in 10,000, given NASA's assumptions. On the other hand, the probability of at least one catastrophe under the Air Force's assumptions is .5155, or slightly more than 50–50. Given the events of January 28, 1986, which risk assessment—NASA's or the Air Force's—appears to be more appropriate? The probability of one or more disasters in 25 missions is so remote using NASA's assessment that it casts serious doubt on the risk assessment practices used by NASA prior to the *Challenger's* fatal mission (McKean, 1986; Biddle, 1986; Robinson, 1986; *Minneapolis Star and Tribune*, Feb. 11, 1986).

Calculating binomial probabilities becomes tedious when n is large. For some values of n and p the binomial probabilities have been tabulated in Table II of Appendix B. Part of Table II is shown in Table 5.3; a graph of the binomial probability distribution for $n = 10$ and $p = .10$ is shown in Figure 5.8. Table II actually contains a total of nine tables, labeled (**a**) through (**i**), one each corresponding to $n = 5, 6, 7, 8, 9, 10, 15, 20,$ and 25. In each of these tables the columns correspond to values of p, and the rows correspond to values of the random

TABLE 5.3   Reproduction of Part of Table II of Appendix B: Cumulative Binomial Probabilities for $n = 10$

f. n = 10

k \ p	.01	.05	.10	.20	.30	.40	.50	.60	.70	.80	.90	.95	.99
0	.904	.599	.349	.107	.028	.006	.001	.000	.000	.000	.000	.000	.000
1	.996	.914	.736	.376	.149	.046	.011	.002	.000	.000	.000	.000	.000
2	1.000	.988	.930	.678	.383	.167	.055	.012	.002	.000	.000	000	.000
3	1.000	.999	.987	.879	.650	.382	.172	.055	.011	.001	.000	.000	.000
4	1.000	1.000	.998	.967	.850	.633	.377	.166	.047	.006	.000	.000	.000
5	1.000	1.000	1.000	.994	.953	.834	.623	.367	.150	.033	.002	.000	.000
6	1.000	1.000	1.000	.999	.989	.945	.828	.618	.350	.121	.013	.001	.000
7	1.000	1.000	1.000	1.000	.998	.988	.945	.833	.617	.322	.070	.012	.000
8	1.000	1.000	1.000	1.000	1.000	.998	.989	.954	.851	.624	.264	.086	.004
9	1.000	1.000	1.000	1.000	1.000	1.000	.999	.994	.972	.893	.651	.401	.096

FIGURE 5.8   Binomial Probability Distribution for $n = 10$ and $p = .10$; $P(x \le 2)$ Shaded

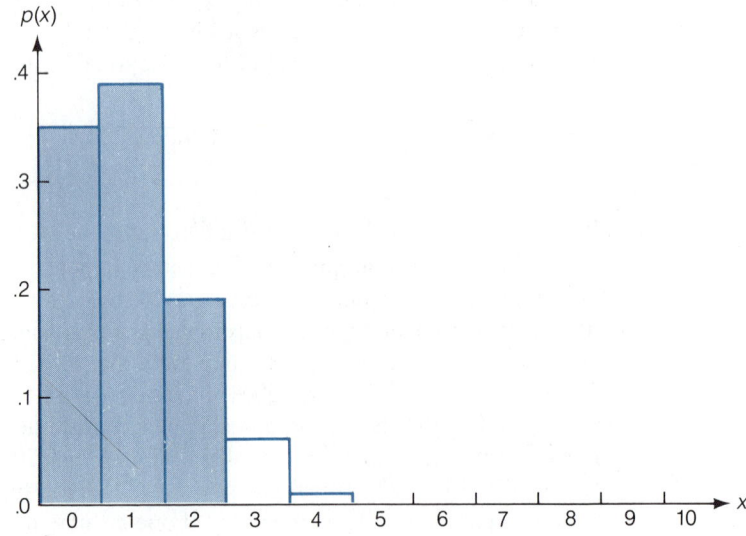

variable x. The entries in the table represent **cumulative** binomial probabilities. Thus, for example, the entry in the column corresponding to $p = .10$ and the row corresponding to $x = 2$ is .930 (shaded), and its interpretation is

$$P(x \leq 2) = P(x = 0) + P(x = 1) + P(x = 2) = .930$$

This probability is also shaded in the graphical representation of the binomial distribution with $n = 10$ and $p = .10$ in Figure 5.8.

You can also use Table II to find the probability that x equals a specific value. For example, suppose you want to find the probability that $x = 2$ in the binomial distribution with $n = 10$ and $p = .10$. This is found by subtraction as follows:

$$P(x = 2) = [P(x = 0) + P(x = 1) + P(x = 2)] - [P(x = 0) + P(x = 1)]$$
$$= P(x \leq 2) - P(x \leq 1)$$
$$= .930 - .736 = .194$$

The probability that a binomial random variable exceeds a specified value can be found using Table II and the notion of complementary events. For example, to find the probability that x exceeds 2 when $n = 10$ and $p = .10$, we use

$$P(x > 2) = 1 - P(x \leq 2) = 1 - .930 = .070$$

Note that this probability is represented by the unshaded portion of the graph in Figure 5.8.

All probabilities in Table II are rounded to three decimal places. Thus, although none of the binomial probabilities in the table is exactly zero, some are small enough (less than .0005) to round to .000. For example, using the formula to find $P(x = 0)$ when $n = 10$ and $p = .6$, we obtain

$$P(x = 0) = \binom{10}{0}(.6)^0(.4)^{10-0} = .4^{10} = .00010486$$

but this is rounded to .000 in Table II of Appendix B (see Table 5.3).

Similarly, none of the table entries is exactly 1.0, but when the cumulative probabilities exceed .9995, they are rounded to 1.000. The row corresponding to the largest possible value for x, $x = n$, is omitted, because all the cumulative probabilities in that row are equal to 1.0 (exactly). For example, in Table 5.3 with $n = 10$, $P(x \leq 10) = 1.0$, no matter what the value of p.

The following example further illustrates the use of Table II.

EXAMPLE 5.8

Suppose a poll of 20 employees is taken in a large company. The purpose is to determine x, the number who favor unionization. Suppose that 60% of all the company's employees favor unionization.

a. Find the mean and standard deviation of x.
b. Use Table II of Appendix B to find the probability that $x < 10$.
c. Use Table II to find the probability that $x > 12$.
d. Use Table II to find the probability that $x = 11$.

SOLUTION

a. The number of employees polled is presumably small compared with the total number of employees in this company. Thus, we may treat x, the number of the 20 who favor unionization, as a binomial random variable. The value of p is the fraction of the total employees who favor unionization; i.e., $p = .6$. Therefore, we calculate the mean and variance:

$$\mu = np = 20(.6) = 12 \qquad \sigma^2 = npq = 20(.6)(.4) = 4.8$$

The standard deviation is then

$$\sigma = \sqrt{4.8} = 2.19$$

b. The tabulated value is

$$P(x \le 9) = .128$$

c. To find the probability

$$P(x > 12) = \sum_{x=13}^{20} p(x)$$

we use the fact that for all probability distributions, $\sum_{\text{All } x} p(x) = 1$. Therefore,

$$P(x > 12) = 1 - P(x \le 12)$$
$$= 1 - \sum_{x=0}^{12} p(x)$$

Consulting Table II, we find the entry in row $k = 12$, column $p = .6$ to be .584. Thus,

$$P(x > 12) = 1 - .584 = .416$$

d. To find the probability that exactly 11 employees favor unionization, recall that the entries in Table II are cumulative probabilities and use the relationship

$$P(x = 11) = [p(0) + p(1) + \cdots + p(10) + p(11)]$$
$$- [p(0) + p(1) + \cdots + p(9) + p(10)]$$
$$= P(x \le 11) - P(x \le 10)$$

Then

$$P(x = 11) = .404 - .245 = .159$$

The probability distribution for x in this example is shown in Figure 5.9. Note that the interval $\mu \pm 2\sigma$ is (7.6, 16.4). ∎

CASE STUDY 5.4

EVALUATING CUSTOMER RESPONSE TO A NEW SALES PROGRAM

Arthur A. Brown, Frank T. Hulswit, and John D. Kettelle (1956) were asked by a large firm to study, and perhaps determine reasons for and solutions to, its lack of growth over the prior 5 years. In their article, Brown et al. refer to the firm as "Penstock Press, a large commercial printing company."

The primary concern of the study was Penstock's sales operations. Accordingly, Brown et al. conducted an experiment to study the sales effectiveness of Penstock's salespeople. The salespeople were instructed to increase their sales efforts toward

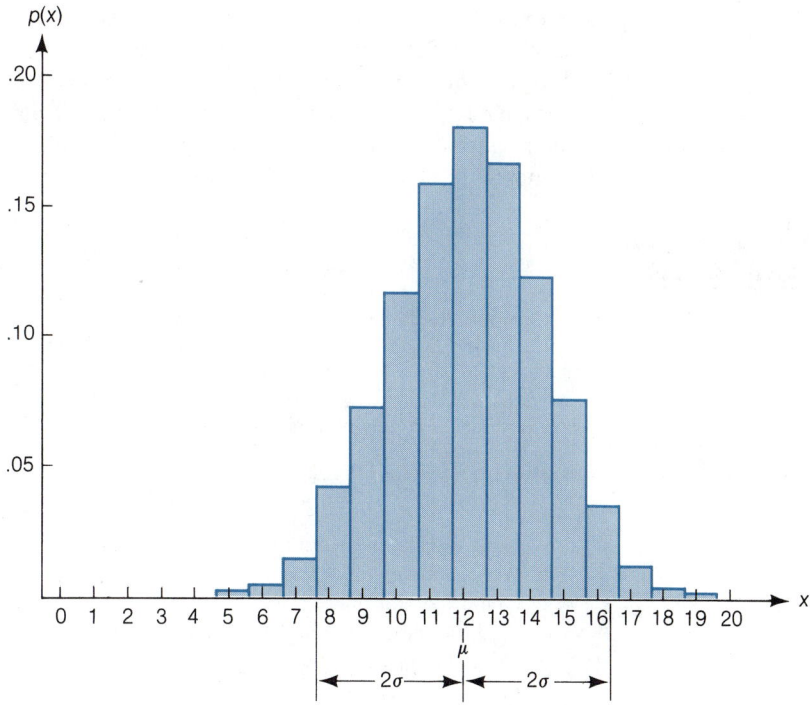

all of Penstock's customers, but in particular toward 60 of the larger customers, for a 4-month experimental period. At the end of the 4-month period it was determined that the probability of a customer making a genuinely positive response to the increased sales effort merely by chance was .25. Of Penstock's 60 large customers, it was noted that 24 made what appeared to be genuinely positive responses. But before concluding that the increased sales effort toward the 60 large customers had paid off, Brown et al. felt it was important to determine how likely it would be for 24 or more of the 60 customers to make positive responses merely by chance. Assuming that the probability of a positive response occurring by chance (.25) is the same for each of the 60 customers and that the response of one customer does not affect that of another, the number of positive responses observed has a binomial probability distribution. Accordingly, the probability of 24 or more positive responses from the 60 customers can be determined as follows:

$$P(x \geq 24) = \sum_{x=24}^{60} \binom{60}{x}.25^x(.75)^{60-x} = .004$$

(Due to the large value of *n*, the number of customers in the experiment, it would be unrealistic to try to compute the above probability by hand. If you had access to binomial tables for *n* = 60 and *p* = .25, the probability could be found using the tables. Otherwise, it would be necessary to use a computer or an approximation such as the one we will discuss in Section 6.5.) The fact that the probability of observing 24 or more genuinely positive responses from the 60 customers merely by chance is only .004 indicates that in actually observing 24 such responses either a very rare event has occurred or the responses were in fact genuine and the increased sales effort did influence the increase in sales.

Brown et al. (1956) concluded that the number of positive responses could not be explained by chance, but that they in fact "implied deliberate continuing business from the customers," i.e., genuine responses. The authors noted that this conclusion was supported by Penstock's salespeople.

·| | | | | | | | | | | | |

EXERCISES 5.35–5.53

LEARNING THE MECHANICS

5.35 Compute the following:

a. $\dfrac{5!}{3!(5-3)!}$ b. $\dbinom{6}{3}$ c. $\dbinom{8}{0}$ d. $\dbinom{5}{5}$ e. $\dbinom{6}{1}$

5.36 Consider the following probability distribution:

$$p(x) = \binom{6}{x}(.4)^x(.6)^{6-x} \qquad (x = 0, 1, 2, \ldots, 6)$$

a. Is x a discrete or a continuous random variable? Explain.
b. What is the name of this probability distribution?
c. Graph the probability distribution.
d. Find the mean and standard deviation of x.
e. Show the mean and the 2 standard deviation interval on each side of the mean on the graph you drew in part **c**.

5.37 Suppose x is a binomial random variable with $n = 5$ and $p = .2$.
a. Use the formula for the binomial probability distribution to calculate the values of $p(x)$ for $x = 0, 1, 2, 3, 4, 5$.
b. Graph $p(x)$.

5.38 Given that x is a binomial random variable, compute $p(x)$ for each of the following cases:
a. $n = 6, \quad x = 3, \quad p = .3$
b. $n = 4, \quad x = 2, \quad q = .6$
c. $n = 2, \quad x = 0, \quad p = .8$

5.39 Suppose x is a binomial random variable with $n = 11$ and $p = .5$.
a. Display $p(x)$ in tabular form.
b. Compute the mean and variance of x.
c. Graph $p(x)$ and locate $E(x)$ and the interval $\mu \pm 2\sigma$ on the graph.
d. What is the probability that x falls within the interval $\mu \pm 2\sigma$?

5.40 Use the results of Exercise 5.39 to find the following probabilities:
a. $P(x \le 3)$ b. $P(x \le 5)$ c. $P(x < 2)$

5.41 Given that x is a binomial random variable with $n = 15$ and $p = .4$, use Table II of Appendix B to find the following probabilities:
a. $P(x \le 1)$ b. $P(x \ge 3)$ c. $P(x \le 5)$
d. $P(x < 10)$ e. $P(x > 10)$ f. $P(x = 6)$

5.42 The binomial probability distribution is a family of probability distributions, with each single distribution depending on the values of n and p. Assume that x is a binomial random variable with $n = 4$.

a. Determine the value of p such that the probability distribution of x is symmetric.

b. Determine a value of p such that the probability distribution of x is skewed to the right.

c. Determine a value of p such that the probability distribution of x is skewed to the left.

d. Graph each of the binomial distributions you obtained in parts a, b, and c. Locate the mean for each distribution on its graph.

e. In general, for what values of p will a binomial distribution be symmetric? Skewed to the right? Skewed to the left?

APPLYING THE CONCEPTS

5.43 Your firm's accountant believes that 10% of the company's invoices contain arithmetic errors. To check this theory, the accountant randomly samples 25 invoices and finds that 7 contain errors. What is the probability that of the 25 invoices written, 7 or more would contain errors if the accountant's theory was valid? What assumptions do you have to make to solve this problem using the methodology of this chapter?

5.44 According to the Internal Revenue Service (IRS), the chances of your tax return being audited are about 2 in 100 if your income is less than $50,000; they increase to about 8 in 100 if your income is $50,000 or more (*Minneapolis Star and Tribune*, Feb. 19, 1986, p. 1M).

a. What is the probability that a taxpayer with income less than $50,000 will be audited by the IRS? With income $50,000 or more?

b. If five taxpayers with incomes under $50,000 are randomly selected, what is the probability that exactly one will be audited? That more than one will be audited?

c. Repeat part b assuming that five taxpayers with incomes of $50,000 or more are randomly selected.

d. If two taxpayers with incomes under $50,000 are randomly selected and two with incomes more than $50,000 are randomly selected, what is the probability that none of these taxpayers will be audited by the IRS?

e. What assumptions did you have to make in order to answer these questions using the methodology presented in this section?

5.45 A particular system in a space vehicle must work properly in order for the space ship to reenter Earth's atmosphere. One particular component of the system operates successfully only 85% of the time. To increase the reliability of the system, four of the components will be installed in such a way that the system will operate successfully if at least one component is working successfully. What is the probability that the system will fail? Assume that the components operate independently.

5.46 An automobile manufacturer has determined that 30% of all gas tanks that were installed on its 1988 compact model are defective.

a. If 15 of the cars are recalled, what is the probability that more than 10 of the 15 will need new gas tanks?

b. If 10,000 of the cars are recalled, what is the probability that fewer than 3,000 will need new gas tanks? Set up the solution but do not perform the calculations. [*Note:* In Section 6.5 we discuss a procedure that can be used to obtain an approximate answer to this question without having to perform the tedious calculations required by the binomial distribution.] Calculate the mean and standard deviation of the distribution, and give an "intuitive" approximation to this probability.

c. In answering parts a and b, what assumptions must be made?

5.47 According to the "January" theory, if the stock market is up in January, it will be up for the whole year (and vice versa). Believe it or not, this indicator of stock market behavior has been correct for 29 of the last 34 years and 100% correct in odd-numbered years ("Heard on the Street," *Wall Street Journal*, Feb. 1, 1984). Suppose there is no truth whatever in this theory and that stock prices are just as likely to move up or down in any given year, regardless of the direction of movement in January.

   a. Find the probability of perfect agreement between the January and annual movements in stock prices over a period of 15 years.

   b. What is the probability of perfect agreement between the January and annual movements in stock prices in at least 10 of 15 years?

5.48 A problem of considerable economic impact on the economy is the burgeoning cost of Medicare and other public-funded medical services. One aspect of this problem concerns the high percentage of people seeking medical treatment who, in fact, have no physical basis for their ailments. One conservative estimate is that the percentage of people who seek medical assistance and who have no real physical ailment is 10%, and some doctors believe that it may be as high as 40%. Suppose we were to randomly sample the records of a doctor and found that 5 of 15 patients seeking medical assistance were physically healthy.

   a. What is the probability of observing 5 or more physically healthy patients in a sample of 15 if the proportion, p, that the doctor normally sees is 10%?

   b. What is the probability of observing 5 or more physically healthy patients in a sample of 15 if the proportion, p, that the doctor normally sees is 40%?

   c. Why might your answer to part **a** make you believe that p is larger than .10?

5.49 According to the U.S. Golf Association (USGA), "The weight of the [golf] ball shall not be greater than 1.620 ounces avoirdupois, and the size not less than 1.680 inches in diameter. The velocity of the ball shall be not greater than 250 feet per second" (USGA, 1982). The USGA periodically checks the specifications of golf balls sold in the United States by randomly sampling balls from pro shops around the country. Two dozen of each kind are sampled, and if more than three do not meet size and/or velocity requirements, that kind of ball is removed from the USGA's approved-ball list (*Golf World*, Sept. 10, 1982).

   a. What assumptions must be made and what information must be known in order to use the binomial probability distribution to calculate the probability that the USGA will remove a particular kind of golf ball from its approved-ball list?

   b. Suppose 10% of all balls produced by a particular manufacturer are less than 1.680 inches in diameter, and assume that the number of such balls, x, in a sample of two dozen balls can be adequately characterized by a binomial probability distribution. Find the mean and standard deviation of the binomial distribution.

   c. Refer to part **b**. If x has a binomial distribution, then so does the number, y, of balls in the sample that meet the USGA's minimum diameter. [*Note:* $x + y = 24$.] Describe the distribution of y. In particular, what are p, q, and n? Also, find $E(y)$ and the standard deviation of y.

5.50 Suppose you are a purchasing officer for a company. You have purchased 50,000 electrical switches and have been guaranteed by the supplier that the shipment will contain no more than .1% defectives. To check the shipment, you randomly sample 500 switches, test them, and find that four are defective. Assuming the supplier's claim is true, compute μ and σ for the number of defectives in a sample of 500 switches. If the supplier's claim is true, is it likely that you would have found four defective switches in the sample? Based on this sample, what inference would you make concerning the supplier's guarantee?

5.51 Many firms utilize sampling plans to control the quality of manufactured items ready for shipment or the quality of items that have been purchased. To illustrate the use of a sampling plan, suppose you are shipping electrical fuses in lots, each containing 10,000 fuses. The plan specifies that you will randomly sample 25 fuses from each lot and accept (and ship) the lot if the number of defective fuses, x, in the sample is less than 3. If $x \geq 3$, you will reject the lot and hold it for a complete reinspection. What is the probability of accepting a lot ($x = 0$, 1, or 2) if the actual fraction defective in the lot is:

a. 1   **b.** .8   **c.** .5   **d.** .2   **e.** .05   **f.** 0

Construct a graph showing $P(A)$, the probability of lot acceptance, as a function of the lot fraction defective, p. This graph is called the **operating characteristic curve** for the sampling plan.

5.52 Refer to Exercise 5.51. Suppose the sampling plan called for sampling $n = 25$ fuses and accepting a lot of $x \leq 3$. Calculate the quantities specified in Exercise 5.51, and construct the operating characteristic curve for this sampling plan. Compare this curve with the curve obtained in Exercise 5.51. (Note how the curve characterizes the ability of the plan to screen bad lots from shipment.)

5.53 After a costly study, a market analyst claims that 12% of all consumers in a particular sales region prefer a certain noncarbonated beverage. To check the validity of this figure, you decide to conduct a survey in the region. You randomly sample $n = 400$ customers and find that $x = 31$ prefer the beverage. Compute μ and σ for the random variable x. Based on a sample of 400, is it likely that you would observe a value of $x \leq 31$ if the market analyst's claim was true? Explain. Do the results of your survey agree with the 12% estimate given by the market analyst?

| | | | | | | | | | | | |

SECTION 5.5

THE POISSON RANDOM VARIABLE (OPTIONAL)

A type of probability distribution useful in describing the number of events that will occur in a specific period of time or in a specific area or volume is the **Poisson distribution** (named after the eighteenth-century physicist and mathematician, Siméon Poisson). The following are typical examples of random variables for which the Poisson probability distribution provides a good model:

1. The number of industrial accidents in a given manufacturing plant per month observed by a plant safety supervisor
2. The number of noticeable surface defects (scratches, dents, etc.) found by quality inspectors on a new automobile (or any manufactured product)
3. The parts per million of some toxicant found in water or air emission from a manufacturing plant (a random variable of great interest to both the business community and the Environmental Protection Agency)
4. The number of arithmetic errors per 100 invoices (or per 1,000 invoices, etc.) in the accounting records of a company
5. The number of customer arrivals per unit time at a service counter (a service station, a hospital clinic, a supermarket checkout counter, etc.)
6. The number of death claims per day received by an insurance company
7. The number of breakdowns of an electronic computer per month

The characteristics of the Poisson random variable are usually difficult to verify for practical examples. The previous examples satisfy them well enough that the Poisson distribution provides a good model in many instances. As with all probability models, the real test of the adequacy of the Poisson model is whether it provides a reasonable approximation to reality—that is, whether empirical data support it.

CHARACTERISTICS OF A POISSON RANDOM VARIABLE

1. The experiment consists of counting the number of times a particular event occurs during a given unit of time or in a given area or volume (or weight, distance, or any other unit of measurement).
2. The probability that an event occurs in a given unit of time, area, or volume is the same for all the units.
3. The number of events that occur in one unit of time, area, or volume is independent of the number that occur in other units.
4. The mean (or expected) number of events in each unit will be denoted by the Greek letter lambda, λ.

The Poisson probability distribution also provides a good approximation to a binomial probability distribution with mean

$$\lambda = np$$

when n is large and p is small (say, $np \leq 7$). (See Exercise 5.59.)

The probability distribution, mean, and variance for a Poisson random variable are shown in the box.

PROBABILITY DISTRIBUTION, MEAN, AND VARIANCE FOR A POISSON RANDOM VARIABLE

$$p(x) = \frac{\lambda^x e^{-\lambda}}{x!} \qquad (x = 0, 1, 2, \ldots)$$

$$\mu = \lambda \qquad \sigma^2 = \lambda$$

where

λ = Mean number of events during a given time period, or over a specific area or volume

$e = 2.71828 \ldots$

The calculation of Poisson probabilities is made easier by the use of Table III in Appendix B. There the cumulative probabilities $P(x \leq k)$ for various values of λ are given. The use of Table III is illustrated in Example 5.9.

EXAMPLE 5.9

Suppose the number, x, of a company's employees who are absent on Mondays has (approximately) a Poisson probability distribution. Furthermore, assume that the average number of Monday absentees is 2.6.

a. Find the mean and standard deviation of x, the number of employees absent on Monday.
b. Use Table III of Appendix B to find the probability that fewer than two employees are absent on a given Monday.
c. Use Table III to find the probability that more than five employees are absent on a given Monday.
d. Use Table III to find the probability that exactly five employees are absent on a given Monday.

SOLUTION

a. The mean and variance of a Poisson random variable are both equal to λ. Thus, for this example,

$$\mu = \lambda = 2.6 \qquad \sigma^2 = \lambda = 2.6$$

Then the standard deviation of x is

$$\sigma = \sqrt{2.6} = 1.61$$

Remember that the mean measures the central tendency of the distribution, and does not necessarily equal a possible value of x. In this example, the mean is 2.6 absences, and although there cannot be 2.6 absences on a given Monday, the average number of Monday absences is 2.6. Similarly, the standard deviation of 1.61 measures the variability of the number of Monday absences. Perhaps a more helpful measure is the interval $\mu \pm 2\sigma$, which in this case stretches from $-.62$ to 5.82. We expect the number of absences to fall in this interval most of the time—with at least 75% relative frequency, and probably with more than 90% relative frequency. The mean and the 2 standard deviation interval around it are shown in Figure 5.10.

FIGURE 5.10
Probability Distribution for Number of Monday Absences

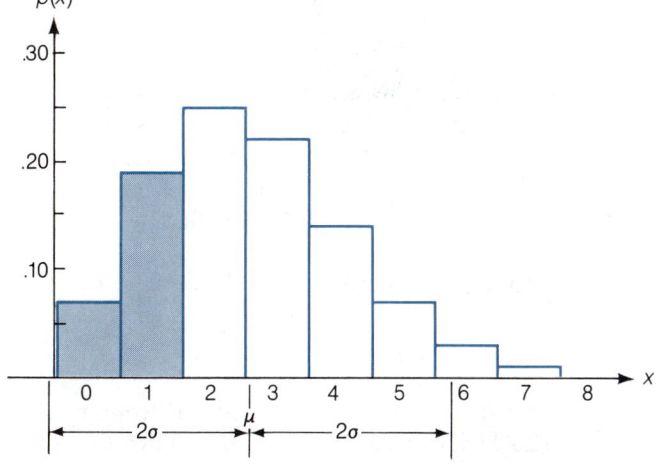

b. A partial reproduction of Table III is shown in Table 5.4. The rows of the table correspond to different values of λ, and the columns correspond to different values of the Poisson random variable x. The entries in the table are cumulative probabilities (much like the binomial probabilities in Table II).

TABLE 5.4

Reproduction of Part of Table III of Appendix B

λ \ x	0	1	2	3	4	5	6	7	8	9
2.2	.111	.335	.623	.819	.928	.975	.993	.998	1.000	
2.4	.091	.308	.570	.779	.904	.964	.988	.997	.999	1.000
2.6	.074	.267	.518	.736	.877	.951	.983	.995	.999	1.000
2.8	.061	.231	.469	.692	.848	.935	.976	.992	.998	.999
3.0	.050	.199	.423	.647	.815	.916	.966	.988	.996	.999
3.2	.041	.171	.380	.603	.781	.895	.955	.983	.994	.998
3.4	.033	.147	.340	.558	.744	.871	.942	.977	.992	.997
3.6	.027	.126	.303	.515	.706	.844	.927	.969	.988	.996
3.8	.022	.107	.269	.473	.668	.816	.909	.960	.984	.994
4.0	.018	.092	.238	.433	.629	.785	.889	.949	.979	.992
4.2	.015	.078	.210	.395	.590	.753	.867	.936	.972	.989
4.4	.012	.066	.185	.359	.551	.720	.844	.921	.964	.985
4.6	.010	.056	.163	.326	.513	.686	.818	.905	.955	.980
4.8	.008	.048	.143	.294	.476	.651	.791	.887	.944	.975
5.0	.007	.040	.125	.265	.440	.616	.762	.867	.932	.968
5.2	.006	.034	.109	.238	.406	.581	.732	.845	.918	.960
5.4	.005	.029	.095	.213	.373	.546	.702	.822	.903	.951
5.6	.004	.024	.082	.191	.342	.512	.670	.797	.886	.941
5.8	.003	.021	.072	.170	.313	.478	.638	.771	.867	.929
6.0	.002	.017	.062	.151	.285	.446	.606	.744	.847	.916

To find the probability that fewer than two employees are absent on a given Monday, we first note that

$$P(x < 2) = P(x \le 1)$$

This probability is a cumulative probability, and is therefore the entry in Table III in the row corresponding to $\lambda = 2.6$ and the column corresponding to $x = 1$. The entry is .267, shown shaded in Table 5.4. This probability corresponds to the shaded area in Figure 5.10, and may be interpreted as meaning that there is a 26.7% chance that fewer than two employees will be absent on a given Monday.

c. To find the probability that more than five employees are absent on a given Monday, we consider the complementary event:

$$P(x > 5) = 1 - P(x \le 5) = 1 - .951 = .049$$

where .951 is the entry in Table III corresponding to $\lambda = 2.6$ and $x = 5$ (see Table 5.4). Note from Figure 5.10 that this is the area in the interval $\mu \pm 2\sigma$, or $-.62$ to 5.82. Then the number of absences should exceed 5 or, equivalently, should be more than 2 standard deviations from the mean, on only about 4.9% of all Mondays. Note that this percentage agrees remarkably well with that given by the Empirical Rule for mound-shaped distributions, which tells us to expect approximately 5% of the measurements (values of the random variable) to lie further than 2 standard deviations from the mean.

d. To use Table III to find the probability that *exactly* five employees are absent on a given Monday, we must write the probability as the difference between two cumulative probabilities:

$$P(x = 5) = P(x \le 5) - P(x \le 4) = .951 - .877$$
$$= .074 \qquad \blacksquare$$

Note that probabilities in Table III are all rounded to three decimal places. Thus, although in theory a Poisson random variable can assume infinitely large values, the values of x in Table III are extended only until the cumulative probability is 1.000. This does not mean that x *cannot* assume larger values, but only that the likelihood is less than $.001$ (in fact, less than $.0005$) that it will do so.

Finally, you may need to calculate Poisson probabilities for values of λ not found in Table III. You may be able to obtain an adequate approximation by interpolation, but if not, consult more extensive tables for the Poisson distribution (see the references at the end of this chapter).

EXERCISES 5.54–5.68

LEARNING THE MECHANICS

5.54 Consider the following probability distribution:

$$p(x) = \frac{2^x e^{-2}}{x!} \qquad (x = 0, 1, 2, \ldots)$$

a. Is x a discrete or continuous random variable? Explain.
b. What is the name of this probability distribution?
c. Graph the probability distribution.
d. Find the mean and standard deviation of x.
e. Find the mean and standard deviation of the probability distribution.

5.55 Given that x is a random variable for which a Poisson probability distribution provides a good approximation, use Table III of Appendix B to compute the following:
a. $P(x \le 2)$ when $\lambda = 1$ **b.** $P(x \le 2)$ when $\lambda = 2$
c. $P(x \le 2)$ when $\lambda = 3$
d. What happens to the probability of the event $\{x \le 2\}$ as λ increases from 1 to 3? Is this intuitively reasonable?

5.56 Assume that x is a random variable having a Poisson probability distribution with a mean of 1.5. Use Table III of Appendix B to find the following probabilities:
a. $P(x \le 2)$ **b.** $P(x \ge 2)$ **c.** $P(x = 2)$
d. $P(x = 0)$ **e.** $P(x > 0)$ **f.** $P(x > 5)$

5.57 Given that x is a random variable for which a Poisson probability distribution with $\lambda = 2$ provides a good characterization:
a. Graph $p(x)$ for $x = 0, 1, 2, \ldots, 9, 10$.
b. Find μ and σ for x, and locate μ and the interval $\mu \pm 2\sigma$ on the graph.
c. What is the probability that x will fall within the interval $\mu \pm 2\sigma$?

5.58 Given that x is a random variable for which a Poisson probability distribution with $\lambda = 4$ provides a good characterization:
a. Graph $p(x)$ for $x = 0, 1, 2, \ldots, 9, 10$.
b. Find μ and σ for x, and locate μ and the interval $\mu \pm 2\sigma$ on the graph.
c. What is the probability that x will fall within the interval $\mu \pm 2\sigma$?

5.59 When n is large and p is small (say $np \leq 7$), the Poisson probability distribution provides a good approximation to the binomial probability distribution. You can investigate the adequacy of this approximation for small values of n by using the exact binomial probabilities provided in Table II of Appendix B. Use Table II to find $p(0)$, $p(1)$, and $p(2)$ for a binomial random variable with $n = 25$ and $p = .05$. Calculate the corresponding Poisson approximations using $\lambda = \mu = np$. (Note that these approximations are reasonably good for n as small as 25. We prefer to use the approximation only for much larger n—say, $n \geq 100$.)

APPLYING THE CONCEPTS

5.60 The Federal Deposit Insurance Corporation (FDIC), established in 1933, insures deposits of up to $100,000 in banks that are members of the Federal Reserve System (and others that voluntarily join the insurance fund) against losses due to bank failure or theft. From 1977 through 1981 the average number of bank failures per year among insured banks was (approximately) 8.5 (*Statistical Abstract of the United States: 1986*, p. 496). Assume that x, the number of bank failures per year among insured banks, can be adequately characterized by a Poisson probability distribution with mean 8.5.
a. Find the expected value and standard deviation of x.
b. In 1980, ten insured banks failed. Find $P(x = 10)$.
c. How far (in standard deviations) does $x = 10$ lie above the mean of the Poisson distribution? That is, find the z-score for $x = 10$.
d. In 1978, seven insured banks failed. Find $P(x \leq 7)$.

5.61 The Environmental Protection Agency (EPA) was established in 1970 as part of the executive branch of the federal government. Its mission is to abate and control pollution in the areas of air, water, solid waste, pesticides, radiation, and toxic substances. The EPA issues pollution standards that vitally affect the safety of consumers and the operations of industry (*The United States Government Manual 1985–1986*). For example, the EPA states that manufacturers of vinyl chloride and similar compounds must limit the amount of these chemicals in plant air emissions to no more than 10 parts per million. Suppose the mean emission of vinyl chloride for a particular plant is 4 parts per million. Assume that the number of parts per million of vinyl chloride in air samples, x, follows a Poisson probability distribution.
a. What is the standard deviation of x for the plant?
b. Is it likely that a sample of air from the plant would yield a value of x that would exceed the EPA limit? Explain.

5.62 The National Transportation Safety Board is responsible for investigating aviation acci-dents. In 1982, 210.1 billion passenger-miles were flown by commercial airlines in the United States. During this period there were 235 fatalities. In 1984, 243.1 billion passenger-miles were flown and there were four fatalities. Based on data from 1979 to 1984, it is known that the average number of fatalities per 100 million passenger-miles flown is approximately .048 (*Statistical Abstract of the United States: 1986*, p. 617). Assuming that airlines fly approximately 18 billion passenger-miles per month, the mean number of fatalities for a 1-month period is 8.64 (180 times the mean per 100 million miles). Suppose the probability distribution for x, the number of fatalities per month, can be approximated by a Poisson probability distribution.

a. What is the probability that no fatalities will occur during any given month? [*Hint:* Either use Table III of Appendix B and interpolate to approximate the probability, or use a calculator or computer to calculate the probability exactly.]

b. Find $E(x)$ and the standard deviation of x.

c. Use your answers to part **b** to describe the probability that as many as 20 fatalities will occur in any given month.

5.63 The safety supervisor at a large manufacturing plant believes the expected number of industrial accidents per month to be 3.4. What is the probability of exactly two accidents occurring next month? Three or more? What assumptions do you need to make to solve this problem using the methodology of this chapter?

5.64 As a check on the quality of the wooden doors produced by a company, its owner requested that each door undergo inspection for defects before leaving the plant. The plant's quality control inspector found that 1 square foot of door surface contains, on the average, .5 minor flaw. Subsequently, 1 square foot of each door's surface was examined for flaws. The owner decided to have all doors reworked that were found to have two or more minor flaws in the square foot of surface that was inspected. What is the probability that a door will fail inspection and be sent back for reworking? What is the probability that a door will pass inspection?

5.65 A can company reports that the number of breakdowns per 8-hour shift on its machine-operated assembly line follows a Poisson distribution with a mean of 1.5. What is the probability of exactly two breakdowns during the midnight shift? What is the probability of fewer than two breakdowns during the afternoon shift? Of no breakdowns during three consecutive 8-hour shifts? (Assume the machine operates independently across shifts.)

5.66 The random variable x, the number of people who arrive at a cashier's counter in a bank during a specified period of time, often possesses (approximately) a Poisson prob-ability distribution. If the mean arrival rate, λ, is known, the Poisson probability dis-tribution can be used to aid in the design of the customer service facility. Suppose you estimate that the mean number of arrivals per minute for cashier service at a bank is one person per minute. What is the probability that in a given minute, the number of arrivals will equal three or more? Can you tell the bank manager that the number of arrivals will rarely exceed three per minute?

5.67 The probability that a health insurance company must pay a major medical claim for a policy is .001. If a group of 1,000 policyholders represents a random sample of all possible policyholders, what is the probability that the insurance company will have to pay at least one major medical claim in this sample? [*Hint:* See Exercise 5.59.]

5.68 The Department of Commerce in a particular state has determined that the number of small businesses that declare bankruptcy per month has approximately a Poisson distribution with a mean equal to 6.4.
  a. Find the probability of at least five bankruptcies occurring next month.
  b. Find the probability of exactly four bankruptcies occurring next month.

S E C T I O N   5.6

THE HYPERGEOMETRIC RANDOM VARIABLE (OPTIONAL)

The **hypergeometric probability distribution** provides a realistic model for some types of enumerative (count) data. The characteristics of the hypergeometric distribution are listed in the box.

CHARACTERISTICS OF A HYPERGEOMETRIC RANDOM VARIABLE

1. The experiment consists of randomly drawing n elements without replacement from a set of N elements, r of which are S's (for Success) and $(N - r)$ of which are F's (for Failure).

2. The hypergeometric random variable x is the number of S's in the draw of n elements.

Note that both the hypergeometric and binomial characteristics stipulate that each draw or trial results in one of two outcomes. The basic difference between these random variables is that the hypergeometric trials are dependent, while the binomial trials are independent. The draws are dependent because the probability of drawing an S (or an F) is dependent on what occurred on preceding draws.

To illustrate the dependence between trials, we note that the probability of drawing an S on the first draw is r/N. Then, the probability of drawing an S on the second draw depends on the outcome of the first. It will be either $(r - 1)/(N - 1)$ or $r/(N - 1)$, depending on whether the first draw was an S or an F. Consequently, the results of the draws represent dependent events.

For example, suppose we define x as the number of women hired in a random selection of three applicants from a total of six men and four women. This random variable satisfies the characteristics of a hypergeometric random variable with $N = 10$ and $n = 3$. The possible outcomes on each trial are either selection of a female (S) or selection of a male (F). Another example of a hypergeometric random variable is the number, x, of defective television picture tubes in a random selection of $n = 4$ from a shipment of $N = 8$ tubes. And, as a third example, suppose $n = 5$ stocks are randomly selected from a list of $N = 15$ stocks. Then, the number x of the five selected companies that pay regular dividends to stockholders is a hypergeometric random variable.

EXAMPLE 5.10

Suppose, as we mentioned earlier, an employer randomly selects three new employees from a total of ten applicants, six men and four women. Let x be the number of women who are hired.

a. Find the mean and standard deviation of x.
b. Find the probability that no women are hired.

PROBABILITY DISTRIBUTION, MEAN, AND VARIANCE
OF THE HYPERGEOMETRIC RANDOM VARIABLE

$$p(x) = \frac{\binom{r}{x}\binom{N-r}{n-x}}{\binom{N}{n}}$$

$$[x = \text{Maximum}[0, n - (N - r)], \ldots, \text{Minimum}(r, n)]$$

$$\mu = \frac{nr}{N} \qquad \sigma^2 = \frac{r(N-r)n(N-n)}{N^2(N-1)}$$

where

$N = $ Total number of elements

$r = $ Number of S's in the N elements

$n = $ Number of elements drawn

$x = $ Number of S's drawn in the n elements

SOLUTION

a. Since x is a hypergeometric random variable with $N = 10$, $n = 3$, and $r = 4$, the mean and variance are

$$\mu = \frac{nr}{N} = \frac{(3)(4)}{10} = 1.2$$

$$\sigma^2 = \frac{r(N-r)n(N-n)}{N^2(N-1)} = \frac{4(10-4)3(10-3)}{(10)^2(10-1)}$$

$$= \frac{(4)(6)(3)(7)}{(100)(9)} = .56$$

The standard deviation is

$$\sigma = \sqrt{.56} = .75$$

b. The probability that no women are hired by the employer, assuming the selection is truly random, is

$$P(x = 0) = p(0) = \frac{\binom{4}{0}\binom{10-4}{3-0}}{\binom{10}{3}}$$

$$= \frac{\dfrac{4!}{0!(4-0)!}\dfrac{6!}{3!(6-3)!}}{\dfrac{10!}{3!(10-3)!}} = \frac{(1)(20)}{120} = \frac{1}{6}$$

The entire probability distribution for x is shown in Figure 5.11. The mean $\mu = 1.2$ and the interval $\mu \pm 2\sigma = (-.3, 2.7)$ are indicated. You can see that if this random variable were to be observed over and over again a large number of times, most of the values of x would fall within the interval $\mu \pm 2\sigma$.

FIGURE 5.11
Probability Distribution for
x in Example 5.10

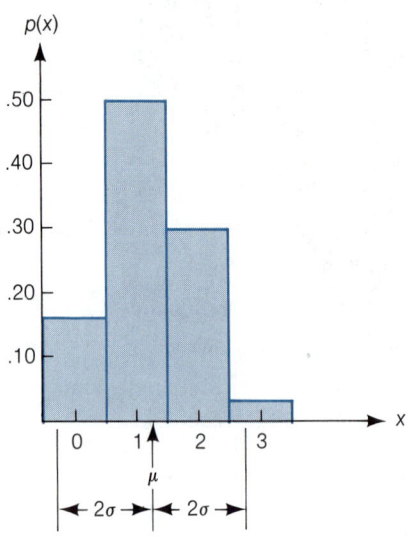

EXERCISES 5.69–5.83

LEARNING THE MECHANICS

5.69 Explain the difference between sampling with replacement and sampling without replacement.

5.70 How do binomial and hypergeometric random variables differ? In what respects are they similar?

5.71 Consider the following probability distribution:

$$p(x) = \frac{\binom{7}{x}\binom{3}{4 - x}}{\binom{10}{4}} \qquad (x = 1, 2, 3, 4)$$

a. Is x a discrete or continuous random variable? Explain.
b. What is the name of this probability distribution?
c. Graph the probability distribution.
d. Find the mean and standard deviation of x.
e. Find the mean and standard deviation of the probability distribution.

5.72 Given that x is a hypergeometric random variable with $N = 8$, $n = 3$, and $r = 5$, compute the following:
a. $P(x = 1)$ **b.** $P(x = 0)$ **c.** $P(x = 3)$ **d.** $P(x \geq 4)$

5.73 Given that x is a hypergeometric random variable with $N = 10$, $n = 5$, and $r = 7$:
  a. Display the probability distribution for x in tabular form.
  b. Compute the mean and variance of x.
  c. Graph $p(x)$ and locate μ and the interval $\mu \pm 2\sigma$ on the graph.
  d. What is the probability that x will fall within the interval $\mu \pm 2\sigma$?

5.74 Given that x is a hypergeometric random variable with $N = 12$, $n = 8$, and $r = 6$:
  a. Display the probability distribution for x in tabular form.
  b. Compute μ and σ for x.
  c. Graph $p(x)$ and locate μ and the interval $\mu \pm 2\sigma$ on the graph.
  d. What is the probability that x will fall within the interval $\mu \pm 2\sigma$?

5.75 Use the results of Exercise 5.74 to find the following probabilities:
  a. $P(x = 1)$ **b.** $P(x = 4)$ **c.** $P(x \le 4)$
  d. $P(x \ge 5)$ **e.** $P(x < 3)$ **f.** $P(x \ge 8)$

5.76 Suppose you plan to sample 10 items from a population of 100 items and would like to determine the probability of observing 4 defective items in the sample. Which probability distribution should you use to compute this probability under the following conditions? Justify your answers.
  a. The sample is drawn without replacement.
  b. The sample is drawn with replacement.

APPLYING THE CONCEPTS

5.77 Suppose you are purchasing cases of wine (12 bottles per case) and that periodically, you select a test case to determine the adequacy of the sealing process. To do this, you randomly select and test three bottles in the case. If a case contains one spoiled bottle of wine, what is the probability it will appear in your sample?

5.78 A television commercial for a particular brand of sugarless chewing gum (call it brand T) claims that "three out of four dentists who recommend sugarless gum to their patients recommend brand T." In order to test this claim, a competitor questioned eight dentists randomly selected from a group of 20 dentists who were known to recommend sugarless gum to their patients. Of the eight dentists, two preferred brand T.
  a. What is the probability of obtaining two or fewer recommendations for brand T in the sample of eight if in fact three out of four dentists in the original group of 20 recommend brand T?
  b. Does the sample result strengthen or weaken the gum manufacturer's claim? Explain.

5.79 If you are purchasing small lots of a manufactured product and it is very costly to test a single item, it may be desirable to test a sample of items from the lot rather than every item in the lot. Such a sampling plan would be based on a hypergeometric probability distribution. For example, suppose each lot contains ten items. You decide to sample four items per lot and reject the lot if you observe one or more defectives. If the lot contains one defective item, what is the probability that you will accept the lot? What is the probability that you will accept the lot if it contains two defective items? Three? Four?

5.80 Construct an operating characteristic curve for the sampling plan in Exercise 5.79 by plotting the probability of lot acceptance, $P(A)$, versus the lot fraction defective. (Use the probabilities of lot acceptance calculated in Exercise 5.79.)

5.81 Refer to Exercises 5.79 and 5.80. Suppose we were to change the *acceptance number* for the sampling plan to 2 (i.e., we will accept the lot if the number, x, of defectives in the sample is $x \le 2$). Construct an operating characteristic curve for this sampling plan, and compare it with the operating characteristic curve in Exercise 5.80. Which sampling plan is more likely to detect lots containing defectives? Explain how this is apparent in the comparison of the operating characteristic curves.

5.82 A marketing manager wants to fill the vacancies for four district managers from among the company's existing sales personnel. Twelve salespersons are judged to be suitable for the district managerial positions, eight men and four women. If the marketing manager randomly selects four persons from this group of 12, find the probability distribution for the number, x, of women selected for the district managerial positions. Present the probability distribution in both graphical and tabular form.

5.83 A curious event was recently described in the *Minneapolis Star and Tribune* (May 27, 1983, p. 1). The Minneapolis Community Development Agency (MCDA) makes home improvement grants each year to homeowners in depressed neighborhoods within the city. Of the $708,000 granted in 1983, $233,000 was awarded by the city council using a "random selection" of 140 homeowners' applications from among a total of 743 applications—601 from the north side and 142 from the south side of Minneapolis. Oddly, all 140 grants awarded were from the north side, clearly a highly improbable outcome if, in fact, the 140 winners were randomly selected from among the 743 applicants.

    a. Suppose the 140 winning applications were randomly selected from among the total of 743, and let x equal the number in the sample from the north side. Find the mean and standard deviation of x.

    b. Use the results of part **a** to support a contention that the grant winners were not randomly selected.

SECTION 5.7

THE GEOMETRIC RANDOM VARIABLE (OPTIONAL)

Another common discrete random variable that has many business applications is the **geometric random variable**. Like the binomial random variable, it arises naturally from a discussion of a coin-tossing experiment (whether the coin is balanced or unbalanced). But instead of tossing the coin a fixed number of times and observing the number, x, of heads, we toss the coin and count the number, x, of tosses until the first head appears. Like the binomial experiment, we assume that the tosses are independent of each other. The geometric random variable has the characteristics listed in the box.

CHARACTERISTICS OF THE GEOMETRIC RANDOM VARIABLE

1. The experiment consists of a sequence of independent trials.
2. Each trial results in one of two outcomes. We denote one of them by S and the other by F.
3. The probability of S remains the same from trial to trial. We will denote this probability by p.
4. The geometric random variable x is defined to be the number of trials until the first S is observed.

The probability distribution for the geometric random variable provides a good model for the length of time a customer must wait for some type of servicing. For this application, the time must be measured in whole units (minutes, hours, etc.). Then, x is equal to the number of time units a customer must wait until being served.

PROBABILITY DISTRIBUTION, MEAN, AND VARIANCE OF A GEOMETRIC RANDOM VARIABLE

$$p(x) = q^{x-1}p \qquad (x = 1, 2, 3, 4, \ldots)$$

$$\mu = \frac{1}{p} \qquad \sigma^2 = \frac{q}{p^2}$$

where

p = Probability of an S outcome

$q = 1 - p$

x = Number of trials until the first S is observed

The number of job applicants interviewed by an employer until the first suitable prospect is found is another discrete random variable that might be modeled by a geometric probability distribution. Or, for a sequence of independent drillings for oil, x could represent the number of drillings until the first successful well is hit.

EXAMPLE 5.11

Let x be the number of days until the closing price of a certain stock shows a gain over the previous day's closing price. Assume that x is a geometric random variable, with p, the probability of a gain in price from one day to the next, equal to .5.

a. Find the mean and standard deviation of x.
b. Find the probability that more than 2 days pass before a gain in price from one day to the next is observed.

SOLUTION

a. The mean and variance for this geometric random variable are

$$\mu = \frac{1}{p} = \frac{1}{.5} = 2 \qquad \sigma^2 = \frac{q}{p^2} = \frac{.5}{(.5)(.5)} = 2$$

Then the standard deviation is

$$\sigma = \sqrt{\sigma^2} = \sqrt{2} = 1.41$$

b. To find the probability that more than 2 days pass before a gain in price is observed, we must find

$$P(x > 2) = p(3) + p(4) + p(5) + \cdots$$

Since the sum is never-ending, we use the complementary relationship:

$$P(x > 2) = 1 - P(x \le 2)$$
$$= 1 - [p(1) + p(2)]$$

Now,

$$p(1) = q^{1-1}p = (.5)^0(.5) = .5$$
$$p(2) = q^{2-1}p = (.5)^1(.5) = .25$$

Thus,

$$P(x > 2) = 1 - (.5 + .25) = .25$$

There is a .25 probability that more than 2 days will pass before the stock shows a gain in its closing price from one day to the next.

The probability distribution for x is shown in Figure 5.12. The expected value of x and the interval $\mu \pm 2\sigma$ are indicated. Note that the geometric probabilities will always decrease as x increases.

FIGURE 5.12
Probability Distribution for
x in Example 5.11

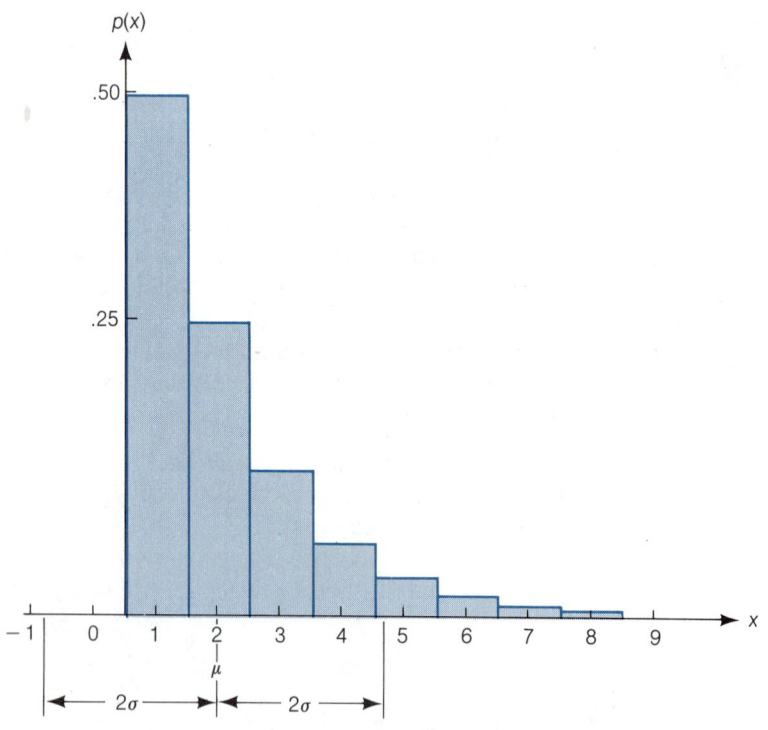

EXERCISES 5.84–5.93

LEARNING THE MECHANICS

5.84 Consider the following probability distribution:

$$p(x) = (.9)^{x-1}(.1) \qquad (x = 1, 2, 3, \ldots)$$

a. Is x a discrete or continuous random variable? Explain.
b. What is the name of this probability distribution?
c. Graph the probability distribution.

d. Find the mean and standard deviation of x.
e. Find the mean and standard deviation of the probability distribution.

5.85 Given that x is a geometric random variable with $p = .2$, compute the following:
   a. $P(x = 2)$ b. $P(x = 3)$ c. $P(x \leq 3)$
   d. $P(x \geq 1)$ e. $P(x = 4)$ f. $P(x > 2)$

5.86 Given that x is a geometric random variable with $p = .7$:
   a. Graph $p(x)$, $x = 1, 2, \ldots$.
   b. Compute the mean and variance of x. Locate μ and the interval $\mu \pm 2\sigma$ on the graph.
   c. What is the probability that x will fall within the interval $\mu \pm 2\sigma$?

5.87 Given that x is a geometric random variable with $p = .3$:
   a. Graph $p(x)$, $x = 1, 2, \ldots$.
   b. Compute the expected value and variance of x. Locate μ and the interval $\mu \pm \sigma$ on your graph.
   c. What is the probability that x will fall within the interval $\mu \pm \sigma$?

APPLYING THE CONCEPTS

5.88 In 1982, the median sale price of existing one-family homes sold to buyers in the United States was $67,800. By 1984, the median price had risen to $72,400. Suppose a large number of the 1984 buyers are going to be randomly selected and interviewed. Let x be the number of interviews conducted until a home buyer who paid more than $72,400 is found.
   a. What is the probability that $x = 4$?
   b. What is the probability that $x \leq 4$?
   c. Find the expected value and standard deviation of x.
   d. Using your answers to part c, describe the likelihood that x will be as large as 7.

5.89 The explosion of the space shuttle *Challenger* on January 28, 1986, occurred on the shuttle program's twenty-fifth mission. Recall from Case Study 5.3 that it had been determined in a risk assessment study conducted for the Air Force that the probability of a catastrophic accident involving the solid-fuel booster rockets of the shuttle was $\frac{1}{35}$. Let x be the number of shuttle missions until the first catastrophe occurs.
   a. Under what conditions would x be a geometric random variable? In the remainder of this exercise assume that these conditions are satisfied.
   b. What is the probability that the first catastrophic accident involving the boosters would occur on the fifth shuttle mission? The twenty-fifth mission? (If you do not have a calculator or computer available to do the necessary calculations for the second question, simply set up the solution.)
   c. What is the probability that the first catastrophe would occur before the twenty-sixth mission? Set up the solution but do not perform the calculations.
   d. Find the mean and standard deviation for x. Use your answers to estimate an upper bound before which the first catastrophe is likely to occur.

5.90 A company that produces food products has determined that the probability that a new cold breakfast cereal obtains more than a 3% market share in its first year on the market is .09. If the company markets one new cold cereal every other year, what is the probability that it will have to wait more than 6 years before marketing a cereal that acquires a market share of more than 3% in its first year?

5.91 The manufacturer of a price-reading optical scanner claims that the probability of it misreading the price of a product (i.e., misreading the bar code on a product's label) is .001. At the time one of the scanners was installed in a supermarket, the store manager tested the performance of the scanner.

a. If the manufacturer's claim is true, what is the probability that the scanner would not misread a price until after the fifth price was read?

b. If in fact the third price was misread, what inference can be made about the manufacturer's claim? Explain.

c. What assumptions must you make in arriving at the solutions for parts **a** and **b**?

5.92 Ten percent of the light bulbs produced by a company are defective. If an inspector tests light bulbs randomly selected from the production line, what is the probability that the first defective bulb will be observed on or after the fourth test? What is the expected number of tests the inspector will have to perform before the first defective bulb is observed?

5.93 An oil company has determined that the probability of striking oil on any particular drilling is .2. Accordingly, what is the probability that it would drill four dry wells before striking oil on the fifth drilling?

| | | | | | | | | | | | | |

S U M M A R Y

In the business world, observations taken on discrete random variables (those that can assume a countable number of values) often have the characteristics of a **binomial, Poisson, hypergeometric,** or **geometric random variable.** In this chapter, we gave the identifying characteristics for each of these random variables, indicated some business data for which the probability models would be appropriate, and gave the formulas for their probability distributions, means, and variances.

Using the probability distribution for a random variable, we were able to calculate the probabilities of specific sample observations. When the probabilities were difficult to calculate, the means and standard deviations provided numerical descriptive measures that enabled us to visualize the probability distributions and thereby to make some approximate probability statements about sample observations.

| | | | | | | | | | | | |

SUPPLEMENTARY EXERCISES 5.94–5.127

[Note: Starred (*) exercises refer to optional sections in this chapter.]

*5.94 Identify the type of random variable—binomial, Poisson, hypergeometric, or geometric—described by each of the following probability distributions:

a. $p(x) = \dfrac{(.5)^x e^{-.5}}{x!}$ $(x = 0, 1, 2, \ldots)$

b. $p(x) = (.4)^{x-1}(.6)$ $(x = 1, 2, 3, \ldots)$

c. $p(x) = \dfrac{\binom{10}{x}\binom{15}{5-x}}{\binom{25}{5}}$ $(x = 0, 1, 2, 3, 4, 5)$

d. $p(x) = \binom{6}{x}(.2)^x(.8)^{6-x}$ $(x = 0, 1, 2, \ldots, 6)$

e. $p(x) = \dfrac{10!}{x!(10-x)!}(.9)^x(.1)^{10-x}$ $(x = 0, 1, 2, \ldots, 10)$

5.95 Given that x is a binomial random variable, compute $p(x)$ for each of the following cases:

a. $n = 4$, $x = 2$, $p = .2$ **b.** $n = 5$, $x = 4$, $p = .4$
c. $n = 3$, $x = 0$, $p = .5$

***5.96** Given that x is a hypergeometric random variable, compute $p(x)$ for each of the following cases:

a. $N = 8$, $n = 5$, $r = 3$, $x = 2$
b. $N = 6$, $n = 2$, $r = 2$, $x = 2$
c. $N = 5$, $n = 4$, $r = 4$, $x = 3$

***5.97** Given that x is a geometric random variable, compute $p(x)$ for each of the following cases:

a. $p = .3$, $x = 3$ **b.** $p = .6$, $x = 4$ **c.** $p = .8$, $x = 2$

***5.98** Given that x is a Poisson random variable, compute $p(x)$ for each of the following cases:

a. $\lambda = 3$, $x = 2$ **b.** $\lambda = 2$, $x = 3$ **c.** $\lambda = .5$, $x = 3$

***5.99** Given that x is a random variable for which a Poisson probability distribution with $\lambda = 4$ provides a good characterization, compute the following:

a. $P(x = 0)$ **b.** $P(x = 3)$ **c.** $P(x = 1)$
d. $P(x = 5)$ **e.** $P(x \le 2)$ **f.** $P(x \ge 2)$

***5.100** Given that x is a geometric random variable with $p = .4$:

a. Graph $p(x)$ for $x = 1, 2, \ldots, 7$.
b. Compute μ and σ for $p(x)$. Locate μ and the interval $\mu \pm 2\sigma$ on the graph.
c. What is the probability that x will fall within the interval $\mu \pm 2\sigma$?

5.101 Which of the following describe discrete random variables, and which describe continuous random variables?

a. The number of damaged inventory items
b. The average monthly sales revenue generated by a salesperson over the past year
c. The number of square feet of warehouse space a company rents
d. The length of time a firm must wait before its copying machine is fixed

5.102 Variables, x, y, and z are three discrete random variables with the same mean and the same range. Their probability distributions are shown in the figures. Which has the largest variance? The next largest? The smallest? Explain.

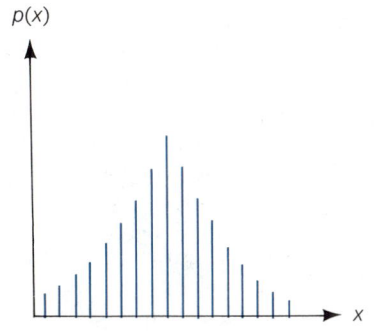

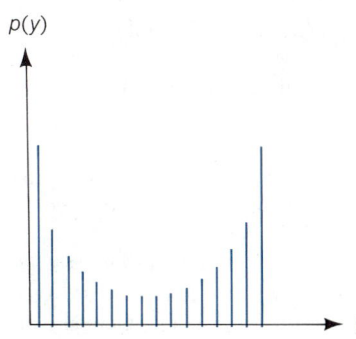

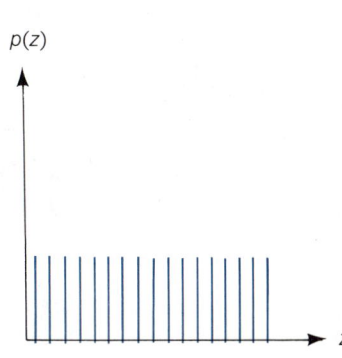

5.103 It is known that for each of two stocks it is equally likely for the stocks to increase, remain the same, or decrease in price by the close of business tomorrow. If only these two stocks are observed and x represents the number of stocks that increase in price by the close of business tomorrow, find the probability distribution for x and display it in tabular form. (Assume the stocks are independent.)

5.104 A shipment of 200 circuit boards contains 20 defectives. One board is randomly selected from the shipment, examined for defects, and placed back in the crate with the rest of the shipment. This sampling procedure is repeated four more times. (This type of sampling is referred to as sampling with replacement.) Let x be the number of defective boards found in the sample of size 5.

    a. Determine the probability distribution for x and display it graphically.

    b. Compute the mean and variance for x. Locate the mean on your graph of part **a**.

5.105 As part of a study of how pricing decisions are made by fabricare firms (the dry cleaning and laundry industry), Joe F. Goetz, Jr. (1985) contacted and questioned 103 fabricare firms. The accompanying table describes the sales volumes of these firms. Assume that the sales volume, x, for each of the six intervals can be adequately approximated by the midpoint of the interval. These midpoints are shown in the second column of the table.

SALES VOLUME	x	RELATIVE FREQUENCY
Over $0 to $75,000	$ 37,500	.14
Over $75,000 to $150,000	112,500	.30
Over $150,000 to $300,000	225,000	.20
Over $300,000 to $500,000	400,000	.17
Over $500,000 to $1,000,000	750,000	.10
Over $1,000,000 to $5,000,000	3,000,000	.09

Source: Goetz (1985), p. 63.

One of the 103 firms is to be randomly selected for more intensive questioning. In this situation the relative frequencies in the table can be interpreted as probabilities that describe the approximate likelihood of the sampled firm having sales volume x.

    a. Find $E(x)$.

    b. Find $E[(x - \mu)^2]$.

    c. Interpret the values you obtained in parts **a** and **b** in the context of the problem.

5.106 The Environmental Protection Agency (EPA) tested 842 in-use automobiles to determine whether there were differences between the cars' actual gas mileage and the mileage projected in the EPA's mileage guide. For highway driving, they found that only 68% of the cars tested had fuel economies within 2 miles per gallon of the mileage guide's projection (*Environmental News*, 1978). Assume this figure holds for the population of cars currently in use for which the EPA has determined projected gas mileages. Suppose the EPA is planning to select 20 cars at random from this population and test their fuel economies to determine how many are within 2 miles per gallon of their EPA projections.

    a. Decide whether the following statement is true or false and explain: The number of cars in the sample of 20 that have mileages within 2 miles per gallon of their EPA projections is not a binomial random variable but, for convenience, could be treated as a binomial random variable.

b. What is the probability (approximately) that fewer than 10 of the 20 cars selected will be within 2 miles per gallon of their EPA projections? (For convenience, use 70% as an approximation to 68%.)

5.107 Refer to Exercise 5.18.

a. Find the expected value of the prize you have won.

b. Suppose you take your winning number to the sales presentation and find that you have won an inflatable raft—a prize worth $49.95. Explain why the expected value of the prize and the actual value of the prize differ.

*5.108 If 20% of the finished products coming off an assembly line are defective, what is the probability that more than three randomly selected finished products would have to be inspected before a defective product is found?

5.109 The owner of construction company A makes bids on jobs so that, if it is awarded the job, company A will make a $10,000 profit. The owner of construction company B makes bids on jobs so that, if it is awarded the job, company B will make a $15,000 profit. Each company describes the probability distribution of the number of jobs the company is awarded per year as shown in the table.

COMPANY A		COMPANY B	
x	p(x)	x	p(x)
2	.05	2	.15
3	.15	3	.30
4	.20	4	.30
5	.35	5	.20
6	.25	6	.05

a. Find the expected number of jobs each will be awarded in a year.

b. What is the expected profit for each company?

c. Find the variance and standard deviation of the distribution of the number of jobs awarded per year for each company.

d. Graph $p(x)$ for both companies A and B. For each company, what proportion of the time will x fall in the interval $\mu \pm 2\sigma$?

5.110 An experiment is to be conducted to determine whether an acclaimed stock market analyst has extrasensory perception (ESP). Five different cards are shuffled and one is chosen at random. The analyst will try to identify which card was drawn without seeing it. The experiment is repeated 20 times and x, the number of correct decisions, is recorded. (Assume that the 20 trials are independent.)

a. If the stock market analyst is guessing (i.e., if the analyst does not possess ESP), what is the value of p, the probability of a correct decision on each trial?

b. If the analyst is guessing, what is the probability of a correct decision on at least half the trials?

c. Suppose the analyst makes the correct decision on 10 of the 20 trials. Do you think this performance is good enough to suggest that the analyst possesses ESP? Support or refute the proposition that the analyst possesses ESP using your results from part **b.**

5.111 The state highway patrol has determined that one out of every six calls for help originating from roadside call boxes is a hoax. Five calls for help have been received and five tow trucks dispatched. What is the probability that none of the calls was a hoax? That only three of the callers really needed assistance? What assumptions do you have to make to solve this problem?

5.112 Refer to Exercise 5.111. If the highway patrol answers 10,000 calls for help next year, and each call costs the patrol about $20 (labor, gas, etc.), approximately how much money will be wasted answering false alarms?

*5.113 Refer to Exercise 5.112. The highway patrol has determined that the expected number of calls for help per hour is 1.1. What is the probability that in the next hour more than two calls for help will be received? Exactly three calls?

5.114 An advertisement for a laundry soap claims that the soap is preferred over all others by 30% of American women. Assuming this claim is true, what is the probability that fewer than 4 women in a random sample of 25 prefer the advertiser's brand? What is the probability that the number of women, x, preferring the brand takes a value in the interval $3 \leq x \leq 13$? If a sample of 25 American women was taken and only 3 preferred the brand, what inference would you make? Why?

*5.115 By mistake, a manufacturer of tape recorders includes three defective recorders in a shipment of ten going out to a small retailer. The retailer has decided to accept the shipment of recorders only if none are found to be defective. Upon receipt of the shipment, the retailer examines only five of the recorders. What is the probability that the shipment will be rejected? If the retailer inspects six of the recorders, what is the probability the shipment will be accepted?

5.116 A manufacturer considers a production lot unacceptable if 10% or more of the units in the lot are defective. In such cases the company wants to scrap (not ship) the entire lot. A company quality control inspector has proposed the following criterion for determining whether to reject a lot: In a sample of ten units from a lot, if two or more are defective, reject the entire lot. If the lot currently under examination is 11% defective, what is the probability that this decision rule will lead the quality control inspector to the correct decision?

5.117 When the price of grain is low, many farmers participate in government-financed, on-farm storage programs rather than selling their grain. But storage invites insect infestations, and grain elevators penalize farmers who sell them insect-infested grain. In 1982, the U.S. Grain Marketing Research Laboratory estimated that 80% of the storage bins of corn in the country were infested with insects, and it has been estimated that the economic loss to farmers in the state of Minnesota is $12.6 million annually (*Minneapolis Tribune*, Aug. 8, 1982, p. 1A). Suppose 20 storage bins of corn are randomly selected and examined for insect infestation.
  a. What is the probability (approximately) that less than one-half of the bins are infested?
  b. What assumptions did you make in answering part **a**?
  c. Why is your answer to part **a** an approximation?
  d. Would you be surprised if all 20 of the bins were infested? Explain.

*5.118 A wholesale office equipment outlet claims that on an average it sells 2.4 typewriters per day. If it has only 5 typewriters in stock at the close of business today and does not expect to receive a shipment of new typewriters until some time after the close of business tomorrow, what is the probability that the outlet's current supply of typewriters will not be sufficient to meet tomorrow's demand?

5.119 If the probability of a customer responding to one of your marketing department's mail questionnaires is .6, what is the probability that, of 20 questionnaires mailed, more than 15 will be returned?

*5.120 Large bakeries typically have fleets of delivery trucks. It was determined by one such bakery that the expected number of delivery truck breakdowns per day was 1.5. What is the probability that there will be exactly two breakdowns today and exactly three tomorrow? Less than two today and more than two tomorrow? Assume that the number of breakdowns is independent from day to day.

*5.121 Of the 38 numbers on the roulette wheel, 18 are colored black. A gambler decides to place money on a black number. If the gambler does not win in either of the first two spins of the wheel, he will leave the casino; otherwise, he will continue playing. Let x equal the number of spins until the gambler achieves the first win.
  a. Find the probability that the gambler loses at the roulette wheel and hence leaves the casino (i.e., the probability that $x > 2$).
  b. Find the probability that the gambler remains betting at the roulette table after two spins of the wheel.
  c. If the gambler always plays black, what is the expected number of spins before he achieves the first win [i.e., find $E(x)$]? Find σ^2, the variance of x.
  d. Is it likely that x would exceed five? Explain. (Assume the gambler will play the game until he achieves the first win.)

5.122 A sales manager has determined that a salesperson makes a sale to 70% of the retailers visited.
  a. Suppose the salesperson visits five retailers today and 20 tomorrow. What is the probability that she makes exactly four sales today *and* more than ten tomorrow?
  b. If the salesperson visits four retailers today and five tomorrow, what is the probability that in these two days she will make exactly two sales?

5.123 If it is known that 5% of the finished products coming off an assembly line are defective, what is the probability that one of the next four products coming off the line is defective? What assumptions do you have to make to solve this problem using the methodology of this chapter?

*5.124 A small life insurance company has determined that on the average it receives five death claims per day. What is the probability that the company will receive three claims or less on a particular day? Exactly five claims? What assumptions must you make to find these probabilities?

5.125 In recent years, the use of the telephone as a data collection instrument for public opinion polls has been steadily increasing. However, one of the major factors bearing on the extent to which the telephone will become an acceptable data collection tool in the future is the *refusal rate*—i.e, the percentage of the eligible subjects actually contacted who refuse to take part in the poll. Suppose that past records indicate a refusal rate of 20% in a large city. A poll of 25 city residents is to be taken, and x is the number of residents contacted by telephone who refuse to take part in the poll.
  a. Find the mean and variance of x.
  b. Find $P(x \leq 5)$.
  c. Find $P(x > 10)$.

5.126 Suppose you are an airport manager. In looking over your records for the past year, you note that 60% of the time the 8:10 P.M. flight from Atlanta is 20 or more minutes late. If you assume that the probability of the 8:10 P.M. flight being 20 or more minutes late on each day during the upcoming 5-day period is .6, what is the probability that the plane will be 20 or more minutes late exactly three times in the next 5 days? At least three times in the next 5 days?

5.127 [*Warning: This exercise is realistic and the calculations involved are tedious.*] According to the Orlando, Florida, *Sentinel Star* (June 1, 1977), the Red Lobster Inns of America decided to take the state of Florida to court over sales taxes. The dispute concerned the 4% sales tax levied on most purchases in the state and focused mainly on the state's "bracket tax collection system." According to the bracket system, a merchant collected 1¢ sales tax for sales between 10¢ and 25¢, 2¢ for sales between 26¢ and 50¢, 3¢ for sales between 51¢ and 75¢, and 4¢ for sales between 76¢ and 99¢. Red Lobster contended that if this system were followed, merchants would always collect more than 4%. That is, if a sale were made for $10.41, 4% would be collected on the $10, but more than 4% would be collected on the 41¢. In fact, 4.878% (2 cents/41 cents) sales tax would be collected on the 41¢. Let $x = 0, 1, 2, \ldots, 99$ be the number of cents (exceeding whole dollars) involved in a sale. Suppose that x has a probability distribution $p(x) = .01$ for $x = 0, 1, 2, \ldots, 99$ (an assumption that might be fairly accurate for restaurant sales). Find the expected value of the percentage of tax paid on the cents portion of a sale. [*Hint:* You can find the percentage tax— call it y—for each value of x. You also know the probabilities associated with each value of x.]

ON YOUR OWN . . .

To control the quality of incoming or outgoing large lots of manufactured items, manufacturers use a lot acceptance sampling plan for quality control. For example, if each lot consists of 1,000 items, the plan will call for the selection of a random sample of n items (n is usually small) from each lot. The items in each sample are carefully inspected, the number of defectives recorded, and the lot considered to be of acceptable quality if the number of defectives is less than or equal to some specified number, a. The number a is called the **acceptance number** for the plan.

To illustrate, simulate sampling from a large lot of items that contains 10% defectives. Place ten poker chips (or marbles, etc.) in a bowl and mark one of the ten as defective. Randomly select a sample of five items from a lot by selecting a chip from the ten, replacing it, and repeating the process four more times. Count the number of times, x, that you observe the defective chip. This process is equivalent to selecting a random sample of $n = 5$ items from a large lot containing 10% defectives.

If you choose $a = 1$ as the acceptance number for the plan, then you will accept only lots for which $x \leq 1$.

By choosing n and a, you change the ability of a sampling plan to screen out bad lots. To investigate the properties of a sampling plan with $n = 5$ and $a = 1$, simulate the process of sampling from 100 lots.

a. Collect the 100 values of x obtained from the simulation, and construct a relative frequency histogram for x. Note that this histogram is an approximation to $p(x)$. Estimate the proportion of lots that will be accepted by the plan by dividing the number of lots accepted by 100.

b. Calculate the exact values of $p(x)$ for $n = 5$ and $p = .1$, and compare these with the results of part **a**.

USING THE COMPUTER . . .

Calculate the mean percentage of college graduates over all the zip codes for one of the census regions listed in Appendix C. Round the mean to the nearest integer, and use the result as an estimate of the percentage of college graduates among all people at least 25 years old in the region.

a. Suppose that 20 individuals from the region respond to an advertisement by an employment agency. If these people represent a random sample of all individuals at least 25 years of age in the region, what is the probability that more than half of them have college degrees?

b. If 2,000 individuals respond, find the mean μ and standard deviation σ of the number of them who have college degrees. Calculate $\mu \pm 2\sigma$ and $\mu \pm 3\sigma$, and use Chebyshev's theorem and the Empirical Rule to estimate the probability that the number of respondents with college degrees falls in each of the intervals. Based on your answers, assess the likelihood that more than half the applicants will have college degrees.

c. What are the potential problems with using the mean percentage of college graduates as an estimate of the percentage for the region? How could the estimate be improved?

REFERENCES

Alexander, G. J. and Francis, J. C. *Portfolio Analysis.* Englewood Cliffs, N.J.: Prentice-Hall, 1986.

Biddle, W. "What destroyed *Challenger?*" *Discover*, Apr. 1986, 40–47.

Brigham, E. F. *Financial Management Theory and Practice*, 2nd ed. Hinsdale, Ill.: Dryden Press, 1979. Chapter 5.

Brown, A. A., Hulswit, F. T., and Kettelle, J. D. "A study of sales operations." *Operations Research*, June 1956, *4*, 296–308.

Chervany, N. L., Anderson, J. C., Benson, P. G., and Hill, A. V. "A management science approach to a Dutch elm disease sanitation program." *Interfaces*, Apr. 1980, *10*, 108.

"'83 report put booster accident as most likely," *Minneapolis Star and Tribune*, Feb. 11, 1986, p. 1.

Elton, E. J. and Gruber, M. J. *Modern Portfolio Theory and Investment Analysis.* New York: Wiley, 1981.

Environmental Protection Agency. *Environment News.* New England Regional Office, Boston, Jan. 1978, 11–12.

Goetz, J. F., Jr. "The pricing decision: A service industry's experience." *Journal of Small Business Management*, Apr. 1985, *23*, 61–67.

Hogg, R. V. and Craig, A. T. *Introduction to Mathematical Statistics*, 4th ed. New York: Macmillan, 1978.

Johnson, R. W. and Sullivan, A. C. "Segmentation of the consumer loan market." *Journal of Retail Banking*, Sept. 1981, *3*, 1–7.

Markowitz, H. M. "Portfolio selection." *Journal of Finance*, *6*, Mar. 1952.

McKean, K. "They fly in the face of danger." *Discover*, Apr. 1986, 48–58.

Mendenhall, W. *Introduction to Probability and Statistics*, 7th ed. Boston: Duxbury, 1987. Chapters 5 and 6.

Misshauk, M. J. *Management Theory and Practice*. Boston: Little, Brown, 1979. Chapter 13.

Murphy, A. H. and Winkler, R. L. "Probability forecasting in meteorology." *Journal of the American Statistical Association*, Sept. 1984, 79, 489–500.

Parzen, E. *Modern Probability Theory and Its Applications*. New York: Wiley, 1960. Chapters 3, 4, 6, and 7.

Robinson, W. V. "NASA blamed for shuttle disaster," *Boston Globe*, June 10, 1986, p. 1.

U.S. Golf Association. *The Rules of Golf*. 1982, p. 10.

The United States Government Manual 1985–1986. Office of the Federal Register, revised July 1, 1985, pp. 485–487.

Webster, A. H. "Straight answers about Dutch elm disease." *Flower and Garden*, May 1978, 28–33, 46–47.

Williams, C. A., Jr., and Heins, R. M. *Risk Management and Insurance*. New York: McGraw-Hill, 1976, pp. 10, 65–66.

Willis, R. E. and Chervany, N. L. *Statistical Analysis and Modeling for Management Decision-Making*. Belmont, Calif.: Wadsworth, 1974. Chapter 5.

WHERE WE'VE BEEN . . .

Because sample data represent observed values of random variables, we needed to find the probabilities associated with specific random variables. As noted in Chapter 5, the appropriate method depends on whether a random variable is discrete or continuous. The probability theory of Chapter 4 provided the mechanism for finding the probabilities associated with discrete random variables. Finding and describing this set of probabilities—the probability distribution for a discrete random variable—was the subject of Chapter 5.

WHERE WE'RE GOING . . .

Since business data are derived from observations on continuous as well as discrete random variables, we need to know probability distributions associated with continuous random variables and also how to use the mean and standard deviation to describe these distributions. Chapter 6 addresses this problem and, in particular, introduces the normal probability distribution. As you will see, the normal probability distribution is one of the most useful distributions in business statistics.

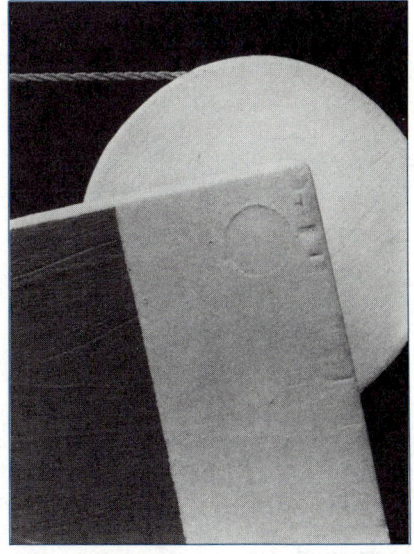

CONTINUOUS RANDOM VARIABLES

CONTENTS

In this chapter we will consider some continuous random variables that are prevalent in business. Recall that a continuous random variable is one that can assume any value within some interval or intervals. For example, the length of time between a consumer's purchase of new automobiles, the thickness of sheets of steel produced in a rolling mill, and the length of time between consumer complaints are all continuous random variables.

The methodology we use to describe continuous random variables will necessarily be somewhat different from that used to describe discrete random variables. We will first discuss the general form of continuous probability distributions, and then we will present three specific types that are used in making business decisions: (1) The uniform distribution is simple but useful, and serves as a good introduction to continuous probability distributions. (2) The normal probability distribution, which plays a basic and important role in both the theory and applications of statistics, is essential to the study of most of the subsequent chapters. (3) The exponential probability distribution has some specialized application to business problems, and we present it in an optional section.

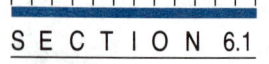

SECTION 6.1

CONTINUOUS PROBABILITY DISTRIBUTIONS

The graphical form of the probability distribution for a continuous random variable, x, will be a smooth curve that might appear as shown in Figure 6.1. This curve, a function of x, is denoted by the symbol $f(x)$ and is variously called a **probability density function**, a **frequency function**, or a **probability distribution**.

The areas under a probability distribution correspond to probabilities for x. For example, the area A between the two points a and b, as shown in Figure 6.1, is the probability that x assumes a value between a and b ($a < x < b$). Because areas over intervals represent probabilities, it follows that the total area under a probability distribution, the probability assigned to all values of x, should equal 1. Note that probability distributions for continuous random variables will have different shapes depending on the relative frequency distributions of real data that the probability distribution is supposed to model. Note too that the events $\{a < x < b\}$ and $\{a \le x \le b\}$ are equivalent for continuous random variables, because no probability is assigned to individual points. In terms of the probability distribution $f(x)$, no area is added by the addition of a single point (or two points) to the interval.

The areas under most probability distributions are obtained by the use of calculus* or other numerical methods. Because this is often a difficult procedure, we give the areas for some of the most common probability distributions in tabular form in Appendix B. Then, to find the area between two values of x, say $x = a$ and $x = b$, you can simply consult the appropriate table.

For each of the continuous random variables presented in this chapter, we will give the formula for the probability distribution along with its mean and standard deviation. These two numbers, μ and σ, will enable you to make some approximate probability statements about a random variable even when you do not have access to a table of areas under the probability distribution.

*Students with knowledge of calculus should note that the probability that x assumes a value in the interval $a < x < b$ is $P(a < x < b) = \int_a^b f(x)\,dx$, assuming the integral exists. Similar to the requirements for a discrete probability distribution, we require $f(x) \ge 0$ and $\int_{-\infty}^{\infty} f(x)\,dx = 1$.

FIGURE 6.1

A Probability Distribution
$f(x)$ for a Continuous
Random Variable x

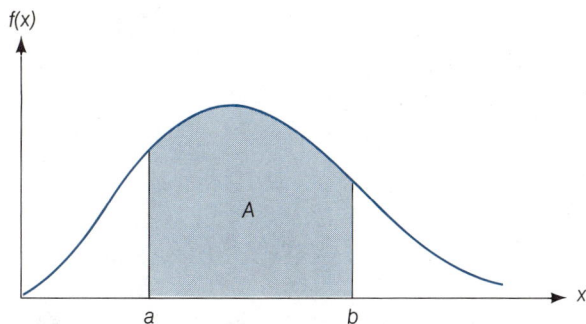

$f(x)$

A

x

a b

S E C T I O N   6.2

**THE UNIFORM
DISTRIBUTION**

Perhaps the simplest of all the continuous probability distributions is the **uniform distribution**. The frequency function has a rectangular shape, as shown in Figure 6.2. Note that the possible values of x consist of all points on the real line between point c and point d. The height of $f(x)$ is constant in that interval and equals $1/(d - c)$. Therefore, the total area under $f(x)$ is given by

$$\text{Total area of rectangle} = (\text{Base})(\text{Height}) = (d - c)\left(\frac{1}{d - c}\right) = 1$$

FIGURE 6.2

The Uniform Probability
Distribution

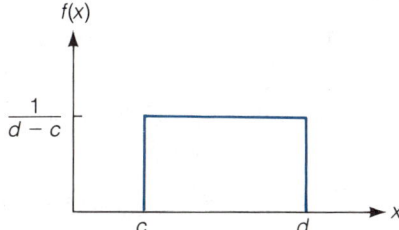

$f(x)$

$\dfrac{1}{d - c}$

c d x

The uniform probability distribution provides a model for continuous random variables that are *evenly distributed* over a certain interval. That is, a uniform random variable is one that is just as likely to assume a value in one interval as it is to assume a value in any other interval of equal size. There is no clustering of values around any value; instead, there is an even spread over the entire region of possible values.

The uniform distribution is sometimes referred to as the **randomness distribution**, since one way of generating a uniform random variable is to perform an experiment in which a point is *randomly* selected on the horizontal axis between the points c and d. If we were to repeat this experiment infinitely often, we would create a uniform probability distribution like that shown in Figure 6.2. The random selection of points on a line can also be used to generate random numbers such as those in Table I of Appendix B. Recall that random numbers are selected in such a way that every digit has an equal probability of selection. Therefore, random numbers are realizations of a uniform random variable. (Random numbers were used to draw random samples in Section 4.6.)

The formulas for the uniform probability distribution and its mean and standard deviation are shown in the box on page 262.

> **PROBABILITY DISTRIBUTION, MEAN, AND STANDARD DEVIATION OF A UNIFORM RANDOM VARIABLE x**
>
> $$f(x) = \frac{1}{d - c} \qquad (c \le x \le d)$$
>
> $$\mu = \frac{c + d}{2} \qquad \sigma = \frac{d - c}{\sqrt{12}}$$

Suppose the interval $a < x < b$ lies within the domain of x; i.e., it falls within the larger interval $c < x < d$. Then the probability that x assumes a value within the interval $a < x < b$ is the area of the rectangle over the interval—namely,[*]

$$\frac{b - a}{d - c}$$

EXAMPLE 6.1

Suppose the research department of a steel manufacturer believes that one of the company's rolling machines is producing sheets of steel of varying thickness. The thickness is a uniform random variable with values between 150 and 200 millimeters. Any sheets less than 160 millimeters thick must be scrapped because they are unacceptable to buyers.

a. Calculate the mean and standard deviation of x, the thickness of the sheets produced by this machine. Then graph the probability distribution and show the mean on the horizontal axis. Also show 1 and 2 standard deviation intervals around the mean.

b. Calculate the fraction of steel sheets produced by this machine that have to be scrapped.

SOLUTION

a. To calculate the mean and standard deviation for x, we substitute 150 and 200 millimeters for c and d, respectively, in the formulas. Thus,

$$\mu = \frac{c + d}{2} = \frac{150 + 200}{2} = 175 \text{ millimeters}$$

and

$$\sigma = \frac{d - c}{\sqrt{12}} = \frac{200 - 150}{\sqrt{12}} = \frac{50}{3.464} = 14.43$$

The uniform probability distribution is

$$f(x) = \frac{1}{d - c} = \frac{1}{200 - 150} = \frac{1}{50}$$

[*]The student who has knowledge of calculus should note that

$P(a < x < b) = \int_a^b f(x)\, dx = \int_a^b 1/(d - c)\, dx = (b - a)/(d - c)$

The graph of this function is shown in Figure 6.3. The mean and the 1 and 2 standard deviation intervals around the mean are shown on the horizontal axis.

FIGURE 6.3

Distribution for x in Example 6.1

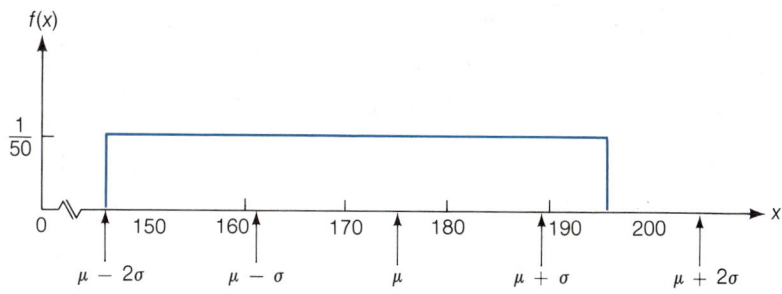

b. To find the fraction of steel sheets produced by the machine that have to be scrapped, we must find the probability that x, the thickness, is less than 160 millimeters. As indicated in Figure 6.4, we need to calculate the area under the frequency function $f(x)$ between the points $x = 150$ and $x = 160$. This is the area of a rectangle with base $160 - 150 = 10$ and height $\frac{1}{50}$. The fraction that has to be scrapped is then

$$P(x < 160) = (\text{Base})(\text{Height}) = (10)\left(\frac{1}{50}\right) = \frac{1}{5}$$

That is, 20% of all the sheets made by this machine must be scrapped.

FIGURE 6.4

Probability That Sheet Thickness, x, Is Between 150 and 160 Millimeters

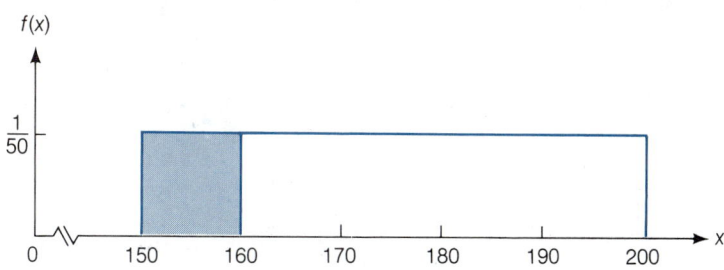

EXERCISES 6.1–6.12

LEARNING THE MECHANICS

6.1 Suppose x is a random variable best described by a uniform probability distribution with $c = 20$ and $d = 45$.
a. Find $f(x)$.
b. Find the mean and variance of x.
c. Graph $f(x)$ and locate μ and the interval $\mu \pm 2\sigma$ on the graph. Find the probability that x assumes a value within the interval $\mu \pm 2\sigma$.

6.2 Refer to Exercise 6.1. Find the following probabilities:

a. $P(20 \leq x \leq 35)$ 　　　　b. $P(20 < x < 35)$
c. $P(x \geq 35)$ 　　　　　　　d. $P(x \leq 20)$
e. $P(x \leq 25)$ 　　　　　　　f. $P(10 \leq x \leq 40)$
g. $P(x \geq 36)$ 　　　　　　　h. $P(x \geq 35.5)$
i. $P(20.5 \leq x \leq 35.5)$ 　　j. $P(x < 20.5)$

6.3 Suppose x is a random variable best described by a uniform probability distribution with $c = 2$ and $d = 5$.

a. Find $f(x)$.
b. Find the mean and variance of x.
c. Graph $f(x)$ and locate μ and the interval $\mu \pm \sigma$. Find the probability that x assumes a value within the interval $\mu \pm \sigma$.

6.4 Use the probability distribution of Exercise 6.3 to find the value of a that makes each of the following probability statements true:

a. $P(x \geq a) = .5$ 　　　　b. $P(x \leq a) = .2$
c. $P(x \leq a) = 0$ 　　　　　d. $P(2.5 \leq x \leq a) = .5$
e. $P(x > a) = .1$

6.5 The random variable x is best described by a uniform probability distribution with $c = 100$ and $d = 200$. Find the probability that x assumes a value:

a. More than 2 standard deviations from μ.
b. Less than 3 standard deviations from μ.
c. Within 2 standard deviations of μ.

6.6 The random variable x is best described by a uniform probability distribution with mean 10 and standard deviation 1. Find c, d, and $f(x)$. Graph the probability distribution.

APPLYING THE CONCEPTS

6.7 As we noted in this section, random numbers are values of a uniform random variable. Construct a relative frequency histogram for the data set listed here. (It was created by the random-number generator of the Minitab computer package.) Except for the expected variation in relative frequencies among the class intervals, does your histogram suggest that the data are observations on a uniform random variable with $c = 0$ and $d = 100$? Explain.

38.8759	35.6438	55.6267	87.4506	60.8422	11.2159	69.2875
88.3734	95.3660	57.8870	50.7119	.7434	12.0605	71.6829
12.4337	98.0716	38.6584	78.3936	94.1727	.8413	23.0278
47.0121	31.8792	21.5478	71.8318	88.2612	93.3017	96.1009
62.6626	11.7828	64.5788	46.7404	28.6777	23.0892	75.1015
44.0466	43.3629	32.9847	87.7819	28.9622		

6.8 Probability distributions and event probabilities can be used to express uncertainty about future events. For example, security analysts use probability distributions to forecast stock prices and quarterly earnings; weather forecasters use event probabilities to forecast precipitation, temperature ranges, and the location of a hurricane's landfall. Such forecasts are known as **probability forecasts** (Murphy and Winkler, 1984). Probability forecasts can be derived from sample data, expert judgment, or a combination of both. When a forecaster possesses relatively little information about the future value of a random variable, the uniform distribution is frequently used to make a probability forecast. Suppose a security analyst believes that the closing price of a particular stock 3 months from today will be between $30 and $40, but has no idea where within the interval the price will be.

a. Graph the uniform distribution the security analyst should use as a probability forecast.

b. According to the analyst's forecast, what is the probability that the stock will close higher than $35? Higher than $38? Between $34 and $36, inclusive? Higher than $50?

6.9 Rapid advances in technology in recent years have led to the development of extremely complex equipment and, consequently, to the need to evaluate the equipment's reliability. The *reliability* of a piece of equipment is frequently defined to be the probability, p, that the equipment performs its intended function successfully for a given period of time under specific conditions (Martz and Waller, 1982). Because p varies from one point in time to another, some reliability analysts treat p as if it were a random variable. Suppose an analyst characterizes the uncertainty about the reliability of a particular robotic device used in an automobile assembly line using the following distribution:

$$f(p) = \begin{cases} 1 & 0 \le p \le 1 \\ 0 & \text{otherwise} \end{cases}$$

a. Graph the analyst's probability distribution for p.

b. Find the mean and variance of p.

c. According to the analyst's probability distribution for p, what is the probability that p is greater than .95? Less than .95?

d. Suppose the analyst receives the additional information that p is definitely between .90 and .95, but that there is complete uncertainty about where it lies between these values. Describe the probability distribution the analyst should now use to describe p.

6.10 The manager of a large department store with three floors reports that the time a customer on the second floor must wait for an elevator has a uniform distribution ranging from 0 to 4 minutes. Find the mean and variance of x, the time a customer on the second floor waits for an elevator. If it takes the elevator 15 seconds to go from floor to floor, find the probability that a hurried customer can reach the first floor in less than 1.5 minutes after pushing the second floor elevator button.

6.11 A bus is scheduled to stop at a certain bus stop every half hour on the hour and half hour. At the end of the day, buses still stop about every 30 minutes, but due to delays earlier in the day, they are equally likely to stop at any time during any given half hour. If you arrive at a bus stop at the end of the day, what is the probability that you will have to wait more than 20 minutes for the bus (no matter when you show up)? How long do you expect to wait for the bus?

6.12 The manager of a local soft-drink bottling company believes that when a new beverage-dispensing machine is set to dispense 7 ounces, it in fact dispenses an amount at random anywhere between 6.5 and 7.5 ounces.

a. Is the amount dispensed by the beverage machine a discrete or continuous random variable? Explain.

b. Graph the frequency function for x, the amount of beverage the manager believes is dispensed by the new machine when it is set to dispense 7 ounces.

c. Find the mean and standard deviation for the distribution graphed in part **b**, and locate the mean and the interval $\mu \pm 2\sigma$ on the graph.

THE NORMAL DISTRIBUTION

One of the most useful and frequently encountered continuous random variables has a **bell-shaped** probability distribution, as shown in Figure 6.5. It is known as a **normal random variable**, and its probability distribution is called a **normal distribution**.

FIGURE 6.5

A Normal Probability Distribution

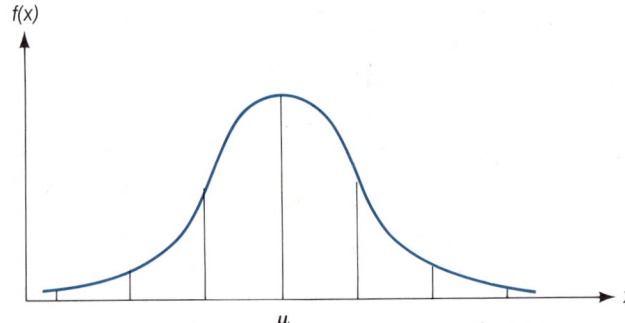

You will see during the remainder of this text that the normal distribution plays a very important role in the science of statistical inference. Many business phenomena generate random variables with probability distributions that are very well approximated by a normal distribution. For example, it has been demonstrated that the monthly rate of return (defined in the footnote on p. 81) for a particular stock is approximately a normal random variable, and the probability distribution for the weekly sales of a corporation might be approximated by a normal probability distribution. The normal distribution might also provide an accurate model for the probability distribution of the weights of loads of produce shipped to a supermarket. You can determine the adequacy of the normal approximation to an existing population of data by comparing the relative frequency distribution of a large sample of the data to the normal probability distribution.

The normal distribution is perfectly symmetric about its mean, μ, as can be seen in the examples in Figure 6.6. Its spread is determined by the value of its standard deviation, σ.

FIGURE 6.6

Several Normal Distributions, with Different Means and Standard Deviations

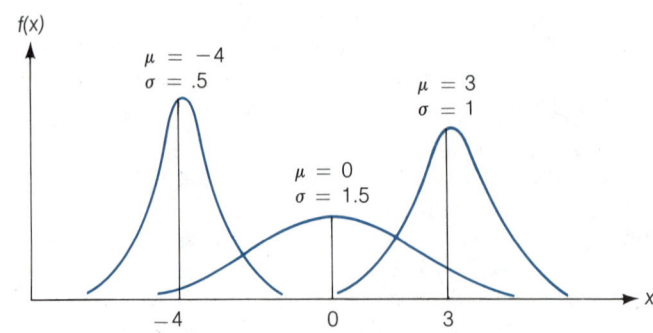

The formula for the normal probability distribution is shown in the next box. Note that the mean μ and the variance σ^2 appear in this formula, so that no separate formulas for μ and σ^2 are necessary. To graph the normal curve we will have to know the numerical values of μ and σ.

PROBABILITY DISTRIBUTION FOR A NORMAL VARIABLE x

$$f(x) = \frac{1}{\sigma\sqrt{2\pi}} e^{-(1/2)[(x-\mu)/\sigma]^2}$$

where

μ = Mean of the normal random variable, x

σ^2 = Variance of the normal random variable, x

π = 3.1416 . . .

e = 2.71828 . . .

Computing the area over intervals under the normal probability distribution is a difficult task.* Consequently, we will use the computer areas listed in Table IV of Appendix B. Although there is an infinitely large number of normal curves— one for each pair of values for μ and σ—we have formed a single table that will apply to any normal curve. This was done by constructing the table of areas as a function of the z-score (presented in Section 3.9). The population z-score for a measurement was defined as the *distance* between the measurement and the population mean, divided by the population standard deviation. Thus, the z-score gives the distance between a measurement and the mean in units equal to the standard deviation. In symbolic form, the z-score for the measurement x is

$$z = \frac{x - \mu}{\sigma}$$

To illustrate the use of Table IV, suppose we know that the length of time between charges of a pocket calculator has a normal distribution, with a mean of 50 hours and a standard deviation of 15 hours. If we were to observe the length of time that elapses before the need for the next charge, what is the probability that this measurement would assume a value between 50 hours and 70 hours? This probability is the area under the normal probability distribution between 50 and 70, as shown in the shaded area, A, of Figure 6.7 on page 268.

---

*The student with knowledge of calculus should note that there is not a closed-form expression for $P(a < x < b) = \int_a^b f(x)\, dx$ for the normal probability distribution. However, the value of this definite integral can be obtained to any desired degree of accuracy by approximation procedures. The areas in Table IV of Appendix B were obtained by using such a procedure.

FIGURE 6.7

Normal Distribution:
$\mu = 50$, $\sigma = 15$

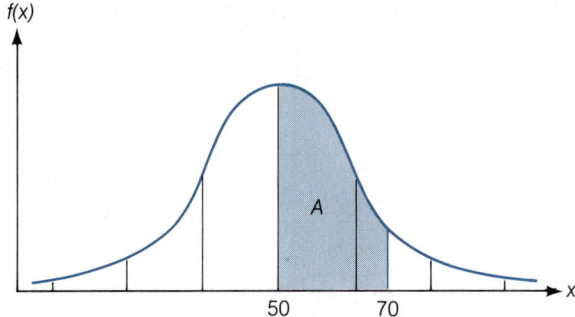

The first step in finding the area A is to calculate the z-score corresponding to the measurement 70. We calculate

$$z = \frac{x - \mu}{\sigma} = \frac{70 - 50}{15} = \frac{20}{15} = 1.33$$

Thus, the measurement 70 is 1.33 standard deviations above the mean, 50. The second step is to refer to Table IV (a partial reproduction of this table is shown in Table 6.1). Note that z-scores are listed in the left-hand column of the table. To find the area corresponding to a z-score of 1.33, we first locate the value 1.3 in the left-hand column. Since this column lists z values to one decimal place only, we refer to the top row of the table to get the second decimal place, .03. Finally, we locate the number where the row labeled $z = 1.3$ and the column labeled .03 meet. This number represents the area between the mean, μ, and the measurement that has a z-score of 1.33:

A = .4082

Thus, the probability that the calculator operates between 50 and 70 hours before needing a charge is .4082.

The use of the z-score simplifies the calculation of normal probabilities because if x is normally distributed with any mean and standard deviation, z is *always* a normal random variable with a mean of 0 and a standard deviation of 1. For this reason z is often referred to as the **standard normal random variable**.

DEFINITION 6.1

The **standard normal random variable** z is defined by the formula

$$z = \frac{x - \mu}{\sigma}$$

where x is a normal random variable with mean μ and standard deviation σ. The standard normal random variable z is normally distributed with mean 0 and standard deviation 1, and can be described as the number of standard deviations between x and μ.

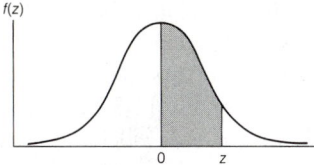

TABLE 6.1 Reproduction of Part of Table IV of Appendix B

z	.00	.01	.02	.03	.04	.05	.06	.07	.08	.09
.0	.0000	.0040	.0080	.0120	.0160	.0199	.0239	.0279	.0319	.0359
.1	.0398	.0438	.0478	.0517	.0557	.0596	.0636	.0675	.0714	.0753
.2	.0793	.0832	.0871	.0910	.0948	.0987	.1026	.1064	.1103	.1141
.3	.1179	.1217	.1255	.1293	.1331	.1368	.1406	.1443	.1480	.1517
.4	.1554	.1591	.1628	.1664	.1700	.1736	.1772	.1808	.1844	.1879
.5	.1915	.1950	.1985	.2019	.2054	.2088	.2123	.2157	.2190	.2224
.6	.2257	.2291	.2324	.2357	.2389	.2422	.2454	.2486	.2517	.2549
.7	.2580	.2611	.2642	.2673	.2704	.2734	.2764	.2794	.2823	.2852
.8	.2881	.2910	.2939	.2967	.2995	.3023	.3051	.3078	.3106	.3133
.9	.3159	.3186	.3212	.3238	.3264	.3289	.3315	.3340	.3365	.3389
1.0	.3413	.3438	.3461	.3485	.3508	.3531	.3554	.3577	.3599	.3621
1.1	.3643	.3665	.3686	.3708	.3729	.3749	.3770	.3790	.3810	.3830
1.2	.3849	.3869	.3888	.3907	.3925	.3944	.3962	.3980	.3997	.4015
1.3	.4032	.4049	.4066	.4082	.4099	.4115	.4131	.4147	.4162	.4177
1.4	.4192	.4207	.4222	.4236	.4251	.4265	.4279	.4292	.4306	.4319
1.5	.4332	.4345	.4357	.4370	.4382	.4394	.4406	.4418	.4429	.4441

Since we will convert all normal random variables to standard normal in order to find probabilities based on Table IV of Appendix B, it is important that you learn to use Table IV well. The following examples illustrate its use.

EXAMPLE 6.2

Find the probability that the standard normal random variable z falls between -1.33 and $+1.33$.

SOLUTION

The standard normal distribution is shown in Figure 6.8. Since all probabilities associated with standard normal random variables can be depicted as areas under the standard normal curve, you should always draw the curve and then equate the desired probability to an area.

FIGURE 6.8
A Distribution of z-Scores
(A Standard Normal
Distribution)

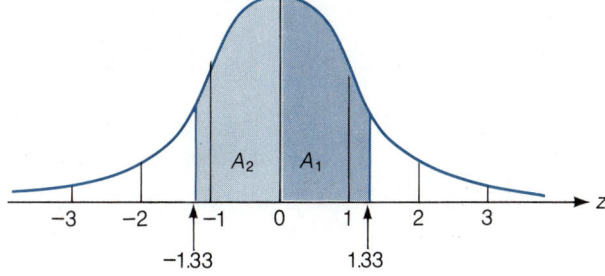

In this example we want to find the probability that z falls between -1.33 and $+1.33$, which is equivalent to the area between -1.33 and $+1.33$, shown shaded in Figure 6.8. Table IV of Appendix B provides the area between $z = 0$ and any value of z looked up, so that if we look up $z = 1.33$ we find that the area between $z = 0$ and $z = 1.33$ is .4082. This is the area labeled A_1 in Figure 6.8. To find the area A_2 between $z = 0$ and $z = -1.33$, we note that the symmetry of the normal distribution implies that the area between $z = 0$ and any point to the left is equal to the area between $z = 0$ and the point equidistant to the right. Thus, in this example the area between $z = 0$ and $z = -1.33$ is equal to the area between $z = 0$ and $z = +1.33$. That is,

$$A_1 = A_2 = .4082$$

The probability that z falls between -1.33 and $+1.33$ is the sum of the areas A_1 and A_2. Using probabilistic notation, we summarize as follows:

$$P(-1.33 < z < +1.33) = P(-1.33 < z < 0) + P(0 \leq z < 1.33)$$
$$= A_1 + A_2 = .4082 + .4082$$
$$= .8164$$

Remember that "<" and "$\leq$" are equivalent in events involving z, because the inclusion (or exclusion) of a single point does not alter the probability of an event involving a continuous random variable. ∎

EXAMPLE 6.3

Find the probability that a standard normal random variable exceeds 1.64, i.e., find $P(z > 1.64)$.

SOLUTION

The area under the standard normal distribution to the right of 1.64 is the shaded area labeled A_1 in Figure 6.9. This area represents the desired probability that z exceeds 1.64. However, when we look up $z = 1.64$ in Table IV of Appendix B, we must remember that the probability given in the table corresponds to the area between $z = 0$ and $z = 1.64$ (the area labeled A_2 in Figure 6.9). From Table IV we find that $A_2 = .4495$. To find the area A_1 to the right of 1.64, we make use of two facts:

1. The standard normal distribution is symmetric about its mean, $z = 0$.
2. The total area under the standard normal probability distribution equals 1.

FIGURE 6.9
Standard Normal
Distribution:
$\mu = 0, \sigma = 1$

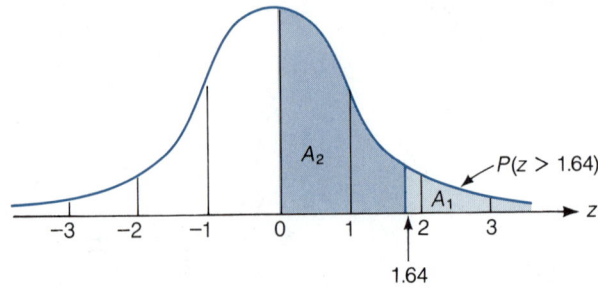

Taken together, these two facts imply that the areas on either side of the mean $z = 0$ equal .5; thus, the area to the right of $z = 0$ in Figure 6.9 is $A_1 + A_2 = .5$. Then

$$P(z > 1.64) = A_1 = .5 - A_2 = .5 - .4495 = .0505$$

To attach some practical significance to this probability, note that the implication is that the chance of a standard normal random variable exceeding 1.64 is approximately .05. Or, since z represents the number of standard deviations between *any* normal random variable and its mean, a normal random variable will exceed its mean by more than 1.64 standard deviations only about 5% of the time. ■

EXAMPLE 6.4

Find the probability that a normal random variable lies to the right of a point $-.74$ standard deviation from its mean.

SOLUTION

We must first interpret the event in terms of a standard normal random variable, so that we can use Table IV of Appendix B to find the event's probability. Since the standard normal random variable z is simply the number of standard deviations between a normal random variable and its mean, the event that a normal random variable lies to the right of a point $-.74$ standard deviation from the mean is equivalent to the event that the standard normal random variable z exceeds $-.74$. The event is shown as the shaded area in Figure 6.10, and we want to find $P(z > -.74)$.

FIGURE 6.10

Standard Normal
Distribution: $\mu = 0$, $\sigma = 1$

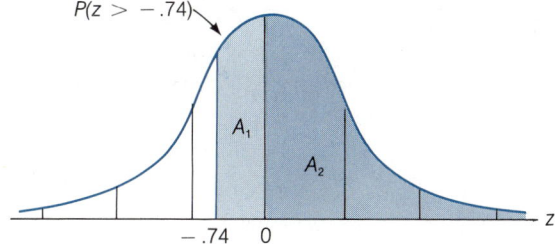

We divide the shaded area into two parts: the area A_1 between $z = -.74$ and $z = 0$, and the area A_2 to the right of $z = 0$. We must always make such a division when the desired area lies on both sides of the mean ($z = 0$), because Table IV contains areas between $z = 0$ and the point you look up. To find A_1, we remember that the sign of z is unimportant when determining the area, because the standard normal distribution is symmetric about its mean. We look up $z = .74$ in Table IV to find that $A_1 = .2704$. The symmetry also implies that half the distribution lies on each side of the mean, so the area A_2 to the right of $z = 0$ is .5. Then

$$P(z > -.74) = A_1 + A_2 = .2704 + .5 = .7704$$ ■

EXAMPLE 6.5

Find the probability that a normal random variable lies more than 1.96 standard deviations from its mean *in either direction*.

SOLUTION

The event that a normal random variable lies beyond 1.96 standard deviations in either direction from its mean is equivalent to the event that the standard normal random variable z exceeds 1.96 in absolute value. That is, we want to find

$$P(z > |1.96|) = P(z < -1.96 \text{ or } z > 1.96)$$

This probability is the shaded area in Figure 6.11. Note that the total shaded area is the sum of two areas, A_1 and A_2—areas that are equal because of the symmetry of the normal distribution.

FIGURE 6.11
Standard Normal
Distribution: $\mu = 0$, $\sigma = 1$

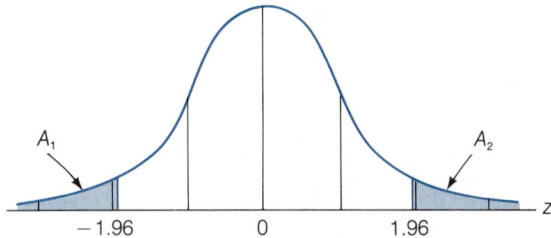

We look up $z = 1.96$ and find the area between $z = 0$ and $z = 1.96$ to be .4750. Then the area to the right of 1.96 is $.5 - .4750 = .0250$, so that

$$P(z > |1.96|) = .0250 + .0250 = .05$$

The implication is that any normal random variable lies more than 1.96 standard deviations from its mean 5% of the time. Recall (Chapter 3) that the Empirical Rule tells us that about 5% of the measurements in mound-shaped distributions will lie beyond 2 standard deviations from the mean; the normal distribution, which is certainly mound-shaped, has 5% of its area beyond 1.96 standard deviations. In fact, the normal distribution provides the model on which the Empirical Rule is based, along with much "empirical" experience with real data that often approximately obey the Rule, whether drawn from a normal distribution or not. ∎

EXAMPLE 6.6

Assume that the length of time, x, between charges of a pocket calculator is normally distributed with a mean of 50 hours and a standard deviation of 15 hours. Find the probability that the calculator will last between 30 and 70 hours between charges.

SOLUTION

The normal distribution with mean $\mu = 50$ and $\sigma = 15$ is shown in Figure 6.12; the desired probability that the calculator lasts between 30 and 70 hours is shaded. In order to find the probability, we must first convert the distribution to standard normal, which we do by calculating the z-score:

$$z = \frac{x - \mu}{\sigma}$$

The z-scores corresponding to the important values of x are shown just under the x values on the horizontal axis in Figure 6.12. Note that $z = 0$ corresponds to the mean of $\mu = 50$ hours, while the x values 30 and 70 yield z-scores of -1.33 and

FIGURE 6.12
Normal Probability
Distribution: $\mu = 50$,
$\sigma = 15$

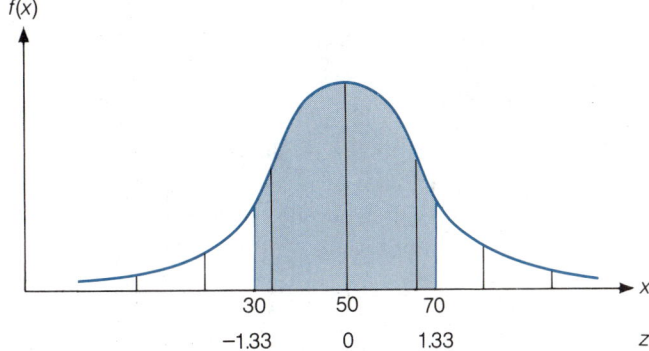

+1.33, respectively. Thus, the event that the calculator lasts between 30 and 70 hours is equivalent to the event that a standard normal random variable lies between −1.33 and +1.33. We found this probability in Example 6.2 (see Figure 6.8) by doubling the area corresponding to $z = 1.33$ in Table IV. That is,

$$P(30 \leq x \leq 70) = P(-1.33 \leq z \leq 1.33) = 2(.4082)$$
$$= .8164$$

The steps to follow when calculating a probability corresponding to a normal random variable are summarized in the box.

STEPS FOR FINDING A PROBABILITY CORRESPONDING TO A NORMAL RANDOM VARIABLE

1. Sketch the normal distribution and indicate the mean of the random variable x. Then shade the area corresponding to the probability you want to find.
2. Convert the boundaries of the shaded area from x values to standard normal random variable z values using the formula

$$z = \frac{x - \mu}{\sigma}$$

Show the z values under the corresponding x values on your sketch.
3. Use Table IV of Appendix B to find the areas corresponding to the z values. If necessary, use the symmetry of the normal distribution to find areas corresponding to negative z values, and the fact that the total area on each side of the mean equals .5 to convert the areas from Table IV to the probabilities of the event you have shaded.

EXAMPLE 6.7

Suppose an automobile manufacturer introduces a new model that has an advertised mean in-city mileage of 27 miles per gallon. Although such advertisements seldom report any measure of variability, suppose you write the manufacturer for the details of the tests, and you find that the standard deviation is 3 miles per gallon. This information leads you to formulate a probability model for the random variable x, the in-city mileage for this car model. You believe that the probability distribution of x can be approximated by a normal distribution with a mean of 27 and a standard deviation of 3.

a. If you were to buy this model of automobile, what is the probability you would purchase one that averages less than 20 miles per gallon for in-city driving?
b. Suppose you purchase one of these new models, and it does get less than 20 miles per gallon for in-city driving. Should you conclude that your probability model is incorrect?

SOLUTION

a. The probability model proposed for x, the in-city mileage, is shown in Figure 6.13. We are interested in finding the area, A, to the left of 20, since this area corresponds to the probability that a measurement chosen from this distribution falls below 20. Or in other words, if this model is correct, the area A represents the fraction of cars that can be expected to get less than 20 miles per gallon for in-city driving. To find A, we first calculate the z value corresponding to $x = 20$. That is,

$$z = \frac{x - \mu}{\sigma} = \frac{20 - 27}{3} = -\frac{7}{3} = -2.33$$

This z value is shown in Figure 6.13 under the corresponding x value.

FIGURE 6.13
Normal Probability
Distribution for x
in Example 6.7:
$\mu = 27$ Miles per Gallon,
$\sigma = 3$ Miles per Gallon

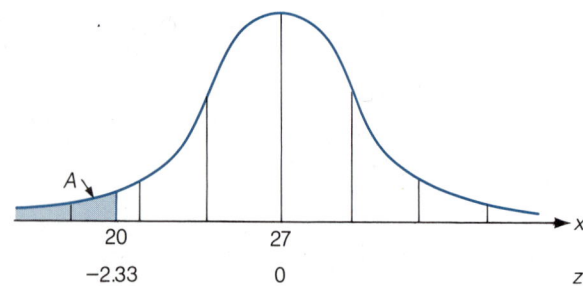

Because of the symmetry of the normal distribution about its mean and because Table IV provides only areas to the right of the mean, we look up 2.33 in Table IV, and find that the corresponding area is .4901. This is the area between $z = 0$ and $z = -2.33$, so we find

$$A = .5 - .4901 = .0099 \approx .01$$

According to this probability model, you should have only about a 1% chance of purchasing a car of this make with an in-city mileage under 20 miles per gallon.

b. Now you are asked to make an inference based on a sample—the car you purchased. You are getting less than 20 miles per gallon for in-city driving. What do you infer? We think you will agree that one of two possibilities is true:

The probability model is correct, and you simply were unfortunate to have purchased one of the cars in the 1% that get less than 20 miles per gallon in the city.

The probability model is incorrect. Perhaps the assumption of a normal distribution is unwarranted, or the mean of 27 is an overestimate, or the standard deviation of 3 is an underestimate, or some combination of these errors was made. At any rate, the form of the actual probability model certainly merits further investigation.

You have no way of knowing with certainty which possibility is the correct one, but the evidence points to the second one. We are again relying on the rare event approach to statistical inference that we introduced earlier. The basic idea is that the sample (one measurement in this case) was so unlikely to have been drawn from the proposed probability model that it casts serious doubt on the model. We would be inclined to believe that the model is somehow in error.

■

Occasionally you will be given a probability and want to find the values of the normal random variable that correspond to the probability. For example, suppose the average daily production of a manufacturer is known to be normally distributed, and management wants to pay an incentive bonus to the crew when their level of production exceeds the 90th percentile of the daily production distribution. To determine the production level at which bonuses will be paid, you will need to be able to use Table IV of Appendix B in reverse, as demonstrated in the following example.

EXAMPLE 6.8

Find the value of z, call it z_0, in the standard normal distribution that will be exceeded only 10% of the time. That is, find z_0 such that $P(z \geq z_0) = .10$.

SOLUTION

In this case we are given a probability, or an area, and asked to find the value of the standard normal random variable that corresponds to the area. Specifically, we want to find the value z_0 such that only 10% of the standard normal distribution exceeds z_0 (see Figure 6.14).

FIGURE 6.14
Standard Normal Distribution for Example 6.8

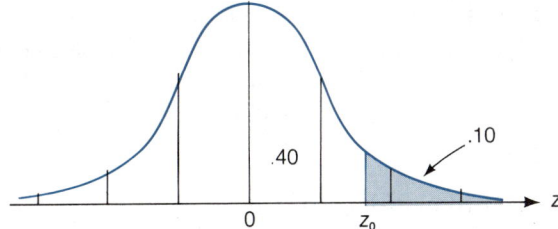

We know that the total area to the right of the mean $z = 0$ is .5, which implies that z_0 must lie to the right of (above) 0. To pinpoint the value, we use the fact that the area to the right of z_0 is .10, which implies that the area between $z = 0$

and z_0 is $.5 - .1 = .4$. But areas between $z = 0$ and some other z value are exactly the types given in Table IV of Appendix B. Therefore, we look up the *area* .4000 in the body of Table IV, and find that the corresponding z value is (to the closest approximation) $z_0 = 1.28$. The implication is that the point 1.28 standard deviations above the mean is the 90th percentile of a normal distribution. ∎

EXAMPLE 6.9

Find the value of z_0 such that 95% of the standard normal z values lie between $-z_0$ and $+z_0$, i.e.,

$$P(-z_0 \leq z \leq z_0) = .95$$

SOLUTION

Here we wish to move an equal distance z_0 in the positive and negative direction from the mean $z = 0$ until 95% of the standard normal distribution is enclosed. This means that the area on each side of the mean will be equal to $\frac{1}{2}(.95) = .475$, as shown in Figure 6.15. Since the area between $z = 0$ and z_0 is .475, we look up .475 in the body of Table IV and find the corresponding z value $z_0 = 1.96$. Thus, as we found in the reverse order in Example 6.5, 95% of a normal distribution lies between plus and minus 1.96 standard deviations of the mean.

FIGURE 6.15
Standard Normal
Distribution:
$\mu = 0, \sigma = 1$

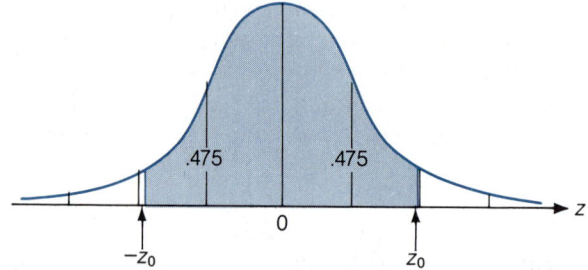

Now that you have learned to use Table IV to find a standard normal z value that corresponds to a specified probability, we will demonstrate a practical application in Example 6.10.

EXAMPLE 6.10

Suppose a paint manufacturer has a daily production, x, that is normally distributed with a mean of 100,000 gallons and a standard deviation of 10,000 gallons. Management wants to create an incentive bonus for the production crew when the daily production exceeds the 90th percentile of the distribution, in hopes that the crew will, in turn, become more productive. At what level of production should management pay the incentive bonus?

SOLUTION

First, we need to find the 90th percentile in the standard normal distribution. We did this in Example 6.8 (see Figure 6.14), and found that $z = 1.28$ is the standard normal value that will be exceeded only 10% of the time.

Next, we want to convert the standard normal value to an x value, a production level in this example. We know that the production level at which the incentive bonus will be paid is $z = 1.28$ standard deviations above the mean level. To determine the production level, remember that

$$z = \frac{x - \mu}{\sigma}$$

If we solve this equation for x, we find

$$x = \mu + z\sigma$$

In this example, $\mu = 100,000$, $z = 1.28$, and $\sigma = 10,000$, so

$$x = 100,000 + 1.28(10,000) = 100,000 + 12,800$$
$$= 112,800$$

Thus, the 90th percentile of the production distribution is $112,800$ gallons. Management should pay an incentive bonus when a day's production exceeds this level if its objective is to pay only when the production is in the top 10% of the current production distribution. ∎

CASE STUDY 6.1

EVALUATING AN
INVESTMENT'S RISK

Frederick S. Hillier (1963) described several ways a business firm can easily, but effectively, evaluate risky investment projects. He noted that, up to the time of his writing,

> such procedures as have been suggested for dealing with risk have tended to be either quite simplified or somewhat theoretical. Thus, these procedures have tended to provide management with only a portion of the information required for a sound decision, or they have assumed the availability of information which is almost impossible to obtain.

In one of his approaches to handling risk, Hillier assumes that the cash flow from an investment to the firm in the ith future year after the investment is made is normally distributed, and shows that the present worth, P, of the proposed investment is therefore normally distributed with mean μ_P and variance σ_P^2. He points out that by describing P with a probability distribution as he has done, management is provided with information about P and, as a result, with some basis upon which to evaluate the risk of the investment decision.

Hillier provides an example of how management can evaluate the risk of an investment by assuming its present worth is normally distributed:

> Suppose that, on the basis of the forecasts regarding prospective cash flow from a proposed investment of $10,000, it is determined that $\mu_P = \$1,000$ and $\sigma_P = \$2,000$. Ordinarily, the current procedure would be to approve the investment since $\mu_P > 0$. However, with additional information available ($\sigma_P = \$2,000$) regarding the considerable risk of the investment, the executive can analyze the situation further. Using widely available tables for the normal distribution, he could note that the probability that $P < 0$, so that the investment won't pay, is 0.31. Furthermore, the probability is 0.16, 0.023, and 0.0013, respectively, that the investment will lose the present worth equivalent of at least $1,000, $3,000, and $5,000, respectively. Considering the financial status of the firm, the executive can use this and similar information to make his decision. Suppose, instead, that the executive is attempting to choose between this investment and a second investment with

μ_P = \$500 and σ_P = \$500. By conducting a similar analysis for the second investment, the executive can decide whether the greater expected earnings of the first investment justifies the greater risk. A useful technique for making this comparison is to superimpose the drawing of the probability distribution of P for the second investment upon the corresponding drawing for the first investment. This same approach generalizes to the comparison of more than two investments.

| | | | | | | | | | | | | |

EXERCISES 6.13–6.37

LEARNING THE MECHANICS

6.13 Use Table IV in Appendix B to calculate the area under the standard normal distribution between the following pairs of z-scores:
a. $z = 0$ and $z = 2$ **b.** $z = 0$ and $z = 3.0$
c. $z = 0$ and $z = 1.5$ **d.** $z = 0$ and $z = .80$

6.14 Repeat Exercise 6.13 for each of the following pairs of standard normal z values:
a. $z = -1$ and $z = 1.5$ **b.** $z = -2$ and $z = 2$
c. $z = -.5$ and $z = 1.5$ **d.** $z = -3$ and $z = 3$

6.15 Repeat Exercise 6.13 for each of the following pairs of standard normal z values:
a. $z = -1.5$ and $z = -1.0$ **b.** $z = -.45$ and $z = .90$
c. $z = -2$ and $z = 0$ **d.** $z = -3$ and $z = 1.4$
e. $z = .5$ and $z = 1.5$ **f.** $z = -2$ and $z = -.5$

6.16 Find each of the following probabilities for the standard normal random variable z:
a. $P(z \geq 3)$ **b.** $P(z \leq -1.6)$ **c.** $P(z \geq 1.645)$
d. $P(z \geq 0)$ **e.** $P(z \leq -1.0)$ **f.** $P(z \leq -1.645)$

6.17 Find each of the following probabilities for the standard normal random variable z:
a. $P(z = 1)$ **b.** $P(z \leq 1)$
c. $P(z < 1)$ **d.** $P(z > 1)$

6.18 Find each of the following probabilities for the standard normal random variable z:
a. $P(-1 \leq z \leq 1)$ **b.** $P(-1.96 \leq z \leq 1.96)$
c. $P(-1.645 \leq z \leq 1.645)$ **d.** $P(-2 \leq z \leq 2)$

6.19 Find each of the following probabilities for the standard normal random variable z, and compare your answers to those of Exercise 6.18:
a. $P(-1 \leq z < 1)$ **b.** $P(-1.96 < z < 1.96)$
c. $P(-1.645 < z \leq 1.645)$ **d.** $P(-2 < z < 2)$

6.20 Find a value of the standard normal random variable z, call it z_0, such that:
a. $P(z \geq z_0) = .05$ **b.** $P(z \geq z_0) = .025$
c. $P(z \leq z_0) = .025$ **d.** $P(z \geq z_0) = .10$
e. $P(z > z_0) = .10$

6.21 Find a value of the standard normal random variable z, call it z_0, such that:
a. $P(z \leq z_0) = .0301$ **b.** $P(-z_0 \leq z \leq z_0) = .95$
c. $P(-z_0 \leq z < z_0) = .90$ **d.** $P(-z_0 \leq z \leq z_0) = .6826$
e. $P(z_0 \leq z \leq 0) = .1628$ **f.** $P(-.75 < z < z_0) = .7026$

6.22 Suppose the random variable x is best described by a normal distribution with $\mu = 30$ and $\sigma = 4$. Find the value of the standard normal random variable z that corresponds to each of the following x values:
  a. $x = 20$ **b.** $x = 30$ **c.** $x = 27.5$
  d. $x = 15$ **e.** $x = 35$ **f.** $x = 25$

6.23 Refer to Exercise 6.22. How many standard deviations away from the mean of x are each of the following x values?
  a. $x = 25$ **b.** $x = 37.5$ **c.** $x = 30$ **d.** $x = 36$

6.24 Suppose the continuous random variable x has a normal probability distribution with mean 120 and variance 36. Draw a rough sketch (i.e., a graph) of the frequency function of x. Locate μ and the interval $\mu \pm 2\sigma$ on the graph. Find the following probabilities:
  a. $P(\mu - 2\sigma \leq x \leq \mu + 2\sigma)$ **b.** $P(x \geq 128)$ **c.** $P(x \leq 108)$
  d. $P(112 \leq x \leq 130)$ **e.** $P(114 \leq x \leq 116)$ **f.** $P(115 \leq x \leq 128)$

6.25 The random variable x has a normal distribution with $\mu = 1,000$ and $\sigma = 10$.
  a. Find the probability that x assumes a value more than 2 standard deviations from its mean. More than 3 standard deviations from μ.
  b. Find the probability that x assumes a value within 1 standard deviation of its mean. Within 2 standard deviations of μ.

6.26 The random variable x has a normal distribution with mean 50 and variance 9. Find the value of x, call it x_0, such that:
  a. $P(x \leq x_0) = .8413$
  b. $P(x > x_0) = .025$
  c. $P(x > x_0) = .95$
  d. $P(41 \leq x < x_0) = .8630$

6.27 The random variable x has a normal distribution with standard deviation 25. It is known that the probability that x exceeds 150 is .90. Find the mean μ of the probability distribution.

APPLYING THE CONCEPTS

6.28 Personnel tests are designed to test a job applicant's cognitive and/or physical abilities. An IQ test is an example of the former; a speed test involving the arrangement of pegs on a peg board is an example of the latter. During the 1970's, the proportion of employers using personnel tests dropped from more than 90% to less than 50%, in part due to concerns with equal-rights laws. As a result of improved testing procedures and the realization that such tests could hold down the costs associated with hiring the wrong person, the use of personnel tests increased dramatically during the 1980's (Dessler, 1986).

    A particular dexterity test is administered nationwide by a private testing service. It is known that for all tests administered last year the distribution of scores was approximately normal with mean 75 and standard deviation 7.5.
  a. A particular employer requires job candidates to score at least 80 on the dexterity test. Approximately what percentage of the test scores during the past year exceeded 80?

b. The testing service reported to a particular employer that the score of one of its job candidates fell at the 98th percentile of the distribution (i.e., approximately 98% of the scores were lower than the candidate's, and only 2% were higher). What was the candidate's score?

6.29 Ideally, a worker seeking a new job in a particular industry should acquire information about wage rates offered by all firms in the industry. However, this information-search could be time-consuming and costly. In particular, the longer an unemployed worker searches for a higher wage, the greater will be the loss in income. Therefore, workers may not find it worthwhile to search until they find the highest available wage rate. The result is that managers may not have to pay top dollar to attract workers. These factors help explain the existing disparity in wage rates among firms (Blair and Kenny, 1982). Suppose the distribution of wage rates nationwide that would be offered to a particular skilled worker can be approximated by a normal distribution with $\mu = \$10.50$ per hour and $\sigma = \$1.25$ per hour. In addition, assume that the worker is offered $12.00 per hour by the first firm contacted.

a. Suppose the worker were to undertake a nationwide job search. What proportion of the wage rates that would be offered to the worker would be greater than $12.00 per hour?

b. If the worker were to complete a nationwide job search and then randomly select one of the many job offers received, what is the probability that the wage rate would be more than $10.00 per hour?

c. The **median**, call it x_m, of a continuous random variable x is the value such that $P(x \geq x_m) = P(x \leq x_m) = .5$. That is, the median is the value x_m such that half the area under the probability distribution lies above x_m and half lies below it. Find the median of the random variable corresponding to the wage rate and compare it to the mean wage rate.

6.30 According to recent studies in the banking industry, "insider loans" were the principal cause of more than half of all bank failures over the past 15 years. (An *insider loan* is money borrowed by directors and stockholders from the nationally chartered bank with which they are affiliated.) If it is known that the amount of money tied up in outstanding insider loans of nationally chartered banks across the country is approximately normally distributed with a mean of $500,000 and a standard deviation of $142,000, find the probability that a nationally chartered bank selected at random has insider loans totaling at least $900,000.

6.31 An important quality characteristic for soft-drink bottlers is the amount of soft drink injected into each bottle. This volume is determined (approximately) by measuring the height of the soft drink in the neck of the bottle and comparing it to a scale that converts the height measurement to a volume measurement (Montgomery, 1985). In a particular filling process, the number of ounces injected into 8-ounce bottles is approximately normally distributed with mean 8.00 ounces and standard deviation .05 ounce. Bottles that contain less than 7.9 ounces do not meet the bottler's quality standard and are sold at a substantial discount.

a. If 20,000 bottles are filled, approximately how many will fail to meet the quality standard?

b. Suppose that, due to the failure of one of the filling system's components, the mean of the filling process shifts to 7.95 ounces. (Assume that the standard deviation remains .05 ounce.) If 20,000 bottles are filled, approximately how many will fail to meet the quality standard?

c. Suppose that a different component fails and, although the mean of the filling process remains 8.00 ounces, the standard deviation increases to .1 ounce. If 20,000 bottles are filled, approximately how many will fail to meet the quality standard?

6.32 The *monthly rate of return* of a stock is a measure investors frequently use for evaluating the behavior of a stock over time. A stock's monthly rate of return generally reflects the amount of money an investor makes (or loses if the return is negative) for every dollar invested in the stock in a given month. Thus, stocks with high average monthly rates of return typically offer more lucrative investment opportunities than stocks with low average monthly rates of return. Eugene Fama (1976) has demonstrated that the probability distribution for the monthly rate of return of a stock can be approximated by a normal probability distribution. Suppose the monthly rates of return to stock ABC are normally distributed with mean .05 and standard deviation .03, and the monthly rates of return to stock XYZ are normally distributed with mean .07 and standard deviation .05. Assume that you have $100 invested in each stock.

a. Over the long run, which stock will yield the higher average monthly rate of return? Why?

b. Suppose you plan to hold each stock for only 1 month. What is the expected value of each investment at the end of 1 month?

c. Which stock offers greater protection against incurring a loss on your investment next month? Why?

6.33 Do security analysts do a good job of forecasting corporate earnings growth and advising their clientele? David Dreman, a *Forbes* columnist, addresses this question in an article titled, "Astrology Might Be Better" (*Forbes*, Mar. 26, 1984). The basis of Dreman's article is a study by Professors Michael Sandretto of Harvard and Sudhir Milkrishna-murthi of the Massachusetts Institute of Technology. The study surveys security analysts' forecasts of annual earnings for the (then) current year for more than 769 companies with five or more forecasts per company per year. The average forecast error for this large number of forecasts was plus or minus 31.3%. To apply this information to a practical situation, suppose the population of analysts' forecast errors is normally distributed with a mean of 31.3% and a standard deviation of 10%.

a. If you obtain a security analyst's forecast for a particular company, what is the probability that it will be in error by more than 50%?

b. If three analysts make the forecast, what is the probability that at least one of the analysts will err by more than 50%?

6.34 In a survey of 2,000 long-distance telephone calls reported in the *Orlando Sentinel* (Mar. 12, 1984), it was found that seven of eight long-distance phone companies were overcharging (charging for additional time) for a given call and that six were charging for unconnected calls. Suppose that the additional time being charged to a long-distance phone call has a normal distribution with a mean of 25 seconds and a standard deviation of 8 seconds.

a. Find the probability that a given long-distance call will be overcharged by at least 40 seconds.

b. By no more than 10 seconds.

6.35 A company that sells annuities must base the annual payout on the probability distribution of the length of life of the participants in the plan. Suppose the probability distribution of the lifetimes of the participants in the plan is approximately a normal distribution with $\mu = 68$ years and $\sigma = 3.5$ years.

a. What proportion of the plan participants would receive payments beyond age 70?

b. Beyond age 75?

6.36 A machine used to regulate the amount of dye dispensed for mixing shades of paint can be set so that it discharges an average of μ milliliters of dye per can of paint. The amount of dye discharged is known to have a normal distribution with variance equal to .160. If more than 6 milliliters of dye are discharged when making a particular shade of blue paint, the shade is unacceptable. Determine the setting for μ so that no more than 1% of the cans of paint will be unacceptable.

6.37 What relationship exists between the standard normal distribution and the box-plot methodology (Section 3.10) for describing distributions of data using quartiles? The answer depends on the true underlying probability distribution of the data. Assume for the remainder of this exercise that the distribution is normal.

a. Calculate the values of the standard normal random variable z, call them z_L and z_U, that correspond to the hinges of the box plot, i.e., the lower and upper quartiles, Q_L and Q_U, of the probability distribution.

b. Calculate the z values that correspond to the inner fences of the box plot for a normal probability distribution.

c. Calculate the z values that correspond to the outer fences of the box plot for a normal probability distribution.

d. What is the probability that an observation lies beyond the inner fences of a normal probability distribution? The outer fences?

e. Can you better understand why the inner and outer fences of a box plot are used to detect outliers in a distribution? Explain.

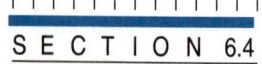

SECTION 6.4

THE EXPONENTIAL DISTRIBUTION (OPTIONAL)

Another important probability distribution that is useful for describing business data is the **exponential probability distribution**. Two business phenomena with frequency functions that might be well approximated by the exponential distribution are the length of time between arrivals at a fast-food drive-through restaurant and the length of time between the filing of claims in a small insurance office. Note that in each of these examples, the measurements are the lengths of time between certain events. For this reason, the exponential distribution is sometimes called the **waiting time distribution**.

The formula for the exponential probability distribution is shown in the box, along with the mean and standard deviation of this frequency function.

PROBABILITY DISTRIBUTION, MEAN, AND STANDARD DEVIATION FOR AN EXPONENTIAL RANDOM VARIABLE x

$$f(x) = \lambda e^{-\lambda x} \qquad (x > 0)$$

$$\mu = \frac{1}{\lambda} \qquad \sigma = \frac{1}{\lambda}$$

Unlike the normal distribution, which has a shape and location determined by the values of the two quantities μ and σ, the shape of the exponential distribution is governed by a single quantity, λ. Further, it is a probability distribution with the property that its mean equals its standard deviation. Exponential distributions corresponding to $\lambda = .5$, 1, and 2 are shown in Figure 6.16.

FIGURE 6.16
Exponential Distributions

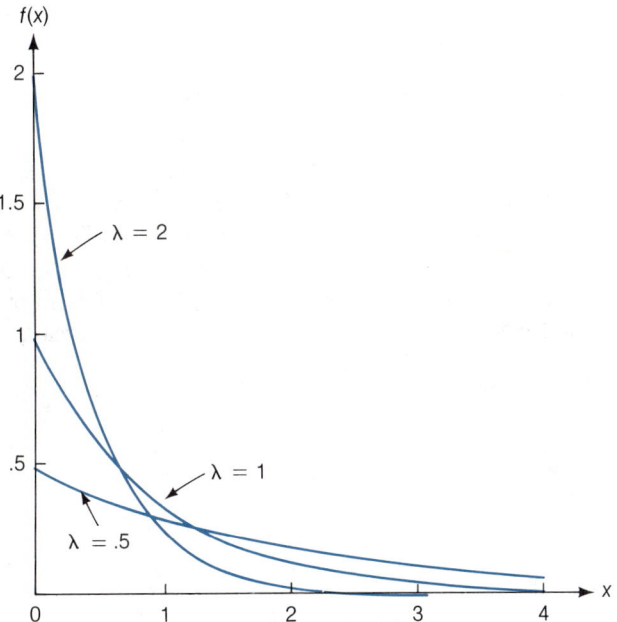

To calculate probabilities for exponential random variables, we need to be able to find areas under the exponential probability distribution. Suppose we want to find the area, A, to the right of some number, a, as shown in Figure 6.17 (page 284). This area can be calculated by using the following formula:

FINDING THE AREA, A, TO THE RIGHT OF A NUMBER, a, FOR AN EXPONENTIAL DISTRIBUTION*

$$A = P(x \geq a) = e^{-\lambda a}$$

Use Table V in Appendix B to find the value of $e^{-\lambda a}$ after substituting the appropriate numerical values for λ and a.

*This area is calculated by integration:

$$\int_a^\infty \lambda e^{-\lambda x}\, dx = -e^{-\lambda x}\Big|_a^\infty = e^{-\lambda a}$$

FIGURE 6.17
The Area, A, to the Right
of a Number, a, for an
Exponential Distribution

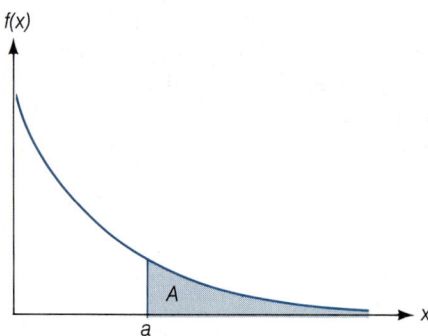

$f(x)$

EXAMPLE 6.11

Suppose the length of time (in days) between sales for an automobile salesperson is modeled as an exponential random variable with $\lambda = .5$. What is the probability that the salesperson goes more than 5 days without a sale?

SOLUTION

The probability we want is the area, A, to the right of $a = 5$ in Figure 6.18. To find this probability, use the formula given for area:

$$A = e^{-\lambda a} = e^{-(.5)(5)} = e^{-2.5}$$

FIGURE 6.18
Exponential Distribution
for Example 6.11: $\lambda = .5$

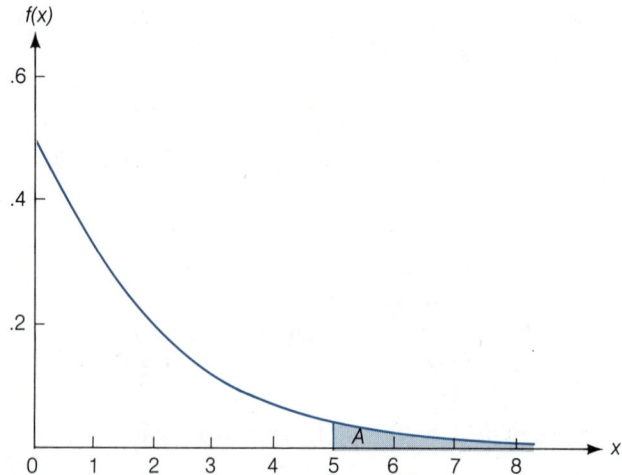

Referring to Table V, we find

$$A = e^{-2.5} = .082085$$

That is, our model indicates that the automobile salesperson has a probability of about .08 of going more than 5 days without a sale. ∎

EXAMPLE 6.12

A microwave oven manufacturer is trying to determine the length of warranty period it should attach to its magnetron tube, the most critical component in the oven. Preliminary testing has shown that the length of life (in years), x, of a magnetron tube has an exponential probability distribution with $\lambda = .16$.

a. Find the mean and standard deviation of x.
b. If a warranty period of 5 years is attached to the magnetron tube, what fraction of tubes must the manufacturer plan to replace (assuming the exponential model with λ = .16 is correct)?
c. Find the probability that the length of life of a magnetron tube will fall within the interval $\mu \pm 2\sigma$.

SOLUTION

a. Using the formulas for the mean and standard deviation for an exponential random variable, we find

$$\mu = \frac{1}{\lambda} = \frac{1}{.16} = 6.25 \text{ years}$$

Also, since $\mu = \sigma$, $\sigma = 6.25$ years.

b. To find the fraction of tubes that will have to be replaced before the 5-year warranty period expires, we need to find the area between 0 and 5 under the distribution. This area, A, is shown in Figure 6.19. To find the required probability, we recall the formula

$$P(x > a) = e^{-\lambda a}$$

FIGURE 6.19
Exponential Distribution
for Example 6.12: λ = .16

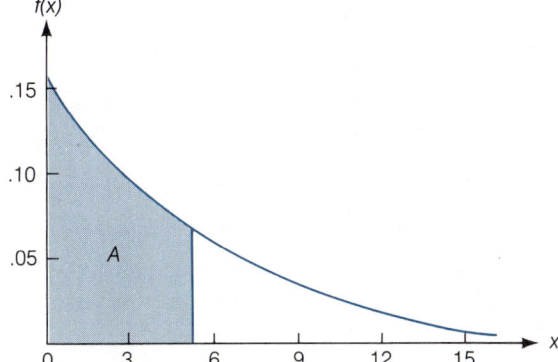

Using this formula and Table V of Appendix B, we find

$$P(x > 5) = e^{-\lambda(5)} = e^{-(.16)(5)} = e^{-.80} = .449329$$

To find the area A, we use the complementary relationship:

$$P(x \leq 5) = 1 - P(x > 5) = 1 - .449329 = .550671$$

So, approximately 55% of the magnetron tubes will have to be replaced during the 5-year warranty period.

c. We would expect the probability that the life of a magnetron tube, x, falls within the interval $\mu \pm 2\sigma$ to be quite large (near .95 if we think the Empirical Rule in Table 3.8 might apply). A graph of the exponential distribution showing the interval from $\mu - 2\sigma$ to $\mu + 2\sigma$ is shown in Figure 6.20 (page 286). Since the point $\mu - 2\sigma$ lies below x = 0, we need to find only the area between x =

0 and $a = \mu + 2\sigma = 6.25 + 2(6.25) = 18.75$. This area, A, which is shaded in Figure 6.20, is

$$A = 1 - P(x > 18.75)$$
$$= 1 - e^{-\lambda(18.75)} = 1 - e^{-(.16)(18.75)} = 1 - e^{-3}$$

FIGURE 6.20

Exponential Distribution for Example 6.12: $\lambda = .16$

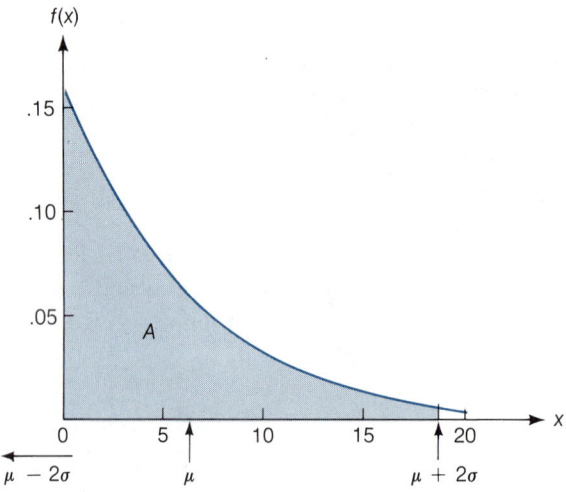

Checking Table V (Appendix B) for the value of e^{-3}, we find $e^{-3} = .049787$. Therefore, the probability that the life, x, of a magnetron tube falls within the interval $\mu \pm 2\sigma$ is

$$A = 1 - e^{-3}$$
$$= 1 - .049787 = .950213$$

You can see that this probability agrees very well with the Empirical Rule even though this probability distribution is not mound-shaped (it is strongly skewed to the right). ∎

CASE STUDY 6.2

QUEUEING THEORY

The formation of waiting lines, or *queues*, is a phenomenon that occurs whenever the demand for a service exceeds its supply. We see this daily at bank-teller windows, supermarket checkout counters, traffic lights, etc. If long queues develop, it may be an indication that not enough service is being provided. If no queues develop, it may be an indication that too much service is being provided. Either situation can prove costly to the service provider. To assist in planning service capacity, an area of study known as **queueing theory** has been employed to model the characteristics of waiting lines.

In the basic structure assumed by most queueing models, *customers* seeking service are generated over time by an *input source*. These customers enter the *queueing system*, join a queue, and await service. Members of the queue are selected for service by some rule (for example, first-come, first-served or random sampling) known as the *service discipline*. A chosen customer is served by the *service mechanism* and then leaves the queueing system. This process is illustrated in Figure 6.21.

FIGURE 6.21

Basic Queueing System

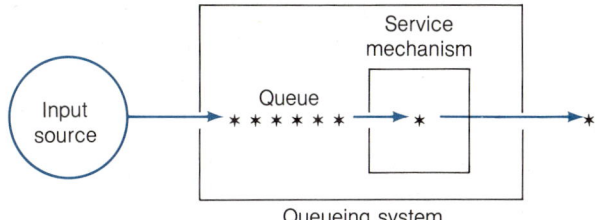

Queueing system

To complete the basic queueing model, certain assumptions must be made to model probabilistically the arrivals to and departures from the queueing system. It has been found that the *interarrival time* (the time between arrivals) to many real queues can be reasonably approximated by an exponential probability distribution. Furthermore, when the specific service required differs among individual customers, the exponential distribution has proved to provide an adequate approximation to the time required to service a customer (i.e., the time that elapses between when service begins and when it ends). Thus, the exponential distribution can be used to describe both the input source and the service mechanism. A more detailed description of the basic queueing theory model and its various assumptions can be found in Hillier and Lieberman (1983).

After developing a queueing model, we can answer such questions as: (1) What is the expected number of customers in the queueing system? (2) What is the expected length of the queue? (3) What is the expected waiting time in the system for an individual customer? (4) What is the expected waiting time in the queue for an individual customer? (5) Are the rates of arrival and departure such that the queue will continue to grow without bound?

W. Blaker Bolling (1972) describes how queueing theory was used to model the emergency room of the Richmond Memorial Hospital in Richmond, Virginia. The input source was the local population of Richmond (including visitors). The service mechanism consisted of eight fully staffed treatment tables. Thus, since it was possible to service eight patients simultaneously, the system was more complex than that described in Figure 6.21. In general, the service discipline was first-come, first-served, unless a serious emergency occurred. However, in modeling the system, a first-come, first-served discipline was assumed for simplicity. Historical data indicated that both interarrival times and service times could be adequately modeled with exponential distributions.

One use of the model was to project the capacity (number of staffed tables) needed to prevent the queue from growing without bound. For example, for the period between 8 and 9 P.M. in August 1972, the interarrival-time distribution was projected to be exponential with mean 5.96 minutes (.0993 hour), and the service-time distribution was projected to be exponential with mean 58 minutes (.9666 hour). Accordingly, the arrival rate was projected to be 10.07 per hour (i.e., $\lambda = 10.07$ for the interarrival-time distribution), and the service rate was projected to be 1.0345 patients per table per hour (i.e., $\lambda = 1.0345$ for the service-time distribution). Since there were just eight service tables, arrivals would exceed departures and, theoretically, the waiting room queue could grow without bound as long as rates of arrival and departure remained unchanged. It was determined through further analysis with the queueing model that it would be necessary to staff at least two more treatment tables to prevent this situation from occurring.

In a discussion about the reliability of computer software, G. J. Schick (1974) says the following:

> Custom software . . . is expensive to develop and requires extensive testing—the goal being to certify that the software is in error-free condition, ready to support the mission for which it was designed. Similar economies should also be expected from an integrated statistical software test program. Traditionally, there are never enough time or resources to test all possible branches and data combinations in a computer program of reasonable size.
>
> Current practice is to design and develop a software system and then to test it to detect errors, until the amount of time and expense required to discover remaining errors is too great to justify further testing. . . . In principle, few large real-time computer programs ever have been tested completely and unequivocally in the sense that every logical data path has been successfully executed under every logical combination for the data at hand for all possible options. One management objective would be to test every logical path in the computer program at least once with some kind of numerical check. At the present state of the art, such a degree of testing is neither feasible nor realistic. In practice, the contractor must be willing to release and the customer willing to accept a level of risk associated with a program that has been less than completely checked.

In finding and correcting errors in a computer program (*debugging*) and determining the program's reliability, Schick and others have noted the importance of the distribution of the time until the next program error is found. If this distribution is assumed to be exponential, with

$$f(x) = \lambda e^{-\lambda x} \qquad (x > 0, \quad \lambda > 0)$$

then, as Schick points out, its mean, $1/\lambda$, would be the average time required to find the next error.

In his article, Schick describes a method relevant to software reliability for estimating the parameter, λ, of the exponential distribution. Using computer debugging data supplied by the U.S. Navy, Schick demonstrates how this estimation procedure and the exponential distribution can be used to estimate the reliability of a computer program. [*Note:* The model used by Schick to represent the distribution for the time until the next error is based on the exponential distribution, but it is slightly more complicated because he assumes that λ varies. For our purposes, however, nothing is lost by assuming the distribution to be exponential.]

After 26 program errors were found, Schick estimated λ to be .042. Accordingly, $1/\lambda = 23.8$ days. This means that the average time it would take to find the next (twenty-seventh) error would be about 24 days. Thus, the probability of it taking, say, 60 or more days to find the next error is

$$P(x \geq 60) = e^{-(.042)(60)} = .08046$$

Over the next 290 days, five more errors were detected. Since this is a rate of about one error every 60 days and since $P(x \geq 60) \approx .08$, it seems unlikely that an exponential distribution with $\lambda = .042$ is an appropriate representation of the distribution for the time until the next error. Based on the number of new errors found and the length of time it took to find them, Schick reestimated λ. He found $\lambda = .0036$. Thus, $1/\lambda = 278$ days, meaning that on average the next error (thirty-second) would not occur for 278 days. At this point, the length of time and, therefore, the cost required to find any remaining program errors may be prohibitive. Debugging should probably be discontinued.

EXERCISES 6.38–6.50

APPLYING THE MECHANICS

6.38 The random variables x and y have exponential distributions with $\lambda = 3$ and $\lambda = .75$, respectively. Using Table V in Appendix B, carefully plot both distributions on the same set of axes.

6.39 Use Table V in Appendix B to determine the value of $e^{-\lambda a}$ for each of the following cases:
a. $\lambda = 1$, $a = 1$ **b.** $\lambda = 1$, $a = 2.5$
c. $\lambda = 2.5$, $a = 3$ **d.** $\lambda = 5$, $a = .3$

6.40 Suppose x has an exponential distribution with $\lambda = 3$. Find the following probabilities:
a. $P(x > 2)$ **b.** $P(x > 1.5)$ **c.** $P(x > 3)$ **d.** $P(x > .45)$

6.41 Suppose x has an exponential distribution with $\lambda = 2.5$. Find the following probabilities:
a. $P(x \leq 3)$ **b.** $P(x \leq 4)$ **c.** $P(x \leq 1.6)$ **d.** $P(x \leq .4)$

6.42 Suppose the random variable x is best approximated by an exponential probability distribution with $\lambda = 2$. Find the mean and variance of x. Find the probability that x will assume a value within the interval $\mu \pm 2\sigma$.

6.43 The random variable x can be adequately approximated by an exponential probability distribution with $\lambda = 1$. Find the probability that x assumes a value:
a. More than 3 standard deviations from μ.
b. Less than 2 standard deviations from μ.
c. Within .5 standard deviation of μ.

APPLYING THE CONCEPTS

6.44 A taxi service based at an airport can be characterized as a transportation system with one source terminal and a fleet of vehicles that take passengers from the terminal to different destinations. Each vehicle returns to the terminal after some random trip time and makes another trip. In order to improve the vehicle-dispatching decisions involved in such a system (e.g., how many passengers should be allocated to a waiting taxi?), Sims and Templeton (1985) used queueing theory (see Case Study 6.2) to model and to evaluate the system. In their model, they assumed travel times of successive trips are independent exponential random variables. Assume $\lambda = .05$.
a. What is the mean trip time for the taxi service?
b. What is the probability that a particular trip will take more than 30 minutes?
c. Two taxis have just been dispatched. What is the probability that both will be gone for more than 30 minutes? That at least one of the taxis will return within 30 minutes?

6.45 The shelf-life of a product is a random variable that is related to consumer acceptance and, ultimately, to sales and profit. Suppose the shelf-life of bread is best approximated by an exponential distribution with mean equal to 2 days. What fraction of the loaves stocked today would you expect to still be saleable (i.e., not stale) 3 days from now?

6.46 An article in the Jacksonville, Florida, *Times Union* (Mar. 11, 1984) reports on the unexplained crash of a small plane on takeoff and the resulting injury to its 24-year-old student pilot. The article notes that Shields Aviation, the renter of the plane, inspects (and presumably performs maintenance) on the aircraft every 100 flight hours. The plane had flown 30 hours since its last inspection. Suppose that x, the time between malfunctions for this particular plane, has an exponential distribution with mean equal to 300 hours. What is the probability that a plane of this type will malfunction within 30 hours after the last inspection?

6.47 As discussed in Case Study 6.2, Bolling (1972) used an exponential distribution with mean 58 minutes to model the service-time distribution of each of the eight treatment tables in the emergency room of Richmond Memorial Hospital.
   a. Using this distribution, find the probability that it will take more than 58 minutes to treat a patient in the emergency room. More than 1.5 hours.
   b. What is the probability that each of the next three patients will require more than 58 minutes for treatment?
   c. Recall from part c of Exercise 6.29 that the median, x_m, of a continuous random variable x is the value such that $P(x \geq x_m) = P(x \leq x_m) = .5$. Is the median time required to treat a patient more or less than 58 minutes?
   d. Using Table V in Appendix B, approximate the median of the service-time distribution.

6.48 In Case Study 6.3, the exponential distribution was used to help evaluate the reliability of a computer program. After 26 program errors were found, the time (in days) required to find the next error was determined to have an exponential distribution with $\lambda = .042$.
   a. Graph this exponential distribution, and locate its mean and the interval $\mu \pm \sigma$ on your graph.
   b. What is the mean time required to find the twenty-seventh program error?
   c. What is the probability that it will take less than 30 days to find the twenty-seventh error?
   d. Find the probability that the time required to find the twenty-seventh error is within the interval $\mu \pm \sigma$.
   e. Find the probability that the time required to find the twenty-seventh error is within the interval $\mu \pm 2\sigma$. How does your answer compare with the approximate probability provided by the Empirical Rule of Chapter 3?

6.49 The following is a sample of the lengths of time between arrivals (rounded to the nearest minute) at the emergency room of a hospital:

8	9	28	3	22	7	5	3	4	23
6	4	5	7	10	5	1	9	14	4
2	6	12	4	1	10	5	3	3	14
18	13	18	15	12	1	5	9	26	37
8	1	19	1	8	11	16	3	31	21

    a. Draw a relative frequency histogram for the data. Does it appear that an exponential distribution could be used to characterize the length of time between arrivals? Explain.

    b. If you were asked to model the length of time between arrivals for this emergency room using an exponential distribution, discuss how you would estimate λ.

6.50 Product *reliability* has been defined as the probability that a product will perform its intended function satisfactorily for its intended life when operating under specified conditions (Lamberson, 1985). The *reliability function*, $R(x)$, for a product indicates the probability of the product's life exceeding x time periods. When the time until failure of a product can be adequately modeled by an exponential distribution, the product's reliability function is $R(x) = e^{-\lambda x}$. Suppose that the time to failure (in years) of a particular product is modeled by an exponential distribution with $\lambda = .5$.

    a. What is the product's reliability function?

    b. What is the probability that the product will perform satisfactorily for at least 4 years?

    c. What is the probability that a particular product will survive longer than the mean life of the product?

    d. If λ changes, will the probability that you calculated in part **c** change? Explain.

    e. If 10,000 units of the product are sold, approximately how many will perform satisfactorily for more than 5 years? About how many will fail within 1 year?

    f. How long should the length of the warranty period be for the product if the manufacturer wants to replace no more than 5% of the units sold while under warranty?

S E C T I O N   6.5

APPROXIMATING A BINOMIAL DISTRIBUTION WITH A NORMAL DISTRIBUTION

When a binomial random variable can assume a large number of values, the calculation of its probabilities may become very tedious. To contend with this problem, we provide tables in Appendix B to give the probabilities for some values of n and p, but these tables are by necessity incomplete. For example, the binomial table (Table II) can be used only for $n = 5, 6, 7, 8, 9, 10, 15, 20$, or 25. To get around this limitation, we seek approximation procedures for calculating the probabilities associated with binomial random variables.

    When n is large, a normal probability distribution can provide a good approximation to the probability histogram of a binomial random variable. To show how the approximation works, we refer to Example 5.8, in which we used the binomial distribution to model the number x of twenty employees who favor unionization. We assumed that 60% of all the company's employees favored unionization. The mean and standard deviation of x were found to be $\mu = 12$ and $\sigma = 2.19$. The binomial distribution for $n = 20$ and $p = .6$ is shown in Figure 6.22 (page 292), and the approximating normal distribution with mean $\mu = 12$ and standard deviation $\sigma = 2.19$ is superimposed.

    As part of Example 5.8, we used Table II to find the probability that $x < 10$. This probability, which is shaded in Figure 6.22, was found to equal .128. To find the normal approximation, we first note that $P(0 \leq x < 10)$ corresponds to the area to the left of $x = 9.5$ on the normal curve in the figure. We use $x = 9.5$ rather than $x = 9$ or $x = 10$ so that all the binomial probability corresponding to $x = 9$ is included in the approximating normal curve area, but none of that corresponding to $x = 10$ is included. Because we are approximating a discrete

FIGURE 6.22
Binomial Distribution for
$n = 20, p = .6$ and
Normal Distribution with
$\mu = 12, \sigma = 2.19$

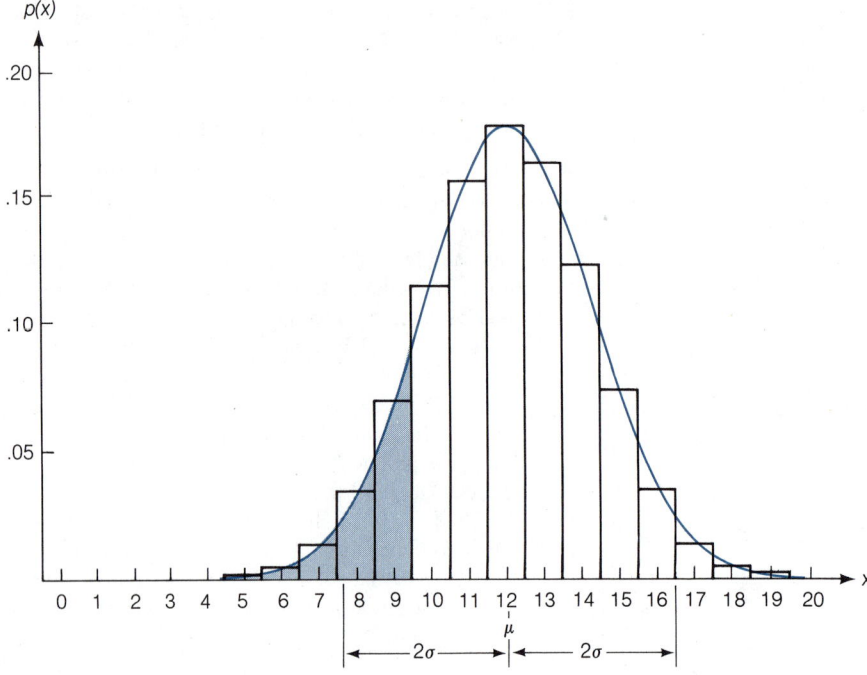

distribution (the binomial) with a continuous distribution (the normal), we call the use of 9.5 (instead of 9 or 10) a **correction for continuity**. That is, we are correcting the discrete distribution so that it can be approximated by the continuous one. The use of the correction for continuity leads to the calculation of the following standard normal z value:

$$z = \frac{x - \mu}{\sigma} = \frac{9.5 - 12}{2.19} = -1.14$$

Using Table IV of Appendix B, we find the area between $z = 0$ and $z = -1.14$ to be .3729. Then the probability that x is less than 10 is approximated by the area under the normal distribution to the left of 9.5, shown shaded in Figure 6.22. That is,

$$P(x \leq 9) \approx P(z \leq -1.14) = .5 - P(-1.14 < z \leq 0)$$
$$= .5 - .3729 = .1271$$

The approximation differs only slightly from the exact binomial probability, .128. Of course, when tables of exact binomial probabilities are available, we will use the exact value rather than a normal approximation.

Use of the normal distribution will not always provide a good approximation for binomial probabilities. The following is a useful rule of thumb: the interval

$\mu \pm 3\sigma$ should lie within the range of the binomial random variable x (i.e., 0 to n) in order for the normal approximation to be adequate. The rule works well because almost all of the normal distribution falls within 3 standard deviations of the mean, so if this interval is contained within the range of x values, there is "room" for the normal approximation to work.

As shown in Figure 6.23(a) for the example above with $n = 20$ and $p = .6$, the interval $\mu \pm 3\sigma = 12 \pm 3(2.19) = (5.43, 18.57)$ lies within the range 0 to 20. However, if we were to try to use the normal approximation with $n = 10$ and $p = .1$, the interval $\mu \pm 3\sigma$ is $1 \pm 3(.95)$, or $(-1.85, 3.85)$. As shown in Figure 6.23(b), this interval is not contained within the range of x, since $x = 0$ is a lower bound for the binomial distribution. Note in Figure 6.23(b) that the normal distribution will not "fit" in the range of x, and therefore will not provide a good approximation to the binomial probabilities.

FIGURE 6.23

Rule of Thumb for Normal Approximation to Binomial Probabilities

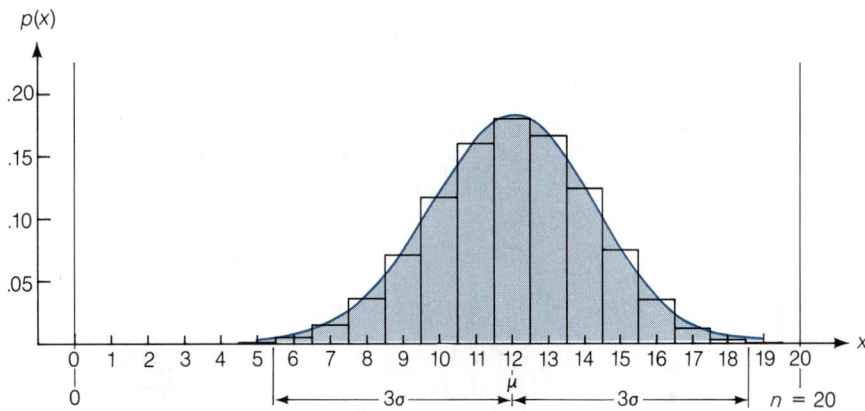

(a) $n = 20$, $p = .6$: Normal approximation is good

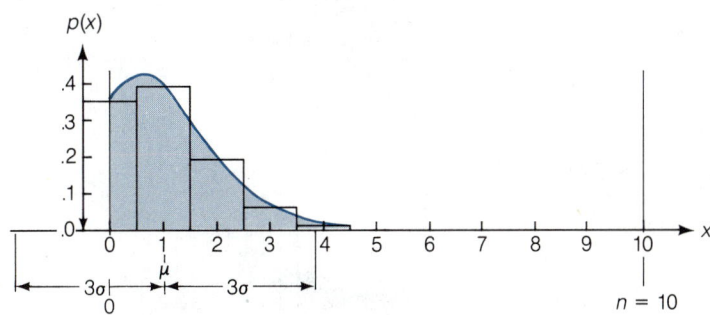

(b) $n = 10$, $p = .1$: Normal approximation is poor

The steps for approximating a binomial probability by a normal probability are given in the box.

USING A NORMAL DISTRIBUTION TO APPROXIMATE BINOMIAL PROBABILITIES

1. After you have determined n and p for the binomial distribution, calculate the interval

   $$\mu \pm 3\sigma = np \pm 3\sqrt{npq}$$

   If the interval lies in the range 0 to n, the normal distribution will provide a reasonable approximation to the probabilities of most binomial events.

2. If the binomial probability to be approximated is of the form $P(x \le a)$ or $P(x > a)$, the correction for continuity is $(a + .5)$, and the approximating standard normal z value is

   $$z = \frac{(a + .5) - \mu}{\sigma}$$

   See Figure 6.24(a).

3. If the binomial probability to be approximated is of the form $P(x \ge a)$ or $P(x < a)$, the correction for continuity is $(a - .5)$, and the approximating standard normal z value is

   $$z = \frac{(a - .5) - \mu}{\sigma}$$

   See Figure 6.24(b).

4. If the binomial probability to be approximated is an interval, such as $P(a \le x < b)$, treat the ends of the interval separately, calculating two distinct z values according to either step 2 or step 3, whichever is appropriate.

5. Sketch the approximating normal distribution and shade the area corresponding to the probability of the event of interest, as in Figure 6.24. Using Table IV of Appendix B and the z value(s) you calculated in steps 2–4, find the shaded area. This is the approximate probability of the binomial event.

FIGURE 6.24

Approximating Binomial Probabilities by Normal Probabilities

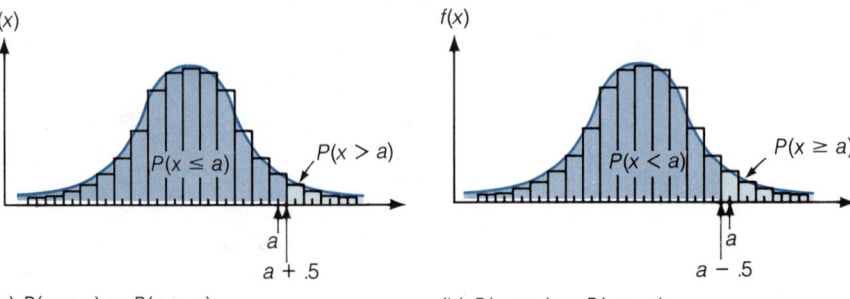

(a) $P(x \le a)$ or $P(x > a)$

(b) $P(x < a)$ or $P(x \ge a)$

EXAMPLE 6.13

The pocket calculator has become relatively inexpensive because its solid-state circuitry is stamped by machine, thus making mass production feasible. One problem with anything that is mass-produced is quality control. The process must be monitored or audited to be sure the output of the process conforms to requirements.

One method of dealing with this problem is **lot acceptance sampling**, in which items being produced are sampled at various stages of the production process and are carefully inspected. The lot of items from which the sample is drawn is then accepted or rejected, based on the number of defectives in the sample. Lots that are accepted may be sent forward for further processing or may be shipped to customers; lots that are rejected may be reworked or scrapped. For example, suppose a manufacturer of calculators chooses 200 stamped circuits from the day's production and determines x, the number of defective circuits in the sample. Suppose that up to a 6% rate of defectives is considered acceptable for the process.

a. Find the mean and standard deviation of x, assuming the defective rate is 6%.
b. Use the normal approximation to determine the probability that 20 or more defectives are observed in the sample of 200 circuits—i.e., find the approximate probability that $x \geq 20$.

SOLUTION

a. The random variable x is binomial, with $n = 200$ and the fraction defective $p = .06$. Thus,

$$\mu = np = 200(.06) = 12$$
$$\sigma = \sqrt{npq} = \sqrt{200(.06)(.94)} = \sqrt{11.28} = 3.36$$

b. We first note that

$$\mu \pm 3\sigma = 12 \pm 3(3.36) = 12 \pm 10.08 = (1.92, 22.08)$$

lies completely within the range from 0 to 200, so a normal probability distribution should provide an adequate approximation to this binomial distribution.

To find the approximating area corresponding to $x \geq 20$, refer to Figure 6.25. Note that we want to include all of the binomial probability histogram from 20 to 200, inclusive. But in order to include the entire rectangle corresponding to $x = 20$, we must begin the approximating area at $20 - .5 = 19.5$. In other

FIGURE 6.25

Normal Approximation to the Binomial Distribution with $n = 200$, $p = .06$

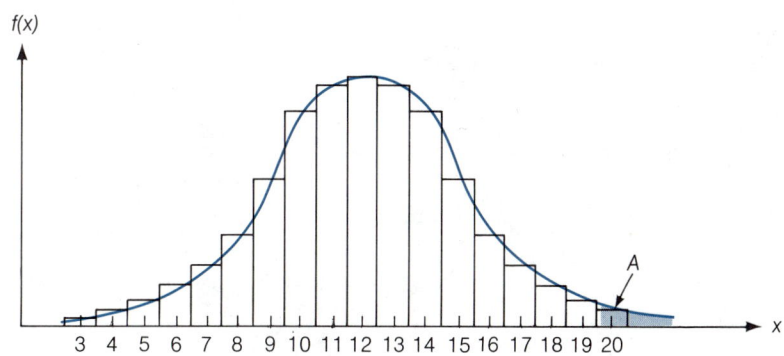

words, since the event is of the form $x \geq a$ with $a = 20$, the correction for continuity is $a - .5 = 20 - .5 = 19.5$. Thus, the z value is

$$z = \frac{(a - .5) - \mu}{\sigma} = \frac{19.5 - 12}{3.36} = \frac{7.5}{3.36} = 2.23$$

Referring to Table IV of Appendix B, we observe that the area to the right of the mean corresponding to $z = 2.23$ (see Figure 6.26) is .4871. So, the area A is

$$A = .5 - .4871 = .0129$$

FIGURE 6.26

Standard Normal Distribution

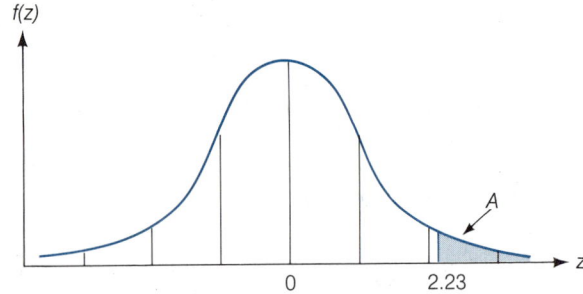

Thus, the normal approximation to the binomial probability is

$$P(x \geq 20) \approx .0129$$

In other words, the probability is extremely small that 20 or more defectives would be observed in a sample of 200 circuits, *if in fact the true defective rate is 6%*. If the manufacturer were to observe $x \geq 20$, the likely reason is that the process is producing more than the acceptable 6% defectives.   ■

EXERCISES 6.51–6.64

LEARNING THE MECHANICS

6.51 Why might you want to use a normal distribution to approximate a binomial distribution?

6.52 Under what circumstances is it appropriate to approximate a binomial distribution with a normal distribution?

6.53 Assume that x is a binomial random variable with n and p as specified in parts **a–f**. For which cases would it be appropriate to use a normal distribution to approximate the binomial distribution?
  a. $n = 50, \quad p = .01$ **b.** $n = 20, \quad p = .45$
  c. $n = 10, \quad p = .4$ **d.** $n = 1{,}000, \quad p = .1$
  e. $n = 200, \quad p = .8$ **f.** $n = 35, \quad p = .7$

6.54 Suppose that x is a binomial random variable with $p = .4$ and $n = 25$.
  a. Would it be appropriate to approximate the probability distribution of x with a normal distribution? Explain.

b. Assuming that a normal distribution provides an adequate approximation to the distribution of x, what are the mean and variance of the approximating normal distribution?

c. Use Table II of Appendix B to find the exact value of $P(x \geq 9)$.

d. Use the normal approximation to find $P(x \geq 9)$.

6.55 Assume that x is a binomial random variable with $n = 25$ and $p = .5$. Use Table II of Appendix B and the normal approximation to find the exact and approximate values, respectively, for the following probabilities:

a. $P(x \leq 12)$ **b.** $P(x \geq 15)$ **c.** $P(9 \leq x \leq 15)$

6.56 Assume that x is a binomial random variable with $n = 100$ and $p = .45$. Use a normal approximation to find the following:

a. $P(x \leq 45)$ **b.** $P(40 \leq x \leq 50)$ **c.** $P(x \geq 38)$

6.57 Assume that x is a binomial random variable with $n = 1,000$ and $p = .50$. Find each of the following probabilities:

a. $P(x > 500)$
b. $P(490 \leq x < 500)$
c. $P(x > 1,000)$

APPLYING THE CONCEPTS

6.58 The *Statistical Abstract of the United States: 1986* reports that 23.4% of the country's 85,407,000 households are inhabited by one person. If 1,000 randomly selected homes are to participate in a Nielsen survey to determine television ratings, find the approximate probability that no more than 250 of these homes are inhabited by one person.

6.59 In 1982, General Motors Corporation's auto assembly plant in Fremont, California, was shut down and turned over to Toyota Motor Corporation as part of a joint venture with the Japanese firm. Before the shutdown, 5,000 workers produced 240,000 cars a year and had an absentee rate of 20%. After the Japanese took over and introduced their distinctive management style, the same number of cars were produced by only 2,500 workers and the absentee rate dropped to 2% ("The Difference Japanese Management Makes," *Business Week*, July 14, 1986, p. 47).

a. With an absentee rate of 20%, what is the probability that at least 90% of a random sample of 50 workers will be on the job on a particular day?

b. What assumption must we make in order to use the normal distribution to approximate the probability required in part **a**?

c. If the absentee rate were 2%, should the probability of the event in part **a** be approximated using a normal distribution? Explain.

6.60 In Case Study 5.3, the number of shuttle catastrophes due to booster failure in n missions was treated as a binomial random variable. Using the binomial distribution and the probability of catastrophe determined by the Air Force's risk assessment study $\left(\frac{1}{35}\right)$, we determined the probability of at least 1 shuttle catastrophe in 25 missions to be .5155.

a. Based on the guidelines presented in this section, would it have been advisable to approximate this probability using the normal approximation to the binomial distribution? Explain.

b. Regardless of your answer to part **a**, use the normal distribution to approximate the binomial probability. Comment on the difference between the exact and approximate probabilities.

c. Refer to part **a**. Would the normal approximation be advisable if $n = 100$? $n = 500$? $n = 1,000$?

d. Approximate the probability that more than 25 catastrophes occur in 1,000 flights, assuming that the probability of a catastrophe in any given flight remains $\frac{1}{35}$.

6.61 In May 1983, after an extensive investigation by the Consumer Product Safety Commission, Honeywell agreed to recall 770,000 potentially defective smoke detectors. The commission suggested that about 40% of the Honeywell detectors were defective. However, Honeywell found only four defectives in a random sample of 2,000 detectors and claimed the recall was not justified (Gross, 1983). Let x be the number of defective smoke detectors found in a random sample of 2,000 detectors.

a. What assumptions must be made in order to characterize x as a binomial random variable? Do these assumptions appear to be satisfied?

b. Assume that the conditions of part **a** hold. Determine the approximate probability of finding four or fewer defective smoke detectors in a random sample of 2,000 if, in fact, 40% of all detectors are defective.

c. Assume that Honeywell's sample data have been reported accurately. Is it *likely* that 40% of their detectors are defective? Explain.

d. Refer to part **c**. Is it *possible* that 40% of Honeywell's detectors are defective? Explain.

6.62 The percentage of fat in the bodies of American men is an approximate normal random variable with mean equal to 15% and standard deviation equal to 2%. If these values were used to describe the body fat of men in the U.S. Army and if 20% or more body fat is characterized as obese, what is the approximate probability that a random sample of 10,000 Army men will contain fewer than 50 who would be characterized as obese? Suppose the Army actually were to check the percentage of body fat for a random sample of 10,000 men. If only 30 contained 20% or more body fat, would you conclude that the Army was successful in reducing the percentage of obese men below the percentage in the general population? Explain your reasoning.

6.63 An advertising agency was hired to introduce a new product. It claimed that after its campaign, 30% of all consumers were familiar with the product. To check the claim, the manufacturer of the product surveyed 2,000 consumers. Of this number, 527 consumers had learned about the product through sources attributable to the campaign. What is the approximate probability that as few as 527 (i.e., 527 or fewer) would have learned about the product if the campaign was really 30% effective?

6.64 To check on the effectiveness of a new production process, 700 photoflash devices were randomly selected from a large number that had been produced. If the process actually produces 6% defectives, what is the approximate probability that:

a. More than 50 defectives appear in the sample of 700?

b. The number of defectives in the sample of 700 is 45 or less?

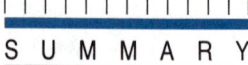

S U M M A R Y

Many **continuous random variables** in business applications have probability distributions that are well approximated by the **normal**, **uniform**, or **exponential probability distributions**. In this chapter we showed the graphical shape of each probability distribution, gave its mean and variance, and pointed out some practical applications of each probability model. In addition, we showed that the normal probability distribution provides a good approximation for the binomial distribution when n is sufficiently large.

| | | | | | | | | | | | | |

SUPPLEMENTARY EXERCISES 6.65–6.89

[*Note: Starred (*) exercises refer to the optional section in this chapter.*]

6.65 Assume that x is a random variable best described by a uniform distribution with $c = 10$ and $d = 90$.
  a. Find $f(x)$.
  b. Find the mean and standard deviation of x.
  c. Graph the probability distribution for x, and locate its mean and the interval $\mu \pm 2\sigma$ on the graph.
  d. Find $P(x \leq 60)$.
  e. Find $P(x \geq 90)$.
  f. Find $P(x \leq 80)$.
  g. Find $P(\mu - \sigma \leq x \leq \mu + \sigma)$.
  h. Find $P(x > 75)$.

6.66 Use Table IV of Appendix B to calculate the area under the standard normal distribution between the following pairs of z-scores:
  a. -1.96 and 1.96 **b.** -1.645 and 1.645 **c.** -3 and 3
  d. -3 and 2.5 **e.** 1.5 and 2.5 **f.** -1.5 and 2.5

6.67 Use Table IV of Appendix B to find the following probabilities:
  a. $P(z \geq .4)$ **b.** $P(z \leq .3)$
  c. $P(z \geq -3.05)$ **d.** $P(-1.75 \leq z \leq -.25)$

6.68 The random variable x has a normal distribution with $\mu = 75$ and $\sigma = 10$. Find the following probabilities:
  a. $P(x \leq 80)$ **b.** $P(x \geq 85)$ **c.** $P(70 \leq x \leq 75)$
  d. $P(x > 80)$ **e.** $P(x = 78)$ **f.** $P(x \leq 110)$

6.69 Find a value of z, call it z_0, such that:
  a. $P(z \geq z_0) = .5517$ **b.** $P(z \leq z_0) = .5080$
  c. $P(z \geq z_0) = .1492$ **d.** $P(z_0 \leq z \leq .59) = .4773$

6.70 Suppose x is a normal random variable with mean 40 and standard deviation 6. Find the value of x, call it x_0, such that:
  a. $P(\mu \leq x < x_0) = .40$
  b. $P(x \geq x_0) = .10$
  c. $P(x_0 \leq x < \mu) = .45$
  d. $P(x < x_0) = .05$
  e. $P(x > x_0) = .40$
  f. $P(x \leq x_0) = .40$

***6.71** Suppose x has an exponential distribution with $\lambda = .3$. Find the following probabilities:
  a. $P(x \leq 2)$ **b.** $P(x > 3)$ **c.** $P(x = 1)$
  d. $P(x \leq 7)$ **e.** $P(4 \leq x \leq 12)$ **f.** $P(x = 2.5)$

6.72 Assume that x is a binomial random variable with $n = 50$ and $p = .6$. Find approximate values for the following probabilities:
  a. $P(x \leq 35)$ **b.** $P(25 \leq x \leq 40)$
  c. $P(x \geq 20)$ **d.** $P(40 \leq x \leq 50)$

6.73 The metropolitan airport commission is considering the establishment of limitations on noise pollution around a local airport. At the present time, the noise level per jet takeoff in one neighborhood near the airport is approximately normally distributed with a mean of 100 decibels and a standard deviation of 6 decibels.

a. What is the probability that a randomly selected jet will generate a noise level greater than 108 decibels in this neighborhood?

b. What is the probability that a randomly selected jet will generate a noise level of exactly 100 decibels?

c. Suppose a regulation is passed that requires jet noise in this neighborhood to be lower than 105 decibels 95% of the time. Assuming the standard deviation of the noise distribution remains the same, how much will the mean level of noise have to be lowered to comply with the regulation?

6.74 Suppose the present value of a risky investment is approximately normally distributed with mean $10,000 and standard deviation $4,000. What is the probability that the present value of the investment is less than $1,000? Greater than $20,000?

6.75 The *tolerance limits* for a particular quality characteristic (e.g., length, weight, or strength) of a product are the minimum and/or maximum values at which the product will operate properly. Tolerance limits are set by the engineering design function of the manufacturing operation (Juran and Gryna, 1980). The tensile strength of a particular metal part can be characterized as being normally distributed with a mean of 25 pounds and a standard deviation of 2 pounds. The upper and lower tolerance limits for the part are 30 pounds and 21 pounds, respectively. A part that falls within the tolerance limits results in a profit of $10. A part that falls below the lower tolerance limit costs the company $2; a part that falls above the upper tolerance limit costs the company $1. Find the company's expected profit per metal part produced.

*6.76 Blending feeders are used to break up tobacco that has been aged in tightly packed hogsheads. One cigarette manufacturer determined that the time between breakdowns for each of its blending feeders is best represented by an exponential distribution with mean equal to 100 hours of operation. Suppose a particular feeder was just repaired and put back into service. What is the probability that it will not break down for at least 50 more hours? What is the probability that it will break down within the next 100 hours?

6.77 E. Brewer and P. Kaeser (1963) conducted a study of the factors that affect the level of production of workers paid on a piecework basis (i.e., paid according to the number of items they produce or process). Their study involved observing the performance of 36 quality control inspectors at a paper mill in England over an 8-week period. The inspectors were responsible for detecting and sorting out paper with defects such as holes, spots, creases, and rust marks. Part of the study entailed computing and analyzing the average hourly earnings for each inspector for each week of the 8-week observation period. Brewer and Kaeser constructed a frequency histogram of the average earnings data and noted that the histogram could be approximated by a normal distribution.

a. How many observations are there in the average earnings data set?

b. Suppose the average earnings histogram can be approximated by a normal distribution with $\mu = \$7.65$ and $\sigma = \$1.25$. Approximately what proportion of the weekly average earnings are over $8.50 per hour?

c. Using the normal distribution of part **b**, is it possible to determine approximately how many inspectors averaged over $8.50 per hour for the 8-week period? If so, how many? If not, why not?

6.78 On December 28, 1980, before millions of television viewers, the Schlitz Brewing Co. conducted a "live" taste test between its beer, Schlitz, and Anheuser-Busch's Budweiser. The taste test was conducted using a panel of 100 "loyal Budweiser drinkers." The panelists were required to sign affidavits stating that they drink at least two six-packs of Budweiser per week. The two beers were served to the tasters without labels and in identical opaque mugs, making it virtually impossible to identify a brand by sight. Since it is against broadcasting industry regulations, the panel was not shown tasting the beers. Instead, the commercial began after both beers had been tasted but before the tasters had indicated their preferences. When the master of ceremonies, a former National Football League referee, signaled the tasters to reveal their preferences, a scoreboard indicated the percentage of the panel that preferred the taste of Schlitz. According to *Fortune* magazine, "there really is no great difference between its beer [Schlitz] and those of other American brewers. Most American beers—unlike most European brands—are subtly flavored, and it requires a trained palate to distinguish among them. Probably no more than one person in a hundred has such a palate" ("Schlitz's Crafty Taste Test," 1981). The scoreboard indicated that 46% of the panel preferred Schlitz. In light of the *Fortune* quote, is the fact that 46% preferred Schlitz surprising? Was Schlitz taking much of a chance by conducting the taste test on "live television"? Explain.

*6.79 Based on sample data collected in the Denver area, Nicholas Kiefer (1985) found that in some cases the exponential distribution is an adequate approximation for the distribution of the time (in weeks) an individual is unemployed. In particular, he found the exponential distribution to be appropriate for white and black workers, but not for Hispanics. Use $\lambda = .075$ to answer the following questions.
a. What is the mean time workers are unemployed according to the exponential distribution?
b. What is the probability that a white worker who just lost her job will be unemployed for at least 2 weeks? More than 6 weeks?
c. What is the probability that an unemployed worker will find a new job within 12 weeks?

6.80 It is quite common for the standard deviation of a random variable to increase proportionally as the mean increases. When this occurs, the **coefficient of variation**,

$$CV = \frac{\sigma}{\mu}$$

the ratio of σ to μ, is the *proportionality constant*. To illustrate, the error (in dollars) in assessing the value of a house increases as the house increases in value. Suppose that long experience with assessors in your part of the country has shown that the coefficient of variation is .08 and that the probability distribution of assessed valuations on the same house by many different assessors is approximately normal with a mean we will call the *true value* of the house. Suppose the true value of your house is $50,000, and it is being assessed for taxation purposes. What is the probability that the assessor will assess your house in excess of $55,000?

6.81 As noted in Exercise 6.80, it is sometimes true that the larger the mean of a random variable, the larger will be its standard deviation. A measure of the relative variability of different data sets can be obtained by computing the coefficient of variation ($s/\bar{x}$) for each. The data sets at the top of page 302 reflect the numbers of checks (in thousands) processed per week by three different banks over the last 6 weeks.

BANK 1	BANK 2	BANK 3
20	60	100
10	63	92
18	58	81
25	65	110
31	70	105
12	54	129

Rank each of these data sets in terms of variability using the:
a. Range   **b.** Standard deviation   **c.** Coefficient of variation

*6.82 Assume that the length of the active life of baking yeast has an exponential distribution with a mean equal to 6 months. If the expiration date marked on a package of yeast is based on a life of 6 months, what is the probability that a package of the yeast will lose its potency before its expiration date?

6.83 On the average, the main chute fails in one of every 1,000 parachutes. Suppose that during a lifetime, a professional parachutist makes 4,000 jumps, and let x equal the number of times the main chute fails. What is the approximate probability that the parachutist's main chute fails on at least one jump? [*Note:* Because n is large and p is so small, the Poisson probability distribution will also provide a good approximation to this probability (see Exercise 5.59). If you covered Section 5.5, find the Poisson approximation to $P(x > 0)$.]

*6.84 The Poisson probability distribution, like the binomial, can be approximated by a normal probability distribution in some situations. The approximation, using $\mu = \lambda$ and $\sigma = \sqrt{\lambda}$, will be good when λ is large (large enough so that the distance between $x = 0$ and λ is at least $3\sigma = 3\sqrt{\lambda}$; i.e., $\lambda \geq 9$). The number of union complaints per month at a particular manufacturing plant has a Poisson probability distribution with $\mu = 40$ complaints per month. Use the normal approximation to the Poisson probability distribution to find:
a. The approximate probability that the number of complaints in a given month will be less than 35.
b. The approximate probability that the number of complaints in a given month will exceed 40.
c. What is the approximate probability that in each of 3 consecutive months, the number of complaints will exceed 40?

*6.85 The number of serious accidents per month in a manufacturing plant has (approximately) a Poisson probability distribution with a mean of 2.
a. If an accident occurs today, what is the probability that the next serious accident will not occur within the next month? [*Note:* If x, the number of events per unit time, has a Poisson distribution with mean λ, then it can be shown that the time between adjacent pairs of events has an exponential probability distribution with mean $1/\lambda$.]
b. What is the probability that more than one accident will occur within the next month?

6.86 A company has a lump-sum incentive plan for salespeople that is dependent on their level of sales. If they sell less than $100,000 per year, they receive a $1,000 bonus; from $100,000 to $200,000, they receive $5,000; and above $200,000, they receive

$10,000. If the annual sales per salesperson has approximately a normal distribution with $\mu = \$180,000$ and $\sigma = \$50,000$:

a. Find p_1, the proportion of salespeople who receive a $1,000 bonus.

b. Find p_2, the proportion of salespeople who receive a $5,000 bonus.

c. Find p_3, the proportion of salespeople who receive a $10,000 bonus.

d. What is the mean value of the bonus payout for the company? [*Hint:* Review the definition for the expected value of a random variable in Chapter 5.]

6.87 Contrary to our intuition, very reliable decisions concerning the proportion of a large group of consumers who favor a particular product or a particular social issue can be based on relatively small samples. For example, suppose the target population of consumers contains 50,000,000 people and we want to decide whether the proportion of consumers, p, in the population who favor some product (or issue) is as large as some value, say .2. Suppose you randomly select a sample as small as 1,600 from the 50,000,000 and you observe the number, x, of consumers in the sample who favor the new product. Assuming that $p = .2$, find the mean and standard deviation of x. Suppose that 400 (or 25%) of the sample of 1,600 consumers favor the new product. Why might this sample result lead you to conclude that p (the proportion of consumers who favor the product in the population of 50,000,000) is larger than .2? [*Hint:* Compare the observed value of x with the values of μ and σ calculated on the assumption that $p = .2$.]

6.88 To help highway planners anticipate the need for road repairs and design future construction projects, data are collected on the volume and weight of truck traffic on specific roadways. Recently, equipment has been developed that can be built into road surfaces to measure traffic volumes and to weigh trucks without requiring them to stop at roadside weigh stations. As with any measuring device, however, the "weigh-in-motion" equipment does not always record truck weights accurately. In an experiment performed by the Minnesota Department of Transportation involving repeated weighing of a 27,907-pound truck, it was found that the weights recorded by the weigh-in-motion equipment were approximately normally distributed with mean 27,315 and a standard deviation of 628 pounds (Dahlin, 1982; Wright, Owen, and Pena, 1983). It follows that the difference between the actual weight and recorded weight, the error of measurement, is normally distributed with mean 592 pounds and standard deviation 628 pounds.

a. What is the probability that the weigh-in-motion equipment understates the actual weight of the truck?

b. If a 27,907-pound truck were driven over the weigh-in-motion equipment 100 times, approximately how many times would the equipment overstate the truck's weight?

c. What is the probability that the error in the weight recorded by the weigh-in-motion equipment for a 27,907-pound truck exceeds 400 pounds?

d. It is possible to adjust (or *calibrate*) the weigh-in-motion equipment to control the mean error of measurement. At what level should the mean error be set so the equipment will understate the weight of a 27,907-pound truck 50% of the time? Only 40% of the time?

6.89 *Simulation* has been defined as "the process of designing a model of a real system and conducting experiments with this model for the purpose either of understanding the behavior of the system or of evaluating various strategies (within the limits imposed by a criterion or set of criteria) for the operation of the system" (Shannon, 1975). In Case Study 6.2, we saw that the Richmond Memorial Hospital used queueing theory to simulate the flow of patients in its emergency room. In that case, an exponential probability distribution was used to model the interarrival times of patients arriving at the emergency room. In general, after we have modeled the arrival process, we can use a computer to simulate arrivals. For Case Study 6.2, this is conceptually equivalent to drawing a random sample of interarrival times from a population of interarrival times whose relative frequency distribution is characterized by an exponential distribution. Many statistical computer packages are available for performing such simulations. For this exercise, Minitab was used to simulate a random sample of size $n = 48$ from each of the three continuous probability distributions described in this chapter. Plot a frequency or relative frequency histogram for each data set, and describe the probability distribution from which it appears the data were sampled. [Note: Statistical tests are available to test whether sample data have been selected from a specified population probability distribution. However, these tests are beyond the scope of this text.]

a.

.49809	.12027	.62005	.22538	.37436	.04148
.67047	.51957	.28542	.02330	.06551	.17087
.19405	.05575	1.83394	.01557	.01218	.05462
.08064	.81027	.45986	.29073	.14429	1.54523
.39153	.94809	.03960	.40591	.16163	.61685
.09053	.40320	.34805	.20736	1.15085	.52765
.29120	.09520	.16720	.49599	.08638	.00107
.32537	.16684	.04859	.29705	.18021	.05824

b.

11.9430	10.4727	13.4021	10.2837	11.5317	12.5550
11.1934	11.2110	10.2613	11.1970	10.5620	12.5638
12.8020	11.0463	10.7942	10.4239	12.8067	10.1791
13.7000	13.9958	11.4107	10.4990	12.4114	12.0499
11.7649	10.5271	13.0135	13.3110	12.9863	13.8829
11.6737	12.0269	12.5397	10.6642	11.5691	11.3977
12.2147	12.4337	12.7399	12.1405	10.4660	13.8000
12.2485	12.1577	10.1007	10.7129	10.8231	13.6000

c.

84.4507	83.4281	59.2906	72.3982	70.9882	76.1517
89.7030	71.9061	66.2978	86.6549	78.8585	90.4534
73.0715	73.7856	71.6502	76.1163	67.2840	63.3765
82.3677	72.8783	77.9567	73.2113	74.2898	72.0788
84.9135	63.5424	70.5931	60.4805	66.4123	86.7108
64.8896	79.4812	68.7999	60.9720	73.8562	83.1790
67.9300	75.5962	66.6784	50.3200	77.7831	79.1616
76.5933	73.4935	66.8368	59.2256	47.0316	69.7614

ON YOUR OWN . . .

For large values of n the computational effort involved in working with the binomial probability distribution is considerable. Fortunately, in many instances the normal distribution provides a good approximation to the binomial distribution. This exercise was designed to demonstrate how well the normal distribution approximates the binomial distribution.

a. Suppose the random variable x has a binomial probability distribution with $n = 10$ and $p = .5$. Using the binomial distribution, find the probability that x takes on a value in each of the following intervals: $\mu \pm \sigma$, $\mu \pm 2\sigma$, and $\mu \pm 3\sigma$.

b. Approximate the probabilities requested in part **a** using a normal approximation to the given binomial distribution.

c. Determine the magnitude of the difference between each of the three probabilities as determined by the binomial distribution and by the normal approximation.

USING THE COMPUTER . . .

Refer to Using the Computer in Chapter 5. Again use the mean percentage of college graduates as an estimate of the percentage of college graduates among all people at least 25 years old in the region you selected.

a. Use the normal approximation to the binomial to estimate the probability that fewer than 10% of 20 applicants (still assuming they represent a random sample) have a college education. Compare your answer with the exact binomial probability of the same event.

b. Use the normal approximation to the binomial to estimate the probability that fewer than 10% of 200 applicants have a college education. If your statistical software package has a function that calculates exact binomial probabilities, use it to calculate the same probability you approximated, and compare the results.

REFERENCES

Blair, R. D. and Kenny, L. W. *Microeconomics for Managerial Decision Making.* New York: McGraw-Hill, 1982. Chapter 10.

Bolling, W. B. "Queuing model of a hospital emergency room." *Industrial Engineering,* Sept. 1972, 26–31.

Brewer, E. and Kaeser, P. "A comparative analysis of incentive plans." *Journal of Industrial Relations,* July 1963, *11*, 183–198.

Dahlin, C. *Minnesota's Experience with Weighing Trucks in Motion.* St. Paul: Minnesota Department of Transportation, 1982.

Dessler, G. "Personnel tests gain in popularity." *St. Paul Pioneer Press and Dispatch,* Mar. 17, 1986, p. 12.

Fama, E. F. *Foundations of Finance.* New York: Basic Books, 1976. Chapter 1.

Gross, S. "Honeywell smoke detectors recalled after 18-month probe." *Minneapolis Star and Tribune,* May 25, 1983, p. 1A.

Hillier, F. S. "The derivation of probabilistic information for the evaluation of risky investments." *Management Science,* Apr. 1963, 9, 443–457.

Hillier, F. S. and Lieberman, G. J. *Introduction to Operations Research*, 3rd ed. San Francisco: Holden-Day, 1983. Chapter 10.

Hogg, R. V. and Craig, A. T. *Introduction to Mathematical Statistics*, 4th ed. New York: Macmillan, 1978. Chapter 1.

Juran, J. M. and Gryna, F. M., Jr. *Quality Planning and Analysis*, 2nd ed. New York: McGraw-Hill, 1980.

Kiefer, N. M. "Specification diagnostics based on Laguerre alternatives for econometric models of duration." *Journal of Econometrics*, 28, 1985, 135–154.

Lamberson, L. R. "Reliability tutorial." *American Society for Quality Control Quality Congress Transaction*, Baltimore, 1985, pp. 88–99.

Lindgren, B. W. *Statistical Theory*, 3rd ed. New York: Macmillan, 1976. Chapters 2 and 3.

Martz, H. F. and Waller, R. A. *Bayesian Reliability Analysis*. New York: Wiley, 1982. pp. 1, 256.

Montgomery, D. C. *Introduction to Statistical Quality Control*. New York: Wiley, 1985.

Mood, A. M., Graybill, F. A., and Boes, D. C. *Introduction to the Theory of Statistics*, 3rd ed. New York: McGraw-Hill, 1974. Chapter 3.

Murphy, A. H. and Winkler, R. L. "Probability forecasting in meteorology." *Journal of the American Statistical Association*, 79, 1984, 489–500.

Neter, J., Wasserman, W., and Whitmore, G. A. *Applied Statistics*, 2nd ed. Boston: Allyn & Bacon, 1982. Chapter 7.

Schick, G. J. "The search for a software reliability model." *Decision Sciences*, Oct. 1974, 5, 529.

"Schlitz's crafty taste test." *Fortune*, Jan. 26, 1981, 32–34.

Shannon, R. E. *Systems Simulation: The Art and Science*. Englewood Cliffs, N.J.: Prentice-Hall, 1975. Chapter 1.

Sims, S. H. and Templeton, J. G. C. "Steady state results for the M/M $(a, b)/c$ batch-service system." *European Journal of Operational Research*, 21, 1985, 260–267.

Winkler, R. L. and Hays, W. *Statistics: Probability, Inference, and Decision*, 2nd ed. New York: Holt, Rinehart and Winston, 1975. Chapter 3.

Wright, J. L., Owen, F., and Pena, D. *Status of Mn/DOT's Weigh-in-Motion Program*. St. Paul: Minnesota Department of Transportation, Jan. 1983.

C H A P T E R 7

WHERE WE'VE BEEN...

We have learned in earlier chapters that the objective of most statistical investigations is inference—that is, making decisions about or estimating some numerical descriptive measure (a parameter) of a population based on sample data. To make the decision or to estimate the population parameter, we use sample data to compute sample statistics (Chapter 3) such as the sample mean or variance.

WHERE WE'RE GOING...

Because sample measurements are observed values of random variables, the value we compute for a sample statistic will vary in a random manner from sample to sample. In other words, since sample statistics are computed from random variables, they themselves are random variables, and they have probability distributions that are either discrete or continuous, as discussed in Chapters 5 and 6. The probability distribution of a sample statistic is called a **sampling distribution** because it characterizes the distribution of values of the statistic over a very large number of samples. Sampling distributions are the topic of this chapter. We will discuss why many sampling distributions tend to be approximately normal, and you will see how sampling distributions can be used to evaluate the accuracy of parameter estimates.

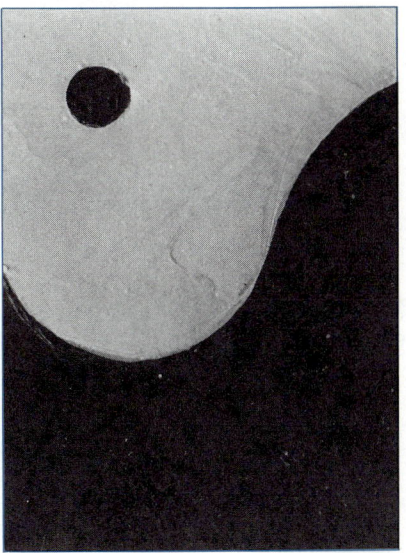

SAMPLING DISTRIBUTIONS

CONTENTS

In Chapters 5 and 6 we assumed that we knew the probability distribution of a random variable, and based on this knowledge, we were able to compute the mean, variance, and probability that the random variable assumed specific values. However, in most practical business applications, this information will not be available. To illustrate, in Example 5.8, we calculated the probability that the binomial random variable x (the number of 20 polled employees who favor unionization) assumed specific values. To do this, it was necessary to assume some value for p, the proportion of the employees in the population who favor unionization. Thus, for the purpose of illustration we assumed $p = .6$, but in all likelihood, the exact value of p would be unknown. In fact, the probable purpose of taking the poll was to estimate p. Similarly, when we modeled the in-city gas mileage of a certain automobile model in Example 6.7, we used the normal probability distribution with an *assumed* mean and standard deviation of 27 and 3 miles per gallon, respectively. In reality, the true mean and standard deviation are unknown quantities that would have to be estimated.

Numerical quantities that describe probability distributions are called **parameters**. Thus, p (the probability of a success in a binomial experiment) and μ and σ (the mean and standard deviation of a normal distribution) are examples of parameters. Since probability distributions are used to characterize populations, it follows that parameters are also numerical descriptive measures of populations.

DEFINITION 7.1

A **parameter** is a numerical descriptive measure of a population. It is calculated from the observations in the population.

We often use the information contained in a sample to make inferences about the parameters of a population. In order to make such inferences, we must compute **sample statistics** that will aid in making these inferences.

DEFINITION 7.2

A **sample statistic** is a numerical descriptive measure of a sample. It is calculated from the observations in the sample.

Some examples of useful sample statistics we have already discussed are the sample mean, $\bar{x}$; sample median; sample variance, s^2; and the sample standard deviation, s. Before we can use these and other sample statistics to make inferences about population parameters, we have to be able to evaluate their properties. How can we decide which sample statistic contains the most information about a population parameter? One purpose of this chapter is to answer this question.

SECTION 7.1

**INTRODUCTION
TO SAMPLING
DISTRIBUTIONS**

If we want to estimate a parameter of a population—say, the population mean, μ—there are a number of sample statistics that we could use for the estimate. Two possibilities are the sample mean, $\bar{x}$, and the sample median, m. Which of these do you think will provide a better estimate of μ?

Before answering this question, consider the following example: Toss a fair die and let x equal the number of dots showing on the up face. Suppose the die is tossed three times, producing the sample measurements 2, 2, 6. The sample mean is $\bar{x} = 3.33$ and the sample median is $m = 2$. Since the population mean of x is $\mu = 3.5$, you can see that for this sample of three measurements, the sample mean $\bar{x}$ provides an estimate that falls closer to μ than does the sample median [see Figure 7.1(a)].

FIGURE 7.1
Comparing the Sample
Mean ($\bar{x}$) and Sample
Median (m) as Estimators
of the Population
Mean (μ)

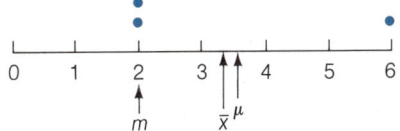

(a) Sample 1: $\bar{x}$ is closer than m to μ

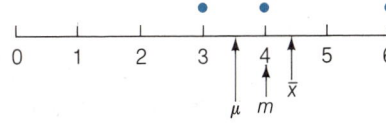

(b) Sample 2: m is closer than $\bar{x}$ to μ

Now suppose we toss the die three more times and obtain the sample measurements 3, 4, 6. The mean and median of this sample are $\bar{x} = 4.33$ and $m = 4$, respectively. This time m is closer to μ [see Figure 7.1(b)].

This simple example illustrates an important point: Neither the sample mean nor the sample median will *always* fall closer to the population mean. Consequently, we cannot compare these two sample statistics or, in general, any two sample statistics on the basis of their performance for a single sample. Instead, we need to recognize that sample statistics are themselves random variables because different samples can lead to different values for the sample statistics. As random variables, sample statistics must be judged and compared on the basis of their probability distributions—i.e., the collection of values and associated probabilities of each statistic that would be obtained if the sampling experiment were repeated a *very large number of times*. We will illustrate this concept with an example.

Suppose it is known that in a certain part of Canada the daily high temperature recorded for all past months of January has a mean of $\mu = 10°F$ and a standard deviation of $\sigma = 5°F$. Consider an experiment consisting of randomly selecting 25 daily high temperatures from the records of past months of January and calculating the sample mean, $\bar{x}$. If this experiment were repeated a very large number of times, the value of $\bar{x}$ would vary from sample to sample. For example, the first sample of 25 temperature measurements might have a mean of $\bar{x} = 9.8$; the second sample, a mean of $\bar{x} = 11.4$; the third sample, a mean of $\bar{x} = 10.5$; etc. If the sampling experiment were repeated a very large number of times and the resulting values of $\bar{x}$ were displayed in a histogram, the histogram would be the approximate probability distribution of $\bar{x}$. If $\bar{x}$ is a good estimator of μ, we would expect the values of $\bar{x}$ to cluster around μ as shown in Figure 7.2 (page 310). This probability distribution is called a **sampling distribution** because it describes the potential outcomes of $\bar{x}$ in repeated sampling.

FIGURE 7.2

Sampling Distribution for $\bar{x}$ Based on a Sample of $n = 25$ Measurements: $\mu = 10$, $\sigma = 5$

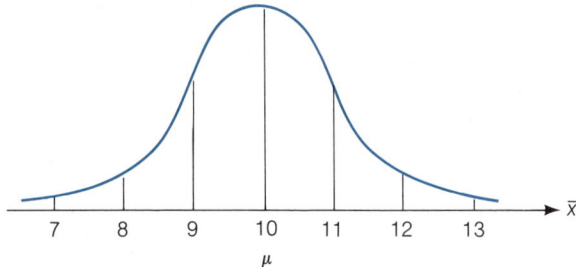

In actual practice, the sampling distribution of a statistic is obtained mathematically or (approximately) by simulating the sampling on a computer using the procedure described above.

DEFINITION 7.3

The **sampling distribution** for a sample statistic calculated from a sample of n measurements is the probability distribution of the statistic.

If $\bar{x}$ has been calculated from a sample of $n = 25$ measurements selected from a population with mean $\mu = 10$ and standard deviation $\sigma = 5$, the sampling distribution (Figure 7.2) provides information about the behavior of $\bar{x}$ in repeated sampling. For example, the probability that you will draw a sample of 25 measurements and obtain a value of $\bar{x}$ in the interval $9 \leq \bar{x} \leq 10$ will be the area under the sampling distribution over that interval.

Since the properties of a statistic are typified by its sampling distribution, it follows that, to compare two statistics, you compare their sampling distributions. For example, if you have two statistics, A and B, for estimating the same parameter (for purposes of illustration, suppose the parameter is the population variance σ^2) and if their sampling distributions are as shown in Figure 7.3, you would choose statistic A in preference to statistic B. You would make this choice because the sampling distribution for statistic A centers over σ^2 and has less spread (variation) than the sampling distribution for statistic B. When you draw a single sample in a practical sampling situation, the probability is higher that statistic A will fall closer to σ^2.

FIGURE 7.3

Two Sampling Distributions for Estimating the Population Variance, σ^2

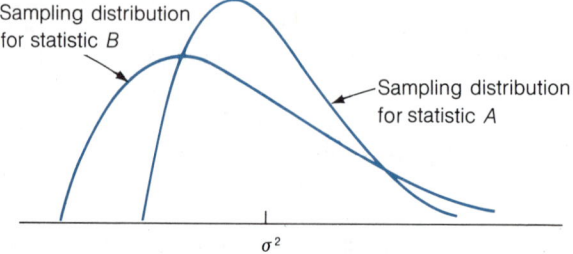

Remember that in practice we will not know the numerical value of the unknown parameter, σ^2, so we will not know whether statistic A or statistic B is closer to σ^2 for particular samples. We have to rely on our theoretical knowledge of the sampling distributions to choose the better sample statistic and then use it sample after sample.

EXAMPLE 7.1

Consider a population consisting of the measurements 0, 3, and 12 and described by the following probability distribution:

x	0	3	12
$p(x)$	$\frac{1}{3}$	$\frac{1}{3}$	$\frac{1}{3}$

A random sample of $n = 3$ measurements is selected from the population.

a. Find the sampling distribution of the sample mean $\bar{x}$.
b. Find the sampling distribution of the sample median m.

SOLUTION

Every possible sample of $n = 3$ measurements is listed in Table 7.1 along with the sample mean and median. Also, because any one sample is as likely to be selected as any other (random sampling), the probability of observing any particular sample is $\frac{1}{27}$.

TABLE 7.1
Possible Samples of $n = 3$ Measurements

POSSIBLE SAMPLES	$\bar{x}$	m	PROBABILITY	POSSIBLE SAMPLES	$\bar{x}$	m	PROBABILITY
0, 0, 0	0	0	$\frac{1}{27}$	3, 3, 12	6	3	$\frac{1}{27}$
0, 0, 3	1	0	$\frac{1}{27}$	3, 12, 0	5	3	$\frac{1}{27}$
0, 0, 12	4	0	$\frac{1}{27}$	3, 12, 3	6	3	$\frac{1}{27}$
0, 3, 0	1	0	$\frac{1}{27}$	3, 12, 12	9	12	$\frac{1}{27}$
0, 3, 3	2	3	$\frac{1}{27}$	12, 0, 0	4	0	$\frac{1}{27}$
0, 3, 12	5	3	$\frac{1}{27}$	12, 0, 3	5	3	$\frac{1}{27}$
0, 12, 0	4	0	$\frac{1}{27}$	12, 0, 12	8	12	$\frac{1}{27}$
0, 12, 3	5	3	$\frac{1}{27}$	12, 3, 0	5	3	$\frac{1}{27}$
0, 12, 12	8	12	$\frac{1}{27}$	12, 3, 3	6	3	$\frac{1}{27}$
3, 0, 0	1	0	$\frac{1}{27}$	12, 3, 12	9	12	$\frac{1}{27}$
3, 0, 3	2	3	$\frac{1}{27}$	12, 12, 0	8	12	$\frac{1}{27}$
3, 0, 12	5	3	$\frac{1}{27}$	12, 12, 3	9	12	$\frac{1}{27}$
3, 3, 0	2	3	$\frac{1}{27}$	12, 12, 12	12	12	$\frac{1}{27}$
3, 3, 3	3	3	$\frac{1}{27}$				

a. From Table 7.1 you can see that $\bar{x}$ can assume the values, 0, 1, 2, 3, 4, 5, 6, 8, 9, and 12. Because $\bar{x} = 0$ occurs in only one sample, $P(\bar{x} = 0) = \frac{1}{27}$. Similarly, $\bar{x} = 1$ occurs in three samples: (0, 0, 3), (0, 3, 0), and (3, 0, 0). Therefore, $P(\bar{x} = 1) = \frac{3}{27} = \frac{1}{9}$. Calculating the probabilities of the remaining values of $\bar{x}$ and arranging them in a table, we obtain the following probability distribution:

$\bar{x}$	0	1	2	3	4	5	6	8	9	12
$p(\bar{x})$	$\frac{1}{27}$	$\frac{3}{27}$	$\frac{3}{27}$	$\frac{1}{27}$	$\frac{3}{27}$	$\frac{6}{27}$	$\frac{3}{27}$	$\frac{3}{27}$	$\frac{3}{27}$	$\frac{1}{27}$

b. In Table 7.1 you can see that the median m can assume one of the three values 0, 3, or 12. The value $m = 0$ occurs in seven different samples. Therefore, $P(m = 0) = \frac{7}{27}$. Similarly, $m = 3$ occurs in thirteen samples and $m = 12$ occurs in seven samples. Therefore, the probability distribution for the median m is as follows:

m	0	3	12
$p(m)$	$\frac{7}{27}$	$\frac{13}{27}$	$\frac{7}{27}$

■

Example 7.1 demonstrates the procedure for finding the exact sampling distribution of a statistic when the number of different samples that could be selected from the population is relatively small. In the real world, populations are often generated by random variables that can assume a very large number of values; the samples are difficult (or impossible) to enumerate. When such a situation occurs, we may choose to obtain the approximate sampling distribution for a statistic by simulating the sampling over and over again and recording the proportion of times different values of the statistic occur. Example 7.2 illustrates this procedure.

EXAMPLE 7.2

Suppose we perform the following experiment: Take a sample of 11 measurements from the uniform distribution shown in Figure 7.4. Calculate the two sample statistics

FIGURE 7.4
Uniform Distribution from 0 to 1

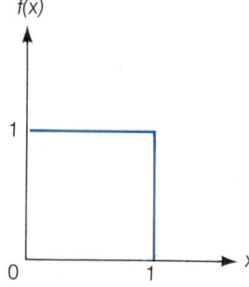

$$\bar{x} = \text{Sample mean} = \frac{\sum_{i=1}^{11} x_i}{11}$$

$m = \text{Median} = \text{Sixth sample measurement when the 11 measurements are arranged in ascending order}$

In this particular example we *know* that the population mean is $\mu = .5$. The objective will be to find out which sample statistic contains more information about μ. We use a computer to generate 1,000 samples, each with $n = 11$ observations. Then, we compute $\bar{x}$ and m for each sample. Our goal is to find the resulting approximate sampling distributions for $\bar{x}$ and m.

SOLUTION

The first ten of the 1,000 samples generated are presented in Table 7.2. For each of the 1,000 samples we compute the sample mean, $\bar{x}$, and the sample median, m. For example, the first computer-generated sample from the uniform distribution (arranged in ascending order) contained the following measurements: .125, .138, .139, .217, .419, .506, .516, .757, .771, .786, .919. The sample mean, $\bar{x}$, and median, m, computed for this sample are

$$\bar{x} = \frac{.125 + .138 + \cdots + .919}{11} = .481$$

$$m = \text{Sixth ordered measurement} = .506$$

TABLE 7.2 First Ten Samples of $n = 11$ Measurements from a Uniform Distribution

SAMPLE	MEASUREMENTS										
1	.217	.786	.757	.125	.139	.919	.506	.771	.138	.516	.419
2	.303	.703	.812	.650	.848	.392	.988	.469	.632	.012	.065
3	.383	.547	.383	.584	.098	.676	.091	.535	.256	.163	.390
4	.218	.376	.248	.606	.610	.055	.095	.311	.086	.165	.665
5	.144	.069	.485	.739	.491	.054	.953	.179	.865	.429	.648
6	.426	.563	.186	.896	.628	.075	.283	.549	.295	.522	.674
7	.643	.828	.465	.672	.074	.300	.319	.254	.708	.384	.534
8	.616	.049	.324	.700	.803	.399	.557	.975	.569	.023	.072
9	.093	.835	.534	.212	.201	.041	.889	.728	.466	.142	.574
10	.957	.253	.983	.904	.696	.766	.880	.485	.035	.881	.732

The relative frequency histograms for $\bar{x}$ and m for the 1,000 samples of size $n = 11$ are shown in Figure 7.5.

FIGURE 7.5 Approximate Sampling Distributions for $\bar{x}$ and m, Example 7.2

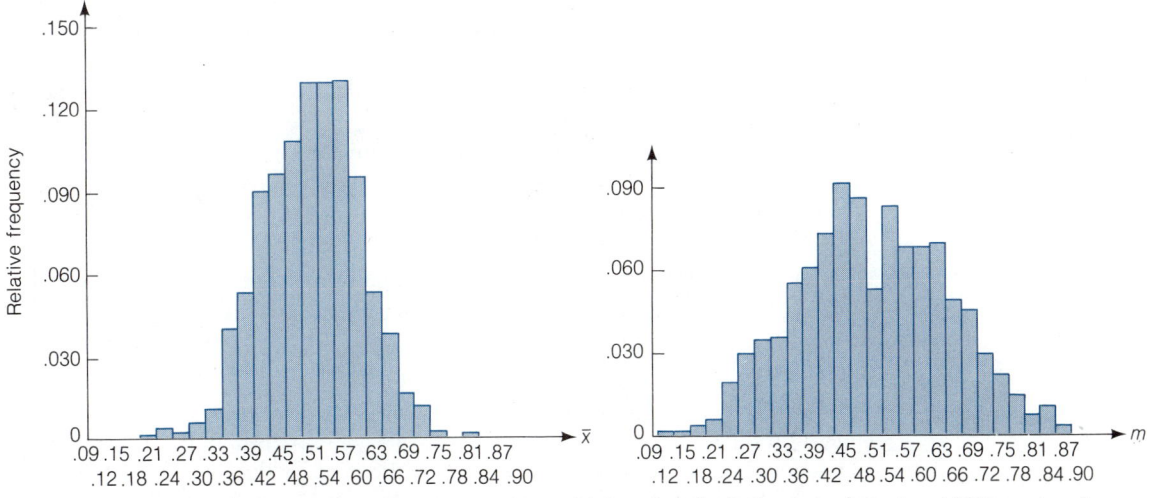

(a) Sampling distribution for $\bar{x}$ (based on 1,000 samples of $n = 11$ measurements)

(b) Sampling distribution for m (based on 1,000 samples of $n = 11$ measurements)

You can see that the values of $\bar{x}$ tend to cluster around μ to a greater extent than do the values of m. Thus, on the basis of the observed sampling distributions, we conclude that $\bar{x}$ contains more information about μ than m does—at least for samples of $n = 11$ measurements from the uniform distribution. ∎

We will not always have to simulate repeated sampling on a computer to find sampling distributions. Many sampling distributions can be derived mathematically, but the theory necessary to do this is beyond the scope of this text. Consequently, when we need to know the properties of a statistic, we will present its sampling distribution and simply describe its properties. Several of the important properties of sampling distributions are discussed in the next section.

||||||||||||||

S E C T I O N 7.2

PROPERTIES OF SAMPLING DISTRIBUTIONS: UNBIASEDNESS AND MINIMUM VARIANCE

The simplest type of statistic used to make inferences about a population parameter is a **point estimator**. A point estimator is a rule or formula that tells us how to use the sample data to calculate a single number that can be used as an estimate of the value of some population parameter. For example, the sample mean, $\bar{x}$, is a point estimator of the population mean, μ. Similarly, the sample variance, s^2, is a point estimator of the population variance, σ^2.

Often, many different point estimators can be found to estimate the same parameter. Each will have a sampling distribution that provides information about the point estimator. By examining the sampling distribution, we can determine how large the difference between an estimate and the true value of the parameter—called the **error of estimation**—is likely to be.

DEFINITION 7.4

A **point estimator** of a population parameter is a rule or formula that tells us how to use the sample data to calculate a single number that can be used as an **estimate** of the population parameter.

We can also determine whether an estimator is likely to overestimate or to underestimate a parameter.

Since the sampling distribution of a point estimator (or of any statistic) describes its behavior in repeated sampling, we look to the sampling distribution to identify characteristics or properties that we would want an estimator to have. As a first consideration, we would like the sampling distribution to center over the parameter we want to estimate. One way to express centrality is in terms of the mean of the sampling distribution. Consequently, we say that a statistic is **unbiased** if its sampling distribution has a mean equal to the parameter it is intended to estimate. When this occurs, the sampling distribution of the statistic will be centered over the parameter as shown in Figure 7.6(a). If the mean of a sampling distribution is not equal to the parameter it is intended to estimate, the statistic is said to be **biased**. The amount of bias is the difference between the mean of the sampling distribution and the value of the parameter you wish to estimate. The sampling distribution for a biased statistic is shown in Figure 7.6(b).

DEFINITION 7.5

If a sample statistic has a sampling distribution with a mean equal to the population parameter the statistic is intended to estimate, the statistic is said to be an **unbiased** estimator of the parameter.

If the mean of the sampling distribution is not equal to the parameter, the statistic is said to be a **biased** estimator of the parameter.

FIGURE 7.6 Sampling Distributions for Unbiased and Biased Estimators

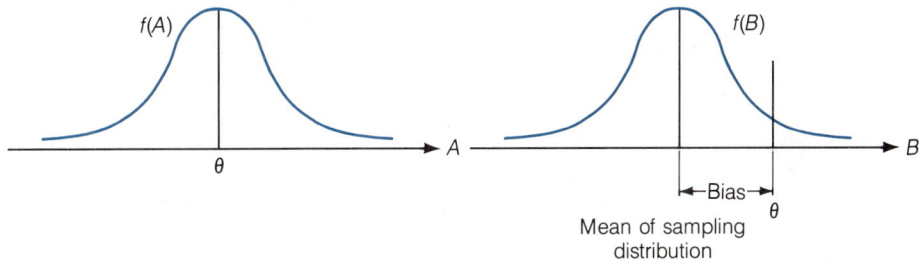

(a)   Unbiased sample statistic for the parameter θ (b)   Biased sample statistic for the parameter θ

The standard deviation of a sampling distribution measures another important property of a statistic—the spread of the estimates generated by repeated sampling. Suppose two statistics, A and B, are both unbiased estimators of the population parameter called θ (theta). Note that θ could be any parameter, such as μ, σ^2, or σ. Since the means of the two sampling distributions are the same, we turn to their standard deviations to decide which will provide estimates that fall closer to the unknown population parameter we are estimating. Naturally, we will choose the sample statistic that has the smaller standard deviation. Figure 7.7 depicts sampling distributions for A and B. Note that the standard deviation of the distribution of A is smaller than the standard deviation for B, indicating that over a large number of samples, the values of A cluster more closely around the unknown population parameter than do the values of B. Therefore, we would choose statistic A instead of statistic B as an estimator of θ.

FIGURE 7.7

Sampling Distributions for
Two Unbiased Estimators

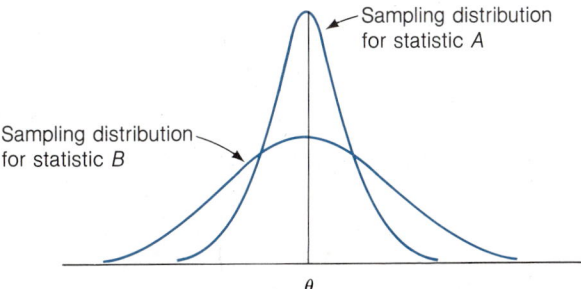

In summary, to make an inference about a population parameter, use the sample statistic with a sampling distribution that is unbiased and has a small standard deviation (usually smaller than the standard deviation of other unbiased sample statistics). How to find this sample statistic will not concern us, because the "best" statistic for estimating a particular parameter is a matter of record. We will simply present an unbiased estimator with its standard deviation for each population parameter we consider. [*Note:* The standard deviation of the sampling distribution for a statistic is usually called the **standard error** of the statistic.]

EXAMPLE 7.3

In Example 7.1, we found the sampling distributions of the sample mean $\bar{x}$ and the sample median m for random samples of $n = 3$ measurements from a population defined by the following probability distribution:

x	0	3	12
$p(x)$	$\frac{1}{3}$	$\frac{1}{3}$	$\frac{1}{3}$

The sampling distributions of $\bar{x}$ and m were found to be the following:

$\bar{x}$	0	1	2	3	4	5	6	8	9	12
$p(\bar{x})$	$\frac{1}{27}$	$\frac{3}{27}$	$\frac{3}{27}$	$\frac{1}{27}$	$\frac{3}{27}$	$\frac{6}{27}$	$\frac{3}{27}$	$\frac{3}{27}$	$\frac{3}{27}$	$\frac{1}{27}$

m	0	3	12
$p(m)$	$\frac{7}{27}$	$\frac{13}{27}$	$\frac{7}{27}$

a. Show that $\bar{x}$ is an unbiased estimator of μ in this situation.
b. Show that m is a biased estimator of μ in this situation.

SOLUTION

a. The expected value of a discrete random variable x (see Section 4.3) is defined to be $E(x) = \Sigma\, xp(x)$, where the summation is over all values of x. Then

$$E(x) = \mu = \sum xp(x) = (0)(\tfrac{1}{3}) + (3)(\tfrac{1}{3}) + (12)(\tfrac{1}{3}) = 5$$

The expected value of the discrete random variable $\bar{x}$ is

$$E(\bar{x}) = \sum (\bar{x})p(\bar{x})$$

summed over all values of $\bar{x}$. Or

$$E(\bar{x}) = (0)(\tfrac{1}{27}) + (1)(\tfrac{3}{27}) + 2(\tfrac{3}{27}) + \cdots + (12)(\tfrac{1}{27}) = 5$$

Since $E(\bar{x}) = \mu$, we see that $\bar{x}$ is an unbiased estimator of μ.

b. The expected value of the sample median m is

$$E(m) = \sum mp(m) = (0)(\tfrac{7}{27}) + (3)(\tfrac{13}{27}) + (12)(\tfrac{7}{27}) = 4.56$$

Since the expected value of m is not equal to μ ($\mu = 5$), the sample median m is a biased estimator of μ. ∎

EXAMPLE 7.4

Refer to Example 7.3 and find the standard deviations of the sampling distributions of $\bar{x}$ and m. Which statistic would appear to be a better estimator for μ?

SOLUTION

The variance of the sampling distribution of $\bar{x}$ (we will denote it by the symbol $\sigma_{\bar{x}}^2$) is found to be

$$\sigma_{\bar{x}}^2 = E\{[\bar{x} - E(\bar{x})]^2\} = \sum (\bar{x} - \mu)^2 p(\bar{x})$$

where, from Example 7.3,

$$E(\bar{x}) = \mu = 5$$

Then

$$\sigma_{\bar{x}}^2 = (0 - 5)^2(\tfrac{1}{27}) + (1 - 5)^2(\tfrac{3}{27}) + (2 - 5)^2(\tfrac{3}{27}) + \cdots + (12 - 5)^2(\tfrac{1}{27})$$
$$= 8.6667$$

and

$$\sigma_{\bar{x}} = \sqrt{8.6667} = 2.94$$

Similarly, the variance of the sampling distribution of m (we will denote it by σ_m^2) is

$$\sigma_m^2 = E\{[m - E(m)]^2\}$$

where, from Example 7.3, the expected value of m is $E(m) = 4.56$. Then

$$\sigma_m^2 = E\{[m - E(m)]^2\} = \sum [m - E(m)]^2 p(m)$$
$$= (0 - 4.56)^2(\tfrac{7}{27}) + (3 - 4.56)^2(\tfrac{13}{27}) + (12 - 4.56)^2(\tfrac{7}{27}) = 20.9136$$

and

$$\sigma_m = \sqrt{20.9136} = 4.57$$

Which statistic appears to be the better estimator for the population mean μ: the sample mean $\bar{x}$ or the median m? To answer this question, we compare the sampling distribution of the two statistics. The sampling distribution of the sample median m is biased (i.e., it is shifted to the left of the mean μ) and its standard deviation $\sigma_m = 4.57$ is much larger than the standard deviation of the sampling distribution of $\bar{x}$, $\sigma_{\bar{x}} = 2.94$. Consequently, the sample mean $\bar{x}$ would be a better estimator of the population mean μ, for the population in question, than would the sample median m. ∎

SECTION 7.3

THE SAMPLING DISTRIBUTION OF THE SAMPLE MEAN

Estimating the mean useful life of automobiles, the mean monthly sales for all automobile dealers in a large city, and the mean breaking strength of a new plastic are practical problems with something in common. In each, we are interested in making an inference about the mean, μ, of some population. Because many practical business problems involve estimating μ, it is particularly important to have a sample statistic that is a good estimator of μ. As we mentioned in Chapter 3, the sample mean, $\bar{x}$, is generally a good choice as an estimator of μ. The mean and standard deviation of the sampling distribution of this useful statistic are related to the mean, μ, and standard deviation, σ, of the sampled population as described in the box.

THE MEAN AND STANDARD DEVIATION OF THE SAMPLING DISTRIBUTION OF $\bar{x}$

Regardless of the shape of the population relative frequency distribution,

1. The mean of the sampling distribution of $\bar{x}$ will equal μ, the mean of the sampled population; i.e., $\mu_{\bar{x}} = \mu$.
2. The standard deviation of the sampling distribution of $\bar{x}$ will equal σ, the standard deviation of the sampled population, divided by the square root of the sample size, n; i.e.,

$$\sigma_{\bar{x}} = \frac{\sigma}{\sqrt{n}} \,^*$$

The standard deviation $\sigma_{\bar{x}}$ is often referred to as the **standard error of the mean**.

Thus, $\bar{x}$ is an unbiased estimator of the population mean μ [i.e., $E(\bar{x}) = \mu$], and the standard deviation of $\bar{x}$ is inversely related to the (square root of the) sample size (i.e., $\sigma_{\bar{x}} = \sigma/\sqrt{n}$).

For example, suppose the sampled population has the uniform probability distribution shown in Figure 7.8(a). The mean and standard deviation of this probability distribution are $\mu = .5$ and $\sigma = .29$ (refer to Section 6.2 for the formulas for μ and σ). Now suppose a sample of 11 measurements is selected from this population. The sampling distribution of the sample mean for samples of size 11 will also have a mean of .5, with a standard deviation

$$\sigma_{\bar{x}} = \frac{\sigma}{\sqrt{n}} = \frac{.29}{\sqrt{11}} = .09$$

(That is, the standard error of $\bar{x}$ is $\sigma/\sqrt{n} = .09$.)

What can be said about the shape of the sampling distribution of $\bar{x}$? Two important theorems provide this information.

THEOREM 7.1

If a random sample of n observations is selected from a population with a normal distribution, the sampling distribution of $\bar{x}$ will be a normal distribution.

---

*If the sample size, n, is large relative to the number, N, of elements in the population, $\sigma/\sqrt{n}$ must be multiplied by a finite population correction factor, $\sqrt{(N - n)/(N - 1)}$. For most sampling situations, this correction factor will be close to 1 and can be ignored.

> **THEOREM 7.2: CENTRAL LIMIT THEOREM**
>
> If a random sample of n observations is selected from a population (any population), then, when n is sufficiently large, the sampling distribution of $\bar{x}$ will be approximately a normal distribution. The larger the sample size, n, the better will be the normal approximation to the sampling distribution of $\bar{x}$.*

FIGURE 7.8

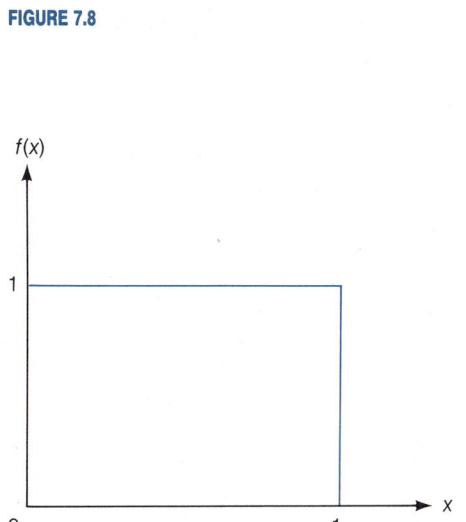

(a) Relative frequency distribution of the sampled population

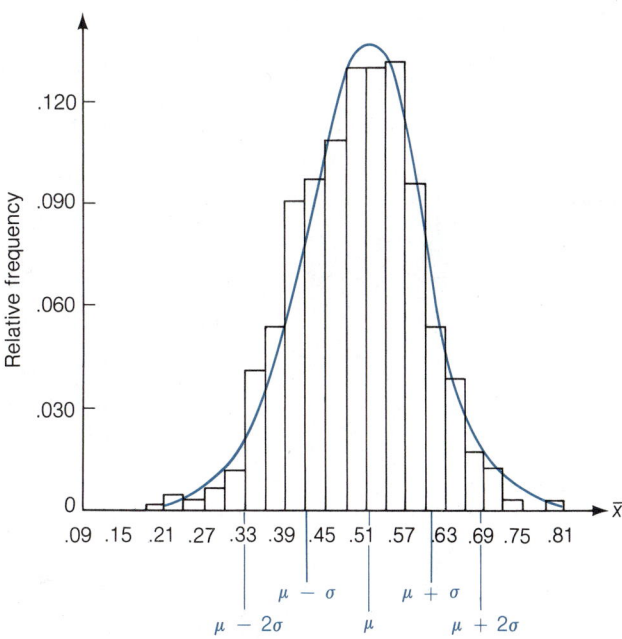

(b) Relative frequency histogram for $\bar{x}$ in 1,000 samples from a uniform distribution

Thus, for sufficiently large samples, the sampling distribution of $\bar{x}$ will be approximately normal. How large must the sample size, n, be so that $\bar{x}$ has a normal sampling distribution? The answer depends on the shape of the relative frequency distribution of the sampled population, as shown in Figure 7.9 (page 320). Generally speaking, the greater the skewness of the sampled population distribution, the larger the sample size must be to obtain an adequate normal approximation to the sampling distribution of $\bar{x}$. But for some populations, particularly those with symmetric distributions, n may be fairly small and the sampling distribution of $\bar{x}$ will be approximately normal.

---

*Also, because of the Central Limit Theorem, the sum of a random sample of n observations, $\sum_{i=1}^{n} x_i$, will have a sampling distribution that will be approximately normal for large samples. This distribution will have a mean equal to $n\mu$ and a variance equal to $n\sigma^2$.

FIGURE 7.9

Sampling Distributions of
$\bar{x}$ for Different Populations
and Different Sample
Sizes

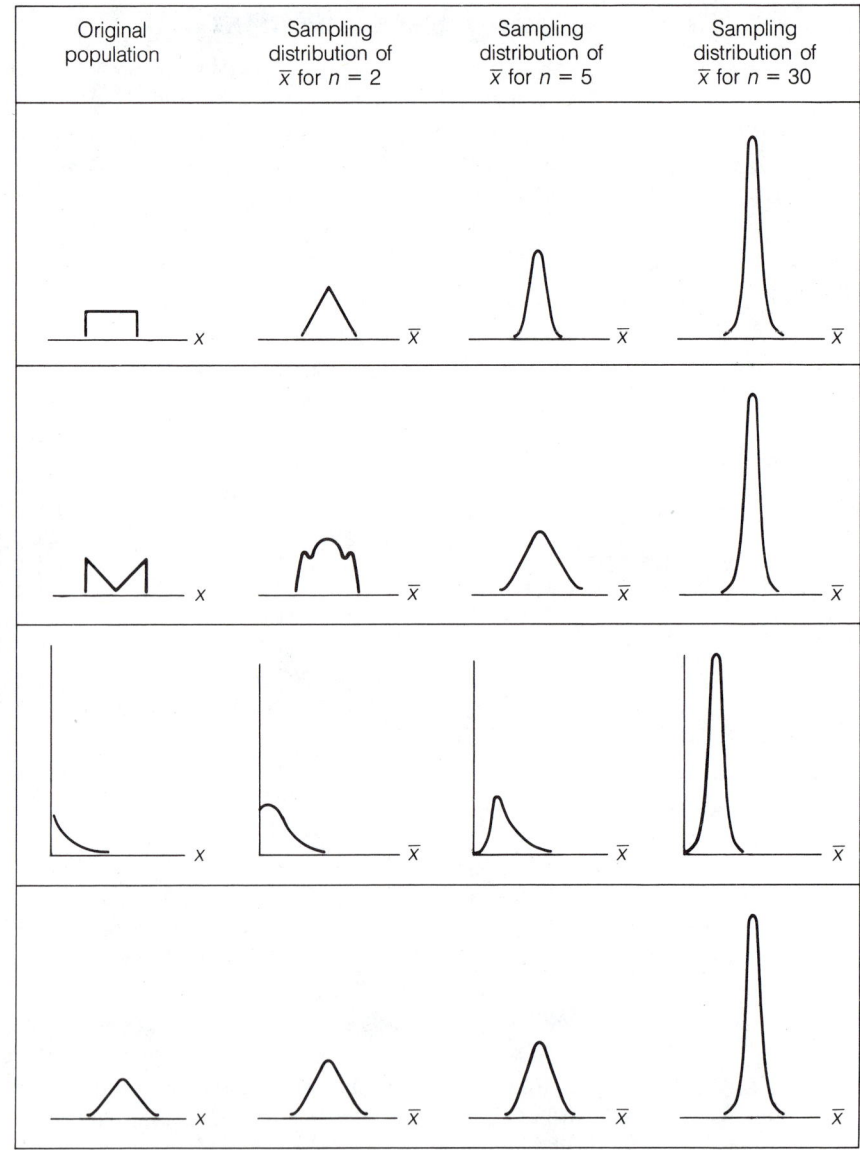

Original population	Sampling distribution of $\bar{x}$ for $n = 2$	Sampling distribution of $\bar{x}$ for $n = 5$	Sampling distribution of $\bar{x}$ for $n = 30$

To demonstrate how small n can be and still achieve approximate normality for the sampling distribution of $\bar{x}$, recall Example 7.2, in which we generated 1,000 samples of $n = 11$ measurements from a uniform distribution. The relative frequency histogram for the 1,000 sample means is shown in Figure 7.8(b), and the normal probability distribution with a mean of .5 and a standard deviation of .09 is superimposed. You can see that this normal probability distribution approximates the computer-generated sampling distribution very well, even though the sample size is only $n = 11$.

The implications of the Central Limit Theorem are apparent by comparing the sampling distribution of $\bar{x}$ in Figure 7.8(b) with the distribution of the sampled population in Figure 7.8(a). In this situation, the population relative frequency distribution is not skewed and, as a result, a very small sample size proved large enough to apply the Central Limit Theorem. The sampling distribution of $\bar{x}$ is approximately normal, even though the distribution of the sampled population is decidedly nonnormal. You will also note that the mean of the sampling distribution is equal to the mean of the distribution of the sampled population, but that the variability of the sampling distribution is substantially less than the variability of the sampled population.

In most real-life applications, the shape of the population distribution will *not* be known. In such cases, we typically require $n \geq 30$ in order to invoke the Central Limit Theorem.

EXAMPLE 7.5

Suppose we have selected a random sample of $n = 25$ observations from a population with mean equal to 80 and standard deviation equal to 5. It is known that the population is not extremely skewed.

a. Sketch the relative frequency distributions for the population and for the sampling distribution of the sample mean, $\bar{x}$.

b. Find the probability that $\bar{x}$ will be larger than 82.

SOLUTION

a. We do not know the exact shape of the population relative frequency distribution, but we do know that it should be centered about $\mu = 80$, its spread should be measured by $\sigma = 5$, and it is not highly skewed. One possibility is shown in Figure 7.10(a). From the Central Limit Theorem, we know that the sampling distribution of $\bar{x}$ will be approximately normal since the sampled population distribution is not extremely skewed. We also know that the sampling distribution will have mean and standard deviation

$$\mu_{\bar{x}} = \mu = 80 \quad \text{and} \quad \sigma_{\bar{x}} = \frac{\sigma}{\sqrt{n}} = \frac{5}{\sqrt{25}} = 1$$

The sampling distribution of $\bar{x}$ is shown in Figure 7.10(b).

FIGURE 7.10 A Population Relative Frequency Distribution and the Sampling Distribution for $\bar{x}$

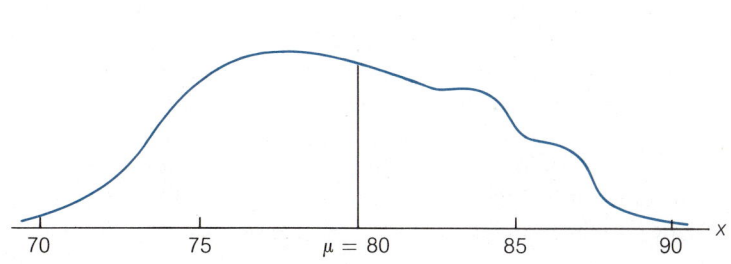

(a)   Population relative frequency distribution

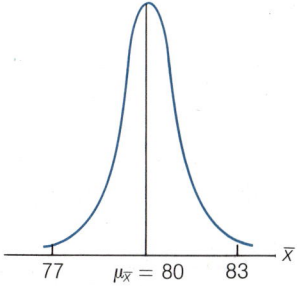

(b)   Sampling distribution of $\bar{x}$

b. The probability that $\bar{x}$ will exceed 82 is equal to the darker shaded area in Figure 7.11. To find this area, we need to find the z value corresponding to $\bar{x} = 82$. Recall that the standard normal random variable z is the difference between any normally distributed random variable and its mean, expressed in units of its standard deviation. Since $\bar{x}$ is a normally distributed random variable with mean $\mu_{\bar{x}} = \mu$ and standard deviation $\sigma_{\bar{x}} = \sigma/\sqrt{n}$, it follows that the standard normal z value corresponding to the sample mean, $\bar{x}$, is

$$z = \frac{(\text{Normal random variable}) - (\text{Mean})}{\text{Standard deviation}} = \frac{\bar{x} - \mu_{\bar{x}}}{\sigma_{\bar{x}}}$$

Therefore, for $\bar{x} = 82$, we have

$$z = \frac{\bar{x} - \mu_{\bar{x}}}{\sigma_{\bar{x}}} = \frac{82 - 80}{1} = 2$$

The area A in Figure 7.11 corresponding to $z = 2$ is given in the table of areas under the normal curve (see Table IV of Appendix B) as .4772. Therefore, the tail area corresponding to the probability that $\bar{x}$ exceeds 82 is

$$P(\bar{x} > 82) = P(z > 2) = .5 - .4772 = .0228$$

FIGURE 7.11

The Sampling Distribution of $\bar{x}$

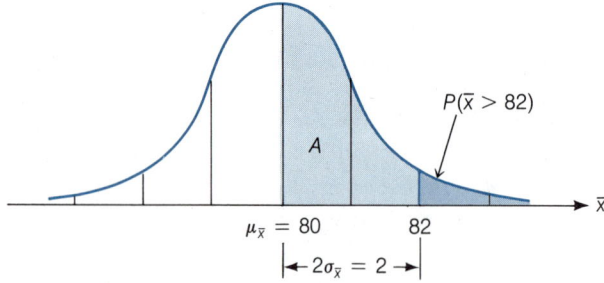

EXAMPLE 7.6

A manufacturer of automobile batteries claims that the distribution of the lifetimes of its best battery has a mean of 54 months and a standard deviation of 6 months. Suppose a consumer group decides to check the claim by purchasing a sample of 50 of these batteries and subjecting them to tests that determine their lifetimes.

a. Assuming the manufacturer's claim is true, describe the sampling distribution of the mean lifetime of a sample of 50 batteries.

b. Assuming the manufacturer's claim is true, what is the probability that the consumer group's sample has a mean lifetime of 52 months or less?

SOLUTION

a. Even though we have no information about the shape of the probability distribution of the lifetimes of the batteries, we can use the Central Limit Theorem to deduce that the sampling distribution for a sample mean lifetime of 50 batteries is approximately normally distributed. Furthermore, the mean of this sampling distribution is the same as the mean of the sampled population, which is $\mu = 54$ months, according to the manufacturer's claim. Finally, the standard deviation of the sampling distribution is given by

$$\sigma_{\bar{x}} = \frac{\sigma}{\sqrt{n}} = \frac{6}{\sqrt{50}} = .85 \text{ month}$$

Thus, if we assume the claim is true, the sampling distribution of the mean lifetime of the 50 batteries is approximately normal with mean 54 months and standard deviation .85 month. The sampling distribution is shown in Figure 7.12.

FIGURE 7.12

Sampling Distribution of $\bar{x}$ in Example 7.6

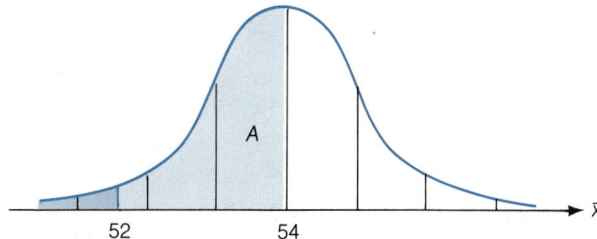

b. If the manufacturer's claim is true, the probability that the consumer group observes a mean battery lifetime of 52 months or less for their sample of 50 batteries, $P(\bar{x} \leq 52)$, is equivalent to the darker shaded area in Figure 7.12. Since the sampling distribution is approximately normal, we can find this area by computing the standard normal z value:

$$z = \frac{\bar{x} - \mu_{\bar{x}}}{\sigma_{\bar{x}}} = \frac{\bar{x} - \mu}{\sigma_{\bar{x}}} = \frac{52 - 54}{.85} = -2.35$$

where $\mu_{\bar{x}}$, the mean of the sampling distribution of $\bar{x}$, is equal to μ, the mean of the sampled population.

The area A shown in Figure 7.12 between $\bar{x} = 52$ and 54 is found in Table IV of Appendix B to be .4906, so the area to the left of 52 is

$$P(\bar{x} \leq 52) = .5 - A = .5 - .4906 = .0094$$

Thus, the probability that the consumer group will observe a sample mean of 52 or less is only .0094 if the manufacturer's claim is true. If the 50 tested batteries do result in a mean of 52 months or less, the consumer group will have strong evidence that the manufacturer's claim is untrue, because such an event is very unlikely to occur if the claim is true. (This is still another application of the rare event approach to statistical inference.)   ∎

In addition to providing a very useful approximation for the sampling distribution of a sample mean, the Central Limit Theorem offers an explanation for the fact that many relative frequency distributions of data are mound-shaped. Many of the macroscopic measurements we take in business research are really means or sums of many microscopic phenomena. For example, a year's sales of a company is the total of the many individual sales the company made during the year. Thus, the year's sales for a sample of similar companies may have a mound-shaped relative frequency distribution. Similarly, the length of time a construction company takes to complete a house might be viewed as the total of the time each of the large number of distinct jobs necessary to build the house takes to complete. The monthly profit of a firm can be viewed as the sum of the profits of all the transactions of the firm for that month. If we adopt viewpoints like these, the Central Limit Theorem offers some explanation for the frequent occurrence of mound-shaped distributions in nature.

CASE STUDY 7.1

EVALUATING THE
CONDITION OF RENTAL
CARS

In the summer of 1986, National Car Rental Systems, Inc., commissioned USAC Properties, Inc. [the performance testing/endorsement arm of the United States Automobile Club (USAC)] to conduct a survey of the general condition of the cars rented to the public by Hertz, Avis, National, and Budget Rent-a-Car.* National was interested in comparing the conditions of the cars it rented with those of the other leading car-rental companies.

Teams of USAC officials would evaluate each company's cars on appearance and cleanliness, accessory performance, mechanical functions, and vehicle safety using a demerit point system designed specifically for this survey. Each car would start with a perfect score of 0 points and would incur demerit points for each discrepancy noted by the inspectors. The number of demerits associated with a discrepancy would be based on the seriousness of the discrepancy.

If all cars in each company's fleet could be inspected and graded, one measure of the overall condition of a company's cars would be the mean of all scores received by the company, i.e., the company's *fleet mean score*. Such a census, however, besides being virtually impossible to conduct logistically, would be prohibitively expensive. It was therefore decided that the fleet mean score would have to be estimated. Accordingly, ten major airports were randomly selected, and ten cars from each company were randomly rented for inspection from each airport by USAC officials, i.e., a sample of size $n = 100$ cars from each company's fleet was drawn and inspected. In the analysis of USAC's inspection results, each company's mean score was used to estimate the company's unknown fleet mean score. (The use of a sample mean to estimate a population mean will be discussed in detail in Chapter 8.) As we have seen in this chapter, $\bar{x}$ is a random variable with a sampling distribution that has a mean equal to the mean of the population from which the sample was drawn. Thus, in the context of this case study, the mean of the sampling distribution of $\bar{x}$ is the unknown fleet mean score. Since the sample size used by USAC was 100, the statisticians who evaluated USAC's inspection results were able to invoke the Central Limit Theorem and assume the sampling distribution of $\bar{x}$ to be approximately normally distributed. This assumption enabled comparisons of fleet mean scores for Hertz, Avis, National, and Budget Rent-a-Car to be made using

*Information by personal communication with Rajiv Tandon, Corporate Vice President and General Manager of the Car Rental Division, National Car Rental Systems, Inc., Minneapolis, Minn.

the conventional large-sample testing procedures that will be presented in Chapters 8 and 9.

CASE STUDY 7.2

REDUCING INVESTMENT RISK THROUGH DIVERSIFICATION

In Case Study 5.2, it was noted that the variance of the monthly rate of return of a security is used by many investors as a measure of the risk or uncertainty involved in investing in the security. In this case study, we demonstrate that an investor can reduce investment risk by investing in more than one security—that is, by *diversifying* investments.

A number of studies by financial analysts have shown that the total risk (total variation) of a stock, as measured by the variance of the stock's rates of return over time, is comprised of two components: *systematic risk* and *unsystematic risk*. Systematic risk (systematic variability) is the portion of total risk caused by factors that simultaneously influence the prices of all stocks. Examples of such factors are changes in federal economic policies and changes in the national political climate. These factors explain why the prices of all stocks tend to move together over time (i.e., generally upward or generally downward). Unsystematic risk (unsystematic variability) is the portion of the total risk of a particular stock due to factors that influence the firm in question but generally do not influence other firms. Examples of such factors are labor strikes, management errors, and lawsuits (Francis, 1980). Although the proportions of systematic and unsystematic risk vary from firm to firm, it has been determined that, for many of the stocks listed on the New York Stock Exchange, systematic risk comprises about 25% and unsystematic risk comprises about 75% of the stock's total risk (Blume, 1971). As is demonstrated in this case study, an investor can use diversification to reduce the unsystematic portion of the total investment risk.

Suppose an investor is considering investing a total of $5,000 in one or more of five different stocks. We will denote the monthly rates of return of these stocks by r_1, r_2, r_3, r_4, and r_5. For simplicity, we will assume these monthly returns are independent and identically distributed random variables with mean $\mu = 10\%$ and standard deviation $\sigma = 4\%$. Suppose the investor has narrowed the choice to two options: (1) invest $5,000 in stock 1 or (2) invest $1,000 in each of the five stocks.

Under the first option, the investor's monthly rate of return is r_1. Under the second option, since equal amounts of money were invested in each stock, the investor's monthly rate of return is $\bar{r} = \sum_{i=1}^{5} r_i/5$. If the first option is chosen, the investor's expected monthly rate of return, $E(r_i)$, is $\mu = 10\%$ and the risk, as measured by the variance of the stock's rate of return, is $\sigma^2 = 16$. If the second option is chosen, the investor's expected monthly rate of return, $E(\bar{r})$, can be shown to be the same as in the first option. However, since the numerator of $\bar{r}$ is the sum of $n = 5$ independent and identically distributed random variables, each with mean μ and variance σ^2, the variance of $\bar{r}$ is $\sigma_{\bar{r}}^2 = \sigma^2/n$. Accordingly, the risk faced by the investor is $\frac{16}{5} = 3.2$ and is lower than under the first option. Thus, the uncertainty faced by the investor is lower if the investor diversifies, rather than putting "all the eggs in one basket." For a more detailed discussion of risk reduction through diversification, see Sharpe (1981) and Elton and Gruber (1981), or Alexander and Francis (1986).

**THE RELATIONSHIP
BETWEEN SAMPLE
SIZE AND A SAMPLING
DISTRIBUTION**

Suppose you draw two random samples from a population—one sample containing $n = 5$ observations and the second containing $n = 10$—and you want to compute $\bar{x}$ for each sample and use these statistics to estimate the population mean, μ. Intuitively, it would seem that the $\bar{x}$ based on the sample of ten measurements would contain more information about μ than the $\bar{x}$ based on five measurements. But how is this larger sample size reflected in the sampling distribution of a statistic?

For the statistics you will encounter in this text, the variance of the sampling distribution of a statistic will be inversely proportional to the sample size.* Or, since the standard deviation of $\bar{x}$ is equal to $\sigma/\sqrt{n}$, you can say that the standard deviation of the sampling distribution is proportional to $1/\sqrt{n}$. So, to reduce the standard deviation of the sampling distribution of a statistic to $\frac{1}{2}$ its original value, you will need 4 times as many observations in your sample $(1/\sqrt{n} = 1/\sqrt{4} = \frac{1}{2})$. Or to reduce the standard deviation to $\frac{1}{3}$ its original value, you will need 9 times as many observations.

The sampling distributions for the sample mean, $\bar{x}$, based on random samples from a normally distributed population, are shown in Figure 7.13 for $n = 1, 4,$ and 16 observations. The curve for $n = 1$ represents the probability distribution of the population. Those for $n = 4$ and $n = 16$ are sampling distributions of $\bar{x}$. Note how the distributions contract (variation decreases) for $n = 4$ and $n = 16$. The standard deviation of $\bar{x}$ based on $n = 16$ measurements is half the corresponding standard deviation of the distribution based on $n = 4$ measurements.

FIGURE 7.13
Three Sampling
Distributions of $\bar{x}$

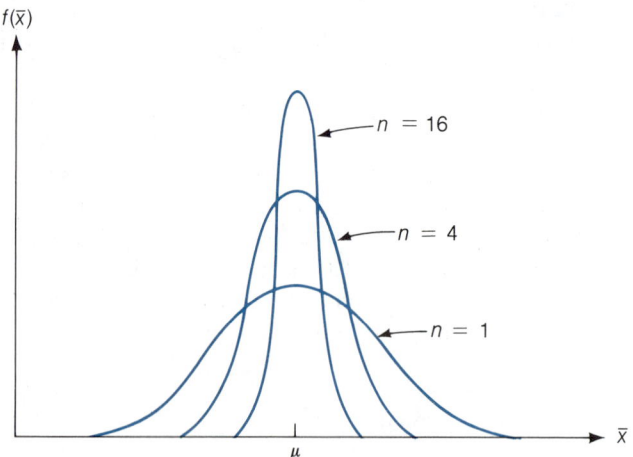

EXAMPLE 7.7

Consider the **Bernoulli** random variable x that can assume the values 0 and 1 with probabilities p and $q = (1 - p)$, respectively. The distribution is summarized in Table 7.3, where we have associated the term *Success* with the outcome $x = 1$, and *Failure* with the outcome $x = 0$. The Bernoulli random variable is just a binomial random variable with $n = 1$ trial.

*Note that this is not true of all statistics, but it is true for most.

TABLE 7.3
Bernoulli Distribution

Outcome	x	p(x)
Failure	0	q
Success	1	p

Suppose a random sample of n measurements is drawn from this Bernoulli distribution, and the sample mean $\bar{x}$ is calculated. Note that

$$\bar{x} = \frac{\sum\limits_{i=1}^{n} x_i}{n}$$

and that $\sum\limits_{i=1}^{n} x_i$ is the total number of Successes (the number of 1's in the sample) in the random sample of n measurements (or *trials*). This sum is therefore a binomial random variable, with n trials and probability of Success p. For example, the Bernoulli random variable might be the status of a single computer microchip (good or defective), x the number of n such chips that are good, and $\bar{x}$ the fraction of good chips in a set of n.

a. Find the mean and standard deviation of the sampling distribution of $\bar{x}$.
b. Simulate the distribution of $\bar{x}$ using $p = .8$, and $n = 1, 10, 25$, and 100 by generating 1,000 samples for each sample size and creating a histogram of the 1,000 sample means.

SOLUTION

a. The mean and variance of the Bernoulli random variable are

$$\mu = E(x) = 0(q) + 1(p) = p$$
$$\sigma^2 = E[(x - \mu)^2] = (0 - p)^2(q) + (1 - p)^2(p)$$
$$= p^2q + q^2p = pq(p + q) = pq$$

Note that these are the mean and standard deviation of a binomial random variable with $n = 1$, which is another description of a Bernoulli random variable. We know that the mean $\bar{x}$ of a random sample is unbiased, so that

$$E(\bar{x}) = \mu = p$$

and the standard error is

$$\sigma_{\bar{x}} = \frac{\sigma}{\sqrt{n}} = \frac{\sqrt{pq}}{\sqrt{n}} = \sqrt{\frac{pq}{n}}$$

Because $\bar{x}$ provides an unbiased estimate of the probability of Success p, and has a standard error that decreases as the sample size increases, we will use it to estimate p in subsequent chapters, where we will refer to it as the sample fraction of Successes, $\hat{p}$.

$$\bar{x} = \hat{p}$$

b. The mean and standard deviation of $\bar{x}$ when $p = .8$ are

$$\mu_{\bar{x}} = p = .8$$

$$\sigma_{\bar{x}} = \sqrt{\frac{pq}{n}} = \sqrt{\frac{(.8)(.2)}{n}} = \frac{.4}{\sqrt{n}}$$

You can see how the standard error of $\bar{x}$ decreases as n increases in Table 7.4.

TABLE 7.4
Mean and Standard Error
of $\bar{x}$ for Bernoulli Random
Variable with $p = .8$

n	$\mu_{\bar{x}}$	$\sigma_{\bar{x}}$
1	.8	$\frac{.4}{\sqrt{1}} = .4000$
10	.8	$\frac{.4}{\sqrt{10}} = .1265$
25	.8	$\frac{.4}{\sqrt{25}} = .0800$
100	.8	$\frac{.4}{\sqrt{100}} = .0400$

The simulation of 1,000 sample means with each of the sample sizes given in Table 7.4 resulted in the sampling distributions shown in Figure 7.14. A relative frequency histogram is used to display each sampling distribution. Note that each sampling distribution more closely resembles a normal distribution than the previous one. As the Central Limit Theorem promises, the distribution of $\bar{x}$ becomes approximately normal for large n, and the approximation improves as n increases. Note too that the values of $\bar{x}$ cluster more closely around their mean (.8) as n is increased. We will make use of these properties to estimate p for binomial random variables in Chapter 8.   ∎

EXAMPLE 7.8

Refer to Example 7.7. Suppose $n = 400$ microchips are to be sampled from a very large batch, of which a proportion p are good. The proportion of good chips in the sample will be determined by assigning a 1 to each good chip, a 0 to each defective chip, and calculating the sample mean of the 400 Bernoulli observations. Assuming that the approximate value of p is .8, what is the probability that $\bar{x}$ will fall within .03 of the exact value of p?

SOLUTION

We first note (see Example 7.7) that the mean and standard deviation of $\bar{x}$ are

$$E(\bar{x}) = p$$

$$\sigma_{\bar{x}} = \sqrt{\frac{pq}{n}} = \sqrt{\frac{pq}{400}}$$

Using the approximate value of p to obtain an approximation for the standard deviation of the sampling distribution of $\bar{x}$, we find

$$\sigma_{\bar{x}} \approx \sqrt{\frac{(.8)(.2)}{400}} = .02$$

FIGURE 7.14   Sampling Distributions for Bernoulli Sample Means

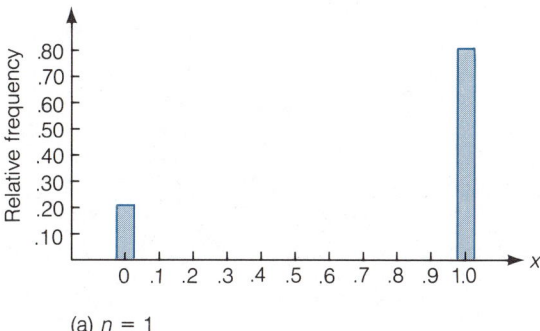

(a) $n = 1$

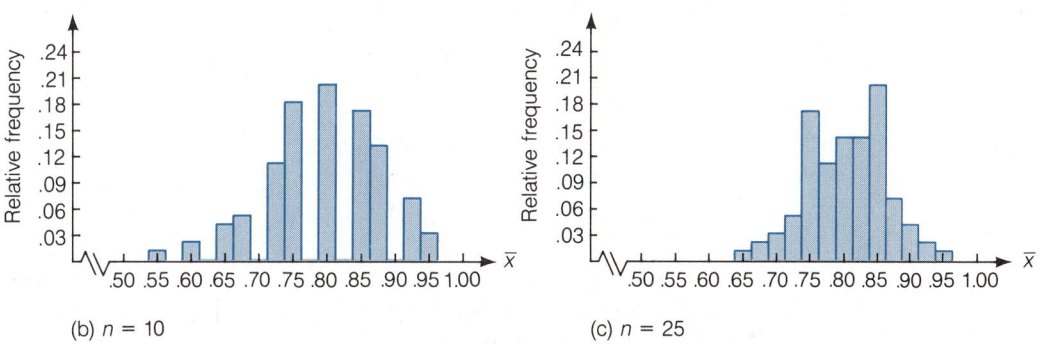

(b) $n = 10$ (c) $n = 25$

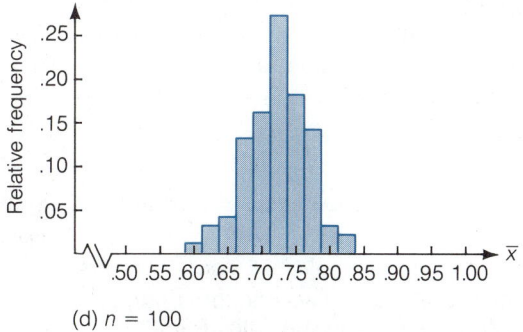

(d) $n = 100$

Next, the Central Limit Theorem implies that the distribution of $\bar{x}$ based on a sample of size 400 is approximately normal. The properties are summarized in Figure 7.15 (page 330). Note that the mean of the sampling distribution is p, so that the standard normal z value is given by the formula

$$z = \frac{\bar{x} - \mu_{\bar{x}}}{\sigma_{\bar{x}}} = \frac{\bar{x} - p}{.02}$$

FIGURE 7.15
Sampling Distribution of $\bar{x}$

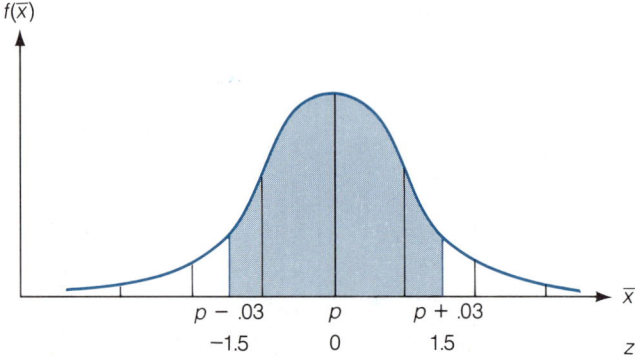

Even though the value of p is unknown, we can find the probability that $\bar{x}$ is within .03 of p by first calculating the z value corresponding to the $\bar{x}$ value that is .03 greater than p:

$$z = \frac{(p + .03) - p}{.02} = \frac{.03}{.02} = 1.5$$

Similarly, the value .03 less than p is 1.5 standard deviations *below* the mean, and has a z value of -1.5. Thus, the event that $\bar{x}$ falls within .03 of p is equivalent to the event that the standard normal random variable z is between -1.5 and 1.5. Using Table IV of Appendix B, we find the probability that $\bar{x}$ falls within 1.5 standard deviations of p is $2(.4332) = .8664$, the area shown shaded in Figure 7.15. The interpretation of this probability is that there is about an 87% chance that the proportion of good chips in the sample of 400 will fall within .03 of the exact proportion of good chips in the entire batch.   ■

For most sampling distributions, the standard deviation of the distribution decreases as the sample size increases. We will use this result in Chapters 8 and 9 to help us determine the sample size needed to obtain a specified accuracy of estimation.

S U M M A R Y

Many practical business problems require that an inference be made about some population **parameter** (call it θ). If we want to make this inference on the basis of information in a sample, we need to compute a **sample statistic** that contains information about θ. The amount of information a sample statistic contains about θ is reflected by its **sampling distribution**, the probability distribution of the sample statistic. The sampling distribution describes the behavior of the statistic in repeated sampling. In particular, we want a sample statistic that is an **unbiased** estimator of θ and has a smaller variance than any other unbiased sample statistic.

When the population parameter of interest is the mean, μ, the sample mean, $\bar{x}$, provides an unbiased estimator with a standard deviation of $\sigma/\sqrt{n}$. In addition, the **Central Limit Theorem** assures us that the sampling distribution of the mean of a large sample will be approximately normally distributed, no matter what the shape of the relative frequency distribution of the sampled population.

The amount of information in a sample that is relevant to some population parameter is related to the sample size. For example, the standard deviation of the sampling distribution of the sample mean, $\bar{x}$, will be inversely proportional to the square root of the sample size (i.e., $1/\sqrt{n}$).

The sampling distributions for all the many statistics that can be computed from sample data could be discussed in detail, but this would delay discussion of the practical objective of this course—the role of statistical inference in business decision-making. Consequently, we will comment further on the sampling distributions of particular statistics when we use them as estimators or decision-makers in the following chapters.

| | | | | | | | | | | | | |

EXERCISES 7.1–7.34   [Note: Starred (*) exercises require the use of a computer.]

LEARNING THE MECHANICS

7.1 In each of the following cases, find the mean and standard deviation of the sampling distribution of the sample mean, $\bar{x}$, for a random sample of size n drawn from a population with mean μ and standard deviation σ:

   a. $n = 10$, $\mu = 20$, $\sigma = 3$ **b.** $n = 40$, $\mu = 100$, $\sigma = 10$
   c. $n = 12$, $\mu = 25$, $\sigma = 2$ **d.** $n = 100$, $\mu = 400$, $\sigma = 9$

7.2 A random sample of $n = 25$ observations is drawn from a normal population with mean equal to 15 and standard deviation equal to 3.

   a. Give the mean and standard deviation of the (repeated) sampling distribution of $\bar{x}$.
   b. Describe the shape of the sampling distribution of $\bar{x}$. Does your answer depend on the sample size?
   c. Calculate the z-score corresponding to a value of $\bar{x}$ equal to 15.5.
   d. Calculate the z-score corresponding to $\bar{x} = 14$.

7.3 Refer to Exercise 7.2. Find the probability that:

   a. $\bar{x}$ is larger than 16   **b.** $\bar{x}$ is less than 16
   c. $\bar{x}$ is larger than 14.2   **d.** $\bar{x}$ falls between 14 and 16
   e. $\bar{x}$ is less than 14

7.4 A random sample of 50 observations is to be drawn from a large population with mean 50 and variance 25.

   a. Give the mean and standard deviation of the (repeated) sampling distribution of $\bar{x}$.
   b. Describe the shape of the sampling distribution of $\bar{x}$. Does your answer depend on the sample size?

7.5 A random sample of 49 observations is to be drawn from a large population with mean 200 and standard deviation 20. Find the probability that:

   a. $\bar{x} \leq 200$ **b.** $\bar{x} < 200$ **c.** $\bar{x} < 205$
   d. $\bar{x} > 190$ **e.** $\bar{x} > 209$ **f.** $193 \leq \bar{x} \leq 200$
   g. $197.1 \leq \bar{x} \leq 202.9$ **h.** $205 \leq \bar{x} \leq 210$

7.6 The table contains 50 random samples of random digits, $x = 0, 1, 2, 3, \ldots, 9$, where $p(x) = \frac{1}{10}$. Each sample contains $n = 6$ measurements.

SAMPLE	SAMPLE	SAMPLE	SAMPLE
8, 1, 8, 0, 6, 6	7, 6, 7, 0, 4, 3	4, 4, 5, 2, 6, 6	0, 8, 4, 7, 6, 9
7, 2, 1, 7, 2, 9	1, 0, 5, 9, 9, 6	2, 9, 3, 7, 1, 3	5, 6, 9, 4, 4, 2
7, 4, 5, 7, 7, 1	2, 4, 4, 7, 5, 6	5, 1, 9, 6, 9, 2	4, 2, 3, 7, 6, 3
8, 3, 6, 1, 8, 1	4, 6, 6, 5, 5, 6	8, 5, 1, 2, 3, 4	1, 2, 0, 6, 3, 3
0, 9, 8, 6, 2, 9	1, 5, 0, 6, 6, 5	2, 4, 5, 3, 4, 8	1, 1, 9, 0, 3, 2
0, 6, 8, 8, 3, 5	3, 3, 0, 4, 9, 6	1, 5, 6, 7, 8, 2	7, 8, 9, 2, 7, 0
7, 9, 5, 7, 7, 9	9, 3, 0, 7, 4, 1	3, 3, 8, 6, 0, 1	1, 1, 5, 0, 5, 1
7, 7, 6, 4, 4, 7	5, 3, 6, 4, 2, 0	3, 1, 4, 4, 9, 0	7, 7, 8, 7, 7, 6
1, 6, 5, 6, 4, 2	7, 1, 5, 0, 5, 8	9, 7, 7, 9, 8, 1	4, 9, 3, 7, 3, 9
9, 8, 6, 8, 6, 0	4, 4, 6, 2, 6, 2	6, 9, 2, 9, 8, 7	5, 5, 1, 1, 4, 0
3, 1, 6, 0, 0, 9	3, 1, 8, 8, 2, 1	6, 6, 8, 9, 6, 0	4, 2, 5, 7, 7, 9
0, 6, 8, 5, 2, 8	8, 9, 0, 6, 1, 7	3, 3, 4, 6, 7, 0	8, 3, 0, 6, 9, 7
8, 2, 4, 9, 4, 6	1, 3, 7, 3, 4, 3		

a. Use the 300 random digits to construct a relative frequency histogram for the data. This relative frequency distribution should approximate $p(x)$.

b. Use the methods of Section 5.3 to show that $E(x) = \mu = 4.5$.

c. Use the methods of Section 5.3 to show that $\sigma^2 = E[(x - \mu)^2] = 8.25$.

d. Suppose you intend to make an inference about the mean, μ, using the median of a sample of $n = 6$ measurements. To see how well the sample median will estimate μ, calculate the median, m, for each of the 50 samples. Construct a relative frequency histogram for the sample medians to see how close they lie to $\mu = 4.5$. Calculate the mean and standard deviation of the 50 medians.

7.7 Calculate $\bar{x}$ for each of the 50 samples in Exercise 7.6.

a. Construct a relative frequency histogram for the sample means to see how close they lie to $\mu = 4.5$. This will be a rough approximation to the sampling distribution of $\bar{x}$ for $n = 6$ observations.

b. Use the results of parts **b** and **c** in Exercise 7.6 to find $\mu_{\bar{x}}$ and $\sigma_{\bar{x}}$. Do these numbers describe the central tendency and variability of your histogram for the 50 sample means?

c. Compare your approximation of the sampling distribution of the sample mean in part **a** with the sampling distribution of the sample median in part **d** of Exercise 7.6. Do the sample means or the sample medians appear to fall closer to $\mu = 4.5$? Or is there very little difference for samples of $n = 6$ observations?

7.8 To see the effect of sample size on the standard deviation of the sampling distribution of a statistic, refer to Exercise 7.6 and combine pairs of samples (moving down the columns of the table) to obtain 25 samples of $n = 12$ measurements. Calculate the median for each sample.

a. Construct a relative frequency histogram for the 25 medians. Compare this with the histogram prepared for Exercise 7.6, part **d**, which is based on samples of $n = 6$ digits.

b. Calculate the mean and standard deviation of the 25 medians. Compare the standard deviation of this sampling distribution with the standard deviation of the sampling distribution in Exercise 7.6, part **d**. What relationship would you expect to exist between the two standard deviations?

7.9 Refer to Exercise 7.8. Repeat the exercise, but use the means of the samples rather than the medians, and compare the results to those obtained in Exercise 7.7.

7.10 A random sample of 40 observations is to be drawn from a large population of measurements. It is known that 30% of the measurements in the population are 1's, 20% of the measurements are 2's, 20% are 3's, and 30% are 4's.
   a. Give the mean and standard deviation of the (repeated) sampling distribution of $\bar{x}$, the sample mean of the 40 observations.
   b. Describe the shape of the sampling distribution of $\bar{x}$. Does your answer depend on the sample size?

*7.11 Using a statistical software package, generate 100 random samples of size $n = 40$ from the population described in Exercise 7.10.
   a. Compute $\bar{x}$ for each sample and plot a frequency distribution for the 100 values of $\bar{x}$.
   b. Compute the mean and standard deviation for the 100 values of $\bar{x}$.
   c. How does the approximation for the sampling distribution of $\bar{x}$ developed in parts **a** and **b** compare with the approximation you obtained in Exercise 7.10? Explain why differences may exist.

*7.12 Use a statistical software package to generate 100 random samples of size $n = 2$ from a population characterized by a uniform probability distribution (Section 6.2) with $c = 0$ and $d = 10$. Compute $\bar{x}$ for each sample and plot a frequency distribution for the 100 values of $\bar{x}$. Repeat this process for $n = 5, 10, 30,$ and 50. Explain how your plots illustrate the Central Limit Theorem.

*7.13 Use a statistical software package to generate 100 random samples of size $n = 2$ from a population characterized by a normal probability distribution with mean 100 and standard deviation 10. Compute $\bar{x}$ for each sample and plot a frequency distribution for the 100 values of $\bar{x}$. Repeat this process for $n = 5, 10, 30,$ and 50. Explain how your plots illustrate Theorem 7.2.

7.14 Suppose a sample of $n = 50$ items is drawn from a population of manufactured products and the weight, x, of each item is recorded. Prior experience has shown that the weight has a probability distribution with $\mu = 6$ ounces and $\sigma = 2.5$ ounces. Then $\bar{x}$, the sample mean, will be approximately normally distributed (because of the Central Limit Theorem).
   a. Calculate $\mu_{\bar{x}}$ and $\sigma_{\bar{x}}$.
   b. What is the probability that the manufacturer's sample has a mean weight of between 5.75 and 6.25 ounces?
   c. What is the probability that the manufacturer's sample has a mean weight of less than 5.5 ounces?
   d. How would the sampling distribution of $\bar{x}$ change if the sample size, n, were increased from 50 to, say, 100?

7.15 A population consists of four numbers, 1, 2, 2, and 3, marked on poker chips.
   a. How many different samples of $n = 2$ chips could be selected (without replacement) from the population? List the possible samples.
   b. Give the probability of selecting any one of these samples (assume that the sampling is random).

    c. Calculate $\bar{x}$ for each of the samples in part **b**.

    d. Calculate the probability associated with each of the possible values of $\bar{x}$.

    e. Graph the population probability distribution and the sampling distribution of $\bar{x}$ (obtained in part **d**).

7.16 Suppose x equals the number of heads observed when a single coin is tossed (i.e., $x = 0$ or $x = 1$). The population corresponding to x is the set of 0's and 1's generated when the coin is tossed repeatedly a large number of times. Suppose we select $n = 2$ observations from this population (i.e., we toss the coin twice and observe two values of x).

    a. List the three different samples (combinations of 0's and 1's) that could be obtained.

    b. Calculate the value of $\bar{x}$ for each of the samples.

    c. List the values that $\bar{x}$ can assume, and find the probabilities of observing these values.

    d. Construct a graph of the sampling distribution of $\bar{x}$.

7.17 Suppose x equals the number of heads observed when a single coin is tossed (i.e., $x = 0$ or $x = 1$). The population corresponding to x is the set of 0's and 1's generated when the coin is tossed repeatedly a large number of times. Suppose we select $n = 3$ observations from this population (i.e., we toss the coin three times and observe the three values of x).

    a. List the four different samples (combinations of 0's and 1's) that could be obtained.

    b. Calculate the value of $\bar{x}$ for each of the samples.

    c. List the values that $\bar{x}$ can assume, and find the probabilities of observing these values.

    d. Construct a graph of the sampling distribution of $\bar{x}$.

7.18 A random sample of size n is to be drawn from a large population with mean 100 and standard deviation 10, and the sample mean, $\bar{x}$, is to be calculated. To see the effect of different sample sizes on the standard deviation of the sampling distribution of $\bar{x}$, plot $\sigma/\sqrt{n}$ against n for $n = 1, 5, 10, 20, 30, 40,$ and 50.

7.19 Refer to Exercise 7.18. If you increase the sample size from $n = 5$ to $n = 25$, does the information in the sample mean, $\bar{x}$, pertinent to μ increase by the same amount as it does for an increase in sample size from $n = 30$ to $n = 50$? How is this answer shown in the graph you constructed for Exercise 7.18?

7.20 A random sample of size $n = 30$ is to be drawn from a large population with $\mu = 500$ and $\sigma = 200$.

    a. What is the standard deviation of the sampling distribution of $\bar{x}$?

    b. In order to reduce the standard deviation of $\bar{x}$ to 50% of the value in part **a**, how much larger would n need to be?

    c. In order to reduce $\sigma_{\bar{x}}$ to 75% of the value in part **a**, how much larger would n need to be?

APPLYING THE CONCEPTS

7.21 Purchased materials, parts, and services account for over 50% of the manufacturing costs of many companies. As a result, it is important for a company's overall quality program to extend to the vendors (i.e., suppliers) from whom the purchases are made. In recent years, many firms have begun including vendor surveillance as part of their quality programs (Juran and Gryna, 1980).

For example, suppose a soft-drink bottler requires bottles with an internal pressure strength of at least 150 pounds per square inch (psi). A prospective bottle vendor claims that its production process yields bottles with a mean internal strength of 157 psi and a standard deviation of 3 psi. As part of its vendor surveillance program, the bottler strikes an agreement with the vendor that permits the bottler to sample from the vendor's production process to verify the vendor's claim. The bottler randomly selected 40 bottles from the last 10,000 produced, measured the internal pressure strength of each, and found the mean strength for the sample to be 1.3 psi below the process mean cited by the vendor.

a. Assuming the vendor's claim to be true, what is the probability of obtaining a sample mean this far or farther below the process mean? What does your answer suggest about the validity of the vendor's claim?

b. If the process standard deviation were 3 psi as claimed by the vendor, but the mean were 156 psi, would the observed sample result be more or less likely than in part **a**? What if the mean were 158 psi?

c. If the process mean were 157 psi as claimed, but the process standard deviation were 2 psi, would the observed sample result be more or less likely than in part **a**? What if the standard deviation were 6 psi?

7.22 Marketing managers face many decision-making situations in which accurate information is needed about future sales; examples include setting quotas for sales representatives and assessing the need for advertising. In retail organizations, the organization's buyers provide marketing managers with information about future sales. Typically, buyers combine historical information on past sales with assessments of future consumer demand to make subjective 6-month forecasts of sales for specific products.

In a recent study, Steven Hartley (1983) conducted an experiment to evaluate the forecasting skills of 140 retail buyers of two large midwestern retail organizations. In one part of the experiment, 61 of the buyers were given historical sales data for the previous 30 months and asked to forecast sales 6 months from now. For each buyer, Hartley calculated the difference between the actual number of units sold 6 months later and the buyer's forecast. This difference is sometimes called *forecast error* and is denoted here as x. In order to characterize the accuracy of this group of 61 buyers, Hartley calculated $\bar{x}$, the mean forecast error for the sample.

a. Assume the sample of 61 buyers was randomly selected from a large population of buyers whose forecast errors have a distribution with mean 10 and standard deviation 16. Describe the sampling distribution of $\bar{x}$.

b. Hartley found $\bar{x} = 13.49$. Given the information in part **a** about the population of buyers, is this a likely result? Explain.

c. Suppose $\bar{x} = 0$ and $s^2 = 1$. What do these statistics tell you about the accuracy of the forecasts of the 61 buyers?

7.23 In Case Study 7.2, an individual was considering investing $1,000 in each of $n = 5$ different stocks. The monthly rate of return on each stock had mean $\mu = 10\%$ and standard deviation $\sigma = 4\%$. The investor's monthly rate of return for the portfolio of five stocks was $\bar{r} = \Sigma r_i / 5$. It was shown that the variance of the investor's monthly rate of return was $\sigma_{\bar{r}}^2 = \sigma^2/n = 3.2$ and that this number is a measure of the risk faced by the investor.

a. If instead the individual were to invest $1,000 in only three of the five stocks, would the risk faced by the investor increase or decrease? Explain.

b. Suppose $1,000 was invested in each of ten stocks with rate of return characteristics identical to those described above. Measure the risk faced by the investor and compare it to the risk associated with investing in just five of the stocks.

7.24 It was not until the 1950's that manufacturing companies began to measure and account for the costs of their quality functions. In general, quality costs are those associated with producing, identifying, avoiding, or repairing products that do not meet requirements. They are often divided into four categories: prevention costs, appraisal costs, internal failure costs, and external failure costs (Montgomery, 1985). This exercise is concerned with an inspection and evaluation activity that would generate appraisal costs.

A particular manufacturing process requires steel rods that are at least 3 meters in length. The rods are purchased in lots of 50,000. To determine whether the lot meets the required quality standards, 100 rods are randomly sampled from each incoming lot and the mean length of rods in the sample is calculated. The quality manager has decided to accept lots whose sample mean is 3.005 meters or more. Assume that the standard deviation of the rod lengths in a lot is .03 meter.

a. If in fact each lot has a mean length of 3 meters, what percentage of the lots received by the manufacturer will be returned to the vendor (i.e., the supplier)?

b. If in fact all of the rods in all of the lots received by the manufacturer are between 2.999 and 3.004 meters in length, what percentage of the lots will be returned to the vendor?

7.25 A local bank reported to the federal government that its 5,246 savings accounts have a mean balance of $1,000 and a standard deviation of $240. Government auditors have asked to randomly sample 64 of the bank's accounts in order to assess the reliability of the mean balance reported by the bank. The auditors say they will certify the bank's report only if the sample mean balance is within $60 of the reported mean balance. What is the probability that the auditors will *not* certify the bank's report, even if the mean balance really is $1,000? (Assume the standard deviation reported by the bank is accurate.)

NUMBER OF FLAWS PER DISK	
x	$p(x)$
0	.75
1	.15
2	.10

7.26 Suppose a manufacturing process that produces floppy disks for microcomputers can be characterized by the probability distribution in the table.

a. Compute the mean and standard deviation of the number of flaws per disk.

b. Describe the approximate sampling distribution of the mean number of flaws per disk in a random sample of 400 disks selected from the production process.

c. Compute the mean and standard deviation of $\bar{x}$.

d. Find the probability that the mean number of flaws per disk in a random sample of 400 disks is less than .30.

7.27 The distribution of starting salaries of men who received college degrees last year had a mean equal to $15,000 and a standard deviation equal to $1,200. A random sample of 36 starting salaries of women who received college degrees last year had a sample mean equal to $15,600.

a. What is the probability that a random sample of 36 starting salaries of men who graduated last year would have a sample mean at least as large as $15,600?

b. Based on your answer to part **a**, does it appear that the starting salaries of women should be characterized by the same probability distribution that characterizes men's starting salaries? Explain.

7.28 To determine whether a metal lathe that produces machine bearings is properly adjusted, a random sample of 36 bearings is collected and the diameter of each is measured. If the standard deviation of the diameter of the machine bearings measured over a long period of time is .001 inch, what is the probability that the mean diameter $\bar{x}$ of the sample will lie within .0001 inch of the population mean diameter of the bearings?

7.29 Refer to Exercise 7.28. Suppose the mean diameter of the bearings produced by the machine is supposed to be .5 inch. The company decides to use the sample mean (from Exercise 7.28) to decide whether the process is in control—i.e., whether it is producing bearings with a mean diameter of .5 inch. The machine will be considered out of control if the mean of the sample of $n = 36$ diameters is less than .4994 inch or larger than .5006 inch. If the true mean diameter of the bearings produced by the machine is .501 inch, what is the probability that the test will fail to imply that the process is out of control?

7.30 Northern States Power (NSP), a private utility in Minnesota, is required by federal and state law to offer in-home energy audits to its customers. The purpose of the law is to encourage households to conserve energy. In early 1981, NSP mailed many of its customers an energy-audit offer, but only 3% of the households responded with an audit request. The State of Minnesota had expected a response rate of 30%–50%. In an attempt to improve this response rate, Richard Weijo (1983) evaluated alternative means of making NSP customers aware of the audit program and of motivating them to make inexpensive energy-saving changes to their homes.

As part of the study, two independent random samples were selected from among the 9,900 households in St. Paul and its suburbs that were scheduled to receive energy-audit offers by NSP in April 1982. The 280 households in the first sample received an inexpensive waterflow controller along with their energy-audit offer. The 348 households in the second sample received only the energy-audit offer. Each household in both samples was contacted by telephone and questioned about the mailing it had received. Five questions were designed to measure how much the head of a household could recall about the information contained in the energy-audit offer. Each household received a score from 0 to 5, corresponding to the number of questions correctly answered. Let x denote the score received by a household whose mailing contained the waterflow controller and y denote the score received by a household whose mailing consisted only of the energy-audit offer. To compare the effectiveness of the two types of mailings with respect to information recall, Weijo calculated $\bar{x}$ and $\bar{y}$ for the two samples of households.

a. The scores received by the 280 households that were mailed the waterflow controller can be viewed as a sample from the population of scores that would exist if all 9,900 households had been sent the waterflow controller and then been interviewed. Suppose this hypothetical population of scores has standard deviation 1.2. Describe the sampling distribution of $\bar{x}$.

b. Suppose all 9,900 households had received only the energy-audit offer and were later scored with respect to their recall of information about the mailing. Assume that the standard deviation of this population of scores is 1.0, and describe the sampling distribution of $\bar{y}$.

c. How large would the sample of households receiving the waterflow controller have to be in order to have $\sigma_{\bar{x}}^2 = \sigma_{\bar{y}}^2$?

7.31 The primary responsibility of government tax assessors is to estimate the market values of all properties in their respective jurisdictions. These estimates form the basis upon which property tax bills are determined. In order to estimate the mean market value for single-family homes in a particular county in 1987, a county tax assessor uses the mean sale price for all single-family homes that sold during 1987. Such a sample is typically treated by tax assessors as if it were a random sample from the population of properties in question (*Standard on Assessment-Ratio Studies*, 1980). Suppose 400 properties sold during 1987 and assume that the standard deviation of the market value for the population of properties is about $50,000.

  a. What is the population in question?
  b. What is the probability that the assessor's sample mean will overestimate the actual mean market value of the population?
  c. What is the probability that the assessor's sample mean will fall within $4,000 of the actual mean market value for the population?

7.32 In any production process, some variation in the quality of the product is unavoidable. Variation in product can be divided into two categories: variation due to special causes and variation due to common causes. The former includes variation due to a specific worker or group of workers, a specific machine, or a specific local condition. The latter includes variation due to faults of the overall production system, such as poor working conditions for workers, use of raw materials of inferior quality, poor design of the product, etc. The discovery and removal of special causes is generally the responsibility of a person directly involved with the production operation in question. In contrast, common causes are the responsibility of management.

  A production process in which all special causes of variation have been eliminated is said to be *stable* or *in statistical control*. The variation that remains is simply random variation. If the magnitude of the random variation is unacceptable to management, it can be reduced through the elimination of common causes (Deming, 1982, 1986).

  It is common practice to monitor the variation in the quality characteristic of a product over time by plotting the quality characteristic on a *control chart*. For example, the amount of alkali in soap might be monitored each hour by randomly selecting from the production process and measuring the quantity of alkali in $n = 5$ test specimens of soap. The mean, $\bar{x}$, of the sample alkaline measurements would be plotted against time, as shown in the accompanying figure. If the process is in statistical control, $\bar{x}$ should assume a distribution with a mean equal to the process mean, μ, with standard deviation equal to the process standard deviation divided by the square root of the sample size, $\sigma_{\bar{x}} = \sigma/\sqrt{n}$. The control chart (see next page) includes a horizontal line to locate the process mean and two lines, called *control limits*, located $3\sigma_{\bar{x}}$ above and below μ. If $\bar{x}$ falls within the control limits, the process is deemed to be in control. If $\bar{x}$ is outside the limits, there is strong evidence that special causes of variation are present, and the process is deemed to be out of control.

  When the production process is in statistical control, the percentage of alkali in a test specimen of soap follows approximately a normal distribution with $\mu = 2\%$ and $\sigma = 1\%$.

  a. If $n = 5$, how far from μ should the upper and lower control limits be located?
  b. If the process is in control, what is the probability that $\bar{x}$ will fall outside the control limits?
  c. If the process mean shifts to $\mu = 3\%$ before the sample is drawn, what is the probability that the sample will lead to the (correct) conclusion that the process is out of control?

Control Chart for
Exercise 7.32

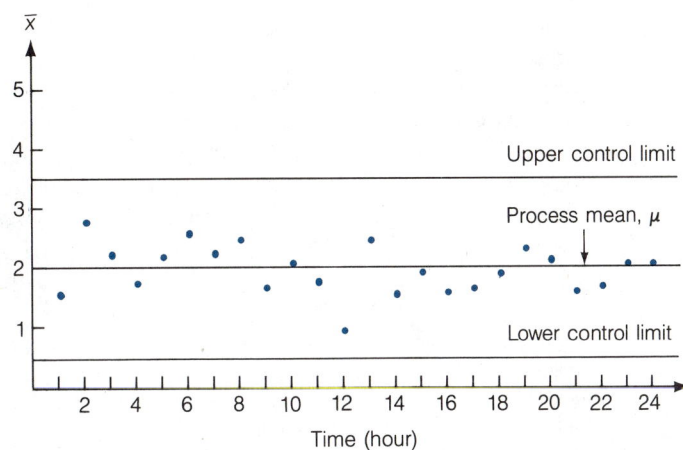

7.33 Refer to Exercise 7.32. The soap company has decided to set tighter control limits than the $3\sigma_{\bar{x}}$ limits described in Exercise 7.32. In particular, when the process is in control, the company is willing to risk a .10 probability that $\bar{x}$ falls outside the control limits.

  a. The company still wants the control limits to be located at equal distances above and below the process mean and is still planning to use $n = 5$ measurements in each hourly sample. Where should the control limits be located?

  b. Suppose the control limits developed in part **a** are implemented, but unknown to the company, μ is currently 3% (not 2%). What is the probability that $\bar{x}$ will fall outside the control limits if $n = 5$? If $n = 10$?

7.34 Refer to Exercise 7.32. In order to improve the sensitivity of control charts, *warning limits* are sometimes included on the chart along with the control limits. These limits are typically set at $\mu \pm 1.96\sigma_{\bar{x}}$. If two successive data points fall outside the warning limits, the process is deemed to be out of control (Wetherill, 1977).

  a. If the soap process is in control (i.e., it follows a normal distribution with $\mu = 2\%$ and $\sigma = 1\%$), what is the probability that the next value of $\bar{x}$ will fall outside the warning limits?

  b. If the soap process is in control, how many of the next 40 values of $\bar{x}$ plotted on the control chart would be expected to fall above the upper warning limit?

  c. If the soap process is in control, what is the probability that the next two values of $\bar{x}$ will fall below the lower warning limit?

USING THE COMPUTER . . .

Calculate the mean and standard deviation for the 1,000 zip codes' median household incomes. We will treat these quantities as the population mean μ and standard deviation σ for this exercise.

  a. Draw 100 random samples of $n = 20$ observations from the 1,000 zip codes' median household incomes. Select the samples with replacement, i.e., replace each measurement before selecting the next.* Calculate the 100 sample means. Generate a stem and leaf display or a relative frequency histogram for the 100 means. Then count the number of the 100 sample means that fall in the intervals $\mu \pm \sigma/\sqrt{n}$, $\mu \pm 2\sigma/\sqrt{n}$, and $\mu \pm 3\sigma/\sqrt{n}$. How do the graphical description and the percentage of means falling in the intervals agree with a normal distribution having mean μ and standard deviation $\sigma/\sqrt{n}$?

b. Repeat part **a** using a sample size of $n = 50$. Is the sampling distribution of the sample means closer to normal for the larger sample size?

REFERENCES

Alexander, G. J. and Francis, J. C. *Portfolio Analysis*. Englewood Cliffs, N.J.: Prentice-Hall, Inc., 1986. Chapters 4 and 5.

Blume, M. "On the assessment of risk," *Journal of Finance*, Mar. 1971, 26, 1–10.

Deming, W. E. *Out of the Crisis*. Cambridge, Mass.: MIT Center for Advanced Engineering Study, 1986.

Deming, W. E. *Quality, Productivity, and Competitive Position*. Cambridge, Mass.: MIT Center for Advanced Engineering Study, 1982. Chapter 7.

Elton, E. J. and Gruber, M. J. *Modern Portfolio Theory and Investment Analysis*. New York: Wiley, 1981.

Francis, J. C. *Investments: Analysis and Management*, 3rd ed. New York: McGraw-Hill, 1980.

Hartley, S. W. *Judgmental Sales Forecasting: An Experimental Investigation of Task Structure and Environmental Complexity*. Unpublished Ph.D. dissertation, University of Minnesota, 1983.

Hogg, R. V. and Craig, A. T. *Introduction to Mathematical Statistics*, 4th ed. New York: Macmillan, 1978. Chapter 4.

Juran, J. M. and Gryna, F. M., Jr. *Quality Planning and Analysis*. New York: McGraw-Hill, 1980.

Lindgren, B. W. *Statistical Theory*, 3rd ed. New York: Macmillan, 1976. Chapter 2.

Montgomery, D. C. *Introduction to Statistical Quality Control*. New York: Wiley, 1985. Chapter 1.

Neter, J., Wasserman, W., and Whitmore, G. A. *Applied Statistics*, 2nd ed. Boston: Allyn & Bacon, 1983. Chapters 8 and 9.

Sharpe, W. F. *Investments*, 2nd ed. Englewood Cliffs, N.J.: Prentice-Hall, 1981. Chapter 5.

Standard on Assessment-Ratio Studies. Chicago: International Association of Assessing Officers, 1980.

Weijo, R. O. *Evaluating Information, Incentive, and Door-to-door Interventions for the MECS Energy Audit Program: A Theoretical Application of the Petty and Cacippo Elaboration Likelihood Model*. Unpublished Ph.D. dissertation, University of Minnesota, 1983.

Wetherill, G. B. *Sampling Inspection and Quality Control*, 2nd ed. New York: Chapman and Hall, 1977. Chapter 3.

Winkler, R. L. and Hays, W. *Statistics: Probability, Inference, and Decision*, 2nd ed. New York: Holt, Rinehart and Winston, 1975. Chapter 5.

*In this and future exercises we will specify sampling with replacement to simulate the sampling from very large or infinite populations. This avoids the use of finite population correction factors (Chapter 20) for samples from (relatively) small populations.

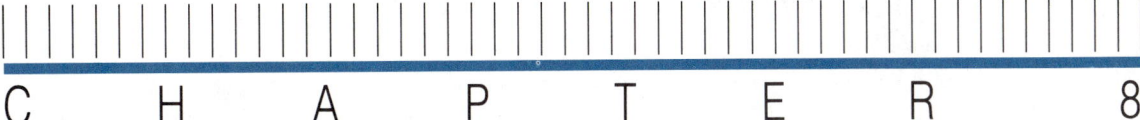

C H A P T E R 8

WHERE WE'VE BEEN . . .

In the preceding chapters we learned that populations are characterized by numerical descriptive measures (called **parameters**), and that decisions about their values are based on sample statistics computed from sample data. Since statistics vary in a random manner from sample to sample, inferences based on them will be subject to uncertainty. This property is reflected in the sampling (probability) distribution of a statistic.

WHERE WE'RE GOING . . .

This chapter puts all the preceding material into practice; that is, we will estimate or make decisions about population means or proportions based on a single sample selected from a population. Most important, we will use the sampling distribution of a sample statistic to assess the uncertainty associated with an inference.

CONTENTS

ESTIMATION AND A TEST OF HYPOTHESIS: SINGLE SAMPLE

The estimation of the mean gas mileage for a new car model, the test of a claim that a certain brand of television tube has a mean life of 5 years, and the estimation of the mean yearly sales for companies in the steel industry are problems with a common element. In each case, we are interested in making an inference about the mean of a population. This important problem constitutes the primary topic of this chapter.

We will concentrate on two types of inferences about a population parameter: **estimation of the parameter** and **tests of hypotheses, or claims, about the parameter.** You will see that different techniques are used for making inferences, depending on whether a sample contains a large or small number of measurements. Regardless, our objectives remain the same. We want to make the best use of the information in the sample to make an inference and to assess its reliability.

In Sections 8.1 and 8.2 we consider a method of estimating a population mean using a large sample, and develop a formula to determine just how large the sample must be to achieve a specified degree of reliability. In Sections 8.3, 8.4, and 8.5, we show how to test hypotheses about population means with large samples. Small-sample inferences about population means are treated in Section 8.6. Finally, inferences about binomial probabilities and the sample sizes necessary to make reliable inferences are covered in Sections 8.7 and 8.8, respectively.

SECTION 8.1

LARGE-SAMPLE ESTIMATION OF A POPULATION MEAN

Suppose a large credit corporation wants to estimate the average amount of money owed by its delinquent debtors; i.e., debtors who are more than 2 months behind in payment. To accomplish this objective, the company plans to sample 100 of its delinquent accounts and to use the sample mean, $\bar{x}$, of the amounts overdue to estimate μ, the mean for *all* delinquent accounts. The sample mean $\bar{x}$ represents a **point estimator** of the population mean μ (Definition 7.4). How can we assess the accuracy of this point estimator?

Recall that for sufficiently large samples the sampling distribution of the sample mean is approximately normal, as shown in Figure 8.1. Now, suppose you plan to take a sample of $n = 100$ measurements and calculate the following interval:

$$\bar{x} \pm 2\sigma_{\bar{x}} = \bar{x} \pm \frac{2\sigma}{\sqrt{n}}$$

FIGURE 8.1
Sampling Distribution of $\bar{x}$

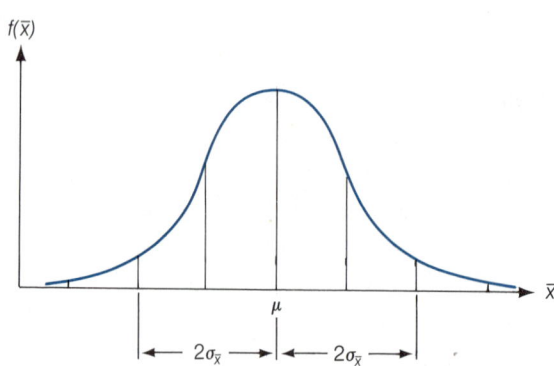

That is, you will form an interval 4 standard deviations wide: from 2 standard deviations below the sample mean to 2 standard deviations above the sample mean. What is the chance, before we have drawn the sample, that this interval will enclose μ, the population mean?

To answer this question, refer to Figure 8.1. If the 100 measurements yield a value of $\bar{x}$ that falls between the two lines on either side of μ, i.e., within 2 standard deviations of μ, then the interval $\bar{x} \pm 2\sigma_{\bar{x}}$ will contain μ. If $\bar{x}$ falls outside either of these boundaries, then the interval $\bar{x} \pm 2\sigma_{\bar{x}}$ will not contain μ. Since the area under the normal curve (the sampling distribution of $\bar{x}$) between these boundaries is approximately .95 (more precisely, from Table IV of Appendix B, the area is .9544), we know that the interval $\bar{x} \pm 2\sigma_{\bar{x}}$ will contain μ with a probability approximately equal to .95.

To illustrate, suppose that the sum and the sum of squared deviations for the sample of debits of 100 delinquent accounts are

$$\sum_{i=1}^{n} x_i = \$23,300 \qquad \sum_{i=1}^{n} (x_i - \bar{x})^2 = 801,900$$

First calculate the sample statistics:

$$\bar{x} = \frac{\sum\limits_{i=1}^{100} x_i}{n} = \frac{23,300}{100} = \$233 \qquad s = \sqrt{\frac{\sum\limits_{i=1}^{100} (x_i - \bar{x})^2}{n-1}} = \sqrt{\frac{801,900}{99}} = \$90$$

Then, we calculate the interval

$$\bar{x} \pm 2\sigma_{\bar{x}} = 233 \pm 2\left(\frac{\sigma}{\sqrt{100}}\right)$$

But now we face a problem. You can see that without knowing the standard deviation, σ, of the original population—i.e., the standard deviation of the amounts of *all* delinquent accounts—we cannot calculate this interval. However, since we have a large sample ($n = 100$ measurements), we can approximate the interval by using the sample standard deviation, s, to approximate σ. Thus,

$$\bar{x} \pm 2\left(\frac{\sigma}{\sqrt{100}}\right) \approx \bar{x} \pm 2\left(\frac{s}{\sqrt{100}}\right)$$

$$= 233 \pm 2\left(\frac{90}{10}\right) = 233 \pm 18$$

That is, we estimate the mean amount of delinquency for all accounts to fall within the interval ($215, $251).

Can we be sure that μ, the true mean amount due, is within the interval ($215, $251)? We cannot be certain, but we can be reasonably confident that it is. This confidence is derived from the knowledge that if we were to repeatedly draw samples of 100 accounts from this group of delinquent accounts and form the interval $\bar{x} \pm 2\sigma_{\bar{x}}$ each time, approximately 95% of the intervals would contain μ. We have no way of knowing (without looking at *all* the delinquent accounts) whether our sample interval is one of the 95% that contain μ or one of the 5% that do not, so we simply state that we are 95% confident our interval ($215, $251) contains μ. Thus, we have given an interval estimate of the mean delinquency per account and a measure of the reliability of the estimate.

The formula that tells us how to calculate an interval estimate based on sample data is called an **interval estimator**. The probability, .95, that measures the confidence that we can place in the interval estimate is called a **confidence coefficient**. The percentage, 95%, is called the **confidence level** for the interval estimate. It is not usually possible to assess the reliability of point estimators precisely because they are single points rather than intervals. Since we prefer to use the estimators for which a measure of reliability can be calculated, interval estimators will usually be used.

DEFINITION 8.1

An **interval estimator** is a formula that tells us how to use sample data to calculate an interval that estimates a population parameter.

DEFINITION 8.2

The **confidence coefficient** is the probability that an interval estimator encloses the population parameter if the estimator is used repeatedly a very large number of times. The **confidence level** is the confidence coefficient expressed as a percentage.

FIGURE 8.2

Interval Estimators for θ: Ten Samples

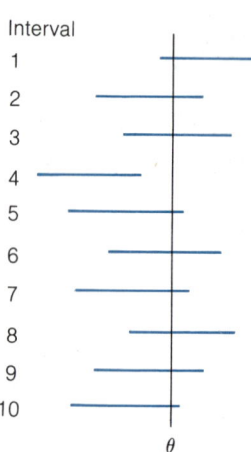

The foregoing is an example of how an interval can be used to estimate a population parameter. When we use an interval estimator, we can usually calculate the probability that the interval estimation *process* will result in an interval that contains the true value of the population parameter. That is, the probability that the interval contains the parameter in repeated usage is usually known. Figure 8.2 shows that happens when a number of samples are drawn from a population and a confidence interval for a parameter, say θ, is calculated from each. The location of θ is indicated by the vertical line in the figure. Ten confidence intervals, one based on each of ten samples, are shown as horizontal line segments. Note that the confidence intervals move from sample to sample—sometimes containing θ and other times missing θ. If our confidence level is 95%, then in the long run, 95% of our sample confidence intervals will contain θ.

Suppose you wish to choose a **confidence coefficient** other than .95. Notice that in Figure 8.1 the confidence coefficient .95 is equal to the total area under the sampling distribution, less .05 of the area; this .05 is divided equally between the two tails. Using this idea, we can construct a confidence interval with any desired confidence coefficient by increasing or decreasing the area (call it α) assigned to the tails of the sampling distribution (see Figure 8.3). For example, if we place

$\alpha/2$ in each tail and if $z_{\alpha/2}$ is the z value such that the area $\alpha/2$ will lie to its right, then the confidence interval with confidence coefficient $(1 - \alpha)$ is

$$\bar{x} \pm z_{\alpha/2}\sigma_{\bar{x}}$$

FIGURE 8.3
Locating $z_{\alpha/2}$ on the
Standard Normal Curve

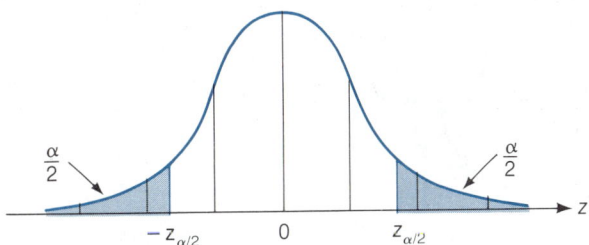

LARGE-SAMPLE 100$(1 - \alpha)$% CONFIDENCE INTERVAL FOR μ

$$\bar{x} \pm z_{\alpha/2}\sigma_{\bar{x}}$$

where $z_{\alpha/2}$ is the z value with an area $\alpha/2$ to its right (see Figure 8.3) and $\sigma_{\bar{x}} = \sigma/\sqrt{n}$. The parameter σ is the standard deviation of the sampled population and n is the sample size. When σ is unknown (as is almost always the case) and n is large (say, $n \geq 30$), the value of σ can be approximated by the sample standard deviation, s.

To illustrate, for a confidence coefficient of .90, $(1 - \alpha) = .90$, $\alpha = .10$, $\alpha/2 = .05$, and $z_{.05}$ is the z value that locates .05 in one tail of the sampling distribution. Recall that Table IV of Appendix B gives the areas between the mean and a specified z value. Since the total area to the right of the mean is .50, $z_{.05}$ will be the z value corresponding to the tabulated area to the right of the mean equal to .4500. This z value is $z_{.05} = 1.645$. Confidence coefficients used in practice (in reports and published articles) usually range from .90 to .99. The most commonly used confidence coefficients with corresponding values of α and $z_{\alpha/2}$ are shown in Table 8.1.

TABLE 8.1
Commonly Used
Values of $z_{\alpha/2}$

CONFIDENCE LEVEL 100$(1 - \alpha)$	α	$\alpha/2$	$z_{\alpha/2}$
90%	.10	.05	1.645
95%	.05	.025	1.96
99%	.01	.005	2.575

EXAMPLE 8.1

Unoccupied seats on flights cause the airlines to lose revenue. Suppose a large airline wants to estimate its average number of unoccupied seats per flight over the past year. To accomplish this, the records of 225 flights are randomly selected and the number of unoccupied seats is noted for each of the sampled flights. The sample mean and standard deviation are

$$\bar{x} = 11.6 \text{ seats} \qquad s = 4.1 \text{ seats}$$

Estimate μ, the mean number of unoccupied seats per flight during the past year, using a 90% confidence interval.

SOLUTION

The general form of the large-sample 90% confidence interval for a population mean is

$$\bar{x} \pm z_{\alpha/2}\sigma_{\bar{x}} = \bar{x} \pm z_{.05}\sigma_{\bar{x}}$$

$$= \bar{x} \pm 1.645\left(\frac{\sigma}{\sqrt{n}}\right)$$

For the 225 records sampled, we have

$$11.6 \pm 1.645\left(\frac{\sigma}{\sqrt{225}}\right)$$

Since we do not know the value of σ (the standard deviation of the number of unoccupied seats per flight for all flights of the year), we use our best approximation, the sample standard deviation s. Then the 90% confidence interval is, approximately,

$$11.6 \pm 1.645\left(\frac{4.1}{\sqrt{225}}\right) = 11.6 \pm .45$$

or, from 11.15 to 12.05. That is, the airline can be 90% confident that the mean number of unoccupied seats per flight was between 11.15 and 12.05 during the sampled year.

Remember that the 90% confidence comes from the knowledge that the interval estimation process will result in an interval that contains μ 90% of the time in repeated sampling. We do not know whether this particular interval (11.15, 12.05) is one of the 90% that contain μ, or one of the 10% that do not. ■

The interpretation of confidence intervals for a population mean is summarized in the box at the top of the next page.

CASE STUDY 8.1

DANCING TO THE CUSTOMER'S TUNE: THE NEED TO ASSESS CUSTOMER PREFERENCES

The following quotations have been extracted from the Dec. 13, 1976, issue of *Business Week*:

"We're dancing to the tune of the customer as never before," says J. Janvier Wetzel, vice-president for sales promotion at Los Angeles–based Broadway Department Stores. "With population growth down to a trickle compared with its previous level, we're no longer spoiled with instant success every time we open a new store. Traditional department stores are locked in the biggest competitive battle in their history."

> **INTERPRETATION OF A CONFIDENCE INTERVAL FOR A POPULATION MEAN**
>
> When we form a $100(1 - \alpha)\%$ confidence interval for μ, we usually express our confidence in the interval with a statement such as, "We can be $100(1 - \alpha)\%$ confident that μ lies between the lower and upper bounds of the confidence interval," where for a particular application we substitute the appropriate numerical values for the confidence, and the lower and upper bounds. The statement reflects our confidence in the process rather than in the particular interval formed. We know that repeated application of the same procedure will result in different lower and upper bounds on the interval. Furthermore, we know that $100(1 - \alpha)\%$ of the resulting intervals will contain μ. There is (usually) no way to determine whether a particular interval is one of those that contain μ, or one that does not. However, unlike point estimators, confidence intervals have some measure of reliability, the confidence coefficient, associated with them, and for that reason are generally preferred to point estimators.

The nation's retailers are becoming uncomfortably aware that today's operating environment is vastly different from that of the 1960s. Population growth is slowing, a growing singles market is emerging, family formations are coming at later ages, and more women are embarking on careers. Of the 71 million households in the U.S. today, the dominant consumer buying segment is families headed by persons over 45. But by 1980 this group will have lost its majority status to the 25 to 40 year-old group. Merchants must now reposition their stores to attract these new customers.

To do so retailers are using market research to ferret out new purchasing attitudes and lifestyles and then translating this into customer buying segments. . . . department stores are taking a hard look at some of the basics of their business by . . . spending heavily for far more elaborate market research. Data on demographics, psychographics (measurement of attitudes), and lifestyles are being fed into retailers' computers so they can make marketing decisions based on actual spending patterns and estimate their inventory needs with less risk.

In order to stock its various departments with the type and style of goods that appeal to its potential group of customers, a downtown department store should be interested in estimating the average age of downtown shoppers, not shoppers in general. Suppose a downtown department store questions 49 downtown shoppers concerning their age (the offer of a small gift certificate may help convince shoppers to respond to such questions). The sample mean and standard deviation are found to be 40.1 and 8.6, respectively. The store could then estimate μ, the mean age of all downtown shoppers, with a 95% confidence interval as follows:

$$\bar{x} \pm 1.96\left(\frac{s}{\sqrt{n}}\right) = 40.1 \pm 1.96\left(\frac{8.6}{\sqrt{49}}\right)$$

$$= 40.1 \pm 2.4$$

Thus, the department store should gear its sales to the segment of consumers with average age between 37.7 and 42.5.

LEARNING THE MECHANICS

8.1 Find $z_{\alpha/2}$ for each of the following:
 a. $\alpha = .10$ **b.** $\alpha = .01$
 c. $\alpha = .05$ **d.** $\alpha = .20$

8.2 What is the confidence level of each of the following confidence intervals for μ?
 a. $\bar{x} \pm 1.96\sigma/\sqrt{n}$ **b.** $\bar{x} \pm 1.645\sigma/\sqrt{n}$ **c.** $\bar{x} \pm 2.575\sigma/\sqrt{n}$
 d. $\bar{x} \pm 1.282\sigma/\sqrt{n}$ **e.** $\bar{x} \pm .99\sigma/\sqrt{n}$

8.3 A random sample of 64 observations from a population produced the following summary statistics:

$$\Sigma x = 500 \qquad \Sigma (x_i - \bar{x})^2 = 3{,}566$$

 a. Find a 95% confidence interval for μ.
 b. Interpret the confidence interval you found in part **a**.

8.4 A random sample of 80 observations from a normally distributed population produced a sample mean of 14.1 and a standard deviation of 2.6.
 a. Find a 95% confidence interval for μ.
 b. Find a 99% confidence interval for μ.
 c. What happens to the width of a confidence interval as the sample size is held fixed and the value of the confidence coefficient is increased?
 d. Would your confidence intervals of parts **a** and **b** be valid if the distribution of the original population were not normal? Explain.

8.5 Refer to Exercise 8.4. Suppose the sample contained only 32 observations.
 a. Recompute the 95% confidence interval for μ.
 b. What is the effect on the width of a confidence interval of reducing the sample size as the confidence coefficient remains fixed? Of increasing the sample size as the confidence coefficient remains fixed?

8.6 Explain what is meant by the statement "We are 95% confident that our interval estimate contains μ."

8.7 Explain the difference between an interval estimator and a point estimator for μ.

8.8 Describe the relationship between the size of the confidence coefficient and the width of the confidence interval.

8.9 Describe the relationship between the sample size and the width of the confidence interval.

APPLYING THE CONCEPTS

8.10 To assess the magnitude of recent rent increases in metropolitan Minneapolis, an apartment referral service randomly sampled and interviewed 32 apartment building owners. They obtained the following data concerning the change in rent for two-bedroom apartments between May 1985 and May 1986:

1.13%	3.32%	.62%	2.03%	1.82%
2.61	1.79	3.59	2.27	1.80
−1.22	2.06	−.45	.05	.34
7.33	5.20	1.25	1.42	.68
2.03	.68	.00	1.41	1.39
2.42	1.00	.90	.61	.00
1.00	−1.07			

a. Carefully describe the population from which the sample was drawn.

b. Use a 95% confidence interval to estimate the mean percentage change in rent for two-bedroom apartments.

c. In constructing the confidence interval of part **b**, was it necessary to assume that the population was normally distributed? Explain.

8.11 A brief article in the *Wall Street Journal* (Mar. 1, 1984) states that of 48 sources surveyed, Prudential Bache Securities gave the lowest estimate (3%) of the rise in the Consumer Price Index for 1984 and Prudential Insurance (Prudential Bache's owner) gave the highest (5.6%).

a. Find an approximate value for the sample standard deviation of the sample of 48 estimates. [*Hint:* Assume that the range is approximately equal to $4s$.]

b. Assume that the 48 estimates represent a random sample of estimates from a large number of estimate sources. If the mean of the sample was 4.3%, find a 90% confidence interval for the mean estimated increase in the Consumer Price Index for this population of estimates.

8.12 At the end of 1982, 1983, and 1984, the average price of a share of stock traded on the New York Stock Exchange (NYSE) was $33.03, $35.11, and $32.31, respectively (*Statistical Abstract of the United States: 1986*, p. 509). To investigate the average share price at the end of 1985, a random sample of 36 NYSE stocks was drawn. Their closing prices on December 31, 1985, are listed (by their NYSE abbreviations) in the table.

STOCK	PRICE	STOCK	PRICE	STOCK	PRICE	STOCK	PRICE
Alcan	$28\frac{7}{8}$	XTRA	$22\frac{5}{8}$	Foxmyr	$22\frac{3}{4}$	Textron	$48\frac{7}{8}$
WinDix	$38\frac{1}{8}$	AmExp	$53\frac{7}{8}$	NiaMP	$20\frac{1}{8}$	Dallas	18
Elgin	$13\frac{3}{8}$	Litton	$83\frac{1}{4}$	CTS	$32\frac{7}{8}$	Traveler	$47\frac{3}{8}$
Unitlnd	$24\frac{7}{8}$	Enserch	$21\frac{1}{2}$	Exxon	$54\frac{3}{8}$	Coachm	$13\frac{3}{8}$
Avon	$27\frac{1}{2}$	Kyocer	46	JoyMfg	$23\frac{1}{2}$	MacMl	$36\frac{1}{2}$
CinBell	$55\frac{5}{8}$	Gensco	$3\frac{1}{8}$	LaLand	$30\frac{7}{8}$	C Psyc	28
Oneida	$16\frac{1}{2}$	SeaLnd	$23\frac{3}{8}$	Kellogg	$68\frac{1}{4}$	Becor	$14\frac{5}{8}$
BosEd	46	Intrfst	$10\frac{1}{4}$	Pansph	$24\frac{7}{8}$	CtData	$20\frac{5}{8}$
Delta Ar	$39\frac{5}{8}$	Pueblo	$18\frac{1}{2}$	Rowan	$7\frac{1}{2}$	SupMkt	$50\frac{1}{2}$

a. Using an 80% confidence interval, estimate the average price of a share of stock at the end of 1985.

b. Use the latest edition of the *Statistical Abstract of the United States* to determine the actual average price of a stock on the NYSE at the end of 1985. Is this figure in agreement with your confidence interval? Explain. If not, provide a possible explanation for the disagreement.

8.13 Automotive engineers are continually improving their products. Suppose a new type of brake light has been developed by General Motors. As part of a product safety evaluation program, General Motors' engineers wish to estimate the mean driver response time to the new brake light. (Response time is the length of time from the point that the brake is applied until the driver in the following car takes some corrective action.) Fifty drivers are selected at random and the response time (in seconds) for each driver is recorded, yielding the following results: $\bar{x} = .72$, $s^2 = .0936$. Estimate the mean driver response time to the new brake light using a 99% confidence interval.

8.14 A new process that has recently been developed can transform ordinary iron into a kind of super-iron called metallic glass ("One Answer to Imports," 1981). Metallic glass is three to four times as strong as the toughest steel alloys and up to 100 times as resistant to corrosion as the best stainless steel. One of the problems with metallic glass, however, is its tendency to become brittle at very high temperatures. In order to estimate the mean temperature, μ, at which a particular type of metallic glass becomes brittle, 36 pieces of the metallic glass were randomly sampled from a recent production run. Each piece was independently subjected to higher and higher temperatures until it became brittle. The temperature at which brittleness was first noticed was recorded for each piece in the sample. The following results were obtained: $\bar{x} = 480°F$, $s = 11°F$. Use a 90% confidence interval to estimate μ. Interpret your confidence interval.

8.15 Nasser Arshadi and Edward Lawrence (1984) investigated the profiles (i.e., career patterns, social backgrounds, and so forth) of the top executives in the U.S. banking industry. They sampled 96 executives and, among other things, found that 80% studied business or economics and that 45% had a graduate degree. With respect to the number of years of service, x, at the same bank, the group had a mean of 23.43 years and a standard deviation of 10.82 years.

a. Construct an 80% confidence interval for $E(x) = \mu$.

b. Interpret your interval in the context of the problem.

c. What assumption(s) was it necessary to make in order to construct the confidence interval of part a?

d. Is your interval estimate for $E(x)$ also an interval estimate for $E(\bar{x})$? Explain. [*Hint:* See Chapter 7.]

RANK	NUMBER OF SURVEY RESPONDENTS
1	24
2	101
3	430
4	355

8.16 [*Note: This exercise requires the use of methods discussed in optional Section 3.8.*] In an article in the *Harvard Business Review*, Rosen, Rynes, and Mahoney (1983) reported the results of a survey conducted to gather the views of managers on issues relating to the salary gap between males and females. A sample of 910 *Harvard Business Review* subscribers was polled. Eighty-three percent of the women surveyed believed that, among the factors that contribute to the salary gap, discrimination in setting wage rates (men earn more than women in similar jobs) was a very important factor. Only 40% of the men surveyed thought this was a very important factor. Concerning their power to influence compensation policy, the respondents were asked to rank each of the following groups from 1 to 4, with 1 indicating the group with the most clout: the federal government, organized labor, personnel administrators, and women's political groups.

Suppose the rankings given in the table were obtained for women's political groups. The actual average (mean) rank for women's political groups obtained by the survey was 3.2, which is approximately the same as the average rank of the hypothetical data given in the table.

a. Use a 90% confidence interval to estimate the average rank for women's political groups that would be obtained if all *Harvard Business Review* subscribers responded to the survey questionnaire.

b. What assumptions had to be made when constructing the confidence interval?

c. In the context of this problem, carefully explain what it means to be 90% confident that your confidence interval includes the population mean.

| | | | | | | | | | | | | |

SECTION 8.2

DETERMINING THE SAMPLE SIZE NECESSARY FOR MAKING INFERENCES ABOUT A POPULATION MEAN

Sometimes the analyst must plan the sampling experiment that generates the data used to make inferences about the population. Such data are generated by **designed experiments**, and perhaps the most important design decision faced by the analyst is to determine the size of the sample. We will show in this section that the appropriate sample size for making an inference about a population mean depends on the desired reliability.

To see this, consider the illustrative example from Section 8.1, in which we estimated the mean overdue amount for all delinquent accounts in a large credit corporation. A sample of 100 delinquent accounts produced an estimate $\bar{x}$ that was within \$18 of the true mean amount due, μ, for all delinquent accounts at the 95% confidence level. That is, the 95% confidence interval for μ was \$36 wide when 100 accounts were sampled. This is illustrated in Figure 8.4(a).

FIGURE 8.4 Relationship Between Sample Size and Width of Confidence Interval for Delinquent Creditors Example

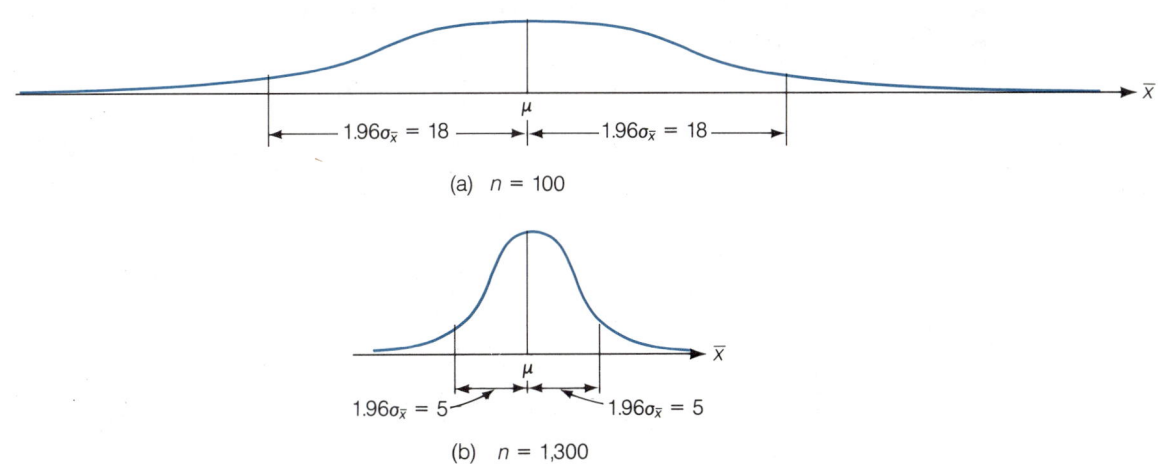

(a) $n = 100$

(b) $n = 1,300$

Now suppose that we want to estimate μ to within \$5 with 95% confidence. That is, we want to narrow the width of the confidence interval from \$36 to \$10, as shown in Figure 8.4(b). How much will the sample size have to be increased to accomplish this? If we want the estimator $\bar{x}$ to be within \$5 of μ, we must have

$$1.96\sigma_{\bar{x}} = 5$$

or, equivalently,

$$1.96\left(\frac{\sigma}{\sqrt{n}}\right) = 5$$

The necessary sample size is obtained by solving this equation for n. To do this we need an approximation for σ. We have an approximation from the initial sample of 100 accounts—namely, the sample standard deviation, $s = 90$. Thus,

$$1.96\left(\frac{\sigma}{\sqrt{n}}\right) \approx 1.96\left(\frac{s}{\sqrt{n}}\right)$$

$$= 1.96\left(\frac{90}{\sqrt{n}}\right)$$

$$= 5$$

$$\sqrt{n} = \frac{1.96(90)}{5} = 35.28$$

$$n = (35.28)^2 = 1,244.68$$

Approximately 1,245 accounts will have to be sampled to estimate the mean overdue amount μ to within \$5 with (approximately) 95% confidence. The confidence interval resulting from a sample of this size will be approximately \$10 wide [see Figure 8.4(b)].

In general, we can express the reliability associated with a confidence interval for the population mean μ in one of two equivalent ways: we can specify the bound, B, within which we want to estimate μ with $100(1 - \alpha)\%$ confidence. This bound B then is equal to the half-width of the confidence interval, as shown in Figure 8.5. Equivalently, we can specify the total width, W, of the $100(1 - \alpha)\%$ confidence interval for μ, also shown in Figure 8.5. Note that $W = 2B$.

FIGURE 8.5

Specifying the Bound B or Total Width W for a Confidence Interval

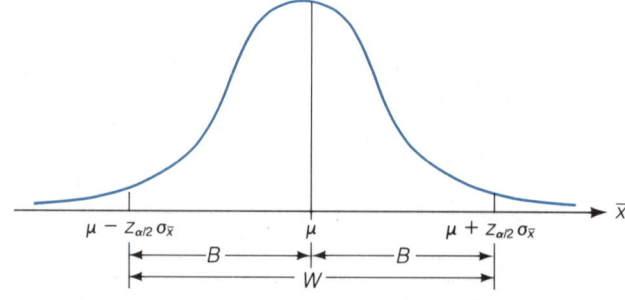

The procedure for finding the sample size necessary to estimate μ to within a given bound B or with a total interval width W is given in the box.

SAMPLE SIZE DETERMINATION FOR $100(1 - \alpha)\%$ CONFIDENCE INTERVALS FOR μ

In order to estimate μ to within a bound B or, equivalently, with a confidence interval of total width W, with $100(1 - \alpha)\%$ confidence, the required sample size is found by solving one of the following equations for n:

$$z_{\alpha/2}\left(\frac{\sigma}{\sqrt{n}}\right) = B \qquad z_{\alpha/2}\left(\frac{\sigma}{\sqrt{n}}\right) = \frac{W}{2}$$

The solution can be written in terms of either B or W:

$$n = \frac{(z_{\alpha/2})^2\sigma^2}{B^2} \qquad n = \frac{4(z_{\alpha/2})^2\sigma^2}{W^2}$$

The value of σ is usually unknown. It can be estimated by the standard deviation, s, from a prior sample. Alternatively, we may approximate the range R of observations in the population, and (conservatively) estimate $\sigma \approx R/4$. In any case, you should round the value of n you obtain *upward* to ensure that the sample size will be sufficient to achieve the specified reliability.

EXAMPLE 8.2

Suppose the manufacturer of official NFL footballs uses a machine to inflate the new balls to a pressure of 13.5 lbs. When the machine is properly calibrated, the mean inflation pressure is 13.5, but uncontrollable factors cause the pressures of individual footballs to vary randomly from about 13.3 to 13.7 lbs. For quality control purposes, the manufacturer wishes to estimate the true mean inflation pressure with a 99% confidence interval that is only .05 lb. wide. What sample size should be specified for the experiment?

SOLUTION

For a 99% confidence interval we have $z_{\alpha/2} = z_{.005} = 2.575$, and to estimate σ we note that the range of observations is $R = 13.7 - 13.3 = .4$, and use $\sigma \approx R/4 = .1$. Thus, in order to have a confidence interval of width $W = .05$, we use the formula derived in the box to find the sample size n:

$$n = \frac{4(z_{\alpha/2})^2\sigma^2}{W^2}$$

$$\approx \frac{4(2.575)^2(.1)^2}{(.05)^2} = 106.09$$

We round this up to $n = 107$ and, realizing that σ was approximated by $R/4$, we might even advise that the sample size be specified as $n = 110$ to be more certain of attaining the objective of a 99% confidence interval with width $W = .05$ lb. ∎

Sometimes the formulas will lead to a solution that indicates a small sample size is sufficient to achieve the confidence interval goal. As we will see in Section 8.6, the procedures and assumptions for small samples differ from those for large samples.

Therefore, if the formulas yield a small sample size ($n < 30$), one simple strategy is to select a sample size $n \geq 30$. Of course, the cost of sampling must also be considered when the sample size is being determined. Although more complex formulas can be derived to take sampling costs into account, these are beyond the scope of this text. For our purposes, it is sufficient to realize that a sampling budget may prove to be a restriction on the sample size, and therefore on the reliability of the confidence interval.

EXERCISES 8.17–8.26

LEARNING THE MECHANICS

8.17 If you wish to estimate a population mean correct to within a bound $B = .2$ with probability .95 and you know from prior sampling that σ^2 is approximately equal to 6.1, how many observations would have to be included in your sample?

8.18 If you wish to estimate a population mean with a 95% confidence interval of width $W = .2$ and you know from prior sampling that σ^2 is approximately equal to 6.1, how many observations would have to be included in your sample? Compare your answer to that for Exercise 8.17. Explain the difference.

8.19 Suppose you wish to estimate the mean of a normal population using a 95% confidence interval and you know from prior information that $\sigma^2 \approx 1$.
   a. To see the effect of the sample size on the width of the confidence interval, calculate the width W of the confidence interval for $n = 16, 25, 49, 100, 400$.
   b. Plot the width as a function of sample size n on graph paper. Connect the points by a smooth curve and note how the width decreases as n increases.

8.20 Suppose you wish to estimate a population mean correct to within a bound $B = .15$ with probability equal to .90. You do not know σ^2, but you know that the observations will range in value between 24 and 27.
   a. Find the approximate sample size that will produce the desired accuracy of the estimate. You wish to be conservative to ensure that the sample size will be ample to achieve the desired accuracy of the estimate. [*Hint:* Using your knowledge of data variation from Section 3.7, assume that the range of the observations will equal 4σ.]
   b. Calculate the approximate sample size making the less conservative assumption that the range of the observations is equal to 6σ.

8.21 It costs you $10 to draw a sample of size $n = 1$ and measure the attribute of interest. You have a budget of $1,200.
   a. Do you have sufficient funds to estimate the population mean for the attribute of interest with a 95% confidence interval 4 units in width? Assume $\sigma = 12$.
   b. If a 90% confidence level were used, would your answer to part **a** change? Explain.

APPLYING THE CONCEPTS

8.22 It costs more to produce defective items—since they must be scrapped or reworked—than it does to produce nondefective items. This simple fact suggests that manufacturers should ensure the quality of their products by perfecting their production processes rather than through inspection of finished products (Deming, 1982). In order to better understand a particular metal-stamping process, a manufacturer wishes to estimate the mean length of items produced by the process during the past 24 hours.

a. How many parts should be sampled in order to estimate the population mean to within .1 mm with 80% confidence? Previous studies of this machine have indicated that the standard deviation of lengths produced by the stamping operation is about 2 mm.

b. Time permits the use of a sample size no larger than 100. If an 80% confidence interval for μ is constructed using $n = 100$, will it be wider or narrower than the interval that would have been obtained using the sample size determined in part **a**? Explain.

c. If management requires that μ be estimated to within .1 mm and that a sample of size no more than 100 be used, what is (approximately) the maximum confidence level that could be attained for a confidence interval that meets management's specifications?

8.23 The EPA standard on the amount of suspended solids that can be discharged into rivers and streams is a maximum of 60 milligrams per liter daily, with a maximum monthly average of 30 milligrams per liter. Suppose you want to test a randomly selected sample of n water specimens and to estimate the mean daily rate of pollution produced by a mining operation. If you want a 95% confidence interval estimate of width 2 milligrams, how many water specimens would you have to include in your sample? Assume prior knowledge indicates that pollution readings in water samples taken during a day are approximately normally distributed with a standard deviation equal to 5 milligrams.

8.24 Suppose a department store wants to estimate μ, the average age of the customers in its contemporary apparel department, correct to within 2 years with probability equal to .95. Approximately how large a sample would be required? [*Note:* Management does not know the standard deviation σ but guesses that the ages of its customers range from 15 to 45. Use a conservative approximation for σ to calculate n.]

8.25 According to a Food and Drug Administration (FDA) study, a cup of coffee contains an average of 115 milligrams of caffeine, with the amount per cup ranging from 60 to 180 milligrams. In contrast, sugar-free Mr. Pibb tested at 58.8 milligrams of caffeine per 12-ounce serving, Coca-Cola and Diet Coke at 45.6 milligrams, and Pepsi at 38.4 milligrams. Suppose you want to repeat the FDA experiment to obtain an estimate of the mean caffeine content in a cup of coffee correct to within 5 milligrams with 95% confidence. How many cups of coffee would have to be included in your sample?

8.26 The United States Golf Association (USGA) tests all new brands of golf balls to assure that they meet USGA specifications. One test conducted is intended to measure the average distance travelled when the ball is hit by a machine called "Iron Byron," a name inspired by the swing of the famous golfer Byron Nelson. Suppose the USGA wants to estimate the mean distance for a new brand with a 90% confidence interval of width 2 yards. Assume that past tests have indicated that the standard deviation of the distances "Iron Byron" hits golf balls is approximately 10 yards. How many golf balls should be hit by "Iron Byron" to achieve the desired accuracy in estimating the mean?

**LARGE-SAMPLE TEST
OF HYPOTHESIS
ABOUT A POPULATION
MEAN**

Suppose building specifications in a certain city require that the average breaking strength of residential sewer pipe be more than 2,400 pounds per foot of length (that is, per lineal foot). Each manufacturer who wants to sell pipe in this city must demonstrate that its product meets the specification. Note that we are again interested in making an inference about the mean, μ, of a population. However, in this example, we are less interested in estimating the value of μ than we are in testing a **hypothesis** about its value. That is, we want to decide whether the mean breaking strength of the pipe exceeds 2,400 pounds per lineal foot.

The method used to reach a decision is based on the rare event concept explained in earlier chapters. We define two hypotheses: (1) The **null hypothesis** is that which represents the status quo to the party performing the sampling experiment—the hypothesis that will be accepted unless the data provide convincing evidence that it is false. (2) The **research** or **alternative hypothesis** is that which will be accepted only if the data provide convincing evidence of its truth. From the point of view of the city conducting the tests, the null hypothesis is that the manufacturer's pipe does *not* meet specifications unless the tests provide convincing evidence otherwise. The null and alternative hypotheses are therefore:

Null hypothesis (H_0): $\mu \le 2,400$ (i.e., the manufacturer's pipe does not meet specifications)

Alternative (research) hypothesis (H_a): $\mu > 2,400$ (i.e., the manufacturer's pipe meets specifications)

How can the city decide when enough evidence exists to conclude that the manufacturer's pipe meets specifications? Since the hypotheses concern the value of the population mean μ, it is reasonable to use the sample mean $\bar{x}$ to make the inference, just as we did when forming confidence intervals for μ in Sections 8.1 and 8.2. The city will conclude that the pipe meets specifications only when the sample mean $\bar{x}$ convincingly indicates that the population mean exceeds 2,400 pounds per lineal foot.

"Convincing" evidence in favor of the alternative hypothesis will exist when the value of $\bar{x}$ exceeds 2,400 by an amount that cannot be readily attributed to sampling variability. The decision-maker, or **test statistic**, will be the z value that measures the distance between the value of $\bar{x}$ and the value of μ specified in the null hypothesis. When the null hypothesis contains more than one value of μ, as in this case (H_0: $\mu \le 2,400$), we use the value of μ closest to the values specified in the alternative hypothesis. The idea is that if the hypothesis that μ *equals* 2,400 can be rejected in favor of $\mu > 2,400$, then μ *less than or equal to* 2,400 can certainly be rejected. Thus, the test statistic is

$$z = \frac{\bar{x} - 2,400}{\sigma_{\bar{x}}} = \frac{\bar{x} - 2,400}{\sigma/\sqrt{n}}$$

Note that a value of $z = 1$ means that $\bar{x}$ is 1 standard deviation above $\mu = 2,400$; a value of $z = 1.5$ means that $\bar{x}$ is 1.5 standard deviations above $\mu = 2,400$, etc. How large must z be before the city can be convinced that the null hypothesis can be rejected in favor of the alternative, and conclude that the pipe meets specifications?

If you examine Figure 8.6, you will note that the chance of observing a value of $\bar{x}$ more than 1.645 standard deviations above 2,400 is only .05, if in fact the true mean μ is 2,400. Thus, if the sample mean is more than 1.645 standard deviations above 2,400, either H_0 is true and a relatively rare event has occurred (probability .05 or less) or H_a is true and the population mean exceeds 2,400. Since we would most likely reject the notion that a rare event has occurred, we would reject the null hypothesis ($\mu \leq 2,400$) and conclude that the alternative hypothesis ($\mu > 2,400$) is true. What is the probability that this procedure will lead us to a wrong decision?

FIGURE 8.6

The Sampling Distribution of $\bar{x}$, Assuming $\mu = 2,400$

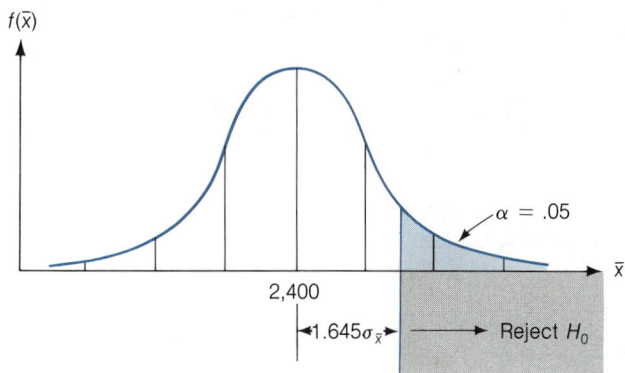

Deciding that the alternative hypothesis is true when in fact it is false is called a **Type I error**. As indicated in Figure 8.6, the probability of making a Type I error—that is, deciding in favor of the alternative hypothesis when in fact the null hypothesis is true—is only $\alpha = .05$. That is,

$$\alpha = P(\text{Type I error})$$
$$= P(\text{Rejecting the null hypothesis when in fact the null hypothesis is true})$$

In our example,

$$\alpha = P(z > 1.645 \text{ when in fact } \mu = 2,400) = .05$$

Therefore, the test can be summarized as follows:

Null and alternative hypotheses: $H_0: \mu = 2,400$ $H_a: \mu > 2,400$

Test statistic: $z = \dfrac{\bar{x} - 2,400}{\sigma_{\bar{x}}}$

Rejection region: $z > 1.645$ for $\alpha = .05$

To illustrate the use of the test, suppose we tested 50 sections of sewer pipe and found the mean and standard deviation for these 50 measurements of breaking strength to be

$\bar{x} = 2,460$ pounds per lineal foot $s = 200$ pounds per lineal foot

As in the case of estimation, we can use s to approximate σ when s is calculated from a large set of sample measurements.

The test statistic is

$$z = \frac{\bar{x} - 2{,}400}{\sigma_{\bar{x}}} = \frac{\bar{x} - 2{,}400}{\sigma/\sqrt{n}} \approx \frac{\bar{x} - 2{,}400}{s/\sqrt{n}}$$

Substituting $\bar{x} = 2{,}460$, $n = 50$, and $s = 200$, we have

$$z \approx \frac{2{,}460 - 2{,}400}{200/\sqrt{50}} = \frac{60}{28.28} = 2.12$$

Therefore, the sample mean lies $2.12\sigma_{\bar{x}}$ above the hypothesized value of μ, 2,400, as shown in Figure 8.7. Since this value of z exceeds 1.645, it falls in the rejection region. That is, we reject the null hypothesis that $\mu = 2{,}400$ and accept the alternative hypothesis, $\mu > 2{,}400$. Thus, it appears that the company's pipe has a mean strength that exceeds 2,400 pounds per lineal foot.

FIGURE 8.7

Location of Test Statistic
When $\bar{x} = 2{,}460$

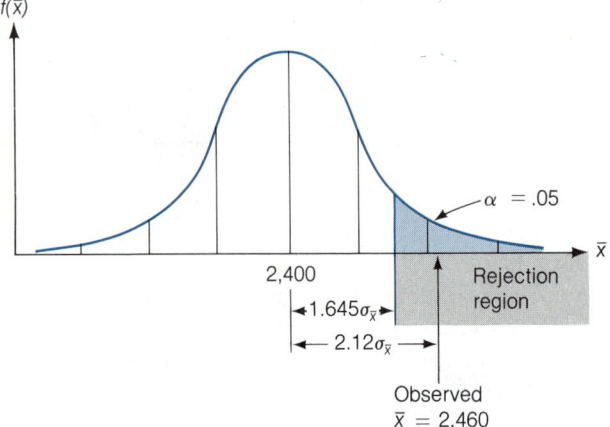

How much faith can be placed in this conclusion? What is the probability that our statistical test would lead us to reject the null hypothesis (and conclude that the company's pipe met the city's specifications) when in fact the null hypothesis was true? The answer is "$\alpha = .05$." That is, we selected the level of risk, α, of making a Type I error when we constructed the test. Thus, the chance is only 1 in 20 that our test would lead us to conclude the manufacturer's pipe satisfied the city's specifications when in fact this conclusion was false.

Now, suppose the sample mean breaking strength for the 50 sections of sewer pipe turned out to be $\bar{x} = 2{,}430$ pounds per lineal foot. Assuming the sample standard deviation is still $s = 200$, the test statistic is

$$z = \frac{2{,}430 - 2{,}400}{200/\sqrt{50}} = \frac{30}{28.28} = 1.06$$

Therefore, the sample mean $\bar{x} = 2{,}430$ is only 1.06 standard deviations above the null hypothesized value of $\mu = 2{,}400$. As shown in Figure 8.8, this value does not fall in the rejection region ($z > 1.645$). Therefore, we know that we cannot reject H_0 using $\alpha = .05$. Even though the sample mean exceeds the city's specification of 2,400 by 30 pounds per lineal foot, it does not exceed the specification by enough to provide *convincing* evidence that the *population mean* exceeds 2,400.

FIGURE 8.8

Location of Test Statistic
When $\bar{x} = 2,430$

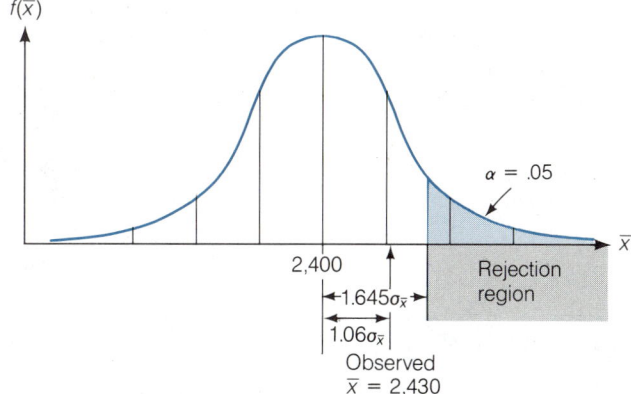

Should we accept the null hypothesis $H_0: \mu \leq 2,400$, and conclude that the manufacturer's pipe does not meet specifications? To do so would be to risk a **Type II error**—that of concluding that the null hypothesis is true (the pipe does not meet specifications) when in fact it is false (the pipe does meet specifications). We will denote the probability of committing a Type II error by β, and we will show in optional Section 8.4 that β is often difficult to specify precisely. Rather than make a decision (accept H_0) for which the probability of error (β) is not precisely specified, we avoid the potential Type II error by avoiding the conclusion that the null hypothesis is true. Instead, we will simply state that *the sample evidence is insufficient to reject H_0 at $\alpha = .05$*. Since the null hypothesis is the "status-quo" hypothesis, the effect of not rejecting H_0 is to maintain the status quo. In our pipe-testing example, the effect of having insufficient evidence to reject the null hypothesis that the pipe does not meet specifications is probably to prohibit the utilization of the manufacturer's pipe unless and until there is sufficient evidence that the pipe does meet specifications. That is, until the data indicate convincingly that the null hypothesis is false, we usually maintain the status quo implied by its truth.

Table 8.2 summarizes the four possible outcomes of a test of hypothesis. The "true state of nature" columns in Table 8.2 refer to the fact that either the null hypothesis H_0 is true, or the alternative hypothesis H_a is true. Note that the true state of nature is unknown to the researcher conducting the test. The "decision" rows in Table 8.2 refer to the action of the researcher, assuming that he or she will either conclude that H_0 is true or that H_a is true based on the results of the sampling experiment. Note that a Type I error can be made *only* when the alternative hypothesis is accepted (equivalently, when the null hypothesis is rejected) and a Type II error can be made *only* when the null hypothesis is accepted. Our policy will be to make a decision only when we know the probability of making the error

TABLE 8.2

Decisions and
Consequences for
a Test of Hypothesis

		TRUE STATE OF NATURE	
		H_0 true	H_a true
DECISION	H_0 true	Correct decision	Type II error (probability β)
	H_a true	Type I error (probability α)	Correct decision

that corresponds to that decision. Since α is usually specified, we will generally be able to reject H_0 (accept H_a) when the sample evidence supports that decision. However, since β is usually not specified, we will generally avoid the decision to accept H_0, preferring instead to state that the sample evidence is insufficient to reject H_0 when the test statistic is not in the rejection region.

The elements of a test of hypothesis are summarized in the box. Note that the first four elements are all specified *before* the sampling experiment is performed. In no case will the results of the sample be used to determine the hypotheses: the data are collected to test the (predetermined) hypotheses, not to formulate them.

ELEMENTS OF A TEST OF HYPOTHESIS

1. *Null hypothesis* (H_0):   A theory about the values of one or more population parameters. The theory generally represents the status quo, which we accept until proven false.
2. *Alternative (research) hypothesis* (H_a):   A theory that contradicts the null hypothesis. The theory generally represents that which we will accept only when sufficient evidence exists to establish its truth.
3. *Test statistic:*   A sample statistic used to decide whether to reject the null hypothesis.
4. *Rejection region:*   The numerical values of the test statistic for which the null hypothesis will be rejected. The rejection region is chosen so that the probability is α that it will contain the test statistic when the null hypothesis is true, thereby leading to a Type I error. The value of α is usually chosen to be small (e.g., .01, .05, or .10), and is referred to as the **level of significance** of the test.
5. *Assumptions:*   Any assumptions made about the population(s) being sampled should be clearly stated.
6. *Experiment and calculation of test statistic:*   The sampling experiment is performed, and the numerical value of the test statistic is determined.
7. *Conclusion:*
   a. If the numerical value of the test statistic falls in the rejection region, we reject the null hypothesis and conclude that the alternative hypothesis is true. We know that the hypothesis-testing process will lead to this conclusion incorrectly (Type I error) only $100\alpha\%$ of the time when H_0 is true.
   b. If the test statistic does not fall in the rejection region, we reserve judgment about which hypothesis is true. We do not conclude that the null hypothesis is true, because we do not (in general) know the probability β that our test procedure will lead to an incorrect acceptance of H_0 (Type II error).*

*In many practical business applications of hypothesis testing, nonrejection leads management to behave as if the null hypothesis were accepted. Accordingly, the distinction between acceptance and nonrejection is frequently blurred in practice. We will discuss the issues connected with the acceptance of the null hypothesis and the calculation of β in more detail in (optional) Section 8.4.

CASE STUDY 8.2

STATISTICS IS MURDER!

The jury trial of an accused murderer is analogous to the statistical hypothesis-testing process. Each of the elements of a test of hypothesis applies to the jury system of deciding the guilt or innocence of the accused:

1. H_0:   The null hypothesis in a jury trial is that the accused is innocent. The status-quo hypothesis in the American system of justice is innocence, which is assumed to be true until proven otherwise.

2. H_a: The alternative hypothesis is guilt, which is accepted only when sufficient evidence exists to establish its truth.

3. *Test statistic:*   The test statistic in a trial is the final vote of the jury, i.e., the number of the jury members who vote "guilty."

4. *Rejection region:*   In a murder trial the jury vote must be unanimous in favor of guilt before the null hypothesis of innocence is rejected in favor of the alternative hypothesis of guilt. Thus, for a 12-member jury trial, the rejection region is $x = 12$, where x is the number of "guilty" votes.

5. *Assumption:*   The primary assumption made in jury trials concerns the method of selecting the jury. The jury is assumed to represent a random sample of citizens who have no prejudice concerning the case.

6. *Experiment and calculation of the test statistic:*   The sampling experiment is analogous to the jury selection, the trial, and the jury deliberations. The final vote of the jury is analogous to the calculation of the test statistic.

7. *Conclusion:*

   a. If the vote of the jury is unanimous in favor of guilt, the null hypothesis of innocence is rejected and the court concludes that the accused murderer is guilty. Although the court does not, in general, know the probability α that the conclusion is in error, the system relies on the belief that the value is made very small by requiring a unanimous vote before guilt is concluded.

   b. Any vote other than a unanimous one for guilt results in the court reserving judgment about the hypotheses, either by declaring the accused "not guilty," or by declaring a mistrial and repeating the "test" with a new jury. (The latter is analogous to collecting more data and repeating a statistical test of hypothesis.) The court never accepts the null hypothesis by declaring the accused "innocent," perhaps recognizing both that innocence is the "status-quo" hypothesis and does not need to be proved, and that the probability β of incorrectly concluding innocence may not be as small as α.

As in the case of tests of statistical hypotheses, we may never know whether the verdict in a murder trial is correct. Instead, we rely on the knowledge that the trial procedure will lead to incorrect conclusions (especially, guilt when the accused is in fact innocent) in only a very small percentage of trials.

The null and alternative hypotheses may take one of several forms. In the sewer pipe example, we tested the null hypothesis that the population mean strength of the pipe is less than or equal to 2,400 pounds per lineal foot, against the alternative hypothesis that the mean strength exceeds 2,400. That is, we tested

H_0: $\mu \leq 2,400$

H_a: $\mu > 2,400$

This is a **one-tailed** or (**one-sided**) statistical test because the alternative hypothesis specifies that the population parameter (the population mean μ, in this example) is strictly greater than a specified value (2,400, in this example). If the alternative hypothesis had been H_a: $\mu < 2,400$, the test would still be one-sided, because the parameter is still specified to be on "one side" of the null hypothesis value. Some statistical investigations seek to show that the population parameter is *either larger or smaller* than some specified value. Such an alternative hypothesis is called a **two-tailed** (or **two-sided**) hypothesis.

While alternative hypotheses are always specified as strict inequalities, such as $\mu < 2,400$, $\mu > 2,400$, or $\mu \neq 2,400$, null hypotheses are usually specified as equalities, such as $\mu = 2,400$. Even when the null hypothesis is an inequality, such as $\mu \leq 2,400$, we specify H_0: $\mu = 2,400$, reasoning that if sufficient evidence exists to show that H_a: $\mu > 2,400$ is true when tested against H_0: $\mu = 2,400$, then surely sufficient evidence exists to reject $\mu < 2,400$ as well. Therefore, the null hypothesis is specified as the value of μ closest to a one-sided alternative hypothesis, and as the only value *not* specified in a two-tailed alternative hypothesis. The steps for selecting the null and alternative hypotheses are summarized in the box.

STEPS FOR SELECTING THE NULL AND ALTERNATIVE HYPOTHESES

1. Select the **alternative hypothesis** as that which the sampling experiment is intended to establish. The alternative hypothesis will assume one of three forms:

FORM	EXAMPLE
One-tailed, upper-tailed	H_a: $\mu > 2,400$
One-tailed, lower-tailed	H_a: $\mu < 2,400$
Two-tailed	H_a: $\mu \neq 2,400$

2. Select the **null hypothesis** as the status quo, that which will be presumed true unless the sampling experiment conclusively establishes the alternative hypothesis. The null hypothesis will be specified as the parameter value closest to the alternative in one-tailed tests, and as the complementary (or only unspecified) value in two-tailed tests.

   Example: H_0: $\mu = 2,400$

The rejection region for a two-tailed test differs from that for a one-tailed test. When we are trying to detect departure from the null hypothesis value in *either* direction, we must establish a rejection region in both tails of the sampling distribution of the test statistic. Figures 8.9(a) and (b) show the one-tailed rejection regions for lower- and upper-tailed tests, respectively. The two-tailed rejection region is illustrated in Figure 8.9(c). Note that a rejection region is established in each tail of the sampling distribution for a two-tailed test.

FIGURE 8.9

Rejection Regions
Corresponding to One-
and Two-Tailed Tests

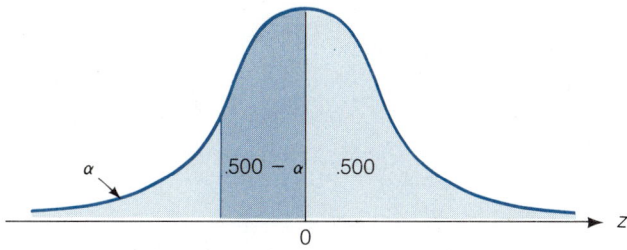

(a) Form of H_a: $<$

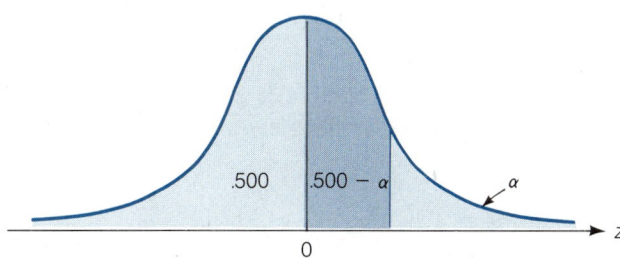

(b) Form of H_a: $>$

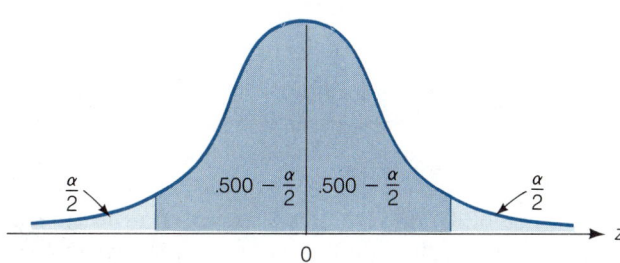

(c) Form of H_a: $\neq$

The rejection regions corresponding to typical values selected for α are shown in Table 8.3 for one- and two-tailed tests. Note that the smaller α you select, the more evidence (the larger z) is required before you can reject H_0.

TABLE 8.3

Rejection Regions for
Common Values of α

| | ALTERNATIVE HYPOTHESIS | | |
	Lower-tailed	Upper-tailed	Two-tailed
$\alpha = .10$	$z < -1.28$	$z > 1.28$	$z < -1.645$ or $z > 1.645$
$\alpha = .05$	$z < -1.645$	$z > 1.645$	$z < -1.96$ or $z > 1.96$
$\alpha = .01$	$z < -2.33$	$z > 2.33$	$z < -2.575$ or $z > 2.575$

EXAMPLE 8.3

A manufacturer of cereal wants to test the performance of its filling machine. The machine is designed to discharge a mean amount of $\mu = 12$ ounces per box, and the manufacturer wants to detect any departure from this setting. The quality control experiment calls for sampling 100 boxes to determine whether the machine is performing to specifications. Set up a test of hypothesis for this quality control experiment, using $\alpha = .01$.

SOLUTION

Since the manufacturer wishes to detect a departure from the setting of $\mu = 12$ in either direction, $\mu < 12$ or $\mu > 12$, we will conduct a two-tailed statistical test. Following the procedure for selecting the null and alternative hypotheses, we specify as the alternative hypothesis that the mean differs from 12 ounces, since detecting the machine's departure from specifications is the purpose of the quality control experiment. The null hypothesis is the presumption that the fill machine is operating properly unless the sampling experiment indicates otherwise. Thus,

$H_0: \quad \mu = 12$

$H_a: \quad \mu \neq 12$ (i.e., $\mu < 12$ or $\mu > 12$)

The test statistic measures the number of standard deviations between the observed value of $\bar{x}$ and the null hypothesized value $\mu = 12$:

$$\text{Test statistic:} \quad z = \frac{\bar{x} - 12}{\sigma_{\bar{x}}}$$

The rejection region must be designated to detect a departure from $\mu = 12$ in *either* direction, so we will reject H_0 for values of z that are either too small (negative) or too large (positive). To determine the precise values of z that comprise the rejection region, we first select α, the probability that the test will lead to incorrect rejection of the null hypothesis. Then we divide α equally between the lower and upper tails of the distribution of z, as shown in Figure 8.10. In this example, $\alpha = .01$, so $\alpha/2 = .005$ is placed in each tail. This area in the tails corresponds to $z = -2.575$ and $z = 2.575$, respectively (from Table 8.3):

FIGURE 8.10
Two-Tailed Rejection
Region: $\alpha = .01$

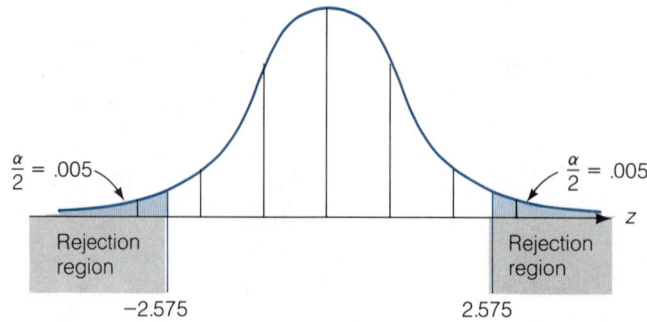

$\frac{\alpha}{2} = .005$

$\frac{\alpha}{2} = .005$

Rejection region

Rejection region

-2.575

2.575

Rejection region: $z < -2.575$ or $z > 2.575$ (Figure 8.10)

Assumptions: Since the sample size of the experiment is large enough ($n > 30$), the Central Limit Theorem will apply, and no assumptions need to be made about the population of fill measurements. The sampling distribution of the sample mean fill of 100 boxes will be approximately normal regardless of the distribution of the individual boxes' fills. ∎

Note that the test in Example 8.3 is set up *before* the sampling experiment is conducted; the data are not used to develop the test. Evidently, the manufacturer does not want to disrupt the filling process to adjust the machine unless the sampling experiment provides very convincing evidence that it is not meeting specifications, because the value of α has been set at .01. If the sampling experiment results in the rejection of H_0, the manufacturer can be 99% confident that the machine needs adjustment.

Once the test is set up, we are ready to perform the sampling experiment and conduct the test. The test is performed in Example 8.4.

EXAMPLE 8.4

Refer to the quality control test set up in Example 8.3. Suppose the sampling experiment is conducted with the following results:

$n = 100$ observations

$\bar{x} = 11.85$ ounces

$s = .5$ ounce

Use the results of the sampling experiment to conduct the test of hypothesis.

SOLUTION

Since the test is completely specified in Example 8.3, we simply substitute the sample statistics into the test statistic:

$$z = \frac{\bar{x} - 12}{\sigma_{\bar{x}}} = \frac{\bar{x} - 12}{\sigma/\sqrt{n}} = \frac{11.85 - 12}{\sigma/\sqrt{100}}$$

$$\approx \frac{11.85 - 12}{s/10} = \frac{-.15}{.5/10} = -3.0$$

The implication is that the sample mean, 11.85, is (approximately) 3 standard deviations below the null hypothesized value of 12.0 in the sampling distribution of $\bar{x}$. You can see in Figure 8.10 that this value of z is in the lower-tail rejection region, which consists of all values of $z < -2.575$. This sampling experiment provides sufficient evidence to reject H_0 and conclude, at the $\alpha = .01$ level of significance, that the mean fill differs from the specification of $\mu = 12$ ounces. It appears that the machine is, on average, underfilling the boxes. ∎

Two final points about the test of hypothesis in Example 8.4 apply to all statistical tests:

1. Although it is tempting to make our conclusion with more than 99% confidence (or, equivalently, at less than the .01 level of significance), we resist the temptation because the level of α is determined *before* the sampling experiment is performed. If we decide that we are willing to tolerate a 1% Type I error rate, the result of the sampling experiment should have no effect on that decision. **In general, the same data should not be used both to set up and to conduct the test.**

2. When we state our conclusion at the .01 level of significance, or at the 99% confidence level, we are referring to the success or failure rate of the *procedure*, not the result of this particular test. We know that the test procedure will lead to the rejection of the null hypothesis only 1% of the time when in fact $\mu = 12$. **Therefore, when the test statistic falls in the rejection region, we infer that the alternative $\mu \neq 12$ is true, and express our confidence in the procedure by quoting the α level of significance, or the $100(1 - \alpha)\%$ confidence level.**

The set-up of a large-sample test of hypothesis about a population mean is summarized in the box. Both the one- and two-tailed tests are shown.

LARGE-SAMPLE TEST OF HYPOTHESIS ABOUT μ

ONE-TAILED TEST	TWO-TAILED TEST
H_0: $\mu = \mu_0{}^*$	H_0: $\mu = \mu_0{}^*$
H_a: $\mu < \mu_0$	H_a: $\mu \neq \mu_0$
(or H_a: $\mu > \mu_0$)	
Test statistic: $z = \dfrac{\bar{x} - \mu_0}{\sigma_{\bar{x}}}$	Test statistic: $z = \dfrac{\bar{x} - \mu_0}{\sigma_{\bar{x}}}$
Rejection region: $z < -z_\alpha$ (or $z > z_\alpha$)	Rejection region: $z < -z_{\alpha/2}$ or $z > z_{\alpha/2}$
where z_α is chosen so that $P(z > z_\alpha) = \alpha$	where $z_{\alpha/2}$ is chosen so that $P(z > z_{\alpha/2}) = \alpha/2$

Assumptions: No assumptions need to be made about the probability distribution of the population, because the Central Limit Theorem assures us that, for large samples, the test statistic will be approximately normally distributed regardless of the shape of the underlying probability distribution of the population.

Once the test has been set up, the sampling experiment is performed and the test statistic is calculated. The next box contains possible conclusions for a test of hypothesis, depending on the result of the sampling experiment.

*Note: μ_0 is the symbol for the numerical value assigned to μ under the null hypothesis.

> **POSSIBLE CONCLUSIONS FOR A TEST OF HYPOTHESIS**
>
> 1. If the calculated test statistic falls in the rejection region, reject H_0 and conclude that the alternative hypothesis H_a is true. State that you are rejecting H_0 at the α level of significance, or with $100(1 - \alpha)\%$ confidence. Remember that the confidence is in the process, not in the particular result of a single test.
> 2. If the test statistic does not fall in the rejection region, conclude that the sampling experiment does not provide sufficient evidence to reject H_0 at the α level of significance. [Generally, we will not "accept" the null hypothesis unless the probability β of a Type II error has been calculated (see optional Section 8.4).]

CASE STUDY 8.3

STATISTICAL QUALITY CONTROL, PART I

In Exercise 7.32 we described a graphical device, known as a control chart, that can be used in manufacturing operations to monitor the variation over time in the quality of products being produced. Although a relatively recent invention (the control chart was developed by Walter A. Shewhart of Bell Telephone Laboratories in 1924), this device has become a basic tool of quality control engineers and operations managers the world over. Japan's emergence as an industrial superpower is due in part to their early adoption and refinement of quality control techniques, such as the control chart, that were developed in the United States (Duncan, 1974). In this case study and Case ·Study 8.4, we expand the discussion of control charts and demonstrate that they are simply vehicles for conducting hypothesis tests.

Suppose it is desired to monitor the pitch diameter of the threads on a particular aircraft fitting. When the process is in control, the pitch diameters follow a normal distribution with mean μ_0 and standard deviation σ_0. Recall from Exercise 7.32 that such monitoring can be accomplished by (1) randomly sampling n items from the production process at regular time intervals, (2) measuring the pitch diameter of each item sampled, and (3) plotting the mean diameter of each sample, $\bar{x}$, on a control chart like that in Figure 8.11. Such a control chart is called an $\bar{x}$-chart.

FIGURE 8.11

$\bar{x}$-Chart

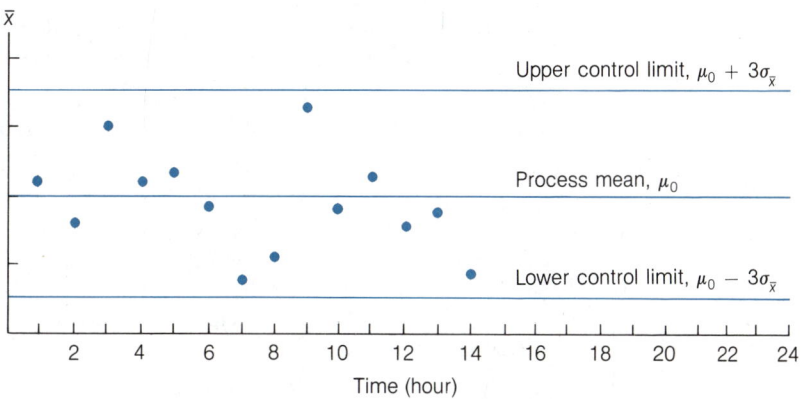

Time (hour)

If a value of $\bar{x}$ falls above the upper control limit or below the lower control limit, there is strong evidence that the process is out of control—i.e., that the quality of the product being produced does not meet established standards. Otherwise, the process is deemed to be in control.

In the language of hypothesis testing, this decision process can be described as follows:

1. There are two hypotheses of interest:

H_0:   Process is in control, $\mu = \mu_0$

H_a: Process is out of control, $\mu \neq \mu_0$

2. The test statistic used to investigate these hypotheses is $\bar{x}$.
3. The upper and lower control limits define the rejection region for the test.
4. Since the control limits are located at $\mu_0 \pm 3\sigma_{\bar{x}}$, the probability of committing a Type I error is $\alpha = .0026$. (Why?)

Thus, each time a quality control engineer plots a sample mean on an $\bar{x}$-chart and observes where it falls in relation to the control limits, the engineer is conducting a two-tailed hypothesis test.

EXERCISES 8.27–8.42

LEARNING THE MECHANICS

8.27 What is the difference between the null hypothesis and the alternative hypothesis?

8.28 Define each of the following:
  a. Type I error   **b.** Type II error   **c.** α **d.** β

8.29 When do you risk making a Type I error? A Type II error?

8.30 In testing a hypothesis, who or what determines the size of the rejection region?

8.31 For each of the following rejection regions, sketch the sampling distribution of z and indicate the location of the rejection region:
  a. $z > 1.96$ **b.** $z > 1.645$
  c. $z > 2.575$ **d.** $z < -1.29$
  e. $z < -1.645$ or $z > 1.645$ **f.** $z < -2.575$ or $z > 2.575$
  g. For each of the rejection regions specified in parts **a–f**, what is the probability that a Type I error will be made when the null hypothesis is true?

8.32 If you test a hypothesis and reject the null hypothesis in favor of the alternative hypothesis, does your test prove that the alternative hypothesis is correct? Explain.

8.33 A random sample of 100 observations from a population with standard deviation 60 yielded a sample mean of 110.
  a. Test the null hypothesis that $\mu = 100$ against the alternative hypothesis that $\mu > 100$ using $\alpha = .05$. Interpret the results of the test.
  b. Test the null hypothesis that $\mu = 100$ against the alternative hypothesis that $\mu \neq 100$ using $\alpha = .05$. Interpret the results of the test.
  c. Compare the results of the two tests you conducted. Explain why the results differ.

8.34 A random sample of 49 observations produced the following sums:

$$\sum x = 20.7 \qquad \sum (x_i - \bar{x})^2 = 2.155$$

a. Test the null hypothesis that $\mu = .47$ against the alternative hypothesis that $\mu < .47$ using $\alpha = .10$. Interpret the results of the test.

b. Test the null hypothesis that $\mu = .47$ against the alternative hypothesis that $\mu \neq .47$ using $\alpha = .10$. Interpret the results.

8.35 Suppose you are interested in conducting the following statistical test:

$$H_0: \quad \mu = 200$$

$$H_a: \quad \mu > 200$$

and you have decided to use the following decision rule: "Reject H_0 if the sample mean of a random sample of 100 items is more than 212." Assume that the standard deviation of the population is 80.

a. Express the decision rule in terms of z.

b. Find α, the probability of making a Type I error, using this decision rule.

8.36 A random sample of 60 observations produced the following sums:

$$\sum x = 27.4 \qquad \sum x^2 = 14.3$$

a. Test the null hypothesis that $\mu = .40$ against the alternative hypothesis that $\mu > .40$. Test using $\alpha = .05$. Interpret the results.

b. Test the null hypothesis that $\mu = .40$ against the alternative hypothesis that $\mu \neq .40$. Test using $\alpha = .05$. Interpret the results.

APPLYING THE CONCEPTS

8.37 In 1895 an Italian criminologist, Cesare Lombroso, proposed that blood pressure be used to test for truthfulness. In the 1930's, William Marston added the measurements of respiration and perspiration to the process and called his machine the *polygraph*—or lie detector. Today, the federal court system will not consider polygraph results as evidence, but nearly half of the state courts do permit polygraph tests under certain circumstances. In addition, its use in screening job applicants is on the rise in both industry and government (Dujack, 1986). But how well does it work? Physicians Michael Phillips, Allan Brett, and John Beary subjected the polygraph to the same careful testing given to medical diagnostic tests. They found that if 1,000 people were subjected to the polygraph and 500 told the truth and 500 lied, the polygraph would indicate that approximately 185 of the truth-tellers were liars and that approximately 120 of the liars were truth-tellers ("Lie Detectors Can Make a Liar of You," *Discover*, June 1986, p. 7).

a. In the application of a polygraph test, an individual is presumed to be a truth-teller (H_0) until "proven" a liar (H_a). In this context, what is a Type I error? A Type II error?

b. According to Phillips, Brett, and Beary, what is the probability (approximately) that a polygraph test will result in a Type I error? A Type II error?

8.38 Small increases in the mean charge for monthly long-distance telephone calls produce substantial increases in the profits for telephone companies. A telephone company's records indicate that private customers pay an average of $17.10 per month for long-distance telephone calls; the standard deviation of the amounts paid for long-distance calls is $9.80.

a. If a random sample of 50 bills is taken, what is the probability that the sample mean is greater than $20?

b. If a random sample of 100 bills is taken, what is the probability that the sample mean is greater than $20?

c. The telephone company suspects that the mean amount paid per month per customer for long-distance service has increased. A random sample of 100 customers' bills during a given month produced a sample mean of $21.25 expended for long-distance calls. Do the data indicate that the mean level of the amounts billed per month for long-distance telephone calls has increased from $17.10? Test using $\alpha = .05$.

8.39 In quality control applications of hypothesis testing, the null and alternative hypotheses are frequently specified as

H_0:   The production process is performing satisfactorily.

H_a: The process is performing in an unsatisfactory manner.

Accordingly, α is sometimes referred to as the *producer's risk*, while β is called the *consumer's risk* (Montgomery, 1985). An injection molder produces plastic golf tees. The process is designed to produce tees with a mean weight of .250 ounce. To investigate whether the injection molder is operating satisfactorily, 40 tees were randomly sampled from the last hour's production. Their weights (in ounces) are listed below:

.247	.252	.248	.251	.248	.252	.250	.252
.253	.253	.254	.253	.256	.254	.253	.252
.253	.250	.253	.254	.256	.251	.256	.251
.249	.254	.256	.253	.255	.255	.253	.255
.251	.253	.251	.251	.253	.251	.254	.253

a. Do the data provide sufficient evidence to conclude that the process is not operating satisfactorily? Test using $\alpha = .05$.

b. In the context of this problem, explain why it makes sense to call α the producer's risk and β the consumer's risk.

8.40 The Environmental Protection Agency (EPA) estimated that the 1988 G-car obtains a mean of 35 miles per gallon on the highway, and the company that manufactures the car claims that it exceeds the EPA estimate in highway driving. To support its assertion, the company randomly selects thirty-six 1988 G-cars and records the mileage obtained for each car over a driving course similar to that used by the EPA. The following data resulted:

$$\bar{x} = 36.8 \text{ miles per gallon} \qquad s = 6.0 \text{ miles per gallon}$$

a. If the auto manufacturer wishes to show that the mean miles per gallon for 1988 G-cars is greater than 35 miles per gallon, what should it choose for the alternative hypothesis? The null hypothesis?

b. Do the data provide sufficient evidence to support the auto manufacturer's claim? Test using $\alpha = .05$.

8.41 In 1979, basic cable television service cost an average of $7.37 per month in the United States (*Statistical Abstract of the United States: 1981*, p. 565). In March 1983, the Federal Communications Commission (FCC) noted that the cost of basic cable service had risen only about 8% since 1979 and that it cost on average no more than $8.00 per month. Suppose a consumer advocacy group doubts the FCC's claim. In order to investigate the claim, the group randomly samples 33 of the more than 4,000 cable systems in the United States and asks each what its basic service charge was in early 1983. The following results are obtained:

$ 8.53	$8.41	$7.80	$8.20	$8.14	$7.89	$7.92
20.01	8.96	6.50	8.79	7.73	8.15	8.00
7.66	7.76	7.63	8.16	7.50	7.64	6.99
9.83	7.86	7.97	6.96	8.63	7.64	
8.13	8.35	7.83	7.88	7.65	7.75	

a. Specify the null and alternative hypotheses that should be used by the consumer advocacy group in investigating the FCC's claim.

b. With respect to the hypotheses you specified in part **a**, explain the practical implications of making a Type I error and a Type II error.

c. Conduct the hypothesis test you described in part **a**, and interpret the test's results in the context of this exercise. Use $\alpha = .10$.

8.42 The introduction of printed circuit boards (PCBs) in the 1950's revolutionized the electronics industry. However, solder-joint defects on PCBs have plagued electronics manufacturers since the introduction of the PCB. A single PCB may contain hundreds of solder joints. Until the 1980's, the only means of checking the quality of solder joints was by visual inspection. Because of the low reliability of visual inspection, some manufacturers required each of their joints to be inspected four times by four different people. Now both X-ray and laser technologies are available for use in inspection (Streeter, 1986). A particular manufacturer of laser-based inspection equipment claims that its product can inspect on average at least 10 solder joints per second when the joints are spaced .1 inch apart. The equipment was tested by a potential buyer on 48 different PCBs. In each case, the equipment was operated for exactly 1 second. The numbers of solder joints inspected on each run are listed below:

10	9	10	10	11	9	12	8
8	9	6	10	7	10	11	9
9	13	9	10	11	10	12	8
9	9	9	7	12	6	9	10
10	8	7	9	11	12	10	0
10	11	12	9	7	9	9	10

a. The potential buyer wants to know whether the sample data refute the manufacturer's claim. Specify the null and alternative hypotheses that the buyer should test.

b. In the context of this exercise, what is a Type I error? A Type II error?

c. Conduct the hypothesis test you described in part **a** and interpret the results in the context of this exercise. Use $\alpha = .05$.

S E C T I O N 8.4

CALCULATING TYPE II ERROR PROBABILITIES: MORE ABOUT β (OPTIONAL)

In our introduction to hypothesis testing in Section 8.3, we showed that the probability of committing a Type I error, α, can be controlled by the selection of the rejection region for the test. Thus, when the test statistic falls in the rejection region and we make the decision to reject the null hypothesis, we do so knowing the error rate for incorrect rejections of H_0. The situation corresponding to accepting the null hypothesis, and thereby risking a Type II error, is not generally as controllable. For that reason, we adopted a policy of nonrejection of H_0 when the test statistic does not fall in the rejection region, rather than risking an error of unknown magnitude.

To see how β, the probability of a Type II error, can be calculated for a test of hypothesis, recall the example in Section 8.3 in which a city tests a manufacturer's pipe to see whether it meets the requirement that the mean strength exceeds 2,400 pounds per lineal foot. The set-up for the test is as follows:

H_0: $\mu = 2{,}400$

H_a: $\mu > 2{,}400$

Test statistic: $z = \dfrac{\bar{x} - 2{,}400}{\sigma/\sqrt{n}}$

Rejection region: $z > 1.645$ for $\alpha = .05$

Figure 8.12(a) shows the rejection region for the **null distribution**—that is, the distribution of the test statistic assuming the null hypothesis is true. The area in the rejection region is .05, and this area represents α, the probability that the test statistic leads to rejection of H_0 when, in fact, H_0 is true.

The Type II error probability β is calculated assuming that the null hypothesis is false, because it is defined as the *probability of accepting H_0 when it is false.* Since H_0 is false for any value of μ exceeding 2,400, one value of β exists for each possible value of μ greater than 2,400 (an infinite number of possibilities). Figures 8.12(b), (c), and (d) show three of the possibilities, corresponding to alternative hypothesis values of μ equal to 2,425, 2,450, and 2,475, respectively. Note that β is the area in the *nonrejection* (or *acceptance*) *region* in each of these distributions, and that β decreases as the true value of μ moves farther from the null hypothesized value of $\mu = 2{,}400$. This is sensible, because the probability of incorrectly accepting the null hypothesis should decrease as the distance between the null and alternative values of μ increases.

In order to calculate the value of β for a specific value of μ in H_a, we proceed as follows:

1. First, calculate the value of $\bar{x}$ that corresponds to the border between the acceptance and rejection regions. For the sewer pipe example, that is the value of $\bar{x}$ that lies 1.645 standard deviations above $\mu = 2{,}400$ in the sampling distribution of $\bar{x}$. Denoting this value by $\bar{x}_0$, corresponding to the largest value of $\bar{x}$ that

FIGURE 8.12

Values of α and β for Various Values of μ

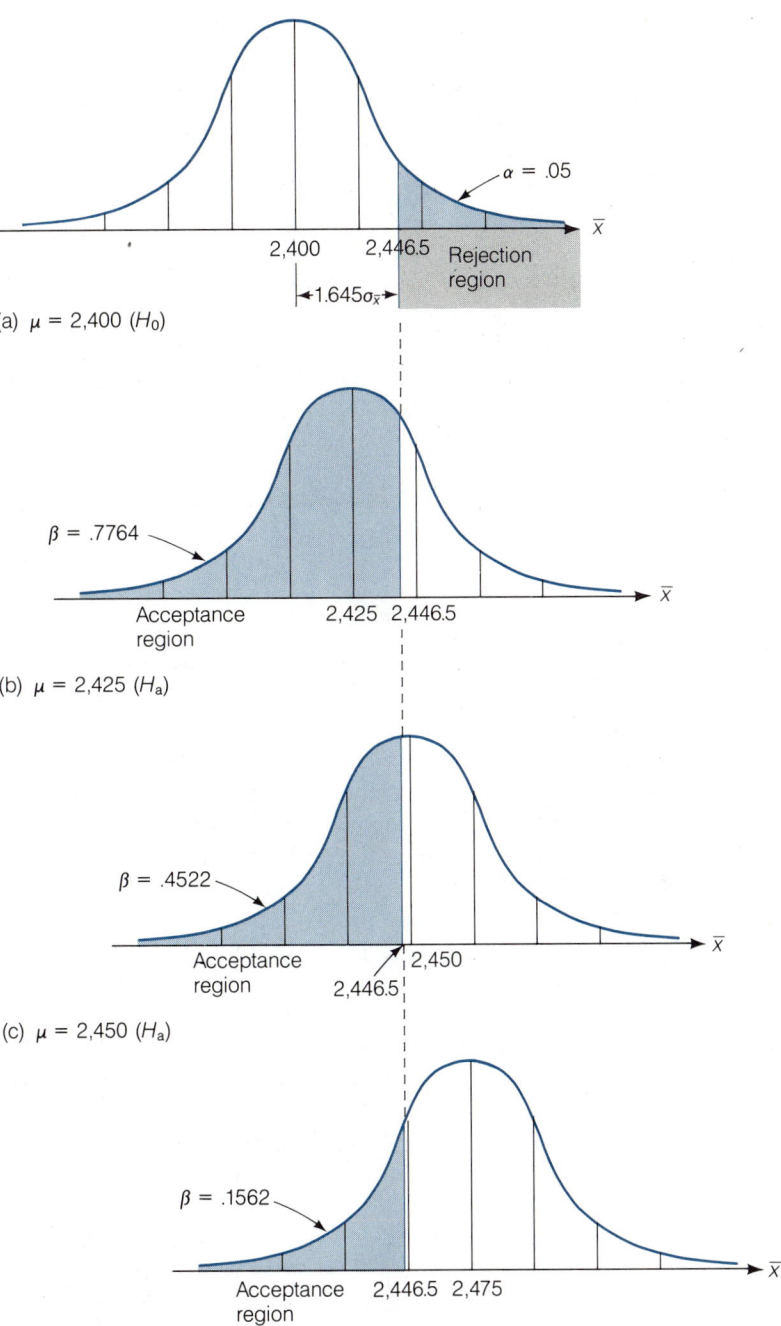

(a) $\mu = 2,400$ (H_0)

(b) $\mu = 2,425$ (H_a)

(c) $\mu = 2,450$ (H_a)

(d) $\mu = 2,475$ (H_a)

supports the null hypothesis, we find (recalling that $s = 200$ and $n = 50$):

$$\bar{x}_0 = \mu_0 + 1.645\sigma_{\bar{x}} = 2,400 + 1.645\left(\frac{\sigma}{\sqrt{n}}\right)$$

$$\approx 2,400 + 1.645\left(\frac{s}{\sqrt{n}}\right) = 2,400 + 1.645\left(\frac{200}{\sqrt{50}}\right)$$

$$= 2,400 + 1.645(28.28) = 2,446.5$$

2. Next, for a particular alternative distribution corresponding to a value of μ, μ_a, we calculate the z value corresponding to $\bar{x}_0$, the border between the rejection and acceptance regions. We then use this z value and Table IV of Appendix B to determine the area in the *acceptance* region. This area is the value of β corresponding to the particular alternative μ_a. For example, for the alternative $\mu_a = 2,425$, we calculate

$$z = \frac{\bar{x}_0 - 2,425}{\sigma_{\bar{x}}} = \frac{\bar{x}_0 - 2,425}{\sigma/\sqrt{n}}$$

$$\approx \frac{\bar{x}_0 - 2,425}{s/\sqrt{n}} = \frac{2,446.5 - 2,425}{28.28} = .76$$

Note in Figure 8.12(b) that the area in the acceptance region is the area to the left of $z = .76$. This area is

$$\beta = .5 + .2764 = .7764$$

Thus, the probability that the test procedure will lead to an incorrect acceptance of the null hypothesis $\mu = 2,400$ when in fact $\mu = 2,425$ is about .78. As the average strength of the pipe increases to 2,450, the value of β decreases to .4522 [Figure 8.12(c)]. If the mean strength is further increased to 2,475, the value of β is further decreased to .1562 [Figure 8.12(d)]. Thus, even if the true mean strength of the pipe exceeds the minimum specification by 75 pounds per lineal foot, the test procedure will lead to an incorrect acceptance of the null hypothesis (rejection of the pipe) approximately 16% of the time. The upshot is that the pipe must be manufactured so that the mean strength well exceeds the minimum requirement if the manufacturer wants the probability of its acceptance by the city to be large (i.e., β to be small).

The steps for calculating β for a large-sample test about a population mean are summarized in the box at the top of the next page.

Following the calculation of β for a particular value of μ_a, you should interpret the value in the context of the hypothesis-testing application. It is often useful to interpret the value of $1 - \beta$, which is known as the **power of the test**, corresponding to a particular alternative, μ_a. Since β is the probability of accepting the null hypothesis when the alternative hypothesis is true with $\mu = \mu_a$, $1 - \beta$ is the probability of the complementary event, or the probability of rejecting the null hypothesis when the alternative $\mu = \mu_a$ is true. That is, the power $1 - \beta$ measures the likelihood that the test procedure will lead to the *correct* decision (reject H_0) for a particular value of the mean in the alternative hypothesis.

STEPS FOR CALCULATING β FOR A LARGE-SAMPLE TEST ABOUT μ

1. Calculate the value(s) of $\bar{x}$ corresponding to the border(s) of the rejection region. There will be one border value for a one-tailed test, and two for a two-tailed test. The formula is one of the following, corresponding to a test with level of significance α:

   Upper-tailed test:

   $$\bar{x}_0 = \mu_0 + z_\alpha \sigma_{\bar{x}} \approx \mu_0 + z_\alpha\left(\frac{s}{\sqrt{n}}\right)$$

   Lower-tailed test:

   $$\bar{x}_0 = \mu_0 - z_\alpha \sigma_{\bar{x}} \approx \mu_0 - z_\alpha\left(\frac{s}{\sqrt{n}}\right)$$

   Two-tailed test:

   $$\bar{x}_{0,L} = \mu_0 - z_{\alpha/2} \sigma_{\bar{x}} \approx \mu_0 - z_{\alpha/2}\left(\frac{s}{\sqrt{n}}\right)$$

   $$\bar{x}_{0,U} = \mu_0 + z_{\alpha/2} \sigma_{\bar{x}} \approx \mu_0 + z_{\alpha/2}\left(\frac{s}{\sqrt{n}}\right)$$

2. Specify the value of μ, μ_a, in the alternative hypothesis for which the value of β is to be calculated. Then convert the border value(s) of $\bar{x}_0$ to z value(s) using the alternative distribution with mean μ_a. The general formula for the z value is

   $$z = \frac{\bar{x}_0 - \mu_a}{\sigma_{\bar{x}}}$$

   Sketch the alternative distribution (centered at μ_a), and shade the area in the acceptance (nonrejection) region. Use Table IV of Appendix B to convert the z statistic(s) to the area β in the acceptance region.

DEFINITION 8.3

The **power** of a test is the probability that the test will lead to the (correct) rejection of the null hypothesis for a particular value of μ in the alternative hypothesis. The power is equal to $1 - \beta$ for the particular alternative considered.

For example, in the sewer pipe example we found that $\beta = .7764$ when $\mu = 2{,}425$. This is the probability that the test leads to the (incorrect) acceptance of the null hypothesis when $\mu = 2{,}425$. Or, equivalently, the power of the test is

$1 - .7764 = .2236$, which means that the test will lead to the (correct) rejection of the null hypothesis only 22% of the time when the pipe exceeds specifications by 25 pounds per lineal foot. When the manufacturer's pipe has a mean strength of 2,475 (that is, 75 pounds per lineal foot in excess of specifications), the power of the test increases to $1 - .1562 = .8438$. That is, the test will lead to the acceptance of the manufacturer's pipe 84% of the time if $\mu = 2,475$.

EXAMPLE 8.5

Recall the quality control experiment in Examples 8.3 and 8.4, in which we tested to determine whether a cereal box filling machine was deviating from the specified mean fill of $\mu = 12$ ounces. The test set-up is repeated below:

H_0: $\mu = 12$

H_a: $\mu \neq 12$ (i.e., $\mu < 12$ or $\mu > 12$)

Test statistic: $z = \dfrac{\bar{x} - 12}{\sigma_{\bar{x}}}$

Rejection region:

$z < -1.96$ or $z > 1.96$ for $\alpha = .05$

$z < -2.575$ or $z > 2.575$ for $\alpha = .01$

Note that two rejection regions have been specified corresponding to values of $\alpha = .05$ and $\alpha = .01$, respectively. Assume that $n = 100$ and $s = .5$.

a. Suppose the machine is underfilling the boxes by an average of .1 ounce, i.e., $\mu = 11.9$. Calculate the values of β corresponding to the two rejection regions. Discuss the relationship between the values of α and β.

b. Calculate the power of the test for each of the rejection regions when $\mu = 11.9$.

SOLUTION

a. We first consider the rejection region corresponding to $\alpha = .05$. The first step is to calculate the border values of $\bar{x}$ corresponding to the two-tailed rejection region, $z < -1.96$ or $z > 1.96$:

$$\bar{x}_{0,L} = \mu_0 - 1.96\sigma_{\bar{x}} \approx \mu_0 - 1.96\left(\frac{s}{\sqrt{n}}\right)$$

$$= 12.0 - 1.96\left(\frac{.5}{10}\right) = 11.902$$

$$\bar{x}_{0,U} = \mu_0 + 1.96\sigma_{\bar{x}} \approx \mu_0 + 1.96\left(\frac{s}{\sqrt{n}}\right)$$

$$= 12.0 + 1.96\left(\frac{.5}{10}\right) = 12.098$$

These border values are shown in Figure 8.13(a).

FIGURE 8.13

Calculation of β for
Example 8.5

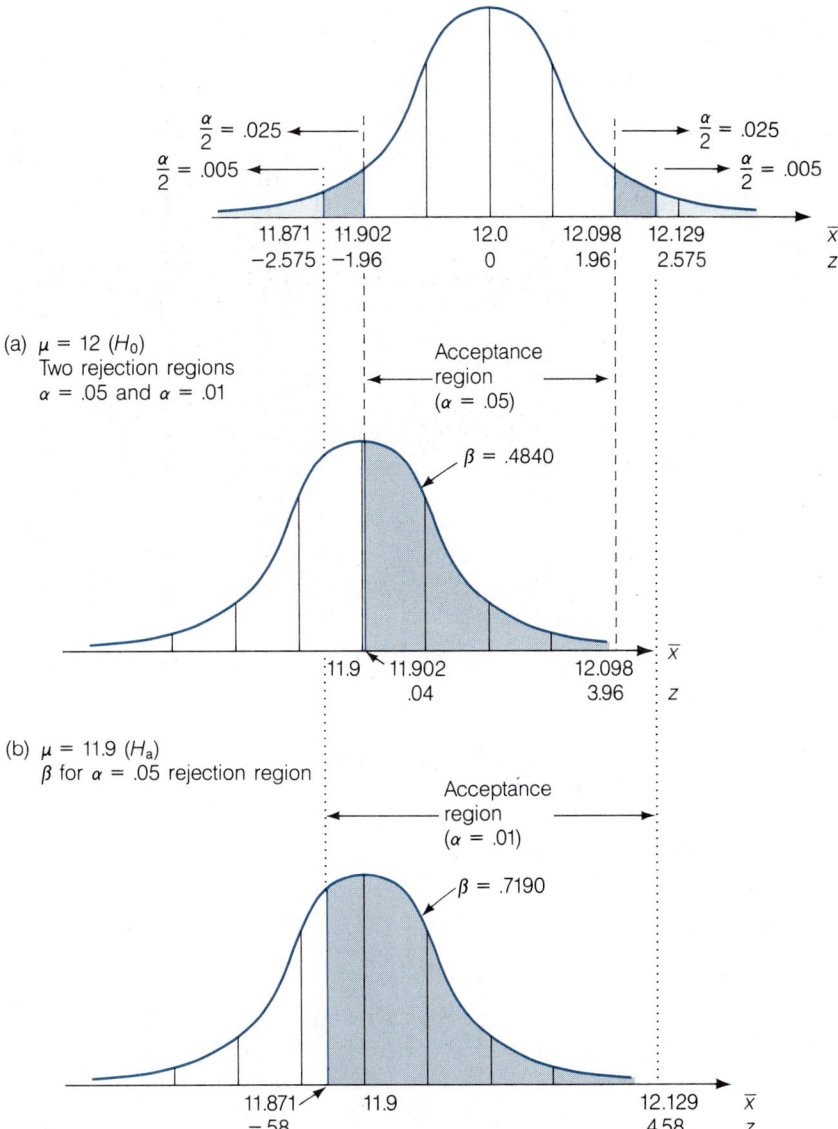

(a) $\mu = 12$ (H_0)
Two rejection regions
$\alpha = .05$ and $\alpha = .01$

(b) $\mu = 11.9$ (H_a)
β for $\alpha = .05$ rejection region

(c) $\mu = 11.9$ (H_a)
β for $\alpha = .01$ rejection region

Next, we convert these values to z values in the alternative distribution with $\mu_a = 11.9$:

$$z_L = \frac{\bar{x}_{0,L} - \mu_a}{\sigma_{\bar{x}}} \approx \frac{11.902 - 11.9}{.05} = .04$$

$$z_U = \frac{\bar{x}_{0,U} - \mu_a}{\sigma_{\bar{x}}} \approx \frac{12.098 - 11.9}{.05} = 3.96$$

These z values are shown in Figure 8.13(b), and you can see that the acceptance (or nonrejection) region is the area between them. Using Table IV of Appendix B, we find that the area between $z = 0$ and $z = .04$ is .0160, and the area between $z = 0$ and $z = 3.96$ is (approximately) .5 (since $z = 3.96$ is off the scale of Table IV). Then the area between $z = .04$ and $z = 3.96$ is, approximately,

$$\beta = .5 - .0160 = .4840$$

Thus, the test with $\alpha = .05$ will lead to a Type II error about 48% of the time when the machine is underfilling by .1 ounce.

For the rejection region corresponding to $\alpha = .01$, $z < -2.575$ or $z > 2.575$, we find

$$\bar{x}_{0,L} = 12.0 - 2.575\left(\frac{.5}{10}\right) = 11.871$$

$$\bar{x}_{0,U} = 12.0 + 2.575\left(\frac{.5}{10}\right) = 12.129$$

These border values of the rejection region are shown in Figure 8.13(c).

Converting these to z values in the alternative distribution with $\mu_a = 11.9$, we find $z_L = -.58$ and $z_U = 4.58$. The area between these values is, approximately,

$$\beta = .2190 + .5 = .7190$$

Thus, the chance that the test procedure with $\alpha = .01$ will lead to an incorrect acceptance of H_0 is about 72%.

Note that the value of β increases from .4840 to .7190 when we decrease the value of α from .05 to .01. This is a general property of the relationship between α and β: **as α is decreased (increased), β is increased (decreased)**.

b. The power is defined to be the probability of (correctly) rejecting the null hypothesis when the alternative is true. When $\mu = 11.9$ and $\alpha = .05$, we find

$$\text{Power} = 1 - \beta = 1 - .4840 = .5160$$

When $\mu = 11.9$ and $\alpha = .01$,

$$\text{Power} = 1 - \beta = 1 - .7190 = .2810$$

You can see that the power of the test is decreased as the level of α is decreased. This means that as the probability of incorrectly rejecting the null hypothesis is decreased, the probability of correctly accepting the null hypothesis for a given alternative is also decreased. Thus, the value of α must be selected carefully, with the realization that a test is made less powerful to detect departures from the null hypothesis when the value of α is decreased.

∎

We have shown that the probability of committing a Type II error, β, is inversely related to α (Example 8.5), and that the value of β decreases as the value of μ moves farther from the null hypothesis value (sewer pipe example). The sample size n also affects β. Remember that the standard deviation of the sampling distribution of $\bar{x}$ is inversely proportional to the (square root of) the sample size: $\sigma_{\bar{x}} = \sigma/\sqrt{n}$. Thus, as illustrated in Figure 8.14, the variability of both the null and alternative sampling distributions is decreased as n is increased. The effect is to decrease both α and β for a fixed rejection region, as illustrated in Figure 8.14(b). If the value of α is specified and remains fixed as n is increased, the value of β is made even smaller, as illustrated in Figure 8.14(c). Conversely, the power of the test for a given alternative hypothesis is increased as the sample size is increased.

FIGURE 8.14 Relationship Between α, β, and n

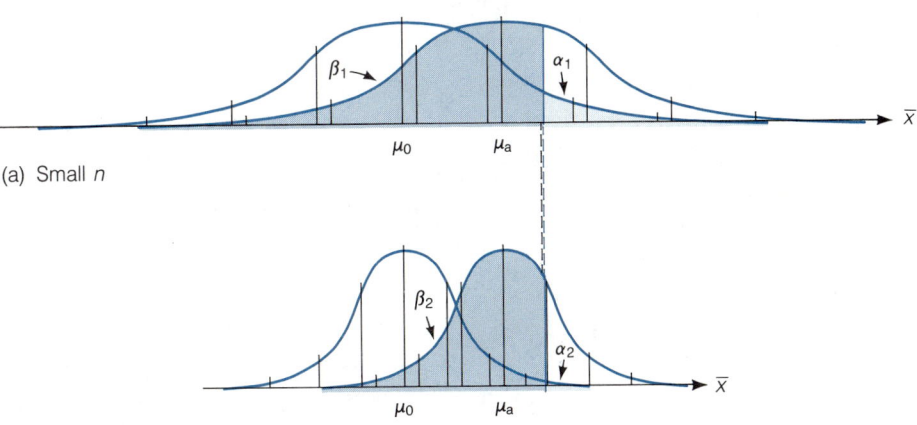

(a) Small n

(b) Large n, fixed rejection region
$(\alpha_2 < \alpha_1)$

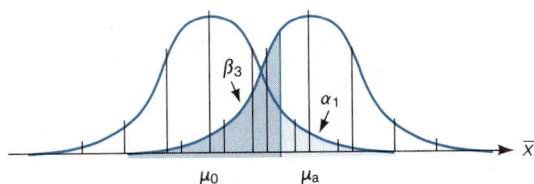

(c) Large n, fixed α
$(\beta_3 < \beta_1)$

The properties of β and power are summarized in the box.

PROPERTIES OF β AND POWER

1. The value of β decreases and the power increases as the distance between the null and alternative values of μ increases (see Figure 8.12).
2. The value of β increases and the power decreases as the value of α is decreased (see Figure 8.13).
3. The value of β decreases and the power increases as the sample size is increased (see Figure 8.14).

EXERCISES 8.43–8.54

LEARNING THE MECHANICS

8.43 What is the relationship between β, the probability of committing a Type II error, and the power of a test?

8.44 List three factors that will increase the power of a test.

8.45 Suppose you want to test H_0: $\mu = 1,000$ against H_a: $\mu > 1,000$ using $\alpha = .05$. The population in question is normally distributed with standard deviation 120. A random sample of size $n = 36$ will be used.
  a. Sketch the sampling distribution of $\bar{x}$ assuming that H_0 is true.
  b. Find the value of $\bar{x}_0$, that value of $\bar{x}$ above which the null hypothesis will be rejected. Indicate the rejection region on your graph of part **a**. Shade the area above the rejection region and label it α.
  c. On your graph of part **a**, sketch the sampling distribution of $\bar{x}$ if $\mu = 1,020$. Shade the area under this distribution that corresponds to the probability that $\bar{x}$ falls in the nonrejection region when $\mu = 1,020$. Label this area β.
  d. Find β.
  e. Compute the power of this test for detecting the alternative H_a: $\mu = 1,020$.

8.46 Refer to Exercise 8.45.
  a. If $\mu = 1,040$ instead of 1,020, what is the probability that the hypothesis test will incorrectly fail to reject H_0? That is, what is β?
  b. If $\mu = 1,040$, what is the probability that the test will correctly reject the null hypothesis? That is, what is the power of the test?
  c. Compare β and the power of the test when $\mu = 1,040$ to the values you obtained in Exercise 8.45 for $\mu = 1,020$. Explain the differences.

8.47 It is desired to test H_0: $\mu = 50$ against H_a: $\mu < 50$ using $\alpha = .10$. The population in question is uniformly distributed with standard deviation 20. A random sample of size 64 will be drawn from the population.
  a. Describe the (approximate) sampling distribution of $\bar{x}$ under the assumption that H_0 is true.
  b. Describe the (approximate) sampling distribution of $\bar{x}$ under the assumption that the population mean is 45.
  c. If μ were really equal to 45, what is the probability that the hypothesis test would lead the investigator to commit a Type II error?
  d. What is the power of this test for detecting the alternative H_a: $\mu = 45$?

8.48 Refer to Exercise 8.47. If the true value of the population mean is $\mu = 48$, what is the power of the test? How does it compare with the power when $\mu = 45$?

8.49 Suppose you want to conduct the two-tailed test of $H_0: \mu = 10$ against $H_a: \mu \neq 10$ using $\alpha = .05$. A random sample of size 100 will be drawn from the population in question. Assume the population has a standard deviation equal to 1.0.

  a. Describe the sampling distribution of $\bar{x}$ under the assumption that H_0 is true.
  b. Describe the sampling distribution of $\bar{x}$ under the assumption that $\mu = 9.9$.
  c. If μ were really equal to 9.9, find the value of β associated with the test.
  d. Find the value of β for the alternative $H_a: \mu = 10.1$.

8.50 Refer to Exercises 8.47 and 8.48.

  a. Find β for each of the following values of the population mean: 49, 47, 45, 43, 41.
  b. Plot each value of β you obtained in part **a** against its associated population mean. Show β on the vertical axis and μ on the horizontal axis. Draw a curve through the five points on your graph.
  c. Use your graph of part **b** to find the approximate probability that the hypothesis test will lead to a Type II error when $\mu = 48$. Compare your answer to the result you obtained in Exercise 8.48.
  d. Convert each of the β values you calculated in part **a** to the power of the test at the specified value of μ. Plot the power on the vertical axis against μ on the horizontal axis. Compare the β graph of part **b** to the *power curve* of this part.
  e. Examine the graphs of parts **b** and **d**. Explain what they reveal about the relationships among the distance between the true mean μ and the null hypothesized mean μ_0, the value of β, and the power.

APPLYING THE CONCEPTS

8.51 If a manufacturer (the vendee) buys all items of a particular type from a particular vendor, the manufacturer is practicing *sole sourcing*. Sole sourcing is a purchasing policy that is generally recognized as an important component of a firm's quality system. One of the major benefits of sole sourcing for the vendee is the improved communication that results from the closer vendee/vendor relationship (Treleven, 1986). As part of a sole sourcing arrangement, a vendor agreed to periodically supply its vendee with sample data from its production process. The vendee uses the data to investigate whether the mean length of rods produced by the vendor's production process is truly 5.0 mm or more, as claimed by the vendor and desired by the vendee.

  a. If the production process has a standard deviation of .01 mm, the vendor supplies $n = 100$ items to the vendee, and the vendee uses $\alpha = .05$ in testing $H_0: \mu = 5.0$ mm against $H_a: \mu < 5.0$ mm, what is the probability that the vendee's test will fail to reject the null hypothesis when in fact $\mu = 4.9975$ mm? What is the name given to this type of error?
  b. Refer to part **a**. What is the probability that the vendee's test will reject the null hypothesis when in fact $\mu = 5.0$? What is the name given to this type of error?
  c. What is the power of the test to detect a departure of .0025 mm below the specified mean rod length of 5.0 mm?

8.52 Refer to Exercise 8.42, in which the performance of a particular type of laser-based inspection equipment was investigated. Assume that the standard deviation of the number of solder joints inspected on each run is 1.2. If $\alpha = .05$ is used in conducting the hypothesis test of interest using a sample of 48 circuit boards, and if the true mean number of solder joints that can be inspected is really equal to 9.5, what is the probability that the test will result in a Type II error?

8.53 Refer to Exercise 8.40, in which the alternative hypothesis that the mean miles per gallon achieved by 1988 G-cars exceeds 35 is tested against the null hypothesis that the mean is 35 (or less). A sample of 36 automobiles were tested; assume that the resulting standard deviation of $s = 6$ is a good estimate of the true standard deviation.

   a. Calculate the power of the test for the mean values of 35.5, 36.0, 36.5, 37.0, and 37.5.

   b. Plot the power of the test on the vertical axis against the mean on the horizontal axis. Draw a curve through the points.

   c. Use the power curve of part **b** to estimate the power for the mean value $\mu = 36.75$. Calculate the power for this value of μ, and compare it to your approximation.

   d. Use the power curve to approximate the power of the test when $\mu = 40$. If the true value of the mean mpg for this model is really 40, what (approximately) are the chances that the test will fail to reject the null hypothesis that the mean is 35?

8.54 Refer to Exercise 8.53. Show what happens to the power curve when the sample size is increased from $n = 36$ to $n = 100$. Assume that the standard deviation is $\sigma = 6$.

SECTION 8.5

OBSERVED SIGNIFICANCE LEVELS: *p*-VALUES

According to the statistical test procedure described in Section 8.3, the value of α and, correspondingly, the rejection region are selected prior to conducting the test, and the conclusion is stated in terms of rejecting or not rejecting the null hypothesis. A second method of presenting the results of a statistical test reports the extent to which the test statistic disagrees with the null hypothesis and leaves to the reader the task of deciding whether to reject the null hypothesis. This measure of disagreement is called the **observed significance level** (or *p***-value**) for the test.

DEFINITION 8.4

The **observed significance level**, or *p*-value, for a specific statistical test is the probability (assuming H_0 is true) of observing a value of the test statistic that is at least as contradictory to the null hypothesis, and as supportive of the alternative hypothesis, as the one computed from the sample data.

For example, the value of the test statistic computed for the sample of $n = 50$ sections of sewer pipe was $z = 2.12$. Since the test was one-tailed—i.e., H_a: $\mu > 2,400$—values of the test statistic even more contradictory to H_0 than the one observed would be values larger than $z = 2.12$. Therefore, the observed significance level (*p*-value) for this test is

$$p\text{-value} = P(z \geq 2.12)$$

or, equivalently, the area under the standard normal curve to the right of $z = 2.12$ (see Figure 8.15).

FIGURE 8.15

Finding the *p*-Value for
an Upper-Tail Test When
z = 2.12

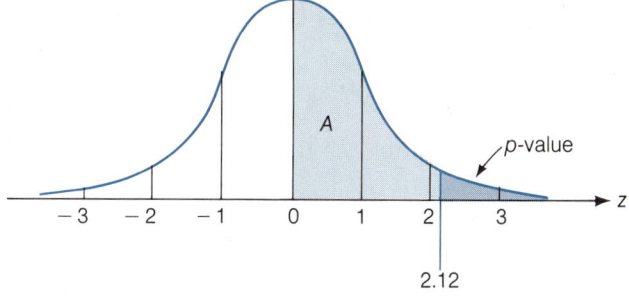

The area A in Figure 8.15 is given in Table IV of Appendix B as .4830. Therefore, the upper-tail area corresponding to $z = 2.12$ is

$$p\text{-value} = .5 - .4830 = .0170$$

Consequently, we say these test results are highly significant; that is, they disagree rather strongly with the null hypothesis, $H_0: \mu = 2,400$, and favor $H_a: \mu > 2,400$. The probability of observing a z value at least as large as 2.12 is only .0170, if in fact H_0 is true.

If you are inclined to select $\alpha = .05$ for this test, then you would reject the null hypothesis because the *p*-value for the test, .0170, is less than .05. In contrast, if you choose $\alpha = .01$, you would not reject the null hypothesis because the *p*-value for the test is larger than .01. Thus, the use of the observed significance level is identical to the test procedure described in the preceding sections except that the choice of α is left to the reader.

The steps for calculating the *p*-value corresponding to a test statistic for a population mean are given in the box.

STEPS FOR CALCULATING THE *p*-VALUE FOR A TEST OF HYPOTHESIS

1. Determine the value of the test statistic z corresponding to the result of the sampling experiment.
2. **a.** If the test is one-tailed, the *p*-value is equal to the tail area beyond z in the same direction as the alternative hypothesis. Thus, if the alternative hypothesis is of the form ">," the *p*-value is the area to the right of, or above, the observed z value. Conversely, if the alternative is of the form "<," the *p*-value is the area to the left of, or below, the observed z value.
   b. If the test is two-tailed, the *p*-value is equal to twice the tail area beyond the observed z value in the direction of the sign of z. That is, if z is positive, the *p*-value is twice the area to the right of, or above, the observed z value. Conversely, if z is negative, the *p*-value is twice the area to the left of, or below, the observed z value.

EXAMPLE 8.6

Find the observed significance level for the test of the mean filling weight in Examples 8.3 and 8.4.

SOLUTION

Example 8.3 presented a two-tailed test of the hypothesis

$$H_0: \quad \mu = 12 \text{ ounces}$$

against the alternative hypothesis

$$H_a: \quad \mu \neq 12 \text{ ounces}$$

The observed value of the test statistic in Example 8.4 was $z = -3.0$, and any value of z less than -3.0 or greater than $+3.0$ (because this is a two-tailed test) would be even more contradictory to H_0. Therefore, the observed significance level for the test is

$$p\text{-value} = P(z < -3.0 \text{ or } z > +3.0)$$

Thus, we calculate the area below the observed z value, $z = -3.0$, and double it. Consulting Table IV of Appendix B, we find that $P(z < -3.0) = .5 - .4987 = .0013$. Therefore, the p-value for this two-tailed test is

$$2P(z < -3.0) = 2(.0013) = .0026$$

We can interpret this p-value as a strong indication that the machine is not filling the boxes according to specifications ($\mu \neq 12$), since we would observe a test statistic this extreme or more extreme only 26 in 10,000 times if the machine were meeting specifications ($\mu = 12$). The extent to which the mean differs from 12 could be better determined by calculating a confidence interval for μ. ∎

When publishing the results of a statistical test of hypothesis in journals, case studies, reports, etc., many researchers make use of p-values. Instead of selecting α a priori and then conducting a test as outlined in this chapter, the researcher will compute and report the value of the appropriate test statistic and its associated p-value. It is left to the reader of the report to judge the significance of the result—i.e., the reader must determine whether to reject the null hypothesis in favor of the alternative hypothesis, based on the reported p-value. This p-value is often referred to as the **attained significance level** of the test. Usually, the null hypothesis will be rejected if the observed significance level is *less* than the fixed significance level, α, chosen by the reader. The inherent advantages of reporting test results in this manner are twofold: (1) readers are able to draw their own conclusions about the reported hypothesis test by choosing α themselves and comparing it to the reported p-value and (2) a measure of the degree of significance of the test result (i.e., the p-value) is provided.

HOW TO DECIDE WHETHER TO REJECT H_0 USING REPORTED p-VALUES

1. Choose the maximum value of α you are willing to tolerate.
2. If the observed significance level (p-value) of the test is less than the chosen value of α, then reject the null hypothesis.

| | | | | | | | | | | | | |

EXERCISES 8.55–8.68

LEARNING THE MECHANICS

8.55 If a hypothesis test were conducted using $\alpha = .05$, for which of the following p-values would the null hypothesis be rejected?
a. .06   **b.** .10   **c.** .01
d. .001   **e.** .251   **f.** .042

8.56 For each α and observed significance level (p-value) pair, indicate whether the null hypothesis would be rejected.
a. $\alpha = .05$, p-value $= .10$
b. $\alpha = .10$, p-value $= .05$
c. $\alpha = .01$, p-value $= .001$
d. $\alpha = .025$, p-value $= .05$
e. $\alpha = .10$, p-value $= .45$

8.57 Explain the difference between statistical significance and practical significance.

8.58 An analyst tested the null hypothesis $\mu \geq 20$ against the alternative hypothesis that $\mu < 20$. The analyst reported a p-value of .06. What is the smallest value of α for which the null hypothesis would be rejected?

8.59 In a test of H_0: $\mu \leq 100$ against H_a: $\mu > 100$, the sample data yielded the test statistic $z = 2.26$. Find the p-value for the test.

8.60 In a test of H_0: $\mu = 100$ against H_a: $\mu \neq 100$, the sample data yielded the test statistic $z = 2.08$. Find the p-value for the test.

8.61 In a test of H_0: $\mu \geq 100$ against H_a: $\mu < 100$, the sample data yielded the test statistic $z = -1.11$. Find the observed significance level of the test.

8.62 In a test of the hypothesis H_0: $\mu = 50$ versus H_a: $\mu > 50$, a sample of $n = 100$ observations possessed mean $\bar{x} = 50.5$ and standard deviation $s = 3.3$. Find and interpret the p-value for this test.

8.63 In a test of the hypothesis H_0: $\mu = 10$ versus H_a: $\mu \neq 10$, a sample of $n = 50$ observations possessed mean $\bar{x} = 9.5$ and standard deviation $s = 2.1$. Find and interpret the p-value for this test.

APPLYING THE CONCEPTS

8.64 The manufacturer of an over-the-counter pain reliever claims that its product brings pain relief to headache sufferers in less than 3.5 minutes, on average. In order to be able to make this claim in its television advertisements, the manufacturer was required by a particular television network to present statistical evidence in support of the claim. The manufacturer reported that for a random sample of 50 headache sufferers, the mean time to relief was 3.3 minutes and the standard deviation was 66 seconds.
a. Do these data support the manufacturer's claim? Test using $\alpha = .05$.
b. Report the p-value of the test.
c. In general, do large p-values or small p-values support the manufacturer's claim? Explain.

8.65 Refer to Exercise 8.39, in which the performance of an injection molder that produces plastic golf tees was investigated. Find the p-value for the test and interpret its value.

8.66 *USA Today* (Aug. 17, 1983) reported that for the 1983–1984 academic year, 4-year private colleges charged students an average of \$4,627 for tuition and fees, while at 4-year public colleges the average was \$1,105. Suppose that for 1984–1985 a random sample of 30 private colleges yielded the following data on tuition and fees: $\bar{x} = \$5,000$ and $s = \$1,643$. Assume that \$4,627 is the population mean for 1983–1984.

a. Specify the null and alternative hypotheses you would use to investigate whether the mean amount for tuition and fees in 1984–1985 was significantly larger (in the statistical sense) than it was in 1983–1984.

b. Calculate the p-value for the hypothesis test you described in part **a**, and explain what the p-value indicates about the statistical significance of the test results.

c. Explain the difference between statistical significance and practical significance in the context of this exercise.

8.67 Florida's housing market remains strong due to the steady stream of new residents fleeing harsh northern winters. This year, the state association of realtors claims that the mean cost of a new home in Florida is \$80,380. One realtor who claims that this figure is too low obtained a random sample of 30 sale prices from a list of all homes sold in Florida during the last 6 months. The sample mean and standard deviation were:

$$\bar{x} = \$83,290 \qquad s = \$6,500$$

a. The realtor wishes to conduct a hypothesis test to substantiate her claim. Identify the null and alternative hypotheses of interest to her.

b. Find the observed significance level for this test, and interpret its value.

8.68 Refer to Exercise 8.40. Find the observed significance level for the test, and interpret its value.

| | | | | | | | | | | | | | |

SECTION 8.6

SMALL-SAMPLE INFERENCES ABOUT A POPULATION MEAN

One of the items of interest to an investor in the stock market is the amount a company's annual earnings per share will increase or decrease over the next year. Recall that earnings per share is computed by dividing the total annual earnings of the company by the total number of shares of stock outstanding. One way of trying to project the change in earnings per share is to ask the opinion of several experts, thus obtaining a sample of projections for the particular company. Then, this sample of opinions can be used to make an inference about the mean projected earnings per share, μ, of all stock analysts. However, time and cost restrictions would probably limit the sample of opinions to a small number, and thus the large-sample inferential techniques of Sections 8.1 and 8.3 may not be applicable.

Many inferences in business must be made on the basis of very limited information—i.e., **small samples**. When an inference is to be made about a population mean, μ, small samples have two immediate problematic effects:

PROBLEM 1

The shape of the sampling distribution of the sample mean $\bar{x}$ now depends on the shape of the population that is sampled. Because the Central Limit Theorem applies only to large samples, we can no longer assume that the sampling distribution of $\bar{x}$ is approximately normal.

PROBLEM 2

Although it is still true that $\sigma_{\bar{x}} = \sigma/\sqrt{n}$, the sample standard deviation s may provide a poor approximation of the population standard deviation σ when the sample size is small.

SOLUTION TO PROBLEM 1

According to Theorem 7.1, the sampling distribution of $\bar{x}$ will be normal (approximately normal) even for small samples *if the population being sampled is normal (approximately normal)*.

SOLUTION TO PROBLEM 2

Instead of using the statistic

$$z = \frac{\bar{x} - \mu}{\sigma_{\bar{x}}} = \frac{\bar{x} - \mu}{\sigma/\sqrt{n}}$$

which requires knowledge of, or a good approximation to, σ, we use the statistic

$$t = \frac{\bar{x} - \mu}{s/\sqrt{n}}$$

which replaces the population standard deviation, σ, by the sample standard deviation, s.

The distribution of the **t statistic** in repeated sampling was discovered by W. S. Gosset, a scientist in the Guinness brewery, who published his discovery in 1908 under the pen name of Student. The main result of Gosset's work is that if we are sampling from a normal distribution, the t statistic will have a sampling distribution very much like that of the z statistic: mound-shaped, symmetric, and with mean 0. The primary difference between the sampling distributions of t and z is that the t distribution is more variable than the z, which follows intuitively when you realize that t contains two random quantities ($\bar{x}$ and s), while z contains only one ($\bar{x}$).

The actual increase in variability in the sampling distribution of t depends on the sample size, n. In particular, the smaller the value of n, the more variable will be the sampling distribution of t. A convenient way of expressing this dependence is to say that the t statistic has $(n - 1)$ **degrees of freedom (df)**. Recall that the quantity $(n - 1)$ is the divisor that appears in the formula for s^2. This number plays a key role in the sampling distribution of s^2 and will appear in discussions of other statistics in later chapters.

In Figure 8.16, we show both the sampling distribution of z and the sampling distribution of a t statistic with 4 degrees of freedom (df). You can see that the increased variability of the t statistic means that the t value, t_α, that locates an area α in the upper tail of the t distribution will be larger than the corresponding value z_α. Values of t that will be used in forming small-sample confidence intervals for μ and rejection regions for small-sample tests of hypotheses about μ are given in Table VI of Appendix B. A partial reproduction of this table is shown in Table 8.4.

FIGURE 8.16

Standard Normal (z) Distribution and t Distribution with 4 df

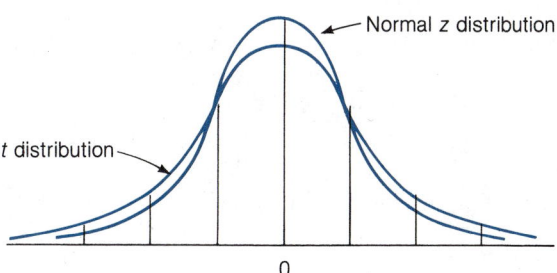

TABLE 8.4

Reproduction of Part of
Table VI of Appendix B

DEGREES OF FREEDOM	$t_{.100}$	$t_{.050}$	$t_{.025}$	$t_{.010}$	$t_{.005}$	$t_{.001}$	$t_{.005}$
1	3.078	6.314	12.706	31.821	63.657	318.31	636.62
2	1.886	2.920	4.303	6.965	9.925	22.326	31.598
3	1.638	2.353	3.182	4.541	5.841	10.213	12.924
4	1.533	2.132	2.776	3.747	4.604	7.173	8.610
5	1.476	2.015	2.571	3.365	4.032	5.893	6.869
6	1.440	1.943	2.447	3.143	3.707	5.208	5.959
7	1.415	1.895	2.365	2.998	3.499	4.785	5.408
8	1.397	1.860	2.306	2.896	3.355	4.501	5.041
9	1.383	1.833	2.262	2.821	3.250	4.297	4.781
10	1.372	1.812	2.228	2.764	3.169	4.144	4.587
11	1.363	1.796	2.201	2.718	3.106	4.025	4.437
12	1.356	1.782	2.179	2.681	3.055	3.930	4.318
13	1.350	1.771	2.160	2.650	3.012	3.852	4.221
14	1.345	1.761	2.145	2.624	2.977	3.787	4.140
15	1.341	1.753	2.131	2.602	2.947	3.733	4.073

Note that t_α values are listed for degrees of freedom from 1 to 29, where α refers to the tail area to the right of t_α. For example, if we want the t value with an area of .025 to its right and 4 df, we look in the table under the column $t_{.025}$ for the entry in the row corresponding to 4 df. This entry is $t_{.025} = 2.776$, as shown in Figure 8.17. The corresponding standard normal z-score is $z_{.025} = 1.96$.

FIGURE 8.17

The $t_{.025}$ Value in a t
Distribution with 4 df and
the Corresponding $z_{.025}$
Value

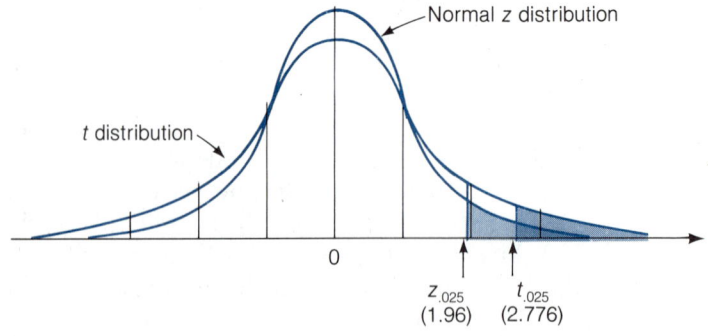

Note that the last row of Table VI, where df = infinity, contains the standard normal z values. This follows from the fact that as the sample size n grows very large, s becomes closer to σ, and thus t becomes closer in distribution to z. In fact, when df = 29, there is little difference between corresponding tabulated values of z and t. Thus, we choose the arbitrary cutoff of $n = 30$ (df = 29) to distinguish between the large- and small-sample inferential techniques.

Returning to the projected earnings per share example, suppose we can get a sample of five expert opinions about next year's earnings per share for a stock. We calculate the mean and standard deviation of these five projections to be

$$\bar{x} = \$2.63 \qquad s = \$.72$$

If we know that last year's earnings were $2.01 per share, is there enough evidence to indicate that the mean expert projection, μ, exceeds last year's figure?

The type of inference desired is a test of hypothesis. Since we want to show that the mean expert projection for this year exceeds last year's earnings per share, we will test the null hypothesis that $\mu = \$2.01$ against the alternative hypothesis that $\mu > \$2.01$. Thus, the elements of the test are

Null hypothesis H_0: $\mu = \$2.01$

Alternative hypothesis H_a: $\mu > \$2.01$

Since σ is unknown and the sample is small ($n = 5$), we use the t statistic:

Test statistic: $t = \dfrac{\bar{x} - \mu_0}{s/\sqrt{n}} = \dfrac{\bar{x} - 2.01}{s/\sqrt{n}}$

Assumption: The relative frequency distribution of the population of projected earnings per share is approximately normal.

Note that we must assume the normality of the population in order to use the t statistic. If we want to test at the $\alpha = .05$ level, the rejection region will be

Rejection region: $t > t_{.05} = 2.132$ where df $= n - 1 = 4$

This rejection region is shown in Figure 8.18.

FIGURE 8.18

Rejection Region for Projected Earnings per Share Test

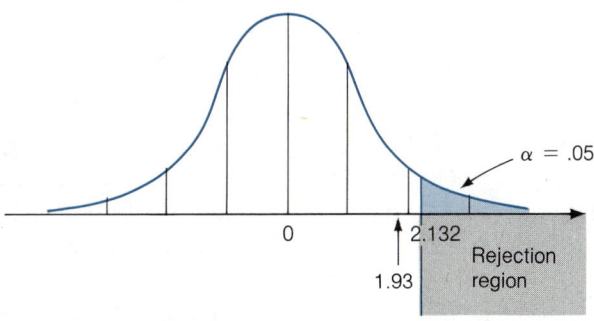

We now calculate

$$t = \frac{\bar{x} - 2.01}{s/\sqrt{n}} = \frac{2.63 - 2.01}{.72/\sqrt{5}}$$

$$= \frac{.62}{.72/2.24} = 1.93$$

Since the value of t, 1.93, calculated from the sample data does not exceed the tabulated value of 2.132, we cannot conclude that the mean projection of all experts exceeds last year's earnings of $2.01. Even though the sample mean $\bar{x} = 2.63$ falls 1.93 standard deviations above the null hypothesized value of $\mu = 2.01$, the small sample provides insufficient evidence to conclude that the population mean μ exceeds 2.01 with 95% confidence.

We summarize the technique for conducting a small-sample test of hypothesis about a population mean in the box.

SMALL-SAMPLE TEST OF HYPOTHESIS ABOUT μ

ONE-TAILED TEST	TWO-TAILED TEST
H_0: $\mu = \mu_0$	H_0: $\mu = \mu_0$
H_a: $\mu < \mu_0$	H_a: $\mu \neq \mu_0$
(or H_a: $\mu > \mu_0$)	

One-tailed test statistic:
$$t = \frac{\bar{x} - \mu_0}{s/\sqrt{n}}$$

Two-tailed test statistic:
$$t = \frac{\bar{x} - \mu_0}{s/\sqrt{n}}$$

Rejection region (one-tailed): $t < -t_\alpha$
(or $t > t_\alpha$ when H_a: $\mu > \mu_0$)

Rejection region (two-tailed): $t < -t_{\alpha/2}$ or $t > t_{\alpha/2}$

where t_α and $t_{\alpha/2}$ are based on $(n - 1)$ degrees of freedom.

Assumption: A random sample is selected from a population with a relative frequency distribution that is approximately normal.

Remember, the basic assumption necessary for the use of the t statistic is that the sampled population has a relative frequency distribution that is approximately normal. What can be done if you know that the population relative frequency distribution is decidedly nonnormal, say highly skewed?

WHAT CAN BE DONE IF THE POPULATION RELATIVE FREQUENCY DISTRIBUTION DEPARTS GREATLY FROM NORMAL?

Answer: Use the nonparametric statistical methods of Chapter 16.

EXAMPLE 8.7

A major car manufacturer wants to test a new engine to determine whether it meets new air pollution standards. The mean emission, μ, of all engines of this type must be less than 20 parts per million of carbon. Ten engines are manufactured for testing purposes, and the mean and standard deviation of the emissions for this sample of engines are determined to be

$$\bar{x} = 17.1 \text{ parts per million} \qquad s = 3.0 \text{ parts per million}$$

Do the data supply sufficient evidence to allow the manufacturer to conclude that this type of engine meets the pollution standard? Assume that the manufacturer is willing to risk a Type I error with probability equal to $\alpha = .01$.

SOLUTION

The manufacturer wants to establish the alternative hypothesis that the mean emission level, μ, for all engines of this type is less than 20 parts per million. The elements of this small-sample one-tailed test are

H_0: $\mu = 20$

H_a: $\mu < 20$

Test statistic: $t = \dfrac{\bar{x} - 20}{s/\sqrt{n}}$

Assumption: The relative frequency distribution of the population of emission levels for all engines of this type is approximately normal.

Rejection region: For $\alpha = .01$ and df $= n - 1 = 9$, the one-tailed rejection region (see Figure 8.19) is $t < -t_{.01} = -2.821$.

FIGURE 8.19
Rejection Region for
Example 8.7

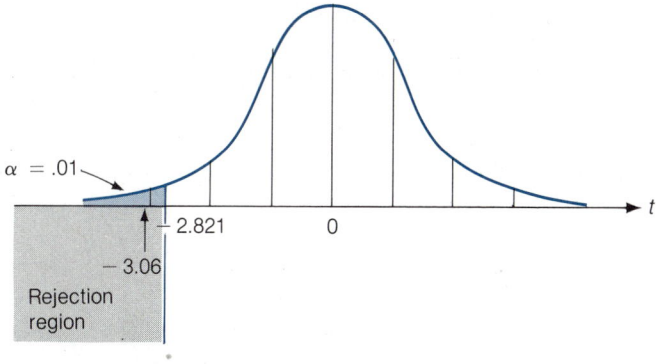

We now calculate the test statistic:

$$t = \frac{\bar{x} - 20}{s/\sqrt{n}} = \frac{17.1 - 20}{3.0/\sqrt{10}} = -3.06$$

Since the calculated t falls in the rejection region (see Figure 8.19), the manufacturer concludes that $\mu < 20$ parts per million and the new engine type meets the pollution standard. Are you satisfied with the reliability associated with this inference? The probability is only $\alpha = .01$ that the test would support the alternative hypothesis if in fact it was false. ∎

EXAMPLE 8.8

Find the observed significance level for the test in Example 8.7.

SOLUTION

The test performed in Example 8.7 was a one-tailed test, in which H_0: $\mu = 20$ would be rejected in favor of H_a: $\mu < 20$ for values of t in the lower tail of the t distribution. Since the value of t computed from the sample data was $t = -3.06$, the observed significance level (or p-value) for the test is equal to the probability that t would assume a value less than or equal to -3.06, if in fact H_0 was true. This is equal to the area in the lower tail of the t distribution (shaded in Figure 8.20). To find this area—i.e., the p-value for the test—we consult the t table in

FIGURE 8.20

The Observed
Significance Level for
the Test in Example 8.7

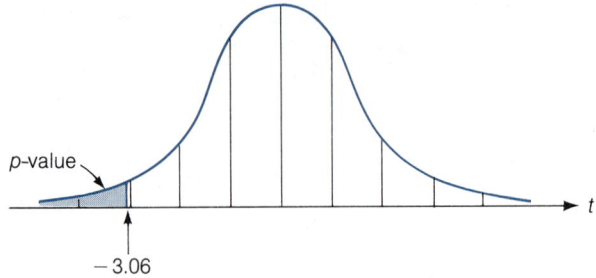

p-value

-3.06

Table VI of Appendix B. Unlike the table of areas under the normal curve, Table VI gives only the t values corresponding to the areas .100, .050, .025, .010, .005, .001, and .0005. Therefore, we can only approximate the p-value for the test. Since the observed t value was based on 9 degrees of freedom, we use the df $= 9$ row in Table VI and move across the row until we reach the t value that is closest to the observed $t = -3.06$. [*Note:* We ignore the minus sign.] The t values corresponding to p-values of .010 and .005 are 2.821 and 3.250, respectively. Since the observed t value falls between $t_{.010}$ and $t_{.005}$, the p-value for the test lies between .010 and .005. We could interpolate to more accurately locate the p-value for the test, but it is easier and adequate for our purposes to choose the larger area as the p-value and thus report it as .010.   ■

We may also use the t distribution to form a small-sample confidence interval for a population mean μ, **if the population is approximately normally distributed.** Recall that the large-sample confidence interval for μ is

$$\bar{x} \pm z_{\alpha/2}\sigma_{\bar{x}} = \bar{x} \pm z_{\alpha/2}\left(\frac{\sigma}{\sqrt{n}}\right)$$

where $100(1 - \alpha)\%$ is the desired confidence level. To form the small-sample confidence interval, replace σ by s and $z_{\alpha/2}$ by $t_{\alpha/2}$ (remember, the degrees of freedom must be specified for the tabulated t value).

SMALL-SAMPLE CONFIDENCE INTERVAL FOR μ

$$\bar{x} \pm t_{\alpha/2}\left(\frac{s}{\sqrt{n}}\right)$$

where $t_{\alpha/2}$ is based on $(n - 1)$ degrees of freedom.

Assumption: The relative frequency distribution of the sampled population is approximately normal.

EXAMPLE 8.9

Some quality control experiments require *destructive sampling* in order to measure some particular characteristic of the product. For example, suppose a manufacturer of printers for personal computers wishes to estimate the mean number of characters printed before the printhead fails. The cost of destructive sampling often dictates small samples. Suppose the printer manufacturer tests $n = 15$ printheads and calculates the following statistics:

$$\bar{x} = 1.23 \text{ million characters} \qquad s = .27 \text{ million characters}$$

Form a 99% confidence interval for the mean number of characters printed before the head fails.

SOLUTION

If we assume that the number of characters printed before printhead failure is normally distributed, we can use the t statistic to form the confidence interval. We use a confidence coefficient of .99 and degrees of freedom $n - 1 = 14$ to find in Table VI of Appendix B:

$$t_{\alpha/2} = t_{.005} = 2.977$$

Thus, the small sample forces us to assume normality and extend the interval almost 3 standard deviations (of $\bar{x}$) on each side of the sample mean in order to form the 99% confidence interval. For these data, the interval is

$$\bar{x} \pm t_{.005}\left(\frac{s}{\sqrt{n}}\right) = 1.23 \pm 2.977\left(\frac{.27}{\sqrt{15}}\right)$$
$$= 1.23 \pm .21 \quad \text{or} \quad (1.02, 1.44)$$

Thus, the manufacturer can be 99% confident that the printhead has a mean life between 1.02 and 1.44 million characters. If the manufacturer were to advertise that the mean life of its printheads is (at least) 1 million characters, the interval would support such a claim. Our confidence is derived from the fact that 99% of the intervals formed in repeated applications of this procedure will contain μ. ∎

We have emphasized throughout this section that the assumption of a normally distributed population is necessary for making small-sample inferences about μ when using the t statistic. While many business phenomena do have approximately normal distributions, it is also true that many business phenomena have distributions that are not normal or even mound-shaped. Empirical evidence acquired over the years has shown that the t distribution is rather insensitive to moderate departures from normality. That is, the use of the t statistic when sampling from mound-shaped populations generally produces credible results; however, for cases in which the distribution is distinctly nonnormal, **nonparametric methods** should be used. Nonparametric statistics are the subject of Chapter 16.

EXERCISES 8.69–8.90

LEARNING THE MECHANICS

8.69 In what ways are the distributions of the z statistic and t statistic alike? How do they differ?

8.70 Under what circumstances should you use the t distribution in testing a hypothesis about a population mean?

8.71 Let t_0 be a particular value of t. Use Table VI of Appendix B to find t_0 values such that the following statements are true:

a. $P(t \geq t_0) = .025$ where df $= 8$ **b.** $P(t \geq t_0) = .01$ where df $= 10$
c. $P(t \leq t_0) = .005$ where df $= 17$ **d.** $P(t \leq t_0) = .05$ where df $= 14$

8.72 Let t_0 be a particular value of t. Use Table VI of Appendix B to find t_0 values such that the following statements are true:

a. $P(-t_0 < t < t_0) = .95$ where df $= 11$
b. $P(t \leq -t_0 \text{ or } t \geq t_0) = .05$ where df $= 11$
c. $P(t \leq t_0) = .05$ where df $= 11$
d. $P(t \leq -t_0 \text{ or } t \geq t_0) = .10$ where df $= 20$
e. $P(t \leq -t_0 \text{ or } t \geq t_0) = .01$ where df $= 6$

8.73 For each of the following rejection regions, sketch the sampling distribution of t and indicate the location of the rejection region on your sketch:

a. $t > 1.440$ where df $= 6$
b. $t < -1.782$ where df $= 12$
c. $t < -2.060$ or $t > 2.060$ where df $= 25$

8.74 For each of the rejection regions defined in Exercise 8.73, what is the probability that a Type I error will be made?

8.75 The following sample of five measurements was randomly selected from a normally distributed population: 4, 7, 3, 4, 6

a. Test the null hypothesis that the mean of the population is 6 against the alternative hypothesis, $\mu < 6$. Use $\alpha = .05$.
b. Test the null hypothesis that the mean of the population is 6 against the alternative hypothesis, $\mu \neq 6$. Use $\alpha = .05$.

8.76 Find the observed significance level for each test in Exercise 8.75.

8.77 Refer to Exercise 8.75.

a. Find a 95% confidence interval for μ.
b. Give the value of t that would be used to form a 90% confidence interval for μ.
c. Describe a practical situation that would motivate you to form a confidence interval for μ rather than testing a hypothesis about μ.

8.78 The following sample of six measurements was randomly selected from a normally distributed population: 1, 3, −1, 5, 1, 2

a. Test the null hypothesis that the mean of the population is 3 against the alternative hypothesis, $\mu < 3$. Use $\alpha = .05$.
b. Test the null hypothesis that the mean of the population is 3 against the alternative hypothesis, $\mu \neq 3$. Use $\alpha = .05$.

8.79 Refer to Exercise 8.78.

a. Find a 95% confidence interval for μ.
b. Give the value of t that would be used to form a 90% confidence interval for μ.

8.80 The following sample of 24 measurements was selected from a population that is approximately normally distributed:

91	80	99	110	95	106	78	121
106	100	97	82	100	83	115	104
114	118	96	101	79	130	94	101

a. Construct an 80% confidence interval for the population mean.
b. Construct a 95% confidence interval for the population mean and compare the width of this interval with that of part **a**.
c. Carefully interpret each of the confidence intervals, and explain why the 80% confidence interval is narrower.

APPLYING THE CONCEPTS

8.81 A *mortgage* is a type of loan that is secured by a designated piece of property. If the borrower defaults on the loan, the lender can sell the property to recover the outstanding debt. The home mortgage is the most important type of personal loan in the United States. In a home mortgage, the borrower pledges the home in question as security for the loan (Sharpe, 1985). A federal bank examiner is interested in estimating the mean outstanding principal balance of all home mortgages foreclosed by the bank due to default by the borrower during the last 3 years. A random sample of 12 foreclosed mortgages yielded the following data (in dollars):

95,982   81,422   39,888   46,836   66,899   69,110
59,200   62,331   105,812   55,545   56,635   72,123

a. Describe the population from which the bank examiner collected the sample data. What characteristic must this population possess to enable us to construct a confidence interval for the mean outstanding principal balance using the method described in this section?
b. Construct a 90% confidence interval for the mean of interest.
c. Carefully interpret your confidence interval in the context of the problem.

8.82 In any bottling process, a manufacturer will lose money if the bottles contain either more or less than is claimed on the label. Accordingly, bottlers pay close attention to the amount of their product being dispensed by bottle-filling machines. Suppose a quality control inspector for a catsup company is interested in testing whether the mean number of ounces of catsup per family-size bottle differs from the labeled amount of 20 ounces. The inspector samples nine bottles, measures the weight of their contents, and finds that $\bar{x} = 19.7$ ounces and $s = .3$ ounce.
a. Does the sample evidence indicate that the catsup dispensing machine needs adjustment? Test at $\alpha = .05$.
b. What is the p-value for the hypothesis test you conducted in part **a**?
c. What assumptions are necessary so that the procedure used in part **a** is valid?
d. Find a 90% confidence interval for the mean number of ounces of catsup being dispensed.

8.83 A recent study indicated that the cost of hiring an employee (excluding salary) ranges from about $1,500 for a secretary to more than $40,000 for a manager (Dessler, 1986). In order to estimate its mean cost of hiring an entry-level secretary, a large corporation randomly selected eight of the entry-level secretaries it hired during the last 2 years and determined the costs (in dollars) involved in hiring each. The following data were obtained:

2,100   1,650   1,315   2,035
2,245   1,980   1,700   2,190

Assume that the population from which these data were sampled is approximately normally distributed.

a. Describe the population from which the corporation collected the sample data.

b. Use a 90% confidence interval to estimate the mean of interest to the corporation.

c. How wide is the confidence interval you constructed in part **b**? Would a 95% confidence interval be wider or narrower? Explain.

8.84 In an article entitled "Huge Phone Bills Look Like Mobster Fraud," the *Orlando Sentinel* (Mar. 15, 1984) comments on the rash of huge telephone bills received by some AT&T customers in early 1984. Unexplained huge bills received during this brief period of time (often by private individuals) possessed the following dollar values: $109,500, $61,180, $125,883, $35,236, $26,337, $93,315, and $36,063. Suppose these bills represent a random sample of the sizes of the thefts of telephone services that AT&T might expect in the future. Use the data to obtain an estimate of the mean size of a theft in the future. Use a 90% confidence interval. List any assumptions you make.

8.85 A company purchases large quantities of naphtha in 50-gallon drums. Because the purchases are ongoing, small shortages in the drums can represent a sizable loss to the company. The weights of the drums vary slightly from drum to drum, so the weight of the naphtha is measured after removing it from the drums. Suppose the company samples the contents of 20 drums, measures the naphtha in each, and calculates $\bar{x} = 49.70$ gallons and $s = .32$ gallon. Do the sample statistics provide sufficient evidence to indicate that the mean fill per 50-gallon drum is less than 50 gallons? Use $\alpha = .10$. List your assumptions.

8.86 A cigarette manufacturer advertises that its new low-tar cigarette "contains on average no more than 4 milligrams of tar." You have been asked to test the claim using the following sample information: $n = 25$, $\bar{x} = 4.16$ milligrams, $s = .30$ milligram. Does the sample information disagree with the manufacturer's claim? Test using $\alpha = .05$. List any assumptions you make.

8.87 In an effort to offset Russia's growing armored force, the U.S. Defense Department has selected a new Army tank designed by Chrysler Corp. The tank, called M-1, can reach an average top speed of 45 miles per hour, a speed the Defense Department believes is faster than the Soviets' fastest and most powerful tank, the T-72. To estimate the mean top speed of the T-72, suppose the Defense Department gained access to data on three of Russia's tanks and found the top speed for the three tanks had an average of 43.5 miles per hour and a standard deviation of 2.5 miles per hour. Find a 95% confidence interval for the mean top speed of the T-72 tanks. What assumptions must you make to form this confidence interval? Interpret the interval estimate.

8.88 Before the tax reform legislation of 1986, interest payments were a deductible expense. That is, persons who itemized their tax deductions (as opposed to taking the standard deduction) could reduce their taxable income by the amount of interest they paid (e.g., mortgage interest and finance charges) during the year. The average interest deduction claimed in 1980 by taxpayers of various income levels is shown in the accompanying table.

   Suppose 12 tax returns are randomly sampled by the Internal Revenue Service from the population of 1986 tax returns with adjusted gross incomes between $25,000 and $30,000. The interest deduction claimed on each return is listed below:

$3,050	$3,101	$3,415	$2,910	$3,333	$3,002
3,872	3,102	3,222	2,806	2,851	2,999

1980

ADJUSTED GROSS INCOME ($ thousands)	AVERAGE INTEREST DEDUCTION ($)
15–20	2,604
20–25	2,792
25–30	3,011
30–50	3,527
50–100	5,626
100–200	10,384
200–500	20,815

Source: *Wall Street Journal*, Dec. 8, 1982, p. 1.

a. Assume that the population of interest deductions from which the sample was drawn is approximately normally distributed. Do the sample data provide sufficient evidence to conclude that in 1986 the average interest deduction claimed by taxpayers in the $25,000–$30,000 adjusted gross income bracket is significantly greater than in 1980? Use $\alpha = .05$.

b. Find and interpret the p-value for the test.

8.89 In a recent nationwide survey of 1,000 men and women conducted by Caldwell Davis Partners (an advertising agency), it was found that two-thirds of the people surveyed perceived themselves as younger than their actual chronological age. These findings may help explain the recent failures of a line of food advertised as being for senior citizens and a shampoo directed at "hair over 40." The survey also indicated that, on average, men and women perceived themselves to be 6 years younger and 7 years younger, respectively, than their actual ages. However, men and women under 30 generally perceived themselves as older than their actual age (Nemy, 1982). A researcher randomly sampled ten college students under the age of 30 and asked them how old they were and how old they perceived themselves to be. The results are shown in the table.

CHRONOLOGICAL AGE	PERCEIVED AGE
20	22
19	21
25	30
22	25
26	22
19	19
18	20
20	18
20	21
21	21

a. Do the sample data support the survey's findings with respect to the perceptions of men and women under 30? Test using $\alpha = .10$.

b. What assumption must hold in order for the procedure you used in part **a** to be valid?

8.90 The Occupational Safety and Health Act of 1970 (OSHA) allows issuance of engineering standards to assure safe workplaces for all Americans. In 1975, the standards for exposure to arsenic in smelters, herbicide production facilities, and other places where arsenic is used were reviewed, and the previous maximum allowable level of .5 milligram per cubic meter of air was reduced to .004. Suppose smelters at two plants are being investigated to determine whether they are meeting OSHA standards. Two analyses of the air are made at each plant, and the results (in milligrams per cubic meter of air) are shown in the table.

PLANT 1		PLANT 2	
Observation	Arsenic level	Observation	Arsenic level
1	.01	1	.05
2	.005	2	.09

a. Do the data provide sufficient evidence to indicate that plant 1 fails to meet the OSHA standard? Specify the null and alternative hypotheses. Then conduct the test using $\alpha = .05$. Interpret your results.

b. Repeat the instructions of part **a** for plant 2.

c. Find the p-values for the tests in parts **a** and **b** and interpret them.

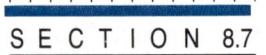

SECTION 8.7

LARGE-SAMPLE INFERENCES ABOUT A BINOMIAL PROBABILITY

Many market studies are conducted by companies with the objective of determining the fraction of buyers of a particular product who prefer the company's brand. For example, a tobacco company may conduct a market study by sampling and interviewing 1,000 smokers to determine their brand preference. The objective of the survey is to estimate the proportion of all smokers who smoke the company's brand. The number, x, of the 1,000 sampled who smoke the company's brand is a binomial random variable (see Section 5.4 for a description of the binomial experiment). The probability, p, that a smoker prefers the company's brand is the parameter to be estimated. That is, we want to estimate the proportion p of all smokers who prefer the company's brand.

How would you estimate the probability, p, of success in a binomial experiment? One logical answer is to use the proportion of successes in the sample. That is, we can estimate p by calculating

$$\hat{p} \text{ (read "}p\text{ hat")} = \frac{\text{Number of successes in the sample}}{\text{Number of trials}} = \frac{x}{n}$$

Thus, if 313 of the 1,000 smokers were found to smoke the company's brand, we would estimate the proportion p of all smokers who prefer the brand to be

$$\hat{p} = \frac{x}{n} = \frac{313}{1,000} = .313$$

To determine the reliability of the estimator $\hat{p}$, we need to know its sampling distribution. That is, if we were to draw samples of 1,000 smokers over and over again, each time calculating a new estimate $\hat{p}$, what would be the frequency distribution of all the $\hat{p}$ values? The answer lies in viewing $\hat{p}$ as the average or mean number of successes per trial over the n trials. Thus, if each success in the sample is assigned a value equal to 1 and each failure is assigned a 0, then the sum of all n sample observations is x, the total number of successes; and $\hat{p} = x/n$ is the average or mean number of successes per trial in the n trials. The Central Limit Theorem tells us that the relative frequency distribution of the sample mean for any population is approximately normal for sufficiently large samples. We demonstrated the properties of $\hat{p}$ in Example 7.7. The (repeated) sampling distribution of $\hat{p}$ has the characteristics indicated in Figure 8.21 and listed in the next box.

FIGURE 8.21

Sampling Distribution of $\hat{p}$

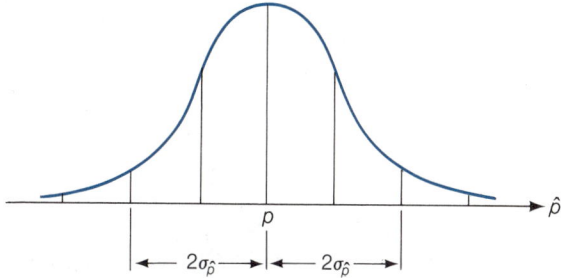

SAMPLING DISTRIBUTION OF $\hat{p}$

1. The mean of the sampling distribution of $\hat{p}$ is p; i.e., $\hat{p}$ is an unbiased estimator of p.
2. The standard deviation of the sampling distribution of $\hat{p}$ is $\sqrt{pq/n}$; i.e., $\sigma_{\hat{p}} = \sqrt{pq/n}$, where $q = 1 - p$.
3. For large samples, the sampling distribution of $\hat{p}$ is approximately normal. A sample size will be considered large if the interval $\hat{p} \pm 3\sigma_{\hat{p}}$ does not include 0 or 1. [Note: This requirement is almost equivalent to that given in Section 6.5 for approximating a binomial distribution with a normal one. The main difference is that we assumed p to be known in Section 6.5; now we are trying to make inferences about an unknown p, so we use $\hat{p}$ to estimate p in checking the adequacy of the normal approximation.]

The fact that $\hat{p}$ is a "sample mean fraction of successes" allows us to form confidence intervals and test hypotheses about p in a manner that is completely analogous to that used for large-sample inferences about μ:

LARGE-SAMPLE CONFIDENCE INTERVAL FOR p

$$\hat{p} \pm z_{\alpha/2}\sigma_{\hat{p}} = \hat{p} \pm z_{\alpha/2}\sqrt{pq/n} \approx \hat{p} \pm z_{\alpha/2}\sqrt{\hat{p}\hat{q}/n}$$

where $\hat{p} = \dfrac{x}{n}$ and $\hat{q} = 1 - \hat{p}$

[Note: When n is large, we can use $\hat{p}$ to approximate the value of p in the formula for $\sigma_{\hat{p}}$.]

Thus, if 313 of 1,000 smokers smoke the company's brand, a 95% confidence interval for the proportion of all smokers who prefer the company's brand is

$$\hat{p} \pm z_{\alpha/2}\sigma_{\hat{p}} = .313 \pm 1.96\sqrt{pq/1,000}$$

where $q = 1 - p$. Just as we needed an approximator for σ in calculating a large-sample confidence interval for μ, we now need an approximation for p. As Table 8.5 shows, the approximation for p need not be especially accurate, because the value of pq needed for the confidence interval is relatively insensitive to changes in p. Therefore, we can use $\hat{p}$ to approximate p. Keeping in mind that $\hat{q} = 1 - \hat{p}$, we substitute these values into the formula for the confidence interval:

$$\hat{p} \pm 1.96\sqrt{pq/1,000} \approx \hat{p} \pm 1.96\sqrt{\hat{p}\hat{q}/1,000}$$
$$= .313 \pm 1.96\sqrt{(.313)(.687)/1,000} = .313 \pm .029$$
$$= (.284, .342)$$

TABLE 8.5

Values of pq for Several Different p Values

p	pq
.5	.25
.6 or .4	.24
.7 or .3	.21
.8 or .2	.16
.9 or .1	.09

The company can be 95% confident that the interval from 28.4% to 34.2% contains the true percentage of all smokers who prefer its brand. That is, in repeated construction of confidence intervals, approximately 95% of all samples would produce confidence intervals that enclose p. Note that the guidelines for interpreting a confidence interval about μ also apply to interpreting a confidence interval for p, because p is the "population mean fraction of successes" in a binomial experiment.

Tests of hypotheses concerning p are also analogous to those for population means (large samples).

LARGE-SAMPLE TEST OF HYPOTHESIS ABOUT p

ONE-TAILED TEST	TWO-TAILED TEST
H_0: $p = p_0$	H_0: $p = p_0$
(p_0 = hypothesized value of p)	H_a: $p \neq p_0$
H_a: $p < p_0$	
(or H_a: $p > p_0$)	

Test statistic: $z = \dfrac{\hat{p} - p_0}{\sigma_{\hat{p}}}$ *Test statistic:* $z = \dfrac{\hat{p} - p_0}{\sigma_{\hat{p}}}$

where $\sigma_{\hat{p}} = \sqrt{[p_0(1 - p_0)]/n}$, where, as usual, we assume H_0 is true until convinced otherwise

Rejection region: $z < -z_{\alpha}$ *Rejection region:* $z < -z_{\alpha/2}$
        (or $z > z_{\alpha}$ or $z > z_{\alpha/2}$
        when H_a: $p > p_0$)

Assumption: The experiment is binomial, and the sample size is large enough that the interval $p_0 \pm 3\sigma_{\hat{p}}$ does not include 0 or 1.

EXAMPLE 8.10

The reputations (and hence, sales) of many businesses can be severely damaged by shipments of manufactured items that contain an unusually large percentage of defectives. For example, a manufacturer of alkaline batteries may want to be reasonably certain that less than 5% of the batteries are defective. Suppose 300 batteries are randomly selected from a very large shipment, each is tested, and 10 defective batteries are found. Does this provide sufficient evidence for the manufacturer to conclude that the fraction defective in the entire shipment is less than .05? Use $\alpha = .01$.

SOLUTION

Before conducting the test of hypothesis, we check to determine whether the sample size is large enough to use the normal approximation for the sampling distribution of $\hat{p}$. The criterion is tested by the interval

$$p_0 \pm 3\sigma_{\hat{p}} = p_0 \pm 3\sqrt{\frac{p_0 q_0}{n}} = .05 \pm 3\sqrt{\frac{(.05)(.95)}{300}}$$

$$= .05 \pm .04 \quad \text{or} \quad (.01, .09)$$

Since the interval lies within the interval $(0, 1)$, the normal approximation will be adequate.

The objective of the sampling is to determine whether there is sufficient evidence to indicate that p is less than .05. Consequently, we will test the null hypothesis that $p = .05$ against the alternative hypothesis that $p < .05$. The elements of the test are

H_0: $p = .05$

H_a: $p < .05$

Test statistic: $z = \dfrac{\hat{p} - .05}{\sigma_{\hat{p}}}$

Rejection region: $z < -z_{.01} = -2.33$ (see Figure 8.22)

FIGURE 8.22

Rejection Region for Example 8.10

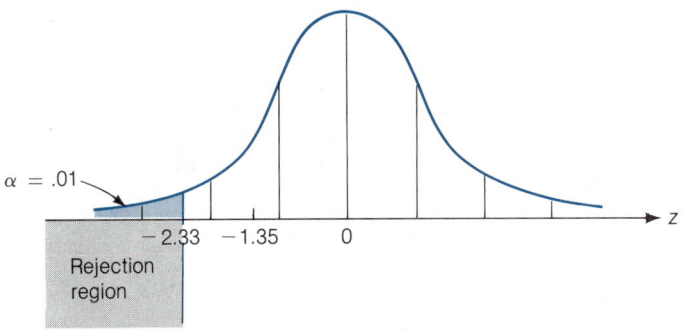

We now calculate the test statistic:

$$z = \frac{\hat{p} - .05}{\sigma_{\hat{p}}} = \frac{(10/300) - .05}{\sqrt{p_0 q_0/n}} = \frac{.033 - .05}{\sqrt{p_0 q_0/300}}$$

Notice that we use p_0 to calculate $\sigma_{\hat{p}}$ because the test statistic is computed on the assumption that the null hypothesis is true; i.e., $p = p_0$. Therefore, substituting the values for p_0 and q_0 into the z statistic, we obtain

$$z \approx \frac{-.017}{\sqrt{(.05)(.95)/300}} = \frac{-.017}{.0126} = -1.35$$

As shown in Figure 8.22, the calculated z value does not fall in the rejection region. Therefore, based on this test, there is insufficient evidence at the .01 level of significance to indicate that the shipment contains fewer than 5% defective batteries.

■

EXAMPLE 8.11

In Example 8.10 we found that we did not have sufficient evidence, at the $\alpha = .01$ level of significance, to indicate that the fraction defective, p, of alkaline batteries was less than $p = .05$. How strong was the weight of evidence favoring the alternative hypothesis (H_a: $p < .05$)? Find the observed significance level for the test.

SOLUTION

The computed value of the test statistic was $z = -1.35$. Therefore, for this one-tailed test,

Observed significance level $= P(z \le -1.35)$

This lower-tail area is shown in Figure 8.23. The area A between $z = 0$ and $z = 1.35$ is given in Table IV of Appendix B as .4115. Therefore, the observed significance level is $.5 - .4115 = .0885$. Note that this probability is quite small. We may not have rejected H_0: $p = .05$ for $\alpha = .01$, but the probability of observing a z value as small as or smaller than -1.35 is only .0885. Therefore, we would reject H_0 if we choose $\alpha = .10$ (since the observed significance level is less than .10), but we would not reject H_0 (the conclusion of Example 8.10) if we choose $\alpha = .05$ or $\alpha = .01$.

FIGURE 8.23

The Observed Significance Level for Example 8.10

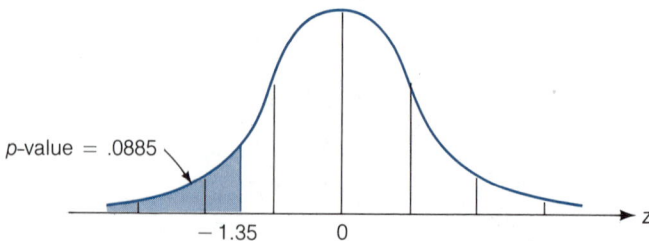

p-value $= .0885$

■

Small-sample estimators and test procedures are also available for p. These are omitted from our discussion because most surveys conducted in business use samples that are large enough to employ the large-sample estimators and tests presented in this section.

CASE STUDY 8.4

STATISTICAL QUALITY
CONTROL, PART 2

In complicated assembly operations (such as railway car assembly), many quality variables could be measured (e.g., strength of welds, degree of corrosion, and number of paint flaws), and in principle, each could be monitored over time using control charts, as described in Case Study 8.3. In some situations, however, an alternative, simpler procedure may be more appropriate. For example, n finished products could be randomly sampled at regular time intervals, inspected for defects, and simply classified as being defective or nondefective products. Then $\hat{p}$, the proportion of defectives in each sample, could be determined and plotted on a control chart called a ***p*-chart**, as illustrated in Figure 8.24. In this way, the proportion of defective products produced and, therefore, product quality and the current capability of the production process could be monitored over time (Wetherill, 1977).

FIGURE 8.24

p-Chart for Proportion
Defective

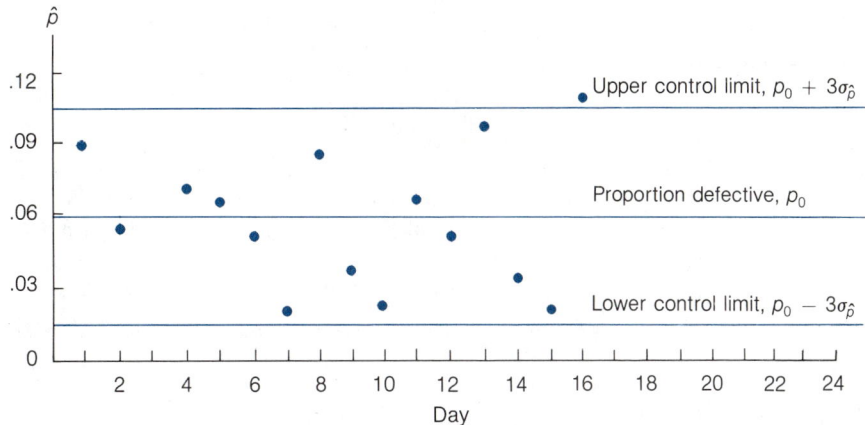

In order to construct the control chart shown in Figure 8.24, it is necessary to know (i.e., have a good estimate for) p_0, the proportion of defectives produced when the process is operating "normally" (i.e., in control). Then, assuming n is large enough to use the normal distribution to approximate the sampling distribution of $\hat{p}$, the control limits are located $3\sigma_{\hat{p}}$ above and below p_0. If a value of $\hat{p}$ falls above the upper limit, it is a signal that the process is turning out more defectives than usual and may be out of control. But what is the significance of the lower control limit? Why should the manufacturer be concerned if fewer defectives than usual are being produced? Two important reasons follow (Caplen, 1970):

1. It may be an indication that the inspector is not performing his or her job carefully and may be missing defectives that normally would be identified. As a result, defective products may be sold to customers.
2. If the inspector is performing adequately, it may be an indication that the production process really is temporarily better. If so, the low $\hat{p}$ value signals management to begin a search for the causes for the improvement. If found, it may be possible to improve the production process permanently.

As in Case Study 8.3, this control chart procedure for monitoring the proportion of defective products is nothing more than a two-tailed hypothesis test with the following elements:

1. H_0: Process is in control, $p = p_0$
   H_a: Process is out of control, $p \neq p_0$
2. The test statistic is $\hat{p}$.
3. The rejection region is defined by the upper and lower control limits.
4. Since the control limits are located at $p_0 \pm 3\sigma_{\hat{p}}$, the probability of committing a Type I error is approximately $\alpha = .0026$. (Why?)

| | | | | | | | | | | | | |

EXERCISES 8.91–8.108   ## LEARNING THE MECHANICS

8.91 Explain the meaning of the phrase "$\hat{p}$ is an unbiased estimator of p."

8.92 A random sample of size $n = 400$ yielded $\hat{p} = .42$.
   a. Is the sample size large enough to use the methods of this section to construct a confidence interval for p? Explain.
   b. Construct a 95% confidence interval for p.

8.93 A random sample of size $n = 144$ yielded $\hat{p} = .76$.
   a. Is the sample size large enough to use the methods of this section to construct a confidence interval for p? Explain.
   b. Construct a 90% confidence interval for p.

8.94 For the binomial sample information summarized in each part, indicate whether the sample size is large enough to use the methods of this chapter to construct a confidence interval for p.
   a. $n = 500,\quad \hat{p} = .05$
   b. $n = 100,\quad \hat{p} = .05$
   c. $n = 10,\quad \hat{p} = .5$
   d. $n = 10,\quad \hat{p} = .3$

8.95 Suppose a random sample of 100 observations from a binomial population gave a value of $\hat{p} = .63$ and you wish to test the null hypothesis that the population parameter p is equal to .70 against the alternative hypothesis that $p < .70$.
   a. Noting that $\hat{p} = .63$, what does your intuition tell you? Does the value of $\hat{p}$ appear to contradict the null hypothesis?
   b. Use the large-sample z test to test H_0: $p = .70$ against the alternative hypothesis H_a: $p < .70$ at the $\alpha = .05$ significance level. How do the test results compare with your intuitive decision from part **a**?
   c. Find and interpret the p-value for the hypothesis test you conducted in part **b**.

8.96 Suppose the sample in Exercise 8.95 has produced $\hat{p} = .83$ and we wish to test H_0: $p = .9$ against the alternative hypothesis H_a: $p < .9$.
   a. Calculate the value of the z statistic for this test.
   b. Note that $\hat{p} - p_0 = .83 - .9 = -.07$ is the same as for Exercise 8.95. Considering this, why is the absolute value of z for this exercise larger than that calculated in Exercise 8.95?
   c. Complete the test using $\alpha = .05$ and interpret the results.
   d. Find the observed significance level for your hypothesis test, and interpret its value.

8.97 A random sample of 50 consumers taste-tested a new snack food. Their responses were coded (0: do not like; 1: like; 2: indifferent) and listed below:

1	0	0	1	2	0	1	1	0	0
0	1	0	2	0	2	2	0	0	1
1	0	0	0	0	1	0	2	0	0
0	1	0	0	1	0	0	1	0	1
0	2	0	0	1	1	0	0	0	1

a. Use an 80% confidence interval to estimate the proportion of consumers who like the snack food.
b. Provide a statistical interpretation for the confidence interval constructed in part a.

8.98 Refer to Exercise 8.97.
a. Test H_0: $p = .5$ against H_a: $p > .5$, where p is the proportion of customers who do not like the snack food. Use $\alpha = .10$.
b. Report the observed significance level of your test.

8.99 Following a year-long investigation begun in 1985, a congressional subcommittee concluded that ineptitude and fraud are present at unacceptable levels within the real-estate appraisal profession. The report indicates that inflated home appraisals are responsible, in part, for many defaulted home mortgages. The subcommittee reported that at least 10%–15% of the $1.3 billion in losses experienced by private mortgage insurers during 1984–1985 was directly attributable to faulty and fraudulent appraisals. One insurer sampled 300 defaulted home mortgages and found 120 of them involved defective appraisals (Harney, 1986).
a. Assume that the 300 defaulted mortgages were randomly sampled from all defaulted mortgages during 1985. Use a 90% confidence interval to estimate the proportion of all defaulted home mortgages in 1985 that involved a defective home appraisal.
b. How wide is the confidence interval you constructed in part a? Would an 80% confidence interval be wider or narrower than your interval of part a? Explain.

8.100 According to a spokesperson for General Mills, the company's "cents-off" coupon offers are designed to get people to buy its products and its refund offers (money returned with proof of repeated purchases) are designed to encourage people to continue buying its products. In a national survey conducted by the Nielsen Clearing House in 1975, 65% of the respondents indicated that they used cents-off coupons when grocery shopping. In a 1980 survey, the Nielsen organization found that 76% of those surveyed used cents-off coupons (*Minneapolis Star*, Nov. 29, 1981, p. 7F). Suppose the 1980 survey consisted of a random sample of 100 shoppers, of whom 76 indicated that they used cents-off coupons.
a. Is the sample size large enough to use the inferential procedures presented in this section? Explain.
b. Does the 1980 sample provide sufficient evidence that the percentage of shoppers using cents-off coupons exceeds 65%? Test using $\alpha = .05$.
c. Find the observed significance level for the test you conducted in part b, and interpret its value.

8.101 In February and March 1985, the Gallup Organization conducted telephone interviews with a random sample of 258 owners and managers of small U.S. businesses (i.e., firms with 20 or more employees, but with less than $50 million in sales). Forty percent of those interviewed were under 45 years of age and 30% have worked at four or more companies (Graham, 1985).

a. Describe the population of interest to the Gallup Organization.

b. Is the sample size large enough to construct a confidence interval for the proportion of owners and managers who are under 45 years of age? Explain.

c. Construct a 95% confidence interval for the proportion of interest in part b.

d. If a 95% confidence interval were used to estimate the proportion of owners and managers who have worked at four or more companies, would the interval be wider, narrower, or the same width as the interval estimate of part c? Explain.

8.102 Standard Oil of California used a sample survey to determine whether people's attitudes toward Standard's corporate image tended to be favorable or unfavorable. The sample results indicated that, for the first time in 30 years, more people had unfavorable than favorable attitudes. Standard Oil responded by initiating an institutional advertising campaign to help improve its image (*Marketing News*, 1976). Fearful that its corporate image had also suffered in recent years, another large oil corporation conducted a similar survey with the following results:

Unfavorable opinions   3,465

Favorable opinions   2,502

No opinions   821

a. Examine the data. Based on your intuition, does it appear that more than 50% of the general public possess an unfavorable attitude toward the company?

b. Do the sample data support the hypothesis that more than 50% of the general public hold unfavorable opinions about the company? Test at $\alpha = .05$.

c. Construct a 90% confidence interval for the proportion of individuals with no opinion.

d. List any assumptions that you made in answering parts b and c.

8.103 Refer to Exercise 8.102. Find the observed significance level for the test you conducted in part b and interpret its value.

8.104 Marketing research has been defined by the American Marketing Association as the "systematic gathering, recording, and analyzing of data about problems relating to the marketing of goods and services" (American Marketing Association, 1961). Companies may have their own marketing research departments, or they may contract the services of a marketing research firm. The marketing research department of a large West Coast manufacturer of facial tissue paper was charged with the responsibility of determining consumer preferences regarding the softness of its newly developed product (brand A) relative to the industry leader (brand B). A random sample of 205 consumers was selected and asked to rank the softness of brands A and B. In the results, 119 ranked brand A as softer, and 86 ranked brand B as softer.

a. Do the data indicate that brand A is perceived by consumers as being superior to brand B in terms of softness? Test using $\alpha = .05$.

b. Find the *p*-value for the test and interpret its value.

8.105 Shoplifting is an escalating problem for retailers. According to *U.S. News and World Report* (Feb. 21, 1977), one New York City store randomly selected 500 shoppers and observed them while they were in the store. Two in twenty-five were seen stealing. How accurate is this estimate? To help you answer this question, construct a 95% confidence interval for *p*, the proportion of all the store's customers who are shoplifters.

8.106 Interested in how well a new computer billing operation is working, a company statistician samples 400 bills that are ready for mailing and checks them for errors. Twenty-four are found to contain at least one error. Find a 90% confidence interval for p, the true proportion of bills that contain errors.

8.107 A producer of frozen orange juice claims that 20% of all orange juice drinkers prefer its product. To test the validity of this claim, a competitor samples 200 orange juice drinkers and finds that only 33 prefer the producer's brand.
  a. Does the sample evidence indicate that the proportion of orange juice drinkers who prefer the producer's brand is significantly less than .20? Test at $\alpha = .10$.
  b. Find the p-value of the hypothesis test you conducted in part **a**. Interpret its value.

8.108 The following is a very useful result concerning the mean and variance of some sampling distributions: Suppose c is a constant and x is a statistic with mean μ and variance σ^2. Then it can be shown (the proof is omitted here) that the mean and variance of cx are

$$E(cx) = cE(x) = c\mu \qquad \sigma_{cx}^2 = c^2\sigma^2$$

Application: If you draw a random sample of n people from a large population of consumers and x is the number in the sample who favor some proposal (favor a particular product, etc.), then x is a binomial random variable with mean $\mu = np$ and variance $\sigma^2 = npq$ (from Chapter 5). The proportion of people in the sample who favor the proposal, x/n, is used to estimate the population proportion p.
  a. Use the information above to show that the sampling distribution of the sample proportion, x/n, has a mean equal to p and a standard deviation equal to $\sqrt{p(1 - p)/n}$.
  b. Let each person in the population who favors the proposal be represented by a 1 and each person who does not by a 0. Then the entire population of consumers can be viewed as a collection of 1's and 0's; x will equal the sum of the 1's and 0's in the sample of n, and x/n will be the sample average. What will be the approximate form of the sampling distribution of the sample proportion when the sample size n is large? Why?
  c. Suppose you select a random sample of 1,600 consumers from a large population that (unknown to you) contains 20% ($p = .2$) who favor the proposal. What is the probability that your sample proportion will differ from the population proportion ($p = .2$) by more than .01?

S E C T I O N   8.8

DETERMINING THE SAMPLE SIZE NECESSARY TO MAKE INFERENCES ABOUT A BINOMIAL PROBABILITY

We showed in Section 8.2 that experiments can be designed to estimate a population mean μ with a specified degree of reliability. An analogous situation exists for estimating a binomial probability p. For example, in Section 8.7 a tobacco company used a sample of 1,000 smokers to calculate a 95% confidence interval for the proportion of smokers who preferred its brand, obtaining the interval .284 to .342. Note that the total width of the interval is about .06 (.342 − .284 ≈ .06). Suppose the company wants to estimate its market share more precisely, say with a 95% confidence interval having a width of .03.

The company wants a confidence interval width W of .03. This corresponds to a half-width, or bound B on the estimate of p, of $B = W/2 = .015$. The sample size n to generate such an interval is found by solving the equation

$$z_{\alpha/2}\sigma_{\hat{p}} = B \quad \text{or} \quad z_{\alpha/2}\sqrt{\frac{pq}{n}} = .015 \quad \text{(see Figure 8.25)}$$

FIGURE 8.25

Specifying the Total Width
W (or Bound B) of a
Confidence Interval for a
Binomial Probability p

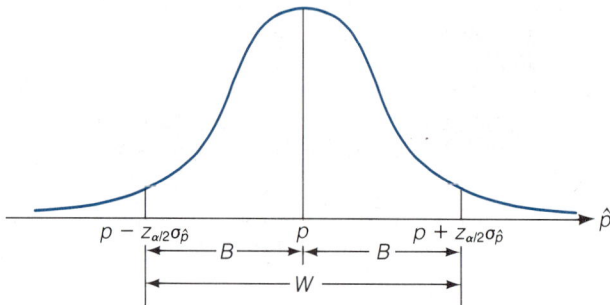

Since a 95% confidence interval is desired, the appropriate z value is $z_{\alpha/2} = z_{.025} = 1.96$. We must approximate the value of the product pq before we can solve the equation for n. As shown in Table 8.5, the closer the value of p and q to .5, the larger the product pq. Thus, to find a conservatively large sample size that will generate a confidence interval with the specified reliability, we generally choose an approximation of p close to .5. In the case of the tobacco company, however, we have an initial sample estimate of $\hat{p} = .313$. A conservatively large estimate of pq can therefore be obtained by using $p = .35$. We now substitute into the equation and solve for n:

$$1.96\sqrt{\frac{(.35)(.65)}{n}} = .015$$

$$n = \frac{(1.96)^2(.35)(.65)}{(.015)^2}$$

$$n = 3,884.28 \approx 3,885$$

The company must sample about 3,885 smokers to estimate the percentage who prefer its brand with a 95% confidence interval of width .03.

The procedure for finding the sample size necessary to estimate a binomial probability p to within a given bound B or with a total interval width W is given in the box at the top of the next page.

EXAMPLE 8.12

Since the deregulation of the telephonic communications industry, many firms have begun manufacturing telephones. Suppose one large manufacturer that entered the market quickly has an initial problem with excessive customer complaints and consequent returns of the phones for repair or replacement. To deal with the problem, the manufacturer wants to put a quality control program in place. How many telephones should be sampled and checked in order to estimate the fraction defective p to within .01 with 90% confidence?

SAMPLE SIZE DETERMINATION FOR 100(1 − α)% CONFIDENCE INTERVAL FOR p

In order to estimate a binomial probability p to within a bound B or, equivalently, with a confidence interval of total width W with $100(1 - \alpha)\%$ confidence, the required sample size is found as follows:

$$z_{\alpha/2}\sqrt{\frac{pq}{n}} = B \qquad z_{\alpha/2}\sqrt{\frac{pq}{n}} = \frac{W}{2}$$

The solution can be written in terms of either B or W:

$$n = \frac{(z_{\alpha/2})^2(pq)}{B^2} \qquad n = \frac{4(z_{\alpha/2})^2(pq)}{W^2}$$

The value of the product pq is usually unknown. It can be estimated by using the sample fraction of successes, $\hat{p}$, from a prior sample. Remember that the value of pq increases as p approaches .5, so that you can obtain conservatively large values of n by approximating p by .5, or values close to .5. In any case, you should round the value of n you obtain *upward* to ensure that the sample size will be sufficient to achieve the specified reliability.

SOLUTION

In order to estimate p to within a bound of .01, we set the half-width of the confidence interval equal to $B = .01$, as shown in Figure 8.26.

FIGURE 8.26

Specified Reliability for Estimate of Fraction Defective in Example 8.12

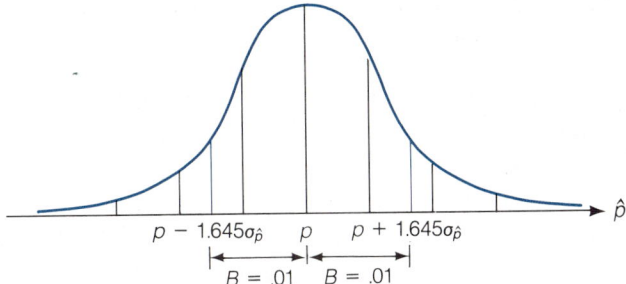

The equation for the sample size n requires an estimate of the product pq. We could most conservatively estimate $pq = .25$ (i.e., use $p = .5$), but this may be overly conservative when estimating a fraction defective (as consumers, we hope so!). A value of .1, corresponding to 10% defective, will probably be conservatively large for this application. The solution is therefore

$$n = \frac{(z_{\alpha/2})^2(pq)}{B^2} \approx \frac{(1.645)^2(.1)(.9)}{(.01)^2}$$

$$= 2{,}435.4 \approx 2{,}436$$

Thus, the manufacturer should sample 2,436 telephones in order to estimate the fraction defective p to within .01 with 90% confidence. Remember that this answer depends on our approximation for pq, where we used .09. If the fraction defective is closer to .05 than .10, we can use a sample of 1,286 telephones (check this) to estimate p to within .01 with 90% confidence. ∎

The cost of sampling will play an important role in the final determination of the sample size to be selected to estimate a binomial probability p. Although more complex formulas can be derived to balance the reliability and cost considerations, we will solve for the necessary sample size and note that the sampling budget may be a limiting factor. Consult the references for a more complete treatment of this problem.

EXERCISES
8.109–8.118

LEARNING THE MECHANICS

8.109 If nothing is known about p, .5 can be substituted for p in the sample size formula. But when this is done, the resulting sample size may be larger than needed. Under what circumstances will using $p = .5$ in the sample size formula yield a sample size larger than needed to construct a confidence interval for p with a specified bound and a specified confidence level?

8.110 In each case, find the approximate sample size necessary to estimate a binomial proportion p correct to within .02 with probability equal to .90.
   a. Assume you know p is near .8.
   b. Assume you have no knowledge of the value of p, but you want to be certain that your sample is large enough to achieve the specified accuracy for the estimate.

8.111 In each case, find the approximate sample size required to construct a 95% confidence interval for p that has width .12.
   a. Assume p is near .3.
   b. Assume you have no prior knowledge about p.

8.112 The following is a 90% confidence interval for p: (.26, .54). How large was the sample used to construct this interval?

APPLYING THE CONCEPTS

8.113 According to estimates made by the General Accounting Office, the Internal Revenue Service (IRS) answered 18.3 million telephone inquiries during the 1986 tax season and 17% of the IRS offices provided answers that were wrong. These estimates were based on data collected from sample calls to numerous IRS offices (Roper, 1986). How many IRS offices should be randomly selected and contacted in order to estimate the proportion of IRS offices that fail to correctly answer questions about gift taxes with a 90% confidence interval of width .06?

8.114 While corporate executives are probably not as highly stressed as air traffic controllers or inner-city police, research has indicated that they are among the more highly pressured work groups. In order to estimate p, the proportion of managers who perceive themselves to be frequently under stress, Hall and Savery (1986) sampled 532 managers in Western Australian corporations. One hundred ninety of these managers fell into the "high stress" group. Assume that random sampling was used in this study. Was the sample size large enough to estimate p to within .03 with 95% confidence? Explain.

8.115 A marketing research organization wishes to estimate the proportion of television viewers who watch a particular prime-time situation comedy on May 24. The proportion is expected to be approximately .30. At a minimum, how many viewers should be randomly selected to ensure that a 95% confidence interval for the true proportion of viewers will have a width of .01 or less?

8.116 Before a bill to increase federal price supports for farmers comes before the U.S. Congress, a congressman would like to know how nonfarmers feel about the issue. Approximately how many nonfarmers should the congressman survey in order to estimate the true proportion favoring this bill to within .05 with probability equal to .90?

8.117 Suppose you are a retailer and you want to estimate the proportion of your customers who are shoplifters. You decide to select a random sample of shoppers and check closely to determine whether they steal any merchandise while in the store. Suppose experience suggests that the percentage of customers who are shoplifters is near 5%. How many customers should you include in your sample if you want to estimate the proportion of shoplifters in your store correct to within .02 with probability equal to .90?

8.118 Some quality control testing involves destructive sampling, i.e., the test to determine whether the item is defective destroys the product. This type of sampling is generally expensive, and high costs often prohibit large sample sizes. For example, suppose the National Highway Safety Administration (NHSA) wishes to determine the proportion of new tires that will fail when subjected to hard braking at a speed of 60 mph. NHSA can obtain the tires for $25 (wholesale price) each. Suppose the budget for the experiment is $10,000, and NHSA wishes to estimate the percentage that will fail to within .02 with 95% confidence. If the entire $10,000 can be spent on tires (i.e., ignoring other costs), and assuming that the true fraction that will fail is approximately .05, can NHSA attain its goal while staying within the budget? Explain.

SUMMARY

The objective of statistics is to make inferences about a population based on information in a sample. In this chapter, we have presented several methods for accomplishing this objective.

The inference-making techniques we discussed are **estimation** and **hypothesis testing**. Estimation of a population parameter is accomplished by using an interval estimate with a probability of coverage (**confidence coefficient**) that is fixed by the researcher at a high level (usually .90, .95, or .99). On the other hand, when a specific **alternative (research) hypothesis** about a parameter is tested, the probability α of falsely rejecting the **null hypothesis** and accepting the alternative hypothesis is chosen to be small. Thus, we try to minimize the chance of error in both of these inference-making procedures.

One of the most important parameters about which inferences are made is the population mean, μ. The sample mean, $\bar{x}$, is used for making the inference, but the inferential procedure depends on the **sample size**. When the sample size is large (we have specified $n > 30$ as large), the **standard normal z statistic** is used. The *t* **statistic** is used when σ is unknown and a small sample is drawn from a normally (or approximately normally) distributed population.

| | | | | | | | | | | | | |

**SUPPLEMENTARY
EXERCISES
8.119–8.139**

[Note: Starred (*) exercises refer to the optional section.]

8.119 Let t_0 be a particular value of t. Use Table VI of Appendix B to find the values such that the following statements are true:
a. $P(t \leq t_0) = .05$ where df $= 20$ **b.** $P(t \geq t_0) = .005$ where df $= 9$
c. $P(t \leq -t_0$ or $t \geq t_0) = .10$ where df $= 8$
d. $P(t \leq -t_0$ or $t \geq t_0) = .01$ where df $= 17$

8.120 This exercise is designed to give you practice computing p-values. Find approximate values for each of the following:
a. For df $= 10$, find $P(t \geq 1.95)$.   **b.** For df $= 25$, find $P(t \leq -2.60)$.
c. For df $= 15$, find $P(t \leq -1.45$ or $t \geq 1.45)$.
d. For df $= 7$, find $P(t \leq -3.33$ or $t \geq 3.33)$.

8.121 If the rejection of the null hypothesis of a particular test would cause your firm to go out of business, would you want α to be small or large? Explain.

8.122 A large New York City bank is interested in estimating (1) the proportion of weeks in which it processes more than 100,000 checks and (2) the average number of checks it processes per week. The bank maintains records of x, the number of checks processed each week. Suppose the bank records the number of checks, x, processed per week for 50 weeks randomly sampled from among the past 6 years. Define each of the following *in the context of the problem:*
a. $\bar{x}$ **b.** $\hat{p}$ **c.** σ_x **d.** μ_x **e.** n **f.** $\sigma_{\bar{x}}$ **g.** p **h.** s_x

8.123 In 1978, only 110 U.S. companies offered any form of child-care assistance to their employees. By 1985, 2,500 companies offered day-care advice, day-care programs, or financial assistance (*Minneapolis Star and Tribune*, Nov. 30, 1986, p. 11A). A large corporation randomly sampled 60 of its employees with children under 5 years of age and asked them whether they would prefer to have the company pay for day-care at a program of the parents' choosing or to have the company supply free on-site day care. The responses were coded (1: prefer choice; 2: prefer on-site program) and appear below:

1	1	2	1	2	1	1	2	2	2
2	1	2	2	2	1	2	1	1	2
1	2	1	2	2	2	2	2	1	2
1	1	2	2	2	2	2	2	2	1
1	2	1	1	1	2	2	2	2	1
2	2	1	2	2	2	1	2	2	2

a. Prior to questioning the sample of employees, management believed that the employees would prefer a program of their own choosing. Do the data support their prior belief? Conduct the appropriate hypothesis test. Use the observed significance level of the test to answer the question.
b. Construct a 90% confidence interval for the true proportion of employees with children under 5 years of age who prefer an on-site program. Interpret the interval, and explain how the interpretation relates to the decision you reached in the test of hypothesis in part **a.**

8.124 A firm's president, vice-presidents, department managers, and others use financial data generated by the firm's accounting system to help them make decisions regarding such things as pricing, budgeting, and plant expansion. To provide reasonable certainty that the system provides reliable data, internal auditors periodically perform various checks of the system (Taylor and Glezen, 1979). Suppose an internal auditor is interested in determining the proportion of sales invoices in a population of 5,000 sales invoices for which the "total sales" figure is in error. She plans to estimate the true proportion of invoices in error based on a random sample of size 100.

   a. Assume that the population of invoices is numbered from 1 to 5,000 and that every invoice ending with a 0 is in error (i.e., 10% are in error). Use the random number table (Table I in Appendix B) to draw a random sample of 100 invoices from the population of 5,000 invoices. For example, random number 456 stands for invoice number 456. List the invoice numbers in your sample and indicate which of your sampled invoices are in error (i.e., those ending in a 0).

   b. Use the results of your sample of part **a** to construct a 90% confidence interval for the true proportion of invoices in error.

   c. Recall that the true population proportion of invoices in error is equal to .1. Compare the true proportion with the estimate of the true proportion you developed in part **b**. Does your confidence interval include the true proportion?

8.125 A company is interested in estimating μ, the mean number of days of sick leave taken by all its employees. The firm's statistician selects at random 100 personnel files and notes the number of sick days taken by each employee. The following sample statistics are computed:

$$\bar{x} = 12.2 \text{ days} \qquad s = 10 \text{ days}$$

   a. Estimate μ using a 90% confidence interval.

   b. How many personnel files would the statistician have to select in order to estimate μ to within 2 days with 99% confidence?

   c. Do the data support the alternative hypothesis that μ, the mean number of sick days taken by the employees, is greater than 10.9 days? Test at $\alpha = .05$. Report the observed significance level of the test.

8.126 According to the *Minneapolis Star* (May 2, 1980, p. 11A), the federal government requires states to certify that they are enforcing the 55-miles-per-hour speed limit and that motorists are driving at that speed. A state is in jeopardy of losing millions of dollars in federal road funds if more than 60% of its vehicles on 55-mph highways are exceeding the speed limit. The Minnesota Highway Patrol conducts 70 radar surveys each year at a total of 50 sites to estimate the proportion p of vehicles exceeding 55 miles per hour. Each sample survey involves at least 400 vehicles.

   a. How large a sample should be selected at site #42 on Interstate 35W to estimate p to within 3% with 90% confidence? Last year approximately 60% of all vehicles exceeded 55 mph.

   b. The highway patrol also estimates μ, the average speed of vehicles on state highways. Accordingly, it wants to know whether the sample size determined in part **a** is large enough to also estimate μ to within .25 mile per hour with 90% confidence. Assume that the standard deviation of vehicle speeds is approximately 2 miles per hour. How large a sample should be taken at site #42 to estimate μ with the desired reliability?

8.127 The EPA sets a limit of 5 parts per million on PCB (a dangerous substance) in water. A major manufacturing firm producing PCB for electrical insulation discharges small amounts from the plant. The company management, attempting to control the amount of PCB in its discharge, has given instructions to halt production if the mean amount of PCB in the effluent exceeds 3 parts per million. A random sampling of 50 water specimens produced the following statistics: $\bar{x} = 3.1$ parts per million, $s = .5$ part per million.

    a. Do these statistics provide sufficient evidence to halt the production process? Use $\alpha = .01$.

    b. If you were the plant manager, would you want to use a large or a small value for α for the test in part **a**? Explain.

    c. Find the p-value for the test and interpret its value.

*8.128 Refer to Exercise 8.127.

    a. In the context of the problem, define a Type II error.

    b. Calculate β for the test described in part **a** of Exercise 8.127 assuming that the true mean is $\mu = 3.1$ ppm.

    c. What is the power of the test to detect the effluent's departure from the standard of 3.0 ppm when the mean is 3.1 ppm?

    d. Repeat parts **b** and **c** assuming that the true mean is 3.2 ppm. What happens to the power of the test as the plant's mean PCB departs further from the standard?

*8.129 Refer to Exercises 8.127 and 8.128.

    a. Suppose an α value of .05 is used to conduct the test. Does this change favor the manufacturer? Explain.

    b. Determine the value of β and the power for the test when $\alpha = .05$ and $\mu = 3.1$.

    c. What happens to the power of the test when α is increased?

8.130 A large mail-order company has placed an order for 5,000 electric can openers with a supplier on condition that no more than 2% of the can openers will be defective. To check the shipment, the company tests a random sample of 400 of the can openers and finds 11 are defective. Does this provide sufficient evidence to indicate that the proportion of defective can openers in the shipment exceeds 2%? Test using $\alpha = .05$.

8.131 Find and interpret the significance level for the hypothesis test in Exercise 8.130.

8.132 Refer to Exercise 8.130. Suppose the company wants to estimate the proportion, p, of defective can openers in the shipment using a 95% confidence interval of width .08. Approximately how large a sample would be required?

8.133 The Internal Revenue Service is conducting an audit of the 10,000 outlets of a large fast-food chain. It is interested in determining the average error in reported income last year for all outlets in the chain. The size of the chain precludes a census (an audit of all 10,000 outlets), so 100 outlets are randomly selected and audited. Let $x =$ Error in reported income $=$ (Actual income $-$ Reported income) for a given firm. The audits yielded the following statistics:

$$\bar{x} = \$12,522 \qquad s = \$4,000$$

    a. Construct a 95% confidence interval for the mean error in reported income per outlet.

b. What does the confidence interval from part **a** reflect regarding the chain's income-reporting behavior last year?

8.134 In the past, a chemical company produced 880 pounds of a certain type of plastic per day. Now, using a newly developed and less expensive process, the mean daily yield of plastic for the first 50 days of production was 871 pounds, the standard deviation was 21 pounds.

a. Do the data provide sufficient evidence to indicate that the mean daily yield for the new process is less than for the old procedure? (Test using $\alpha = .01$.)

b. What assumptions must you make in order to use the statistical test you employed?

8.135 Refer to Exercise 8.134. Find and interpret the p-value for the test conducted.

*8.136 Refer to Exercise 8.134. Calculate the probability β that the test fails to reject the null hypothesis that the new process has a mean daily yield of $\mu = 880$ when in fact the true mean is $\mu = 875$. What is the power of the test to determine that $\mu < 880$ when $\mu = 875$?

8.137 A market researcher wants to select one sample to estimate both μ, the average age of people living within 5 miles of a proposed shopping mall site, and p, the proportion of people within that 5-mile radius who are between 20 and 40 years of age. He wants to estimate μ with a 95% confidence interval that is no more than 6 years wide and p with a 90% confidence interval of width no greater than .1. It is known from previous studies of this population that the standard deviation of the ages in the population is 10 years, and it is believed that p is near .4. How large a sample does the researcher need to draw in order to construct confidence intervals for both μ and p that satisfy the above specifications?

8.138 A discount store claims that its steel-belted radial tires are more resistant to wear than those of a major tire company. The following experiment was performed to test this claim. On each of 40 cars, one discount tire and one rubber company tire were mounted on the rear axle. After each car was driven 8,000 miles, the tires were inspected for wear. Suppose the tires of the discount store show less wear on 32 of the cars. What would you conclude about the discount store's claim? Why?

8.139 [Note: This exercise uses material discussed in optional Section 5.5.] A survey of 2,000 Americans reported in the *Gainesville Sun* (Mar. 19, 1984) contains both good and bad news for the nation's pharmacists and physicians. The good news is that there is plenty of business. Americans have a minor physical ailment once every 3 days, on the average. The bad news is that the respondents handle 90% of the problems themselves by treating themselves with over-the-counter drugs, home remedies, or simply ignoring their ailments. The estimate of the time between ailments is based on the reported number, x, of ailments a respondent might expect to encounter in a typical 2-week period. The average for the survey of 2,000 respondents was 4.5. How accurate is this sample estimate of the mean number, μ, of ailments per person per 2-week period for the population of all adult Americans? To answer this question, find a 99% confidence interval for μ and interpret your results. [Hint: The probability distribution of the number of ailments per person in a 2-week period can be approximated by a Poisson probability distribution. For a Poisson random variable, $\sigma^2 = \mu$ and, therefore, an estimate of σ^2 is provided by the sample mean $\bar{x}$.]

ON YOUR OWN . . .

Choose a population pertinent to your major area of interest that has an unknown mean (or, if the population is binomial, that has an unknown proportion of success). For example, a marketing major may be interested in the proportion of consumers who prefer a particular product. An advertising major might want to estimate the proportion of the television viewing audience who regularly watch a particular program. An economics major may want to estimate the mean monthly expenditure of college students on food.

Define the parameter you want to estimate and conduct a **pilot study** to obtain an initial estimate of the parameter of interest and, more important, an estimate of the variability associated with the estimator. A pilot study is a small experiment (perhaps 20 to 30 observations) used to gain some information about the population of interest. The purpose is to help plan more elaborate future experiments. Based on the results of your pilot study, determine the sample size necessary to estimate the parameter to within a reasonable bound (of your choice) with a 95% confidence interval.

USING THE COMPUTER . . .

Refer to **Using the Computer** in Chapter 7. Recall the values of the "population" mean μ and standard deviation σ for the 1,000 zip code income measurements. Suppose our objective is to sample from this population and to estimate the mean μ using a 95% confidence interval.

a. Determine the sample size n_1 necessary to estimate μ to within $2,000 with 95% confidence. Then generate 100 95% confidence intervals by repeatedly drawing samples of size n_1 (with replacement) from the 1,000 measurements, and using the sample statistics to form a confidence interval. Treat σ as unknown when forming the confidence intervals. What percentage of the confidence intervals contain μ?

b. Determine the sample size n_2 necessary to estimate μ to within $500 with 95% confidence. Then generate 100 95% confidence intervals by repeatedly drawing samples of size n_2 (with replacement) from the 1,000 measurements, and using the sample statistics to form a confidence interval. Treat σ as unknown when forming the confidence intervals. What percentage of the confidence intervals contain μ?

c. Repeat parts **a** and **b** using an 80% confidence interval. Compare the results.

REFERENCES

Arshadi, N. and Lawrence, E. C. "A characteristic appraisal of top bank executives." *Journal of Retail Banking*, Vol. 5, No. 4, Winter 1983–1984, 19–25.

Caplen, R. A *Practical Approach to Quality Control*. London: Business Books, 1970. Chapter 15.

Deming, W. E. *Quality, Productivity, and Competitive Position*. Cambridge, Mass.: M.I.T. Center for Advanced Engineering Study, 1982.

Dessler, G. "Personnel tests gain in popularity," *St. Paul Pioneer Press and Dispatch*, Mar. 17, 1986, p. 12.

Dujack, S. R. "Science leaves no doubt: The polygraph lies," *Minneapolis Star and Tribune*, July 31, 1986, p. 11A.

Duncan, A. J. *Quality Control and Industrial Statistics*, 4th ed. Homewood, Ill.: Richard D. Irwin, 1974. Chapter 1.

Graham, E. "The entrepreneurial mystique," *Wall Street Journal*, May 20, 1985, p. 3c.

Hall, K. and Savery, L. K. "Tight rein, more stress." *Harvard Business Review*, Jan.–Feb. 1986, pp. 160–164.

Harney, K. "The nation's housing," *Minneapolis Star and Tribune*, Sept. 27, 1986, p. 1S.

Mendenhall, W., Reinmuth, J. E., Beaver, R., and Duhan, D. *Statistics for Management and Economics*, 5th ed. Boston: Duxbury, 1986. Chapters 8 and 9.

Montgomery, D. C. *Introduction to Statistical Quality Control*. New York: Wiley, 1985. Chapter 3.

Nemy, E. "Survey says two-thirds of Americans see themselves as younger than they are," *Minneapolis Tribune*, Dec. 19, 1982, p. 2F.

"One answer to imports: Wonder-iron." *Fortune*, Feb. 9, 1981, p. 71.

Report of Definitions Committee of the American Marketing Association. Chicago: American Marketing Association, 1961.

Roper, J. E. "Survey: 17% of IRS phone answers wrong," *Minneapolis Star and Tribune*, July 26, 1986, page 4A.

Rosen, B., Rynes, S., and Mahoney, T. A. "Compensation, jobs, and gender." *Harvard Business Review*, July–Aug. 1983, 170–190.

Sharpe, W. F. *Investments*, 3rd cd. Englewood Cliffs, N.J.: Prentice-Hall, Inc., 1985. Chapter 11.

Streeter, J. P. "Solder joint inspection using a laser inspector," *Quality Congress Transactions*. Milwaukee: American Society for Quality Control, 1986. pp. 507–515.

Taylor, D. H. and Glezen, G. W. *Auditing, Integrated Concepts and Procedures*. New York: Wiley, 1979. p. 3.

Treleven, M. "Sole sourcing from the vendor side," *Quality Congress Transactions*. Milwaukee: American Society for Quality Control, 1986. pp. 584–590.

U.S. Bureau of the Census. *Statistical Abstract of the United States: 1986*. Washington, D.C.: U.S. Government Printing Office, 1986.

Wetherill, G. B. *Sampling, Inspection, and Quality Control*, 2nd ed., London: Chapman and Hall, 1977. Chapter 3.

Willis, R. E. and Chervany, N. L. *Statistical Analysis and Modeling for Management Decision-making*. Belmont, Calif.: Wadsworth, 1974. Chapters 8 and 11.

C H A P T E R 9

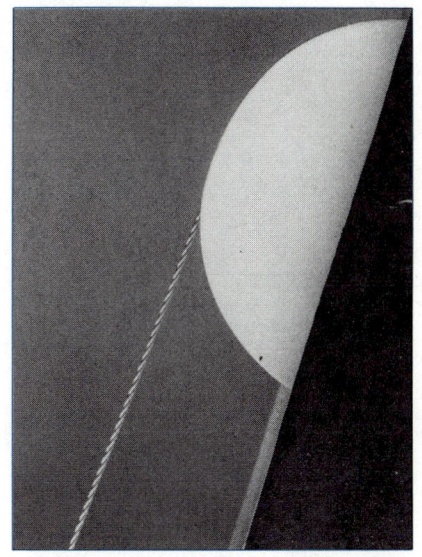

WHERE WE'VE BEEN . . .

The two methods for making statistical inferences, estimation and tests of hypotheses, were presented in Chapter 8. Confidence intervals and tests of hypotheses based on single samples were used to make inferences about sampled populations. We gave confidence intervals and tests of hypotheses concerning a population mean, μ, and a binomial proportion, p, and learned how to select the sample size necessary to obtain a specified amount of information concerning a parameter.

WHERE WE'RE GOING . . .

Now that we have learned to make inferences about a single population, it is natural that we would want to compare two populations. We may want to compare the mean costs per pound in the manufacture of two drugs or the mean lives of two industrial products. We may also wish to compare two population proportions, say the proportions of consumers who prefer a product before and after an advertising campaign. How to decide whether differences exist in population means or proportions and how to estimate these differences will be the subject of this chapter.

TWO SAMPLES: ESTIMATION AND TESTS OF HYPOTHESES

CONTENTS

SECTION 9.1

**LARGE-SAMPLE
INFERENCES ABOUT
THE DIFFERENCE
BETWEEN TWO
POPULATION MEANS:
INDEPENDENT
SAMPLING**

Suppose a chain of department stores is considering two suburbs of a large city as alternatives for locating a new store. The final decision about which location to choose is to be based on a comparison of the mean incomes of families living in the two suburbs.* The store is to be located in the suburb that has the higher mean income per household.

Let μ_1 represent the mean income of families in suburb 1 and μ_2 represent the mean income of families in suburb 2. Then our objective is to make an inference about $(\mu_1 - \mu_2)$, the difference between the mean incomes for the two suburbs.

Suppose independent samples of 100 households are randomly selected from each suburb, and the mean incomes, $\bar{x}_1$ and $\bar{x}_2$, are calculated for the two samples. An intuitively appealing estimator for $(\mu_1 - \mu_2)$ is the difference between the sample means, $(\bar{x}_1 - \bar{x}_2)$. The performance of this estimator in repeated sampling is summarized by the properties of its (repeated) sampling distribution (see Figure 9.1).

FIGURE 9.1

Sampling Distribution of $(\bar{x}_1 - \bar{x}_2)$

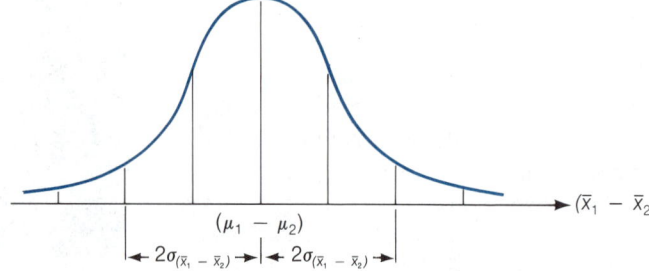

PROPERTIES OF THE SAMPLING DISTRIBUTION OF $(\bar{x}_1 - \bar{x}_2)$

1. The sampling distribution of $(\bar{x}_1 - \bar{x}_2)$ is approximately normal for *large* samples.
2. The mean of the sampling distribution of $(\bar{x}_1 - \bar{x}_2)$ is $(\mu_1 - \mu_2)$.
3. If the two samples are independent, the standard deviation of the sampling distribution is

$$\sigma_{(\bar{x}_1 - \bar{x}_2)} = \sqrt{\frac{\sigma_1^2}{n_1} + \frac{\sigma_2^2}{n_2}}$$

where σ_1^2 and σ_2^2 are the variances of the two populations being sampled, and n_1 and n_2 are the respective sample sizes.

Since the shape of the sampling distribution is approximately normal for large samples, we can use the z statistic to make inferences about $(\mu_1 - \mu_2)$, just as we did for a single mean. The procedures for forming confidence intervals and testing

---

*Assume that the incomes within a suburb are moderately homogeneous and hence the distributions are not heavily skewed. For this case, the mean would be a satisfactory measure of central tendency for the data.

hypotheses are summarized in the boxes. Note the similarity of these procedures to their counterparts for a single mean (Sections 8.1 and 8.3).

LARGE-SAMPLE CONFIDENCE INTERVAL FOR $(\mu_1 - \mu_2)$

$$(\bar{x}_1 - \bar{x}_2) \pm z_{\alpha/2}\sigma_{(\bar{x}_1 - \bar{x}_2)} = (\bar{x}_1 - \bar{x}_2) \pm z_{\alpha/2}\sqrt{\frac{\sigma_1^2}{n_1} + \frac{\sigma_2^2}{n_2}}$$

Assumptions: The two samples are randomly selected in an independent manner from the two populations. The sample sizes, n_1 and n_2, are large enough so that $\bar{x}_1$ and $\bar{x}_2$ each have approximately normal sampling distributions and so that s_1^2 and s_2^2 provide good approximations to σ_1^2 and σ_2^2. This will generally be true if $n_1 \geq 30$ and $n_2 \geq 30$.

LARGE-SAMPLE TEST OF HYPOTHESIS FOR $(\mu_1 - \mu_2)$

ONE-TAILED TEST

H_0: $(\mu_1 - \mu_2) = D_0$

H_a: $(\mu_1 - \mu_2) < D_0$
   [or H_a: $(\mu_1 - \mu_2) > D_0$]

TWO-TAILED TEST

H_0: $(\mu_1 - \mu_2) = D_0$

H_a: $(\mu_1 - \mu_2) \neq D_0$

where D_0 = Hypothesized difference between the means (this is often 0)

Test statistic: $\quad z = \dfrac{(\bar{x}_1 - \bar{x}_2) - D_0}{\sigma_{(\bar{x}_1 - \bar{x}_2)}}$

Test statistic: $\quad z = \dfrac{(\bar{x}_1 - \bar{x}_2) - D_0}{\sigma_{(\bar{x}_1 - \bar{x}_2)}}$

where $\sigma_{(\bar{x}_1 - \bar{x}_2)} = \sqrt{\dfrac{\sigma_1^2}{n_1} + \dfrac{\sigma_2^2}{n_2}}$

Rejection region: $\quad z < -z_\alpha$
   [or $z > z_\alpha$ when
   H_a: $(\mu_1 - \mu_2) > D_0$]

Rejection region: $\quad z < -z_{\alpha/2}$
   or $z > z_{\alpha/2}$

Assumptions: Same as for the large-sample confidence interval above.

For example, suppose the means and standard deviations of the incomes of the sampled households from the two suburbs are as follows:

SUBURB 1	SUBURB 2
$\bar{x}_1 = \$18{,}750$	$\bar{x}_2 = \$15{,}150$
$s_1 = \$3{,}200$	$s_2 = \$2{,}700$
$n_1 = 100$	$n_2 = 100$

Then to form a 95% confidence interval for the difference $(\mu_1 - \mu_2)$ between the true mean suburban incomes, we calculate

$$(\bar{x}_1 - \bar{x}_2) \pm 1.96 \sqrt{\frac{\sigma_1^2}{n_1} + \frac{\sigma_2^2}{n_2}} = (18{,}750 - 15{,}150) \pm 1.96 \sqrt{\frac{\sigma_1^2}{100} + \frac{\sigma_2^2}{100}}$$

To complete the calculations for this confidence interval, we must estimate σ_1^2 and σ_2^2. Since the samples are both relatively large, the sample variances s_1^2 and s_2^2 will provide reasonable approximations. Thus, our interval is approximately

$$3{,}600 \pm 1.96 \sqrt{\frac{(3{,}200)^2}{100} + \frac{(2{,}700)^2}{100}} = 3{,}600 \pm 821 = (2{,}779,\ 4{,}421)$$

When this estimation procedure is used, confidence intervals of this type will enclose the difference in population means, $(\mu_1 - \mu_2)$, 95% of the time. Therefore, we are reasonably confident that the mean income of households in suburb 1 is between $2,779 and $4,421 higher than the mean income of households in suburb 2. Based on this information, the department store chain should build the new store in suburb 1.

EXAMPLE 9.1

The management of a restaurant wants to determine whether a new advertising campaign has increased its mean daily income (gross). The daily incomes for 50 business days prior to the campaign's beginning were recorded. After conducting the advertising campaign and allowing a 20-day period for the advertising to take effect, the restaurant management recorded the income for 30 business days. These two samples will allow the management to make an inference about the effect of the advertising campaign on the restaurant's daily income. A summary of the results of the two samples is shown below:

BEFORE CAMPAIGN	AFTER CAMPAIGN
$n_1 = 50$	$n_2 = 30$
$\bar{x}_1 = \$1{,}255$	$\bar{x}_2 = \$1{,}330$
$s_1 = \$215$	$s_2 = \$238$

Do these samples provide sufficient evidence for the management to conclude that the mean income has been increased by the advertising campaign? Test using $\alpha = .05$.

SOLUTION

We can best answer this question by performing a test of hypothesis. Defining μ_1 as the mean daily income before the campaign and μ_2 as the mean daily income after the campaign, we will attempt to support the alternative (research) hypothesis that $\mu_2 > \mu_1$ [i.e., that $(\mu_1 - \mu_2) < 0$]. Thus, we will test the null hypothesis, $(\mu_1 - \mu_2) = 0$, rejecting this hypothesis if $(\bar{x}_1 - \bar{x}_2)$ equals a large negative value.

The elements of the test are as follows:

H_0: $(\mu_1 - \mu_2) = 0$ (i.e., $\mu_1 = \mu_2$; note that $D_0 = 0$ for this hypothesis test)

H_a: $(\mu_1 - \mu_2) < 0$ (i.e., $\mu_1 < \mu_2$)

Test statistic: $z = \dfrac{(\bar{x}_1 - \bar{x}_2) - D_0}{\sigma_{(\bar{x}_1 - \bar{x}_2)}} = \dfrac{(\bar{x}_1 - \bar{x}_2) - 0}{\sigma_{(\bar{x}_1 - \bar{x}_2)}}$

Rejection region: $z < -z_\alpha = -1.645$ (see Figure 9.2)

FIGURE 9.2
Rejection Region for
Example 9.1

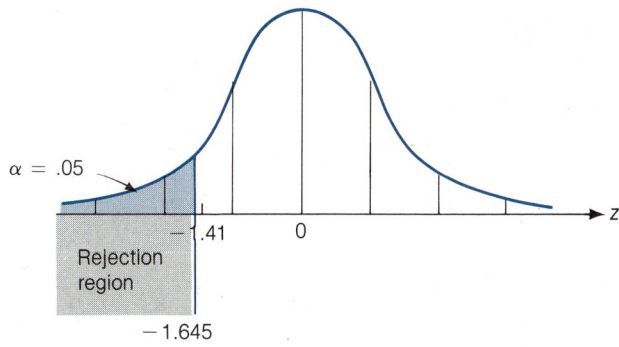

Assuming the samples before and after the campaign are independent, we now calculate

$$z = \frac{(\bar{x}_1 - \bar{x}_2) - 0}{\sigma_{(\bar{x}_1 - \bar{x}_2)}} = \frac{(1{,}255 - 1{,}330)}{\sqrt{\dfrac{\sigma_1^2}{n_1} + \dfrac{\sigma_2^2}{n_2}}}$$

$$\approx \frac{-75}{\sqrt{\dfrac{s_1^2}{n_1} + \dfrac{s_2^2}{n_2}}} = \frac{-75}{\sqrt{\dfrac{(215)^2}{50} + \dfrac{(238)^2}{30}}} = \frac{-75}{53.03} = -1.41$$

As you can see in Figure 9.2, the calculated z value does not fall in the rejection region. The samples do not provide sufficient evidence, at $\alpha = .05$, for the restaurant management to conclude that the advertising campaign has increased the mean daily income. ∎

EXAMPLE 9.2

Find the observed significance level for the test from Example 9.1.

SOLUTION

The alternative hypothesis in Example 9.1, H_a: $(\mu_1 - \mu_2) < 0$, required a lower one-tailed test using

$$z = \frac{\bar{x}_1 - \bar{x}_2}{\sigma_{(\bar{x}_1 - \bar{x}_2)}}$$

as a test statistic. Since the value of z calculated from the sample data was -1.41, the observed significance level (p-value) for the test is the probability of observing a value of z at least as contradictory to the null hypothesis as $z = -1.41$; i.e.,

$$p\text{-value} = P(z \leq -1.41)$$

This probability is computed assuming H_0 is true and is equal to the shaded area shown in Figure 9.3.

The tabulated area corresponding to $z = 1.41$ in Table IV of Appendix B is .4207. Therefore, the observed significance level for the test is

$$p\text{-value} = .5 - .4207 = .0793$$

FIGURE 9.3

The Observed Significance Level for Example 9.1

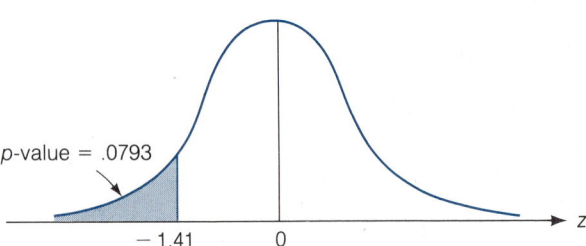

You will recall that in Example 9.1, we chose $\alpha = .05$ as the probability of a Type I error and consequently did not reject H_0; that is, we did not find sufficient evidence to indicate that $(\mu_1 - \mu_2) < 0$. If the results of the test had been presented in terms of an observed significance level and left for us to interpret, we might not have reached such an inflexible conclusion. Observing a value of z as small as $z = -1.41$ is an improbable event (rare event) if, in fact, $\mu_1 = \mu_2$. Since the probability is fairly small (.0793)—in fact, quite close to .05—we would conclude that there is some evidence to suggest that $\mu_1 < \mu_2$. Naturally, we would be more certain of this conclusion if the observed significance level were smaller, say, .05, .01, or, better yet, .001. However, the practical question to be answered is not whether the test results are statistically significant but whether the difference between μ_1 and μ_2 is large enough to have a practical business significance. To shed light on this question, we will wish to estimate the difference, $(\mu_1 - \mu_2)$.

EXAMPLE 9.3

Find a 95% confidence interval for the difference in mean daily incomes before and after the advertising campaign of Example 9.1 and discuss the implications of the confidence interval.

SOLUTION

The 95% confidence interval for $(\mu_1 - \mu_2)$ is

$$(\bar{x}_1 - \bar{x}_2) \pm z_{\alpha/2} \sqrt{\frac{\sigma_1^2}{n_1} + \frac{\sigma_2^2}{n_2}}$$

Once again, we will substitute s_1^2 and s_2^2 for σ_1^2 and σ_2^2 because these quantities will provide good approximations to σ_1^2 and σ_2^2 for samples as large as $n_1 = 50$ and $n_2 = 30$. Then, the 95% confidence interval for $(\mu_1 - \mu_2)$ is

$$(1{,}255 - 1{,}330) \pm 1.96 \sqrt{\frac{(215)^2}{50} + \frac{(238)^2}{30}} = -75 \pm 103.94$$

Thus, we estimate the difference in mean daily income to fall in the interval $-\$178.94$ to $\$28.94$. In other words, we estimate that μ_2, the mean daily income *after* the advertising campaign, could be larger than μ_1, the mean daily income *before* the campaign, by as much as $\$178.94$ per day, or it could be less than μ_1 by $\$28.94$ per day.

Now what should the restaurant management do? You can see that the sample sizes collected in the experiment were not large enough to detect a difference between μ_1 and μ_2. To be able to detect a difference (if in fact a difference exists), the management will have to repeat the experiment and increase the sample sizes. This will reduce the width of the confidence interval for $(\mu_1 - \mu_2)$. The restaurant management's best estimate of $(\mu_1 - \mu_2)$ is the point estimate $(\bar{x}_1 - \bar{x}_2) = -\$75$. Thus, the management must decide whether the cost of conducting the advertising campaign is overshadowed by a possible gain in mean daily income estimated at $\$75$ (but which might be as large as $\$178.94$ or could be as low as $-\$28.94$). Based on this analysis, the management will decide whether to continue the experiment or reject the new advertising program as a poor investment.   ■

EXERCISES 9.1–9.17

LEARNING THE MECHANICS

9.1 The purpose of this exercise is to show how taking the difference between two sample means affects the variability of the statistic.
  a. Suppose the first sample is selected from a population with mean $\mu_1 = 150$ and variance $\sigma_1^2 = 900$. Within what range should the sample mean vary about 95% of the time in repeated samples of 100 measurements from this distribution? That is, construct an interval extending 2 standard deviations of $\bar{x}_1$ on each side of μ_1.
  b. Suppose the second sample is selected independently of the first from a second population with mean $\mu_2 = 150$ and variance $\sigma_2^2 = 1{,}600$. Within what range should the sample mean vary about 95% of the time in repeated samples of 100 measurements from this distribution? That is, construct an interval extending 2 standard deviations of $\bar{x}_2$ on each side of μ_2.
  c. Now consider the difference between the two sample means, $(\bar{x}_1 - \bar{x}_2)$. What are the mean and standard deviation of the sampling distribution of $(\bar{x}_1 - \bar{x}_2)$?
  d. Within what range should the difference in sample means vary about 95% of the time in repeated independent samples of 100 measurements each from the two populations? That is, construct an interval extending 2 standard deviations of $(\bar{x}_1 - \bar{x}_2)$ on each side of $(\mu_1 - \mu_2)$.
  e. What, in general, can be said about the variability of the difference between independent sample means relative to the variability of the individual sample means?

9.2 Independent random samples of 64 observations each are chosen from two normal populations with the following means and standard deviations:

POPULATION 1	POPULATION 2
$\mu_1 = 12$	$\mu_2 = 10$
$\sigma_1 = 4$	$\sigma_2 = 3$

Let $\bar{x}_1$ and $\bar{x}_2$ denote the two sample means.
  a. Give the mean and standard deviation of the sampling distribution of $\bar{x}_1$.
  b. Give the mean and standard deviation of the sampling distribution of $\bar{x}_2$.
  c. Find the mean and standard deviation of the sampling distribution of $(\bar{x}_1 - \bar{x}_2)$.
  d. Will the statistic $(\bar{x}_1 - \bar{x}_2)$ be normally distributed? Explain.

9.3 Refer to Exercise 9.2.
  a. Give the z-score corresponding to $(\bar{x}_1 - \bar{x}_2) = 1.5$.
  b. Find the probability that $(\bar{x}_1 - \bar{x}_2)$ is larger than 1.5.
  c. Find the probability that $(\bar{x}_1 - \bar{x}_2)$ is less than 1.5.
  d. Find the probability that $(\bar{x}_1 - \bar{x}_2)$ is larger than 1 or less than -1.

9.4 Two independent random samples have been selected, 100 observations from population 1 and 100 from population 2. Sample means $\bar{x}_1 = 70$ and $\bar{x}_2 = 50$ were obtained. From previous experience with these populations, it is known that the variances are $\sigma_1^2 = 100$ and $\sigma_2^2 = 64$.
  a. Find $\sigma_{(\bar{x}_1 - \bar{x}_2)}$.
  b. Sketch the approximate sampling distribution for $(\bar{x}_1 - \bar{x}_2)$ assuming $(\mu_1 - \mu_2) = 5$.
  c. Locate the observed value of $(\bar{x}_1 - \bar{x}_2)$ on the graph you drew in part **b**. Does it appear that this value contradicts the null hypothesis $H_0: (\mu_1 - \mu_2) = 5$?
  d. Use Table IV in Appendix B to determine the rejection region for the test of H_0: $(\mu_1 - \mu_2) = 5$ against $H_a: (\mu_1 - \mu_2) \neq 5$. Use $\alpha = .05$.
  e. Conduct the hypothesis test of part **d** and interpret your results.

9.5 Refer to Exercise 9.4. Construct a 95% confidence interval for $(\mu_1 - \mu_2)$. Interpret the interval. Which inference provides more information about the value of $(\mu_1 - \mu_2)$—the test of hypothesis in Exercise 9.4 or the confidence interval in this exercise?

9.6 Two independent random samples have been selected, 100 from population 1 and 150 from population 2. Sample means $\bar{x}_1 = 1,025$ and $\bar{x}_2 = 1,039$ were obtained. The sample standard deviations are $s_1 = 10$ and $s_2 = 12$.
  a. Describe the sampling distribution of $(\bar{x}_1 - \bar{x}_2)$. Assume that $(\mu_1 - \mu_2) = -5$.
  b. Test $H_0: (\mu_1 - \mu_2) = -5$ against $H_a: (\mu_1 - \mu_2) < -5$, using $\alpha = .01$. Interpret the results of your test.
  c. Report the p-value of your test.

APPLYING THE CONCEPTS

9.7 Wansley, Roenfeldt, and Cooley (1983) compared the profiles of a sample of 44 firms that merged during 1975–1976 with those of a sample of 44 firms that did not merge.

The table displays information obtained on the firms' price–earnings ratios.

	MERGED FIRMS	NONMERGED FIRMS
Sample mean	7.295	14.666
Sample standard deviation	7.374	16.089

a. The analysis performed by Wansley, Roenfeldt, and Cooley indicated that "merged firms generally have smaller price–earnings ratios." Do you agree? Test using $\alpha = .05$.
b. Report the p-value of the test you conducted in part **a**.
c. What assumption(s) was it necessary to make in order to perform the test in part **a**?
d. Do you think that the distributions of the price–earnings ratios for the populations from which these samples were drawn are normally distributed? Why or why not? [*Hint:* Note the relative values of the sample means and standard deviations.]

9.8 A paper company conducted an experiment to compare the mean time to unload shipments of logs for two different unloading procedures. Random samples of 50 trucks each were unloaded using a new method and the company's current method. The objective of the experiment is to determine whether the new method will reduce the mean unloading time. The sample means and standard deviations are shown in the table.

NEW METHOD	CURRENT METHOD
$n_1 = 50$	$n_2 = 50$
$\bar{x}_1 = 25.4$ minutes	$\bar{x}_2 = 27.3$ minutes
$s_1 = 3.1$ minutes	$s_2 = 3.7$ minutes

a. Do the data provide sufficient evidence to indicate that the mean unloading time for the new method is less than the mean unloading time for the method currently in use? Test using $\alpha = .05$.
b. Give the observed significance level for the test.

9.9 Refer to Exercise 9.8. Find a 90% confidence interval for the difference in mean unloading times between the two methods.

9.10 Tennant Co., a Minnesota manufacturer of industrial floor-cleaning machines, recently began using "quality circles" to help improve the quality of its product. The term *quality circles* describes a process in which groups comprised of both white-collar and blue-collar employees attempt to solve quality, productivity, and/or work environment problems. According to a recent survey of 6,800 companies by the New York Stock Exchange's Office of Economic Research, 74% had quality circle programs.
In 1979, at the time Tennant began its quality circles program, a sample of finished machines was found to have an average of 4.2 defects per machine. Because of such defects, the company had to pay its employees $16.6 million for the 39,600 hours of labor required to rework the defective machines. In 1982, a sample of machines revealed a substantial improvement in quality. The mean number of defectives was reduced to 1.3 flaws per machine, and rework was down to 3,500 hours of labor (Marcotty, 1983a, 1983b). Assume that each sample of machines was randomly selected and was of size 100. Further, assume that the sample standard deviations for 1979 and 1982 were 2.0 and 1.1, respectively.

a. While the decline in the average number of defects per machine between 1979 and 1982 appears to be significant from a managerial perspective, is it statistically significant? To answer the question, conduct the appropriate hypothesis test and report and interpret the observed significance level of the test.

b. In the context of the problem, describe the Type I and Type II errors associated with your hypothesis test of part **a**.

9.11 Thirty-six stocks were randomly selected from those listed on the New York Stock Exchange (NYSE), and 36 stocks were randomly selected from those listed on the American Stock Exchange (ASE). The closing prices of all 72 stocks on December 15, 1986, and their abbreviations (as listed in the *Wall Street Journal*) are listed in the table.

NYSE		NYSE		ASE		ASE	
Firm	Closing Price	Firm	Closing Price	Firm	Closing Price	Firm	Closing Price
Algln	$14\frac{1}{8}$	Fuqua	$24\frac{1}{4}$	Andal	$8\frac{1}{2}$	IntBknt	5
BCelts	$15\frac{7}{8}$	BwnSh	18	BSD	$2\frac{5}{8}$	BinkMf	$24\frac{1}{8}$
Glenfed	$24\frac{1}{8}$	BellInd	$21\frac{1}{8}$	Fstcrp	$11\frac{1}{4}$	AmIntl	$6\frac{1}{4}$
Newell	28	TranEx	$14\frac{3}{4}$	UnvPat	$14\frac{1}{8}$	FreqE	$21\frac{5}{8}$
CartWl	$77\frac{1}{2}$	CinMil	$19\frac{7}{8}$	Endvco	$5\frac{5}{8}$	Duplex	$17\frac{7}{8}$
HCA	$30\frac{3}{4}$	Radice	8	Telsci	$4\frac{7}{8}$	Ducom	$16\frac{1}{4}$
FstChic	32	Dynlct	$15\frac{1}{2}$	Stepan	$30\frac{3}{4}$	Devlcp	$14\frac{3}{4}$
Becor	$11\frac{5}{8}$	ThmBet	$43\frac{3}{4}$	Zentec Corp	$4\frac{3}{8}$	Baker	$16\frac{3}{4}$
HeclaM	$11\frac{1}{8}$	ACyan	$80\frac{1}{2}$	Nantck	$6\frac{5}{8}$	Damson	$\frac{3}{8}$
Tonka	$20\frac{1}{8}$	Hecks	$11\frac{7}{8}$	Pantast	$10\frac{1}{4}$	CitFst	86
Chmpln	$31\frac{7}{8}$	Berkey	$4\frac{1}{2}$	Pentron	$1\frac{5}{8}$	Unimar	$6\frac{1}{4}$
UtdMM	13	OhMatr	$16\frac{1}{8}$	PropCT	$24\frac{1}{8}$	MMed	$26\frac{1}{2}$
HazLab	23	Purolat	25	VST	$8\frac{1}{2}$	Digicon	$\frac{1}{2}$
Upjohn	95	Wendy	$10\frac{3}{4}$	AmBld	$3\frac{1}{4}$	MSR	$1\frac{1}{2}$
Navistr	$5\frac{3}{8}$	Amern	$30\frac{3}{8}$	Valspar	$41\frac{7}{8}$	Nichols	$7\frac{7}{8}$
Wrld Ar	$4\frac{1}{8}$	KimCl	$83\frac{5}{8}$	WangB	$12\frac{5}{8}$	Gull	$13\frac{1}{4}$
Conrac	$12\frac{3}{4}$	SonyCp	$21\frac{1}{4}$	Hlthch	$9\frac{1}{2}$	Frnkln	$12\frac{3}{4}$
OklaGE	$35\frac{1}{4}$	KeysCo	$6\frac{7}{8}$	Lionel	$5\frac{5}{8}$	GrtLKC	$35\frac{1}{4}$

Source: *Wall Street Journal*, Dec. 16, 1986, pp. 54–56. Reprinted by permission of *Wall Street Journal*, © Dow Jones & Company, Inc. (1986). All rights reserved.

a. Use a 95% confidence interval to estimate the difference between the mean price of a share of stock traded on the NYSE and the mean price of a share of stock traded on the ASE. Interpret your result.

b. Carefully define the populations about which you can make inferences using the confidence interval you constructed in part **a**.

c. Suppose ten more stocks were randomly selected from each stock exchange. If the confidence interval of part **a** were recalculated using $n_1 = 46$ and $n_2 = 46$, would the width of the resulting confidence interval necessarily be narrower? Explain.

d. Test the null hypothesis that the mean price on the NYSE exceeds that on the ASE by $20 per share against the alternative that the difference is not $20 per share. Use $\alpha = .05$.

e. Which of the procedures, the confidence interval of part **a** or the test of hypothesis of part **d**, provides more information about the value of the true difference between the mean earnings per share on the two exchanges? Justify your answer.

9.12 As part of a study in participative management, George H. Hines (1974) sampled workers from two types of New Zealand sociocultural backgrounds: those who believe in the existence of a class system and those who believe that they live and work in a classless society. Each worker in the sampling was selected from a work environment with a high degree of participatory management. Do workers who consider themselves to be social equals with their management superiors possess different levels of job satisfaction than those workers who see themselves as socially different from management? Each worker in the independent random samples was asked this question by rating his or her job satisfaction on a scale of 1 (poor) to 7 (excellent). Based on the results of this study (shown in the table), what can you say about differences in job satisfaction for the two different sociocultural types of workers? Use $\alpha = .10$.

	BELIEF IN EXISTENCE OF A CLASS SYSTEM	
	Yes	No
Sample size	175	277
Mean	5.42	5.19
Standard deviation	1.24	1.17

9.13 An experiment has been conducted to compare the productivity of two machines. Machine 1 produced an average of 51.4 items per hour and a standard deviation of $s_1 = 2.1$ for 35 randomly selected hours during the past 2 weeks. Machine 2 produced an average of 49.5 items per hour and a standard deviation of $s_2 = 1.8$ for 45 randomly selected hours during the past 2 weeks.

a. Describe the populations being compared.

b. Do the samples provide sufficient evidence at $\alpha = .10$ to conclude that machine 1 produces more items per hour, on the average, than machine 2?

c. Report the p-value for the test you conducted in part **b**.

9.14 Refer to Exercise 9.13. Construct a 95% confidence interval for $(\mu_1 - \mu_2)$. Would a 99% confidence interval be narrower or wider than the one you constructed? Why?

A	B
$\bar{x}_1 = 365$	$\bar{x}_2 = 352$
$s_1 = 23$	$s_2 = 41$

9.15 Two manufacturers of corrugated fiberboard each claim that the strength of their product tests at more than 360 pounds per square inch on the average. As a result of consumer complaints, a consumer products testing firm believes that firm A's product is stronger than firm B's. To test its belief, 100 fiberboards were chosen randomly from firm A's inventory and 100 were chosen from firm B's inventory. The strength of each fiberboard was tested and the results (in pounds per square inch) are summarized in the table.

a. Does the sample information support the consumer products testing firm's belief? Test at $\alpha = .05$.

b. What assumptions did you make in conducting the test in part a? Do you think such assumptions could comfortably be made in practice? Why or why not?

9.16 Refer to Exercise 9.15. Does the sample information support firm A's claim that the mean strength of its corrugated fiberboard is more than 360 pounds per square inch? Test at $\alpha = .10$.

9.17 The mean or median value of a community's single-family homes is sometimes used as an indicator of the community's wealth or "buying power." Such information may be useful to retail merchants looking for communities in which to locate their businesses and/or in deciding where to concentrate their advertising. A retailer believes that the mean price of a single-family home in Edina, Minnesota, is substantially more than in St. Louis Park, Minnesota. If this is confirmed by a statistical hypothesis test, the retailer will increase her advertising budget for the Edina area. The table contains the asking prices of a random sample of 36 homes that were for sale in Edina in late October 1986 and a random sample of 36 homes that were on the market at the same time in St. Louis Park.

EDINA		ST. LOUIS PARK	
$386,000	$165,000	$ 63,500	$ 85,900
89,900	151,900	79,900	79,900
345,900	149,500	83,900	64,900
219,000	143,000	198,800	148,900
159,900	128,500	103,900	139,900
364,000	98,500	59,900	79,900
190,000	179,500	64,500	99,900
349,500	299,000	139,500	73,900
177,000	99,900	109,900	63,900
379,000	111,000	71,000	147,500
159,500	229,000	69,900	28,900
225,000	139,900	189,900	62,900
119,500	106,900	45,900	69,900
98,500	93,900	67,900	83,900
189,900	99,900	79,900	119,900
228,500	675,000	89,900	89,900
218,000	84,500	138,800	75,500
98,500	109,900	139,500	74,800

Source: MLS Compilation of Listings, Division II, No. 44, Oct. 31–Nov. 6, 1986. Greater Minneapolis Area Board of Realtors.

a. Describe the populations from which the sample data were selected.

b. What are the parameters of interest to the retailer?

c. Do the data indicate that the mean price of a home in Edina is more than $50,000 greater than the mean price in St. Louis Park? Test using $\alpha = .05$.

d. Use a 95% confidence interval to estimate the true difference in the mean prices for Edina and St. Louis Park.

| | | | | | | | | | | | | | |

S E C T I O N 9.2

SMALL-SAMPLE INFERENCES ABOUT THE DIFFERENCE BETWEEN TWO POPULATION MEANS: INDEPENDENT SAMPLING

Suppose a television network wanted to determine whether major sports events or first-run movies attract more viewers in the prime-time hours. It selected 28 prime-time evenings; of these, 13 had programs devoted to major sports events, and the remaining 15 had first-run movies. The number of viewers (estimated by a television viewer rating firm) was recorded for each program. If μ_1 is the mean number of sports viewers per evening of sports programming and μ_2 is the mean number of movie viewers per evening, we want to detect a difference between μ_1 and μ_2—if such a difference exists. Therefore, we want to test the null hypothesis

$$H_0: \quad (\mu_1 - \mu_2) = 0$$

against the alternative hypothesis

$$H_a: \quad (\mu_1 - \mu_2) \neq 0 \quad \text{(i.e., either } \mu_1 > \mu_2 \text{ or } \mu_2 > \mu_1)$$

Since the sample sizes are small, s_1^2 and s_2^2 will be unreliable estimates of σ_1^2 and σ_2^2 and the z test statistic will be inappropriate for the test. But, as in the case of a single mean (Section 8.6), we can construct a Student's t statistic. This statistic (formula to be given subsequently) has the familiar t distribution described in Chapter 8. **To use the t statistic, both sampled populations must be approximately normally distributed with equal population variances, and the random samples must be selected independently of each other.** The normality and equal variances assumptions would imply relative frequency distributions for the populations that would appear as shown in Figure 9.4. We will assume that the distributions of the two populations of numbers of television viewers will approximately satisfy these assumptions.

FIGURE 9.4

Assumptions for the Two-Sample t:
(1) Normal Populations
(2) Equal Variances

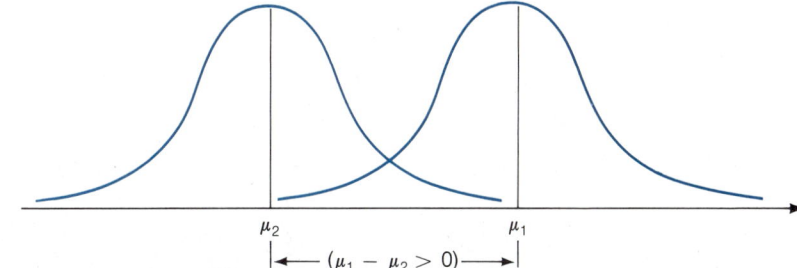

Since we assume the two populations have equal variances ($\sigma_1^2 = \sigma_2^2 = \sigma^2$), it is reasonable to combine the sum of squares of deviations from the two samples to construct a pooled sample estimator of σ^2 for use in the t statistic. Thus, if s_1^2 and s_2^2 are the two sample variances (both estimating the variance σ^2 common to both populations), the pooled estimator of σ^2, denoted as s_p^2, is

$$s_p^2 = \frac{\overbrace{\sum_{i=1}^{n_1} (x_i - \bar{x}_1)^2}^{\text{From sample 1}} + \overbrace{\sum_{i=1}^{n_2} (x_i - \bar{x}_2)^2}^{\text{From sample 2}}}{n_1 + n_2 - 2}$$

or

$$s_{\mathrm{p}}^2 = \frac{(n_1 - 1)s_1^2 + (n_2 - 1)s_2^2}{(n_1 - 1) + (n_2 - 1)}$$

$$= \frac{(n_1 - 1)s_1^2 + (n_2 - 1)s_2^2}{n_1 + n_2 - 2}$$

where x_1 represents a measurement from sample 1 and x_2 represents a measurement from sample 2. Recall that the term *degrees of freedom* was defined in Section 8.6 as 1 less than the sample size for each sample—i.e., $(n_1 - 1)$ for sample 1 and $(n_2 - 1)$ for sample 2. Since we are pooling the information on σ^2 obtained from both samples, the degrees of freedom associated with the pooled variance s_{p}^2 is equal to the sum of the degrees of freedom for the two samples, namely, the denominator of s_{p}^2—i.e., $(n_1 - 1) + (n_2 - 1) = n_1 + n_2 - 2$.

Note that the second formula given for s_{p}^2 shows that the pooled variance is simply a *weighted average* of the two sample variances, s_1^2 and s_2^2. The weight given each variance is proportional to its degrees of freedom. The result is an average, or "pooled," variance that is closer to the sample variance with larger degrees of freedom. If the two variances have the same number of degrees of freedom (i.e., if the sample sizes are equal), then the pooled variance is a simple average of the two sample variances.

To obtain the small-sample test statistic for testing $H_0: (\mu_1 - \mu_2) = D_0$, substitute the pooled estimate of σ^2 into the formula for the two-sample z statistic (Section 9.1) to obtain

$$t = \frac{(\bar{x}_1 - \bar{x}_2) - D_0}{\sqrt{s_{\mathrm{p}}^2 \left(\dfrac{1}{n_1} + \dfrac{1}{n_2} \right)}}$$

It can be shown that this statistic, like the t statistic of Chapter 8, has a t distribution with $(n_1 + n_2 - 2)$ degrees of freedom.

We will use the television viewer example to outline the final steps for this t test. The hypothesized difference in mean number of viewers is $D_0 = 0$. The rejection region will be two-tailed and will be based on a t distribution with $(n_1 + n_2 - 2)$ or $(13 + 15 - 2) = 26$ df. Letting $\alpha = .05$, the rejection region for the test would be

$$t < -t_{\alpha/2} \quad \text{or} \quad t > t_{\alpha/2}$$

The value for $t_{.025}$ with df $= 26$ given in Table VI of Appendix B is 2.056. Thus, the rejection region for the television example is

$$t < -2.056 \quad \text{or} \quad t > 2.056$$

This rejection region is shown in Figure 9.5.

FIGURE 9.5
Rejection Region for
Two-Tailed t Test:
$\alpha = .05$, df $= 26$

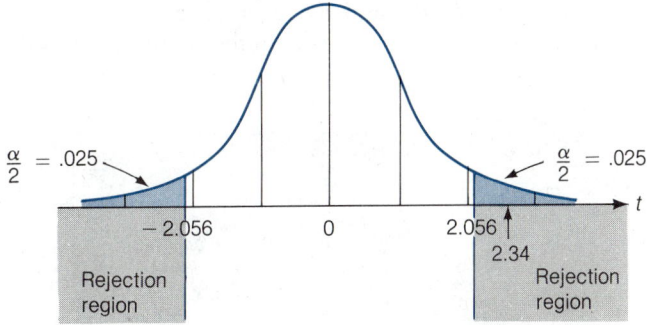

Now, suppose the television network's samples produce the results shown in the table:

SPORTS	MOVIE
$n_1 = 13$	$n_2 = 15$
$\bar{x}_1 = 6.8$ million	$\bar{x}_2 = 5.3$ million
$s_1 = 1.8$ million	$s_2 = 1.6$ million

We must assume that these independent samples were selected from normal distributions (i.e., the number of viewers is normally distributed for each type of event) with equal variances. [*Note:* Although the sample estimates of variance are not equal, the assumption that the population variances are equal may still be valid. We will present a method for checking this assumption statistically in Section 9.3.] We now calculate

$$s_p^2 = \frac{(n_1 - 1)s_1^2 + (n_2 - 1)s_2^2}{n_1 + n_2 - 2} = \frac{(13 - 1)(1.8)^2 + (15 - 1)(1.6)^2}{13 + 15 - 2}$$

$$= \frac{74.72}{26} = 2.87$$

Then,

$$t = \frac{(\bar{x}_1 - \bar{x}_2) - D_0}{\sqrt{s_p^2\left(\frac{1}{n_1} + \frac{1}{n_2}\right)}} = \frac{(6.8 - 5.3) - 0}{\sqrt{2.87\left(\frac{1}{13} + \frac{1}{15}\right)}}$$

$$= \frac{1.5}{.64} = 2.34$$

Since the observed value of t, $t = 2.34$, falls in the rejection region (see Figure 9.5), the samples provide sufficient evidence to indicate that the mean numbers of viewers differ for major sports events and first-run movies shown in prime time. Or, we say that the test results are statistically significant at the $\alpha = .05$ level of significance. Because the rejection was in the positive or upper tail of the t distribution, the indication is that the mean number of viewers for sports events exceeds that for movies.

The t statistic can also be used to construct confidence intervals for the difference between population means. Both the confidence interval and the test of hypothesis procedures are summarized in the boxes.

SMALL-SAMPLE CONFIDENCE INTERVAL FOR $(\mu_1 - \mu_2)$

$$(\bar{x}_1 - \bar{x}_2) \pm t_{\alpha/2} \sqrt{s_p^2 \left(\frac{1}{n_1} + \frac{1}{n_2} \right)}$$

where

$$s_p^2 = \frac{(n_1 - 1)s_1^2 + (n_2 - 1)s_2^2}{n_1 + n_2 - 2}$$

and $t_{\alpha/2}$ is based on $(n_1 + n_2 - 2)$ df.

Assumptions: 1. Both sampled populations have relative frequency distributions that are approximately normal.
2. The population variances are equal.
3. The samples are randomly and independently selected from the populations.

SMALL-SAMPLE TEST OF HYPOTHESIS FOR $(\mu_1 - \mu_2)$ (INDEPENDENT SAMPLES)

ONE-TAILED TEST

H_0: $(\mu_1 - \mu_2) = D_0$
H_a: $(\mu_1 - \mu_2) < D_0$
$\quad$ [or H_a: $(\mu_1 - \mu_2) > D_0$]

Test statistic:

$$t = \frac{(\bar{x}_1 - \bar{x}_2) - D_0}{\sqrt{s_p^2 \left(\frac{1}{n_1} + \frac{1}{n_2} \right)}}$$

Rejection region:

$t < -t_\alpha$
[or $t > t_\alpha$ when
H_a: $(\mu_1 - \mu_2) > D_0$]
where t_α is based on
$(n_1 + n_2 - 2)$ df.

TWO-TAILED TEST

H_0: $(\mu_1 - \mu_2) = D_0$
H_a: $(\mu_1 - \mu_2) \neq D_0$

Test statistic:

$$t = \frac{(\bar{x}_1 - \bar{x}_2) - D_0}{\sqrt{s_p^2 \left(\frac{1}{n_1} + \frac{1}{n_2} \right)}}$$

Rejection region:

$t < -t_{\alpha/2}$ or $t > t_{\alpha/2}$
where $t_{\alpha/2}$ is based on
$(n_1 + n_2 - 2)$ df.

Assumptions: Same as for the small-sample confidence interval for $(\mu_1 - \mu_2)$ above.

EXAMPLE 9.4

Suppose you want to estimate the difference in annual operating costs for automobiles with rotary engines and those with standard engines. You randomly select 8 owners of cars with rotary engines and 12 owners of cars with standard engines who have purchased their cars within the last 2 years and are willing to participate in the experiment. Each of the 20 owners keeps accurate records of the amount spent on operating his or her car (including gasoline, oil, repairs, etc.) for a 12-month period. All costs are recorded on a per-thousand-mile basis to adjust for differences in mileage driven during the 12-month period. The results are summarized in the table. Estimate the true difference $(\mu_1 - \mu_2)$ in the mean operating costs per thousand miles between cars with rotary and cars with standard engines. Use a 90% confidence interval.

ROTARY	STANDARD
$n_1 = 8$	$n_2 = 12$
$\bar{x}_1 = \$56.96$	$\bar{x}_2 = \$52.73$
$s_1 = \$4.85$	$s_2 = \$6.35$

SOLUTION

The objective of this experiment is to obtain a 90% confidence interval for $(\mu_1 - \mu_2)$. To use the small-sample confidence interval for $(\mu_1 - \mu_2)$, the following assumptions must be satisfied:

1. The operating cost per thousand miles is normally distributed for cars with both rotary and standard engines. Since these costs are averages (because we observe them on a per-thousand-mile basis), the Central Limit Theorem lends credence to this assumption.
2. The variance in cost is the same for the two types of cars. Under these circumstances, we might expect the variation in costs from automobile to automobile to be about the same for both types of engines.
3. The samples are randomly and independently selected from the two populations. We have randomly chosen 20 different owners for the two samples in such a way that the cost measurement for one owner is not dependent on the cost measurement for any other owner. Therefore, this assumption would be valid.

The first step in performing the test is to calculate the pooled estimate of variance:

$$s_p^2 = \frac{(n_1 - 1)s_1^2 + (n_2 - 1)s_2^2}{n_1 + n_2 - 2}$$

$$= \frac{(8 - 1)(4.85)^2 + (12 - 1)(6.35)^2}{8 + 12 - 2}$$

$$= 33.7892$$

where s_p^2 possesses $(n_1 + n_2 - 2) = (8 + 12 - 2) = 18$ df. Then, the 90% confidence interval for $(\mu_1 - \mu_2)$, the difference in mean operating costs for the two types of automobiles, is

$$(\bar{x}_1 - \bar{x}_2) \pm t_{\alpha/2}\sqrt{s_p^2\left(\frac{1}{n_1} + \frac{1}{n_2}\right)} = (56.96 - 52.73) \pm t_{.05}\sqrt{33.7892\left(\frac{1}{8} + \frac{1}{12}\right)}$$

$$= 4.23 \pm 1.734(2.653) = 4.23 \pm 4.60$$

$$= (-.37, 8.83)$$

This means that, with 90% confidence, we estimate the difference in mean operating costs per thousand miles between cars with rotary engines and those with standard engines to fall in the interval from $-\$.37$ to $\$8.83$. In other words, we estimate the mean operating costs for rotary engines to be anywhere from $\$.37$ less than to $\$8.83$ more than the operating costs per thousand miles for standard engines. Although the sample means seem to suggest that rotary cars cost more to operate, there is insufficient evidence to indicate that $(\mu_1 - \mu_2)$ differs from 0 because the interval includes 0 as a possible value for $(\mu_1 - \mu_2)$. To show a difference in mean operating costs (if it exists), it would be necessary to increase the sample sizes and thereby narrow the width of the confidence interval for $(\mu_1 - \mu_2)$. ■

The two-sample t statistic is a powerful tool for comparing population means when the assumptions are satisfied. It has also been shown to retain its usefulness when the sampled populations are only approximately normally distributed. And when the sample sizes are equal, the assumption of equal population variances can be relaxed. That is, when $n_1 = n_2$, σ_1^2 and σ_2^2 can be quite different and the test statistic will still have (approximately) a Student's t distribution. When the experimental situation does not satisfy the assumptions, other statistical tests are available. These nonparametric statistical tests are described in Chapter 16.

WHAT CAN BE DONE IF THE ASSUMPTIONS ARE NOT SATISFIED?

Answer: If you are concerned that your assumptions are not satisfied, use the Wilcoxon rank sum test for independent samples to test for a shift in population distributions. (See Chapter 16.)

| | | | | | | | | | | | | | |

EXERCISES 9.18–9.33   LEARNING THE MECHANICS

9.18 To use the t statistic to test for differences in the means of two populations, what assumptions must be made about the two populations? About the two samples?

9.19 Two populations are described in each of the following cases. In which cases would it be appropriate to apply the small-sample t test to investigate the difference between the population means?

a. Population 1:   Normal distribution with variance σ_1^2
   Population 2:   Skewed to the right with variance $\sigma_2^2 = \sigma_1^2$

b. Population 1:   Normal distribution with variance σ_1^2
   Population 2:   Normal distribution with variance $\sigma_2^2 \neq \sigma_1^2$

c. Population 1:   Skewed to the left with variance σ_1^2
   Population 2:   Skewed to the left with variance $\sigma_2^2 = \sigma_1^2$

d. Population 1:   Normal distribution with variance σ_1^2
   Population 2:   Normal distribution with variance $\sigma_2^2 = \sigma_1^2$

e. Population 1:   Uniform distribution with variance σ_1^2
   Population 2:   Uniform distribution with variance $\sigma_2^2 = \sigma_1^2$

9.20 In the t tests of this section, σ_1^2 and σ_2^2 are assumed to be equal. Thus, we say $\sigma_1^2 = \sigma_2^2 = \sigma^2$. Why is a pooled estimator of σ^2 used instead of either s_1^2 or s_2^2?

9.21 Assume that $\sigma_1^2 = \sigma_2^2 = \sigma^2$. Calculate the pooled estimator of σ^2 for each of the following cases:

a. $s_1^2 = 100,\quad s_2^2 = 90,\quad n_1 = n_2 = 25$
b. $s_1^2 = 12,\quad s_2^2 = 16,\quad n_1 = 20,\quad n_2 = 10$
c. $s_1^2 = .15,\quad s_2^2 = .20,\quad n_1 = 8,\quad n_2 = 12$
d. $s_1^2 = 2,500,\quad s_2^2 = 2,000,\quad n_1 = 16,\quad n_2 = 17$
e. Note that the pooled estimate is a weighted average of the sample variances. Which of the variances does the pooled estimate fall nearer in each of the above cases?

9.22 Independent random samples from normal populations produced the results shown in the table.
a. Calculate the pooled estimate of σ^2.
b. Do the data provide sufficient evidence to indicate that $\mu_2 > \mu_1$? Test using $\alpha = .10$.
c. Find the approximate observed significance level for the test, and interpret its value.

SAMPLE 1	SAMPLE 2
2.1	3.4
3.6	3.0
1.4	4.1
3.0	3.9
2.9	3.5
3.2	

2.70   3.58
.0?0?9   .43243
.90409

9.23 Refer to Exercise 9.22.
a. Find a 90% confidence interval for $(\mu_1 - \mu_2)$. Interpret the confidence interval.
b. Which of the two inferential procedures, the test of hypothesis in Exercise 9.22 or the confidence interval in this exercise, provides more information about $(\mu_1 - \mu_2)$? Justify your answer.

9.24 Independent random samples from two normal populations produced the results shown in the table.
a. Calculate the pooled estimate of σ^2.
b. Do these data provide sufficient evidence to indicate that $\mu_1 \neq \mu_2$? Test using $\alpha = .05$.
c. Find the approximate p-value for the test, and interpret its value.

SAMPLE 1	SAMPLE 2
$n_1 = 12$	$n_2 = 16$
$\bar{x}_1 = 35$	$\bar{x}_2 = 43$
$s_1 = 4.2$	$s_2 = 3.7$

9.25 Refer to Exercise 9.24. Find a 95% confidence interval for $(\mu_1 - \mu_2)$. Interpret the confidence interval.

9.26 Independent random samples from approximately normal populations produced the results shown in the table.

SAMPLE 1				SAMPLE 2			
52	33	42	44	52	43	47	56
41	50	44	51	62	53	61	50
45	38	37	40	56	52	53	60
44	50	43		50	48	60	55

a. Do the data provide sufficient evidence to conclude that $(\mu_2 - \mu_1) > 10$? Test using $\alpha = .01$.
b. Construct a 98% confidence interval for $(\mu_2 - \mu_1)$. Interpret your result.

APPLYING THE CONCEPTS

9.27 An industrial plant wants to determine which of two types of fuel—gas or electric—will produce more useful energy at the lower cost. One measure of economical energy production, called the *plant investment per delivered quad*, is calculated by taking the amount of money (in dollars) invested in the particular utility by the plant and dividing by the delivered amount of energy (in quadrillion British thermal units). The smaller this ratio, the less an industrial plant pays for its delivered energy.

Random samples of 11 plants using electrical utilities and 16 plants using gas utilities were taken, and the plant investment per quad was calculated for each. The data produced the results shown in the table.

	ELECTRIC	GAS
Sample size	11	16
Mean investment/quad (billions)	$22.5	$17.5
Variance	17.5	15

a. Do these data provide sufficient evidence at $\alpha = .05$ to indicate a difference in the average investment per quad between the plants using gas and those using electrical utilities?

b. Find a 90% confidence interval for $(\mu_1 - \mu_2)$. Give a practical interpretation of this interval.

OPERATOR 1	OPERATOR 2
$n_1 = 10$	$n_2 = 10$
$\bar{x}_1 = 35$	$\bar{x}_2 = 31$
$s_1^2 = 17.2$	$s_2^2 = 19.1$

9.28 A manufacturing company is interested in determining whether there is a significant difference between the average number of units produced per day by two different machine operators. A random sample of ten daily outputs was selected for each operator from the outputs over the past year. The data on number of items produced per day are summarized in the table.

a. Do the samples provide sufficient evidence at $\alpha = .05$ to conclude that a difference does exist between the mean daily outputs of the machine operators?

b. What assumptions must you make so that this test will be valid?

c. Find a 90% confidence interval for $(\mu_1 - \mu_2)$. Explain clearly the meaning of your confidence interval.

9.29 Marketing strategists would like to be able to predict consumer response to new products and their accompanying promotional schemes. To this end, studies that examine the differences between buyers and nonbuyers of a product are of interest. One such study conducted by Shuchman and Riesz (1975) was aimed at characterizing the purchasers and nonpurchasers of Crest toothpaste. Purchasers were defined as households that converted to Crest following its endorsement by the Council on Dental Therapeutics of the American Dental Association on August 1, 1960, and remained "loyal" to Crest until at least April 1963. Nonpurchasers were defined as households that did not convert to Crest during the same time period. Using demographic data collected from a sample of 499 purchasers and 499 nonpurchasers, Shuchman and Riesz demonstrated that both the mean household size (number of persons) and mean household income were significantly larger for purchasers than for nonpurchasers. A similar study utilized random samples of size 20 and yielded the data in the table on the age of the householder primarily responsible for making toothpaste purchases.

PURCHASERS				NONPURCHASERS			
34	35	23	44	28	22	44	33
52	46	28	48	55	63	45	31
28	34	33	52	60	54	53	58
41	32	34	49	52	52	66	35
50	45	29	59	25	48	59	61

a. Do the data present sufficient evidence to conclude there is a difference in the mean age of purchasers and nonpurchasers? Use $\alpha = .10$.

b. What assumptions are necessary in order to answer part **a**?

c. Give the observed significance level for the test, and interpret its value.

d. Find a 90% confidence interval for the difference between the mean ages of purchasers and nonpurchasers.

9.30 One way corporations raise money for expansion is to issue *bonds*, which are loan agreements to repay the purchaser a specified amount of money with a fixed rate of interest paid periodically over the life of the bond. The sale of the bonds is usually handled by an underwriting firm. In a study described in the *Harvard Business Review* (July–Aug. 1979), D. Logue and R. Rogalski ask the question, "Does It Pay to Shop for Your Bond Underwriter?" The reason for the question is that the price of a bond may rise or fall after its issuance. Therefore, whether a corporation receives the market price for a bond depends on the skill of the underwriter. The mean change in the prices of 27 bonds handled over a 12-month period by one underwriter and in the prices of 23 bonds handled by another are given in the table.

	UNDERWRITER 1	UNDERWRITER 2
Sample size	27	23
Sample mean	−.0491	−.0307
Sample variance	.009800	.002465

a. Do the data provide sufficient evidence to indicate a difference in the mean change in bond prices handled by the two underwriters? Test using $\alpha = .05$.

b. Find a 95% confidence interval for the mean difference for the two underwriters, and interpret it.

U.S. PLANTS	JAPANESE PLANTS
7.11%	3.52%
6.06%	2.02%
8.00%	4.91%
6.87%	3.22%
4.77%	1.92%

.0656   .0312

.0122   .0123

9.31 With the emergence of Japan as an industrial superpower, U.S. businesses have begun taking a close look at Japanese management styles and philosophies. Some of the credit for the high quality of Japanese products has been attributed to the Japanese system of permanent employment for their workers. In the United States, high job turnover rates are common in many industries and are associated with high product defect rates. High turnover rates mean U.S. plants are more highly populated with inexperienced workers who are unfamiliar with the company's product lines than is the case in Japan. In a recent study of the air-conditioner industry in Japan and the United States, David Garvin (1983) reported that the difference in the average annual turnover rate of workers between U.S. plants and Japanese plants was 3.1%. In another study, five Japanese and five U.S. plants that manufacture air-conditioners were randomly sampled, and their turnover rates were found to be those listed in the table.

a. Do the data provide sufficient evidence to indicate that the mean annual percentage turnover for U.S. plants exceeds the corresponding mean percentage for Japanese plants? Test using $\alpha = .05$.

b. Report the observed significance level of the test you conducted in part **a**.

c. List any assumptions you made in conducting the hypothesis test of part **a**.

9.32 Suppose you are personnel manager for a company, and you suspect a difference in the mean length of work time lost due to sickness for two types of employees: those who work at night versus those who work during the day. In particular, you suspect that the mean time lost for the night shift exceeds the mean for the day shift. To check your theory, you randomly sample the records for ten employees for each shift category and record the number of days lost due to sickness within the past year. The data are shown in the table.

NIGHT SHIFT, 1		DAY SHIFT, 2	
21	2	13	18
10	19	5	17
14	6	16	3
33	4	0	24
7	12	7	1
$\bar{x}_1 = 12.8$		$\bar{x}_2 = 10.4$	
$\sum_{i=1}^{n} x_i^2 = 2{,}436$		$\sum_{i=1}^{n} x_i^2 = 1{,}698$	

a. Calculate s_1^2 and s_2^2.

b. Show that the pooled estimate of the common population standard deviation, σ, is 8.86. Look at the range of the observations within each of the two samples. Does it appear that the estimate, 8.86, is a reasonable value for σ?

c. If μ_1 and μ_2 represent the mean number of days per year lost due to sickness for the night and day shifts, respectively, test the null hypothesis $H_0: \mu_1 = \mu_2$ against the alternative $H_a: \mu_1 > \mu_2$. Use $\alpha = .05$. Do the data provide sufficient evidence to indicate that $\mu_1 > \mu_2$?

d. What assumptions must be satisfied so that the t test from part c is valid?

e. Suppose you were concerned that the assumptions of part d might not be satisfied. What alternative to a Student's t test do you have?

9.33 Sales quotas are sales volume objectives assigned to specific sales units, such as regions, districts, or salespersons' territories. They are usually expressed in terms of dollar sales volume. Sometimes, in order to achieve manufacturing efficiency or long-term goals, sales managers set quotas for specific products at challenging levels. The underlying idea is that, by setting challenging quotas and attaching significant rewards to their achievement, it is possible to direct salespersons' efforts along desired paths (Winer, 1973). The Universal Products Company (real company, fictitious name) manufactures and markets electronic and electromechanical industrial equipment. It has a sales force of over 1,000 salespersons, organized in ten districts and 135 branch offices. Salespersons have sales quotas on two specific products, Dataprinters and Micromagnetics, as well as an overall sales volume quota. Many salespersons have complained that having to make the existing quota on Dataprinters takes an inordinate amount of time and keeps them from generating a higher overall sales volume. In order to determine how a relaxation of the Dataprinter quota would affect total sales volume, Winer compared the sales volumes of a sample of branch offices whose salespersons all worked under the standard quota with the sales volumes of a sample of branch offices whose salespersons all were given a lower Dataprinter quota. Data were collected for a 7-month period and are reported in the accompanying table in terms of total sales per worker-month (in thousands of dollars).

a. Do the data present sufficient evidence to indicate a difference between mean sales per worker-month for the two types of sales quotas? Use $\alpha = .10$.

BRANCH	LOWER QUOTA	STANDARD QUOTA
1		17.7
2	15.6	
3		15.1
4	14.0	
5		12.3
6		12.0
7	11.2	
8	11.0	
9		10.5
10	10.3	
11		10.0
12	9.4	

b. What assumptions, if any, was it necessary to make in order to carry out the hypothesis test required in part **a**?

c. Estimate the difference between the mean sales per worker-month for the two types of quotas using a 90% confidence interval.

d. Combine the information provided by the test of hypothesis in part **a** and the confidence interval in part **c**, and summarize the results of this study in words that a manager who is not familiar with the language of statistics could understand.

SECTION 9.3

COMPARING TWO POPULATION VARIANCES: INDEPENDENT RANDOM SAMPLES

Suppose you want to use the two-sample t statistic to compare the mean productivity of two paper mills. However, you are concerned that the assumption of equal variances of the productivity for the two plants may be unrealistic. It would be helpful to have a statistical procedure to check the validity of this assumption.

The common statistical procedure for comparing population variances σ_1^2 and σ_2^2 makes an inference about the ratio, σ_1^2/σ_2^2, based on the ratio of the sample variances, s_1^2/s_2^2. Thus, we will attempt to support the alternative hypothesis that the ratio σ_1^2/σ_2^2 differs from 1 (i.e., the variances are unequal) by testing the null hypothesis that the ratio equals 1 (i.e., the variances are equal).

$$H_0: \quad \frac{\sigma_1^2}{\sigma_2^2} = 1 \quad (\sigma_1^2 = \sigma_2^2)$$

$$H_a: \quad \frac{\sigma_1^2}{\sigma_2^2} \neq 1 \quad (\sigma_1^2 \neq \sigma_2^2)$$

We will use the test statistic $F = s_1^2/s_2^2$.

To establish a rejection region for the test statistic, we need to know how s_1^2/s_2^2 is distributed in repeated samples. That is, we need to know the sampling distribution of s_1^2/s_2^2. As you will subsequently see, the sampling distribution of s_1^2/s_2^2 is based on two of the assumptions already required for the t test, namely:

1. The two sampled populations are normally distributed.
2. The samples are randomly and independently selected from their respective populations.

When these assumptions are satisfied and when the null hypothesis is true (i.e., $\sigma_1^2 = \sigma_2^2$), the sampling distribution of $F = s_1^2/s_2^2$ is the **F distribution** with $\nu_1 = (n_1 - 1)$ numerator degrees of freedom and $\nu_2 = (n_2 - 1)$ denominator degrees of freedom. The shape of the F distribution will depend on the degrees of freedom associated with s_1^2 and s_2^2—i.e., $(n_1 - 1)$ and $(n_2 - 1)$. An F distribution with $\nu_1 = 7$ and $\nu_2 = 9$ df is shown in Figure 9.6. As you can see, the distribution is skewed to the right.

FIGURE 9.6

An F Distribution with 7 and 9 df

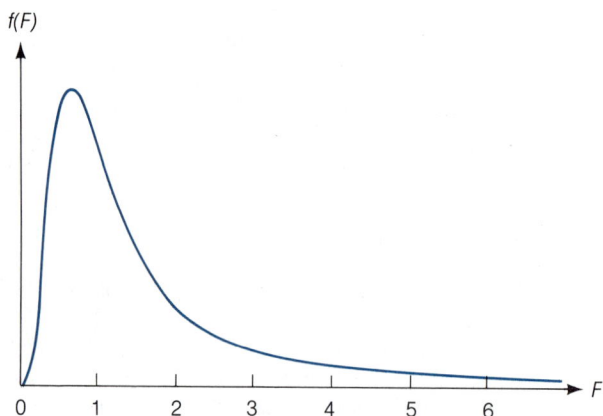

We need to be able to find F values corresponding to the tail areas of this distribution in order to establish the rejection region for our test of hypothesis because, when the population variances are unequal, we expect the ratio F of the sample variances to be either very large or very small. The upper-tail F values can be found in Tables VII, VIII, IX, and X of Appendix B. Table VIII is partially reproduced in Table 9.1. It gives F values that correspond to $\alpha = .05$ upper-tail areas for different degrees of freedom. The columns of Tables VII, VIII, IX, and X correspond to various degrees of freedom for the numerator sample variance, s_1^2, while the rows correspond to the degrees of freedom for the denominator sample variance, s_2^2. Thus, if the numerator degrees of freedom is 7 and the denominator degrees of freedom is 9, we look in the seventh column and ninth row of Table VIII to find $F_{.05} = 3.29$. As shown in Figure 9.7, $\alpha = .05$ is the tail area to the right of 3.29 in the F distribution with 7 and 9 df. That is, if $\sigma_1^2 = \sigma_2^2$, the probability that the F statistic will exceed 3.29 is $\alpha = .05$.

Suppose we want to compare the variability in production for two paper mills and we have obtained the results shown in the table.

SAMPLE 1	SAMPLE 2
$n_1 = 13$ days	$n_2 = 18$ days
$\bar{x}_1 = 26.3$ production units	$\bar{x}_2 = 19.7$ production units
$s_1 = 8.2$ production units	$s_2 = 4.7$ production units

TABLE 9.1 Reproduction of Part of Table VIII of Appendix B: $\alpha = .05$

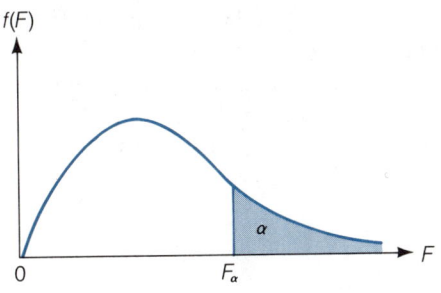

	ν_1	NUMERATOR DEGREES OF FREEDOM								
	ν_2	1	2	3	4	5	6	7	8	9
DENOMINATOR DEGREES OF FREEDOM	1	161.4	199.5	215.7	224.6	230.2	234.0	236.8	238.9	240.5
	2	18.51	19.00	19.16	19.25	19.30	19.33	19.35	19.37	19.38
	3	10.13	9.55	9.28	9.12	9.01	8.94	8.89	8.85	8.81
	4	7.71	6.94	6.59	6.39	6.26	6.16	6.09	6.04	6.00
	5	6.61	5.79	5.41	5.19	5.05	4.95	4.88	4.82	4.77
	6	5.99	5.14	4.76	4.53	4.39	4.28	4.21	4.15	4.10
	7	5.59	4.74	4.35	4.12	3.97	3.87	3.79	3.73	3.68
	8	5.32	4.46	4.07	3.84	3.69	3.58	3.50	3.44	3.39
	9	5.12	4.26	3.86	3.63	3.48	3.37	3.29	3.23	3.18
	10	4.96	4.10	3.71	3.48	3.33	3.22	3.14	3.07	3.02
	11	4.84	3.98	3.59	3.36	3.20	3.09	3.01	2.95	2.90
	12	4.75	3.89	3.49	3.25	3.11	3.00	2.91	2.85	2.80
	13	4.67	3.81	3.41	3.18	3.03	2.92	2.83	2.77	2.71
	14	4.60	3.74	3.34	3.11	2.96	2.85	2.76	2.70	2.65

FIGURE 9.7

An F Distribution for 7 and 9 df: $\alpha = .05$

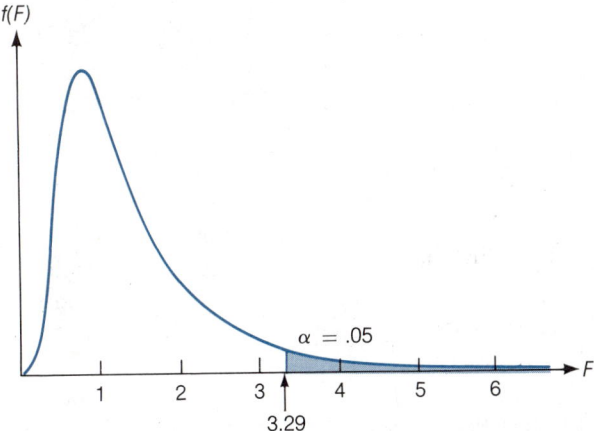

To form the rejection region for a two-tailed F test we want to make certain that the upper tail is used because only the upper-tail values of F are shown in Tables VII, VIII, IX, and X. To accomplish this, *we will always place the larger sample*

variance in the numerator of the F test statistic. This has the effect of doubling the tabulated value for α, since we double the probability that the F ratio will fall in the upper tail by always placing the larger sample variance in the numerator. That is, we make the test two-tailed by putting the larger variance in the numerator rather than establishing rejection regions in both tails.

Thus, for our production example, we have a numerator s_1^2 with $\nu_1 = n_1 - 1 = 12$ and a denominator of s_2^2 with $\nu_2 = n_2 - 1 = 17$. Therefore, the test statistic will be

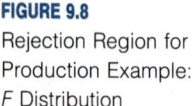

$$F = \frac{\text{Larger sample variance}}{\text{Smaller sample variance}} = \frac{s_1^2}{s_2^2}$$

and we will reject H_0: $\sigma_1^2 = \sigma_2^2$ for $\alpha = .10$ when the calculated value of F exceeds the tabulated value:

$$F_{.05} = 2.38 \qquad \text{(see Figure 9.8)}$$

FIGURE 9.8

Rejection Region for Production Example: *F* Distribution

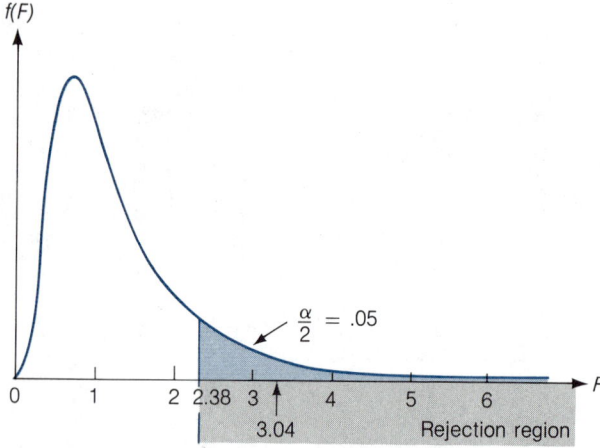

Now, what do the data tell us? We calculate

$$F = \frac{s_1^2}{s_2^2} = \frac{(8.2)^2}{(4.7)^2} = 3.04$$

and compare it to the rejection region shown in Figure 9.8. You can see that the F value 3.04 falls in the rejection region, and therefore, the data provide sufficient evidence to indicate that the population variances differ. Consequently, we would be reluctant to use the two-sample t statistic to compare the population means because the assumption of equal population variances is questionable.

What would you have concluded if the value of F calculated from the samples had not fallen in the rejection region? Would you have concluded that the null hypothesis of equal variances was true? No, because then you risk the possibility of a Type II error (accepting H_0 when H_a is true) without knowing the value of β, the probability of accepting H_0: $\sigma_1^2 = \sigma_2^2$ if in fact it is false. Since we will not consider the calculation of β for specific alternatives in this text, when the F statistic does not fall in the rejection region, we simply conclude that insufficient sample evidence exists to refute the null hypothesis that $\sigma_1^2 = \sigma_2^2$.

The F test for equal population variances is summarized in the box.

F TEST FOR EQUAL POPULATION VARIANCES

ONE-TAILED TEST

H_0: $\sigma_1^2 = \sigma_2^2$

H_a: $\sigma_1^2 < \sigma_2^2$
  (or H_a: $\sigma_1^2 > \sigma_2^2$)

Test statistic:

$$F = \frac{s_2^2}{s_1^2} \text{ when } H_a: \ \sigma_1^2 < \sigma_2^2$$

$$\left(\text{or } F = \frac{s_1^2}{s_2^2} \text{ when } H_a: \ \sigma_1^2 > \sigma_2^2 \right)$$

Rejection region:

$F > F_\alpha$
where F_α is based on $\nu_1 = n_2 - 1$ and $\nu_2 = n_1 - 1$ df
(or $F > F_\alpha$ when H_a: $\sigma_1^2 > \sigma_2^2$,
where F_α is based on $\nu_1 = n_1 - 1$ and $\nu_2 = n_2 - 1$ df)

TWO-TAILED TEST

H_0: $\sigma_1^2 = \sigma_2^2$

H_a: $\sigma_1^2 \neq \sigma_2^2$

Test statistic:

$$F = \frac{\text{Larger sample variance}}{\text{Smaller sample variance}}$$

$$= \frac{s_1^2}{s_2^2} \text{ when } s_1^2 > s_2^2$$

$$\left(\text{or } F = \frac{s_2^2}{s_1^2} \text{ when } s_2^2 > s_1^2 \right)$$

Rejection region:

$F > F_{\alpha/2}$ when $s_1^2 > s_2^2$
where $F_{\alpha/2}$ is based on $\nu_1 = n_1 - 1$ and $\nu_2 = n_2 - 1$ df)
(or $F > F_{\alpha/2}$ when $s_2^2 > s_1^2$,
where $F_{\alpha/2}$ is based on $\nu_1 = n_2 - 1$ and $\nu_2 = n_1 - 1$ df)

Assumptions: 1. Both sampled populations are normally distributed.
  2. The samples are random and independent.

EXAMPLE 9.5

Refer to Example 9.4, in which we used the two-sample t statistic to compare the mean operating costs of cars with rotary and standard engines. Use the F test to check the assumption that the population variances are equal. Use $\alpha = .10$.

SOLUTION

The elements of the test are as follows:

$$H_0: \ \frac{\sigma_1^2}{\sigma_2^2} = 1$$

$$H_a: \ \frac{\sigma_1^2}{\sigma_2^2} \neq 1$$

Test statistic: $F = \dfrac{\text{Larger sample variance}}{\text{Smaller sample variance}} = \dfrac{s_2^2}{s_1^2}$

To find the rejection region, we proceed as follows: The numerator degrees of freedom is $v_1 = n_2 - 1 = 11$; the denominator degrees of freedom is $v_2 = n_1 - 1 = 7$. Thus, from Table VIII of Appendix B we find the rejection region:

$$F > F_{\alpha/2} = F_{.05} \approx 3.60$$

(Since no tabled value is given for $v_1 = 11$, we average the entries at 10 and 12 to obtain $F_{.05} \approx 3.60$.)

We now calculate

$$F = \frac{s_2^2}{s_1^2} = \frac{(6.35)^2}{(4.85)^2} = 1.71$$

This F value is not in the rejection region. Therefore, there is insufficient evidence at the $\alpha = .10$ level to refute the assumption of equal population variances. ∎

The following example shows that the F statistic is sometimes used to compare population variances in their own right rather than just to check the validity of an assumption.

EXAMPLE 9.6

STOCK 1	STOCK 2
$n_1 = 25$	$n_2 = 25$
$\bar{x}_1 = .250$	$\bar{x}_2 = .125$
$s_1 = .76$	$s_2 = .46$

Suppose an investor wants to compare the risks associated with two different stocks, where the risk of a given stock is measured by the variation in daily price changes. Suppose we obtain a random sample of 25 daily price changes for stock 1 and 25 for stock 2. The sample results are summarized in the table. Compare the risks associated with the two stocks by testing the null hypothesis that the variances of the price changes for the stocks are equal. Use $\alpha = .10$.

SOLUTION

Since we wish to detect a difference in population variances, we will want to detect either $\sigma_1^2 > \sigma_2^2$ or $\sigma_2^2 > \sigma_1^2$. Therefore, we choose as the alternative (research) hypothesis, H_a: $\sigma_1^2 \neq \sigma_2^2$, and will conduct the following two-tailed test:

$$H_0: \quad \frac{\sigma_1^2}{\sigma_2^2} = 1$$

$$H_a: \quad \frac{\sigma_1^2}{\sigma_2^2} \neq 1$$

Test statistic: $\quad F = \dfrac{\text{Larger sample variance}}{\text{Smaller sample variance}} = \dfrac{s_1^2}{s_2^2}$

Assumptions: 1. The changes in daily prices for each stock have relative frequency distributions that are approximately normal.
         2. The samples are randomly and independently selected from a set of daily stock reports.

Rejection region: $\quad F > F_{\alpha/2} = F_{.05} = 1.98$
where $F_{.05}$ possesses $v_1 = n_1 - 1 = 24$ and $v_2 = n_2 - 1 = 24$ df.

We calculate

$$F = \frac{s_1^2}{s_2^2} = \frac{(.76)^2}{(.46)^2} = 2.73$$

The calculated F exceeds the rejection value of 1.98. Therefore, we conclude that the variances of daily price changes differ for the two stocks. It appears that the risk, as measured by the variance of daily price changes, is greater for stock 1 than for stock 2. How much reliability can we place in this inference? Only one time in ten (since $\alpha = .10$), on the average, would this statistical test lead us to conclude erroneously that σ_1^2 and σ_2^2 were different when in fact they were equal.

∎

EXAMPLE 9.7

Find the approximate observed significance level for the F test in Example 9.6.

SOLUTION

Since the observed value of the F statistic in Example 9.6 was 2.73, the observed significance level for the test would equal the probability of observing a value of F at least as contradictory to $H_0: \sigma_1^2 = \sigma_2^2$ as $F = 2.73$, if in fact H_0 was true. Since we give the F tables in Appendix B only for values of α equal to .10, .05, .025, and .01, we can only approximate the observed significance level. Checking Tables VII, VIII, IX, and X, we find $F_{.05} = 1.98$, $F_{.025} = 2.27$, and $F_{.01} = 2.66$. Since the observed value of $F = 2.73$ slightly exceeds $F_{.01}$, the observed significance level for the test will be slightly less than

Approximate p-value $= 2(.01) = .02$

Note that we doubled the α value shown in Table X because this was a two-tailed test.

∎

We have presented the F test as a test of hypothesis of the equality of variances—i.e., $\sigma_1^2 = \sigma_2^2$. Although this is the most common application of the test, it can also be used to test a hypothesis that the ratio of the population variances is equal to some specified value, $H_0: \sigma_1^2 / \sigma_2^2 = k$. The test would be conducted exactly the same way as a test of hypothesis concerning the equality of variances except that we would use the test statistic

$$F = \frac{s_1^2}{s_2^2}\left(\frac{1}{k}\right)$$

> **WHAT DO YOU DO IF THE ASSUMPTION OF NORMAL POPULATION DISTRIBUTIONS IS NOT SATISFIED?**
>
> *Answer:* The F test is much more sensitive to departures from normality than the t test for comparing population means discussed in Section 9.2. If you have doubts about the normality of the population frequency distributions, you can use a nonparametric method for comparing the two population variances. A method can be found in the references listed at the end of this chapter.

| | | | | | | | | | | | | | |

EXERCISES 9.34–9.46

LEARNING THE MECHANICS

9.34 Under what conditions is the sampling distribution of s_1^2/s_2^2 an F distribution?

9.35 Use Tables VII, VIII, IX, and X of Appendix B to find each of the following F values:
  a. $F_{.05}$ where $\nu_1 = 9$ and $\nu_2 = 6$ **b.** $F_{.01}$ where $\nu_1 = 18$ and $\nu_2 = 14$
  c. $F_{.025}$ where $\nu_1 = 11$ and $\nu_2 = 4$ **d.** $F_{.10}$ where $\nu_1 = 20$ and $\nu_2 = 5$

9.36 Given ν_1 and ν_2, find the following probabilities:
  a. $\nu_1 = 2$, $\nu_2 = 30$, $P(F \geq 5.39)$ **b.** $\nu_1 = 24$, $\nu_2 = 10$, $P(F < 2.74)$
  c. $\nu_1 = 7$, $\nu_2 = 1$, $P(F \leq 236.8)$ **d.** $\nu_1 = 40$, $\nu_2 = 40$, $P(F > 2.11)$

9.37 For each of the following cases, identify the rejection region that should be used to test $H_0: \sigma_1^2 = \sigma_2^2$ against $H_a: \sigma_1^2 > \sigma_2^2$. Use $\nu_1 = 30$ and $\nu_2 = 20$.
  a. $\alpha = .10$ **b.** $\alpha = .05$
  c. $\alpha = .025$ **d.** $\alpha = .01$

9.38 For each of the following cases, identify the rejection region that should be used to test $H_0: \sigma_1^2 = \sigma_2^2$ against $H_a: \sigma_1^2 \neq \sigma_2^2$. Use $\nu_1 = 10$ and $\nu_2 = 12$.
  a. $\alpha = .20$ **b.** $\alpha = .10$
  c. $\alpha = .05$ **d.** $\alpha = .02$

SAMPLE 1	SAMPLE 2
$n_1 = 12$	$n_2 = 27$
$\bar{x}_1 = 31.7$	$\bar{x}_2 = 37.4$
$s_1^2 = 3.87$	$s_2^2 = 8.75$

9.39 Independent random samples were selected from each of two normally distributed populations, $n_1 = 12$ from population 1 and $n_2 = 27$ from population 2. The means and variances for the two samples are shown in the table.
  a. Do the data provide sufficient evidence to indicate a difference between the population variances? Test using $\alpha = .10$.
  b. Find the interpret the approximate p-value for the test.

SAMPLE 1	SAMPLE 2
3.1	2.3
4.4	1.4
1.2	3.7
1.7	8.9
.7	5.5
3.4	

9.40 Independent random samples were selected from each of two normally distributed populations, $n_1 = 6$ from population 1 and $n_2 = 5$ from population 2. The data are shown in the table.
  a. Do these data provide sufficient evidence to indicate a difference between the population variances? Use $\alpha = .05$.
  b. Find and interpret the approximate observed significance level for the test.

APPLYING THE CONCEPTS

9.41 Tests of product quality can be completely automated or can be conducted using human inspectors or human inspectors aided by mechanical devices. While human inspection is frequently the most economical alternative, it can lead to serious inspection error problems (Benson and Ohta, 1986). Numerous studies have demonstrated that inspectors rarely are able to detect as many as 85% of the defective items that they inspect and that performance varies across inspectors (Sinclair, 1978). To evaluate the performance of inspectors in a new company, a quality manager had a sample of 12 novice inspectors evaluate 200 finished products. The same 200 items were evaluated by 12 experienced inspectors. The quality of each item—whether defective or nondefective—was known to the manager. The table lists the number of inspection errors (classifying a defective item as nondefective, or vice versa) made by each inspector.
  a. Prior to conducting this experiment the manager believed the variance in inspection errors was lower for experienced inspectors than for novice inspectors. Do the sample data support her belief? Test using $\alpha = .05$.

NOVICE INSPECTORS	EXPERIENCED INSPECTORS
30 45	31 19
35 31	15 18
26 33	25 24
40 29	19 10
36 21	28 20
20 48	17 ·21

b. What is the approximate p-value of the test you conducted in part **a**?

9.42 Suppose your firm has been experimenting with two different physical arrangements of its assembly line. It has been determined that both arrangements yield approximately the same average number of finished units per day. To obtain an arrangement that produces greater process control you suggest that the arrangement with the smaller variance in the number of finished units produced per day be permanently adopted. Two independent random samples yield the results shown in the table. Do the samples provide sufficient evidence at $\alpha = .10$ to conclude that the variances of the number of units produced per day differ for the two arrangements? If so, which arrangement would you choose? If not, what would you suggest the firm do?

ASSEMBLY LINE 1	ASSEMBLY LINE 2
$n_1 = 21$ days	$n_2 = 21$ days
$s_1^2 = 1{,}432$	$s_2^2 = 3{,}761$

INSTRUMENT 1	INSTRUMENT 2
29	26
28	34
30	30
28	32
30	28

9.43 The quality control department of a paper company measures the brightness (a measure of reflectance) of finished paper on a periodic basis throughout the day. Two instruments that are available to measure the paper specimens are subject to error, but they can be adjusted so the mean readings for a control paper specimen are the same for both instruments. Suppose you are concerned about the precision of the two instruments— namely, that instrument 2 is less precise than instrument 1. To check this theory, five measurements of a single paper sample are made on both instruments. The data are shown in the table. Do the data provide sufficient evidence to indicate that instrument 2 is less precise than instrument 1? Test using $\alpha = .05$.

9.44 Recent technological advances have turned the telephone operator's workplace into what some experts call "the office of the future." For example, directory-assistance operators use computers to look up telephone numbers, and then by flicking a key have a recorded voice give the number to the customer. The computerization of the operator's workplace has significantly shortened the time necessary to service a customer. Thus, while the number of telephone calls made in this country per year increased from 67 billion in 1950 to 310 billion in 1980, the number of telephone operators decreased from 244,000 to 128,000. However, the computer not only speeds the operator's work, it monitors the operator's work as well—somewhat like a foreman. This monitoring has led some telephone company offices to issue a guideline of 30 seconds as the maximum amount of time an operator should spend completing operator-assisted calls (Serrin, 1983). To study the effect of this guideline on the mean time to complete a call, a researcher planned to estimate the difference in the mean time between offices that do and offices that do not issue the guideline to its operators. The completion times (in seconds) shown in the table were sampled from the computer-collected data that results from monitoring the operators.

TIME GUIDELINE ISSUED		TIME GUIDELINE NOT ISSUED	
31.2	26.3	30.0	21.4
33.5	27.3	22.6	39.8
27.2	25.9	35.9	45.3
23.5	20.3	41.3	37.1
28.8	26.8	28.9	25.6

a. The researcher knows that in order to be able to use the two-sample t statistic in constructing a confidence interval, the assumption of equal population variances should first be examined. Test the equality of the population variances using $\alpha = .05$. What does your test indicate about the appropriateness of using the two-sample t in this situation?

b. List any assumptions you made in conducting the hypothesis test of part **a**.

c. In the context of this problem, describe the Type I and Type II errors associated with your hypothesis test of part **a**.

d. What is the approximate p-value of the test you conducted in part **a**?

9.45 Refer to Exercise 9.33 in Section 9.2, in which a t test was used to determine whether sales volume was affected by the type of sales quota assigned to salespersons. In conducting the t test, it was assumed that the variance of total sales per worker-month was the same for both populations being studied. Do the data provide sufficient evidence to indicate that this assumption may be violated? Test using $\alpha = .05$.

9.46 In Exercise 9.28 a manufacturing company was interested in determining whether a significant difference existed between the mean number of units produced per day by two different machine operators. Two independent random samples yielded the data shown in the table.

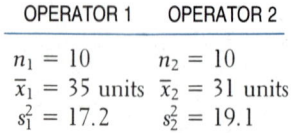

OPERATOR 1	OPERATOR 2
$n_1 = 10$	$n_2 = 10$
$\bar{x}_1 = 35$ units	$\bar{x}_2 = 31$ units
$s_1^2 = 17.2$	$s_2^2 = 19.1$

a. In order to conduct the hypothesis test required in Exercise 9.28 you had to assume $\sigma_1^2 = \sigma_2^2$. Use an F test with $\alpha = .10$ to determine whether the data violate this assumption. Explain the significance of your result.

b. If your conclusion in part **a** were incorrect, would you have committed a Type I error or a Type II error? Explain.

| | | | | | | | | | | | | |

SECTION 9.4

INFERENCES ABOUT THE DIFFERENCE BETWEEN TWO POPULATION MEANS: PAIRED DIFFERENCE EXPERIMENTS

Suppose you want to compare the mean daily sales of two restaurants located in the same city. If you were to record the restaurants' total sales for each of 12 days (2 work weeks), the results might appear as shown in Table 9.2.

Test the null hypothesis that the mean daily sales, μ_1 and μ_2, for the two restaurants are equal against the alternative hypothesis that they differ; i.e.,

$$H_0: (\mu_1 - \mu_2) = 0$$

$$H_a: (\mu_1 - \mu_2) \neq 0$$

TABLE 9.2   Daily Sales for Two Restaurants

DAY	RESTAURANT 1	RESTAURANT 2
1 (Monday)	$ 759	$ 678
2 (Tuesday)	981	933
3 (Wednesday)	1,005	918
4 (Thursday)	1,449	1,302
5 (Friday)	1,905	1,782
6 (Saturday)	2,073	1,971
7 (Monday)	693	639
8 (Tuesday)	873	825
9 (Wednesday)	1,074	999
10 (Thursday)	1,338	1,281
11 (Friday)	1,932	1,827
12 (Saturday)	2,106	2,049
	$\bar{x}_1 = \$1,349.00$	$\bar{x}_2 = \$1,267.00$
	$s_1 = \$530.07$	$s_2 = \$516.03$

Using the two-sample t statistic (Section 9.2) we would calculate

$$s_p^2 = \frac{(n_1 - 1)s_1^2 + (n_2 - 1)s_2^2}{n_1 + n_2 - 2}$$

$$= \frac{(12 - 1)(530.07)^2 + (12 - 1)(516.03)^2}{12 + 2 - 2}$$

$$= 273,630.6$$

and

$$t = \frac{(\bar{x}_1 - \bar{x}_2) - 0}{\sqrt{s_p^2\left(\frac{1}{n_1} + \frac{1}{n_2}\right)}} = \frac{(1,349.00 - 1,267.00)}{\sqrt{273,630.6\left(\frac{1}{12} + \frac{1}{12}\right)}}$$

$$= \frac{82.0}{213.54} = .38$$

This small t value will not lead to rejection of H_0 when compared to the t distribution with $n_1 + n_2 - 2 = 22$ df, even if α were chosen as large as .20 ($t_{\alpha/2} = t_{.10} = 1.321$). Thus, we might conclude that insufficient evidence exists to infer that there is a difference in mean daily sales for the two restaurants.

However, if you examine the data in Table 9.2 more closely, you will find this conclusion difficult to accept. The sales of restaurant 1 exceed those of restaurant 2 *for every one of the 12 days*. This, in itself, is strong evidence to indicate that μ_1 differs from μ_2, and we will subsequently confirm this fact. Why, then, was the t test unable to detect this difference?

The cause of this apparent inconsistency with the test result is that the two-sample t is inappropriate, because the assumption of independent samples is invalid.

If you examine the pairs of daily sales, you will note that the sales of the two restaurants tend to rise and fall together over the days of the week. This pattern suggests a very strong daily dependence between the two samples and a violation of the assumption of independence required for the two-sample t test of Section 9.2. In this particular situation, note the *large variation within samples* (reflected by the large value of s_p^2) in comparison to the *small difference between the sample means*. Because s_p^2 was so large, the t test was unable to detect a possible difference between μ_1 and μ_2.

TABLE 9.3

Daily Sales and Differences for Two Restaurants

DAY	RESTAURANT 1	RESTAURANT 2	(RESTAURANT 1 − RESTAURANT 2)
1 (Monday)	$ 759	$ 678	$ 81
2 (Tuesday)	981	933	48
3 (Wednesday)	1,005	918	87
4 (Thursday)	1,449	1,302	147
5 (Friday)	1,905	1,782	123
6 (Saturday)	2,073	1,971	102
7 (Monday)	693	639	54
8 (Tuesday)	873	825	48
9 (Wednesday)	1,074	999	75
10 (Thursday)	1,338	1,281	57
11 (Friday)	1,932	1,827	105
12 (Saturday)	2,106	2,049	57

$$\bar{x}_D = \$82.00$$
$$s_D = \$32.00$$

Now, consider a valid method to analyze the data of Table 9.1. We add to this table a column of differences between the daily sales of the restaurants, and thus form Table 9.3. We can regard these daily differences in sales as a random sample of all daily differences, past and present. Then we can use this sample to make inferences about the mean, μ_D, of the population of differences, *which is equal to the difference* ($\mu_1 - \mu_2$): i.e., the mean of the population (sample) of differences equals the difference between the population (sample) means. Thus, our test becomes

$$H_0: \quad \mu_D = 0 \quad (\text{i.e., } \mu_1 - \mu_2 = 0)$$
$$H_a: \quad \mu_D \neq 0 \quad (\text{i.e., } \mu_1 - \mu_2 \neq 0)$$

The test statistic is a one-sample t (Section 8.6) since we are now analyzing a single sample of differences:

Test statistic: $\quad t = \dfrac{\bar{x}_D - 0}{s_D/\sqrt{n_D}}$

where

$\bar{x}_D$ = Sample mean of differences

s_D = Sample standard deviation of differences

n_D = Number of differences

Assumptions: 1. The population of differences in daily sales is approximately normally distributed.
2. The sample differences are randomly selected from a population of differences.

To find the rejection region, we first choose $\alpha = .05$. Then we will reject H_0 if

$$t < -t_{.025} \quad \text{or} \quad t > t_{.025}$$

where $t_{.025}$ is based on $(n_D - 1)$ degrees of freedom.

Referring to Table VI of Appendix B, we find the t value corresponding to $\alpha/2 = .025$ and $n_D - 1 = 12 - 1 = 11$ df to be $t_{.025} = 2.201$. Thus, the null hypothesis will be rejected if $t < -2.201$ or $t > 2.201$. Note that the number of degrees of freedom has decreased from $n_1 + n_2 - 2 = 22$ to 11 by using the **paired difference experiment** rather than the two independent random samples design.

Now calculate

$$t = \frac{\bar{x}_D - 0}{s_D/\sqrt{n_D}} = \frac{82.00}{32.00/\sqrt{12}} = 8.88$$

Because this value of t falls in the rejection region, we conclude that the difference in mean daily sales for the two restaurants differs from 0. The fact that $\bar{x}_1 - \bar{x}_2 = \bar{x}_D = \$82.00$ strongly suggests that the mean daily sales for restaurant 1 exceeds the mean daily sales for restaurant 2.

This kind of experiment, in which observations are paired and the differences analyzed, is called a **paired difference experiment**. In many cases a paired difference experiment can provide more information about the difference between population means than an independent samples experiment. The differencing removes the variability due to the dimension on which the observations are paired. For instance, in the restaurant example, the day-to-day variability in daily sales is removed by analyzing the differences between the restaurants' daily sales. The removal of the variability due to this extra dimension is called **blocking**, and the paired difference experiment is a simple example of a **randomized block experiment**. In our example, the days represent the blocks. [*Note:*   Randomized block experiments are discussed in greater detail in Chapter 15.]

Some other examples for which the paired difference experiment might be appropriate are the following:

1. To compare the performance of two automobile salespeople, we might test a hypothesis about the difference $(\mu_1 - \mu_2)$ in their respective mean monthly sales. If we randomly choose n_1 months of salesperson 1's sales and independently choose n_2 months of salesperson 2's sales, the month-to-month variability caused by the seasonal nature of new car sales might inflate s_p^2 and prevent the two-sample t statistic from detecting a difference between μ_1 and μ_2, if such a difference actually exists. However, by taking the difference in monthly sales for the two salespeople for each of n months, the month-to-month variability (seasonal variation) in sales can be eliminated and the probability of detecting a difference between μ_1 and μ_2, if a difference exists, is increased.

2. Suppose you want to estimate the difference $(\mu_1 - \mu_2)$ in mean price between two major brands of premium gasoline. If you were to choose two independent random samples of stations for each brand, the variability in price due to geographic location may be large. To eliminate this source of variability, you could choose pairs of stations, one station for each brand, in close geographic proximity and use the sample of differences between the prices of the brands to make an inference about $(\mu_1 - \mu_2)$.

3. Suppose a college placement center wants to estimate the difference $(\mu_1 - \mu_2)$ in mean starting salaries for men and women graduates who seek jobs through the center. If it independently samples men and women, the starting salaries may vary due to their different college majors and differences in grade-point averages. To eliminate these sources of variability, the placement center could match male and female job-seekers according to their majors and grade-point averages. Then the difference between the starting salaries of each pair in the sample could be used to make an inference about $(\mu_1 - \mu_2)$.

The hypothesis-testing and confidence interval procedures based on a paired difference experiment are summarized in the next two boxes.

PAIRED DIFFERENCE TEST OF HYPOTHESIS

ONE-TAILED TEST	**TWO-TAILED TEST**
H_0: $(\mu_1 - \mu_2) = D_0$, i.e., $(\mu_D = D_0)$	H_0: $(\mu_1 - \mu_2) = D_0$, i.e., $(\mu_D = D_0)$
H_a: $(\mu_1 - \mu_2) < D_0$, i.e., $(\mu_D < D_0)$ [or H_a: $(\mu_1 - \mu_2) > D_0$, i.e., $(\mu_D > D_0)$]	H_a: $(\mu_1 - \mu_2) \neq D_0$, i.e., $(\mu_D \neq D_0)$
Test statistic: $t = \dfrac{\bar{x}_D - D_0}{s_D / \sqrt{n_D}}$	*Test statistic:* $t = \dfrac{\bar{x}_D - D_0}{s_D / \sqrt{n_D}}$
Rejection region: $t < -t_\alpha$ [or $t > t_\alpha$ when H_a: $(\mu_1 - \mu_2) > D_0$]	*Rejection region:* $t < -t_{\alpha/2}$ or $t > t_{\alpha/2}$
where t_α has $(n_D - 1)$ df.	where $t_{\alpha/2}$ has $(n_D - 1)$ df.

Assumptions: 1. The relative frequency distribution of the population of differences is normal.
2. The differences are randomly selected from the population of differences.

PAIRED DIFFERENCE CONFIDENCE INTERVAL

$$\bar{x}_D \pm t_{\alpha/2}\frac{s_D}{\sqrt{n_D}}$$

where $t_{\alpha/2}$ has $(n_D - 1)$ df.

Assumption: Same as for the paired difference test (previous box).

EXAMPLE 9.8

A paired difference experiment is conducted to compare the starting salaries of male and female college graduates who find jobs. Pairs are formed by choosing a male and a female with the same major and similar grade-point averages. Suppose a random sample of ten pairs is formed in this manner, and the starting annual salary of each person is recorded. The results are shown in Table 9.4. Test to see whether there is evidence that the mean starting salary, μ_1, for males exceeds the mean starting salary, μ_2, for females. Use $\alpha = .05$.

TABLE 9.4

PAIR	MALE	FEMALE	DIFFERENCE (Male – Female)
1	$14,300	$13,800	$ 500
2	16,500	16,600	−100
3	15,400	14,800	600
4	13,500	13,500	0
5	18,500	17,600	900
6	12,800	13,000	−200
7	14,500	14,200	300
8	16,200	15,100	1,100
9	13,400	13,200	200
10	14,200	13,500	700

SOLUTION

Since we are interested in determing whether the data indicate that μ_1 exceeds μ_2—i.e., whether the mean starting salary for men exceeds the mean starting salary for women—we will choose a one-sided alternative (research) hypothesis. Then the elements of the paired difference test are

H_0: $\mu_D = 0$ $(\mu_1 - \mu_2 = 0)$

H_a: $\mu_D > 0$ $(\mu_1 - \mu_2 > 0)$

Test statistic: $t = \dfrac{\bar{x}_D - 0}{s_D/\sqrt{n_D}}$

Assumption: The relative frequency distribution for the population of differences is normal.

Since the test is upper-tailed, we will reject H_0 if

$$t > t_\alpha = t_{.05} = 1.833$$

where t_α is based on $n_D - 1 = 9$ df. The rejection region is shown in Figure 9.9.

FIGURE 9.9

Rejection Region
for Example 9.8

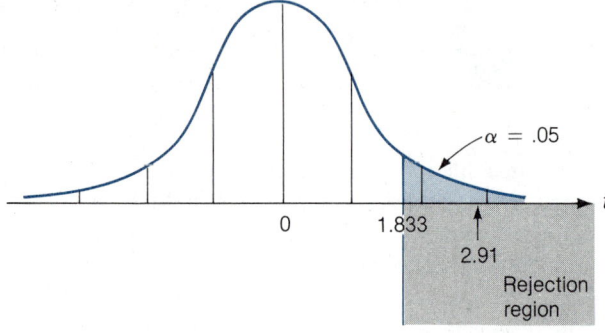

Using x_{Di} to represent the ith difference measurement, we now calculate

$$\sum_{i=1}^{10} x_{Di} = 500 + (-100) + \cdots + 700 = 4{,}000$$

and

$$\sum_{i=1}^{10} x_{Di}^2 = 3{,}300{,}000$$

Then,

$$\bar{x}_D = \frac{\sum_{i=1}^{10} x_{Di}}{10} = \frac{4{,}000}{10} = 400$$

$$s_D^2 = \frac{\sum_{i=1}^{n_D} (x_{Di} - \bar{x}_D)^2}{n_D - 1} = \frac{\sum_{i=1}^{10} x_{Di}^2 - \left(\sum_{i=1}^{10} x_{Di}\right)^2 \Big/ 10}{9}$$

$$= \frac{3{,}300{,}000 - (4{,}000)^2 / 10}{9} = 188{,}888.89$$

$$s_D = \sqrt{s_D^2} = 434.61$$

Substituting these values into the formula for the test statistic, we find that

$$t = \frac{\bar{x}_D - 0}{s_D / \sqrt{n_D}} = \frac{400}{434.61 / \sqrt{10}} = \frac{400}{137.44} = 2.91$$

As you can see in Figure 9.9, the calculated t falls in the rejection region. Thus, we conclude at $\alpha = .05$ that the mean starting salary for males exceeds the mean starting salary for females. ∎

One measure of the amount of information about $(\mu_1 - \mu_2)$ gained by using a paired difference experiment rather than an independent samples experiment in Example 9.8 is the relative widths of the confidence intervals obtained by the two methods. A 95% confidence interval for $(\mu_1 - \mu_2)$ using the paired difference experiment is

$$\bar{x}_D \pm t_{\alpha/2} \frac{s_D}{\sqrt{n_D}}$$

where t_α has $n_D - 1 = 9$ df. Substituting into this formula, we obtain

$$400 \pm t_{.025} \frac{434.61}{\sqrt{10}} = 400 \pm 2.262 \frac{434.61}{\sqrt{10}}$$

$$= 400 \pm 310.88 \approx 400 \pm 311 = (\$89, \$711)$$

If we analyzed the same data as though they were from an independent samples experiment,* we would first calculate the following quantities:

MALES	FEMALES
$\bar{x}_1 = \$14,930$	$\bar{x}_2 = \$14,530$
$s_1^2 = 3,009,000$	$s_2^2 = 2,331,222.22$

Then

$$s_p^2 = \frac{(n_1 - 1)s_1^2 + (n_2 - 1)s_2^2}{n_1 + n_2 - 2} = \frac{9(3,009,000) + 9(2,331,222.22)}{18}$$

$$= 2,670,111.11$$

where s_p^2 is based on $(n_1 + n_2 - 2) = (10 + 10 - 2) = 18$ df. The 95% confidence interval is

$$(\bar{x}_1 - \bar{x}_2) \pm t_{\alpha/2} \sqrt{s_p^2 \left(\frac{1}{n_1} + \frac{1}{n_2}\right)} = 400 \pm t_{.025} \sqrt{2,670,111.11\left(\frac{1}{10} + \frac{1}{10}\right)}$$

$$= 400 \pm 2.101 \sqrt{2,670,111.11\left(\frac{1}{10} + \frac{1}{10}\right)}$$

$$= 400 \pm 1,535.35$$

$$\approx 400 \pm 1,535 = (-\$1,135, \$1,935)$$

*This is done only to provide a measure of the increase in the amount of information obtained by a paired design in comparison to an unpaired design. Actually, if an experiment is designed using pairing, an unpaired analysis would be invalid because the assumption of independent samples would not be satisfied.

The confidence interval for the independent sampling experiment is about five times wider than the corresponding paired difference confidence interval. Blocking out the variability due to differences in majors and grade-point averages significantly increases the information about the difference in mean male and female starting salaries by providing a much more accurate (smaller confidence interval for the same confidence coefficient) estimate of $(\mu_1 - \mu_2)$.

You may wonder whether a paired difference experiment is always superior to an independent samples experiment. The answer is: Most of the time, but not always. We sacrifice half the degrees of freedom in the t statistic when a paired difference design is used instead of an independent samples design. This is a *loss* of information, and unless this loss is more than compensated for by the reduction in variability obtained by blocking (pairing), the paired difference experiment will result in a net loss of information about $(\mu_1 - \mu_2)$. Thus, we should be convinced that the pairing will significantly reduce variability before performing the paired difference experiment. Most of the time this will happen.

One final note: The pairing of the observations is determined *before* the experiment is performed (that is, by the *design* of the experiment). *A paired difference experiment is never obtained by pairing the sample observations after the measurements have been acquired.* Such is the stuff of which statistical lies are made!

WHAT DO YOU DO WHEN THE ASSUMPTION OF A NORMAL DISTRIBUTION FOR THE POPULATION OF DIFFERENCES IS NOT SATISFIED?

Answer: Use the Wilcoxon signed rank test for the paired difference design. (See Chapter 16.)

CASE STUDY 9.1

COMPARING SALARIES FOR EQUIVALENT WORK

Procedures to maintain comparable salaries between federal white-collar workers and those in the private sector are mandated by federal law. William M. Smith (1976) has discussed one statistical mechanism used in complying with this law. An annual survey, the *National Survey for Professional, Administrative, Technical, and Clerical Pay,* is conducted by the Bureau of Labor Statistics. Salary information is collected for approximately 85 work-level categories ranging from clerical to administrative positions in 20 occupations.

The design of the survey resembles a paired difference design because the occupation and experience of employees in the private and government sectors are matched as closely as possible before a comparison is made. Test statistics like the paired difference t can be used to compare the mean salaries for the two sectors. Then, if the data indicate that the mean salaries differ for certain levels, the need for an adjustment is indicated.

Smith points out that presidential or legislative intervention often prevents the adjustments indicated by the data from being enacted. However, the results of the survey are valuable as a salary guide for the private sector and to those performing general economic analyses.

LEARNING THE MECHANICS

9.47 A paired difference experiment yielded n_D pairs of observations. In each case, what is the rejection region for testing H_0: $\mu_D = 2$ against H_a: $\mu_D > 2$?

a. $n_D = 10,\quad \alpha = .05$ **b.** $n_D = 20,\quad \alpha = .10$

c. $n_D = 5,\quad \alpha = .025$ **d.** $n_D = 9,\quad \alpha = .01$

9.48 A paired difference experiment yielded the data shown in the table.

PERSON	BEFORE x_1	AFTER x_2
1	83	92
2	60	71
3	55	56
4	99	104
5	77	89

a. Compute $\bar{x}_D$ and s_D.
b. Demonstrate that $\bar{x}_D = \bar{x}_1 - \bar{x}_2$.
c. Is there sufficient evidence to conclude that $\mu_1 \neq \mu_2$? Test using $\alpha = .05$.
d. Find and interpret the approximate p-value for the paired difference test.
e. What assumptions are necessary so the paired difference test will be valid?

9.49 Refer to Exercise 9.48. Construct a 95% confidence interval for μ_D.

9.50 A paired difference experiment produced the following data:

$$n_D = 18 \qquad \bar{x}_1 = 92 \qquad \bar{x}_2 = 95.5 \qquad \bar{x}_D = -3.5 \qquad s_D^2 = 21$$

a. Determine the values of t for which the null hypothesis, $(\mu_1 - \mu_2) = 0$, would be rejected in favor of the alternative hypothesis $(\mu_1 - \mu_2) < 0$. Use $\alpha = .10$.
b. Conduct the paired difference test described in part **a**. Draw the appropriate conclusions.
c. What assumptions are necessary so that the paired difference test will be valid?
d. Construct a 90% confidence interval for the mean difference μ_D.
e. Which of the two inferential procedures, the confidence interval of part **d** or the test of hypothesis of part **b**, provides more information about the difference between the population means?

9.51 Frequently, a paired difference experiment provides more information about the difference between two population means than an independent random samples experiment. Explain. Also explain when it may not.

9.52 A paired difference experiment yielded the data shown in the table.
a. Test H_0: $\mu_D = 10$ against H_a: $\mu_D \neq 10$, where $\mu_D = (\mu_1 - \mu_2)$. Use $\alpha = .05$.
b. Report the p-value for the test you conducted in part **a**. Interpret the p-value.

PAIR	x	y
1	55	44
2	68	55
3	40	25
4	55	56
5	75	62
6	52	38
7	49	31

APPLYING THE CONCEPTS

9.53 Do people generally receive the asking price when they sell their home? The accompanying table lists the asking price and sale price for a random sample of ten 1986 single-family home sales in Eden Prairie, Minnesota.

SINGLE-FAMILY HOME	ASKING PRICE	SALE PRICE
1	$ 98,000	$ 94,300
2	122,500	119,800
3	109,900	105,000
4	98,000	100,000
5	79,900	79,900
6	275,000	265,000
7	99,700	96,000
8	97,900	99,200
9	209,000	209,000
10	173,900	169,000

Source: *MLS Compilation of Listings, Division II*, No. 44, Oct. 31–Nov. 6, 1986. Greater Minneapolis Area Board of Realtors.

a. Explain why the samples are not independent.

b. What is the dimension on which the samples are paired or matched?

c. Conduct a test of hypothesis that will help answer the question of interest for people in Eden Prairie who sold their homes in 1986. Use $\alpha = .05$.

9.54 *Time* (Dec. 7, 1981) reports that a growing number of corporations are encouraging their employees to "get things off their chests" by communicating directly to management on hot-line phones, through the mail, or in face-to-face meetings. For example, American Express Corp. has instituted a mail program that guarantees their employees anonymity (if desired) and an answer from the responsible person (including the company chairperson) within 10 days. A similar program at IBM generates an average of 13,000 letters a year from IBM's 195,000 U.S. employees. Such programs reflect a growing concern on the part of American business for ways to improve worker productivity.

Before instituting an employee complaint/suggestion program in its manufacturing plant, a company randomly sampled eight workers and measured their productivity in terms of the number of items produced per day. A year after the start of the complaint program the productivity of seven of these workers was reevaluated. (The eighth had been promoted in the interim.) The productivity data are shown in the table.

EMPLOYEE ID NUMBER	AUGUST 1986	AUGUST 1987
1011	10	9
0033	9	11
0998	12	14
0006	8	9
1802	10	9
0246	11	14
0777	14	–
1112	11	13

a. Do the data provide evidence that the complaint/suggestion program has helped to increase worker productivity? Test using $\alpha = .10$, and clearly state any assumptions you make in conducting the test.

b. Discuss how the 1-year gap between productivity evaluations could weaken the results of the study.

9.55 A large corporation is considering hiring a consultant to conduct a yearly in-house assertiveness training course for its incoming group of management trainees. Before signing a contract with the consultant, the corporation wishes to assess the effectiveness of the course. Ten trainees were randomly selected to participate in the course and each was given tests designed to measure assertiveness before the course and after the course was completed. Higher scores on the test indicate higher levels of assertiveness. The test scores are shown in the table.

TRAINEE	BEFORE	AFTER	TRAINEE	BEFORE	AFTER
1	50	62	6	56	65
2	61	63	7	47	42
3	51	49	8	66	69
4	43	44	9	42	35
5	52	48	10	50	43

a. Do the data provide sufficient evidence to conclude that the trainees are more assertive after taking the course? Test using $\alpha = .05$.

b. Report the approximate p-value for the test and interpret its value.

c. Based on the results of the test, would you recommend that the corporation hire the consultant? Explain.

d. What assumptions must you make so that the paired t test will be valid?

FIRM	1960's	1970's
1	10	10
2	8	12
3	9	8
4	7	16
5	8	14
6	7	6
7	6	11
8	9	12
9	8	11
10	7	12

9.56 Among the better known and most frequently enforced antitrust laws are those against price fixing (Sherman Act, Section One) and monopolization or attempt to monopolize (Sherman Act, Section Two). Other antitrust violations include price discrimination, retail price maintenance, and tying the sale of a product to the purchase of another product. Alan R. Beckenstein, H. Landis Gabel, and Karlene Roberts (1983) surveyed 188 *Fortune 500* industrial companies and found that, on average (without adjusting for inflation), these companies "spent $3\frac{1}{2}$ times the amount on antitrust investigations, legal fees, fines, damages, court costs, and out-of-court settlements in the 1970's as they did in the 1960's." Further, they claim that, on average, these companies faced 3.22 more antitrust litigations in the 1970's than in the 1960's. As a result, Beckenstein, Gabel, and Roberts argue that companies need to give more attention to the possible legal consequences of their actions than they have in the past. They conclude that today's legal risks make it imperative that managers at all levels in the organization be involved with designing and implementing strategies to comply with the antitrust laws. Another survey of ten *Fortune 500* companies yielded the data in the table about the number of litigations faced by the firms in the 1960's and 1970's.

a. Do the data provide sufficient evidence to reject Beckenstein, Gabel, and Roberts' claim? Conduct a test of hypothesis and use the p-value of your test to help you answer this question.

b. Use a 90% confidence interval to estimate the difference in the mean number of antitrust litigations per firm in the 1970's and the 1960's. Compare your results to Beckenstein, Gabel, and Roberts' result.

9.57 A manufacturer of automobile shock absorbers was interested in comparing the durability of its shocks with that of the shocks produced by its biggest competitor. To make the comparison, one of the manufacturer's and one of the competitor's shocks were randomly selected and installed on the rear wheels of each of six cars. After the cars had been driven 20,000 miles, the strength of each test shock was measured, coded, and recorded. Results of the examination are shown in the table.

CAR NUMBER	MANUFACTURER'S SHOCK	COMPETITOR'S SHOCK
1	8.8	8.4
2	10.5	10.1
3	12.5	12.0
4	9.7	9.3
5	9.6	9.0
6	13.2	13.0

  a. Do that data present sufficient evidence to conclude that there is a difference in the mean strength of the two types of shocks after 20,000 miles of use? Use $\alpha = .05$.

  b. Find the approximate observed significance level for the test, and interpret its value.

  c. What assumptions are necessary to apply a paired difference analysis to the data?

  d. Construct a 95% confidence interval for $(\mu_1 - \mu_2)$. Interpret the confidence interval.

9.58 Suppose the data in Exercise 9.57 are based on independent random samples.

  a. Do the data provide sufficient evidence to indicate a difference between the mean strengths for the two types of shocks? Use $\alpha = .05$.

  b. Construct a 95% confidence interval for $(\mu_1 - \mu_2)$. Interpret your result.

  c. Compare the confidence intervals you obtained in Exercise 9.57 and in part **b** of this exercise. Which is wider? To what do you attribute the difference in width? Assuming in each case that the appropriate assumptions are satisfied, which interval provides you with more information about $(\mu_1 - \mu_2)$? Explain.

  d. Are the results of an unpaired analysis valid when the data have been collected from a paired experiment?

9.59 In Case Study 9.1 we discussed the *National Survey for Professional, Administrative, Technical, and Clerical Pay*, which is conducted annually to determine whether federal pay scales are commensurate with private sector salaries. Recall that the government and private workers in the study are matched as closely as possible before the salaries are compared. Suppose that the table shows the annual salaries for 12 pairs of individuals in the sample, matched on job level and experience.

PAIR	PRIVATE	GOVERNMENT	PAIR	PRIVATE	GOVERNMENT
1	$12,500	$11,750	7	$15,800	$14,500
2	22,300	20,900	8	17,500	17,900
3	14,500	14,800	9	23,300	21,400
4	32,300	29,900	10	42,100	43,200
5	20,800	21,500	11	16,800	15,200
6	19,200	18,400	12	14,500	14,200

a. Use these data to construct a 99% confidence interval for the difference between the mean salaries of the private and government sectors.

b. What assumptions are necessary for the validity of the procedure you used in part **a**?

9.60 A pupillometer is a device used to observe changes in an individual's pupil dilations as he or she is exposed to different visual stimuli. Since there is a direct correlation between the amount an individual's pupil dilates and his or her interest in the stimuli, marketing organizations sometimes use pupillometers to help them evaluate potential consumer interest in new products, alternative package designs, and other factors. The Design and Market Research Laboratories of the Container Corporation of America used a pupillometer to evaluate consumer reaction to different silverware patterns for one of its clients (McGuire, 1973). Suppose 15 consumers were chosen at random, and each was shown two different silverware patterns. The pupillometer readings for each consumer are shown in the table (in millimeters).

CONSUMER	PATTERN 1	PATTERN 2	CONSUMER	PATTERN 1	PATTERN 2
1	1.00	.80	9	.98	.91
2	.97	.66	10	1.46	1.10
3	1.45	1.22	11	1.85	1.60
4	1.21	1.00	12	.33	.21
5	.77	.81	13	1.77	1.50
6	1.32	1.11	14	.85	.65
7	1.81	1.30	15	.15	.05
8	.91	.32			

a. Use a 90% confidence interval to estimate the difference in the mean amount of pupil dilation per consumer for silverware patterns 1 and 2.

b. Give a practical interpretation of the interval.

c. Does the confidence interval in part **a** support the inference that the mean dilation for pattern 1 exceeds that for pattern 2 by more than .1 millimeter?

d. Conduct a test of hypothesis to determine whether the data indicate that the mean dilation for pattern 1 exceeds that for pattern 2 by more than .1 millimeter. Use $\alpha = .05$.

|||||||||||||
SECTION 9.5

INFERENCES ABOUT THE DIFFERENCE BETWEEN POPULATION PROPORTIONS: INDEPENDENT BINOMIAL EXPERIMENTS

Suppose a manufacturer of campers wants to compare the potential market for its products in the northeastern United States to the market in the southeastern United States. Such a comparison would help the manufacturer decide where to concentrate sales efforts. The company randomly chooses 1,000 households in the northeastern United States (NE) and 1,000 households in the southeastern United States (SE) and determines whether each household plans to buy a camper within the next 5 years. The objective is to use this sample information to make an inference about the difference $(p_1 - p_2)$ between the proportion p_1 of *all* households in the NE and the proportion p_2 of *all* households in the SE that plan to purchase a camper within 5 years.

The two samples represent independent binomial experiments (see Section 5.4 for the characteristics of binomial experiments), with the binomial random variables

x_1 and x_2 being the numbers of the 1,000 sampled households in each area that indicate they will purchase a camper within 5 years. The results of the sampling can be summarized as follows:

NE	SE
$n_1 = 1{,}000$	$n_2 = 1{,}000$
$x_1 = 42$	$x_2 = 24$

We can now calculate the *sample* proportions $\hat{p}_1$ and $\hat{p}_2$ of the households in the NE and SE, respectively, that are prospective buyers:

$$\hat{p}_1 = \frac{x_1}{n_1} = \frac{42}{1{,}000} = .042$$

$$\hat{p}_2 = \frac{x_2}{n_2} = \frac{24}{1{,}000} = .024$$

The difference between the sample proportions, $(\hat{p}_1 - \hat{p}_2)$, is an intuitively appealing point estimator of the difference between the population parameters, $(p_1 - p_2)$. For our example, the estimate is

$$(\hat{p}_1 - \hat{p}_2) = .042 - .024 = .018$$

To judge the reliability of the estimator $(\hat{p}_1 - \hat{p}_2)$, we must observe its performance in repeated sampling from the two populations. That is, we need to know the sampling distribution of $(\hat{p}_1 - \hat{p}_2)$. Properties of the sampling distribution are given in the box. Remember that $\hat{p}_1$ and $\hat{p}_2$ can be viewed as means of the number of successes in the respective samples so that the Central Limit Theorem will apply when the sample sizes are large.

PROPERTIES OF THE SAMPLING DISTRIBUTION OF $(\hat{p}_1 - \hat{p}_2)$

1. If the sample sizes n_1 and n_2 are large (see Section 8.7 for a guideline), the sampling distribution of $(\hat{p}_1 - \hat{p}_2)$ is approximately normal.
2. The mean of the sampling distribution of $(\hat{p}_1 - \hat{p}_2)$ is $(p_1 - p_2)$; i.e.,

$$E(\hat{p}_1 - \hat{p}_2) = p_1 - p_2$$

Thus, $(\hat{p}_1 - \hat{p}_2)$ is an unbiased estimator of $(p_1 - p_2)$.
3. The standard deviation of the sampling distribution of $(\hat{p}_1 - \hat{p}_2)$ is

$$\sigma_{(\hat{p}_1 - \hat{p}_2)} = \sqrt{\frac{p_1 q_1}{n_1} + \frac{p_2 q_2}{n_2}}$$

Since the distribution of $(\hat{p}_1 - \hat{p}_2)$ in repeated sampling is approximately normal, we can use the z statistic to derive confidence intervals for $(p_1 - p_2)$ or to test a hypothesis about $(p_1 - p_2)$. For the camper example, a 95% confidence interval for the difference $(p_1 - p_2)$ is

$$(\hat{p}_1 - \hat{p}_2) \pm 1.96\sigma_{(\hat{p}_1 - \hat{p}_2)} = (\hat{p}_1 - \hat{p}_2) \pm 1.96\sqrt{\frac{p_1 q_1}{n_1} + \frac{p_2 q_2}{n_2}}$$

The quantities $p_1 q_1$ and $p_2 q_2$ must be estimated in order to complete the calculation of the standard deviation, $\sigma_{(\hat{p}_1 - \hat{p}_2)}$, and hence of the confidence interval. In Section 8.7, we showed that the value of pq is relatively insensitive to the value chosen to approximate p. Therefore, $\hat{p}_1\hat{q}_1$ and $\hat{p}_2\hat{q}_2$ will provide satisfactory estimates of $p_1 q_1$ and $p_2 q_2$, respectively. Then

$$(\hat{p}_1 - \hat{p}_2) \pm 1.96\sqrt{\frac{p_1 q_1}{n_1} + \frac{p_2 q_2}{n_2}} \approx (\hat{p}_1 - \hat{p}_2) \pm 1.96\sqrt{\frac{\hat{p}_1\hat{q}_1}{n_1} + \frac{\hat{p}_2\hat{q}_2}{n_2}}$$

$$= (.042 - .024) \pm 1.96\sqrt{\frac{(.042)(.958)}{1,000} + \frac{(.024)(.976)}{1,000}}$$

$$= .018 \pm .016 = (.002, .034)$$

Thus, we are 95% confident that the interval $(.002, .034)$ contains $(p_1 - p_2)$. It appears that there are between .2% and 3.4% more households in the NE than in the SE that plan to purchase campers in the next 5 years.

LARGE-SAMPLE $100(1 - \alpha)$% CONFIDENCE INTERVAL FOR $(p_1 - p_2)$

$$(\hat{p}_1 - \hat{p}_2) \pm z_{\alpha/2}\ \sigma_{(\hat{p}_1 - \hat{p}_2)} = (\hat{p}_1 - \hat{p}_2) \pm z_{\alpha/2}\sqrt{\frac{p_1 q_1}{n_1} + \frac{p_2 q_2}{n_2}}$$

$$\approx (\hat{p}_1 - \hat{p}_2) \pm z_{\alpha/2}\sqrt{\frac{\hat{p}_1\hat{q}_1}{n_1} + \frac{\hat{p}_2\hat{q}_2}{n_2}}$$

Assumptions: The two samples are independent random samples from binomial distributions. Both samples should be large enough that the normal distribution provides an adequate approximation to the sampling distribution of $\hat{p}_1$ and $\hat{p}_2$ (see Section 8.7).

The z statistic,

$$z = \frac{(\hat{p}_1 - \hat{p}_2) - (p_1 - p_2)}{\sigma_{(\hat{p}_1 - \hat{p}_2)}}$$

is used to test the null hypothesis that $(p_1 - p_2)$ equals some specified difference—say, D_0. For the special case where $D_0 = 0$—i.e., where we want to test the null hypothesis that $(p_1 - p_2) = 0$ (or equivalently, that $p_1 = p_2$)—the best estimate of $p_1 = p_2 = p$ is obtained by dividing the total number of successes $(x_1 + x_2)$ for the two samples by the total number of observations $(n_1 + n_2)$; i.e.,

$$\hat{p} = \frac{x_1 + x_2}{n_1 + n_2}$$

or,

$$\hat{p} = \frac{n_1\hat{p}_1 + n_2\hat{p}_2}{n_1 + n_2}$$

The second equation shows that $\hat{p}$ is a weighted average of $\hat{p}_1$ and $\hat{p}_2$, with the larger sample receiving more weight. If the sample sizes are equal, than $\hat{p}$ is a simple average of the two sample proportions of successes.

We now substitute the weighted average $\hat{p}$ for both p_1 and p_2 in the formula for the standard deviation of $(\hat{p}_1 - \hat{p}_2)$:

$$\sigma_{(\hat{p}_1 - \hat{p}_2)} = \sqrt{\frac{p_1 q_1}{n_1} + \frac{p_2 q_2}{n_2}} \approx \sqrt{\frac{\hat{p}\hat{q}}{n_1} + \frac{\hat{p}\hat{q}}{n_2}} = \sqrt{\hat{p}\hat{q}\left(\frac{1}{n_1} + \frac{1}{n_2}\right)}$$

The test is summarized in the box.

LARGE-SAMPLE TEST OF HYPOTHESIS ABOUT $(p_1 - p_2)$

ONE-TAILED TEST

H_0: $(p_1 - p_2) = 0$*

H_a: $(p_1 - p_2) < 0$
   [or H_a: $(p_1 - p_2) > 0$]

Test statistic: $z = \dfrac{(\hat{p}_1 - \hat{p}_2)}{\sigma_{(\hat{p}_1 - \hat{p}_2)}}$

Rejection region: $z < -z_\alpha$
   [or $z > z_\alpha$ when
   $H_a:(p_1 - p_2) > 0$]

TWO-TAILED TEST

H_0: $(p_1 - p_2) = 0$

H_a: $(p_1 - p_2) \neq 0$

Test statistic: $z = \dfrac{(\hat{p}_1 - \hat{p}_2)}{\sigma_{(\hat{p}_1 - \hat{p}_2)}}$

Rejection region:
   $z < -z_{\alpha/2}$ or $z > z_{\alpha/2}$

Note: $\sigma_{(\hat{p}_1 - \hat{p}_2)} = \sqrt{\dfrac{p_1 q_1}{n_1} + \dfrac{p_2 q_2}{n_2}} \approx \sqrt{\hat{p}\hat{q}\left(\dfrac{1}{n_1} + \dfrac{1}{n_2}\right)}$

where $\hat{p} = \dfrac{x_1 + x_2}{n_1 + n_2} = \dfrac{n_1\hat{p}_1 + n_2\hat{p}_2}{n_1 + n_2}$

Assumption: Same as for large-sample confidence interval for $(p_1 - p_2)$. (See previous box.)

---

*The test can be adapted to test for a difference $D_0 \neq 0$. Because most applications call for a comparison of p_1 and p_2, implying $D_0 = 0$, we will confine our attention to this case.

EXAMPLE 9.9

A consumer agency wants to determine whether there is a difference between the proportions of the two leading automobile models that need major repairs (more than \$300) within 2 years of their purchase. A sample of 400 2-year owners of model 1 are contacted, and a sample of 500 2-year owners of model 2 are contacted. The numbers x_1 and x_2 of owners who report that their cars needed major repairs within the first 2 years are 53 and 78, respectively. Test the null hypothesis that no difference exists between the proportions in populations 1 and 2 needing major repairs against the alternative that a difference does exist. Use $\alpha = .10$.

SOLUTION

If we define p_1 and p_2 as the true proportions of model 1 and model 2 owners, respectively, whose cars need major repairs within 2 years, the elements of the test are

$$H_0: \quad (p_1 - p_2) = 0$$
$$H_a: \quad (p_1 - p_2) \neq 0$$

Test statistic: $\quad z = \dfrac{(\hat{p}_1 - \hat{p}_2)}{\sigma_{(\hat{p}_1 - \hat{p}_2)}}$

Rejection region $(\alpha = .10)$: $\quad z > z_{\alpha/2} = z_{.05} = 1.645$
$\quad\quad$ or $z < -z_{\alpha/2} = -z_{.05} = -1.645$
$\quad\quad$ (see Figure 9.10)

FIGURE 9.10
Rejection Region for Example 9.9

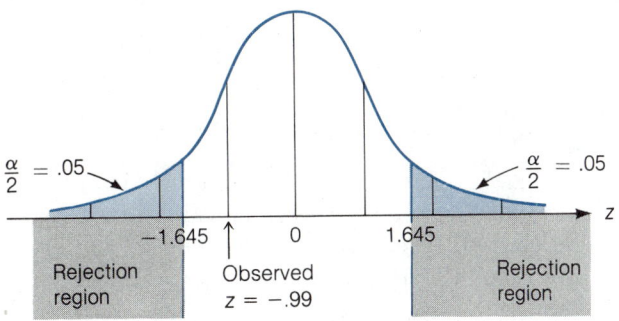

$\dfrac{\alpha}{2} = .05$ $\dfrac{\alpha}{2} = .05$

-1.645 $\quad$ 0 $\quad$ 1.645 $\quad$ z

Rejection region $\quad$ Observed $z = -.99$ $\quad$ Rejection region

We now calculate

$$z = \frac{(\hat{p}_1 - \hat{p}_2)}{\sigma_{(\hat{p}_1 - \hat{p}_2)}} = \frac{(\hat{p}_1 - \hat{p}_2)}{\sqrt{\dfrac{p_1 q_1}{n_1} + \dfrac{p_2 q_2}{n_2}}}$$

$$\approx \frac{(\hat{p}_1 - \hat{p}_2)}{\sqrt{\hat{p}\hat{q}\left(\dfrac{1}{n_1} + \dfrac{1}{n_2}\right)}}$$

where

$$\hat{p}_1 = \frac{x_1}{n_1} = \frac{53}{400} = .1325$$

$$\hat{p}_2 = \frac{x_2}{n_2} = \frac{78}{500} = .1560$$

and

$$\hat{p} = \frac{n_1\hat{p}_1 + n_2\hat{p}_2}{n_1 + n_2} = \frac{400(.1325) + 500(.1560)}{400 + 500} = .1456$$

Thus, $\hat{p}$ is a weighted average of $\hat{p}_1$ and $\hat{p}_2$, with more weight given the larger sample of model 2 owners. We substitute to obtain the (approximate) z test statistic:

$$z \approx \frac{.1325 - .1560}{\sqrt{(.1456)(.8544)\left(\frac{1}{400} + \frac{1}{500}\right)}} = \frac{-.0235}{.0237} = -.99$$

The samples provide insufficient evidence at $\alpha = .10$ to detect a difference between the proportions of the two models that need repairs within 2 years. Even though 2.35% more sampled owners of model 2 found major repairs, this difference is only .99 standard deviation ($z = -.99$) from the hypothesized zero difference between the true proportions. ∎

EXAMPLE 9.10

Find the observed significance level for the test in Example 9.9.

SOLUTION

The observed value of z for this two-tailed test was $z = -.99$. Therefore, the observed significance level is

p-value $= P(z < -.99$ or $z > .99)$

This probability is equal to the shaded area shown in Figure 9.11. The area corresponding to $z = .99$ is given in Table IV of Appendix B as .3389. Therefore, the observed significance level for the test, the sum of the two shaded tail areas under the standard normal curve, is

p-value $= 2(.5 - .3389) = .3222$

FIGURE 9.11
The Observed
Significance Level
for the Test of
Example 9.9

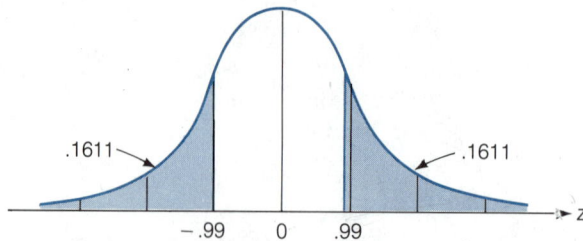

The probability of observing a z as large as .99 or less than $-.99$ if in fact $p_1 = p_2$ is .3222. This large p-value indicates that there is little or no evidence of a difference between p_1 and p_2. ∎

CASE STUDY 9.2

HOTEL ROOM
INTERVIEWING—
ANXIETY AND
SUSPICION

Writing in the *Sloan Management Review*, Lois Kaufman and John Wolf (1982) describe the typical approach used by sales managers for attracting and interviewing prospective sales representatives:

> To maintain product distribution and enhance customer service, corporate sales personnel are typically assigned to a district sales office that is tied to a regional office by telephone or telex. The managers who are responsible for staffing sales territories usually operate with considerable autonomy. When hiring sales representatives, they usually advertise for these positions in local newspapers, and traditionally invite applicants to interviews which are held in hotel rooms.*

The purpose of the article was to examine the effects of the hotel interview site on prospective sales representatives, and on women in particular.

As part of their study, Kaufman and Wolf asked a sample of 74[†] female college students from Rutgers University, Montclair State College, and Union College whether they would agree to a job interview in a room at a local hotel. Sixty-two percent said they would. A sample of 74 college women were asked whether they would agree to a job interview in a room of a local office building. Ninety-eight percent said they would. The authors used a one-tailed hypothesis test to examine the following hypotheses:

$$H_0: \quad p_1 - p_2 = 0$$
$$H_a: \quad p_1 - p_2 > 0$$

where

p_1 = Proportion of female college students who, if offered a job interview in a room of a local office building, would say they would attend the interview

p_2 = Proportion of female college students who, if offered a job interview in a room of a local hotel, would say they would attend the interview

They obtained an observed significance level of less than .05 for their z statistic and concluded that the proportion of women who would agree to an interview in an office building is significantly greater than the proportion willing to interview in a hotel room.

A similar but less extreme result was obtained when college men were asked the same questions. For the women $\hat{p}_1 - \hat{p}_2 = .36$, while for the men $\hat{p}_1 - \hat{p}_2 = .09$. Based on these results and other information supplied by the men and women who participated in their study, Kaufman and Wolf explained the study's findings as follows:

> Both men and women find hotel rooms more stressful than offices, but women are even more anxious than men about the prospect of interviewing in hotels. Most respondents said they would take precautions if they interviewed in hotels. Hotel room hiring was described with such words as "shady," "suspicious," "secretive," and "fishy."

---

*Reprinted from "Hotel room interviewing—anxiety and suspicion" by L. Kaufman and J. Wolf, *Sloan Management Review*, Vol. 23, Spring 1982, by permission of the publisher. Copyright © 1982 by the Sloan Management Review Association. All rights reserved.
†Sample sizes were obtained via personal communication from Professor Kaufman.

Kaufman and Wolf summarized the implications of their findings by saying:

> We believe that companies should reexamine their hiring practices and recognize that hotel room interviewing may not be the best approach to recruiting and hiring talented employees. Companies should consider using college campuses, local placement agencies, civic centers, libraries, or government buildings as possible interview sites. Although companies may not intend to discourage applicants when they interview in hotels, our survey showed that many women as well as some men will not attend an interview held in a hotel.

EXERCISES 9.61–9.80

LEARNING THE MECHANICS

9.61 What are the characteristics of a binomial experiment?

9.62 The quantities $\hat{p}_1$ and $\hat{p}_2$ have been defined as x_1/n_1 and x_2/n_2, respectively. What assumptions do we make about x_1 and x_2?

9.63 In each case, determine whether the sample sizes are large enough to conclude that the sampling distribution of $(\hat{p}_1 - \hat{p}_2)$ is approximately normal.
  a. $n_1 = 10, \quad n_2 = 12, \quad \hat{p}_1 = .50, \quad \hat{p}_2 = .42$
  b. $n_1 = 10, \quad n_2 = 12, \quad \hat{p}_1 = .10, \quad \hat{p}_2 = .08$
  c. $n_1 = n_2 = 30, \quad \hat{p}_1 = .20, \quad \hat{p}_2 = .30$
  d. $n_1 = 100, \quad n_2 = 200, \quad \hat{p}_1 = .05, \quad \hat{p}_2 = .09$
  e. $n_1 = 100, \quad n_2 = 200, \quad \hat{p}_1 = .95, \quad \hat{p}_2 = .91$

9.64 In each case, find the values of z for which $H_0: (p_1 - p_2) = 0$ would be rejected in favor of $H_a: (p_1 - p_2) < 0$.
  a. $\alpha = .01$ **b.** $\alpha = .025$
  c. $\alpha = .05$ **d.** $\alpha = .10$

9.65 Random samples of $n_1 = 200$ and $n_2 = 220$ from two binomial populations, 1 and 2, produced $x_1 = 38$ and $x_2 = 71$ successes, respectively.
  a. Compute the sample proportions, $\hat{p}_1$ and $\hat{p}_2$.
  b. Find the values of z for which the null hypothesis $H_0: (p_1 - p_2) = 0$ would be rejected in favor of the alternative hypothesis $H_a: (p_1 - p_2) \neq 0$. Use $\alpha = .10$.
  c. Test the hypotheses described in part **b**. Interpret your results.
  d. Find and interpret the p-value for the test.
  e. What assumptions must be satisfied so the test will be valid?

9.66 Refer to Exercise 9.65. Construct a 98% confidence interval for $(p_1 - p_2)$. Interpret your confidence interval.

9.67 A random sample of size $n_1 = 500$ from population 1 and a random sample of size $n_2 = 500$ from population 2 yielded $x_1 = 140$ and $x_2 = 192$ successes, respectively.
  a. Given

$$H_0: (p_1 - p_2) = 0 \qquad H_a: (p_1 - p_2) < 0$$

  find the values of z for which the null hypothesis would be rejected in favor of the alternative hypothesis. Use $\alpha = .025$.
  b. Conduct the test described above. Interpret the results.
  c. Find the observed significance level of the hypothesis test, and interpret its value.
  d. What assumptions must be satisfied so the test will be valid?

9.68 Refer to Exercise 9.67. Construct an 80% confidence interval for $(p_1 - p_2)$. Interpret your confidence interval.

9.69 Random samples of size $n_1 = 50$ and $n_2 = 60$ were drawn from populations 1 and 2, respectively. The samples yielded $\hat{p}_1 = .4$ and $\hat{p}_2 = .3$. Test $H_0: (p_1 - p_2) = 0$ against $H_a: (p_1 - p_2) > 0$ using $\alpha = .05$.

APPLYING THE CONCEPTS

9.70 What makes entrepreneurs different from chief executive officers (CEOs) of *Fortune 500* companies? The *Wall Street Journal* hired the Gallup Organization to investigate this question. For the study, entrepreneurs were defined as chief executive officers of companies listed by *Inc.* magazine as among the 500 fastest-growing smaller companies in the United States. The Gallup Organization sampled 207 CEOs of *Fortune 500* companies and 153 entrepreneurs. They obtained the results shown in the table.

	FORTUNE 500 CEOs	ENTREPRENEURS
Age: Under 45 years old	.092 19 .09179	96 .627 .62745
Education: Completed 4 years of college	195	116
Employment record: Have been fired or dismissed from a job	19	47

Source: Graham (1985). Reprinted by permission of *Wall Street Journal*, © Dow Jones & Company, Inc. (1985). All rights reserved.

a. In each of the three areas—age, education, and employment record—are the sample sizes large enough to use the inferential methods of this section to investigate the differences between *Fortune 500* CEOs and entrepreneurs? Justify your answer.

b. Do the data indicate that *Fortune 500* CEOs and entrepreneurs differ in terms of education? Test using $\alpha = .05$.

c. What assumption(s) must be satisfied in order for your test of part **b** to be valid?

9.71 Refer to Exercise 9.70. Use a 95% confidence interval to estimate the difference in the proportions of *Fortune 500* CEOs and entrepreneurs who are under 45 years of age.

9.72 Refer to Exercise 9.70.
a. Test to determine whether the data indicate that the fractions of CEOs and entrepreneurs who have been fired or dismissed from a job differ at the $\alpha = .01$ level of significance.
b. Construct a 99% confidence interval for the difference between the fractions of CEOs and entrepreneurs who have been fired or dismissed from a job.
c. Which inferential procedure provides more information about the difference between employment records, the test of hypothesis of part **a** or the confidence interval of part **b**? Explain.

9.73 To aid in the development of new products and/or to guide their marketing programs, producers of food products attempt to identify the taste preferences of various segments of the population. Accordingly, a recent study by Professor Susan Schiffman of the Duke University Center for the Study of Aging and Human Development should be of interest to the food industry. Professor Schiffman conducted an experiment that showed that a person's ability to identify food by smell and taste decreases with increasing age. As a result, she recommends adding simulated odors to the food of older people to improve its flavor. Part of Professor Schiffman's experiment involved asking a random sample of older persons and a random sample of college students to smell, taste, and identify a variety of foods that had been blended to prevent identification by "feel." Subjects were blindfolded during the experiment (Meier, 1980). Suppose that blended apple was correctly identified by 81 of 100 students and by 51 of 100 older people.

a. Would these data support Professor Schiffman's conclusion that the ability to identify food decreases with age? Test using $\alpha = .05$.

b. Report the p-value of the hypothesis test, and interpret its value.

c. What assumption must be satisfied so the hypothesis test of part **a** will be valid? Why was this assumption necessary?

9.74 Suppose a firm switches its table salt container from a cylinder (expensive) to a rectangular box (inexpensive). The firm samples 1,000 households nationwide, both before and after the switch, to estimate the percentage of households that purchase its brand of salt. The results obtained are shown in the table.

	BEFORE	AFTER
Sample size	1,000	1,000
Number of households using firm's brand	475	305

a. Estimate the true difference in the percentage of households that use the firm's salt before and after the packaging switch. Use a 90% confidence interval.

b. Interpret the confidence interval of part **a**. Express clearly the reliability of the interval.

9.75 Refer to Exercise 9.74. The firm's vice-president in charge of sales claims the switch to the box has seriously hurt the firm's market share. Does the sample evidence support this claim at the .05 significance level?

9.76 Moving companies (home movers, etc.) are required by the government to publish a Carrier Performance Report each year. One of the descriptive statistics they must include in this report is the percentage of shipments on which a \$50 or greater claim for loss or damage was filed in the previous year. Suppose company A and company B each decide to estimate this figure by sampling their records, and they obtain the data shown in the table.

	COMPANY A	COMPANY B
Total shipments delivered	9,542	6,631
Number of shipments on which a claim of \$50 or greater was filed	1,653	501

a. Estimate the true proportion of shipments on which a claim of \$50 or greater was made against company A. Use an estimate that reflects its reliability.

b. Repeat part **a** for company B.

c. Use a 95% confidence interval to estimate the true difference in the proportions of shipments that result in claims being made against company A and company B.

9.77 Refer to Exercise 9.76. Test the null hypothesis that no difference exists between the true percentage of shipments resulting in claims made against company A and company B against the alternative hypothesis that a difference does exist. Use $\alpha = .05$. Do your test results indicate that one carrier is superior to the other? If so, which one? Explain how you arrived at your conclusion.

9.78 If α were set at .01 in Exercise 9.77, would you be less likely or more likely to reject the null hypothesis if in fact it is true? Explain.

9.79 The Reserve Mining Company of Minnesota commissioned a team of physicians to study the breathing patterns of its miners who were exposed to taconite dust. The physicians compared the breathing of 307 miners who had been employed in Reserve's Babbit, Minnesota, mine for more than 20 years with that of 35 Duluth area men with no history of exposure to taconite dust. The physicians concluded that "there is no significant difference in respiratory symptoms or breathing ability between the group of men who have worked in the taconite industry for more than 20 years and a group of men of similar smoking habits but without exposure to taconite dust" (*Minneapolis Tribune*, Feb. 20, 1977). Using the statistical procedures you have learned in this chapter, design a hypothesis test (give H_0, H_a, test statistic, etc.) that would have been appropriate for use in the physicians' study.

9.80 Refer to Exercise 9.79. Suppose the physicians determined that 61 of the 307 miners had breathing irregularities and that 5 of the 35 Duluth men had breathing irregularities. Test to determine whether the data indicate that a higher proportion of breathing irregularities exist among those who have been exposed to taconite dust than among those who have not been exposed.

SECTION 9.6

DETERMINING THE SAMPLE SIZE

You can find the appropriate sample size to estimate the difference between two population parameters with a specified degree of reliability by using the method described in Sections 8.2 and 8.8. That is, to estimate the difference between two parameters correct to within B units with probability $(1 - \alpha)$, set $z_{\alpha/2}$ standard deviations of the sampling distribution of the estimator equal to B. Then solve for the sample size. To do this, you have to specify a particular ratio between n_1 and n_2. Most often, you will want to have equal sample sizes—i.e., $n_1 = n_2 = n$. We will illustrate the procedure with two examples.

EXAMPLE 9.11

The sales manager for a chain of supermarkets wants to determine whether store location, management, and other factors produce a difference in the mean meat purchase per customer (zero purchases to be excluded) at two different stores. The estimate of the difference in mean meat purchase per customer is to be correct to within $2.00 with probability equal to .95. If the two sample sizes are to be equal, find $n_1 = n_2 = n$, the number of customer meat sales to be randomly selected from each store.

SOLUTION

To solve the problem, you have to know something about the variation in the dollar amount of meat sales per customer. Suppose you know that the sales have a range of approximately $30 at each store. Then you could approximate $\sigma_1 = \sigma_2 = \sigma$ by letting the range equal 4σ, and

$$4\sigma \approx \$30$$
$$\sigma \approx \$7.50$$

The next step is to solve the equation

$$z_{\alpha/2} \sqrt{\frac{\sigma_1^2}{n_1} + \frac{\sigma_2^2}{n_2}} = B$$

for n, where $n = n_1 = n_2$. Since we want the estimate to lie within $B = \$2.00$ of $(\mu_1 - \mu_2)$ with probability equal to .95, $z_{\alpha/2} = z_{.025} = 1.96$. Then, letting $\sigma_1 = \sigma_2 = 7.5$ and solving for n, we have

$$1.96 \sqrt{\frac{(7.5)^2}{n} + \frac{(7.5)^2}{n}} = 2.00$$

$$1.96 \sqrt{\frac{2(7.5)^2}{n}} = 2.00$$

$$n = 108.05 \approx 108$$

Consequently, you will have to randomly sample 108 meat sales per store to estimate the difference in mean meat sales per customer correct to within $2.00 with probability approximately equal to .95. ∎

EXAMPLE 9.12

A production supervisor suspects a difference exists between the proportions of defective items produced by two different machines. Experience has shown that the proportion defective for the two machines is in the neighborhood of .03. If the supervisor wants to estimate the difference in the proportions using a 95% confidence interval of width .01, how many items must be randomly sampled from the production of each machine? (Assume that you want $n_1 = n_2 = n$.)

SOLUTION

For the specified level of reliability, $z_{\alpha/2} = z_{.025} = 1.96$. Then, letting $p_1 = p_2 = .03$ and $n_1 = n_2 = n$, we find the required sample size per machine by solving the following equation for n:

$$z_{\alpha/2} \sqrt{\frac{p_1 q_1}{n_1} + \frac{p_2 q_2}{n_2}} = \frac{W}{2}$$

where W is the desired width of the confidence interval, i.e., $W = .01$. Substituting and solving for n, we obtain

$$1.96 \sqrt{\frac{(.03)(.97)}{n} + \frac{(.03)(.97)}{n}} = .005$$

$$1.96 \sqrt{\frac{2(.03)(.97)}{n}} = .005$$

$$n = 8,943.2$$

You can see that this may be a tedious sampling procedure. If the supervisor insists on estimating $(p_1 - p_2)$ correct to within .005 with probability equal to .95, approximately 9,000 items will have to be inspected for each machine. ∎

You can see from the calculations in Example 9.12 that $\sigma_{(\hat{p}_1 - \hat{p}_2)}$ (and hence the solution, $n_1 = n_2 = n$) depends on the actual (but unknown) values of p_1 and p_2.

In fact, the solution for $n_1 = n_2 = n$ is largest when $p_1 = p_2 = .5$. Therefore, if we have no prior information on the approximate values of p_1 and p_2, we use $p_1 = p_2 = .5$ in the formula for $\sigma_{(\hat{p}_1 - \hat{p}_2)}$. If p_1 and p_2 really are close to .5, then the values of n_1 and n_2 that you have calculated will be appropriate. If p_1 and p_2 differ substantially from .5, then your solutions for n_1 and n_2 will be larger than needed. Consequently, using $p_1 = p_2 = .5$ when solving for n_1 and n_2 is a conservative procedure because the sample sizes n_1 and n_2 will be at least as large as (and probably larger than) needed.

The procedures for determining sample sizes necessary for estimating $(\mu_1 - \mu_2)$ or $(p_1 - p_2)$ for the case $n_1 = n_2$ are given in the box.

DETERMINATION OF SAMPLE SIZE FOR TWO-SAMPLE PROCEDURES

1. To estimate $(\mu_1 - \mu_2)$ to within a given bound B with probability $(1 - \alpha)$ or, equivalently, with a $100(1 - \alpha)\%$ confidence interval of width $W = 2B$, use the following formula to solve for equal sample sizes that will achieve the desired reliability:

$$n_1 = n_2 = \frac{(z_{\alpha/2})^2(\sigma_1^2 + \sigma_2^2)}{B^2} = \frac{4(z_{\alpha/2})^2(\sigma_1^2 + \sigma_2^2)}{W^2}$$

You will need to substitute estimates for the values of σ_1^2 and σ_2^2 before solving for the sample size. These estimates might be sample variances s_1^2 and s_2^2 from prior sampling (e.g., a pilot sample), or from an educated (and conservatively large) guess based on the range, i.e., $s \approx R/4$.

2. To estimate $(p_1 - p_2)$ to within a given bound B with probability $(1 - \alpha)$ or, equivalently, with a $100(1 - \alpha)\%$ confidence interval of width $W = 2B$, use the following formula to solve for equal sample sizes that will achieve the desired reliability:

$$n_1 = n_2 = \frac{(z_{\alpha/2})^2(p_1q_1 + p_2q_2)}{B^2} = \frac{4(z_{\alpha/2})^2(p_1q_1 + p_2q_2)}{W^2}$$

You will need to substitute estimates for the values of p_1 and p_2 before solving for the sample size. These estimates might be based on prior samples, obtained from educated guesses, or, most conservatively, specified as $p_1 = p_2 = .5$.

EXERCISES 9.81–9.89

LEARNING THE MECHANICS

9.81 Suppose you want to estimate the difference between two population means correct to within 1.5 with probability .95. If prior information suggests that the population variances are approximately equal to

$$\sigma_1^2 = \sigma_2^2 = 12$$

and you want to select independent random samples of equal size from the populations, how large should the sample sizes, n_1 and n_2, be?

9.82 A pollster wants to estimate the difference between the proportions of men and women who favor a particular national candidate using a 90% confidence interval of width .06. Suppose the pollster has no prior information about the proportions. If equal numbers of men and women are to be polled, how large should the samples be?

9.83 Suppose you want to estimate the difference between two population proportions correct to within .04 with confidence coefficient equal to .90. Also suppose you think both p_1 and p_2 are near .4 and you want to select samples of equal size from the two populations. Find the required sample sizes, n_1 and n_2.

9.84 Enough money has been budgeted to collect independent random samples of size $n_1 = n_2 = 100$ from populations 1 and 2 in order to estimate $(\mu_1 - \mu_2)$. Prior information indicates that $\sigma_1 = \sigma_2 = 10$. Have sufficient funds been allocated to construct a 90% confidence interval for $(\mu_1 - \mu_2)$ of width 4 or less? Justify your answer.

APPLYING THE CONCEPTS

9.85 Nationally televised home shopping was introduced in 1985. Overnight it became the hottest craze in television programming. By December 1986 there were 34 home shopping cable services (Covert, 1986). Who uses these home shopping services? Are the shoppers primarily men or women? Suppose you want to estimate the difference in the proportions of men and women who say they have used or expect to use televised home shopping using an 80% confidence interval of width .06 or less.
   a. Approximately how many people should be included in your samples?
   b. Suppose you want to obtain individual estimates for the two proportions of interest. Will the sample size found in part **a** be large enough to provide estimates of each proportion correct to within .02 with probability equal to .90? Justify your response.

9.86 In Exercise 9.44 you should have rejected the hypothesis that $\sigma_1^2 = \sigma_2^2$ and concluded that it is inappropriate to use the two-sample t statistic in making inferences about $(\mu_1 - \mu_2)$. In such cases, inferences about $(\mu_1 - \mu_2)$ can still be made if data are plentiful enough so that large-sample procedures such as those in Section 9.1 can be used. Accordingly, how many additional completion times should be sampled in order that the difference in the mean completion times for operators working with and without the 30-second guideline can be estimated to within 2 seconds with probability .80? Assume equal sample sizes are desired.

9.87 Refer to Exercise 9.73 in Section 9.5 and the experiment conducted to investigate whether a person's ability to identify food by smell and taste decreases with increasing age. How large should Professor Schiffman's samples be if she wishes to estimate the difference in the proportion of students and the proportion of older people who are able to identify blended apple to within .05 with probability .90?

9.88 Rat damage creates a large financial loss in the production of sugarcane. One aspect of the problem that has been investigated by the U.S. Department of Agriculture concerns the optimal place to locate rat poison. To be most effective in reducing rat damage, should the poison be located in the middle of the field or on the outer perimeter? One way to answer this question is to determine where the greater amount of damage occurs. If damage is measured by the proportion of cane stalks that have been damaged by rats, how many stalks from each section of the field should be sampled in order to estimate the true difference between the proportions of stalks damaged in the two sections to within .02 with probability .95?

9.89 Suppose you are interested in the growth rate of dividends. Consider investing $1,000 in a stock, and suppose you want to estimate the dividend rate on your $1,000 investment at the end of 5 years. Particularly, you want to compare two types of stocks, electrical utilities and oil companies. To conduct your study, you plan to randomly select n oil stocks and n electrical utility stocks. For each stock, you will check the records, calculate the number of shares of stock you could have purchased 5 years ago for $1,000, and then calculate the dividend rate (in percent) that the stock would be paying today on your $1,000 investment. Suppose you think the dividend rates will vary over a range of roughly 25%. To obtain an approximate value of σ_1 and σ_2, let $\sigma_1 = \sigma_2 = \sigma$ and let the range be 4σ. Then the range is $25 \approx 4\sigma$ and $\sigma \approx 6.25$. How large should n be if you want to estimate the difference in mean rates of dividend return with a 95% confidence interval of width 6%?

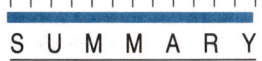

S U M M A R Y

We have presented various techniques for using the information in two samples to make inferences about the difference between population parameters. As you would expect, we are able to make reliable inferences with fewer assumptions about the sampled populations when the sample sizes are large. When we cannot take large samples from the populations, the **two-sample t statistic** permits us to use the limited sample information to make inferences about the **difference between means** when the assumptions of normality and equal population variances are at least approximately true. The **paired difference experiment** offers the possibility of increasing the information about $(\mu_1 - \mu_2)$ by pairing similar observational units to control variability. In designing a paired difference experiment, we expect that the reduction in variability will more than compensate for the loss in degrees of freedom.

Two other inferential procedures for making comparisons between population parameters were presented in this chapter. The **F test** was used to compare two population variances, σ_1^2 and σ_2^2. This test is useful in checking the assumption of equal population variances, an assumption that is essential to the independent samples t test (and confidence interval) for a comparison of two population means. The F test can also be used to compare the variances of two populations when these variances assume practical importance as a measure of risk, error, etc.

This chapter concluded with a comparison of two binomial parameters, p_1 and p_2. Practical examples of such comparisons are numerous; they frequently appear in the analysis of business surveys. A company might want to compare the proportion of consumers who prefer a new product A to a new (or old) product B. Or, the comparison might occur in a production setting when a manufacturer wants to compare the fractions of defectives that emerge from two production lines.

SUPPLEMENTARY
EXERCISES 9.90–9.115

[*Note: In each problem, state the assumptions necessary for the procedure to be valid.*]

9.90 Was the average amount spent by firms in the electronics industry on company-sponsored research and development (R&D) higher in 1985 than it was in 1984? The table lists R&D expenditures (in millions of dollars) for a sample of firms in the electronics industry.

FIRM	1984	1985
Kollmorgen	18.7	23.0
Harris	105.2	115.7
Equatorial Comm.	3.8	3.4
Andrew	13.0	15.9
Timeplex	5.7	8.7
Rogers	6.1	7.4
Hazeltine	10.0	9.3
Siltec	4.3	4.6
General Instr.	48.2	52.5
Singer	57.0	57.0

a. A securities analyst who follows the electronics industry believes R&D expenditures have increased. Do the data support the analyst's beliefs? Test using $\alpha = .10$.

b. In the context of this exercise, what are the Type I and Type II errors associated with the hypothesis test of part **a**?

c. What assumptions must hold in order for your test of part **a** to be valid?

9.91 Refer to Exercise 9.90. Use a 95% confidence interval to estimate the mean difference between 1985 and 1984 R&D expenditures. Interpret the interval.

9.92 To market a new cigarette, a tobacco company decides to use two different advertising agencies, one operating in the East and one in the West. After the cigarette has been on the market for 6 months, random samples of smokers are taken from each of the two regions and questioned concerning their cigarette preference. The numbers in the samples favoring the new brand are shown in the table. Do the data provide sufficient evidence to indicate a difference in the proportions preferring the new brand between the two regions? Use $\alpha = .05$. Based on your test, what inference can be made about the difference in the effectiveness of the two advertising agencies?

	SAMPLE SIZE	NUMBER PREFERRING NEW BRAND
East	500	12
West	450	15

A	B
43	46
48	49
37	43
52	41
45	48

9.93 When new instruments are developed to perform chemical analyses of products (food, medicine, etc.), they are usually evaluated with respect to two criteria: accuracy and precision. *Accuracy* refers to the ability of the instrument to identify correctly the nature and amounts of a product's components. *Precision* refers to the consistency with which the instrument will identify the components of the same material in repeated analyses. Thus, a large variability in the identification of the components of a single sample of a product indicates a lack of precision. Suppose a pharmaceutical firm is considering two brands of an instrument designed to identify the components of certain drugs. As part of a comparison of precision, ten test-tube samples of a well-mixed batch of a drug are selected and then five are analyzed by instrument A and five by instrument B. The data shown in the table are the percentages of the primary component of the drug given by the instruments. Do these data provide evidence of a difference in the precision of the two machines? Use $\alpha = .10$.

9.94 The procedure outlined in this exercise was developed by Tele-Research, Inc., for evaluating the effectiveness of newly developed television commercials prior to their release. The exercise describes part of an actual study in which the procedure was used (Jenssen, 1966). Three hundred ninety-two shoppers were randomly selected as they entered a large Los Angeles supermarket and were asked to describe their preferences for several product brands. One of these was a brand, XYZ, whose new television commercial was the object of the study. Ostensibly in exchange for their time, the shoppers were given a packet of ten different cents-off coupons for products sold in the supermarket. Included were coupons for XYZ. The coupons could be used only in that particular store and only on that particular day. A second sample of 387 shoppers was given the same interview, but was also asked to watch four television commercials in a trailer parked outside the supermarket. One of the commercials was a newly developed ad for XYZ. Following the viewing, the shoppers were asked for their reactions to the commercials. The shoppers were then given the same packet of coupons. Of the 392 shoppers not exposed to the television commercials, 57 redeemed the coupon for XYZ. Of the 387 shoppers who were exposed to XYZ's commercials, 84 redeemed the XYZ coupon.

a. Do the sample data provide sufficient evidence to conclude that the new XYZ commercial motivates shoppers to purchase the XYZ brand? Use $\alpha = .05$.

b. Find and interpret the observed significance level for the test.

9.95 List the assumptions necessary for each of the following inferential techniques:

a. Large-sample inferences about the difference $(\mu_1 - \mu_2)$ between population means using a two-sample z statistic

b. Small-sample inferences about $(\mu_1 - \mu_2)$ using an independent samples design and a two-sample t statistic

c. Small-sample inferences about $(\mu_1 - \mu_2)$ using a paired difference design and a single-sample t statistic to analyze the differences

d. Large-sample inferences about the difference $(p_1 - p_2)$ between binomial proportions using a two-sample z statistic

9.96 Advertising companies often try to characterize the average user of a client's product so advertisements can be targeted at particular segments of the buying community. Suppose a new movie is about to be released and an advertising company wants to determine whether to aim the advertisement at people under 25 years old or those over 25. It plans to arrange an advance showing of the movie to a number of individuals from each group and then to obtain an opinion about the movie from each individual. How many individuals should be included in each sample if the advertising company wants to estimate the difference in the proportions of viewers in each age group who will like the movie to within .05 with 90% confidence? Assume the sample size for each group will be the same and about half of each group will like the movie.

9.97 To compare the rate of return an investor can expect on tax-free municipal bonds with the rate of return on taxable bonds, an investment advisory firm randomly samples ten bonds of each type and computes the annual rate of return over the past 3 years for each bond. The rate of return is then adjusted for taxes, assuming the investor is in a 30% tax bracket. The means and standard deviations for the adjusted returns are shown in the table at the top of page 480.

a. Test to determine whether there is a difference in the mean rates of return between tax-free and taxable bonds for investors in the 30% tax bracket. Use $\alpha = .05$.

b. Find the approximate observed significance level for the test of part **a**, and interpret its value.

TAX-FREE BONDS	TAXABLE BONDS
$\bar{x}_1 = 7.8\%$	$\bar{x}_2 = 7.3\%$
$s_1 = 1.1\%$	$s_2 = 1.0\%$

    c. What assumptions were necessary for the validity of the testing procedure you used in part **a**?

9.98 Refer to Exercise 9.97. Do the sample data cast doubt on the assumption of equal population variances? Use $\alpha = .10$.

9.99 A major interface between consumers and retailers occurs when the consumer is dissatisfied with the product purchased and returns to the retailer to obtain satisfaction. The actions taken by retailers in such situations, however, may not conform to the expectations of consumers. The resulting frustration and ill will benefit neither the consumer nor the retailer. Ronald Dornoff and Clint Tankersley (1975) conducted a study to test the hypothesis that differences exist between retailers' and consumers' perceptions regarding actions taken by retailers in market transactions. A random sample of 300 consumers was selected from the Cincinnati Metropolitan Area Telephone Directory and asked via mail questionnaire to react to scenarios like the following:

> A customer calls the retailer to report that her refrigerator purchased 2 weeks ago is not cooling properly and that all the food has spoiled.

> Action that should be taken by the retailer: The customer should be reimbursed for the value of the spoiled food.

One hundred usable questionnaires were returned. The same questionnaire was presented in person to 100 managers and assistant managers of a random sample of 40 retail establishments drawn from the yellow pages of the Cincinnati Telephone Directory. For the above scenario, 89 consumers agreed with the action prescribed for the retailer, 3 disagreed, and 8 had no opinion. Thirty-seven retailers (managers and/or assistant managers) agreed with the prescribed action, 54 disagreed, and 9 had no opinion.

    a. Use a 95% confidence interval to estimate the difference in the proportions of consumers and retailers who agree with the action prescribed in the scenario. Draw appropriate conclusions regarding the hypothesis of interest to Dornoff and Tankersley.

    b. What assumption(s), if any, must be made in constructing the confidence interval?

    c. Discuss the implications of the composition of the sample of retailers for the validity of the conclusions you made in part **a**. How would you improve the sampling procedure?

9.100 An automobile manufacturer wants to estimate the difference in the mean miles per gallon rating for two models it produces. If the range of ratings is expected to be about 6 miles per gallon for each model, how many cars of each model must be tested in order to estimate the difference in mean rating with a 90% confidence interval with a width of 1 mile per gallon?

9.101 In 1986, the American Society for Quality Control commissioned the Gallup Organization to explore the attitudes, beliefs, and experiences of upper-level executives in U.S. businesses with respect to the quality and quality practices related to their companies' products and services. The following is one of the questions asked of the executives: "Poor quality—as measured by repair, rework and scrap costs, lost sales, and so on—is said to cost American business billions of dollars annually. How much does poor quality cost your company, as a percent of gross sales?" The table describes the responses (rounded to the nearest percent) of 387 service company executives and 311 industrial company executives.

	SERVICE COMPANIES	INDUSTRIAL COMPANIES
Less than 5%	45%	47%
5%–10%	24	23
11%–19%	5	12
20%–29%	3	6
30%–49%	2	2
50% or more	1	0
Don't know	20	10

Source: "Gallup survey: Top executives talk quality," *Quality Progress*, Dec. 1986, pp. 49–54. © 1986 American Society for Quality Control. Reprinted by permission.

a. Use a 90% confidence interval to estimate the proportion of all executives in U.S. service companies who believe poor quality costs their firm 10% or less of gross sales.

b. Use a 90% confidence interval to estimate the difference between the proportion of all executives in service companies and the proportion of all executives in industrial companies who believe poor quality costs their companies 10% or less of gross sales.

c. What assumptions must hold in order for your confidence intervals of parts **a** and **b** to be valid?

9.102 Refer to Exercise 9.101. Do the data provide sufficient evidence to indicate that a difference exists between the proportion of service company executives and the proportion of industrial company executives who do not know how much poor quality costs their companies? Test using $\alpha = .10$.

9.103 Management training programs are often instituted to teach supervisory skills and thereby increase productivity. Suppose a company psychologist administers a set of examinations to each of ten supervisors before such a training program begins and then administers similar examinations at the end of the program. The examinations are designed to measure supervisory skills, with higher scores indicating increased skill. The results of the tests are shown in the table.

SUPERVISOR	BEFORE TRAINING PROGRAM	AFTER TRAINING PROGRAM
1	63	78
2	93	92
3	84	91
4	72	80
5	65	69
6	72	85
7	91	99
8	84	82
9	71	81
10	80	87

a. Do the data provide evidence that the training program is effective in increasing supervisory skills, as measured by the examination scores? Use $\alpha = .10$.

b. Find and interpret the approximate p-value for the test.

BANK 1	BANK 2
$\bar{x}_1 = 2.2$	$\bar{x}_2 = 1.8$
$s_1 = 1.15$	$s_2 = 1.10$

9.104 Two banks, bank 1 and bank 2, independently sampled 40 and 50 of their business accounts, respectively, and determined the number of the bank's services (loans, checking, savings, investment counseling, etc.) each sampled business was using. Both banks offer the same services. A summary of the data supplied by the samples is listed in the table. Do the samples provide sufficient evidence to conclude that the average number of services used by bank 1's business customers is significantly greater (at $\alpha = .10$) than the average number of services used by bank 2's business customers?

9.105 Find a 99% confidence interval for $(\mu_1 - \mu_2)$ in Exercise 9.104. Does the interval include 0? Interpret the confidence interval.

9.106 Suppose you have been offered similar jobs in two different locales. To help in deciding which job to accept, you would like to compare the cost of living in the two cities. One of your primary concerns is the cost of housing, so you obtain a copy of a newspaper from each locale and begin to study the housing prices in the classified advertisements. One convenient method for getting a general idea of prices is to compute the prices on a per-square-foot basis. This is done by dividing the price of the house by the heated area (in square feet) of the house. Random samples of 63 advertisements in locale 1 and 78 in locale 2 produce the results shown in the table. Is there evidence that the mean housing price per square foot differs in the two locales? Use $\alpha = .01$.

LOCALE 1	LOCALE 2
$\bar{x}_1 = \$50.40$ per square foot	$\bar{x}_2 = \$53.70$ per square foot
$s_1 = \$4.50$ per square foot	$s_2 = \$5.30$ per square foot

9.107 A number of computer programs are available to conduct two-sample tests of hypotheses to compare the means of two populations, for both independent and pooled samples. Most of these report both the test statistic and the observed significance level for the test, but some report only the observed significance level of the test. Suppose you use one of these programs to test the null hypothesis $H_0: (\mu_1 - \mu_2) = 0$ versus the alternative $H_a: (\mu_1 - \mu_2) \neq 0$ with independent samples of size 12 and 10, respectively. Assuming you are using $\alpha = .05$, what conclusions would you reach in each of the following instances of an observed significance level reported by the program?

a. P-VALUE $= .0429$ **b.** P-VALUE $= .1984$
c. P-VALUE $= .0001$ **d.** P-VALUE $= .0344$
e. P-VALUE $= .0545$ **f.** P-VALUE $= .9633$

g. You should always be sure that the program is performing the calculations correctly, especially programs with which you do not have much experience. Even programs that perform the calculations correctly usually do not remind you of the assumptions necessary for the validity of the procedure. What assumptions are necessary for this experiment?

9.108 Some power plants are located near rivers or oceans so the water can be used for cooling their condensers. As part of an environmental impact study, suppose a power company wants to estimate the mean difference in water temperature between the discharge of its plant and the offshore waters. How many sample measurements must be taken at each site to obtain a 95% confidence interval of width .4°C? Assume the range in readings will be about 4°C at each site and the same number of readings will be taken at each site.

9.109 The use of preservatives by food processors has become a controversial issue. Suppose two preservatives are extensively tested and determined safe for use in meats. A processor wants to compare the preservatives for their effects on retarding spoilage. Suppose 15 cuts of fresh meat are treated with preservative A and 15 with B, and the number of hours until spoilage begins is recorded for each of the 30 cuts of meat. The results are summarized in the table.

PRESERVATIVE A	PRESERVATIVE B
$\bar{x}_1 = 106.4$ hours	$\bar{x}_2 = 96.5$ hours
$s_1 = 10.3$ hours	$s_2 = 13.4$ hours

a. Is there evidence of a difference in the mean time until spoilage begins for the two preservatives at $\alpha = .05$?
b. Can you recommend an experimental design that the processor could have used to reduce the variability in the data?

9.110 Refer to Exercise 9.109. Construct a 95% confidence interval for the difference between the mean times until spoilage for the two preservatives.

9.111 An economist wants to investigate the difference in unemployment rates between an urban industrial community and a university community in the same state. She interviews 525 potential members of the work force in the industrial community and 375 in the university community. Of these, 47 and 22, respectively, are unemployed. Use a 95% confidence interval to estimate the difference in unemployment rates in the two communities.

9.112 A large department store plans to renovate one of its floors, with one of the results being an increase in floor space for one department. The management has narrowed the decision about which department to enlarge to two departments: men's clothing and sporting goods. The final decision will be based on mean sales, with the department having the greater mean to be enlarged. The last 12 months' sales data are shown in the table.

MONTH	MEN'S CLOTHING	SPORTING GOODS	MONTH	MEN'S CLOTHING	SPORTING GOODS
1	$15,726	$17,533	7	$15,525	$16,774
2	11,243	10,895	8	15,799	16,223
3	22,325	19,449	9	16,449	16,135
4	23,494	21,500	10	16,993	17,834
5	12,676	18,925	11	19,832	18,429
6	13,492	21,426	12	32,434	34,565

a. Use these data to form a 95% confidence interval for the mean difference in monthly sales for the two departments.
b. On the basis of the confidence interval formed in part a, can you make a recommendation to the store management as to which department should be enlarged?
c. What assumptions are necessary to make valid the procedure you used in part a?

9.113 Smoke detectors are highly recommended safety devices for early fire detection in homes and businesses. It is extremely important that the devices be nondefective. Suppose 100 brand A smoke detectors are tested and 12 fail to emit a warning signal. Subjected to the same test, 15 out of 90 brand B detectors fail to operate. Form a 90% confidence interval to estimate the difference in the fractions of defective smoke detectors produced by the two companies. Interpret this confidence interval.

9.114 The federal government is interested in determining whether salary discrimination exists between men and women in the private sector. Suppose random samples of 15 women and 22 men are drawn from the population of first-level managers in the private sector. The information on their annual salaries is summarized in the table.

WOMEN	MEN
$\bar{x}_1 = \$23,400$	$\bar{x}_2 = \$24,700$
$s_1 = \$2,300$	$s_2 = \$3,100$
$n_1 = 15$	$n_2 = 22$

a. Do the data provide sufficient evidence at $\alpha = .05$ to indicate that the mean salary of male first-level managers exceeds the mean salary of females in that position?

b. What assumptions are necessary for the validity of the test used in part **a**?

9.115 Refer to Exercise 9.114. Conduct a test to determine whether the data indicate that the assumption of equal salary variances is false. Use $\alpha = .10$.

ON YOUR OWN . . .

Many stock market indexes, such as the Dow Jones Average, act both as indicators of stock market trends and as economic indicators. One way of comparing economic conditions at the end of two consecutive years would be to estimate the difference in the mean closing prices of all stocks on the New York Stock Exchange. Below, we have outlined two methods of sampling to estimate the difference in mean closing prices on the last day of market operations for two consecutive years, say 1986 and 1987.

METHOD 1

TWO INDEPENDENT SAMPLES

STEP 1   Obtain lists of the closing prices of all stocks on the New York Stock Exchange for the last operating days of 1986 and 1987. (Any library will have these available.)

STEP 2   Using a table of random numbers, randomly choose 15 stocks from the 1986 list and record the closing price of each.

STEP 3   Again refer to a table of random numbers and choose a second (independent) sample of 15 closing prices from the 1987 list.

STEP 4   Using the two samples of closing prices, form a 95% confidence interval for the true difference in mean closing prices for the two years.

METHOD 2

PAIRED SAMPLES

STEP 1   Same as for method 1.

STEP 2   Same as for method 1.

STEP 3 Obtain the 1987 closing prices for the *same stocks as those used in 1986*.

STEP 4 Using this set of paired observations, form a 95% confidence interval for the true mean difference in closing prices for the two years.

Before actually collecting any data, state which method you think will provide more information (and why). Then, to compare the two methods, first perform the entire experiment outlined in method 1. After you have completed this, obtain the 1987 closing prices for the *same* stocks as the 1986 stocks analyzed and complete step 4 of method 2.

Which method provided a narrower confidence interval and thus more information on this performance of the experiment? Does this agree with your preliminary answer?

USING THE COMPUTER . . .

Select two of the census regions given in Appendix C, and consider the sports purchasing index. Suppose a marketing firm wants to target one of the two regions for a sports magazine marketing campaign.

a. Treat the sports index measurements for the zip codes in the regions as a random sample of sports index measurements from all the zip codes for the regions. Test the null hypothesis that the populations' mean sports purchasing indexes are equal using $\alpha = .01$, and place a 99% confidence interval on the true difference between the mean purchasing index for the two regions.

b. Repeat part **a** using the following pairs of α and confidence levels for the tests and confidence intervals: (.05, 95%), (.10, 90%), and (.20, 80%). Describe what happens to the tests and confidence intervals as α is increased and the confidence level is decreased. Which do you think is more informative, the tests or the confidence intervals? Explain.

REFERENCES

Beckenstein, A. R. Gabel, H. L., and Roberts, K. "An executive's guide to antitrust compliance." *Harvard Business Review*, Sept.–Oct. 1983, 94–102.

Benson, P. G. and Ohta, H. "Classifying sensory inspectors with heterogeneous inspection-error probabilities." *Journal of Quality Technology*, Vol. 18, No. 2, Apr. 1986, 79–90.

Covert, C. "Television viewers snapping up Home Shopping Network," *Minneapolis Star and Tribune*, Dec. 16, 1986, p. 1C.

Dornoff, R. J. and Tankersley, C. B. "Perceptual differences in market transactions: A source of customer frustration." *Journal of Consumer Affairs*, Summer 1975, 9, 97–103.

Garvin, D. A. "Quality on the line." *Harvard Business Review*, Sept.–Oct. 1983, 65–75.

Gibbons, J. D. *Nonparametric Statistical Inference*. New York: McGraw-Hill, 1971.

Graham, E. "The entrepreneurial mystique." *Wall Street Journal*, May 20, 1985, p. 1C.

Hines, G. H. "Sociocultural influences on employee expectancy and participative management." *Academy of Management Journal*, 1974, 17(2).

Hollander, M. and Wolfe, D. A. *Nonparametric Statistical Methods*. New York: Wiley, 1973.

Jenssen, W. J. "Sales effects of TV, radio, and print advertising." *Journal of Advertising Research*, June 1966, 6, 2–7.

Kaufman, L. and Wolf, J. "Hotel room interviewing—anxiety and suspicion." *Sloan Management Review*, Spring 1982, 23(3).

Marcotty, J. "Quantity of quality circles proves they're no fad." *Minneapolis Tribune*, Nov. 20, 1983a, p. 1D.

Marcotty, J. "Tennant tightens up loose screws." *Minneapolis Tribune*, Nov. 20, 1983b, p. 1D.

McGuire, E. P. *Evaluating New Product Proposals*. New York: National Industrial Conference Board, 1973. pp. 54–55.

Meier, P. "Taste: It's in the buds—and they don't improve with age." *Minneapolis Tribune*, June 8, 1980.

Mendenhall, W. *Introduction to Probability and Statistics*, 7th ed. Boston: Duxbury, 1986. Chapter 8.

Neter, J., Wasserman, W., and Whitmore, G. A. *Applied Statistics*, 2nd ed. Boston: Allyn & Bacon, 1982. Chapter 13.

Serrin, W. "Technology takes toll on operators." *Minneapolis Tribune*, Nov. 27, 1983, p. 1D.

Shuchman, A. and Riesz, P. C. "Correlates of persuasibility: The Crest case." *Journal of Marketing Research*, Feb. 1975, *12*, 7–11.

Siegel, S. *Nonparametric Statistics for the Behavioral Sciences*. New York: McGraw-Hill, 1956.

Sinclair, M. A. "A collection of modelling approaches for visual inspection in industry." *International Journal of Production Research*, Vol. 16, No. 4, 1978, 275–292.

Smith, W. M. "Federal pay procedures and the comparability survey." *Monthly Labor Review*, Aug. 1976, 27–31.

Wansley, J. W., Roenfeldt, R. L., and Cooley, P. L. "Abnormal returns from merger profiles." *Journal of Financial and Quantitative Analysis*, Vol. 18, No. 2, June 1983, 149–162.

Winer, L. "The effect of product sales quotas on sales force productivity." *Journal of Marketing Research*, May 1973, *10*, 180.

Winkler, R. L. and Hays, W. L. *Statistics: Probability, Inference, and Decision*, 2nd ed., New York: Holt, Rinehart and Winston, 1975. Chapter 6.

WHERE WE'VE BEEN . . .

The answers to many questions that arise in business require knowledge about the mean of a population or about the difference between two population means. Estimating and testing hypotheses about means, or the difference between two means, were the subjects of Chapters 8 and 9.

WHERE WE'RE GOING . . .

Suppose you want to predict the assessed value of a house in a particular community. Using the methods of Chapter 8, you could select a random sample of houses from the community and use the mean of their assessed values to predict the assessed value of the house of interest to you. But using this procedure would ignore the information contained in easily observed variables that are related to assessed house value—namely, the square feet of floor space, number of bathrooms, age of the house, etc. In this chapter we will consider the problem of relating the mean value of a single dependent variable y (for example, assessed house value) to a single independent variable x (say, square feet of floor space) using a linear relationship. The more complex problem of relating y to many independent variables will be the topic of Chapter 11.

SIMPLE LINEAR REGRESSION

CONTENTS

Much business research is devoted to the topic of **modeling**—i.e., trying to describe how variables are related. For example, an econometrician might be interested in modeling the relationship between the level of consumption expenditure and disposable personal income. An advertising agency might want to know the relationship between a firm's sales revenue and the amount spent on advertising. And an investment firm may be interested in relating the performance of the stock market to the current discount rate of the Federal Reserve Board.

The simplest graphical model for relating a variable y to a single independent variable x is a straight line. In this chapter we discuss **simple linear (straight-line) models** and show how to fit them to a set of data points using the **method of least squares**. We then show how to judge whether a relationship exists between y and x, and how to use the model to estimate $E(y)$, the mean value of y, and to predict a future value of y for a given value of x. The totality of these methods is called a **simple linear regression analysis**.

Most models for business variables are much more complicated than implied by a straight-line relationship. Nevertheless, the methods of this chapter are very useful, and they set the stage for the formulation and fitting of more complex models in succeeding chapters. Thus, this chapter provides an intuitive justification for the techniques used in a regression analysis and identifies most of the types of inferences we will want to make using a **multiple linear regression analysis** later in this book.

SECTION 10.1

PROBABILISTIC MODELS.

An important consideration in merchandising a product is the amount of money spent on advertising. Suppose you want to model the monthly sales revenue of an appliance store as a function of the monthly advertising expenditure. The first question to be answered is this: Do you think an exact relationship exists between these two variables? That is, can the exact value of sales revenue be predicted if the advertising expenditure is specified? We think you will agree this is not possible for several reasons. Sales depend on many variables other than advertising expenditure—for example, time of year, state of general economy, inventory, and price structure. However, even if many variables are included in the model (the topic of Chapter 11), it is still unlikely that we can predict the monthly sales *exactly*. There will almost certainly be some variation in sales due strictly to **random phenomena** that cannot be modeled or explained. We will refer to all unexplained variations in sales—caused by important but unincluded variables or by unexplainable random phenomena—as **random error**.

If we construct a model that hypothesizes an exact relationship between variables, it is called a **deterministic model**. For example, if we believe that monthly sales revenue y will be exactly 10 times the monthly advertising expenditure x, we write

$$y = 10x$$

This represents a **deterministic** relationship between the variables y and x.

On the other hand, if we believe that the model should be constructed to allow for random error, then we hypothesize a **probabilistic model**. This includes both a deterministic component and a random error component. For example, if we hypothesize that the sales y is related to advertising x by

$$y = 10x + \text{Random error}$$

we are hypothesizing a **probabilistic** relationship between y and x. Note that the deterministic component of this probabilistic model is $10x$.

GENERAL FORM OF A PROBABILISTIC MODEL

y = Deterministic component + Random error

where y is the variable to be predicted.

As you will see, the random error plays an important role in testing hypotheses and finding confidence intervals for the deterministic portion of the model and enables us to estimate the magnitude of the error of prediction when the model is used to predict some value of y to be observed in the future.

We begin with the simplest of probabilistic models—**a first-order linear model**, which graphs as a straight line. The elements of the straight-line model are summarized in the box.

A FIRST-ORDER (STRAIGHT-LINE) MODEL

$$y = \beta_0 + \beta_1 x + \varepsilon$$

where

$\quad\quad y$ = **Dependent** or **response variable** (variable to be modeled)

$\quad\quad x$ = **Independent*** or **predictor variable** (variable used as a predictor of y)

$\quad\quad \varepsilon$ (epsilon) = Random error component

$\quad\quad \beta_0$ (beta zero) = y-intercept of the line—i.e., point at which the line intercepts or cuts through the y-axis (see Figure 10.1)

$\quad\quad \beta_1$ (beta one) = Slope of the line—i.e., amount of increase (or decrease) in the deterministic component of y for every 1-unit increase in x (see Figure 10.1)

FIGURE 10.1

The Straight-Line Model

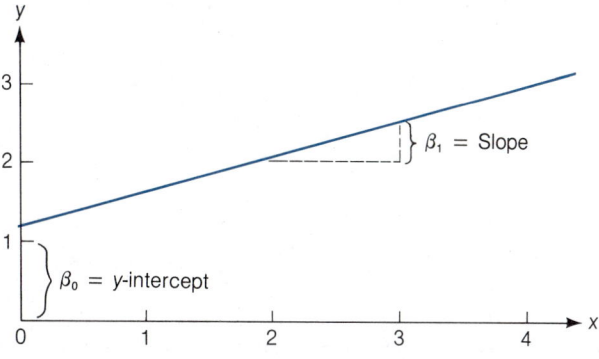

*The word *independent* should not be interpreted in a probabilistic sense. The phrase *independent variable* is used in regression analysis to refer to a predictor variable for the response y.

Note that we use Greek symbols β_0 and β_1 to represent the y-intercept and slope of the model. They are population parameters with numerical values that will be known only if we have access to the entire population of (x, y) measurements.

It is helpful to think of regression modeling as a five-step procedure:

STEP 1   Hypothesize the deterministic component of the probabilistic model.
STEP 2   Use sample data to estimate unknown parameters in the model.
STEP 3   Specify the probability distribution of the random error term, and estimate any unknown parameters of this distribution.
STEP 4   Statistically check the usefulness of the model.
STEP 5   When satisfied that the model is useful, use it for prediction, estimation, and other purposes.

In this chapter we will skip step 1 and deal with steps 2–5 for only the straight-line model. Chapters 11 and 12 will discuss how to build more complex models.

EXERCISES 10.1–10.6   LEARNING THE MECHANICS

10.1 In each case, graph the line that passes through the given points.
  a. (0, 0) and (4, 4)   **b.** (0, 2) and (2, 0)
  c. (−2, 2) and (6, 3)   **d.** (−4, −1) and (3, 4)

10.2 The equation for a straight line (deterministic) is

$$y = \beta_0 + \beta_1 x$$

If the line passes through the point $(-1, 3)$, then $x = -1$, $y = 3$ must satisfy the equation; i.e.,

$$3 = \beta_0 + \beta_1(-1)$$

Similarly, if the line passes through the point $(3, 4)$, then $x = 3$, $y = 4$ must satisfy the equation; i.e.,

$$4 = \beta_0 + \beta_1(3)$$

Use these two equations to solve for β_0 and β_1, and find the equation of the line that passes through the points $(-1, 3)$ and $(3, 4)$.

10.3 Refer to Exercise 10.2. Find the equations of the lines that pass through the points listed in Exercise 10.1.

10.4 Plot the following lines:
  a. $y = 3 + 2x$ **b.** $y = 3 - 2x$ **c.** $y = -3 + 2x$
  d. $y = -x$ **e.** $y = 2x$ **f.** $y = .50 + 1.25x$

10.5 Give the slope and y-intercept for each of the lines defined in Exercise 10.4.

10.6 Why do we generally prefer a probabilistic model to a deterministic model? Give examples for which the two types of models might be appropriate.

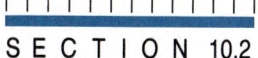

SECTION 10.2

FITTING THE MODEL: THE METHOD OF LEAST SQUARES

Suppose an appliance store conducts a 5-month experiment to determine the effect of advertising on sales revenue. The results are shown in Table 10.1. (The number of measurements is small, and the measurements themselves are unrealistically simple to avoid arithmetic confusion in this initial example.) The relationship between sales revenue, y, and advertising expenditure, x, is hypothesized to follow a first-order linear model, that is,

$$y = \beta_0 + \beta_1 x + \varepsilon$$

The question is this: How can we best use the information in the sample of five observations in Table 10.1 to estimate the unknown y-intercept β_0 and slope β_1?

TABLE 10.1

Advertising–Sales Data

MONTH	ADVERTISING EXPENDITURE x ($ hundreds)	SALES REVENUE y ($ thousands)
1	1	1
2	2	1
3	3	2
4	4	2
5	5	4

To gain information on the approximate values of these parameters, it is helpful to plot the sample data. Such a plot, called a **scattergram**, locates each of the five data points on a graph, as in Figure 10.2. Note that the scattergram suggests a general tendency for y to increase as x increases. If you place a ruler on the scattergram, you will see that a line may be drawn through three of the five points, as shown in Figure 10.3 (page 492). To obtain the equation of this visually fitted line, note that the line intersects the y-axis at $y = -1$, so the y-intercept is -1. Also, y increases exactly 1 unit for every 1-unit increase in x, indicating that the slope is $+1$. Therefore, the equation is

$$\tilde{y} = -1 + 1(x) = -1 + x$$

where $\tilde{y}$ is used to denote the value of y predicted from the visually fitted model.

FIGURE 10.2

Scattergram for Data in Table 10.1

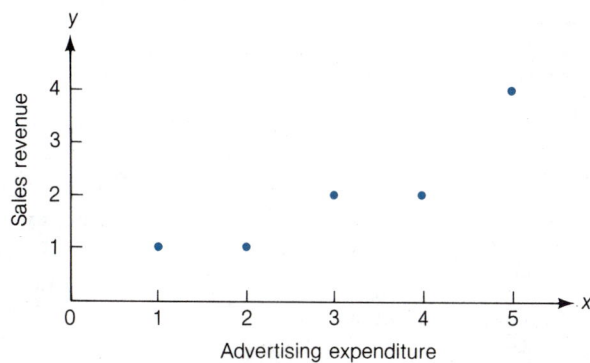

FIGURE 10.3

Visual Straight-Line
Fit to the Data

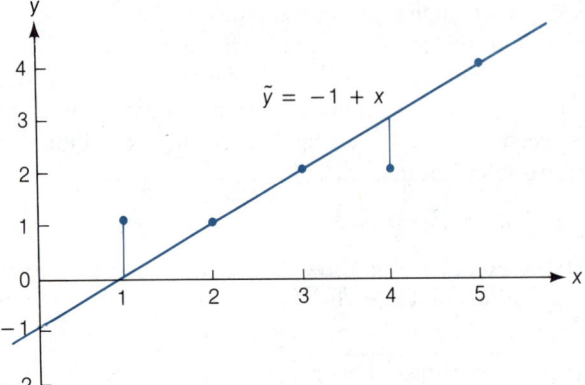

One way to decide quantitatively how well a straight line fits a set of data is to note the extent to which the data points deviate from the line. For example, to evaluate the visual model in Figure 10.3 we calculate the **deviations**—i.e., the differences between the observed and the predicted values of y. These deviations, or **errors**, are the vertical distances between observed and predicted values (see Figure 10.3). The observed and predicted values of y, their differences, and their squared differences are shown in Table 10.2. Note that the **sum of errors** equals 0 and the **sum of squares of the errors (SSE)** is equal to 2.

TABLE 10.2

Comparing Observed
and Predicted Values
for the Visual Model

x	y	$\hat{y} = -1 + x$	$(y - \hat{y})$	$(y - \hat{y})^2$
1	1	0	$(1 - 0) =$ 1	1
2	1	1	$(1 - 1) =$ 0	0
3	2	2	$(2 - 2) =$ 0	0
4	2	3	$(2 - 3) = -1$	1
5	4	4	$(4 - 4) =$ 0	0
			Sum of errors = 0	Sum of squared errors (SSE) = 2

You can see by shifting the ruler around the graph that it is possible to find many lines for which the sum of the errors is equal to 0, but it can be shown that there is one (and only one) line for which the *SSE is a minimum*. This line is called the **least squares line**, the **regression line**, the **least squares prediction equation**, or the **fitted line**.

To find the least squares line for a set of data, assume that we have a sample of n data points that can be identified by corresponding values of x and y, say (x_1, y_1), (x_2, y_2), . . . , (x_n, y_n). For example, the $n = 5$ data points shown in Table 10.2

are $(1, 1)$, $(2, 1)$, $(3, 2)$, $(4, 2)$, and $(5, 4)$. The straight-line model for the response, y, in terms of x, is

$$y = \beta_0 + \beta_1 x + \varepsilon$$

The equation of the line that relates the mean value $E(y)$ to x (called the **line of means**) is

$$E(y) = \beta_0 + \beta_1 x$$

The fitted line, which we will calculate using the five data points, is represented as

$$\hat{y} = \hat{\beta}_0 + \hat{\beta}_1 x$$

The "hats" can be read as "estimator of." Thus, $\hat{y}$ is an estimator of the mean value of y, $E(y)$, and a predictor of some future value of y; and $\hat{\beta}_0$ and $\hat{\beta}_1$ are estimators of β_0 and β_1, respectively.

For a given data point, say the point (x_i, y_i), the observed value of y is y_i and the predicted value of y would be obtained by substituting x_i into the prediction equation:

$$\hat{y}_i = \hat{\beta}_0 + \hat{\beta}_1 x_i$$

And the deviation of the ith value of y from its predicted value is

$$y_i - \hat{y}_i = y_i - (\hat{\beta}_0 + \hat{\beta}_1 x_i)$$

Then the sum of squares of the deviations of the y values about their predicted values for all the n data points is

$$SSE = \sum_{i=1}^{n} [y_i - (\hat{\beta}_0 + \hat{\beta}_1 x_i)]^2$$

The quantities $\hat{\beta}_0$ and $\hat{\beta}_1$ that make the SSE a minimum are called the **least squares estimates** of the population parameters β_0 and β_1, and the prediction equation $\hat{y} = \hat{\beta}_0 + \hat{\beta}_1 x$ is called the **least squares line**.

DEFINITION 10.1

The **least squares line** is one that has a smaller SSE than any other straight-line model.

The values of $\hat{\beta}_0$ and $\hat{\beta}_1$ that minimize the SSE are (proof omitted) given by the formulas in the following box.*

*Students who are familar with calculus should note that the values of $\hat{\beta}_0$ and $\hat{\beta}_1$ that minimize $SSE = \sum_{i=1}^{n} (y_i - \hat{y}_i)^2$ are obtained by setting the two partial derivatives $\partial SSE/\partial \beta_0$ and $\partial SSE/\partial \beta_1$ equal to 0. Furthermore, we denote the *sample* solutions to the equations by $\hat{\beta}_0$ and $\hat{\beta}_1$, where the ^ (hat) denotes that these are sample estimates of the true population intercept β_0 and slope β_1. The solutions to these two equations yield the formulas shown in the box.

FORMULAS FOR THE LEAST SQUARES ESTIMATES

Slope: $\hat{\beta}_1 = \dfrac{SS_{xy}}{SS_{xx}}$

y-intercept: $\hat{\beta}_0 = \bar{y} - \hat{\beta}_1\bar{x}$

where

$$SS_{xy} = \sum_{i=1}^{n} x_i y_i - \frac{\left(\sum\limits_{i=1}^{n} x_i\right)\left(\sum\limits_{i=1}^{n} y_i\right)}{n}$$

$$SS_{xx} = \sum_{i=1}^{n} x_i^2 - \frac{\left(\sum\limits_{i=1}^{n} x_i\right)^2}{n}$$

n = Sample size

TABLE 10.3

Preliminary Computations for the Advertising–Sales Example

x_i	y_i	x_i^2	$x_i y_i$
1	1	1	1
2	1	4	2
3	2	9	6
4	2	16	8
5	4	25	20
Totals $\Sigma x_i = 15$	$\Sigma y_i = 10$	$\Sigma x_i^2 = 55$	$\Sigma x_i y_i = 37$

Preliminary computations for finding the least squares line for the advertising–sales example are contained in Table 10.3. We can now calculate*

$$SS_{xy} = \sum x_i y_i - \frac{\left(\sum x_i\right)\left(\sum y_i\right)}{5} = 37 - \frac{(15)(10)}{5}$$

$$= 37 - 30 = 7$$

$$SS_{xx} = \sum x_i^2 - \frac{\left(\sum x_i\right)^2}{5} = 55 - \frac{(15)^2}{5}$$

$$= 55 - 45 = 10$$

Then, the slope of the least squares line is

$$\hat{\beta}_1 = \frac{SS_{xy}}{SS_{xx}} = \frac{7}{10} = .7$$

*Since summations are used extensively from this point on, we will omit the limits on Σ when the summation includes all the measurements in the sample, i.e., when the symbol is $\sum\limits_{i=1}^{n}$, we will write Σ.

and the y-intercept is

$$\hat{\beta}_0 = \bar{y} - \hat{\beta}_1\bar{x} = \frac{\sum y_i}{5} - \hat{\beta}_1\frac{\left(\sum x_i\right)}{5}$$

$$= \frac{10}{5} - (.7)\frac{15}{5} = 2 - (.7)(3) = 2 - 2.1 = -.1$$

The least squares line is thus

$$\hat{y} = \hat{\beta}_0 + \hat{\beta}_1 x = -.1 + .7x$$

The graph of this line is shown in Figure 10.4.

FIGURE 10.4

The Line
$\hat{y} = -.1 + .7x$
Fit to the Data

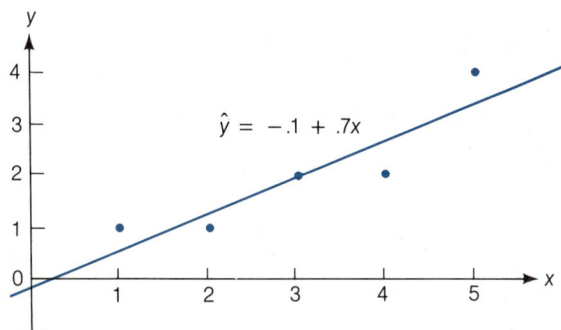

The observed and predicted values of y, the deviations of the y values about their predicted values, and the squares of these deviations are shown in Table 10.4. Note that the sum of squares of the deviations (SSE) is 1.10, and (as we would expect) this is less than the SSE = 2.0 obtained in Table 10.2 for the visually fitted line.

TABLE 10.4

Comparing Observed and Predicted Values for the Least Squares Model

x	y	$\hat{y} = -.1 + .7x$	$(y - \hat{y})$	$(y - \hat{y})^2$
1	1	.6	$(1 - .6) = $ .4	.16
2	1	1.3	$(1 - 1.3) = -.3$	.09
3	2	2.0	$(2 - 2.0) = $ 0	.00
4	2	2.7	$(2 - 2.7) = -.7$	.49
5	4	3.4	$(4 - 3.4) = $ .6	.36
			Sum of errors = 0	SSE = 1.10

To summarize, we have defined the best-fitting straight line to be the one that satisfies the least squares criterion; that is, the sum of the squared errors will be smaller than for any other straight-line model.

| | | | | | | | | | | | | | |

EXERCISES 10.7–10.16   **LEARNING THE MECHANICS**

10.7 The following table is similar to Table 10.3. It is used for making the preliminary computations for finding the least squares line for the given pairs of x and y values.

x_i	y_i	x_i^2	$x_i y_i$
7	2		
4	4		
6	2		
2	5		
1	7		
1	6		
3	5		
Totals $\Sigma x_i =$	$\Sigma y_i =$	$\Sigma x_i^2 =$	$\Sigma x_i y_i =$

a. Complete the table.   **b.** Find SS_{xy}.   **c.** Find SS_{xx}.
d. Find $\hat{\beta}_1$.   **e.** Find $\bar{x}$ and $\bar{y}$.   **f.** Find $\hat{\beta}_0$.
g. Find the least squares line.

10.8 Refer to Exercise 10.7. After the least squares line has been obtained, the following table (which is similar to Table 10.4) can be used for (1) comparing the observed and the predicted values of y, and (2) computing SSE.

x	y	$\hat{y} =$	$(y - \hat{y})$	$(y - \hat{y})^2$
7	2			
4	4			
6	2			
2	5			
1	7			
1	6			
3	5			
			$\Sigma(y - \hat{y}) =$	$SSE = \Sigma(y - \hat{y})^2 =$

a. Complete the table.
b. Plot the least squares line on a scattergram of the data. Plot the following line on the same graph:

$$\hat{y} = 14 - 2.5x$$

c. Show that SSE is larger for the line of part **b** that it is for the least squares line.

10.9 Construct a scattergram for the data in the table.

x	.5	1	1.5
y	2	1	3

a. Plot the following two lines on your scattergram:

$$y = 3 - x$$
$$y = 1 + x$$

b. Which of these lines would you choose to characterize the relationship between x and y? Explain.

c. Show that the sum of errors for both of these lines equals 0.

d. Which of these lines has the smaller SSE?

e. Find the least squares line for the data, and compare it to the two lines described in part **a**.

10.10 Consider the following pairs of measurements:

x	5	3	−1	2	7	6	4
y	4	3	0	1	8	5	3

a. Construct a scattergram for the data.

b. What does the scattergram suggest about the relationship between x and y?

c. What are the least squares estimates of β_0 and β_1?

d. Plot the least squares line on your scattergram. Does the line appear to fit the data well? Explain.

APPLYING THE CONCEPTS

10.11 Individuals who report perceived wrongdoing of a corporation or public agency are known as *whistle blowers*. Janet P. Near and Marcia P. Miceli (1986) used regression analysis to study factors that affect the extent of the retaliation by the organization against the whistle blower. Among the factors they studied were the pay rate and education level of the whistle blower, the seriousness of the organization's wrongdoing, and the external whistle-blowing channel used. Near and Miceli developed an index to measure the extent of retaliation. The index was based on the number of forms of reprisal experienced by the whistle blower, the number of forms of reprisal with which the whistle blower was threatened, and the number of different types of people within the organization (i.e., coworkers, immediate supervisor, and so forth) who retaliated against them. The table lists the retaliation index (higher numbers indicate more extensive retaliation) and salary for a sample of 15 whistle blowers from federal agencies.

RETALIATION INDEX	SALARY	RETALIATION INDEX	SALARY
301	$62,000	535	$15,800
550	36,500	455	44,000
755	17,600	615	46,600
327	20,000	700	12,100
500	30,100	650	62,000
377	35,000	630	21,000
290	47,500	360	11,900
452	54,000		

Source: Data adapted from results presented by Near and Miceli (1986).

a. Construct a scattergram for the data. Does it appear that the extent of retaliation increases, decreases, or stays the same with an increase in salary? Explain.

b. Use the method of least squares to fit a straight line to the data.

c. Graph the least squares line on your scattergram. Does the least squares line support your answer to the question in part **a**? Explain.

10.12 Due primarily to the price controls of the Organization of Petroleum Exporting Countries (OPEC), a cartel of crude oil suppliers, the price of crude oil rose dramatically from the mid-1970's to the early 1980's. As a result, motorists were confronted with a similar upward spiral of gasoline prices. The data in the table are typical prices for a gallon of regular leaded gasoline and a barrel of crude oil (refiner acquisition cost) for the indicated years.

YEAR	GASOLINE y (cents/gallon)	CRUDE OIL x ($/bbl)
1973	38.8	3.89
1975	56.7	7.67
1976	59.0	8.19
1977	62.2	8.57
1978	62.6	9.00
1979	85.7	12.64
1980	119.1	21.59
1981	133.1	31.77
1982	122.2	28.52
1983	115.7	26.19
1984	112.9	25.88

Source: *Statistical Abstract of the United States: 1986*, pp. 698, 703.

Given that $\Sigma y = 968$, $\Sigma x = 183.91$, $\Sigma y^2 = 96,299.58$, $\Sigma x^2 = 4,099.7935$, and $\Sigma xy = 19,485.868$:

a. Use the data to calculate the least squares line that describes the relationship between the price of a gallon of gasoline and the price of a barrel of crude oil.

b. Plot your least squares line on a scattergram of the data. Does your least squares line appear to be an appropriate characterization of the relationship between y and x? Explain.

c. If the price of crude oil fell to $8 per barrel, to what level (approximately) would the price of regular gasoline fall? Justify your response.

10.13 A company that developed a new type of fertilizer is interested in investigating the relationship between the yield of potatoes, y, and the amount of the new fertilizer that is applied to the potato plants, x. An agronomist divided a field into eight plots of equal size and applied differing amounts of fertilizer to each. The yield of potatoes (in pounds) and the fertilizer application (in pounds) were recorded for each plot. The data are shown in the table.

x	1	1.5	2	2.5	3	3.5	4	4.5
y	25	31	27	28	36	35	32	34

a. Construct a scattergram for the data.

b. Find the least squares estimates of β_0 and β_1.

c. According to your least squares line, approximately how many pounds of potatoes would you expect from a plot to which 3.75 pounds of fertilizer had been applied? [*Note:* A measure of the reliability of these predictions will be discussed in Section 10.8.]

10.14 A car dealer is interested in modeling the relationship between the number of cars sold by the firm each week and the average number of salespeople who work on the showroom floor per day during the week. The dealer believes the relationship between the two variables can best be described by a straight line. The sample data shown in the table were supplied by the car dealer.

WEEK OF	NUMBER OF CARS SOLD y	AVERAGE NUMBER OF SALESPEOPLE ON DUTY x
January 30	20	6
June 29	18	6
March 2	10	4
October 26	6	2
February 7	11	3

a. Construct a scattergram for the data.
b. Assuming the relationship between the variables is best described by a straight line, use the method of least squares to estimate the y-intercept and the slope of the line.
c. Plot the least squares line on your scattergram.
d. According to your least squares line, approximately how many cars should the dealer expect to sell in a week if an average of five salespeople are kept on the showroom floor each day? [*Note:* A measure of the reliability of these predictions will be discussed in Section 10.8.]

10.15 Is the number of games won by a major league baseball team in a season related to the team's batting average? The accompanying table shows the number of games won and the batting averages for the 14 teams in the American League for the 1986 season.

TEAM	NUMBER OF GAMES WON y	TEAM BATTING AVERAGE x
Cleveland	84	.284
New York	90	.271
Boston	95	.271
Toronto	86	.269
Texas	87	.267
Detroit	87	.263
Minnesota	71	.261
Baltimore	73	.258
California	92	.255
Milwaukee	77	.255
Seattle	67	.253
Kansas City	76	.252
Oakland	76	.252
Chicago	72	.247

Source: *Official American League Averages*, 1986. The American League of Professional Baseball Clubs, New York, pp. 2, 24–27.

a. If you were to model the relationship between the number of games won, y, by a major league team and the team's batting average, x, using a straight line, would you expect the slope of the line to be positive or negative? Explain.

b. Construct a scattergram for the data. Does the pattern revealed by the scattergram agree with your answer to part a?

c. Given that $\Sigma y = 1,133$, $\Sigma x = 3.658$, $\Sigma y^2 = 92,703$, $\Sigma x^2 = .957118$, and $\Sigma xy = 296.734$, fit a simple linear regression model to the data.

d. Graph the least squares line on your scattergram. Does your least squares line seem to fit the points on your scattergram?

e. Can you explain why the number of games won does not appear to be strongly related to a team's batting average?

10.16 An appliance company is interested in relating the sales rate of 17-inch color television sets to the price per set. To do this, the company randomly selected 15 weeks in the past year and recorded the number of sets sold each week and the price at which the sets were being sold during that week. The data are shown in the table.

WEEK	NUMBER OF 17-INCH COLOR TELEVISION SETS SOLD PER WEEK y	PRICE x ($)	WEEK	NUMBER OF 17-INCH COLOR TELEVISION SETS SOLD PER WEEK y	PRICE x ($)
1	55	350	9	20	400
2	54	360	10	45	340
3	25	385	11	50	350
4	18	400	12	35	335
5	51	370	13	30	330
6	20	390	14	30	325
7	45	375	15	53	365
8	19	390			

a. Given that $SS_{xx} = 9,043.3333$, $SS_{xy} = -2,393.3333$, $\bar{y} = 36.6667$, and $\bar{x} = 364.3333$, find the least squares line relating y to x.

b. Plot the data and graph the least squares line as a check on your calculations.

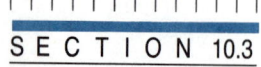

SECTION 10.3

MODEL ASSUMPTIONS

In Section 10.2, we assumed that the probabilistic model relating the firm's sales revenue y to advertising dollars x is

$$y = \beta_0 + \beta_1 x + \varepsilon$$

and recall that the least squares estimate of the deterministic component of the model $\beta_0 + \beta_1 x$ is

$$\hat{y} = \hat{\beta}_0 + \hat{\beta}_1 x = -.1 + .7x$$

Now we turn our attention to the random component ε of the probabilistic model and its relation to the errors in estimating β_0 and β_1. In particular, we will see how the probability distribution of ε determines how well the model describes the true relationship between the dependent variable y and the independent variable x.

We make four basic assumptions about the general form of the probability distribution of ε:

ASSUMPTION 1 The mean of the probability distribution of ε is 0. That is, the average of the errors over an infinitely long series of experiments is 0 for each setting of the independent variable x. This assumption implies that the mean value of y, $E(y)$, for a given value of x is $E(y) = \beta_0 + \beta_1 x$.

ASSUMPTION 2 The variance of the probability distribution of ε is constant for all values of the independent variable, x. For our straight-line model, this assumption means that the variance of ε is equal to a constant, say σ^2, for all values of x.

ASSUMPTION 3 The probability distribution of ε is normal.

ASSUMPTION 4 The errors associated with any two different observations are independent. That is, the error associated with one value of y has no effect on the errors associated with other y values.

The implications of the first three assumptions can be seen in Figure 10.5, which shows distributions of errors for three particular values of x—namely, x_1, x_2, and x_3. Note that the relative frequency distributions of the errors are normal, with a mean of 0 and a constant variance σ^2 (all the distributions shown have the same amount of spread or variability). The straight line shown in Figure 10.5 plots the mean value $E(y)$ for a given value of x. Then, the line of means is given by the equation

$$E(y) = \beta_0 + \beta_1 x$$

FIGURE 10.5
The Probability
Distribution of ε

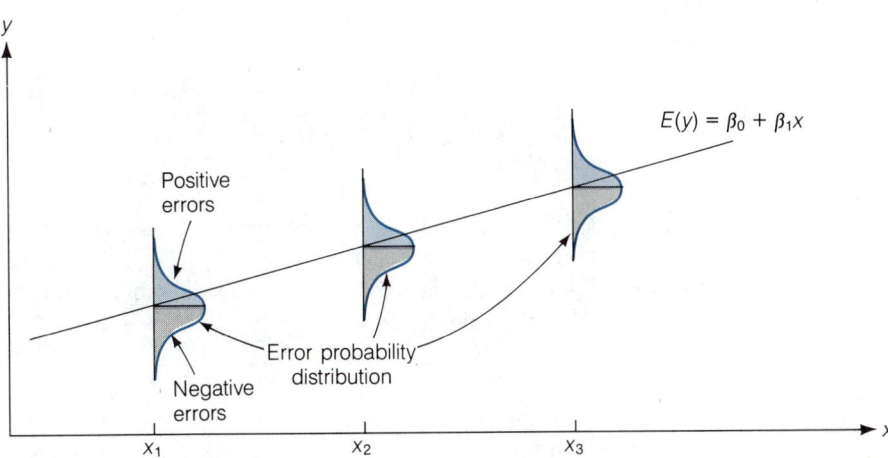

Various techniques exist for checking the validity of these assumptions, and there are remedies to be applied when they appear to be invalid. We will discuss some of these techniques and remedies in Chapters 11 and 12.

In actual practice, the assumptions need not hold exactly in order for least squares estimators and test statistics (to be described subsequently) to possess the measures of reliability that we would expect from a regression analysis. The assumptions will be satisfied adequately for many applications encountered in business.

| | | | | | | | | | | | |

S E C T I O N 10.4

AN ESTIMATOR OF σ^2

S is variance of ε

It seems reasonable to assume that the greater the variability of the random error ε (which is measured by its variance σ^2), the greater will be the errors in the estimation of the model parameters β_0 and β_1 and in the error of prediction when $\hat{y}$ is used to predict y for some value of x. Consequently, you should not be surprised, as we proceed through this chapter, to find that σ^2 appears in the formulas for all confidence intervals and test statistics that we use.

In most practical situations, σ^2 will be unknown, and we must use our data to estimate its value. The best (proof omitted) estimate, s^2, of σ^2 is obtained by dividing the sum of squares of deviations,

$$SSE = \sum (y_i - \hat{y}_i)^2$$

by the number of degrees of freedom (df) associated with this quantity. We use 2 df to estimate the y-intercept and slope in the straight-line model, leaving $(n - 2)$ df for the error variance estimation (see the formulas in the box).

ESTIMATION OF σ^2

$$s^2 = \frac{SSE}{\text{Degrees of freedom for error}} = \frac{SSE}{n - 2}$$

where

$$SSE = \sum (y_i - \hat{y}_i)^2 = SS_{yy} - \hat{\beta}_1 SS_{xy}$$

$$SS_{yy} = \sum (y_i - \bar{y})^2 = \sum y_i^2 - \frac{\left(\sum y_i\right)^2}{n}$$

Warning: When performing these calculations, you may be tempted to round the calculated values of SS_{yy}, $\hat{\beta}_1$, and SS_{xy}. Be certain to carry at least six significant figures for each of these quantities to avoid substantial errors in the calculation of SSE.

In the advertising–sales example, we previously calculated $SSE = 1.10$ for the least squares line $\hat{y} = -.1 + .7x$. Recalling that there were $n = 5$ data points, we have $n - 2 = 5 - 2 = 3$ df for estimating σ^2. Thus,

$$s^2 = \frac{SSE}{n - 2} = \frac{1.10}{3} = .367$$

is the estimated variance, and

$$s = \sqrt{.367} = .61$$

is the estimated standard deviation of ε.

You may be able to obtain an intuitive feeling for s by recalling the interpretation given to a standard deviation in Chapter 3 and remembering that the least squares line estimates the mean value of y for a given value of x. Since s measures the spread of the distribution of y values about the least squares line, we should not be surprised to find that most of the observations lie within $2s$ or $2(.61) = 1.22$ of the least squares line. For this simple example (only five data points), all five data points fall within $2s$ of the least squares line. In Section 10.8, we will use s to evaluate the error of prediction when the least squares line is used to predict a value of y to be observed for a given value of x.

**EXERCISES
10.17–10.25**

LEARNING THE MECHANICS

10.17 Suppose you fit a least squares line to 12 data points and calculate SSE = .507. Find s^2, the estimator of σ^2, the variance of the random error term ε.

10.18 Calculate SSE and s^2 for each of the following cases:
  a. $n = 18$, $SS_{yy} = 95$, $SS_{xy} = 50$, $\hat{\beta}_1 = .75$
  b. $n = 35$, $\Sigma y^2 = 860$, $\Sigma y = 50$, $SS_{xy} = 2{,}700$, $\hat{\beta}_1 = .2$
  c. $n = 20$, $\Sigma(y_i - \bar{y})^2 = 58$, $SS_{xy} = 91$, $SS_{xx} = 170$

10.19 Refer to Exercises 10.7 and 10.10. Calculate SSE, s^2, and s for the least squares lines obtained in those exercises.

10.20 Visually compare the following scattergrams. If a least squares line were determined for each data set, which do you think would have the smallest variance, s^2? Explain.

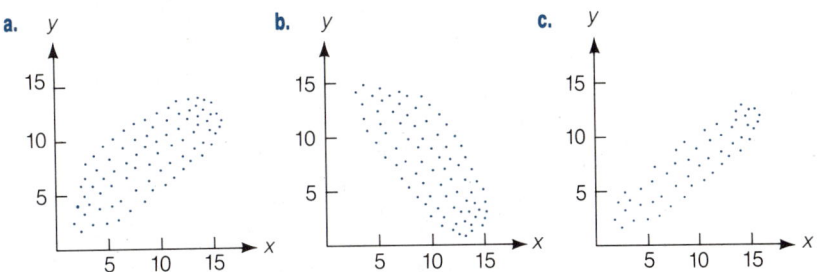

APPLYING THE CONCEPTS

10.21 In order to improve the quality of the output of any production process, it is necessary first to understand the capabilities of the process (Deming, 1982). In a particular manufacturing process, the useful life of a cutting tool is related to the speed at which the tool is operated. It is necessary to understand this relationship in order to predict when the tool should be replaced and how many spare tools should be available. The data in the table were derived from life tests for the two different brands of cutting tools currently used in the production process.

CUTTING SPEED (Meters per Minute)	USEFUL LIFE (HOURS) Brand A	USEFUL LIFE (HOURS) Brand B	CUTTING SPEED (Meters per Minute)	USEFUL LIFE (HOURS) Brand A	USEFUL LIFE (HOURS) Brand B
30	4.5	6.0	50	1.0	3.7
30	3.5	6.5	60	4.0	3.8
30	5.2	5.0	60	2.0	3.0
40	5.2	6.0	60	1.1	2.4
40	4.0	4.5	70	1.1	1.5
40	2.5	5.0	70	.5	2.0
50	4.4	4.5	70	3.0	1.0
50	2.8	4.0			

a. Construct a scattergram for each brand of cutting tool.
b. For each brand, use the method of least squares to model the relationship between useful life and cutting speed.
c. Find SSE, s^2, and s for each least squares line.
d. For a cutting speed of 70 meters per minute, find $\hat{y} \pm 2s$ for each least squares line.
e. For which brand would you feel more confident in using the least squares line to predict useful life for a given cutting speed? Explain.

10.22 In Exercise 10.12, the price of a gallon of gasoline, y, was modeled as a function of the price of a barrel of crude oil, x, using the following equation: $y = \beta_0 + \beta_1 x + \varepsilon$. Estimates of β_0 and β_1 were found to be 34.143 and 3.2213, respectively.
a. Estimate the variance and standard deviation of ε.
b. Suppose the price of a barrel of crude oil is \$15. Estimate the mean and standard deviation of the price of a gallon of gasoline under these circumstances.
c. Repeat part **b** for a crude oil price of \$30 per barrel.
d. What assumptions about ε did you make in answering parts **b** and **c**?

10.23 Although the cable television industry could provide viewers with 100 or more channels, a study by the A. C. Nielsen Co. suggests that that may be many more than viewers want or will ever watch. The Nielsen survey indicates that as the number of television channels increases, the *percentage* of channels viewed for 10 minutes a week or more declines (Landro and Mayer, 1982). In a similar study, 20 households were sampled, and the number of channels available to each household was recorded. In addition, each household was asked to monitor its television viewing for 1 week and report the number of channels watched for 10 minutes or more. The results appear in the accompanying table.
a. Do these data tend to support the Nielsen findings? Find the appropriate least squares line and use it to justify your answer.
b. Plot your least squares line on a scattergram of the data.
c. Calculate SSE, s^2, and s. For the given number of channels available to a particular household, within what approximate bounds would you expect this least squares line to be able to predict the percentage of channels watched for 10 minutes?

HOUSEHOLD	NUMBER OF CHANNELS AVAILABLE	NUMBER OF CHANNELS WATCHED FOR 10 MINUTES	HOUSEHOLD	NUMBER OF CHANNELS AVAILABLE	NUMBER OF CHANNELS WATCHED FOR 10 MINUTES
1	12	6	11	25	10
2	29	10	12	8	6
3	4	3	13	5	4
4	20	8	14	10	4
5	40	12	15	16	9
6	5	3	16	4	4
7	6	5	17	5	1
8	4	4	18	45	13
9	14	8	19	35	5
10	20	6	20	50	10

10.24 A breeder of thoroughbred horses wishes to model the relationship between the gestation period and the length of life of a horse. The breeder believes that the two variables may follow a linear trend. The information in the table was supplied to the breeder from various thoroughbred stables across the state.

HORSE	GESTATION PERIOD x (days)	LIFE LENGTH y (years)
1	416	24
2	279	25.5
3	298	20
4	307	21.5
5	356	22
6	403	23.5
7	265	21

a. Fit a least squares line to the data. Plot the data points and graph the least squares line as a check on your calculations.
b. According to your least squares line, approximately how long would you expect a horse to live whose gestation period was 400 days?
c. Calculate SSE and s^2.
d. Give an interpretation of the standard deviation s in the context of this problem.

10.25 A company keeps extensive records on its new salespeople on the premise that sales should increase with experience. A random sample of seven new salespeople produced the data on experience and sales shown in the table at the top of page 506.
a. Fit a least squares line to the data.
b. Plot the data and graph the least squares line.
c. Predict the sales that a new salesperson would be expected to generate after 6 months on the job. After 9 months.
d. Calculate SSE, s^2, and s.
e. Within what approximate distance do you expect your predictions in part c to fall from the true number of sales generated by the new salesperson? [*Note:* A more precise measure of reliability for these predictions will be discussed in Section 10.8.]

MONTHS ON JOB x	MONTHLY SALES y ($ thousands)
2	2.4
4	7.0
8	11.3
12	15.0
1	.8
5	3.7
9	12.0

SECTION 10.5

ASSESSING THE USEFULNESS OF THE MODEL: MAKING INFERENCES ABOUT THE SLOPE β_1

Refer again to the data of Table 10.1 and suppose that the appliance store's sales revenue is *completely unrelated* to the advertising expenditure. What could be said about the values of β_0 and β_1 in the hypothesized probabilistic model

$$y = \beta_0 + \beta_1 x + \varepsilon$$

if x contributes no information for the prediction of y? The implication is that the mean of y—i.e., the deterministic part of the model $E(y) = \beta_0 + \beta_1 x$—does not change as x changes. Regardless of the value of x, you always predict the same value of y. In the straight-line model, this means that the true slope, β_1, is equal to 0. Therefore, to test the null hypothesis that the linear model contributes no information for the prediction of y against the alternative hypothesis that the linear model is useful for predicting y, we test

$$H_0: \quad \beta_1 = 0$$
$$H_a: \quad \beta_1 \neq 0$$

If the data support the alternative hypothesis, we conclude that x does contribute information for the prediction of y using the straight-line model [although the true relationship between $E(y)$ and x could be more complex than a straight line]. Thus, to some extent, this is a test of the usefulness of the hypothesized model.

The appropriate test statistic is found by considering the sampling distribution of $\hat{\beta}_1$, the least squares estimator of the slope β_1.

SAMPLING DISTRIBUTION OF $\hat{\beta}_1$

If the four assumptions about ε (see Section 10.3) are satisfied, then the sampling distribution of $\hat{\beta}_1$, the least squares estimator of slope, will be normal, with mean β_1 (the true slope) and standard deviation

$$\sigma_{\hat{\beta}_1} = \frac{\sigma}{\sqrt{SS_{xx}}} \qquad \text{(see Figure 10.6)}$$

FIGURE 10.6

Sampling Distribution of $\hat{\beta}_1$

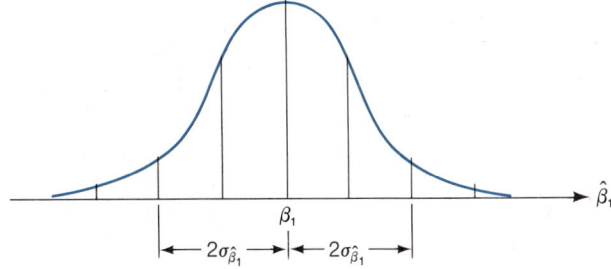

Since σ will usually be unknown, the appropriate test statistic will generally be a Student's t statistic formed as follows:

$$t = \frac{\hat{\beta}_1 - \text{Hypothesized value of } \beta_1}{s_{\hat{\beta}_1}} \quad \text{where} \quad s_{\hat{\beta}_1} = \frac{s}{\sqrt{SS_{xx}}}$$

$$= \frac{\hat{\beta}_1 - 0}{s/\sqrt{SS_{xx}}}$$

Note that we have substituted the estimator s for σ, and then formed $s_{\hat{\beta}_1}$ by dividing s by $\sqrt{SS_{xx}}$. The number of degrees of freedom associated with this t statistic is the same as the number of degrees of freedom associated with s. Recall that this will be $(n-2)$ df when the hypothesized model is a straight line (see Section 10.4).

The set-up of our test of the usefulness of the model is summarized in the box:

A TEST OF MODEL USEFULNESS

ONE-TAILED TEST	**TWO-TAILED TEST**
H_0: $\beta_1 = 0$	H_0: $\beta_1 = 0$
H_a: $\beta_1 < 0$	H_a: $\beta_1 \neq 0$
(or H_a: $\beta_1 > 0$)	

Test statistic: $t = \dfrac{\hat{\beta}_1}{s_{\hat{\beta}_1}} = \dfrac{\hat{\beta}_1}{s/\sqrt{SS_{xx}}}$ Test statistic: $t = \dfrac{\hat{\beta}_1}{s_{\hat{\beta}_1}} = \dfrac{\hat{\beta}_1}{s/\sqrt{SS_{xx}}}$

Rejection region: $t < -t_\alpha$ Rejection region: $t < -t_{\alpha/2}$
                 (or $t > t_\alpha$)   or $t > t_{\alpha/2}$

where t_α is based on $(n-2)$ df.   where $t_{\alpha/2}$ is based on $(n-2)$ df.

Assumptions: The four assumptions about ε listed in Section 10.3.

For the advertising–sales example, we will choose $\alpha = .05$ and, since $n = 5$, df $= (n-2) = 5 - 2 = 3$. Then the rejection region for the two-tailed test is

$$t < -t_{.025} = -3.182 \quad \text{or} \quad t > t_{.025} = 3.182$$

We previously calculated $\hat{\beta}_1 = .7$, $s = .61$, and $SS_{xx} = 10$. Thus,

$$t = \frac{\hat{\beta}_1}{s/\sqrt{SS_{xx}}} = \frac{.7}{.61/\sqrt{10}} = \frac{.7}{.19} = 3.7$$

Since this calculated t value falls in the upper-tail rejection region (see Figure 10.7), we reject the null hypothesis and conclude that the slope β_1 is not 0. The sample evidence indicates that x contributes information for the prediction of y using a linear model for the relationship between sales revenue and advertising.

FIGURE 10.7
Rejection Region and
Calculated t Value for
Testing Whether the
Slope $\beta_1 = 0$

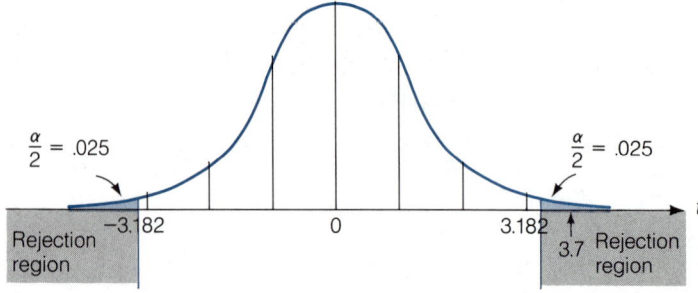

What conclusion can be drawn if the calculated t value does not fall in the rejection region? We know from previous discussions of the philosophy of hypothesis testing that such a t value does *not* lead us to accept the null hypothesis. That is, we do not conclude that $\beta_1 = 0$. Additional data might indicate that β_1 differs from 0, or a more complex relationship may exist between y and x, requiring the fitting of a model other than the straight-line model. We discuss several such models in Chapter 11.

Another way to make inferences about the slope β_1 is to estimate it using a confidence interval. This interval is formed as shown in the box.

A 100($1 - \alpha$)% CONFIDENCE INTERVAL FOR THE SLOPE β_1

$$\hat{\beta}_1 \pm t_{\alpha/2} s_{\hat{\beta}_1} \qquad \text{where} \quad s_{\hat{\beta}_1} = \frac{s}{\sqrt{SS_{xx}}}$$

and $t_{\alpha/2}$ is based on $(n - 2)$ df.

For the advertising–sales example, a 95% confidence interval for the slope β_1 is

$$\hat{\beta}_1 \pm t_{.025} s_{\hat{\beta}_1} = .7 \pm 3.182 \left(\frac{s}{\sqrt{SS_{xx}}} \right) = .7 \pm 3.182 \left(\frac{.61}{\sqrt{10}} \right) = .7 \pm .61$$

Thus, we estimate with 95% confidence that the interval from .09 to 1.31 includes the slope parameter β_1.

Since all the values in this interval are positive, it appears that β_1 is positive and that the mean of y, $E(y)$, increases as x increases. However, the rather large width of the confidence interval reflects the small number of data points (and, consequently, a lack of information) in the experiment. We would expect a narrower interval if the sample size were increased.

| | | | | | | | | | | | | |

**EXERCISES
10.26–10.39**

LEARNING THE MECHANICS

10.26 Construct both a 95% and a 90% confidence interval for β_1 for each of the following cases:
  a. $\hat{\beta}_1 = 31$, $s = 3$, $SS_{xx} = 35$, $n = 10$
  b. $\hat{\beta}_1 = 64$, $SSE = 1,960$, $SS_{xx} = 30$, $n = 14$
  c. $\hat{\beta}_1 = -8.4$, $SSE = 146$, $SS_{xx} = 64$, $n = 20$

10.27 Consider the following pairs of observations:

x	1	4	3	2	5	6	0
y	1	3	3	1	4	7	2

  a. Construct a scattergram for the data.
  b. Use the method of least squares to fit a straight line to the seven data points in the table.
  c. Plot the least squares line on your scattergram of part **a**.
  d. Specify the null and alternative hypotheses you would use to test whether the data provide sufficient evidence to indicate that x contributes information for the (linear) prediction of y.
  e. What is the test statistic that should be used in conducting the hypothesis test of part **d**? Specify the degrees of freedom associated with the test statistic.
  f. Conduct the hypothesis test of part **d** using $\alpha = .05$.

10.28 Refer to Exercise 10.27. Construct an 80% and a 98% confidence interval for β_1.

10.29 Do the accompanying data provide sufficient evidence to conclude that a straight line is useful for characterizing the relationship between x and y?

y	4	2	4	3	2	4
x	1	6	5	3	2	4

APPLYING THE CONCEPTS

10.30 Based on an observational study of five chief executives, Mintzberg (1973) identified ten managerial roles that can be found in all managerial jobs: figurehead, leader, liaison, monitor, disseminator, spokesperson, entrepreneur, disturbance handler, resource allocator, and negotiator. In a recent observational study of 19 managers from a medium-sized manufacturing plant, Luthans, Rosenkrantz, and Hennessey (1985) extended Mintzberg's work by investigating which activities *successful* managers actually perform. Each manager was observed during eighty 10-minute intervals over a 2-week period and their activities were recorded. The authors used regression analysis to investigate which of the recorded activities were related to managerial success. To measure success, Luthans et al. devised an index based on the manager's length of time in the organization and his or her level within the firm; the higher the index, the more successful the manager. The table presents data (which are representative of the data collected by Luthans et al.) that can be used to determine whether managerial success can in part be explained by extensiveness of a manager's network-building interactions with people outside the manager's work unit. Such interactions include phone and face-to-face meetings with customers and suppliers, attending external meetings, and doing public relations work.

MANAGER	MANAGER SUCCESS INDEX y	NUMBER OF INTERACTIONS WITH OUTSIDERS x	MANAGER	MANAGER SUCCESS INDEX y	NUMBER OF INTERACTIONS WITH OUTSIDERS x
1	40	12	11	70	20
2	73	71	12	47	81
3	95	70	13	80	40
4	60	81	14	51	33
5	81	43	15	32	45
6	27	50	16	50	10
7	53	42	17	52	65
8	66	18	18	30	20
9	25	35	19	42	21
10	63	82			

a. Construct a scattergram for the data.

b. Given $SS_{yy} = 7,006.6316$, $SS_{xx} = 10,824.5263$, $SS_{xy} = 2,561.2632$, $\bar{y} = 54.5789$, and $\bar{x} = 44.1579$, use the method of least squares to find a prediction equation for managerial success.

c. Find SSE, s^2, and s for your prediction equation. Interpret the standard deviation s in the context of this problem.

d. Plot the least squares line on your scattergram of part a. Does it appear that the number of interactions with outsiders contributes information for the prediction of managerial success? Explain.

e. Conduct a formal statistical hypothesis test to answer the question posed in part d. Use $\alpha = .05$.

f. Construct a 95% confidence interval for β_1. Interpret the interval in the context of the problem.

10.31 The expenses involved in a manufacturing operation may be categorized as being for *raw material*, *direct labor*, and *overhead*. The term *direct labor* refers to the persons employed to transform the raw materials into the finished product. *Overhead* refers to all expenses other than those for raw materials and direct labor that are involved with running the factory (e.g., supervisory labor, maintenance of equipment, and office supplies) (Gray and Johnston, 1977). A manufacturer of ten-speed racing bicycles is interested in estimating the relationship between its monthly factory overhead and the total number of bicycles produced per month. The estimate will be used to help develop the manufacturing budget for next year. The data in the table have been collected for the previous 12 months.

MONTH	PRODUCTION LEVEL (Thousands of units)	OVERHEAD ($ thousands)	MONTH	PRODUCTION LEVEL (Thousands of units)	OVERHEAD ($ thousands)
1	16.9	41.4	7	16.3	37.5
2	15.6	35.0	8	15.5	37.0
3	17.4	38.3	9	23.4	47.9
4	11.6	29.5	10	28.4	55.6
5	17.7	39.6	11	27.1	53.1
6	17.6	37.4	12	19.2	40.6

a. Find the least squares prediction equation relating monthly overhead y to monthly production level x.

b. Test to determine whether the straight-line model contributes information for the prediction of overhead costs. Use $\alpha = .05$.

c. Which of the four assumptions we make about ε may be inappropriate in this problem? Explain.

10.32 During June, July, and early August of 1981, a total of ten bids were made by DuPont, Seagram, and Mobil to take over Conoco. Finally, on August 5, DuPont announced that it had succeeded. The total value of the offer accepted by Conoco was $7.54 billion, making it the largest takeover in the history of American business. As part of an analysis of the Conoco takeover, Richard S. Ruback (1982) used regression analysis to examine whether movements in the rate of return of each of the above-mentioned companies' common stock could be explained by movements in the rate of return of the stock market as a whole. He used the following model: $y = \beta_0 + \beta_1 x + \varepsilon$, where y is the daily rate of return of a stock, x is the daily rate of return of the stock market as a whole as measured by the daily rate of return of Standard & Poor's 500 Composite Index, and ε is believed to satisfy the assumptions of Section 10.3. (This model is known in the finance literature as the *market model*. Note that the parameter β_1 reflects the sensitivity of the stock's rate of return to movements in the stock market as a whole.) Using daily data from the beginning of 1979 through the end of 1980 ($n = 504$), Ruback obtained the least squares lines shown in the table for the four firms in question. The t statistics associated with values of $\hat{\beta}_1$ are shown to the right of each least squares prediction equation.

FIRM	ESTIMATED MARKET MODEL	
Conoco	$\hat{y} = .0010 + 1.40x$	$(t = 21.93)$
DuPont	$\hat{y} = -.0005 + 1.21x$	$(t = 18.76)$
Mobil	$\hat{y} = .0010 + 1.62x$	$(t = 16.21)$
Seagram	$\hat{y} = .0013 + .76x$	$(t = 6.05)$

Reprinted from "The Conoco takeover and stockholder returns" by R. S. Ruback, *Sloan Management Review*, Vol. 23, Winter 1982, pp. 13–33, by permission of the publisher. Copyright © 1982 by the Sloan Management Review Association. All rights reserved.

a. For each of the models, test $H_0: \beta_1 = 0$ versus $H_a: \beta_1 \neq 0$. Use $\alpha = .01$. Draw the appropriate conclusion regarding the usefulness of the market model in each case.

b. If the rate of return of Standard & Poor's 500 Composite Index increased by .10, how much change would occur in the mean rate of return of Conoco's common stock? How much change would occur in the mean rate of return of Seagram's common stock?

c. Which of the two stocks, Conoco or Seagram, appears to be more responsive to changes in the market as a whole? Explain.

10.33 Refer to Exercise 10.11, in which the extent of retaliation against whistle blowers was investigated. Since an individual's salary is a reasonably good indicator of his or her power within an organization, the data of Exercise 10.11 can be used to investigate whether the extent of retaliation is related to the power of the whistle blower in the organization. Near and Miceli (1986) were unable to reject the hypothesis that the extent of retaliation is unrelated to the whistle blower's power. Do you agree? Test using $\alpha = .05$.

10.34 In Exercise 10.12, the following least squares line was developed to describe the relationship between the price of a gallon of regular gasoline, y, and the price of a barrel of crude oil, x:

$$\hat{y} = 34.143 + 3.2213x$$

Construct a 90% confidence interval to estimate the mean increase in the price of gasoline (in cents) per $1.00 increase per barrel in the price of crude oil.

10.35 A large car rental agency sells its cars after using them for a year. Among the records kept for each car are mileage and maintenance costs for the year. To evaluate the performance of a particular car model in terms of maintenance costs, the agency wants to use a 95% confidence interval to estimate the mean increase in maintenance costs for each additional 1,000 miles driven. Assume the relationship between maintenance cost and miles driven is linear. Use the data in the table to accomplish the objective of the rental agency.

CAR	MILES DRIVEN x (thousands)	MAINTENANCE COST y (dollars)
1	54	326
2	27	159
3	29	202
4	32	200
5	28	181
6	36	217

10.36 Buyers are often influenced by bulk advertising of a particular product. For example, suppose you have a product that sells for 25¢. If it is advertised at 2/50¢, 3/75¢, or 4/$1, some people may think they are getting a bargain. To test this theory, a store manager advertised an item for equal periods of time at five different bulk rates and observed the data listed in the table. Do the data provide sufficient evidence to indicate that sales increase as the number in the bulk increases?

ADVERTISED NUMBER IN BULK SALE	VOLUME SOLD
x	$\cdot y$
1	27
2	36
3	34
4	63
5	52

10.37 The precision of $\hat{\beta}_1$ as an estimator of β_1 is generally measured by its standard deviation $\sigma_{\hat{\beta}_1}$. In general, the larger the value of $\sigma_{\hat{\beta}_1}$, the wider (less precise) are confidence intervals for β_1; the smaller the value of $\sigma_{\hat{\beta}_1}$, the narrower (more precise) are confidence intervals for β_1.

a. Examine the formula for $\sigma_{\hat{\beta}_1}$ and explain how the observed values of the independent variable influence the size of $\sigma_{\hat{\beta}_1}$.

b. Sometimes it is possible to obtain data for a regression study by setting the independent variable, x, at different levels and observing the resulting values of the dependent variable, y. For example, suppose a supermarket chain was interested in studying the relationship between the sales of a product and the number of square feet of display space devoted to the product. Data for such a study would be generated by utilizing display areas of different sizes in different stores and observing the resulting sales. If you were designing such a study, how would your answer to part **a** influence the choice of display area sizes?

10.38 Do the data in Exercise 10.23 support the theory that the percentage of channels watched for 10 or more minutes decreases as the number of channels available increases? Test using $\alpha = .10$. Comment on the assumptions necessary for the validity of the test.

10.39 Will driving at or below the 55-mile-per-hour speed limit provide a substantial savings in fuel? To investigate the relationship between automobile gasoline consumption and driving speed, a small economy car was driven twice over the same stretch of an interstate freeway at each of six different speeds. The numbers of miles per gallon measured for each of the 12 trips are shown in the table.

MILES PER HOUR	50	55	60	65	70	75
MILES PER GALLON	34.8, 33.6	34.6, 34.1	32.8, 31.9	32.6, 30.0	31.6, 31.8	30.9, 31.7

a. Fit a least squares line to the data.

b. Is there sufficient evidence to conclude that a straight-line model provides useful information about the relationship between gasoline consumption and speed? Test using $\alpha = .05$.

c. Construct a 90% confidence interval for β_1 and interpret it.

**CORRELATION:
ANOTHER MEASURE
OF THE USEFULNESS
OF THE MODEL**

The claim is often made that the crime rate and the unemployment rate are "highly correlated." Another popular belief is that the Gross National Product (GNP) and the rate of inflation are "correlated." Some people even believe that the Dow Jones Industrial Average and the lengths of fashionable skirts are "correlated." Thus, the term **correlation** implies a relationship between two variables.

The **Pearson product moment correlation coefficient r**, defined in the box, provides a quantitative measure of the strength of the linear relationship between x and y, just as does the least squares slope $\hat{\beta}_1$. However, unlike the slope, the correlation coefficient r is *scaleless*. The value of r is always between -1 and $+1$, no matter what the units of x and y are.

DEFINITION 10.2

The **Pearson product moment coefficient of correlation r** is a measure of the strength of the linear relationship between two variables x and y. It is computed (for a sample of n measurements on x and y) as follows:

$$r = \frac{SS_{xy}}{\sqrt{SS_{xx}SS_{yy}}}$$

Note that r is computed from the same quantities used in fitting the least squares line. Since both r and $\hat{\beta}_1$ provide information about the utility of the model, it is not surprising that there is a similarity in their computational formulas. In particular, note that SS_{xy} appears in the numerators of both expressions and, since both denominators are always positive, r and $\hat{\beta}_1$ will always be of the same sign (either both positive or both negative). A value of r near or equal to 0 implies little or no linear relationship between the values of y and x that were observed in the sample.

FIGURE 10.8

Values of r and
Their Implications

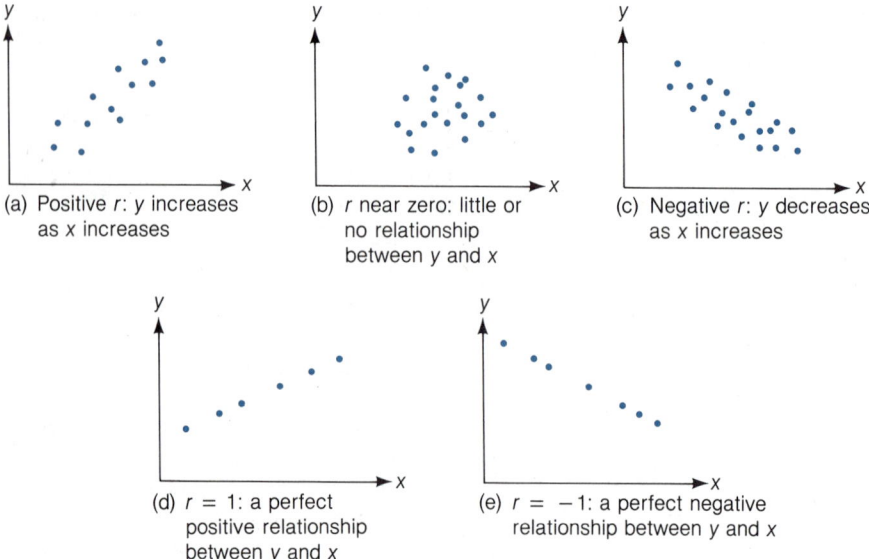

(a) Positive r: y increases as x increases

(b) r near zero: little or no relationship between y and x

(c) Negative r: y decreases as x increases

(d) $r = 1$: a perfect positive relationship between y and x

(e) $r = -1$: a perfect negative relationship between y and x

In contrast, the closer r is to 1 or -1, the stronger the linear relationship between y and x. And, if $r = 1$ or $r = -1$, all the points fall exactly on the least squares line. Positive values of r imply that y increases as x increases; negative values imply that y decreases as x increases. Each of these situations is portrayed in Figure 10.8.

EXAMPLE 10.1

A firm wants to know the correlation between the size of its sales force and its yearly sales revenue. The records for the past 10 years are examined, and the results listed in Table 10.5 are obtained. Calculate the coefficient of correlation r for the data.

TABLE 10.5

Sales Force–Revenue Data, Example 10.1

YEAR	NUMBER OF SALESPEOPLE x	SALES y ($ hundred thousands)	YEAR	NUMBER OF SALESPEOPLE x	SALES y ($ hundred thousands)
1978	15	1.35	1983	29	2.93
1979	18	1.63	1984	30	3.41
1980	24	2.33	1985	32	3.26
1981	22	2.41	1986	35	3.63
1982	25	2.63	1987	38	4.15

SOLUTION

We need to calculate SS_{xy}, SS_{xx}, and SS_{yy}:

$$SS_{xy} = \sum x_i y_i - \frac{\left(\sum x_i\right)\left(\sum y_i\right)}{10} = 800.62 - \frac{(268)(27.73)}{10} = 57.456$$

$$SS_{xx} = \sum x_i^2 - \frac{\left(\sum x_i\right)^2}{10} = 7,668 - \frac{(268)^2}{10} = 485.6$$

$$SS_{yy} = \sum y_i^2 - \frac{\left(\sum y_i\right)^2}{10} = 83.8733 - \frac{(27.73)^2}{10} = 6.97801$$

Then, the coefficient of correlation is

$$r = \frac{SS_{xy}}{\sqrt{SS_{xx}SS_{yy}}} = \frac{57.456}{\sqrt{(485.6)(6.97801)}} = \frac{57.456}{58.211} = .99$$

Thus, the size of the sales force and sales revenue are very highly correlated—at least over the past 10 years. The implication is that a strong positive linear relationship exists between these variables (see Figure 10.9 on page 516). We must be careful, however, not to jump to any unwarranted conclusions. For instance, the firm may be tempted to conclude that the best thing it can do to increase sales is to hire a large number of new salespeople. The implication of such a conclusion is that there is a *causal* relationship between the two variables. However, **high correlation does not imply causality**. The fact is that many things have probably contributed both to the increase in the size of the sales force and to the increase

FIGURE 10.9

Scattergram for
Example 10.1

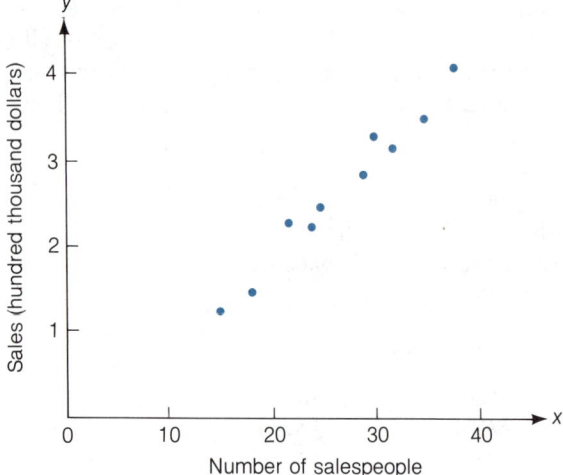

in sales revenue. The firm's expertise has undoubtedly grown, the rate of inflation has increased (so that 1987 dollars are not worth as much as 1978 dollars), and perhaps the scope of products and services sold by the firm has widened. We must be careful not to infer a causal relationship on the basis of high sample correlation. The only safe conclusion when a high correlation is observed in the sample data is that a linear trend may exist between x and y.   ∎

Keep in mind that the correlation coefficient r measures the correlation between x values and y values in the sample, and that a similar coefficient of correlation exists for the population from which the data points were selected. The **population correlation coefficient** is denoted by the symbol ρ (rho). As you might expect, ρ is estimated by the corresponding sample statistic, r. Or, rather than estimating ρ, we might want to test the hypothesis $H_0: \rho = 0$ against $H_a: \rho \neq 0$—i.e., test the hypothesis that x contributes no information for the prediction of y using the straight-line model against the alternative that the two variables are at least linearly related. However, we have already performed this identical test in Section 10.5 when we tested $H_0: \beta_1 = 0$ against $H_a: \beta_1 \neq 0$. That is, the null hypothesis $H_0: \rho = 0$ is equivalent to the hypothesis $H_0: \beta_1 = 0.$* When we tested the null hypothesis $H_0: \beta_1 = 0$ in connection with the advertising–sales example, the data led to a rejection of the hypothesis for $\alpha = .05$. This implies that the null hypothesis of a zero linear correlation between the two variables (advertising and sales) can also be rejected at $\alpha = .05$. The only real difference between the least squares slope $\hat{\beta}_1$ and the coefficient of correlation r is the measurement scale. Therefore, the information they provide about the usefulness of the least squares model is to some extent redundant. For this reason, we will use the slope to make inferences about the existence of a positive or negative linear relationship between the two variables.

---

*The correlation test statistic equivalent to $t = \hat{\beta}_1 / s_{\hat{\beta}_1}$ is

$$t = \frac{r}{\sqrt{(1 - r^2)/(n - 2)}}$$

SECTION 10.7

THE COEFFICIENT OF DETERMINATION

Another way to measure the contribution of x in predicting y is to consider how much the errors of prediction of y were reduced by using the information provided by x. To illustrate, suppose a sample of data has the scattergram shown in Figure 10.10(a). If we assume that x contributes no information for the prediction of y, the best prediction for a value of y is the sample mean $\bar{y}$, which is shown as the horizontal line in Figure 10.10(b). The vertical line segments in Figure 10.10(b) are the deviations of the points about the mean $\bar{y}$. Note that the sum of squares of deviations for the model $\hat{y} = \bar{y}$ is

$$SS_{yy} = \sum (y_i - \bar{y})^2$$

FIGURE 10.10 A Comparison of the Sum of Squares of Deviations for Two Models

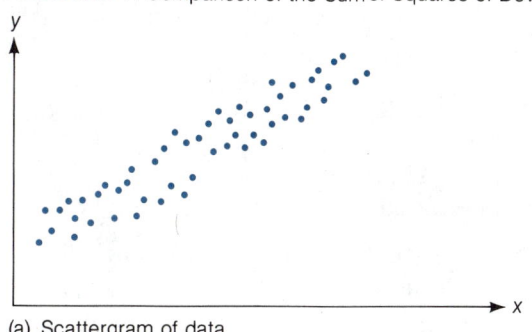

(a) Scattergram of data

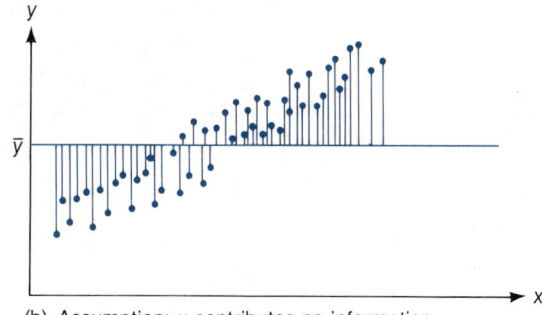

(b) Assumption: x contributes no information for predicting y, $\hat{y} = \bar{y}$

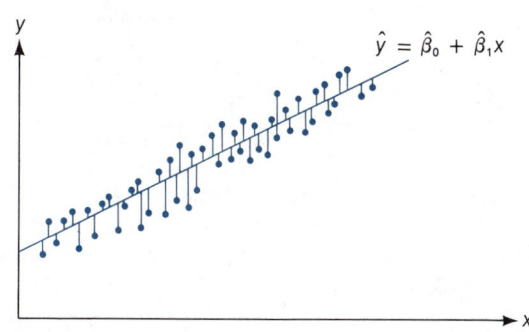

(c) Assumption: x contributes information for predicting y, $\hat{y} = \hat{\beta}_0 + \hat{\beta}_1 x$

Now suppose you fit a least squares line to the same set of data and locate the deviations of the points about the line as shown in Figure 10.10(c). Compare the deviations about the prediction lines in parts (b) and (c) of Figure 10.10. You can see that:

1. If x contributes little or no information for the prediction of y, the sums of squares of deviations for the two lines,

$$SS_{yy} = \sum (y_i - \bar{y})^2 \quad \text{and} \quad SSE = \sum (y_i - \hat{y}_i)^2$$

will be nearly equal.

2. If x does contribute information for the prediction of y, the SSE will be smaller than SS_{yy}. In fact, if all the points fall on the least squares line, then SSE $= 0$.

Then, the reduction in the sum of squares of deviations that can be attributed to x, expressed as a proportion of SS_{yy}, is

$$\frac{SS_{yy} - SSE}{SS_{yy}}$$

Note that SS_{yy} is the "total sample variation" of the observations around the sample mean $\bar{y}$, and that SSE is the remaining "unexplained sample variability" after fitting the line $\hat{y}$, so that the difference ($SS_{yy} - SSE$) is the "explained sample variability" attributable to the linear relationship with x. Then a verbal description of the proportion is as follows:

$$\frac{SS_{yy} - SSE}{SS_{yy}} = \frac{\text{Explained sample variability}}{\text{Total sample variability}}$$

$$= \text{Proportion of total sample variability}$$
$$\text{explained by the linear relationship}$$

It can be shown that this proportion is equal to the square of the simple linear coefficient of correlation r (the Pearson product moment coefficient of correlation).

DEFINITION 10.3

The **coefficient of determination** is the square of the coefficient of correlation. It represents the proportion of the total sample variability around $\bar{y}$ that is explained by the linear relationship between y and x. It may be computed as

$$r^2 = \frac{SS_{yy} - SSE}{SS_{yy}} = 1 - \frac{SSE}{SS_{yy}}$$

Note that r^2 is always between 0 and 1, because r is between -1 and $+1$. Thus, an r^2 of .60 means that the sum of squares of deviations of the y values about their predicted values has been reduced 60% by the use of the least squares equation $\hat{y}$, instead of $\bar{y}$, to predict y.

EXAMPLE 10.2   Calculate the coefficient of determination for the advertising–sales example. The data are repeated in Table 10.6.

TABLE 10.6

ADVERTISING EXPENDITURE x ($ hundreds)	SALES REVENUE y ($ thousands)
1	1
2	1
3	2
4	2
5	4

SOLUTION

We first calculate

$$SS_{yy} = \sum y_i^2 - \frac{\left(\sum y_i\right)^2}{5} = 26 - \frac{(10)^2}{5} = 26 - 20 = 6$$

From previous calculations,

$$SSE = \sum (y_i - \hat{y}_i)^2 = 1.10$$

Then, the coefficient of determination is given by

$$r^2 = \frac{SS_{yy} - SSE}{SS_{yy}} = \frac{6.0 - 1.1}{6.0} = \frac{4.9}{6.0} = .82$$

So we know that by using the advertising expenditure x to predict y with the least squares line

$$\hat{y} = -.1 + .7x$$

the total sum of squares of deviations of the five sample y values about their predicted values has been reduced 82% by the use of $\hat{y}$, instead of $\bar{y}$, to predict y. ∎

CASE STUDY 10.1

ESTIMATING THE COST
OF A CONSTRUCTION
PROJECT

As evidenced by the cost overruns of public building projects, the initial estimate of the ultimate cost of a structure is often rather poor. These estimates usually rely on a precise definition of the proposed building in terms of working drawings and specifications. However, cost estimators do not take random error into account, so no measure of reliability is possible for their deterministic estimates. Crandall and Cedercreutz (1976) propose the use of a probabilistic model to make cost estimates. They use regression models to relate cost to independent variables like volume, amount of glass, and floor area. Crandall and Cedercreutz's rationale for choosing this approach is that "one of the principal merits of the least squares regression model, for the purpose of preliminary cost estimating, is the method of dealing with anticipated error." They go on to point out that when random error is anticipated, "statistical methods, such as regression analysis, attack the problem head on."

Crandall and Cedercreutz intially focused on the cost of mechanical work (heating, ventilating, and plumbing), since this part of the total cost is generally difficult to predict. Conventional cost estimates rely heavily on the amount of ductwork and piping used in construction, but this information is not precisely known until too late to be of use to the cost estimator. One of several models discussed was a simple linear model relating mechanical cost to floor area. Based on the data associated with 26 factory and warehouse buildings, the least squares prediction equation given in Figure 10.11 was found. It was concluded that floor area and mechanical cost are linearly related, since the t statistic (for testing H_0: $\beta_1 = 0$) was found to equal 3.61, which is significant with an α as small as .002. Thus, floor area should be useful when predicting the mechanical cost of a factory or warehouse. In addition, the regression model enables the reliability of the predicted cost to be assessed.

FIGURE 10.11

Simple Linear Model
Relating Cost to
Floor Area

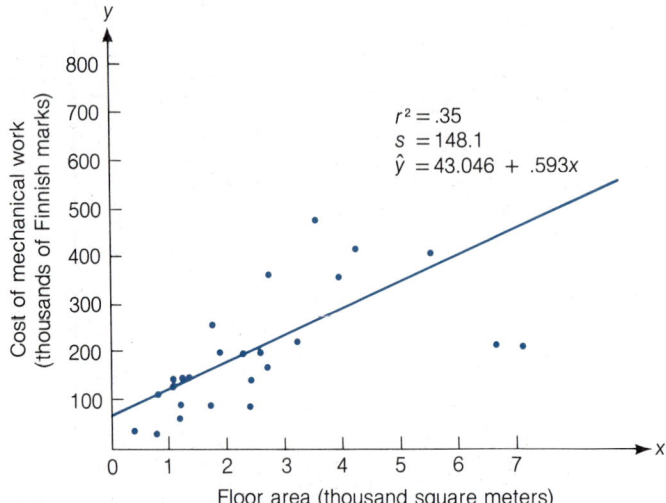

The value of the coefficient of determination r^2 was found to be .35. This tells us that only 35% of the variation among mechanical costs is accounted for by the differences in floor areas. Since there is only one independent variable in the model, this relatively small value of r^2 should not be too surprising. If other variables related to mechanical cost were included in the model, they would probably account for a significant portion of the remaining 65% of the variation in mechanical cost not explained by floor area. In the next chapter, we discuss this important aspect of relating a response to more than one independent variable.

EXERCISES
10.40–10.53

LEARNING THE MECHANICS

10.40 Explain what each of the following sample correlation coefficients tells you about the relationship between the x and y values in the sample:

a. $r = 1$ **b.** $r = -1$ **c.** $r = 0$
d. $r = .90$ **e.** $r = .10$ **f.** $r = -.88$

10.41 Describe the slope of the least squares line if:

a. $r = .7$ **b.** $r = -.7$ **c.** $r = 0$
d. $r^2 = .64$

10.42 Construct a scattergram for each data set. Then calculate r and r^2 for each data set. Interpret their values.

a.

x	−2	−1	0	1	2
y	−2	1	2	5	6

b.

x	−2	−1	0	1	2
y	6	5	3	2	0

c.

x	1	2	2	3	3	3	4
y	2	1	3	1	2	3	2

d.

x	0	1	3	5	6
y	0	1	2	1	0

10.43 Calculate r^2 for the least squares line in each of the following exercises. Interpret their values.

a. Exercise 10.7   **b.** Exercise 10.10   **c.** Exercise 10.27

APPLYING THE CONCEPTS

10.44 Find the correlation coefficient and the coefficient of determination for the sample data listed in the table and interpret your results; discuss whether a causal relationship can be inferred.

YEAR	NUMBER OF 18-HOLE AND LARGER GOLF COURSES IN THE U.S.	NUMBER OF DIVORCES IN THE U.S. (in millions)
1960	2,725	2.9
1965	3,769	3.5
1970	4,845	4.3
1975	6,282	6.5
1977	6,551	8.0
1978	6,699	8.6
1979	6,787	8.8
1980	6,856	9.9
1981	6,944	10.8
1982	7,059	11.5
1983	7,125	11.6
1984	7,230	12.3

Source: U.S. Bureau of the Census, *Statistical Abstract of the United States: 1986*, pp. 35, 230.

10.45 Find the correlation coefficient and the coefficient of determination for the sample data listed in the table and interpret your results. For these data, $SS_{xx} = 13,143,088.00$, $SS_{yy} = 1,552,011.334$, $SS_{xy} = 49,153.00$, $\bar{x} = 2,023.00$, and $\bar{y} = 1,543.67$.

YEAR	GROSS NATIONAL PRODUCT x (billions of dollars)	NEW HOUSING STARTS y (thousands)	YEAR	GROSS NATIONAL PRODUCT x (billions of dollars)	NEW HOUSING STARTS y (thousands)
1960	507	1,296	1978	2,164	2,036
1965	691	1,510	1979	2,418	1,760
1970	993	1,469	1980	2,632	1,313
1973	1,326	2,057	1981	2,958	1,100
1974	1,434	1,353	1982	3,069	1,072
1975	1,549	1,171	1983	3,305	1,712
1976	1,718	1,548	1984	3,663	1,756
1977	1,918	2,002			

Source: U.S. Bureau of the Census, *Statistical Abstract of the United States: 1986*, pp. 431, 724.

10.46 Data on monthly sales, y, price per unit during the month, x_1, and amount spent on advertising, x_2, for a product are shown for a 5-month period in the table at the top of page 522. Based on this sample, which variable—price or advertising expenditure—appears to provide more information about sales? Explain.

MONTH	TOTAL MONTHLY SALES y (thousands)	PRICE PER UNIT x_1	AMOUNT SPENT ON ALL FORMS OF ADVERTISING x_2 (hundreds)
June	$40	$ .85	$6.0
July	50	.76	5.0
August	55	.75	8.0
September	30	1.00	7.5
October	45	.80	5.5

10.47 In Exercise 10.15, we gave the number of games won, y, and the batting average, x, for the 14 American League baseball teams at the end of the 1986 season. They are repeated here in the table.

TEAM	NUMBER OF GAMES WON y	TEAM BATTING AVERAGE x
Cleveland	84	.284
New York	90	.271
Boston	95	.271
Toronto	86	.269
Texas	87	.267
Detroit	87	.263
Minnesota	71	.261
Baltimore	73	.258
California	92	.255
Milwaukee	77	.255
Seattle	67	.253
Kansas City	76	.252
Oakland	76	.252
Chicago	72	.247

Source: *Official American League Averages*, 1986. The American League of Professional Baseball Clubs, New York, pp. 2, 24–27.

Recall that $\Sigma y = 1{,}133$, $\Sigma x = 3.658$, $\Sigma y^2 = 92{,}703$, $\Sigma x^2 = .957118$, and $\Sigma xy = 296.734$.

a. Calculate the correlation coefficient, r, and the coefficient of determination, r^2, for the data. Interpret their values.

b. Do the data provide sufficient evidence to conclude that a correlation exists between a team's number of wins and its batting average? Test using $\alpha = .05$. [*Hint:* See the last paragraph of Section 10.6.]

10.48 A *negotiable certificate of deposit* is a marketable receipt for funds deposited in a bank for a specified period of time at a specified rate of interest (Cook, 1977). The accompanying table lists the end-of-quarter interest rate for 3-month certificates of deposit during the period January 1979 through June 1986. The table also lists end-of-quarter values of Standard & Poor's 500 Stock Composite Average (an indicator of stock market activity) for the same time period. Find the coefficient of determination and the correlation coefficient for the data, and interpret your results.

YEAR	QUARTER	INTEREST RATE	S&P 500	YEAR	QUARTER	INTEREST RATE	S&P 500
1979	I	10.13	101.59	1983	I	8.69	152.96
	II	9.95	102.91		II	9.20	168.11
	III	11.89	109.32		III	9.39	166.07
	IV	13.43	107.94		IV	9.69	164.93
1980	I	17.57	104.69	1984	I	10.08	159.18
	II	8.49	114.24		II	11.34	153.18
	III	11.29	125.46		III	11.29	166.10
	IV	18.65	135.76		IV	8.60	167.24
1981	I	14.43	136.00	1985	I	9.02	180.66
	II	16.90	131.21		II	7.44	191.85
	III	16.84	116.18		III	7.93	182.08
	IV	12.49	122.55		IV	7.80	211.28
1982	I	14.21	111.96	1986	I	7.24	238.90
	II	14.46	109.61		II	6.73	250.84
	III	10.66	120.42				
	IV	8.66	135.28				

Source: *Standard & Poor's Statistical Service, Current Statistics,* 1986, Standard & Poor's Corporation.

10.49 In the summer of 1981, the Minnesota Department of Transportation installed a state-of-the-art weigh-in-motion scale in the concrete surface of the eastbound lanes of Interstate 494 in Bloomington, Minnesota. The system is computerized and monitors traffic continuously. It is capable of distinguishing among 13 different types of vehicles (car, five-axle semi, five-axle twin trailer, etc.). The primary purpose of the system is to provide traffic counts and weights for use in the planning and design of future roadways. After installation, a study was undertaken to determine whether the scale's readings correspond with the static weights of the vehicles being monitored. (Studies of this type are known as *calibration studies.*) After some preliminary comparisons using a two-axle-six-tire truck carrying different loads (see table), calibration adjustments were made in the software of the weigh-in-motion system and the scales were re-evaluated (Wright, Owen, and Pena, 1983).

TRIAL NUMBER	STATIC WEIGHT OF TRUCK x (thousand pounds)	WEIGH-IN-MOTION READING PRIOR TO CALIBRATION ADJUSTMENT y_1 (thousand pounds)	WEIGH-IN-MOTION READING AFTER CALIBRATION ADJUSTMENT y_2 (thousand pounds)
1	27.9	26.0	27.8
2	29.1	29.9	29.1
3	38.0	39.5	37.8
4	27.0	25.1	27.1
5	30.3	31.6	30.6
6	34.5	36.2	34.3
7	27.8	25.1	26.9
8	29.6	31.0	29.6
9	33.1	35.6	33.0
10	35.5	40.2	35.0

Source: Adapted from data in Wright, Owen, and Pena (1983).

a. Construct two scattergrams, one of y_1 versus x and the other of y_2 versus x.

b. Use the scattergram of part **a** to evaluate the performance of the weigh-in-motion scale both before and after the calibration adjustment.

c. Calculate the correlation coefficient for both sets of data, and interpret their values. Explain how these correlation coefficients can be used to evaluate the weigh-in-motion scale.

d. Suppose the sample correlation coefficient for y_2 and x were 1. Could this happen if the static weights and the weigh-in-motion readings disagreed? Explain.

10.50 A problem of economic and social concern in the United States is the importation and sale of illicit drugs. The data shown in the table are part of a larger body of data collected by the Florida attorney general's office in an attempt to relate the incidence of drug seizures and drug arrests to the characteristics of the Florida counties. Given are the number, y, of drug arrests per county in 1982, the density, x_1, of the county (population per square mile), and the number, x_2, of law enforcement employees. In order to simplify the calculations, we show data for only ten counties.

	COUNTY									
	1	2	3	4	5	6	7	8	9	10
POPULATION DENSITY, x_1	169	68	278	842	18	42	112	529	276	613
NUMBER OF LAW ENFORCEMENT EMPLOYEES, x_2	498	35	772	5,788	18	57	300	1,762	416	520
NUMBER OF ARRESTS IN 1982, y	370	44	716	7,416	25	50	189	1,097	256	432

a. Fit a least squares line to relate the number, y, of drug arrests per county in 1982 to the county population density, x_1.

b. We might expect the mean number of arrests to increase as the population density increases. Do the data support this theory? Test using $\alpha = .05$.

c. Calculate the coefficient of determination for this regression analysis and interpret its value.

10.51 Repeat parts **a**, **b**, and **c** of Exercise 10.50 using the number, x_2, of county law enforcement employees as the independent variable. Then answer the following questions.

d. Which least squares line has the lower SSE?

e. Which independent variable explains more of the variation in y? Explain.

10.52 Refer to Exercise 10.50.

a. Calculate the correlation coefficient, r, between the county population density, x_1, and the number of law enforcement employees, x_2.

b. Does the correlation between x_1 and x_2 differ significantly from 0? Test using $\alpha = .05$.

10.53 A firm's *demand curve* describes the quantity of its product that can be sold at different possible prices, other things being equal (Leftwich, 1973). Over the period of a year, a tire company varied the price of one of its radial tires in order to estimate the demand curve for the tire. They observed that when the price was set very low or very high, they sold few tires. The latter result they understood; the former they determined was due to consumer misperception that the tire's low price must be linked to poor quality. The data in the table describe the tire's sales over the experimental period.

TIRE PRICE x ($)	NUMBER SOLD y (hundreds)
20	13
35	57
45	85
60	43
70	17

a. Calculate a least squares line to approximate the firm's demand curve.
b. Construct a scattergram and plot your least squares line as a check on your calculations.
c. Test $H_0: \beta_1 = 0$ using a two-tailed test and $\alpha = .05$. Draw the appropriate conclusion in the context of the problem.
d. Does the nonrejection of H_0 in part c imply that no relationship exists between tire price and sales volume? Explain.
e. Calculate the coefficient of determination for the least squares line of part a and interpret its value in the context of the problem.

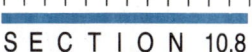

SECTION 10.8

USING THE MODEL FOR ESTIMATION AND PREDICTION

If we are satisfied that a useful model has been found to describe the relationship between sales revenue and advertising, we are ready to accomplish the original objectives for building the model: using it to estimate or to predict sales on the basis of advertising dollars spent.

The most common uses of a probabilistic model can be divided into two categories. The first is the use of the model for estimating the mean value of y, E(y), for a specific value of x.

For our example, we may want to estimate the mean sales revenue for *all* months during which $400 ($x = 4$) is expended on advertising.

The second use of the model entails predicting a particular y value for a given value of x.

That is, if we decide to expend $400 next month, we want to predict the firm's sales revenue for that month.

In the case of estimating a mean value of y, we are attempting to estimate the mean result of a very large number of experiments at the given x value. In the second case, we are trying to predict the outcome of a single experiment at the given x value.

In which of these model uses do you expect to have more success; i.e., which value—the mean or individual value of y—can we estimate (or predict) with more accuracy?

Before answering this question, we first consider the problem of choosing an estimator (or predictor) of the mean (or individual) y value. We will use the least squares model

$$\hat{y} = \hat{\beta}_0 + \hat{\beta}_1 x$$

both to estimate the mean value of y and to predict a particular value of y for a given value of x. For our example, we found

$$\hat{y} = -.1 + .7x$$

so that the estimated mean value of sales revenue for all months when $x = 4$ (advertising $= \$400$) is

$$\hat{y} = -.1 + .7(4) = 2.7$$

or \$2,700 (the units of y are thousands of dollars). The identical value is used to predict the y value when $x = 4$. That is, both the estimated mean value and the predicted value of y equal $\hat{y} = 2.7$ when $x = 4$, as shown in Figure 10.12.

The difference in these two model uses lies in the relative accuracy of the estimate and the prediction. These accuracies are best measured by the repeated sampling errors of the least squares line when it is used as an estimator and as a predictor, respectively. These errors are given in the box.

FIGURE 10.12

Estimated Mean Value and Predicted Individual Value of Sales Revenue y for $x = 4$

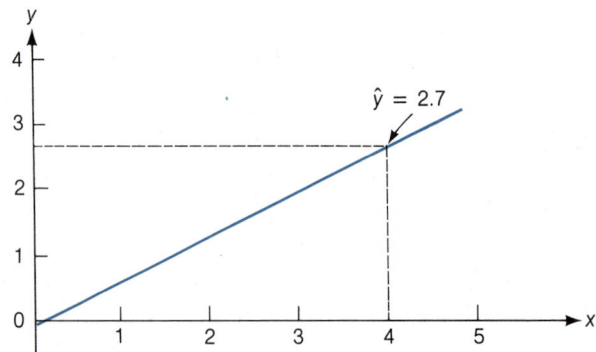

SAMPLING ERRORS FOR THE ESTIMATOR OF THE MEAN OF y AND THE PREDICTOR OF AN INDIVIDUAL y

1. The standard deviation of the sampling distribution of the estimator $\hat{y}$ of the mean value of y at a particular value of x, say x_{p}, is

$$\sigma_{\hat{y}} = \sigma\sqrt{\frac{1}{n} + \frac{(x_{\mathrm{p}} - \bar{x})^2}{SS_{xx}}}$$

where σ is the standard deviation of the random error ε.

2. The standard deviation of the prediction error for the predictor $\hat{y}$ of an individual y value for $x = x_{\mathrm{p}}$ is

$$\sigma_{(y-\hat{y})} = \sigma\sqrt{1 + \frac{1}{n} + \frac{(x_{\mathrm{p}} - \bar{x})^2}{SS_{xx}}}$$

where σ is the standard deviation of the random error ε.

The true value of σ will rarely be known. Thus, we estimate σ by s and calculate the estimation and prediction intervals as shown in the following boxes.

A 100(1 − α)% CONFIDENCE INTERVAL FOR THE MEAN VALUE OF y FOR x = x_p

$$\hat{y} \pm t_{\alpha/2}(\text{Estimated standard deviation of } \hat{y})$$

or

$$\hat{y} \pm t_{\alpha/2}\, s \sqrt{\frac{1}{n} + \frac{(x_p - \bar{x})^2}{SS_{xx}}}$$

where $t_{\alpha/2}$ is based on $(n - 2)$ df.

A 100(1 − α)% PREDICTION INTERVAL FOR AN INDIVIDUAL y FOR x = x_p

$$\hat{y} \pm t_{\alpha/2}\big[\text{Estimated standard deviation of } (y - \hat{y})\big]$$

or

$$\hat{y} \pm t_{\alpha/2}\, s \sqrt{1 + \frac{1}{n} + \frac{(x_p - \bar{x})^2}{SS_{xx}}}$$

where $t_{\alpha/2}$ is based on $(n - 2)$ df.

EXAMPLE 10.3

Find a 95% confidence interval for mean monthly sales when the appliance store spends $400 on advertising.

SOLUTION

For a $400 advertising expenditure, $x_p = 4$ and, since $n = 5$, df $= n - 2 = 3$. Then the confidence interval for the mean value of y is

$$\hat{y} \pm t_{\alpha/2}\, s \sqrt{\frac{1}{n} + \frac{(x_p - \bar{x})^2}{SS_{xx}}}$$

or

$$\hat{y} \pm t_{.025}\, s \sqrt{\frac{1}{5} + \frac{(4 - \bar{x})^2}{SS_{xx}}}$$

Recall that $\hat{y} = 2.7$, $s = .61$, $\bar{x} = 3$, and $SS_{xx} = 10$. From Table VI in Appendix B, $t_{.025} = 3.182$. Thus, we have

$$2.7 \pm (3.182)(.61) \sqrt{\frac{1}{5} + \frac{(4 - 3)^2}{10}} = 2.7 \pm (3.182)(.61)(.55)$$

$$= 2.7 \pm 1.1 \quad \text{or} \quad (1.6, 3.8)$$

We estimate that the interval from $1,600 to $3,800 encloses the mean sales revenue when the store expends $400 a month on advertising. Note that we used a small amount of data for purposes of illustration in fitting the least squares line and that the width of the interval could be decreased by using a larger number of data points.

EXAMPLE 10.4

Predict the monthly sales for next month if a $400 expenditure is to be made on advertising. Use a 95% prediction interval.

SOLUTION

To predict the sales for a particular month for which $x_p = 4$, we calculate the 95% prediction interval as

$$\hat{y} \pm t_{\alpha/2} \, s \sqrt{1 + \frac{1}{n} + \frac{(x_p - \bar{x})^2}{SS_{xx}}} = 2.7 \pm (3.182)(.61)\sqrt{1 + \frac{1}{5} + \frac{(4 - 3)^2}{10}}$$

$$= 2.7 \pm (3.182)(.61)(1.14)$$

$$= 2.7 \pm 2.2 \text{ or } (.5, 4.9)$$

Therefore, we predict that the sales next month will fall in the interval from $500 to $4,900. ∎

As in the case for the confidence interval for the mean value of y (Example 10.3), the prediction interval for y (Example 10.4) is quite large. This is because we have chosen a small number of data points to fit the least squares line. The width of the prediction interval could be reduced by using a larger number of data points.

A comparison of the confidence interval for the mean value of y and the prediction interval for some future value of y for a $400 advertising expenditure ($x = 4$) is illustrated in Figure 10.13. It is important to note that the prediction interval for an individual value of y will always be wider than the confidence interval for a mean value of y. You can see this by examining the formulas for the two intervals and you can see it in Figure 10.13.

FIGURE 10.13

A 95% Confidence Interval for Mean Sales and a Prediction Interval for Sales When $x = 4$

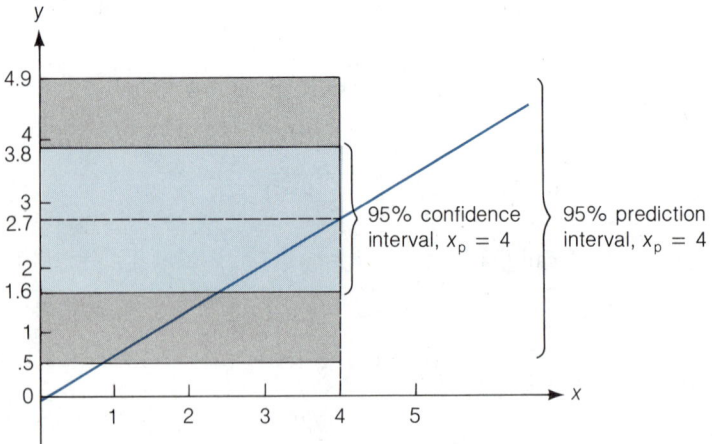

The error in estimating the mean value of y, $E(y)$, for a given value of x, say x_p, is the distance between the least squares line and the true line of means, $E(y) = \beta_0 + \beta_1 x$. This error, $[\hat{y} - E(y)]$, is shown in Figure 10.14. In contrast, the error $(y_p - \hat{y})$ in predicting some future value of y is the sum of two errors—

FIGURE 10.14

Error of Estimating the
Mean Value of y for a
Given Value of x

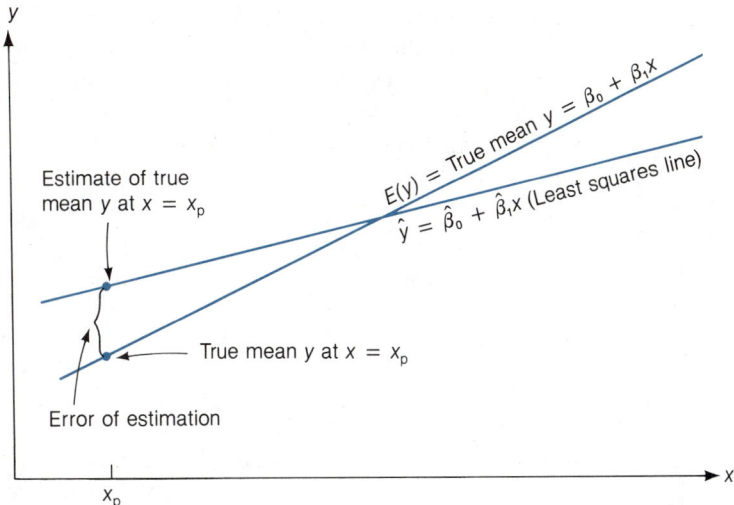

the error of estimating the mean of y, $E(y)$, shown in Figure 10.14, plus the random error that is a component of the value of y to be predicted (see Figure 10.15). Consequently, the error of predicting a particular value of y will be larger than the error of estimating the mean value of y for a particular value of x. Note from their formulas that both the error of estimation and the error of prediction take their smallest values when $x_p = \bar{x}$. The farther x_p lies from $\bar{x}$, the larger will be the errors of estimation and prediction. You can see why this is true by noting the deviations for different values of x_p between the line of means $E(y) = \beta_0 + \beta_1 x$ and the predicted line of means $\hat{y} = \hat{\beta}_0 + \hat{\beta}_1 x$ shown in Figure 10.15. The deviation is larger at the extremities of the interval where the largest and smallest values of x in the data set occur.

FIGURE 10.15

Error of Predicting a
Future Value of y for
a Given Value of x

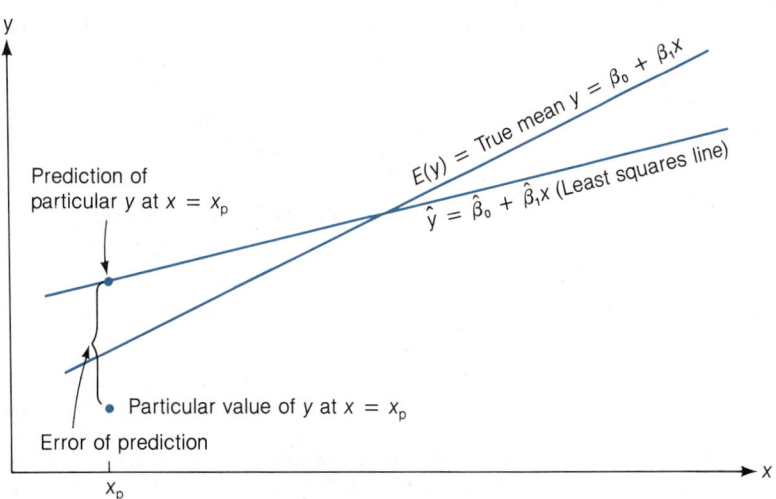

Both the confidence intervals for the mean values and the prediction intervals for individual values are depicted over the entire range of the regression line in Figure 10.16. You can see that the confidence interval is always narrower than the prediction interval, and that they are both narrowest at the mean $\bar{x}$, increasing steadily as the distance $|x - \bar{x}|$ increases.

FIGURE 10.16

Confidence Intervals for Mean Values and Prediction Intervals for Individual Values

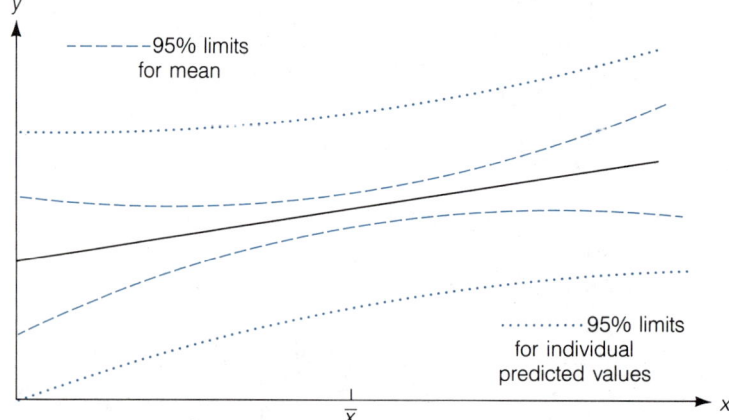

The confidence interval width grows smaller as n is increased; thus, in theory, you can obtain as precise an estimate of the mean value of y as desired (at any given x) by selecting a large enough sample. The prediction interval for a particular value of y also grows smaller as n increases, but there is a lower limit on its width. If you examine the formula for the prediction interval (page 527), you will see that the interval can get no smaller than $\hat{y} \pm z_{\alpha/2}\sigma$.* Thus, the only way to obtain more accurate predictions for individual values of y is to reduce the standard deviation of the regression, σ. This can be accomplished only by improving the model, either by using a curvilinear (rather than linear) relationship with x, or by adding new independent variables to the model, or both. Methods of improving the model are discussed in Chapters 11 and 12.

EXERCISES
10.54–10.63

LEARNING THE MECHANICS

10.54 Consider the following pairs of measurements:

x	1	2	3	4	5	6	7
y	3	5	4	6	7	7	10

a. Construct a scattergram for these data.
b. Find the least squares line and plot it on your scattergram.

*The result follows from the facts that, for large n, $t_{\alpha/2} \approx z_{\alpha/2}$, $s \approx \sigma$, and the last two terms under the radical in the standard error of the predictor are approximately 0.

c. Find s^2.
d. Find a 90% confidence interval for the mean value of y when $x_p = 4$. Plot the upper and lower bounds of the confidence interval on your scattergram.
e. Find a 90% prediction interval for y when $x_p = 4$. Plot the upper and lower bounds of the prediction interval on your scattergram.
f. Compare the widths of the intervals you constructed in parts **d** and **e**. Which is wider and why?

10.55 Consider the following pairs of measurements:

x	1	−1	2	0	4	−2	4	−2	5	1
y	−1	−5	1	−3	6	−6	4	−5	4	0

For these data, $SS_{xx} = 57.6$, $SS_{yy} = 162.5$, $SS_{xy} = 94.0$, and $\hat{y} = -2.4583 + 1.6319x$.
a. Construct a scattergram for the data.
b. Plot the least squares line on your scattergram.
c. Use a 95% confidence interval to estimate the mean value of y when $x_p = 5$. Plot the upper and lower bounds of the interval on your scattergram.
d. Repeat part **c** for $x_p = 1.2$ and $x_p = -2$.
e. Compare the widths of the three confidence intervals you constructed in parts **c** and **d** and explain why they differ.

10.56 Refer to Exercise 10.55.
a. Using no information about x, estimate and calculate a 95% confidence interval for the mean value of y. [*Hint:* Use the one-sample t methodology of Section 8.6.]
b. Plot the estimated mean value and the confidence interval as horizontal lines on your scattergram of Exercise 10.55.
c. Compare the confidence intervals you calculated in parts **c** and **d** of Exercise 10.55 with that you calculated in part **a** of this exercise. Does x appear to contribute information about the mean value of y?
d. Check the answer you gave in part **c** with a statistical test of the null hypothesis $H_0: \beta_1 = 0$ versus $H_a: \beta_1 \neq 0$. Use $\alpha = .05$.

10.57 In fitting a least squares line to $n = 12$ data points, the following quantities were computed:

$$SS_{xx} = 25 \quad \bar{x} = 2 \quad SS_{yy} = 17 \quad \bar{y} = 3 \quad SS_{xy} = 20$$

a. Find the least squares line.   **b.** Graph the least squares line.
c. Calculate SSE.   **d.** Calculate s^2.
e. Find a 95% confidence interval for the mean value of y when $x_p = 1$.
f. Find a 95% prediction interval for y when $x_p = 1.5$.

APPLYING THE CONCEPTS

10.58 The reasons given by workers for quitting their jobs generally fall into one of two categories: (1) worker quits to seek or take a different job; and (2) worker quits to withdraw from the labor force. Economic theory suggests that wages and quit rates are related. Evidence regarding this relationship is important for understanding how labor markets operate (Koch and Ragan, 1986). The table lists quit rates (quits per 100 employees) and the average hourly wage in a sample of 15 manufacturing industries during 1983.

INDUSTRY	QUIT RATE	AVERAGE WAGE	INDUSTRY	QUIT RATE	AVERAGE WAGE
1	1.4	$ 8.20	9	2.0	$ 7.99
2	.7	10.35	10	3.8	5.54
3	2.6	6.18	11	2.3	7.50
4	3.4	5.37	12	1.9	6.43
5	1.7	9.94	13	1.4	8.83
6	1.7	9.11	14	1.8	10.93
7	1.0	10.59	15	2.0	8.80
8	.5	13.29			

a. Construct a scattergram for these data.

b. Use the method of least squares to model the quit rate as a function of average hourly wage.

c. Do the data present sufficient evidence to conclude that average hourly wage rate contributes useful information for the prediction of quit rates? What does your model suggest about the relationship between quit rates and wages?

d. Find a 95% prediction interval for the quit rate in an industry with an average hourly wage of $9.00.

e. Estimate the mean quit rate for industries with an average hourly wage of $6.50. Use a 95% confidence interval.

10.59 Refer to Exercise 10.12. Find a 95% confidence interval for the mean price of a gallon of gasoline when crude oil costs $30 per barrel.

10.60 Managers are an important part of any organization's resource base. Accordingly, the organization should be just as concerned about forecasting its future managerial needs as it is with forecasting its needs for, say, the natural resources used in its production processes. According to William F. Glueck (1977), one commonly used procedure for forecasting the demand for managers is to model the relationship between sales and the number of managers needed. The theory is that "the demand for managers is the result of the increases and decreases in the demand for products and services that an enterprise offers its customers and clients" (p. 274). In order to develop this relationship, data such as those shown in the table can be collected from a firm's records.

a. Use simple linear regression to model the relationship between the number of managers and the number of units sold.

b. Plot your least squares line on a scattergram of the data. Does it appear that the relationship between y and x is linear? If not, does it appear that your least squares model will provide a useful approximation to the relationship? Explain.

c. Test the usefulness of your model. Use $\alpha = .05$. State your conclusion in the context of the problem.

d. The company projects that next May it will sell 39 units. Use your least squares model to construct (1) a 90% confidence interval for the mean number of managers needed when the sales level is 39 and (2) a 90% prediction interval for the number of managers needed next May.

e. Compare the widths and interpret the two intervals you constructed in part d.

DATE	MONTHLY SALES x (units)	NUMBER OF MANAGERS y	DATE	MONTHLY SALES x (units)	NUMBER OF MANAGERS y
3/83	5	10	9/85	30	22
6/83	4	11	12/85	31	25
9/83	8	10	3/86	36	30
12/83	7	10	6/86	38	30
3/84	9	9	9/86	40	31
6/84	15	10	12/86	41	31
9/84	20	11	3/87	51	32
12/84	21	17	6/87	40	30
3/85	25	19	9/87	48	32
6/85	24	21	12/87	47	32

10.61 Refer to Exercise 10.21.

a. Use a 90% confidence interval to estimate the mean useful life of brand A cutting tools when the cutting speed is 45 meters per minute. Repeat for brand B. Compare the width of the two intervals, and comment on the reasons for any difference.

b. Use a 90% prediction interval to predict the useful life of a brand A cutting tool when the cutting speed is 45 meters per minute. Repeat for brand B. Compare the widths of the two intervals to each other, and to the two intervals you calculated in part **a.** Comment on the reasons for any differences.

c. Note that the estimation and prediction you performed in parts **a** and **b** were for a value of x that was not included in the original sample; that is, the value $x = 45$ was not part of the sample. However, the value is within the range of x values in the sample, so that the regression model spans the x value for which the estimation and prediction was made. In such situations, estimation and prediction represent **interpolations**.

Suppose you were asked to predict the useful life for a cutting speed of $x = 100$ meters per minute. Since the given value of x is outside the range of the sample x values, the prediction is an example of **extrapolation**. Predict the useful life of a brand A cutting tool that is operated at 100 meters per minute, and construct a 95% prediction interval for the actual useful life of the tool. What additional assumption do you have to make in order to assure the validity of an extrapolation?

TOTAL EXPENDITURES y ($ billions)	AVERAGE NIGHTS x
2.03	18.6
1.50	17.6
.88	8.3
1.03	9.2
2.48	19.4
3.34	28.9
2.50	18.0
3.52	27.8
1.67	12.1
2.88	20.8

10.62 Explain why the confidence interval for the mean value of y for $x = x_p$ gets wider the farther x_p is from $\bar{x}$. What are the implications of this phenomenon for estimation and prediction?

10.63 Many variables affect the total dollar expenditure by quarter for tourists visiting the state of Florida. One of these is the average number of nights stayed in Florida. Shown in the table are the total expenditures, y, per quarter and the average number of nights, x, stayed in Florida for 10 quarters.

a. Plot the data points on graph paper.

b. Find the least squares line relating y to x. As a check on your calculations, graph the line in your plot from part **a** to see if the line appears to model the relationship between y and x.

c. Do the data provide sufficient evidence to indicate that the number of nights stayed in Florida contributes information for the prediction of total quarterly tourist expenditures? Test using $\alpha = .05$.

d. Calculate r and r^2, and interpret their values.

e. Find a 90% confidence interval for the mean quarterly tourist expenditures if the average number of nights stayed in Florida is 15.

f. Find a 90% prediction interval for the mean quarterly tourist expenditures if the average number of nights stayed in Florida is 15.

g. Explain the difference in the widths of the intervals found in parts e and f.

SECTION 10.9

SIMPLE LINEAR REGRESSION: AN EXAMPLE

In the previous sections we have presented the basic elements necessary to fit and use a straight-line regression model. In this final section we assemble these elements by applying them in an example.

Suppose a fire insurance company wants to relate the amount of fire damage in major residential fires to the distance between the residence and the nearest fire station. The study is to be conducted in a large suburb of a major city; a sample of 15 recent fires in this suburb is selected. The amount of damage, y, and the distance, x, between the fire and the nearest fire station are recorded for each fire. The results are given in Table 10.7.

TABLE 10.7

Fire Damage Data

DISTANCE FROM FIRE STATION x (miles)	FIRE DAMAGE y ($ thousands)	DISTANCE FROM FIRE STATION x (miles)	FIRE DAMAGE y ($ thousands)
3.4	26.2	2.6	19.6
1.8	17.8	4.3	31.3
4.6	31.3	2.1	24.0
2.3	23.1	1.1	17.3
3.1	27.5	6.1	43.2
5.5	36.0	4.8	36.4
.7	14.1	3.8	26.1
3.0	22.3		

STEP 1 First, we hypothesize a model to relate fire damage, y, to the distance, x, from the nearest fire station. We will hypothesize a straight-line probabilistic model:

$$y = \beta_0 + \beta_1 x + \varepsilon$$

STEP 2 Next, we use the data to estimate the unknown parameters in the deterministic component of the hypothesized model. We make some preliminary calculations:

$$SS_{xx} = \sum x_i^2 - \frac{\left(\sum x_i\right)^2}{15} = 196.16 - \frac{(49.2)^2}{15}$$

$$= 196.160 - 161.376 = 34.784$$

$$SS_{yy} = \sum y_i^2 - \frac{\left(\sum y_i\right)^2}{15} = 11,376.48 - \frac{(396.2)^2}{15}$$

$$= 11,376.480 - 10,464.96267 = 911.517334$$

$$SS_{xy} = \sum x_i y_i - \frac{\left(\sum x_i\right)\left(\sum y_i\right)}{15} = 1{,}470.65 - \frac{(49.2)(396.2)}{15}$$

$$= 1{,}470.650 - 1{,}299.536 = 171.114$$

Then the least squares estimate of the slope β_1 and intercept β_0 are

$$\hat{\beta}_1 = \frac{SS_{xy}}{SS_{xx}} = \frac{171.114}{34.784} = 4.919331$$

$$\hat{\beta}_0 = \bar{y} - \hat{\beta}_1\bar{x} = \frac{396.2}{15} - 4.919\left(\frac{49.2}{15}\right)$$

$$= 26.413333 - (4.919331)(3.28) = 26.413333 - 16.135406$$

$$= 10.277927$$

And the least squares equation is

$$\hat{y} = 10.278 + 4.919x$$

This prediction equation is graphed in Figure 10.17, along with a plot of the data points.

FIGURE 10.17

Least Squares Model for
the Fire Damage Data

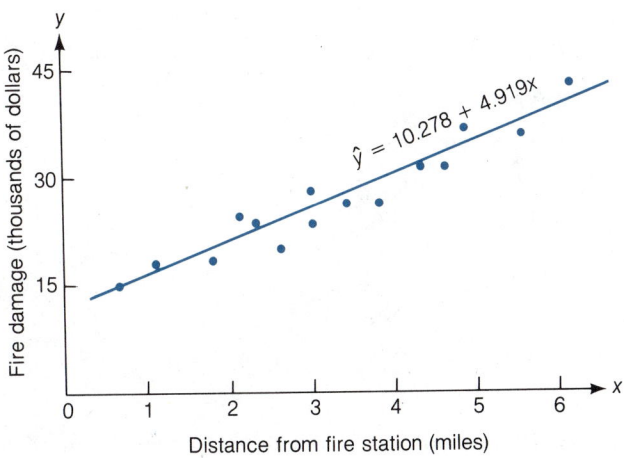

STEP 3 Now, we specify the probability distribution of the random error component, ε. The assumptions about the distribution will be identical to those listed in Section 10.3. Although we know that these assumptions are not completely satisfied (they rarely are for any practical problem), we are willing to assume they are approximately satisfied for this example. We have to estimate the variance σ^2 of ε, so we calculate

$$SSE = \sum (y_i - \hat{y}_i)^2 = SS_{yy} - \hat{\beta}_1 SS_{xy}$$

where the last expression represents a shortcut formula for SSE. Thus,

$$SSE = 911.517334 - (4.919331)(171.114)$$
$$= 911.517334 - 841.766405 = 69.750929*$$

To estimate σ^2, we divide SSE by the degrees of freedom available for error, $n - 2$. Thus,

$$s^2 = \frac{SSE}{n - 2} = \frac{69.750929}{15 - 2} = 5.3655$$
$$s = \sqrt{5.3655} = 2.32$$

STEP 4 We can now check the usefulness of the hypothesized model—that is, whether x really contributes information for the prediction of y using the straight-line model. First test the null hypothesis that the slope β_1 is zero—i.e., that there is no linear relationship between fire damage and the distance from the nearest fire station—against the alternative that the slope β_1 differs from zero. We test

H_0: $\beta_1 = 0$

H_a: $\beta_1 \neq 0$

Test statistic: $t = \dfrac{\hat{\beta}_1 - 0}{s_{\hat{\beta}_1}} = \dfrac{\hat{\beta}_1}{s/\sqrt{SS_{xx}}}$

Assumptions: Those made about ε in Section 10.3

For $\alpha = .05$, we will reject H_0 if

$$t > t_{\alpha/2} \quad \text{or} \quad t < -t_{\alpha/2}$$

where for $n = 15$, df $= n - 2 = 15 - 2 = 13$ and $t_{.025} = 2.160$. We then calculate the t statistic:

$$t = \frac{\hat{\beta}_1}{s_{\hat{\beta}_1}} = \frac{\hat{\beta}_1}{s/\sqrt{SS_{xx}}}$$
$$= \frac{4.919}{2.32/\sqrt{34.784}} = \frac{4.919}{.393} = 12.5$$

This large t value leaves little doubt that the distance between the fire and the fire station contributes information for the prediction of fire damage. Particularly, it appears (as we might suspect) that fire damage increases as the distance increases.

We gain additional information about the relationship by forming a confidence interval for the slope β_1. A 95% confidence interval is

$$\hat{\beta}_1 \pm t_{.025}\, s_{\hat{\beta}_1} = 4.919 \pm (2.160)(.393)$$
$$= 4.919 \pm .849 = (4.070, 5.768)$$

---

*For problems where rounding is necessary, at least six significant figures should be carried for these quantities. Otherwise, the calculated value of SSE may be substantially in error.

We estimate that the interval from \$4,070 to \$5,768 encloses the mean increase (β_1) in fire damage per additional mile distance from the fire station.

Another measure of the usefulness of the model is the coefficient of correlation r:

$$r = \frac{SS_{xy}}{\sqrt{SS_{xx}\ SS_{yy}}}$$

$$= \frac{171.114}{\sqrt{(34.784)(911.517)}} = \frac{171.114}{178.062} = .96$$

The high correlation provides further support for our conclusion that β_1 differs from 0; it appears that fire damage and distance from the fire station are highly correlated.

The coefficient of determination is

$$r^2 = (.96)^2 = .92$$

which implies that 92% of the sum of squares of deviations of the y values about $\bar{y}$ is explained by the distance, x, between the fire and the fire station. All signs point to a strong linear relationship between y and x.

STEP 5 We are now prepared to use the least squares model. Suppose the insurance company wants to predict the fire damage if a major residential fire were to occur 3.5 miles from the nearest fire station; i.e., $x_p = 3.5$. The predicted value is

$$\hat{y} = \hat{\beta}_0 + \hat{\beta}_1 x_p$$
$$= 10.278 + (4.919)(3.5)$$
$$= 10.278 + 17.216 = 27.5$$

(we round to the nearest tenth to be consistent with the units of the original data in Table 10.7). If we want a 95% prediction interval, we calculate

$$\hat{y} \pm t_{.025}\ s\sqrt{1 + \frac{1}{n} + \frac{(x_p - \bar{x})^2}{SS_{xx}}} = 27.5 \pm (2.16)(2.32)\sqrt{1 + \frac{1}{15} + \frac{(3.5 - 3.28)^2}{34.784}}$$

$$= 27.5 \pm (2.16)(2.32)\sqrt{1.0681}$$

$$= 27.5 \pm 5.2 \quad \text{or} \quad (22.3,\ 32.7)$$

The model yields a 95% prediction interval for fire damage in a major residential fire 3.5 miles from the nearest station of \$22,300 to \$32,700.

One caution before closing: We would not use this prediction model to make predictions for homes less than .7 mile or more than 6.1 miles from the nearest fire station. A look at the data in Table 10.7 reveals that all the x values fall between .7 and 6.1. It is dangerous to use the model to make predictions outside the region in which the sample data fall. A straight line might not provide a good model for the relationship between the mean value of y and x for values of x beyond the range of the sample data.

**USING THE
COMPUTER FOR
SIMPLE LINEAR
REGRESSION**

All of the examples of simple linear regression that we have presented thus far have required rather tedious calculations involving SS_{yy}, SS_{xx}, SS_{xy}, and so forth. Even with the use of a pocket calculator, the process is laborious and susceptible to error. Fortunately, the use of computers can significantly reduce the labor involved in regression calculations. In this section we introduce the regression output from one statistical software package, the SAS System. Though this is just one of the many statistical packages available that provide simple linear regression output, most produce essentially the same quantities, differing only in format and labeling. The regression output of other packages will be presented in Chapter 11.

The SAS output for the fire damage example of Section 10.9 is presented in Figure 10.18. We have shaded the parts of the printout corresponding to most of the key simple linear regression quantities introduced in this chapter.

First, the estimates of the y-intercept and the slope are found about halfway down the printout on the left-hand side, under the column labeled Parameter Estimate and in the rows labeled INTERCEP and X, respectively. The values are $\hat{\beta}_0 = 10.277929$ and $\hat{\beta}_1 = 4.919331$. When rounded to three decimal places, these quantities agree with our calculations in Section 10.9.

Next, we find the measures of variability: SSE, s^2, and s. They are shaded in the upper portion of the printout. SSE is found under the column heading Sum of Squares and in the row labeled Error: SSE = 69.75098. The estimate of the error variance σ^2 is under the column heading Mean Square and in the row labeled Error: $s^2 = 5.36546$. The estimate of the standard deviation σ is directly to the right of the heading Root MSE: $s = 2.31635$. Again, all values (after rounding) agree with the corresponding quantities we calculated in Section 10.9.

The coefficient of determination is shown (shaded) under the heading R-Square in the upper portion of the printout: $r^2 = .9235$. Again, to two decimal places, this agrees with our calculation in Section 10.9. The coefficient of correlation r is not given on the printout.

The t statistic for testing H_0: $\beta_1 = 0$ versus H_a: $\beta_1 \neq 0$ is given (shaded) in the center of the page under the column heading T for H0: Parameter = 0 in the row corresponding to X. The value $t = 12.525$ agrees with our computed value when we used the formula

$$t = \frac{\hat{\beta}_1}{s_{\hat{\beta}_1}} = \frac{\hat{\beta}_1}{s/\sqrt{SS_{xx}}}$$

To determine which hypothesis this test statistic supports, we can establish a rejection region using the t table (Table VI of Appendix B), just as we did in Section 10.9. However, the printout makes this unnecessary, because the observed significance level, or p-value, is shown (shaded) immediately to the right of the t statistic, under the column heading Prob > |T|. Remember that if the observed significance level is less than the α value you select, then the test statistic supports the alternative hypothesis at that level. For example, if we select $\alpha = .05$ in this example, the observed significance level of .0001 given on the printout indicates that we should

FIGURE 10.18   SAS Printout for Fire Damage Regression Analysis

FIRE DAMAGE EXAMPLE
STRAIGHT-LINE MODEL WITH PREDICTION INTERVALS

Model: MODEL1
Dep Variable: Y

Analysis of Variance

Source	DF	Sum of Squares	Mean Square	F Value	Prob>F
Model	1	841.76636	841.76636	156.886	0.0001
Error	13	69.75098	5.36546		
C Total	14	911.51733			

Root MSE	2.31635	R-Square	0.9235	
Dep Mean	26.41333	Adj R-Sq	0.9176	
C.V.	8.76961			

Parameter Estimates

Variable	DF	Parameter Estimate	Standard Error	T for H0: Parameter=0	Prob > |T|
INTERCEP	1	10.277929	1.42027781	7.237	0.0001
X	1	4.919331	0.39274775	12.525	0.0001

Obs	X	Y	Predict Value	Residual	Lower95% Predict	Upper95% Predict
1	3.4	26.2000	27.0037	-0.8037	21.8344	32.1729
2	1.8	17.8000	19.1327	-1.3327	13.8141	24.4514
3	4.6	31.3000	32.9068	-1.6068	27.6186	38.1951
4	2.3	23.1000	21.5924	1.5076	16.3577	26.8271
5	3.1	27.5000	25.5279	1.9721	20.3573	30.6984
6	5.5	36.0000	37.3342	-1.3342	31.8334	42.8351
7	0.7	14.1000	13.7215	0.3785	8.1087	19.3342
8	3	22.3000	25.0359	-2.7359	19.8622	30.2097
9	2.6	19.6000	23.0682	-3.4682	17.8678	28.2686
10	4.3	31.3000	31.4311	-0.1311	26.1908	36.6713
11	2.1	24.0000	20.6085	3.3915	15.3442	25.8729
12	1.1	17.3000	15.6892	1.6108	10.1999	21.1785
13	6.1	43.2000	40.2858	2.9142	34.5906	45.9811
14	4.8	36.4000	33.8907	2.5093	28.5640	39.2175
15	3.8	26.1000	28.9714	-2.8714	23.7843	34.1585
16 *	3.5	.	27.4956	.	22.3239	32.6672

Sum of Residuals   -3.73035E-14
Sum of Squared Residuals   69.7510
Predicted Resid SS (Press)   93.2117

reject H_0. We can conclude that there is sufficient evidence at $\alpha = .05$ to infer that a linear relationship between fire damage and distance from the station is useful for predicting damage.

If you wish to conduct a one-tailed test, the observed significance level is half that given on the printout (assuming the sign of the test statistic agrees with the alternative hypothesis). Thus, if we were testing $H_0: \beta_1 = 0$ versus $H_a: \beta_1 > 0$ in this example, the observed significance level would be $\frac{1}{2}(.0001) = .00005$.

Predicted y values and the corresponding prediction intervals are given in the lower portion of the SAS printout. To find the 95% prediction interval for the fire damage y when the distance from the fire station is $x = 3.5$ miles, first locate the value 3.5 in the column labeled X (the last value in the column). The prediction is given in the center column labeled Predict Value in the row corresponding to 3.5:

$$\hat{y} = 27.4956$$

The lower and upper confidence bounds are given in the columns headed Lower and Upper 95% Predict, respectively:

Lower = 22.3239

Upper = 32.6672

Again, all values agree after rounding with our calculations in Section 10.9.

Although much more information is given on the SAS printout, we have discussed only those aspects that have been presented in the chapter. In the next two chapters we will expand our discussion to include other important components of the SAS printout, as well as other statistical packages. The point here is that the computer can alleviate much of the burden of calculation involved in a regression analysis, and enable us to spend more time on the interpretation of the model. The time spent in learning how to read computer regression output is a good investment.

EXERCISES 10.64–10.65

LEARNING THE MECHANICS

10.64 Refer to Exercise 10.11, in which the extent of retaliation (as measured by the retaliation index) is related to the salary of 15 whistle blowers using a least squares model. A computer printout for this regression model is given here.

a. Find the least squares estimates of the intercept β_0 and the slope β_1.
b. Identify the values of SSE, s^2, and s. Based on the standard deviation, how accurately do you expect to be able to predict the retaliation index using the salary of the whistle blower?
c. Find and interpret the coefficient of determination r^2.
d. Find the t statistic and its observed significance level for testing the usefulness of the model. Interpret the values.
e. Find and interpret the 95% confidence interval for the mean retaliation index of whistle blowers with salaries of $35,000.

SAS Simple Linear
Regression Printout
for Exercise 10.64

RETALIATION INDEX VS. SALARY

Model: MODEL1
Dep Variable: R_INDEX

Analysis of Variance

Source	DF	Sum of Squares	Mean Square	F Value	Prob>F
Model	1	20853.93900	20853.93900	0.925	0.3538
Error	13	293208.46100	22554.49700		
C Total	14	314062.40000			

Root MSE	150.18155	R-Square	0.0664	
Dep Mean	499.80000	Adj R-Sq	-0.0054	
C.V.	30.04833			

Parameter Estimates

Variable	DF	Parameter Estimate	Standard Error	T for H0: Parameter=0	Prob > ¦T¦
INTERCEP	1	575.028672	87.31829499	6.585	0.0001
SALARY	1	-0.002186	0.00227386	-0.962	0.3538

Obs	SALARY	R_INDEX	Predict Value	Residual	Lower95% Mean	Upper95% Mean
1	62000	301.0	439.5	-138.5	280.1	598.8
2	36500	550.0	495.2	54.7770	410.8	579.6
3	17600	755.0	536.5	218.5	418.9	654.2
4	20000	327.0	531.3	-204.3	421.6	641.0
5	30100	500.0	509.2	-9.2163	422.8	595.6
6	35000	377.0	498.5	-121.5	414.7	582.3
7	47500	290.0	471.2	-181.2	365.6	576.8
8	54000	452.0	457.0	-4.9600	329.4	584.6
9	15800	535.0	540.5	-5.4827	416.5	664.5
10	44000	455.0	478.8	-23.8246	382.7	574.9
11	46600	615.0	473.1	141.9	370.2	576.1
12	12100	700.0	548.6	151.4	410.6	686.5
13	62000	650.0	439.5	210.5	280.1	598.8
14	21000	630.0	529.1	100.9	422.6	635.7
15	11900	360.0	549.0	-189.0	410.3	687.7

10.65 Refer to Exercise 10.30, in which an index of managerial success was related to the number of interactions he or she has with outsiders. The simple linear regression printout for these data is shown at the top of page 542.

a. Find the least squares equation, and compare it to that you calculated in Exercise 10.30.

b. Find the standard deviation s, and interpret the value in terms of this exercise.

c. Find and interpret r^2.

d. Is there sufficient evidence to indicate that the model is useful for predicting y? What is the observed significance level of the test?

SAS Simple Linear
Regression Printout
for Exercise 10.65

Source	DF	Sum of Squares	Mean Square	F Value	Prob>F
Model	1	606.03751	606.03751	1.610	0.2216
Error	17	6400.59407	376.50553		
C Total	18	7006.63158			

Root MSE	19.40375	R-Square	0.0865
Dep Mean	54.57895	Adj R-Sq	0.0328
C.V.	35.55171		

Parameter Estimates

Variable	DF	Parameter Estimate	Standard Error	T for H0: Parameter=0	Prob > |T|
INTERCEP	1	44.130454	9.36159293	4.714	0.0002
INTERACT	1	0.236617	0.18650103	1.269	0.2216

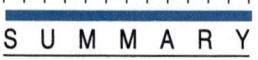

S U M M A R Y

We have introduced an extremely useful tool in this chapter—**the method of least squares** for fitting a prediction equation to a set of data. The application of this methodology, along with the associated inferential procedures, is referred to as **regression analysis**. In five steps we showed how to use sample data to build a model relating a dependent variable y to a single independent variable x.

1. Hypothesize a **probabilistic model**. In this chapter, we confined our attention to the **first-order (straight-line) model**, $y = \beta_0 + \beta_1 x + \varepsilon$.
2. Use the method of least squares to estimate the unknown parameters in the **deterministic component**, $\beta_0 + \beta_1 x$. The least squares estimates yield a model $\hat{y} = \hat{\beta}_0 + \hat{\beta}_1 x$ with a **sum of squared errors (SSE)** that is smaller than the SSE for any other straight-line model.
3. Specify the probability distribution of the **random error component**, ε.
4. Assess the usefulness of the hypothesized model. Included here are making inferences about the **slope**, β_1; calculating the **coefficient of correlation**, r; and calculating the **coefficient of determination**, r^2.
5. Finally, if we are satisfied with the model, we can use it to **estimate the mean y value**, $E(y)$, for a given x value and/or to **predict an individual y value** for a specific value of x.

SUPPLEMENTARY EXERCISES
10.66–10.80

10.66 In fitting a least squares line to $n = 15$ data points, the following quantities were computed:

$$SS_{xx} = 50 \qquad SS_{yy} = 25 \qquad SS_{xy} = -30 \qquad \bar{x} = 1.3 \qquad \bar{y} = 27$$

a. Find the least squares line.
b. Graph the least squares line.
c. Calculate SSE.
d. Calculate s^2.
e. Find a 90% confidence interval for β_1. Interpret this interval.
f. Find a 90% confidence interval for the mean value of y when $x_p = 1.8$.
g. Find a 90% prediction interval for y when $x_p = 1.8$.

10.67 Consider the following sample data:

y	5	1	3
x	5	1	3

a. Construct a scattergram for the data.
b. It is possible to find many lines for which $\Sigma (y - \hat{y}) = 0$. For this reason, the criterion $\Sigma (y - \hat{y}) = 0$ is not used for identifying the "best"-fitting straight line. Find two lines that have $\Sigma (y - \hat{y}) = 0$.
c. Find the least squares line.
d. Compare the value of SSE for the least squares line to that of the two lines you found in part b. What principle of least squares is demonstrated by this comparison?

10.68 Emotional exhaustion, or *burnout*, is a significant problem for people with careers in the field of human services. It seriously affects productivity and feelings of job satisfaction. Michael P. Leiter and Kimberly Ann Meechan (1986) used regression analysis to investigate the relationship between burnout and different aspects of the human services professional's job and job-related behavior. To measure emotional exhaustion, they used the Maslach Burnout Inventory (a questionnaire from which an index of exhaustion can be developed). One of the independent variables they considered was the proportion of the person's social contacts with individuals who belong to the person's work group. They called this variable *concentration*. The table lists the values of the emotional exhaustion index (higher values indicate greater emotional exhaustion) and concentration for a sample of 25 human services professionals who work in a large public hospital.

EXHAUSTION INDEX y	CONCENTRATION x (%)	EXHAUSTION INDEX y	CONCENTRATION x (%)
100	20	493	86
525	60	892	83
300	38	527	79
980	88	600	75
310	79	855	81
900	87	709	75
410	68	791	77
296	12	718	77
120	35	684	77
501	70	141	17
920	80	400	85
810	92	970	96
506	77		

For this data set, $SS_{yy} = 1,800,417.44$, $SS_{xx} = 14,026.16$, $SS_{xy} = 124,348.520$, $\bar{y} = 578.32$, and $\bar{x} = 68.560$.

a. Construct a scattergram for the data. Do the variables x and y appear to be related?
b. Calculate the correlation coefficient for the data and interpret its value.
c. Use the method of least squares to estimate the straight-line model relating emotional exhaustion and concentration, and plot it on your scattergram.

d. Calculate the coefficient of determination for your least squares line and interpret it.

e. Test the usefulness of the straight-line relationship with concentration for predicting burnout. Use $\alpha = .05$.

f. Is there evidence that the correlation between emotional exhaustion and concentration differs from 0? Does your conclusion mean that concentration causes emotional exhaustion? Explain.

g. Use a 95% prediction interval to forecast the level of emotional exhaustion for a human services professional who has 80% of her social contacts within her work group. Interpret the prediction in the context of the problem.

h. Use a 95% confidence interval to estimate the mean exhaustion level for all professionals who have 80% of their social contacts within their work groups. Interpret the interval, and compare its width with that of the prediction interval in part **g**.

10.69 Spiraling energy costs have generated interest in energy conservation in businesses of all sizes. Consequently, firms planning to build new plants or make additions to existing facilities have become very conscious of the energy efficiency of proposed new structures. As a result, such firms are interested in knowing the relationship between a building's yearly energy consumption and the factors that influence heat loss. Some of these factors are the number of building stories above and below ground, the materials used in the construction of the building shell, the number of square feet of building shell, and climatic conditions (ASHRAE, 1968). The table lists the energy consumption in British thermal units (a BTU is the amount of heat required to raise 1 pound of water 1° Fahrenheit) for 1987 for 22 buildings that were all subjected to the same climatic conditions. The SAS printout that fits the straight-line model relating BTU consumption, y, to building shell area, x, is also given.

BTU/YEAR (Thousands)	SHELL AREA (Square feet)	BTU/YEAR (Thousands)	SHELL AREA (Square feet)
3,870,000	30,001	2,680,000	23,680
1,371,000	13,530	337,500	5,650
2,422,000	26,060	567,500	8,001
672,200	6,355	555,300	6,147
233,100	4,576	239,400	2,660
218,900	24,680	2,629,000	19,240
354,000	2,621	1,102,000	10,700
3,135,000	23,350	423,500	9,125
1,470,000	18,770	423,500	6,510
1,408,000	12,220	1,691,000	13,530
2,201,000	25,490	1,870,000	18,860

a. Find the least squares estimates of the intercept β_0 and the slope β_1.

b. Investigate the usefulness of the model you developed in part **a**. Is yearly energy consumption positively linearly related to the shell area of the building? Test using $\alpha = .10$.

c. Show how to calculate the observed significance level of the test from part **b** by using that given on the printout. Interpret its value.

d. Find the coefficient of determination r^2 and interpret its value.

e. A company wishes to build a new warehouse that will contain 8,000 square feet of shell area. Find the predicted value and a 95% prediction interval on the printout. Comment on the usefulness of this interval.

SAS Printout for
Exercise 10.69

Source	DF	Sum of Squares	Mean Square	F Value	Prob>F
Model	1	1.658498E+13	1.658498E+13	42.028	0.0001
Error	20	7.89232E+12	394616010047		
C Total	21	2.44773E+13			

Root MSE	628184.69422	R-Square	0.6776
Dep Mean	1357904.54545	Adj R-Sq	0.6614
C.V.	46.26133		

Parameter Estimates

Variable	DF	Parameter Estimate	Standard Error	T for H0: Parameter=0	Prob > |T|
INTERCEP	1	-99045	261617.65980	-0.379	0.7090
AREA	1	102.814048	15.85924082	6.483	0.0001

Obs	AREA	BTU	Predict Value	Residual	Lower95% Predict	Upper95% Predict
1	30001	3870000	2985479	884521	1546958	4424000
2	13530	1371000	1292029	78971.2	-47949.3	2632007
3	26060	2422000	2580289	-158289	1183940	3976637
4	6355	672200	554338	117862	-810192	1918868
5	4576	233100	371432	-138332	-1005463	1748327
6	24680	218900	2438405	-2219505	1054223	3822588
7	2621	354000	170430	183570	-1222796	1563657
8	23350	3135000	2301663	833337	927871	3675455
9	18770	1470000	1830774	-360774	482352	3179196
10	12220	1408000	1157342	250658	-184021	2498706
11	25490	2201000	2521685	-320685	1130530	3912840
12	23680	2680000	2335591	344409	959345	3711838
13	5650	337500	481854	-144354	-887287	1850995
14	8001	567500	723570	-156070	-631698	2078838
15	6147	555300	532953	22347.3	-832898	1898804
16	2660	239400	174440	64959.9	-1218433	1567313
17	19240	2629000	1879097	749903	528832	3229362
18	10700	1102000	1001065	100935	-343656	2345786
19	9125	423500	839133	-415633	-511035	2189301
20	6510	423500	570274	-146774	-793294	1933842
21	13530	1691000	1292029	398971	-47949.3	2632007
22	18860	1870000	1840028	29972.3	491266	3188789
23	8000	.	723467	.	-631806	2078740

f. The application of the model you developed in part **a** to the warehouse problem of part **e** is appropriate only if certain assumptions can be made about the new warehouse. What are these assumptions?

10.70 Refer to Exercise 10.30, in which managerial success, y, was modeled as a function of the number, x, of contacts a manager makes with people outside his or her work unit during a specific period of time. The data are given in the table on page 510. Recall that $SS_{yy} = 7,006.6316$, $SS_{xx} = 10,824.5263$, $SS_{xy} = 2,561.2632$, $\bar{y} = 54.5789$, and $\bar{x} = 44.1579$.

a. A particular manager was observed for 2 weeks as in the Luthans, Rosenkrantz, and Hennessey (1985) study. She made 55 contacts with people outside her work unit. Predict the value of the manager's success index. Use a 90% prediction interval.

b. A second manager was observed for 2 weeks. This manager made 110 contacts with people outside his work unit. Give two reasons why caution should be exercised in using the least squares model developed from the given data set to construct a prediction interval for this manager's success index.

c. In the context of this problem, determine the value of x whose associated prediction interval for y is the narrowest.

10.71 The table shows the number of passengers carried by scheduled airlines and railroads in the United States over the period from 1950 to 1980.

YEAR	PASSENGERS CARRIED BY SCHEDULED AIR CARRIERS IN THE U.S. y (millions)	PASSENGERS CARRIED BY RAILROAD IN THE U.S. x (millions)
1950	19	488
1955	42	433
1960	62	327
1965	103	306
1970	169	289
1974	208	275
1975	205	270
1976	223	272
1977	240	276
1978	275	262
1979	317	274
1980	297	281

Source: U.S. Bureau of the Census, *Statistical Abstract of the United States: 1982.*

a. Find the correlation coefficient and the coefficient of determination for the data in the table and interpret their values.

b. Do the data provide sufficient evidence to indicate that x and y are correlated? Test using $\alpha = .05$.

10.72 A large supermarket chain has its own store brand for many grocery items. These tend to be priced lower than other brands. For a particular item, the chain wants to study the effect of varying the price for the major competing brand on the sales of the store brand item, while the prices for the store brand and all other brands are held fixed. The experiment is conducted at one of the chain's stores over a 7-week period and the results are shown in the table.

WEEK	MAJOR COMPETITOR'S PRICE x	STORE BRAND SALES y
1	37¢	122
2	32	107
3	29	99
4	35	110
5	33	113
6	31	104
7	35	116

a. Find the least squares line relating store brand sales, y, to major competitor's price, x.

b. Plot the data and graph the line as a check on your calculations.

c. Does x contribute information for the prediction of y?

d. Calculate r and r^2 and interpret their values.

e. Find a 90% confidence interval for mean store brand sales when the competitor's price is 33¢.

f. Suppose you were to set the competitor's price at 33¢. Find a 90% prediction interval for next week's sales.

10.73 Sometimes it is known from theoretical considerations that the straight-line relationship between two variables, x and y, passes through the origin of the xy-plane. Consider the relationship between the total weight of a shipment of 50-pound bags of flour, y, and the number of bags in the shipment, x. Since a shipment containing $x = 0$ bags (i.e., no shipment at all) has a total weight of $y = 0$, a straight-line model of the relationship between x and y should pass through the point $x = 0$, $y = 0$. In such a case you could assume $\beta_0 = 0$ and characterize the relationship between x and y with the following model:

$$y = \beta_1 x + \varepsilon$$

The least squares estimate of β_1 for this model is

$$\hat{\beta}_1 = \frac{\sum x_i y_i}{\sum x_i^2}$$

From the records of past flour shipments, 15 shipments were randomly chosen and the data shown in the table were recorded.

WEIGHT OF SHIPMENT	NUMBER OF 50-POUND BAGS IN SHIPMENT	WEIGHT OF SHIPMENT	NUMBER OF 50-POUND BAGS IN SHIPMENT
5,050	100	7,162	150
10,249	205	24,000	500
20,000	450	4,900	100
7,420	150	14,501	300
24,685	500	28,000	600
10,206	200	17,002	400
7,325	150	16,100	400
4,958	100		

a. Find the least squares line for the given data under the assumption that $\beta_0 = 0$. Plot the least squares line on a scatttergram of the data.

b. Find the least squares line for the given data using the model

$$y = \beta_0 + \beta_1 x + \varepsilon$$

(i.e., do not restrict β_0 to equal 0). Plot this line on the same scatterplot you constructed in part **a**.

c. Refer to part **b**. Why might $\hat{\beta}_0$ be different from 0 even though the true value of β_0 is known to be 0?

d. The estimated standard error of $\hat{\beta}_0$ is equal to

$$s\sqrt{\frac{1}{n} + \frac{\bar{x}^2}{SS_{xx}}}$$

Use the t statistic

$$t = \frac{\hat{\beta}_0 - 0}{s\sqrt{\frac{1}{n} + \frac{\bar{x}^2}{SS_{xx}}}}$$

to test the null hypothesis H_0: $\beta_0 = 0$ against the alternative H_a: $\beta_0 \neq 0$. Use $\alpha = .10$. Should you include β_0 in your model?

10.74 As a result of the increase in the number of suburban shopping centers, many center-city stores are suffering financially. A downtown department store thinks that increased advertising might help lure more shoppers into the area. To study the effect of advertising on sales, records were obtained for several mid-year months during which the store varied advertising expenditures. Those records are shown in the table.

ADVERTISING EXPENSE x ($ thousands)	SALES y ($ thousands)
.9	30
1.1	34
.8	32
1.2	37
.7	31

a. Estimate the coefficient of correlation between sales and advertising expenditures.
b. Do the data provide sufficient evidence to indicate a nonzero correlation between sales, y, and advertising expense, x?
c. Can you use the results of parts **a** and **b** to conclude that additional advertising expense will *cause* sales to increase? Why or why not?

10.75 A certain manufacturer evaluates the sales potential for a product in a new marketing area by selecting several stores within the area to sell the product on a trial basis for a 1-month period. The sales figures for the trial period are then used to project sales for the entire area. [*Note:* The same number of trial stores is used each time.]

TOTAL SALES DURING TRIAL PERIOD x ($ hundreds)	TOTAL SALES FOR FIRST MONTH FOR THE ENTIRE AREA y ($ hundreds)
16.8	48.2
14.0	46.8
18.3	54.3
22.1	59.7
14.9	48.3
23.2	67.5

a. Use the data in the table to develop a simple linear model for predicting first-month sales for the entire area based on sales during the trial period.
b. Plot the data, and graph the line as a check on your calculations.
c. Do the data provide sufficient evidence to indicate that total sales during the trial period contribute information for predicting total sales during the first month? Test using $\alpha = .10$.
d. Find the approximate p-value for the test in part c, and interpret its value.
e. Use a 90% prediction interval to predict total sales for the first month for the entire area if the trial sales are $2,000.

10.76 As part of the first-year evaluation for new salespeople, a large food-processing firm projects the second-year sales for each salesperson based on his or her sales for the first year.

FIRST-YEAR SALES x ($ thousands)	SECOND-YEAR SALES y ($ thousands)
75.2	99.3
91.7	125.7
100.3	136.1
64.2	108.6
81.8	102.0
110.2	153.7
77.3	108.8
80.1	105.4

a. Use the data in the table on eight salespeople for this firm to fit a simple linear prediction model for second-year sales based on the first year's sales. Assume the data have been adjusted in terms of a base year to discount inflation effects.
b. Plot the data, and graph the line as a check on your calculations.
c. Do the data provide sufficient information to indicate that x contributes information for the prediction of y?
d. Find the approximate p-value for the test in part c, and interpret its value.
e. Calculate r^2, and interpret its value.
f. If a salesperson has first-year sales of $90,000, find a 90% prediction interval for next year's sales.

10.77 In the late 1970's and early 1980's, the prices of single-family homes in the United States rose faster than the rate of inflation. As a result, many investors directed their funds to the housing market as a hedge against inflation. One way an investor can assess the value of a specific house is to compare it to the sale prices of similar houses that have recently been sold. Another popular approach, according to Cho and Reichert (1980), involves the use of a regression analysis to model the relationship between price and the variables that influence price. Independent variables that could be utilized are total living area, number of rooms, number of baths, age of property, and so on. Of these factors, Cho and Reichert indicate that total living area provides the most information for determining the worth of a house. The table in the margin on page 550 lists the final selling price and total living area for a sample of 24 homes in the same geographic area that were sold during the last 3 months of 1987. The regression printout using a straight-line model to relate price to area for these data is also shown on page 550.

AREA (square feet)	PRICE ($)
2,100	150,000
1,455	114,900
1,630	106,500
2,600	195,000
1,210	75,500
1,857	126,600
2,000	135,400
2,400	178,650
2,256	145,100
1,290	62,600
2,332	168,200
1,725	138,100
3,000	205,000
1,400	79,400
2,750	200,000
2,900	215,100
2,500	180,800
1,535	120,900
1,333	70,000
2,455	165,200
3,010	185,000
2,180	160,000
1,870	119,900
1,582	99,900

Source	DF	Sum of Squares	Mean Square	F Value	Prob>F
Model	1	42825909688	42825909688	247.857	0.0001
Error	22	3801265207.6	172784782.16		
C Total	23	46627174896			

Root MSE	13144.76254	R-Square	0.9185	
Dep Mean	141572.91667	Adj R-Sq	0.9148	
C.V.	9.28480			

Parameter Estimates

Variable	DF	Parameter Estimate	Standard Error	T for H0: Parameter=0	Prob > |T|
INTERCEP	1	−15124	10308.485043	−1.467	0.1565
AREA	1	76.174547	4.83848421	15.743	0.0001

Obs	AREA	PRICE	Predict Value	Residual	Lower95% Mean	Upper95% Mean
1	2100	150000	144842	5157.9	139261	150423
2	1455	114900	95709	19190.5	87495.9	103923
3	1630	106500	109040	−2540.0	102017	116064
4	2600	195000	182929	12070.7	175142	190717
5	1210	75500.0	77046.7	−1546.7	66887.4	87206.1
6	1857	126600	126332	268.3	120416	132247
7	2000	135400	137225	−1824.6	131631	142819
8	2400	178650	167694	10955.6	161152	174237
9	2256	145100	156725	−11625.3	150814	162637
10	1290	62600.0	83140.7	−20540.7	73642.8	92638.6
11	2332	168200	162515	5685.4	156304	168725
12	1725	138100	116277	21823.4	109791	122763
13	3000	205000	213399	−8399.2	202423	224376
14	1400	79400.0	91519.9	−12119.9	82892.2	100148
15	2750	200000	194356	5644.5	185450	203261
16	2900	215100	205782	9318.3	195657	215906
17	2500	180800	175312	5488.1	168190	182433
18	1535	120900	101803	19096.5	94160.9	109446
19	1333	70000.0	86416.2	−16416.2	77264.5	95568
20	2455	165200	171884	−6684.0	165035	178733
21	3010	185000	214161	−29160.9	203098	225224
22	2180	160000	150936	9064.0	145236	156636
23	1870	119900	127322	−7421.9	121449	133195
24	1582	99900	105384	−5483.7	98056	112711
25	2200	.	152460	.	146713	158206

a. Construct a scattergram for the data.

b. Find the least squares line and plot it on your scattergram.

c. Find r^2 and interpret its value in the context of the problem.

d. Do the data provide evidence that living area contributes information for predicting the price of a home? Use $\alpha = .05$.

e. Find a 95% confidence interval for β_1. Does your confidence interval support the conclusion you reached in part d? Explain. (Note that the standard error of $\hat{\beta}_1$ is given on the printout in the column headed Standard Error in the row corresponding to AREA.)

f. Find the observed significance level for the test in part **d**, and interpret its value.

g. Estimate the mean selling price for homes with a total living area of 2,200 square feet. Use a 95% confidence interval.

10.78 The table lists the 1985 sales (y, in millions of dollars) and number of employees (x, in thousands) for a random sample of 20 *Fortune 500* companies.

COMPANY	SALES y	NUMBER OF EMPLOYEES x	COMPANY	SALES y	NUMBER OF EMPLOYEES x
Procter & Gamble	$13,552.0	62.2	Telex	$ 591.0	6.6
H. B. Fuller	457.9	4.4	Westinghouse Electric	10,700.2	125.6
J. P. Stevens	1,858.2	27.2	Tribune	1,937.8	15.0
Mattel	1,050.9	20.0	Allied-Signal	9,115.0	143.8
Intel	1,365.0	21.3	Polaroid	1,295.2	12.9
Nashua	621.6	5.3	Lancaster Colony	440.6	5.8
GAF	732.0	4.3	Magic Chef	1,061.8	9.7
Coleco	776.0	4.4	Georgia Kraft	557.0	3.1
General Mills	5,654.1	63.2	Compaq Computer	503.9	1.9
ITT	12,714.0	232.0	Merck	3,547.5	30.9

Source: "The Fortune 500," *Fortune*, Apr. 28, 1986, pp. 175–232.

For this data set, $SS_{yy} = 368,426,099.57$, $SS_{xx} = 69,382.432$, $SS_{xy} = 4,352,045.534$, $\bar{y} = 3,426.585$, and $\bar{x} = 39.980$. Perform a regression analysis that follows the five steps presented in this chapter; be sure to state all assumptions you make. Give a prediction interval for a *Fortune 500* company with 100 employees.

10.79 When a sample is drawn from a population with correlation coefficient, ρ, equal to 0, the sample correlation may still turn out to be significantly different from 0. In such cases the correlation is said to be *spurious* or *irrelevant*. As more measurements are included in the sample, the spurious correlation will shrink toward 0. Calculate the sample correlation coefficient for the data in the table and comment on whether you believe the correlation is spurious.

DATE	PRIME INTEREST RATE	RATE OF GROWTH IN MONEY SUPPLY (M1[a])	DATE	PRIME INTEREST RATE	RATE OF GROWTH IN MONEY SUPPLY (M1[a])
2/84	11.0	.55	1/85	10.75	1.48
3/84	11.5	.43	2/85	10.5	1.19
4/84	12.0	.04	3/85	10.5	.47
5/84	12.5	1.06	4/85	10.5	.49
6/84	13.0	.94	5/85	10.5	1.17
7/84	13.0	−.09	6/85	10.0	1.65
8/84	13.0	.15	7/85	9.5	.78
9/84	13.0	.42	8/85	9.5	1.70
10/84	12.75	−.62	9/85	9.5	.99
11/84	12.0	.71	10/85	9.5	−.13
12/84	11.25	.93	11/85	9.5	1.11
			12/85	9.5	1.10

[a]M1 = Currency + Demand deposits + Traveler's checks + Other checkable deposits.
Source: *Economic Report of the President, 1985, 1986*, pp. 303, 327, 333.

10.80 In Exercise 10.79, you considered the correlation between the prime interest rate and the rate of growth of the money supply for the years 1984–1985.

   a. Do the data of Exercise 10.79 indicate that the population correlation coefficient, ρ, is less than 0? Test using $\alpha = .05$ and report the p-value of your test.

   b. What do the results of your hypothesis test suggest about the form of the relationship between the prime interest rate and the rate of growth of the money supply?

   c. Construct a scattergram of the data in Exercise 10.79. Does the pattern of points in your scattergram support the conclusion you made in part **a**? Explain.

ON YOUR OWN . . .

The Gross National Product (GNP) is one of the nation's best-known economic indicators. Many economists have developed models to forecast future values of the GNP. There is surely a large number of variables that should be included if an accurate prediction is to be made. For the moment, however, consider the simple case of choosing one important variable to include in a simple straight-line model for GNP.

First, list three independent variables (x_1, x_2, and x_3) that you think might be (individually) strongly related to the GNP. Next, obtain ten yearly values (preferably for the last 10 years) of the three independent variables and the GNP.*

a. Use the least squares formulas given in this chapter to fit three straight-line models—one for each independent variable—for predicting the GNP.

b. Interpret the sign of the estimated slope coefficient, $\hat{\beta}_1$, in each case, and test the utility of the model by testing $H_0: \beta_1 = 0$ against $H_a: \beta_1 \neq 0$.

c. Calculate the coefficient of determination r^2 for each model. Which of the independent variables predicts the GNP best over the 10 sample years when a straight-line model is used? Is this variable necessarily best in general (i.e., for all years)? Explain.

USING THE COMPUTER . . .

Suppose we want to model the relationship between the median household income y and the percentage of college graduates x in a zip code using a straight-line model.

a. Draw a random sample of 100 zip codes from the 1,000 described in Appendix C, and extract y and x for each zip code sampled. Use a software package that includes a regression program to obtain the least squares fit for the model of interest.

   (1) Graph the fitted model.

   (2) Interpret the estimated intercept and slope.

   (3) Interpret the estimated standard deviation of the error term.

   (4) Evaluate the usefulness of the model.

---

*The assumption that the random errors are independent is debatable for time series data. For the purposes of illustration, we assume they are approximately independent. The problem of dependent errors is discussed in Chapter 14.

(5) Estimate the mean median income for all zip codes having 15% college graduates using a 95% confidence interval.

(6) Predict the median income for a particular zip code having 15% college graduates using a 95% prediction interval.

b. Repeat part **a** using the entire set of 1,000 zip codes. Compare the results to those you obtained in part **a**.

REFERENCES

ASHRAE Guide and Data Book: Applications. New York: American Society of Heating, Refrigerating, and Air-Conditioning Engineers, Inc., 1968. Section IV.

Cho, C. C. and Reichert, A. "An application of multiple regression analysis for appraising single-family housing values." *Business Economics*, Jan. 1980, *15*, 47–52.

Cook, T. Q. (ed.) *Instruments of the Money Market*, 4th ed. Richmond, Va.: Federal Reserve Bank of Richmond, 1977.

Crandall, J. S. and Cedercreutz, M. "Preliminary cost estimates for mechanical work." *Building Systems Design*, Oct.–Nov. 73, 35–51.

Deming, W. E. *Quality, Productivity, and Competitive Position.* Cambridge, Mass.: MIT Center For Advanced Engineering Study, 1982.

Draper, N. and Smith, H. *Applied Regression Analysis*, 2nd ed. New York: Wiley, 1981.

Glueck, W. F. *Management.* Hinsdale, Ill.: Dryden Press, 1977.

Gray, J. and Johnston, K. S. *Accounting and Management Action*, 2nd ed. New York: McGraw-Hill, 1977. pp. 267–268.

Hartwig, F. and Dearing, B. E. *Exploratory Data Anaylsis.* Beverly Hills, Calif.: Sage, 1979. pp. 23–24.

Koch, P. D. and Ragan, J. F., Jr. "Investigating the causal relationship between quits and wages: An exercise in comparative dynamics." *Economic Inquiry*, Vol. 24 (January 1986), 61–83.

Landro, L. and Mayer, J. "Cable-TV viewing study dims prospect of large increase in number of channels." *Wall Street Journal*, Nov. 16, 1982, 10.

Leftwich, R. H. *The Price System and Resource Allocation*, 5th ed. Hinsdale, Ill.: Dryden Press, 1973. p. 123.

Leiter, M. P. and Meechan, K. A. "Role structure and burnout in the field of human services." *Journal of Applied Behavioral Science*, Vol. 22, No. 1, 1986, 47–52.

Luthans, F., Rosenkrantz, S. A., and Hennessey, H. W. "What do successful managers really do? An observational study of managerial activities." *Journal of Applied Behavioral Science*, Vol. 21, No. 3 (August 1985), 255–270.

Mansfield, E. *Economics*, 3rd ed. New York: Norton, 1980. p. 166.

Mendenhall, W. and McClave, J. T. *A Second Course in Business Statistics: Regression Analysis*, 2nd ed. San Francisco: Dellen, 1986.

Miller, R. B. and Wichern, D. W. *Intermediate Business Statistics: Analysis of Variance, Regression, and Time Series.* New York: Holt, Rinehart and Winston, 1977. Chapter 5.

Mintzberg, H. *The Nature of Managerial Work.* New York: Harper and Row, 1973.

Near, J. P. and Miceli, M. P. "Retaliation against whistle blowers: Predictors and effects." *Journal of Applied Psychology*, Vol. 71, No. 1, 1986, 137–145.

Neter, J., Wasserman, W., and Kutner, M. H. *Applied Linear Regression Models.* Homewood, Ill.: Richard D. Irwin, 1983.

Ruback, R. S. "The Conoco takeover and stockholder returns." *Sloan Management Review*, Winter 1982, 23, 13–33.

Tukey, J. W. *Exploratory Data Analysis.* Reading, Mass.: Addison-Wesley, 1977. Chapter 2.

Weisberg, S. *Applied Linear Regression.* New York: Wiley, 1980.

Wright, J. L., Owen, F., and Pena, D. "Status of MN/DOT's weigh-in-motion program." St. Paul: Minnesota Department of Transportation, Jan. 1983.

Younger, M. S. *A Handbook for Linear Regression.* North Scituate, Mass.: Duxbury, 1979.

WHERE WE'VE BEEN . . .

In Chapter 10 we demonstrated how to model the relationship between a dependent variable, y, and an independent variable, x, using a straight line. We fit the straight line to the data points, used r and r^2 to measure the strength of the relationship between y and x, and used the resulting prediction equation to estimate the mean value of y or to predict some future value of y for a given value of x.

WHERE WE'RE GOING . . .

This chapter converts the basic concept of Chapter 10 into a powerful estimation and prediction device by modeling the mean value of y as a function of two or more independent variables. This will enable you to model a response, y (say, the assessed value of a house), as a function of quantitative variables (such as floor space and age of the house) or as a function of qualitative variables (such as type of construction and location). As in the case of simple linear regression, multiple regression analysis includes fitting the model to a data set, testing the utility of the model, and using it for the estimation of the mean value of y for given values of the independent variables. We also use the model to predict some particular value of y to be observed in the future.

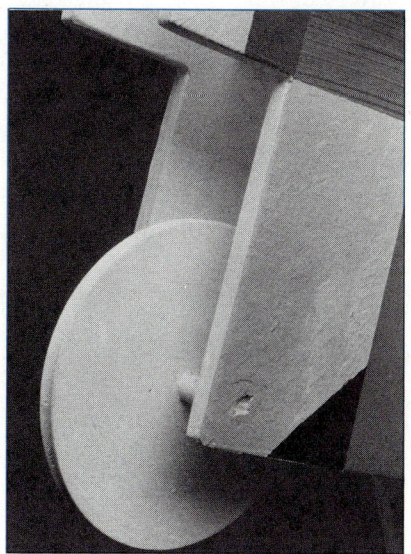

MULTIPLE REGRESSION

CONTENTS

A MULTIPLE REGRESSION ANALYSIS: THE MODEL AND THE PROCEDURE

Most practical applications of regression use models that are more complex than the first-order (straight-line) model. For example, a realistic probabilistic model for monthly sales revenue would include more than just the advertising expenditure discussed in Chapter 10 in order to provide a good predictive model for sales. Factors such as season, inventory on hand, sales force, and price are a few of the many variables that might influence sales. Thus, we would want to incorporate these and other potentially important independent variables into the model if we need to make accurate predictions.

Probabilistic models that include terms involving x^2, x^3 (or higher-order terms), or more than one independent variable are called **multiple regression models**. The general form of these models is

$$y = \beta_0 + \beta_1 x_1 + \beta_2 x_2 + \cdots + \beta_k x_k + \varepsilon$$

The dependent variable, y, is now written as a function of k independent variables, $x_1, x_2, \ldots, x_k$. The random error term is added to make the model probabilistic rather than deterministic. The value of the coefficient β_i determines the contribution of the independent variable x_i, given that the other x variables are held constant, and β_0 is the y-intercept. The coefficient $\beta_0, \beta_1, \ldots, \beta_k$ will usually be unknown, because they represent population parameters.

At first glance it might appear that the regression model shown above would not allow for anything other than straight-line relationships between y and the independent variables, but this is not true. Actually, $x_1, x_2, \ldots, x_k$ can be functions of variables as long as the functions do not contain unknown parameters. For example, the dollar sales, y, of new housing in a region could be a function of the independent variables

$x_1 =$ Mortgage interest rate

$x_2 =$ (Mortgage interest rate)$^2 = x_1^2$

$x_3 =$ Unemployment rate in the region

and so on. You could even insert a cyclical term (if it would be useful) of the form $x_4 = \sin t$, where t is a time variable. The multiple regression model is quite versatile and can be made to model many different types of response variables.

The same steps we followed in developing a straight-line model are applicable to the multiple regression model.

STEP 1 Hypothesize the form of the model. This involves the choice of the independent variables to be included in the model.

STEP 2 Estimate the unknown parameters $\beta_0, \beta_1, \ldots, \beta_k$.

STEP 3 Specify the probability distribution of the random error component ε and estimate its variance, σ^2.

STEP 4 Check the utility of the model.

STEP 5 Use the fitted model to estimate the mean value of y or to predict a particular value of y for given values of the independent variables.

The initial step—hypothesizing the form of the model—is the subject of Chapter 12. In this chapter we assume that the form of the model is known, and we will discuss steps 2–5 for a given model.

CASE STUDY 11.1

PREDICTING CORPORATE
EXECUTIVE
COMPENSATION

Towers, Perrin, Forster & Crosby (TPF&C), an international management consulting firm, has developed a unique and interesting application of multiple regression analysis. Many firms are interested in evaluating their management salary structure, and TPF&C uses multiple regression models to accomplish this salary evaluation. The Compensation Management Service, as TPF&C calls it, measures both the internal and external consistency of a company's pay policies to determine whether they reflect management's intent.

The dependent variable, y, used to represent executive compensation is annual salary. The independent variables used to explain salary structure include the executive's age, education, rank, and bonus eligibility; number of employees under the executive's direct supervision; as well as variables that describe the company for which the executive works, such as annual sales, profit, and total assets.

The initial step in developing models for executive compensation is to obtain a sample of executives from various client firms, which TPF&C calls the Compensation Data Bank. The data for these executives are used to estimate the model coefficients (the β parameters), and these estimates are then substituted into the linear model to form a prediction equation. To predict a particular executive's compensation, TPF&C substitutes into the prediction equation the values of the independent variables that pertain to the executive (the executive's age, rank, etc.). This application of multiple regression analysis is developed more fully in Section 11.8.

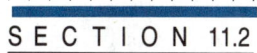

S E C T I O N 11.2

MODEL ASSUMPTIONS

We noted in Section 11.1 that the multiple regression model is of the form

$$y = \beta_0 + \beta_1 x_1 + \beta_2 x_2 + \cdots + \beta_k x_k + \varepsilon$$

where y is the response variable that you wish to predict; $\beta_0, \beta_1, \ldots, \beta_k$ are parameters with unknown values; $x_1, x_2 \ldots, x_k$ are information-contributing variables that are measured without error; and ε is a random error component. Since $\beta_0, \beta_1, \ldots, \beta_k$ and $x_1, x_2, \ldots, x_k$ are nonrandom, the quantity

$$\beta_0 + \beta_1 x_1 + \beta_2 x_2 + \cdots + \beta_k x_k$$

represents the deterministic portion of the model. Therefore, y is composed of two components—one fixed and one random—and, consequently, y is a random variable.

$$y = \underbrace{\beta_0 + \beta_1 x_1 + \cdots + \beta_k x_k}_{\substack{\text{Deterministic} \\ \text{portion of model}}} + \underbrace{\varepsilon}_{\substack{\text{Random} \\ \text{error}}}$$

We will assume (as in Chapter 10) that the random error can be positive or negative and that for any setting of the x values, $x_1, x_2, \ldots, x_k$, the random error ε has a normal probability distribution with mean equal to 0 and variance equal to σ^2. Further, we assume that the random errors associated with any (and every) pair of y values are probabilistically independent. That is, the error, ε, associated with any one y value is independent of the error associated with any other y value.

These assumptions are summarized as follows:

ASSUMPTIONS FOR RANDOM ERROR ε

1. For any given set of values of x_1, x_2, . . . , x_k, the random error ε has a normal probability distribution with mean equal to 0 and variance equal to σ^2.
2. The random errors are independent (in a probabilistic sense).

The assumptions that we have described for a multiple regression model imply that the mean value, $E(y)$, for a given set of values of x_1, x_2, . . . , x_k is equal to

$$E(y) = \beta_0 + \beta_1 x_1 + \beta_2 x_2 + \cdots + \beta_k x_k$$

Models of this type are called **linear** statistical models because $E(y)$ is a **linear function** of the unknown parameters β_0, β_1, . . . , β_k.

All the estimation and statistical test procedures described in this chapter depend on the data satisfying the assumptions described in this section. Since we will rarely, if ever, know for certain whether this occurs, we will want to know how well a regression analysis works and how much faith we can place in our inferences when certain assumptions are not satisfied. We will have more to say on this topic after we discuss the methods of a regression analysis more thoroughly and have shown how they are used in a practical situation.

S E C T I O N   11.3

FITTING THE MODEL: THE METHOD OF LEAST SQUARES

The method of fitting multiple regression models is identical to that of fitting the first-order (straight-line) model—namely, the method of least squares. That is, we choose the estimated model

$$\hat{y} = \hat{\beta}_0 + \hat{\beta}_1 x_1 + \cdots + \hat{\beta}_k x_k$$

that minimizes

$$\text{SSE} = \sum (y_i - \hat{y}_i)^2$$

As in the case of the straight-line model, the sample estimates $\hat{\beta}_0$, $\hat{\beta}_1$, . . . , $\hat{\beta}_k$ will be obtained as solutions to a set of simultaneous linear equations.*

The primary difference between fitting the simple and multiple regression models is computational difficulty. The $(k + 1)$ simultaneous linear equations that must be solved to find the $(k + 1)$ estimated coefficients $\hat{\beta}_0$, $\hat{\beta}_1$, . . . , $\hat{\beta}_k$ are often difficult (sometimes impossible) to solve with a pocket or desk calculator. Consequently, we resort to the use of computers. Many computer packages have been developed to fit a multiple regression model by the method of least squares. We

---

*Students who are familiar with calculus should note that $\hat{\beta}_0$, $\hat{\beta}_1$, . . . , $\hat{\beta}_k$ are the solutions to the set of equations $\partial\text{SSE}/\partial\hat{\beta}_0 = 0$, $\partial\text{SSE}/\partial\hat{\beta}_1 = 0$, . . . , $\partial\text{SSE}/\partial\hat{\beta}_k = 0$. The solution, given in matrix notation, is presented in Mendenhall and Sincich (1986) as well as other texts listed in the references at the end of the chapter.

will present output from several of the more popular computer packages, commencing, in our first example, with the computer output from the SAS System. Since the SAS regression output is similar to that of most other package regression programs, you should have little trouble interpreting regression output from other packages as you encounter them in future examples and exercises, at your computer center, or in using a microcomputer.

To illustrate, suppose we theorize that monthly electrical usage, y, in all-electric homes is related to the size, x, of the home by the model $y = \beta_0 + \beta_1 x + \beta_2 x^2 + \varepsilon$. To estimate the unknown parameters β_0, β_1, and β_2, values of y and x were collected for each of ten homes during a particular month. The data are shown in Table 11.1.

TABLE 11.1

SIZE OF HOME x (square feet)	MONTHLY USAGE y (kilowatt-hours)	SIZE OF HOME x (square feet)	MONTHLY USAGE y (kilowatt-hours)
1,290	1,182	1,840	1,711
1,350	1,172	1,980	1,804
1,470	1,264	2,230	1,840
1,600	1,493	2,400	1,956
1,710	1,571	2,930	1,954

Notice that we include a term involving x^2 in the model above because we expect curvature in the graph of the response model relating y to x. The term involving x^2 is called a **second-order**, or **quadratic**, term. Figure 11.1 illustrates that the electrical usage appears to increase in a curvilinear manner with the size of the home. This provides some support for the inclusion of the second-order term, x^2, in the model.

FIGURE 11.1

Scattergram of the Home Size–Electrical Usage Data

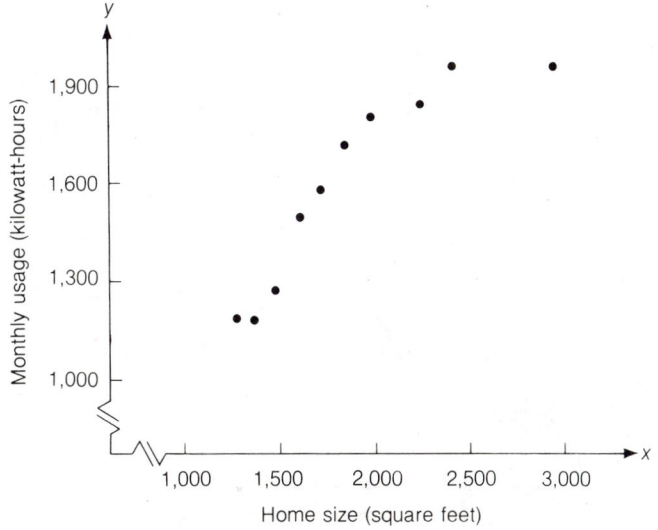

Part of the output from the SAS multiple regression routine for the data in Table 11.1 is reproduced in Figure 11.2. The least squares estimates of the β parameters appear in the column labeled ESTIMATE. You can see that $\hat{\beta}_0 = -1,216.1$, $\hat{\beta}_1 = 2.3989$, and $\hat{\beta}_2 = -.00045$. Therefore, the equation that minimizes the SSE for the data is

$$\hat{y} = -1,216.1 + 2.3989x - .00045x^2$$

The minimum value of SSE, 15,332.6, also appears in the printout. [*Note:* Much detail on the printout has not yet been discussed. We will continue throughout this chapter to shade the aspects of the printout that are under discussion.]

FIGURE 11.2   SAS Computer Printout for the Home Size–Electrical Usage Data

SOURCE	DF	SUM OF SQUARES	MEAN SQUARE	F VALUE	PR > F
MODEL	2	831069.54637065	415534.77318533	189.71	0.0001
ERROR	7	15332.55362935	2190.36480419		ROOT MSE
CORRECTED TOTAL	9	846402.10000000		R-SQUARE	46.8013333
				0.981885	

PARAMETER	ESTIMATE	T FOR HO: PARAMETER = 0	PR > ¦T¦	STD ERROR OF ESTIMATE
INTERCEPT	-1216.14388700	-5.01	0.0016	242.80636850
X	2.39893018	9.76	0.0001	0.24583560
X*X	-0.00045004	-7.62	0.0001	0.00005908

Note that the graph of the multiple regression model (Figure 11.3, a response curve) provides a good fit to the data of Table 11.1. Furthermore, the small value of $\hat{\beta}_2$ does *not* imply that the curvature is insignificant, since the numerical value of $\hat{\beta}_2$ is dependent on the scale of the measurements. We will test the contribution of the second-order coefficient β_2 in Section 11.5.

FIGURE 11.3

Least Squares Model for the Home Size–Electrical Usage Data

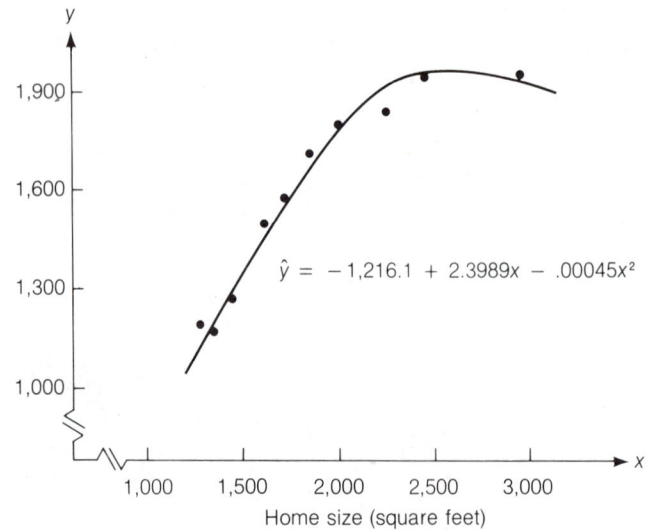

$\hat{y} = -1,216.1 + 2.3989x - .00045x^2$

Home size (square feet)

The ultimate goal of this multiple regression analysis is to use the fitted model to predict electrical usage, y, for a home of a specific size (area), x. And, of course, we will want to give a prediction interval for y so we will know how much faith we can place in the prediction. That is, if the prediction model is used to predict electrical usage, y, for a given size of home, x, what will be the error of prediction? To answer this question, we need to estimate σ^2, the variance of ε.

S E C T I O N 11.4

ESTIMATION OF σ^2, THE VARIANCE OF ε

You will recall that σ^2 is the variance of the random error, ε. If $\sigma^2 = 0$, all the random errors will equal 0 and the predicted values, $\hat{y}$, will be identical to $E(y)$; that is $E(y)$ will be estimated without error. In contrast, a large value of σ^2 implies large (absolute) values of ε and larger deviations between the predicted values, $\hat{y}$, and the mean value, $E(y)$. Consequently, the larger the value of σ^2, the greater will be the error in estimating the model parameters $\beta_0, \beta_1, \ldots, \beta_k$ and the error in predicting a value of y for a specific set of values of $x_1, x_2, \ldots, x_k$. Thus, σ^2 plays a major role in making inferences about $\beta_0, \beta_1, \ldots, \beta_k$, in estimating $E(y)$, and in predicting y for specific values of $x_1, x_2, \ldots, x_k$.

Since the variance, σ^2, of the random error, ε, will rarely be known, we must use the results of the regression analysis to estimate its value. You will recall that σ^2 is the variance of the probability distribution of the random error, ε, for a given set of values for $x_1, x_2, \ldots, x_k$, and hence that it is the mean value of the squares of the deviations of the y values (for given values of $x_1, x_2, \ldots, x_k$) about the mean value $E(y)$.* Since the predicted value, $\hat{y}$, estimates $E(y)$ for each of the data points, it seems natural to use

$$SSE = \sum (y_i - \hat{y}_i)^2$$

to construct an estimator of σ^2.

For example, in the second-order model describing electrical usage as a function of home size, we found that $SSE = 15,332.6$. We now want to use this quantity to estimate the variance of ε. Recall that the estimator for the straight-line model was $s^2 = SSE/(n - 2)$ and note that the denominator is $(n - $ Number of estimated β parameters), which is $(n - 2)$ in the first-order (straight-line) model. Since we must estimate one more parameter, β_2, for the second-order model, the estimator of σ^2 is

$$s^2 = \frac{SSE}{n - 3}$$

That is, the denominator becomes $(n - 3)$ because there are now three β parameters in the model. The numerical estimate for this example is

$$s^2 = \frac{SSE}{10 - 3} = \frac{15,332.6}{7} = 2,190.36$$

---

*Recall from Section 11.2 that $y = E(y) + \varepsilon$. Therefore, ε is equal to the deviation $y - E(y)$. Also, by definition, the variance of a random variable is the expected value of the square of the deviation of the random variable from its mean. According to our model, $E(\varepsilon) = 0$. Therefore, $\sigma^2 = E(\varepsilon^2)$.

In many computer printouts and textbooks, s^2 is called the **mean square for error (MSE)**. This estimate of σ^2 is shown in the column titled MEAN SQUARE in the SAS printout in Figure 11.2.

For the general multiple regression model

$$y = \beta_0 + \beta_1 x_1 + \beta_2 x_2 + \cdots + \beta_k x_k + \varepsilon$$

we must estimate the $(k + 1)$ parameters $\beta_0, \beta_1, \beta_2, \ldots, \beta_k$. Thus, the estimator of σ^2 is SSE divided by the quantity $(n -$ Number of estimated β parameters$)$.

We will use the estimator of σ^2 both to check the utility of the model (Sections 11.5 and 11.6) and to provide a measure of reliability of predictions and estimates when the model is used for those purposes (Section 11.7). Thus, you can see that the estimation of σ^2 plays an important part in the development of a regression model.

ESTIMATOR OF σ^2 FOR MULTIPLE REGRESSION MODEL WITH k INDEPENDENT VARIABLES

$$\text{MSE} = \frac{\text{SSE}}{n - \text{Number of estimated } \beta \text{ parameters}} = \frac{\text{SSE}}{n - (k + 1)}$$

SECTION 11.5

ESTIMATING AND TESTING HYPOTHESES ABOUT THE β PARAMETERS

Sometimes the individual β parameters in a model have particular practical significance, and we want to estimate their values or test hypotheses about them. For example, if electrical usage, y, is related to home size, x, by the straight-line relationship

$$y = \beta_0 + \beta_1 x_1 + \varepsilon$$

then β_1 has a very practical interpretation. That is, you saw in Chapter 10 that β_1 is the increase in mean kilowatt-hours of electrical usage, y, for a 1-square-foot increase in home size, x.

As proposed in the preceding sections, suppose that the electrical usage, y, is related to home size, x, by the quadratic model

$$y = \beta_0 + \beta_1 x + \beta_2 x^2 + \varepsilon$$

Then the mean value of y for a given value of x is

$$E(y) = \beta_0 + \beta_1 x + \beta_2 x^2$$

What is the practical interpretation of β_2? As noted earlier, the parameter β_2 measures the curvature of the response curve shown in Figure 11.3. If $\beta_2 > 0$, the slope of the curve will increase as x increases, as shown in Figure 11.4(a). If $\beta_2 < 0$, the slope of the curve will decrease as x increases, as shown in Figure 11.4(b).

Intuitively, we would expect the electrical usage, y, to rise almost proportionally to home size, x. Then, eventually, as the size of the home increases, the increase

FIGURE 11.4

The Interpretation of β_2 for a Second-Order Model

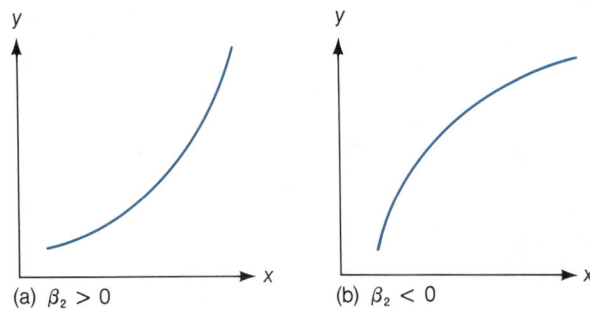

(a) $\beta_2 > 0$ (b) $\beta_2 < 0$

in electrical usage for a 1-unit increase in home size might begin to decrease. Thus, a forecaster of electrical usage would want to determine whether this type of curvature actually was present in the response curve, or, equivalently, the forecaster would want to test the null hypothesis

H_0: $\beta_2 = 0$ (No curvature in the response curve)

against the alternative hypothesis

H_a: $\beta_2 < 0$ (Downward curvature exists in the response curve)

A test of this hypothesis can be performed using a Student's t test.

The t test utilizes a test statistic analogous to that used to make inferences about the slope of the straight-line model (Section 10.5). The t statistic is formed by dividing the sample estimate, $\hat{\beta}_2$, of the parameter, β_2, by the estimated standard deviation of the sampling distribution of $\hat{\beta}_2$:

Test statistic: $t = \dfrac{\hat{\beta}_2}{s_{\hat{\beta}_2}}$

We use the symbol $s_{\hat{\beta}_2}$ to represent the estimated standard deviation of $\hat{\beta}_2$. The formula for computing $s_{\hat{\beta}_2}$ is very complex, but its computation is performed automatically as part of most standard multiple regression computer analyses. Thus, most computer packages list the estimated standard deviation $s_{\hat{\beta}_i}$ for each estimated model coefficient $\hat{\beta}_i$. In addition, they usually give the calculated t values for testing H_0: $\beta_i = 0$ for each coefficient in the model.

The rejection region for the test is found in exactly the same way as the rejection regions for the t tests in Chapters 8 and 9. That is, we consult Table VI in Appendix B to obtain an upper-tail value of t. This is a value, t_α, such that $P(t > t_\alpha) = \alpha$. We can then use this value to construct rejection regions for either one- or two-tailed tests. To illustrate, in the electrical usage example, the error degrees of freedom is $(n - 3) = 7$, the denominator of the estimate of σ^2. Then the rejection region for a one-tailed test with $\alpha = .05$ is

Rejection region: $t < -t_\alpha$: $\alpha = .05$,   df $= 7$
$t < -1.895$ (see Figure 11.5 on the next page)

FIGURE 11.5
Rejection Region for Test
of β_2

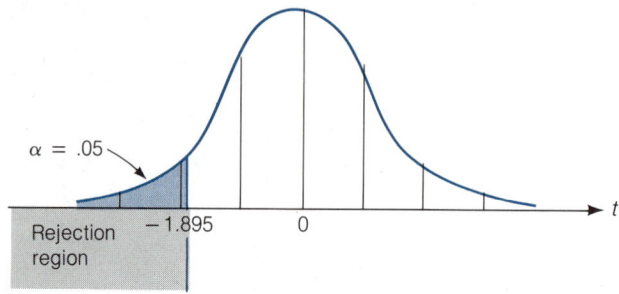

$\alpha = .05$

Rejection region -1.895 0 t

In Figure 11.6 we again show a portion of the SAS printout for the electrical usage example. The following quantities are shaded:

1. The estimated coefficients, $\hat{\beta}_0$, $\hat{\beta}_1$, and $\hat{\beta}_2$
2. The SSE
3. The MSE (estimate of σ^2, the variance of ε)

FIGURE 11.6   SAS Output for the Home Size–Electrical Usage Data

SOURCE	DF	SUM OF SQUARES	MEAN SQUARE	F VALUE	PR > F
MODEL	2	831069.54637065	415534.77318533	189.71	0.0001
ERROR	7	15332.55362935	2190.36480419		ROOT MSE
CORRECTED TOTAL	9	846402.10000000		R-SQUARE	46.8013333
				0.981885	

PARAMETER	ESTIMATE	T FOR H0: PARAMETER = 0	PR > ¦T¦	STD ERROR OF ESTIMATE
INTERCEPT	-1216.14388700	-5.01	0.0016	242.80636850
X	2.39893018	9.76	0.0001	0.24583560
X*X	-0.00045004	-7.62	0.0001	0.00005908

The estimated standard deviations for the estimated model coefficients appear under the column labeled STD ERROR OF ESTIMATE. The t statistics for testing the null hypotheses that the coefficients β_0, β_1, . . . , β_k individually equal 0 appear under the column headed T FOR H0: PARAMETER = 0. The t value corresponding to the test of the null hypothesis H_0: $\beta_2 = 0$ is the last one in the column; i.e., $t = -7.62$. Since this value falls in the rejection region (i.e., it is less than -1.895), we conclude that the second-order term $\beta_2 x^2$ makes an important contribution to the prediction model of electrical usage.

The SAS printout shown in Figure 11.6 also lists the two-tailed observed significance levels (or p-values) for each t value. These values appear under the column headed PR > |T|. The observed significance level .0001 corresponds to the quadratic term, and this implies that we would reject H_0: $\beta_2 = 0$ in favor of H_a: β_2

$\neq 0$ at any α level larger than .0001. Since our alternative hypothesis was one-sided, H_a: $\beta_2 < 0$, the observed significance level is half that given in the printout; i.e., $\frac{1}{2}(.0001) = .00005$. Thus, there is very strong evidence that the mean electrical usage increases more slowly per square foot for large houses than for small houses.

We can also form a 95% confidence interval for the parameter β_2 as follows:

$$\hat{\beta}_2 \pm t_{\alpha/2}s_{\hat{\beta}_2} = -.000450 \pm (2.365)(.0000591)$$

or $(-.000590, -.000310)$. Note that the t value 2.365 corresponds to $\alpha/2 = .025$ and $(n - 3) = 7$ df. This interval constitutes a 95% confidence interval for β_2, the rate of change in curvature in mean electrical usage as home size is increased. Note that all values in the interval are negative, providing strong support for the conclusion of our test although the test was one-tailed, while the confidence interval is two-tailed.

Testing a hypothesis about a single β parameter that appears in any multiple regression model is accomplished in exactly the same manner as described for the second-order electrical usage model. The form of the t test is shown in the box.

TEST OF AN INDIVIDUAL PARAMETER COEFFICIENT IN THE MULTIPLE REGRESSION MODEL

$$y = \beta_0 + \beta_1 x_1 + \beta_2 x_2 + \cdots + \beta_k x_k + \varepsilon$$

ONE-TAILED TEST

H_0: $\beta_i = 0^*$

H_a: $\beta_i > 0$
(or $\beta_i < 0$)

Test statistic: $t = \dfrac{\hat{\beta}_i}{s_{\hat{\beta}_i}}$

Rejection region: $t > t_\alpha$
(or $t < -t_\alpha$)

TWO-TAILED TEST

H_0: $\beta_i = 0^*$

H_a: $\beta_i \neq 0$

Test statistic: $t = \dfrac{\hat{\beta}_i}{s_{\hat{\beta}_i}}$

Rejection region:
$t > t_{\alpha/2}$ or $t < -t_{\alpha/2}$

where

 n = Number of observations

 k = Number of independent variables in the model

and $t_{\alpha/2}$ is based on $[n - (k + 1)]$ df.

Assumptions: See Section 11.2 for the assumptions about the probability distribution of the random error component ε.

---

*To test the null hypothesis that a parameter, β_i, equals some value other than 0, say H_0: $\beta_i = \beta_{i0}$, use the test statistic $t = (\hat{\beta}_i - \beta_{i0})/s_{\hat{\beta}_i}$. All other aspects of the test will be as described in the box.

EXAMPLE 11.1

A collector of antique grandfather clocks believes that the price received for the clocks at an antique auction increases with the age of the clocks and with the number of bidders. Thus, the following model is hypothesized:

$$y = \beta_0 + \beta_1 x_1 + \beta_2 x_2 + \varepsilon$$

where

y = Auction price

x_1 = Age of clock (years) x_2 = Number of bidders

A sample of 32 auction prices of grandfather clocks, along with their age and the number of bidders, is given in Table 11.2. The model $y = \beta_0 + \beta_1 x_1 + \beta_2 x_2 + \varepsilon$ is fit to the data, and a portion of the SAS printout is shown in Figure 11.7. Test the hypothesis that the auction price increases as the number of bidders increases (and age is held constant), i.e., $\beta_2 > 0$. Use $\alpha = .05$.

TABLE 11.2
Auction Price Data

AGE x_1	NUMBER OF BIDDERS x_2	AUCTION PRICE y	AGE x_1	NUMBER OF BIDDERS x_2	AUCTION PRICE y
127	13	$1,235	170	14	$2,131
115	12	1,080	182	8	1,550
127	7	845	162	11	1,884
150	9	1,522	184	10	2,041
156	6	1,047	143	6	854
182	11	1,979	159	9	1,483
156	12	1,822	108	14	1,055
132	10	1,253	175	8	1,545
137	9	1,297	108	6	729
113	9	946	179	9	1,792
137	15	1,713	111	15	1,175
117	11	1,024	187	8	1,593
137	8	1,147	111	7	785
153	6	1,092	115	7	744
117	13	1,152	194	5	1,356
126	10	1,336	168	7	1,262

FIGURE 11.7 SAS Printout for Example 11.1

SOURCE	DF	SUM OF SQUARES	MEAN SQUARE	F VALUE	PR > F
MODEL	2	4277159.70340504	2138579.85170252	120.65	0.0001
ERROR	29	514034.51534496	17725.32811534		ROOT MSE
CORRECTED TOTAL	31	4791194.21875000	R-SQUARE		133.13650181
			0.892713		

PARAMETER	ESTIMATE	T FOR H0: PARAMETER = 0	PR > ITI	STD ERROR OF ESTIMATE
INTERCEPT	-1336.72205214	-7.71	0.0001	173.35612607
X1	12.73619884	14.11	0.0001	0.90238049
X2	85.81513260	9.86	0.0001	8.70575681

SOLUTION

The hypothesis of interest concerns the parameter β_2. Specifically,

$$H_0: \quad \beta_2 = 0$$
$$H_a: \quad \beta_2 > 0$$

Test statistic: $t = \dfrac{\hat{\beta}_2}{s_{\hat{\beta}_2}}$

Rejection region: For $\alpha = .05$, $t > t_{.05}$
where df $= n - (k + 1) = 32 - 3 = 29$
or $t > 1.699$ (see Figure 11.8)

FIGURE 11.8
Rejection Region for
$H_0: \beta_2 = 0$

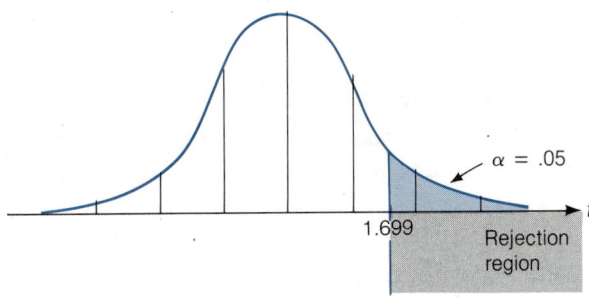

The calculated t value, $t = 9.86$, is indicated in Figure 11.7. This value exceeds 1.699 and therefore falls in the rejection region. Thus, the collector can conclude that the mean auction price of the clocks increases as the number of bidders increases, when age is held constant.

Note that the values $\hat{\beta}_1 = 12.74$ and $\hat{\beta}_2 = 85.82$ (shaded in Figure 11.7) are easily interpreted. We estimate that the mean auction price increases \$12.74 per year of age of the clock, for a fixed number of bidders, and the mean price increases by \$85.82 per additional bidder for a fixed age clock.   ■

Be careful not to try to interpret the estimated intercept $\hat{\beta}_0 = -1,336.72$ in the same way we interpreted $\hat{\beta}_1$ and $\hat{\beta}_2$ in Example 11.1. You might think that this implies a negative price for clocks 0 years of age with 0 bidders. However, these zeros are meaningless numbers in this example, since the ages range from 108 to 194 and the number of bidders ranges from 5 to 15. Keep in mind that we are modeling y within the range of values observed for the predictor variables and that interpretations of the models for values of the independent variables outside their sampled ranges can be very misleading.

Some computer programs use an F test to test hypotheses concerning the individual β parameters. If you conduct a two-tailed t test and reject the hypothesis if $t > t_{\alpha/2}$ or $t < -t_{\alpha/2}$, then the corresponding F test will imply rejection if the computed value of F (which is equal to the square of the computed t statistic) is larger than F_α, because the square of the Student's t with ν degrees of freedom is equal to an F statistic with 1 df in the numerator and ν df in the denominator.

Thus, $t_{\alpha/2}^2 = F_\alpha$, where t is based on ν df and F has 1 numerator and ν denominator degrees of freedom, respectively. As an example, when we tested the curvature parameter β_2 in the second-order model relating electrical usage to home size, the computed t value was -7.62 (see Figure 11.6). The equivalent F statistic yields

$$F = t^2 = (-7.62)^2 = 58.06$$

Suppose we wanted to conduct a two-tailed statistical test of H_0: $\beta_2 = 0$ against H_a: $\beta_2 \neq 0$. The upper-tail rejection region for a two-tailed test with $\alpha = .05$ is

$$F > F_{.05} \quad \text{where } F_{.05} \text{ is based on } \nu_1 = 1 \text{ df, } \nu_2 = 7 \text{ df}$$

or

$$F > 5.59$$

Note that the F value, 5.59, is equal to the square of 2.365, the value of t that corresponds to $t_{.025}$ with 7 df. In other words, you can conduct a two-tailed test of the null hypothesis H_0: $\beta_i = 0$ using either a two-tailed t test or a one-tailed F test. If you want to conduct a one-tailed test to detect H_a: $\beta_i > 0$ (or H_a: $\beta_i < 0$), the F test is not appropriate. You will have to conduct the test using a t statistic.

EXERCISES 11.1–11.11   LEARNING THE MECHANICS

[Note: Starred (*) exercises require the use of a computer.]

11.1 The SAS System was used to fit the model $y = \beta_0 + \beta_1 x_1 + \beta_2 x_2 + \varepsilon$ to $n = 20$ data points and the accompanying printout was obtained.

SAS Printout for
Exercise 11.1

Dep Variable: Y

Analysis of Variance

Source	DF	Sum of Squares	Mean Square	F Value	Prob>F
Model	2	128329.27624	64164.63812	7.223	0.0054
Error	17	151015.72376	8883.27787		
C Total	19	279345.00000			

Root MSE	94.25114	R-Square	0.4594
Dep Mean	360.50000	Adj R-Sq	0.3958
C.V.	26.14456		

Parameter Estimates

Variable	DF	Parameter Estimate	Standard Error	T for H0: Parameter=0	Prob > |T|
INTERCEP	1	506.346067	45.16942487	11.210	0.0001
X1	1	-941.900226	275.08555975	-3.424	0.0032
X2	1	-429.060418	379.82566485	-1.130	0.2743

a. What are the sample estimates of β_0, β_1, and β_2?
b. What is the least squares prediction equation?

c. Find SSE, MSE, and s. Interpret the standard deviation in the context of the problem.

d. Test H_0: $\beta_1 = 0$ against H_a: $\beta_1 \neq 0$. Use $\alpha = .05$.

e. Use a 95% confidence interval to estimate β_2.

11.2 Suppose you fit the multiple regression model

$$y = \beta_0 + \beta_1 x_1 + \beta_2 x_2 + \beta_3 x_3 + \varepsilon$$

to $n = 30$ data points and obtained the following result:

$$\hat{y} = 3.4 - 4.6x_1 + 2.7x_2 + .93x_3$$

The estimated standard errors of $\hat{\beta}_2$ and $\hat{\beta}_3$ are 1.86 and .29, respectively.

a. Test the null hypothesis H_0: $\beta_2 = 0$ against the alternative hypothesis H_a: $\beta_2 \neq 0$. Use $\alpha = .05$.

b. Test the null hypothesis H_0: $\beta_3 = 0$ against the alternative hypothesis H_a: $\beta_3 \neq 0$. Use $\alpha = .05$.

c. The null hypothesis H_0: $\beta_2 = 0$ is not rejected. In contrast, the null hypothesis H_0: $\beta_3 = 0$ is rejected. Explain how this can happen even though $\hat{\beta}_2 > \hat{\beta}_3$.

11.3 Suppose you fit the second-order model,

$$y = \beta_0 + \beta_1 x + \beta_2 x^2 + \varepsilon$$

to $n = 25$ data points. Your estimate of β_2 is $\hat{\beta}_2 = .47$, and the estimated standard error of the estimate is $s_{\hat{\beta}_2} = .15$.

a. Test the null hypothesis that the mean value of y is related to x by the (*first-order*) linear model

$$E(y) = \beta_0 + \beta_1 x$$

(H_0: $\beta_2 = 0$) against the alternative hypothesis that the true relationship is given by the quadratic model (a *second-order* linear model),

$$E(y) = \beta_0 + \beta_1 x + \beta_2 x^2$$

(H_a: $\beta_2 \neq 0$). Use $\alpha = .05$.

b. Suppose you wanted to determine only whether the quadratic curve opens upward; i.e., as x increases, the slope of the curve increases. Give the test statistic and the rejection region for the test for $\alpha = .05$. Do the data support the theory that the slope of the curve increases as x increases? Explain.

c. What is the value of the F statistic for testing the null hypothesis H_0: $\beta_2 = 0$ against H_a: $\beta_2 \neq 0$?

d. Could the F statistic in part c be used to conduct the test in part b? Explain.

11.4 How is the number of degrees of freedom available for estimating σ^2, the variance of ε, related to the number of variables in a regression model?

APPLYING THE CONCEPTS

11.5 Economists have two major types of data available to them: **time series data** and **cross-sectional data**. For example, an economist estimating a consumption function, say, household food consumption as a function of household income and household size, might measure the variables of interest for a particular sample of households at a particular point in time. In this case, the economist is using *cross-sectional data*. If instead, the economist is interested in how total consumption in the United States is related to national income, the economist probably would track these variables over time. In this case, the economist is using *time series data* (Wonnacott and Wonnacott, 1979). The cross-sectional data in the table have been collected for a random sample of 25 households in Washington, D.C.

HOUSEHOLD	FOOD CONSUMPTION DURING 1987 ($ thousands)	1987 HOUSEHOLD INCOME ($ thousands)	NUMBER OF PERSONS IN HOUSEHOLD AT END OF 1987	HOUSEHOLD	FOOD CONSUMPTION DURING 1987 ($ thousands)	1987 HOUSEHOLD INCOME ($ thousands)	NUMBER OF PERSONS IN HOUSEHOLD AT END OF 1987
1	3.2	31.1	4	14	3.1	85.2	2
2	2.4	20.5	2	15	4.5	35.6	9
3	3.8	42.3	4	16	3.5	68.5	3
4	1.9	18.9	1	17	4.0	10.5	5
5	2.5	26.5	2	18	3.5	21.6	4
6	3.0	29.8	4	19	1.8	29.9	1
7	2.6	24.3	3	20	2.9	28.6	3
8	3.2	38.1	4	21	2.6	20.2	2
9	3.9	52.0	5	22	3.6	38.7	5
10	1.7	16.0	1	23	2.8	11.2	3
11	2.9	41.9	3	24	4.5	14.3	7
12	1.7	9.9	1	25	3.5	16.9	5
13	4.5	33.1	7				

a. It has been hypothesized that household food consumption, y, is related to household income, x_1, and to the size of the household, x_2, as follows:

$$y = \beta_0 + \beta_1 x_1 + \beta_2 x_2 + \varepsilon$$

The SAS computer printout for fitting the model to the data is shown here. Give the least squares prediction equation.

SAS Printout for Exercise 11.5

DEPENDENT VARIABLE: FOOD

SOURCE	DF	SUM OF SQUARES	MEAN SQUARE	F VALUE
MODEL	2	15.46228509	7.73114255	100.80
ERROR	22	1.68731491	0.07669613	
CORRECTED TOTAL	24	17.14960000		PR > F
				0.0001

R-SQUARE	C.V.	ROOT MSE	FOOD MEAN
0.901612	8.9221	0.27694067	3.10400000

PARAMETER	ESTIMATE	T FOR H0: PARAMETER=0	PR > ¦T¦	STD ERROR OF ESTIMATE
INTERCEPT	1.43260377	9.76	0.0001	0.14673751
INCOME	0.00999062	3.15	0.0046	0.00316806
SIZE	0.37928986	13.68	0.0001	0.02772472

b. Do the data provide sufficient evidence to conclude that food consumption increases with household income? Test using $\alpha = .01$.

c. As a check on your conclusion of part b, construct a scattergram of household food consumption versus household income. Does the plot support your conclusion in part b? Explain.

d. In Chapter 10, we used the method of least squares to fit a straight line to a set of data points that were plotted in two dimensions. In this exercise, we are fitting a plane to a set of points plotted in three dimensions. We are attempting to determine the plane, $\hat{y} = \hat{\beta}_0 + \hat{\beta}_1 x_1 + \hat{\beta}_2 x_2$, that, according to the principle of least squares, best fits the data points. Sketch the least squares plane you developed in part a. Be sure to label all three axes of your graph.

11.6 Henry and Haynes (1978) report that in the middle 1970's Data Resources, Inc. (DRI), a firm that supplies economic information analyses and advice to government, industry, and financial institutions, used multiple regression to develop a model that characterized a particular bank's demand for mortgage loans. Working with quarterly time series data on the variables described here, DRI obtained the following least squares model:

$$\hat{y} = 37,350.40 + .61x_1 - 155.74x_2 + 19,934.7x_3 - 5,354.9x_4 + 5,317.61x_5$$
$$\quad\quad (3.5) \quad\quad (5.5) \quad\quad (-2.8) \quad\quad (4.5) \quad\quad (-3.9) \quad\quad (2.7)$$

where

y = Mortgage loan demand (in dollars)

x_1 = Seasonally adjusted mortgage loans outstanding during previous period

x_2 = Deposits at mutual savings banks and savings and loan associations

x_3 = Average number of housing starts per month

x_4 = State's rate of unemployment

x_5 = Conventional mortgage loan interest rate

The numbers in parentheses are the t statistics associated with the estimates of the model coefficients above them. Assume $n = 28$. The above model, along with one for the demand for commercial loans and another for the demand for installment loans, became the heart of the bank's loan-demand forecasting system.

a. Find the estimated standard deviation of $\hat{\beta}_2$.

b. Prior to fitting the model, DRI hypothesized that β_2 should be negative because x_2 represents an alternative source of mortgage money available to the consumer. Do the data support this hypothesis? Test using $\alpha = .01$. State your conclusion in the context of the problem.

c. Report the approximate p-value of your test.

d. Provide an economic explanation for why it is reasonable to expect β_4 to be negative.

11.7 To run a manufacturing operation efficiently, it is necessary to know how long it takes employees to manfacture the product. Without such information, the cost of making the product cannot be determined. Furthermore, management would not be able to establish an effective incentive plan for its employees because it would not know how to set work standards (Chase and Aquilano, 1979). Estimates of production time are frequently obtained using time studies. The data in the table at the top of page 572 were obtained from a recent time study of a sample of 15 employees on an automobile assembly line.

TIME TO COMPLETE TASK y (minutes)	MONTHS OF EXPERIENCE x	TIME TO COMPLETE TASK y (minutes)	MONTHS OF EXPERIENCE x
10	24	17	3
20	1	18	1
15	10	16	7
11	15	16	9
11	17	17	7
19	3	18	5
11	20	10	20
13	9		

a. The SAS computer printout for fitting the model, $y = \beta_0 + \beta_1 x + \beta_2 x^2 + \varepsilon$, is shown here. Find the least squares prediction equation.

```
DEPENDENT VARIABLE: Y

SOURCE                        DF    SUM OF SQUARES    MEAN SQUARE    F VALUE

MODEL                          2    156.11947722      78.05973861     65.59
ERROR                         12     14.28052278       1.19004356    PR > F
CORRECTED TOTAL               14    170.40000000                     0.0001


R-SQUARE                   C.V.             ROOT MSE          Y MEAN

0.916194                 7.3709          1.09089118       14.80000000


                                 T FOR HO:       PR > !T!     STD ERROR OF
PARAMETER        ESTIMATE     PARAMETER=0                      ESTIMATE

INTERCEPT      20.09110757         27.72        0.0001         0.72470507
X              -0.67052219         -4.33        0.0010         0.15470634
X*X             0.00953474          1.51        0.1576         0.00632580
```

b. Plot the fitted equation on a scattergram of the data. Is there sufficient evidence to support the inclusion of the quadratric term in the model? Explain.

c. Test the null hypothesis that $\beta_2 = 0$ against the alternative that $\beta_2 \neq 0$. Use $\alpha = .01$. Does the quadratic term make an important contribution to the model?

d. Your conclusion in part **c** should have been to drop the quadratic term from the model. Do so and fit the "reduced model," $y = \beta_0 + \beta_1 x + \varepsilon$, to the data.

e. Define β_1 in the context of this exercise. Find a 90% confidence interval for β_1 in the reduced model of part **d**.

11.8 A researcher wished to investigate the effects of several factors on production line supervisors' attitudes toward handicapped workers. A study was conducted involving 40 randomly selected supervisors. The response y, a supervisor's attitude toward handicapped workers, was measured with a standardized attitude scale. Independent variables used in the study were

$$x_1 = \begin{cases} 1 & \text{if the supervisor is female} \\ 0 & \text{if the supervisor is male} \end{cases}$$

x_2 = Number of years of experience in a supervisory job

The researcher fit the model

$$y = \beta_0 + \beta_1 x_1 + \beta_2 x_2 + \beta_3 x_2^2 + \varepsilon$$

to the data with the following results:

$$\hat{y} = 50 + 5x_1 + 5x_2 - .1x_2^2 \qquad s_{\hat{\beta}_3} = .03$$

a. Is there sufficient evidence to indicate that the quadratic term in years of experience, x_2^2, is useful for predicting attitude score? Use $\alpha = .05$.

b. Sketch the predicted attitude score, $\hat{y}$, as a function of the number of years of experience, x_2, for male supervisors ($x_1 = 0$). Next, substitute x_1 into the least squares equation and thereby obtain a plot of the prediction equation for female supervisors. [Note: For both males and females, plotting $\hat{y}$ for $x_2 = 0$, 2, 4, 6, 8, and 10 will produce a good picture of the prediction equations. The vertical distance between the males' and females' prediction curves is the same for all values of x_2.]

11.9 An employer has found that factory workers who are with the company longer tend to invest more in a company investment program per year than workers with less time in the company. The following model is believed to be adequate in modeling the relationship of annual amount invested, y, to years working for the company, x:

$$y = \beta_0 + \beta_1 x + \beta_2 x^2 + \varepsilon$$

The employer checks the records for a sample of 50 factory employees for a previous year, and fits the above model to get $\hat{\beta}_2 = .0015$ and $s_{\hat{\beta}_2} = .000712$. Test to determine whether the employer can conclude that $\beta_2 > 0$. Use $\alpha = .05$.

11.10 To project personnel needs for the Christmas shopping season, a department store wants to project sales for the season. The sales for the previous Christmas season are an indication of what to expect for the current season. However, the projection should also reflect the current economic environment by taking into consideration sales for a more recent period. The following model might be appropriate:

$$y = \beta_0 + \beta_1 x_1 + \beta_2 x_2 + \varepsilon$$

where

x_1 = Previous Christmas sales

x_2 = Sales for August of current year

y = Sales for current Christmas

Data for 10 previous years were used to fit the prediction equation, and the following were calculated (all units in thousands of dollars):

$$\hat{\beta}_1 = .62 \qquad s_{\hat{\beta}_1} = .273$$
$$\hat{\beta}_2 = .55 \qquad s_{\hat{\beta}_2} = .181$$

and the regression model standard deviation is $s = 2.5$.

a. Test to determine whether evidence exists to indicate that August sales contribute information for predicting Christmas sales.

b. If this model were used to predict Christmas sales, what would the approximate precision of the prediction be? [Note: We will show how to determine the precision of a multiple regression prediction in Section 11.7.]

*11.11 The owner of an apartment building in Minneapolis believed that her 1982 property tax bill was too high due to an overassessment of the property's value by the city tax assessor. The owner hired an independent real estate appraiser to investigate the appropriateness of the city's assessment. The appraiser used regression analysis to explore the relationship between the sale prices of apartments sold in Minneapolis during 1982 and various characteristics of the properties. Twenty-five apartment buildings were randomly sampled from all apartment buildings that were sold during 1982. The table lists the data collected by the appraiser. The real estate appraiser hypothesized that the sale price (i.e., market value) of an apartment building is related to the other variables in the table according to the following model:

$$y = \beta_0 + \beta_1 x_1 + \beta_2 x_2 + \beta_3 x_3 + \beta_4 x_4 + \beta_5 x_5 + \varepsilon$$

CODE NO.	SALE PRICE y ($)	NUMBER OF APARTMENT UNITS x_1	AGE OF STRUCTURE x_2 (years)	LOT SIZE x_3 (sq. ft.)	NUMBER OF ON-SITE PARKING SPACES x_4	GROSS BUILDING AREA x_5
0229	90,300	4	82	4,635	0	4,266
0094	384,000	20	13	17,798	0	14,391
0043	157,500	5	66	5,913	0	6,615
0079	676,200	26	64	7,750	6	34,144
0134	165,000	5	55	5,150	0	6,120
0179	300,000	10	65	12,506	0	14,552
0087	108,750	4	82	7,160	0	3,040
0120	276,538	11	23	5,120	0	7,881
0246	420,000	20	18	11,745	20	12,600
0025	950,000	62	71	21,000	3	39,448
0015	560,000	26	74	11,221	0	30,000
0131	268,000	13	56	7,818	13	8,088
0172	290,000	9	76	4,900	0	11,315
0095	173,200	6	21	5,424	6	4,461
0121	323,650	11	24	11,834	8	9,000
0077	162,500	5	19	5,246	5	3,828
0060	353,500	20	62	11,223	2	13,680
0174	134,400	4	70	5,834	0	4,680
0084	187,000	8	19	9,075	0	7,392
0031	155,700	4	57	5,280	0	6,030
0019	93,600	4	82	6,864	0	3,840
0074	110,000	4	50	4,510	0	3,092
0057	573,200	14	10	11,192	0	23,704
0104	79,300	4	82	7,425	0	3,876
0024	272,000	5	82	7,500	0	9,542

Source: Robinson Appraisal Co., Inc., Mankato, Minnesota.

a. Fit the real estate appraiser's model to the data in the table. Report the least squares prediction equation.

b. Find the standard deviation of the regression model and interpret context of this problem.

c. Do the data provide sufficient evidence to conclude that value inc ___ ___ number of units in an apartment building? Report the observed significance level, and reach a conclusion using $\alpha = .05$.

d. Interpret the value of $\hat{\beta}_1$ in terms of these data. Remember that your interpretation must recognize the presence of the other variables in the model.

e. Construct a scattergram of sale price versus age. What does your scattergram suggest about the relationship between these variables?

f. Test $H_0: \beta_2 = 0$ against $H_a: \beta_2 < 0$ using $\alpha = .01$. Interpret the result in the context of the problem. Does the result agree with your observation in part e? Why is it reasonable to conduct a one-tailed rather than a two-tailed test of this null hypothesis?

g. What is the observed significance level of the hypothesis test of part f?

SECTION 11.6

CHECKING THE USEFULNESS OF A MODEL: R^2 AND THE ANALYSIS OF VARIANCE F TEST

Conducting t tests on each β parameter in a model is not a good way to determine whether a model is contributing information for the prediction of y. If we were to conduct a series of t tests to determine whether the independent variables are contributing to the predictive relationship, it is very likely that we would make one or more errors in deciding which terms to retain in the model and which to exclude. For example, suppose that all the β parameters (except β_0) are in fact equal to 0. Although the probability of concluding that any *single* β parameter differs from 0 is only α, the probability of rejecting *at least one* of a set of null hypotheses when each is true is much higher. You can see why this is true by considering the following analogy. The probability of observing a head on a single toss of a coin is .5, but the probability of observing *at least one* head in five tosses of a coin is .97. Thus, in multiple regression models for which a large number of independent variables are being considered, conducting a series of t tests may include a large number of insignificant variables and exclude some useful ones. If we want to test the utility of a multiple regression model, we will need a global test (one that encompasses all the β parameters). We would also like to find some statistical quantity that measures how well the model fits the data.

We commence with the easier problem—finding a measure of how well a linear model fits a set of data. For this we use the multiple regression equivalent of r^2, the coefficient of determination for the straight-line model (Chapter 10). Thus, we define the **sample multiple coefficient of determination, R^2,** as

$$R^2 = 1 - \frac{\sum (y_i - \hat{y}_i)^2}{\sum (y_i - \bar{y})^2} = 1 - \frac{\text{SSE}}{\text{SS}_{yy}} = \frac{\text{SS}_{yy} - \text{SSE}}{\text{SS}_{yy}} = \frac{\text{Explained variability}}{\text{Total variability}}$$

where $\hat{y}_i$ is the predicted value of y_i for the model. Just as for the simple linear model, R^2 is a sample statistic that represents the fraction of the sample variation of the y values (measured by SS_{yy}) that is attributable to the regression model. Thus, $R^2 = 0$ implies a complete lack of fit of the model to the data, and $R^2 = 1$ implies a perfect fit, with the model passing through every sample data point. In general, the larger the value of R^2, the better the model fits the data.

To illustrate, the value $R^2 = .982$ for the electrical usage example is indicated in Figure 11.9. This very high value of R^2 implies that 98.2% of the sample variation in electrical usage is attributable to, or explained by, the independent variable (home size) x. Thus, R^2 is a sample statistic that tells how well the model fits the data, and thereby represents a measure of the usefulness of the model.

FIGURE 11.9   SAS Printout for Electrical Usage Example

SOURCE	DF	SUM OF SQUARES	MEAN SQUARE	F VALUE	PR > F
MODEL	2	831069.54637065	415534.77318533	189.71	0.0001
ERROR	7	15332.55362935	2190.36480419		ROOT MSE
CORRECTED TOTAL	9	846402.10000000		R-SQUARE	46.8013333
				0.981885	

PARAMETER	ESTIMATE	T FOR HO: PARAMETER = 0	PR > ¦T¦	STD ERROR OF ESTIMATE
INTERCEPT	-1216.14388700	-5.01	0.0016	242.80636850
X	2.39893018	9.76	0.0001	0.24583560
X*X	-0.00045004	-7.62	0.0001	0.00005908

The fact that R^2 is a sample statistic implies that it can be used to make inferences about the usefulness of the model for predicting y values for specific settings of the independent variables. In particular, for the electrical usage data, the test

H_0: $\beta_1 = \beta_2 = 0$

H_a: At least one of the parameters β_1 and β_2 is nonzero

would formally test the global usefulness of the model.

The test statistic used to test this hypothesis is an F statistic, and several equivalent versions of the formula can be used (although we will usually rely on the computer to calculate the F statistic):

$$\text{Test statistic:} \quad F = \frac{(SS_{yy} - SSE)/k}{SSE/[n - (k + 1)]} = \frac{R^2/k}{(1 - R^2)/[n - (k + 1)]}$$

Both these formulas indicate that the F statistic is the ratio of the *explained* variability divided by the model degrees of freedom to the *unexplained* variability divided by the error degrees of freedom. Thus, the larger the proportion of the total variability accounted for by the model, the larger the F statistic.

To determine when the ratio becomes large enough that we can confidently reject the null hypothesis and conclude that the model is more useful than no model at all for predicting y, we compare the calculated F statistic to a tabled F value with k df in the numerator and $[n - (k + 1)]$ df in the denominator. Tables of the F distribution for various values of α are given in Tables VII, VIII, IX, and X of Appendix B.

Rejection region: $F > F_\alpha$ where $\nu_1 = k$ df, $\nu_2 = n - (k + 1)$ df

For the electrical usage example, $n = 10$, $k = 2$, $n - (k + 1) = 7$, and $\alpha = .05$. Consequently, we will reject H_0: $\beta_1 = \beta_2 = 0$ if

$$F > F_{.05} \quad \text{where } \nu_1 = 2, \; \nu_2 = 7$$

or

$$F > 4.74 \quad \text{(see Figure 11.10)}$$

FIGURE 11.10

Rejection Region for the F Statistic with $\nu_1 = 2$, $\nu_2 = 7$, and $\alpha = .05$

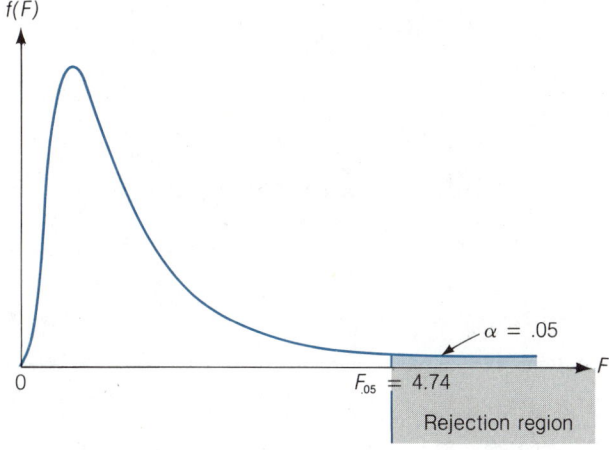

From the computer printout (Figure 11.9), we find that the computed F is 189.71. Since this value greatly exceeds the tabulated value of 4.74, we conclude that at least one of the model coefficients β_1 and β_2 is nonzero. Therefore, this global F test indicates that the second-order model $y = \beta_0 + \beta_1 x + \beta_2 x^2 + \varepsilon$ is useful for predicting electrical usage.

The F statistic is also given as a part of most regression printouts, usually in a portion of the printout called "Analysis of Variance." This is an appropriate descriptive term, since the F statistic relates the explained and unexplained portions of the total variance of y. For example, the elements of the SAS computer printout in Figure 11.9 that lead to the calculation of the F value are:

$$\text{F VALUE} = \frac{\text{SUM OF SQUARES(MODEL)}/\text{DF(MODEL)}}{\text{SUM OF SQUARES(ERROR)}/\text{DF(ERROR)}}$$
$$= \frac{\text{MEAN SQUARE(MODEL)}}{\text{MEAN SQUARE(ERROR)}}$$

From Figure 11.9 we see that F VALUE = 189.71. Note, too, that the observed significance level for the F statistic is given under the heading PR > F as .0001, which means that we would reject the null hypothesis H_0: $\beta_1 = \beta_2 = 0$ at any α value greater than .0001.

The analysis of variance F test for testing the usefulness of the model is summarized in the box.

TESTING GLOBAL USEFULNESS OF THE MODEL: THE ANALYSIS OF VARIANCE F TEST

H_0: $\beta_1 = \beta_2 = \cdots = \beta_k = 0$ (All model terms are unimportant for predicting y)

H_a: At least one $\beta_i \neq 0$ (At least one model term is useful for predicting y)

Test statistic:

$$F = \frac{(SS_{yy} - SSE)/k}{SSE/[n - (k + 1)]} = \frac{R^2/k}{(1 - R^2)/[n - (k + 1)]}$$

$$= \frac{\text{MEAN SQUARE(MODEL)}}{\text{MEAN SQUARE(ERROR)}}$$

where n is the sample size and k is the number of terms in the model.

Rejection region: $F > F_\alpha$, with k numerator degrees of freedom and $[n - (k + 1)]$ denominator degrees of freedom.

Assumptions: The standard regression assumptions about the random error component (Section 11.2).

Caution: A rejection of the null hypothesis leads to the conclusion [with $100(1 - \alpha)\%$ confidence] that the model is useful. However, "useful" does not necessarily mean "best." Another model may prove even more useful in terms of providing more reliable estimates and predictions. This global F test is usually regarded as a test that the model *must* pass to merit further consideration.

EXAMPLE 11.2

Refer to Example 11.1, in which an antique collector modeled the auction price, y, of grandfather clocks as a function of the age of the clock, x_1, and the number of bidders, x_2. The hypothesized model was

$$y = \beta_0 + \beta_1 x_1 + \beta_2 x_2 + \varepsilon$$

A sample of 32 observations was obtained, with the results summarized in the SAS printout repeated in Figure 11.11. Discuss the coefficient of determination R^2 for this example and then conduct the global F test of model usefulness using $\alpha = .05$.

FIGURE 11.11   SAS Printout for Example 11.2

SOURCE	DF	SUM OF SQUARES	MEAN SQUARE	F VALUE	PR > F
MODEL	2	4277159.70340504	2138579.85170252	120.65	0.0001
ERROR	29	514034.51534496	17725.32811534		ROOT MSE
CORRECTED TOTAL	31	4791194.21875000	R-SQUARE		133.13650181
			0.892713		

PARAMETER	ESTIMATE	T FOR H0: PARAMETER = 0	PR > !T!	STD ERROR OF ESTIMATE
INTERCEPT	-1336.72205214	-7.71	0.0001	173.35612607
X1	12.73619884	14.11	0.0001	0.90238049
X2	85.81513260	9.86	0.0001	8.70575681

SOLUTION

The R^2 value is .89 (see Figure 11.11). This implies that 89% of the variation of the sample y values (the auction prices) about their mean can be explained by the least squares model. We now test:

H_0: $\beta_1 = \beta_2 = 0$ [Note: $k = 2$]

H_a: At least one of the two model coefficients is nonzero

Test statistic: $F = \dfrac{R^2/k}{(1 - R^2)/[n - (k + 1)]}$

Rejection region: $F > F_\alpha$ where $\nu_1 = k$ df, $\nu_2 = n - (k + 1)$ df

For this example, $n = 32$, $k = 2$, and $n - (k + 1) = 32 - 3 = 29$. Then, for $\alpha = .05$, we will reject H_0: $\beta_1 = \beta_2 = 0$ if $F > F_{.05}$—i.e., if $F > 3.33$ (obtained from Table VIII in Appendix B). The computed value of the F test statistic is 120.65 (see Figure 11.11). Since this value of F falls in the rejection region ($F = 120.65$ greatly exceeds $F_{.05} = 3.33$), the data provide strong evidence that at least one of the model coefficients is nonzero. The model appears to be useful for predicting auction prices. ∎

Can we be sure that the best prediction model has been found if the global F test indicates that a model is useful? Unfortunately, we cannot. There is no way of knowing whether the addition of other independent variables will further improve the usefulness of the model, as Example 11.3 indicates.

EXAMPLE 11.3

Refer to Examples 11.1 and 11.2. Suppose the collector, having observed many auctions, believes that the *rate of increase* of the auction price with age will be driven upward by a large number of bidders. Thus, instead of a relationship like that shown in Figure 11.12(a), in which the rate of increase in price with age is the same for any number of bidders, the collector believes the relationship is like that shown in Figure 11.12(b). Note that as the number of bidders increases from 5 to 15, the slope of the price versus age line increases. When the slope of the relationship between y and one independent variable (x_1) depends on the value of

FIGURE 11.12

Examples of No
Interaction and
Interaction Models

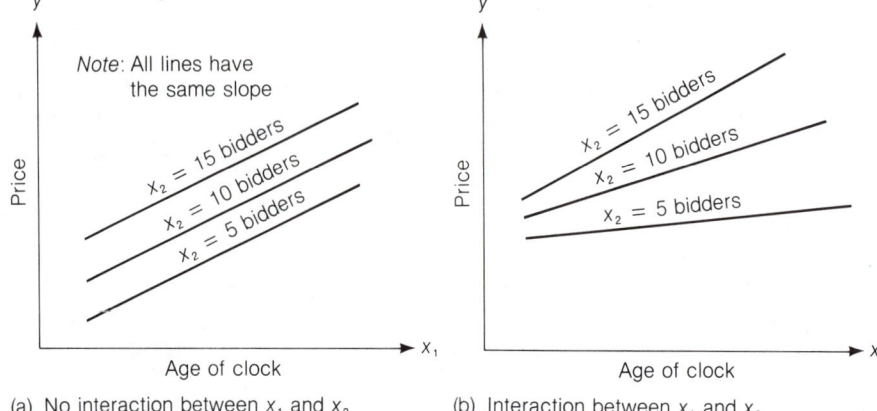

(a) No interaction between x_1 and x_2

(b) Interaction between x_1 and x_2

a second independent variable (x_2), as is the case here, we say that x_1 and x_2 **interact**.* A model that accounts for this type of interaction is written

$$y = \beta_0 + \beta_1 x_1 + \beta_2 x_2 + \beta_3 x_1 x_2 + \varepsilon$$

Note that the increase in the mean price, $E(y)$, for each 1-year increase in age, x_1, is no longer given by the constant, β_1, but is now $\beta_1 + \beta_3 x_2$. That is, the amount $E(y)$ increases for each 1-unit increase in x_1 is *dependent on the number of bidders*, x_2. Thus, the two variables x_1 and x_2 interact to affect y.

The 32 data points listed in Table 11.2 were used to fit the first-order model with interaction. A portion of the SAS printout is shown in Figure 11.13.

FIGURE 11.13 Portion of the SAS Printout for the Model with Interaction

SOURCE	DF	SUM OF SQUARES	MEAN SQUARE	F VALUE	PR > F
MODEL	3	4572547.98717668	1524182.66239223	195.19	0.0001
ERROR	28	218646.23157332	7808.79398476		ROOT MSE
CORRECTED TOTAL	31	4791194.21875000		R-SQUARE	88.36738077
				0.954365	

PARAMETER	ESTIMATE	T FOR H0: PARAMETER = 0	PR > !T!	STD ERROR OF ESTIMATE
INTERCEPT	322.75435309	1.10	0.2806	293.32514660
X1	0.87328775	0.43	0.6688	2.01965115
X2	-93.40991991	-3.14	0.0039	29.70767946
X1*X2	1.29789828	6.15	0.0001	0.21102602

Test the hypothesis that the price–age slope increases as the number of bidders increases–i.e., that age, x_1, and number of bidders, x_2, interact positively.

*Further discussion of interaction is given in Chapter 12.

SOLUTION

The model is

$$y = \beta_0 + \beta_1 x_1 + \beta_2 x_2 + \beta_3 x_1 x_2 + \varepsilon$$

and the hypothesis of interest to the collector concerns the parameter β_3. Specifically,

H_0: $\beta_3 = 0$

H_a: $\beta_3 > 0$

Test statistic: $t = \dfrac{\hat{\beta}_3}{s_{\hat{\beta}_3}}$

Rejection region: For $\alpha = .05$, $t > t_{.05}$, where df $= n - (k + 1)$

In this example, $n = 32$, $k = 3$, df $= n - (k + 1) = 32 - 4 = 28$, and thus, $t_{.05} = 1.701$.

The t value corresponding to $\hat{\beta}_3$ is indicated in Figure 11.13. The value, $t = 6.15$, exceeds 1.701, and therefore falls in the rejection region. Thus, the collector can conclude that the rate of change of the mean price of the clocks with age increases as the number of bidders increases—i.e., x_1 and x_2 interact positively. Thus, it appears that the interaction term should be included in the model. ∎

One note of caution: Although the coefficient of x_2 is negative ($\hat{\beta}_2 = -93.41$), this does *not* imply that the auction price decreases as the number of bidders increases. Since interaction is present, the rate of change (slope) of mean auction price with the number of bidders *depends on* x_1, the age of the clock. Thus, for example, the estimated rate of change of y with x_2 for a 150-year-old clock is

Estimated x_2 slope $= \hat{\beta}_2 + \hat{\beta}_3 x_1$

$$= -93.41 + 1.30(150) = 101.60$$

In other words, we estimate that the auction price of a 150-year-old clock will *increase* by \$101.60 for every additional bidder. Although this rate of increase will vary as x_1 is changed, it will remain positive for the range of values of x_1 included in the sample. Extreme care is needed in interpreting the signs and sizes of coefficients in a multiple regression model.

To summarize the discussion in this section, the value of R^2 is an indicator of how well the prediction equation fits the data. More important, it can be used (in the F statistic) to determine whether the data provide sufficient evidence to indicate that the model contributes information for the prediction of y. Intuitive evaluations of the contribution of the model based on the computed value of R^2 must be examined with care. The value of R^2 will increase as more and more variables are added to the model. Consequently, you could force R^2 to take a value very close to 1 even though the model contributes no information for the prediction of y. In fact, R^2 will equal 1 when the number of terms in the model equals the number of data points. Therefore, you should not rely solely on the value of R^2 to tell you whether the model is useful for predicting y. Use the F test.

LEARNING THE MECHANICS

[*Note:* *Starred (*) exercises require the use of a computer.*]

11.12 The model $y = \beta_0 + \beta_1 x + \beta_2 x^2 + \varepsilon$ was fit to $n = 19$ data points with the results shown in the accompanying printout.

SAS Printout for
Exercise 11.12

Dep Variable: Y

Analysis of Variance

Source	DF	Sum of Squares	Mean Square	F Value	Prob>F
Model	2	24.22335	12.11167	65.478	0.0001
Error	16	2.95955	0.18497		
C Total	18	27.18289			

Root MSE	0.43008	R-Square	0.8911	
Dep Mean	3.56053	Adj R-Sq	0.8775	
C.V.	12.07921			

Parameter Estimates

Variable	DF	Parameter Estimate	Standard Error	T for H0: Parameter=0	Prob > ¦T¦
INTERCEP	1	0.734606	0.29313351	2.506	0.0234
X	1	0.765179	0.08754136	8.741	0.0001
XSQ	1	-0.030810	0.00452890	-6.803	0.0001

a. Find R^2 and interpret its value.
b. Test the null hypothesis that $\beta_1 = \beta_2 = 0$ against the alternative hypothesis that at least one of β_1 and β_2 are nonzero. Calculate the test statistic using the two formulas given in this section, and compare your results to each other and to that given on the printout. Use $\alpha = .05$ and interpret the result of your test.
c. Find the observed significance level for this test on the printout, and interpret it.
d. Test H_0: $\beta_2 = 0$ against H_a: $\beta_2 \neq 0$. Use $\alpha = .05$ and interpret the result of your test. Report and interpret the observed significance level of the test.

11.13 Suppose you fit the model

$$y = \beta_0 + \beta_1 x_1 + \beta_2 x_2 + \beta_3 x_1 x_2 + \beta_4 x_1^2 + \beta_5 x_2^2 + \varepsilon$$

to $n = 30$ data points and you obtain

   SSE $= .42$ $R^2 = .91$

a. Do the values of SSE and R^2 suggest that the model provides a good fit to the data? Explain.
b. Is the model of any use in predicting y? Test the null hypothesis that $E(y) = \beta_0$, i.e.,

   H_0: $\beta_1 = \beta_2 = \cdots = \beta_5 = 0$

against the alternative hypothesis

H_a: At least one of the parameters $\beta_1, \beta_2, \ldots, \beta_5$ is nonzero

Use $\alpha = .05$.

11.14 Suppose you fit the model

$$y = \beta_0 + \beta_1 x_1 + \beta_2 x_2 + \varepsilon$$

to $n = 20$ data points and obtain

$$\sum (y_i - \hat{y}_i)^2 = 1.47 \qquad \sum (y_i - \bar{y})^2 = 2.73$$

a. Construct an analysis of variance table for this regression analysis, using the printout in Exercise 11.12 as a model. Be sure to include the sources of variability, the degrees of freedom, the sums of squares, the mean squares, and the F statistic. Calculate R^2 for the regression analysis.

b. Test the null hypothesis that $\beta_1 = \beta_2 = 0$ against the alternative hypothesis that at least one of the parameters differs from 0. Calculate the test statistic in two different ways and compare the results. Use $\alpha = .05$ to reach a conclusion about whether the model contributes information for the prediction of y.

11.15 If the analysis of variance F test leads to the conclusion that at least one of the model parameters is nonzero, can you conclude that the model is the best predictor for the dependent variable y? Can you conclude that all of the terms in the model are important for predicting y? What is the appropriate conclusion?

APPLYING THE CONCEPTS

11.16 If producers (providers) of goods (services) are able to reduce the unit cost of their goods by increasing the scale of their operation, they are the beneficiaries of an economic force known as *economies of scale*. Economies of scale cause a firm's long-run average costs to decline (Ferguson and Maurice, 1970). The question of whether economies of scale, diseconomies of scale, or neither (i.e., constant economies of scale) exist in the U.S. motor freight, common carrier industry has been debated for years. In an effort to settle the debate within a specific subsection of the trucking industry, Sugrue, Ledford, and Glaskowsky (1982) used regression analysis to model the relationship between each of a number of profitability/cost measures and the size of the operation. In one case, they modeled expense per vehicle-mile, y_1, as a function of the firm's total revenue, x. In another case, they modeled expense per ton-mile, y_2, as a function of x. Data were collected from 264 firms and the least squares results obtained are shown in the table.

DEPENDENT VARIABLE	$\hat{\beta}_0$	$\hat{\beta}_1$	r	F
Expense per vehicle-mile	2.279	$-.00000069$	$-.0783$	1.616
Expense per ton-mile	.1680	$-.000000066$	$-.0902$	2.148

Source: Sugrue, P. K., Ledford, M. H., and Glaskowsky, N. A., Jr. "Operating economies of scale in the U.S. long-haul common carrier, motor freight industry." *Transportation Journal*, Fall 1982, 22, 27–41.

a. Investigate the usefulness of the two models estimated by Sugrue, Ledford, and Glaskowsky. Use $\alpha = .05$. Draw appropriate conclusions in the context of the problem.

b. Are the observed significance levels of the hypothesis tests you conducted in part **a** greater than .10 or less than .10? Explain.

c. What do your hypothesis tests of part **a** suggest about economies of scale in the subsection of the trucking industry investigated by Sugrue, Ledford, and Glaskowsky —the long-haul, heavy-load, intercity-general-freight common carrier sector? Explain.

*11.17   Refer to Exercise 11.11, in which a regression model was used to explore the valuation of apartment buildings in Minneapolis.

    a. Find R^2 for the least squares prediction equation found in part **a** of Exercise 11.11. Interpret its value in the context of the problem.

    b. Use an F test to investigate the usefulness of the model hypothesized by the real estate appraiser. Use $\alpha = .01$, and carefully state your conclusion in the context of this application.

    c. What is the observed significance level of the F test?

11.18   In hopes of increasing the company's share of the fine food market, researchers for a meat-processing firm that prepares meats for exclusive restaurants are working to improve the quality of its hickory-smoked hams. One of their studies concerns the effect of time spent in the smokehouse on the flavor of the ham. Hams that were in the smokehouse for varying amounts of time were each subjected to a taste test by a panel of ten food experts. The following model was thought to be appropriate by the researchers:

$$y = \beta_0 + \beta_1 t + \beta_2 t^2 + \varepsilon$$

where

    $y =$ Mean of the taste scores for the ten experts

    $t =$ Time in the smokehouse (hours)

Assume the least squares model estimated using a sample of 20 hams is

$$\hat{y} = 20.3 + 5.2t - .0025t^2$$

and that $s_{\hat{\beta}_2} = .0011$. The coefficient of determination is $R^2 = .79$.

    a. Is there evidence to indicate that the overall model is useful? Test at $\alpha = .05$.

    b. Is there evidence to indicate that the quadratic term is important in this model? Test at $\alpha = .05$.

11.19   Writing in *Accounting Review*, Benston (1966) describes how multiple regression can be used by accountants in cost analysis. He points out that multiple regression models can be used to shed light on "the factors that cause costs to be incurred and the magnitudes of their effects" (p. 658). The independent variables of such a regression model are the factors believed to be related to cost, the dependent variable. The estimates of the coefficients of the regression model provide measures of the magnitude of the factors' effects on cost. In some instances, however, Benston notes that it may be desirable to use physical units instead of cost as the dependent variable in a cost analysis. Such would be the case if most of the cost associated with the activity of interest is a function of some physical unit, such as hours of labor. The advantage of this approach is that the regression model will provide estimates of the number of labor hours required under different circumstances, and these hours can then be costed at the current labor rate.

The sample data shown in the table have been collected from a firm's accounting and production records to provide cost information about the firm's shipping department. The variables for which data were collected were suggested by Benston.

WEEK	HOURS OF LABOR y	THOUSANDS OF POUNDS SHIPPED x_1	PERCENTAGE OF UNITS SHIPPED BY TRUCK x_2	AVERAGE NUMBER OF POUNDS PER SHIPMENT x_3
1	100	5.1	90	20
2	85	3.8	99	22
3	108	5.3	58	19
4	116	7.5	16	15
5	92	4.5	54	20
6	63	3.3	42	26
7	79	5.3	12	25
8	101	5.9	32	21
9	88	4.0	56	24
10	71	4.2	64	29
11	122	6.8	78	10
12	85	3.9	90	30
13	50	3.8	74	28
14	114	7.5	89	14
15	104	4.5	90	21
16	111	6.0	40	20
17	110	8.1	55	16
18	100	2.9	64	19
19	82	4.0	35	23
20	85	4.8	58	25

The SAS computer printout for fitting the model $y = \beta_0 + \beta_1 x_1 + \beta_2 x_2 + \beta_3 x_3 + \varepsilon$ to the data is shown here.

SAS Printout for Exercise 11.19

DEPENDENT VARIABLE: LABOR

SOURCE	DF	SUM OF SQUARES	MEAN SQUARE	F VALUE
MODEL	3	5158.31382780	1719.43794260	17.87
ERROR	16	1539.88617220	96.24288576	PR > F
CORRECTED TOTAL	19	6698.20000000		0.0001

R-SQUARE	C.V.	ROOT MSE	LABOR MEAN
0.770104	10.5148	9.81034585	93.30000000

PARAMETER	ESTIMATE	T FOR H0: PARAMETER=0	PR > !T!	STD ERROR OF ESTIMATE
INTERCEPT	131.92425208	5.13	0.0001	25.69321439
WEIGHT	2.72608977	1.20	0.2483	2.27500488
TRUCK	0.04721841	0.51	0.6199	0.09334856
AVGSHIP	-2.58744391	-4.03	0.0010	0.64281819

a. Find the least squares prediction equation.

b. Use an F test to investigate the usefulness of the model specified in part **a**. Use $\alpha = .01$, and state your conclusion in the context of the problem.

c. Test $H_0: \beta_2 = 0$ versus $H_a: \beta_2 \neq 0$ using $\alpha = .05$. What do the results of your test suggest about the magnitude of the effects of x_2 on labor costs?

d. Find R^2, and interpret its value in the context of the problem.

e. If shipping department employees are paid \$7.50 per hour, how much less, on average, will it cost the company per week if the average number of pounds per shipment increases from a level of 20 to 21? Assume that x_1 and x_2 remain unchanged. Your answer to this question is an estimate of what is known in economics as the *expected marginal cost* associated with a 1-pound increase in x_3.

f. With what approximate precision can this model be used to predict the hours of labor? [*Note:* The precision of multiple regression predictions is discussed in Section 11.7.]

11.20 Because the coefficient of determination R^2 always increases when a new independent variable is added to the model, it may be tempting to include many variables in a model to force R^2 to be near 1. However, doing so reduces the degrees of freedom available for estimating σ^2, which adversely affects our ability to make reliable inferences. As an example, suppose you want to use 18 economic indicators to predict next year's GNP. You fit the model

$$y = \beta_0 + \beta_1 x_1 + \beta_2 x_2 + \cdots + \beta_{17} x_{17} + \beta_{18} x_{18} + \varepsilon$$

where $y = $ GNP and $x_1, x_2, \ldots, x_{18}$ are indicators. Only 20 years of data ($n = 20$) are used to fit the model, and you obtain $R^2 = .95$. Test to determine whether this impressive-looking R^2 is large enough to infer that the model is useful—i.e., that at least one term in the model is important for predicting GNP. Use $\alpha = .05$.

11.21 Refer to Exercise 11.8. Recall that the dependent variable being modeled was production line supervisors' scores on a test designed to measure attitudes toward handicapped workers. The independent variables were the sex of the supervisor ($x_1 = 1$ if female, 0 if male) and the supervisor's years of experience (x_2). Suppose the same model is proposed as in Exercise 11.8, with the addition of an interaction between x_1 and x_2; i.e.,

$$y = \beta_0 + \beta_1 x_1 + \beta_2 x_2 + \beta_3 x_2^2 + \beta_4 x_1 x_2 + \varepsilon$$

This model is fit to the same data (consisting of 40 observations) as in Exercise 11.8 with the result

$$\hat{y} = 50 + 5x_1 + 6x_2 - .2x_2^2 - x_1 x_2$$

and

$$s_{\hat{\beta}_4} = .02 \qquad R^2 = .87$$

a. Interpret the value of R^2.

b. Is there sufficient evidence to indicate that this model is useful for predicting attitude score? Test $H_0: \beta_1 = \beta_2 = \beta_3 = \beta_4 = 0$ using $\alpha = .05$.

c. Is there evidence that the interaction between sex and years of experience is useful in the prediction model?

d. Sketch the predicted attitude score, $\hat{y}$, as a function of the number of years of experience, x_2, for males ($x_1 = 0$). Next, substitute $x_1 = 1$ into the least squares equation and thereby obtain a plot of the prediction equation for females. Compare these sketches with those obtained when the model without interaction was fit in Exercise 11.8. [*Note:* For both males and females, plotting $\hat{y}$ for $x_2 = 0, 2, 4, 6,$ 8, and 10 will produce a good picture of the prediction equations. The interaction term allows the vertical distance between the males' and females' prediction curves to change as x_2 changes.]

11.22 Carson Bays (1986) used regression analysis to investigate the determinants of survival size of nonprofit hospitals. For a given sample of hospitals, survival size, y, is defined as the largest size hospital (in terms of number of beds) exhibiting growth in market share over a specific time interval. Suppose ten states are randomly selected and the survival size for all nonprofit hospitals in each state is determined for the time period 1981–1982 and for 1984–1985, yielding two observations per state. The 20 survival sizes are listed in the table along with the following data for each state, for the second year in each time interval:

x_1 = Percentage of beds that are in for-profit hospitals

x_2 = Ratio of the number of persons enrolled in health maintenance organizations (HMOs) to the number of persons covered by hospital insurance

x_3 = State population (in thousands)

x_4 = Percent of state that is urban

Bays hypothesized that the following model characterizes the relationship between survival size and the four variables listed above:

$$y = \beta_0 + \beta_1 x_1 + \beta_2 x_2 + \beta_3 x_3 + \beta_4 x_4 + \varepsilon$$

STATE	TIME PERIOD	SURVIVAL SIZE y	x_1	x_2	x_3	x_4
1	1	370	.13	.09	5,800	89
1	2	390	.15	.09	5,955	87
2	1	455	.08	.11	17,648	87
2	2	450	.10	.16	17,895	85
3	1	500	.03	.04	7,332	79
3	2	480	.07	.05	7,610	78
4	1	550	.06	.005	11,731	80
4	2	600	.10	.005	11,790	81
5	1	205	.30	.12	2,932	44
5	2	230	.25	.13	3,100	45
6	1	425	.04	.01	4,148	36
6	2	445	.07	.02	4,205	38
7	1	245	.20	.01	1,574	25
7	2	200	.30	.01	1,560	28
8	1	250	.07	.08	2,471	38
8	2	275	.08	.10	2,511	38
9	1	300	.09	.12	4,060	52
9	2	290	.12	.20	4,175	54
10	1	280	.10	.02	2,902	37
10	2	270	.11	.05	2,925	38

Source: Adapted from Bays, C. W. "The determinants of hospital size: A survival analysis." *Applied Economics*, 1986, *18*, 359–377.

a. The model was fit to the data in the table using the SAS System with the results given in the accompanying printout. Report the least squares prediction equation.

SAS Printout for Exercise 11.22

Dep Variable: Y

Analysis of Variance

Source	DF	Sum of Squares	Mean Square	F Value	Prob>F
Model	4	246537.05939	61634.26485	28.180	0.0001
Error	15	32807.94061	2187.19604		
C Total	19	279345.00000			

Root MSE	46.76747	R-Square	0.8826	
Dep Mean	360.50000	Adj R-Sq	0.8512	
C.V.	12.97295			

Parameter Estimates

Variable	DF	Parameter Estimate	Standard Error	T for H0: Parameter=0	Prob > ¦T¦
INTERCEP	1	295.327091	40.17888737	7.350	0.0001
X1	1	-480.837576	150.39050364	-3.197	0.0060
X2	1	-829.464955	196.47303539	-4.222	0.0007
X3	1	0.007934	0.00355335	2.233	0.0412
X4	1	2.360769	0.76150774	3.100	0.0073

b. Find the regression standard deviation s and interpret its value in the context of the problem.

c. Use an F test to investigate the usefulness of the hypothesized model. Report the observed significance level, and use $\alpha = .025$ to reach your conclusion.

d. Prior to collecting the data it was hypothesized that increases in the number of for-profit hospital beds would decrease the survival size of nonprofit hospitals. Do the data support this hypothesis? Test using $\alpha = .05$.

*11.23 A company that services microcomputers is interested in developing a regression model that will assist them in manpower planning. In particular, they want a model that describes the relationship between the time a service person spends on a preventive maintenance service call to a customer, y, and two independent variables: the number of microcomputers to be serviced, x_1, and the service person's number of months of experience in preventive maintenance, x_2. Company records were sampled and the data in the table were obtained.

a. Fit the model $y = \beta_0 + \beta_1 x_1 + \beta_2 x_2 + \varepsilon$ to the data.

b. Investigate whether the model is useful. Test using $\alpha = .10$.

c. Find R^2 for the fitted model. Interpret your result.

d. Fit the model $y = \beta_0 + \beta_1 x_1 + \beta_2 x_2 + \beta_3 x_1 x_2 + \varepsilon$ to the data.

e. Find R^2 for the model of part **d**.

f. Explain why you should not rely solely on a comparison of the two R^2 values for drawing conclusions about which model is more useful for predicting y.

g. Do the data provide sufficient evidence to indicate that the interaction term, $x_1 x_2$, contributes information for the prediction of y? [Hint: Test H_0: $\beta_3 = 0$.]

MAINTENANCE TIME (hours)	NUMBER OF MICROCOMPUTERS	EXPERIENCE (months)
1.0	1	12
3.1	3	8
17.0	10	5
14.0	8	2
6.0	5	10
1.8	1	1
11.5	10	10
9.3	5	2
6.0	4	6
12.2	10	8

h. Can you be certain that the model you selected in part **f** is the best model to use in predicting maintenance time? Explain.

*11.24 Regression analysis can be used to model the relationship between the selling price of a house and its total living area (for more details, see Exercise 10.77). The table contains the final selling prices and total living areas for a sample of 30 houses in the same geographic area that were sold during 1987.

AREA (sq. ft.)	PRICE	AREA (sq. ft.)	PRICE	AREA (sq. ft.)	PRICE
1,600	$101,900	1,910	$104,000	2,390	$123,300
1,980	108,700	1,810	104,500	2,560	133,000
2,130	114,500	2,035	112,000	2,370	125,000
1,990	110,300	3,000	210,000	2,420	123,300
1,710	100,900	2,500	160,000	2,300	119,200
2,357	121,100	2,120	113,600	2,320	120,000
2,335	118,700	1,880	106,500	2,470	126,800
2,650	137,200	1,775	105,000	2,460	127,100
2,070	109,300	1,650	99,000	3,650	225,100
1,930	108,400	2,250	116,100	2,750	205,600

a. Construct a scattergram for the data. Is there evidence to suggest that it would be inappropriate to represent the relationship between price and area with a straight line? Explain.

b. Fit the following model to the data:

$$y = \beta_0 + \beta_1 x + \beta_2 x^2 + \varepsilon$$

where y = price and x = area. Plot the fitted equation on the scattergram of part **a**.

c. Find the standard deviation for the least squares equation of part **b**. Interpret the value in the context of this application.

d. Use an F test to investigate whether the model is useful for predicting y. Test using $\alpha = .05$.

e. Do the data provide sufficient evidence to indicate that the second-order term (x^2) contributes information for the prediction of y? Test using $\alpha = .05$.

SECTION 11.7

USING THE MODEL FOR ESTIMATION AND PREDICTION

In Section 10.8 we discussed the use of the least squares line for estimating the mean value of y, $E(y)$, for some particular value of x, say $x = x_p$. We also showed how to use the same fitted model to predict, when $x = x_p$, some value of y to be observed in the future. Recall that the least squares line yielded the same value for both the estimate of $E(y)$ and the prediction of some future value of y. That is, both are the result of substituting x_p into the prediction equation, $\hat{y} = \hat{\beta}_0 + \hat{\beta}_1 x$, and calculating $\hat{y}$. There the equivalence ends. The confidence interval for the mean, $E(y)$, was narrower than the prediction interval for y because of the additional uncertainty attributable to the random error, ε, when predicting some future value of y.

These same concepts carry over to the multiple regression model. For example, suppose we want to estimate the mean electrical usage for a given home size, say $x_p = 1,500$ square feet. Assuming the quadratic model represents the true relationship between electrical usage and home size, we want to estimate

$$E(y) = \beta_0 + \beta_1 x_p + \beta_2 x_p^2 = \beta_0 + \beta_1(1,500) + \beta_2(1,500)^2$$

Substituting into the least squares prediction equation yields the following estimate of $E(y)$:

$$\hat{y} = \hat{\beta}_0 + \hat{\beta}_1(1,500) + \hat{\beta}_2(1,500)^2$$
$$= -1,216.144 + 2.3989(1,500) - .00045004(1,500)^2 = 1,369.7$$

To form a confidence interval for the mean, we need to know the standard deviation of the sampling distribution for the estimator $\hat{y}$. For multiple regression models, the form of this standard deviation is rather complex. However, some regression packages allow us to obtain the confidence intervals for mean values of y at any given setting of the independent variables. A portion of the SAS output for the electrical usage example is shown in Figure 11.14. The mean value and corresponding 95% confidence interval for $x_p = 1,500$ are shown in the columns labeled PREDICTED VALUE, LOWER 95% CL FOR MEAN, and UPPER 95% CL FOR MEAN. Note that

$$\hat{y} = 1,369.7$$

FIGURE 11.14

SAS Printout for Estimated Mean Value and Corresponding Confidence Interval for $x_p = 1,500$

X	PREDICTED VALUE	LOWER 95% CL FOR MEAN	UPPER 95% CL FOR MEAN
1500	1369.66088739	1324.98831001	1414.33346477

which agrees with our earlier calculation. The 95% confidence interval for the true mean of y is shown to be 1,325.0 to 1,414.3 (see Figure 11.15).

If we were interested in predicting the electrical usage for a particular 1,500-square-foot home, $\hat{y} = 1,369.7$ would be used as the predicted value. However, the prediction interval for a particular value of y will be wider than the confidence interval for the mean value. This is reflected by the printout shown in Figure 11.16,

FIGURE 11.15
Confidence Interval for
Mean Electrical Usage

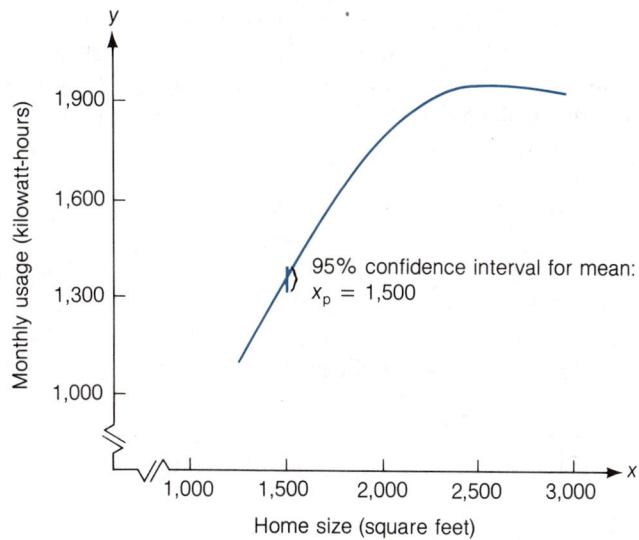

which gives the predicted value of y and corresponding 95% prediction interval for $x_p = 1,500$. This prediction interval, which extends from 1,250.3 to 1,489.0, is shown in Figure 11.17.

FIGURE 11.16
SAS Printout for Predicted
Value and Corresponding
Prediction Interval for
$x_p = 1,500$

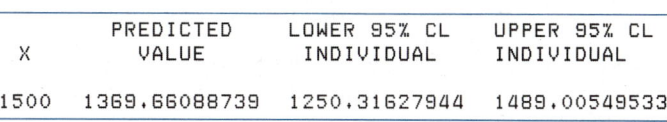

X	PREDICTED VALUE	LOWER 95% CL INDIVIDUAL	UPPER 95% CL INDIVIDUAL
1500	1369.66088739	1250.31627944	1489.00549533

FIGURE 11.17
Prediction Interval for
Electrical Usage

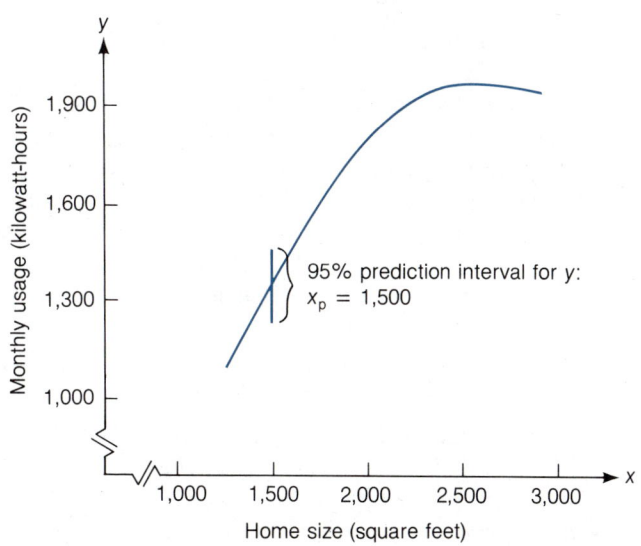

Unfortunately, not all computer packages have the capability to produce confidence intervals for means and prediction intervals for particular y values. This is a rather serious oversight since the estimation of mean values and the prediction of particular values represent the culmination of our model building efforts: using the model to make inferences about the dependent variable y.

CASE STUDY 11.2

PREDICTING THE
SALES OF CREST
TOOTHPASTE

Knowing the determinants of demand for its product would assist a company in focusing its marketing efforts on the appropriate market segments (e.g., middle-income families with school-age children). In addition, knowing the relationship between those determinants and product sales would assist the company in predicting sales and, therefore, in its planning endeavors throughout the organization. Using a model developed by Roger D. Carlson (1981) as her starting point, Carolyn I. Allmon (1982) employed multiple regression analysis to investigate the determinants of demand for Crest toothpaste and to predict Crest's sales.

Allmon modeled sales of Crest as a function of advertising expenditures, personal disposable income, and the ratio of Crest's advertising budget to that for Colgate toothpaste, Crest's closest competitor. More specifically, Allmon hypothesized the following model:

$$y = \beta_0 + \beta_1 x_1 + \beta_2 x_2 + \beta_3 x_3 + \varepsilon$$

where

y = Sales of Crest toothpaste in the current year, computed by multiplying Crest's market share by the total market (thousands of dollars)

x_1 = Advertising budget for Crest in the current year (thousands of dollars)

x_2 = U.S. personal disposable income in the current year (billions of dollars)

x_3 = Ratio of Crest's advertising budget to Colgate's advertising budget in current year (rounded to nearest hundredth)

Allmon hypothesized that the effect of each independent variable on sales is positive—i.e., that β_1, β_2, and β_3 have positive signs. Her belief that β_1 is positive follows from the fact that Procter and Gamble, the firm that manufactures Crest, has continued to advertise Crest heavily over many years. It is unlikely they would do this if advertising did not positively affect sales. That β_2 should be positive follows from a study of toothbrushing frequency in which it was shown that as family income increased, family members brushed their teeth more frequently. Allmon argues that since Colgate and Crest are competitors, sales of Colgate should affect sales of Crest and vice versa. Therefore, the larger is x_3, the higher y should be. Accordingly, β_3 should be positive.

The model was estimated using the method of least squares on data collected for the years 1967–1979. The following results were obtained:

$$\hat{y} = 30{,}626 + 3.8932x_1 + 86.519x_2 - 29{,}607x_3$$
$$\phantom{\hat{y} = }(19{,}808) \quad (2.0812) \quad (18.693) \quad (23{,}822)$$
$$\phantom{\hat{y} = }1.5461 \quad\quad 1.8706 \quad\quad 4.6283 \quad\quad -1.2429$$
$$R^2 = .9575 \quad\quad F = 67.595$$

where the numbers in parentheses are the standard errors of the coefficient estimates and the numbers below them are the associated t statistics.

Since $F = 67.595 > F_{.01} = 6.99$ (with $\nu_1 = 3$, $\nu_2 = 9$), Allmon concluded that at least one of the model coefficients is nonzero and that the model is useful for predicting sales. Next, she conducted one-sided, upper-tailed hypothesis tests to determine whether the data support her expectations about the positive influence of each of the independent variables on sales. Comparing $t_{.05} = 1.833$ (df = 9) to the t statistics given, it can be seen that the data support her hypotheses with respect to β_1 and β_2, but that the null hypothesis H_0: $\beta_3 = 0$ cannot be rejected. Thus, she concluded,

> my model indicates that income is a significant determinant of sales as is advertising expenditure in the current year. The competition of Colgate dental cream via its advertising expenditure appears not to affect Crest sales significantly. This information could be used by Procter and Gamble in the formulation of their advertising policy, for example, in determining the segment of the market to which most advertising should be directed.*

Allmon's primary purpose for developing the model was prediction. As part of her test of the predictive power of the model, Allmon substituted the values for x_1, x_2, and x_3 in 1980 (which were known at the time) into the fitted model and calculated $\hat{y}$. Her model predicted sales of \$251,060,340 and actual sales were \$245,000,000. Allmon's model did substantially better than Carlson's model, the model she had set out to refine. Carlson's model predicted sales of \$275,447,450.

In concluding the report of her findings, Allmon pointed out several limitations of the model. Among them were the following:

1. It was developed from annual data. Accordingly, the results should be viewed cautiously, since evidence exists that suggests the cumulative effect of advertising on sales lasts less than a year.
2. The sample is quite small. Accordingly, even one more data point may substantially change the coefficient estimates and overall model fit.
3. None of the data were obtained from primary sources (such as Procter and Gamble). All data except disposable income were collected from the publication *Advertising Age*, a secondary data source.

S E C T I O N 11.8

MULTIPLE
REGRESSION: AN
EXAMPLE

Let us return to the executive compensation example introduced in Case Study 11.1. Recall that the management consultant firm of Towers, Perrin, Forster & Crosby (TPF&C) uses a multiple regression model to project executive salaries. Suppose the list of independent variables given in Table 11.3 (page 594) is to be used to build a model for the salaries of corporate executives.

*Allmon, C.I. "Advertising and sales relationships for toothpaste: Another look." *Business Economics*, Sept. 1982, *17*, 55–61. Published by the National Association of Business Economists.

TABLE 11.3

List of Independent
Variables for Executive
Compensation Example

INDEPENDENT VARIABLE	DESCRIPTION
x_1	Years of experience
x_2	Years of education
x_3	1 if male; 0 if female
x_4	Number of employees supervised
x_5	Corporate assets ($ millions)
x_6	x_1^2
x_7	$x_3 x_4$

STEP 1   The first step is to hypothesize a model relating executive salary to the independent variables listed in Table 11.3. TPF&C have found that executive compensation models that use the logarithm of salary as the dependent variable are better predictive models than those using the salary as the dependent variable. This is probably because salaries tend to be incremented in *percentages* rather than dollar values. When a dependent variable undergoes percentage changes as the independent variables are varied, the logarithm of the dependent variable will be more suitable as a dependent variable. The model we propose is

$$y = \beta_0 + \beta_1 x_1 + \beta_2 x_2 + \beta_3 x_3 + \beta_4 x_4 + \beta_5 x_5 + \beta_6 x_6 + \beta_7 x_7 + \varepsilon$$

where $y = \log(\text{Executive salary})$, $x_6 = x_1^2$ (second-order term in years of experience), and $x_7 = x_3 x_4$ (cross product or interaction term between sex and number of employees supervised). The variable x_3 is a **dummy variable**; it is used to describe an independent variable that is not measured on a numerical scale, but instead is **qualitative (categorical)** in nature. Sex is such a variable, since its values, male and female, are categories rather than numbers. Thus, we assign the value $x_3 = 1$ if the executive is male, $x_3 = 0$ if the executive is female. (For more detail on the use and interpretation of dummy variables, see Chapter 12.) The interaction term, $x_3 x_4$, allows for the possibility that the relationship between the number of employees supervised, x_4, and corporate salary is dependent on sex, x_3. For example, as the number of supervised employees increases, with all other factors being equal, a woman's salary might rise more rapidly than a man's. (The concept of interaction is also explained in more detail in Chapter 12.)

STEP 2   Now, we estimate the model coefficients $\beta_0, \beta_1, \ldots, \beta_7$. Suppose that a sample of 100 executives is selected, and the variables y and $x_1, x_2, \ldots, x_7$ are recorded (or, in the case of x_6 and x_7, calculated). The sample is then used as input for a computer regression routine; the SAS output is shown in Figure 11.18. The least squares model is

$$\hat{y} = 8.88 + .045x_1 + .033x_2 + .119x_3 + .00033x_4 + .0020x_5 - .00072x_6 + .00031x_7$$

STEP 3   The next step is to specify the probability distribution of ε, the random error component. We assume that ε is normally distributed, with a mean of 0 and a constant variance σ^2. Furthermore, we assume that the errors are independent. The estimate of the variance, σ^2, is given in the SAS printout as

$$s^2 = \text{MSE} = \frac{\text{SSE}}{n - (k + 1)} = \frac{\text{SSE}}{100 - (7 + 1)} = .0021$$

FIGURE 11.18 SAS Printout for Executive Compensation Example

SOURCE	DF	SUM OF SQUARES	MEAN SQUARE	F VALUE	PR > F
MODEL	7	27.06425564	3.86632223	1819.30	0.0001
ERROR	92	0.19551523	0.00212517		ROOT MSE
CORRECTED TOTAL	99	27.25977087		R-SQUARE	0.0460995
				0.992828	

PARAMETER	ESTIMATE	T FOR H0: PARAMETER = 0	PR > ¦T¦	STD ERROR OF ESTIMATE
INTERCEPT	8.87878688	192.49	0.0001	0.04612667
X1 (EXPERIENCE)	0.04460301	26.83	0.0001	0.00166257
X2 (EDUCATION)	0.03326230	12.31	0.0001	0.00270306
X3 (SEX)	0.11892473	6.89	0.0001	0.01724977
X4 (EMPLOYEES SUPERVISED)	0.00033216	19.97	0.0001	0.00001664
X5 (ASSETS)	0.00201021	73.25	0.0001	0.00002744
X6 (= X1*X1)	-0.00071702	-15.11	0.0001	0.00004746
X7 (= X3*X4)	0.00031244	16.16	0.0001	0.00001933

STEP 4 We now want to see how well the model predicts salaries. First, note that $R^2 = .993$. This implies that 99.3% of the variation in y (logarithm of salaries) for these 100 sampled executives is accounted for by the model. The significance of this can be tested:

H_0: $\beta_1 = \beta_2 = \cdots = \beta_7 = 0$

H_a: At least one of the model coefficients is nonzero

Test statistic: $F = \dfrac{R^2/k}{(1 - R^2)/[n - (k + 1)]}$

Rejection region: For $\alpha = .05$, $F > F_{.05}$,
where $\nu_1 = k = 7$ df, $\nu_2 = n - (k + 1) = 92$ df

where from Table VIII in Appendix B, $F_{.05} \approx 2.1$. The test statistic is given on the SAS printout. Since $F = 1,819.3$ exceeds the tabulated value of F and, in fact, has an observed significance level of .0001, we conclude that the model does contribute information for predicting executive salaries. It appears that at least one of the β parameters in the model differs from 0.

We may be particularly interested in whether the data provide evidence that the mean salary of executives increases as the asset value of the company increases, when all other variables (experience, education, etc.) are held constant. In other words, we may want to know whether the data provide sufficient evidence to show that $\beta_5 > 0$. We use the following test:

H_0: $\beta_5 = 0$

H_a: $\beta_5 > 0$

Test statistic: $t = \dfrac{\hat{\beta}_5}{s_{\hat{\beta}_5}}$

For $\alpha = .05$, $n = 100$, $k = 7$, and df $= n - (k + 1) = 92$, we will reject H_0 if $t > t_{.05}$, where (because the degrees of freedom of t are so large) $t_{.05} \approx z_{.05} = 1.645$. Thus, we reject H_0 if

$$t > 1.645$$

The computed t value corresponding to the independent variable, x_5, is 73.25 (see Figure 11.18). Since this value exceeds 1.645 (and has an observed significance level of .0001), we find evidence that the mean salary of executives does increase as the company assets increase, when all other variables are held constant.

STEP 5　The culmination of the modeling effort is to use the model for estimation and prediction. Suppose a firm is trying to determine fair compensation for an executive with the characteristics shown in Table 11.4. The least squares model can be used to obtain a predicted value for the logarithm of salary. That is,

$$\hat{y} = \hat{\beta}_0 + \hat{\beta}_1(12) + \hat{\beta}_2(16) + \hat{\beta}_3(0) + \hat{\beta}_4(400) + \hat{\beta}_5(160.1) + \hat{\beta}_6(144) + \hat{\beta}_7(0)$$

TABLE 11.4
Values of Independent Variables for a Particular Executive

$x_1 = 12$ years of experience
$x_2 = 16$ years of education
$x_3 = 0$ (female)
$x_4 = 400$ employees supervised
$x_5 = \$160.1$ million (the firm's asset value)
$x_6 = x_1^2 = 144$
$x_7 = x_3 x_4 = 0$

This predicted value is given in Figure 11.19, a partial reproduction of the SAS regression printout for this problem: $\hat{y} = 10.298$. The 95% prediction interval is also given: from 10.203 to 10.392. To predict the salary of an executive with these characteristics we take the antilogarithm of these values. That is, the predicted salary is $e^{10.298} = \$29,700$ (rounded to the nearest hundred) and the 95% prediction interval is from $e^{10.203}$ to $e^{10.392}$ (or from \$27,000 to \$32,600). Thus, an executive with the characteristics given in Table 11.4 should be paid between \$27,000 and \$32,600 to be consistent with the sample data.

FIGURE 11.19　SAS Printout for Executive Compensation Problem

X1	X2	X3	X4	X5	X6	X7	PREDICTED VALUE	LOWER 95% CL INDIVIDUAL	UPPER 95% CL INDIVIDUAL
12	16	0	400	160.1	144	0	10.29766682	10.20298295	10.39235070

| | | | | | | | | | |

S E C T I O N 11.9

STATISTICAL COMPUTER PROGRAMS

There are a number of different statistical program packages; some of the most popular are BMDP, Minitab, SAS, and SPSS (see the references at the end of the chapter). You may have access to one or more of these packages at your computer center.

The multiple regression programs for these packages may differ in what they are programmed to do, how they do it, and the appearance of their computer printouts, but all of them print the basic outputs needed for regression analysis. For example, some will compute confidence intervals for $E(y)$ and prediction intervals for y. Others will not. Some test the null hypotheses that the individual β parameters equal 0 using Student's t tests, while others use F tests.* But all give the least squares estimates, the values of SSE, s^2, etc.

To illustrate, the Minitab, SAS, and SPSS regression analysis computer printouts for Example 11.3 are shown in Figure 11.20 (pages 598–599). For that example, we fit the model

$$y = \beta_0 + \beta_1 x_1 + \beta_2 x_2 + \beta_3 x_1 x_2 + \varepsilon$$

to $n = 32$ data points. The variables in the model were

y = Auction price

x_1 = Age of clock (years)

x_2 = Number of bidders

Notice that the Minitab printout gives the prediction equation at the top of the printout. The independent variables, shown in the prediction equation and listed at the left side of the printout, are AGE, BIDDERS, and AGE-BID. Each program treats the product AGE-BID = AGE × BIDDERS as a third independent variable, which must be computed before the fitting begins. For this reason, the prediction equation will always appear on the printout as first-order even though some of the independent variables shown in the prediction equation may actually be the squares or cross products of other independent variables. The SPSS printout in Figure 11.20 lists the independent variables in a different order: AGE_BID, AGE, and BIDDERS. The order is determined statistically rather than by the user, and the user must therefore pay attention to the name of each variable when interpreting the SPSS printout. Note that SPSS also prints the intercept last rather than first, using the label of (Constant). Minitab prints the intercept first, but with no label.

The estimates of the regression coefficients appear opposite the identifying variable in the Minitab column titled COEFFICIENT, in the SAS column titled Parameter Estimate, and in the SPSS column titled B. Compare the estimates given in these three columns. Note that the Minitab printout gives the estimates with a lesser degree of accuracy (fewer decimal places) than the SAS and SPSS printouts. (Ignore the column titled Beta in the SPSS printout. These are standardized estimates and will not be discussed in this text.)

The estimated standard errors of the estimates are given in the Minitab column titled ST. DEV. OF COEF., in the SAS column titled Standard Error, and in the SPSS column titled SE B.

The values of the test statistics for testing H_0: $\beta_i = 0$, where $i = 1, 2, 3$, are shown in the Minitab column titled T-RATIO = COEF/S.D., in the SAS column

---

*A two-tailed Student's t test based on ν df is equivalent to an F test where the F statistic has 1 df in the numerator and ν df in the denominator. See Section 11.5.

FIGURE 11.20   Computer Printouts for Example 11.3

(a) Minitab Regression Printout

```
THE REGRESSION EQUATION IS
PRICE = 323 + 0.87 AGE - 93.4 BIDDERS + 1.30 AGE-BID

                                 ST. DEV.     T-RATIO =
COLUMN        COEFFICIENT        OF COEF.     COEF/S.D.
                 322.8             293.3        1.10
AGE              0.873             2.020        0.43
BIDDERS         -93.41            29.71        -3.14
AGE-BID          1.2979            0.2110       6.15


S = 88.37


R-SQUARED = 95.4 PERCENT
R-SQUARED = 94.9 PERCENT, ADJUSTED FOR D.F.

ANALYSIS OF VARIANCE

DUE TO        DF           SS       MS=SS/DF
REGRESSION     3        4572548      1524183
RESIDUAL      28         218646         7809
TOTAL         31        4791194
```

(b) SAS Regression Printout

```
Dep Variable: PRICE

                        Analysis of Variance

                     Sum of        Mean
Source      DF       Squares       Square       F Value    Prob>F

Model        3  4572547.9872  1524182.6624     195.188    0.0001
Error       28   218646.23157     7808.79398
C Total     31  4791194.2187

        Root MSE      88.36738    R-Square      0.9544
        Dep Mean    1327.15625    Adj R-Sq      0.9495
        C.V.           6.65840

                      Parameter Estimates

                     Parameter      Standard     T for H0:
Variable   DF        Estimate         Error    Parameter=0   Prob > |T|

INTERCEP    1      322.754353   293.32514660      1.100       0.2806
AGE         1        0.873288     2.01965115      0.432       0.6688
BIDDERS     1      -93.409920    29.70767946     -3.144       0.0039
AGE_BID     1        1.297898     0.21102602      6.150       0.0001
```

(c) SPSS Regression Printout
_____

```
Equation Number 1      Dependent Variable..    PRICE


Beginning Block Number  1.  Method:  Enter
    AGE       BIDDERS  AGE_BID



Variable(s) Entered on Step Number
    1..     AGE_BID
    2..     AGE
    3..     BIDDERS



Multiple R               .97692
R Square                 .95436
Adjusted R Square        .94948
Standard Error         88.36738


Analysis of Variance
                         DF      Sum of Squares      Mean Square
Regression                3      4572547.98718    1524182.66239
Residual                 28       218646.23157       7808.79398

F =      195.18797        Signif F =   .0000



------------------- Variables in the Equation -------------------

Variable              B          SE B        Beta       T   Sig T

AGE_BID          1.29790        .21103     1.37032    6.150  .0000
AGE               .87329       2.01965      .06085     .432  .6688
BIDDERS        -93.40992      29.70768     -.67471   -3.144  .0039
(Constant)     322.75435     293.32515               1.100  .2806
```
_____

titled T for H0: Parameter = 0, and in the SPSS column titled T. Note that the computed *t* values shown are identical (except for the number of decimal places), but Minitab does not give the observed significance level of the test. Consequently, to draw conclusions from the Minitab printout, you must compare the computed values of *t* with the critical values given in a *t* table (Table VI in Appendix B). In contrast, the SAS printout gives the observed significance level for each *t* test in the column titled Prob > |T|, and the SPSS printout in the column titled Sig T. Note that these observed significance levels have been computed assuming that the tests are two-tailed. The observed significance levels for one-tailed tests would equal half of these values.

   The Minitab printout gives the value SSE = 218646 under the ANALYSIS OF VARIANCE column headed SS and in the row identified as RESIDUAL. The value of $s^2 = 7809$ is shown in the same row under the column headed MS = SS/DF, and the degrees of freedom, DF, appears in the same row as 28. The corresponding values are shown at the top of the SAS printout in the row labeled Error and in the columns designated as Sum of Squares, Mean Square, and DF, respectively. These quantities appear with similar headings in the SPSS printout, but Error is called Residual in SPSS. The standard deviation s for the regression equation is given on each printout, but with different labels. Minitab shows $s = 88.37$ in the center of the printout; SAS labels it Root MSE in the left center of the printout. SPSS calls the standard deviation the Standard Error and prints it in the center of the printout. All the estimates are identical, except for the number of decimal places.

The value of R^2, as defined in Section 11.6, is given in the Minitab printout as 95.4 PERCENT (we defined this quantity as a ratio where $0 \leq R^2 \leq 1$). It is given in the SAS printout as 0.9544, and shown in the left column of the SPSS printout as 0.95436. (Ignore the quantities shown in the Minitab printout as R^2 ADJUSTED FOR D.F., in the SAS printout as Adj R Sq, and in the SPSS printout as Adjusted R Square. These quantities are adjusted for the degrees of freedom associated with the total SS and SSE and are not used or discussed in this text.)

The F statistic for testing the usefulness of the model (Section 11.6)—i.e., testing the null hypothesis that all model parameters (except β_0) equal zero—is shown under the title F VALUE as 195.188 at the top of the SAS printout. In addition, the SAS printout gives the observed significance level of this F test under Prob > F as 0.0001. This F value, 195.18797, is also printed at the center of the SPSS printout with the observed significance level given as Signif F. The F statistic for testing the usefulness of the model is not given in the Minitab printout. If you are using Minitab and wish to obtain the value of this statistic, you must compute it using the analysis of variance formula given in Section 11.6:

$$F = \frac{\text{MEAN SQUARE(MODEL)}}{\text{MEAN SQUARE(ERROR)}}$$

These quantities are given in the Minitab printout under the column marked MS = SS/DF. The rows are labeled REGRESSION and RESIDUAL, which are synonymous with MODEL and ERROR, respectively. Thus,

$$F = \frac{1,524,183}{7,809} = 195.18$$

a value that agrees with the values given in the SAS and SPSS printouts. The logic behind this test and other tests of hypotheses concerning sets of the β parameters will be further discussed in Section 12.4.

Although we will henceforth use the three popular statistical packages discussed in this section (SAS, Minitab, and SPSS), there are many others available. The rapid expansion of the microcomputer market has been accompanied by significant growth in the availability of statistical software. It is important to evaluate any new package carefully to assure that the results it produces are complete and correct. Fortunately, there is usually much similarity in the computer printouts produced by the packages, so you will find it relatively easy to learn how to read a new regression printout after you have become familiar with those used in this text. We will intermix the different packages in the examples and exercises to help familiarize you with different formats.

| | | | | | | | | | | | | |

S E C T I O N 11.10

RESIDUAL ANALYSIS: CHECKING THE REGRESSION ASSUMPTIONS

When we apply a regression analysis to a set of data, we never know for certain that the assumptions of Section 11.2 are satisfied. How far can we deviate from the assumptions and still expect a multiple regression analysis to yield results that will have the reliability stated in this chapter? How can we detect departures (if they exist) from the assumptions of Section 11.2 and what can we do about them? We provide some partial answers to these questions in this section and will direct you to further discussion in succeeding chapters.

Remember from Section 11.2 that

$$y = E(y) + \varepsilon$$

where the expected value $E(y)$ of y for a given set of values of $x_1, x_2, \ldots, x_k$ is

$$E(y) = \beta_0 + \beta_1 x_1 + \beta_2 x_2 + \cdots + \beta_k x_k$$

and ε is a random error. The first assumption we made was that the mean value of the random error for *any* given set of values of $x_1, x_2, \ldots, x_k$ is $E(\varepsilon) = 0$. One consequence of this assumption is that the mean $E(y)$ for a specific set of values of $x_1, x_2, \ldots, x_k$ is

$$E(y) = \beta_0 + \beta_1 x_1 + \beta_2 x_2 + \cdots + \beta_k x_k$$

That is,

$$y = \underbrace{E(y)}_{\substack{\text{Mean value of } y \\ \text{for specific values} \\ \text{of } x_1, x_2, \ldots, x_k}} + \underbrace{\varepsilon}_{\substack{\text{Random} \\ \text{error}}}$$

The second consequence of the assumption is that the least squares estimators of the model parameters, $\beta_0, \beta_1, \beta_2, \ldots, \beta_k$, will be unbiased regardless of the remaining assumptions that we attribute to the random errors and their probability distributions.

The properties of the sampling distributions of the parameter estimators $\hat{\beta}_0, \hat{\beta}_1, \ldots, \hat{\beta}_k$ will depend on the remaining assumptions that we specify concerning the probability distributions of the random errors. Recall that we assumed that for any given set of values of $x_1, x_2, \ldots, x_k$, ε has a normal probability distribution with mean equal to 0 and variance equal to σ^2. Also, we assumed that the random errors are probabilistically independent.

It is unlikely that these assumptions are ever satisfied exactly in a practical application of regression analysis. Fortunately, experience has shown that least squares regression analysis produces reliable statistical tests, confidence intervals, and prediction intervals as long as the departures from the assumptions are not too great. In this section and in Chapter 12 we present some methods for determining whether the data indicate significant departures from the assumptions.

Because the assumptions all concern the random error component, ε, of the model, the first step is to estimate the random error. Since the actual random error associated with a particular value of y is the difference between the actual y value and its mean, we estimate the error by the difference between the actual y value and the *estimated* mean. This estimated error is called the **regression residual**, or simply the **residual**, and is denoted by $\hat{\varepsilon}$. The actual error ε and residual $\hat{\varepsilon}$ are shown in Figure 11.21.

$$
\begin{aligned}
\text{Actual random error} &= \varepsilon \\
&= (\text{Actual } y \text{ value}) - (\text{Mean of } y) \\
&= y - E(y) = y - (\beta_0 + \beta_1 x_1 + \beta_2 x_2 + \cdots + \beta_k x_k)
\end{aligned}
$$

$$
\begin{aligned}
\text{Estimated random error (residual)} &= \hat{\varepsilon} \\
&= (\text{Actual } y \text{ value}) - (\text{Estimated mean of } y) \\
&= y - \hat{y} = y - (\hat{\beta}_0 + \hat{\beta}_1 x_1 + \hat{\beta}_2 x_2 + \cdots + \hat{\beta}_k x_k)
\end{aligned}
$$

FIGURE 11.21

Actual Random Error and Regression Residual

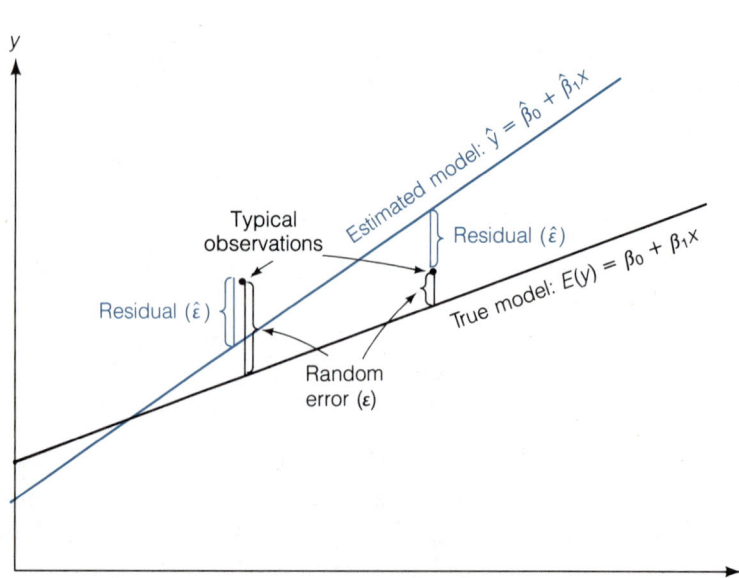

Since the true mean of y (i.e., the true regression model) is not known, the actual random error cannot be calculated. However, because the residual is based on the estimated mean (the least squares regression model), it can be calculated and used to estimate the random error and to check the regression assumptions. Such checks are generally referred to as **residual analyses**. Some useful properties of residuals are given in the box.

PROPERTIES OF REGRESSION RESIDUALS

1. A residual is equal to the difference between the observed y value and its estimated (regression) mean:

Residual $= y - \hat{y}$

2. The mean of the residuals is equal to 0. This property follows from the fact that the sum of the differences between the observed y values and the least squares regression model is equal to 0.

$$\sum(\text{Residuals}) = \sum(y - \hat{y}) = 0$$

3. The standard deviation of the residuals is equal to the standard deviation of the regression model, s. This property follows from the fact that the sum of the squared residuals is equal to SSE, which when divided by the error degrees of freedom is equal to the variance of the regression model, s^2. The square root of the variance is both the standard deviation of the residuals and the standard deviation of the regression model.

$$\sum(\text{Residuals})^2 = \sum(y - \hat{y})^2 = \text{SSE}$$

$$s = \sqrt{\frac{\sum(\text{Residuals})^2}{n - (k + 1)}} = \sqrt{\frac{\text{SSE}}{n - (k + 1)}}$$

The following examples show how the analysis of regression residuals can be used to verify the assumptions associated with the model, and to improve the model when the assumptions do not appear to be satisfied. Although the residuals can be calculated and plotted by hand, we rely on the computer for these tasks in the examples and exercises. Most statistical computer packages now include residual analyses as a standard component of their regression modeling programs.

EXAMPLE 11.4

The data for the home size–electrical usage example used throughout this chapter are repeated in Table 11.5 (page 604). SAS printouts for a straight-line model and a quadratic model fitted to the data are shown in Figures 11.22(a) and 11.22(b), respectively. The residuals from these models are shaded in the printouts. The residuals are then plotted on the vertical axis against the variable x, size of home, on the horizontal axis in Figures 11.23(a) and 11.23(b), respectively (pages 606–607).

a. Verify that each residual is equal to the difference between the observed y value and the estimated mean value.
b. Analyze the residual plots.

TABLE 11.5
Home Size–Electrical Usage Data

SIZE OF HOME x (square feet)	MONTHLY USAGE y (kilowatt-hours)	SIZE OF HOME x (square feet)	MONTHLY USAGE y (kilowatt-hours)
1,290	1,182	1,840	1,711
1,350	1,172	1,980	1,804
1,470	1,264	2,230	1,840
1,600	1,493	2,400	1,956
1,710	1,571	2,930	1,954

FIGURE 11.22
SAS Printouts for Electrical Usage Example

(a) Straight-line Model

Dep Variable: Y

Analysis of Variance

Source	DF	Sum of Squares	Mean Square	F Value	Prob>F
Model	1	703957.18342	703957.18342	39.536	0.0002
Error	8	142444.91658	17805.61457		
C Total	9	846402.10000			

Root MSE	133.43768	R-Square	0.8317	
Dep Mean	1594.70000	Adj R-Sq	0.8107	
C.V.	8.36757			

Parameter Estimates

Variable	DF	Parameter Estimate	Standard Error	T for H0: Parameter=0	Prob > |T|
INTERCEP	1	578.927752	166.96805715	3.467	0.0085
X	1	0.540304	0.08592981	6.288	0.0002

Obs	Y	Predict Value	Residual
1	1182.0	1275.9	-93.9204
2	1172.0	1308.3	-136.3
3	1264.0	1373.2	-109.2
4	1493.0	1443.4	49.5852
5	1571.0	1502.8	68.1517
6	1711.0	1573.1	137.9
7	1804.0	1648.7	155.3
8	1840.0	1783.8	56.1935
9	1956.0	1875.7	80.3417
10	1954.0	2162.0	-208.0

Sum of Residuals   0
Sum of Squared Residuals   142444.9166

(b) Quadratic Model

Dep Variable: Y

Analysis of Variance

Source	DF	Sum of Squares	Mean Square	F Value	Prob>F
Model	2	831069.54637	415534.77319	189.710	0.0001
Error	7	15332.55363	2190.36480		
C Total	9	846402.10000			

Root MSE	46.80133	R-Square	0.9819	
Dep Mean	1594.70000	Adj R-Sq	0.9767	
C.V.	2.93480			

Parameter Estimates

Variable	DF	Parameter Estimate	Standard Error	T for H0: Parameter=0	Prob > |T|
INTERCEP	1	-1216.143887	242.80636850	-5.009	0.0016
X	1	2.398930	0.24583560	9.758	0.0001
XSQ	1	-0.000450	0.00005908	-7.618	0.0001

Obs	Y	Predict Value	Residual
1	1182.0	1129.6	52.4359
2	1172.0	1202.2	-30.2136
3	1264.0	1337.8	-73.7916
4	1493.0	1470.0	22.9586
5	1571.0	1570.1	0.9359
6	1711.0	1674.2	36.7685
7	1804.0	1769.4	34.5998
8	1840.0	1895.5	-55.4654
9	1956.0	1949.1	6.9431
10	1954.0	1949.2	4.8287

Sum of Residuals   -2.27374E-12
Sum of Squared Residuals   15332.5536

SOLUTION

a. For the straight-line model the residual is calculated for the first y value as follows:

Residual = (Observed y value) − (Estimated mean)

$$= y - \hat{y} = 1{,}182 - 1{,}275.9 = -93.9$$

where the estimated mean is the first number (shaded) in the column labeled Predict Value on the SAS printout in Figure 11.22(a). Similarly, the residual for the first y value using the quadratic model is

Residual = $1{,}182 - 1{,}129.6 = 52.4$

Both residuals agree (after rounding) with the first values given in the column labeled Residual in Figures 11.22(a) and 11.22(b), respectively. Although the residuals both correspond to the same observed y value, 1,182, they differ because the estimated mean value changes depending on whether the straight-line model or quadratic model is used. Similar calculations produce the remaining residuals.

b. The plot of the residuals for the straight-line model [Figure 11.23(a)] reveals a nonrandom pattern. The residuals exhibit a mound shape, with the residuals for the small values of x below the 0 (mean) line, the residuals corresponding to the middle values of x above the 0 line, and the residual for the largest value of x again below the 0 line. The indication is that the mean value of the random error (which the residuals estimate) *within* the ranges of x (small, medium, large) may not be equal to 0. Such a pattern usually indicates that curvature needs to be added to the model.

FIGURE 11.23
Residual Plots for
Electrical Usage Example

(a) Straight-line model

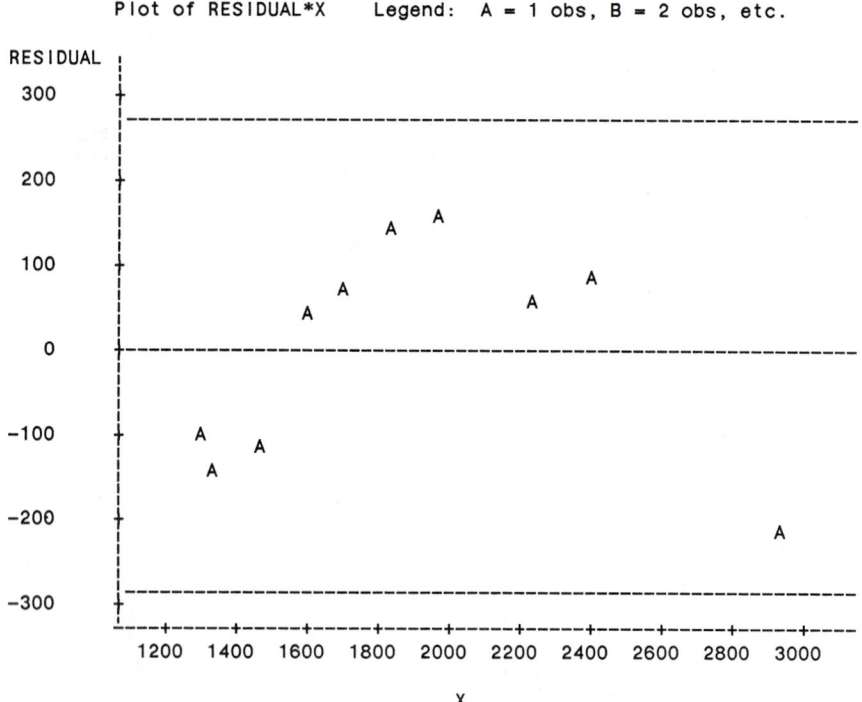

When the second-order term is added to the model, the pattern disappears in Figure 11.23(b), where the residuals appear to be randomly distributed around the 0 line, as expected. Note, too, that the ±2 standard deviation lines are at about ±95 on the quadratic residual plot, compared to (about) ±275 on the straight-line plot. The implication is that the quadratic model provides a considerably better model for predicting electrical usage, verifying our conclusions from previous analyses in this chapter. ∎

FIGURE 11.23 (cont.)
(b) Quadratic Model

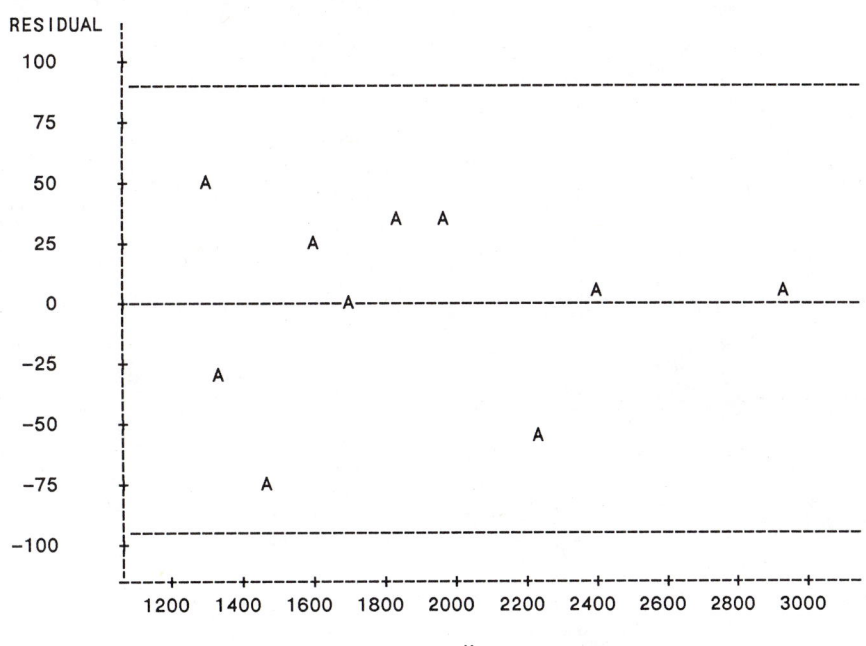

Residual analyses are also useful for detecting one or more observations that deviate significantly from the regression model. We expect approximately 95% of the residuals to fall within 2 standard deviations of the 0 line, and all or almost all of them to lie within 3 standard deviations of their mean of 0. Residuals that are extremely far from the 0 line, and disconnected from the bulk of the other residuals, are called **outliers**, and should receive special attention from the regression analyst.

EXAMPLE 11.5

The data for the grandfather clock example used throughout this chapter are repeated in Table 11.6 (page 608), with one important difference: the auction price of the clock at the top of the second column has been changed from $2,131 to $1,131 (shaded in Table 11.6). The interaction model

$$E(y) = \beta_0 + \beta_1 x_1 + \beta_2 x_2 + \beta_3 x_1 x_2$$

is again fit to these (modified) data, with the printout shown in Figure 11.24. The residuals are shown shaded in the printout, and then plotted against the number of bidders, x_2, in Figure 11.25 (page 610). Analyze the residual plot.

TABLE 11.6
Auction Price Data

AGE x_1	NUMBER OF BIDDERS x_2	AUCTION PRICE y	AGE x_1	NUMBER OF BIDDERS x_2	AUCTION PRICE y
127	13	$1,235	170	14	$1,131
115	12	1,080	182	8	1,550
127	7	845	162	11	1,884
150	9	1,522	184	10	2,041
156	6	1,047	143	6	854
182	11	1,979	159	9	1,483
156	12	1,822	108	14	1,055
132	10	1,253	175	8	1,545
137	9	1,297	108	6	729
113	9	946	179	9	1,792
137	15	1,713	111	15	1,175
117	11	1,024	187	8	1,593
137	8	1,147	111	7	785
153	6	1,092	115	7	744
117	13	1,152	194	5	1,356
126	10	1,336	168	7	1,262

SOLUTION

The residual plot dramatically reveals the one altered measurement. Note that one of the two residuals at $x_2 = 14$ bidders falls more than 3 standard deviations below the 0 line. Note that no other residual lies more than 2 standard deviations from the (mean) 0 line.

What do we do with outliers once we identify them? First, we try to determine the cause. Were the data entered into the computer incorrectly? Was the observation recorded incorrectly when the data were collected? If so, correct the observation and rerun the program. Another possibility is that the observation is not representative of the conditions you are trying to model. For example, in this case the low price may be attributable to extreme damage to the clock, or to a clock of inferior quality compared to the others. In these cases we probably would exclude the observation from the analysis. In many cases you may not be able to determine the cause of the outlier. Even so, you may want to rerun the regression analysis excluding the outlier in order to assess the effect of that observation on the results of the analysis.

Figure 11.26 (page 610) shows the printout when the outlier observation is excluded from the grandfather clock analysis, and Figure 11.27 shows the new plot of the residuals vs. the number of bidders. Now only one of the residuals lies beyond 2 standard deviations from 0, and none of them lie beyond 3 standard deviations. Also, the model statistics indicate a much better model without the outlier. Most notably, the standard deviation (s) has decreased from 200.3 to 85.28, indicating a model that will provide more precise estimates and predictions (narrower confidence and prediction intervals) for clocks that are similar to those in the reduced sample. Remember, though, that if you decide to remove the outlier from the analysis, and in fact it belongs to the same population as the rest of the sample, the resulting model may provide misleading estimates and predictions. ∎

FIGURE 11.24
Printout for Grandfather
Clock Example with
Altered Data

THE REGRESSION EQUATION IS
PRICE = - 511 + 8.16 AGE + 19.7 BIDDERS + 0.320 AGE-BID

COLUMN	COEFFICIENT	ST. DEV. OF COEF.	T-RATIO = COEF/S.D.
	-510.5	664.8	-0.77
AGE	8.160	4.577	1.78
BIDDERS	19.74	67.33	0.29
AGE-BID	0.3197	0.4783	0.67

S = 200.3

R-SQUARED = 73.0 PERCENT
R-SQUARED = 70.1 PERCENT, ADJUSTED FOR D.F.

ANALYSIS OF VARIANCE

DUE TO	DF	SS	MS=SS/DF
REGRESSION	3	3029141	1009714
RESIDUAL	28	1123116	40111
TOTAL	31	4152256	

ROW	AGE	Y PRICE	PRED. Y VALUE	ST.DEV. PRED. Y	RESIDUAL
1	127	1235.0	1310.3	59.2	-75.3
2	115	1080.0	1106.0	62.0	-26.0
3	127	845.0	948.2	61.0	-103.2
4	150	1522.0	1322.8	37.1	199.2
5	156	1047.0	1180.2	60.2	-133.2
6	182	1979.0	1831.8	82.8	147.2
7	156	1822.0	1597.9	61.8	224.1
8	132	1253.0	1186.1	39.7	66.9
9	137	1297.0	1179.3	39.0	117.7
10	113	946.0	914.4	58.5	31.6
11	137	1713.0	1560.6	78.3	152.4
12	117	1024.0	1072.8	53.0	-48.8
13	137	1147.0	1115.8	44.2	31.2
14	153	1092.0	1149.9	58.9	-57.9
15	117	1152.0	1187.1	69.6	-35.1
16	126	1336.0	1117.9	43.4	218.1
17	170	1131.0	1914.0	116.5	-783.0
18	182	1550.0	1598.1	62.7	-48.1
19	162	1884.0	1598.3	56.9	285.7
20	184	2041.0	1776.6	70.6	264.4
21	143	854.0	1049.2	58.8	-195.2
22	159	1483.0	1422.1	40.5	60.9
23	108	1055.0	1130.6	97.8	-75.6
24	175	1545.0	1523.0	55.3	22.0
25	108	729.0	696.4	99.5	32.6
26	179	1792.0	1642.9	57.5	149.1
27	111	1175.0	1223.7	107.1	-48.7
28	187	1593.0	1651.7	68.5	-58.7
29	111	785.0	781.9	80.7	3.1
30	115	744.0	823.5	75.4	-79.5
31	194	1356.0	1481.4	133.3	-125.4
32	168	1262.0	1374.6	57.6	-112.6

FIGURE 11.25

Residual Plot vs. Number of Bidders

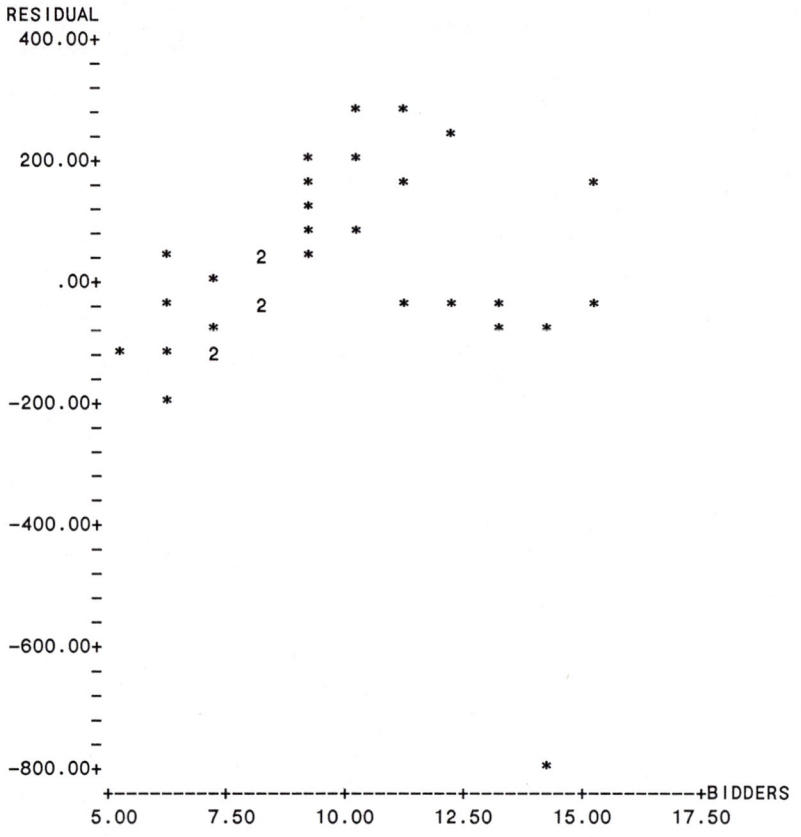

FIGURE 11.26

Minitab Printout with Outlier Excluded

```
THE REGRESSION EQUATION IS
PRICE-2 = 476 - 0.46 AGE-2 - 114 BIDDERS2 + 1.48 AGE-BID2


                                 ST. DEV.      T-RATIO =
COLUMN        COEFFICIENT        OF COEF.      COEF/S.D.
                 475.8            296.2          1.61
AGE-2           -0.465            2.093         -0.22
BIDDERS2      -114.19            31.03          -3.68
AGE-BID2         1.4775           0.2280         6.48


S = 85.28


R-SQUARED = 95.2 PERCENT
R-SQUARED = 94.7 PERCENT, ADJUSTED FOR D.F.


ANALYSIS OF VARIANCE

  DUE TO      DF            SS        MS=SS/DF
REGRESSION     3       3927844       1309282
RESIDUAL      27        196341          7272
TOTAL         30       4124186
```

FIGURE 11.27

Minitab Residual Plot

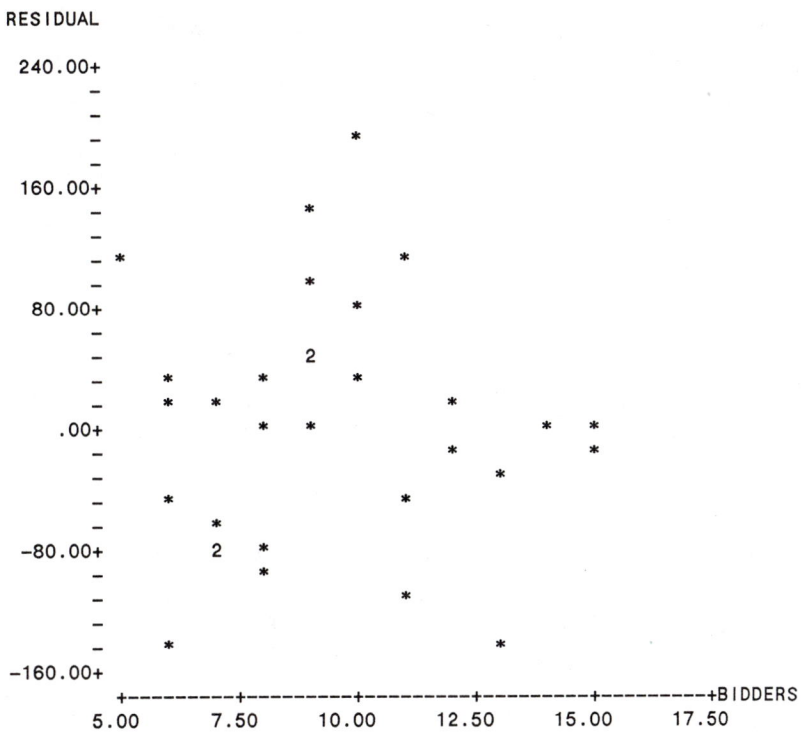

Outlier analysis is another example of testing the assumption that the expected (mean) value of the random error component is 0, since this assumption is in doubt for the error terms corresponding to the outliers. The last example in this section checks the assumption of the normality of the random error component.

EXAMPLE 11.6

Refer to Example 11.5. Use a stem and leaf display (Section 2.4) to plot the frequency distribution of the residuals in the grandfather clock example, both before and after the outlier residual is removed. Analyze the plots and determine whether the assumption of normality of the error distribution is reasonable.

SOLUTION

The stem and leaf displays for the two sets of residuals are constructed using Minitab, and are shown in Figure 11.28 (page 612). Note that the outlier appears to skew the frequency distribution in Figure 11.28(a), while the stem and leaf display in Figure 11.28(b) appears to be more mound-shaped. Although the displays do not provide formal statistical tests of normality, they do provide a descriptive display. Relative frequency histograms can also be used to check the normality assumption. In this example the normality assumption appears to be more plausible after the outlier is removed. Consult the references for methods to conduct statistical tests of normality using the residuals. ∎

FIGURE 11.28*

Stem and Leaf Displays
for Grandfather Clock
Example

(a) Outlier Included

```
STEM-AND-LEAF DISPLAY OF RESIDUAL
LEAF DIGIT UNIT =  10.0000
1 2 REPRESENTS 120.


       STEM  LEAF

    1   -7   8
    1   -6
    1   -5
    1   -4
    1   -3
    1   -2
    6   -1   93210
   16   -0   7775544432
   16    0   0233366
    9    1   14459
    4    2   1268
```

(b) Outlier Excluded

```
STEM-AND-LEAF DISPLAY OF RESIDUAL
LEAF DIGIT UNIT =  10.0000
1 2 REPRESENTS 120.



    3   -1*  331
    9   -0.  987765
  (7)   -0*  4321000
   15   +0*  011223344
    6   +0.  79
    4    1*  004
    1    1.  9
```

A summary of the residual analyses presented in this section to check the assumption that the random error ε is normally distributed with mean 0 is presented in the box. Using the residuals to check the remaining assumptions of constant variance and independence will be presented in Chapters 12 and 14, respectively. Residual analysis is a useful tool for the regression analyst, not only to check the assumptions, but also to provide information about how the model can be improved.

*Recall that the left column of the Minitab printout shows the number of measurements at least as extreme as the stem. In Figure 11.28(a), for example, the 6 corresponding to the STEM = −1 means that six measurements are less than or equal to −100. If one of the numbers in the leftmost column is enclosed in parentheses, the number in parentheses is the number of measurements in that row, and the median is contained in that row.

STEPS IN A RESIDUAL ANALYSIS

1. Calculate and plot the residuals against the independent variables, preferably with the assistance of a computer program.
2. Analyze the plot, looking for curvature—either a mound or bowl shape. Both shapes are distinguished by groups of residuals at the low and high values of the x variable on one side of the 0 line, and the residuals for the medium x values on the opposite side of the 0 line. This shape signals the need for a curvature term in the model. Try a second-order term in the variable against which the residuals are plotted.
3. Examine the residual plots for outliers. Draw lines on the residual plots at 2 and 3 standard deviation distances below and above the 0 line. Examine residuals outside the 3 standard deviation lines as potential outliers, and check to see that approximately 5% of the residuals exceed the 2 standard deviation lines. Determine whether each outlier can be explained as an error in data collection or transcription, or corresponds to a member of a population different from that of the remainder of the sample, or simply represents an unusual observation. If the observation is determined to be an error, fix it or remove it. Even if cause cannot be determined, you may want to rerun the regression analysis without the observation to determine its effect on the analysis.
4. Plot a frequency distribution of the residuals, using a stem and leaf display or a histogram. Check to see if obvious departures from normality exist. Extreme skewness of the frequency distribution may indicate the need for a transformation of the dependent variable, a topic we discuss in Chapter 12.

EXERCISES
11.25–11.34

LEARNING THE MECHANICS

[*Note: Starred* (*) *exercises require the use of a computer.*]

11.25 When a multiple regression model is used for estimating the mean of the dependent variable and for predicting a particular value of y, which will be narrower, the confidence interval for the mean or the prediction interval for the particular y value?

11.26 Refer to Exercise 11.1, in which the model

$$y = \beta_0 + \beta_1 x_1 + \beta_2 x_2 + \varepsilon$$

was fit to $n = 20$ data points. The Minitab regression printout for these data is shown on page 614.
   a. Find the least squares prediction equation.
   b. Find and interpret the standard deviation of the regression model.
   c. Calculate the F statistic. Does the model contribute information for the prediction of y? Test using $\alpha = .05$.
   d. Place a 90% confidence interval on the coefficient β_2.
   e. Find and interpret the coefficient of determination R^2.

Minitab Printout for
Exercise 11.26

```
THE REGRESSION EQUATION IS
Y = 506 - 942 X1 - 429 X2


                                ST. DEV.      T-RATIO =
       COLUMN     COEFFICIENT   OF COEF.      COEF/S.D.
                     506.35        45.17        11.21
       X1           -941.9        275.1         -3.42
       X2           -429.1        379.8         -1.13


       S = 94.25


       R-SQUARED = 45.9 PERCENT
       R-SQUARED = 39.6 PERCENT, ADJUSTED FOR D.F.


       ANALYSIS OF VARIANCE


         DUE TO      DF          SS        MS=SS/DF
         REGRESSION   2       128329         64165
         RESIDUAL    17       151016          8883
         TOTAL       †9       279345
```

11.27 Refer to Exercise 11.26. The SPSS printout for the same data set is given here. Repeat parts a–e of Exercise 11.26 using the SPSS printout.

SPSS Printout for
Exercise 11.27

```
Multiple R              .67779
R Square                .45939
Adjusted R Square       .39579
Standard Error        94.25114


Analysis of Variance
                      DF      Sum of Squares      Mean Square
Regression             2        128329.27624      64164.63812
Residual              17        151015.72376       8883.27787


F =        7.22308      Signif F =   .0054



------------------ Variables in the Equation ------------------


Variable           B          SE B        Beta          T    Sig T


X1          -941.90023    275.08556      -.61728     -3.424   .0032
X2          -429.06042    379.82566      -.20365     -1.130   .2743
(Constant)   506.34607     45.16942                  11.210   .0000
```

11.28 Refer to Exercise 11.12, in which a quadratic model was fit to $n = 19$ data points. Two residual plots are shown here—one corresponding to the fitting of a straight-line model to the data, and the other to the quadratic model fit in Exercise 11.12. Analyze the two plots. Is the need for a quadratic term evident from the residual plot for the straight-line model? Does your conclusion agree with your test of the quadratic term in Exercise 11.12?

Residual Plot for Straight-
line Model

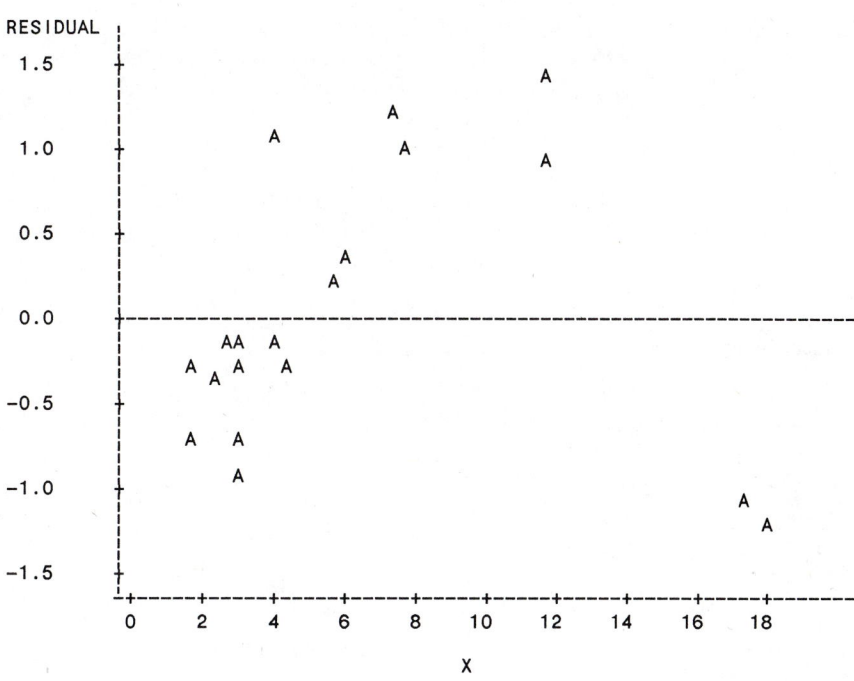

Residual Plot for
Quadratic Model

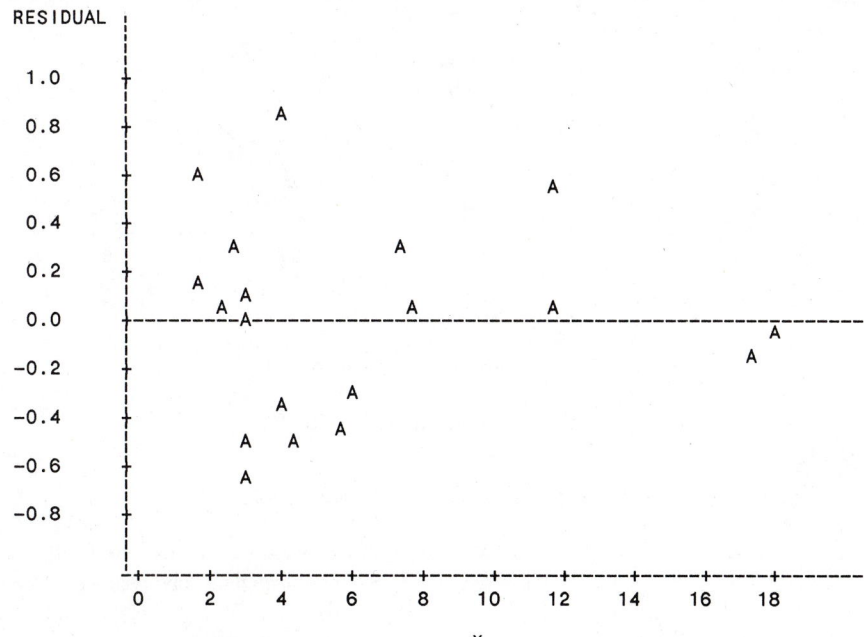

APPLYING THE CONCEPTS

11.29 Refer to Exercise 11.5, in which the 1987 food consumption expenditure, y, of a sample of 25 households in Washington, D.C., was related to the household income, x_1, and size of household, x_2, by the model

$$y = \beta_0 + \beta_1 x_1 + \beta_2 x_2 + \varepsilon$$

Plots of the residuals from this model are shown here and on page 617—one vs. x_1 and one vs. x_2. Analyze the plots. Is there visual evidence of a need for a quadratic term in either x_1 or x_2?

Residual Plot for Exercise 11.29: $\hat{\varepsilon}$ vs. x_1

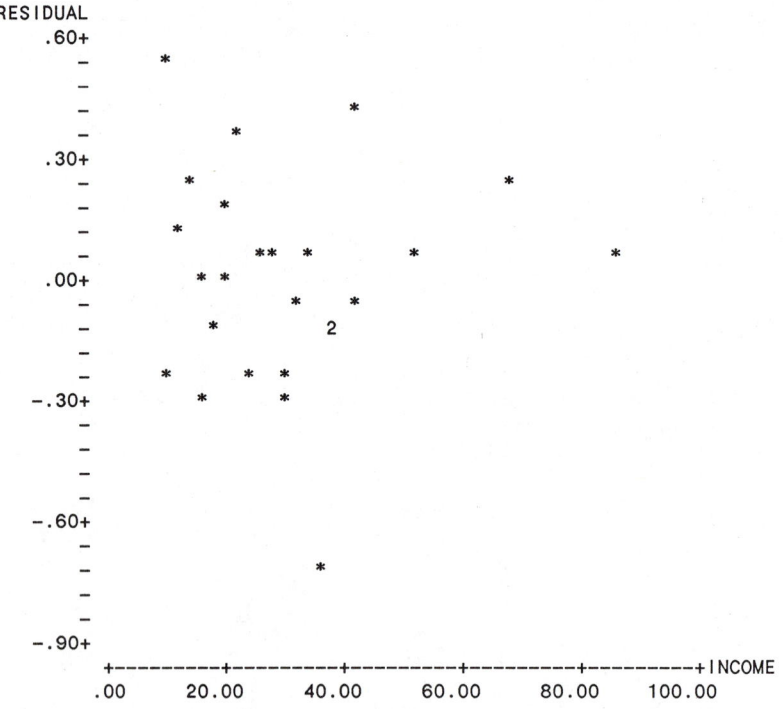

11.30 Refer to Exercise 11.29. Suppose a 26th household is added to the sample, with the following characteristics:

Food consumption: 6.5

Income: 62.3

Persons in household: 5

Two Minitab printouts are shown here—one for the same model fit to the first 25 observations, the second fit to all 26 observations.

Residual Plot for Exercise
11.29: $\hat{\varepsilon}$ vs. x_2

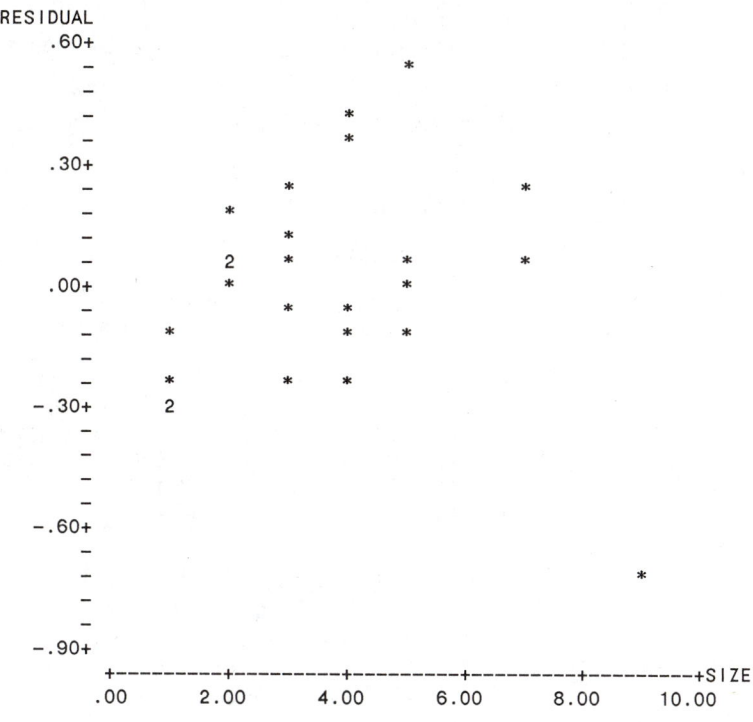

Minitab Printout for
Exercise 11.30: Model Fit
to 25 Observations

```
THE REGRESSION EQUATION IS
FOOD = 1.43 + 0.00999 INCOME + 0.379 SIZE
```

COLUMN	COEFFICIENT	ST. DEV. OF COEF.	T-RATIO = COEF/S.D.
	1.4326	0.1467	9.76
INCOME	0.009991	0.003168	3.15
SIZE	0.37929	0.02772	13.68

```
S = 0.2769
```

```
R-SQUARED = 90.2 PERCENT
R-SQUARED = 89.3 PERCENT, ADJUSTED FOR D.F.
```

```
ANALYSIS OF VARIANCE
```

DUE TO	DF	SS	MS=SS/DF
REGRESSION	2	15.4623	7.7311
RESIDUAL	22	1.6873	0.0767
TOTAL	24	17.1496	

Minitab Printout for
Exercise 11.30: Model Fit
to 26 Observations

```
THE REGRESSION EQUATION IS
FOOD = 1.16 + 0.0187 INCOME + 0.406 SIZE

                                 ST. DEV.      T-RATIO =
   COLUMN       COEFFICIENT      OF COEF.      COEF/S.D.
                   1.1554          0.2882         4.01
   INCOME        0.018735         0.006034        3.11
   SIZE          0.40577          0.05551         7.31


   S = 0.5581


   R-SQUARED = 74.6 PERCENT
   R-SQUARED = 72.4 PERCENT, ADJUSTED FOR D.F.


   ANALYSIS OF VARIANCE

     DUE TO      DF         SS        MS=SS/DF
   REGRESSION     2       21.076       10.538
   RESIDUAL      23        7.163        0.311
   TOTAL         25       28.239
```

a. Record the least squares estimates of the model parameters, and note differences in the estimates. Interpret each estimate.

b. Find and interpret the standard deviation for each model.

c. Conduct the analysis of variance F test for each model using $\alpha = .05$.

d. Place a 95% confidence interval on the mean rate of change in food consumption per additional person in the household for each model, assuming household income is constant.

e. According to the results of parts **a–d**, how much influence does the additional observation have on the model?

11.31 Refer to Exercises 11.29 and 11.30. Residual plots against household income and size of household for the models corresponding to 26 observations are shown on page 619.

a. Analyze the plots. Are there any outliers?

b. What are possible explanations for any outliers you identified in part **a**?

11.32 Refer to Exercises 11.29–11.31. The accompanying stem and leaf displays represent the frequency distributions of the residuals for the two data sets, one with $n = 25$ (below) and one with $n = 26$ (page 620). Analyze the displays, especially with regard to the normality assumption.

Stem and Leaf Display for
25 Residuals

```
STEM-AND-LEAF DISPLAY OF RESIDUAL
LEAF DIGIT UNIT =     .1000
1 2 REPRESENTS 1.2



     1    -0S   7
     1    -0F
     6    -0T   32222
    11    -0*   11100
    (8)   +0*   00000001
     6    +0T   2223
     2    +0F   45
```

Residual Plot for Exercise
11.31: $\hat{\varepsilon}$ vs. x_1

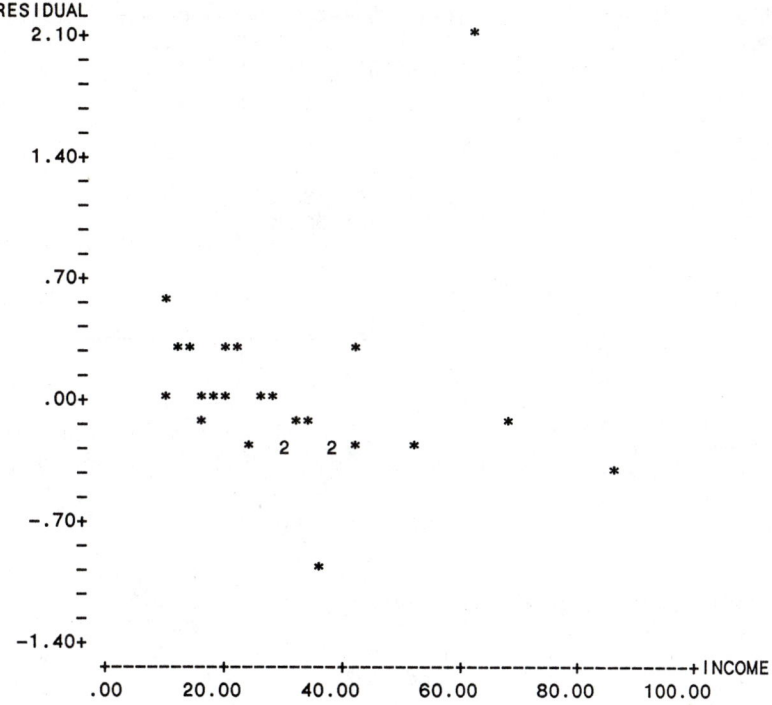

Residual Plot for Exercise
11.31: $\hat{\varepsilon}$ vs. x_2

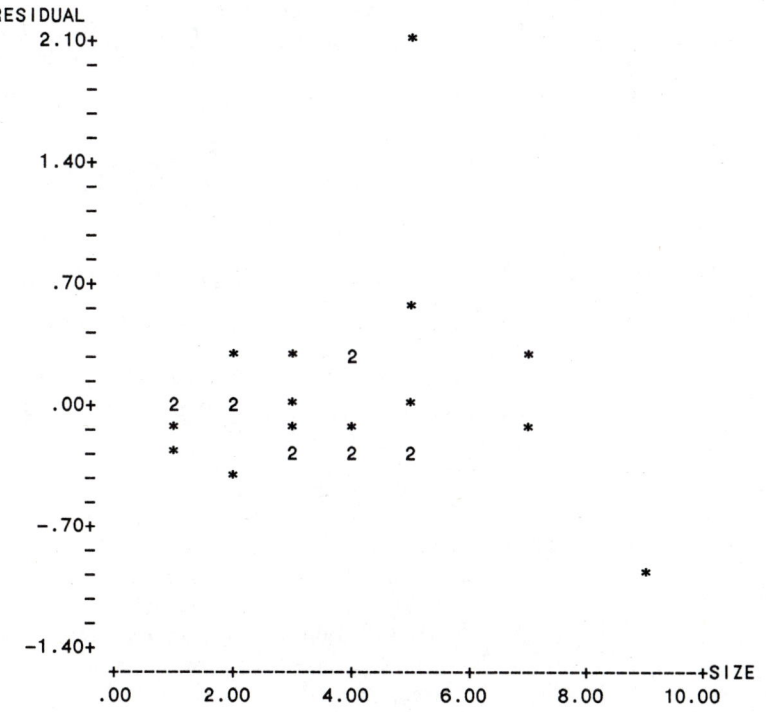

Stem and Leaf Display for
26 Residuals

```
STEM-AND-LEAF DISPLAY OF RESIDUAL
LEAF DIGIT UNIT =     .1000
1 2 REPRESENTS 1.2



     1     -0. 9
   (16)    -0* 4333222211110000
     9     +0* 0022223
     2     +0. 6
     1      1*
     1      1.
     1      2* 1
```

11.33 Refer to Exercise 11.7, in which a quadratic model was used to relate the time to complete a task, y, to the months of experience, x, for a sample of 15 employees on an automobile assembly line. The SPSS printouts for both straight-line and quadratic models are given here.

SPSS Printout for Exercise
11.33: Straight-line Model

```
Multiple R              .94886
R Square                .90033
Adjusted R Square       .89266
Standard Error          1.14301


Analysis of Variance
                     DF      Sum of Squares      Mean Square
Regression            1          153.41583        153.41583
Residual             13           16.98417          1.30647


F =     117.42733      Signif F =  .0000


------------------ Variables in the Equation -------------------

Variable             B           SE B        Beta          T   Sig T

X                -.44494       .04106     -.94886    -10.836   .0000
(Constant)      19.27908       .50788                 37.960   .0000
```

a. Find and interpret the standard deviation of each regression model in the context of this application.

b. Find and interpret R^2 for each model.

c. Conduct the analysis of variance F test for each model. Use $\alpha = .05$ in each case. Interpret the results of the tests.

d. Test the quadratic term in the second-order model. Use $\alpha = .05$. Does the quadratic term for experience appear to provide information for the prediction of time to complete the task?

SPSS Printout for Exercise
11.33: Quadratic Model

```
Multiple R              .95718
R Square                .91619
Adjusted R Square       .90223
Standard Error         1.09089

Analysis of Variance
                    DF      Sum of Squares      Mean Square
Regression           2           156.11948         78.05974
Residual            12            14.28052          1.19004

F =      65.59402       Signif F =  .0000


------------------ Variables in the Equation ------------------

Variable              B           SE B         Beta          T   Sig T

X                 -.67052        .15471     -1.42992     -4.334  .0010
XSQ         9.534744E-03  6.32580E-03       .49728      1.507  .1576
(Constant)      20.09111        .72471                 27.723  .0000
```

11.34 Refer to Exercise 11.33. Residual plots for the straight-line and quadratic models are
shown below and on page 622. Analyze the plots. Does your residual analysis support
the statistical test you conducted in part **d** of Exercise 11.33?

Residual Plot for Exercise
11.34: Straight-line Model

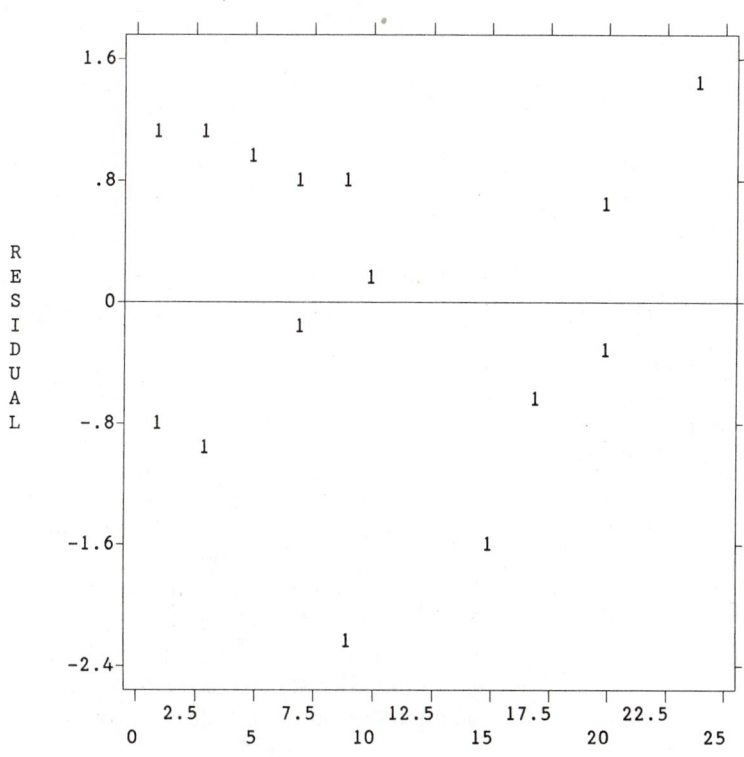

Residual Plot for Exercise
11.34: Quadratic Model

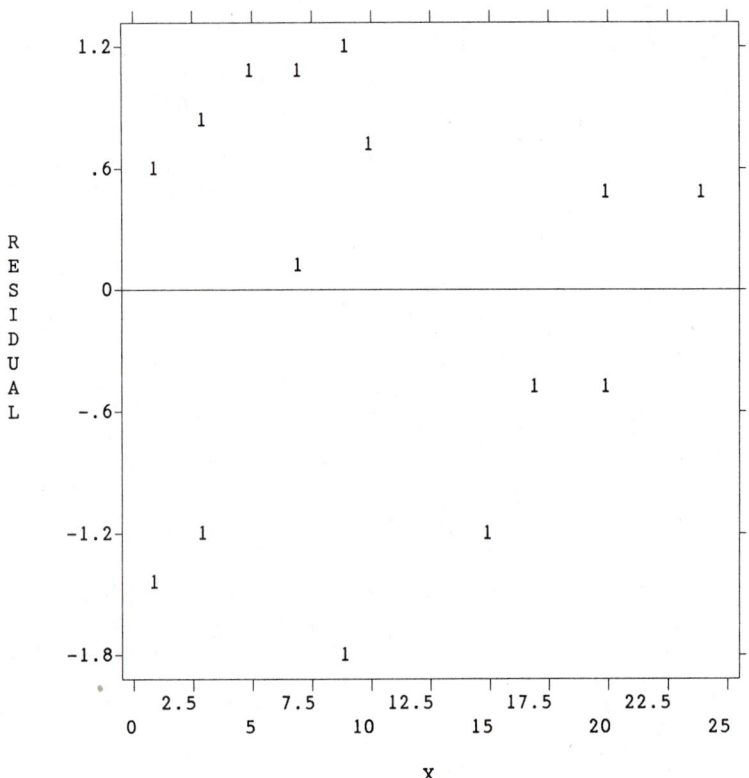

SOME PITFALLS: ESTIMABILITY, MULTICOLLINEARITY, AND EXTRAPOLATION

There are several problems you should be aware of when constructing a prediction model for some response, y. A few of the most important are discussed in this section.

PROBLEM 1
PARAMETER
ESTIMABILITY

Suppose you want to fit a model relating a firm's monthly profit, y, to the advertising expenditure, x. We propose the first-order model $E(y) = \beta_0 + \beta_1 x$. Now, suppose we have 3 months of data, and the firm spent \$1,000 on advertising during each month. The data are shown in Figure 11.29. You can see the problem: The parameters of the line cannot be estimated when all the data are concentrated at a single x value. Recall that it takes two points (x values) to fit a straight line. Thus, the parameters are not estimable when only one x value is observed.

FIGURE 11.29

Profit and Advertising
Expenditure Data:
3 Months

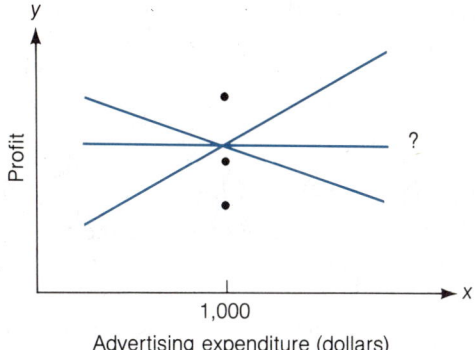

A similar problem would occur if we attempted to fit the second-order model

$$E(y) = \beta_0 + \beta_1 x + \beta_2 x^2$$

to a set of data for which only one *or two* different x values were observed (see Figure 11.30). At least three different x values must be observed before a second-order model can be fit to a set of data (that is, before all three parameters are estimable). In general, the number of levels of x must be at least one more than the order of the polynomial in x that you want to fit.

FIGURE 11.30

Only Two x Values
Observed—Second-Order
Model Is Not Estimable

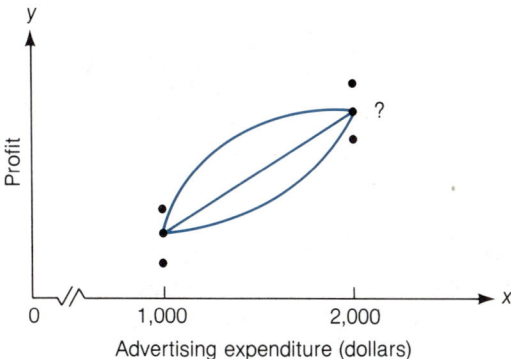

Since most business variables are not controlled by the researcher, the independent variables will almost always be observed at a sufficient number of levels to permit estimation of the model parameters. However, when the computer program you use suddenly refuses to fit a model, the problem is probably inestimable parameters.

PROBLEM 2

MULTICOLLINEARITY

Often, two or more of the independent variables used in the model for $E(y)$ will contribute redundant information. That is, the independent variables will be correlated with each other. For example, suppose we want to construct a model to predict the gasoline mileage rating of a truck as a function of its load, x_1, and the horsepower, x_2, of its engine. In general, you would expect heavy loads to require greater horsepower and to result in lower mileage ratings. Thus, although both x_1 and x_2 contribute information for the prediction of mileage rating, some of the information is overlapping because x_1 and x_2 are correlated.

If the model

$$E(y) = \beta_0 + \beta_1 x_1 + \beta_2 x_2$$

were fit to a set of data, we might find that the t values for both $\hat{\beta}_1$ and $\hat{\beta}_2$ (the least squares estimates) are nonsignificant. However, the F test for $H_0: \beta_1 = \beta_2 = 0$ would probably be highly significant. The tests may seem to be contradictory, but really they are not. The t tests indicate that the contribution of one variable, say $x_1 =$ Load, is not significant after the effect of $x_2 =$ Horsepower has been discounted (because x_2 is also in the model). The significant F test, on the other hand, tells us that at least one of the two variables is making a contribution to the prediction of y (i.e., β_1, β_2, or both differ from 0). In fact, both are probably contributing, but the contribution of one overlaps with that of the other.

When highly correlated independent variables are present in a regression model, the results may be confusing. The researcher may want to include only one of the variables in the final model. One way of deciding which variable to include is by using **stepwise regression**, a topic discussed in Chapter 12. Generally, only one (or a small number) of a set of multicollinear independent variables will be included in the regression model by a stepwise regression procedure. This procedure tests the parameter associated with each variable in the presence of all the variables already in the model. For example, if at one step, the variable Truck load is included as a significant variable in the prediction of the mileage rating, then the variable Horsepower will probably never be added in a future step. Thus, if a set of independent variables is thought to be multicollinear, some screeening by stepwise regression may be helpful.

PROBLEM 3

PREDICTION OUTSIDE THE EXPERIMENTAL REGION

By the late 1960's, many research economists had developed highly technical models to relate the state of the economy to various economic indexes and other independent variables. Many of these models were multiple regression models, where, for example, the dependent variable y might be next year's growth in GNP and the independent variables might include this year's rate of inflation, this year's Consumer Price Index, and other factors. In other words, the model might be constructed to predict next year's economy using this year's knowledge.

Unfortunately, these models were almost unanimously unsuccessful in predicting the recession in the early 1970's. What went wrong? One of the problems was that the regression models were used to predict y for values of the independent variables that were outside the region in which the model was developed. For example, the inflation rate in the late 1960's, when the models were developed, ranged from 6% to 8%. When the double-digit inflation of the early 1970's became a reality, some researchers attempted to use the same models to predict future growth in GNP. As you can see in Figure 11.31, the model may be very accurate for predicting y when x is in the range of experimentation, but the use of the model outside that range is a dangerous practice.

FIGURE 11.31
Using a Regression Model
Outside the Experimental
Region

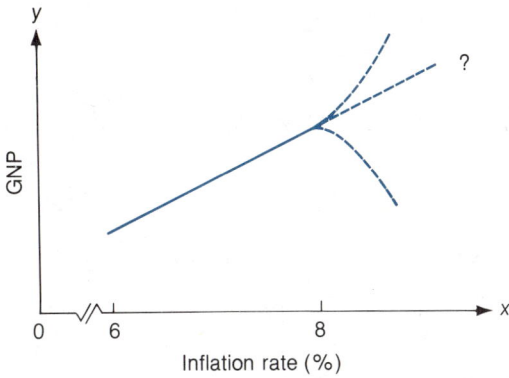

PROBLEM 4
CORRELATED ERRORS

Another problem associated with using a regression model to predict an economic variable y based on independent variables $x_1, x_2, \ldots, x_k$ arises from the fact that the data are frequently time series. That is, the values of both the dependent and independent variables are observed sequentially over a period of time. The observations tend to be correlated over time, which in turn often causes the prediction errors of the regression model to be correlated. Thus, the assumption of independent errors is violated, and the model tests and prediction intervals are no longer valid. One solution to this problem is to construct a time series model; this will be the subject of Chapter 14.

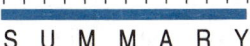

S U M M A R Y

We have discussed some of the methodology of **multiple regression analysis**, a technique for modeling a dependent variable y as a function of several independent variables $x_1, x_2, \ldots, x_k$. The steps we follow in constructing and using multiple regression models are much the same as those for the simple straight-line models:

1. The form of the probabilistic model is hypothesized.
2. The model coefficients are estimated using the method of least squares.
3. The probability distribution of ε is specified and σ^2 is estimated.
4. The utility of the model is checked.
5. If the model is deemed useful, it may be used to make estimates and to predict values of y to be observed in the future.

We have covered steps 2–5 in this chapter, assuming that the model was specified. The most important topic, model building—step 1, is discussed in Chapter 12. Additional material on these topics can be found in the references at the end of the chapter.

**SUPPLEMENTARY
EXERCISES
11.35–11.63**

[*Note: Starred (*) exercises require the use of a computer.*]

11.35 Suppose you used Minitab to fit the model

$$y = \beta_0 + \beta_1 x_1 + \beta_2 x_2 + \varepsilon$$

to $n = 15$ data points and you obtained the printout shown on page 626.

```
THE REGRESSION EQUATION IS
Y =      90.1 -   1.84 X1 +   .285 X2

                                       ST. DEV.      T-RATIO =
             COLUMN   COEFFICIENT    OF COEF.       COEF/S.D.

              --              90.1         23.1          3.90
    X1   C2             -1.836          .367         -5.01
    X2   C3               .285          .231          1.24

THE ST. DEV. OF Y ABOUT REGRESSION LINE IS
S =        10.7
WITH (   15- 3) =   12 DEGREES OF FREEDOM

R-SQUARED = 91.6 PERCENT
R-SQUARED = 90.2 PERCENT, ADJUSTED FOR D.F.

ANALYSIS OF VARIANCE

    DUE TO         DF       SS     MS=SS/DF

REGRESSION       2    14801.       7400.
RESIDUAL        12     1364.        114.
TOTAL           14    16165.
```

a. What is the least squares prediction equation?

b. Find R^2 and interpret its value.

c. Is there sufficient evidence to indicate that the model is useful for predicting y? Conduct an F test using $\alpha = .05$.

d. Test the null hypothesis $H_0: \beta_1 = 0$ against the alternative hypothesis $H_a: \beta_1 \neq 0$. Test using $\alpha = .05$. Draw the appropriate conclusions.

e. Find the standard deviation of the regression model, and interpret it.

11.36 Several states now require all high school seniors to pass an achievement test before they can graduate. On the test, the seniors must demonstrate their familiarity with basic verbal and mathematical skills. Suppose the educational testing company that creates and administers these exams wants to model the score, y, on one of its exams as a function of the student's IQ, x_1, and socioeconomic status (SES). The SES is a **categorical (qualitative) variable** with three levels: low, medium, and high. As we will demonstrate in Chapter 12, two **dummy (indicator) variables** are needed to describe a qualitative independent variable with three levels. Thus, we define

$$x_2 = \begin{cases} 1 & \text{if SES is medium} \\ 0 & \text{if SES is low or high} \end{cases} \qquad x_3 = \begin{cases} 1 & \text{if SES is high} \\ 0 & \text{if SES is low or medium} \end{cases}$$

Data were collected for a random sample of 60 seniors who have taken the test, and the model

$$E(y) = \beta_0 + \beta_1 x_1 + \beta_2 x_2 + \beta_3 x_3$$

was fit to the data, with the results shown in the accompanying SAS printout.

a. Identify the least squares equation.

b. Interpret the value of R^2 and test to determine whether the data provide sufficient evidence to indicate that this model is useful for predicting achievement test scores.

c. Sketch the relationship between predicted achievement test score and IQ for the three levels of SES. [Note: Three graphs of $\hat{y}$ versus x_1 must be drawn: the first

SOURCE	DF	SUM OF SQUARES	MEAN SQUARE	F VALUE	PR > F
MODEL	3	12268.56439492	4089.52146497	188.33	0.0001
ERROR	56	1216.01893841	21.71462390		
CORRECTED TOTAL	59	13484.58333333		R-SQUARE	ROOT MSE
				0.909822	4.65989527

PARAMETER	ESTIMATE	T FOR H0: PARAMETER=0	PR > |T|	STD ERROR OF ESTIMATE
INTERCEPT	-13.06166081	-3.21	0.0022	4.07101383
X1	0.74193946	17.56	0.0001	0.04224805
X2	18.60320572	12.49	0.0001	1.48895324
X3	13.40965415	8.97	0.0001	1.49417069

for the low SES model ($x_2 = x_3 = 0$), the second for the medium SES model ($x_2 = 1$, $x_3 = 0$), and the third for the high SES model ($x_2 = 0$, $x_3 = 1$). The increase in predicted achievement test score per unit increase in IQ is the same for all three levels of SES; i.e., all three lines are parallel.]

11.37 Refer to Exercise 11.36. We now use the same data to fit the model

$$E(y) = \beta_0 + \beta_1 x_1 + \beta_2 x_2 + \beta_3 x_3 + \beta_4 x_1 x_2 + \beta_5 x_1 x_3$$

Thus, we now add the interaction between IQ and SES to the model. The SAS printout for this model is shown here.

SOURCE	DF	SUM OF SQUARES	MEAN SQUARE	F VALUE	PR > F
MODEL	5	12515.10021009	2503.02004202	139.42	0.0001
ERROR	54	969.48312324	17.95339117		
CORRECTED TOTAL	59	13484.58333333		R-SQUARE	ROOT MSE
				0.928104	4.23714422

PARAMETER	ESTIMATE	T FOR H0: PARAMETER = 0	PR > |T|	STD ERROR OF ESTIMATE
INTERCEPT	0.60129643	0.11	0.9096	5.26818519
X1	0.59526252	10.70	0.0001	0.05563379
X2	-3.72536406	-0.37	0.7115	10.01967496
X3	-16.23196444	-1.90	0.0631	8.55429931
X1*X2	0.23492147	2.29	0.0260	0.10263908
X1*X3	0.30807756	3.53	0.0009	0.08739554

a. Identify the least squares prediction equation.
b. Interpret the value of R^2 and test to determine whether the data provide sufficient evidence to indicate that this model is useful for predicting achievement test scores.
c. Sketch the relationship between predicted achievement test score and IQ for the three levels of SES. [Note: The interaction terms in the model allow nonparallelism among the three SES models; i.e., the mean increase in achievement test score per unit increase in IQ differs for the three levels of SES. To determine whether the interaction between IQ and SES is contributing to the prediction of achievement test score, we must test the null hypothesis H_0: $\beta_4 = \beta_5 = 0$ against the alternative that at least one of the coefficients of the interaction terms is nonzero. The method for testing portions of a regression model involving more than one β parameter (but less than all of them) will be discussed in Chapter 12.]

11.38 The recent expansion of U.S. grain exports has intensified the importance of the linkage between the domestic grain transportation system and international transportation. As a first step in evaluating the economics of this interface, Martin and Clement (1982) used multiple regression to estimate ocean transport rates for grain shipped from the lower Columbia River international ports. These ports include Portland, Oregon, and Vancouver, Longview, and Kalama, Washington. Rates per long ton, y, were modeled as a function of the following independent variables:

x_1 = Shipment size (long tons)

x_2 = Distance to destination port (miles)

x_3 = Bunker fuel price ($ per barrel)

$x_4 = \begin{cases} 1 & \text{if American flagship} \\ 0 & \text{if foreign flagship} \end{cases}$

x_5 = Size of port as measured by U.S. Defense Mapping Agency standards

x_6 = Quantity of grain exported from region during year of interest

The method of least squares was used to fit the model to 140 observations from the period 1978–1980. The following results were obtained:

$$\hat{y} = -18.469 - .367x_1 + 6.434x_2 - .2692x_2^2 + 1.7992x_3 + 50.292x_4 + 2.275x_5 - .018x_6$$
$$\phantom{\hat{y} = }(-2.76)\quad(-5.62)\quad(3.64)\quad(-2.25)\quad(12.96)\quad(19.14)\quad(1.71)\quad(-2.69)$$
$$R^2 = .8979$$

The numbers in parentheses are the t statistics associated with the $\hat{\beta}$ values above them.

a. Test H_0: $\beta_1 = \beta_2 = \beta_3 = \beta_4 = \beta_5 = \beta_6 = \beta_7 = 0$. Use $\alpha = .01$. Interpret the results of your test in the context of the problem.

b. Binkley and Harrer (1979) estimated a similar rate function using multiple regression but used different independent variables. The coefficient of determination for their model was .46. Compare the explanatory power of the Binkley and Harrer model with that of Martin and Clement.

c. According to the least squares model, do transport rates increase with distance? Do they increase at an increasing rate? Explain.

*11.39 Refer to Exercises 11.11 and 11.17, in which regression analysis was used to explore the valuation of apartment buildings in Minneapolis.

a. Verify that number of apartment units, x_1, is highly correlated with gross building area, x_5.

b. Eliminate x_1 from the appraiser's model and refit the model to the data. Compare the standard deviations of the two models.

c. Use an F test to investigate the usefulness of this model. Use $\alpha = .05$.

d. Use the least squares equation of part **b** to estimate the value of an apartment building that is 50 years old with gross area 20,000 square feet, lot size 15,000 square feet, and 10 parking spaces. Use a computer program to compute a 95% confidence interval for the model's estimate.

11.40 Suppose you have developed a regression model to explain the relationship between y and x_1, x_2, and x_3. The ranges of the variables you used to develop your model are as follows: $10 \le y \le 100$, $5 \le x_1 \le 55$, $.5 \le x_2 \le 1$, and $1,000 \le x_3 \le 2,000$. Explain why you would have more confidence using your prediction equation to predict y when $x_1 = 30$, $x_2 = .6$, and $x_3 = 1,300$ than when $x_1 = 60$, $x_2 = .4$, and $x_3 = 900$.

11.41 Plastics made under different environmental conditions are known to have differing strengths. A scientist would like to know which combination of temperature and pressure yields a plastic with a high breaking strength. A small preliminary experiment was conducted at two pressure levels and two temperature levels. The following model was proposed:

$$E(y) = \beta_0 + \beta_1 x_1 + \beta_2 x_2 + \beta_3 x_1 x_2$$

where

$y = $ Breaking strength (pounds)

$x_1 = $ Temperature (°F) $x_2 = $ Pressure (pounds per square inch)

A sample of $n = 16$ observations yielded

$$\hat{y} = 226.8 + 4.9x_1 + 1.2x_2 - .7x_1 x_2$$

with

$$s_{\hat{\beta}_1} = 1.11 \qquad s_{\hat{\beta}_2} = .27 \qquad s_{\hat{\beta}_3} = .34$$

Do the data indicate there is an interaction between temperature and pressure? Test using $\alpha = .05$.

11.42 A large government agency would like to predict the number of people it will hire within the next year to fill the 30 positions that are currently open. Historically, the agency has been unable to fill all its job openings. It has been decided to model the number of positions filled in a year, y, as a function of the number of positions open, x_1, and the recruiting budget for the year in dollars, x_2 (e.g., for advertising the positions, paying travel expenses, etc.). A random sample of 10 years of recruiting records was drawn from the agency's 30 years of records. The model

$$E(y) = \beta_0 + \beta_1 x_1 + \beta_2 x_2$$

was fit to the data using the Minitab regression computer program package. The results shown in the printout on page 630 were obtained.

a. Identify the least squares prediction equation.
b. Is there sufficient evidence to indicate that the model contributes information for predicting the number, y, of positions that will be filled? Conduct an F test using $\alpha = .05$.
c. Test the null hypothesis $H_0: \beta_2 = 0$ against the alternative hypothesis $H_a: \beta_2 \ne 0$ using $\alpha = .05$. Interpret the results of your test in the context of the problem.
d. Use the least squares prediction equation to predict how many of the 30 positions the agency will fill next year if the recruiting budget is $10,000.
e. Which (if any) of the assumptions we make about ε in a regression analysis are likely to be violated in this problem? Explain.

```
THE REGRESSION EQUATION IS
Y =    .0562 +    .273 X1 +  .0006 X2


                                    ST. DEV.     T-RATIO =
           COLUMN   COEFFICIENT     OF COEF.     COEF/S.D.


            --              .056        .902          .06
    X1     C2             .2733        .0971         2.81
    X2     C3           .000560      .000129         4.34

THE ST. DEV. OF Y ABOUT REGRESSION LINE IS
S =        1.33
WITH (   10- 3) =     7 DEGREES OF FREEDOM

R-SQUARED = 97.9 PERCENT
R-SQUARED = 97.3 PERCENT, ADJUSTED FOR D.F.

ANALYSIS OF VARIANCE

    DUE TO        DF       SS      MS=SS/DF

    REGRESSION     2    583.18      291.59
    RESIDUAL       7     12.42        1.77
    TOTAL          9    595.60
```

11.43 A metropolitan bus company wants to know whether changes in numbers of bus riders are related to changes in gasoline prices. By using information in the company files and gasoline price information obtained from fuel distributors, the company planned to fit the following model:

$$y = \beta_0 + \beta_1 x_1 + \beta_2 x_2 + \beta_3 x_1 x_2 + \varepsilon$$

where

x_1 = Average wholesale price for regular gasoline in a given month

$x_2 = \begin{cases} 1 & \text{if the bus travels a city route only} \\ 0 & \text{if the bus travels a suburb–city route} \end{cases}$

y = Total number of riders in a bus over the month

a. For the above model, how would you test to determine whether the relationship between the mean number of riders and gasoline price is different for the two types of bus routes?

b. Suppose 12 months of data are kept, and the least squares model is

$$\hat{y} = 500 + 50x_1 + 5x_2 - 10x_1x_2$$

Graph the predicted relationship between number of riders and gasoline price for city buses and for suburb–city buses. Compare the slopes.

c. If $s_{\hat{\beta}_3} = 3.0$, do the data indicate that gasoline price affects the number of riders differently for city and suburb–city buses? Use $\alpha = .05$.

11.44 During the winter months a sample of 100 homes is taken to obtain information concerning the relationship between kilowatt usage, y, and total window and glass area, x (measured as a percentage of the total wall area). A second-order model, $y = \beta_0 + \beta_1 x + \beta_2 x^2 + \varepsilon$, was used to model this relationship. The multiple coefficient of determination for the data was .24. Test whether the data indicate that the model contributes information for the prediction of y. Use $\alpha = .05$. [*Hint:* Use the methods of Section 11.6.]

*11.45 Refer to Exercise 10.24. The breeder of thoroughbred horses has been advised that the prediction model could probably be improved if a quadratic term were added. The following model is therefore proposed:

$$y = \beta_0 + \beta_1 x + \beta_2 x^2 + \varepsilon$$

where, as before

$y =$ Lifetime of horse (years)

$x =$ Gestation period of horse (days)

a. Find the least squares prediction equation and test its adequacy.
b. Has the addition of the quadratic term contributed significant information for the prediction of a thoroughbred horse's lifetime? Test $H_0: \beta_2 = 0$ against the alternative $H_a: \beta_2 \neq 0$ using $\alpha = .05$.
c. Construct residual plots for the straight-line and quadratic models, displaying gestation period on the horizontal axis. Does the result of the residual analysis agree with the test you conducted in part **b**? Which is more reliable, and why?

11.46 Most companies institute rigorous safety programs to assure employee safety. Suppose accident reports over the last year at a company are sampled, and the number of hours the employee had worked before the accident occurred, x, and the amount of time the employee lost from work, y, are recorded. A quadratic model is proposed to investigate a fatigue hypothesis that more serious accidents occur near the end of workdays than near the beginning. Thus, the proposed model is

$$E(y) = \beta_0 + \beta_1 x + \beta_2 x^2$$

A total of 60 accident reports are examined and part of the computer printout appears as shown:

SOURCE	DF	SUM OF SQUARES	MEAN SQUARE	F VALUE
MODEL	2	112.110	56.055	1.28
ERROR	57	2496.201	43.793	R-SQUARE
TOTAL	59	2608.311		.0430

a. Do the data support the fatigue hypothesis? Use $\alpha = .05$ to test whether the proposed model is useful in predicting the lost work time, y.
b. Does the result of the test in part **a** necessarily mean that no fatigue factor exists? Explain.

11.47 Refer to Exercise 11.46. Suppose the company persists in using the quadratic model despite its apparent lack of usefulness. The fitted model is

$$\hat{y} = 12.3 + .25x - .0033x^2$$

where $\hat{y}$ is the predicted time lost (days) and x is the number of hours worked prior to an accident.

a. Use the model to predict the number of days missed by an employee who has an accident after 6 hours of work.
b. Suppose the 95% prediction interval for the predicted value in part **a** is determined to be (1.35, 26.01). Interpret this interval. Does this interval support your conclusion about this model in Exercise 11.46?

11.48 *Operations management* is concerned with planning and controlling those organizational functions and systems that produce goods and services (Schroeder, 1985). One concern of the operations manager of a production process is the level of productivity of the process. An operations manager at a large manufacturing plant is interested in predicting the level of productivity of assembly line A next year (i.e., the number of units that will be produced by the assembly line next year). To do so, she has decided to use regression analysis to model the level of productivity, y, as a function of time, x. The number of units produced by assembly line A was determined for each of the past 15 years ($x = 1, 2, \ldots, 15$). The model

$$E(y) = \beta_0 + \beta_1 x + \beta_2 x^2$$

was fit to the data using Minitab. The results shown in the printout were obtained.

```
THE REGRESSION EQUATION IS
Y = - 1187, - 1333, X1 -  45,6 X2

                               ST, DEV,    T-RATIO =
          COLUMN   COEFFICIENT OF COEF,    COEF/S,D,

            --       -1187        446,       -2,66
   X1   C2           1333,        128,       10,38
   X2   C3          -45,59        7,80       -5,84

THE ST, DEV, OF Y ABOUT REGRESSION LINE IS
S =       501,
WITH (  15- 3) =   12 DEGREES OF FREEDOM

R-SQUARED = 97,3 PERCENT
R-SQUARED = 96,9 PERCENT, ADJUSTED FOR D,F,

ANALYSIS OF VARIANCE

   DUE TO       DF          SS       MS=SS/DF

REGRESSION      2     110578719,    55289359,
RESIDUAL       12       3013365,      251114,
TOTAL          14     113592083,
```

a. Identify the least squares prediction equation.

b. Find R^2 and interpret its value in the context of this problem.

c. Is there sufficient evidence to indicate that the model is useful for predicting the productivity of assembly line A? Test using $\alpha = .05$.

d. Test the null hypothesis $H_0: \beta_2 = 0$ against the alternative hypothesis $H_a: \beta_2 \neq 0$ using $\alpha = .05$. Interpret the results of your test in the context of this problem.

e. Which (if any) of the assumptions we make about ε in a regression analysis are likely to be violated in this problem? Explain.

[*Note:* In this exercise time series data were used to obtain the least squares prediction equation. We will discuss time series and time series models in Chapters 13 and 14.]

11.49 To increase the motivation and productivity of workers, an electronics manufacturer decides to experiment with a new pay incentive structure at one of two plants. The experimental plan will be tried at plant A for 6 months, while workers at plant B will remain on the original pay plan. To evaluate the effectiveness of the new plan, the average assembly time for part of an electronic system was measured for employees at both plants at the beginning and end of the 6-month period. Suppose the following model was proposed:

$$y = \beta_0 + \beta_1 x_1 + \beta_2 x_2 + \varepsilon$$

where

y = Assembly time (hours) at end of 6-month period

x_1 = Assembly time (hours) at beginning of 6-month period

$x_2 = \begin{cases} 1 & \text{if plant A} \\ 0 & \text{if plant B} \end{cases}$ (dummy variable)

A sample of $n = 42$ observations yielded

$$\hat{y} = .11 + .98x_1 - .53x_2$$

where

$$s_{\hat{\beta}_1} = .231 \qquad s_{\hat{\beta}_2} = .48$$

Test to see whether, after allowing for the effect of initial assembly time, plant A had a lower mean assembly time than plant B. Use $\alpha = .01$. [*Note:* When the {0, 1} coding is used to define a dummy variable, the coefficient of the variable represents the difference between the mean response at the two levels represented by the variable. Thus, the coefficient β_2 is the difference in mean assembly time between plant A and plant B at the end of the 6-month period, and $\hat{\beta}_2$ is the sample estimator of that difference.]

11.50 A company that relies on door-to-door sales wants to determine the relationship, if any, between the proportion of customers who buy its product, y, and two independent variables: price, x_1, and years of experience of the salesperson, x_2. Twenty salespeople employed by the company are randomly assigned to sell the products, five to each of four prices, ranging from \$1.98 to \$5.98. Each salesperson makes a sales presentation to 30 prospects, and the percentage of sales is recorded. The 20 observations are used to fit the model

$$y = \beta_0 + \beta_1 x_1 + \beta_2 x_2 + \varepsilon$$

The least squares model is

$$\hat{y} = -.30 - .010x_1 + .10x_2$$

with $s_{\hat{\beta}_1} = .0030$, $s_{\hat{\beta}_2} = .025$, and $R^2 = .86$.

a. Interpret the values of $\hat{\beta}_1$ and $\hat{\beta}_2$.
b. Is there sufficient evidence to conclude that the overall model is useful for predicting y? Use $\alpha = .05$.
c. Do the data support the hypothesis that as the price of the product is increased the mean proportion of buyers will decrease?
d. Is there evidence that as the experience of the salesperson increases the mean proportion of buyers increases?

11.51 Refer to Exercise 11.50. Suppose it is claimed that the least squares model cannot be correct, since $\hat{\beta}_0 = -.30$, and a negative proportion of buyers is clearly impossible. How do you refute this argument?

11.52 The Environmental Protection Agency (EPA) wants to model the gas mileage ratings, y, of automobiles as a function of their engine size, x. A second-order model,

$$E(y) = \beta_0 + \beta_1 x + \beta_2 x^2$$

is proposed. A sample of 50 engines of varying sizes is selected, and the miles per gallon rating of each is determined. The least squares model is

$$\hat{y} = 51.3 - 10.1x + .15x^2$$

The size, x, of the engine is measured in hundreds of cubic inches. Also, $s_{\hat{\beta}_2} = .0037$ and $R^2 = .93$.

a. Sketch the predicted relationship between y and x for values of x between $x = 1$ and $x = 4$.

b. Is there evidence that the quadratic term in the model is contributing to the prediction of the miles per gallon rating, y? Use $\alpha = .05$.

c. Use the model to estimate the mean miles per gallon rating for all cars with 350-cubic-inch engines ($x = 3.5$).

d. Suppose a 95% confidence interval for the quantity estimated in part **c** is determined to be (17.2, 18.4). Interpret this interval.

e. Suppose you purchase an automobile with a 350-cubic-inch engine and determine that the miles per gallon rating is 14.7. Is it surprising that this value lies outside the confidence interval given in part **d**? Explain.

11.53 To determine whether extra personnel are needed for the day, the owners of a water adventure park would like to find a model that would allow them to predict the day's attendance each morning based on the day of the week and weather conditions. The model is of the form

$$E(y) = \beta_0 + \beta_1 x_1 + \beta_2 x_2 + \beta_3 x_3$$

where

$y = $ Daily attendance

$x_1 = \begin{cases} 1 & \text{if weekend} \\ 0 & \text{otherwise} \end{cases}$ (dummy variable)

$x_2 = \begin{cases} 1 & \text{if sunny} \\ 0 & \text{if overcast} \end{cases}$ (dummy variable)

$x_3 = $ Predicted daily high temperature (°F)

After taking 30 days of data, the owners obtained the following least squares model:

$$\hat{y} = -105 + 25x_1 + 100x_2 + 10x_3$$

with $s_{\hat{\beta}_1} = 10$, $s_{\hat{\beta}_2} = 30$, and $s_{\hat{\beta}_3} = 4$. Also, $R^2 = .65$.

a. Interpret the estimated model coefficients.

b. Is there sufficient evidence to conclude that this model is useful in the prediction of daily attendance? Use $\alpha = .05$.

c. Is there sufficient evidence to conclude that mean attendance increases on weekends? Use $\alpha = .10$.

d. Use the model to predict the attendance on a sunny weekday with a predicted high temperature of 95° F.

e. Suppose the 90% prediction interval for part **d** is (645, 1,245). Interpret this interval.

11.54 Refer to Exercise 11.53. The owners of the water adventure park are advised that the prediction model could probably be improved if interaction terms were added. In particular, it is thought that the *rate* of increase in mean attendance with increases in predicted high temperature will be greater on weekends than on weekdays. The following model is therefore proposed:

$$E(y) = \beta_0 + \beta_1 x_1 + \beta_2 x_2 + \beta_3 x_3 + \beta_4 x_1 x_3$$

The same 30 days of data as were used in Exercise 11.53 are used to obtain the least squares model

$$\hat{y} = 250 - 700x_1 + 100x_2 + 5x_3 + 15x_1 x_3$$

with $s_{\hat{\beta}_4} = 3.0$ and $R^2 = .96$.

a. Graph the predicted day's attendance, y, against the day's predicted high temperature, x_3, for a sunny weekday and for a sunny weekend day. Graph both on the same graph for x_3 between 70°F and 100°F. Note that the slope for the weekend day is greater.

b. Do the data indicate that the interaction term is a useful addition to the model? Use $\alpha = .05$.

c. Use this model to predict the attendance for a sunny weekday with a predicted high temperature of 95°F.

d. Suppose the 90% prediction interval for part **c** is (800, 850). Compare this with the prediction interval for the model without interaction in Exercise 11.53, part **e**. Do the relative widths of the prediction intervals support or refute your conclusion about the usefulness of the interaction term (part **b**)?

11.55 Refer to Exercise 11.54. The owners, noting that $\hat{\beta}_1 = -700$, conclude the model is ridiculous because it seems to imply that the mean attendance will be 700 less on weekends than on weekdays. Refute their argument.

11.56 Many students must work part-time to help finance their college education. A survey of 100 students was completed at a university to determine whether the number of hours worked per week, x, was affecting their grade-point averages, y. A quadratic model was proposed:

$$y = \beta_0 + \beta_1 x + \beta_2 x^2 + \varepsilon$$

The 100 observations yielded the least squares model

$$\hat{y} = 2.8 - .005x - .0002x^2$$

with $R^2 = .12$ and $s = .5$.

a. Do these statistics indicate that the model is useful in explaining grade-point averages? Use $\alpha = .05$.

b. Interpret the values of R^2 and s in terms of this application of regression analysis. Do you think the quadratic relationship between y and x is strong? Approximately how precisely would you expect to be able to predict a particular student's grade-point average if you knew the number of hours he or she worked per week?

c. If you changed the model to a straight-line model relating grade-point average to number of hours worked, would the value of R^2 increase, thereby providing a better prediction model for grade-point average? Explain.

11.57 A manufacturer of electronic components for the computer industry believes that the yearly sales for one of its products can best be modeled as a function of the mean price of the product over the year, the mean price of its competitors' products over the year, and the mean number of salespeople per month over the year. Since much of the increase in sales revenue and prices over time may be due simply to inflation, it is advisable to eliminate the effects of inflation on the data before developing a model for annual sales. One way to do this is to express all sales revenue and price data in terms of constant dollars (i.e., inflation-adjusted dollars). We can convert each year's sales and price data to constant (1967) dollars by dividing each year's value by the Producer Price Index and multiplying the result by 100. The Producer Price Index is designed to measure average changes in prices over time of all commodities, at all stages of processing, produced or imported for sale in primary markets in the United States. Data on the four variables of interest to the manufacturer along with the Producer Price Index are presented in the table. [*Note:* Indexes such as the Producer Price Index and the transformation of current dollars to constant dollars are topics discussed in more detail in Chapter 13.]

YEAR	SALES REVENUE y_1 (thousands of \$)	MEAN NUMBER OF SALES PEOPLE x_1	MEAN PRICE OF PRODUCT x_2 (\$)	MEAN PRICE OF COMPETITORS' PRODUCTS x_3 (\$)	PRODUCER PRICE INDEX[a]
1967	50	5	30	25	100.0
1968	120	7	30	26	102.5
1969	140	11	33	28	106.5
1970	135	16	34	30	110.4
1971	163	16	33	31	114.0
1972	233	16	36	34	119.1
1973	241	21	40	37	134.7
1974	255	27	45	42	160.1
1975	286	26	50	48	174.9
1976	330	30	53	54	183.0
1977	389	33	58	58	194.2
1978	425	36	60	61	209.3
1979	445	38	71	72	235.6
1980	472	37	80	81	268.8
1981	501	37	90	93	293.4
1982	510	38	92	92	299.3
1983	490	36	92	90	303.1
1984	505	37	94	94	310.3

[a]Source: *Statistical Abstract of the United States: 1986*, U.S. Bureau of the Census, p. 471.

a. Express the sales revenue data in terms of constant 1967 dollars. Call this new variable y_2. Do the same for the price data for the product of interest and the competitors' price data. Call these new variables x_4 and x_5, respectively.

b. Use the method of least squares to fit the model

$$y_2 = \beta_0 + \beta_1 x_1 + \beta_2 x_4 + \beta_3 x_5 + \varepsilon$$

to the data. Describe what your least squares equation suggests about the relationship between y_2 and each of the three independent variables.

c. Find R^2 and interpret its value in the context of this problem.

d. In using an F test to investigate the usefulness of the model in part b, what are the null and alternative hypotheses?

e. Conduct the hypothesis test of part d. Use $\alpha = .05$ and interpret your results in the context of the problem. What is the observed significance level of your F test?

f. Test H_0: $\beta_2 = 0$ versus H_a: $\beta_2 < 0$ using $\alpha = .05$. Interpret the results of your test in the context of the problem.

g. Use a 95% confidence interval to estimate β_1. Interpret the results in the context of the problem.

h. Predict sales revenue (in 1967 dollars) for the product for a year in which the mean number of salespeople is 35, the mean price of the product is $90, the mean price of competitors' products is $92, and the Producer Price Index is 315.0.

i. Explain why it would not be advisable to use the least squares equation of part b to predict y_2 when $x_1 > 38$, $x_4 > \$31$, and $x_5 > \$32$.

*11.58 A company that services two brands of microcomputers would like to be able to predict the amount of time it takes to perform preventive maintenance on each brand. They believe the following predictive model is appropriate:

$$y = \beta_0 + \beta_1 x_1 + \beta_2 x_2 + \varepsilon$$

where

y = Maintenance time

$$x_1 = \begin{cases} 1 & \text{if brand A} \\ 0 & \text{if brand B} \end{cases}$$

x_2 = Service person's number of months of experience in preventive maintenance

Ten service people were randomly selected, and each was randomly assigned to perform preventive maintenance on either a brand A or brand B microcomputer. The following data were obtained.

MAINTENANCE TIME (Hours)	BRAND	EXPERIENCE (Months)	MAINTENANCE TIME (Hours)	BRAND	EXPERIENCE (Months)
2.0	1	2	1.5	0	2
1.8	1	4	1.7	1	6
.8	0	12	1.2	0	5
1.1	1	12	1.4	1	9
1.0	0	8	1.2	0	7

a. Fit the model to the data.

b. Investigate whether the overall model is useful. Test using $\alpha = .05$.

c. Find R^2 for the fitted model. Does the value of R^2 support your findings in part b? Explain.

d. Find a 90% confidence interval for β_2. Interpret your result in the context of the exercise.

e. Use the fitted model to predict how long it will take a person with 6 months of experience to service a brand B microcomputer.

f. How long would it take the person referred to in part **e** to service ten brand B microcomputers? List any assumptions you made in reaching your prediction.

g. Find a 95% prediction interval for the time required to perform preventive maintenance on a brand A microcomputer by a person with 4 months of experience.

11.59 Many colleges and universities develop regression models for predicting the grade-point average (GPA) of incoming freshmen. This predicted GPA can then be used to make admission decisions. Although most models use many independent variables to predict GPA, we will illustrate by choosing two variables:

x_1 = Verbal score on college entrance examination (percentile)

x_2 = Mathematics score on college entrance examination (percentile)

The data in the table are obtained for a random sample of 40 freshmen at one college.

VERBAL x_1	MATHEMATICS x_2	GPA y	VERBAL x_1	MATHEMATICS x_2	GPA y
81	87	3.49	79	75	3.45
68	99	2.89	81	62	2.76
57	86	2.73	50	69	1.90
100	49	1.54	72	70	3.01
54	83	2.56	54	52	1.48
82	86	3.43	65	79	2.98
75	74	3.59	56	78	2.58
58	98	2.86	98	67	2.73
55	54	1.46	97	80	3.27
49	81	2.11	77	90	3.47
64	76	2.69	49	54	1.30
66	59	2.16	39	81	1.22
80	61	2.60	87	69	3.23
100	85	3.30	70	95	3.82
83	76	3.75	57	89	2.93
64	66	2.70	74	67	2.83
83	72	3.15	87	93	3.84
93	54	2.28	90	65	3.01
74	59	2.92	81	76	3.33
51	75	2.48	84	69	3.06

The SPSS printout corresponding to the model

$$y = \beta_0 + \beta_1 x_1 + \beta_2 x_2 + \varepsilon$$

is shown on page 639.

a. Interpret the least squares estimates $\hat{\beta}_1$ and $\hat{\beta}_2$ in the context of this application.

b. Interpret the standard deviation and the coefficient of determination of the regression model in the context of this application.

c. Is this model useful for predicting GPA? Conduct a statistical test to justify your answer.

d. Sketch the relationship between predicted GPA, $\hat{y}$, and verbal score, x_1, for the following mathematics scores: $x_2 = 60$, 75, and 90.

Residual Plot: $\hat{\varepsilon}$ vs. x_1 for
First-Order Model

```
Multiple R             .82527
R Square               .68106
Adjusted R Square      .66382
Standard Error         .40228

Analysis of Variance
                    DF     Sum of Squares      Mean Square
Regression           2           12.78595          6.39297
Residual            37            5.98755           .16183

F =      39.50530      Signif F =   .0000


------------------- Variables in the Equation -------------------

Variable            B         SE B        Beta          T    Sig T

X1              .02573  4.02357E-03      .59719      6.395   .0000
X2              .03361  4.92751E-03      .63702      6.822   .0000
(Constant)    -1.57054      .49375                  -3.181   .0030
```

11.60 Refer to Exercise 11.59. The residuals from the first-order model are plotted vs. x_1 here and vs. x_2 on page 640. Analyze the two plots, and determine whether visual evidence exists that curvature (a quadratic term) for either x_1 or x_2 should be added to the model.

Residual Plot: $\hat{\varepsilon}$ vs. x_1
for First-Order Model

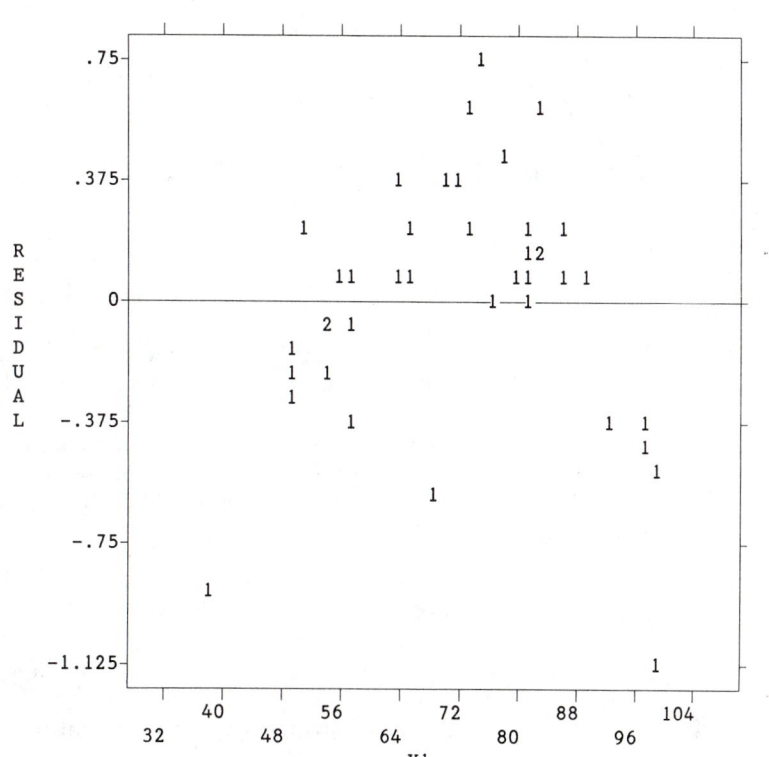

Residual Plot: $\hat{\varepsilon}$ vs. x_2 for
First-Order Model

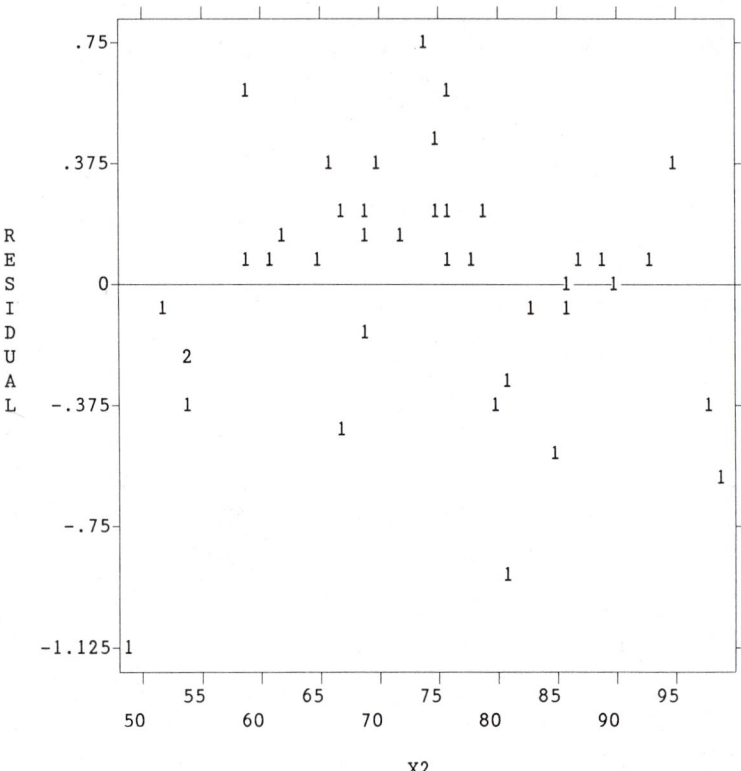

11.61 Refer to Exercises 11.59 and 11.60. The complete second-order model

$$y = \beta_0 + \beta_1 x_1 + \beta_2 x_2 + \beta_3 x_1^2 + \beta_4 x_2^2 + \beta_5 x_1 x_2 + \varepsilon$$

is fit to the data given in Exercise 11.59. The resulting SPSS printout is reproduced on page 641.

a. Compare the standard deviations of the first- and second-order regression models. With what relative precision will these two models predict GPA?

b. Test whether this model is useful for predicting GPA. Use $\alpha = .05$.

c. Test whether the interaction term, $\beta_5 x_1 x_2$, is important for the prediction of GPA. Use $\alpha = .10$. Note that this term permits the distance between three mathematics score curves for GPA versus verbal score to change as the verbal score changes.

11.62 Refer to Exercises 11.59–11.61. The residuals of the second-order model are plotted on pages 641–642 against both x_1 and x_2. Compare the residual plots to those of the first-order model in Exercise 11.60. Given the analyses you have performed in Exercises 11.59–11.62, which of the two models do you think is preferable as a predictor of GPA: the first- or second-order model? [Note: In Chapter 12 we will show how to conduct a statistical test to compare two models.]

SPSS Printout for
Exercise 11.61

```
Multiple R                .96777
R Square                  .93657
Adjusted R Square         .92724
Standard Error            .18714

Analysis of Variance
                     DF        Sum of Squares        Mean Square
Regression            5             17.58274            3.51655
Residual             34              1.19076             .03502

F =     100.40901      Signif F =   .0000
```

```
------------------- Variables in the Equation -------------------

Variable              B           SE B          Beta         T    Sig T

X1              .16681        .02124       3.87132      7.852   .0000
X2              .13760        .02673       2.60754      5.147   .0000
X1SQ       -1.10825E-03  1.17288E-04      -3.71359     -9.449   .0000
X2SQ       -8.43267E-04  1.59423E-04      -2.37284     -5.290   .0000
X1X2       2.410891E-04  1.43974E-04        .49600      1.675   .1032
(Constant)     -9.91676       1.35441                  -7.322   .0000
```

Residual Plot for Exercise
11.62: $\hat{\varepsilon}$ vs. x_1 for
Second-Order Model

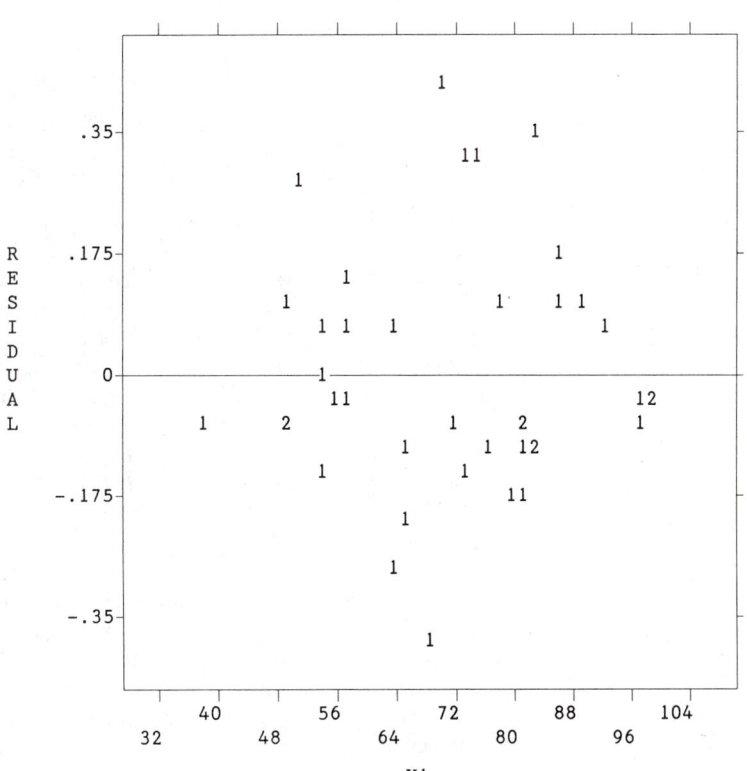

Residual Plot for Exercise
11.62: $\hat{\varepsilon}$ vs. x_2 for
Second-Order Model

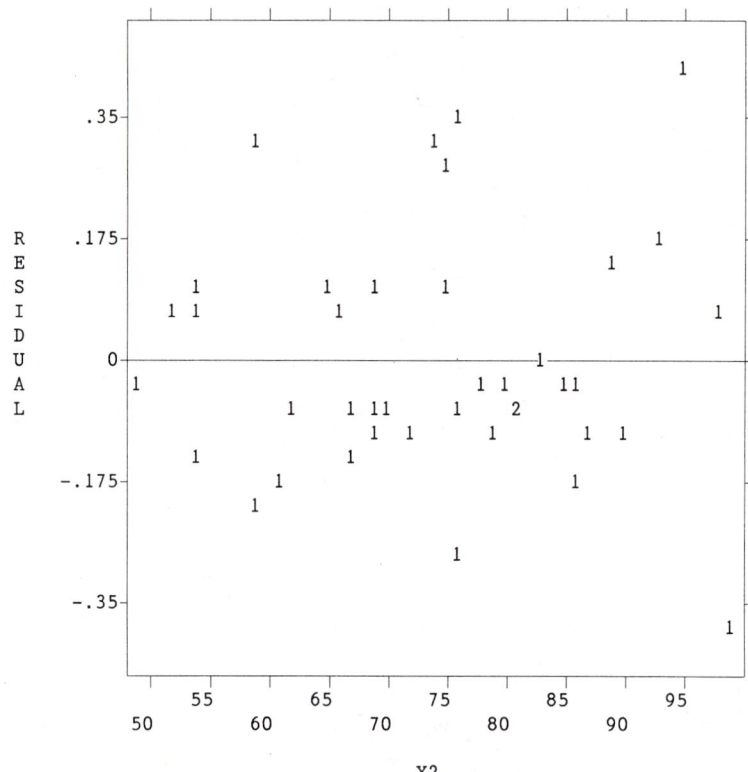

*11.63 An economist is interested in estimating the demand function for passenger car motor fuel in the United States. While demand is clearly a function of many variables, the economist initially wants to model demand as a function of consumer income and the price of gasoline and plans to estimate the model using yearly time series data over the period 1965–1983. As a measure of income, it was decided to use the average gross weekly earnings for each year for production or nonsupervisory workers on private nonagricultural payrolls. These data are available from the Bureau of Labor Statistics (BLS). However, since both earnings and the cost of living increased over this period, the BLS earnings data do not reflect actual purchasing power. Accordingly, the earnings data—originally expressed in *current dollars* (i.e., the number of dollars actually earned)—were converted to 1967 dollars. This was accomplished by dividing each figure by the Consumer Price Index (CPI) for that year and multiplying the result by 100. (This procedure is described in detail in Chapter 13.) The resulting earnings data—called *real earnings*—are the number of dollars that would have to have been earned in 1967 to equal the purchasing power of current year weekly earnings. These data appear in the accompanying table.

Data on the actual price of gasoline and the relative price of gasoline (computed by dividing the CPI for gasoline by the CPI for all items) were also collected and appear in the table. The relative price data reflect the price of gasoline relative to the prices of other consumer goods. The economist believes the relative price may have a substantial

YEAR	MOTOR FUEL CONSUMED BY CARS (billion gallons)	POPULATION OF UNITED STATES (millions)	AVERAGE GROSS REAL WEEKLY EARNINGS (1967 $)	PRICE OF GALLON OF REGULAR GASOLINE ($)	RELATIVE PRICE OF GALLON OF GASOLINE
1965	50.3	194.3	101.01	.32	1.004
1966	53.31	196.6	101.67	.33	.998
1967	55.11	198.7	101.84	.34	1.000
1968	58.52	200.7	103.39	.34	.973
1969	62.45	202.7	104.38	.36	.954
1970	65.8	205.1	103.04	.36	.908
1971	69.51	207.1	104.95	.36	.876
1972	73.5	209.9	109.26	.37	.859
1973	78.0	211.9	109.23	.40	.887
1974	74.2	213.9	104.78	.53	1.083
1975	76.5	216.0	101.45	.57	1.060
1976	78.8	218.0	102.90	.59	1.043
1977	80.7	220.2	104.13	.62	1.037
1978	83.8	222.6	104.25	.63	1.005
1979	80.2	225.1	101.15	.86	1.222
1980	73.7	227.7	95.26	1.19	1.496
1981	71.7	230.0	93.69	1.31	1.508
1982	72.8	232.3	92.45	1.22	1.346
1983	73.4	234.5	94.07	1.16	1.261

Source: All data from *Statistical Abstract of the United States*, various years; except average gross real weekly earnings, which are from *Employment and Earnings*, U.S. Department of Labor, Bureau of Labor Statistics, Dec. 1986, p. 79.

influence on demand. For example, even though actual gasoline prices increase, consumers may actually increase their demand for gasoline if the relative price decreases.

Finally, data have been collected on motor fuel consumed per year and the population of the United States. The population data have been included so that the effects of the growing U.S. population on demand for motor fuel could be removed if desired. This can be accomplished by dividing the total motor fuel consumed in a year by the population that year. The result is called *per capita* motor fuel consumption. By using per capita consumption as the measure of demand, the economist can distinguish the effects on demand of factors such as price and income from the effects of population growth (Blair and Kenny, 1982).

a. Explain what is being measured by the yearly per capita motor fuel consumption variable.

b. Initially, the economist hypothesizes that per capita motor fuel consumption is a linear function of the relative price of gasoline. Use the method of least squares to estimate this model.

c. Investigate the usefulness of the preliminary model. Use $\alpha = .05$. Also, find R^2.

d. Next, the economist would like to expand the model described in part b to include a second independent variable, average gross real weekly earnings. Use the method of least squares to fit this expanded model.

e. Investigate the usefulness of the expanded model of part **d**. Use $\alpha = .05$. Also, find R^2.

f. From an economic (or intuitive) perspective, do the signs of the estimated coefficients of the expanded model seem to be appropriate? Explain. [*Note:* The model will be modified in part **b** of Exercise 12.103 to resolve this problem.]

g. Using the expanded model, estimate the mean per capita demand for motor fuel when average gross weekly earnings are $110.36 and the relative price of a gallon of gasoline is 1.521. What reservations (if any) do you have about the goodness of your estimate?

h. Calculate the residuals of the expanded model, and plot them against both independent variables. Check for evidence of curvature and for outliers.

ON YOUR OWN ...

This is a continuation of the "On Your Own" presented in Chapter 10, in which you selected three independent variables as predictors of the Gross National Product and obtained 10 years of data for each. Now fit the multiple regression model (use an available computer package, if possible)

$$y = \beta_0 + \beta_1 x_1 + \beta_2 x_2 + \beta_3 x_3 + \varepsilon$$

where

y = Gross National Product

x_1 = First variable you chose

x_2 = Second variable you chose

x_3 = Third variable you chose

a. Compare the coefficients $\hat{\beta}_1$, $\hat{\beta}_2$, and $\hat{\beta}_3$ to their corresponding slope coefficients in the Chapter 10 "On Your Own," where you fit three separate straight-line models. How do you account for the differences?

b. Calculate the coefficient of determination R^2, and conduct the F test of the null hypothesis H_0: $\beta_1 = \beta_2 = \beta_3 = 0$. What is your conclusion?

If the independent variables you chose are themselves highly correlated, you may encounter some results that are difficult to explain. For example, the co-efficients $\hat{\beta}_1$, $\hat{\beta}_2$, and $\hat{\beta}_3$ may assume signs that are the opposite of what you expected. Or you may get a highly significant F value in part **b**, but the individual t statistics for the x_1, x_2, and x_3 may all be nonsignificant. This phenomenon—a high correlation between the independent variables in a regression model—is **multicollinearity**. This topic was discussed in Section 11.11.

USING THE COMPUTER ...

Suppose we wish to model the relationship of the median income y to the percent of college graduates x_1 and the percent of women in the work force x_2 in a zip code.

a. Draw a random sample of 100 zip codes from the 1,000 described in Appendix C, and extract y, x_1, and x_2 for each zip code sampled. Use a software package that includes a regression program to obtain the least squares fit of the following models:

$$E(y) = \beta_0 + \beta_1 x_1 + \beta_2 x_2$$
$$E(y) = \beta_0 + \beta_1 x_1 + \beta_2 x_2 + \beta_3 x_1 x_2$$

1. Interpret the estimated β parameters of the models.
2. Interpret the estimated standard deviations of the error terms.
3. Evaluate the usefulness of the models. Is there evidence of interaction between percent of college graduates and percent of women in the work force in their relationship with income?
4. Use the model of preference to estimate the mean median income for all zip codes having 20% college graduates and 40% women in the work force using a 95% confidence interval.
5. Use the model of preference to predict the median income for a particular zip code having 20% college graduates and 40% women in the work force using a 95% prediction interval.
6. Plot the residuals against x_1 and x_2 for each model. Check for evidence that curvature is present for either x_1 or x_2.
7. Plot a histogram or stem and leaf display of the residuals for each model. Comment on the assumption of normality for the random error component.
8. Use the residual plots to determine whether outliers exist.

b. Repeat part **a** using the entire set of 1,000 zip codes. Compare the results to those you obtained in part **a**.

REFERENCES

Allmon, C. I. "Advertising and sales relationships for toothpaste: Another look." *Business Economics*, Sept. 1982, *17*, 55–61.

Benston, G. J. "Multiple regression analysis of cost behavior." *Accounting Review*, Oct. 1966, *41*, 657–672.

Binkley, J. K. and Harrer, B. "Major determinants of ocean freight rates for grains: An econometric analysis." *American Journal of Agricultural Economics*, Feb. 1981, *63*, 47–57.

Blair, R. D. and Kenny, L. W. *Microeconomics for Managerial Decision Making*. New York: McGraw-Hill, 1982, Chapter 3.

Carlson, R. D. "Advertising and sales relationships for toothpaste." *Business Economics*, Sept. 1981, *16*, 36–39.

Chase, R. B. and Aquilano, N. J. *Production and Operations Management*, rev. ed. Homewood, Ill.: Richard D. Irwin, 1979, Chapter 11.

Chatterjee, S. and Price, B. *Regression Analysis by Example*. New York: Wiley, 1977.

Dixon, W. J., Brown, M. B., Engelman, L., Frane, J. W., Hill, M. A., Jennrich, R. I., and Toporek, J. D. *BMDP Statistical Software*. Berkeley: University of California Press, 1983.

Draper, N. R. and Smith, H. *Applied Regression Analysis*, 2nd ed. New York: Wiley, 1981.

Ferguson, C. E. and Maurice, S. C. *Economics Analysis*. Homewood, Ill.: Richard D. Irwin, 1970. Chapter 6.

Henry, W. R. and Haynes, W. W. *Managerial Economics: Analysis and Cases*, 4th ed. Dallas: Business Publications, 1978. Chapter 5.

Martin, M. V. and Clement, D. A. "An analysis of port-specific international grain freight rates: The case of the lower Columbia River port area." *Transportation Journal*, Fall 1982, *22*, 18–26.

Mendenhall, W. and Sincich, T. *A Second Course in Business Statistics: Regression Analysis*, 2nd ed. San Francisco, Dellen, 1986.

Miller, R. B. and Wichern, D. W. *Intermediate Business Statistics: Analysis of Variance, Regression, and Time Series*. New York: Holt, Rinehart and Winston, 1977. Chapters 6–8.

Neter, J., Wasserman, W., and Kutner, M. *Applied Linear Regression Models*. Homewood, Ill.: Richard D. Irwin, 1983.

Nie, N., Hull, C. H., Jenkins, J. G., Steinbrenner, K., and Bent, D. H. *Statistical Package for the Social Sciences*, 2nd ed. New York: McGraw-Hill, 1975.

Ryan, T. A., Joiner, B. L., and Ryan, B. F. *Minitab Student Handbook*, North Scituate, Mass.: Duxbury, 1976.

SAS User's Guide: Statistics, 1982 ed. Ray, A. A. (ed.). SAS Institute, Inc. Box 8000, Cary, N.C. 27511.

Schroeder, R. G. *Operations Management: Decision Making in the Operations Function*, 2nd ed., New York: McGraw-Hill, 1985.

Weisberg, S. *Applied Linear Regression*. New York: Wiley, 1980.

Winkler, R. L. and Hays, W. L. *Statistics: Probability, Inference, and Decision*, 2nd ed. New York: Holt, Rinehart and Winston, 1975. Chapter 10.

Wonnacott, R. J. and Wonnacott, T. H. *Econometrics*, 2nd ed. New York: Wiley, 1979. Chapter 6.

Younger, M. S. A *Handbook for Linear Regression*. North Scituate, Mass.: Duxbury, 1979.

C H A P T E R 12

WHERE WE'VE BEEN . . .

One of the most important topics in applied statistics, regression analysis, was presented in Chapters 10 and 11. Simple linear regression, using a straight line to model the relationship between a dependent variable y and a single independent variable x, was the topic of Chapter 10. Multiple regression, relating a dependent variable to any number of independent variables, was the topic of Chapter 11. In both chapters, we learned how to fit regression models to a set of data and how to use the model to estimate the mean value of y or to predict a future value of y for a given value of x.

WHERE WE'RE GOING . . .

In Chapters 10 and 11, an important problem was circumvented— the selection of a model that is appropriate for the given data. No matter how much you know about regression analysis, or how well you can fit a model to a set of data and interpret the results, the information will be of little value if you choose an ill-fitting model to relate the mean value of y to the independent variables. The process of choosing a reasonable model and using the data to modify and improve it, is called *model building*. We introduce you to this topic in Chapter 12.

INTRODUCTION TO MODEL BUILDING

CONTENTS

We have indicated in Chapters 10 and 11 that the first step in the construction of a regression model is to hypothesize the form of the deterministic portion of the probabilistic model. This **model building**, or model construction, stage is the key to the success (or failure) of the regression analysis. If the hypothesized model does not reflect, at least approximately, the true nature of the relationship between the mean response $E(y)$ and the independent variables $x_1, x_2, \ldots, x_k$, the modeling effort will usually be unrewarded.

By *model building*, we mean developing a model that will provide a good fit to a set of data and that will give good estimates of the mean value of y and good predictions of future values of y for given values of the independent variables. To illustrate, suppose you wish to relate the demand, y, for a given product to advertising expenditure, x, and (unknown to you) the second-order model

$$E(y) = \beta_0 + \beta_1 x + \beta_2 x^2$$

would permit you to predict y with a very small error of prediction [see Figure 12.1(a)]. Unfortunately, you have erroneously chosen the first-order model

$$E(y) = \beta_0 + \beta_1 x$$

to explain the relationship between y and x [see Figure 12.1(b)].

FIGURE 12.1
Two Models for Relating Demand y to Advertising Expenditure x

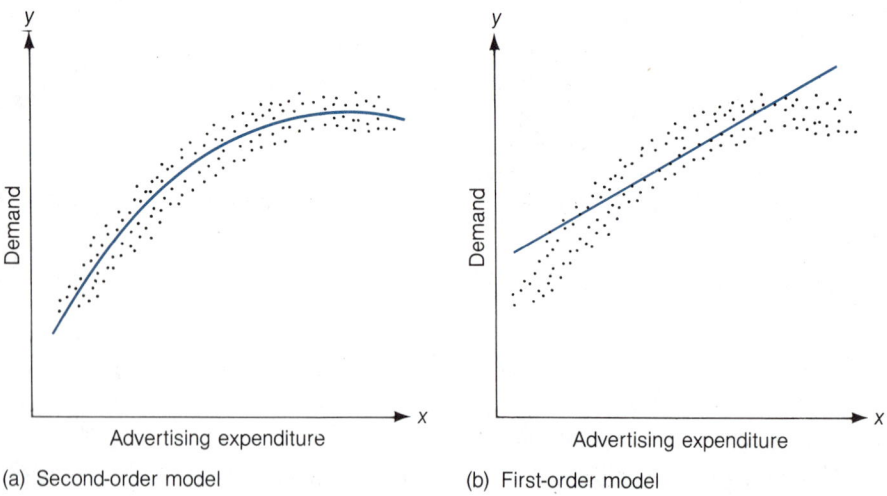

(a) Second-order model   (b) First-order model

The consequence of choosing the wrong model is clearly demonstrated by comparing Figures 12.1(a) and (b). The errors of prediction for the second-order model are relatively small in comparison to those for the first-order model. The lesson to be learned from this simple example is clear. Choosing a good set of independent (predictor) variables, $x_1, x_2, \ldots, x_k$, will not guarantee a good prediction equation. In addition to selecting independent variables that contain information about y, you must specify an equation relating y to $x_1, x_2, \ldots, x_k$ that will provide a good fit to your data.

In the following sections, we will present some useful models for relating a response y to one or more predictor variables.

THE TWO TYPES OF INDEPENDENT VARIABLES: QUANTITATIVE AND QUALITATIVE

In Chapter 2, we defined two types of variables that may arise in business applications: quantitative and qualitative. In regression analysis, the dependent variable will always be quantitative, but the independent variables may be either quantitative or qualitative. As you will see, the way an independent variable enters the model depends on its type.

DEFINITION 12.1

A **quantitative** independent variable is one that assumes numerical values corresponding to the points on a line. An independent variable that is not quantitative is called **qualitative**.

The Gross National Product, prime interest rate, number of defects in a product, and kilowatt-hours of electricity used per day are all examples of quantitative independent variables. On the other hand, suppose three different styles of packaging, A, B, and C, are used by a manufacturer. This independent variable is qualitative, since it is not measured on a numerical scale. Certainly, the style of packaging is an independent variable that may affect the sales of a product, and we would want to include it in a model describing the product's sales, y.

DEFINITION 12.2

The **levels** of an independent variable are its different intensity settings.

For a quantitative independent variable, the levels correspond to the numerical values it assumes. For example, if the number of defects in a product ranges from 0 to 3, the independent variable has four levels: 0, 1, 2, and 3.

The levels of a qualitative variable are not numerical. They can be defined only by describing them. For example, the independent variable for the style of packaging was observed at three levels: A, B, and C. Occasionally, a quantitative variable (e.g., income) will be reported in qualitative levels (e.g., low, medium, and high). Thus, the distinction between quantitative and qualitative variables is often a function of the type of information available.

EXAMPLE 12.1

In Chapter 11 we considered the problem of predicting executive salaries as a function of several independent variables. Consider the following four independent variables that may affect executive salaries:

a. Number of years of experience
b. Sex of the employee
c. Firm's net asset value
d. Rank of the employee

For each of these independent variables, give its type and describe the levels you would expect to observe.

SOLUTION

a. The independent variable for the number of years of experience is quantitative because its values are numerical. We would expect to observe levels ranging from 0 to 40 (approximately) years.

b. The independent variable for sex is qualitative because its levels can be described only by the nonnumerical labels "female" and "male."

c. The independent variable for the firm's net asset value is quantitative, with a large number of possible levels corresponding to the range of dollar values representing various firms' net asset values.

d. Suppose the independent variable for the rank of the employee is observed at three levels: supervisor, assistant vice-president, and vice-president. Since we cannot assign a realistic numerical measure of relative importance to each position, rank is a qualitative independent variable. ■

Quantitative independent variables are treated differently from qualitative variables in regression modeling. In the next section, we will begin our discussion of how quantitative variables are used in the modeling effort.

| | | | | | | | | | | | | |

EXERCISES 12.1–12.5

APPLYING THE CONCEPTS

12.1 The process of quality planning and control requires a continuous interaction between the customer and the manufacturer. In order to be able to design, monitor, and control the quality of a product, the manufacturer needs to know which attributes of the product play an important role in the customer's decision to buy and use the product (Schroeder, 1985). The following quality attributes were identified for a particular product:
a. Weight   **b.** Color   **c.** Age   **d.** Hardness   **e.** Package design
Regression analysis can be used to explore the relationship between these variables and sales of the product. Classify each of the variables as quantitative or qualitative and describe the levels that the variables may assume.

12.2 The marketing department of a large consumer food products company conducted a study to investigate the effect of the following independent variables on the total number of units sold per month of one of the company's new products:
a. Monthly advertising expenditure   **b.** Type of container
c. Color of container   **d.** Medium used for advertising
e. Net weight of the product
Classify each of the variables as quantitative or qualitative and describe the levels that the variables may assume.

12.3 Companies keep personnel files on their employees that contain important information on each individual's background. The data could be used, for example, to predict employee performance ratings. Identify the independent variables listed below as qualitative or quantitative. For qualitative variables, suggest several levels that might be observed. For quantitative variables, give a range of values (levels) for which the variable might be observed.
a. Age   **b.** Years of experience with the company
c. Highest educational degree   **d.** Job classification
e. Marital status   **f.** Religious preference
g. Salary   **h.** Sex

12.4 Which of the assumptions about ε (Section 10.3) prohibit the use of a qualitative variable as a dependent variable?

12.5 Exercise 11.38 described the least squares model developed by Martin and Clement to estimate ocean transport rates for grain shipped from the lower Columbia River international ports. Classify y, x_1, x_2, and x_4 as quantitative or qualitative variables and describe the levels that each variable may assume.

S E C T I O N 12.2

MODELS WITH A SINGLE QUANTITATIVE INDEPENDENT VARIABLE

The most common linear models relating y to a single quantitative independent variable x are those derived from a polynomial expression of the type shown in the box. Specific models, obtained by assigning particular values to p, are listed below.

FORMULA FOR A pTH-ORDER POLYNOMIAL WITH ONE QUANTITATIVE INDEPENDENT VARIABLE

$$E(y) = \beta_0 + \beta_1 x + \beta_2 x^2 + \beta_3 x^3 + \cdots + \beta_p x^p$$

where p is an integer and $\beta_0, \beta_1, \ldots, \beta_p$ are unknown parameters that must be estimated.

---

1. FIRST-ORDER MODEL

$$E(y) = \beta_0 + \beta_1 x$$

Comments on model parameters:

β_0: y-intercept

β_1:   Slope of the line

General comments: The first-order model is used when you expect the rate of change in y per unit change in x to remain fairly stable over the range of values of x for which you wish to predict y (see Figure 12.2). Most relationships between $E(y)$ and x are curvilinear, but the curvature over the range of values of x for which you wish to predict y may be very slight. When this occurs, a first-order (straight-line) model should provide a good fit to your data.

FIGURE 12.2

Graph of a First-Order Model

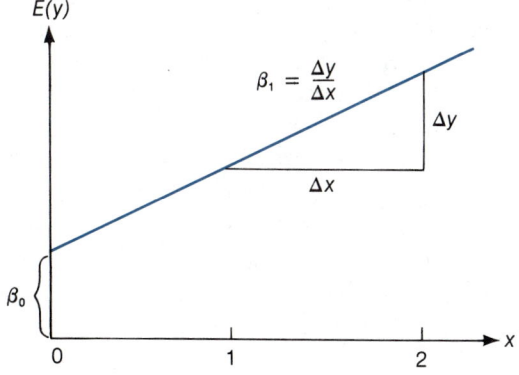

2. SECOND-ORDER MODEL

$$E(y) = \beta_0 + \beta_1 x + \beta_2 x^2$$

Comments on model parameters:

β_0:   *y*-intercept

β_1:   Changing the value of β_1 shifts the parabola to the right or left; increasing the value of β_1 causes the parabola to shift to the left

β_2:   Rate of curvature

General comments: A second-order model traces a parabola, one that opens either downward ($\beta_2 < 0$) or upward ($\beta_2 > 0$), as shown in Figure 12.3. Since most relationships will possess some curvature, a second-order model will often be a good choice to relate *y* to *x*.

FIGURE 12.3

Graphs of Two Second-Order Models

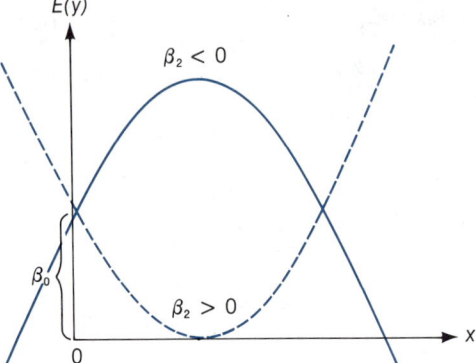

3. THIRD-ORDER MODEL

$$E(y) = \beta_0 + \beta_1 x + \beta_2 x^2 + \beta_3 x^3$$

Comments on model parameters:

β_0:   *y*-intercept

β_3:   The magnitude of β_3 controls the rate of reversal of curvature for the curve

General comments: Reversals in curvature are not common, but such relationships can be modeled by third- and higher-order polynomials. As can be seen in Figure 12.3, a second-order model contains no reversals in curvature. The slope continues to either increase or decrease as *x* increases and produces either a trough or a peak. A third-order model (see Figure 12.4) contains one reversal in curvature and produces one peak and one trough. In general, a graph of a *p*th-order polynomial will contain a total of $(p - 1)$ peaks and troughs.

Most functional relationships in nature seem to be smooth (except for random error), that is, they are not subject to rapid and irregular reversals in direction. Consequently, the second-order polynomial model is perhaps the most useful of those described above. To develop a better understanding of how this model is used, consider Example 12.2.

FIGURE 12.4

Graphs of Two Third-Order Models

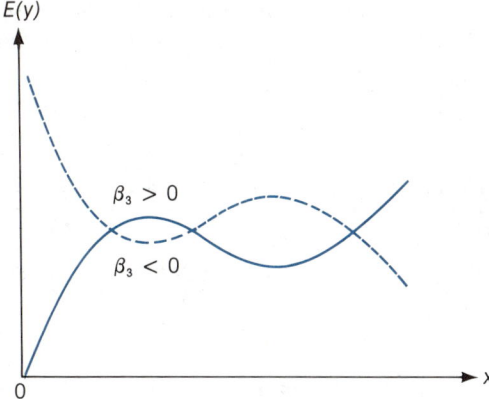

EXAMPLE 12.2

Power companies must be able to predict the peak power load at their various stations in order to operate effectively. The *peak power load* is the maximum amount of power that must be generated each day to meet demand.

Suppose a power company located in the southern part of the United States decides to model daily peak power load, y, as a function of the daily high temperature, x, and the model is to be constructed for the summer months when demand is greatest. Although we would expect the peak power load to increase as the high temperature increases, the *rate* of increase in $E(y)$ might also increase as x increases. That is, a 1-unit increase in high temperature from $100°$ to $101°$ F might result in a larger increase in power demand than would a 1-unit increase from $80°$ to $81°$ F. Therefore, we postulate the second-order model

$$E(y) = \beta_0 + \beta_1 x + \beta_2 x^2$$

and we expect β_2 to be positive.

A random sample of 25 summer days is selected, and the data are shown in Table 12.1. Fit a second-order model using these data, and test the hypothesis that the power load increases at an increasing *rate* with temperature—i.e., that $\beta_2 > 0$.

TABLE 12.1

Power Load Data

TEMPERATURE (°F)	PEAK LOAD (Megawatts)	TEMPERATURE (°F)	PEAK LOAD (Megawatts)	TEMPERATURE (°F)	PEAK LOAD (Megawatts)
94	136.0	106	178.2	76	100.9
96	131.7	67	101.6	68	96.3
95	140.7	71	92.5	92	135.1
108	189.3	100	151.9	100	143.6
67	96.5	79	106.2	85	111.4
88	116.4	97	153.2	89	116.5
89	118.5	98	150.1	74	103.9
84	113.4	87	114.7	86	105.1
90	132.0				

SOLUTION

The SAS printout shown in Figure 12.5 gives the least squares fit of the second-order model using the data in Table 12.1. The prediction equation is

$$\hat{y} = 385.048 - 8.293x + .05982x^2$$

A plot of this equation and the observed values is given in Figure 12.6.

FIGURE 12.5 Portion of the SAS Printout for the Second-Order Model of Example 12.2

SOURCE	DF	SUM OF SQUARES	MEAN SQUARE	F VALUE	PR > F
MODEL	2	15011.77199776	7505.88599888	259.69	0.0001
ERROR	22	635.87840224	28.90356374	R-SQUARE	ROOT MSE
CORRECTED TOTAL	24	15647.65040000		0.959363	5.37620347

PARAMETER	ESTIMATE	T FOR H0: PARAMETER = 0	PR > ¦T¦	STD ERROR OF ESTIMATE
INTERCEPT	385.04809323	6.98	0.0001	55.17243578
TEMP	-8.29252680	-6.38	0.0001	1.29904502
TEMP*TEMP	0.05982337	7.93	0.0001	0.00754855

FIGURE 12.6
Plot of the Observations
and the Second-Order
Least Squares Fit

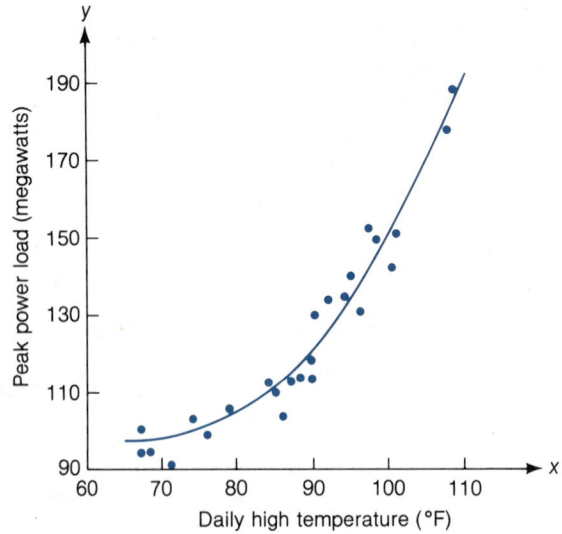

We now test whether the sample value $\hat{\beta}_2 = .05982$ is large enough to conclude *in general* that the power load increases at an increasing rate with temperature:

H_0: $\beta_2 = 0$

H_a: $\beta_2 > 0$

Test statistic: $t = \dfrac{\hat{\beta}_2}{s_{\hat{\beta}_2}}$

For $\alpha = .05$, $n = 25$, and $k = 2$, we reject H_0 if

$$t > t_{.05}$$

where $t_{.05} = 1.717$ (from Table VI of Appendix B) is based on $n - (k + 1) = 22$ degrees of freedom. From Figure 12.5, the calculated value of t is 7.93. Since this value exceeds $t_{.05} = 1.717$, we reject H_0 at $\alpha = .05$ and conclude that the mean power load increases at an increasing rate with temperature. ∎

EXERCISES 12.6–12.20 **LEARNING THE MECHANICS**

12.6 The graphs depict pth-order polynomials with one independent variable. For each graph, identify the order of the polynomial. Find the value of β_0 and β_1 for each.

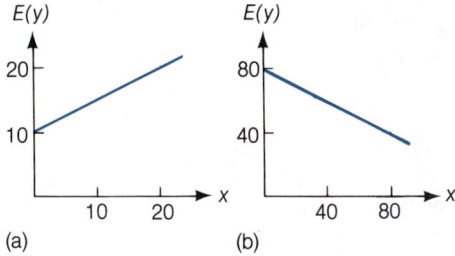

(a)   (b)

12.7 The graphs depict pth-order polynomials for one independent variable. For each graph, identify the order of the polynomial, the value of β_0, and the sign of β_2.

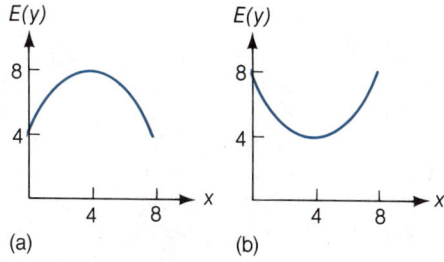

(a)   (b)

12.8 Graph the following polynomials and identify the order of each on your graph:

a. $E(y) = 2 + 3x$ b. $E(y) = 2 + 3x^2$
c. $E(y) = 1 + 2x + 2x^2 + x^3$ d. $E(y) = 2x + 2x^2 + x^3$
e. $E(y) = 2 - 3x^2$ f. $E(y) = -2 + 3x$

12.9 Suppose $E(y)$ can best be modeled by a second-order polynomial in x, where x is a quantitative variable. Write the probabilistic model for y.

12.10 Minitab was used to fit the model

$$y = \beta_0 + \beta_1 x + \beta_2 x^2 + \varepsilon$$

to $n = 17$ data points. The printout at the top of page 656 was obtained. Note that on the printout, C21 is y, C22 is x, and C23 is x^2.

```
THE REGRESSION EQUATION IS
C21 = 63.1 + 1.17 C22 - 0.0374 C23

                                   ST. DEV.       T-RATIO =
COLUMN          COEFFICIENT        OF COEF.       COEF/S.D.
                   63.11             11.92           5.30
C22                 1.1683            0.8276         1.41
C23                -0.03742           0.01234       -3.03

S = 9.289

R-SQUARED = 86.0 PERCENT
R-SQUARED = 84.0 PERCENT, ADJUSTED FOR D.F.

ANALYSIS OF VARIANCE

   DUE TO       DF          SS        MS=SS/DF
REGRESSION      2        7434.4        3717.2
RESIDUAL        14       1208.1          86.3
TOTAL           16       8642.5
```

a. Report the least squares prediction equation.

b. Plot the prediction equation.

c. Specify the null and alternative hypotheses you would use to determine whether y decreases at an increasing rate as x increases.

d. Conduct the hypothesis test of part c using $\alpha = .05$. Interpret your result.

APPLYING THE CONCEPTS

[Note: Starred (*) exercises require the use of a computer.]

12.11 A company that sells and services copy machines conducted a study to relate the number of service calls per month required for a particular brand of table-top copier to the age of the copier (in months). All copiers used in the study were utilized by their owners to produce between 10,000 and 12,000 copies per month. The company suspects that new and old copiers require more service calls than those of middle age.

a. Based on this information, propose an appropriate model relating the mean number of service calls per month to the copier's age. Define all variables in your proposed model.

b. Indicate whether you think β_0 and β_2 assume positive or negative values, and explain the reasons for your decisions.

12.12 Suppose you want to model the appraised value of a house, y, as a function of the number, x, of square feet of living space it contains. Regardless of how the appraisal is to be used, we would expect y to increase as x increases. In some instances, particularly if the appraisal is for tax purposes, the rate of increase in y decreases as x increases.

a. Write a suitable linear model to relate y to x.

b. Specify the signs of the coefficients in your model so they agree with the given verbal explanation of the relationship between y and x. Sketch the relationship on graph paper.

12.13 The strength of a certain plastic is thought to be related to the amount of pressure used to produce the plastic. Researchers believe that, as pressure is increased, the strength of the plastic increases until, at some point, increases in pressure have a detrimental effect on strength. Write a model to relate the strength, y, of the plastic to pressure, x, that would reflect the above beliefs. Sketch the model.

12.14 An economist has proposed the following model to describe the relationship between the number of items produced per day (output) and the number of hours of labor expended per day (input) in a particular production process:

$$y = \beta_0 + \beta_1 x + \beta_2 x^2 + \varepsilon$$

where

y = Number of items produced per day

x = Number of hours of labor per day

A portion of the Minitab computer printout that results from fitting this model to a sample of 25 weeks of production data is reproduced here. Do the data provide sufficient evidence to indicate that the *rate* of increase in output per unit increase of input decreases as the input increases? Test using $\alpha = .05$.

```
THE REGRESSION EQUATION IS
Y =   -6.17  +   2.04 X1   -  .0323 X2

                             ST. DEV.    T-RATIO =
        COLUMN    COEFFICIENT  OF COEF.   COEF/S.D.

          --         -6.173      1.666      -3.71
   X1    C2           2.036       .185      11.02
   X2    C3           -.03231    .00489      -6.60

THE ST. DEV. OF Y ABOUT REGRESSION LINE IS
S =        1.243
WITH (   25- 3) =   22 DEGREES OF FREEDOM

R-SQUARED = 95.5 PERCENT
R-SQUARED = 95.1 PERCENT, ADJUSTED FOR D.F.

ANALYSIS OF VARIANCE

    DUE TO       DF      SS     MS=SS/DF

REGRESSION    2    718.168    359.084
RESIDUAL     22     33.992      1.545
TOTAL        24    752.160
```

12.15 A company is considering having the employees on its assembly line work 4 days per week for 10 hours each instead of 5 days for 8 hours. Management is concerned that the effect of fatigue due to longer afternoons of work might increase assembly times to an unsatisfactory level. An experiment with the 4-day week is planned in which time studies will be conducted on some of the workers during the afternoons. It is believed that an adequate model of the relationship between assembly time, y, and time since lunch, x, should allow for the average assembly time to decrease for a while after lunch (as workers get back in the groove) before it starts to increase as the workers become tired.
  a. Propose a model to relate $E(y)$ and x that would reflect management's belief. Define all terms in your model.
  b. Sketch the shape of the function described by your hypothesized model.

PRESSURE	MILEAGE
x	y
(pounds per square inch)	(miles per gallon)
30	29
31	32
32	36
33	38
34	37
35	33
36	26

12.16 Underinflated or overinflated tires can affect gas mileage. A new brand of tire was tested for its effect on mileage at different pressures with the results shown in the table.
a. Plot the data on a scattergram.
b. If you were given only the information for $x = 30, 31, 32, 33$, what kind of model would you suggest? For $x = 33, 34, 35, 36$? For all the data?

*12.17 Many studies have confirmed that the productivity of manufacturing operations increases as a result of learning. As more units are produced, the work force becomes more proficient at the task and each successive unit takes less time to complete, up to a point. Generally, productivity steadily rises following the initiation of a manufacturing process, then levels off and becomes stationary. However, a recent article presents evidence that suggests this may not be the case in less developed countries. The table presents data collected on the sacking production of Victory Jute Mill, Chittagong, Bangladesh, a jute spinning and weaving plant.

AGE OF MILL (years)	PRODUCTIVITY (kgs. per labor hour)	AGE OF MILL (years)	PRODUCTIVITY (kgs. per labor hour)
2	4.08	12	4.57
3	4.18	13	4.56
4	4.42	14	4.59
5	4.48	15	4.51
6	4.55	16	4.46
7	4.58	22	4.32
8	4.63	23	4.28
9	4.65	24	4.24
10	4.67	25	4.17
11	4.66	26	4.13

Source: Kibria, M. G. and Tisdell, C. A. "International comparisons of learning curves and productivity." *Management International Review*, Vol. 25, No. 4 (1985), 66–72.

a. Construct a scattergram for these data. What does your scattergram reveal about the productivity of the mill as it aged?
b. Based on the pattern revealed in your scattergram, propose a model to relate productivity, y, to the age of the mill, x.
c. Fit your model of part b to the data. Report the least squares prediction equation and plot it on the scattergram of part a.
d. Do the data provide sufficient evidence to conclude that the model proposed in part b provides information for the prediction of productivity? Test using $\alpha = .05$.

*12.18 A veterinarian who works for a large midwestern pig cooperative believes that she has developed a daily vitamin pellet that will substantially increase the weight of mature pigs within 1 month. However, she is uncertain of the relationship between the daily dose (amount of the vitamin) and the percentage gain in weight after 1 month. To better understand this relationship, she randomly selects 16 pigs of the same age and weight and feeds them different doses for a 1-month trial period. The resulting data are given in the table.

a. Plot the data on a scattergram.
b. Fit the model $E(y) = \beta_0 + \beta_1 x + \beta_2 x^2$ to the data.

WEIGHT GAIN (% original weight)	DAILY PELLET DOSE	WEIGHT GAIN (% original weight)	DAILY PELLET DOSE
82	0	90	4
78	0	95	4
80	1	89	5
87	1	93	5
87	2	90	6
95	2	85	6
97	3	84	7
90	3	90	7

c. Is there evidence to support the inclusion of the second-order term in the model of part b? Test using $\alpha = .05$.

d. Plot the fitted model of part b on the scattergram of part a.

*12.19 The Federal Reserve System (FRS) was established in 1913 to provide central banking facilities for the United States. It is often referred to as the "banker's bank." One of its major responsibilities is to control the flow of the nation's money supply in order to facilitate orderly economic growth. One of the ways this is accomplished is through the buying and selling of government securities. The sale of securities to the public draws money from the commercial banking system; the purchase of securities by the FRS from the public increases the money in the commercial banking system. The resulting ebb and flow of the money supply affects the level of interest rates (the prices paid for borrowed money) in the economy (*Federal Reserve System*, 1963). In Exercise 10.79 we investigated the correlation between the rate of growth of the money supply and the prime interest rate. The data used in that exercise are repeated in the table.

DATE	PRIME INTEREST RATE	RATE OF GROWTH IN MONEY SUPPLY (M1[a])	DATE	PRIME INTEREST RATE	RATE OF GROWTH IN MONEY SUPPLY (M1[a])
2/84	11.0	.55	1/85	10.75	1.48
3/84	11.5	.43	2/85	10.5	1.19
4/84	12.0	.04	3/85	10.5	.47
5/84	12.5	1.06	4/85	10.5	.49
6/84	13.0	.94	5/85	10.5	1.17
7/84	13.0	−.09	6/85	10.0	1.65
8/84	13.0	.15	7/85	9.5	.78
9/84	13.0	.42	8/85	9.5	1.70
10/84	12.75	−.62	9/85	9.5	.99
11/84	12.0	.71	10/85	9.5	−.13
12/84	11.25	.93	11/85	9.5	1.11
			12/85	9.5	1.10

[a]M1 = Currency + Demand deposits + Traveler's checks + Other checkable deposits.
Source: *Economic Report of the President*, 1985, 1986, pp. 303, 333, 327.

a. Fit a straight line to the data using the prime interest rate as the dependent variable, y. Compute R^2.

b. Based on the results of part **a**, describe in words the apparent relationship between interest rates and M1 growth.

c. Plot the least squares line on a scattergram of the data. Does it appear that a second-order model might better explain the variation in interest rates?

d. Fit a second-order model to the data and compute R^2.

e. Plot the prediction equation you developed in part **d** on a scattergram of the data.

f. Do the data provide sufficient evidence to conclude that $\beta_2 \neq 0$? Test using $\alpha = .05$. Draw the appropriate conclusions regarding the usefulness of the second-order model relative to the first-order model of part **a** for explaining the variation in interest rates.

12.20 Automobile accidents result in a tragic loss of life and, in addition, they represent a serious dollar loss to the nation's economy. Shown in the table are the number of highway deaths (to the nearest hundred) and the number of licensed vehicles (in hundreds of thousands) for the years 1950–1985. (The years are coded 1–36 for convenience.) During the years 1974–1985 (years 25–36 in the table), the nationwide 55-mile-per-hour speed limit was in effect.

YEAR	DEATHS y	NUMBER OF VEHICLES x_1	YEAR	DEATHS y	NUMBER OF VEHICLES x_1
1	34.8	49.2	19	54.9	103.1
2	37.0	51.9	20	55.8	107.4
3	37.8	53.3	21	54.6	111.2
4	38.0	56.3	22	54.3	116.3
5	35.6	58.6	23	56.3	112.3
6	38.4	62.8	24	55.5	129.8
7	39.6	65.2	25	46.4	134.9
8	38.7	67.6	26	45.9	137.9
9	37.0	68.8	27	47.0	143.5
10	37.9	72.1	28	49.5	148.8
11	38.1	74.5	29	51.5	153.6
12	38.1	76.4	30	51.9	159.4
13	40.8	79.7	31	52.6	164.9
14	43.6	83.5	32	50.8	165.7
15	47.7	87.3	33	46.0	166.5
16	49.1	91.8	34	44.6	167.7
17	53.0	95.9	35	46.2	174.2
18	52.9	98.9	36	45.6	175.7

Source: *Accident Facts*, National Safety Council.

a. Write a second-order model relating the number, y, of highway deaths for a year to the number, x_1, of licensed vehicles.

b. The SAS computer printout for fitting the model to the data is shown on page 661. Is there sufficient evidence to indicate that the model provides information for the prediction of the number of annual highway deaths? Test using $\alpha = .05$.

c. Give the observed significance level for the test of part **b**, and interpret it.

d. Does the second-order term contribute information for the prediction of y? Test using $\alpha = .05$.

e. Give the observed significance level for the test of part **d**, and interpret it.

SAS Printout for Exercise 12.20

Dep Variable: Y

Analysis of Variance

Source	DF	Sum of Squares	Mean Square	F Value	Prob>F
Model	2	1254.34593	627.17296	49.690	0.0001
Error	33	416.51713	12.62173		
C Total	35	1670.86306			

Root MSE	3.55271	R-Square	0.7507	
Dep Mean	45.76389	Adj R-Sq	0.7356	
C.V.	7.76312			

Parameter Estimates

Variable	DF	Parameter Estimate	Standard Error	T for H0: Parameter=0	Prob > |T|
INTERCEP	1	-0.526668	5.30185537	-0.099	0.9215
X1	1	0.824314	0.10408155	7.920	0.0001
X1SQ	1	-0.003201	0.00045741	-6.998	0.0001

SECTION 12.3

MODELS WITH TWO QUANTITATIVE INDEPENDENT VARIABLES

1. FIRST-ORDER MODEL

$$E(y) = \beta_0 + \beta_1 x_1 + \beta_2 x_2$$

Comments on model parameters:

β_0: y-intercept, the value of $E(y)$ when $x_1 = x_2 = 0$

β_1: Change in $E(y)$ for a 1-unit increase in x_1, when x_2 is held fixed

β_2: Change in $E(y)$ for a 1-unit increase in x_2, when x_1 is held fixed

General comments: The graph in Figure 12.7 (page 662) traces a **response surface** (in contrast to the **response curve** used to relate $E(y)$ to a *single* quantitative variable). In particular, a first-order model relating $E(y)$ to two independent quantitative variables, x_1 and x_2, graphs as a plane in three-dimensional space. The plane traces the value of $E(y)$ for every combination of values (x_1, x_2) that correspond to points in the x_1, x_2 plane. Most response surfaces in the real world are well behaved (smooth), and they have curvature. Consequently, a first-order model is appropriate only if the response surface is fairly flat over the x_1, x_2 region that is of interest to you.

The assumption that a first-order model will adequately characterize the relationship between $E(y)$ and the variables x_1 and x_2 is equivalent to assuming that x_1 and x_2 do not interact; that is, you assume that the effect on $E(y)$ of a change in x_1 (for a fixed value of x_2) is the same regardless of the value of x_2 (and vice versa).

FIGURE 12.7
Computer-Generated
Graph of a First-Order
Model

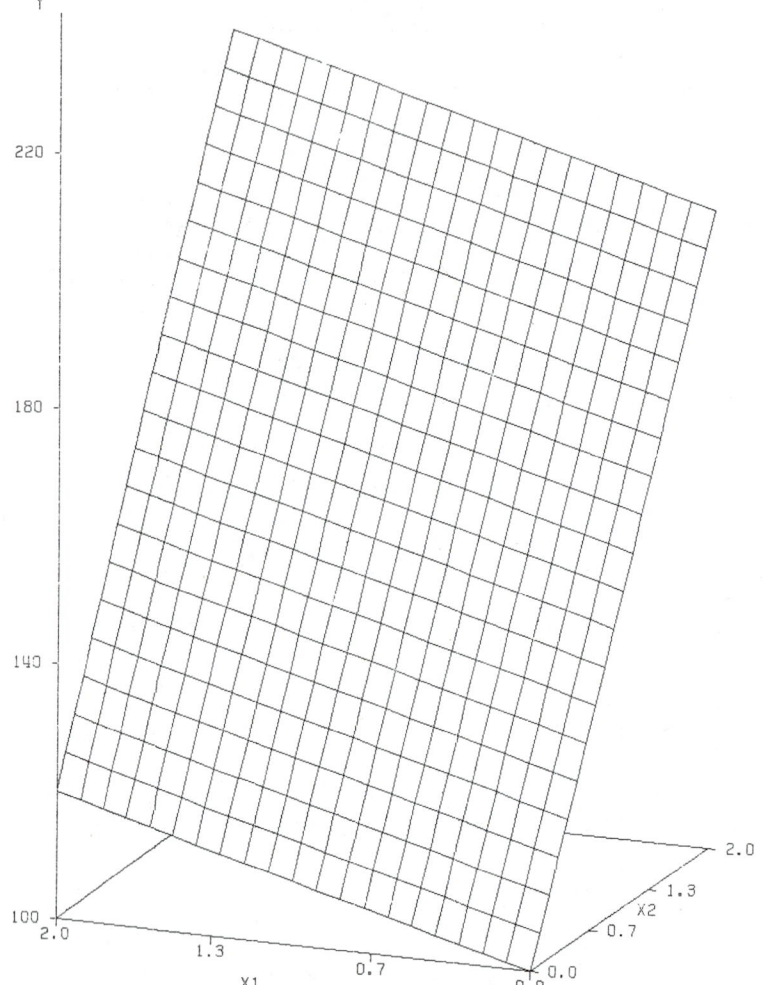

Thus, no interaction implies that the effect of changes in one variable (say x_1) on $E(y)$ is *independent* of the value of the second variable (say x_2). For example, if we assign values to x_2 in a first-order model, the graph of $E(y)$ as a function of x_1 would produce parallel lines as shown in Figure 12.8. These lines, called **contour lines**, show the contours of the surface when it is sliced by three planes, each of which is parallel to the $E(y)$, x_1 plane, at distances $x_2 = 1$, 2, and 3 from the origin.

DEFINITION 12.3

Two variables x_1 and x_2 are said to **interact** if the change in $E(y)$ for a 1-unit change in x_1 (when x_2 is held fixed) is dependent on the value of x_2.

FIGURE 12.8

A Graph Indicating No
Interaction Between
x_1 and x_2

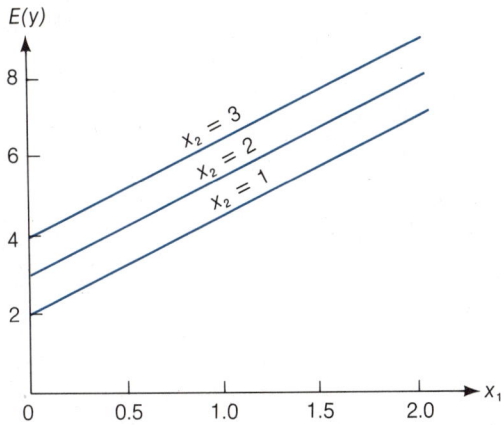

2. AN INTERACTION MODEL (SECOND-ORDER)

$$E(y) = \beta_0 + \beta_1 x_1 + \beta_2 x_2 + \beta_3 x_1 x_2$$

Comments on model parameters:

β_0: y-intercept, the value of $E(y)$ when $x_1 = x_2 = 0$

β_1 and β_2:   Changing β_1 and β_2 causes the surface to shift along the x_1- and x_2-axes

β_3:   Controls the rate of twist in the ruled surface (see Figure 12.9 on page 664)

General comments: This model is said to be second-order because the order of the highest-order term ($x_1 x_2$) in x_1 and x_2 is 2; i.e., the sum of the exponents of x_1 and x_2 equals 2. This interaction model traces a ruled surface in a three-dimensional space (Figure 12.9). You could produce such a surface by placing a pencil perpendicular to a line and moving it along the line, while rotating it around the line. The resulting surface would appear as a twisted plane. A graph of $E(y)$ as a function of x_1 for given values of x_2 (say $x_2 = 1$, 2, and 3) produces nonparallel contour lines (see Figure 12.10), thus indicating that the change in $E(y)$ for a given change in x_1 is dependent on the value of x_2 and, therefore, that x_1 and x_2 interact. Interaction is an extremely important concept because it is easy to get in the habit of fitting first-order models and individually examining the relationships between $E(y)$ and each of a set of independent variables, $x_1, x_2, \ldots, x_k$. Such a procedure is meaningless when interaction exists (which is, at least to some extent, almost always the case), and it can lead to gross errors in interpretation. For example, suppose the relationship between $E(y)$ and x_1 and x_2 is as shown in Figure 12.10 and that you have observed y for each of the $n = 9$ combinations of values of x_1 and x_2 ($x_1 = 1$, 2, 3 and $x_2 = 1$, 2, 3). If you fit a first-order model in x_1 and x_2 to the data, the fitted plane would be (except for random error) approximately parallel to the x_1, x_2 plane, thus suggesting that x_1 and x_2 contribute very little information about $E(y)$. That this is not the case is clearly indicated by the figure.

FIGURE 12.9

Computer-Generated Graph for an Interaction Model (Second-Order)

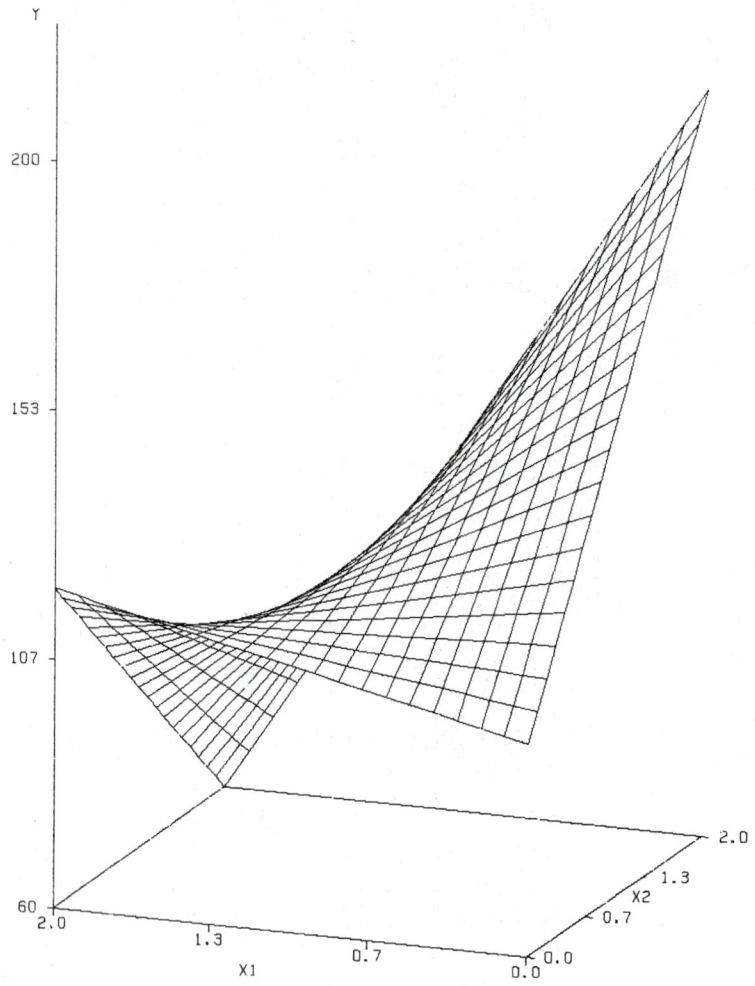

FIGURE 12.10

A Graph Indicating Interaction Between x_1 and x_2

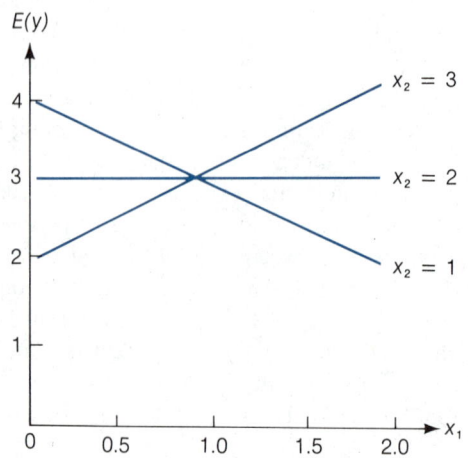

Fitting a first-order model to the data would not allow for the twist in the true surface and would therefore give a false impression of the relationship between $E(y)$ and x_1 and x_2. The procedure for detecting interaction between two independent variables can be seen by examining the model. The interaction model differs from the noninteraction first-order model only in the inclusion of the $\beta_3 x_1 x_2$ term:

Interaction model: $E(y) = \beta_0 + \beta_1 x_1 + \beta_2 x_2 + \beta_3 x_1 x_2$

First-order model: $E(y) = \beta_0 + \beta_1 x_1 + \beta_2 x_2$

Therefore, to test for the presence of interaction, we test

H_0: $\beta_3 = 0$ (no interaction)

against the alternative hypothesis

H_a: $\beta_3 \neq 0$ (interaction)

using the familiar Student's t test of Section 11.5.

3. A COMPLETE SECOND-ORDER MODEL

$$E(y) = \beta_0 + \beta_1 x_1 + \beta_2 x_2 + \beta_3 x_1 x_2 + \beta_4 x_1^2 + \beta_5 x_2^2$$

Comments on model parameters:

β_0: y-intercept, the value of $E(y)$ when $x_1 = x_2 = 0$

β_1 and β_2: Changing β_1 and β_2 causes the surface to shift along the x_1- and x_2-axes

β_3: The value of β_3 controls the rotation of the surface

β_4 and β_5: Signs and values of these parameters control the type of surface and the rates of curvature

The following three types of surfaces may be produced by a second-order model:

β_4 and β_5 positive: A paraboloid that opens upward [Figure 12.11(a)]

β_4 and β_5 negative: A paraboloid that opens downward [Figure 12.11(b)]

β_4 and β_5 differ in sign: A saddle-shaped surface [Figure 12.11(c)]

FIGURE 12.11

Graphs of Three Second-Order Surfaces

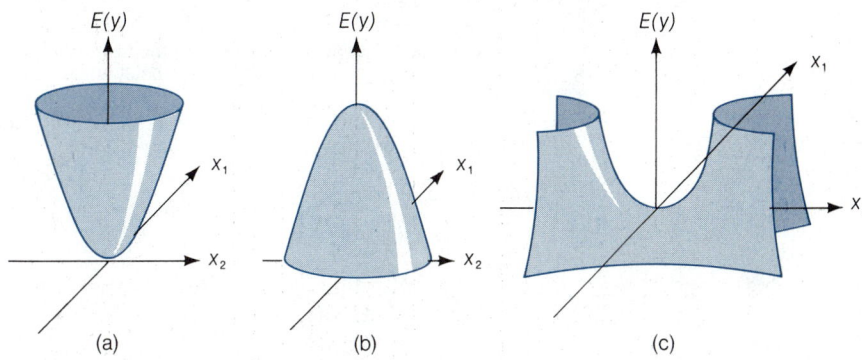

(a) (b) (c)

General comments: A complete second-order model is the three-dimensional equivalent of a second-order model with a single quantitative variable. Instead of tracing parabolas, it traces paraboloids and saddle surfaces. Since you fit only a portion of the complete surface to the data, a complete second-order model provides a large variety of gently curving surfaces. It is a good choice for a model if you expect curvature in the response surface relating $E(y)$ to x_1 and x_2.

EXERCISES
12.21–12.30

LEARNING THE MECHANICS

12.21 Consider an application in which you are trying to relate a response y to two independent variables, x_1 and x_2.
  a. Write a first-order model relating $E(y)$ to x_1 and x_2.
  b. Modify the model you constructed in part **a** to include an interaction term.
  c. Modify the model you constructed in part **b** to make it a complete second-order model.

12.22 Suppose the true relationship between $E(y)$ and the quantitative independent variables x_1 and x_2 is described by the first-order model

$$E(y) = 4 - x_1 + 2x_2$$

  a. Describe the corresponding response surface.
  b. Plot the contour lines of the response surface for $x_1 = 2, 3, 4$, where $0 \le x_2 \le 5$.
  c. Plot the contour lines of the response surface for $x_2 = 2, 3, 4$, where $0 \le x_1 \le 5$.
  d. Use the contour lines you plotted in parts **b** and **c** to explain how changes in the settings of x_1 and x_2 affect $E(y)$.
  e. Use your graph from part **b** to determine how much $E(y)$ changes when x_1 is changed from 4 to 2 and, simultaneously, x_2 is changed from 1 to 2.

12.23 Suppose the true relationship between $E(y)$ and the quantitative independent variables x_1 and x_2 is

$$E(y) = 4 - x_1 + 2x_2 + x_1x_2$$

  a. Identify the order of the model.
  b. Describe the corresponding response surface.
  c. Plot the contour lines of the response surface for $x_1 = 0, 1, 2$, where $0 \le x_2 \le 5$.
  d. Explain why the contour lines you plotted in part **c** are not parallel.
  e. Use the contour lines you plotted in part **c** to explain how changes in the settings of x_1 and x_2 affect $E(y)$.
  f. Use your graph from part **c** to determine how much $E(y)$ changes when x_1 is changed from 2 to 0 and, simultaneously, x_2 is changed from 4 to 5.

12.24 What does it mean to say that two variables affect the mean response, $E(y)$, independently of each other?

12.25 Minitab was used to fit the model

$$y = \beta_0 + \beta_1 x_1 + \beta_2 x_2 + \beta_3 x_1 x_2 + \varepsilon$$

to $n = 15$ data points. The results are shown in the printout, where C1 is y, C2 is x_1, C3 is x_2, and C4 is x_1x_2.

Minitab Printout for
Exercise 12.25

```
THE REGRESSION EQUATION IS
C1 = - 2.55 + 3.82 C2 + 2.63 C3 - 1.29 C4
```

COLUMN	COEFFICIENT	ST. DEV. OF COEF.	T-RATIO = COEF/S.D.
	-2.550	1.142	-2.23
C2	3.8150	0.5286	7.22
C3	2.6300	0.3443	7.64
C4	-1.2850	0.1594	-8.06

```
S = 0.7127
```

```
R-SQUARED = 85.6 PERCENT
R-SQUARED = 81.6 PERCENT, ADJUSTED FOR D.F.
```

```
ANALYSIS OF VARIANCE
```

DUE TO	DF	SS	MS=SS/DF
REGRESSION	3	33.149	11.050
RESIDUAL	11	5.587	0.508
TOTAL	14	38.736	

a. What is the prediction equation for the response surface?
b. Describe the geometric form of the response surface of part a.
c. Plot the prediction equation for the case when $x_2 = 1$. Do this twice more on the same graph for the cases when $x_2 = 3$ and $x_2 = 5$.
d. Explain what it means to say that x_1 and x_2 interact. Explain why your graph of part c suggests that x_1 and x_2 interact.
e. Specify the null and alternative hypotheses you would use to test whether x_1 and x_2 interact.
f. Conduct the hypothesis test of part e using $\alpha = .01$.

APPLYING THE CONCEPTS

[Note: *Starred (*) exercises require the use of a computer.*]

12.26 The Department of Energy wants to develop a regression model to help forecast annual gasoline consumption in the United States, y. They have decided to model $E(y)$ as a function of two independent variables:

x_1 = Number of cars (millions) in use during year

x_2 = Number of trucks (millions) in use during year

a. Identify the independent variables as quantitative or qualitative.
b. Write the first-order model for $E(y)$.
c. Write the complete second-order model for $E(y)$.
d. With respect to the model in part c, specify the null and alternative hypotheses you would use to test whether the second-order terms contribute information for the prediction of annual gasoline consumption.

12.27 Some corporations, instead of owning a fleet of cars, rent cars from a rental agency. A corporation may do this because it is sometimes more economical to rent new cars for a year than to buy new cars each year. A major rental agency wants to develop a model that will allow it to estimate the average annual cost to the prospective customer of renting cars, y, as a function of two independent variables:

x_1 = Number of cars rented

x_2 = Average number of miles driven per car during the year (in thousands)

a. Identify the independent variables as quantitative or qualitative.

b. Write the first-order model for $E(y)$.

c. Write a model for $E(y)$ that contains all first-order and interaction terms. Sketch typical response curves showing the mean cost, $E(y)$, versus the average mileage driven, x_2, for different values of x_1. (Assume that x_1 and x_2 interact.)

d. Write the complete second-order model for $E(y)$.

12.28 Refer to Exercise 12.27. Suppose the model from part **c** is fit, with the following result:

$$\hat{y} = 1 + .05x_1 + x_2 + .05x_1x_2$$

(The units of $\hat{y}$ are thousands of dollars.) Graph the estimated cost $\hat{y}$ as a function of the average number of miles driven, x_2, over the range $x_2 = 10$ to $x_2 = 50$ (10,000–50,000 miles) for $x_1 = 1$, 5, and 10. Do these functions agree (approximately) with the graphs you drew for Exercise 12.27, part **c**?

12.29 An economist is interested in modeling the relationship between quarterly sales of central air-conditioning systems (in thousands) for single-family homes in the United States and two quantitative independent variables, housing starts in the previous quarter (in thousands) and the gross national product (in billions of 1972 dollars). Data were collected, and a model was fit. The displayed portion of the resulting Minitab printout describes the least squares prediction equation. In the prediction equation, $x_3 = x_1^2$ and $x_4 = x_2^2$.

Minitab Printout for Exercise 12.29

```
THE REGRESSION EQUATION IS
Y =    149. + .472 X1 - .0993 X2
     - .0005 X3 + .0000 X4

                              ST. DEV.   T-RATIO =
       COLUMN   COEFFICIENT   OF COEF.   COEF/S.D.
       --            148.5       224.5        .66
X1     C2             .472        .126       3.74
X2     C3            -.099        .162       -.61
X3     C12       -.000535     .000359      -1.49
X4     C13        .0000153    .0000294       .52

THE ST. DEV. OF Y ABOUT REGRESSION LINE IS
S =     6.884
WITH (  16-5) = 11 DEGREES OF FREEDOM

R-SQUARED = 92.0 PERCENT
R-SQUARED = 89.1 PERCENT, ADJUSTED FOR D.F.

ANALYSIS OF VARIANCE

   DUE TO      DF        SS    MS=SS/DF

REGRESSION    4    6002.08    1500.52
RESIDUAL     11     521.36      47.40
TOTAL        15    6523.44
```

a. Write the prediction equation for the response surface.

b. Describe the geometric form of the response surface of part **a**.

c. Do the data provide sufficient evidence to conclude that the model hypothesized by the economist is useful for predicting quarterly sales of central air-conditioning systems? Test using $\alpha = .01$.

d. Does it appear that the variation in air conditioner sales could be adequately explained by a less complex regression model? Explain. [*Note:* In the next section we discuss a formal procedure for making this inference.]

*12.30 A supermarket chain is interested in exploring the relationship between the sales of its store-brand vegetables (y), the amount spent on promotion of the vegetables in local newspapers (x_1), and the amount of shelf space allocated to the brand (x_2). One of the chain's supermarkets was randomly selected and over a 20-week period x_1 and x_2 were varied as reported in the table.

WEEK	SALES ($)	ADVERTISING EXPENDITURES ($)	SHELF SPACE (sq. ft.)	WEEK	SALES ($)	ADVERTISING EXPENDITURES ($)	SHELF SPACE (sq. ft.)
1	2,010	201	75	11	5,005	996	75
2	1,850	205	50	12	2,500	625	50
3	2,400	355	75	13	3,005	860	50
4	1,575	208	30	14	3,480	1,012	50
5	3,550	590	75	15	5,500	1,135	75
6	2,015	397	50	16	1,995	635	30
7	3,908	820	75	17	2,390	837	30
8	1,870	400	30	18	4,390	1,200	50
9	4,877	997	75	19	2,785	990	30
10	2,190	515	30	20	2,989	1,205	30

a. Fit the following model to the data:

$$y = \beta_0 + \beta_1 x_1 + \beta_2 x_2 + \beta_3 x_1 x_2 + \varepsilon$$

b. Conduct an F test to investigate the overall usefulness of this model. Use $\alpha = .05$.

c. Test for the presence of interaction between advertising expenditure and shelf space. Use $\alpha = .05$.

d. Explain what it means to say that advertising expenditures and shelf space interact.

e. Explain how you could be misled by using a first-order model instead of an interaction model to explain how advertising expenditure and shelf space influence sales.

S E C T I O N   12.4

MODEL BUILDING: TESTING PORTIONS OF A MODEL

The presentation of models with one and with two quantitative independent variables raises a very general question. Do certain terms in the model contribute more information than others for the prediction of y?

To illustrate, suppose you have collected data on a response, y, and two quantitative independent variables, x_1 and x_2, and you are considering the use of either a first-order or a second-order model to relate $E(y)$ to x_1 and x_2. Will the second-order model provide better predictions of y than the first-order model? To answer this question, examine the two models and note that the second-order model contains all the terms contained in the first-order model plus three additional terms, those involving β_3, β_4, and β_5.

First-order model: $E(y) = \beta_0 + \beta_1 x_1 + \beta_2 x_2$ Second-order terms

Second-order model: $E(y) = \beta_0 + \beta_1 x_1 + \beta_2 x_2 + \beta_3 x_1 x_2 + \beta_4 x_1^2 + \beta_5 x_2^2$

Therefore, asking whether the second-order model contributes more information for the prediction of y than the first-order model is equivalent to asking whether at least one of the parameters, β_3, β_4, and β_5, differs from 0—i.e., whether the terms involving β_3, β_4, and β_5 should be retained in the model. Therefore, to test whether the second-order terms should be included in the model, we test the null hypothesis

H_0: $\beta_3 = \beta_4 = \beta_5 = 0$

(i.e., the second-order terms do not contribute information for the prediction of y) against the alternative hypothesis

H_a: At least one of the parameters, β_3, β_4, or β_5, differs from 0

(i.e., at least one of the second-order terms contributes information for the prediction of y).

The procedure for conducting this test is intuitive: First, we use the method of least squares to fit the first-order model and calculate the corresponding sum of squares for error, SSE_1 (the sum of squares of the deviations between observed and predicted y values). Next, we fit the second-order model and calculate its sum of squares for error, SSE_2. Then, we compare SSE_1 to SSE_2. If the second-order terms contribute to the model, then SSE_2 should be much smaller than SSE_1, and the difference $(SSE_1 - SSE_2)$ will be large. The larger the difference, the greater the weight of evidence that the second-order model provides better predictions of y than does the first-order model.

The sum of squares for error will always decrease when new terms are added to the model. The question is whether this decrease is large enough to conclude that it is due to more than just an increase in the number of model terms and to chance. To test the null hypothesis that the parameters of the second-order terms β_3, β_4, and β_5 simultaneously equal 0, we use an F statistic calculated as follows:

$$F = \frac{\text{Drop in SSE/Number of } \beta \text{ parameters being tested}}{s^2 \text{ for the complete second-order model}} = \frac{(SSE_1 - SSE_2)/3}{SSE_2 - /[n - (5 + 1)]}$$

When the assumptions listed in Section 11.2 about the error term, ε, are satisfied and the β parameters for the second-order terms are all 0 (i.e., when H_0 is true), this F statistic has an F distribution with $\nu_1 = 3$ and $\nu_2 = n - 6$ degrees of freedom. Note that ν_1 is the number of β parameters being tested and ν_2 is the number of degrees of freedom associated with s^2 in the complete model.

If the second-order terms *do* contribute to the model (i.e., if H_a is true), we expect the F statistic to be large. Thus, we use a one-tailed test and reject H_0 when F exceeds some critical value, F_α, as shown in Figure 12.12. A summary of the steps used in testing the null hypothesis that a set of model parameters are all equal to 0 is shown in the next box.

FIGURE 12.12
Rejection Region for
the F Test of
H_0: $\beta_3 = \beta_4 = \beta_5 = 0$

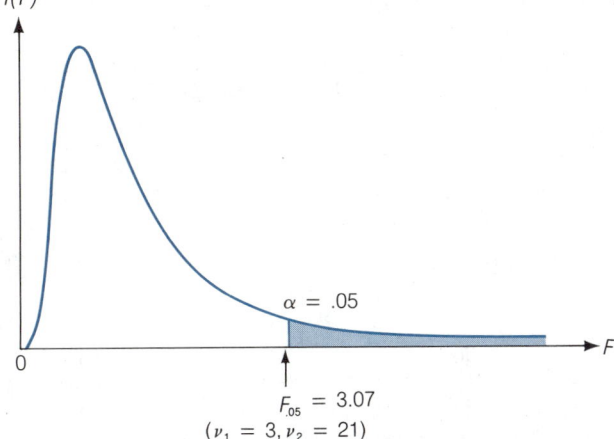

$f(F)$

$\alpha = .05$

F

0

$F_{.05} = 3.07$
$(\nu_1 = 3, \nu_2 = 21)$

F TEST FOR TESTING THE NULL HYPOTHESIS: SET OF β PARAMETERS EQUAL 0

Reduced model: $E(y) = \beta_0 + \beta_1 x_1 + \cdots + \beta_g x_g$

Complete model: $E(y) = \beta_0 + \beta_1 x_1 + \cdots + \beta_g x_g$
$$+ \beta_{g+1} x_{g+1} + \cdots + \beta_k x_k$$

H_0: $\beta_{g+1} = \beta_{g+2} = \cdots = \beta_k = 0$

H_a: At least one of the β parameters under test is nonzero

Test statistic: $F = \dfrac{(\text{SSE}_1 - \text{SSE}_2)/(k - g)}{\text{SSE}_2/[n - (k + 1)]}$

where

SSE_1 = Sum of squared errors for the reduced model

SSE_2 = Sum of squared errors for the complete model

$k - g$ = Number of β parameters specified in H_0

$k + 1$ = Number of β parameters in the complete model

n = Sample size

Rejection region: $F > F_\alpha$

where

$\nu_1 = k - g$ = Degrees of freedom for the numerator

$\nu_2 = n - (k + 1)$ = Degrees of freedom for the denominator

---

EXAMPLE 12.3

Many companies manufacture products (e.g., steel, paint, gasoline) that are at least partially chemically produced. In many instances, the quality of the finished product is a function of the temperature and pressure at which the chemical reactions take place. Suppose you wanted to model the quality, y, of a product as a function of the temperature, x_1, and the pressure, x_2, at which it is produced. Four inspectors independently assign a quality score between 0 and 100 to each product, and then the quality, y, is calculated by averaging the four scores. An experiment is conducted by varying temperature between 80° and 100°F and pressure between 50 and 60 pounds per square inch. The resulting data are given in Table 12.2.

TABLE 12.2
Temperature, Pressure, and Quality of the Finished Product

x_1 (°F)	x_2 (pounds per square inch)	y	x_1 (°F)	x_2 (pounds per square inch)	y	x_1 (°F)	x_2 (pounds per square inch)	y
80	50	50.8	90	50	63.4	100	50	46.6
80	50	50.7	90	50	61.6	100	50	49.1
80	50	49.4	90	50	63.4	100	50	46.4
80	55	93.7	90	55	93.8	100	55	69.8
80	55	90.9	90	55	92.1	100	55	72.5
80	55	90.9	90	55	97.4	100	55	73.2
80	60	74.5	90	60	70.9	100	60	38.7
80	60	73.0	90	60	68.8	100	60	42.5
80	60	71.2	90	60	71.3	100	60	41.4

a. Fit a second-order model to the data.
b. Sketch the response surface.
c. Do the data provide sufficient evidence to indicate that the second-order terms contribute information for the prediction of y?

SOLUTION   a. The complete second-order model is

$$E(y) = \beta_0 + \beta_1 x_1 + \beta_2 x_2 + \beta_3 x_1 x_2 + \beta_4 x_1^2 + \beta_5 x_2^2$$

The data in Table 12.2 were used to fit this model, and a portion of the SAS output is shown in Figure 12.13. The least squares prediction equation is

$$\hat{y} = -5{,}127.90 + 31.10 x_1 + 139.75 x_2 - .146 x_1 x_2 - .133 x_1^2 - 1.14 x_2^2$$

FIGURE 12.13   Portion of the SAS Printout for Example 12.3

SOURCE	DF	SUM OF SQUARES	MEAN SQUARE	F VALUE	PR > F
MODEL	5	8402.26453714	1680.45290743	596.32	0.0001
ERROR	21	59.17842582	2.81802028	R-SQUARE	ROOT MSE
CORRECTED TOTAL	26	8461.44296296		0.993006	1.67869601

PARAMETER	ESTIMATE	T FOR H0: PARAMETER = 0	PR > |T|	STD ERROR OF ESTIMATE
INTERCEPT	-5127.89907417	-46.49	0.0001	110.29601483
X1	31.09638889	23.13	0.0001	1.34441322
X2	139.74722222	44.50	0.0001	3.14005411
X1*X2	-0.14550000	-15.01	0.0001	0.00969196
X1*X1	-0.13338889	-19.46	0.0001	0.00685325
X2*X2	-1.14422222	-41.74	0.0001	0.02741299

b. A three-dimensional graph of this prediction model is shown in Figure 12.14. Note that the mean quality seems to be greatest for temperatures of about 85°–90°F and for pressures of about 55–57 pounds per square inch.* Further experimentation in these ranges might lead to a more precise determination of the optimal temperature–pressure combination.

FIGURE 12.14

Plot of Second-Order Least Squares Model for Example 12.3

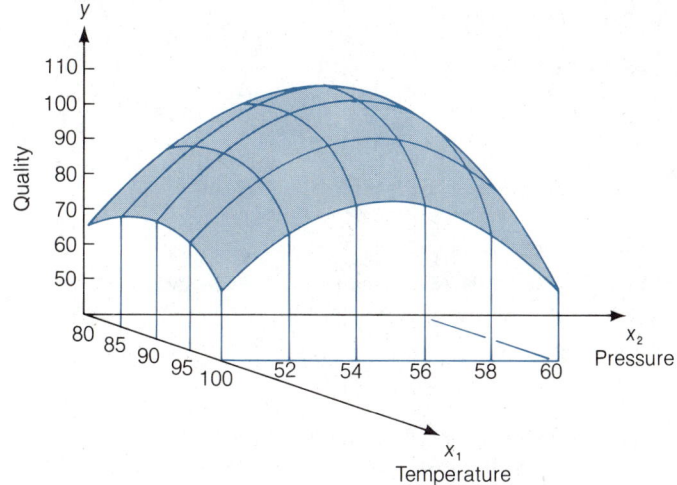

c. To determine whether the data provide sufficient evidence to indicate that the second-order terms contribute information for the prediction of y, we test

$$H_0:\quad \beta_3 = \beta_4 = \beta_5 = 0$$

against the alternative hypothesis

$$H_0:\quad \text{At least one of the parameters, } \beta_3, \beta_4, \text{ or } \beta_5, \text{ differs from } 0$$

The first step in conducting the test is to drop the second-order terms out of the complete (second-order) model and fit the reduced model

$$E(y) = \beta_0 + \beta_1 x_1 + \beta_2 x_2$$

to the data. The SAS computer printout for this procedure is shown in Figure 12.15 (page 674). You can see that the sums of squares for error, given in Figures 12.13 and 12.15 for the complete and reduced models, respectively, are

$$SSE_2 = 59.17842582 \qquad SSE_1 = 6,671.50851852$$

and that s^2 for the complete model is

$$s_2^2 = 2.81802028$$

*Students with knowledge of calculus should note that we can solve for the exact temperature and pressure that maximize quality in the least squares model by solving $\partial\hat{y}/\partial x_1 = 0$ and $\partial\hat{y}/\partial x_2 = .0$ for x_1 and x_2. These estimated optimal values are $x_1 = 86.25°F$ and $x_2 = 55.58$ pounds per square inch. Remember, however, that these represent only sample estimates of the coordinates for the optimal value.

FIGURE 12.15   SAS Computer Printout for the Reduced (First-Order) Model in Example 12.3

DEPENDENT VARIABLE: Y

SOURCE	DF	SUM OF SQUARES	MEAN SQUARE	F VALUE
MODEL	2	1789.93444444	894.96722222	3.22
ERROR	24	6671.50851852	277.97952160	PR > F
CORRECTED TOTAL	26	8461.44296296		0.0577

R-SQUARE	C.V.	ROOT MSE	Y MEAN
0.211540	24.8984	16.67271788	66.96296296

PARAMETER	ESTIMATE	T FOR H0: PARAMETER=0	PR > ¦T¦	STD ERROR OF ESTIMATE
INTERCEPT	106.08518519	1.90	0.0700	55.94500427
X1	-0.91611111	-2.33	0.0285	0.39297973
X2	0.78777778	1.00	0.3262	0.78595946

Recall that $n = 27$, $k = 5$, and $g = 2$. Therefore, the calculated value of the F statistic, based on $\nu_1 = k - g = 3$ and $\nu_2 = n - (k + 1) = 21$ degrees of freedom is

$$F = \frac{(SSE_1 - SSE_2)/(k - g)}{SSE_2/[n - (k + 1)]} = \frac{(SSE_1 - SSE_2)/(k - g)}{s_2^2}$$

where $\nu_1 = k - g$ is equal to the number of parameters involved in H_0 and s_2^2 is the value of s^2 for the complete model. Therefore,

$$F = \frac{(6{,}671.50851852 - 59.17842582)/3}{2.81802028} = 782.1$$

The final step in the test is to compare this computed value of F with the tabulated value based on $\nu_1 = 3$ and $\nu_2 = 21$ degrees of freedom. If we choose $\alpha = .05$, $F_{.05} = 3.07$. Since the computed value of F falls in the rejection region (see Figure 12.12)—i.e., it exceeds $F_{.05} = 3.07$—we reject H_0 and conclude that at least one of the second-order terms contributes information for the prediction of y. In other words, the data support the contention that the curvature we see in the response surface is not due simply to random variation in the data. The second-order model does appear to provide better predictions of y than a first-order model.   ∎

Example 12.3 demonstrates the motivation for testing a hypothesis that each one of a set of β parameters equals 0, and it also demonstrates the procedure. Other applications of this test appear in the following sections.

EXERCISES
12.31–12.41

LEARNING THE MECHANICS

12.31 Suppose you fit the regression model

$$y = \beta_0 + \beta_1 x_1 + \beta_2 x_2 + \beta_3 x_1 x_2 + \beta_4 x_1^2 + \beta_5 x_2^2 + \varepsilon$$

to $n = 30$ data points and you wish to test H_0: $\beta_3 = \beta_4 = \beta_5 = 0$.

a. What would the alternative hypothesis be?

b. Explain in detail how you would find the quantities necessary to compute the F statistic for your hypothesis test.

c. How many numerator and denominator degrees of freedom are associated with your F statistic?

12.32 Refer to Exercise 12.31. Suppose you fit the complete and reduced models for the hypothesis test described in Exercise 12.31 and obtained $SSE_1 = 246.1$ and $SSE_2 = 215.2$. Conduct the hypothesis test and interpret the results of your test. Use $\alpha = .05$.

12.33 Explain why the F test used to compare complete and reduced models is a one-tailed, upper-tailed test.

12.34 Minitab was used to fit the complete model

$$y = \beta_0 + \beta_1 x_1 + \beta_2 x_2 + \beta_3 x_3 + \beta_4 x_4 + \varepsilon$$

to $n = 20$ data points. The following is the resulting printout. [Note that the dependent variable y is represented by C1, and the independent variables x_1–x_4 are represented by C2–C5, respectively.]

Minitab Printout for Complete Model in Exercise 12.34

```
THE REGRESSION EQUATION IS
C1 = 14.6 - 0.611 C2 + 0.439 C3 - 0.080 C4 - 0.064 C5

                                      ST. DEV.      T-RATIO =
     COLUMN        COEFFICIENT        OF COEF.      COEF/S.D.
                      14.575            4.887          2.98
     C2              -0.6113           0.1775         -3.44
     C3               0.4388           0.2199          2.00
     C4              -0.0796           0.1083         -0.74
     C5              -0.0636           0.1247         -0.51

     S = 3.190

     R-SQUARED = 84.5 PERCENT
     R-SQUARED = 80.3 PERCENT, ADJUSTED FOR D.F.

     ANALYSIS OF VARIANCE

         DUE TO      DF          SS        MS=SS/DF
         REGRESSION   4       831.09        207.77
         RESIDUAL    15       152.66         10.18
         TOTAL       19       983.75
```

The independent variables x_3 and x_4 were dropped from the above model and Minitab was used to fit the resulting reduced model. The printout is shown here:

Minitab Printout for Reduced Model in Exercise 12.34

```
THE REGRESSION EQUATION IS
C1 = 14.0 - 0.642 C2 + 0.396 C3

                                  ST. DEV.       T-RATIO =
COLUMN         COEFFICIENT       OF COEF.        COEF/S.D.
                  13.968            4.626            3.02
C2               -0.6422           0.1675           -3.84
C3                0.3959           0.2061            1.92

S = 3.072

R-SQUARED = 83.7 PERCENT
R-SQUARED = 81.8 PERCENT, ADJUSTED FOR D.F.

ANALYSIS OF VARIANCE

DUE TO         DF            SS       MS=SS/DF
REGRESSION      2        823.31        411.66
RESIDUAL       17        160.44          9.44
TOTAL          19        983.75
```

a. Report the least squares prediction equations for the complete and reduced models.
b. Find SSE_1 and SSE_2. Interpret each of these quantities.
c. How many β parameters are in the complete model? The reduced model?
d. Specify the null and alternative hypotheses you would use to investigate whether the complete model contributes more information for the prediction of y than the reduced model.
e. Conduct the hypothesis test of part d. Use $\alpha = .05$.
f. What is the approximate p-value of the test of part e?

APPLYING THE CONCEPTS

[Note: Starred (*) exercises require the use of a computer.]

12.35 A large research and development company rates the performance of each of the members of its technical staff once a year. Each person is rated on a scale of 0 to 100 by his or her immediate supervisor, and this merit rating is used to help determine the size of the person's pay raise for the coming year. The company's personnel department is interested in developing a regression model to help them forecast the merit rating that an applicant for a technical position will receive after he or she has been with the company 3 years. The company proposes to use the following model to forecast the merit ratings of applicants who have just completed their graduate studies and have no prior related job experience:

$$E(y) = \beta_0 + \beta_1 x_1 + \beta_2 x_2 + \beta_3 x_1 x_2 + \beta_4 x_1^2 + \beta_5 x_2^2$$

where

y = Applicant's merit rating after 3 years

x_1 = Applicant's grade-point average (GPA) in graduate school

x_2 = Applicant's verbal score on the Graduate Record Examination (percentile)

A random sample of n = 40 employees who have been with the company more than 3 years was selected. Each employee's merit rating after 3 years, his or her graduate school GPA, and the percentile in which the verbal Graduate Record Exam score fell were recorded. The above model was fit to these data. The following is a portion of the resulting computer printout:

SOURCE	DF	SUM OF SQUARES	MEAN SQUARE
MODEL	5	4911.56	982.31
ERROR	34	1830.44	53.84
TOTAL	39	6742.00	R-SQUARE
			0.729

The reduced model, $E(y) = \beta_0 + \beta_1 x_1 + \beta_2 x_2$, was fit to the same data and the resulting computer printout is partially reproduced here:

SOURCE	DF	SUM OF SQUARES	MEAN SQUARE
MODEL	2	3544.84	1772.42
ERROR	37	3197.16	86.41
TOTAL	39	6742.00	R-SQUARE
			0.526

a. Identify the null and alternative hypotheses for a test to determine whether the complete model contributes information for the prediction of y.

b. Identify the null and alternative hypotheses for a test to determine whether a second-order model contributes more information than a first-order model for the prediction of y.

c. Conduct the hypothesis test you described in part **a**. Test using α = .05. Draw the appropriate conclusions in the context of the problem.

d. Conduct the hypothesis test you described in part **b**. Test using α = .05. Draw the appropriate conclusions in the context of the problem.

*12.36 A firm would like to be able to forecast its yearly sales in each of its sales regions. The firm has decided to base its forecasts on regional population size and its yearly regional advertising expenditures. The population data in the table at the top of page 678 were obtained from the Bureau of the Census, and the advertising and sales data were obtained from the firm's internal records.

a. Fit a complete second-order model to the data.

b. Is the complete second-order model useful for forecasting sales? Test using α = .05.

c. The firm is planning to market its product in a new sales region next year. The region has a population of 400,000, and the firm plans to spend $12,000 on advertising. Use the fitted model you obtained in part **a** to forecast next year's sales in this new region. [Note: We would want to express this estimate as a prediction

SALES REGION	SALES (Thousands of units)	POPULATION OF REGION (Thousands)	ADVERTISING EXPENDITURES ($ thousands)
1	65	200	8
2	80	210	10
3	85	205	9
4	100	300	8.5
5	108	320	12
6	114	290	10
7	40	90	6
8	45	85	8
9	150	450	9
10	42	87	9
11	220	480	13
12	200	500	15

interval, but its computation is beyond the scope of this text. The procedure is described in the references at the end of the chapter. You may also find that it can be obtained using your statistical computer package.]

12.37 Refer to Exercise 12.36, in which a firm would like to develop a regression model to forecast its yearly sales in each of its sales regions. The model under consideration is a complete second-order model:

$$E(y) = \beta_0 + \beta_1 x_1 + \beta_2 x_2 + \beta_3 x_1 x_2 + \beta_4 x_1^2 + \beta_5 x_2^2$$

where

y = Yearly regional sales

x_1 = Population of sales region

x_2 = Yearly regional advertising expenditures

The following is a portion of the computer printout that results from fitting this model to the $n = 12$ data points given in Exercise 12.36:

```
SOURCE   DF   SUM OF SQUARES   MEAN SQUARE

MODEL    5        38638.97        7727.79
ERROR    6          159.94          26.66
TOTAL   11        38798.91      R-SQUARE
                                   0.996
```

The reduced first-order model, $E(y) = \beta_0 + \beta_1 x_1 + \beta_2 x_2$, was fit to the same data and the resulting computer printout is partially reproduced here:

```
SOURCE   DF   SUM OF SQUARES   MEAN SQUARE

MODEL    2         36704.5        18352.2
ERROR    9          2094.4          232.7
TOTAL   11         38798.9      R-SQUARE
                                   0.946
```

Is there sufficient evidence to conclude that a second-order model contributes more information for the prediction of y than a first-order model? Test using $\alpha = .05$.

12.38 Refer to Exercise 12.26, in which the Department of Energy wants to develop a regression model to help forecast annual gasoline consumption in the United States. The complete and reduced models for the test that you described in part **d** of Exercise 12.26 were fit to $n = 25$ data points. The resulting values of SSE_1 and SSE_2 were 1,065.9 and 400.6, respectively.

  a. Conduct the test to determine whether the data present sufficient evidence to indicate interaction between x_1 and x_2. Test using $\alpha = .05$.

  b. Find the approximate observed significance level for the test in part **a**.

12.39 In Exercise 11.5 we found that a first-order model was useful for characterizing the relationship between household food consumption and the two independent variables, household income and size of household. The sample data are repeated in the table and the SAS printout at the top of page 680 summarizes the results of fitting a first-order model to the data.

HOUSEHOLD	FOOD CONSUMPTION DURING 1987 y ($ thousands)	1987 HOUSEHOLD INCOME x_1 ($ thousands)	NUMBER OF PERSONS IN HOUSEHOLD AT END OF 1987 x_2	HOUSEHOLD	FOOD CONSUMPTION DURING 1987 y ($ thousands)	1987 HOUSEHOLD INCOME x_1 ($ thousands)	NUMBER OF PERSONS IN HOUSEHOLD AT END OF 1987 x_2
1	3.2	31.1	4	14	3.1	85.2	2
2	2.4	20.5	2	15	4.5	35.6	9
3	3.8	42.3	4	16	3.5	68.5	3
4	1.9	18.9	1	17	4.0	10.5	5
5	2.5	26.5	2	18	3.5	21.6	4
6	3.0	29.8	4	19	1.8	29.9	1
7	2.6	24.3	3	20	2.9	28.6	3
8	3.2	38.1	4	21	2.6	20.2	2
9	3.9	52.0	5	22	3.6	38.7	5
10	1.7	16.0	1	23	2.8	11.2	3
11	2.9	41.9	3	24	4.5	14.3	7
12	1.7	9.9	1	25	3.5	16.9	5
13	4.5	33.1	7				

  a. Fit a complete second-order model to the data.

  b. Is the complete second-order model useful for explaining the variation in household food consumption? Test using $\alpha = .05$.

  c. Is there sufficient evidence to conclude that at least one of the β parameters associated with the second-order terms differs from 0? Test using $\alpha = .05$.

  d. Do the data provide sufficient evidence to conclude that household income and household size interact? Test using $\alpha = .05$.

  e. Describe the danger involved in conducting a series of hypothesis tests (such as in parts **b**, **c**, and **d**) for determining which terms to retain in a regression model and which to exclude.

SAS Printout for Exercise 12.39

DEPENDENT VARIABLE: FOOD

SOURCE	DF	SUM OF SQUARES	MEAN SQUARE	F VALUE
MODEL	2	15.46228509	7.73114255	100.80
ERROR	22	1.68731491	0.07669613	PR > F
CORRECTED TOTAL	24	17.14960000		0.0001

R-SQUARE	C.V.	ROOT MSE	FOOD MEAN
0.901612	8.9221	0.27694067	3.10400000

PARAMETER	ESTIMATE	T FOR H0: PARAMETER=0	PR > |T|	STD ERROR OF ESTIMATE
INTERCEPT	1.43260377	9.76	0.0001	0.14673751
INCOME	0.00999062	3.15	0.0046	0.00316806
SIZE	0.37928986	13.68	0.0001	0.02772472

12.40 In Exercise 11.19 we found that a first-order model was useful for explaining the variation in the number of hours worked per week in the shipping department of a particular firm. The data used to fit the model are repeated in the accompanying table.

WEEK	HOURS OF LABOR y	THOUSANDS OF POUNDS SHIPPED x_1	PERCENTAGE OF UNITS SHIPPED BY TRUCK x_2	AVERAGE NUMBER OF POUNDS PER SHIPMENT x_3
1	100	5.1	90	20
2	85	3.8	99	22
3	108	5.3	58	19
4	116	7.5	16	15
5	92	4.5	54	20
6	63	3.3	42	26
7	79	5.3	12	25
8	101	5.9	32	21
9	88	4.0	56	24
10	71	4.2	64	29
11	122	6.8	78	10
12	85	3.9	90	30
13	50	3.8	74	28
14	114	7.5	89	14
15	104	4.5	90	21
16	111	6.0	40	20
17	110	8.1	55	16
18	100	2.9	64	19
19	82	4.0	35	23
20	85	4.8	58	25

a. Write a complete second-order model for the data.

b. The SAS computer printout for fitting the second-order model of part **a** is shown here. Find the prediction equation.

SAS Printout for Exercise 12.40

DEPENDENT VARIABLE: LABOR

SOURCE	DF	SUM OF SQUARES	MEAN SQUARE	F VALUE
MODEL	9	6043.40529897	671.48947766	10.25
ERROR	10	654.79470103	65.47947010	PR > F
CORRECTED TOTAL	19	6698.20000000		0.0006

R-SQUARE	C.V.	ROOT MSE	LABOR MEAN
0.902243	8.6730	8.09193859	93.30000000

PARAMETER	ESTIMATE	T FOR H0: PARAMETER=0	PR > !T!	STD ERROR OF ESTIMATE
INTERCEPT	655.80722615	2.90	0.0159	226.28033719
WEIGHT	-57.32665806	-1.73	0.1138	33.08376144
TRUCK	-3.38961628	-1.77	0.1079	1.91983172
AVGSHIP	-28.27072935	-2.66	0.0238	10.61901566
WEIGHT*TRUCK	0.22403392	1.52	0.1583	0.14691511
WEIGHT*AVGSHIP	2.20142243	2.44	0.0348	0.90189596
TRUCK*AVGSHIP	0.08858647	2.23	0.0496	0.03966598
WEIGHT*WEIGHT	0.45327631	0.34	0.7414	1.33580032
TRUCK*TRUCK	0.00371216	0.92	0.3800	0.00404162
AVGSHIP*AVGSHIP	0.20833049	1.64	0.1315	0.12683455

c. Do the data provide sufficient evidence to indicate that the model is useful for predicting y? Test using $\alpha = .05$.

d. Is there sufficient evidence to indicate that the second-order terms are useful for predicting y? Test using $\alpha = .05$.

*12.41 In Exercise 11.11, a real estate appraiser used regression analysis to explore the relationship between the sale prices of apartments and various characteristics of the apartments. The data are repeated in the table on page 682.

a. Fit a first-order model to the data. (You may already have done this in Exercise 11.11.)

b. Do the data provide sufficient evidence to conclude that the model of part **a** is useful for predicting sale price? Test using $\alpha = .05$.

c. Drop x_3 and x_4 from the model of part **a** and refit the model to the data.

d. Do the data provide sufficient evidence to conclude that the model of part **c** is useful for predicting sale price? Test using $\alpha = .05$.

CODE NO.	SALE PRICE y ($)	NUMBER OF APARTMENT UNITS x_1	AGE OF STRUCTURE x_2 (years)	LOT SIZE x_3 (sq. ft.)	NUMBER OF ON-SITE PARKING SPACES x_4	GROSS BUILDING AREA x_5
0229	90,300	4	82	4,635	0	4,266
0094	384,000	20	13	17,798	0	14,391
0043	157,500	5	66	5,913	0	6,615
0079	676,200	26	64	7,750	6	34,144
0134	165,000	5	55	5,150	0	6,120
0179	300,000	10	65	12,506	0	14,552
0087	108,750	4	82	7,160	0	3,040
0120	276,538	11	23	5,120	0	7,881
0246	420,000	20	18	11,745	20	12,600
0025	950,000	62	71	21,000	3	39,448
0015	560,000	26	74	11,221	0	30,000
0131	268,000	13	56	7,818	13	8,088
0172	290,000	9	76	4,900	0	11,315
0095	173,200	6	21	5,424	6	4,461
0121	323,650	11	24	11,834	8	9,000
0077	162,500	5	19	5,246	5	3,828
0060	353,500	20	62	11,223	2	13,680
0174	134,400	4	70	5,834	0	4,680
0084	187,000	8	19	9,075	0	7,392
0031	155,700	4	57	5,280	0	6,030
0019	93,600	4	82	6,864	0	3,840
0074	110,000	4	50	4,510	0	3,092
0057	573,200	14	10	11,192	0	23,704
0104	79,300	4	82	7,425	0	3,876
0024	272,000	5	82	7,500	0	9,542

Source: Robinson Appraisal Co., Inc., Mankato, Minnesota.

SECTION 12.5

MODELS WITH ONE QUALITATIVE INDEPENDENT VARIABLE

Suppose we want to write a model for the mean profit, $E(y)$, per sales dollar of a construction company as a function of the sales engineer who estimates and bids on a job. (For the purpose of explanation, we will ignore other independent variables that might affect the response.) Further, suppose there are three sales engineers: Adams, Brown, and Clark. Then Sales engineer is a single qualitative variable set at three levels corresponding to Adams, Brown, and Clark. Note that with a qualitative independent variable, we cannot attach a quantitative meaning to a given level. All we can do is describe it.

To simplify our notation, let μ_A be the mean profit per sales dollar for Adams, and let μ_B and μ_C be the corresponding mean profits for Brown and Clark. Our objective is to write a single prediction equation that will give the mean value of y for the three sales engineers. This can be done as follows:

$$E(y) = \beta_0 + \beta_1 x_1 + \beta_2 x_2$$

where

$$x_1 = \begin{cases} 1 & \text{if Brown is the sales engineer} \\ 0 & \text{if Brown is not the sales engineer} \end{cases}$$

$$x_2 = \begin{cases} 1 & \text{if Clark is the sales engineer} \\ 0 & \text{if Clark is not the sales engineer} \end{cases}$$

The variables x_1 and x_2 are not meaningful independent variables as for the case of the models with quantitative independent variables. Instead, they are **dummy** (or **indicator**) **variables** that make the model function. To see how they work, let $x_1 = 0$ and $x_2 = 0$. This condition will apply when we are seeking the mean response for Adams (neither Brown nor Clark will be the sales engineer; hence, it must be Adams). Then the mean value of y when Adams is the sales engineer is

$$\mu_A = E(y) = \beta_0 + \beta_1(0) + \beta_2(0) = \beta_0$$

This tells us that the mean profit per sales dollar for Adams is β_0. Or, it means that $\beta_0 = \mu_A$.

Now suppose we want to represent the mean response, $E(y)$, when Brown is the sales engineer. Checking the dummy variable definitions, we see that we should let $x_1 = 1$ and $x_2 = 0$:

$$\mu_B = E(y) = \beta_0 + \beta_1 x_1 + \beta_2 x_2 = \beta_0 + \beta_1(1) + \beta_2(0) = \beta_0 + \beta_1$$

or, since $\beta_0 = \mu_A$,

$$\mu_B = \mu_A + \beta_1$$

Then it follows that the interpretation of β_1 is

$$\beta_1 = \mu_B - \mu_A$$

which is the difference in the mean profit per sales dollar between Brown and Adams.

Finally, if we want the mean value of y when Clark is the sales engineer, we let $x_1 = 0$ and $x_2 = 1$:

$$\mu_C = E(y) = \beta_0 + \beta_1(0) + \beta_2(1) = \beta_0 + \beta_2$$

or, since $\beta_0 = \mu_A$,

$$\mu_C = \mu_A + \beta_2$$

Then it follows that the interpretation of β_2 is

$$\beta_2 = \mu_C - \mu_A$$

Note that we are able to describe *three levels* of the qualitative variable with only *two dummy variables*. This is because the mean of the base level (Adams, in this case) is accounted for by the intercept β_0.

Since Sales engineer is a qualitative variable, we will use a bar graph to show the value of mean profit $E(y)$ for the three levels of Sales engineer (see Figure 12.16 on page 684). In particular, note that the height of the bar, $E(y)$, for each level of

FIGURE 12.16

Bar Chart Comparing $E(y)$
for the Three Sales
Engineers

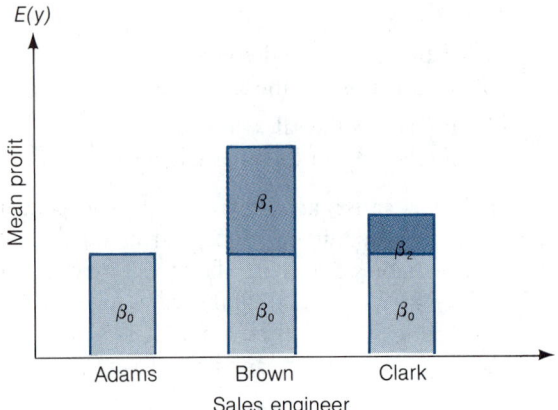

engineer is equal to the sum of the model parameters shown in the preceding equations. You can see that the height of the bar corresponding to Adams is β_0; i.e., $E(y) = \beta_0$. Similarly, the heights of the bars corresponding to Brown and Clark are $E(y) = \beta_0 + \beta_1$ and $E(y) = \beta_0 + \beta_2$, respectively.*

Now, carefully examine the model with a single qualitative independent variable at three levels, because we will use exactly the same pattern for any number of levels. Also, the interpretation of the parameters will always be the same.

One level is selected as the base level (we used Adams as level A). Then for the 1–0 system of coding† for the dummy variables,

$$\mu_A = \beta_0$$

The coding for all dummy variables is as follows: To represent the mean value of y for a particular level, let that dummy variable equal 1; otherwise, the dummy variable is set equal to 0. Using this system of coding, we have

$$\mu_B = \beta_0 + \beta_1$$
$$\mu_C = \beta_0 + \beta_2$$
$$\vdots$$

Because $\mu_A = \beta_0$, any other model parameter will represent the difference in means between that level and the base level:

$$\beta_1 = \mu_B - \mu_A$$
$$\beta_2 = \mu_C - \mu_A$$
$$\vdots$$

*Either β_1 or β_2, or both, could be negative. If, for example, β_1 were negative, the height of the bar corresponding to Brown would be *reduced* (rather than increased) from the height of the bar for Adams by the amount β_1. Figure 12.16 is constructed assuming β_1 and β_2 are positive quantities.

†We do not have to use a 1–0 system of coding for the dummy variables. Any two-value system will work, but the interpretation given to the model parameters will depend on the code. Using the 1–0 system makes the model parameters easy to interpret.

PROCEDURE FOR WRITING A MODEL WITH ONE QUALITATIVE INDEPENDENT VARIABLE AT k LEVELS

Always use one less dummy variable than the number of levels of the qualitative variable. Thus, for a qualitative variable with k levels, we use $k - 1$ dummy variables:

$$y = \beta_0 + \beta_1 x_1 + \beta_2 x_2 + \cdots + \beta_{k-1} x_{k-1} + \varepsilon$$

where x_i is the dummy variable for level i and

$$x_i = \begin{cases} 1 & \text{if } y \text{ is observed at level } i \\ 0 & \text{otherwise} \end{cases}$$

Then, for this system of coding

$$\mu_A = \beta_0 \qquad \text{and} \qquad \beta_1 = \mu_B - \mu_A$$
$$\mu_B = \beta_0 + \beta_1 \qquad\qquad \beta_2 = \mu_C - \mu_A$$
$$\mu_C = \beta_0 + \beta_2 \qquad\qquad \beta_3 = \mu_D - \mu_A$$
$$\mu_D = \beta_0 + \beta_3 \qquad\qquad \vdots$$
$$\vdots$$

EXAMPLE 12.4

Suppose a large chain of department stores wants to compare the mean dollar amounts owed by its delinquent credit card customers in three different annual income groups: under $12,000, $12,000–$25,000, and over $25,000. A sample of ten customers with delinquent accounts is selected from each group and the amount owed by each is recorded, as shown in Table 12.3. Do the data provide sufficient evidence to indicate that the mean dollar amounts owed by customers differ for the three income groups?

TABLE 12.3

Income Class:
Dollars Owed

	CATEGORY		
	1	2	3
	Under $12,000	$12,000–$25,000	Over $25,000
	$148	$513	$335
	76	264	643
	393	433	216
	520	94	536
	236	535	128
	134	327	723
	55	214	258
	166	135	380
	415	280	594
	153	304	465
TOTALS	$2,296	$3,099	$4,278

SOLUTION

Note that Income is ordinarily a quantitative variable, but in this example only the group (low, medium, high) in which the customer's income falls is known. We therefore treat Income as a qualitative variable. The model relating $E(y)$ to this single qualitative variable, Income level, is

$$E(y) = \beta_0 + \beta_1 x_1 + \beta_2 x_2$$

where

$$x_1 = \begin{cases} 1 & \text{if income level 2} \\ 0 & \text{if not} \end{cases} \qquad x_2 = \begin{cases} 1 & \text{if income level 3} \\ 0 & \text{if not} \end{cases}$$

and

$$\beta_1 = \mu_2 - \mu_1 \qquad \beta_2 = \mu_3 - \mu_1$$

where μ_1, μ_2, and μ_3 are the mean responses for income categories 1, 2, and 3, respectively. Testing the null hypothesis that the means for the three income levels are equal, i.e., $\mu_1 = \mu_2 = \mu_3$, is equivalent to testing

$$H_0: \quad \beta_1 = \beta_2 = 0$$

because if $\beta_1 = \mu_2 - \mu_1 = 0$ and $\beta_2 = \mu_3 - \mu_1 = 0$, then μ_1, μ_2, and μ_3 must be equal. The alternative hypothesis is

$$H_a: \quad \text{At least one of the parameters, } \beta_1 \text{ or } \beta_2, \text{ differs from 0}$$

which implies that at least two of the three means (μ_1, μ_2, and μ_3) differ.

There are two ways to conduct this test. We can fit the complete model shown above and the reduced model (deleting the terms involving β_1 and β_2),

$$E(y) = \beta_0$$

and conduct the F test described in the preceding section (we leave this as an exercise for you). Or, we can use the F test of the complete model (Section 11.6), which tests the null hypothesis that all parameters in the model, with the exception of β_0, equal 0. Either way you conduct the test, you will obtain the same computed value of F, the value shown on the SAS printout for a test of the complete model. The SAS printout for fitting the complete model,

$$E(y) = \beta_0 + \beta_1 x_1 + \beta_2 x_2$$

is shown in Figure 12.17 and the value of the F statistic for testing the complete model, $F = 3.48$, is shaded. We will wish to compare this value with the tabulated value of F based on $\nu_1 = 2$ and $\nu_2 = 27$ degrees of freedom. If we choose $\alpha = .05$, we will reject $H_0: \beta_1 = \beta_2 = 0$ if the computed value of F exceeds $F_{.05} = 3.35$. Since the computed value of F, $F = 3.48$, exceeds $F_{.05} = 3.35$, we reject H_0 and conclude that at least one of the parameters, β_1 or β_2, differs from 0. Or, equivalently, we conclude that the data provide sufficient evidence to indicate that the mean indebtedness does vary from one income group to another.

FIGURE 12.17 SAS Computer Printout for Example 12.4

```
DEPENDENT VARIABLE: Y

SOURCE                    DF    SUM OF SQUARES      MEAN SQUARE    F VALUE

MODEL                      2    198772.46666667    99386.23333333    3.48
ERROR                     27    770670.90000000    28543.36666667   PR > F
CORRECTED TOTAL           29    969443.36666667                     0.0452

R-SQUARE              C.V.           ROOT MSE           Y MEAN

0.205038            52.3978        168.94782232      322.43333333

                              T FOR H0:    PR > :T:   STD ERROR OF
PARAMETER      ESTIMATE     PARAMETER=0             ESTIMATE

INTERCEPT    229.60000000      4.30      0.0002     53.42599243
X1            80.30000000      1.06      0.2973     75.55576307
X2           198.20000000      2.62      0.0141     75.55576307
```

We need to make two additional comments about Example 12.4. First, regression analysis is not the only way to analyze these data. Another procedure for calculating the value of the F statistic, known as an **analysis of variance**, is described in Chapter 15. Second, if you choose to analyze the data by fitting complete and reduced models (Section 12.4), you will find that the least squares estimate of β_0 in the reduced model,

$$E(y) = \beta_0$$

is $\bar{y}$, the mean of all $n = 30$ observations, and the sum of squares for error for the reduced model is

$$SSE_1 = \sum (y_i - \hat{y}_i)^2 = \sum (y_i - \bar{y})^2 = 969,443.367$$

This value is shown in the SAS printout in Figure 12.17 as SUM OF SQUARES corresponding to CORRECTED TOTAL. We leave the remaining steps, calculating the drop in SSE and the resulting F statistic, to you. You will find that the value you obtain will be the same as the value of F shown in the SAS printout in Figure 12.17.

EXERCISES 12.47–12.54

LEARNING THE MECHANICS

12.42 Write a regression model relating the mean value of y to a qualitative independent variable that can assume two levels. Interpret all the terms in the model.

12.43 Write a regression model relating $E(y)$ to a qualitative independent variable that can assume three levels. Interpret all the terms in the model.

12.44 The following model was used to relate $E(y)$ to a single qualitative variable with four levels:

$$E(y) = \beta_0 + \beta_1 x_1 + \beta_2 x_2 + \beta_3 x_3$$

where

$$x_1 = \begin{cases} 1 & \text{if level 2} \\ 0 & \text{if not} \end{cases} \qquad x_2 = \begin{cases} 1 & \text{if level 3} \\ 0 & \text{if not} \end{cases} \qquad x_3 = \begin{cases} 1 & \text{if level 4} \\ 0 & \text{if not} \end{cases}$$

This model was fit to $n = 30$ data points and the following result was obtained:

$$\hat{y} = 10.2 - 4x_1 + 12x_2 + 2x_3$$

Find estimates for $E(y)$ when the qualitative independent variable is set at each of the following levels:

a. Level 1   **b.** Level 2   **c.** Level 3   **d.** Level 4

e. Specify the null and alternative hypotheses you would use to test whether $E(y)$ is the same for all four levels of the independent variable.

12.45 Minitab was used to fit the following model to $n = 15$ data points:

$$y = \beta_0 + \beta_1 x_1 + \beta_2 x_2 + \varepsilon$$

where

$$x_1 = \begin{cases} 1 & \text{if level 2} \\ 0 & \text{if not} \end{cases} \qquad x_2 = \begin{cases} 1 & \text{if level 3} \\ 0 & \text{if not} \end{cases}$$

The resulting printout is shown below. (Note that in the printout C1 represents y, and C2 and C3 represent x_1 and x_2, respectively.)

Minitab Printout for
Exercise 12.45

```
THE REGRESSION EQUATION IS
C1 = 80.0 + 16.8 C2 + 40.4 C3

                                   ST. DEV.       T-RATIO =
COLUMN        COEFFICIENT          OF COEF.       COEF/S.D.
                 80.000             4.082           19.60
C2               16.800             5.774            2.91
C3               40.400             5.774            7.00

S = 9.129

R-SQUARED = 80.5 PERCENT
R-SQUARED = 77.2 PERCENT, ADJUSTED FOR D.F.

ANALYSIS OF VARIANCE

DUE TO          DF              SS         MS=SS/DF
REGRESSION       2           4118.9         2059.5
RESIDUAL        12           1000.0           83.3
TOTAL           14           5118.9
```

a. Report the least squares prediction equation.
b. Interpret the values of $\hat{\beta}_1$ and $\hat{\beta}_2$.

c. Interpret the following hypotheses in terms of μ_1, μ_2, and μ_3:

H_0: $\beta_1 = \beta_2 = 0$

H_a: At least one of the parameters β_1 and β_2 differs from 0

d. Conduct the hypothesis test of part c.

APPLYING THE CONCEPTS

[Note: Starred (*) exercises require the use of a computer.]

12.46 In 1983, 4-year private colleges charged an average of $4,627 for tuition and fees for the year; 4-year public colleges charged $1,105 (USA Today, Aug. 17, 1983, p. 1). In order to estimate the difference in the mean amounts charged for the 1985–1986 academic year, random samples of 40 private colleges and 40 public colleges were contacted and questioned about their tuition structures.

   a. Which of the procedures described in Chapter 9 could be used to estimate the difference in mean charges between private and public colleges?

   b. Propose a regression model involving the qualitative independent variable—type of college—that could be used to investigate the difference between the means. Be sure to specify the coding scheme for the dummy variable in the model.

   c. Explain how the regression model you developed in part b could be used to estimate the difference between the population means.

*12.47 In Exercise 9.11 you estimated the difference between the mean price of a share of stock traded on the New York Stock Exchange (NYSE) and the mean price of a share of stock traded on the American Stock Exchange (ASE). The data (except firm name) are repeated in the accompanying table, which also shows price data for 15 stocks randomly selected from those traded over the counter (OTC). All prices are closing prices on December 15, 1986.

NYSE			ASE			OTC	
$14\frac{1}{8}$	23	$15\frac{1}{2}$	$8\frac{1}{2}$	$8\frac{1}{2}$	$14\frac{3}{4}$	$27\frac{1}{4}$	9
$15\frac{7}{8}$	95	$43\frac{3}{4}$	$2\frac{5}{8}$	$3\frac{1}{4}$	$16\frac{3}{4}$	9	$32\frac{1}{2}$
$24\frac{1}{8}$	$5\frac{3}{8}$	$80\frac{1}{2}$	$11\frac{1}{4}$	$41\frac{7}{8}$	$\frac{3}{8}$	3	$8\frac{3}{4}$
28	$4\frac{1}{8}$	$11\frac{7}{8}$	$14\frac{1}{8}$	$12\frac{5}{8}$	86	18	$11\frac{3}{4}$
$77\frac{1}{2}$	$12\frac{3}{4}$	$4\frac{1}{2}$	$5\frac{5}{8}$	$9\frac{1}{2}$	$6\frac{1}{4}$	$8\frac{1}{2}$	17
$30\frac{3}{4}$	$35\frac{1}{4}$	$16\frac{1}{8}$	$4\frac{7}{8}$	$5\frac{5}{8}$	$26\frac{1}{2}$	$11\frac{3}{4}$	$14\frac{1}{2}$
32	$24\frac{1}{4}$	25	$30\frac{3}{4}$	5	$\frac{1}{2}$	32	$21\frac{1}{4}$
$11\frac{5}{8}$	18	$10\frac{3}{4}$	$4\frac{3}{8}$	$24\frac{1}{8}$	$1\frac{1}{2}$	$9\frac{1}{8}$	
$11\frac{1}{8}$	$21\frac{1}{8}$	$30\frac{3}{8}$	$6\frac{5}{8}$	$6\frac{1}{4}$	$7\frac{7}{8}$		
$20\frac{1}{8}$	$14\frac{3}{4}$	$83\frac{5}{8}$	$10\frac{1}{4}$	$21\frac{5}{8}$	$13\frac{1}{4}$		
$31\frac{7}{8}$	$19\frac{7}{8}$	$21\frac{1}{4}$	$1\frac{5}{8}$	$17\frac{7}{8}$	$12\frac{3}{4}$		
13	8	$6\frac{7}{8}$	$24\frac{1}{8}$	$16\frac{1}{4}$	$35\frac{1}{4}$		

Source: Wall Street Journal, Dec. 16, 1986, pp. 51–52, 54–56.

a. Propose a regression model involving one qualitative independent variable that could be used to investigate whether the mean prices of NYSE, ASE, and OTC stocks differ. Be sure to specify the coding scheme for dummy variables in your model.

b. Fit your model to the data. Report the least squares prediction equation.

c. Do the data provide sufficient evidence to conclude that the mean prices differ? Test using $\alpha = .05$.

12.48 An independent testing laboratory has been hired to compare the lifelength (in months) of four different brands of color television picture tubes, A, B, C, and D. Life data have been obtained on ten randomly selected picture tubes of each brand. Propose a regression model involving the qualitative independent variable, Brand of color tube, to estimate the mean lifelength of a tube. Interpret each term in your model.

12.49 The director of marketing of a company that sells business machines is interested in modeling mean monthly sales (in thousands of dollars) per salesperson, $E(y)$, as a function of the type of sales incentive plan that is in effect: commission only, straight salary, or salary plus commission on each sale. The director has proposed the following model:

$$E(y) = \beta_0 + \beta_1 x_1 + \beta_2 x_2$$

where

$$x_1 = \begin{cases} 1 & \text{if salesperson is paid a straight salary} \\ 0 & \text{if otherwise} \end{cases}$$

$$x_2 = \begin{cases} 1 & \text{if salesperson is paid a salary plus commission} \\ 0 & \text{if otherwise} \end{cases}$$

A portion of the computer printout that results from using Minitab to fit this model to the sales data collected from a sample of 15 salespersons (five from each incentive plan) is shown here.

```
THE REGRESSION EQUATION IS
Y =     20.0 -   8.60 X1 +  3.80 X2

                                ST. DEV.    T-RATIO =
         COLUMN   COEFFICIENT   OF COEF.    COEF/S.D.

           --        20.000       2.898       6.90
X1   C21             -8.60        4.10       -2.10
X2   C22              3.80        4.10        .93

THE ST. DEV. OF Y ABOUT REGRESSION LINE IS
S =       6.481
WITH ( 15- 3) = 12 DEGREES OF FREEDOM

R-SQUARED = 44.5 PERCENT
R-SQUARED = 35.2 PERCENT, ADJUSTED FOR D.F.

ANALYSIS OF VARIANCE

    DUE TO       DF     SS     MS=SS/DF

REGRESSION    2    403.60    201.80
RESIDUAL     12    504.00     42.00
TOTAL        14    907.60
```

a. Do the data provide sufficient evidence to conclude that there is a difference in mean monthly sales among the three incentive plans? Test using $\alpha = .05$.

b. Use the least squares prediction equation to estimate the mean sales for salespersons working on a straight salary basis.

c. Use the least squares prediction equation to estimate the mean sales for salespersons working on a commission only basis.

[*Note:* We would prefer to use confidence intervals for the estimates in parts **b** and **c**, but their calculation is beyond the scope of this text. This procedure can be found in the references at the end of the chapter.]

12.50 Refer to Exercise 12.49. Find a 90% confidence interval for the difference between the mean monthly sales for salespersons on salary plus commission versus those on commission only.

12.51 The manager of a supermarket wants to model the total weekly sales of beer, y, as a function of brand. This model will enable the manager to plan the store's inventory. The market carries three brands, B_1, B_2, and B_3.

a. What type of independent variable is Brand of beer?

b. Write the model relating mean weekly beer sales, $E(y)$, as a function of brand of beer. Be sure to explain any dummy variables you use.

c. Interpret the β parameters of your model in part **b**.

d. In terms of the model parameters, what is the mean weekly sales for brand B_3?

12.52 Refer to Exercise 12.51. Suppose the manager uses brand B_1 as the base level and obtains the prediction equation

$$\hat{y} = 450 + 60x_1 - 30x_2$$

where

$$x_1 = \begin{cases} 1 & \text{if brand } B_2 \\ 0 & \text{otherwise} \end{cases} \qquad x_2 = \begin{cases} 1 & \text{if brand } B_3 \\ 0 & \text{otherwise} \end{cases}$$

a. What is the difference between the estimated mean weekly sales for brands B_2 and B_1?

b. What is the estimated mean weekly sales for brand B_2?

[*Note:* We would generally form confidence intervals for the true means in order to assess the reliability of these estimates. Our objective in these exercises is to develop the ability to use the models to obtain the estimates.]

12.53 Five varieties of peas are currently being tested by a large agribusiness cooperative in Ohio to determine which is best suited for production. A field was divided into 20 plots, with each variety of peas planted in four plots. The yields (in bushels of peas) produced from each plot are shown in the table:

VARIETY OF PEAS				
A	B	C	D	E
26.2	29.2	29.1	21.3	20.1
24.3	28.1	30.8	22.4	19.3
21.8	27.3	33.9	24.3	19.9
28.1	31.2	32.8	21.8	22.1

SAS was used to fit the model

$$y = \beta_0 + \beta_1 x_1 + \beta_2 x_2 + \beta_3 x_3 + \beta_4 x_4 + \varepsilon$$

to these data, using the coding $x_1 = 1$ for variety A, $x_2 = 1$ for variety B, $x_3 = 1$ for variety C, and $x_4 = 1$ for variety D. The resulting printout is shown below.

SAS Printout for Exercise 12.53

Dep Variable: Y

Analysis of Variance

Source	DF	Sum of Squares	Mean Square	F Value	Prob>F
Model	4	342.04000	85.51000	23.966	0.0001
Error	15	53.52000	3.56800		
C Total	19	395.56000			

Root MSE	1.88892	R-Square	0.8647
Dep Mean	25.70000	Adj R-Sq	0.8286
C.V.	7.34986		

Parameter Estimates

Variable	DF	Parameter Estimate	Standard Error	T for H0: Parameter=0	Prob > |T|
INTERCEP	1	20.350000	0.94445752	21.547	0.0001
X1	1	4.750000	1.33566463	3.556	0.0029
X2	1	8.600000	1.33566463	6.439	0.0001
X3	1	11.300000	1.33566463	8.460	0.0001
X4	1	2.100000	1.33566463	1.572	0.1367

a. Write the model relating mean yield to pea variety, and interpret all the parameters in the model.

b. Report the least squares model from the SAS printout.

c. What null and alternative hypotheses are tested by the global F test for this model? Interpret the hypotheses both in terms of the β parameters and the mean yields for the five varieties of peas.

d. Test the hypotheses of part c using $\alpha = .05$.

e. Place a 95% confidence interval on the difference between the mean yields of varieties D and E.

*12.54 A firm's *debt-to-equity ratio* is a measure of the extent to which management is using borrowed funds. It is a measure of considerable importance to individuals and organizations that are potential lenders to the firm. A high debt-to-equity ratio signals that in case of default, the lender will likely not recover outstanding loans to the firm since insufficient equity exists to cover all the firm's obligations. In addition, the higher the debt, the more funds that are required to service the debt. As a result, if business turns downward, the firm may not have sufficient operating funds to meet debt service payments (Spiro, 1982). The debt-to-equity ratios for firms in four different industries are given in the table for the fiscal year that ended during 1986.

INSURANCE		PUBLISHING	
Firm	Debt-to-equity	Firm	Debt-to-equity
Chubb	.24	Deluxe Check	.03
Kemper	.09	New York Times	.32
St. Paul Cos.	.12	Times Mirror	.51
Lincoln National	.09	Dow Jones	.15
USF & G	.00	Gannett	.73
Aetna Life & Cas.	.07		

ELECTRIC UTILITIES		BANKING	
Firm	Debt-to-equity	Firm	Debt-to-equity
Pacific G & E	.84	U.S. Bancorp	.54
Houston Ind.	1.00	Sun Trust Banks	.29
Florida Progress	1.03	Mellon Bank	.82
Penn Power & Light	1.11	Michigan National	.13
North States Power	.83	Southeast Banking	.41
Ohio Edison	1.16		
Orange & Rockland	.70		

Source: "39th Annual Report on American Industry," *Forbes*, Jan. 12, 1987.

a. Propose a regression model involving the qualitative independent variable Industry that could be used to investigate whether the mean debt-to-equity ratio varies among the four industries. Be sure to specify the coding scheme for the dummy variables in your model.

b. Test the null hypothesis that the mean debt-to-equity ratios are equal in the four industries. Use $\alpha = .05$.

c. Do the data provide sufficient evidence to conclude that the mean debt-to-equity ratios of the electric utilities industry and the insurance industry differ? Test using $\alpha = .10$.

SECTION 12.6

COMPARING THE SLOPES OF TWO OR MORE LINES

Suppose you wish to relate the mean monthly sales, $E(y)$, of a company to monthly advertising expenditure, x, for three different advertising media—say, newspaper, radio, and television—and you wish to use first-order (straight-line) models to model the responses for all three media. Graphs of these three relationships might appear as shown in Figure 12.18 on page 694.

Since the lines in Figure 12.18 are hypothetical, a number of practical business questions arise. Is one advertising medium as effective as any other; that is, do the three mean sales lines differ for the three advertising media? Do the increases in mean sales per dollar increase in advertising differ for the three advertising media; that is, do the slopes of the three lines differ? Note that each of the two practical business questions has been rephrased into a question about the parameters that define the three lines of Figure 12.18. To answer them, we must write a single linear statistical model that will characterize the three lines of Figure 12.18. Then the practical business questions can be answered by testing hypotheses about the model parameters.

FIGURE 12.18
Graphs of the
Relationship Between
Mean Sales, $E(y)$, and
Advertising Expenditure, x

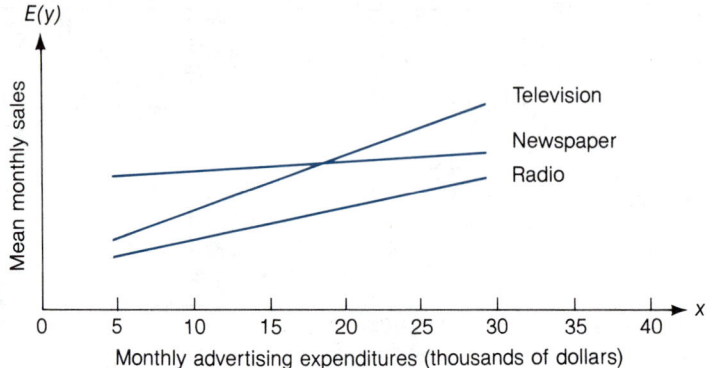

In the preceding example, the response (monthly sales) is a function of *two* independent variables, one quantitative (advertising expenditure, x) and one qualitative (type of medium). We will examine the different models that can be constructed relating $E(y)$ to these two independent variables.

1. The straight-line relationship between mean sales, $E(y)$, and advertising expenditure is the same for all three media; that is, a single line will describe the relationship between $E(y)$ and advertising expenditure, x_1, for all the media (see Figure 12.19).

$$E(y) = \beta_0 + \beta_1 x_1 \quad x_1 = \text{Advertising expenditure}$$

FIGURE 12.19
The Relationship Between
$E(y)$ and x_1 Is the Same
for All Media

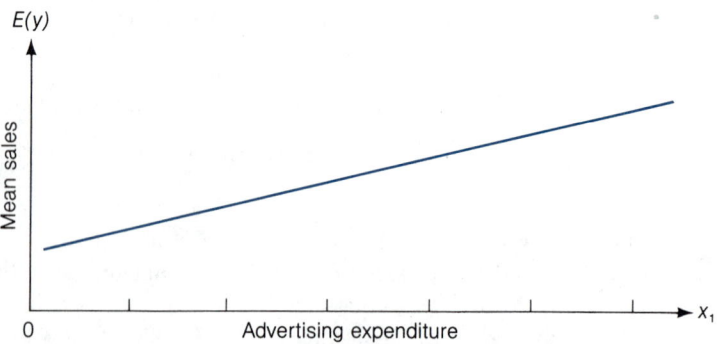

2. The straight lines relating mean sales, $E(y)$, to advertising expenditure, x_1, differ from one medium to another, but the increase in mean sales per unit increase in dollar advertising expenditure, x_1, is the same for all media. That is, the lines are parallel but have different y-intercepts (see Figure 12.20).

$$E(y) = \beta_0 + \beta_1 x_1 + \beta_2 x_2 + \beta_3 x_3$$

$$x_1 = \text{Advertising expenditure}$$

$$x_2 = \begin{cases} 1 & \text{if radio medium} \\ 0 & \text{if not} \end{cases} \qquad x_3 = \begin{cases} 1 & \text{if television medium} \\ 0 & \text{if not} \end{cases}$$

FIGURE 12.20
Parallel Response Lines
for the Three Media

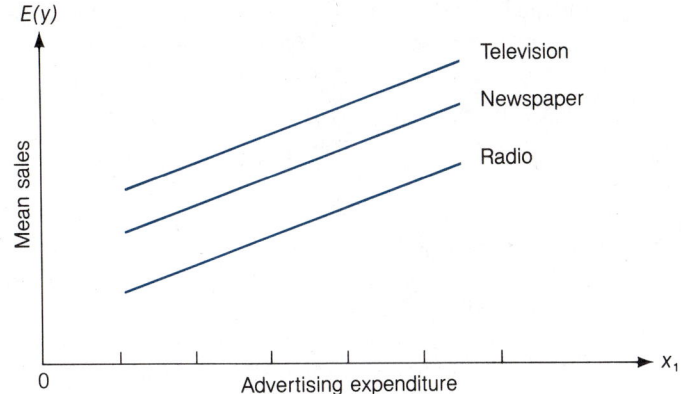

Notice that this model is essentially a combination of a first-order model with a single quantitative variable and the model with a single qualitative variable:

First-order model with a single
quantitative variable: $E(y) = \beta_0 +$ $\boxed{\beta_1 x_1}$

Model with a single
qualitative variable
at three levels: $E(y) = \beta_0 +$ $\boxed{\beta_2 x_2 + \beta_3 x_3}$

where x_1, x_2, and x_3 are defined as above. The model described implies no interaction between the two independent variables, advertising expenditure x_1 and the qualitative variable Type of advertising medium. The change in $E(y)$ for a 1-unit increase in x_1 is identical (i.e., the slopes of the lines are equal) for all three advertising media. The terms corresponding to each of the independent variables are called **main effect terms** because they imply no interaction.

3. The straight lines relating mean sales, $E(y)$, to advertising expenditure, x_1, differ for the three advertising media; that is, the intercepts and slopes differ for the three lines (see Figure 12.21). As you will see, this interaction model is obtained

FIGURE 12.21
Different Response Lines
for the Three Media

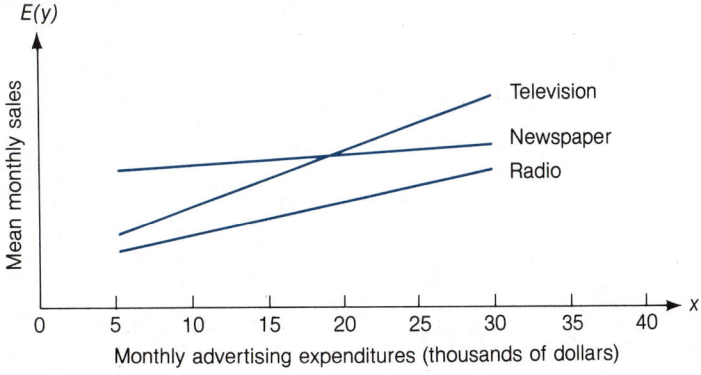

by adding interaction terms (those involving the cross product terms, one each from each of the two independent variables):

$$E(y) = \beta_0 + \underbrace{\beta_1 x_1}_{\substack{\text{Main effect,} \\ \text{advertising} \\ \text{expenditure}}} + \underbrace{\beta_2 x_2 + \beta_3 x_3}_{\substack{\text{Main effect,} \\ \text{type of medium}}} + \underbrace{\beta_4 x_1 x_2 + \beta_5 x_1 x_3}_{\text{Interaction}}$$

Note that each of the preceding models is obtained by adding terms to model 1, the single first-order model used to model the responses for all three media. Model 2 is obtained by adding the main effect terms for the qualitative variable, Type of medium; and model 3 is obtained by adding the interaction terms to model 2.

Will a single line (Figure 12.19) characterize the responses for all three media or do the three response lines differ as shown in Figure 12.21? A test of the null hypothesis that a single first-order model adequately describes the relationship between $E(y)$ and advertising expenditure x_1 for all three media is a test of the null hypothesis that the parameters of model 3, β_2, β_3, β_4, and β_5, equal 0; i.e.,

$$H_0: \quad \beta_2 = \beta_3 = \beta_4 = \beta_5 = 0$$

This hypothesis can be tested by fitting the complete model (model 3) and the reduced model (model 1) and conducting an F test, as described in Section 12.4.

Suppose we assume that the response lines for the three media will differ but wonder whether the data present sufficient evidence to indicate differences in the slopes of the lines. To test the null hypothesis that model 2 adequately describes the relationship between $E(y)$ and advertising expenditure x_1, we wish to test

$$H_0: \quad \beta_4 = \beta_5 = 0$$

that is, that the two independent variables, advertising expenditure x_1 and the qualitative variable, Type of medium, do not interact. This test can be conducted by fitting the complete model (model 3) and the reduced model (model 2), calculating the drop in the sum of squares for error, and conducting an F test.

EXAMPLE 12.5

Substitute the appropriate values of the dummy variables in model 3 to obtain the equations of the three response lines in Figure 12.21.

SOLUTION

The complete model that characterizes the three lines in Figure 12.21 is

$$E(y) = \beta_0 + \beta_1 x_1 + \beta_2 x_2 + \beta_3 x_3 + \beta_4 x_1 x_2 + \beta_5 x_1 x_3$$

where

x_1 = Advertising expenditure

$$x_2 = \begin{cases} 1 & \text{if radio medium} \\ 0 & \text{if not} \end{cases} \qquad x_3 = \begin{cases} 1 & \text{if television medium} \\ 0 & \text{if not} \end{cases}$$

Examining the coding, you can see that $x_2 = x_3 = 0$ when the advertising medium is newspaper. Substituting these values into the expression for $E(y)$, we obtain the newspaper medium line.

Newspaper medium line:

$$E(y) = \beta_0 + \beta_1 x_1 + \beta_2(0) + \beta_3(0) + \beta_4 x_1(0) + \beta_5 x_1(0)$$
$$= \beta_0 + \beta_1 x_1$$

Similarly, we substitute the appropriate values of x_2 and x_3 into the expression for $E(y)$ to obtain the radio medium line and the television medium line.

Radio medium line:

$$E(y) = \beta_0 + \beta_1 x_1 + \beta_2(1) + \beta_3(0) + \beta_4 x_1(1) + \beta_5 x_1(0)$$

$$= \underbrace{\beta_0 + \beta_2}_{y\text{-intercept}} + \underbrace{(\beta_1 + \beta_4)}_{\text{Slope}} x_1$$

Television medium line:

$$E(y) = \beta_0 + \beta_1 x_1 + \beta_2(0) + \beta_3(1) + \beta_4 x_1(0) + \beta_5 x_1(1)$$

$$= \underbrace{\beta_0 + \beta_3}_{y\text{-intercept}} + \underbrace{(\beta_1 + \beta_5)}_{\text{Slope}} x_1$$ ∎

If you were to fit model 3, obtain estimates of $\beta_0, \beta_1, \beta_2, \ldots, \beta_5$, and substitute them into the equations for the three media lines shown in Example 12.5, you would obtain exactly the same prediction equations as you would obtain if you fit three separate straight lines, one to each of the three sets of media data. You may ask why we would not fit the three lines separately. Why fit a model (model 3) that combines all three lines into the same equation? The answer is that you need to use this procedure if you wish to use statistical tests to compare the three media lines. We need to be able to express a practical question about the lines in terms of a hypothesis that a set of parameters in the model equal 0. You could not do this if you were to perform three separate regression analyses and fit a line to each set of media data.

EXAMPLE 12.6

An industrial psychologist conducted an experiment to investigate the relationship between worker productivity and a measure of salary incentive for two manufacturing plants, one, A, with union representation and the other, B, with nonunion representation. The productivity, y, per worker was measured by recording the number of machined castings that a worker could produce in a 4-week, 40-hour-per-week period. The incentive was the amount, x_1, of bonus (in cents per casting) paid for all castings produced in excess of 1,000 per worker for the 4-week period. Nine workers were selected from each plant and three from each group of nine were assigned to receive a 20¢ bonus per casting, three a 30¢ bonus, and three a 40¢ bonus per casting. The productivity data for the 18 workers, three for each plant type and incentive combination, are shown in the table at the top of page 698.

		INCENTIVE						
	20¢/casting			30¢/casting			40¢/casting		
Union plant	1,435	1,512	1,491	1,583	1,529	1,610	1,601	1,574	1,636
Nonunion plant	1,575	1,512	1,488	1,635	1,589	1,661	1,645	1,616	1,689

a. Plot the data points, and graph the prediction equations for the two productivity lines. Assume that the relationship between mean productivity and incentive is first-order.

b. Do the data provide sufficient evidence to indicate a difference in worker response to incentives between the two plants?

SOLUTION

If we assume that a first-order model* is adequate to detect a change in mean productivity, $E(y)$, as a function of incentive, x_1, then the model that produces two productivity lines, one for each plant, is

$$E(y) = \beta_0 + \beta_1 x_1 + \beta_2 x_2 + \beta_3 x_1 x_2$$

where

$$x_1 = \text{Incentive} \qquad x_2 = \begin{cases} 1 & \text{if nonunion plant} \\ 0 & \text{if union plant} \end{cases}$$

a. The SAS printout for the regression analysis is shown in Figure 12.22. The prediction equation obtained by reading the parameter estimates from the printout is

$$\hat{y} = 1,365.833 + 6.217x_1 + 47.778x_2 + .033x_1x_2$$

FIGURE 12.22 SAS Computer Printout for the Complete Model of Example 12.6

```
DEPENDENT VARIABLE: Y

SOURCE                          DF    SUM OF SQUARES      MEAN SQUARE    F VALUE

MODEL                            3    57332.38888889    19110.79629630    11.46
ERROR                           14    23349.22222223     1667.80158730    PR > F
CORRECTED TOTAL                 17    80681.61111112                     0.0005

R-SQUARE              C.V.           ROOT MSE          Y MEAN

0.710600              2.5901         40.83872656       1576.72222222

                              T FOR H0:   PR > ITI    STD ERROR OF
PARAMETER         ESTIMATE    PARAMETER=0              ESTIMATE

INTERCEPT       1365.8333333     26.35      0.0001     51.83641257
X1                 6.21666667     3.73      0.0022      1.66723403
X2                47.77777778     0.65      0.5251     73.30775769
X1*X2              0.03333333     0.01      0.9889      2.35782498
```

*Although the model contains a term involving x_1x_2, it is first-order (that is, it graphs as a straight line) in the quantitative variable x_1. The variable x_2 is a dummy variable that introduces or deletes terms in the model. The order of a term is determined only by the quantitative variables that appear in the term.

The prediction equation for the union plant can be obtained by substituting $x_2 = 0$ into the general prediction equation. Then

$$\hat{y} = \hat{\beta}_0 + \hat{\beta}_1 x_1 + \hat{\beta}_2(0) + \hat{\beta}_3 x_1(0)$$
$$= \hat{\beta}_0 + \hat{\beta}_1 x_1$$
$$= 1,365.833 + 6.217 x_1$$

Similarly, the prediction equation for the nonunion plant is obtained by substituting $x_2 = 1$ into the general prediction equation. Then,

$$\hat{y} = \hat{\beta}_0 + \hat{\beta}_1 x_1 + \hat{\beta}_2 x_2 + \hat{\beta}_3 x_1 x_2$$
$$= \hat{\beta}_0 + \hat{\beta}_1 x_1 + \hat{\beta}_2(1) + \hat{\beta}_3 x_1(1)$$

$$= \underbrace{(\hat{\beta}_0 + \hat{\beta}_2)}_{y\text{-intercept}} + \underbrace{(\hat{\beta}_1 + \hat{\beta}_3)}_{\text{Slope}} x_1$$
$$= (1,365.833 + 47.778) + (6.217 + .033)x_1$$
$$= 1,413.611 + 6.250 x_1$$

The graphs of these prediction equations are shown in Figure 12.23.

FIGURE 12.23

Graphs of the Prediction Equations for the Two Productivity Lines of Example 12.6

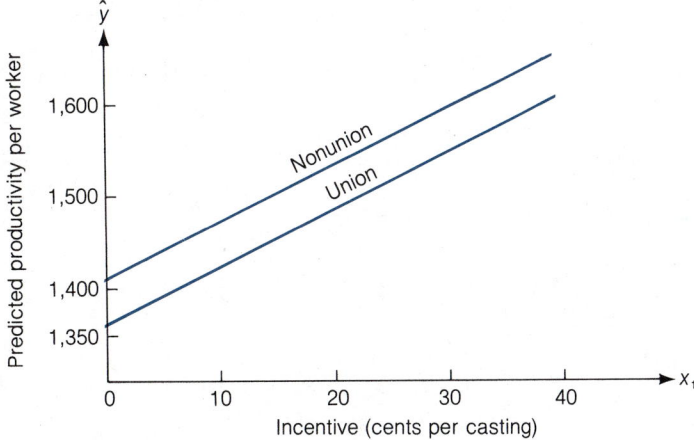

b. To determine whether the data provide sufficient evidence to indicate a difference in worker response to incentives for the two plants, we test the null hypothesis that a *single* line characterizes the relationship between productivity per worker and the amount of incentive, x_1, against the alternative hypothesis that we need two separate lines to characterize the relationship, one for each plant. If there is no difference in mean response $E(y)$ to x_1 between the two plants, then we do not need Type of plant in the model; i.e., we do not need the terms involving x_2. Therefore, we wish to test

$$H_0: \quad \beta_2 = \beta_3 = 0$$

against the alternative hypothesis

H_a: At least one of the two parameters, β_2 or β_3, differs from 0

The SAS computer printout for fitting the reduced model,

$$E(y) = \beta_0 + \beta_1 x_1$$

to the data is shown in Figure 12.24. Reading SSE_2 and SSE_1 from Figures 12.22 and 12.24, respectively, we obtain

Complete model: $SSE_2 = 23,349.22$

Reduced model: $SSE_1 = 34,056.28$

Drop in SSE: $SSE_1 - SSE_2 = 10,707.06$

FIGURE 12.24   SAS Computer Printout for the Reduced Model of Example 12.6

DEPENDENT VARIABLE: Y

SOURCE	DF	SUM OF SQUARES	MEAN SQUARE	F VALUE
MODEL	1	46625.33333333	46625.33333333	21.91
ERROR	16	34056.27777778	2128.51736111	PR > F
CORRECTED TOTAL	17	80681.61111112		0.0003

R-SQUARE	C.V.	ROOT MSE	Y MEAN
0.577893	2.9261	46.13585765	1576.72222222

PARAMETER	ESTIMATE	T FOR H0: PARAMETER=0	PR > :T:	STD ERROR OF ESTIMATE
INTERCEPT	1389.72222222	33.56	0.0001	41.40819949
X1	6.23333333	4.68	0.0003	1.33182749

The value of s^2 for the complete model is obtained from Figure 12.22:

$$s^2 = 1,667.80$$

Substituting these values, along with $k = 3$ and $g = 1$, into the formula for the F statistic yields

$$F = \frac{(SSE_1 - SSE_2)/(k - g)}{s^2} = \frac{10,707.06/2}{1,667.80} = 3.21$$

The numerator degrees of freedom (the number of parameters involved in H_0) is 2 and the denominator degrees of freedom (the number associated with s^2 in the complete model) is 14. If we choose $\alpha = .05$, the tabulated value of $F_{.05}$, given in Table VIII of Appendix B, is 3.74. Since the computed value, $F = 3.21$, is less than the tabulated value, $F_{.05} = 3.74$, there is insufficient evidence (at $\alpha = .05$) to indicate a difference in worker response to incentives between the two plants. Therefore, there is no evidence to indicate that two different lines, one for each plant, are needed to describe the relationship between the mean productivity per worker, $E(y)$, and the amount of incentive, x_1. ∎

EXAMPLE 12.7

Refer to Example 12.6 and explain how you would determine whether the data provide sufficient evidence to indicate that the incentive, x_1, affects mean productivity.

SOLUTION

If incentive did *not* affect mean productivity, we would not need terms involving x_1 in the model. Therefore, we would test the null hypothesis

$$H_0: \quad \beta_1 = \beta_3 = 0$$

against the alternative hypothesis

$$H_a: \quad \text{At least one of the parameters, } \beta_1 \text{ or } \beta_3, \text{ differs from } 0$$

We would fit the reduced model,

$$E(y) = \beta_0 + \beta_2 x_2$$

to the data and find SSE_1. The values of SSE_2 and s^2 for the complete model would be the same as those used in Example 12.6. Finally, you would calculate the value of the F statistic and compare it with a tabulated value of F based on $\nu_1 = 2$ and $\nu_2 = 14$ degrees of freedom. If the test leads to rejection of H_0, you have evidence to indicate that the increase in mean productivity that appears to be present in the graphs in Figure 12.23 is not due to random variation in the data. ∎

| | | | | | | | | | | | | |

**EXERCISES
12.55–12.65**

LEARNING THE MECHANICS

12.55 Suppose you are interested in a business application with a response y, one quantitative independent variable x_1, and one qualitative variable at three levels.
  a. Write a first-order model that relates the mean response $E(y)$ to the quantitative independent variable.
  b. Add the main effect terms for the qualitative independent variable to the model of part **a**. Specify the coding scheme you use.
  c. Add terms to the model of part **b** to allow for interaction between the quantitative and qualitative independent variables.
  d. Under what circumstances will the response lines of the model in part **c** be parallel?
  e. Under what circumstances will the model in part **c** have only one response line?

12.56 SAS was used to fit the following model to $n = 15$ data points:

$$y = \beta_0 + \beta_1 x_1 + \beta_2 x_2 + \beta_3 x_3 + \varepsilon$$

where x_1 is a quantitative variable, and x_2 and x_3 are dummy variables describing a qualitative variable at three levels using the coding scheme

$$x_2 = \begin{cases} 1 & \text{if level 2} \\ 0 & \text{otherwise} \end{cases} \qquad x_3 = \begin{cases} 1 & \text{if level 3} \\ 0 & \text{if otherwise} \end{cases}$$

The resulting printout is shown below:

SAS Printout for Complete Model of Exercise 12.56

Dep Variable: Y

Analysis of Variance

Source	DF	Sum of Squares	Mean Square	F Value	Prob>F
Model	3	4747.70480	1582.56827	46.894	0.0001
Error	11	371.22853	33.74805		
C Total	14	5118.93333			

Root MSE	5.80931	R-Square	0.9275	
Dep Mean	99.06667	Adj R-Sq	0.9077	
C.V.	5.86404			

Parameter Estimates

Variable	DF	Parameter Estimate	Standard Error	T for H0: Parameter=0	Prob > ¦T¦
INTERCEP	1	44.802703	8.55817652	5.235	0.0003
X1	1	2.172673	0.50335248	4.316	0.0012
X2	1	9.412913	4.05315974	2.322	0.0404
X3	1	15.631532	6.81368981	2.294	0.0425

a. What is the response line (equation) for $E(y)$ when $x_2 = x_3 = 0$? When $x_2 = 1$ and $x_3 = 0$? When $x_2 = 0$ and $x_3 = 1$?

b. What is the general least squares prediction equation for the above model?

c. What is the least squares prediction equation associated with level 1? Level 2? Level 3? Plot these on the same graph.

d. Specify the null and alternative hypotheses that should be used to investigate whether a difference exists between the response lines of the qualitative variable levels 1, 2, and 3.

e. Use the model with only the quantitative variable x_1 (printout shown at the top of page 703) to conduct the hypothesis test of part d. Use $\alpha = .05$.

APPLYING THE CONCEPTS

[Note: Starred (*) exercises require the use of a computer.]

12.57 The Florida Citrus Commission is interested in evaluating the performance of two orange juice extractors, brand A and brand B. It is believed that the size of the fruit used in the test may influence the juice yield (amount of juice per pound of oranges) obtained by the extractors. The commission wants to develop a regression model relating

SAS Printout for Reduced Model of Exercise 12.56

Dep Variable: Y

Analysis of Variance

Source	DF	Sum of Squares	Mean Square	F Value	Prob>F
Model	1	4522.90741	4522.90741	98.650	0.0001
Error	13	596.02592	45.84815		
C Total	14	5118.93333			

Root MSE	6.77113	R-Square	0.8836	
Dep Mean	99.06667	Adj R-Sq	0.8746	
C.V.	6.83492			

Parameter Estimates

Variable	DF	Parameter Estimate	Standard Error	T for H0: Parameter=0	Prob > |T|
INTERCEP	1	33.904568	6.78960378	4.994	0.0002
X1	1	3.083380	0.31044102	9.932	0.0001

the mean juice yield, $E(y)$, to the type of orange juice extractor (brand A or brand B) and the size of orange (diameter), x_1.

a. Identify the independent variables as qualitative or quantitative.

b. Write a model that describes the relationship between $E(y)$ and size of orange as two parallel lines, one for each brand of extractor.

c. Modify the model of part b to permit the slopes of the two lines to differ.

d. Sketch typical response lines for the model of part b. Do the same for the model of part c. Carefully label your graphs.

e. Specify the null and alternative hypotheses you would use to determine whether the model in part c provides more information for predicting yield than does the model in part b.

f. Explain how you would obtain the quantities necessary to compute the F statistic that would be used in testing the hypotheses you described in part e.

12.58 An economist is interested in modeling the mean monthly demand $E(y)$ (in thousands of units) for a particular product as a function of the product's price (in dollars) and the season of the year. The following model has been proposed:

$$E(y) = \beta_0 + \beta_1 x_1 + \beta_2 x_2 + \beta_3 x_3 + \beta_4 x_4$$

where

x_1 = Price

$$x_2 = \begin{cases} 1 & \text{if spring} \\ 0 & \text{otherwise} \end{cases} \qquad x_3 = \begin{cases} 1 & \text{if summer} \\ 0 & \text{otherwise} \end{cases} \qquad x_4 = \begin{cases} 1 & \text{if fall} \\ 0 & \text{otherwise} \end{cases}$$

A portion of the Minitab computer printout that results from fitting this model to a sample of 16 months of sales data selected from the last 2 years is shown below.

```
THE REGRESSION EQUATION IS
Y =     12.1 -   1.11 X1 +   3.94 X2
    +   7.17 X3 +   3.72 X4

                               ST. DEV.    T-RATIO =
          COLUMN COEFFICIENT   OF COEF.    COEF/S.D.

          --               12.067    1.473       8.19
X1   C2                     -1.113     .187      -5.93
X2   C3                      3.942     .858       4.59
X3   C4                      7.166     .815       8.80
X4   C5                      3.724     .819       4.55

THE ST. DEV. OF Y ABOUT REGRESSION LINE IS
S =      1.135
WITH ( 16 - 5) =   11 DEGREES OF FREEDOM

R-SQUARED = 93.1 PERCENT
R-SQUARED = 90.6 PERCENT, ADJUSTED FOR D.F.

ANALYSIS OF VARIANCE

 DUE TO       DF        SS    MS=SS/DF

REGRESSION    4    191.590    47.897
RESIDUAL     11     14.160     1.287
TOTAL        15    205.750
```

The reduced model, $E(y) = \beta_0 + \beta_1 x_1$, was also fit to the same data, and the resulting computer printout is partially reproduced here:

```
THE REGRESSION EQUATION IS
Y =    17.4 -   1.35 X1

                               ST. DEV.    T-RATIO =
          COLUMN  COEFFICIENT  OF COEF.    COEF/S.D.

          --               17.394    2.863       6.08
X1   C2                     -1.348     .403      -3.34

THE ST. DEV. OF Y ABOUT REGRESSION LINE IS
S =      2.859
WITH ( 16- 2) =    14 DEGREES OF FREEDOM

R-SQUARED = 44.4 PERCENT
R-SQUARED = 40.4 PERCENT, ADJUSTED FOR D.F.

ANALYSIS OF VARIANCE

 DUE TO       DF        SS    MS=SS/DF

REGRESSION    1     91.345    91.345
RESIDUAL     14    114.405     8.172
TOTAL        15    205.750
```

a. Is there sufficient evidence to conclude that mean monthly demand depends on the season of the year? Test using $\alpha = .05$.

b. Find an estimate for $E(y)$ when the price is $8.00 and it is summer. Interpret your result.

[*Note:*   We prefer to use a confidence interval for the estimate of part **b** but its calculation is beyond the scope of this text. This procedure can be found in the references at the end of the chapter.]

*12.59 In Exercises 11.11 and 12.41 a real estate appraiser used regression analysis to explore the relationship between the sale prices of apartments and various characteristics of the apartments. Some of the data from those exercises are reproduced in the table, which also contains data on the physical condition of each apartment building (E: excellent; G: good; F: fair).

CODE NO.	SALE PRICE	NUMBER OF APARTMENT UNITS	CONDITION OF APARTMENT BLDG.
0229	$ 90,300	4	F
0094	384,000	20	G
0043	157,500	5	G
0079	676,200	26	E
0134	165,000	5	G
0179	300,000	10	G
0087	108,750	4	G
0120	276,538	11	G
0246	420,000	20	G
0025	950,000	62	G
0015	560,000	26	G
0131	268,000	13	F
0172	290,000	9	E
0095	173,200	6	G
0121	323,650	11	G
0077	162,500	5	G
0060	353,500	20	F
0174	134,400	4	E
0084	187,000	8	G
0031	155,700	4	E
0019	93,600	4	F
0074	110,000	4	G
0057	573,200	14	E
0104	79,300	4	F
0024	272,000	5	E

Source: Robinson Appraisal Co., Inc., Mankato, Minnesota.

a. Write a model that describes the relationship between sale price and number of apartment units as three parallel lines, one for each level of physical condition. Be sure to specify the dummy variable coding scheme you use.

b. Plot y against x_1 (number of apartment units) for all buildings in excellent condition. On the same graph, plot y against x_1 for all buildings in good condition. Do this again for all buildings in fair condition. Does it appear that the model you specified in part **a** is appropriate? Explain.

c. Fit the model from part **a** to the data. Report the least squares prediction equation for each of the three building condition levels.

d. Plot the three prediction equations of part **c** on a scattergram of the data.

e. Do the data provide sufficient evidence to conclude that the relationship between sale price and number of units differs depending on the physical condition of the apartments? Test using $\alpha = .05$.

*12.60 A consumer advocacy organization conducted an experiment to compare the effectiveness of three commercially available weight-reducing diets. Ten people were assigned to each of the three diets for a 1-month period of time. Their weights (in pounds) were recorded at the beginning of the month and again at the end of the month. The results obtained are shown in the table.

DIET A		DIET B		DIET C	
x_1, weight before	y, weight loss	x_1, weight before	y, weight loss	x_1, weight before	y, weight loss
227	14	255	19	206	7
286	16	193	8	222	9
180	−2	186	4	168	2
176	8	145	15	132	0
204	15	219	16	173	−3
155	5	273	19	210	8
303	17	289	25	269	10
146	7	168	6	275	15
215	15	194	12	241	8
187	6	248	21	219	5

a. Construct a first-order regression model to relate weight loss, y, to weight before the program, x_1, and the type of diet. Be sure to specify the dummy variable coding scheme you use.

b. Suppose the effect of initial weight on weight loss varies from diet to diet. Write the appropriate regression model for this case. Sketch typical response curves depicting this situation.

c. Fit the models in parts **a** and **b** to the data given in the table.

d. Do the data provide sufficient evidence to indicate an interaction between initial weight and type of diet? That is, is the model of part **b** preferable to the model of part **a**? Test using $\alpha = .05$.

e. With respect to the model of part **b**, specify the null and alternative hypotheses you would use to test whether a difference exists among the mean weight losses for the three weight-reducing programs.

f. Conduct the test of part **e** using $\alpha = .05$.

12.61 Researchers for a dog food company have developed a new puppy food they hope will compete with the major brands. One premarketing test involved comparing the new food with two competitors in terms of weight gain. Fifteen 8-week-old German shepherd puppies, each from a different litter, were divided into three groups of five puppies each. Each group was fed one of the three brands of food.

a. Set up a model that assumes the initial weight, x_1, is linearly related to final weight, y, but does not allow for differences among the three brands; i.e., assume the response curve is the same for the three brands of dog food. Sketch the response curve as it might appear.

b. Set up a model that assumes the effect of initial weight is linearly related to final weight, and allows the intercepts of the lines to differ for the three brands. In other words, assume the initial weight and brand both affect final weight, but in an independent fashion. Sketch typical response curves.

c. Now write the main effects plus interaction model. For this model we assume the initial weight is linearly related to final weight, but both the slope and the intercept of the line depend on the brand. Sketch typical response curves.

12.62 A company is studying three different safety programs, A, B, and C, in an attempt to reduce the number of work-hours lost due to accidents. Each program is to be tried at three of the company's nine factories, and the plan is to monitor the lost work-hours, y, for a 1-year period beginning 6 months after the new safety program is instituted.

a. Write a main effects model relating $E(y)$ to the lost work-hours, x_1, the year before the plan is instituted and to the type of program that is instituted.

b. In terms of the model parameters from part **a**, what hypothesis would you test to determine whether the mean work-hours lost differ for the three safety programs?

12.63 Refer to Exercise 12.62. After the three safety programs have been in effect for 18 months, the complete main effects model is fit to the $n = 9$ data points. With safety program A as the base level, the following results were obtained:

$$\hat{y} = -2.1 + .88x_1 - 150x_2 + 35x_3 \qquad \text{SSE} = 1,527.27$$

Then the reduced model $E(y) = \beta_0 + \beta_1 x_1$ is fit, with the result

$$\hat{y} = 15.3 + .84x_1 \qquad \text{SSE} = 3,113.14$$

Test whether the mean work-hours lost differ for the three programs. Use $\alpha = .05$.

12.64 An insurance company is experimenting with three different training programs, A, B, and C, for its salespeople. The following main effects model is proposed:

$$E(y) = \beta_0 + \beta_1 x_1 + \beta_2 x_2 + \beta_3 x_3$$

where

y = Monthly sales (in thousands of dollars)

x_1 = Number of months experience

$$x_2 = \begin{cases} 1 & \text{if training program B used} \\ 0 & \text{otherwise} \end{cases} \qquad x_3 = \begin{cases} 1 & \text{if training program C used} \\ 0 & \text{otherwise} \end{cases}$$

Training program A is the base level.

a. What hypothesis would you test to determine whether the mean monthly sales differ for salespeople trained by the three programs?

b. After experimenting with 50 salespeople over a 5-year period, the complete model is fit, with the result

$$\hat{y} = 10 + .5x_1 + 1.2x_2 - .4x_3 \qquad \text{SSE} = 140.5$$

Then the reduced model $E(y) = \beta_0 + \beta_1 x_1$ is fit to the same data, with the result

$$\hat{y} = 11.4 + .4x_1 \qquad SSE = 183.2$$

Test the hypothesis you formulated in part **a**. Use $\alpha = .05$.

*12.65 In Exercise 11.63, an economist modeled the per capita demand for passenger car motor fuel in the United States as a function of two quantitative independent variables, personal income and the relative price of a gallon of gasoline. In this exercise we explore the relationship between per capita demand for motor fuel, y, and only one of those variables, the relative price of a gallon of gasoline, x. The data are repeated in the table.

YEAR	MOTOR FUEL CONSUMED BY CARS (Billion gallons)	POPULATION OF UNITED STATES (Millions)	RELATIVE PRICE OF GALLON OF GASOLINE
1965	50.3	194.3	1.004
1966	53.31	196.6	.998
1967	55.11	198.7	1.000
1968	58.52	200.7	.973
1969	62.45	202.7	.954
1970	65.8	205.1	.908
1971	69.51	207.1	.876
1972	73.5	209.9	.859
1973	78.0	211.9	.887
1974	74.2	213.9	1.083
1975	76.5	216.0	1.060
1976	78.8	218.0	1.043
1977	80.7	220.2	1.037
1978	83.8	222.6	1.005
1979	80.2	225.1	1.222
1980	73.7	227.7	1.496
1981	71.7	230.0	1.508
1982	72.8	232.3	1.346
1983	73.4	234.5	1.261

Source: *Statistical Abstract of the United States*, various years.

a. Construct a scattergram of the data.
b. In 1973, the Organization of Petroleum Exporting Countries (OPEC) began manipulating the supply of oil—and therefore gasoline—which subsequently caused prices to climb to record heights. This accounts for the unusual pattern you should have observed in the scattergram. Propose a first-order regression model that allows for differences in the relationship between y and x before and after OPEC began manipulating oil prices. [*Note:* Allow for differences in the slopes of the two regression lines.]
c. Fit your proposed model to the data.
d. Plot the two prediction equations associated with the least squares model obtained in part **c** on the scattergram of the data constructed in part **a**. Interpret these equations in the context of the problem.

e. Do the data provide sufficient evidence to conclude that the relationship between demand for motor fuel and the relative price of gasoline changed after 1973? Test using $\alpha = .05$.

f. Is there sufficient evidence to conclude that the slopes of the demand functions prior to 1973 and after 1973 differ? Test using $\alpha = .05$.

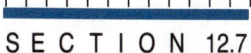

SECTION 12.7

COMPARING TWO OR MORE RESPONSE CURVES

Suppose we think that the relationship between mean monthly sales, $E(y)$, and advertising expenditure, x_1 (Section 12.6), is second-order. The scenario for writing the models for this situation is as follows.

1. The mean sales curves are identical for all three advertising media; that is, a single second-order curve will suffice to describe the relationship between $E(y)$ and x_1 for all the media (see Figure 12.25).

$$E(y) = \beta_0 + \beta_1 x_1 + \beta_2 x_1^2 \qquad x_1 = \text{Advertising expenditure}$$

FIGURE 12.25

The Relationship Between $E(y)$ and x_1 Is the Same for All Media

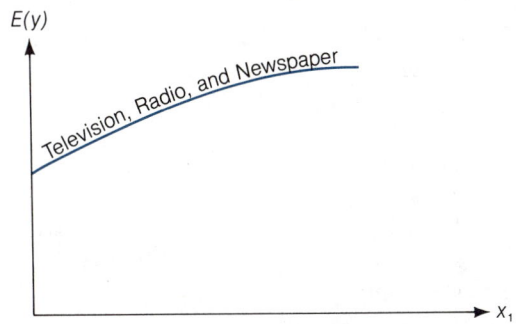

2. The response curves have the same shapes but different y-intercepts (see Figure 12.26).

$$E(y) = \beta_0 + \beta_1 x_1 + \beta_2 x_1^2 + \beta_3 x_2 + \beta_4 x_3$$

$x_1 = \text{Advertising expenditure}$

$$x_2 = \begin{cases} 1 & \text{if radio medium} \\ 0 & \text{if not} \end{cases} \qquad x_3 = \begin{cases} 1 & \text{if television medium} \\ 0 & \text{if not} \end{cases}$$

FIGURE 12.26

The Response Curves Have the Same Shapes but Different y-intercepts

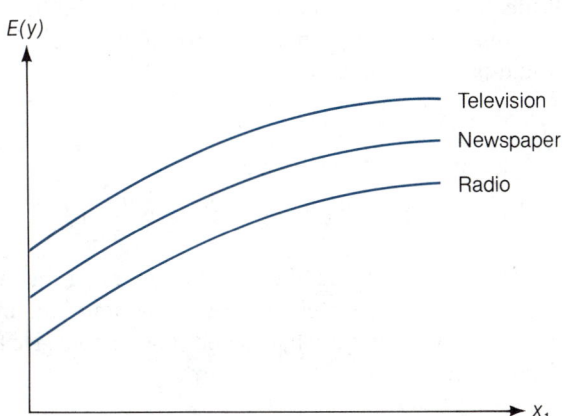

3. The response curves for the three advertising media are different—i.e., Advertising expenditure and Type of medium interact (see Figure 12.27).

$$E(y) = \beta_0 + \beta_1 x_1 + \beta_2 x_1^2 + \beta_3 x_2 + \beta_4 x_3 + \beta_5 x_1 x_2 + \beta_6 x_1 x_3 + \beta_7 x_1^2 x_2 + \beta_8 x_1^2 x_3$$

FIGURE 12.27

The Response Curves Differ for the Three Media

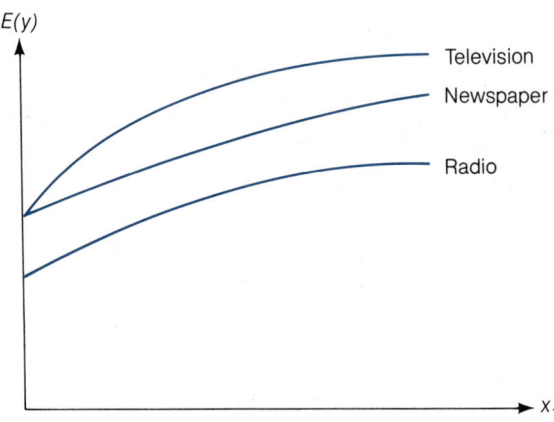

EXAMPLE 12.8

Give the equation of the second-order model for the radio advertising medium.

SOLUTION

Model 3 characterizes the relationship between $E(y)$ and x_1 for the radio advertising medium (see the coding) when $x_2 = 1$ and $x_3 = 0$. Substituting these values into model 3, we obtain

$$\begin{aligned}
E(y) &= \beta_0 + \beta_1 x_1 + \beta_2 x_1^2 + \beta_3 x_2 + \beta_4 x_3 + \beta_5 x_1 x_2 + \beta_6 x_1 x_3 + \beta_7 x_1^2 x_2 + \beta_8 x_1^2 x_3 \\
&= \beta_0 + \beta_1 x_1 + \beta_2 x_1^2 + \beta_3(1) + \beta_4(0) + \beta_5 x_1(1) + \beta_6 x_1(0) + \beta_7 x_1^2(1) + \beta_8 x_1^2(0) \\
&= (\beta_0 + \beta_3) + (\beta_1 + \beta_5) x_1 + (\beta_2 + \beta_7) x_1^2
\end{aligned}$$

∎

EXAMPLE 12.9

What null hypothesis about the parameters of model 3 would you test if you wished to determine whether the second-order curves for the three media differ?

SOLUTION

If the curves were identical, we would not need the independent variable Type of medium in the model; that is, we would delete all terms involving x_2 and x_3. This would produce model 1,

$$E(y) = \beta_0 + \beta_1 x_1 + \beta_2 x_1^2$$

and the null hypothesis would be

$$H_0: \quad \beta_3 = \beta_4 = \beta_5 = \beta_6 = \beta_7 = \beta_8 = 0$$

∎

EXAMPLE 12.10

Suppose we assume that the response curves for the three media differ but we want to know whether the second-order terms contribute information for the prediction of y. Or, equivalently, will a second-order model give better predictions than a first-order model?

SOLUTION

The only difference between model 3 and a first-order model are those terms involving x_1^2. Therefore, the null hypothesis, "the second-order terms contribute no information for the prediction of y," is equivalent to

$$H_0: \quad \beta_2 = \beta_7 = \beta_8 = 0 \qquad \blacksquare$$

Examples 12.9 and 12.10 identify two tests that answer practical questions concerning a collection of second-order models. Other comparisons among the curves can be made by testing appropriate sets of model parameters (see the exercises).

The models described in the preceding sections provide only an introduction to statistical modeling. Models can be constructed to relate $E(y)$ to any number of quantitative and/or qualitative independent variables. You can compare response curves and surfaces for different levels of a qualitative variable or for different combinations of levels of two or more qualitative independent variables. A general explanation of how to write linear statistical models can be found in Chapter 6 of Mendenhall and Sincich (1986).

CASE STUDY 12.1

FORECASTING PEAK-HOUR TRAFFIC VOLUME*

In designing future metropolitan roadways or redesigning existing roads, highway engineers rely heavily on traffic volume forecasts. Should the road have two, three, or four lanes in each direction? How should the on- and off-ramps be configured? Should the on-ramps be metered? Should one or more lanes be reversible to accommodate rush hour traffic? Should one or more lanes be restricted to carpools? The answers to such questions depend primarily on the traffic volumes the road must be able to handle during the peak travel times of the day—morning and evening rush hours.

Traffic forecasters at the Minnesota Department of Transportation use regression analysis to estimate weekday peak-hour traffic volumes on existing and proposed roadways. In particular, they model the traffic volume for the hour with the highest volume (this is the peak hour, typically 7–8 A.M.), y, as a function of the road's total traffic volume for the day, x_1. The model is developed using traffic data from existing roadways. Upon obtaining a forecast for the total weekday traffic volume for the road in question, the forecasters use the regression model to estimate (forecast) the mean peak-hour volume for the road.

For a recent project involving the redesign of a section of Interstate 494 in Bloomington, Minnesota, the forecasters collected data on peak-hour traffic volumes and weekday traffic volumes (using electronic sensors that count vehicles) at eight locations in the Minneapolis area believed to be similar to the one being redesigned. One of the characteristics that was common to the Interstate 494 site and the eight other locations was that the peak-hour volume at each was nearing its theoretical upper bound. A sample of $n = 72$ measurements was obtained and a scattergram of the data was constructed (see Figure 12.28 on page 712).

*Personal communications from John Sem, Director; Allan E. Pint, State Traffic Forecast Engineer; and James Page, Sr., Transportation Planner, Traffic and Commodities Studies Section, Minnesota Department of Transportation, St. Paul, Minnesota.

FIGURE 12.28

Scattergram: Peak-Hour
Traffic Volume Versus
Total Traffic Volume for
the Day

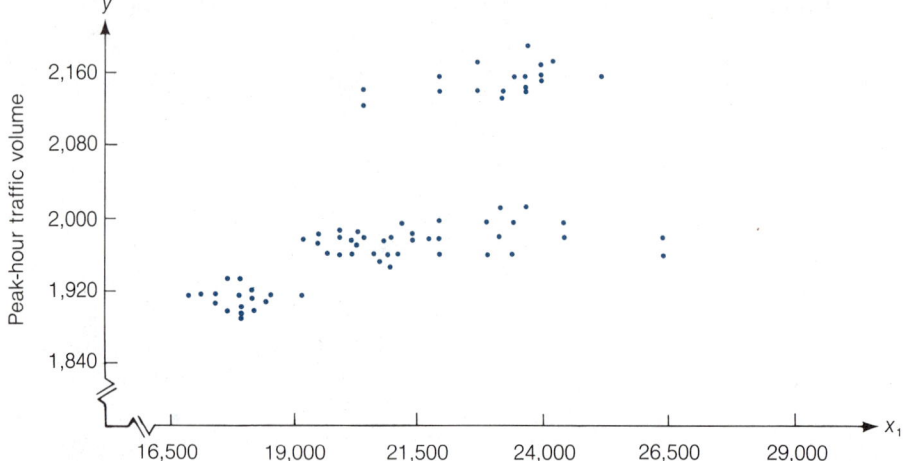

The traffic forecasters were surprised to see the isolated group of observations at the top of the scattergram. They investigated and found that all these data points were collected on Interstate 35W at the intersection of 46th Street, on the south side of Minneapolis. It turned out that, while all locations in the sample were three-lane highways (including this location), this location was unique because the highway widens to four lanes just a short distance north of the electronic sensor. Accordingly, this location was *not* similar to the section of Interstate 494 being redesigned, as was originally thought. It was decided that the 18 measurements collected at this location should be kept in the sample (in order to maintain a large sample size for the estimation of σ^2), but that a separate peak-hour-volume response curve should be estimated for this location through the use of a dummy variable in the regression model.

Knowing that peak-hour traffic volumes have a theoretical upper bound and that the locations sampled were nearing their bounds, the forecasters hypothesized that a second-order model should be used to explain the variation in peak-hour volume. Since this hypothesis was also supported by the scattergram, they fit the following model to the $n = 72$ data points:

$$E(y) = \beta_0 + \beta_1 x_1 + \beta_2 x_1^2 + \beta_3 x_2 + \beta_4 x_1 x_2$$

where

$$x_2 = \begin{cases} 1 & \text{if Interstate 35W and 46th Street} \\ 0 & \text{otherwise} \end{cases}$$

The results are shown in the Minitab printout in Figure 12.29. In order to formally investigate whether the relationship between peak-hour volume and total traffic volume for the day at the Interstate 35W location is different from that at the other locations, they eliminated the terms $\beta_3 x_2$ and $\beta_4 x_1 x_2$ from the model and fit the resulting reduced model to the same data, with the results shown in Figure 12.30. From the information contained on the two printouts, the forecasters calculated

FIGURE 12.29 Peak-Hour Traffic Volume: Complete Model

```
THE REGRESSION EQUATION IS
Y =    779. + .104 X1 -   .0000 X2
   +    98.0 X3 +   .0029 X4

                                 ST. DEV.    T-RATIO=
          COLUMN   COEFFICIENT   OF COEF.    COEF/S.D.

          --            779.0      141.8       5.49
   X1  C2              .1039       .0136       7.63
   X2  C3         -.000002220  .000000324     -6.84
   X3  C4             98.0        75.6         1.30
   X4  C5             .00291      .00332        .88

THE ST. DEV. OF Y ABOUT REGRESSION LINE IS
S =      15.47
WITH (   72- 5) =   67 DEGREES OF FREEDOM

R-SQUARED = 97.2 PERCENT
R-SQUARED = 97.0 PERCENT, ADJUSTED FOR D.F.

ANALYSIS OF VARIANCE

   DUE TO      DF        SS    MS=SS/DF

REGRESSION     4     555747.6   138936.9
RESIDUAL      67      16031.3      239.3
TOTAL         71     571778.9
```

FIGURE 12.30 Peak-Hour Traffic Volume: Reduced Model

```
THE REGRESSION EQUATION IS
Y =    197. +   .149 X1 - .0000 X2

                                 ST. DEV.    T-RATIO =
          COLUMN   COEFFICIENT   OF COEF.    COEF/S.D.

          --            197.4      578.9        .34
   X1  C2              .1492       .0555       2.69
   X2  C3          -.00000295   .00000132     -2.24

THE ST. DEV. OF Y ABOUT REGRESSION LINE IS
S =      65.45
WITH (   72- 3) = 69 DEGREES OF FREEDOM

R-SQUARED = 48.3 PERCENT
R-SQUARED = 46.8 PERCENT, ADJUSTED FOR D.F.

ANALYSIS OF VARIANCE

   DUE TO      DF        SS    MS=SS/DF

REGRESSION     2     276177    138088
RESIDUAL      69     295602      4284
TOTAL         71     571779
```

$$F = \frac{(295{,}602 - 16{,}031.3)/2}{16{,}031.3/67} = 584.2$$

and compared it to $F_{.01} \approx 4.96$, where $\nu_1 = 2$ and $\nu_2 = 67$. They concluded that the two response curves were different. In addition, the forecasters noted that β_1 was significantly greater than 0 and that β_2 was significantly less than 0—as would be expected, since peak-hour volume has an upper bound. Accordingly, for estimating and predicting peak-hour volumes at the Interstate 494 site, the forecasters chose the following prediction equation:

$$\hat{y} = 779 + .1039x_1 - .00000222x_1^2$$

EXERCISES 12.66–12.77

LEARNING THE MECHANICS

12.66 Consider an application for which you want to relate the response variable y to one quantitative variable and one qualitative variable at three levels.
  a. Write a complete second-order model that relates $E(y)$ to the quantitative variable.
  b. Add the main effect terms for the qualitative variable (at three levels) to the model of part **a**.
  c. Add terms to the model of part **b** to allow for interaction between the quantitative and qualitative independent variables.

12.67 Refer to Exercise 12.66, part **c**.
  a. Under what circumstances will the response curves of the model have the same shape but different y-intercepts?
  b. Under what circumstances will the response curves of the model be parallel lines?
  c. Under what circumstances will the response curves of the model be identical?

12.68 Write a model for $E(y)$ in which there are two independent variables, one quantitative and one qualitative with four levels. Construct the model so the associated response curves are second-order and the independent variables do not interact.

12.69 Minitab was used to fit the following model to $n = 25$ data points:

$$y = \beta_0 + \beta_1 x_1 + \beta_2 x_1^2 + \beta_3 x_2 + \beta_4 x_3 + \beta_5 x_1 x_2$$
$$+ \beta_6 x_1 x_3 + \beta_7 x_1^2 x_2 + \beta_8 x_1^2 x_3 + \varepsilon$$

where x_1 is a quantitative variable and

$$x_2 = \begin{cases} 1 & \text{if level 2} \\ 0 & \text{otherwise} \end{cases} \qquad x_3 = \begin{cases} 1 & \text{if level 3} \\ 0 & \text{otherwise} \end{cases}$$

The resulting printout is shown at the top of page 715. (Note that C1 represents y, and C2–C9 represent the eight independent variable terms in the model, in the same order as specified above.)

Minitab Printout for
Complete Model in
Exercise 12.69

```
THE REGRESSION EQUATION IS
C1 = 48.8 - 3.36 C2 + 0.0749 C3 - 2.36 C4 - 7.60 C5 + 3.71 C6
         + 2.66 C7 - 0.0183 C8 - 0.0372 C9
```

COLUMN	COEFFICIENT	ST. DEV. OF COEF.	T-RATIO = COEF/S.D.
	48.780	4.186	11.65
C2	-3.3618	0.4708	-7.14
C3	0.07492	0.01088	6.89
C4	-2.364	6.922	-0.34
C5	-7.595	6.102	-1.24
C6	3.7144	0.8295	4.48
C7	2.6590	0.6469	4.11
C8	-0.01826	0.02175	-0.84
C9	-0.03724	0.01453	-2.56

```
S = 3.190

R-SQUARED = 98.9 PERCENT
R-SQUARED = 98.4 PERCENT, ADJUSTED FOR D.F.
CONTINUE?

ANALYSIS OF VARIANCE
```

DUE TO	DF	SS	MS=SS/DF
REGRESSION	8	15256.9	1907.1
RESIDUAL	16	162.9	10.2
TOTAL	24	15419.8	

Minitab was also used to fit the model

$$y = \beta_0 + \beta_1 x_1 + \beta_2 x_1^2 + \varepsilon$$

to the same data set. The printout is shown below:

Minitab Printout for
Reduced Model in
Exercise 12.69

```
THE REGRESSION EQUATION IS
C1 = 33.9 + 0.86 C2 - 0.0110 C3
```

COLUMN	COEFFICIENT	ST. DEV. OF COEF.	T-RATIO = COEF/S.D.
	33.93	20.74	1.64
C2	0.862	2.218	0.39
C3	-0.01102	0.05103	-0.22

```
S = 26.10

R-SQUARED =  2.8 PERCENT
R-SQUARED =   .0 PERCENT, ADJUSTED FOR D.F.

ANALYSIS OF VARIANCE
```

DUE TO	DF	SS	MS=SS/DF
REGRESSION	2	433.7	216.9
RESIDUAL	22	14986.0	681.2
TOTAL	24	15419.8	

a. What is the general least squares prediction equation for the complete model?

b. What is the equation of the response curve for $E(y)$ when $x_2 = 0$ and $x_3 = 0$? When $x_2 = 1$ and $x_3 = 0$? When $x_2 = 0$ and $x_3 = 1$?

c. On the same graph, plot the least squares prediction equations associated with level 1, with level 2, and with level 3.

d. Specify the null and alternative hypotheses that should be used to investigate whether the second-order response curves for the three levels differ.

e. Conduct the hypothesis test of part **d**. Use $\alpha = .05$.

APPLYING THE CONCEPTS

[*Note: Starred (*) exercises require the use of a computer.*]

12.70 A pharmaceutical company wants to develop a regression model that will predict the mean time to relief after the administration of a new pain-killing drug. Two variables that are believed to be good predictors of relief time are the age of the patient and the method of drug administration. There are three methods of administration: (1) orally in liquid form; (2) orally in pill form; (3) intravenously. The following model was proposed by the pharmaceutical company:

$$E(y) = \beta_0 + \beta_1 x_1 + \beta_2 x_2 + \beta_3 x_3 + \beta_4 x_1 x_2 + \beta_5 x_1 x_3$$

where

y = Time to relief (in minutes)

x_1 = Age of patient (in years)

$x_2 = \begin{cases} 1 & \text{if drug administered orally in pill form} \\ 0 & \text{if not} \end{cases}$

$x_3 = \begin{cases} 1 & \text{if drug administered intravenously} \\ 0 & \text{if not} \end{cases}$

The data in the table, obtained on 12 patients, were used to fit the above model.

METHOD 1		METHOD 2		METHOD 3	
Age	Time to relief	Age	Time to relief	Age	Time to relief
51	22	46	28	37	19
36	25	40	24	60	17
31	20	26	23	25	21
20	25	32	25	38	20

a. Multiple regression analysis produced the following results:

SOURCE	DF	SUM OF SQUARES	MEAN SQUARE	F	PR > F	R-SQUARE
MODEL	5	87.473	17.495	4.90	0.0394	0.803120
ERROR	6	21.443	3.574			
CORRECTED TOTAL	11	108.916				

Test whether the model is useful in predicting mean time to relief. Use $\alpha = .05$.

b. What hypothesis would you use to test whether the age–administration method interaction terms contribute to the prediction of mean time to relief?

c. The reduced model

$$E(y) = \beta_0 + \beta_1 x_1 + \beta_2 x_2 + \beta_3 x_3$$

was fit to the data and produced SSE = 38.289. Does this provide sufficient evidence at $\alpha = .05$ to indicate that the interaction terms should be kept in the model?

12.71 An operations manager is interested in modeling $E(y)$, the expected length of time per month (in hours) that a machine will be shut down for repairs as a function of the type of machine (001 or 002) and the age of the machine (in years). He has proposed the following model:

$$E(y) = \beta_0 + \beta_1 x_1 + \beta_2 x_1^2 + \beta_3 x_2$$

where

$$x_1 = \text{Age of machine} \qquad x_2 = \begin{cases} 1 & \text{if machine type 001} \\ 0 & \text{if machine type 002} \end{cases}$$

Data were obtained on $n = 20$ machine breakdowns and were used to estimate the parameters of the above model. A portion of the regression analysis computer printout is shown here:

SOURCE	DF	SUM OF SQUARES	MEAN SQUARE
MODEL	3	2396.364	798.788
ERROR	16	128.586	8.037
TOTAL	19	2524.950	R-SQUARE
			0.949

The reduced model, $E(y) = \beta_0 + \beta_1 x_1 + \beta_2 x_2$, was fit to the same data. The regression analysis computer printout is partially reproduced here:

SOURCE	DF	SUM OF SQUARES	MEAN SQUARE
MODEL	2	2342.42	1171.21
ERROR	17	182.53	10.74
TOTAL	19	2524.95	R-SQUARE
			0.928

Is there sufficient evidence to conclude that the second-order (x_1^2) term in the model proposed by the operations manager is necessary? Test using $\alpha = .05$.

*12.72 Refer to Exercise 12.71. The data that were used to fit the operations manager's complete and reduced models are displayed in the table at the top of page 718.

a. Use the data to test the null hypothesis that $\beta_1 = \beta_2 = 0$. Test using $\alpha = .10$.

b. Carefully interpret the results of the test in the context of the problem.

DOWN TIME PER MONTH (hours)	MACHINE TYPE	MACHINE AGE (years)	DOWN TIME PER MONTH (hours)	MACHINE TYPE	MACHINE AGE (years)
10	001	1.0	10	002	2.0
20	001	2.0	20	002	4.0
30	001	2.7	30	002	5.0
40	001	4.1	44	002	8.0
9	001	1.2	9	002	2.4
25	001	2.5	25	002	5.1
19	001	1.9	20	002	3.5
41	001	5.0	42	002	7.0
22	001	2.1	20	002	4.0
12	001	1.1	13	002	2.1

12.73 An equal rights group has charged that women are being discriminated against in terms of the salary structure in a state university system. It is thought that a complete second-order model will be adequate to describe the relationship between salary and years of experience for both groups. A sample is to be taken from the records for faculty members (all of equal status) within the system and the following model is to be fit:

$$E(y) = \beta_0 + \beta_1 x_1 + \beta_2 x_1^2 + \beta_3 x_2 + \beta_4 x_1 x_2 + \beta_5 x_1^2 x_2$$

where

$y = $ Annual salary (in thousands of dollars)

$x_1 = $ Experience (years) $x_2 = \begin{cases} 1 & \text{if female} \\ 0 & \text{if male} \end{cases}$

a. What hypothesis would you test to determine whether the *rate* of increase of mean salary with experience is different for males and females?
b. What hypothesis would you test to determine whether there are differences in mean salaries that are attributable to sex?

[*Note*:   In practice, we would include other variables in the model. We include only two here to simplify the exercise.]

12.74 Refer to Exercise 12.20, where we presented data on the number of highway deaths and the number of licensed vehicles on the road for the years 1950–1985. We mentioned that the number of deaths, y, may also have been affected by the existence of the national 55-mile-per-hour speed limit during the years 1974–1985 (i.e., years 25–36). Define the dummy variable

$x_2 = \begin{cases} 1 & \text{if 55-mile-per-hour speed limit was in effect} \\ 0 & \text{if not} \end{cases}$

a. Introduce the variable x_2 into the second-order model of Exercise 12.20 to account for the presence or absence of the 55-mile-per-hour speed limit in a given year. Include terms involving the interaction between x_2 and x_1.
b. Refer to your model for part **a**. Sketch on a single piece of graph paper your visualization of the two response curves, the second-order curves relating y to x_1 before and after the imposition of the 55-mile-per-hour speed limit.

c. Suppose that x_1 and x_2 do not interact. How would that affect the graphs of the two response curves of part **b**?

d. Refer to part **c**. Suppose that x_1 and x_2 do interact. How would this affect the graphs of the two response curves?

12.75 In Exercise 12.20, we fit a second-order model to data relating the number, y, of U.S. highway deaths per year to the number, x_1, of licensed vehicles on the road. In Exercise 12.74 we added a qualitative variable, x_2, to account for the presence or absence of the 55-mile-per-hour national speed limit. The accompanying SAS computer printout gives the results of fitting the model

$$E(y) = \beta_0 + \beta_1 x_1 + \beta_2 x_1^2 + \beta_3 x_2 + \beta_4 x_1 x_2 + \beta_5 x_1^2 x_2$$

to the data. Use this printout and the printout for Exercise 12.20 to determine whether the data provide sufficient evidence to indicate that the qualitative variable (speed limit) contributes information for the prediction of the annual number of highway deaths. Test using $\alpha = .05$. Discuss the practical implications of your test results.

SAS Printout for
Exercise 12.75

Dep Variable: Y

Analysis of Variance

Source	DF	Sum of Squares	Mean Square	F Value	Prob>F
Model	5	1299.14564	259.82913	20.970	0.0001
Error	30	371.71742	12.39058		
C Total	35	1670.86306			

Root MSE	3.52003	R-Square	0.7775	
Dep Mean	45.76389	Adj R-Sq	0.7405	
C.V.	7.69171			

Parameter Estimates

Variable	DF	Parameter Estimate	Standard Error	T for HO: Parameter=0	Prob > |T|
INTERCEP	1	0.216246	7.14583612	0.030	0.9761
X1	1	0.812026	0.15649688	5.189	0.0001
X1SQ	1	-0.003177	0.00076431	-4.156	0.0002
X2	1	-61.651263	34.60299050	-1.782	0.0849
X1X2	1	1.060693	0.60068353	1.766	0.0876
X1SQX2	1	-0.004076	0.00237521	-1.716	0.0965

*12.76 In order to divide a set of existing and/or potential sales accounts into sales territories, it is helpful to know how many sales calls should be made to each account during the year (Churchill, Ford, and Walker, 1985). The data in the table on page 720 were collected for ten wholesale (W) accounts of similar size (i.e., the maximum potential sales at each account are similar) and ten retail (R) accounts of similar size.

a. Construct a scattergram of the data with 1987 sales on the vertical axis and number of sales calls on the horizontal axis. Use "R" as a plotting symbol to represent retail accounts, and "W" as a plotting symbol to represent wholesale accounts.

ACCOUNT NO.	1987 SALES (thousands of $)	NUMBER OF SALES CALLS MADE IN 1987	TYPE OF ACCOUNT
005	92	5	R
192	144	9	W
083	60	10	R
791	130	12	W
483	48	1	R
029	116	5	W
632	87	9	R
191	110	3	W
086	100	9	R
333	90	1	W
246	107	6	R
130	147	10	W
055	151	9	W
128	127	5	W
115	103	5	R
201	133	7	W
099	90	10	R
111	82	2	W
200	85	4	R
076	19	2	R

b. Based on your scattergram, write a second-order model that describes the relationship between sales revenue and the number of sales calls made for both wholesale and retail accounts.

c. Fit the model you proposed in part **b** to the data. Plot the resulting prediction equations on your scattergram of part **a**.

d. Based on your graph of part **c** and the value of R^2 that you obtained in fitting the model, comment on the goodness of fit of your model to the data.

e. What null hypothesis about the parameters of your model would you test to determine whether the sales response curves for the two types of accounts differ?

f. Conduct the hypothesis test of part **e**. Use $\alpha = .01$.

g. What null hypothesis about the parameters of your model would you test to determine whether the second-order terms in your model contribute information for the prediction of sales?

h. Conduct the hypothesis test of part **g**. Use $\alpha = .05$.

i. Based on the fitted model you obtained in part **c**, estimate the optimal number of sales calls that should be made to wholesale accounts.

*12.77 *Productivity* has been defined as the relationship between inputs and outputs of a productive system. In order to manage a system's productivity, it is necessary to measure it. Productivity is typically measured by dividing a measure of system output by a measure of the inputs to the system. Some examples of productivity measures are Sales/Salesperson, (Yards of carpet laid)/(Number of carpet layers), Shipments/[(Direct labor) + (Indirect labor) + (Materials)]. Notice that productivity can be improved by producing greater output with the same inputs or by producing the same output with fewer inputs. In manufacturing operations, productivity ratios like the shipments example just cited generally vary with the volume of output produced (Schroeder, 1985). The production data in the table have been collected for a random sample of months for the three regional plants of a particular manufacturing firm. Each plant manufactures the same product.

NORTH PLANT		SOUTH PLANT		WEST PLANT	
Productivity ratio	Number of units produced	Productivity ratio	Number of units produced	Productivity ratio	Number of units produced
1.30	1,000	1.43	1,015	1.61	501
.90	400	1.50	925	.74	140
1.21	650	.91	150	1.19	303
.75	200	.99	222	1.88	930
1.32	850	1.33	545	1.72	776
1.29	600	1.15	402	1.39	400
1.18	756	1.51	709	1.86	810
1.10	500	1.01	176	.99	220
1.26	925	1.24	392	.79	160
.93	300	1.49	699	1.59	626
.81	258	1.37	800	1.82	640
1.12	590	1.39	660	.91	190

a. Construct a scattergram for the data. Plot the data from the north plant using dots, from the south plant using small circles, and from the west plant using small triangles.

b. Visually fit each plant's response curve to the scattergram.

c. Based on the results of part b, propose a second-order regression model that could be used to estimate the relationship between productivity and volume for the three plants.

d. Fit the model you proposed in part c to the data.

e. Is there sufficient evidence to conclude that the productivity response curves for the three plants differ? Test using $\alpha = .05$.

f. Do the data provide sufficient evidence to conclude that the second-order model contributes more information for the prediction of productivity than a first-order model? Test using $\alpha = .05$.

g. Next month, 890 units are scheduled to be produced at the west plant. Use the model you developed in part d to predict next month's productivity ratio at the west plant.

| | | | | | | | | | | | | | |

SECTION 12.8

MODEL BUILDING: STEPWISE REGRESSION

The problem of predicting executive salaries was discussed in Chapter 11. Perhaps the biggest problem in building a model to describe executive salaries is choosing the important independent variables to be included in the model. The list of potentially important independent variables is extremely long, and we need some objective method of screening out those that are not important.

The problem of deciding which of a large set of independent variables to include in a model is common. Trying to determine which variables influence the profit of a firm, affect product quality, or are related to the state of the economy are only a few examples.

A systematic approach to building a model with a large number of independent variables is difficult because the interpretation of multivariable interactions and higher-order polynomials is tedious. We therefore turn to a screening procedure known as **stepwise regression**.

The most commonly used stepwise regression procedure, available in most popular computer packages, works as follows: The user first identifies the response, y, and the set of potentially important independent variables, $x_1, x_2, \ldots, x_k$, where k will generally be large. (Note that this set of variables could represent both first- and higher-order terms, as well as any interaction terms that might be important information contributors.) The response and independent variables are then entered into the computer, and the stepwise procedure begins.

STEP 1   The computer fits all possible one-variable models of the form

$$E(y) = \beta_0 + \beta_1 x_i$$

to the data. For each model, the test of the null hypothesis

$$H_0: \quad \beta_1 = 0$$

against the alternative hypothesis

$$H_a: \quad \beta_1 \neq 0$$

is conducted using the t (or the equivalent F) test for a single β parameter. The independent variable that produces the largest (absolute) t value is declared the best one-variable predictor of y. Call this independent variable x_1.

STEP 2   The stepwise program now begins to search through the remaining $(k-1)$ independent variables for the best two-variable model of the form

$$E(y) = \beta_0 + \beta_1 x_1 + \beta_2 x_i$$

This is done by fitting all two-variable models containing x_1 and each of the other $(k-1)$ options for the second variable x_i. The t values for the test $H_0: \beta_2 = 0$ are computed for each of the $(k-1)$ models (corresponding to the remaining independent variables x_i, $i = 2, 3, \ldots, k$), and the variable having the largest t is retained. Call this variable x_2.

At this point, some computer packages diverge in methodology. The better packages now go back and check the t value of $\hat{\beta}_1$ *after $\hat{\beta}_2 x_2$ has been added to the model.* If the t value has become nonsignificant at some specified α level (say $\alpha = .10$), the variable x_1 is removed and a search is made for the independent variable with a β parameter that will yield the most significant t value in the presence of $\hat{\beta}_2 x_2$. Other packages do not recheck $\hat{\beta}_1$, but proceed directly to step 3.

The best-fitting plane may yield a different value for $\hat{\beta}_1$ than that obtained in step 1, because $\hat{\beta}_1$ and $\hat{\beta}_2$ may be correlated. Thus, both the value of $\hat{\beta}_1$ and, therefore, its significance will usually change from step 1 to step 2. For this reason, the computer packages that recheck the t values at each step are preferred.

STEP 3 The stepwise procedure now checks for a third independent variable to include in the model with x_1 and x_2. That is, we seek the best model of the form

$$E(y) = \beta_0 + \beta_1 x_1 + \beta_2 x_2 + \beta_3 x_i$$

To do this, we fit all the $(k - 2)$ models using x_1, x_2, and each of the $(k - 2)$ remaining variables, x_i, as a possible x_3. The criterion is again to include the independent variable with the largest t value. Call this best third variable x_3.

The better programs now recheck the t values corresponding to the x_1 and x_2 coefficients, replacing the variables with t values that have become nonsignificant. This procedure is continued until no additional independent variables can be found that yield significant t values (at the specified α level) in the presence of the variables already in the model.

The result of the stepwise procedure is a model containing only those terms with t values that are significant at the specified α level. Thus, in most practical situations, only several of the large number of independent variables will remain. However, it is very important *not* to jump to the conclusion that all the independent variables important for predicting y have been identified or that the unimportant independent variables have been eliminated. Remember, the stepwise procedure is using only *sample estimates* of the true model coefficients (β's) to select the important variables. An extremely large number of single β parameter t tests have been conducted, and the probability is very high that one or more errors have been made in including or excluding variables. That is, we have very probably included some unimportant independent variables in the model (Type I errors) and eliminated some important ones (Type II errors).

There is a second reason why we might not have arrived at a good model. When we choose the variables to be included in the stepwise regression, we may often omit high-order terms (to keep the number of variables manageable). Consequently, we may have initially omitted several important terms from the model. Thus, we should recognize stepwise regression for what it is—an objective screening procedure.

Now, we will consider interactions and quadratic terms (for quantitative variables) among variables screened by the stepwise procedure. It would be best to develop this response surface model with a second set of data independent of that used for the screening, so the results of the stepwise procedure can be partially verified with new data. However, this is not always possible because in many business modeling situations only a small amount of data is available.

Remember, do not be deceived by the impressive looking t values that result from the stepwise procedure—it has retained only the independent variables with the largest t values. Also, if you have used a main effects model for your stepwise procedure, remember that it may be greatly improved by the addition of interaction and quadratic terms.

EXAMPLE 12.11

In Section 11.8, we fit a multiple regression model for executive salaries as a function of experience, education, sex, and other factors. A preliminary step in the construction of this model was the determination of the most important independent variables. Ten independent variables were considered, as shown in Table 12.4. It would be very difficult to construct a second-order model with ten independent variables. Therefore, use the sample of 100 executives from Section 11.8 to decide which of the ten variables should be included in the construction of the final model for executive salaries.

TABLE 12.4

Independent Variables in the Executive Salary Example

INDEPENDENT VARIABLE	DESCRIPTION
x_1	Experience (years)—quantitative
x_2	Education (years)—quantitative
x_3	Sex (1 if male, 0 if female)—qualitative
x_4	Number of employees supervised—quantitative
x_5	Corporate assets (millions of dollars)—quantitative
x_6	Board member (1 if yes, 0 if no)—qualitative
x_7	Age (years)—quantitative
x_8	Company profits (past 12 months, millions of dollars)—quantitative
x_9	Has international responsibility (1 if yes, 0 if no)—qualitative
x_{10}	Company's total sales (past 12 months, millions of dollars)—quantitative

SOLUTION

We will use stepwise regression with the main effects of the ten independent variables to identify the most important variables. The dependent variable, y, is the natural logarithm of the executive salaries. The SAS stepwise regression printout is shown in Figure 12.31 (pages 725–727). Note that the first variable included in the model is x_4, Number of employees supervised. At the second step, x_5, Corporate assets, enters the model. At the sixth step, x_6, a dummy variable for the qualitative variable Board member or not, is brought into the model. However, because the significance (.2295) of the F statistic (SAS uses the $F = t^2$ statistic in the stepwise procedure rather than the t statistic) for x_6 is greater than the preassigned $\alpha = .10$, x_6 is removed from the model. Thus, at step 7 the procedure indicates that the five-variable model including x_1, x_2, x_3, x_4, and x_5 is best. That is, none of the other independent variables can meet the $\alpha = .10$ criterion for admission to the model.

Thus, in our final modeling effort (Section 11.8) we concentrated on these five independent variables, and determined that several second-order terms were important in the prediction of executive salaries. ∎

CASE STUDY 12.2

A STATISTICAL MODEL FOR LAND APPRAISAL

New factors and a lack of knowledge about the importance of factors that affect value continue to complicate the job of the rural appraiser. In order to provide knowledge on the subject, this article reports on and evaluates a study in which multiple linear regression equations were used to evaluate and quantify factors affecting value. . . . It is believed that the findings obtained with these equations, and the relationships they indicate, will be of value to the appraiser.

FIGURE 12.31 SAS Stepwise Regression Computer Printout for Example 12.11

STEP 1
VARIABLE X4 ENTERED   R-SQUARE = 0.42071677

	DF	SUM OF SQUARES	MEAN SQUARE	F	PROB>F
REGRESSION	1	11.46864285	11.46864285	71.17	0.0001
ERROR	98	15.79112802	0.16113396		
TOTAL	99	27.25977087			

	B VALUE	STD ERROR	F	PROB>F
INTERCEPT	10.20077500			
X4 (EMPLOYEES SUPERVISED)	0.00057284	0.00006790	71.17	0.0001

STEP 2
VARIABLE X5 ENTERED   R-SQUARE = 0.78299675

	DF	SUM OF SQUARES	MEAN SQUARE	F	PROB>F
REGRESSION	2	21.34431198	10.67215599	175.00	0.0001
ERROR	97	5.91545889	0.06098411		
TOTAL	99	27.25977087			

	B VALUE	STD ERROR	F	PROB>F
INTERCEPT	9.87702903			
X4 (EMPLOYEES SUPERVISED)	0.00058353	0.00004178	195.06	0.0001
X5 (ASSETS)	0.00183730	0.00014438	161.94	0.0001

STEP 3
VARIABLE X4 ENTERED   R-SQUARE = 0.89667614

	DF	SUM OF SQUARES	MEAN SQUARE	F	PROB>F
REGRESSION	3	24.44318616	8.14772872	277.71	0.0001
ERROR	96	2.81658471	0.02933942		
TOTAL	99	27.25977087			

(continued)

FIGURE 12.31 (continued)

	B VALUE	STD ERROR	F	PROB>F
INTERCEPT	9.66449288			
X1 (EXPERIENCE)	0.01870784	0.00182032	105.62	0.0001
X4 (EMPLOYEES SUPERVISED)	0.00055251	0.00002914	359.59	0.0001
X5 (ASSETS)	0.00191195	0.00010041	362.60	0.0001

STEP 4
VARIABLE X3 ENTERED

R-SQUARE = 0.94815717

	DF	SUM OF SQUARES	MEAN SQUARE	F	PROB>F
REGRESSION	4	25.84654710	6.46163678	434.37	0.0001
ERROR	95	1.41322377	0.01487604		
TOTAL	99	27.25977087			

	B VALUE	STD ERROR	F	PROB>F
INTERCEPT	9.40077349			
X1 (EXPERIENCE)	0.02074868	0.00131310	249.68	0.0001
X3 (SEX)	0.30011726	0.03089939	94.34	0.0001
X4 (EMPLOYEES SUPERVISED)	0.00055288	0.00002075	710.15	0.0001
X5 (ASSETS)	0.00190876	0.00007150	712.74	0.0001

STEP 5
VARIABLE X2 ENTERED

R-SQUARE = 0.96039323

	DF	SUM OF SQUARES	MEAN SQUARE	F	PROB>F
REGRESSION	5	26.18009940	5.23601988	455.87	0.0001
ERROR	94	1.07967147	0.01148587		
TOTAL	99	27.25977087			

	B VALUE	STD ERROR	F	PROB>F
INTERCEPT	8.85387930			
X1 (EXPERIENCE)	0.02141724	0.00116047	340.61	0.0001
X2 (EDUCATION)	0.03315807	0.00615303	29.04	0.0001
X3 (SEX)	0.31927842	0.02738298	135.95	0.0001
X4 (EMPLOYEES SUPERVISED)	0.00056061	0.00001829	939.84	0.0001
X5 (ASSETS)	0.00193684	0.00006304	943.98	0.0001

FIGURE 12.31   (continued)

STEP 6
VARIABLE X6 ENTERED   R-SQUARE = 0.96100666

	DF	SUM OF SQUARES	MEAN SQUARE	F	PROB>F
REGRESSION	6	26.19682148	4.36613691	382.00	0.0001
ERROR	93	1.06294939	0.01142956		
TOTAL	99	27.25977087			

	B VALUE	STD ERROR	F	PROB>F
INTERCEPT	8.87509152			
X1 (EXPERIENCE)	0.02133460	0.00115963	338.48	0.0001
X2 (EDUCATION)	0.03272195	0.00614851	28.32	0.0001
X3 (SEX)	0.31093801	0.02817264	121.81	0.0001
X4 (EMPLOYEES SUPERVISED)	0.00055820	0.00001835	925.32	0.0001
X5 (ASSETS)	0.00193764	0.00006289	949.31	0.0001
X6 (BOARD)	0.03866226	0.03196369	1.46	0.2295

STEP 7
VARIABLE X6 REMOVED   R-SQUARE = 0.96039323

	DF	SUM OF SQUARES	MEAN SQUARE	F	PROB>F
REGRESSION	5	26.18009940	5.23601988	455.87	0.0001
ERROR	94	1.07967147	0.01148587		
TOTAL	99	27.25977087			

	B VALUE	STD ERROR	F	PROB>F
INTERCEPT	8.85387930			
X1 (EXPERIENCE)	0.02141724	0.00116047	340.61	0.0001
X2 (EDUCATION)	0.03315807	0.00615303	29.04	0.0001
X3 (SEX)	0.31927842	0.02738298	135.95	0.0001
X4 (EMPLOYEES SUPERVISED)	0.00056061	0.00001829	939.84	0.0001
X5 (ASSETS)	0.00193684	0.00006304	943.98	0.0001

The authors of this statement, James O. Wise and H. Jackson Dover (1974), use stepwise regression to identify a number of important factors (variables) that can be used to predict rural property values. They obtained their results by analyzing a sample of 105 cases from seven counties in Georgia. Part of their findings are duplicated in Table 12.5. The variable names are listed in the order in which the stepwise regression procedure identified their importance, and the t values found at each step are given for each variable. Note that both qualitative and quantitative variables have been included. Since each qualitative variable is at two levels, only one main effect term could be included in the model for each factor.

Since there were 105 cases used in the study, a large number of degrees of freedom are associated with each t statistic (first 103, then 102, etc.). Thus, we should compare the value of the test statistic to a corresponding z value (1.645 for $\alpha = .10$ and the two-sided alternative hypothesis H_a: $\beta_i \neq 0$) when we judge the importance of each variable. Although Wise and Dover imply that the variable Size is important, we might not include it, since the t value is only 1.142.

TABLE 12.5

Stepwise Regression Analysis of Price per Acre

VARIABLE NAME	t VALUE
Residential land (yes–no)	10.466
Seedlings and saplings (number)	6.692
Percent ponds (percent)	4.141
Distance to state park (miles)	3.985
Branches or springs (yes–no)	3.855
Site index (ratio)	3.160
Size (acres)	1.142
Farmland (yes–no)	2.288

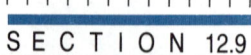

SECTION 12.9

MODEL BUILDING: RESIDUAL ANALYSIS AND TRANSFORMATIONS

In Section 11.10 we discussed the use of residual analysis to check the regression assumption that the random error component of the model is normally distributed with mean 0. In this section we return to the topic of residual analysis to check another of the assumptions—namely, the assumption of constant error variance over the range of the independent variable(s). You will see that this analysis sometimes leads to an extension of the model building process: transformation of the dependent variable.

Recall from Section 11.10 that a residual is defined as the difference between the actual and predicted values of y:

$$\text{Residual} = y - \hat{y}$$

Also, we showed that the mean of the residuals is equal to 0, and the standard deviation of the residuals is equal to the standard deviation of the least squares regression model, s. This sample standard deviation is used as an estimate of the standard deviation, σ, of the random error component, ε. The validity of our tests and confidence intervals based on the least squares estimators depends on the assumption that the standard deviation σ is constant over the range of the independent variable(s).

The variance of the error component is unlikely to be exactly constant in real applications of regression analysis, but moderate departures will not adversely affect the inferences based on the least squares model. However, it is important to detect significant departures, and to modify the model when we do. Residual plots are useful for checking this assumption. Because the variance of the dependent variable y often depends on its mean value, we plot the residuals against the estimated mean value, $\hat{y}$. Some common patterns of nonconstant variances and possible explanations for them are discussed in detail below.

Probably the most common nonconstant variance pattern in business applications is one in which the standard deviation of the residuals is proportional to the mean, as shown in Figure 12.32.

FIGURE 12.32
Residual Standard Deviation Proportional to $\hat{y}$

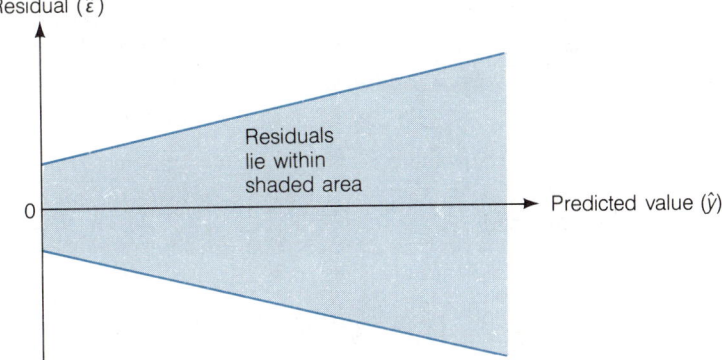

The pattern in Figure 12.32 occurs when the error component tends to increase as the mean increases. The **multiplicative model** is one that accounts for this tendency, where

$$y = E(y)\varepsilon \qquad \text{Multiplicative model}$$

rather than the **additive model** we have been using,

$$y = E(y) + \varepsilon \qquad \text{Additive model}$$

Assuming the standard deviation of ε is σ, the standard deviation of the additive model response is the constant σ, but the standard deviation of the *multiplicative model* response is

$$\text{Standard deviation of } y: \quad E(y)\sigma \qquad \text{Multiplicative model}$$

Thus, the standard deviation of the response increases in direct proportion to the mean $E(y)$ for the multiplicative model.

Applications involving responses measured in dollar units often require multiplicative models. For example, salary increases in organizations are often based on a percentage of the average salary within a rank or level of experience, which implies that both the salaries and the variation in salaries increase as the mean salary in the organization increases. Or, both the gross sales and the variation in the gross sales of a firm may tend to grow proportionally with the number of years the firm has been in business. In short, many of the responses in business applications exhibit a growth in variability proportional to the growth in the mean level of the response. If the plot of the residuals of the additive model for such a response looks like Figure 12.32, the multiplicative model may be more appropriate.

How should we proceed if we find a pattern of increasing residual variation and want to use a multiplicative model? Since the logarithm of a product is the sum of the logarithms, the multiplicative model can be written as

$$\log(y) = \log[E(y)] + \log(\varepsilon)$$

where we use "log" to refer to the natural logarithm. (However, any base can be used without affecting the conclusions of the analysis.) This **transformation** of the dependent variable puts the model back into additive form, so that we can use the least squares methodology. Different interpretations of parameter estimates are required, and the error assumptions now must be made about the random component $\log(\varepsilon)$. The following examples illustrate the use of a residual plot to detect the need for a multiplicative model, and the use of the log transform to fit it.

EXAMPLE 12.12

The data in Table 12.6 are executive salaries, y, and years of experience, x, for a sample of 50 mid-management executives from a major industry. Fit the second-order additive model

$$y = \beta_0 + \beta_1 x + \beta_2 x^2 + \varepsilon$$

and plot the residuals versus the estimated mean salary, $\hat{y}$.

TABLE 12.6
Salary and Experience Data for 50 Executives from a Major Industry

YEARS OF EXPERIENCE x	SALARY y	YEARS OF EXPERIENCE x	SALARY y	YEARS OF EXPERIENCE x	SALARY y
7	$26,075	21	$43,628	28	$99,139
28	79,370	4	16,105	23	52,624
23	65,726	24	65,644	17	50,594
18	41,983	20	63,022	25	53,272
19	62,309	20	47,780	26	65,343
15	41,154	15	38,853	19	46,216
24	53,610	25	66,537	16	54,288
13	33,697	25	67,447	3	20,844
2	22,444	28	64,785	12	32,586
8	32,562	26	61,581	23	71,235
20	43,076	27	70,678	20	36,530
21	56,000	20	51,301	19	52,745
18	58,667	18	39,346	27	67,282
7	22,210	1	24,833	25	80,931
2	20,521	26	65,929	12	32,303
18	49,727	20	41,721	11	38,371
11	33,233	26	82,641		

SOLUTION

We used Minitab to obtain the computer printout for the second-order additive model shown in Figure 12.33. The least squares prediction equation is

$$\hat{y} = 20{,}242 + 522x + 53x^2$$

The positive sign for the coefficient x^2 indicates that salaries are increasing at an increasing rate with experience. (Plot the equation over the relevant range of experience to see this.)

FIGURE 12.33
Minitab Regression
Printout: Second-Order
Additive Model

```
THE REGRESSION EQUATION IS
Y =   20242, +   522, X1 +   53,0 X2

                                  ST, DEV,   T-RATIO =
             COLUMN   COEFFICIENT  OF COEF,   COEF/S,D,

               --        20242,      4422,      4,58
       X1      C1          522,       617,      0,85
       X2      C3          53,0        19,6     2,71

THE ST, DEV, OF Y ABOUT REGRESSION LINE IS
S =       8123,
WITH (  50- 3) =   47 DEGREES OF FREEDOM

R-SQUARED = 81,6 PERCENT
R-SQUARED = 80,8 PERCENT, ADJUSTED FOR D,F,

ANALYSIS OF VARIANCE

   DUE TO       DF            SS        MS=SS/DF

REGRESSION     2    13723521024,   6861758464,
RESIDUAL      47     3101292032,     65984928,
TOTAL         49    16824811520,
```

The model appears to provide useful information for the prediction of salaries. The F statistic, computed as the ratio of MS(Regression) to MS(Residual), is

$$F = \frac{6{,}861{,}758{,}464}{65{,}984{,}928} = 103.99$$

This provides strong evidence of the model's usefulness. Further confirmation is given by $R^2 = .816$, the significance of the quadratic term ($t = 2.71$), and the estimate of standard deviation, $s = \$8{,}123$. The latter indicates that we would expect to be able to predict an executive's salary based on his or her experience to within (roughly) $\pm 2s \approx \$16{,}000$.

An experienced analyst might regard the increasing rate of increase in salaries and the concept that the prediction intervals are of about the same width ($\approx \$32{,}000$) for low- and high-salaried executives as tip-offs that an additive model is inappropriate. However, an examination of the residual plot in Figure 12.34 (page 732) is, perhaps, the first obvious indication of a potential problem. Note that the Minitab program plots **standardized residuals**—residuals divided by the standard deviation,

$s = \$8,123$. This makes it easy to determine the deviation of the residuals from their mean of 0, because the vertical axis is simply the number of standard deviations from the mean (much like a z value).

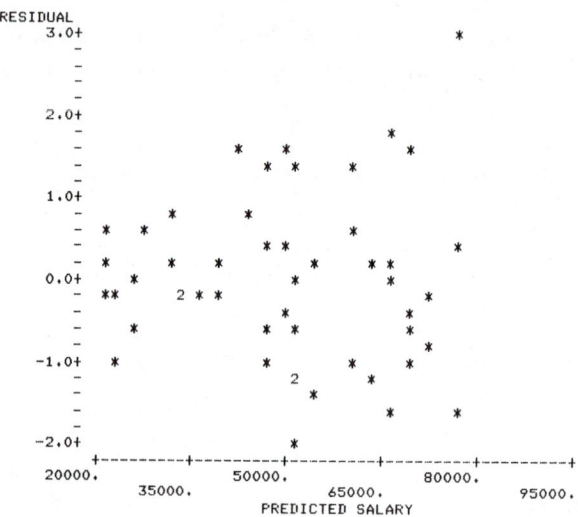

Note the funnel shape of the residuals: the residuals increase in variability as the predicted salary increases. Note that the residuals are all less than 1 standard deviation from the 0 mean when the estimated mean salary, $\hat{y}$, is low, but the range of residual variability stretches to ± 2 standard deviations (or more) for high estimated mean salaries. The residual plot indicates that a multiplicative model may be more appropriate.

EXAMPLE 12.13

Use a multiplicative model to relate salary to years of experience for the executive salary data in Table 12.6. Interpret and evaluate the adequacy of the model.

SOLUTION

We first hypothesize a second-order multiplicative model, so that the logarithmic transform of salary is related to the years of experience by

$$\log(y) = \beta_0 + \beta_1 x + \beta_2 x^2 + \varepsilon$$

where the random component ε represents a logarithmic transform of the multiplicative random error. The Minitab printout for this model is shown in Figure 12.35. The least squares equation is

$$\log(\hat{y}) = 9.8429 + .0497x + .000009x^2$$

This model appears to be useful for predicting log(salary), since the global F statistic is

$$F = \frac{3.6061}{.0243} = 148.40$$

FIGURE 12.35
Minitab Printout for
Second-Order
Multiplicative Model

```
THE REGRESSION EQUATION IS
Y =   9.84 +  0.0497 X1 + 0.0000 X2

                                  ST. DEV.   T-RATIO =
            COLUMN    COEFFICIENT   OF COEF.   COEF/S.D.
            --          9.8429       0.0848      116.08
    X1      C1          0.0497       0.0118        4.20
    X2      C3          0.000009     0.000375      0.03

THE ST. DEV. OF Y ABOUT REGRESSION LINE IS
S =       0.156
WITH (   50- 3) =   47 DEGREES OF FREEDOM

R-SQUARED = 86.4 PERCENT
R-SQUARED = 85.8 PERCENT, ADJUSTED FOR D.F.
ANALYSIS OF VARIANCE

   DUE TO       DF          SS       MS=SS/DF
REGRESSION      2        7.2123       3.6061
RESIDUAL       47        1.1400       0.0243
TOTAL          49        8.3523
```

and $R^2 = .864$, but the test statistic for the quadratic term is not significant ($t = .03$), which means that we cannot reject the null hypothesis that $\beta_2 = 0$. Before we try a simple straight-line multiplicative model, we first note that the logarithmic transformation appears to have solved the problem of increasing residual variation. The residual plot for the multiplicative model is shown in Figure 12.36, and the variation of the residuals is more uniform over the range of the estimated mean log(salary).

FIGURE 12.36
Minitab Standardized
Residual Plot for Second-
Order Multiplicative Model

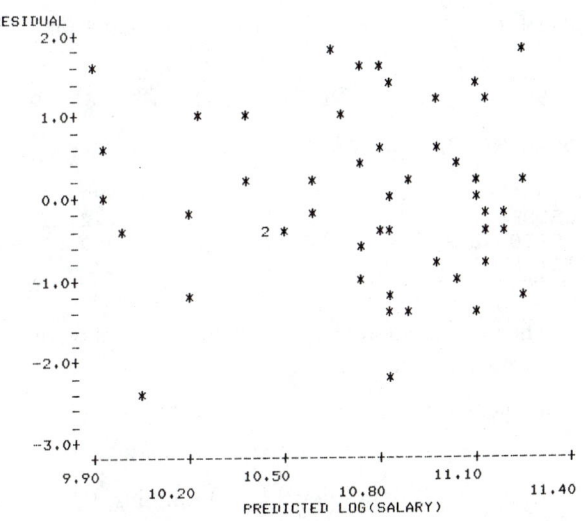

Since the quadratic term does not appear to contribute information for the prediction of log(salary), we fit the straight-line multiplicative model

$$\log(y) = \beta_0 + \beta_1 x + \varepsilon$$

The Minitab regression printout for this model is given in Figure 12.37. The least squares prediction equation is

$$\log(\hat{y}) = 9.8413 + .04998x$$

The F statistic to test the usefulness of the model

$$F = \frac{7.2122}{.0238} = 303.03$$

which indicates that the straight-line multiplicative model contributes information for the prediction of log(salary). The value of R^2, .864, implies that 86.4% of the variation in log(salary) around its mean is accounted for by the straight-line relationship with years of experience. This value of R^2 is not comparable to that for the additive model, because the measure of total variability around the mean, SS_{yy}, differs for the salary and transformed salary. Thus, the coefficients of determination are measuring proportions of different quantities.

FIGURE 12.37

Minitab Printout for First-Order Multiplicative Model

```
THE REGRESSION EQUATION IS
Y =     9.84 +0.0500 X1

                                          ST. DEV.    T-RATIO =
              COLUMN       COEFFICIENT    OF COEF.    COEF/S.D.
              --                9.8413      0.0564      174.63
     X1       C1              0.04998     0.00287       17.43

THE ST. DEV. OF Y ABOUT REGRESSION LINE IS
S =       0.154
WITH (   50- 2) =   48 DEGREES OF FREEDOM

R-SQUARED = 86.4 PERCENT
R-SQUARED = 86.1 PERCENT, ADJUSTED FOR D.F.

ANALYSIS OF VARIANCE

  DUE TO        DF         SS        MS=SS/DF
  REGRESSION     1      7.2122        7.2122
  RESIDUAL      48      1.1401        0.0238
  TOTAL         49      8.3523
```

The interpretation of a multiplicative model requires taking antilogarithms of the least squares equation:

$$\text{antilog}[\log(\hat{y})] = \hat{y}$$

$$= \text{antilog}(9.8413 + .04998x)$$

$$= \text{antilog}(9.8413) \cdot \text{antilog}(.04998x)$$

$$= (18{,}794)(1.0513)^x$$

This multiplicative equation tells us that the estimated salary for an executive with 0 years experience ($x = 0$) is \$18,794 (this value may have no practical meaning since no executives in the sample have 0 years of experience). More important, the second term tells us that each additional year of experience is worth an estimated *multiple* of 1.0513 of the previous year's salary. In other words, an executive's salary is estimated to increase at a compounded rate of 5.13% per year of experience.

If we use the model to estimate the mean salary for an executive with 10 years experience, we find

$$\hat{y} = 18,794(1.0513)^{10} = \$30,995$$

This is the antilog of the logarithmic mean, and is called the **geometric mean**; it differs from the arithmetic mean we are accustomed to estimating with additive models. If we also obtain a confidence interval for the log(salary) mean, and then take the antilog, the result is a confidence interval for the geometric mean. In general, the geometric mean is preferred as a measure of central tendency for distributions that are skewed to the right, since it is not affected as much as the arithmetic mean by extreme observations.

We can also use the multiplicative model to obtain a prediction of a particular executive's salary. The geometric mean $\hat{y}$ is the predicted value, and we can assess the reliability by placing a prediction interval on the estimate of log(y), and then taking antilogs of the two endpoints of the interval.

To obtain a rough idea of the multiplicative model's predictive reliability, note that the standard deviation of the model is $s = .154$. Then $\pm 2s = \pm .308$, and the antilog is

$$\text{antilog}(\pm .308) = (.73,\ 1.36)$$

The interpretation is that the actual salary of an executive will generally (with about 95% confidence) fall between 73% and 136% of the model's predicted value. This translates to an interval of (\$14,600, \$27,200) when the predicted salary is \$20,000, and an interval of (\$58,400, \$108,800) when the predicted salary is \$80,000. Notice how the width of the interval increases as the predicted salary increases, which is exactly the premise of and reason for the selection of a multiplicative rather than an additive model. Of course, these intervals are only rough approximations, and the precise prediction intervals are obtained by using the exact formulas. ∎

Although the multiplicative model, with the corresponding logarithmic transformation, is probably the most common solution to nonconstant residual variation in business applications, several others deserve mention. When the dependent variable is a *count* of something, such as the number of defects in workmanship an automobile dealer must repair while a new car is under warranty, or the number of customers per day at the drive-in window of a bank, the Poisson distribution (Section 5.5) often provides the best description of the response variable. Since the Poisson random variable has a standard deviation equal to the square root of its mean, the residual plot of a regression model fit to a count variable may appear like that in Figure 12.38 (page 736). The appropriate transformation for such a pattern is to use the square root of the response as the dependent variable. If the original response is Poisson, the square-root transform will have (approximately) constant variance.

FIGURE 12.38
Residual Pattern for Count
Response Variable:
Standard Deviation
Proportional to Square
Root of $\hat{y}$

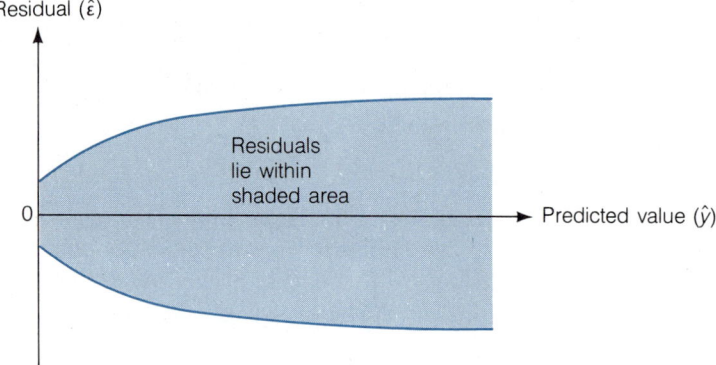

When the response is a proportion or percentage, such as the proportion of defective computer chips in a shipment, or the percentage of women employed on a university's business faculty, the binomial distribution is often appropriate. Recall that the standard deviation of a binomial random variable is proportional to $\sqrt{pq}$, where p represents the true mean proportion for the binomial random variable. Thus, if we plot the residual standard deviation against the mean proportion p over the interval 0 to 1, the result will have the "football" shape shown in Figure 12.39, with peak variability at $p = .5$, and minimal variability near 0 and 1. Unfortunately, a rather complex transformation is required to make the variability relatively uniform over the range of p. If the response variable is denoted by $\hat{p}$, the transformation is

$$y^* = \text{Transform}(\hat{p}) = \sin^{-1}(\sqrt{\hat{p}})$$

Because of the complexity, this transformation is rarely used in practice. If the range of the proportion response is relatively narrow (say, from .3 to .5, .7 to .9, or the like), then the residual variation will be relatively uniform (look at Figure 12.39 over these narrow ranges), and the transformation can be avoided. When the range of the response spans most of the interval from 0 to 1, the transformation may be necessary.

FIGURE 12.39
Risidual Pattern for
Binomial Response
Variable: Standard
Deviation Proportional to
$\sqrt{pq}$

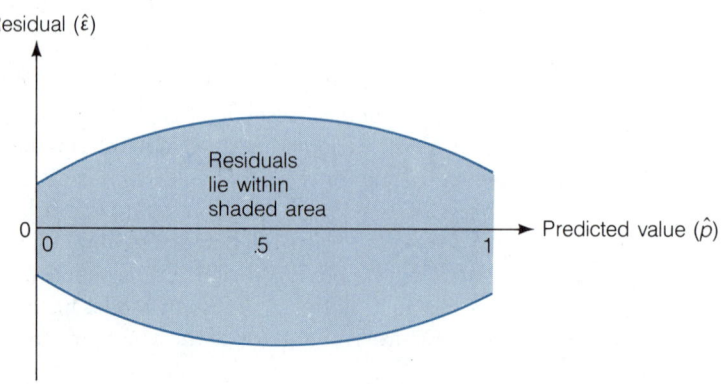

A summary of the patterns of residual variation and the corresponding transformation that will stabilize the residual variance is given in Table 12.7. Remember that transforming the dependent variable may stabilize the residual variation, but you will need to take special care with the interpretation of the least squares estimators and the corresponding inferences for which they are utilized.

TABLE 12.7 Stabilizing Transformations for Responses with Unequal Variances

TYPE OF RESPONSE	STANDARD DEVIATION PROPORTIONAL TO:	RESIDUAL PATTERN	STABILIZING TRANSFORMATION
Multiplicative	Mean response	$\hat{\varepsilon}$ vs $\hat{y}$	$\log(y)$
Count (Poisson)	Square root of mean response	$\hat{\varepsilon}$ vs $\hat{y}$	$\sqrt{y}$
Proportion (Binomial)	$\sqrt{(\text{Mean})(1 - \text{Mean})}$	$\hat{\varepsilon}$ vs $\hat{y}$	$\sin^{-1}\sqrt{y}$

LEARNING THE MECHANICS

12.78 Assume that a mean response $E(y)$ obeys the following multiplicative model:

$$E(y) = 1{,}000(.9)^x$$

a. Plot the model over the range 0 to 10.
b. Interpret the two parameters: 1,000 and .9.
c. The general form of such a model might be written

$$E(y) = \beta_0(\beta_1)^x$$

Interpret the parameters β_0 and β_1.

12.79 Give several reasons why the multiplicative model might be useful in business applications, and give several practical applications for which it might be useful.

12.80 What is the difference between the shapes of residual plots corresponding to a multiplicative response and a count (Poisson) response?

12.81 Suppose $\hat{p}$ is the sample proportion of successes for a binomial experiment with $n = 100$ trials.

a. What is the formula for the standard deviation of $\hat{p}$ in terms of the true probability of success p?

b. Calculate the standard deviation for values of p equal to .1, .2, . . . , .9.

c. Using the horizontal axis to represent the values of p between 0 and 1, plot a ± 2 standard deviation interval for the estimate $\hat{p}$ on each side of the 0 line on the vertical axis. Connect the points using a smooth curve. The plot should have the "football" shape discussed in this section.

d. Why do we often need to transform responses that are binomial sample proportions?

APPLYING THE CONCEPTS

[Note: Starred (*) exercises require the use of a computer.]

12.82 State highway agencies regularly invite contractors to submit bids for the construction of new highways. Typical contracts consist of a number of items, and each bidder must quote a cost to the state on an item-by-item basis. One item on most bid specifications is HBP (hot bituminous pavement). The accompanying table contains the winning HBP unit price bids (in dollars/ton) for 30 contracts let over the past 6 months. The total quantity of HBP specified and the number of bidders are also shown for each contract.

UNIT PRICE (per ton)	QUANTITY (thousands of tons)	NUMBER OF BIDDERS	UNIT PRICE (per ton)	QUANTITY (thousands of tons)	NUMBER OF BIDDERS
$153.32	1.0	4	$151.13	1.6	7
74.11	7.2	10	79.18	7.3	5
29.72	16.7	5	204.94	.2	9
54.67	11.9	5	81.06	6.8	4
68.39	9.3	4	37.62	11.4	8
119.04	3.7	5	17.13	20.0	3
116.14	1.7	6	37.81	13.4	4
146.49	.1	9	130.72	1.8	2
81.81	7.8	5	26.07	18.5	2
19.58	18.4	9	39.59	14.7	5
141.08	2.9	5	66.20	9.1	4
101.72	4.7	10	160.25	2.4	5
24.88	17.4	10	19.39	17.3	9
19.43	18.4	4	86.60	4.5	9
39.63	11.2	9	47.27	11.6	9

a. Assuming that economies of scale are likely to be present in the supply of HBP for specific contracts, what is the expected relationship between unit price and quantity?

b. Let y represent the unit price, x_1 the quantity, and x_2 the number of bidders. SAS was used to fit the additive model

$$y = \beta_0 + \beta_1 x_1 + \varepsilon$$

with the results shown in the accompanying printout. Interpret the least squares estimates, and evaluate the model's usefulness for predicting unit price. Be sure to interpret the estimated standard deviation of the model.

SAS Printout for Additive Model in Exercise 12.82

Dep Variable: Y

Analysis of Variance

Source	DF	Sum of Squares	Mean Square	F Value	Prob>F
Model	1	69039.20554	69039.20554	196.615	0.0001
Error	28	9831.89259	351.13902		
C Total	29	78871.09814			

Root MSE	18.73870	R-Square	0.8753
Dep Mean	79.16567	Adj R-Sq	0.8709
C.V.	23.67024		

Parameter Estimates

Variable	DF	Parameter Estimate	Standard Error	T for H0: Parameter=0	Prob > |T|
INTERCEP	1	148.055230	5.98682026	24.730	0.0001
X1	1	-7.570282	0.53988803	-14.022	0.0001

c. The residual plot for this model, with the predicted unit price on the horizontal axis, is shown below. What does the pattern of residuals suggest about the assumption that the standard deviation of the error component is constant?

Residual Plot for Additive Model in Exercise 12.82

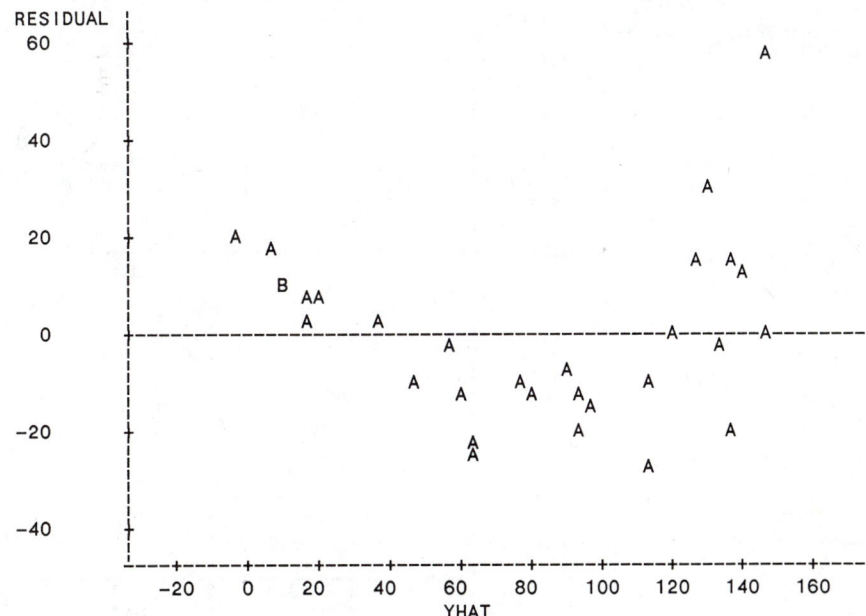

Plot of RESIDUAL*YHAT   Legend: A = 1 obs, B = 2 obs, etc.

12.83 Refer to Exercise 12.82. We now fit a multiplicative model by using the logarithmic transformation of unit price in the following model:

$$\log(y) = \beta_0 + \beta_1 x + \varepsilon$$

The results are reproduced in the accompanying SAS printout.

SAS Printout for Multiplicative Model in Exercise 12.83

Dep Variable: LOG_Y

Analysis of Variance

Source	DF	Sum of Squares	Mean Square	F Value	Prob>F
Model	1	15.59165	15.59165	799.634	0.0001
Error	28	0.54596	0.01950		
C Total	29	16.13761			

Root MSE	0.13964	R-Square	0.9662	
Dep Mean	4.12870	Adj R-Sq	0.9650	
C.V.	3.38210			

Parameter Estimates

Variable	DF	Parameter Estimate	Standard Error	T for H0: Parameter=0	Prob > |T|
INTERCEP	1	5.163967	0.04461255	115.751	0.0001
X1	1	-0.113765	0.00402313	-28.278	0.0001

Residual Plot for Multiplicative Model in Exercise 12.83

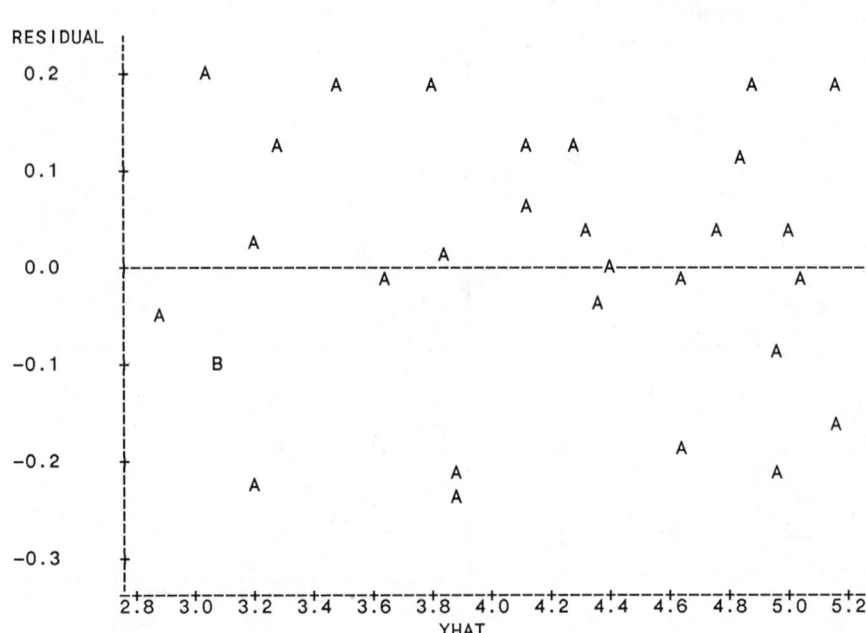

Plot of RESIDUAL*YHAT   Legend: A = 1 obs, B = 2 obs, etc.

a. Interpret the least squares estimates, and evaluate the usefulness of the model for predicting unit price. Be sure to interpret the estimated standard deviation of the model.

b. The residual plot of the multiplicative model is also shown, with the predicted value of log(unit price) on the horizontal axis. Evaluate the assumption that the error component of the transformed model has constant standard deviation.

12.84 Refer to Exercises 12.82 and 12.83. We now want to add the number of bidders, x_2, to the model for unit price.

a. Assuming that x_2 is a measure of competition for a contract, what relationship is expected between unit price and number of bidders?

b. The following model is fit to the data:

$$\log(y) = \beta_0 + \beta_1 x_1 + \beta_2 x_2 + \varepsilon$$

and the SAS printout is shown below. Interpret the least squares estimates, and test the usefulness of the model for predicting unit price.

SAS Printout for Multiplicative Model in Exercise 12.84

Dep Variable: LOG_Y

Analysis of Variance

Source	DF	Sum of Squares	Mean Square	F Value	Prob>F
Model	2	15.67976	7.83988	462.333	0.0001
Error	27	0.45784	0.01696		
C Total	29	16.13761			

Root MSE	0.13022	R-Square	0.9716	
Dep Mean	4.12870	Adj R-Sq	0.9695	
C.V.	3.15401			

Parameter Estimates

Variable	DF	Parameter Estimate	Standard Error	T for H0: Parameter=0	Prob > |T|
INTERCEP	1	5.304182	0.07425955	71.428	0.0001
X1	1	-0.114461	0.00376421	-30.408	0.0001
X2	1	-0.021711	0.00952441	-2.280	0.0308

c. Test to determine whether the data support the hypothesis that the mean unit price decreases as the number of bidders increases. Use $\alpha = .05$.

d. With what approximate precision will this model predict the unit price of HBP?

*12.85 In Exercises 11.11 and 12.41, a real estate appraiser used regression analysis to explore the relationship between the sale prices of apartments and various characteristics of the apartments. A portion of the data is repeated in the table on page 742.

a. Fit a first-order additive model relating sale price to gross area of the building. Interpret the least squares estimates, and evaluate the usefulness of the model.

CODE NO.	SALE PRICE y ($)	GROSS BUILDING AREA sq. ft.	CODE NO.	SALE PRICE y ($)	GROSS BUILDING AREA sq. ft.
0229	90,300	4,266	0095	173,200	4,461
0094	384,000	14,391	0121	323,650	9,000
0043	157,500	6,615	0077	162,500	3,828
0079	676,200	34,144	0060	353,500	13,680
0134	165,000	6,120	0174	134,400	4,680
0179	300,000	14,552	0084	187,000	7,392
0087	108,750	3,040	0031	155,700	6,030
0120	276,538	7,881	0019	93,600	3,840
0246	420,000	12,600	0074	110,000	3,092
0025	950,000	39,448	0057	573,200	23,704
0015	560,000	30,000	0104	79,300	3,876
0131	268,000	8,088	0024	272,000	9,542
0172	290,000	11,315			

Source: Robinson Appraisal Co., Inc., Mankato, Minnesota.

b. Plot the residuals of the model you fit in part **a**, with the estimated sale price, $\hat{y}$, on the horizontal axis. Evaluate the assumption that the error component has constant standard deviation.

c. Use the logarithmic transformation of sale price to fit the first-order multiplicative model

$$\log(y) = \beta_0 + \beta_1 x + \varepsilon$$

Interpret the least squares estimates, and evaluate the usefulness of the model.

d. Plot the residuals of the multiplicative model against the predicted log(sale price). Evaluate the assumption that the error component of the multiplicative model has a constant standard deviation.

e. Compare the approximate precision with which the additive and multiplicative models can be used to predict sale price.

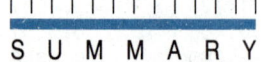

S U M M A R Y

Although this chapter provides only an introduction to the very important topic of **model building**, it enables you to construct many interesting and useful models for business phenomena. You can build on this foundation and, with experience, develop competence in this fascinating area of statistics. Successful model building requires a delicate blend of knowledge of the process being modeled, geometry, and formal statistical testing.

The first step in model building is to identify the response variable, y, and a set of independent variables. Each independent variable is then classified as either **quantitative** or **qualitative**, and **dummy variables** are defined to represent the qualitative independent variables. If the total number of independent variables is large, you may want to use **stepwise regression** to screen out those that do not seem important for the prediction of y.

When the number of independent variables is manageable, the model builder is ready to begin a systematic effort. At least **second-order models**, those containing **two-way interactions** and **quadratic terms** in the quantitative variables, should be considered. Remember that a model with no interaction terms implies that each of the independent variables affects the response independently of the other independent variables. Quadratic terms add curvature to the contour curves when $E(y)$ is plotted as a function of the independent variable. The F test for testing a set of β parameters aids in deciding the final form of the prediction model.

Many problems can arise in regression modeling, and the intermediate steps are often tedious and frustrating. However, the end result of a careful and determined modeling effort is very rewarding—you will have a better understanding of the process generating the dependent variable y and a predictive model for y.

SUPPLEMENTARY EXERCISES
12.86–12.104

[*Note: Starred (*) exercises require the use of a computer.*]

12.86 Investors are interested in knowing the relationship between the behavior of a mutual fund and the behavior of the stock market as a whole. Researchers in finance have hypothesized that the model that appropriately characterizes this relationship is

$$E(y) = \beta_0 + \beta_1 x$$

where

  $y = $ Monthly rate of return of a mutual fund

  $x = $ Monthly rate of return of the stock market as a whole as measured by the monthly rate of return to a market index such as Standard & Poor's 500 Composite Index

The value of β_1 in the above model is referred to as the mutual fund's *beta coefficient*. Assuming the above model is true, investors can predict how the returns of an individual mutual fund will react to changes in the behavior of the market. For example, if $\beta_1 > 1$, the implication is that the return to the mutual fund will be greatly influenced by the behavior of the market and will move in the same direction as the change in the market return. If $0 \le \beta_1 < 1$, the return to the mutual fund will be less sensitive to changes in market behavior but will also move in the same direction as the change in the market return.

In a recent study, Alexander and Stover (1980) included a dummy variable in the above model to determine whether the beta coefficient for an individual mutual fund depends on whether the market is moving generally upward (a *bull market*) or generally downward (a *bear market*).

a. Modify the above regression model (as Alexander and Stover did) to reflect the possibility that $E(y)$ may depend on whether the market is bullish or bearish. Include an interaction term in your model and carefully define the dummy variable coding scheme you use.

b. Using the model you developed in part **a**, describe the differences that may exist between the response curves of $E(y)$ under bull and bear markets.

c. Specify the hypothesis you would test to determine whether a mutual fund's beta coefficient is different during bull and bear markets.

d. Specify the hypothesis you would test to determine whether $E(y)$ should be characterized as $E(y) = \beta_0 + \beta_1 x$ or as in the modified model you developed in part **a**.

12.87 The audience for a product's advertising can be divided into four segments according to the degree of exposure received as a result of the advertising. These segments are groups of consumers who receive very high (VH), high (H), medium (M), or low (L) exposure to the advertising. A company is interested in exploring whether its advertising effort affects its product's market share. Accordingly, the company identifies 24 sample groups of consumers who have been exposed to its advertising, six groups at each exposure level. Then, the company determines its product's market share within each group.

    a. Write a regression model that expresses the company's market share as a function of advertising exposure level. Define all terms in your model, and list any assumptions you make about them.

    b. Did you include interaction terms in your model? Why or why not?

    c. How many degrees of freedom are associated with the F test you would use to test the overall usefulness of the model you constructed in part **a**?

*12.88 In Exercise 10.60 the number of managers in a firm was related to the monthly sales using simple regression. The data are repeated in the table.

DATE	MONTHLY SALES x (units)	NUMBER OF MANAGERS y	DATE	MONTHLY SALES x (units)	NUMBER OF MANAGERS y
3/83	5	10	9/85	30	22
6/83	4	11	12/85	31	25
9/83	8	10	3/86	36	30
12/83	7	10	6/86	38	30
3/84	9	9	9/86	40	31
6/84	15	10	12/86	41	31
9/84	20	11	3/87	51	32
12/84	21	17	6/87	40	30
3/85	25	19	9/87	48	32
6/85	24	21	12/87	47	32

    a. Plot the estimated first-order model on a scattergram of the data. Does it appear that a second-order model would provide more information for the prediction of y than the first-order model? Explain.

    b. Fit the following second-order model to the data:

$$y = \beta_0 + \beta_1 x + \beta_2 x^2 + \varepsilon$$

    c. Test the usefulness of the model described in part **b**. Use $\alpha = .10$. Draw the appropriate conclusions in the context of the problem.

    d. In stating your conclusions in part **c**, are you risking a Type I error, a Type II error, or both? Explain.

    e. Plot the least squares model of part **b** on a scattergram of the data. Using this plot, the plot of part **a**, and any other information or tests you believe are relevant, comment on the appropriateness of using the first-order model versus the second-order model for predicting demand for managers.

12.89 One of the distinguishing features of banks as compared with other financial service institutions is their well-developed system for permitting convenient, personal access to the services they offer. One indication of the demand for these services is the number of trips individuals make to their bank each year. Murphy and Stock (1983) employed multiple regression analysis to investigate the determinants of household trips to the bank in the state of Oklahoma. In the summer of 1979, personal interviews were conducted with a random sample of 597 residents of Oklahoma. The interviewees were asked questions about their yearly banking activities, including number of trips to the bank per year and number of miles to the bank. At the same time, data were collected on other variables, such as number of cars in the household, kind of work (if any) done by the interviewee, etc. The method of least squares was used to develop the following model:

$$\hat{y} = 18.40 + 2.02x_1 - .254x_2 + 1.65x_3 + 1.124x_4 - .06x_5 + .00017x_6 - 8.679x_7 - 3.21x_8 - 1.29x_9 - 11.59x_{10} + .092x_{11}$$
$$\quad\quad (3.696)\ (-3.788)\ (1.637)\ (1.138)\ (-.63)\ (2.15)\ (-3.052)\ (-.958)\ (-.407)\ (-2.705)\ (3.239)$$

$$R^2 = .1836$$
$$F = 11.96$$

where

y = Number of trips to the bank per year

x_1 = Number of people in household

x_2 = Miles to bank

x_3 = Number of cars in household

x_4 = Years of education

x_5 = Miles to work

x_6 = Total family income

$x_7 = \begin{cases} 1 & \text{if not employed} \\ 0 & \text{otherwise} \end{cases}$ $x_8 = \begin{cases} 1 & \text{if white-collar job} \\ 0 & \text{otherwise} \end{cases}$

$x_9 = \begin{cases} 1 & \text{if blue-collar job} \\ 0 & \text{otherwise} \end{cases}$ $x_{10} = \begin{cases} 1 & \text{if farm-related job} \\ 0 & \text{otherwise} \end{cases}$

x_{11} = Number of shopping trips per year for purposes other than banking

The numbers in parentheses are the t statistics associated with the coefficient estimates above them.

a. Identify which independent variables in Murphy and Stock's model are qualitative and which are quantitative.

b. Murphy and Stock use four dummy variables to describe the kind of work done by the sample of Oklahomans. How many different levels can the variable, kind of work, assume? List them.

c. Interpret the value of R^2 in the context of the problem.

d. Test the usefulness of Murphy and Stock's model for explaining the variation in y. Use $\alpha = .01$.

e. Specify the null and alternative hypotheses you would use to test whether the trips-to-the-bank response surface is the same regardless of the kind of work done by the interviewee.

12.90 A fast-food restaurant chain is interested in modeling the mean weekly sales of a restaurant, $E(y)$, as a function of the weekly traffic flow on the street where the restaurant is located and the city in which the restaurant is located. The table contains data that were collected on 24 restaurants in four cities. The model that has been proposed is

$$E(y) = \beta_0 + \beta_1 x_1 + \beta_2 x_2 + \beta_3 x_3 + \beta_4 x_4$$

where

$x_1 = $ Traffic flow

$$x_2 = \begin{cases} 1 & \text{if city 1} \\ 0 & \text{otherwise} \end{cases} \qquad x_3 = \begin{cases} 1 & \text{if city 2} \\ 0 & \text{otherwise} \end{cases} \qquad x_4 = \begin{cases} 1 & \text{if city 3} \\ 0 & \text{otherwise} \end{cases}$$

CITY	TRAFFIC FLOW (thousands of cars)	WEEKLY SALES y ($ thousands)	CITY	TRAFFIC FLOW (thousands of cars)	WEEKLY SALES y ($ thousands)
1	59.3	6.3	3	75.8	8.2
1	60.3	6.6	3	48.3	5.0
1	82.1	7.6	3	41.4	3.9
1	32.3	3.0	3	52.5	5.4
1	98.0	9.5	3	41.0	4.1
1	54.1	5.9	3	29.6	3.1
1	54.4	6.1	3	49.5	5.4
1	51.3	5.0	4	73.1	8.4
1	36.7	3.6	4	81.3	9.5
2	23.6	2.8	4	72.4	8.7
2	57.6	6.7	4	88.4	10.6
2	44.6	5.2	4	23.2	3.3

A portion of the computer printout that results from fitting the model to the data in the table is shown at the top of page 747. The reduced model, $E(y) = \beta_0 + \beta_1 x_1$, was also fit to the same data and the resulting computer printout is partially reproduced below:

SOURCE	DF	SUM OF SQUARES	MEAN SQUARE
MODEL	1	111.3423	111.3423
ERROR	22	7.8073	0.3549
CORRECTED TOTAL	23	119.1496	R-SQUARE
			0.934

a. Test the null hypothesis that $\beta_1 = \beta_2 = \beta_3 = \beta_4 = 0$ using $\alpha = .05$. Interpret the results of your test.

b. Is mean weekly sales, $E(y)$, dependent on the city where a restaurant is located? Test using $\alpha = .05$. Interpret the results of your test.

c. Describe the nature of the response lines that the complete model, $E(y) = \beta_0 + \beta_1 x_1 + \beta_2 x_2 + \beta_3 x_3 + \beta_4 x_4$, would generate. Does the model imply interaction between city and traffic flow?

d. Use the prediction equation based on the complete model to graph the response lines that relate predicted weekly sales, $\hat{y}$, to traffic flow, x_1 (for each of the four cities). Do the graphed response lines suggest an interaction between city and traffic flow?

SAS Printout for Exercise 12.90

Dep Variable: Y

Analysis of Variance

Source	DF	Sum of Squares	Mean Square	F Value	Prob>F
Model	4	116.65552	29.16388	222.173	0.0001
Error	19	2.49407	0.13127		
C Total	23	119.14958			

Root MSE	0.36231	R-Square	0.9791	
Dep Mean	5.99583	Adj R-Sq	0.9747	
C.V.	6.04265			

Parameter Estimates

Variable	DF	Parameter Estimate	Standard Error	T for H0: Parameter=0	Prob > ¦T¦
INTERCEP	1	1.083388	0.32100795	3.375	0.0032
X1	1	0.103673	0.00409449	25.320	0.0001
X2	1	-1.215762	0.20538681	-5.919	0.0001
X3	1	-0.530757	0.28481946	-1.863	0.0779
X4	1	-1.076525	0.22650014	-4.753	0.0001

e. Write a model that includes interaction between city and traffic flow.

*f. Fit the model of part e to the data.

*g. Do the data provide sufficient evidence to indicate that the slopes of the lines differ for at least two of the four cities? Test using $\alpha = .05$.

12.91 One factor that must be considered in developing a shipping system that is beneficial to both the customer and the seller is time of delivery. A manufacturer of farm equipment can ship its products by either rail or truck. Quadratic models are thought to be adequate in relating time of delivery to distance traveled for both modes of transportation. Consequently, it has been suggested that the following model be fit:

$$E(y) = \beta_0 + \beta_1 x_1 + \beta_2 x_1^2 + \beta_3 x_2 + \beta_4 x_1 x_2 + \beta_5 x_1^2 x_2$$

where

$y = $ Shipping time

$x_1 = $ Distance to be shipped $x_2 = \begin{cases} 1 & \text{if rail} \\ 0 & \text{if truck} \end{cases}$

a. What hypothesis would you test to determine whether the data indicate that the quadratic distance terms are useful in the model—i.e., whether curvature is present in the relationship between mean delivery time and distance?

b. What hypothesis would you test to determine whether there is a difference in mean delivery time by rail and by truck?

12.92 Refer to Exercise 12.91. Suppose the model is fit to a total of 50 observations on delivery time. The sum of squared errors is SSE = 226.12. Then, the reduced model

$$E(y) = \beta_0 + \beta_1 x_1 + \beta_2 x_1^2$$

is fit to the same data, and SSE = 259.34. Test whether the data indicate that the mean delivery time differs for rail and truck deliveries. Use $\alpha = .05$.

*12.93 In Exercise 12.87, a company was concerned about the relationship between its market share and its advertising effort. The data in the table were obtained by the company.

MARKET SHARE WITHIN SAMPLE GROUP	ADVERTISING EXPOSURE LEVEL	MARKET SHARE WITHIN SAMPLE GROUP	ADVERTISING EXPOSURE LEVEL	MARKET SHARE WITHIN SAMPLE GROUP	ADVERTISING EXPOSURE LEVEL
10.1	L	11.2	M	11.9	H
10.3	L	10.9	M	12.9	H
10.0	L	10.8	M	10.7	VH
10.3	L	11.0	M	10.8	VH
10.2	L	12.2	H	11.0	VH
10.5	L	12.1	H	10.5	VH
10.6	M	11.8	H	10.8	VH
11.0	M	12.6	H	10.6	VH

a. Fit the model you constructed in part **a** of Exercise 12.87 to the data.
b. Is there evidence to suggest that the firm's expected market share differs for different levels of advertising exposure? Test using $\alpha = .05$.

12.94 To make a product more appealing to the consumer, an automobile manufacturer is experimenting with a new type of paint that is supposed to help the car maintain its new-car look. The durability of this paint depends on the length of time the car body is in the oven after its has been painted. In the initial experiment, three groups of ten car bodies each were baked for three different lengths of time—12, 24, and 36 hours— at the standard temperature setting. Then, the paint finish of each of the 30 cars was analyzed to determine a durability rating, y.
a. Write a second-order model relating the mean durability, $E(y)$, to the length of baking.
b. Could a third-order model be fit to the data? Explain.

12.95 Refer to Exercise 12.94. Suppose the Research and Development Department develops three new types of paint to be tested. Thus, 90 cars are to be tested—30 for each type of paint—in the manner described in Exercise 12.94. Write a model that describes $E(y)$ as a function of the type of paint and bake time. Assume that the independent variables interact.

*12.96 In Exercise 12.86, we discussed the relationship between the behavior of an individual mutual fund and the behavior of the stock market as a whole. The table lists the monthly rates of return for the Dreyfus Fund (a mutual fund) and the monthly rates of return for Standard & Poor's 500 Composite Index (S&P) for the period January 1966 to December 1971. The bear market periods were from January 1966 through September 1966 and from December 1968 through May 1970. The bull market periods were from October 1966 through November 1968 and from June 1970 through December 1971 (Alexander and Stover, 1980).

TIME PERIOD	RETURNS Dreyfus	S&P	TIME PERIOD	RETURNS Dreyfus	S&P	TIME PERIOD	RETURNS Dreyfus	S&P	TIME PERIOD	RETURNS Dreyfus	S&P
1/66	.008	.006	7/67	.073	.047	1/69	−.001	−.007	7/70	.051	.075
2/66	.067	−.013	8/67	−.019	−.007	2/69	−.070	−.043	8/70	.051	.051
3/66	−.008	−.021	9/67	.010	.034	3/69	.015	.036	9/70	.047	.035
4/66	.021	.022	10/67	−.029	−.028	4/69	.014	.023	10/70	−.026	−.010
5/66	−.074	−.049	11/67	.011	.007	5/69	−.003	.003	11/70	.040	.054
6/66	.024	−.015	12/67	.025	.028	6/69	−.066	−.054	12/70	.046	.058
7/66	−.011	−.012	1/68	−.063	−.043	7/69	−.059	−.059	1/71	.044	.042
8/66	.099	−.073	2/68	−.042	−.026	8/69	.057	.045	2/71	.025	.014
9/66	−.011	−.005	3/68	.022	.011	9/69	−.001	−.024	3/71	.031	.038
10/66	.015	.049	4/68	.100	.083	10/69	.050	.046	4/71	.035	.038
11/66	.079	.010	5/68	.021	.016	11/69	−.027	−.030	5/71	−.034	−.037
12/66	.008	.000	6/68	.001	.011	12/69	−.027	−.018	6/71	−.008	.002
1/67	.086	.080	7/68	−.038	−.017	1/70	−.083	−.074	7/71	−.026	−.040
2/67	.010	.007	8/68	.021	.016	2/70	−.044	.059	8/71	.045	.041
3/67	.041	.041	9/68	.056	.040	3/70	−.007	.003	9/71	−.028	−.006
4/67	.048	.044	10/68	.010	.009	4/70	−.110	−.089	10/71	−.039	−.040
5/67	−.039	−.048	11/68	.062	.053	5/70	−.060	−.055	11/71	.019	.003
6/67	.027	.019	12/68	−.025	−.040	6/70	−.042	−.048	12/71	.075	.090

Sources: Standard & Poor's Composite Index returns from Ibbotson, R. G. and Singuefield, R. A., *Stocks, Bonds, Bills, and Inflation: The Past (1926–1976) and the Future (1977–2000)*, Financial Analysts Research Foundation, 1977; and Dreyfus returns from *The Wall Street Journal*.

a. Fit the model you developed in part a of Exercise 12.86 to the data shown in the table.

b. Using the fitted model, estimate the Dreyfus Fund's beta coefficient for bull markets. Estimate the corresponding parameter for bear markets. Describe the relative responsiveness of the mutual fund to the market during bullish and bearish periods.

c. Test the hypothesis you specified in part c of Exercise 12.86. Draw the appropriate conclusions. Test using $\alpha = .05$.

d. Test the hypothesis you specified in part d of Exercise 12.86. Draw the appropriate conclusions. Test using $\alpha = .05$.

12.97 To model the relationship between y, a dependent variable, and x, an independent variable, a researcher has taken one measurement on y at each of five different x values. Drawing on his mathematical expertise, the researcher realizes that he can fit the fourth-order polynomial model

$$E(y) = \beta_0 + \beta_1 x + \beta_2 x^2 + \beta_3 x^3 + \beta_4 x^4$$

and it will pass exactly through all five points, yielding SSE = 0. The researcher, delighted with the "excellent" fit of the model, eagerly sets out to use it to make inferences. What problems will he encounter in attempting to make inferences?

12.98 Due to an increase in gasoline prices, many service stations are offering self-service gasoline at reduced prices. Suppose an oil company wants to model the mean monthly gasoline sales, $E(y)$, of its affiliated stations as a function of the type of service they offer: self-service, full service, or both.

a. How many dummy variables will be needed to describe the qualitative independent variable Type of service?

b. Write the main effects model relating $E(y)$ of the type of service. Describe the coding of the dummy variables.

*12.99 Refer to Exercises 12.20, 12.74, and 12.75, in which we related the number of annual highway deaths, y, to the number of licensed vehicles, x_1, and the presence or absence of the 55-mile-per-hour national speed limit ($x_2 = 1$ for years in which the law was in effect, $x_2 = 0$ for years in which it was not in effect). The data are given on page 660.

a. Fit the complete second-order model from Exercise 12.75 to the data, and plot the residuals against the predicted value of y. Evaluate the assumption that the standard deviation of the error component is constant.

b. Since the response variable, the number of highway deaths per year, is a "count" variable, the square-root transformation might be utilized to stabilize the variance of the random error. Perform this transformation and fit the second-order model to the transformed data.

c. Fit a first-order model to these data, and test to determine whether the second-order terms are useful for predicting the number of highway deaths. Use $\alpha = .05$.

d. Fit a model that does not include any terms involving the speed-limit dummy variable, x_2. Is there evidence that the change in the speed-limit law is associated with a change in the number of highway deaths? Use $\alpha = .05$.

e. Evaluate the approximate precision with which the transformed and untransformed models can be used to predict the number of highway deaths in a given year. Remember that the standard deviation of the transformed regression model must be squared ("untransformed") before you can interpret it.

f. Plot the residuals from the complete second-order model against the predicted square root of y. How does this residual plot compare with that in part **a**? Do you think the transformation is advisable in this case?

*12.100 In Case Study 11.2, we described the first-order regression model developed by Carolyn I. Allmon for predicting the sales of Crest toothpaste. The data she used to estimate the model are presented in the table. Using these data and the procedures you learned in this chapter, attempt to develop a second-order model that, according to the appropriate F test, explains more of the variation in sales than Allmon's model. Describe each step of your model building process.

YEAR	CREST SALES ($ thousand)	ADVERTISING BUDGET ($ thousand)	ADVERTISING RATIO	INCOME ($ billion)	YEAR	CREST SALES ($ thousand)	ADVERTISING BUDGET ($ thousand)	ADVERTISING RATIO	INCOME ($ billion)
1966	86,250	—	—	—	1974	126,000	18,250	1.27	998.3
1967	105,000	16,300	1.25	547.9	1975	162,000	17,300	1.07	1,096.1
1968	105,000	15,800	1.34	593.4	1976	191,625	23,000	1.17	1,194.4
1969	121,600	16,000	1.22	638.9	1977	189,000	19,300	1.07	1,311.5
1970	113,750	14,200	1.00	695.3	1978	210,000	23,056	1.54	1,462.9
1971	113,750	15,000	1.15	751.8	1979	224,250	26,000	1.59	1,641.7
1972	128,925	14,000	1.13	810.3	1980	245,000	28,000	1.56	1,821.7
1973	142,500	15,400	1.05	914.5					

Source: Allmon, C. I. "Advertising and sales relationships for toothpaste: Another look." *Business Economics*, Sept. 1982, 17, 58. Published by the National Association of Business Economists.

12.101 Many companies must accurately estimate their costs before a job is begun in order to acquire a contract and make a profit. A heating and plumbing contractor, for example, may base cost estimates for new homes on the total area of the house and whether central air conditioning is to be installed.

   a. Write a main effects model relating the mean cost of material and labor, $E(y)$, to the area and central air conditioning variables.

   b. Write a complete second-order model for the mean cost as a function of the same two variables.

   c. What hypothesis would you test to determine whether the second-order terms are useful for predicting mean cost?

   d. Explain how you would compute the F statistic needed to test the hypothesis of part **c**.

*12.102 In Exercises 11.63 and 12.65, models for characterizing the demand for passenger car motor fuel were investigated. The data used in those exercises are repeated in the table. Using the procedures you learned in this chapter, build a regression model that includes both qualitative and quantitative independent variables and provides a better explanation of the variation in motor fuel demand than either of the models developed in Exercises 11.63 and 12.65. Use the appropriate F tests to establish your model's superiority.

YEAR	MOTOR FUEL CONSUMED BY CARS (billion gallons)	POPULATION OF UNITED STATES (millions)	AVERAGE GROSS REAL WEEKLY EARNINGS (1967 $)	RELATIVE PRICE OF GALLON OF GASOLINE
1965	50.3	194.3	101.01	1.004
1966	53.31	196.6	101.67	.998
1967	55.11	198.7	101.84	1.000
1968	58.52	200.7	103.39	.973
1969	62.45	202.7	104.38	.954
1970	65.8	205.1	103.04	.908
1971	69.51	207.1	104.95	.876
1972	73.5	209.9	109.26	.859
1973	78.0	211.9	109.23	.887
1974	74.2	213.9	104.78	1.083
1975	76.5	216.0	101.45	1.060
1976	78.8	218.0	102.90	1.043
1977	80.7	220.2	104.13	1.037
1978	83.8	222.6	104.25	1.005
1979	80.2	225.1	101.15	1.222
1980	73.7	227.7	95.26	1.496
1981	71.7	230.0	93.69	1.508
1982	72.8	232.3	92.45	1.346
1983	73.4	234.5	94.07	1.261

Source: All data from *Statistical Abstract of the United States*, various years; except average gross real weekly earnings, which are from *Employment and Earnings*, U.S. Department of Labor, Bureau of Labor Statistics, Dec. 1986, p. 79.

12.103 Refer to Exercise 12.101. The contractor samples 25 recent jobs and fits both the complete second-order model (part **b**) and the reduced main effects model (part **a**), so that a test can be conducted to determine whether the additional complexity of the second-order model is necessary. The resulting SSE and R^2 values are:

Main effects: SSE = 8.548 and R^2 = .950

Second-order: SSE = 6.133 and R^2 = .964

a. Is there sufficient evidence to conclude that the second-order terms are important for predicting the mean cost? Use $\alpha = .05$.

b. Suppose the contractor decides to use the main effects model to predict costs. Use the global F test (Section 11.6) to determine whether the main effects model is useful for predicting costs.

*12.104 The table lists the $n = 72$ observations used by the Minnesota Department of Transportation to develop the peak-hour traffic volume model described in Case Study 12.1. Observations 55–72 are from Interstate 35W at 46th Street.

OBSERVATION NUMBER	PEAK-HOUR VOLUME	24-HOUR VOLUME	OBSERVATION NUMBER	PEAK-HOUR VOLUME	24-HOUR VOLUME	OBSERVATION NUMBER	PEAK-HOUR VOLUME	24-HOUR VOLUME
1	1,990.94	20,070	25	1,923.87	18,184	49	1,978.72	24,249
2	1,989.63	21,234	26	1,922.79	16,926	50	1,975.29	23,321
3	1,986.96	20,633	27	1,917.64	19,062	51	1,973.55	22,842
4	1,986.96	20,676	28	1,916.17	18,043	52	1,973.91	20,626
5	1,983.78	19,818	29	1,916.17	18,043	53	1,972.92	26,166
6	1,983.13	19,931	30	1,916.13	16,691	54	1,966.65	21,755
7	1,982.47	19,266	31	1,912.49	17,339	55	2,120.00	20,250
8	1,981.53	19,658	32	1,912.49	17,339	56	2,140.00	20,251
9	1,979.83	19,203	33	1,909.98	17,867	57	2,160.00	21,852
10	1,979.83	19,958	34	1,907.04	17,773	58	2,186.52	23,511
11	1,978.40	19,152	35	1,907.46	17,678	59	2,180.29	22,431
12	1,978.90	21,651	36	1,905.14	18,024	60	2,174.03	23,734
13	1,977.38	20,198	37	1,902.37	17,405	61	2,174.03	23,734
14	1,972.87	20,508	38	2,017.76	23,517	62	2,167.97	23,387
15	1,964.45	19,783	39	2,009.38	23,017	63	2,160.02	24,885
16	1,962.85	20,815	40	2,007.10	22,808	64	2,160.54	23,332
17	1,964.26	20,105	41	2,007.28	23,152	65	2,159.72	23,838
18	1,961.85	20,500	42	2,004.17	24,352	66	2,155.61	23,662
19	1,961.26	19,593	43	1,997.58	20,939	67	2,147.93	22,948
20	1,958.97	20,818	44	1,994.53	21,822	68	2,147.93	22,948
21	1,943.78	17,480	45	1,984.70	22,918	69	2,147.85	23,551
22	1,927.83	17,768	46	1,984.01	21,129	70	2,144.23	21,637
23	1,928.36	17,659	47	1,983.17	21,674	71	2,142.41	23,543
24	1,925.65	18,357	48	1,982.02	26,148	72	2,137.39	22,594

a. Use the data to replicate the model building process described in Case Study 12.1.

b. Compare your results of part **a** with those presented in Figures 12.28–12.30.

c. Calculate and plot the residuals of the model in a stem and leaf display or a box plot. Evaluate the assumption of normality of the random error component, and determine whether any apparent outliers exist.

    d. Plot the residuals against the predicted fuel consumed. Evaluate the assumption that the error component has a constant standard deviation. Is any transformation of the response suggested by the plot?

ON YOUR OWN . . .

We continue our "On Your Own" theme from Chapters 10 and 11. Remember that you selected three independent variables related to the annual GNP. Now, increase your list of three variables to include approximately ten that you feel would be useful in predicting the GNP. Obtain data for as many years as possible for the new list of variables and the GNP. With the aid of a computer analysis package, use a stepwise regression program to choose the important variables among those you have listed. To test your intuition, list the variables in the order you think they will be selected before you conduct the analysis. How does your list compare with the stepwise regression results?

    After the group of ten variables has been narrowed to a smaller group of variables by the stepwise analysis, try to improve the model by including interactions and quadratic terms. Be sure to consider the meaning of each interaction or quadratic term before adding it to the model—a quick sketch can be very helpful. See if you can systematically construct a useful model for predicting the GNP. You might want to hold out the last several years of data to test the predictive ability of your model after it is constructed. (As noted in Section 12.8, using the same data to construct *and* to evaluate predictive ability can lead to invalid statistical tests and a false sense of security.)

USING THE COMPUTER . . .

Consider the relationship between the mean median household income y of a zip code, the percent of college graduates x_1 in the zip code, and the census region of the country to which the zip code belongs. Randomly select 50 zip codes from each region.

a. Fit each of the following models to your sample:
    1. Complete second-order model
    2. Complete second-order model *without* the qualitative variable census region
    3. Main effects plus interaction (complete first-order) model
    4. Main effects model including both percent college graduates and census region
    5. Main effects model including only percent college graduates

b. Test the complete second-order model (1) against each of the reduced models, (2) and (3). Be sure to write the hypotheses you are testing in terms of this exercise. Which of the models is preferred as a result of your testing? If model (1) is preferred, proceed to part **c**. If model (2) is preferred, test it against model (5). If model (3) is preferred, test it against model (4). Select the preferred model among the five as a result of the testing.

c. Interpret the preferred model. Discuss the value of R^2 and the standard deviation of the model. Plot the predicted value of mean median income against the percent of college graduates, using a different plotting symbol for each census region (assuming the variable census region is in your preferred model). Interpret your plot.

d. Use a prediction interval to predict the mean median income for a zip code in the South region with 20% college graduates.

e. What assumptions are necessary to ensure the validity of the inferences in parts **b**–**d**? Plot the residuals of the preferred model against the predicted value of mean income. Interpret the pattern. Does the plot indicate a need to transform the dependent variable?

f. Repeat parts **a**–**e** using a multiplicative model—that is, using the logarithmic transformation of mean median income as the dependent variable. Be careful to recognize the transformation when you interpret the model and perform the prediction. Which model is to be preferred, the additive or multiplicative? Explain.

g. Repeat this exercise using all the measurements in the data base, rather than your sample. Compare the results to those based on your sample.

REFERENCES

Alexander, G. J. and Stover, R. D. "Consistency of mutual fund performance during varying market conditions." *Journal of Economics and Business,* Spring 1980, 32, 219–226.

Allmon, C. I. "Advertising and sales relationships for toothpaste: Another look." *Business Economics,* Sept. 1982, 17, 58.

Chatterjee, S. and Price, B. *Regression Analysis by Example.* New York: Wiley, 1977.

Churchill, G. A., Jr., Ford, N. M., and Walker, O. C., Jr. *Sales Force Management,* 2nd ed. Homewood, Ill.: Richard D. Irwin, 1985.

Draper, N. and Smith, H. *Applied Regression Analysis,* 2nd ed., New York: Wiley, 1981.

Federal Reserve System: Purpose and Functions. Washington, D.C.: Board of Governors of the Federal Reserve System, 1963.

Graybill, F. A. *Theory and Application of the Linear Model.* North Scituate, Mass.: Duxbury, 1976.

Mendenhall, W. *Introduction to Linear Models and the Design and Analysis of Experiments.* Belmont, Calif.: Wadsworth, 1968.

Mendenhall, W. and Sincich, T. *A Second Course in Business Statistics: Regression Analysis,* 2nd ed. San Francisco: Dellen, 1986.

Miller, R. B. and Wichern, D. W. *Intermediate Business Statistics: Analysis of Variance, Regression, and Time Series.* New York: Holt, Rinehart and Winston, 1977. Chapters 6–8.

Murphy, N. B. and Stock, D. R. "Determinants of the use of banking facilities: Trips to the bank in Oklahoma." *Review of Regional Economics and Business,* Oct. 1983, 8, 33–35.

Neter, J., Wasserman, W., and Kutner, M. *Applied Linear Regression Models.* Homewood, Ill.: Richard D. Irwin, 1983.

Schroeder, R. G. *Operations Management: Decision Making in the Operations Function,* 2nd ed. New York: McGraw-Hill, 1985. Chapters 20 and 22.

Spiro, H. T. *Finance for the Non-financial Manager,* 2nd ed. New York: Wiley, 1982. Chapter 16.

Weisberg, S. *Applied Linear Regression,* 2nd ed. New York: Wiley, 1985.

Winkler, R. L. and Hays, W. L. *Statistics: Probability, Inference and Decision,* 2nd ed. New York: Holt, Rinehart and Winston, 1975. Chapter 10.

Wise, J. O. and Dover, H. J. "An evaluation of a statistical method of appraising rural property." *Appraisal Journal,* Jan. 1974, 42, 103–113.

Younger, M. S. *A Handbook for Linear Regression.* North Scituate, Mass.: Duxbury, 1979.

CHAPTER 13

WHERE WE'VE BEEN . . .

In Chapters 10, 11, and 12, we discussed the construction, estimation, and use of regression models. We saw that regression models provide very powerful tools for analyzing and exploiting the relationships among variables. However, when the data are collected sequentially over time, the assumption of independent random errors (essential for the valid use of regression models) is probably not satisfied.

WHERE WE'RE GOING . . .

In this chapter, we consider data that are collected sequentially over time, i.e., time series data. We begin with a type of time series data often used to characterize some aspect of the economy—namely, index numbers. The remainder of the chapter is devoted to a study of analytical and graphical methods that help us understand the behavior of time series data. Methods for forecasting future values of a time series will be discussed in Chapter 14.

CONTENTS

TIME SERIES: INDEX NUMBERS AND DESCRIPTIVE ANALYSES

If you turn to the financial section of a newspaper, you are very likely to see a graph of the Dow Jones Average* over the past several months or years. The Dow Jones Average is a number based on the daily stock prices of 30 large corporations listed on the New York Stock Exchange and is calculated at the close of each day's trading. Many people believe the Dow Jones Average characterizes the present status of the stock market, which explains the predisposition of the news media to report it and graph its values. Numerical variables, such as the Dow Jones Average, that are calculated, measured, or observed sequentially on a regular chronological basis are called **time series**. The rate of inflation, Consumer Price Index, balance of trade, Producer Price Index, and annual proft of a firm are other examples of business and economic time series.

Time series data, like other types of data we have discussed in previous chapters, are subjected to two kinds of analyses: **descriptive** and **inferential**. Descriptive analyses, the topic of this chapter, use graphical and numerical techniques to provide a clear understanding of the time series. After graphing the data, you will often want to use it to make inferences about the future values of the time series; i.e., you will want to **forecast** future values. For example, once you understand the past and present trends of the Dow Jones Average, you would probably want to forecast its future trend before making decisions about buying and selling stocks. Since significant amounts of money may be riding on the accuracy of your forecasts, you would be interested in measures of their reliability. Forecasts and their measures of reliability are examples of **inferential techniques** in time series analysis. Inferential techniques will be the topic of Chapter 14.

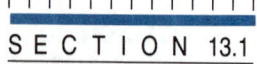

S E C T I O N   13.1

INDEX NUMBERS: AN INTRODUCTION

The most common technique for characterizing a business or economic time series is to compute **index numbers**. Index numbers measure how a time series changes over time. Change is measured relative to a preselected time period, called the **base period**.

DEFINITION 13.1

An **index number** is a number that measures the change in a variable over time relative to the value of the variable during a specific **base period**.

Two types of indexes dominate business and economic applications: **price** and **quantity indexes**. Price indexes measure changes in the price of a commodity or group of commodities over time. The CPI is a price index because it measures price changes of a group of commodities that are intended to reflect typical purchases of American consumers. On the other hand, an index constructed to measure the change in the total number of automobiles produced annually by American manufacturers would be an example of a quantity index.

*We are referring to the Dow Jones Industrial Average.

Methods of calculating index numbers range from very simple to extremely complex, depending on the numbers and types of commodities represented by the index. The next two sections provide details on the calculation and interpretation of several important types of index numbers.

| | | | | | | | | | | |

S E C T I O N   13.2

SIMPLE INDEX NUMBERS

When an index number is based on the price or quantity of a single commodity, it is called a **simple index number**.

> **DEFINITION 13.2**
>
> A **simple index number** is based on the relative changes (over time) in the price or quantity of a single commodity.

TABLE 13.1

Silver Prices, 1970–1985

YEAR	PRICE ($/ounce)
1970	1.771
1971	1.546
1972	1.684
1973	2.558
1974	4.708
1975	4.419
1976	4.353
1977	4.620
1978	5.440
1979	11.090
1980	20.633
1981	10.481
1982	7.950
1983	11.439
1984	8.141
1985	6.192

Source: *1986 CRB Commodity Yearbook*, Commodity Research Bureau, p. 228.

For example, consider the price of silver between 1970 and 1985, shown in Table 13.1. To construct a simple index to describe the relative changes in silver prices, we must first choose a **base period**. The choice is important because the price for all other periods will be compared with the price during the base period. We will select 1972 as the base period, a time just preceding the period of rapid economic inflation associated with dramatic oil price increases.

To calculate the simple index number for a particular year, we divide that year's price by the price during the base year and multiply the result by 100. Thus, for the 1975 silver price index number, we calculate

$$1975 \text{ index number} = \left(\frac{1975 \text{ silver price}}{1972 \text{ silver price}}\right)100 = \left(\frac{4.419}{1.684}\right)100$$
$$= 262.4$$

Similarly, the index number for 1985 is

$$1985 \text{ index number} = \left(\frac{1985 \text{ silver price}}{1972 \text{ silver price}}\right)100 = \left(\frac{6.192}{1.684}\right)100$$
$$= 367.70$$

The index number for the base period is always 100. In our example, we have

$$1972 \text{ index number} = \left(\frac{1972 \text{ silver price}}{1972 \text{ silver price}}\right)100 = 100$$

Thus, the silver price has risen by 162.4% (the difference between the 1975 and 1972 index numbers) between 1972 and 1975, and by 267.70% between 1972 and 1985. The simple index numbers for silver between 1970 and 1985 are given in Table 13.2 (page 758), and are portrayed graphically in Figure 13.1. The steps for calculating simple index numbers are summarized in the next box.

FIGURE 13.1

Graph of Silver Price
Index, 1970–1985

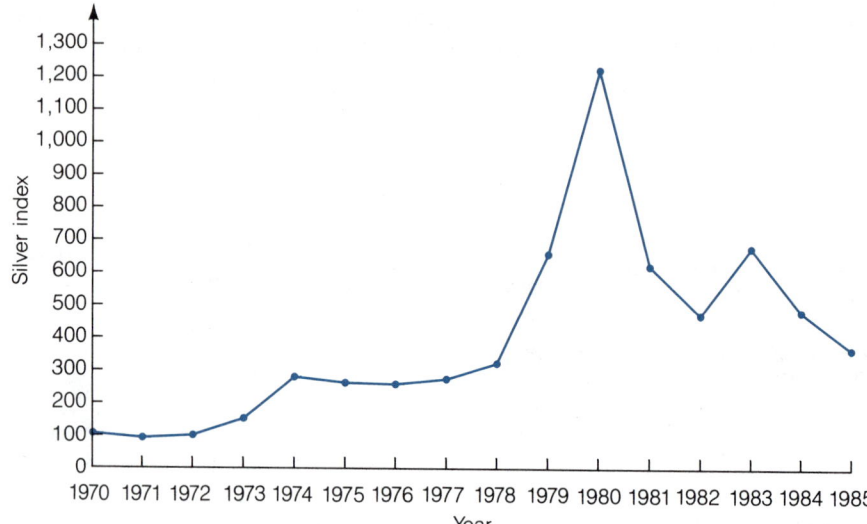

TABLE 13.2

Simple Index
Numbers for
Silver Prices
(Base 1972)

YEAR	INDEX
1970	105.17
1971	91.81
1972	100.00
1973	151.90
1974	279.57
1975	262.41
1976	258.49
1977	274.35
1978	323.04
1979	658.55
1980	1,225.24
1981	622.39
1982	472.09
1983	679.28
1984	483.43
1985	367.70

STEPS FOR CALULATING A SIMPLE INDEX NUMBER

1. Obtain the prices or quantities for the commodity over the time period of interest.
2. Select a base period.
3. Calculate the index number for each period according to the formula:

$$\text{Index number at time } t = \left(\frac{\text{Time series value at time } t}{\text{Time series value at base period}} \right) 100$$

Symbolically,

$$I_t = \left(\frac{Y_t}{Y_0} \right) 100$$

where I_t is the index number at time t, Y_t is the time series value at time t, and Y_0 is the time series value at the base period.

EXAMPLE 13.1

Foreign crude oil prices between 1970 and 1985 are shown in Table 13.3. Construct a simple index for foreign crude oil prices using 1972 as the base period and portray the index on the same graph as the silver price index (Table 13.2).

SOLUTION

Represent the foreign crude oil price at time t by Y_t, the price during the base period (1972) by Y_0, and the simple index number at time t by I_t. Then

$$I_t = \left(\frac{Y_t}{Y_0} \right) 100$$

TABLE 13.3
Foreign Crude Oil Prices,
1970–1985

YEAR	PRICE ($/barrel)	YEAR	PRICE ($/barrel)	YEAR	PRICE ($/barrel)
1970	1.80	1976	11.51	1981	32.00
1971	2.18	1977	12.70	1982	34.00
1972	2.48	1978	15.40	1983	30.00
1973	5.18	1979	18.00	1984	26.00
1974	10.46	1980	28.00	1985	26.00
1975	11.51				

Source: 1986 CRB Commodity Year Book, Commodity Research Bureau.

For example, the index number for 1975 is

$$I_{1975} = \left(\frac{Y_{1975}}{Y_0}\right)100 = \left(\frac{11.51}{2.48}\right)100 = 464.11$$

The interpretation is that crude oil prices increased by 364.11% between 1972 and 1975. All the index numbers for foreign crude oil prices are similarly calculated and are given in Table 13.4. The graphs of the silver and crude oil price indexes

FIGURE 13.2
Simple Indexes for Silver
and Crude Oil Prices
(Base 1972)

TABLE 13.4
Simple Index
Numbers for
Crude Oil Prices
(Base 1972)

YEAR	INDEX
1970	72.58
1971	87.90
1972	100.00
1973	208.87
1974	421.77
1975	464.11
1976	464.11
1977	512.10
1978	620.97
1979	725.81
1980	1,129.03
1981	1,290.32
1982	1,370.97
1983	1,209.68
1984	1,048.39
1985	1,048.39

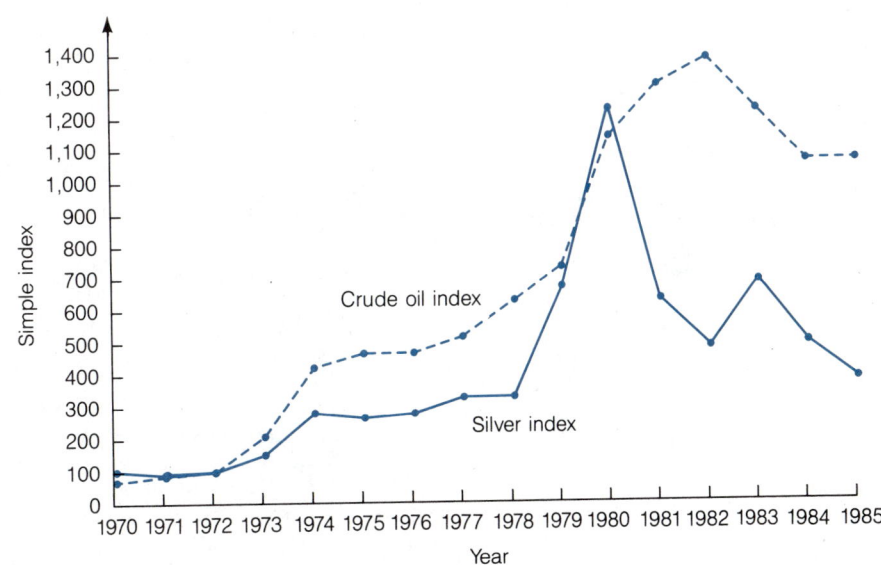

are combined in Figure 13.2. Since the same base period was used for both simple indexes, the two graphs intersect at the 1972 base period, where both indexes have a value of 100.

Portraying two indexes on the same graph (as we did in Example 13.1) indicates another reason indexes are calculated: Two or more commodities' relative price (or quantity) changes can be compared, even though the units of measurement are different (dollars per ounce and dollars per barrel, in our examples). In other words, apples and oranges *can* be compared, so long as you use index numbers to represent them. Figure 13.2 reveals that, during the period 1972–1978, the price of crude oil escalated more rapidly than that of silver, relative to the 1972 price. Also, the very high silver prices in 1980 are obvious on the graph. Note that silver prices began to fall in 1981, but the crude oil prices did not begin their decline until 1983. Of course, these index numbers provide only a *descriptive* comparison of the two time series. Any inferential implications of such a comparison will require a blend of inferential statistical analysis and economic theory.

EXERCISES 13.1–13.8

LEARNING THE MECHANICS

13.1 Explain in words how to construct a simple index.

13.2 The table lists the value of manufactured goods exported by the United States during the period 1975–1984. It also contains several values for each of two simple indexes for exported manufactured goods

YEAR	EXPORTS OF MANUFACTURED GOODS (millions of $)	SIMPLE INDEX (Base year = 1975)	SIMPLE INDEX (Base year = 1980)
1975	10,919.2		49.06
1976	11,206.1	102.63	
1977	10,857.0	99.43	
1978	12,416.8		55.79
1979	16,234.8		
1980	22,255.3	203.82	
1981	20,633.0		
1982	16,739.2	153.30	75.21
1983	14,852.0		66.73
1984	15,139.9	138.65	68.03

Source: *Business Statistics: 1984*, p. 79.

a. Calculate the missing values of each simple index.
b. Interpret the 1984 value for each index.

13.3 The table describes beer production in the United States for the period 1970–1984. Use 1977 as the base period to compute the simple index for this time series.

YEAR	U.S. BEER PRODUCTION (million barrels)	YEAR	U.S. BEER PRODUCTION (million barrels)
1970	133.1	1978	179.1
1971	137.4	1979	184.2
1972	141.3	1980	194.1
1973	148.6	1981	193.7
1974	156.2	1982	196.2
1975	160.6	1983	195.38
1976	163.7	1984	192.23
1977	170.5		

Source: Standard & Poor's Statistical Service, annual, Standard & Poor's Corporation.

13.4 Refer to Exercise 13.3. Is this an example of a quantity index or a price index? Explain.

13.5 Refer to Exercise 13.3. Recompute the simple index using 1980 as the base period. Plot the two indexes on the same graph.

APPLYING THE CONCEPTS

13.6 The stock of Abbott Laboratories has had the yearly closing prices shown in the table.

YEAR	CLOSING PRICE	YEAR	CLOSING PRICE
1973	$49.875	1980	$56.500
1974	50.750	1981	27.000
1975	41.250	1982	38.750
1976	49.125	1983	45.250
1977	56.500	1984	41.750
1978	33.750	1985	68.375
1979	41.125	1986	45.625

Source: *Daily Stock Price Record*, NYSE, 1973–1986; *Security Owner's Stock Guide*, Standard & Poor's, Jan. 1987.

a. Using 1973 as the base period, calculate the simple index for this stock's yearly closing price between 1973 and 1986.
b. By what percentage did the stock price increase or decrease between 1973 and 1986? Between 1978 and 1986?

13.7 The table lists the average retail price (in cents per gallon, including taxes) of motor gasoline for a sample of 85 urban areas in the United States for each month during the period January 1980 through September 1986.

YEAR	JAN.	FEB.	MAR.	APR.	MAY	JUNE	JULY	AUG.	SEPT.	OCT.	NOV.	DEC.
1980	111.0	118.6	123.0	124.2	124.4	124.6	124.7	124.3	123.1	122.3	122.2	123.1
1981	126.9	135.3	138.8	138.1	137.0	136.2	135.3	134.8	135.8	135.3	135.1	134.8
1982	134.1	131.8	126.8	121.0	122.4	129.6	131.8	131.0	129.5	128.0	126.8	124.4
1983	121.3	117.0	113.5	119.8	124.3	126.1	127.2	126.9	125.7	123.9	122.4	121.5
1984	120.0	119.3	119.4	121.1	122.1	121.4	119.7	118.4	118.9	119.5	119.3	117.9
1985	114.5	112.8	115.5	119.9	122.3	123.3	123.3	122.2	120.9	119.8	120.1	120.3
1986	119.0	111.9	98.3	89.5	92.7	95.8	89.5	84.8	86.7			

Source: *Monthly Energy Review*, Energy Information Administration, 1983, 1984, 1986.

a. Using January 1980 as the base period, calculate and plot the simple index for monthly retail gasoline prices between January 1980 and September 1986.
b. Interpret the value of the index you obtained for March 1983 and for August 1986.
c. Use the simple index to interpret the change in the price of gasoline between March 1981 and August 1986.
d. Is the index you constructed in part **a** a price or quantity index? Explain.

13.8 Civilian employment is broadly classified by the federal government into two categories—agricultural and nonagricultural. Employment figures (in thousands of workers) for these categories for the years 1974–1984 are given in the table at the top of page 762.

YEAR	AGRICULTURAL	NONAGRICULTURAL	YEAR	AGRICULTURAL	NONAGRICULTURAL
1974	3,515	83,279	1980	3,364	95,938
1975	3,408	82,438	1981	3,368	97,030
1976	3,331	85,421	1982	3,401	96,125
1977	3,283	88,734	1983	3,383	97,450
1978	3,387	92,661	1984	3,321	101,685
1979	3,347	95,477			

Source: *Business Statistics: 1984*, Department of Commerce, Bureau of Economic Analysis, 1985, p. 22.

a. Compute simple indexes for each of the two time series using 1974 as the base period.

b. Which segment has shown the greater percentage change in employment over the period shown?

c. Are these indexes price or quantity indexes? Explain.

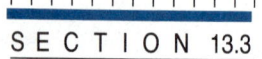

SECTION 13.3

COMPOSITE INDEX NUMBERS

A **composite index number** represents combinations of the prices or quantities of several commodities. For example, suppose you want to construct an index for the total number of sales of the three major automobile manufacturers in the United States. The first step is to collect data on the sales of each manufacturer during the period in which you are interested. The total sales of automobiles by each manufacturer between 1972 and 1985 are shown in Table 13.5. To summarize the information from all three time series in a single index, we add the sales of each manufacturer for each year. That is, we form a new time series consisting of the total number of automobiles sold by the three manufacturers.

TABLE 13.5 Sales of Automobiles by Three Manufacturers (Thousands)

YEAR	GENERAL MOTORS	FORD	CHRYSLER	YEAR	GENERAL MOTORS	FORD	CHRYSLER
1972	7,790.52	5,593.04	2,192.00	1979	8,993.00	5,810.30	1,796.00
1973	8,683.80	5,871.00	2,423.00	1980	7,101.00	4,328.45	1,225.00
1974	6,690.00	5,258.93	2,015.00	1981	6,762.00	4,313.18	1,283.00
1975	6,629.00	4,577.77	1,773.00	1982	6,244.00	4,254.90	1,182.00
1976	8,568.00	5,304.44	2,371.00	1983	7,769.00	4,934.23	1,493.96
1977	9,068.00	6,422.30	2,328.00	1984	8,256.35	5,584.65	2,034.35
1978	9,482.00	6,462.06	2,212.00	1985	9,305.00	5,550.50	2,157.37

Source: *Moody's Industrial Manual*, 1986.

We now construct a simple index for the *total* of the three series. Selecting 1977 as the base year, we divide each total by the 1977 total sales. The resulting **simple composite index** is shown in Table 13.6.

DEFINITION 13.3

A **simple composite index** is a simple index for a time series consisting of the total price or total quantity of two or more commodities.

TABLE 13.6

Simple Composite Index for Total Automobiles Sold by Three Manufacturers

YEAR	INDEX	YEAR	INDEX
1972	87.41	1979	93.16
1973	95.28	1980	71.02
1974	78.37	1981	69.36
1975	72.85	1982	65.56
1976	91.16	1983	79.68
1977	100.00	1984	89.10
1978	101.90	1985	95.48

EXAMPLE 13.2

One of the primary uses of index numbers is to characterize changes in stock prices over time. Stock market indexes have been constructed for many different types of companies and industries, and several composite indexes have been developed to characterize all stocks. These indexes are reported on a daily basis in the news media (e.g., Standard and Poor's 500 Stocks Index and Dow Jones 65 Stocks Index).

Consider the monthly prices given in Table 13.7 (page 764) for four high-technology company stocks listed on the New York Stock Exchange between 1984 and 1986. To see how this type of stock fared as the market began to rally in the mid 1980's, construct a simple composite index using January 1984 as the base period. Graph the index, and comment on its implications.

SOLUTION

First, we calculate the total of the four stock prices each month. These totals are shown in Table 13.7. Then the simple composite index is calculated by dividing each monthly total by the January 1984 total. The index values are given in Table 13.8 (page 765), and a graph of the simple composite index is shown in Figure 13.3.

FIGURE 13.3

Graph of Simple Composite Index of Four Stocks

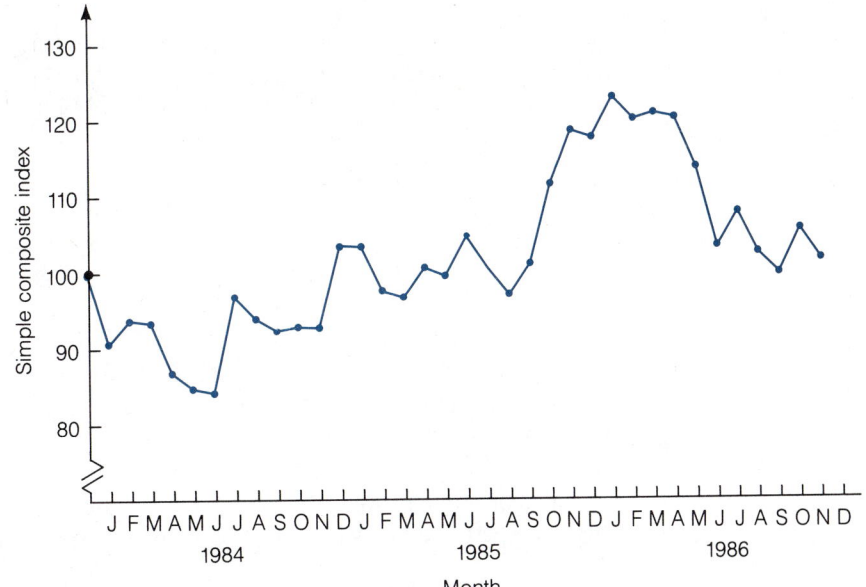

TABLE 13.7 Monthly Prices of Four High-technology Company Stocks

	BELL INDUSTRIES	XEROX	HARRIS	IBM	TOTAL
Jan. 1984	$29.875	$44.000	$38.500	$114.125	$226.500
Feb.	24.750	41.125	29.375	110.250	205.500
Mar.	26.500	41.250	31.000	114.000	212.750
Apr.	25.625	40.500	32.000	113.750	211.875
May	24.000	37.250	28.000	107.750	197.000
June	23.000	38.375	25.125	105.750	192.250
July	21.375	33.750	25.125	110.750	191.000
Aug.	27.125	38.375	30.375	123.750	219.625
Sept.	24.375	37.625	26.750	124.250	213.000
Oct.	23.500	35.500	25.750	124.625	209.375
Nov.	22.625	37.250	28.875	121.750	210.500
Dec.	22.125	37.875	27.125	123.125	210.250
Jan. 1985	22.875	43.375	32.250	136.375	234.875
Feb.	24.625	45.375	30.500	134.000	234.500
Mar.	22.500	43.375	28.250	127.000	221.125
Apr.	21.750	45.500	25.500	126.500	219.250
May	22.250	50.000	27.250	128.625	228.125
June	20.875	52.625	28.375	123.750	225.625
July	23.500	53.875	28.625	131.375	237.375
Aug.	23.375	51.750	26.125	126.625	227.875
Sept.	22.250	50.375	23.500	123.875	220.000
Oct.	24.500	50.375	24.500	129.875	229.250
Nov.	26.875	60.125	26.125	139.750	252.875
Dec.	26.500	59.750	27.250	155.500	269.000
Jan. 1986	24.500	64.250	26.625	151.500	266.875
Feb.	26.625	70.625	30.750	150.875	278.875
Mar.	26.875	68.000	28.250	149.125	272.250
Apr.	25.625	60.000	32.250	156.250	274.125
May	26.125	61.250	33.000	152.375	272.750
June	22.000	56.125	33.250	146.500	257.875
July	19.375	53.250	28.875	132.500	234.000
Aug.	18.750	57.000	29.750	138.750	244.250
Sept.	17.625	51.500	28.625	134.500	232.250
Oct.	18.000	54.750	29.750	123.625	226.125
Nov.	20.250	60.500	31.500	127.125	239.375
Dec.	20.500	60.000	29.750	120.000	230.250

Source: *Daily Stock Price Record*, NYSE 1983–1986, *Standard & Poor's Security Owner's Stock Guide*, Aug. 1986–Jan. 1987.

The plot of the simple composite index for these high-technology stocks shows that their performance was rather "flat" through 1984 and most of 1985. The index

TABLE 13.8   Simple Composite Index of Stock Prices

1984	INDEX	1985	INDEX	1986	INDEX
Jan.	100.00	Jan.	103.70	Jan.	117.83
Feb.	90.73	Feb.	103.53	Feb.	123.12
Mar.	93.93	Mar.	97.63	Mar.	120.20
Apr.	93.54	Apr.	96.80	Apr.	121.03
May	86.98	May	100.72	May	120.42
June	84.88	June	99.61	June	113.85
July	84.33	July	104.80	July	103.31
Aug.	96.96	Aug.	100.61	Aug.	107.84
Sept.	94.04	Sept.	97.13	Sept.	102.54
Oct.	92.44	Oct.	101.21	Oct.	99.83
Nov.	92.94	Nov.	111.64	Nov.	105.68
Dec.	92.83	Dec.	118.76	Dec.	101.66

begins a dramatic increase in November 1985, peaking at 23.12% over the January 1984 value in February 1986. However, while the rest of the market continued to enjoy new highs throughout 1986, these high-technology stocks receded to about the same level as January 1984 by the end of 1986.   ∎

A simple composite price index has a major drawback: The quantity of the commodity that is purchased during each period is not taken into account when only the price totals are used to calculate the index. We can remedy this situation by constructing a **weighted composite price index**.

DEFINITION 13.4

A **weighted composite price index** weights the prices by quantities purchased prior to calculating totals for each time period. The weighted totals are then used to compute the index in the same way that the unweighted totals are used for simple composite indexes.

Since the quantities purchased change from time period to time period, the choice of which time period quantities to use as the basis for the weighted composite index is an important one. A **Laspeyres index** uses the base period quantities as weights. The rationale is that the prices at each time period should be compared as if the same quantities were purchased each period as were purchased during the base period. This method measures price inflation (or deflation) by fixing the purchase quantities at their base period values. The method for calculating a Laspeyres index is given in the box at the top of page 766.

STEPS FOR CALCULATING A LASPEYRES INDEX

1. Collect price information for each of the k price series to be used in the composite index. Denote these series by $P_{1t}, P_{2t}, \ldots, P_{kt}$.
2. Select a base period. Call this time period t_0.
3. Collect purchase quantity information for the base period. Denote the k quantities by $Q_{1t_0}, Q_{2t_0}, \ldots, Q_{kt_0}$.
4. Calculate the weighted totals for each time period according to the formula

$$\sum_{i=1}^{k} Q_{it_0} P_{it}$$

5. Calculate the Laspeyres index, I_t, at time t by taking the ratio of the weighted total at time t to the base period weighted total and multiplying by 100. That is,

$$I_t = \frac{\sum_{i=1}^{k} Q_{it_0} P_{it}}{\sum_{i=1}^{k} Q_{it_0} P_{it_0}} \times 100$$

EXAMPLE 13.3

The January prices for the four high-technology company stocks are given in Table 13.9. Suppose that, in January 1984, an investor purchased the quantities shown in the table. [*Note:* Only January prices and quantities are used to simplify the example. The same methods can be applied to calculate the index for other months.] Calculate the Laspeyres index for the investor's portfolio of high-technology stocks using January 1984 as the base period.

TABLE 13.9

January Prices of High-technology Stocks with Quantities Purchased

	BELL INDUSTRIES	XEROX	HARRIS	IBM
Shares purchased	500	100	100	1,000
January 1984 price	$29.875	$44.000	$38.500	$114.125
January 1985 price	$22.875	$43.375	$32.250	$136.375
January 1986 price	$24.500	$64.250	$26.625	$151.500

SOLUTION

First, we calculate the weighted price totals for each time period, using the January 1984 quantities as weights. Thus,

$$\text{January 1984 weighted total} = \sum_{i=1}^{4} Q_{i,1984} P_{i,1984}$$

$$= 500(29.875) + 100(44.000) + 100(38.500) + 1{,}000(114.125)$$

$$= 137{,}312.5$$

$$\text{January 1985 weighted total} = \sum_{i=1}^{4} Q_{i,1984} P_{i,1985}$$

$$= 500(22.875) + 100(43.375) + 100(32.250) + 1,000(136.375)$$

$$= 155,375.0$$

$$\text{January 1986 weighted total} = \sum_{i=1}^{4} Q_{i,1984} P_{i,1986}$$

$$= 500(24.500) + 100(64.250) + 100(26.625) + 1,000(151.500)$$

$$= 172,837.5$$

Then the Laspeyres index is calculated by multiplying the ratio of each weighted total to the base period weighted total by 100. Thus,

$$I_{1984} = \frac{\sum_{i=1}^{4} Q_{i,1984} P_{i,1984}}{\sum_{i=1}^{4} Q_{i,1984} P_{i,1984}} \times 100 = 100$$

$$I_{1985} = \frac{\sum_{i=1}^{4} Q_{i,1984} P_{i,1985}}{\sum_{i=1}^{4} Q_{i,1984} P_{i,1984}} \times 100 = \frac{155,375.0}{137,312.5} = 113.2$$

$$I_{1986} = \frac{\sum_{i=1}^{4} Q_{i,1984} P_{i,1986}}{\sum_{i=1}^{4} Q_{i,1984} P_{i,1984}} \times 100 = \frac{172,837.5}{137,312.5} = 125.9$$

The implication is that these stocks were worth 13.2% more to the investor in January 1985 than in January 1984 and 25.9% more in January 1986. ■

The Laspeyres index is appropriate when the base period quantities are reasonable weights to apply to all time periods. This is the case in applications such as that described in Example 13.3, where the base period quantities represent actual quantities of stock purchased and held for some period of time. Laspeyres indexes are also appropriate when the base period quantities remain reasonable approximations of purchase quantities in subsequent periods. However, it can be misleading when the relative purchase quantities change significantly from those in the base period.

Probably the best known Laspeyres index is the all-items Consumer Price Index (CPI). This monthly composite index is made up of hundreds of item prices, and the U.S. Bureau of Labor Statistics (BLS) sampled over 30,000 families' purchases in 1982–1984 to determine the base period quantities. Thus, beginning in 1987, the all-items CPI published each month reflects quantities purchased in 1982–1984 by a sample of families across the United States. However, as prices increase for some commodities more quickly than for others, consumers tend to substitute less expensive commodities where possible. For example, as automobile and gasoline prices rapidly inflated in the mid-1970's, consumers began to purchase smaller cars. The net effect of using the base period quantities for the CPI is to overestimate the effect of inflation on consumers, because the quantities are fixed at levels that will actually change in response to price changes.

There are several solutions to the problem of purchase quantities that change relative to those of the base period. One is to change the base period regularly, so that the quantities are regularly updated. A second solution is to compute the index at each time period by using the purchase quantities of that period, rather than those of the base period. A **Paasche index** is calculated by using price totals weighted by the purchase quantities of the period the index value represents. The steps for calculating a Paasche index are given in the box.

STEPS FOR CALCULATING A PAASCHE INDEX

1. Collect price information for each of the k price series to be used in the composite index. Denote these series by $P_{1t}, P_{2t}, \ldots, P_{kt}$.
2. Select a base period. Call this time period t_0.
3. Collect purchase quantity information for every period. Denote the k quantities for period t by $Q_{1t}, Q_{2t}, \ldots, Q_{kt}$.
4. Calculate the Paasche index for time t by multiplying the ratio of the weighted total at time t to the weighted total at time t_0 (base period) by 100, where the weights used are the purchase quantities for time period t. Thus,

$$I_t = \frac{\sum_{i=1}^{k} Q_{it} P_{it}}{\sum_{i=1}^{k} Q_{it} P_{it_0}} \times 100$$

EXAMPLE 13.4

The January prices and volumes (actual quantities purchased) in thousands of shares for the four high-technology company stocks are shown for 1984, 1985, and 1986 in Table 13.10. Calculate and interpret the Paasche index, using January 1984 as the base period.

TABLE 13.10   January Prices and Volumes of High-technology Company Stocks

	BELL INDUSTRIES		XEROX		HARRIS		IBM	
	Price	Volume	Price	Volume	Price	Volume	Price	Volume
January 1984	$29.875	229.7	$44.000	4,843.2	$38.500	1,377.5	$114.125	27,213.2
January 1985	22.875	487.4	43.375	11,869.0	32.250	2,834.7	136.375	36,521.8
January 1986	24.500	167.3	64.250	7,772.9	26.625	1,651.6	151.500	31,936.9

Source: *Daily Stock Price Record*, NYSE, Jan. 1984–1986.

SOLUTION

The key to calculating a Paasche index is to remember that the weights (purchase quantities) change for each time period. Thus,

$$
I_{1984} = \frac{\sum\limits_{i=1}^{4} Q_{i,1984} P_{i,1984}}{\sum\limits_{i=1}^{4} Q_{i,1984} P_{i,1984}} \times 100 = 100
$$

$$
I_{1985} = \frac{\sum\limits_{i=1}^{4} Q_{i,1985} P_{i,1985}}{\sum\limits_{i=1}^{4} Q_{i,1985} P_{i,1984}} \times 100 = \frac{5,598,047}{4,813,983} \times 100 = 116.3
$$

$$
I_{1986} = \frac{\sum\limits_{i=1}^{4} Q_{i,1986} P_{i,1986}}{\sum\limits_{i=1}^{4} Q_{i,1986} P_{i,1984}} \times 100 = \frac{5,385,922}{4,055,391} \times 100 = 132.8
$$

The implication is that 1985 prices represent a 16.3% increase over 1984 prices, assuming the purchase quantities were at January 1985 levels for *both* periods. Similarly, the 1986 index value of 132.8 implies a 32.8% increase when purchase quantities are at the January 1986 level. ∎

The Paasche index is most appropriate when you want to compare current prices to base period prices at *current* purchase levels. However, there are several major problems associated with the Paasche index. First, it requires that purchase quantities be known for every time period. This rules out a Paasche index for applications such as the CPI because the time and monetary resource expenditures required to collect quantity information are considerable. (Recall that more than 30,000 families were sampled to estimate purchase quantities in 1982–1984.) A second problem is that, although each period is compared to the base period, it is difficult to compare the index at two other periods because the quantities used are different for each period. For example, for the four high-technology stocks in Example 13.4, we calculated index values of 116.3 in 1985 and 132.8 in 1986. Although this apparently represents an increase of 16.5% from 1985 to 1986, these two index values are determined using different quantities, and therefore, the change in the index is affected by changes in both prices *and* quantities. This fact makes it difficult to interpret the change in a Paasche index between periods when neither is the base period.

Although there are other types of indexes that use different weighting factors, the Laspeyres and Paasche indexes are the most popular composite indexes. Depending on the primary objective in constructing an index, one of them will probably be suitable for most purposes.

The Consumer Price Index (CPI), first published by the U.S. Bureau of Labor Statistics (BLS) in 1919, is the country's principal measure of price changes. One major use of the CPI is as an indicator of inflation, through which the success or failure of government economic policies can be monitored. A second major use of the CPI is to escalate income payments. Millions of workers have escalator clauses in their collective bargaining contracts that call for increases in wage rates based on increases in the CPI. In addition, the incomes of Social Security beneficiaries and retired military and federal civil service employees are tied to the CPI. It has been estimated that a 1% increase in the CPI can trigger an increase of over $1 billion in income payments.

Since 1978, the BLS has published two national, all-items indexes: the new CPI-U and the traditional CPI-W. The CPI-U measures the price change of a constant market basket of goods and services that are representative of the purchases of all urban residents—approximately 80% of the U.S. population. The CPI-W measures the price change of a constant market basket of goods and services that are representative of the purchases of urban wage earners and clerical workers—approximately 50% of all urban residents. The base period for both indexes is 1967. The CPI-U is the index typically reported by the press and broadcast media. The CPI-W is the index used in the escalator clauses of most labor contracts and government benefit programs. In addition to these two national indexes, the BLS publishes CPI-U and CPI-W indexes for each of 27 metropolitan areas. The national indexes and the metropolitan indexes are reported monthly in the BLS's *CPI Detailed Report*.

The market basket of goods priced by both the CPI-U and the CPI-W includes a homeownership component. Accounting for over 30% of the overall weight of the indexes, the homeownership component influences the indexes more than food, energy, or medical care. This component includes the costs associated with purchasing a home (the price of the home and mortgage interest), as well as the cost of property taxes, property insurance, and maintenance and repairs. During the 1970's and early 1980's, the use of these quantities to measure the cost of homeownership met with much criticism. The following two arguments were made by critics:

1. Since the CPI is used to measure the change in purchasing power for the purpose of escalating income or determining the rate of inflation, it "should not include the impact of rising prices on the value of assets such as houses. Just as the CPI excludes changes in the value of stocks and bonds, . . . the change in the asset value of the house (appreciation or depreciation) and the cost of equity in holding that asset should be distinguished from the change in the cost of the shelter provided by the house. It is the cost of consuming the shelter provided by the house—not the investment aspects of homeownership—which should be reflected in an index used to keep real income constant" (*CPI Issues*, p. 2).

2. The CPI overstates the rate of inflation because "it uses *current* house prices and *current* mortgage interest rates . . . the CPI should not measure the costs of purchasing the base period houses in today's prices and today's mortgage interest rates, but rather the CPI should measure what people are actually paying for housing" (*CPI Issues*, p. 2).

In response to these criticisms, the BLS developed and experimented with an entirely new approach to measuring the cost of housing. As a result, instead of explicitly including in the market basket the homeownership costs described above, the BLS now recommends that a *rental equivalency* component be included. This approach assumes that a household's cost of consuming the flow of services from the housing unit can be represented by the income that the household could receive by renting the home to someone else. This rental equivalency approach to measuring homeownership costs was implemented in an experimental version of the CPI-U called the CPI-U-X1.

Figure 13.4 shows the movement that the CPI-U and CPI-U-X1 would have displayed over the period 1970–1981. Notice that the two indexes generally move together, but that the CPI-U-X1 tends to stay below the CPI-U, particularly during the periods of high mortgage interest rates in 1970, 1974–1975, and 1978–1981. If, as some critics charge, the CPI-U overstates inflation during periods of high mortgage interest rates, it appears that the experimental CPI-U-X1 should provide a better measure of inflation. In January 1983, the BLS changed the official CPI-U to include the rental equivalency approach to measuring homeownership costs and made a similar change in the CPI-W in 1985.

FIGURE 13.4

Changes in the Consumer Price Index for All Urban Consumers: Official (CPI-U) and Experimental Rental Equivalence (CPI-U-X1) Measures [*Note:* Percent changes are calculated using 12 months of unadjusted data.]
Source: Gillingham and Lane, 1982, p. 13.

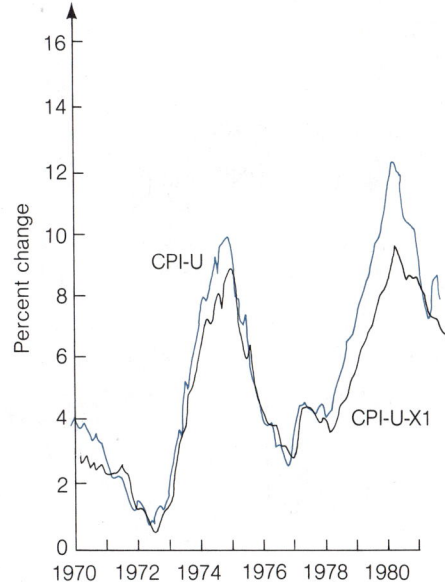

EXERCISES 13.9–13.17

LEARNING THE MECHANICS

13.9 Explain in words how to calculate the following types of indexes:
  a. Simple composite index
  b. Weighted composite index
  c. Laspeyres index
  d. Paasche index

13.10 Using May 1984 as the base period, compute and plot the simple composite index for the commodities listed in the table.

DATE 1984	PRICE OF EGGS (cents/lb.)	PRICE OF COCOA BEANS (cents/lb.)	PRICE OF GASOLINE (cents/gallon)
Jan.	96.1	1.34	113.1
Feb.	92.9	1.34	112.5
Mar.	79.4	1.34	112.5
Apr.	91.4	1.28	114.5
May	68.9	1.35	115.4
June	61.0	1.29	114.7
July	59.9	1.22	112.9
Aug.	58.6	1.17	111.6
Sept.	58.4	1.21	112.0
Oct.	55.3	1.25	112.7
Nov.	61.3	1.26	112.4
Dec.	58.4	1.13	110.9

Source: *Standard & Poor's Basic Statistics: Metals,* 1985, pp. 20, 26.

13.11 Explain in words the difference between Laspeyres and Paasche indexes.

APPLYING THE CONCEPTS

13.12 The gross national product (GNP), which is used as an indicator of the health of the U.S. economy, is the sum of several components. One of these is personal consumption expenditures, which is itself the sum of expenditures for durable goods, nondurable goods, and services. Consider the personal consumption expenditures data listed in the table.

YEAR	DURABLE GOODS ($ billion)	NONDURABLE GOODS ($ billion)	SERVICES ($ billion)	YEAR	DURABLE GOODS ($ billion)	NONDURABLE GOODS ($ billion)	SERVICES ($ billion)
1961	41.6	155.3	138.1	1973	122.9	334.4	351.3
1962	46.7	161.6	147.0	1974	121.9	375.7	388.3
1963	51.4	167.1	156.1	1975	132.2	407.3	437.0
1964	56.3	176.9	167.1	1976	156.5	440.4	482.8
1965	62.8	188.6	178.7	1977	178.2	478.8	547.4
1966	67.7	204.7	192.4	1978	200.2	528.2	618.0
1967	69.6	212.6	208.1	1979	213.4	600.0	693.7
1968	80.0	230.4	225.6	1980	214.7	668.8	784.5
1969	85.5	247.0	247.2	1981	235.4	730.7	883.0
1970	85.2	265.7	270.8	1982	245.1	757.5	982.2
1971	97.1	277.7	293.4	1983	279.8	801.7	1,074.4
1972	111.2	299.3	322.4	1984	318.8	856.9	1,166.2

Source: *Statistical Abstract of the United States: 1986,* Bureau of the Census, p. 435.

a. Using these three component values for the years 1961–1984, construct a simple composite index for personal consumption, using 1967 as the base year.

b. Suppose we want to update the index by making 1974 the base year. Update the index using only the index values you calculated in part **a**, without referring to the original data.

13.13 Refer to Exercise 13.12, in which a personal consumption expenditure index was constructed. Graph the personal consumption expenditure index for the years 1961–1984, first using 1967 as the base year and then using 1974 as the base year. What effect does changing the base year have on the graph of this index?

13.14 Refer to Exercise 13.12. Suppose the output quantities in 1967, measured in billions of units purchased, are as follows:

Durable goods: 8.5

Nondurable goods: 120.1

Services: 30.7

Use the outputs to calculate the Laspeyres index from 1961 to 1984 with 1967 as the base period.

13.15 Refer to Exercises 13.12 and 13.14. Plot the simple composite index and Laspeyres index on the same graph. Comment on the differences between the two indexes.

13.16 The level of price and production of metals in the United States is one measure of the strength of the industrial economy. The table lists the 1984 prices (in dollars per ton) and production (in tons) for three metals important to U.S. industry.

MONTH	COPPER		PIG IRON		LEAD	
	Price	Production	Price	Production	Price	Production
Jan.	$1,361.6	100.7	$213	4,311	$530.0	46.1
Feb.	1,399.0	95.1	213	4,497	520.0	47.0
Mar.	1,483.6	104.0	213	5,083	529.0	51.0
Apr.	1,531.6	95.6	213	5,077	540.0	23.0
May	1,431.2	103.3	213	5,166	531.0	26.5
June	1,383.8	106.9	213	4,565	580.0	13.5
July	1,326.8	95.9	213	4,329	642.8	27.4
Aug.	1,328.8	96.7	213	4,057	602.6	25.8
Sept.	1,307.8	95.7	213	3,473	513.6	20.5
Oct.	1,278.4	89.1	213	3,739	480.8	24.6
Nov.	1,354.2	100.5	213	3,817	528.4	21.5
Dec.	1,305.2	96.9	213	3,694	462.2	27.9

Source: *Standard & Poor's Basic Statistics: Metals*, 1985.

a. Compute simple composite price and quantity indexes for the 12-month period, using January as the base period.

b. Compute the Laspeyres price index for the 12-month period using January as the base period.

c. Plot the simple composite and Laspeyres indexes on the same graph. Comment on the differences.

13.17 Refer to Exercise 13.16.

    a. Compute the Paasche price index for metals for the 12-month period using January as the base period.

    b. Plot the Laspeyres and Paasche indexes on the same graph. Comment on the differences.

    c. Compare the Laspeyres and Paasche index values for September and December. Which index is more appropriate for describing the change in this 4-month period? Explain.

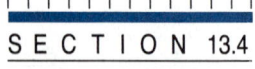

SECTION 13.4

SMOOTHING WITH MOVING AVERAGES

As you have seen in the previous sections, index numbers are useful for describing trends and changes in time series. However, time series often have such irregular fluctuations that trends are difficult to describe. Index numbers can be misleading in such cases because the series is changing so rapidly. Methods for removing the rapid fluctuations in a time series so the general trend can be seen are called **smoothing** techniques.

Probably the simplest smoothing technique is the **moving average method**. The **N-point moving average** of a time series is the average of the time series values at N adjacent time periods. For example, consider the time series consisting of the annual closing-day Dow Jones Average (DJA) from 1961 to 1986 (see Table 13.11). As shown in Figure 13.5, there are time periods during which the DJA oscillates rather rapidly.

TABLE 13.11

Dow Jones Average and 7-Point Moving Average

YEAR	DJA	7-POINT MOVING AVERAGE	YEAR	DJA	7-POINT MOVING AVERAGE
1961	731.14	—	1974	759.37	875.90
1962	652.10	—	1975	802.49	864.50
1963	762.95	—	1976	974.92	848.51
1964	874.13	811.48	1977	835.15	854.24
1965	969.26	841.86	1978	805.01	874.19
1966	785.69	863.04	1979	838.74	909.05
1967	905.11	873.89	1980	963.99	949.58
1968	943.75	875.41	1981	899.01	1,003.36
1969	800.36	872.76	1982	1,046.54	1,109.31
1970	838.92	892.50	1983	1,258.64	1,260.34
1971	884.76	871.68	1984	1,211.57	—
1972	950.71	851.50	1985	1,546.67	—
1973	923.88	876.44	1986	1,895.95	—

Source: *The Dow Jones Investor's Handbook*, 1986; *Wall Street Journal*, Jan. 2, 1987.

FIGURE 13.5   Dow Jones Average, 1961–1986

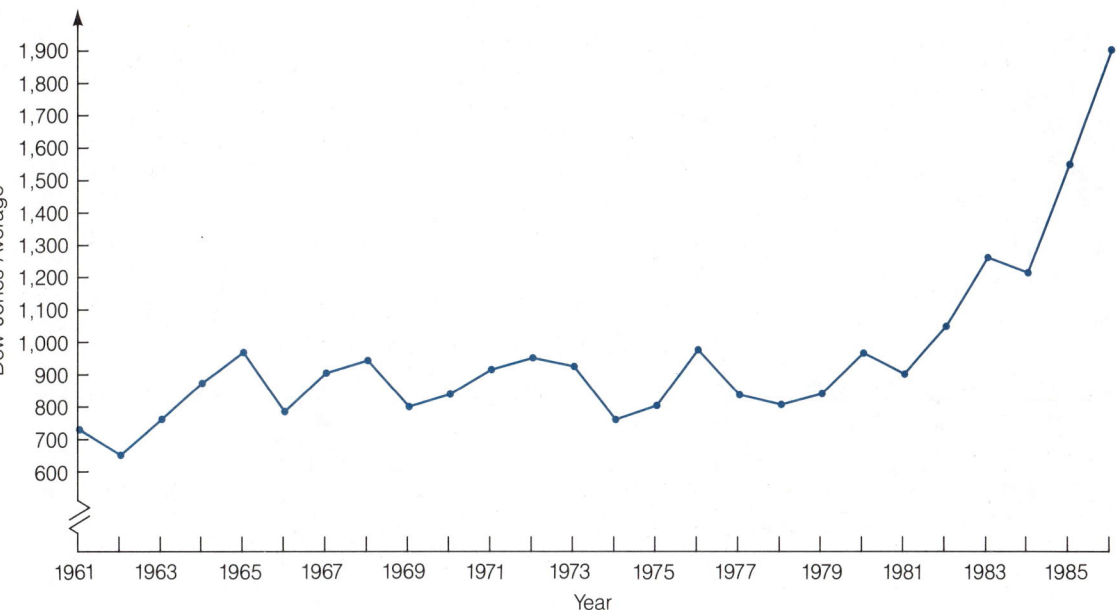

To smooth the DJA series, we will use a 7-point moving average. The value of the 7-point moving average for a particular year is the mean of the DJA for seven periods: the three previous time periods, the time period of interest, and the three subsequent time periods. Denoting the DJA time series by Y_t and the 7-point moving average by M_t, we can write the calculation formula for M_t as

$$M_t = \frac{Y_{t-3} + Y_{t-2} + Y_{t-1} + Y_t + Y_{t+1} + Y_{t+2} + Y_{t+3}}{7}$$

For example,

$$M_{1964} = \frac{Y_{1961} + Y_{1962} + Y_{1963} + Y_{1964} + Y_{1965} + Y_{1966} + Y_{1967}}{7}$$

$$= \frac{731.14 + 652.10 + 762.95 + 874.13 + 969.26 + 785.69 + 905.11}{7}$$

$$= 811.48$$

The complete 7-point moving average is given in Table 13.11. Note that the first three and last three values for Y_t have no corresponding moving average, M_t, because each 7-point moving average calculation requires values for Y_t for three preceding and three subsequent time periods. The 7-point moving average is plotted along with the DJA values in Figure 13.6 on page 776.

FIGURE 13.6   Moving Average for the Dow Jones Average

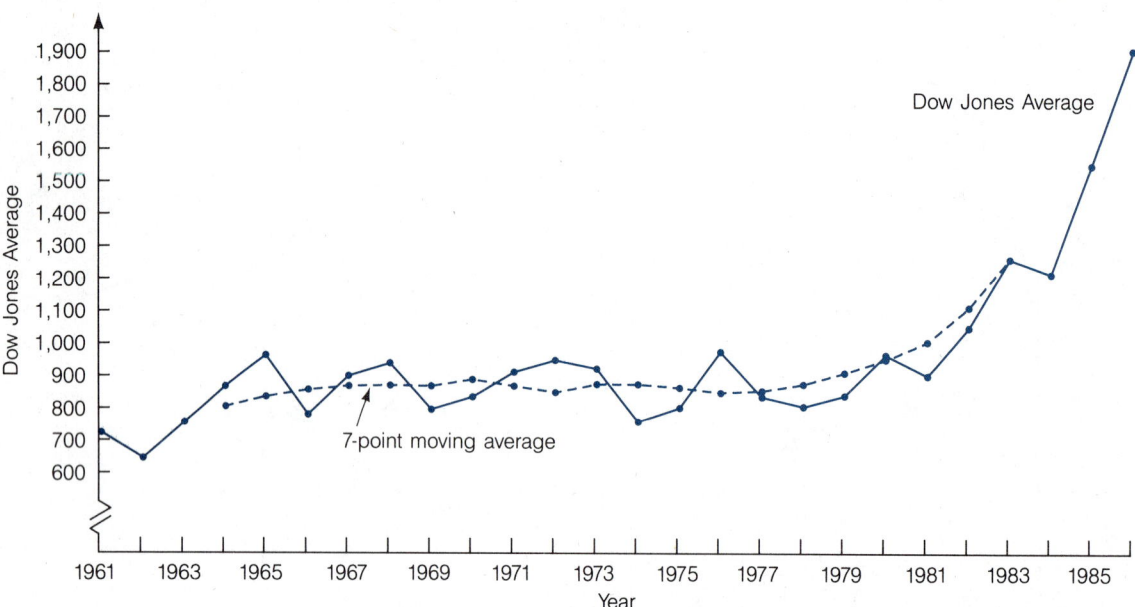

Note that the trend of the DJA is easier to follow with the moving average than with the series itself. The 7-point moving average removes most of the short-term fluctuation, making it possible to see the long-term trend. The main disadvantage of the 7-point moving average is the loss of three data points at each end of the series.

The choice of N in calculating an N-point moving average is an important one. Large values of N produce a smoother moving average, but more points on each end of the series are lost and the moving average may wind up so smooth that important changes in the time series are lost. Small values of N retain more points on each end of the series, but do not yield the same degree of smoothing. We usually try several values of N to determine one that yields a smooth series without missing important changes or losing too many points at the beginning and end of the series.

To see the effect of changing N, we will consider a 4-point moving average for the DJA series. The selection of an even value of N produces a special problem: The average of an even number of points is not "centered" at any specific time period. For example, the average of four time period values is centered between the second and third time periods. Thus, for the DJA data (Table 13.11), the first 4-point moving average is

$$M_{1962.5} = \frac{Y_{1961} + Y_{1962} + Y_{1963} + Y_{1964}}{4}$$

$$= \frac{731.14 + 652.10 + 762.95 + 874.13}{4}$$

$$= 755.08$$

TABLE 13.12

Uncentered 4-Point Moving Average for DJA Series

YEAR	UNCENTERED 4-POINT MOVING AVERAGE
1961	
1962	
1963	755.08
1964	814.61
1965	848.01
1966	883.55
1967	900.95
1968	858.73
1969	872.04
1970	866.95
1971	868.69
1972	899.57
1973	879.68
1974	859.11
1975	865.17
1976	842.98
1977	854.39
1978	863.46
1979	860.72
1980	876.69
1981	937.07
1982	1,042.05
1983	1,103.94
1984	1,265.86
1985	1,478.21
1986	

The complete uncentered 4-point moving average for the DJA series is given in Table 13.12.

It is often inconvenient to use an uncentered moving average because its time periods do not correspond to the time periods of the original time series. For this reason a **centered moving average** is defined, for even N-point moving averages, as the mean of each consecutive pair of uncentered moving average values. For example, the first value of the centered 4-point moving average for the DJA series is

$$M_{1963} = \frac{M_{1962.5} + M_{1963.5}}{2}$$

$$= \frac{755.08 + 814.61}{2}$$

$$= 784.85$$

The general formula for the centered moving average for an even value of N is

$$M_t = \frac{M_{t-.5} + M_{t+.5}}{2}$$

The centered 4-point moving average for the DJA series is given in Table 13.13, and shown graphically with the 7-point moving average in Figure 13.7 (page 778). Note that the 4-point moving average is not as smooth as the 7-point moving average, but it more quickly reflects changes in the series. Also, only two points are lost on each end of the centered 4-point moving average. The method for constructing moving averages is summarized in the box on page 779.

TABLE 13.13 Centered 4-Point Moving Average

YEAR	CENTERED 4-POINT MOVING AVERAGE	YEAR	CENTERED 4-POINT MOVING AVERAGE
1961	—	1974	862.14
1962	—	1975	854.08
1963	784.85	1976	848.69
1964	831.31	1977	858.92
1965	865.78	1978	862.09
1966	892.25	1979	868.71
1967	879.84	1980	906.88
1968	865.39	1981	989.56
1969	869.50	1982	1,073.00
1970	867.82	1983	1,184.90
1971	884.13	1984	1,372.04
1972	889.63	1985	—
1973	869.40	1986	—

FIGURE 13.7   4-Point and 7-Point Moving Averages for Dow Jones Average Time Series

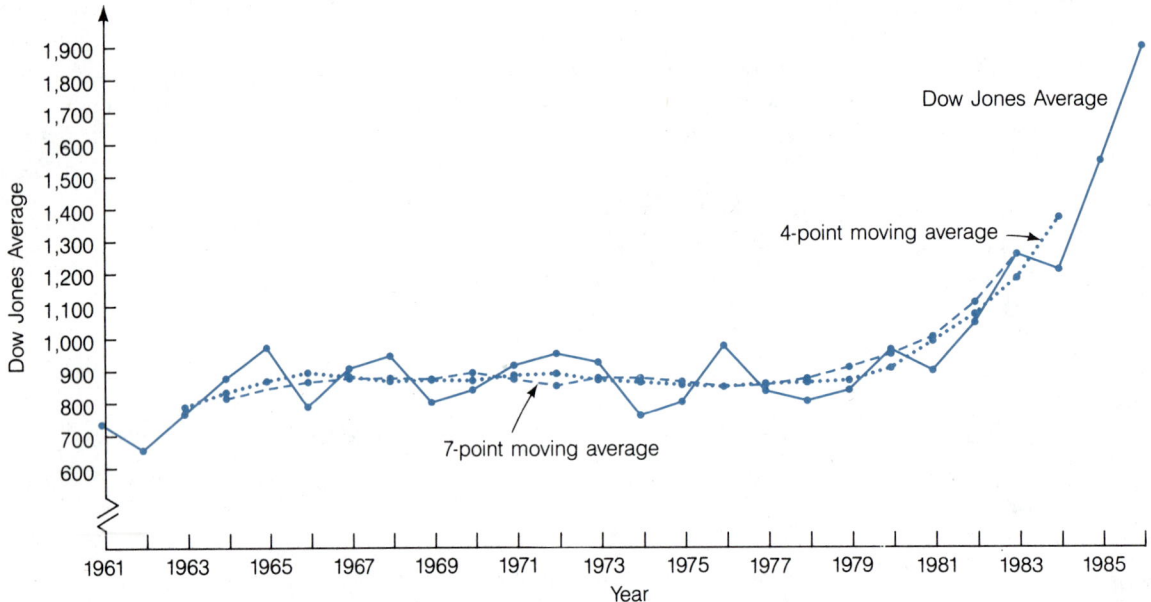

You may have noticed that the N-point moving average for an odd value of N assigns equal weight to each of the N values of the time series. Thus, for a 3-point moving average,

$$M_t = \frac{Y_{t-1} + Y_t + Y_{t+1}}{3} = \tfrac{1}{3}Y_{t-1} + \tfrac{1}{3}Y_t + \tfrac{1}{3}Y_{t+1}$$

Every value used to calculate M_t is given an equal weight of $\tfrac{1}{3}$. On the other hand, the centered 4-point moving average is

$$M_t = \frac{M_{t-.5} + M_{t+.5}}{2}$$

$$= \frac{\dfrac{Y_{t-2} + Y_{t-1} + Y_t + Y_{t+1}}{4} + \dfrac{Y_{t-1} + Y_t + Y_{t+1} + Y_{t+2}}{4}}{2}$$

$$= \tfrac{1}{8}Y_{t-2} + \tfrac{1}{4}Y_{t-1} + \tfrac{1}{4}Y_t + \tfrac{1}{4}Y_{t+1} + \tfrac{1}{8}Y_{t+2}$$

Thus, for N even, the centered N-point moving average assigns one-half the weight to the first and last time periods as to the middle time periods used in the calculation of M_t.

STEPS FOR CONSTRUCTING A MOVING AVERAGE

1. Select N, the number of consecutive time series values that will be averaged to form the N-point moving average. Remember that the larger is the value of N, the smoother is the moving average, and the more points that are lost on each end of the series. Usually you will need to try several values of N before deciding which is most appropriate.

2. If N is odd, the N-point moving average is the mean of N consecutive values of the time series:

$$M_t = \frac{Y_{t-(N-1)/2} + \cdots + Y_t + \cdots + Y_{t+(N-1)/2}}{N}$$

Note that $(N - 1)/2$ points are lost on each end of the series.

3. If N is even, first calculate the uncentered moving average as the mean of N consecutive values of the time series:

$$M_{t-.5} = \frac{Y_{t-N/2} + \cdots + Y_t + \cdots + Y_{t+N/2-1}}{N}$$

and

$$M_{t+.5} = \frac{Y_{t-N/2+1} + \cdots + Y_t + \cdots + Y_{t+N/2}}{N}$$

Then calculate the centered moving average by computing the mean of each adjacent pair of uncentered values:

$$M_t = \frac{M_{t-.5} + M_{t+.5}}{2}$$

Note that $N/2$ values are lost at each end of the series.

Both are specific examples of **weighted moving averages**, where M_t is defined as the weighted average of T values of the time series; i.e.,

$$M_t = \sum_{k=1}^{T} W_k Y_k$$

with the sum of the weights equal to 1:

$$\sum_{k=1}^{T} W_k = 1$$

Many other types of moving averages have been developed, using a variety of different weighting schemes to accomplish specific objectives. A scheme that has received much attention is one that assigns positive weight to current and past values of the time series, and zero weight to future values. This application of weighted moving average is called **exponential smoothing**, and is the topic of Section 13.5.

| | | | | | | | | | | | |

**EXERCISES
13.18–13.25**

LEARNING THE MECHANICS

13.18 Which will produce a smoother description of the trend of a time series, a 3- or 7-point moving average? Why?

13.19 Compute the missing values in the 3- and 5-point moving average columns of the table.

YEAR	PRICE OF GOLD T_t ($ per troy ounce)	3-POINT MOVING AVERAGE	5-POINT MOVING AVERAGE
1970	36		
1971	41	45.33	
1972	59		
1973	98		103.80
1974	160	139.66	120.60
1975	161	148.66	
1976	125		157.60
1977	148	155.66	
1978	194		277.60
1979	308	371.66	344.60
1980	613	460.33	
1981	460	483.00	436.20
1982	376	420.00	
1983	424		
1984	361		
1985	318		

Source: *Standard & Poor's Statistics: Metals,* Standard & Poor's Statistical Service, 1985, p. 236; *Standard & Poor's Current Statistics,* Nov. 1986.

13.20 Plot the gold-price time series and the 3-point moving average of Exercise 13.19 on a graph. Examine the differences between the time series and its 3-point moving average [i.e., examine the distances $(Y_t - M_t)$]. Can you identify any cyclical patterns in the time series?

13.21 Refer to Exercise 13.3. Calculate a 3-point and a 5-point moving average for U.S. beer production between 1970 and 1984.

APPLYING THE CONCEPTS

13.22 Consider the data on quarterly housing starts given in the table.
  a. Compute the missing values for the uncentered and centered 4-point moving average columns of the table.
  b. Plot the time series, Y_t, and the centered 4-point moving average on the same graph. Comment on the smoothing achieved by the moving average.

YEAR	QUARTER	HOUSING STARTS Y_t (thousands of dwellings)	UNCENTERED 4-POINT MOVING AVERAGE	CENTERED 4-POINT MOVING AVERAGE M_t
1980	I	218.7		
	II	269.7		
	III	341.4	297.78	300.36
	IV	361.3	302.93	
1981	I	239.3	283.63	261.42
	II	300.4	239.20	
	III	233.5	220.73	229.97
	IV	183.6	214.13	
1982	I	165.4	226.83	220.48
	II	274.0	252.48	239.66
	III	284.3		
	IV	286.2	327.43	
1983	I	295.5	368.20	347.82
	II	443.7	389.93	379.07
	III	447.4	402.90	
	IV	373.1		
1984	I	347.4	407.83	
	II	492.4	401.90	
	III	418.4	396.58	399.24
	IV	349.4	400.78	398.68
1985	I	326.1	412.75	
	II	509.2		
	III	466.3		

Source: *Standard & Poor's Basic Statistics: Building and Building Materials*,
1984; *Standard & Poor's Current Statistics*, Nov. 1986.

13.23 The table lists the total number of cars and trucks sold by General Motors each year from 1970 to 1985. Calculate moving averages for $N = 3$ and $N = 7$. Plot both on the same graph along with the original data. Which moving average series tracks more closely to the original sales series? Explain.

YEAR	SALES (millions of units)	YEAR	SALES (millions of units)
1970	5.308	1978	9.482
1971	7.779	1979	8.993
1972	7.791	1980	7.101
1973	8.684	1981	6.762
1974	6.690	1982	6.244
1975	6.629	1983	7.769
1976	8.568	1984	8.256
1977	9.068	1985	9.305

Source: *Moody's Industrial Manual*, 1986, p. 1401.

13.24 The total number of commercial and industrial business failures in the United States is an indicator of the general business climate. The table lists the monthly failures for the years 1981–1983.

	JAN.	FEB.	MAR.	APR.	MAY	JUNE	JULY	AUG.	SEPT.	OCT.	NOV.	DEC.
1981	1,109	1,133	1,212	1,557	1,464	1,408	1,432	1,172	1,777	1,604	1,368	1,558
1982	1,535	1,854	2,088	1,882	1,941	2,452	1,631	2,441	2,507	2,457	2,102	2,018
1983	2,455	2,397	2,881	2,471	2,292	2,841	2,313	3,218	2,384	2,511	3,287	2,484

Source: *Business Statistics: 1984,* Department of Commerce, Bureau of Economic Analysis, 1985, p. 22.

a. Plot the time series.

b. Calculate moving averages for $N = 3$, 5, and 7. Plot each moving average series on the same graph. Which moving average series best characterizes the long-run trend? Explain.

13.25 Refer to Exercise 13.7, where the average retail prices of gasoline for 85 urban areas in the United States are given for January 1980 through September 1986. To see the trend in gasoline prices, compute the 5-point moving average for the series, and then plot the retail prices and the moving average on the same graph.

S E C T I O N 13.5

EXPONENTIAL SMOOTHING

As we saw in Section 13.4, moving averages smooth time series data by creating a weighted average of past, current, and future values of the series. However, the objective of smoothing is often to provide a means for forecasting future values of the time series. Moving averages are not useful for forecasting because their calculation requires the use of future values. (Recall that values at the end of the series are lost because the future values are unavailable.) To be useful for forecasting, a weighted moving average that assigns no weight to future values must be defined.

Exponential smoothing is one type of weighted average that assigns positive weights to past and current values only. A single weight, w, called the **exponential smoothing constant**, is selected so that w is between 0 and 1. Then the exponentially smoothed series, E_t, is calculated as follows:

$$E_1 = Y_1$$
$$E_2 = wY_2 + (1 - w)E_1$$
$$E_3 = wY_3 + (1 - w)E_2$$
$$\vdots$$
$$E_t = wY_t + (1 - w)E_{t-1}$$

Thus, the exponentially smoothed value at time t assigns the weight w to the current series value and the weight $(1 - w)$ to the previous smoothed value.

For example, consider the Dow Jones Average time series in Table 13.11. Suppose we want to calculate the exponentially smoothed series using a smoothing constant of $w = .3$. The calculations proceed as follows:

$$E_{1961} = Y_{1961} = 731.14$$

$$E_{1962} = .3Y_{1962} + (1 - .3)E_{1961}$$
$$= .3(652.10) + .7(731.14) = 707.43$$

$$E_{1963} = .3Y_{1963} + (1 - .3)E_{1962}$$
$$= .3(762.95) + .7(707.43) = 724.09$$

$$\vdots$$

All the exponentially smoothed values corresponding to $w = .3$ are given in Table 13.14. Note that no values are lost at either end of the smoothed series.

TABLE 13.14

Dow Jones Industrial Average with Exponential Smoothing (1961–1986)

YEAR	DJA	EXPONENTIALLY SMOOTHED DJA ($w = .3$)	YEAR	DJA	EXPONENTIALLY SMOOTHED DJA ($w = .3$)
1961	731.14	731.14	1974	759.37	856.12
1962	652.10	707.43	1975	802.49	840.03
1963	762.95	724.09	1976	974.92	880.50
1964	874.13	769.10	1977	835.15	866.90
1965	969.26	829.15	1978	805.01	848.33
1966	785.69	816.11	1979	838.74	845.45
1967	905.11	842.81	1980	963.99	881.01
1968	943.75	873.09	1981	899.01	886.41
1969	800.36	851.27	1982	1,046.54	934.45
1970	838.92	847.57	1983	1,258.64	1,031.71
1971	884.76	858.73	1984	1,211.57	1,085.67
1972	950.71	886.32	1985	1,546.67	1,223.97
1973	923.88	897.59	1986	1,895.95	1,425.56

Source: *The Dow Jones Investor's Handbook,* 1986; *Wall Street Journal,* Jan. 2, 1987.

The DJA and exponentially smoothed DJA are graphed in Figure 13.8 (page 784). Like many averages, the exponentially smoothed series changes less rapidly than the time series itself. The choice of w affects the smoothness of E_t. The smaller (closer to 0) is the value of w, the smoother is E_t. Since small values of w give more weight to the past values of the time series, the smoothed series is not affected by rapid changes in the current values and, therefore, appears smoother than the original series. Conversely, choosing w near 1 yields an exponentially smoothed series that is much like the original series. That is, large values of w give more weight to the current value of the time series so the smoothed series looks like the original series. This concept is illustrated in Figure 13.9. The steps for calculating an exponentially smoothed series are given in the box on page 785.

FIGURE 13.8 Exponentially Smoothed Values ($w = .3$) for the Dow Jones Average

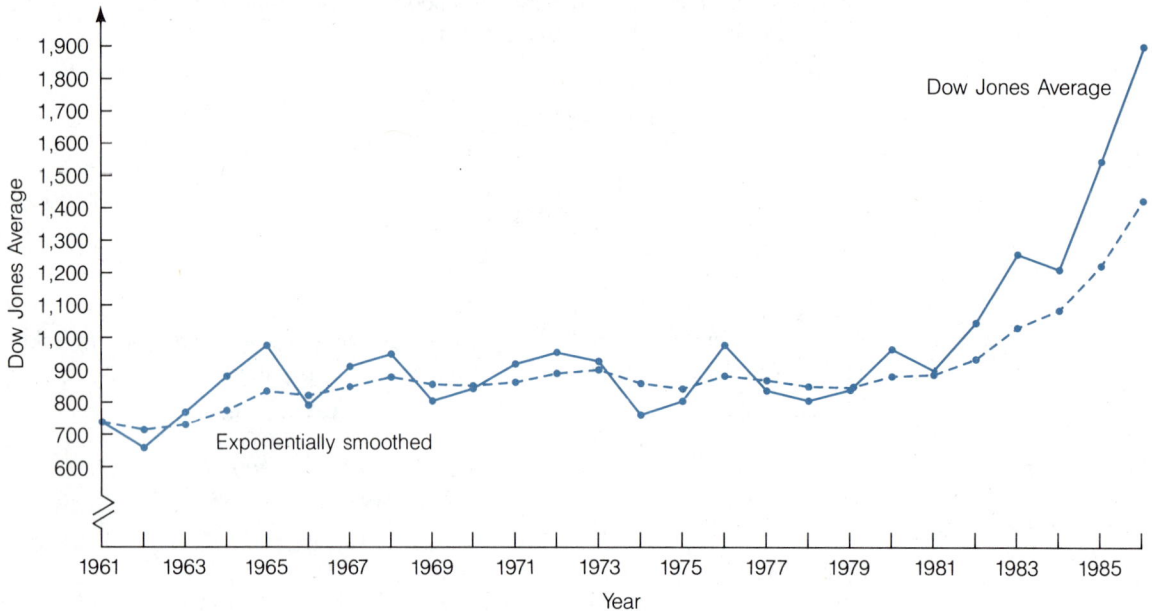

FIGURE 13.9 Exponentially Smoothed Values ($w = .3$ and $w = .7$) for the Dow Jones Average

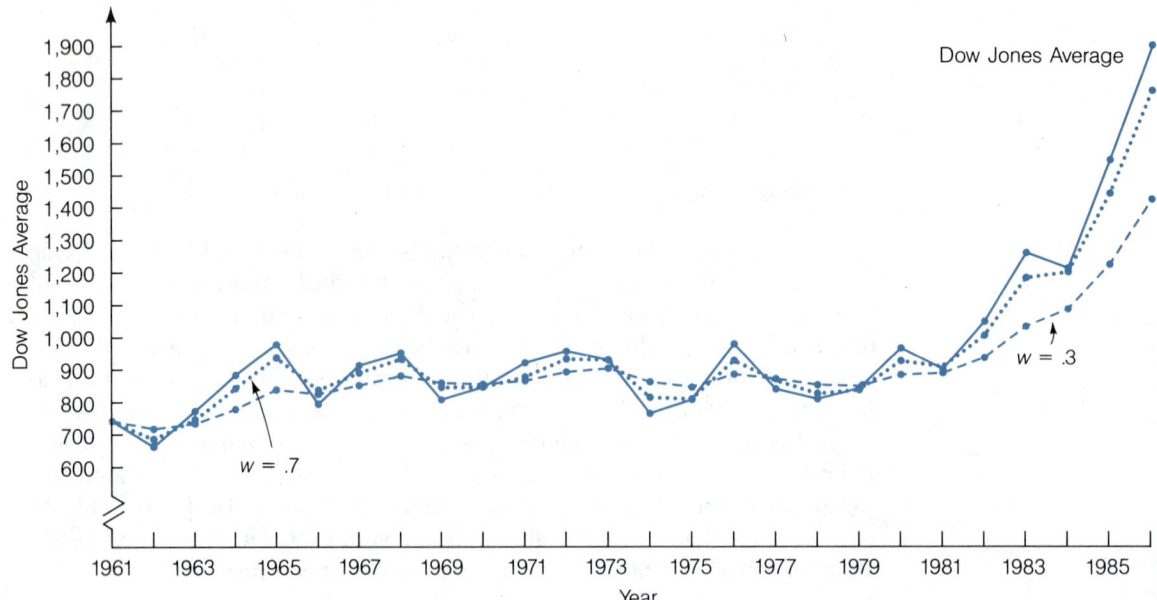

STEPS FOR CALCULATING AN EXPONENTIALLY SMOOTHED SERIES

1. Select an exponential smoothing constant, w, between 0 and 1. Remember that small values of w give less weight to the current value of the series and yield a smoother series. Larger choices of w assign more weight to the current value of the series and yield a more variable series.

2. Calculate the exponentially smoothed series E_t from the original time series Y_t as follows:

$$E_1 = Y_1$$
$$E_2 = wY_2 + (1 - w)E_1$$
$$E_3 = wY_3 + (1 - w)E_2$$
$$\vdots$$
$$E_t = wY_t + (1 - w)E_{t-1}$$

EXAMPLE 13.5

Consider the IBM common stock price from January 1984 to December 1986, shown in Table 13.15. Create the exponentially smoothed series using $w = .5$, and plot both series.

TABLE 13.15 IBM Stock Prices and Exponentially Smoothed Series

1984	IBM STOCK PRICE	EXPONENTIALLY SMOOTHED STOCK PRICE ($w = .5$)	1985	IBM STOCK PRICE	EXPONENTIALLY SMOOTHED STOCK PRICE ($w = .5$)	1986	IBM STOCK PRICE	EXPONENTIALLY SMOOTHED STOCK PRICE ($w = .5$)
Jan.	$114.125	$114.125	Jan.	$136.375	$129.503	Jan.	$151.500	$148.071
Feb.	110.250	112.187	Feb.	134.000	131.752	Feb.	150.875	149.473
Mar.	114.000	113.094	Mar.	127.000	129.376	Mar.	149.125	149.299
Apr.	113.750	113.422	Apr.	126.500	127.938	Apr.	156.250	152.775
May	107.750	110.586	May	128.625	128.281	May	152.375	152.575
June	105.750	108.168	June	123.750	126.016	June	146.500	149.537
July	110.750	109.459	July	131.375	128.695	July	132.500	141.019
Aug.	123.750	116.604	Aug.	126.625	127.660	Aug.	138.750	139.884
Sept.	124.250	120.427	Sept.	123.875	125.768	Sept.	134.500	137.192
Oct.	124.625	122.526	Oct.	129.875	127.821	Oct.	123.625	130.409
Nov.	121.750	122.138	Nov.	139.750	133.786	Nov.	127.125	128.767
Dec.	123.125	122.632	Dec.	155.500	144.643	Dec.	120.000	124.383

Source: *Daily Stock Price Record*, NYSE, 1984–1986; *Standard & Poor's Security Owner's Stock Guide*, Aug. 1986–Jan. 1987.

SOLUTION

To create the exponentially smoothed series with $w = .5$, we calculate

$$E_1 = Y_1 = 114.125$$
$$E_2 = wY_2 + (1 - w)E_1$$
$$= .5(110.250) + .5(114.125) = 112.187$$
$$\vdots$$
$$E_{36} = wY_{36} + (1 - w)E_{35}$$
$$= .5(120.000) + .5(128.767) = 124.383$$

The plot of the original and exponentially smoothed series is shown in Figure 13.10.

FIGURE 13.10
IBM Stock Price and
Exponentially Smoothed
Price ($w = .5$)

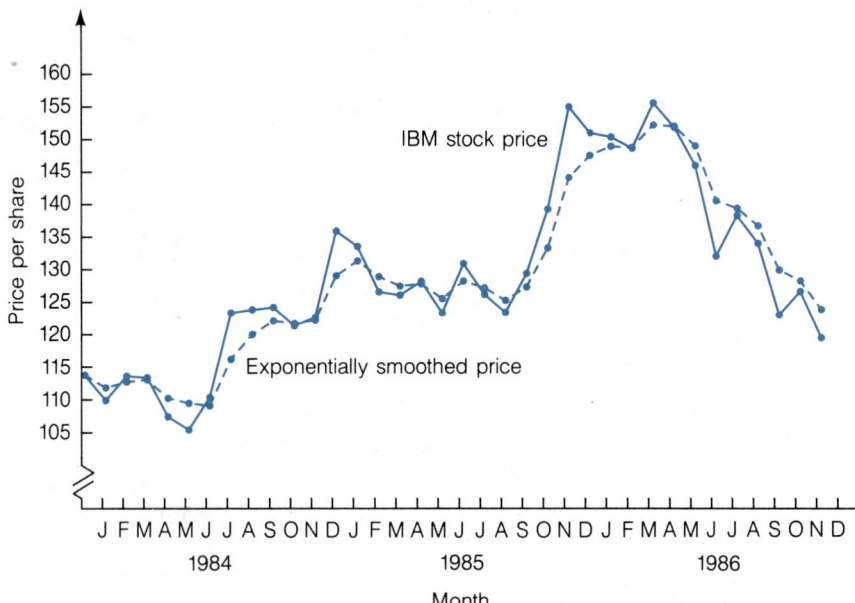

The smoothed series provides a good picture of the general trend of the original series. Notice, however, that the exponentially smoothed series and the original price series are very similar in appearance. This is because the IBM price series is itself relatively smooth over this time period. Nevertheless, the exponentially smoothed series will be less sensitive to any short-term deviations of the prices from the trend and therefore, may be useful even for relatively smooth series.   ■

One of the primary uses of exponential smoothing is for forecasting future values of a time series. Because only current and past values of the time series are used in exponential smoothing, it is easily adapted to forecasting. We will demonstrate this application of exponentially smoothed series in Chapter 15.

EXERCISES
13.26–13.33

LEARNING THE MECHANICS

13.26 Describe the effect of selecting an exponential constant of $w = .2$. Of $w = .8$. Which will produce a smoother trend?

13.27 The table lists the number of Chevrolet passenger cars sold by General Motors to automotive dealers in the United States and Canada from 1977 to 1985.

YEAR	SALES (millions of units)	EXPONENTIALLY SMOOTHED SALES ($w = .5$)
1977	2.133	
1978	2.349	2.241
1979	2.233	2.237
1980	1.740	
1981	1.444	
1982	.986	
1983	1.289	
1984	1.455	
1985	4.882	

Source: *Moody's Industrial Manual*, 1986, p. 1401.

a. Calculate the missing values in the exponentially smoothed series using $w = .5$.
b. Graph the time series and the exponentially smoothed series on the same graph.

13.28 Refer to Exercise 13.22.
a. Calculate the exponentially smoothed series for housing starts using a smoothing constant of $w = .5$.
b. Plot on a graph both the centered 4-point moving average from Exercise 13.22 and the exponentially smoothed series in part **a** of this exercise. Comment on the differences between the two smoothed series.

13.29 Refer to Exercise 13.3.
a. Calculate the exponentially smoothed series for U.S. beer production for the period 1970–1984 using $w = .2$.
b. Calculate the exponentially smoothed series using $w = .8$.
c. Plot the two exponentially smoothed series ($w = .2$ and $w = .8$) on the same graph.

APPLYING THE CONCEPTS

13.30 Standard & Poor's 500 Stock Composite Average (S&P 500) is a stock market index. Like the Dow Jones Average, it is an indicator of stock market activity. The table at the top of page 788 contains end-of-quarter values of the S&P 500 for the years 1971–1985.
a. Compute an exponentially smoothed series for the S&P 500 data for the period 1971–1985 using a smoothing coefficient of $w = .7$.
b. Plot the original time series and the exponentially smoothed series on the same graph.

YEAR	QUARTER	S&P 500	YEAR	QUARTER	S&P 500	YEAR	QUARTER	S&P 500
1971	I	100.31	1976	I	102.77	1981	I	134.94
	II	99.20		II	104.28		II	129.06
	III	98.34		III	105.24		III	118.77
	IV	102.09		IV	107.46		IV	119.13
1972	I	107.20	1977	I	98.42	1982	I	117.09
	II	107.14		II	100.48		II	109.82
	III	110.55		III	96.53		III	121.79
	IV	118.06		IV	95.10		IV	138.91
1973	I	111.52	1978	I	89.21	1983	I	152.96
	II	104.26		II	95.53		II	168.11
	III	108.43		III	102.54		III	166.07
	IV	97.55		IV	96.11		IV	164.93
1974	I	93.98	1979	I	101.59	1984	I	159.18
	II	86.00		II	102.91		II	153.18
	III	63.54		III	109.32		III	166.10
	IV	68.56		IV	107.94		IV	167.24
1975	I	83.36	1980	I	105.36	1985	I	180.66
	II	95.19		II	113.72		II	191.85
	III	83.87		III	127.14		III	182.08
	IV	98.19		IV	131.44		IV	211.28

Source: Standard & Poor's, *The Analysts Handbook*, Jan. 1987, p. 5.

13.31 Refer to Exercise 13.19. Use the exponential smoothing constant $w = .5$ to smooth the gold price series between 1970 and 1985.

13.32 Refer to Exercise 13.8. Using $w = .4$, compute an exponentially smoothed series for each of the two time series: agricultural and nonagricultural employment. Plot each smoothed series.

13.33 There has been phenomenal growth in the transportation sector of the economy since 1960. The personal consumption expenditure figures are given in the table.

YEAR	PERSONAL CONSUMPTION EXPENDITURE ON TRANSPORTATION ($ billions)	YEAR	PERSONAL CONSUMPTION EXPENDITURE ON TRANSPORTATION ($ billions)	YEAR	PERSONAL CONSUMPTION EXPENDITURE ON TRANSPORTATION ($ billions)
1960	42.4	1969	75.7	1977	179.3
1961	44.8	1970	80.6	1978	198.1
1962	47.4	1971	92.3	1979	219.4
1963	49.5	1972	105.4	1980	236.6
1964	54.3	1973	114.6	1981	261.5
1965	58.4	1974	117.9	1982	267.3
1966	60.4	1975	129.4	1983	291.9
1967	63.3	1976	155.2	1984	319.5
1968	69.3				

Source: *Statistical Abstract of the United States: 1986*, U.S. Bureau of the Census.

a. Compute exponentially smoothed values of this personal consumption time series using the smoothing constants $w = .2$ and $w = .8$.

b. Plot the actual series and the two smoothed series on the same graph. Comment on the trend in personal consumption expenditure on transportation in the 1970's and early 1980's as compared to the 1960's.

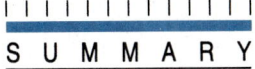

SUMMARY

Time series are observations made sequentially over time. **Index numbers** measure the changes in a time series or group of time series. **Simple index numbers** are based on a single series, while **composite index numbers** measure changes in several series simultaneously. Price indexes that are weighted by purchase quantities are **weighted composite indexes**. **Laspeyres indexes** use weights that are base period purchase quantities, while **Paasche indexes** use the current period purchase quantities as weights.

Smoothing techniques are used to make it easier to discern trends in time series. **Moving averages** combine past, current, and future values of the time series. **Exponential smoothing** combines past and current values of the series.

SUPPLEMENTARY EXERCISES 13.34–13.45

13.34 The U.S. steel industry was the object of much economic attention in the 1970's and early 1980's due to the increasing market share of imported steel, the effects of several recessions, and other economic woes. Prices of different varieties of U.S. steel are given in the table for the period 1971–1984.

YEAR	COLD ROLLED STEEL (¢/pound)	HOT ROLLED STEEL (¢/pound)	GALVANIZED STEEL (¢/pound)	YEAR	COLD ROLLED STEEL (¢/pound)	HOT ROLLED STEEL (¢/pound)	GALVANIZED STEEL (¢/pound)
1971	10.00	7.48	9.61	1978	23.11	15.53	20.47
1972	10.77	8.40	10.88	1979	25.55	17.05	22.32
1973	11.08	8.40	10.59	1980	26.50	18.46	23.88
1974	12.78	9.10	12.39	1981	31.50	20.15	26.88
1975	16.03	11.13	14.80	1982	33.25	20.80	26.75
1976	18.16	12.20	16.07	1983	36.17	22.23	28.43
1977	20.39	13.79	18.10	1984	28.15	23.75	30.30

Source: *Standard & Poor's Basic Statistics: Metals, 1985.*

a. Compute 3-point moving averages for each of the three price series.

b. Plot the three price series and their smoothed 3-point moving averages on the same graph.

13.35 Refer to Exercise 13.34.

a. Compute the exponentially smoothed series corresponding to each of the price series using the smoothing constant $w = .5$.

b. Plot the prices and their exponentially smoothed series on the same graph.

c. What is the main advantage associated with using exponential smoothing instead of moving averages for relatively short series like these?

13.36 Refer to Exercise 13.34.

  a. Calculate a simple composite index for the three steel price series using 1977 as the base period.
  b. Is the index a price index or a quantity index?
  c. What information would you need in order to calculate a Laspeyres index with a base period of 1977? A Paasche index with a base period of 1977?

13.37 Foreign exchange rates, the values of foreign currency in U.S. dollars, are important to investors and international travellers. The table lists the monthly foreign exchange rates of the British pound in 1984 and 1985.

MONTH	1984 ($/£)	1985 ($/£)	MONTH	1984 ($/£)	1985 ($/£)
Jan.	1.41	1.13	July	1.32	1.38
Feb.	1.44	1.10	Aug.	1.31	1.39
Mar.	1.45	1.13	Sept.	1.26	1.36
Apr.	1.43	1.23	Oct.	1.23	1.42
May	1.39	1.25	Nov.	1.24	1.44
June	1.37	1.28	Dec.	1.18	1.44

Source: *Standard & Poor's Statistical Service, Current Statistics*, 1986.

  a. Calculate a simple index for the foreign exchange rate series using January 1984 as the base period.
  b. Plot the index, and use the plot to identify the best time for a U.S. traveller to visit Britain during this period.

13.38 Refer to Exercise 13.33. Using 1975 as the base period, compute a simple index for the personal consumption series.

13.39 A major portion of total consumer credit is extended in the categories of automobile loans, mobile home loans, and revolving credit. Amounts outstanding (in billions of dollars) for the period 1974–1984 are given in the table.

YEAR	AUTOMOBILE	MOBILE HOME	REVOLVING CREDIT	YEAR	AUTOMOBILE	MOBILE HOME	REVOLVING CREDIT
1974	54.3	14.6	13.7	1980	112.2	18.8	58.5
1975	57.2	14.4	15.0	1981	119.8	19.9	64.5
1976	67.7	14.6	17.2	1982	126.3	22.4	69.6
1977	82.9	15.0	39.3	1983	143.1	23.9	82.0
1978	101.6	15.2	48.3	1984	172.6	24.6	101.6
1979	116.4	16.8	57.0				

Source: *Statistical Abstract of the United States: 1986*, U.S. Bureau of the Census, p. 502.

  a. Calculate a simple composite index using 1974 as the base period.
  b. Compute a simple composite index for the series using 1980 as the base period.
  c. Are the indexes constructed in parts **a** and **b** price or quantity indexes?
  d. Compute a simple index for automobile loans using 1980 as the base. Plot the simple index and the composite index from part **b** on the same graph.

13.40 Refer to Exercise 13.39.

    a. Calculate a 3-point moving average of the simple composite index generated in part **b** of that exercise.

    b. Using a smoothing constant of $w = .3$, calculate an exponentially smoothed series corresponding to the simple composite index.

    c. Plot the simple composite index, the 3-point moving average, and the exponentially smoothed series on the same graph. Comment on the relative smoothness of the three series.

13.41 Refer to Exercise 13.39. Assume that in 1980 the number of outstanding loans of each type are as follows:

              Automobile: 40,000

           Mobile home: 10,000

      Revolving credit: 100,000

    a. Calculate a Laspeyres index for 1980–1984 using 1980 as the base and the quantities given above.

    b. Which category of credit is given most weight in the calculation of the Laspeyres index?

13.42 Refer to Exercise 13.39. Suppose the numbers of outstanding loans in each category from 1980 to 1984 are as shown in the table.

YEAR	AUTOMOBILE	MOBILE HOME	REVOLVING CREDIT
1980	40,000	10,000	100,000
1981	45,000	11,000	90,000
1982	50,000	15,000	80,000
1983	53,000	17,000	82,000
1984	60,000	21,000	89,000

    a. Calculate the Paasche index for 1980–1984 using 1980 as a base and the quantities given in the table.

    b. Compare the simple composite index (from Exercise 13.39), the Laspeyres index (from Exercise 13.41), and the Paasche index. Explain why the 1982 values are different for each index, and interpret each.

13.43 Three of many indicators used for measuring the level of economic activity are the index of net business formation, the index of industrial production, and the index of new private housing units authorized by local building permits. End-of-year values of these indicators for the period 1970–1984 are given in the table at the top of page 792.

    a. Calculate a simple composite index for the three indicator series, using 1970 as the base period.

    b. Compute the simple index for each of the three indicator series, using a base period of 1970 for each. Plot the three simple indexes and the simple composite index on the same graph.

YEAR	INDEX OF NET BUSINESS FORMATION	INDEX OF INDUSTRIAL PRODUCTION	INDEX OF NEW PRIVATE HOUSING UNITS AUTHORIZED	YEAR	INDEX OF NET BUSINESS FORMATION	INDEX OF INDUSTRIAL PRODUCTION	INDEX OF NEW PRIVATE HOUSING UNITS AUTHORIZED
1970	106.4	107.8	111.0	1978	128.2	144.1	156.3
1971	108.5	109.6	158.8	1979	128.3	152.5	135.1
1972	115.9	119.7	182.4	1980	122.4	147.0	100.0
1973	114.9	129.8	158.4	1981	118.6	151.0	83.9
1974	109.2	129.3	103.6	1982	113.2	138.6	82.2
1975	107.0	117.8	89.8	1983	114.8	147.6	131.8
1976	115.6	130.5	119.0	1984	117.1	163.3	135.4
1977	123.2	138.2	153.8				

Source: *Statistical Abstract of the United States: 1986*, pp. 521, 725; *Annual U.S. Economic Data*, Federal Reserve Bank of St. Louis, 1985, p. 15.

13.44 Refer to Exercise 13.43.

a. Calculate an exponentially smoothed series corresponding to the business formation index using $w = .2$. Using $w = .8$.

b. Plot on a graph both the business formation index and the two exponentially smoothed series from part **a**. Which exponential smoothing constant yields a smoother series? Explain.

13.45 The number of dollars a person receives in a year is referred to as his or her **monetary** (or **money**) **income**. This figure can be adjusted to reflect the purchasing power of the dollars received relative to the purchasing power of dollars in some base period. The result is called a person's **real income**. Monetary income and real income can be compared to determine, for example, whether an increase in a person's monetary income truly reflects an increase in his or her purchasing power. The Consumer Price Index (CPI) can be used to adjust monetary income to obtain real income (in terms of 1967 dollars). To compute your real income for a specific year, simply divide your monetary income for the year by that year's CPI and multiply by 100. The table lists the CPI for each year during the period 1970–1985.

YEAR	CPI	YEAR	CPI	YEAR	CPI
1970	116.3	1976	170.5	1981	272.4
1971	121.3	1977	181.5	1982	289.1
1972	125.3	1978	195.4	1983	298.4
1973	133.1	1979	217.4	1984	311.1
1974	147.7	1980	246.8	1985	322.2
1975	161.2				

Source: *Statistical Abstract of the United States: 1986*, U.S. Bureau of the Census.

a. Suppose your monetary income increased from $20,000 in 1970 to $46,000 in 1985. What were your real incomes in 1970 and 1985? Were you able to buy more goods and services in 1970 or in 1985? Explain.

b. What monetary income would have been required in 1985 to provide equivalent purchasing power to a 1970 monetary income of $20,000?

ON YOUR OWN . . .

Select a time series of interest to you. It should have at least 24 consecutive values (years, quarters, months, etc.).

a. Calculate a simple index for the series using the first value as the base. Plot the index. How much does the series change from the first value to the last?
b. Use a 5-point moving average to smooth the index values, and plot the moving average on the graph you constructed in part a.
c. Examine the plot of the simple index relative to the moving average. If in your judgment the index exhibits significantly greater variability than the smoothed series, calculate and plot the exponentially smoothed series with $w = .3$. If the index and moving average are each relatively smooth, use $w = .7$.
d. What do the smoothed values of the index indicate about the long-term trend of the series?

REFERENCES

Box, G. E. P. and Jenkins, G. M. *Time Series Analysis: Forecasting and Control*, 2nd ed. San Francisco: Holden-Day, 1977.

Gillingham, R. and Lane, W., "Changing the treatment of shelter costs for homeowners in the CPI." *Monthly Labor Review*, June 1982, 9–14.

Mendenhall, W. and Sincich, T. *A Second Course in Business Statistics: Regression Analysis*, 2nd ed. San Francisco: Dellen, 1986.

Nelson, C. R. *Applied Time Series Analysis for Managerial Forecasting*. San Francisco: Holden-Day, 1983.

U.S. Department of Labor, *BLS Handbook of Methods*. Vol. II. "The Consumer Price Index." Bureau of Labor Statistics, Bulletin 2134-2, Apr. 1984.

U.S. Department of Labor, *The Consumer Price Index: Concepts and Content over the Years*. Bureau of Labor Statistics, Report 517, May 1978.

U.S. Department of Labor, *CPI Issues*. Bureau of Labor Statistics, Report S93, Feb. 1980.

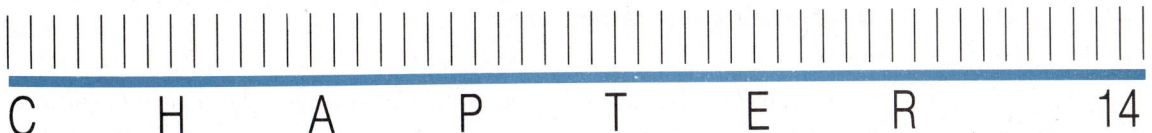

WHERE WE'VE BEEN . . .

In Chapter 13 we discussed methods for describing time series. Index numbers were used to describe changes in a time series; moving averages and exponential smoothing were introduced to describe trends.

WHERE WE'RE GOING . . .

In Chapter 14 we use mathematical models (like the regression models of Chapters 10–12) to describe time series. These models range in complexity from the relatively simple exponential smoothing model to time series models that account for correlation between values observed at different points in time. The primary objective of constructing these models is to use them for forecasting future values of the time series.

CONTENTS

TIME SERIES: MODELS AND FORECASTING

In Chapter 13 we showed how to use various *descriptive* techniques to obtain a picture of the behavior of a time series. Now we want to expand our coverage to include techniques that will enable us to make statistical inferences about the time series. These *inferential* techniques are generally focused on the problem of *forecasting* future values of the time series, and we will discuss a number of methods for predicting the future with something other than a crystal ball. And, unlike fortune tellers, we will show how to provide measures of reliability for the forecasts. Nevertheless, forecasting is one of the more precarious types of statistical inference, because we are trying to predict a value outside the region of the sample data.

We will find that separating the time series into basic components often assists in the modeling and forecasting process. We discuss these time series components in Section 14.1. We then present several of the popular forecasting techniques in Sections 14.2–14.5. In Section 14.6 we show how to use regression residuals to test the assumption that the errors are uncorrelated. Finally, in Section 14.7 we show how to take advantage of residual correlation to improve time series forecasts.

| | | | | | | | | | | | | |

SECTION 14.1

TIME SERIES COMPONENTS

Before forecasts of future values of a time series can be made, some type of model that can be projected into the future must be used to describe the series. Time series models range in complexity from **descriptive models**, such as the simple moving averages and exponential smoothing models discussed in Chapter 13, to **inferential models**, such as the combinations of regression and specialized time series models to be discussed in this chapter. Whether the model is simple or complex, the objective is the same: to produce accurate forecasts of future values of the time series.

Many different algebraic representations of time series models have been proposed. One of the most widely used is an **additive model*** of the form

$$Y_t = T_t + C_t + S_t + R_t$$

The **secular trend**, T_t, also known as the **long-term trend**, is a time series that describes the long-term movements of Y_t. For example, if you want to characterize the secular trend of the production of automobiles since 1930, you would show T_t as an upward-moving time series over the period from 1930 to the present. This does not imply that the automobile production series has always moved upward from month to month and from year to year, but it does mean the long-term trend has been an increasing one over that period of time.

The **cyclical effect**, C_t, generally describes fluctuations of the time series about the secular trend that are attributable to business and economic conditions at the time. For example, the closing Dow Jones Average for the last business day of the year for the years 1961–1986 is given in Table 14.1. You can see in Figure 14.1 that it has a generally increasing secular trend. However, during periods of recession,

---

*Another useful form of the model is **multiplicative**: $Y_t = T_t C_t S_t R_t$. This can be changed to an additive form by taking natural logarithms, i.e., $\ln Y_t = \ln T_t + \ln C_t + \ln S_t + \ln R_t$.

TABLE 14.1
Dow Jones Industrial
Average, 1961–1986

YEAR	DJA	YEAR	DJA	YEAR	DJA
1961	731.14	1970	838.92	1979	838.74
1962	652.10	1971	884.76	1980	936.99
1963	762.95	1972	950.71	1981	899.01
1964	874.13	1973	923.88	1982	1,046.54
1965	969.26	1974	759.37	1983	1,258.64
1966	785.69	1975	802.49	1984	1,211.57
1967	905.11	1976	974.92	1985	1,546.67
1968	943.75	1977	835.15	1986	1,895.95
1969	800.36	1978	805.01		

Source: *The Dow Jones Investor's Handbook*, 1986; *The Wall Street Journal*, Jan. 2, 1987.

FIGURE 14.1 Secular Trend* for the Dow Jones Average

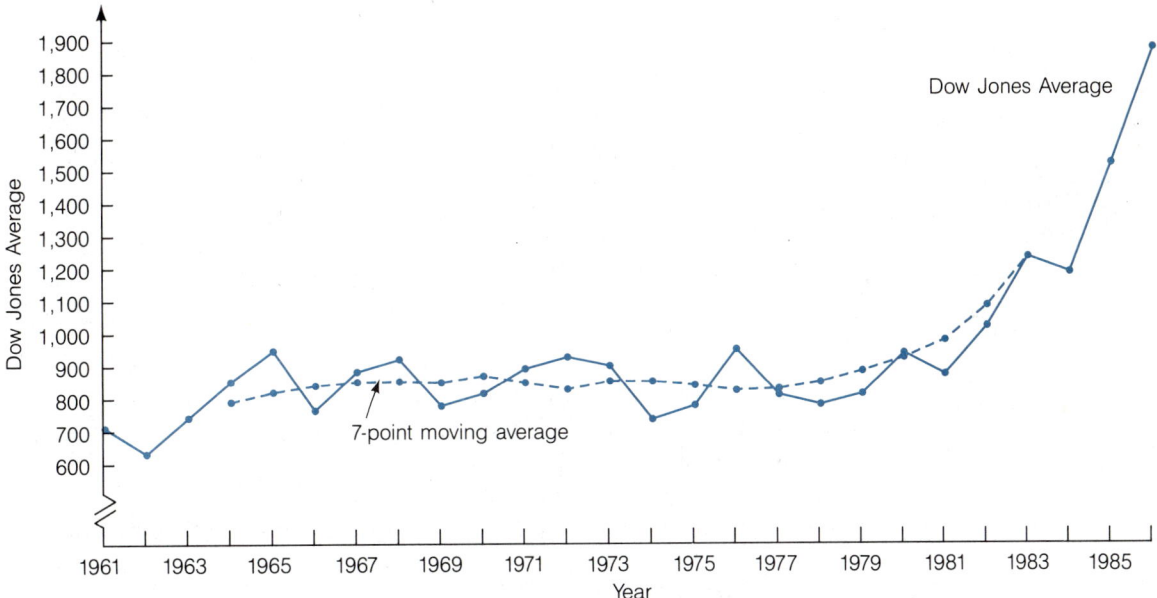

the Dow Jones Average tends to lie below the secular trend, while in times of general economic expansion, it lies above the long-term trend line.

The **seasonal effect**, S_t, describes the fluctuations in the time series that recur during specific time periods. For example, quarterly power loads for a utility company tend to be highest in the summer months (quarter III), with another smaller peak in the winter months (quarter I). The spring and fall (quarters II and IV) seasonal effects are negative, meaning that the series tends to lie below the long-term trend line during those quarters.

*The secular trend shown in Figure 14.1 is a 7-point moving average.

The **residual effect**, R_t, is what remains of Y_t after the secular, cyclical, and seasonal components have been removed. Part of the residual effect may be attributable to unpredictable rare events (earthquake, presidential assassination, people landing on the moon, etc.) and part to the randomness of human actions. In any case, the presence of the residual component makes it impossible to forecast the future values of a time series without error. Thus, the presence of the residual effect emphasizes a point we first made in Chapter 10 in connection with regression models: No business phenomena should be described by deterministic models. All realistic business models, time series or otherwise, should include a residual component.

Each of the four components contributes to the determination of the value of Y_t at each time period. Although it will not always be possible to characterize each component separately, the component model provides a useful theoretical formulation that helps the time series analyst achieve a better understanding of the phenomena affecting the path followed by the time series.

SECTION 14.2

FORECASTING: EXPONENTIAL SMOOTHING

In Chapter 13, we discussed exponential smoothing as a method for describing a time series by removing the irregular fluctuations. In terms of the time series components discussed in the previous section, exponential smoothing tends to deemphasize (or "smooth") most of the residual effects. This, coupled with the fact that exponential smoothing uses only past values of the series, makes it a useful tool for forecasting time series.

Recall that the formula for exponential smoothing is

$$E_t = wY_t + (1 - w)E_{t-1}$$

where w, the **exponential smoothing constant**, ranges from 0 to 1. The selection of w controls the smoothness of E_t. A choice near 0 places more emphasis (weight) on *past* values of the time series and, therefore, yields a smoother series. On the other hand, a choice near 1 gives more weight to *current* values of the series.

Suppose the objective is to forecast the next value of the time series, Y_{t+1}. The **exponentially smoothed forecast** for Y_{t+1} is simply the smoothed value at time t:

$$F_{t+1} = E_t$$

where F_{t+1} is the **forecast** of Y_{t+1}. To help interpret this forecast formula, substitute the smoothing formula for E_t:

$$\begin{aligned} F_{t+1} = E_t &= wY_t + (1 - w)E_{t-1} \\ &= wY_t + (1 - w)F_t \\ &= F_t + w(Y_t - F_t) \end{aligned}$$

Note that we have substituted F_t for E_{t-1}, since the forecast at time t is the smoothed value at time $(t - 1)$. The final equation provides insight into the exponential smoothing forecast: The forecast at time $(t + 1)$ is equal to the forecast at time t, F_t, plus a correction for the error in the forecast at time t, $(Y_t - F_t)$. This is the reason the exponentially smoothed forecast is referred to as an *adaptive forecast*—the forecast at time $(t + 1)$ is explicitly adapted for the error in the forecast at time t.

Because exponential smoothing consists of averaging past and present values, the smoothed values will tend to lag behind the series when a long-term trend exists. In addition, the averaging tends to smooth any seasonal component. Therefore, exponentially smoothed forecasts are appropriate only when the trend and seasonal components are relatively insignificant. Since the exponential smoothing model assumes that the time series has little or no trend or seasonal component, the forecast F_{t+1} is used to forecast not only Y_{t+1}, but also *all* future values of Y_t. That is, the forecast for two time periods ahead is

$$F_{t+2} = F_{t+1}$$

and for three time periods ahead is

$$F_{t+3} = F_{t+2} = F_{t+1}$$

The exponential smoothing forecasting technique is summarized in the box.

CALCULATION OF EXPONENTIALLY SMOOTHED FORECASTS

1. Given the observed time series, $Y_1, Y_2, \ldots, Y_t$, first calculate the exponentially smoothed values $E_1, E_2, \ldots, E_t$ using

   $E_1 = Y_1$
   $E_2 = wY_2 + (1 - w)E_1$

   $\vdots$

   $E_t = wY_t + (1 - w)E_{t-1}$

2. Use the last smoothed value to forecast the next time series value:

   $F_{t+1} = E_t$

3. Assuming that Y_t is relatively free of trend and seasonal components, use the same forecast for all future values of Y_t:

   $F_{t+2} = F_{t+1}$
   $F_{t+3} = F_{t+1}$

   $\vdots$

Two important points must be made about exponentially smoothed forecasts:

1. The choice of w is critical. If you decide that w will be small (near 0), you will obtain a smooth, slowly changing series of forecasts. On the other hand, the

selection of a large value of w (near 1) will yield more rapidly changing forecasts that depend mostly on the current values of the series. In general, several values of w should be tried to determine how sensitive the forecast series is to the choice of w. Forecasting experience will provide the best basis for the choice of w for a particular application.

2. The further into the future you forecast, the less certain you can be of the accuracy of your forecast. Since the exponentially smoothed forecast is constant for all future values, any changes in trend and/or seasonality are not taken into account. However, the uncertainty associated with future forecasts applies not only to exponentially smoothed forecasts, but also to all methods of forecasting. In general, time series forecasting should be confined to the short term.

EXAMPLE 14.1

The annual Dow Jones Averages from 1961 to 1983 are given in Table 14.2, along with the exponentially smoothed values using $w = .3$ and $w = .7$. Use the exponential smoothing technique to forecast the values from 1984 to 1986 using both $w = .3$ and $w = .7$.

TABLE 14.2

Dow Jones Industrial Average (1961–1983) with Exponentially Smoothed Values

YEAR	DJA	EXPONENTIALLY SMOOTHED ($w = .3$)	($w = .7$)	YEAR	DJA	EXPONENTIALLY SMOOTHED ($w = .3$)	($w = .7$)
1961	731.14	731.14	731.14	1973	923.88	897.59	924.75
1962	652.10	707.43	675.81	1974	759.37	856.12	808.98
1963	762.95	724.09	736.81	1975	802.49	840.03	804.44
1964	874.13	769.10	832.93	1976	974.92	880.50	923.78
1965	969.26	829.15	928.36	1977	835.15	866.90	861.74
1966	785.69	816.11	828.49	1978	805.01	848.33	822.03
1967	905.11	842.81	882.12	1979	838.74	845.45	833.73
1968	943.75	873.09	925.26	1980	963.99	881.01	924.91
1969	800.36	851.27	837.83	1981	899.01	886.41	906.78
1970	838.92	847.57	838.59	1982	1,046.54	934.45	1,004.61
1971	884.76	858.73	870.91	1983	1,258.64	1,031.71	1,182.43
1972	950.71	886.32	926.77				

Source: *The Dow Jones Investor's Handbook*, 1986.

SOLUTION

First, we calculate the exponentially smoothed forecasts using $w = .3$. Following the steps outlined in the box, we find:

$$F_{1984} = E_{1983} = 1,031.71$$

$$F_{1985} = F_{1984} = 1,031.71$$

$$F_{1986} = F_{1985} = F_{1984} = 1,031.71$$

The same steps are repeated using $w = .7$, and both sets of forecasts are shown in Table 14.3. Also shown are the actual Dow Jones Averages from 1984 to 1986. The **forecast error**, defined as the actual value minus the forecast value, is given for each exponentially smoothed forecast.

YEAR	ACTUAL DJA	FORECAST (w = .3)	FORECAST ERROR	FORECAST (w = .7)	FORECAST ERROR
1984	1,211.57	1,031.71	179.86	1,182.43	29.14
1985	1,546.67	1,031.71	514.96	1,182.43	364.24
1986	1,895.95	1,031.71	864.24	1,182.43	713.52

TABLE 14.3
Dow Jones Industrial
Forecasts: 1984–1986

Source: *The Dow Jones Investor's Handbook*, 1986; *The Wall Street Journal*, Jan. 2, 1987.

Notice that the one-step-ahead forecasts for 1984 have considerably smaller forecast errors than the two- and three-step-ahead forecasts for 1985 and 1986. Neither the $w = .3$ nor the $w = .7$ forecast projects the 1984–1986 upturn in the DJA, because exponentially smoothed forecasts implicitly assume no trend exists in the time series. This example dramatically illustrates the risk associated with anything other than very short-term forecasting.

Many time series have long-term, or secular, trends. For such series the exponentially smoothed forecast is inappropriate for all but the very short term. In the next section we present an extension of the exponentially smoothed forecast—the **Holt–Winters forecast**—that allows for secular trend in the forecasts.

SECTION 14.3

FORECASTING TRENDS: THE HOLT–WINTERS FORECASTING MODEL

The exponentially smoothed forecasts for the Dow Jones Industrial Average in Section 14.2 have large forecast errors because they do not recognize the trend in the time series. In this section we present an extension of the exponential smoothing method of forecasting that explicitly recognizes the trend in a time series. The **Holt–Winters forecasting model** consists of both an exponentially smoothed component (E_t) and a trend component (T_t). The trend component is used in the calculation of the exponentially smoothed value. The following equations show that both E_t and T_t are weighted averages:

$$E_t = wY_t + (1 - w)(E_{t-1} + T_{t-1})$$
$$T_t = v(E_t - E_{t-1}) + (1 - v)T_{t-1}$$

Note that the equations require *two* smoothing constants, w and v, each of which is between 0 and 1. As before, w controls the smoothness of E_t; a choice near 0 places more emphasis on past values of the time series, while a value of w near 1 gives more weight to current values of the series, and deemphasizes the past.

The trend component of the series is estimated **adaptively**, using a weighted average of the most recent change in the level and the trend estimate from the previous period. A choice of the weight v near 0 places more emphasis on the past estimates of trend (represented by T_{t-1}), while a choice of v near 1 gives more weight to the current change in level [represented by $(E_t - E_{t-1})$].

The calculation of the Holt–Winters components, which proceeds much like the exponential smoothing calculations, is summarized in the next box.

STEPS FOR CALCULATING COMPONENTS OF HOLT–WINTERS MODEL

1. Select an exponential smoothing constant w between 0 and 1. Small values of w give less weight to the current values of the time series, and more weight to the past. Larger choices assign more weight to the current value of the series.
2. Select a trend smoothing constant v between 0 and 1. Small values of v give less weight to the current changes in the level of the series, and more weight to the past trend. Larger values assign more weight to the most recent trend of the series and less to past trends.
3. Calculate the two components, E_t and T_t, from the time series Y_t beginning at time $t = 2$ as follows:[*]

$$E_2 = Y_2$$
$$T_2 = Y_2 - Y_1$$
$$E_3 = wY_3 + (1 - w)(E_2 + T_2)$$
$$T_3 = v(E_3 - E_2) + (1 - v)T_2$$

$$\vdots$$

$$E_t = wY_t + (1 - w)(E_{t-1} + T_{t-1})$$
$$T_t = v(E_t - E_{t-1}) + (1 - v)T_{t-1}$$

[*Note:* E_1 and T_1 are not defined.]

EXAMPLE 14.2

The yearly sales data for a firm's first 35 years of operation are given in Table 14.4. Calculate the Holt–Winters exponential smoothing and trend components for this

TABLE 14.4

A Firm's Yearly Sales Revenue (Thousands of Dollars)

t	Y_t	t	Y_t	t	Y_t
1	4.8	13	48.4	25	100.3
2	4.0	14	61.6	26	111.7
3	5.5	15	65.6	27	108.2
4	15.6	16	71.4	28	115.5
5	23.1	17	83.4	29	119.2
6	23.3	18	93.6	30	125.2
7	31.4	19	94.2	31	136.3
8	46.0	20	85.4	32	146.8
9	46.1	21	86.2	33	146.1
10	41.9	22	89.9	34	151.4
11	45.5	23	89.2	35	150.9
12	53.5	24	99.1		

---

[*]The calculation begins at time $t = 2$ rather than at $t = 1$ because the first two observations are needed to obtain the first estimate of trend, T_2.

time series using $w = .7$ and $v = .5$. Show the data and the exponential smoothing component, E_t, on the same graph.

SOLUTION

Following the formulas for the Holt–Winters components given in the box, we calculate:

$$E_2 = Y_2 = 4.0$$
$$T_2 = Y_2 - Y_1 = 4.0 - 4.8 = -.8$$
$$E_3 = .7Y_3 + (1 - .7)(E_2 + T_2)$$
$$= .7(5.5) + .3(4.0 - .8)$$
$$= 4.8$$
$$T_3 = .5(E_3 - E_2) + (1 - .5)T_2$$
$$= .5(4.8 - 4.0) + .5(-.8)$$
$$= 0$$

$\vdots$

All the E_t and T_t values are given in Table 14.5, and a graph of Y_t and E_t is shown in Figure 14.2 (page 804). Note that the trend component, T_t, measures the general upward trend in Y_t. The choice of $v = .5$ gives equal weight to the most recent trend and to past trends in the sales of the firm. The result is that the exponential smoothing component, E_t, provides a smooth, upward-trending description of the firm's sales. ∎

TABLE 14.5 Holt–Winters Components for Sales Data

MONTH	SALES Y_t	E_t (w = .7)	T_t (v = .5)	MONTH	SALES Y_t	E_t (w = .7)	T_t (v = .5)	MONTH	SALES Y_t	E_t (w = .7)	T_t (v = .5)
1	4.8	—	—	13	48.4	50.5	1.1	25	100.3	100.1	3.8
2	4.0	4.0	-.8	14	61.6	58.6	4.6	26	111.7	109.4	6.5
3	5.5	4.8	.0	15	65.6	64.9	5.4	27	108.2	110.5	3.8
4	15.6	12.4	3.8	16	71.4	71.1	5.8	28	115.5	115.1	4.2
5	23.1	21.0	6.2	17	83.4	81.5	8.1	29	119.2	119.3	4.2
6	23.3	24.5	4.8	18	93.6	92.4	9.5	30	125.2	124.7	4.8
7	31.4	30.8	5.6	19	94.2	96.5	6.8	31	136.3	134.2	7.2
8	46.0	43.1	8.9	20	85.4	90.8	.5	32	146.8	145.2	9.1
9	46.1	47.9	6.9	21	86.2	87.7	-1.2	33	146.1	148.5	6.2
10	41.9	45.8	2.4	22	89.9	88.9	-.1	34	151.4	152.4	5.0
11	45.5	46.3	1.4	23	89.2	89.1	.1	35	150.9	152.9	2.7
12	53.5	51.8	3.5	24	99.1	96.1	3.6				

Our objective is to use the Holt–Winters exponentially smoothed series to forecast the future values of the time series. For the one-step-ahead forecast, this is accomplished by adding the most recent exponentially smoothed component to the most

FIGURE 14.2

Sales Data and Holt–
Winters Exponentially
Smoothed Series

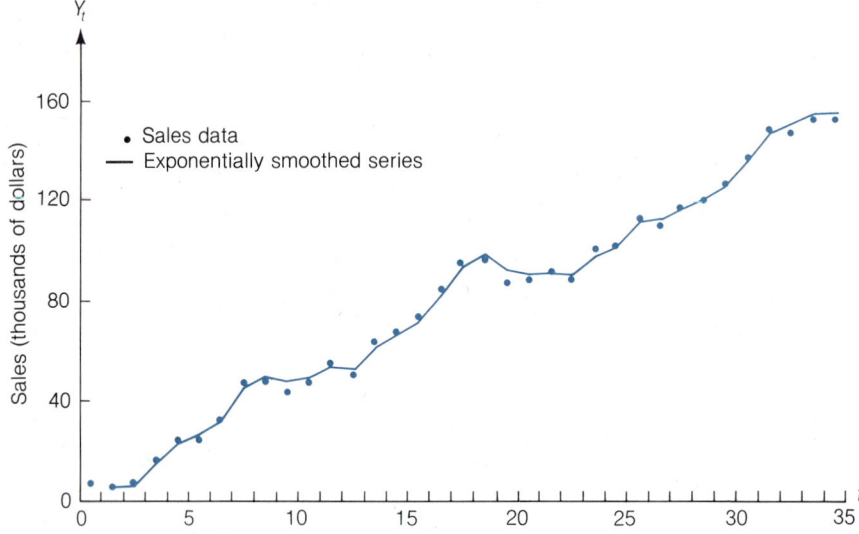

recent trend component. That is, the forecast at time $(t + 1)$, given observed values up to time t, is

$$F_{t+1} = E_t + T_t$$

The idea is that we are constructing the forecast by combining the most recent smoothed estimate, E_t, with the estimate of the expected increase (or decrease) attributable to trend, T_t.

The forecast for two steps ahead is similar, except that we add estimated trend for *two* periods:

$$F_{t+2} = E_t + 2T_t$$

Similarly, for the k-step-ahead forecast, we add the estimated increase (or decrease) in trend over k periods:

$$F_{t+k} = E_t + kT_t$$

The Holt–Winters forecasting methodology is summarized in the box.

HOLT–WINTERS FORECASTING

1. Calculate the exponentially smoothed and trend components, E_t and T_t, for each observed value of Y_t $(t \geq 2)$ using the formulas given in the previous box.

2. Calculate the one-step-ahead forecast using

$$F_{t+1} = E_t + T_t$$

3. Calculate the k-step-ahead forecast using

$$F_{t+k} = E_t + kT_t$$

EXAMPLE 14.3

Refer to Example 14.2 and Table 14.5, where we listed the firm's 35 yearly sales figures, along with the Holt–Winters components using $w = .7$ and $v = .5$. Use the Holt–Winters forecasting technique to forecast the firm's annual sales in years 36–40.

SOLUTION

For year 36 we calculate

$$F_{36} = E_{35} + T_{35}$$
$$= 152.9 + 2.7 = 155.6$$

The forecast 2 years ahead is

$$F_{37} = E_{35} + 2T_{35} = 152.9 + 2(2.7) = 158.3$$

For years 38–40 we find

$$F_{38} = 152.9 + 3(2.7) = 161.0$$
$$F_{39} = 152.9 + 4(2.7) = 163.7$$
$$F_{40} = 152.9 + 5(2.7) = 166.4$$

These forecasts are displayed in Figure 14.3. Note that the upward trend in the forecast is a result of the Holt–Winters estimated trend component.

FIGURE 14.3
Holt–Winters Sales
Forecasts, Years 36–40

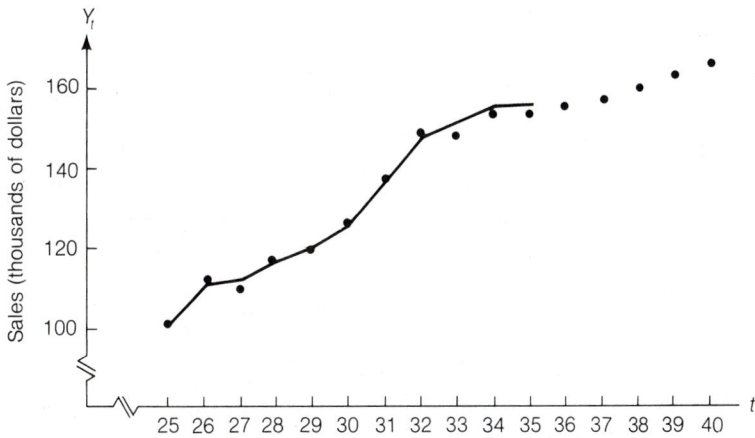

The selection of $w = .7$ and $v = .5$ as the smoothing and trend weights for the sales forecasts in Example 14.3 was based on the objectives of assigning more weight to recent series values in the exponentially smoothed component, and of assigning equal weights to the recent and past trend estimates. However, you may want to try several different combinations of weights when using the Holt–Winters forecasting model so that you can assess the sensitivity of the forecasts to the choice of weights. Experience with the particular time series and Holt–Winters forecasts will help in the selection of w and v in a practical application.

You can use the forecast errors to evaluate the accuracy of the forecast, and to aid in the selection of both the type of forecast to be utilized and the parameters of the forecast formula (e.g., the weights in the exponentially smoothed or Holt–Winters forecasts). Two popular measures of forecast accuracy are the **mean absolute deviation (MAD)** and the **root mean squared error (RMSE)** of the forecasts. Their formulas are given in the box.

MEASURES OF FORECAST ACCURACY

1. The **mean absolute deviation (MAD)** is defined as the mean absolute difference between the forecast and actual values of the time series:

$$\text{MAD} = \frac{\sum\limits_{t=1}^{N} |F_t - Y_t|}{N}$$

where N is the number of forecasts used for evaluation.

2. The **root mean squared error (RMSE)** is defined as the square root of the mean squared difference between the forecast and actual values of the time series:

$$\text{RMSE} = \sqrt{\frac{\sum\limits_{t=1}^{N} (F_t - Y_t)^2}{N}}$$

Note that both measures require one or more actual values of the time series against which to compare the forecasts. Thus, we can either wait several time periods until the observed values are available, or we can hold out several of the values at the end of the time series, not using them to model the time series, but saving them for evaluating the forecasts obtained from the model.

EXAMPLE 14.4

In Section 14.2 we found that the exponentially smoothed forecasts of the Dow Jones Industrial Average were generally unsatisfactory, because they did not recognize the trend in the time series. We now try the Holt–Winters model to account for the trend. Table 14.6 shows the original series through 1983, with two Holt–Winters models: one with a trend weight $v = .2$ and the other with trend weight $v = .8$. Both have smoothing weight $w = .5$. Use the time period 1984–1986 and apply the MAD and RMSE criteria to evaluate the two forecasting models.

SOLUTION

The first step is to calculate the Holt–Winters forecasts for 1984–1986 corresponding to the two models. The formula

$$F_{t+k} = E_t + kT_t$$

is used to obtain the forecasts given in Table 14.7, which also presents the actual DJA values and the forecast errors.

TABLE 14.6

Dow Jones Industrial Average (1961–1983) with Two Holt–Winters Forecasting Models

YEAR	DJA	HOLT–WINTERS MODEL 1		HOLT–WINTERS MODEL 2	
		$E\ (w = .5)$	$T\ (v = .2)$	$E\ (w = .5)$	$T\ (v = .8)$
1961	731.14	—	—	—	—
1962	652.10	652.10	−79.04	652.10	−79.04
1963	762.95	668.01	−60.05	668.01	−3.08
1964	874.13	741.04	−33.43	769.53	80.60
1965	969.26	838.43	−7.27	909.69	128.25
1966	785.69	808.43	−11.82	911.82	27.35
1967	905.11	850.86	−.97	922.14	13.73
1968	943.75	896.82	8.42	939.81	16.88
1969	800.36	852.80	−2.07	878.52	−45.65
1970	838.92	844.83	−3.25	835.90	−43.23
1971	884.76	863.17	1.07	838.71	−6.39
1972	950.71	907.47	9.72	891.51	40.96
1973	923.88	920.53	10.38	928.18	37.52
1974	759.37	845.14	−6.77	862.54	−45.01
1975	802.49	820.43	−10.36	810.01	−51.02
1976	974.92	892.50	6.13	866.95	35.35
1977	835.15	866.89	−.22	868.73	8.49
1978	805.01	835.84	−6.39	841.11	−20.39
1979	838.74	834.10	−5.46	829.73	−13.18
1980	963.99	896.31	8.08	890.27	45.79
1981	899.01	901.70	7.54	917.54	30.97
1982	1,046.54	977.89	21.27	997.52	70.18
1983	1,258.64	1,128.90	47.22	1,163.17	146.56

TABLE 14.7

Forecasts and Forecast Errors for Two Holt–Winters Models; Dow Jones Industrial Average: 1984–1986

YEAR	DJA Actual	MODEL 1 Forecast	Error	MODEL 2 Forecast	Error
1984	1,211.57	1,176.12	35.45	1,309.73	−98.16
1985	1,546.67	1,223.34	323.33	1,456.29	90.38
1986	1,895.95	1,270.56	625.39	1,602.85	293.10

To compare the two models, we first calculate the mean absolute deviations:

$$\text{MAD(model 1)} = \frac{|35.45| + |323.33| + |625.39|}{3}$$

$$= 328.06$$

$$\text{MAD(model 2)} = \frac{|-98.16| + |90.38| + |293.10|}{3}$$

$$= 160.55$$

Next, we calculate the root mean squared errors:

$$\text{RMSE(model 1)} = \sqrt{\frac{(35.45)^2 + (323.33)^2 + (625.39)^2}{3}}$$

$$= 406.99$$

$$\text{RMSE(model 2)} = \sqrt{\frac{(-98.16)^2 + (90.38)^2 + (293.10)^2}{3}}$$

$$= 185.93$$

Note that both criteria lead to the same conclusion: Model 2 provides more accurate predictions of the Dow Jones Industrial Average for the period 1984–1986 than does model 1. Note that the only difference between the two Holt–Winters models is the value of the trend weight: $v = .2$ for model 1 and $v = .8$ for model 2. The higher value in model 2 gives more weight to the short-term trend than to the long-term trend. For the 3 years in question the model sensitive to the short-term trend better forecasts the market's rally in the mid-1980's than does the model that is more sensitive to the long-term trend of the market.   ∎

Criteria such as MAD and RMSE for assessing forecast accuracy require special care in interpretation. The number of time periods included in the evaluation is critical to the decision about which model is preferred. The choice also depends on how many time periods ahead the analyst plans to forecast. Once these decisions are made, the criteria can be used to measure forecast accuracy. Nevertheless, only *inferential* models, such as regression models, that have explicit random error components can be used to estimate the reliability of a forecast *before* the actual value of the time series is observed. We will discuss an inferential model in the next section.

EXERCISES 14.1–14.9

LEARNING THE MECHANICS

14.1 State two criteria for evaluating the accuracy of an exponentially smoothed forecast.

14.2 The table lists the number of new privately owned housing units started each year during the period 1975–1985.
  a. Calculate the missing values of the exponentially smoothed housing-starts series.
  b. Plot the housing-starts series and the exponentially smoothed series on the same graph.

YEAR	HOUSING UNITS STARTED (thousands)	EXPONENTIALLY SMOOTHED ($w = .6$)
1975	1,160	
1976	1,540	1,388.00
1977	1,987	1,747.40
1978	1,692	
1979	1,583	1,635.46
1980	1,292	
1981	957	1,145.95
1982	1,010	1,064.38
1983	1,560	
1984	1,612	
1985	1,735	

Source: *Housing Units Authorized by Building Permits and Public Contracts, Annual: 1985.* Dept. of Commerce, U.S. Bureau of the Census.

14.3 The U.S. beer production for the years 1970–1984 is given in the table.

YEAR	BEER PRODUCTION (million barrels)	YEAR	BEER PRODUCTION (million barrels)
1970	133.1	1978	179.1
1971	137.4	1979	184.2
1972	141.3	1980	194.1
1973	148.6	1981	193.7
1974	156.2	1982	196.2
1975	160.6	1983	195.38
1976	163.7	1984	192.23
1977	170.5		

Source: *Standard & Poor's Current Statistics,* 1986.

a. Use the 1970–1981 values to forecast the 1982–1984 production using simple exponential smoothing with $w = .3$. With $w = .7$. Calculate the forecast errors associated with each.

b. Use the Holt–Winters model with $w = .7$ and $v = .3$ to forecast the 1982–1984 production. Repeat with $w = .3$ and $v = .7$. Calculate the forecast errors associated with each.

c. Compare the two simple exponential smoothing forecasts to the two Holt–Winters forecasts using the MAD and RMSE criteria. Does there appear to be a trend in the beer production time series? Which forecasting model is likely to be more appropriate?

14.4 Refer to part **a** of Exercise 14.3. Use the 1970–1984 values to forecast the 1985 production, using exponential smoothing with $w = .3$. With $w = .7$. Can the errors be measured or estimated for these 1985 forecasts? Explain.

APPLYING THE CONCEPTS

14.5 Standard & Poor's 500 Stock Composite Average (S&P 500) is a stock market index. Like the Dow Jones Industrial Average, it is an indicator of stock market activity. The table contains end-of-quarter values of the S&P 500 for the years 1971–1985.

YEAR	QUARTER	S&P 500	YEAR	QUARTER	S&P 500	YEAR	QUARTER	S&P 500
1971	I	100.31	1976	I	102.77	1981	I	134.94
	II	99.20		II	104.28		II	129.06
	III	98.34		III	105.24		III	118.77
	IV	102.09		IV	107.46		IV	119.13
1972	I	107.20	1977	I	98.42	1982	I	117.09
	II	107.14		II	100.48		II	109.82
	III	110.55		III	96.53		III	121.79
	IV	118.06		IV	95.10		IV	138.91
1973	I	111.52	1978	I	89.21	1983	I	152.96
	II	104.26		II	95.53		II	168.11
	III	108.43		III	102.54		III	166.07
	IV	97.55		IV	96.11		IV	164.93
1974	I	93.98	1979	I	101.59	1984	I	159.18
	II	86.00		II	102.91		II	153.18
	III	63.54		III	109.32		III	166.10
	IV	68.56		IV	107.94		IV	167.24
1975	I	83.36	1980	I	105.36	1985	I	180.66
	II	95.19		II	113.72		II	191.85
	III	83.87		III	127.14		III	182.08
	IV	98.19		IV	131.44		IV	211.28

Source: Standard & Poor's, *The Analysts Handbook*, Jan. 1987, p. 5.

   a. Use $w = .7$ to smooth the series from 1971 through 1984. Then forecast the four quarterly values in 1985 using *only* the information through the fourth quarter of 1984.
   b. Compute the forecast errors for the 1985 forecasts.
   c. Repeat parts **a** and **b** using $w = .3$.
   d. Use the Holt–Winters methodology with $w = .7$ and $v = .5$ to forecast the 1985 quarterly values. Repeat with $w = .3$ and $v = .5$.

14.6 Refer to Exercise 14.5. Use the MAD and RMSE criteria to compare the four models' forecasts of the four 1985 S&P 500 values.

14.7 Refer to Exercise 14.5.
   a. Use the 1980–1985 values to forecast the quarterly 1986 values using simple exponential smoothing with $w = .3$ and $w = .7$.
   b. Calculate the forecasts using the Holt–Winters model with $w = .3$ and $v = .5$. Repeat with $w = .7$ and $v = .5$.
   c. Is there any way to know which of the four sets of forecasts for 1986 is best? On the basis of the criteria you applied to compare the forecasts for 1985, which model would you choose for 1986?

14.8 Gold and other precious metals became attractive investments during the inflationary period of the late 1970's and early 1980's. The table shows monthly gold prices from January 1979 to December 1985.

	1979	1980	1981	1982	1983	1984	1985
Jan.	233.7	691.0	523.5	378.0	479.9	371.3	302.7
Feb.	251.3	636.0	500.5	353.7	490.4	386.4	299.2
Mar.	239.7	500.5	513.7	320.0	419.7	394.7	304.4
Apr.	246.3	518.0	477.2	354.5	432.0	382.0	325.3
May	270.2	526.5	475.5	318.7	437.7	377.7	316.5
June	283.5	661.5	422.0	317.5	412.8	378.1	316.8
July	290.1	629.0	401.5	346.2	423.4	346.8	318.2
Aug.	315.2	636.7	430.0	405.2	416.4	348.1	330.4
Sept.	402.0	682.0	428.7	409.7	411.5	341.3	322.9
Oct.	382.0	644.0	430.8	427.7	393.9	340.6	326.2
Nov.	411.2	627.2	406.7	414.2	382.7	341.5	325.7
Dec.	559.5	589.7	397.5	444.7	388.0	319.5	322.8

Source: *Standard & Poor's Basic Statistics: Metals*, 1985; *Standard & Poor's Current Statistics*, 1986.

a. Use exponential smoothing with $w = .5$ to calculate monthly smoothed values from January 1979 to December 1984. Then forecast the twelve 1985 monthly gold prices.

b. Calculate the forecast errors for 1985. What is the trend in errors as the time distance increases?

c. Calculate 12 one-step-ahead forecasts for 1985 by updating the exponentially smoothed values with each month's actual value, and then forecasting the next month's value.

d. Calculate each month's one-step-ahead forecast error, and compare them with the forecast errors in part **b**. What does the comparison indicate about the difference between short-term and long-term forecasting?

e. Repeat parts **a–d** using the Holt–Winters technique with $w = .5$ and $v = .5$.

14.9 Refer to Exercise 14.8. Two models were used to forecast the monthly 1985 gold prices: an exponential smoothing model with $w = .5$ and a Holt–Winters model with $w = .5$ and $v = .5$.

a. Use the MAD and RMSE criteria to evaluate the two models' accuracy for forecasting the 1985 values using only the 1979–1984 data.

b. Use the MAD and RMSE criteria to evaluate the two models' accuracy when making the 12 one-step-ahead forecasts for 1985, updating the models with each month's actual value before forecasting the next month's value (parts **c** and **d** of Exercise 14.8).

SECTION 14.4

FORECASTING TRENDS: SIMPLE LINEAR REGRESSION

Perhaps the simplest **inferential** forecasting model is one with which you are familiar: the simple linear regression model. A straight-line model is fit relating the time series, Y_t, to the time, t, and the least squares line is used to forecast future values of Y_t.

Suppose a firm is interested in forecasting its sales revenues for each of the next 5 years. To make such forecasts and assess their reliability, a time series model must be constructed. Refer again to the yearly sales data for a firm's 35 years of operation, given in Table 14.4. A plot of the data (Figure 14.4 on page 812) reveals a linearly increasing trend, so the model

$$E(Y_t) = \beta_0 + \beta_1 t$$

FIGURE 14.4
Plot of Sales Data

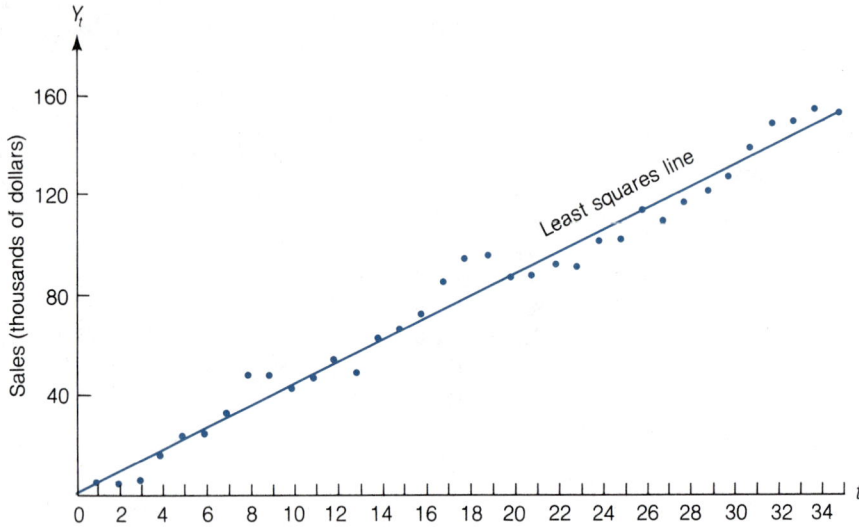

seems plausible for the secular trend. Fitting the model by least squares (see Section 10.2), we find the least squares model

$$\hat{Y}_t = \hat{\beta}_0 + \hat{\beta}_1 t = .4015 + 4.2956t$$

with

$$SSE = 1,345.45$$

This least squares line is shown in Figure 14.4, and the SAS printout is given in Figure 14.5. We can now forecast sales for years 36–40. The forecasts of sales and the corresponding 95% prediction intervals are shown in the printout. For example, for $t = 36$, we have

$$\hat{Y}_{36} = 155.0$$

with the 95% prediction interval (141.3, 168.8). Similarly, we can obtain the forecasts and prediction intervals for years 37–40. The observed sales, forecast sales, and prediction intervals are shown in Figure 14.6. Although it is not easily perceptible in the figure, the prediction intervals widen as we attempt to forecast further into the future (see the printout in Figure 14.5). This agrees with the intuitive notion that short-term forecasts should be more reliable than long-term forecasts.

There are two problems associated with forecasting time series using a least squares model.

PROBLEM 1

We are using the least squares model to forecast values outside the region of observation of the independent variable, t. That is, we are forecasting for values of t between 36 and 40, but the observed sales are for t values between 1 and 35.

FIGURE 14.5   SAS Printout for Least Squares Fit (Straight Line) to Y_t = Sales

SOURCE	DF	SUM OF SQUARES	MEAN SQUARE	F VALUE	PR > F
MODEL	1	65875.20816807	65875.20816807	1615.72	0.0001
ERROR	33	1345.45354622	40.77131958	R-SQUARE	ROOT MSE
CORRECTED TOTAL	34	67220.66171429		0.979985	6.38524233

PARAMETER	ESTIMATE	T FOR HO: PARAMETER = 0	PR > :T:	STD ERROR OF ESTIMATE
INTERCEPT	0.40151261	0.18	0.8567	2.20570829
T	4.29563025	40.20	0.0001	0.10686692

T	PREDICTED VALUE	LOWER 95% CL INDIVIDUAL	UPPER 95% CL INDIVIDUAL
36	155.04420168	141.30017574	168.78822762
37	159.33983193	145.53232286	173.14734101
38	163.63546218	149.76135290	177.50957147
39	167.93109244	153.98731054	181.87487434
40	172.22672269	158.21024159	186.24320379

FIGURE 14.6

Observed (Years 1–35) and Forecast (Years 36–40) Sales Using the Straight-Line Model

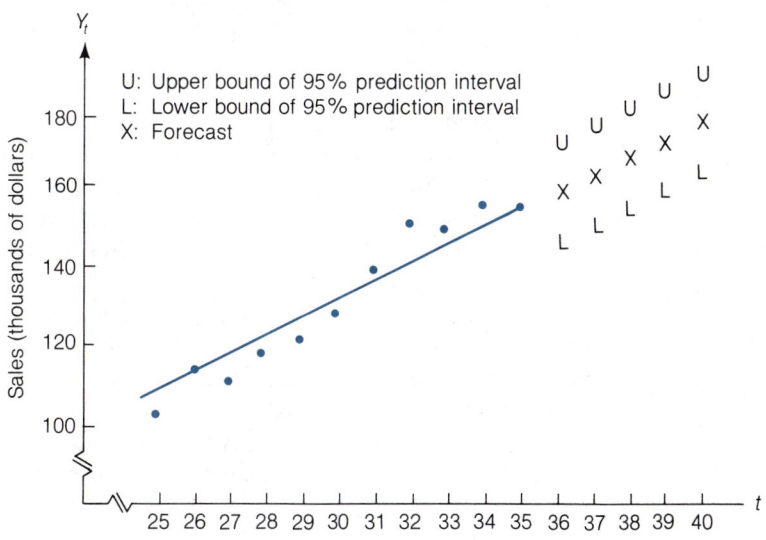

As we noted in Chapters 10–12, it is extremely risky to use a least squares regression model for prediction outside the experimental region.

Problem 1 obviously cannot be avoided. Since forecasting always involves predictions about the future values of a time series, some or all of the independent variables will probably be outside the region of observation on which the model was developed. It is important that the forecaster recognize the dangers of this type of prediction. If underlying conditions change drastically after the model is estimated (e.g., if federal price controls are imposed on the firm's products during the 36th year of operation), the forecasts and their confidence intervals are probably useless.

PROBLEM 2

Although the straight-line model may adequately describe the secular trend of the sales, we have not attempted to build any cyclical effects into the model. Thus, the effect of inflationary and recessionary periods will be to increase the error of the forecasts because the model does not anticipate such periods.

Fortunately, the forecaster often has some degree of control over problem 2, as we will demonstrate in the remainder of the chapter.

In forming the prediction intervals for the forecasts, we made the standard regression assumptions (Chapters 10 and 11) about the random error component of the model. We assume the errors have mean 0, constant variance, normal probability distributions, and are *independent*. The latter assumption is dubious in time series models, especially in the presence of short-term trends. Often, if a year's value lies above the secular trend line, the next year's value has a tendency to be above the line also. That is, the errors tend to be correlated (see Figure 14.4).

We will discuss how to deal with correlated errors in Sections 14.6–14.7. For now, we can characterize the simple linear regression forecasting method as simple and useful for discerning secular trends, but probably too simplistic for most time series. And, as with all forecasting methods, the simple linear regression forecasts should be applied only over the short term.

CASE STUDY 14.1

FORECASTING
THE DEMAND FOR
EMERGENCY ROOM
SERVICES

In Case Study 6.2, we described how queueing theory was used to model the emergency room in the Richmond Memorial Hospital in Richmond, Virginia. In this case study, we describe how a regression model was used to forecast the demand for emergency room services. In particular, we describe the procedure used by the hospital to forecast the average number of emergency room visits per day during the month of August 1970.

Data were collected on the emergency room's operations since its opening in October 1955 (month 1) through January 1970 (month 124). The data on patient visits for each August since the emergency room opened are shown in Table 14.8. A straight-line regression model was used to model the trend in Y_t, the average number of visits per day during August. With time, t (measured in months), as the independent variable, the following least squares model was obtained:

$$\hat{Y}_t = 38.788 + .60990t$$

TABLE 14.8

Emergency Room Data
for the Month of August:
1959–1969

MONTH t	NUMBER OF VISITS	AVERAGE NUMBER OF VISITS PER DAY Y_t	MONTH t	NUMBER OF VISITS	AVERAGE NUMBER OF VISITS PER DAY Y_t
11	1,367	44.09	71	3,019	97.38
23	1,642	52.96	83	2,794	90.12
35	1,780	57.41	95	2,846	91.80
47	2,060	66.45	107	3,001	96.80
59	2,257	72.80	119	3,548	114.45

This least squares line is plotted on a scattergram of the data in Figure 14.7. The plot reveals an upward trend in emergency room visits. Although some seasonal variation was present, a similar upward trend was observed for the other 11 months of the year. It was determined that the increase in the demand for emergency room services was greater than the rise in the Richmond area population. This finding substantiated management's belief that the emergency room was increasingly serving as a replacement for the family physician.

FIGURE 14.7

Least Squares Trend Line for Average Number of Visits per Day During August

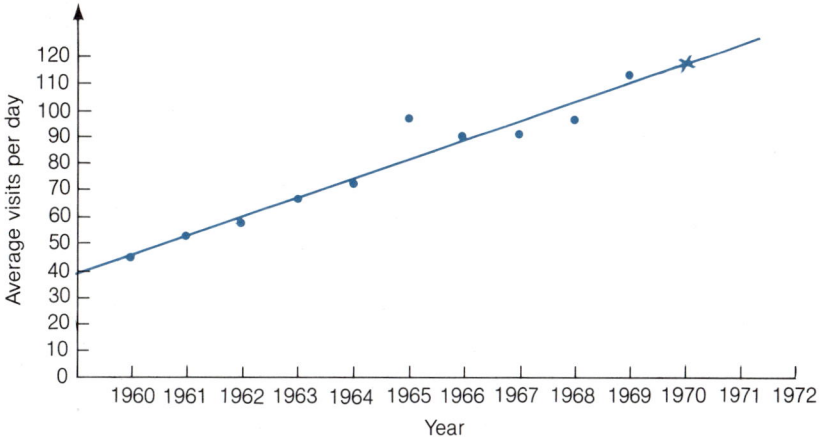

The model yielded a point forecast of $\hat{Y}_{131} = 38.788 + .60990(131) = 118.68$ for the average number of visits per day in August 1970. The associated 95% prediction interval was (100.20, 137.16). The actual demand in August turned out to be 119.20 and is indicated by a cross ($\times$) in the plot of Figure 14.7. Thus, the actual demand fell close to the least squares trend line and well within the bounds of the prediction interval.

Even for projections 18 months into the future, the hospital found this least squares prediction procedure to be superior to previous methods employed by the hospital. As a result, it was adopted for regular use in budgeting for the emergency room and associated services (Bolling, 1972).

SECTION 14.5

SEASONAL REGRESSION MODELS

Many time series have distinct seasonal patterns. Retail sales are usually highest around Christmas, spring, and fall, with lulls in the winter and summer periods. Energy usage is highest in summer and winter, and lowest in spring and fall. Teenage unemployment rises in summer months when schools are not in session, and falls near Christmas when many businesses hire part-time help.

Multiple regression models can be used to forecast future values of a time series with strong seasonal components. To accomplish this, the mean value of the time series, $E(Y_t)$, is given a mathematical form that describes both the secular trend and seasonal components of the time series. Although the seasonal model can assume a wide variety of mathematical forms, the use of dummy variables to describe seasonal differences is common.

For example, consider the power load data for a southern utility company shown in Table 14.9. Data have been obtained on a quarterly basis from 1974 to 1985. A model that combines the expected growth in usage and the seasonal component is

$$E(Y_t) = \beta_0 + \beta_1 t + \beta_2 Q_1 + \beta_3 Q_2 + \beta_4 Q_3$$

TABLE 14.9
Quarterly Power Loads for a Southern Utility Company, 1974–1985

YEAR	QUARTER	POWER LOAD (megawatts)	YEAR	QUARTER	POWER LOAD (megawatts)
1974	1	68.8	1980	1	130.6
	2	65.0		2	116.8
	3	88.4		3	144.2
	4	69.0		4	123.3
1975	1	83.6	1981	1	142.3
	2	69.7		2	124.0
	3	90.2		3	146.1
	4	72.5		4	135.5
1976	1	106.8	1982	1	147.1
	2	89.2		2	119.3
	3	110.7		3	138.2
	4	91.7		4	127.6
1977	1	108.6	1983	1	143.4
	2	98.9		2	134.0
	3	120.1		3	159.6
	4	102.1		4	135.1
1978	1	113.1	1984	1	149.5
	2	94.2		2	123.3
	3	120.5		3	154.4
	4	107.4		4	139.4
1979	1	116.2	1985	1	151.6
	2	104.4		2	133.7
	3	131.7		3	154.5
	4	117.9		4	135.1

where

t = Time period, ranging from $t = 1$ for quarter 1 of 1974, to $t = 48$ for quarter 4 of 1985

$$Q_1 = \begin{cases} 1 & \text{if quarter 1} \\ 0 & \text{if quarter 2, 3, or 4} \end{cases}$$

$$Q_2 = \begin{cases} 1 & \text{if quarter 2} \\ 0 & \text{if quarter 1, 3, or 4} \end{cases}$$

$$Q_3 = \begin{cases} 1 & \text{if quarter 3} \\ 0 & \text{if quarter 1, 2, or 4} \end{cases}$$

The printout in Figure 14.8 shows the least squares fit of this model to the data in Table 14.9.

Note that the model appears to fit well, with $R^2 = .91$, indicating that the model accounts for 91% of the sample variability in power loads over the 12-year period.

FIGURE 14.8   SAS Printout of Least Squares Fit to Power Load Time Series

```
DEPENDENT VARIABLE: POWER LOAD

SOURCE                      DF    SUM OF SQUARES      MEAN SQUARE    F VALUE

MODEL                        4    28374.99250583     7093.74812646   114.88
ERROR                       43     2655.13561917       61.74733998   PR > F
CORRECTED TOTAL             47    31030.12812500                     0.0001

R-SQUARE           C.V.            ROOT MSE          Y3 MEAN

0.914434          6.6766          7.85794757        117.69375000

                                  T FOR H0:      PR > !T!   STD ERROR OF
PARAMETER            ESTIMATE     PARAMETER=0               ESTIMATE

INTERCEPT          70.50852273       22.63        0.0001     3.11552479
T                   1.63621066       19.92        0.0001     0.08213932
QUARTER    1       13.65863199        4.25        0.0001     3.21744388
           2       -3.73591200       -1.16        0.2512     3.21219719
           3       18.46954400        5.76        0.0001     3.20904506
```

The global $F = 114.88$ strongly supports the hypothesis that the model has predictive utility. The standard deviation (ROOT MSE) of 7.86 indicates that the model predictions will usually be accurate to within approximately $\pm 2(7.86)$, or about ± 16 megawatts. Furthermore, $\hat{\beta}_1 = 1.64$ indicates an estimated average growth in load of 1.64 megawatts per quarter. Finally the seasonal dummy variables have the following interpretations (refer to Chapter 12):*

$\hat{\beta}_2 = 13.66$ Quarter 1 loads average 13.66 megawatts more than quarter 4 loads.

$\hat{\beta}_3 = -3.74$ Quarter 2 loads average 3.74 megawatts less than quarter 4 loads.

$\hat{\beta}_4 = 18.47$ Quarter 3 loads average 18.47 megawatts more than quarter 4 loads.

Thus, as expected, winter and summer loads exceed spring and fall loads, with the peak occurring during the summer months.

In order to forecast the 1986 power loads, we calculate the predicted value $\hat{Y}$ for $k = 49$, 50, 51, and 52, at the same time substituting the dummy variable appropriate for each quarter. Thus, for 1986,

$$\hat{Y}_{\text{Quarter 1}} = \hat{\beta}_0 + \hat{\beta}_1(49) + \hat{\beta}_2 = 70.51 + 1.636(49) + 13.66 = 164.3$$
$$\hat{Y}_{\text{Quarter 2}} = \hat{\beta}_0 + \hat{\beta}_1(50) + \hat{\beta}_3 = 148.6$$
$$\hat{Y}_{\text{Quarter 3}} = \hat{\beta}_0 + \hat{\beta}_1(51) + \hat{\beta}_4 = 172.4$$
$$\hat{Y}_{\text{Quarter 4}} = \hat{\beta}_0 + \hat{\beta}_1(52) = 155.6$$

*These interpretations assume a fixed value of t. In practical terms this is unrealistic, since each quarter is associated with a different value of t. Nevertheless, the coefficients of the seasonal dummy variables provide insight into the seasonality of these time series data.

The predicted values and 95% prediction intervals are given in Table 14.10; the data and least squares predicted values are graphed in Figure 14.9. The color line on the graph connects the predicted values. Also shown in Table 14.10 and Figure 14.9 are the actual 1986 quarterly power loads. Notice that all 1986 power loads fall inside the forecast intervals.

TABLE 14.10

Predicted Power Loads, Confidence Intervals, and Actual Power Loads (Megawatts) for 1986

QUARTER	PREDICTED LOAD	LOWER 95% CONFIDENCE LIMIT	UPPER 95% CONFIDENCE LIMIT	ACTUAL LOAD
1	164.3	147.3	181.4	151.3
2	148.6	131.5	165.6	132.9
3	172.4	155.4	189.5	160.5
4	155.6	138.5	172.6	161.0

FIGURE 14.9

Regression Forecasting Model for a Southern Utility Company

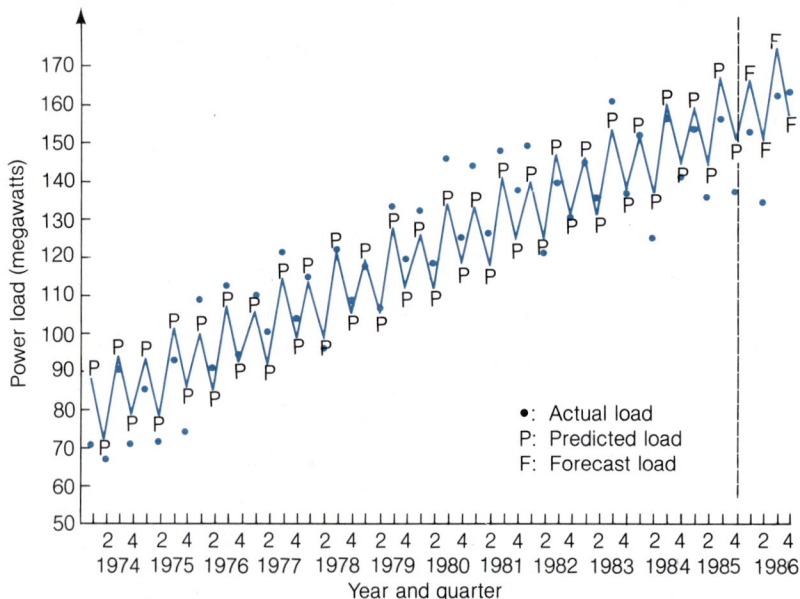

The seasonal model used to forecast the power loads is an **additive model** because the secular trend component ($\beta_1 t$) is added to the seasonal component ($\beta_2 Q_1 + \beta_3 Q_2 + \beta_4 Q_3$) to form the model. A **multiplicative model** would have the same form, except that the dependent variable would be the natural logarithm of power load; i.e.,

$$\ln Y_t = \beta_0 + \beta_1 t + \beta_2 Q_1 + \beta_3 Q_2 + \beta_4 Q_3 + \varepsilon$$

To see the multiplicative nature of this model, we take the antilogarithm of both sides of the equation to get

$$Y_t = \exp\{\beta_0 + \beta_1 t + \beta_2 Q_1 + \beta_3 Q_2 + \beta_4 Q_3 + \varepsilon\}$$
$$= \underbrace{\exp\{\beta_0\}}_{\text{Constant}} \; \underbrace{\exp\{\beta_1 t\}}_{\substack{\text{Secular} \\ \text{trend}}} \; \underbrace{\exp\{\beta_2 Q_1 + \beta_3 Q_2 + \beta_4 Q_3\}}_{\substack{\text{Seasonal} \\ \text{component}}} \; \underbrace{\exp\{\varepsilon\}}_{\substack{\text{Residual} \\ \text{component}}}$$

The multiplicative model often provides a better forecasting model when the time series is changing at an increasing rate over time.

If the time series data are observed monthly, a regression forecasting model would have to use 11 dummy variables to describe monthly seasonality, or three dummy variables could be used (as in the previous models) if the seasonal changes are hypothesized to occur quarterly. In general, this approach to seasonal modeling requires one dummy variable less than the number of seasonal changes expected to occur.

There are approaches in addition to the regression dummy variable method for forecasting seasonal time series. Trigonometric (sine and cosine) terms can be used in regression models to model periodicity. Other time series models (the Holt–Winters exponential smoothing model, for example) do not make use of the regression approach at all, and there are various methods for adding seasonal components to these models. We have chosen to discuss the regression approach because it makes use of the important modeling concepts covered in Chapters 11 and 12, and because the regression forecasts are accompanied by prediction intervals that provide some measure of the forecast reliability. While most other methods do not have explicit measures of reliability, many have proved their merit by providing good forecasts for particular applications. Consult the references at the end of this chapter for details of other seasonal models.

| | | | | | | | | | | | |

**EXERCISES
14.10–14.15**

LEARNING THE MECHANICS

[*Note:* *Starred (*) exercises require the use of a computer.*]

14.10 What is the advantage of regression forecasts as compared with exponentially smoothed forecasts? Does this advantage assure that regression forecasts will prove to be more accurate?

14.11 The annual price of galvanized steel (in cents per pound) from 1971 to 1984 is shown in the table. The SAS printout for the simple linear regression model fit to these data is also shown on page 820. The time variable t begins with $t = 1$ in 1971, and is incremented by 1 for each additional year:

$t = \text{Year} - 1970$

YEAR	PRICE	YEAR	PRICE
1971	9.61	1978	20.47
1972	10.88	1979	22.32
1973	10.59	1980	23.88
1974	12.39	1981	26.88
1975	14.80	1982	26.75
1976	16.07	1983	28.43
1977	18.10	1984	30.30

Source: *Standard & Poor's Basic Statistics: Metals,* 1985.

SAS Printout for Exercise 14.11

Dep Variable: PRICE

Analysis of Variance

Source	DF	Sum of Squares	Mean Square	F Value	Prob>F
Model	1	660.72601	660.72601	936.327	0.0001
Error	12	8.46788	0.70566		
C Total	13	669.19389			

Root MSE	0.84003	R-Square	0.9873	
Dep Mean	19.39071	Adj R-Sq	0.9863	
C.V.	4.33215			

Parameter Estimates

Variable	DF	Parameter Estimate	Standard Error	T for H0: Parameter=0	Prob > |T|
INTERCEP	1	6.609231	0.47421483	13.937	0.0001
T	1	1.704198	0.05569371	30.599	0.0001

Obs	PRICE	Predict Value	Residual	Lower95% Predict	Upper95% Predict
1	9.6100	8.3134	1.2966	6.2613	10.3656
2	10.8800	10.0176	0.8624	8.0090	12.0263
3	10.5900	11.7218	-1.1318	9.7502	13.6935
4	12.3900	13.4260	-1.0360	11.4845	15.3676
5	14.8000	15.1302	-0.3302	13.2116	17.0489
6	16.0700	16.8344	-0.7644	14.9312	18.7377
7	18.1000	18.5386	-0.4386	16.6431	20.4341
8	20.4700	20.2428	0.2272	18.3473	22.1383
9	22.3200	21.9470	0.3730	20.0438	23.8503
10	23.8800	23.6512	0.2288	21.7326	25.5699
11	26.8800	25.3554	1.5246	23.4139	27.2969
12	26.7500	27.0596	-0.3096	25.0880	29.0312
13	28.4300	28.7638	-0.3338	26.7552	30.7724
14	30.3000	30.4680	-0.1680	28.4158	32.5202
15	.	32.1722	.	30.0704	34.2740
16	.	33.8764	.	31.7193	36.0335

a. Find and interpret the least squares estimates.
b. What are the forecasts for 1985 and 1986?
c. Find the 95% forecast intervals for 1985 and 1986.

*14.12 Quarterly retail sales over a 10-year period for a department store are shown (in hundred thousand dollars) in the table.

YEAR	QUARTER			
	1	2	3	4
1	8.3	10.3	8.7	13.5
2	9.8	12.1	10.1	15.4
3	12.1	14.5	12.7	17.1
4	13.7	16.0	14.2	19.2
5	17.4	19.7	18.0	23.1
6	18.2	20.5	18.6	24.0
7	20.0	22.2	20.5	25.1
8	22.3	25.1	22.9	27.7
9	24.7	26.9	25.1	29.8
10	25.8	28.7	26.0	32.2

a. Write a regression model that contains trend and seasonal components to describe the sales data.

b. Use a least squares regression program to fit the model. Evaluate the fit of the model and interpret the coefficients.

c. Use the regression model to forecast the quarterly sales during the 11th year. Give 95% confidence intervals for the forecasts.

APPLYING THE CONCEPTS

14.13 There was phenomenal growth in the transportation sector of the U.S. economy during the 1960's and 1970's. The personal consumption expenditure figures (in billion dollars) for this sector are given in the table for the years 1960–1984.

YEAR	EXPENDITURE	YEAR	EXPENDITURE	YEAR	EXPENDITURE
1960	42.4	1969	75.7	1978	198.1
1961	44.8	1970	80.6	1979	219.4
1962	47.4	1971	92.3	1980	236.6
1963	49.5	1972	105.4	1981	261.5
1964	54.3	1973	114.6	1982	267.3
1965	58.4	1974	117.9	1983	291.9
1966	60.4	1975	129.4	1984	319.5
1967	63.3	1976	155.2		
1968	69.3	1977	179.3		

Source: *Statistical Abstract of the United States: 1986*, U.S. Bureau of the Census.

a. Fit the simple regression model

$$E(Y_t) = \beta_0 + \beta_1 t$$

where t is the number of years since 1960 (i.e., $t = 0, 1, \ldots, 22$).

b. Forecast the personal consumption expenditure from 1985 to 1987. Calculate 95% confidence intervals for these forecasts.

14.14 The table presents the quarterly sales index for a particular brand of calculator at a campus bookstore. The quarters are based on an academic year, so the first quarter represents fall; the second, winter; the third, spring; and the fourth, summer.

YEAR	FIRST QUARTER	SECOND QUARTER	THIRD QUARTER	FOURTH QUARTER
1983	438	398	252	160
1984	464	429	376	216
1985	523	496	425	318
1986	593	576	456	398
1987	636	640	526	498

We defined the time variable as $t = 1$ for the first quarter of 1983, $t = 2$ for the second quarter of 1983, etc. The seasonal dummy variables are as follows:

$$Q_1 = \begin{cases} 1 & \text{if quarter 1} \\ 0 & \text{otherwise} \end{cases} \qquad Q_2 = \begin{cases} 1 & \text{if quarter 2} \\ 0 & \text{otherwise} \end{cases} \qquad Q_3 = \begin{cases} 1 & \text{if quarter 3} \\ 0 & \text{otherwise} \end{cases}$$

The SAS printout for the model

$$E(Y_t) = \beta_0 + \beta_1 t + \beta_2 Q_1 + \beta_3 Q_2 + \beta_4 Q_3$$

is reproduced on page 823.

a. Interpret the least squares estimates, and evaluate the usefulness of the model.
b. Which of the assumptions about the random error component is in doubt when a regression model is fit to time series data?
c. Find the forecasts and the 95% prediction intervals for the 1988 quarterly sales.

14.15 The U.S. labor force is comprised of the civilian labor force and resident military personnel. The table lists the size of the labor force over the period 1970–1984.

YEAR	U.S. LABOR FORCE (in thousands)	YEAR	U.S. LABOR FORCE (in thousands)
1970	84,889	1978	103,882
1971	86,355	1979	106,559
1972	88,847	1980	108,544
1973	91,203	1981	110,315
1974	93,670	1982	111,872
1975	95,453	1983	113,226
1976	97,826	1984	115,241
1977	100,665		

Source: *Business Statistics: 1984*, Dept. of Commerce, Bureau of Economic Analysis, 1985, p. 43.

a. Use the method of least squares to fit a simple regression model to the data.
b. Forecast the labor force for 1985 and 1986.
c. Construct 95% confidence intervals for the forecasts of part **b**.
d. Check the accuracy of your forecasts by looking up the actual labor force size for 1985 and 1986 in either *Business Statistics* or the *Statistical Abstract of the United States*.

SAS Printout for Exercise 14.14

Dep Variable: Y

Analysis of Variance

Source	DF	Sum of Squares	Mean Square	F Value	Prob>F
Model	4	318560.30000	79640.07500	117.817	0.0001
Error	15	10139.50000	675.96667		
C Total	19	328699.80000			

Root MSE	25.99936	R-Square	0.9692
Dep Mean	440.90000	Adj R-Sq	0.9609
C.V.	5.89688		

Parameter Estimates

Variable	DF	Parameter Estimate	Standard Error	T for H0: Parameter=0	Prob > |T|
INTERCEP	1	119.850000	16.94950835	7.071	0.0001
T	1	16.512500	1.02771490	16.067	0.0001
Q1	1	262.337500	16.72998649	15.681	0.0001
Q2	1	222.825000	16.57140484	13.446	0.0001
Q3	1	105.512500	16.47552320	6.404	0.0001

Obs	Y	Predict Value	Residual	Lower95% Predict	Upper95% Predict
1	438.0	398.7	39.3000	335.5	461.9
2	398.0	375.7	22.3000	312.5	438.9
3	252.0	274.9	-22.9000	211.7	338.1
4	160.0	185.9	-25.9000	122.7	249.1
5	464.0	464.7	-0.7500	403.4	526.1
6	429.0	441.7	-12.7500	380.4	503.1
7	376.0	340.9	35.0500	279.6	402.3
8	216.0	251.9	-35.9500	190.6	313.3
9	523.0	530.8	-7.8000	470.1	591.5
10	496.0	507.8	-11.8000	447.1	568.5
11	425.0	407.0	18.0000	346.3	467.7
12	318.0	318.0	0	257.3	378.7
13	593.0	596.9	-3.8500	535.5	658.2
14	576.0	573.9	2.1500	512.5	635.2
15	456.0	473.1	-17.0500	411.7	534.4
16	398.0	384.0	13.9500	322.7	445.4
17	636.0	662.9	-26.9000	599.7	726.1
18	640.0	639.9	0.1000	576.7	703.1
19	526.0	539.1	-13.1000	475.9	602.3
20	498.0	450.1	47.9000	386.9	513.3
21	.	729.0	.	662.8	795.1
22	.	706.0	.	639.8	772.1
23	.	605.1	.	539.0	671.3
24	.	516.1	.	450.0	582.3

S E C T I O N   14.6

AUTOCORRELATION AND THE DURBIN–WATSON TEST

Recall that one of the assumptions we make when using a regression model to make predictions is that the errors are independent. However, with time series data, this assumption is questionable. The cyclical component of a time series may result in deviations from the secular trend that tend to cluster alternately on the positive and negative sides of the trend, as shown in Figure 14.10.

The observed errors between the time series and the regression model for the secular trend (and seasonal component, if present) are called **time series residuals**. Thus, if the time series Y_t has an estimated trend of $\hat{Y}_t$, then the time series residual* is

$$\hat{R}_t = Y_t - \hat{Y}_t$$

Note that time series residuals are defined just as the residuals for any regression model. However, we will usually plot time series residuals versus time to determine whether a cyclical component is apparent.

FIGURE 14.10

Illustration of Cyclical Errors

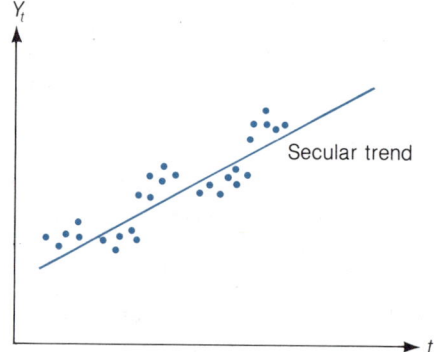

For example, consider the sales forecasting data in Table 14.4 (page 802), to which we fit a simple straight-line regression model. The plot of the data and model is repeated in Figure 14.11, and a plot of the time series residuals is shown in Figure 14.12.

Notice the tendency of the residuals to group alternately into positive and negative clusters. That is, if the residual for year t is positive, there is a tendency for the residual for year $(t + 1)$ to be positive. These cycles are indicative of possible positive correlation between neighboring residuals. The correlation between time series residuals at different points in time is called **autocorrelation**, and the autocorrelation of neighboring residuals (time periods t and $t + 1$) is called **first-order autocorrelation**.

DEFINITION 14.1

The correlation between time series residuals at different points in time is called **autocorrelation**. Correlation between neighboring residuals (at times t and $t + 1$) is called **first-order autocorrelation**. In general, correlation between residuals at times t and $t + d$ is called dth-order autocorrelation.

---

*We use $\hat{R}_t$ rather than $\hat{\varepsilon}$ to denote a time series residual because, as we shall see, time series residuals often do not satisfy the regression assumptions associated with the random component ε.

FIGURE 14.11

Plot of Sales Data

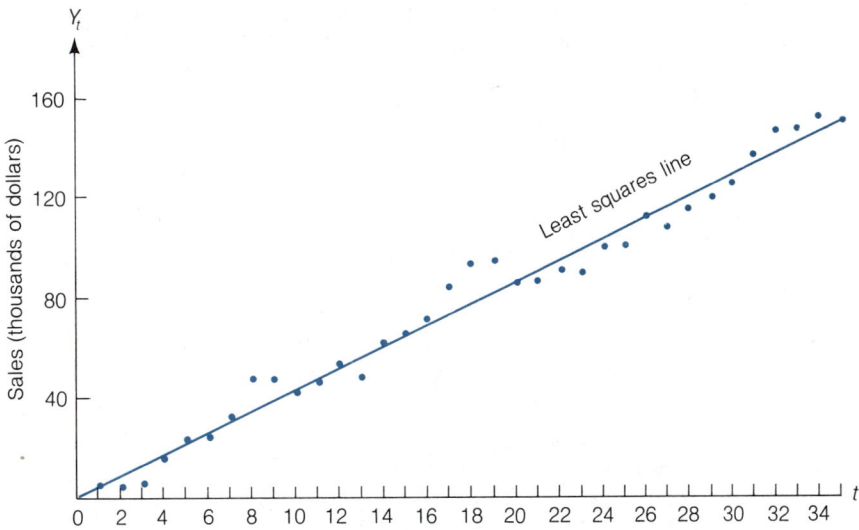

FIGURE 14.12

Plot of Residuals Versus
Time for the Sales Data:
Least Squares Model

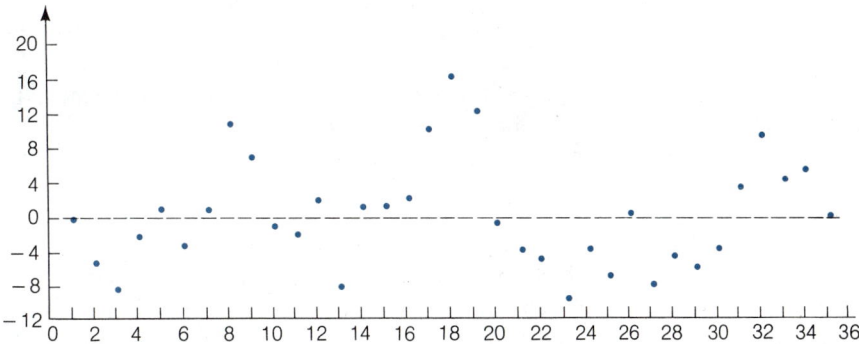

Rather than speculate about the presence of autocorrelation among time series residuals, we prefer to test for it. For most business economic time series, the relevant test is for first-order autocorrelation. Other higher-order autocorrelations may indicate seasonality, e.g., fourth-order autocorrelation in a quarterly time series. However, when we use the term *autocorrelation* in this text we are referring to first-order autocorrelation unless otherwise specified. So, we test

H_0: No first-order autocorrelation of residuals

H_a: Positive first-order autocorrelation of residuals

The **Durbin–Watson d statistic** is used to test for the presence of first-order autocorrelation. The statistic is given by the formula

$$d = \frac{\sum\limits_{t=2}^{n} (\hat{R}_t - \hat{R}_{t-1})^2}{\sum\limits_{t=1}^{n} \hat{R}_t^2}$$

where n is the number of observations (time periods) and $(\hat{R}_t - \hat{R}_{t-1})$ represents the difference between a pair of successive time series residuals. The value of d always falls in the interval from 0 to 4. The interpretations of the values of d are given in the box. Most statistical software packages include a routine that calculates d for time series residuals.

INTERPRETATION OF DURBIN–WATSON d STATISTIC

$$d = \frac{\sum\limits_{t=2}^{n} (\hat{R}_t - \hat{R}_{t-1})^2}{\sum\limits_{t=1}^{n} \hat{R}_t^2}$$

Range of d: $0 \le d \le 4$

1. If the residuals are uncorrelated, then $d \approx 2$.
2. If the residuals are positively autocorrelated, then $d < 2$, and if the autocorrelation is very strong, $d \approx 0$.
3. If the residuals are negatively autocorrelated, then $d > 2$, and if the autocorrelation is very strong, $d \approx 4$.

Durbin and Watson (1951) give tables for the lower-tail values of the d statistic, which we show in Tables XIV ($\alpha = .05$) and XV ($\alpha = .01$) of Appendix B. Part of Table XIV is reproduced in Table 14.11. For the sales example, we have $k = 1$ independent variable and $n = 35$ observations. Using $\alpha = .05$ for the one-tailed test for positive autocorrelation, we obtain the tabled values $d_L = 1.40$ and $d_U = 1.52$. The meaning of these values is illustrated in Figure 14.13. Because of the complexity of the sampling distribution of d, it is not possible to specify a single point that acts as a boundary between the rejection and nonrejection regions, as we did for the z, t, F, and other test statistics. Instead, an upper (d_U) and lower (d_L) bound are specified so that a d value less than d_L *does* provide strong evidence of positive autocorrelation at $\alpha = .05$ (recall that small d values indicate positive autocorrelation), a d value greater than d_U does *not* provide evidence of positive autocorrelation at $\alpha = .05$, and a value of d between d_L and d_U might or might not be significant at the $\alpha = .05$ level. If $d_L < d < d_U$, more information is needed before we can reach any conclusion about the presence of autocorrelation.

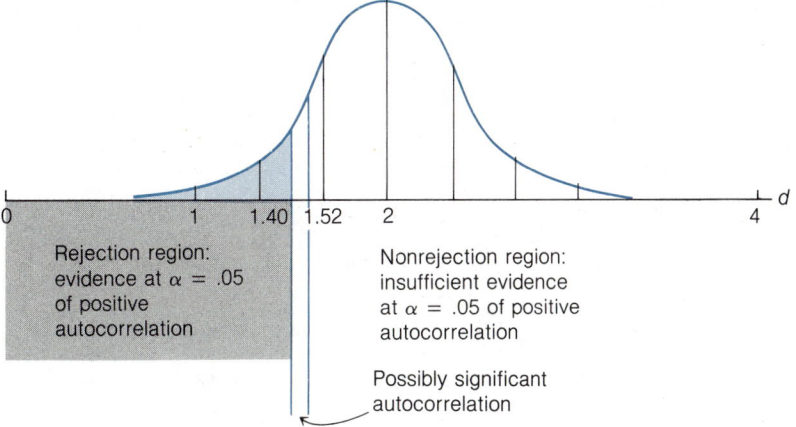

Rejection region:
evidence at $\alpha = .05$
of positive
autocorrelation

Nonrejection region:
insufficient evidence
at $\alpha = .05$ of positive
autocorrelation

Possibly significant
autocorrelation

TABLE 14.11
Reproduction of Part of
Table XIV of Appendix B
($\alpha = .05$)

n	$k = 1$		$k = 2$		$k = 3$		$k = 4$		$k = 5$	
	d_L	d_U	d_L	d_U	d_L	d_U	d_L	d_U	d_L	d_U
31	1.36	1.50	1.30	1.57	1.23	1.65	1.16	1.74	1.09	1.83
32	1.37	1.50	1.31	1.57	1.24	1.65	1.18	1.73	1.11	1.82
33	1.38	1.51	1.32	1.58	1.26	1.65	1.19	1.73	1.13	1.81
34	1.39	1.51	1.33	1.58	1.27	1.65	1.21	1.73	1.15	1.81
35	1.40	1.52	1.34	1.58	1.28	1.65	1.22	1.73	1.16	1.80
36	1.41	1.52	1.35	1.59	1.29	1.65	1.24	1.73	1.18	1.80
37	1.42	1.53	1.36	1.59	1.31	1.66	1.25	1.72	1.19	1.80
38	1.43	1.54	1.37	1.59	1.32	1.66	1.26	1.72	1.21	1.79
39	1.43	1.54	1.38	1.60	1.33	1.66	1.27	1.72	1.22	1.79
40	1.44	1.54	1.39	1.60	1.34	1.66	1.29	1.72	1.23	1.79

Tests for negative autocorrelation and two-tailed tests can be conducted by making use of the symmetry of the sampling distribution of the d statistic about its mean, 2 (see Figure 14.13). The test procedure is summarized in the box on page 829.

The SAS printout for the sales example is presented in Figure 14.14 (page 828). It shows that the computed value of d is .82, which is less than the tabulated value of $d_L = 1.40$. Thus, we conclude that the residuals of the straight-line model for sales are positively autocorrelated.

Once strong evidence of first-order autocorrelation has been established, as in the case of the sales example, doubt is cast on the least squares results and any inferences drawn from them. In the next section, we will present a time series model that accounts for the autocorrelation of the random errors. The residual correlation can be taken into account in a time series model and thereby used to improve both the fit of the model and the reliability of model inferences and forecasts.

FIGURE 14.14 SAS Printout for the Regression Analysis: Annual Sales Data

```
DEPENDENT VARIABLE: Y      SALES (THOUSANDS OF DOLLARS)

SOURCE                  DF      SUM OF SQUARES      MEAN SQUARE      F VALUE

MODEL                    1      65875.20816807   65875.20816807     1615.72
ERROR                   33       1345.45354622      40.77131958      PR > F
CORRECTED TOTAL         34      67220.66171429                       0.0001

R-SQUARE           C.V.              ROOT MSE            Y MEAN

0.979985         8.2154            6.38524233         77.72285714

                          T FOR HO:     PR > :T:     STD ERROR OF
PARAMETER     ESTIMATE    PARAMETER=0                  ESTIMATE

INTERCEPT    0.40151261        0.18       0.8567      2.20570829
T            4.29563025       40.20       0.0001      0.10686692

OBSERVATION     OBSERVED VALUE    PREDICTED VALUE        RESIDUAL

     1            4.80000000         4.69714286        0.10285714
     2            4.00000000         8.99277311       -4.99277311
     3            5.50000000        13.28840336       -7.78840336
     4           15.60000000        17.58403361       -1.98403361
     5           23.10000000        21.87966387        1.22033613
     6           23.30000000        26.17529412       -2.87529412
     7           31.40000000        30.47092437        0.92907563
     8           46.00000000        34.76655462       11.23344538
     9           46.10000000        39.06218487        7.03781513
    10           41.90000000        43.35781513       -1.45781513
    11           45.50000000        47.65344538       -2.15344538
    12           53.50000000        51.94907563        1.55092437
    13           48.40000000        56.24470588       -7.84470588
    14           61.60000000        60.54033613        1.05966387
    15           65.60000000        64.83596639        0.76403361
    16           71.40000000        69.13159664        2.26840336
    17           83.40000000        73.42722689        9.97277311
    18           93.60000000        77.72285714       15.87714286
    19           94.20000000        82.01848739       12.18151261
    20           85.40000000        86.31411765       -0.91411765
    21           86.20000000        90.60974790       -4.40974790
    22           89.90000000        94.90537815       -5.00537815
    23           89.20000000        99.20100840      -10.00100840
    24           99.10000000       103.49663866       -4.39663866
    25          100.30000000       107.79226891       -7.49226891
    26          111.70000000       112.08789916       -0.38789916
    27          108.20000000       116.38352941       -8.18352941
    28          115.50000000       120.67915966       -5.17915966
    29          119.20000000       124.97478992       -5.77478992
    30          125.20000000       129.27042017       -4.07042017
    31          136.30000000       133.56605042        2.73394958
    32          146.80000000       137.86168067        8.93831933
    33          146.10000000       142.15731092        3.94268908
    34          151.40000000       146.45294118        4.94705882
    35          150.90000000       150.74857143        0.15142857

     SUM OF RESIDUALS                                  0.00000000
     SUM OF SQUARED RESIDUALS                       1345.45354622
     SUM OF SQUARED RESIDUALS - ERROR SS              -0.00000000
     FIRST ORDER AUTOCORRELATION                       0.58962415
     DURBIN-WATSON D                                   0.82072679
```

DURBIN–WATSON d TEST

ONE-TAILED TEST	TWO-TAILED TEST
H_0: No first-order autocorrelation	H_0: No first-order autocorrelation
H_a: Positive first-order autocorrelation (or H_a: Negative first-order autocorrelation)	H_a: Positive or negative first-order autocorrelation

Test statistic:

$$d = \frac{\sum\limits_{t=2}^{n} (\hat{R}_t - \hat{R}_{t-1})^2}{\sum\limits_{t=1}^{n} \hat{R}_t^2}$$

Test statistic:

$$d = \frac{\sum\limits_{t=2}^{n} (\hat{R}_t - \hat{R}_{t-1})^2}{\sum\limits_{t=1}^{n} \hat{R}_t^2}$$

Rejection region: $d < d_{L,\alpha}$
[or $(4 - d) < d_{L,\alpha}$ if
H_a: Negative first-order
autocorrelation]

Rejection region:
$d < d_{L,\alpha/2}$ or $(4 - d) < d_{L,\alpha/2}$

where $d_{L,\alpha}$ is the lower tabled value corresponding to k independent variables and n observations. The corresponding upper value, $d_{U,\alpha}$, defines a "possibly significant" region between $d_{L,\alpha}$ and $d_{U,\alpha}$ (see Figure 14.13).

where $d_{L,\alpha/2}$ is the lower tabled value corresponding to k independent variables and n observations. The corresponding upper value, $d_{U,\alpha/2}$ defines a "possibly significant" region between $d_{L,\alpha/2}$ and $d_{U,\alpha/2}$ (see Figure 14.13).

Assumption: The residuals are normally distributed.

**EXERCISES
14.16–14.20**

LEARNING THE MECHANICS

14.16 Define autocorrelation. Explain why it is important in time series modeling and forecasting.

14.17 What do the following Durbin–Watson statistics suggest about the autocorrelation of the time series residuals from which each was calculated?
a. $d = 3.9$ **b.** $d = .2$ **c.** $d = 1.99$

14.18 For each case, indicate the decision regarding the test of the null hypothesis of no first-order autocorrelation against the alternative hypothesis of positive first-order autocorrelation.
a. $k = 2$, $n = 20$, $\alpha = .05$, $d = 1.1$
b. $k = 2$, $n = 20$, $\alpha = .01$, $d = 1.1$
c. $k = 5$, $n = 65$, $\alpha = .05$, $d = .95$
d. $k = 1$, $n = 31$, $\alpha = .01$, $d = 1.35$

APPLYING THE CONCEPTS

14.19 The decrease in the value of the dollar from 1964 to 1984 is illustrated by the data in the table. The buying power of the dollar (compared with 1967) is listed for each year. The first-order model

$$Y_t = \beta_0 + \beta_1 t + \varepsilon$$

was fit to the data using the method of least squares. The following least squares estimates of β_0 and β_1 were calculated:

$$\hat{\beta}_0 = 83.501 \quad \text{and} \quad \hat{\beta}_1 = .0420$$

YEAR t	VALUE Y_t	YEAR t	VALUE Y_t	YEAR t	VALUE Y_t
1964	1.076	1971	.824	1978	.512
1965	1.058	1972	.799	1979	.460
1966	1.029	1973	.751	1980	.405
1967	1.000	1974	.677	1981	.367
1968	.960	1975	.620	1982	.346
1969	.911	1976	.587	1983	.335
1970	.860	1977	.551	1984	.321

Source: *Statistical Abstract of the United States: 1986,* p. 468.

a. Calculate and plot the regression residuals against t. Is there a tendency for the residuals to have long positive and negative runs? To what do you attribute this phenomenon?

b. Calculate the Durbin–Watson d statistic, and test the null hypothesis that the time series residuals are uncorrelated. Use $\alpha = .05$.

c. What assumption(s) must be satisfied in order for the test of part **b** to be valid?

14.20 The table gives the volume of retail sales (in millions of dollars) of passenger cars in the United States by motor vehicle dealers for the years 1984 and 1985.

MONTH	TIME t	1984 SALES Y_t (millions)	TIME t	1985 SALES Y_t (millions)
Jan.	1	$20.76	13	$22.70
Feb.	2	21.07	14	23.19
Mar.	3	19.97	15	22.94
Apr.	4	21.06	16	24.18
May	5	21.26	17	24.15
June	6	21.94	18	24.07
July	7	21.23	19	24.15
Aug.	8	20.40	20	25.30
Sept.	9	20.65	21	27.74
Oct.	10	21.83	22	23.02
Nov.	11	22.20	23	23.19
Dec.	12	22.16	24	24.12

Source: *Standard & Poor's Current Statistics,* Nov. 1986, p. 14.

We used the SAS System to fit the first-order time series model

$$Y_t = \beta_0 + \beta_1 t + \varepsilon$$

to the data using the method of least squares, and obtained the printout shown here.

SAS Printout for Exercise 14.20

Dep Variable: Y

Analysis of Variance

Source	DF	Sum of Squares	Mean Square	F Value	Prob>F
Model	1	50.37956	50.37956	44.726	0.0001
Error	22	24.78078	1.12640		
C Total	23	75.16033			

Root MSE	1.06132	R-Square	0.6703	
Dep Mean	22.63667	Adj R-Sq	0.6553	
C.V.	4.68850			

Parameter Estimates

Variable	DF	Parameter Estimate	Standard Error	T for H0: Parameter=0	Prob > ¦T¦
INTERCEP	1	20.020362	0.44718746	44.770	0.0001
T	1	0.209304	0.03129660	6.688	0.0001

Durbin-Watson D	1.533
(For Number of Obs.)	24
1st Order Autocorrelation	0.211

a. Interpret the least squares estimates.
b. With what approximate precision do you expect this model to predict annual sales (assuming the necessary assumptions about the random component ε are satisfied)?
c. Is there evidence at the $\alpha = .10$ level of significance that the residuals are autocorrelated?

SECTION 14.7

FORECASTING WITH AUTOREGRESSIVE MODELS

The Durbin–Watson d test can be used to test for the presence of first-order autocorrelation. After we are convinced that the time series residuals are autocorrelated, we will want to take advantage of this autocorrelation to obtain better forecasts.

We can take advantage of autocorrelated residuals by modeling the residual component of the time series. A useful model for autocorrelated time series residuals is the **first-order autoregressive model** for the residual component R_t:*

$$R_t = \phi R_{t-1} + \varepsilon_t$$

*We use R_t to denote the true residual, $Y_t - E(Y_t)$, and $\hat{R}_t$ to denote the estimated residual, $Y_t - \hat{Y}_t$.

where ϕ (the Greek letter phi) is a constant coefficient between -1 and 1, and ε_t is a normally distributed and *independent* time series (called **white noise** by time series analysts) with a mean of 0 and a constant variance. This model implies that values of R_t are autocorrelated. The first-order autocorrelation of R_t and R_{t-1} is equal to ϕ, and the dth-order autocorrelation between residuals d time periods apart, R_t and R_{t-d}, is equal to ϕ^d. The autocorrelations corresponding to first-order autoregressive models are shown for several values of ϕ in Figure 14.15.

FIGURE 14.15

Autocorrelation Functions for Several First-Order Autoregressive Models:

$R_t = \phi R_{t-1} + \varepsilon_t$

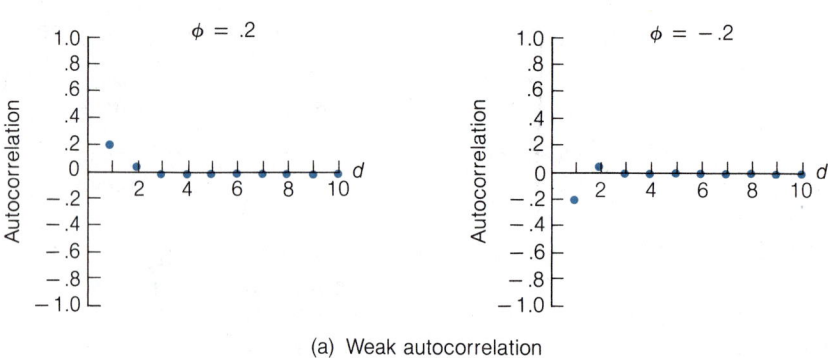

(a) Weak autocorrelation

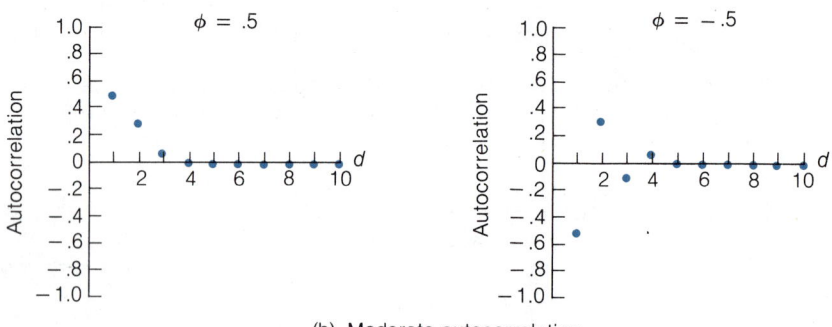

(b) Moderate autocorrelation

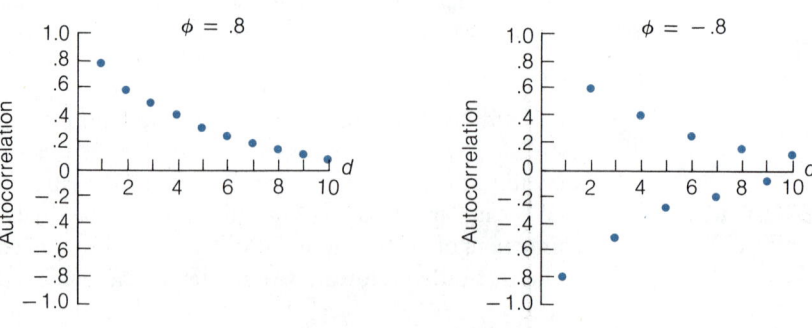

(c) Strong autocorrelation

Note that the first-order autocorrelation is largest, and the autocorrelations diminish rapidly as the order d (distance between time periods) increases* because positive values of ϕ imply positive autocorrelations for all orders d. Since positive autocorrelations generate cyclic behavior in the time series residuals, first-order autoregressive models used for business and economic time series have positive values for ϕ.

In summary, the pair of models

$$Y_t = \beta_0 + \beta_1 t + R_t$$
$$R_t = \phi R_{t-1} + \varepsilon_t$$

provides a description of the yearly sales of the firm that makes use of the autocorrelation of the time series residuals and therefore provides a more realistic description. In order to estimate the parameters of this pair of models (β_0, β_1, and ϕ), a modification of least squares is used. Although the details of this technique are beyond the scope of this text, several statistical software packages include the technique that provides estimates of these parameters. For example, the SAS printout that results from using this modified least squares technique to fit the pair of models to the sales revenue data of Table 14.4 is shown in Figure 14.16. [Caution: The SAS time series model is defined so that ϕ has the opposite sign from the value contained in our model. Consequently, you must multiply the estimate of ϕ shown in the portion of the SAS printout titled ESTIMATES OF THE AUTOREGRESSIVE PARAMETERS (beneath the heading COEFFICIENT) by -1 to obtain the estimate of ϕ for our model.] The fitted models are

$$\hat{Y}_t = .4058 + 4.2959t + \hat{R}_t$$
$$\hat{R}_t = .5896\hat{R}_{t-1}$$

with

$$SSE = 877.69$$

FIGURE 14.16

SAS Printout for Straight-Line Autoregressive Model Fit to Sales Data

```
               ESTIMATES OF THE AUTOREGRESSIVE PARAMETERS
       LAG      COEFFICIENT          STD ERROR           T RATIO
        1       -0.58962415          0.14277861        -4.129639


                    YULE-WALKER ESTIMATES

       SSE        877.6854      DFE                   32
       MSE         27.42767     ROOT MSE        5.237143
       SBC        223.1868      AIC             218.5208
       REG RSQ      0.9412      TOTAL RSQ         0.9869
```

VARIABLE	DF	B VALUE	STD ERROR	T RATIO	APPROX PROB
INTERCPT	1	0.40575699	3.99697517	0.102	0.9198
T	1	4.29593038	0.18983105	22.630	0.0001

*The autoregressive model is called a **stationary model** because the autocorrelations depend only on the distance, d, between residuals, and not on the time, t. Most time series models postulate a stationary residual component.

Note that the SSE has been reduced from 1,345.45 for the least squares fit to 877.69 for the pair of models just shown. Thus, we expect our forecasts to be more reliable for the straight-line autoregressive models.

To obtain forecasts for years 36–40, we find

$$\hat{R}_{35} = Y_{35} - [.4058 + 4.2959(35)] = 150.9 - 150.7623 = .1377$$

We use the estimated value of $\hat{R}_{35} = .1377$ to obtain

$$\hat{R}_{36} = \hat{\phi}\hat{R}_{35} = (.5896)(.1377) = .0812$$

and then

$$\hat{Y}_{36} = \hat{\beta}_0 + \hat{\beta}_1(36) + \hat{R}_{36}$$
$$= .4058 + (4.2959)(36) + .0812 = 155.14$$

Approximate 95% prediction limits (not shown on the SAS printout) are (144.8, 165.5). Note that this interval is narrower than the interval we obtained for the same forecast using the least squares model, (141.3, 168.8). The procedure for forecasting Y_{37}–Y_{40} is similar, and the entire set of forecasts and approximate prediction limits are shown in Figure 14.17.

FIGURE 14.17

Forecasts and Prediction Intervals for Years 36–40 Using Straight-Line Autoregressive Models

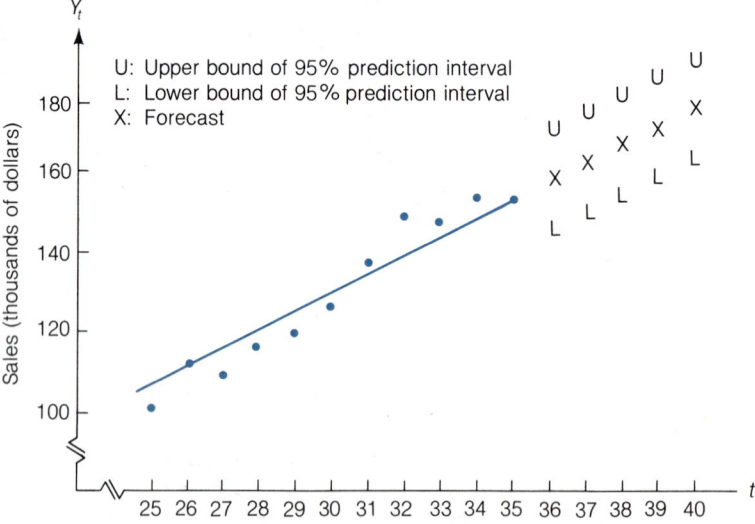

A comparison of Figures 14.6 and 14.17 reveals that the prediction intervals are somewhat narrower when the autoregressive model is used for the random component. It appears that the autoregressive model for R_t helps identify the short-term cyclical effects on sales, thereby making the forecasts more reliable.

The type of modeling exemplified by the combination of regressive–autoregressive models for the sales data is useful for many business time series. First, a regressive model is postulated for the deterministic component to describe the secular trend and, if appropriate, the seasonal effect. Then, the random component is modeled to describe the cyclical and residual effects. The autoregressive model is useful for this random component; it has the general form

$$R_t = \phi_1 R_{t-1} + \phi_2 R_{t-2} + \cdots + \phi_p R_{t-p} + \varepsilon_t$$

and is called an **autoregressive model of order** p. The name *autoregressive* comes from the fact that R_t is regressed on its own past values. As the order, p, is increased, more complex autocorrelation functions can be modeled. An even more flexible model is the **autoregressive–moving average (ARMA) model:**

$$R_t = \phi_1 R_{t-1} + \cdots + \phi_p R_{t-p} + \varepsilon_t + \theta_1 \varepsilon_{t-1} + \cdots + \theta_q \varepsilon_{t-q}$$

**FORECASTING WITH REGRESSION MODELS
WITH FIRST-ORDER AUTOREGRESSIVE RESIDUALS**

1. Postulate a regression model for Y_t as a function of the predictor time series $x_{1t}, x_{2t}, \ldots, x_{kt}$ (e.g., time t, seasonal dummy variables, and so on):

   $$Y_t = \beta_0 + \beta_1 x_{1t} + \beta_2 x_{2t} + \cdots + \beta_k x_{kt} + R_t$$

2. Use the method of least squares to fit the regression model and calculate the Durbin–Watson d statistic for the time series residuals.

3. Assuming the Durbin–Watson test reveals autocorrelation, model the time series residual, R_t, using the first-order autoregressive model

   $$R_t = \phi R_{t-1} + \varepsilon_t$$

4. Use a statistical software package to obtain estimates of the parameters β_0, $\beta_1, \ldots, \beta_k$, and ϕ.

5. To forecast the value for Y_{t+1}, first compute

   $$\hat{R}_t = Y_t - (\hat{\beta}_0 + \hat{\beta}_1 x_{1t} + \cdots + \hat{\beta}_k x_{kt})$$
   $$\hat{R}_{t+1} = \hat{\phi} \hat{R}_t$$
   $$\hat{Y}_{t+1} = \hat{\beta}_0 + \hat{\beta}_1 x_{1,t+1} + \cdots + \hat{\beta}_k x_{k,t+1} + \hat{R}_{t+1}$$

6. Approximate 95% prediction limits for the forecast are given by

   $$\hat{Y}_{t+1} \pm 2\sqrt{\text{MSE}}$$

7. Future forecasts can be calculated recursively, first calculating the residual

   $$\hat{R}_{t+2} = \hat{\phi} \hat{R}_{t+1}$$

   and then

   $$\hat{Y}_{t+2} = \hat{\beta}_0 + \hat{\beta}_1 x_{1,t+2} + \cdots + \hat{\beta}_k x_{k,t+2} + \hat{R}_{t+2}$$

   and so forth.

8. An approximate 95% prediction interval for the m-step-ahead forecast is

   $$\hat{Y}_{t+m} \pm 2\sqrt{\text{MSE}(1 + \hat{\phi}^2 + \cdots + \hat{\phi}^{2(m-1)})}$$

The ARMA model relates the current time series residuals to a linear function of the last p residuals and a linear function of the current and last q white noise errors. Of course, as the ARMA model becomes more complex, there are more parameters to estimate, and the techniques for identifying the model, for estimating the parameters, and for forecasting become more complex. We will leave the details of the general ARMA approach to the references at the end of this chapter. The forecasting technique for the first-order autoregressive model is summarized in the preceding box.

EXAMPLE 14.5

Consider the power load data for a southern utility company given in Table 14.9. In Section 14.5, we fit the model

$$E(Y_t) = \beta_0 + \beta_1 t + \beta_2 Q_1 + \beta_3 Q_2 + \beta_4 Q_3$$

using the method of least squares. The least squares fit for this model is shown in Figure 14.18. Note that the Durbin–Watson d statistic is small ($d = .50$), indicating positively correlated residuals. (Note that $d < d_L \approx 1.37$ from Table XIV, using $\alpha = .05$, $k = 4$, and $n = 48$.)

FIGURE 14.18

Least Squares Model for Power Load Data

DEP VARIABLE: LOAD

ANALYSIS OF VARIANCE

SOURCE	DF	SUM OF SQUARES	MEAN SQUARE	F VALUE	PROB>F
MODEL	4	28374.99251	7093.74813	114.883	0.0001
ERROR	43	2655.13562	61.74733998		
C TOTAL	47	31030.12813			

ROOT MSE	7.857948	R-SQUARE	0.9144	
DEP MEAN	117.6937	ADJ R-SQ	0.9065	
C.V.	6.676606			

PARAMETER ESTIMATES

VARIABLE	DF	PARAMETER ESTIMATE	STANDARD ERROR	T FOR H0: PARAMETER=0	PROB > |T|
INTERCEP	1	70.50852273	3.11552479	22.631	0.0001
T	1	1.63621066	0.08213932	19.920	0.0001
Q1	1	13.65863199	3.21744388	4.245	0.0001
Q2	1	-3.73591200	3.21219719	-1.163	0.2512
Q3	1	18.46954400	3.20904506	5.755	0.0001

DURBIN-WATSON D	0.504
(FOR NUMBER OF OBS.)	48
1ST ORDER AUTOCORRELATION	0.657

Fit the regression–autoregression pair of models given by

$$Y_t = \beta_0 + \beta_1 t + \beta_2 Q_1 + \beta_3 Q_2 + \beta_4 Q_3 + R_t$$
$$R_t = \phi R_{t-1} + \varepsilon_t$$

and use these models to forecast the 1986 quarterly power loads. Give approximate prediction intervals for each forecast.

SOLUTION

Using SAS PROC AUTOREG yields the printout shown in Figure 14.19. The fitted models are

$$\hat{Y}_t = 68.84 + 1.65t + 13.82 Q_1 - 3.44 Q_2 + 18.71 Q_3$$
$$\hat{R}_t = .657 \hat{R}_{t-1}$$

FIGURE 14.19 First-Order Autoregressive Model for Power Load Data

```
                              EXAMPLE 14.5
                      A U T O R E G   P R O C E D U R E
DEPENDENT VARIABLE = LOAD


                      ORDINARY LEAST SQUARES ESTIMATES

            SSE            2655.136    DFE                   43
            MSE            61.74734    ROOT MSE        7.857948
            SBC            348.2005    AIC             338.8445
            REG RSQ          0.9144    TOTAL RSQ         0.9144
            DURBIN-WATSON    0.5038


    VARIABLE  DF       B VALUE       STD ERROR    T RATIO  APPROX PROB

    INTERCPT   1    70.5085227      3.11552479    22.631      0.0001
    T          1     1.6362107      0.08213932    19.920      0.0001
    Q1         1    13.6586320      3.21744388     4.245      0.0001
    Q2         1    -3.7359120      3.21219719    -1.163      0.2512
    Q3         1    18.4695440      3.20904506     5.755      0.0001


               ESTIMATES OF AUTOCORRELATIONS

LAG  COVARIANCE  CORRELATION  -1 9 8 7 6 5 4 3 2 1 0 1 2 3 4 5 6 7 8 9 1
  0    55.3153    1.000000    |                   |*******************|
  1    36.3449    0.657050    |                   |*************      |


                  PRELIMINARY MSE=      31.43491

            ESTIMATES OF THE AUTOREGRESSIVE PARAMETERS
            LAG    COEFFICIENT      STD ERROR      T RATIO
             1     -0.65704967     0.11632116    -5.648582


                     YULE-WALKER ESTIMATES

            SSE            1290.97    DFE                   42
            MSE           30.73737    ROOT MSE         5.54413
            SBC           318.0239    AIC             306.7967
            REG RSQ         0.9170    TOTAL RSQ         0.9584


    VARIABLE  DF       B VALUE       STD ERROR    T RATIO  APPROX PROB

    INTERCPT   1    68.8433212      4.49089158    15.330      0.0001
    T          1     1.6505670      0.15051367    10.966      0.0001
    Q1         1    13.8245278      1.69507481     8.156      0.0001
    Q2         1    -3.4396410      1.90614907    -1.804      0.0783
    Q3         1    18.7128218      1.65184460    11.328      0.0001
```

A comparison of Figures 14.18 and 14.19 is instructive. First, note that the coefficient estimates change when the autoregressive residual model is added, but only slightly. The average increase in load per quarter is now estimated to be $\hat{\beta}_1 = 1.65$ (compared to 1.64), and the seasonal dummy variable estimates also change slightly, but their interpretations remain the same: Peak loads occur in the summer and winter quarters. Of special significance is the change in standard deviation (ROOT MSE) from 7.86 for the regression model to 5.54 for the regression–autoregression pair. Thus, we expect the prediction error associated with the pair of models to be approximately $\pm2(5.54)$, or about 11 megawatts. This represents a reduction of about 30% when compared with the error associated with the regression model. In other words, accounting for the autocorrelation between residuals in this case has reduced estimated prediction (and forecasting) error by about 30%.

The forecasting process begins with the calculation of the residual for the last observation, quarter 4 of 1985 ($t = 48$):

$$\hat{R}_{48} = Y_{48} - [\hat{\beta}_0 + \hat{\beta}_1(48)]$$
$$= 135.1 - [68.84 + 1.65(48)] = -12.94$$
$$\hat{R}_{49} = \hat{\phi}\hat{R}_{48} = (.657)(-12.94) = -8.50$$
$$\hat{Y}_{49} = \hat{\beta}_0 + \hat{\beta}_1(49) + \hat{\beta}_2 + \hat{R}_{49}$$
$$= 68.84 + 1.65(49) + 13.82 - 8.50 = 155.01$$

For quarter 2 of 1986, we forecast

$$\hat{R}_{50} = \hat{\phi}\hat{R}_{49} = (.657)(-8.50) = -5.58$$
$$\hat{Y}_{50} = \hat{\beta}_0 + \hat{\beta}_1(50) + \hat{\beta}_3 + \hat{R}_{50}$$
$$= 68.84 + 1.65(50) - 3.44 - 5.58 = 142.3$$

Similar computations give the forecasts for the last two quarters of 1986. The forecasts are shown in Table 14.12. Also shown are approximate 95% forecast intervals, which are computed using

$$\hat{Y}_{49} \pm 2\sqrt{\text{MSE}} = 155.01 \pm 2\sqrt{30.74}$$
$$= (143.9, 166.1)$$
$$\hat{Y}_{50} \pm 2\sqrt{\text{MSE}(1 + \hat{\phi}^2)} = 142.3 \pm 2\sqrt{30.74[1 + (.657)^2]}$$
$$= (129.0, 155.6)$$

and so forth. Note that the forecast intervals are narrower than those given for the regression model alone (Table 14.10). This reflects the advantage of using the regression–autoregression pair when the residuals are autocorrelated: Forecast accuracy will usually be increased.

TABLE 14.12

1986 Power Load Forecasts (Megawatts) Using the Regression–Autoregression Models

QUARTER	FORECAST	LOWER 95% FORECAST BOUND	UPPER 95% FORECAST BOUND	ACTUAL LOAD
1	155.0	143.9	166.1	151.3
2	142.3	129.0	155.6	132.9
3	168.0	152.7	183.3	160.5
4	152.2	136.1	168.3	161.0

Regression and ARMA models represent an extremely powerful combination for forecasting business and economic time series. All components of the time series— secular, seasonal, cyclical, and residual—can be modeled by the combination. However, the successful and skillful application of this approach requires considerable practice and expertise. Additionally, once the parameters are estimated, the model is relatively inflexible. Unlike the exponential smoothing model, which is adaptive to changes in the time series, the regression–ARMA parameters do not change with time unless the parameters are periodically reestimated. Nevertheless, the regression–ARMA pair is widely applied by forecasters.

| | | | | | | | | | | |

**EXERCISES
14.21–14.26**

LEARNING THE MECHANICS

14.21 Write the regression–autoregression pair of models that would be fit to annual data if only a secular trend were being postulated in the regression model. Explain the role of each of the parameters in the models.

14.22 Repeat Exercise 14.21 assuming that the data are quarterly and that seasonal dummy variables are to be added to the regression model.

APPLYING THE CONCEPTS

14.23 The Dow Jones Industrial Average (DJA) is the most widely followed stock market indicator. The values of the DJA from 1961 to 1983 are given in the table. The results of using SAS PROC AUTOREG to fit the regression–autoregression pair

$$Y_t = \beta_0 + \beta_1 t + R_t$$
$$R_t = \phi R_{t-1} + \varepsilon_t$$

where t is the number of years since 1960 (i.e., $t = 1, 2, \ldots, 23$) are given in the computer printout shown here. Evaluate the fit of the model, and interpret the estimated coefficients.

YEAR	DJA
1961	731.14
1962	652.10
1963	762.95
1964	874.13
1965	969.26
1966	785.69
1967	905.11
1968	943.75
1969	800.36
1970	838.92
1971	884.76
1972	950.71
1973	923.88
1974	759.37
1975	802.49
1976	974.92
1977	835.15
1978	805.01
1979	838.74
1980	963.99
1981	899.01
1982	1,046.54
1983	1,258.64

Source: *The Dow Jones Investor's Handbook*, 1986.

DEPENDENT VARIABLE = Y

ORDINARY LEAST SQUARES ESTIMATES

SSE	231681.1	DFE	21
MSE	11032.44	ROOT MSE	105.0354
SBC	283.5475	AIC	281.2765
REG RSQ	0.3221	TOTAL RSQ	0.3221
DURBIN-WATSON	1.1952		

VARIABLE	DF	B VALUE	STD ERROR	T RATIO	APPROX PROB
INTERCPT	1	753.403755	45.2714457	16.642	0.0001
T	1	10.428745	3.3017595	3.159	0.0048

ESTIMATES OF AUTOCORRELATIONS

LAG	COVARIANCE	CORRELATION	-1 9 8 7 6 5 4 3 2 1 0 1 2 3 4 5 6 7 8 9 1
0	10073.1	1.000000	|**********************|
1	2499.24	0.248110	|***** |

PRELIMINARY MSE= 9453.007

ESTIMATES OF THE AUTOREGRESSIVE PARAMETERS

LAG	COEFFICIENT	STD ERROR	T RATIO
1	-0.24811023	0.21661502	-1.145397

YULE-WALKER ESTIMATES

SSE	212122.4	DFE	20
MSE	10606.12	ROOT MSE	102.986
SBC	284.718	AIC	281.3115
REG RSQ	0.2807	TOTAL RSQ	0.3793

VARIABLE	DF	B VALUE	STD ERROR	T RATIO	APPROX PROB
INTERCPT	1	743.437201	56.9693091	13.050	0.0001
T	1	11.529724	4.1269552	2.794	0.0112

14.24 Refer to Exercise 14.23. Use the estimated models to forecast the Dow Jones Industrial Average from 1984 to 1986, and complete the table. Compare these forecasts to the exponentially smoothed forecasts in Table 14.3.

YEAR	ACTUAL DJA	FORECAST	FORECAST ERROR
1984	1,211.57		
1985	1,546.67		
1986	1,895.95		

14.25 Refer to Exercises 14.23 and 14.24. Calculate approximate 95% forecast bounds. Do these bounds contain the actual DJA values?

14.26 Between 1980 and 1984 foreign exporters of wheat increased their share of the international wheat market at the expense of the United States. Since the wheat industry in the United States relies heavily on exports to maintain its vitality, any loss in market share is reason for concern. Noncompetitive U.S. prices are blamed in part for the decline in market share. Darr and Gribbons (1985) used a regression model with a first-order autoregressive residual term to ascertain the trend in U.S. wheat prices relative to the price trends of other major wheat exporters. In the case of Argentina, they modeled Y_t, the ratio of Argentina's wheat price to the price of U.S. wheat, as a function of time, t. Using monthly data from 1980–1984 (i.e., $t = 1, 2, \ldots , 60$), they obtained the following fitted models:

$$\hat{Y}_t = \begin{array}{c} 1.20 \\ (50.61) \end{array} \begin{array}{c} -.00073t \\ (-9.42) \end{array} + \hat{R}_t$$

$$\hat{R}_t = \begin{array}{c} .267\hat{R}_{t-1} \\ (-2.99) \end{array}$$

where the numbers in parentheses are the t statistics for the coefficient estimates above them.

a. Is the autocorrelation of the residuals weak, moderate, or strong? Explain.

b. Darr and Gribbons concluded that the regression results provide evidence that Argentina's price declined relative to the U.S. price over the period 1980–1984. Do you agree? Explain.

c. Explain how the above models can be used to forecast the price of Argentina's wheat relative to the U.S. price for 1985 and beyond.

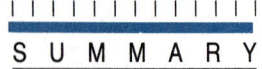

SUMMARY

Time series are often modeled as a combination of four components: **secular, seasonal, cyclical,** and **residual. Exponential smoothing** is an adaptive method for forecasting time series with little or no secular or seasonal trends. The **Holt–Winters model** provides an adaptive forecasting technique for a time series with a significant trend (secular) component. Simple linear regression can be used to forecast long-term trends of time series, and also allows the computation of prediction intervals to evaluate the forecast's reliability. Multiple regression models can be used to describe both long-term and seasonal components.

Since many business and economic time series models exhibit cyclical behavior, the **Durbin–Watson _d_ statistic** is important for testing **residual autocorrelation**. When autocorrelation is present, the combination of regression and autoregression models represents a useful forecasting tool. This pair of models is capable of describing all components of the time series, and of providing forecasts with prediction intervals.

Forecasting is an especially difficult aspect of statistical inference, because, by definition, we are extrapolating out of the range of the time period containing the data. Forecasts should be confined to the short term and, when possible, some measure of forecast reliability should be calculated. However, probably the best measure of a forecasting technique's usefulness is the comparison of the forecasts against the future realization of the time series values.

SUPPLEMENTARY EXERCISES 14.27–14.42

[Note: Starred (*) exercises require the use of a computer.]

14.27 Civilian employment is broadly classified by the federal government into two categories—agricultural and nonagricultural. Employment figures (in thousands of workers) for these categories are given in the table for the years 1974–1984.

YEAR	AGRICULTURAL	NONAGRICULTURAL
1974	3,515	83,279
1975	3,408	82,438
1976	3,331	85,421
1977	3,283	88,734
1978	3,387	92,661
1979	3,347	95,477
1980	3,364	95,938
1981	3,368	97,030
1982	3,401	96,125
1983	3,383	97,450
1984	3,321	101,685

Source: *Business Statistics: 1984*, Dept. of Commerce, Bureau of Economic Analysis, 1985, p. 22.

a. Use $w = .5$ to compute exponential smoothing forecasts for each of the two series for 1985.

b. Use the Holt–Winters model with $w = .5$ and $v = .5$ to compute 1985 forecasts for each of the series.

14.28 Refer to Exercise 14.13. Use $w = .3$ and $v = .7$ to compute the Holt–Winters forecasts for 1985–1987. Compare these to the linear regression forecasts obtained in Exercise 14.13.

14.29 The stock of Abbott Laboratories has had the yearly closing prices shown in the table on page 842.

a. Use exponential smoothing with $w = .8$ to forecast the 1987 and 1988 closing prices. If you buy at the end of 1986 and sell at the end of 1988, what is your expected gain (loss)?

YEAR	CLOSING PRICE	YEAR	CLOSING PRICE
1973	$49.875	1980	$56.500
1974	50.750	1981	27.000
1975	41.250	1982	38.750
1976	49.125	1983	45.250
1977	56.500	1984	41.750
1978	33.750	1985	68.375
1979	41.125	1986	45.625

Source: *Daily Stock Price Record*, NYSE, 1973–1986;
Security Owners' Stock Guide, Standard & Poor's, Jan.
1987.

b. Repeat part **a** using the Holt–Winters model with $w = .8$ and $v = .5$.

c. In which forecast do you have more confidence? Explain.

14.30 Refer to Exercise 14.29.

a. Fit a simple linear regression model to the stock price data.

b. Plot the fitted regression line on a scattergram of the data.

c. Forecast the 1987 and 1988 closing prices using the regression model.

d. Construct 95% prediction intervals for the forecasts of part **c**. Interpret the intervals in the context of the problem.

14.31 Refer to Exercise 14.30. Calculate the time series residuals for the simple linear model, and use the Durbin–Watson d statistic to test for the presence of autocorrelation.

14.32 The Gross National Product (GNP) is a measure of total U.S. output, and is therefore an important indicator of the U.S. economy. The quarterly GNP values (in billions of dollars) from 1971–1985 are given in the table below:

YEAR	QUARTER 1	2	3	4
1971	1,049.3	1,068.9	1,086.6	1,105.8
1972	1,142.4	1,171.7	1,196.1	1,233.5
1973	1,283.5	1,307.6	1,337.7	1,376.7
1974	1,387.7	1,423.8	1,451.6	1,473.8
1975	1,479.8	1,516.7	1,578.5	1,621.8
1976	1,672.0	1,698.6	1,729.0	1,772.5
1977	1,834.8	1,895.1	1,954.4	1,988.9
1978	2,031.7	2,139.5	2,202.5	2,281.6
1979	2,335.5	2,377.9	2,454.8	2,502.9
1980	2,572.9	2,578.8	2,639.1	2,736.0
1981	2,866.6	2,912.5	3,004.9	3,032.2
1982	3,021.4	3,070.2	3,090.7	3,109.6
1983	3,171.5	3,272.0	3,362.2	3,436.2
1984	3,553.3	3,644.7	3,694.6	3,758.7
1985	3,909.3	3,965.0	4,030.5	4,087.7

Source: *Standard & Poor's Current Statistics*, Jan. 1986,
Nov. 1986.

Use $w = .5$ and $v = .5$ to calculate Holt–Winters forecasts for 1986. Then complete the following table:

QUARTER	ACTUAL GNP	FORECAST	FORECAST ERROR
1	4,149.2		
2	4,175.6		
3	4,234.3		
4	4,260.6		

*14.33 Refer to Exercise 14.32.
a. Use the simple linear regression model to forecast the 1986 quarterly GNP. Place 95% confidence limits on the forecasts.
b. The GNP values given are *seasonally adjusted*, which means that an attempt to remove seasonality has been made prior to reporting the figures. Add quarterly dummy variables to the model. Use the partial F test (discussed in Section 12.4) to determine whether the data indicate the significance of the seasonal component. Does the test support the assertion that the GNP figures are seasonally adjusted?
c. Use the seasonal model to forecast the 1986 quarterly GNP values.

*14.34 Refer to Exercise 14.33.
a. Calculate the time series residuals for the seasonal model, and use the Durbin–Watson test to determine whether the residuals are autocorrelated. Use $\alpha = .05$.
b. Use SAS PROC AUTOREG, or a similar procedure, to fit the seasonal regression–autoregression pair of models. Forecast the 1986 quarterly GNP, and calculate approximate 95% forecast bounds. Do these bounds contain the actual GNP values? (See Exercise 14.32 for the actual values.)

*14.35 Refer to Exercises 14.32–14.34. A multiplicative model for GNP can be formed by using the natural logarithm of GNP as the dependent variable in the regression models.
a. Fit the simple linear regression and seasonal regression models using ln(GNP) as the dependent variable.
b. Compute the forecasts and forecast bounds for the quarterly 1986 GNP using the models in part **a**. Take the antilogarithm of the regression forecasts and bounds to express them in the original units (billions of dollars).

14.36 Refer to Exercises 14.32–14.35. For each of the forecasting models apply the MAD and RMSE criteria to evaluate the forecasts for the four quarters of 1986. Which of the forecasting models performs best according to each criterion?

14.37 Commercial banks, finance companies, credit unions, and retail outlets are the principal issuers of consumer installment credit (CIC). Since commercial banks and credit unions typically offer lower interest rates, they are referred to as *primary lenders*. Finance companies and retail outlets are referred to as *secondary lenders*. From 1945 to 1970, CIC in the United States grew from \$2.5 billion to \$101 billion, a growth rate $4\frac{1}{2}$ times greater than that of the Gross National Product. In the process, there was a dramatic increase in the yearly percentage market share of CIC held by primary lenders relative to secondary lenders. Dauten, Apilado, and Warner (1973) used regression analysis on the data given in the table at the top of page 844 to obtain projections of the market share of primary lenders beyond 1970.

| YEAR | PRIMARY LENDERS | | TOTAL | SECONDARY LENDERS | | TOTAL |
	Commercial banks	Credit unions		Finance companies	Retail outlets	
1945	30.2%	4.1%	34.3%	36.9%	27.9%	64.8%
1950	39.4	4.0	43.4	36.1	19.7	55.8
1955	36.7	5.8	42.5	40.9	15.5	56.4
1960	38.9	9.2	48.1	35.4	14.7	50.1
1965	40.6	10.3	50.9	34.0	13.7	47.7
1966	40.4	10.6	51.0	33.6	13.9	47.5
1967	40.4	11.1	51.5	33.0	14.1	47.1
1968	41.1	11.3	52.4	32.4	13.8	46.2
1969	41.1	11.8	52.9	32.3	13.4	45.7
1970	41.4	12.4	53.8	30.8	13.9	44.7

Source: Dauten, J. J., Apilado, V. P., and Warner, D. C. "Consumer credit: Changing patterns in supply and demand variables, and their implications," *Journal of Consumer Affairs*, Winter 1973, 7, 97.

a. Use the method of least squares to model the secular trend in the market share of primary lenders.
b. Plot the least squares model on a scattergram of the data.
c. Use your least squares model to find point forecasts for the market share of primary lenders for each of the years 1971 through 1975. Discuss some of the pitfalls of this forecasting technique.
d. Dauten, Apilado, and Warner concluded that if their ". . . projections are borne out over time, the relative position of primary and secondary lenders with respect to market share will have completely reversed in the 20-year period from 1955 to 1975." Do your projections yield the same conclusion?
e. Check the accuracy of your forecast for 1975 by determining the actual market share for primary lenders in 1975 from data supplied in the *Federal Reserve Bulletin*.

14.38 A major portion of total consumer credit is extended in the categories of automobile loans, mobile home loans, and revolving credit. Amounts outstanding (in billions of dollars) for the period 1974–1984 are given in the table.

YEAR	AUTOMOBILE	MOBILE HOME	REVOLVING CREDIT
1974	54.3	14.6	13.7
1975	57.2	14.4	15.0
1976	67.7	14.6	17.2
1977	82.9	15.0	39.3
1978	101.6	15.2	48.3
1979	116.4	16.8	57.0
1980	112.2	18.8	58.5
1981	119.8	19.9	64.5
1982	126.3	22.4	69.6
1983	143.1	23.9	82.0
1984	172.6	24.6	101.6

Source: *Statistical Abstract of the United States: 1986*, U.S. Bureau of the Census, p. 502.

a. Use simple linear regression models for each credit category to forecast the 1985 and 1986 values. Place 95% confidence bounds on each forecast.

b. Calculate the Holt–Winters forecasts for 1985 and 1986 using $w = .7$ and $v = .7$. Compare the results with the simple linear regression forecasts of part **a**.

14.39 Consider the monthly IBM stock prices from January 1984 to December 1986 shown in the table. Also shown are the exponentially smoothed values for $w = .5$.

1984	STOCK PRICE	EXPONENTIALLY SMOOTHED STOCK PRICE	1985	STOCK PRICE	EXPONENTIALLY SMOOTHED STOCK PRICE	1986	STOCK PRICE	EXPONENTIALLY SMOOTHED STOCK PRICE
Jan.	$114.125	$114.125	Jan.	$136.375	$129.503	Jan.	$151.500	$148.071
Feb.	110.250	112.187	Feb.	134.000	131.752	Feb.	150.875	149.473
Mar.	114.000	113.094	Mar.	127.000	129.376	Mar.	149.125	149.299
Apr.	113.750	113.422	Apr.	126.500	127.938	Apr.	156.250	152.775
May	107.750	110.586	May	128.625	128.281	May	152.375	152.575
June	105.750	108.168	June	123.750	126.016	June	146.500	149.537
July	110.750	109.459	July	131.375	128.695	July	132.500	141.019
Aug.	123.750	116.604	Aug.	126.625	127.660	Aug.	138.750	139.884
Sept.	124.250	120.427	Sept.	123.875	125.768	Sept.	134.500	137.192
Oct.	124.625	122.526	Oct.	129.875	127.821	Oct.	123.625	130.409
Nov.	121.750	122.138	Nov.	139.750	133.786	Nov.	127.125	128.767
Dec.	123.125	122.632	Dec.	155.500	144.643	Dec.	120.000	124.383

Source: *Daily Stock Price Record*, NYSE, 1984–1986; *Standard & Poor's Security Owner's Stock Guide*, Aug. 1986–Jan. 1987.

a. Plot both the stock price and the smoothed stock price on the same graph.

b. Use the exponentially smoothed series to forecast the monthly values of the IBM stock price from January 1987 to March 1987.

14.40 Refer to Exercise 14.39. The SAS printout for the simple linear regression model fit to the IBM stock prices is shown on page 846. The time t ranges from 1 to 36 over the 3-year period of the sample.

a. Interpret the least squares estimates.

b. With what approximate precision do you expect to be able to predict the IBM stock price using this model?

c. Give the forecasts and the 95% forecast intervals for the January–March 1987 prices. How does the precision of these forecasts agree with the approximation obtained in part **b**?

d. What assumptions does the random error component of the model have to satisfy in order to make the model inferences (such as the forecast intervals in part **c**) valid?

e. Test to determine whether there is evidence of first-order autocorrelation in the random error component. Use $\alpha = .05$. What does the result of this test imply about the validity of the model inferences?

SAS Printout for
Exercise 14.40

DEP VARIABLE: PRICE

ANALYSIS OF VARIANCE

SOURCE	DF	SUM OF SQUARES	MEAN SQUARE	F VALUE	PROB>F
MODEL	1	2880.38080	2880.38080	25.590	0.0001
ERROR	34	3826.94342	112.55716		
C TOTAL	35	6707.32422			

ROOT MSE	10.6093	R-SQUARE	0.4294
DEP MEAN	129.4479	ADJ R-SQ	0.4127
C.V.	8.195803		

PARAMETER ESTIMATES

VARIABLE	DF	PARAMETER ESTIMATE	STANDARD ERROR	T FOR H0: PARAMETER=0	PROB > |T|
INTERCEP	1	113.51845	3.61141766	31.433	0.0001
T	1	0.86105212	0.17021234	5.059	0.0001

OBS	ACTUAL	PREDICT VALUE	STD ERR PREDICT	LOWER95% PREDICT	UPPER95% PREDICT	RESIDUAL
1	114.1	114.4	3.4640	91.6988	137.1	-0.2545
2	110.3	115.2	3.3188	92.6497	137.8	-4.9906
3	114.0	116.1	3.1760	93.5957	138.6	-2.1016
4	113.8	117.0	3.0361	94.5366	139.4	-3.2127
5	107.8	117.8	2.8994	95.4725	140.2	-10.0737
6	105.8	118.7	2.7665	96.4032	141.0	-12.9348
7	110.8	119.5	2.6378	97.3288	141.8	-8.7958
8	123.8	120.4	2.5141	98.2492	142.6	3.3431
9	124.3	121.3	2.3961	99.1643	143.4	2.9821
10	124.6	122.1	2.2847	100.1	144.2	2.4960
11	121.8	123.0	2.1809	101.0	145.0	-1.2400
12	123.1	123.9	2.0858	101.9	145.8	-0.7261
13	136.4	124.7	2.0007	102.8	146.7	11.6629
14	134.0	125.6	1.9270	103.7	147.5	8.4268
15	127.0	126.4	1.8659	104.5	148.3	0.5658
16	126.5	127.3	1.8187	105.4	149.2	-0.7953
17	128.6	128.2	1.7866	106.3	150.0	0.4687
18	123.8	129.0	1.7703	107.2	150.9	-5.2674
19	131.4	129.9	1.7703	108.0	151.7	1.4966
20	126.6	130.7	1.7866	108.9	152.6	-4.1145
21	123.9	131.6	1.8187	109.7	153.5	-7.7255
22	129.9	132.5	1.8659	110.6	154.4	-2.5866
23	139.8	133.3	1.9270	111.4	155.2	6.4273
24	155.5	134.2	2.0007	112.2	156.1	21.3163
25	151.5	135.0	2.0858	113.1	157.0	16.4552
26	150.9	135.9	2.1809	113.9	157.9	14.9692
27	149.1	136.8	2.2847	114.7	158.8	12.3581
28	156.1	137.6	2.3961	115.5	159.7	18.4971
29	152.4	138.5	2.5141	116.3	160.6	13.8860
30	146.5	139.4	2.6378	117.1	161.6	7.1500
31	132.5	140.2	2.7665	117.9	162.5	-7.7111
32	138.8	141.1	2.8994	118.7	163.4	-2.3221
33	134.5	141.9	3.0361	119.5	164.4	-7.4332
34	123.6	142.8	3.1760	120.3	165.3	-19.1692
35	127.1	143.7	3.3188	121.1	166.2	-16.5303
36	120.0	144.5	3.4640	121.8	167.2	-24.5163
37	.	145.4	3.6114	122.6	168.2	.
38	.	146.2	3.7608	123.4	169.1	.
39	.	147.1	3.9118	124.1	170.1	.

SUM OF RESIDUALS	2.13163E-13
SUM OF SQUARED RESIDUALS	3826.943
PREDICTED RESID SS (PRESS)	4387.173

DURBIN-WATSON D	0.410
(FOR NUMBER OF OBS.)	36
1ST ORDER AUTOCORRELATION	0.716

14.41   Refer to Exercises 14.39 and 14.40. The result of applying SAS PROC AUTOREG to these data is shown at the top of page 847.

a. Interpret the model estimates after they have been adjusted for first-order autocorrelation.

b. Forecast the January–March 1987 prices, and place approximate 95% forecast intervals on each. In which do you have more confidence, the least squares forecasts of Exercise 14.40, or the forecasts that take first-order autocorrelation into account? Explain.

SAS Printout for
Exercise 14.41

DEPENDENT VARIABLE = PRICE

ORDINARY LEAST SQUARES ESTIMATES

SSE	3826.943	DFE	34
MSE	112.5572	ROOT MSE	10.6093
SBC	277.3175	AIC	274.1505
REG RSQ	0.4294	TOTAL RSQ	0.4294
DURBIN-WATSON	0.4101		

VARIABLE	DF	B VALUE	STD ERROR	T RATIO	APPROX PROB
INTERCPT	1	113.518452	3.61141766	31.433	0.0001
T	1	0.861052	0.17021234	5.059	0.0001

ESTIMATES OF AUTOCORRELATIONS

LAG	COVARIANCE	CORRELATION	-1 9 8 7 6 5 4 3 2 1 0 1 2 3 4 5 6 7 8 9 1
0	106.304	1.000000	| |**********************|
1	76.1561	0.716399	| |************** |

PRELIMINARY MSE= 51.74586

ESTIMATES OF THE AUTOREGRESSIVE PARAMETERS

LAG	COEFFICIENT	STD ERROR	T RATIO
1	-0.71639896	0.12145239	-5.898599

YULE-WALKER ESTIMATES

SSE	1523.419	DFE	33
MSE	46.16421	ROOT MSE	6.794425
SBC	248.4611	AIC	243.7105
REG RSQ	0.1095	TOTAL RSQ	0.7729

VARIABLE	DF	B VALUE	STD ERROR	T RATIO	APPROX PROB
INTERCPT	1	116.144600	6.94125906	16.732	0.0001
T	1	0.636707	0.31611029	2.014	0.0522

*14.42 Refer to Exercises 14.39–14.40. Another method often employed when forecasting economic time series is *first differencing*. This method models the difference between successive values of a time series, $Y_t - Y_{t-1}$. For example, the pair of models

$$Y_t - Y_{t-1} = R_t$$
$$R_t = \phi R_{t-1} + \varepsilon_t$$

is the first difference–autoregressive pair.

a. Fit this pair of models to the IBM price data given in Exercise 14.39. To accomplish this, first compute the differences (there will be 35 differences), and then use SAS PROC AUTOREG or a similar procedure to fit the autoregressive model.

b. Use the estimated models to forecast the January–March 1987 prices. To accomplish this, use

$$\hat{R}_{36} = Y_{36} - Y_{35}$$
$$\hat{R}_{37} = \hat{\phi}\hat{R}_{36}$$
$$\hat{Y}_{37} = Y_{36} + \hat{R}_{37}$$

$$\vdots$$

ON YOUR OWN . . .

Refer to "On Your Own" for Chapter 13. Use the same time series and simple index for this exercise. Hold out the last three values of the time series to evaluate the forecasting techniques utilized.

a. Using all but the last three values of the simple index, use the exponential smoothing forecasting technique with the same value of w you selected for smoothing in part **c** of the Chapter 13 "On Your Own."

b. Use the Holt–Winters technique to forecast the last three values of the simple index, selecting v consistent with the trend you found in part **d** of the Chapter 13 "On Your Own," and using the same value of w used in part **a.**

c. Use a simple linear regression model to forecast the last three values of the simple index. Do not use these last three values in fitting the model.

d. Use the Durbin–Watson test to determine whether the residuals of the simple regression are autocorrelated. Fit a first-order autoregressive model to the residuals, and use the simple regression–autoregressive residual pair to forecast the last three values of the simple index.

e. Use the MAD and RMSE criteria to evaluate the forecasts of parts **a–d**. Which technique appears to provide the best forecasts for the last three index values? Plot all the forecasts and the actual values. Does the plot support your choice of the best technique?

f. Convert the best forecasts of the index values to forecasts of the actual time series. Compare the forecasts to the actual values using the MAD and RMSE criteria. Is your conclusion about the best forecasting technique for your series unchanged?

REFERENCES

Abraham, B. and Ledholter, J. *Statistical Methods for Forecasting*. New York: Wiley, 1983.

Anderson, T. W. *The Statistical Analysis of Time Series*. New York: Wiley, 1971.

Bolling, W. B. "Queuing model of a hospital emergency room." *Industrial Engineering*, Sept. 1972, 26–31.

Box, G. E. P. and Jenkins, G. M. *Timer Series Analysis: Forecasting and Control*, 2nd ed. San Francisco: Holden-Day, 1977.

Darr, T. and Gribbons, G. "How U.S. exports are faring in the world wheat market," *Monthly Labor Review*, Oct. 1985, 10–24.

Durbin, J. and Watson, G. S. "Testing for serial correlation in least squares regression, I." *Biometrika*, 1950, 37, 409–428.

Durbin, J. and Watson, G. S. "Testing for serial correlation in least squares regression, II." *Biometrika*, 1951, 38, 159–178.

Durbin, J. and Watson, G. S. "Testing for serial correlation in least squares regression, III." *Biometrika*, 1971, 58, 1–19.

Fuller, W. A. *Introduction to Statistical Time Series*. New York: Wiley, 1976.

Granger, C. W. J. and Newbold, P. *Forecasting Economic Time Series*. New York: Academic Press, 1977.

Mendenhall, W. and Sincich, T. *A Second Course in Business Statistics: Regression Analysis*, 2nd ed. San Francisco: Dellen, 1986.

Nelson, C. R. *Applied Time Series Analysis for Managerial Forecasting*. San Francisco: Holden-Day, 1973.

WHERE WE'VE BEEN . . .

As we have seen in preceding chapters, the solutions of many business problems are based on inferences about population means. Methods for estimating and testing hypotheses about a single mean and the comparison of two means were presented in Chapters 8 and 9. Chapters 10–12 dealt with linear models for estimating the mean value of a response using regression models, and Chapters 13 and 14 treated the special case where the response measurements represent a time series.

WHERE WE'RE GOING . . .

In this chapter we extend the methodology of Chapters 8–12 in two important ways. First, we discuss the critical elements in the *design* of a sampling experiment. Then we show how to *analyze* the experiment in order to compare more than two populations. We will present several of the more popular experimental designs and show how to use the computer to analyze designed experiments using both analysis of variance and regression programs.

ANALYSIS OF VARIANCE

CONTENTS

Most of the data we have analyzed in previous chapters were collected in **observational** sampling experiments rather than **designed** sampling experiments. In observational experiments the analyst has little or no control over the variables under study and merely observes their values. In contrast, designed experiments are those in which the analyst attempts to control the levels of one or more variables to determine their effect on a variable of interest. Although the opportunity for such control is not always present in real business settings, it is instructive, even for observational experiments, to have a working knowledge of the analysis and interpretation of data that result from designed experiments and to know at least the basics of how to design experiments when the opportunity arises.

We first present the basic elements of an experimental design in Section 15.1. We then discuss two of the simpler, but more popular, experimental designs in Sections 15.2 and 15.3. Slightly more complex experiments are discussed in Section 15.4.

In optional Section 15.5 we will show how the statistical analysis of designed experiments can be accomplished using regression methodology.

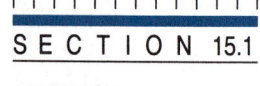

SECTION 15.1

ELEMENTS OF A
DESIGNED
EXPERIMENT

Many of the elements in a designed experiment are the same as or similar to those we introduced in regression analysis. For example, the **response** is the variable of interest in the experiment. The response might be the SAT score of a high school senior, the total sales of a firm last year, or the total income of a particular household this year. In regression analyses we referred to the response as the **dependent variable**, y, and we use them synonymously.

> **DEFINITION 15.1**
>
> The **response** is the variable of interest to be measured in the experiment. We also refer to the response as the **dependent variable**.

The intent of most statistical experiments is to determine the effect of one or more variables on the response. These variables, which we called the **independent variables** in regression analyses, are often referred to as the **factors** in a designed experiment. Like independent variables, factors are either **quantitative** or **qualitative** depending on whether the variable is measured on a numerical scale or not. For example, we might want to explore the effect of the qualitative factor "Gender" on the response "SAT score." In other words, we want to compare the SAT scores of male and female high school seniors. Or, we might wish to determine the effect of the quantitative factor "Number of salespeople" on the response "Total sales" for retail firms. Often two or more factors are of interest. For example, we might want to determine the effect of the quantitative factor "Number of wage earners" and the qualitative factor "Location" on the response "Household income."

> **DEFINITION 15.2**
>
> **Factors** are those variables whose effect on the response is of interest to the experimenter. **Quantitative** factors are measured on a numerical scale, while **qualitative** factors are those that are not (naturally) measured on a numerical scale. We referred to factors as **independent variables** in regression analyses, and we use the terms synonymously.

Levels are the values of the factors that are utilized in the experiment. The levels of qualitative factors are usually nonnumerical. For example, the levels of Gender are Male and Female, and the levels of Location might be North, East, South, and West. * The levels of quantitative factors are the numerical values of the variable utilized in the experiment. The Number of salesmen for each of a set of companies, the Number of wage earners in each of a set of households, and the GPAs for a set of high school seniors all represent levels of the respective quantitative factors.

> **DEFINITION 15.3**
>
> The **levels** of a factor are the values of the factor utilized in the experiment.

When a *single factor* is employed in an experiment, the **treatments** of the experiment are the levels of the factor. For example, if the effect of the factor "Gender" on the response "SAT score" is being investigated, the treatments of the experiment are the two levels of Gender: Female and Male. Or, if the effect of the Number of wage earners on Household income is the subject of the experiment, then the numerical values assumed by the quantitative factor "Number of wage earners" are the treatments. If *two or more factors* are utilized in an experiment, the **treatments** are the factor-level combinations employed. For example, if the effects of the factors "Gender and GPA" on the response "SAT score" are being investigated, then the treatments are the combinations of the levels of Gender and GPA employed; (Female, 2.61), (Male, 3.43), and (Female, 3.82) are all treatments if those particular factor-level combinations are utilized in the experiment.

> **DEFINITION 15.4**
>
> The **treatments** of an experiment are the factor-level combinations utilized.

The objects on which the response variable and factors are observed are the **experimental units**. For example, SAT score, High school GPA, and Gender are all variables that can be observed on the same experimental unit—a high school

---

*The levels of a qualitative variable may bear numerical labels. For example , the Locations could be numbered 1, 2, 3, and 4. However, in such cases the numerical labels for a qualitative variable will usually be codes representing nonnumerical levels.

senior. Or, Total sales, Earnings per share, and Number of salespeople can be measured on a particular firm in a particular year, and the firm–year combination is the experimental unit. As another example, Total income, Number of wage earners, and Location can be observed for a household at a particular point in time, and the household–time combination is the experimental unit. Every experiment, whether observational or designed, has experimental units on which the variables are measured. However, the identification of the experimental units is more important in designed experiments, when the experimenter must actually sample the experimental units and measure the variables.

DEFINITION 15.5

An **experimental unit** is the object on which the response and factors are observed or measured.

When the specification of the treatments and the method of assigning the experimental units to each of the treatments is controlled by the analyst, the experiment is said to be **designed**. In contrast, if the analyst is just an observer of the treatments on a sample of experimental units, the experiment is **observational**. For example, if you give one randomly selected group of employees a particular training program and withhold it from another randomly selected group in order to evaluate the effect of the training on worker productivity, then you are designing an experiment. If, on the other hand, you compare the productivity of employees with college degrees with the productivity of employees without college degrees, the experiment is observational.

DEFINITION 15.6

A **designed experiment** is one for which the analyst controls the specification of the treatments and the method of assigning the experimental units to each treatment. An **observational experiment** is one for which the analyst simply observes the treatments and the response on a sample of experimental units.

The diagram in Figure 15.1 provides an overview of the experimental process and a summary of the terminology we have introduced in this section. Note that the experimental unit is at the core of the process. The method by which it is sampled from the population determines the type of experiment, and the level of every factor (the treatment) and the response are all observed on the experimental unit.

EXAMPLE 15.1

The USGA (United States Golf Association) regularly tests golf equipment to assure that it conforms to USGA standards. Suppose it wishes to compare the mean distance travelled by four different brands of golf balls when struck by a driver (the club with least club loft used to maximize distance). The following experiment is designed: Ten balls of each brand are randomly selected, each is struck by "Iron Byron" (the USGA's golf robot named for the famous golfer, Byron Nelson) using a driver, and the distance travelled is recorded. Identify each of the following elements in this

FIGURE 15.1

Sampling Experiments:
Process and Terminology

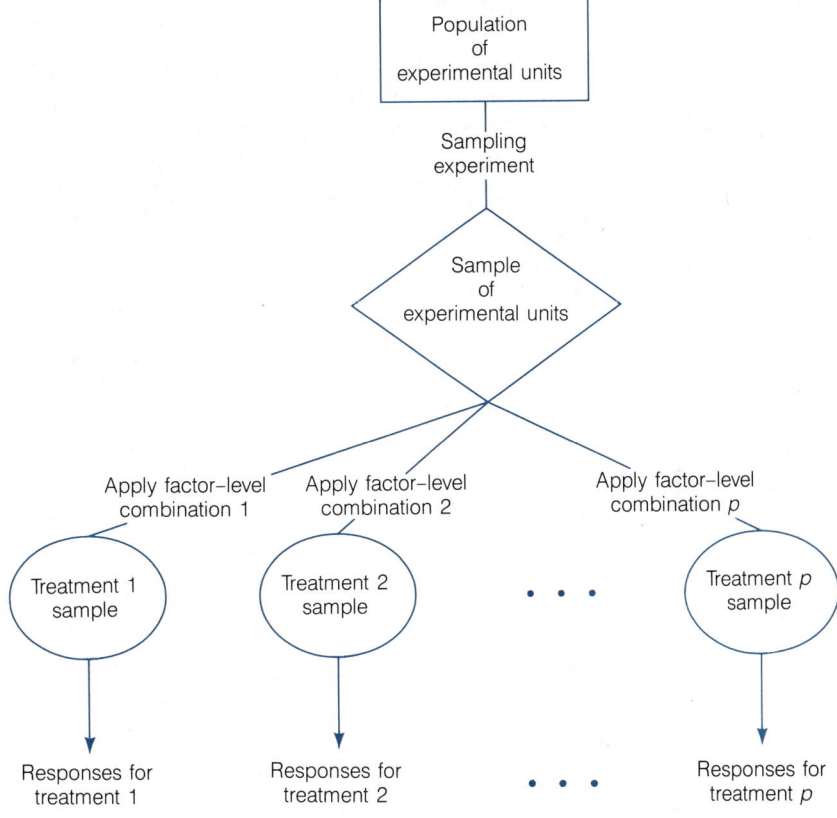

designed experiment: response, factors, factor types, levels, treatments, and experimental units.

SOLUTION

The response is the variable of interest, Distance travelled. The only factor being
investigated is Brand of golf ball, and it is nonnumerical and therefore qualitative.
The four brands (say A, B, C, and D) represent the levels of this factor. Since only
one factor is employed, the treatments are the four levels of this factor—that is,
the four brands. The experimental unit is a golf ball; more specifically, it is a golf
ball at a particular position in the striking sequence, since the distance travelled
can be recorded only when the ball is struck, and we would expect the distance to
be different (due to random factors such as wind resistance, landing place, and so
forth) if the same ball were struck a second time. Note that ten experimental units
are sampled for each treatment, generating a total of 40 observations.   ■

EXAMPLE 15.2

Refer to Example 15.1. Suppose the USGA is also interested in comparing the
mean distances the four brands of golf balls travel when struck by a five-iron. Ten
balls of each brand are randomly selected, five to be struck by the driver and five
by the five-iron. Identify the elements of the experiment, and construct a schematic
diagram similar to Figure 15.1 to provide an overview of this experiment.

The response is unchanged: Distance travelled. The experiment now has two factors, Brand of golf ball and Club utilized. There are four levels of Brand and two of Club (driver and five-iron, or 1 and 5). Treatments are factor-level combinations, so there are $4 \times 2 = 8$ treatments in this experiment: (A, 1), (A, 5), (B, 1), (B, 5), (C, 1), (C, 5), (D, 1), and (D, 5). The experimental units are still the combinations of golf ball and hitting position. Note that five experimental units are sampled per treatment, generating 40 observations. The experiment is summarized in Figure 15.2.

FIGURE 15.2
Two Factor Golf Experiment Summary: Example 15.2

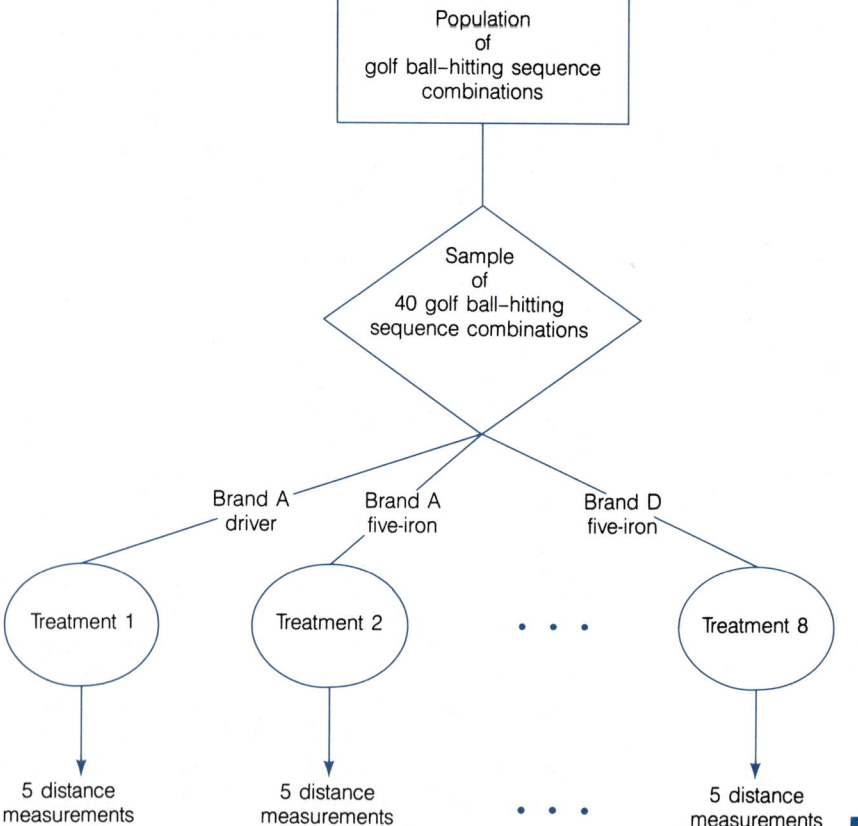

Our objective in designing an experiment is usually to maximize the amount of information obtained about the relationship between the treatments and the response. Of course, we are almost always subject to constraints on budget, time, and even the availability of experimental units. Nevertheless, designed experiments are generally preferred to observational experiments. Not only do we have better control of the amount and quality of the information collected, but also observational experiments are subject to biases in the selection of the experimental units representing each treatment. Inferences based on observational experiments always carry the implicit assumption that the sample has no hidden bias that was not considered in the statistical analysis. Better understanding of the potential problems with observational experiments is a by-product of our study of experimental design in the remainder of this chapter.

EXERCISES 15.1–15.8

LEARNING THE MECHANICS

15.1 What are the treatments for a designed experiment that utilizes one qualitative factor with four levels (A, B, C, and D)?

15.2 What are the treatments for a designed experiment with two factors, one qualitative with two levels (A and B) and one quantitative with five levels (50, 60, 70, 80, and 90)?

15.3 What are the experimental units on which each of the following responses are observed?
   a. College GPA
   b. Statewide unemployment rate in December
   c. Gasoline mileage rating for a model of automobile
   d. Number of defective sectors on a computer diskette

15.4 What is the difference between an observational and a designed experiment?

APPLYING THE CONCEPTS

15.5 Suppose the mean debt-to-equity ratio is to be compared for four types of companies: insurance, publishing, electric utilities, and banking. If random samples of each type of company are to be taken to make the comparison, identify each of the following elements:
   a. Response
   b. Factor(s) and factor type(s)
   c. Treatments
   d. Experimental units

15.6 Brockhaus (1980) compared the risk-taking propensity of three types of managers: entrepreneurs, newly hired managers, and newly promoted managers. Samples of individuals in each group were administered a test that measures one's propensity for taking risks. Identify each of the following elements:
   a. Response
   b. Factor(s) and factor type(s)
   c. Treatments
   d. Experimental units

15.7 A quality control supervisor measures the quality of a steel ingot on a scale from 0 to 10. He designs an experiment in which three different temperatures (ranging from 1100° to 1200°F) and five different pressures (ranging from 500 to 600 psi) are utilized, with 20 ingots produced at each temperature–pressure combination. Identify the following elements of the experiment:
   a. Response
   b. Factor(s) and factor type(s)
   c. Treatments
   d. Experimental units

15.8 Brief descriptions of a number of experiments are given below. Determine whether each is observational or designed, and explain your reasoning.

a. An economist obtains the unemployment rate and gross state product for a sample of states over the past 10 years, with the objective of examining the relationship between the unemployment rate and the gross state product by census region.

b. A manager in a paper production facility installs one of three different incentive programs in each of nine plants to determine the effect of each program on productivity.

c. A marketer of microcomputers runs ads in each of four national publications for one quarter and keeps track of the number of sales that are attributable to each publication's ad.

d. An electric utility engages a consultant to monitor the discharge from its smokestack on a monthly basis over a 1-year period in order to relate the level of sulfur dioxide in the discharge to the load on the facility's generators.

e. Intrastate trucking rates are compared before and after governmental deregulation of prices charged, with the comparison also taking into account distance of haul, goods hauled, and the price of diesel fuel.

S E C T I O N 15.2

THE COMPLETELY RANDOMIZED DESIGN

The simplest experimental design, a **completely randomized design**, consists of the *independent random selection* of experimental units representing each treatment. For example, we could independently select random samples of 20 female and 15 male high school seniors to compare their mean SAT scores. Or, we could independently select random samples of 30 households from each of four census districts in order to compare the mean income per household among the districts. In both examples our objective is to compare treatment means by selecting random, independent samples for each treatment.

> **DEFINITION 15.7**
>
> A **completely randomized design** is one for which independent random samples of experimental units are selected for each treatment.

The objective of a completely randomized design is usually to compare the treatment means. If we denote the true, or population, means of the p treatments as $\mu_1, \mu_2, \ldots, \mu_p$, then we will test the null hypothesis that the treatment means are all identical against the alternative that at least two of the treatment means differ:

H_0: $\mu_1 = \mu_2 = \cdots = \mu_p$

H_a: At least two of the p treatment means differ

The μ's might represent the means of *all* female and male high school seniors' SAT scores or the means of *all* households' income in each of four census regions.

In order to conduct a statistical test of these hypotheses, we will use the means of the independent random samples selected from the treatment populations using the completely randomized design. That is, we compare the p sample means $\bar{y}_1, \bar{y}_2, \ldots, \bar{y}_p$.

For example, suppose you select independent random samples of five female and five male high school seniors, and obtain sample mean SAT scores of 550 and 590, respectively. Can we conclude that males score 40 points higher, on average, than females? To answer the question, we must consider the amount of sampling variability among the experimental units (students). If the scores are as depicted in the dot diagram shown in Figure 15.3, then the difference between the means is small relative to the sampling variability of the scores within the treatments, female and male. We would not be inclined to reject the null hypothesis of equal population means in this case.

FIGURE 15.3
Dot Diagram of SAT
Scores: Difference
Between Means
Dominated by Sampling
Variability

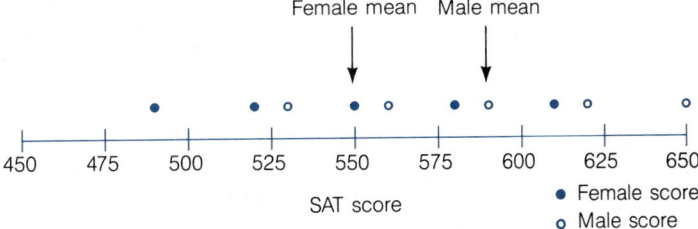

In contrast, if the data are as depicted in the dot diagram of Figure 15.4, then the sampling variability is small relative to the difference between the two means. We would be inclined to favor the alternative hypothesis that the population means differ in this case.

FIGURE 15.4
Dot Diagram of SAT
Scores: Difference
Between Means Large
Relative to Sampling
Variability

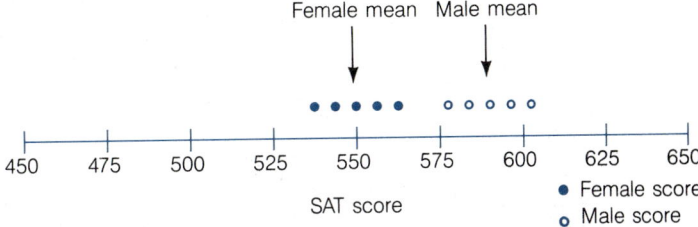

You can see that the key is to compare the difference between the treatment means to the amount of sampling variability. To conduct a formal statistical test of the hypotheses requires numerical measures of the difference between the treatment means and the sampling variability within each treatment. The variation between the treatment means is measured by the **Sum of Squares for Treatment (SST)**, which is calculated by squaring the distance between each treatment mean and the overall mean of *all* sample measurements, multiplying each squared distance by the number of sample measurements for the treatment, and adding the results over all treatments:

$$\text{SST} = \sum_{i=1}^{p} n_i(\bar{y}_i - \bar{y})^2$$
$$= 5(550 - 570)^2 + 5(590 - 570)^2 = 4{,}000$$

where we use $\bar{y}$ to represent the overall mean response of all sample measurements—that is, the mean of the combined samples. The symbol n_i is used to denote the sample size for the ith treatment. You can see that the value of SST is 4,000 for the two samples of five female and five male SAT scores.

Next, we must measure the sampling variability within the treatments. We call this the **Sum of Squares for Error** (SSE) because it measures the variability around the treatment means that is attributed to sampling error. Suppose the ten measurements in the first dot diagram (Figure 15.3) are 490, 520, 550, 580, and 610 for the five females and 530, 560, 590, 620, and 650 for the five males. Then the value of SSE is computed by summing the squared distance between each response measurement and the corresponding treatment mean and then adding the squared differences over all measurements in the entire sample:

$$SSE = \sum_{j=1}^{n_1}(y_{1j} - \bar{y}_1)^2 + \sum_{j=1}^{n_2}(y_{2j} - \bar{y}_2)^2 + \cdots + \sum_{j=1}^{n_p}(y_{pj} - \bar{y}_p)^2$$

where the symbol y_{1j} is the jth measurement in sample 1, y_{2j} is the jth measurement in sample 2, and so on. This rather complex-looking formula is not difficult to interpret in practice. For our samples of SAT scores,

$$
\begin{aligned}
SSE &= [(490 - 550)^2 + (520 - 550)^2 + (550 - 550)^2 + (580 - 550)^2 + (610 - 550)^2] \\
&\quad + [(530 - 590)^2 + (560 - 590)^2 + (590 - 590)^2 + (620 - 590)^2 + (650 - 590)^2] \\
&= 18,000
\end{aligned}
$$

To make the two measures of variability comparable, we divide each by the degrees of freedom to convert the sum of squares to mean squares. First, the **Mean Square for Treatments** (MST), which measures the variability among the treatment means, is equal to

$$MST = \frac{SST}{p - 1} = \frac{4,000}{2 - 1} = 4,000$$

where the number of degrees of freedom for the p treatments is $(p - 1)$. Next, the **Mean Square for Error** (MSE), which measures the sampling variability within the treatments, is

$$MSE = \frac{SSE}{n - p} = \frac{18,000}{10 - 2} = 2,250$$

Finally, we calculate the ratio of MST to MSE, an **F statistic**:

$$F = \frac{MST}{MSE} = \frac{4,000}{2,250} = 1.78$$

Values of the F statistic near 1 indicate that the two sources of variation, that between treatment means and that within treatments, are approximately equal. In this case, the difference between the treatment means may well be attributable to sampling error, which provides little support for the alternative hypothesis that the population treatment means differ. Values of F well in excess of 1 indicate that the variation among treatment means well exceeds that within means and therefore support the alternative hypothesis that the population treatment means differ.

When does F exceed 1 by enough to reject the null hypothesis that the means are equal? This depends on the degress of freedom for treatments and for error, and on the value of α selected for the test. We compare the calculated F value to a tabled F value (Tables VII–X of Appendix B) with $\nu_1 = (p - 1)$ degrees of freedom in the numerator and $\nu_2 = (n - p)$ degrees of freedom in the denominator, and corresponding to a Type I error probability of α. For the SAT example, the F statistic has $\nu_1 = (2 - 1) = 1$ numerator degree of freedom and $\nu_2 = (10 - 2) = 8$ denominator degrees of freedom. Thus, for $\alpha = .05$ we find (Table VIII of Appendix B):

$$F_{.05} = 5.32 \quad \text{for } \nu_1 = 1 \text{ and } \nu_2 = 8$$

The implication is that MST would have to be 5.32 times greater than MSE before we could conclude at the .05 level of significance that the two population treatment means differ. Since the data yielded $F = 1.78$, our initial impressions for the dot diagram in Figure 15.3 are confirmed: There is insufficient information to conclude that the mean SAT scores differ for the populations of female and male high school seniors. The rejection region and the calculated F value are shown in Figure 15.5.

FIGURE 15.5

Rejection Region and Calculated F Values for SAT Score Sample

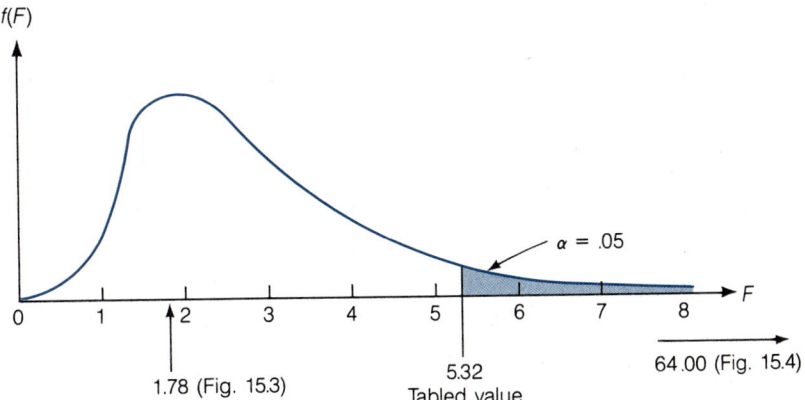

In contrast, consider the dot diagram in Figure 15.4. Since the means are the same as in the first example, 550 and 590, respectively, the variation between the means is the same, MST = 4,000. But the variation within the two treatments appears to be considerably smaller. The observed SAT scores are 540, 545, 550, 555, and 560 for females and 580, 585, 590, 595, and 600 for males. The variation within the treatments is measured by

$$
\begin{aligned}
\text{SSE} &= [(540 - 550)^2 + (545 - 550)^2 + (550 - 550)^2 + (555 - 550)^2 + (560 - 550)^2] \\
&\quad + [(580 - 590)^2 + (585 - 590)^2 + (590 - 590)^2 + (595 - 590)^2 + (600 - 590)^2] \\
&= 500
\end{aligned}
$$

$$\text{MSE} = \frac{\text{SSE}}{n - p} = \frac{500}{8} = 62.5$$

Then the F ratio is

$$F = \frac{MST}{MSE} = \frac{4,000}{62.5} = 64.0$$

Again, our visual analysis of the dot diagram is confirmed statistically: $F = 64.0$ well exceeds the tabled F value, 5.32, corresponding to the .05 level of significance (see Figure 15.5). We would therefore reject the null hypothesis at that level and conclude that the SAT mean score of males differs from that of females.

Recall that we performed a hypothesis test for the difference between two means in Section 9.2 using a two-sample t statistic for two independent samples. When two independent samples are being compared, the t and F tests are equivalent. To see this, consider the second sample, depicted in Figure 15.4, and recall the formula

$$t = \frac{\bar{y}_1 - \bar{y}_2}{\sqrt{s_p^2 \left(\frac{1}{n_1} + \frac{1}{n_2}\right)}} = \frac{590 - 550}{\sqrt{(62.5)\left(\frac{1}{5} + \frac{1}{5}\right)}} = \frac{40}{5} = 8$$

where we used the fact that $s_p^2 = MSE$, which you can verify by comparing the formulas. Note that the calculated F for these samples ($F = 64$) equals the square of the calculated t for the same samples ($t = 8$). Likewise, the tabled F value (5.32) equals the square of the tabled t value at the two-sided .05 level of significance ($t_{.025} = 2.306$ with 8 df). Since both the rejection region and the calculated values are related in the same way, the tests are equivalent. Moreover, the assumptions that must be met to assure the validity of the t and F test are the same:

1. The probability distributions of the populations of responses associated with each treatment must all be normal.
2. The probability distributions of the populations of responses associated with each treatment must have equal variances.
3. The samples of experimental units selected for the treatments must be random and independent.

In fact, the only real difference between the tests is that the F test can be used to compare *more than two* treatment means, while the t test is applicable to two samples only. The F test is summarized in the box on page 861.

Calculation formulas for MST and MSE are given in Appendix D. We will rely on some of the many computer programs available to compute the F statistic, concentrating on the interpretation of the results rather than their calculations.

EXAMPLE 15.3

Suppose the United States Golf Association (USGA) wants to compare the mean distances travelled by four different brands of golf balls when struck with a driver. A completely randomized design is employed, with Iron Byron, the USGA's robotic golfer, using a driver to hit a random sample of ten balls of each brand in a random sequence. The distance is recorded for each hit, and the results are shown in Table 15.1, organized by brand.

TEST TO COMPARE p TREATMENT MEANS FOR A COMPLETELY RANDOMIZED DESIGN

H_0: $\mu_1 = \mu_2 = \cdots = \mu_p$

H_a: At least two treatment means differ

Test statistic: $F = \dfrac{\text{MST}}{\text{MSE}}$

Assumptions: 1. All p population probability distributions are normal.
2. The p population variances are equal.
3. Samples are selected randomly and independently from the respective populations.

Rejection region: $F > F_\alpha$, where F_α is based on $(p - 1)$ numerator degrees of freedom (associated with MST) and $(n - p)$ denominator degrees of freedom (associated with MSE.)

TABLE 15.1
Results of Completely
Randomized Design:
"Iron Byron" Driver

	BRAND A	BRAND B	BRAND C	BRAND D
	251.2	263.2	269.7	251.6
	245.1	262.9	263.2	248.6
	248.0	265.0	277.5	249.4
	251.1	254.5	267.4	242.0
	265.5	264.3	270.5	246.5
	250.0	257.0	265.5	251.3
	253.9	262.8	270.7	262.8
	244.6	264.4	272.9	249.0
	254.6	260.6	275.6	247.1
	248.8	255.9	266.5	245.9
Means	251.3	261.1	270.0	249.4

a. Set up the test to compare the mean distances for the four brands. Use $\alpha = .10$.

b. Use the SAS System Analysis of Variance program to obtain the test statistic. Interpret the results.

SOLUTION

a. To compare the mean distances of the four brands, we first specify the hypotheses to be tested. Denoting the population mean of the ith brand by μ_i, we test

H_0: $\mu_1 = \mu_2 = \mu_3 = \mu_4$

H_a: The mean distances differ for at least two of the brands

The test statistic compares the variation among the four treatment (brand) means to the sampling variability within each of the treatments.

Test statistic: $F = \dfrac{\text{MST}}{\text{MSE}}$

Rejection region: $F > F_\alpha = F_{.10}$
with $\nu_1 = (p - 1) = 3$ df and $\nu_2 = (n - p) = 36$ df

From Table VII of Appendix B, we find $F_{.10} \approx 2.25$ for 3 and 36 df. Thus, we will reject H_0 if $F > 2.25$.

The assumptions necessary to assure the validity of the test are as follows: (1) The probability distributions of the distances for each brand are normal. (2) The variances of the distance probability distributions for the four brands are identical. (3) The samples of ten golf balls for each brand are selected randomly and independently.

b. The SAS printout for the data in Table 15.1 resulting from this completely randomized design is given in Figure 15.6. Note that the top part of the printout is identical to that in a regression analysis. The Total Sum of Squares is designated the "Corrected Total," and it is partitioned into the "Model" and "Error" Sums of Squares. The bottom part of the printout further partitions the "Model" component into the factors that comprise the model. In this single-factor experiment, the Model and Brand sums of squares are the same. The Sum of Squares column is headed "Type I SS," one of four types of sums of squares that SAS will calculate. The distinction becomes important only in multi-factor experiments with unequal numbers of observations per treatment; we will need to utilize only the Type I sum of squares. See the references at the end of this chapter for a more complete discussion of using SAS to analyze more complex experiments.

The values of the mean squares MST and MSE (shaded on the printout) are 903.07 and 25.22, respectively. The F ratio, 35.81, also shaded on the printout, exceeds the tabled value 2.25. We therefore reject the null hypothesis at the .10 level of significance, concluding that at least two of the brands differ with respect to mean distance travelled when struck by the driver.

FIGURE 15.6
SAS Analysis of Variance
Printout for Golf Ball
Distance Data:
Completely Randomized
Design

Dependent Variable: DISTANCE

Source	DF	Sum of Squares	Mean Square	F Value	Pr > F
Model	3	2709.19875	903.06625	35.81	0.0001
Error	36	907.86100	25.21836		
Corrected Total	39	3617.05975			

R-Square	C.V.	Root MSE	DISTANCE Mean
0.749006	1.9469768	5.02179	257.927500

Source	DF	Type I SS	Mean Square	F Value	Pr > F
BRAND	3	2709.20	903.07	35.81	0.0001

The observed significance level of the F test is also given on the printout: .0001. This is the area to the right of the calculated F value and implies that we would reject the null hypothesis that the means are equal at any α level greater than .0001. The result of the test is summarized in Figure 15.7.

FIGURE 15.7
F Test for Completely
Randomized Design: Golf
Ball Experiment

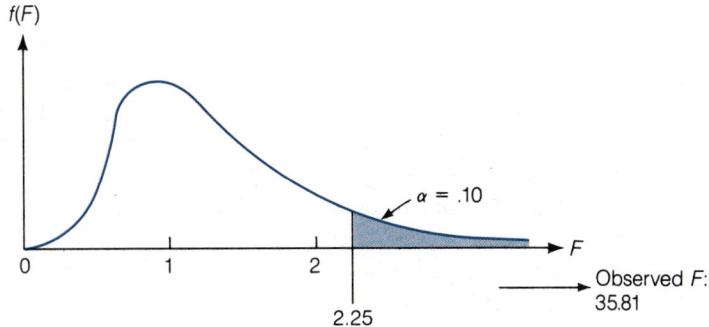

The results of an analysis of variance (ANOVA) can be summarized in a simple tabular format, similar to that obtained from the SAS program in Example 15.3. The general form of the table is shown in Table 15.2, where the symbols df, SS, and MS stand for degrees of freedom, Sum of Squares, and Mean Square, respectively. Note that the two sources of variation, Treatments and Error, add to the Total Sum of Squares, SS_{yy}, that we first introduced in simple linear regression analysis (Chapter 10). The ANOVA summary table for Example 15.3 is given in Table 15.3, and the partitioning of the Total Sum of Squares into its two components is illustrated in Figure 15.8.

TABLE 15.2
ANOVA Summary Table
for a Completely
Randomized Design

SOURCE	df	SS	MS	F
Treatments	$p-1$	SST	$MST = \dfrac{SST}{p-1}$	$\dfrac{MST}{MSE}$
Error	$n-p$	SSE	$MSE = \dfrac{SSE}{n-p}$	
Total	$n-1$	SS(Total)		

TABLE 15.3
ANOVA Summary Table
for Example 15.3

SOURCE	df	SS	MS	F
Brands	3	2,709.20	903.07	35.81
Error	36	907.86	25.22	
Total	39	3,617.06		

FIGURE 15.8
Partitioning of the Total
Sum of Squares for the
Completely Randomized
Design

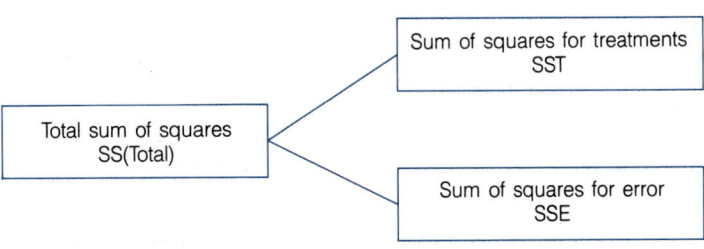

Suppose the F test results in a rejection of the null hypothesis that the treatment means are equal. Is the analysis complete? Usually, the conclusion that at least two of the treatment means differ leads to another question. Which of the means differ, and by how much? For example, the F test in Example 15.3 leads to the conclusion that at least two of the brands of golf balls are associated with different mean distances travelled when struck with a driver. Now the question is, which of the brands differ? How are the brands ranked with respect to mean distance?

We can place confidence intervals on the difference between the various pairs of treatment means in the experiment. If there are p treatment means, then there are $c = p(p - 1)/2$ pairs of means that can be compared. If we want to have $100(1 - \alpha)\%$ confidence that each of the c confidence intervals contains the true difference it is intended to estimate, then each individual confidence interval will have to be formed using a smaller value of α than would a single interval. For example, if we want to compare the $4(3)/2 = 6$ pairs of golf balls Brand means and we want 95% confidence that all six confidence intervals comparing the means contain the true differences between the Brand means, then each individual interval will need to be constructed using a smaller level of significance than .05 in order to have 95% confidence that the six intervals collectively include the true differences. *

There are a number of procedures available for making **multiple comparisons** of a set of treatment means which, under various assumptions, assure that the overall confidence level associated with all the comparisons remains at or above the specified $100(1 - \alpha)\%$ level. Perhaps the simplest of these to apply is the **Bonferroni procedure**, which calls for specifying the level of significance of each comparison at α/c, where α is the overall level of significance desired, and c is the number of pairs of means to be compared. The result is a set of confidence intervals in which we can be *at least* $100(1 - \alpha)\%$ confident. The Bonferroni procedure is therefore conservative, since the confidence level is generally greater than that we specify. Of course, this means that the intervals are somewhat wider than they need to be in order to have the specified confidence level. Procedures that are more exact are also more complex than the Bonferroni procedure. The Bonferroni procedure is summarized in the box. Other multiple comparison procedures can be found in the references at the end of the chapter.

EXAMPLE 15.4

Refer to Example 15.3, in which we concluded that at least two of the four brands of golf balls are associated with different mean distances travelled when struck with a driver. Use the Bonferroni procedure to compare the $4(3)/2 = 6$ pairs of treatment means. Use an overall confidence level of 90%.

*The reason each interval must be formed at a higher confidence level than that specified for the collection of intervals can be demonstrated as follows:

P{At least one of c intervals fails to contain the true difference}

$= 1 - $ P{All c intervals contain the true differences}

$\geq 1 - (1 - \alpha)^c \geq \alpha$

Thus, to make this probability of at least one failure equal to α, we must specify the individual levels of significance to be less than α.

BONFERRONI PROCEDURE FOR MULTIPLE COMPARISONS: COMPLETELY RANDOMIZED DESIGN

1. Specify the overall level of significance α or, equivalently, the overall confidence level $100(1 - \alpha)\%$.
2. If there are c pairs of treatment means to be compared, specify the level of significance for each comparison at α/c.
3. Calculate the $100(1 - \alpha/c)\%$ confidence interval for each of the c pairs of means using the following formula:

$$(\bar{y}_i - \bar{y}_j) \pm t_{\alpha/(2c)} \, s \sqrt{\left(\frac{1}{n_i} + \frac{1}{n_j}\right)}$$

where $s = \sqrt{\text{MSE}}$ and $t_{\alpha/(2c)}$ is the tabulated value of t (Table VI of Appendix B) that locates area $\alpha/(2c)$ in the upper tail of the t distribution with $(n - p)$ degrees of freedom (the number of degrees of freedom associated with error in the ANOVA).
4. Summarize the results of the multiple comparisons by ranking the treatment means, showing which pairs are significantly different.

Assumptions: Same as for the ANOVA.

SOLUTION

The level of significance specified for the multiple comparisons is $\alpha = .10$. The Bonferroni procedure therefore specifies the level of significance for each comparison at $\alpha/c = .10/6 = .0167$. The t statistic to be used in the formation of the six confidence intervals splits this area into two tails, with $\alpha/(2c) = .0167/2 = .0083$ in each tail. The degrees of freedom associated with error are $(n - p) = (40 - 4) = 36$, which is more than the maximum (29) contained in the t table (Table VI of Appendix B). We will use the standard normal z table (Table IV of Appendix B) to find $z_{.0083} \approx 2.4$. That is, as shown in Figure 15.9 (page 866), we will form six confidence intervals using plus and minus 2.4 standard deviations in order to have 90% confidence in the overall results of all six intervals. Contrast this with the fact that we would be using plus and minus $z = 1.645$ standard deviations if we were forming a single 90% confidence interval.*

We are now ready to form the confidence intervals. Calculating $s = \sqrt{\text{MSE}} = \sqrt{25.22} = 5.02$ (also given in the SAS printout shown in Figure 15.6 as "Root MSE"), we begin with a comparison of brands A and B:

$$(\bar{y}_A - \bar{y}_B) \pm t_{.0083} s \sqrt{(\tfrac{1}{10} + \tfrac{1}{10})}$$
$$(251.3 - 261.1) \pm (2.4)(5.02)(.447)$$
$$-9.8 \pm 5.4 \quad \text{or} \quad (-15.2, -4.4)$$

*If the t table is used, we may have to round $\alpha/(2c)$ *down* to enable the use of Table VI. In Example 15.4, we would approximate $\alpha/(2c) = .0083 \approx .005$. This would provide slightly wider intervals than necessary, but assures an overall confidence level of at least $100(1 - \alpha)\%$. Alternatively, many computer programs will provide a more exact tabled t value corresponding to a given level of significance and a specified number of degrees of freedom.

FIGURE 15.9
Confidence Level for
Bonferroni Multiple
Comparisons:
Example 15.4

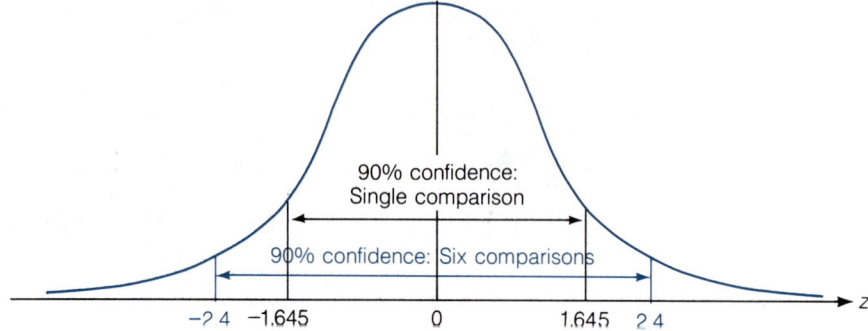

Thus, we conclude that the brand B mean distance exceeds the brand A mean distance by between 4.4 and 15.2 yards. The remainder of the intervals are easier to calculate. The interval half-widths are all the same (5.4), since the sample sizes are the same for each brand. The rest of the comparisons are summarized below:

$$A - C: \quad -18.7 \pm 5.4 \quad \text{or} \quad (-24.1, -13.3)$$
$$A - D: \quad 1.9 \pm 5.4 \quad \text{or} \quad (-3.5, 7.3)$$
$$B - C: \quad -8.9 \pm 5.4 \quad \text{or} \quad (-14.3, -3.5)$$
$$B - D: \quad 11.7 \pm 5.4 \quad \text{or} \quad (6.3, 17.1)$$
$$C - D: \quad 20.6 \pm 5.4 \quad \text{or} \quad (15.2, 26.0)$$

We are 90% confident that the intervals collectively contain all the differences between the true brand mean distances. Note that intervals that contain 0, such as the (brand A − brand D) interval from −3.5 to 7.3, do not support a conclusion that the true brand mean distances differ. If both endpoints of the interval are positive, as with the (brand B − brand D) interval, the implication is that the first brand (B) mean distance exceeds the second (D). Conversely, if both endpoints of the interval are negative, as with the (brand A − brand C) interval, the implication is that the second brand mean distance (C) exceeds the first (A).

A convenient summary of the results of the Bonferroni multiple comparisons is a listing of the brand means from highest to lowest, with a solid line connecting those that are *not* significantly different. This summary is shown in Figure 15.10. The interpretation is that brand C's mean distance exceeds all others; brand B's mean exceeds that of brands A and D; and the means of brands A and D do not differ significantly. All these inferences are made with 90% confidence, the overall confidence level of the Bonferroni multiple comparisons.

FIGURE 15.10
Summary of Bonferroni
Multiple Comparisons,
Example 15.4

BRAND	MEAN
C	270.0
B	261.1
A	251.3
D	249.4

Since the samples were selected independently in a completely randomized design, we can also form a confidence interval for an individual treatment mean. We use the one-sample t confidence interval of Section 8.6 to form the interval, again using the standard deviation, $s = \sqrt{\text{MSE}}$, as the measure of sampling variability for the experiment. For example, if we wish to place a 90% confidence interval on the mean distance travelled by brand C (apparently the "longest ball" of those tested), we calculate

$$\bar{y}_C \pm t_{.05,36} s \sqrt{\tfrac{1}{10}}$$
$$270.0 \pm (1.645)(5.02)(.32)$$
$$270.0 \pm 2.6 \quad \text{or} \quad (267.4, 272.6)$$

Thus, we are 90% confident that the true mean distance travelled for brand C is between 267.4 and 272.6 yards, when hit with a driver by Iron Byron. ∎

The procedure for conducting an analysis of variance for a completely randomized design is summarized in the box. Remember that the hallmark of this design is

---

STEPS FOR CONDUCTING AN ANOVA FOR A COMPLETELY RANDOMIZED DESIGN

1. Be sure the design is truly completely randomized, with independent random samples for each treatment.
2. Create an ANOVA summary table that specifies the variability attributable to Treatments and Error and leads to the calculation of the F statistic to test the null hypothesis that the treatment means are equal in the population. Use either an ANOVA computer program or the calculation formulas in Appendix D to obtain the necessary numerical computations.
3. If the F test leads to the conclusion that the means differ:
   a. Conduct multiple comparisons of as many of the pairs of means as you wish to compare by using the Bonferroni (or some other) multiple comparisons procedure. Use the results to summarize the statistically significant differences among the treatment means.
   b. If desired, form confidence intervals for one or more individual treatment means.
4. If the F test leads to the nonrejection of the null hypothesis that the treatment means differ, several possibilities exist:
   a. The treatment means are equal—that is, the null hypothesis is true.
   b. The treatment means really differ, but other important factors affecting the response are not accounted for by the completely randomized design. These factors inflate the sampling variability, as measured by MSE, resulting in smaller values of the F statistic. Either increase the sample size for each treatment or use a different experimental design (as in Section 15.3) that accounts for the other factors affecting the response.
   Be careful not to automatically reach conclusion a, since the possibility of a Type II error must be considered if you accept H_0.

independent random samples of experimental units associated with each treatment. We will discuss a design with dependent samples in the next section.

In the past 10 years computers and computer training have become integral parts of the curricula of secondary schools and universities. As a result, younger business professionals tend to be more comfortable with computers than their more senior counterparts.

> 'The older the person is, the worse it is,' said Arnold S. Kahn of the American Psychological Association. 'The older they are and the longer they wait before learning, the more dissatisfied they will become.' The computer's unrelenting march into offices and factories is often cited as a chief cause of work-related stress. As computers and robots come into the workplace, many workers fear they'll never master the new skills required (Aplin-Brownlee, 1984).

In 1984, Gary W. Dickson,[*] Professor of Management Information Systems at the University of Minnesota, investigated the computer literacy of middle managers with 10 years or more management experience. As part of his study, Dickson designed a questionnaire to measure a manager's technical knowledge of computers. If the questionnaire were properly designed, the scores received by managers could be used as predictors of their knowledge of computers, with higher scores indicating greater knowledge. To check the design of the questionnaire (i.e., its validity), 19 middle managers from the Minneapolis–St. Paul metropolitan area were randomly sampled and asked to complete the questionnaire. Their scores appear in Table 15.4. (The highest possible score on the questionnaire was 169.)

TABLE 15.4

Questionnaire Scores of Middle Managers

MANAGER	LEVEL OF TECHNICAL EXPERTISE	SCORE	MANAGER	LEVEL OF TECHNICAL EXPERTISE	SCORE
1	1	82	11	1	80
2	1	114	12	1	105
3	1	90	13	2	110
4	1	80	14	2	133
5	2	128	15	3	128
6	2	90	16	2	130
7	3	156	17	2	104
8	1	88	18	3	151
9	1	93	19	3	140
10	2	130			

Prior to completing the questionnaire, the managers were asked to describe their knowledge of and experience with computers. This information was used to classify the managers as possessing a high (3), medium (2), or low (1) level of technical computer expertise. These data also appear in Table 15.4.

[*]Personal communication with Gary W. Dickson, Jan. 1984.

In order to evaluate the questionnaire's design, Dickson used analysis of variance to compare the mean scores of each of the three groups of managers. If the questionnaire were properly designed, the mean scores should differ. In particular, the mean score of managers with a high level of expertise should be greater than the mean score of managers with a medium level of expertise, etc. Dickson used Minitab to obtain the ANOVA table displayed in Figure 15.11.

FIGURE 15.11
ANOVA Table for
Managers' Scores

```
ANALYSIS OF VARIANCE

DUE TO   DF     SS    MS=SS/DF   F-RATIO

FACTOR    2   7634.      3817.     19.19
ERROR    16   3182.       199.
TOTAL    18  10815.

LEVEL   N    MEAN    ST. DEV.

1        8   91.5      12.3
2        7  117.9      16.6
3        4  143.8      12.4

POOLED ST. DEV. =        14.1
```

Since $F = 19.19$ is greater than $F_{.01} = 6.23$ ($\nu_1 = 2$, $\nu_2 = 16$), Dickson concluded that the mean scores differ for at least two of the three groups of managers. However, this result could be obtained even with a poorly designed questionnaire. For example, a significant F statistic could result even if the mean for the high group were lower than the mean for the low group. Accordingly, Dickson used confidence intervals to examine the differences between individual group means. Using the information on the Minitab printout along with the appropriate t value, he developed the following 95% confidence intervals:

$$7.15 \le \mu_3 - \mu_2 \le 44.65$$
$$10.92 \le \mu_2 - \mu_1 \le 41.88$$

Since these confidence intervals indicate that $\mu_3 > \mu_2 > \mu_1$, Dickson concluded that the questionnaire could be used as a predictor of managers' technical knowledge of computers.

EXERCISES 15.9–15.28

LEARNING THE MECHANICS

15.9 Use Tables VII, VIII, IX, and X of Appendix B to find each of the following F values:
    a. $F_{.05}$, $\nu_1 = 2$, $\nu_2 = 2$ **b.** $F_{.01}$, $\nu_1 = 2$, $\nu_2 = 2$
    c. $F_{.10}$, $\nu_1 = 20$, $\nu_2 = 40$ **d.** $F_{.025}$, $\nu_1 = 12$, $\nu_2 = 9$

15.10 Find the following probabilities:
    a. $P(F \le 2.88)$ for $\nu_1 = 20$, $\nu_2 = 21$
    b. $P(F > 3.52)$ for $\nu_1 = 15$, $\nu_2 = 15$
    c. $P(F > 2.40)$ for $\nu_1 = 15$, $\nu_2 = 15$
    d. $P(F \le 1.69)$ for $\nu_1 = 40$, $\nu_2 = 40$

15.11 In which dot diagram is the difference between the sample means small relative to the variability within the sample observations? Justify your answer.

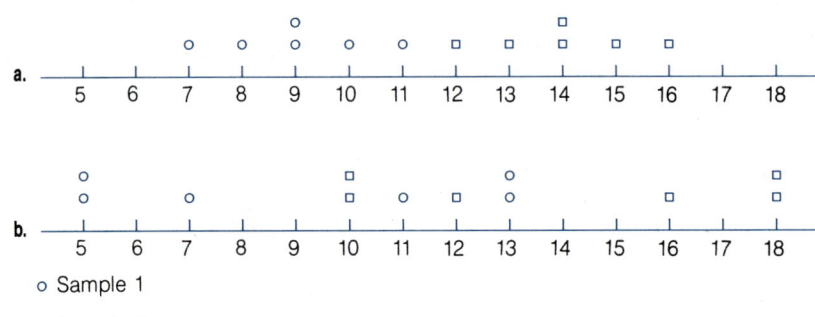

o Sample 1

□ Sample 2

15.12 Refer to Exercise 15.11. Assume that the two samples represent independent, random samples corresponding to two treatments in a completely randomized design.
  a. Calculate the treatment means, i.e., the means of samples 1 and 2, for both dot diagrams.
  b. Use the means to calculate the Sum of Squares for Treatments (SST) for each dot diagram.
  c. Calculate the sum of squared differences between each sample measurement and the corresponding sample mean, and sum the squared differences over the two samples to obtain the Sum of Squares for Error (SSE) for each dot diagram.
  d. Calculate the Total Sum of Squares [SS(Total)] for the two dot diagrams by adding the Sums of Squares for Treatment and Error. What percentage of SS(Total) is accounted for by the treatments—that is, what percentage of the Total Sum of Squares is the Sum of Squares for Treatment—in each case?
  e. Convert the Sums of Squares for Treatment and Error to mean squares by dividing each by the appropriate number of degrees of freedom. Calculate the F ratio of the Mean Square for Treatment (MST) to the Mean Square for Error (MSE) for each dot diagram.
  f. Use the F ratios to test the null hypothesis that the two samples are drawn from populations with identical means. Use $\alpha = .05$.
  g. What assumptions must be made about the probability distributions corresponding to the responses for each treatment in order to assure the validity of the F tests conducted in part f?

15.13 Refer to Exercises 15.11 and 15.12. Conduct a two-sample t test (Section 9.2) of the null hypothesis that the two treatment means are equal for each dot diagram. Use $\alpha = .05$ and two-tailed tests. In the course of the test, compare each of the following with the F tests in Exercise 15.12:
  a. The pooled variances and the MSEs
  b. The t and the F test statistics
  c. The tabled values of t and F that determine the rejection regions
  d. The conclusions of the t and F tests
  e. The assumptions that must be made in order to assure the validity of the t and F tests

15.14 Refer to Exercises 15.11 and 15.12. Complete the following ANOVA table for each of the two dot diagrams:

SOURCE	df	SS	MS	F
Treatments				
Error				
Total				

15.15 Suppose that the Total Sum of Squares for a completely randomized design with $p = 6$ treatments and $n = 36$ total measurements (six per treatment) is equal to 500. In each of the following cases, conduct an F test of the null hypothesis that the six treatment means are the same. Use $\alpha = .10$.
 a. The Sum of Squares for Treatment (SST) is 20% of SS(Total).
 b. SST is 50% of SS(Total).
 c. SST is 80% of SS(Total).
 d. What happens to the F ratio as the percentage of the Total Sum of Squares attributable to treatments is increased?

15.16 A partially completed ANOVA summary for a completely randomized design is shown in the table.

SOURCE	df	SS	MS	F
Treatments	6	16.9		
Error				
Total	41	45.2		

 a. Complete the ANOVA table.
 b. How many treatments are involved in the experiment?
 c. Do the data provide sufficient evidence to indicate a difference among the population means? Test using $\alpha = .10$.
 d. Find the approximate observed significance level for the test in part **c**, and interpret it.
 e. Suppose that $\bar{x}_1 = 3.7$ and $\bar{x}_2 = 4.1$. Do the data provide sufficient evidence to indicate a difference between μ_1 and μ_2? Assume that there are seven observations for each treatment. Test using $\alpha = .10$.
 f. Refer to part **e**. Find a 90% confidence interval for $(\mu_1 - \mu_2)$.
 g. Refer to part **e**. Find a 90% confidence interval for μ_1.

15.17 The Minitab printout for an experiment utilizing a completely randomized design is shown here.

```
ANALYSIS OF VARIANCE
SOURCE      DF          SS         MS         F
FACTOR       3       57258      19086     14.80
ERROR       34       43836       1289
TOTAL       37      101094
```

Note: Minitab uses "FACTOR" instead of "Treatments."

a. How many treatments are involved in the experiment? What is the total sample size?

b. Conduct a test of the null hypothesis that the treatment means are equal. Use $\alpha = .01$.

c. What additional information is needed in order to be able to compare specific pairs of treatment means?

15.18 Refer to Exercise 15.17. Suppose the treatment means are 190.8, 260.1, 191.7, and 279.4, with sample sizes 8, 12, 10, and 8, respectively. Use the Bonferroni technique to compare all pairs of treatment means with an overall level of significance $\alpha = .10$.

15.19 The data in the accompanying table resulted from an experiment that utilized a completely randomized design.

TREATMENT 1	TREATMENT 2	TREATMENT 3
3.8	5.4	1.3
1.2	2.0	.7
4.1	4.8	2.2
5.5	3.8	
2.3		

a. Use the appropriate calculation formulas in Appendix D to complete the following ANOVA table:

SOURCE	df	SS	MS	F
Treatments				
Error				
Total				

b. Test the null hypothesis that $\mu_1 = \mu_2 = \mu_3$, where μ_i represents the true mean for treatment i, against the alternative that at least two of the means differ. Use $\alpha = .01$.

c. Use the Bonferroni technique to compare all pairs of means. Use an $\alpha = .05$ overall level of significance.

APPLYING THE CONCEPTS

15.20 An accounting firm that specializes in auditing the financial records of large corporations is interested in evaluating the appropriateness of the fees it charges for its services. As part of its evaluation it wants to compare the costs it incurs in auditing corporations of different sizes. The accounting firm decided to measure the size of its client corporations in terms of their yearly sales. Accordingly, its population of client corporations was divided into three subpopulations:

A: Those with sales over $250 million

B: Those with sales between $100 million and $250 million

C: Those with sales under $100 million

The firm chose random samples of ten corporations from each of the subpopulations and determined the costs (in thousands of dollars) given in the table from its records.

A	B	C
250	100	80
150	150	125
275	75	20
100	200	186
475	55	52
600	80	92
150	110	88
800	160	141
325	132	76
230	233	200

a. Construct a dot diagram (refer to Figures 15.3 and 15.4) for the sample data using different types of dots for each of the three samples. Indicate the location of each of the sample means. Based on the information reflected in your dot diagram, do you believe that a significant difference exists among the subpopulation means? Explain.

b. SAS was used to conduct the analysis of variance calculations, resulting in the printout shown here. Conduct a test to determine whether the three classes of firms have different mean costs incurred in audits. Use $\alpha = .05$.

SAS Printout for Exercise 15.20

General Linear Models Procedure

Dependent Variable: COST

Source	DF	Sum of Squares	Mean Square	F Value	Pr > F
Model	2	318861.667	159430.833	8.44	0.0014
Error	27	510163.000	18894.926		
Corrected Total	29	829024.667			

R-Square	C.V.	Root MSE	COST Mean
0.384623	72.220043	137.459	190.333333

Source	DF	Type I SS Mean Square	F Value	Pr > F
TREATMNT	2	318861.67	159430.83	8.44	0.0014

c. What is the observed significance level for the test in part b? Interpret it.

d. Use the Bonferroni technique to compare all pairs of means. Use $\alpha = .05$ as the overall level of significance.

e. What assumptions must be met in order to assure the validity of the inferences you made in parts b and d?

15.21 R. H. Brockhaus (1980) of St. Louis University conducted a study to determine whether entrepreneurs, newly hired managers, and newly promoted managers differ in their risk-taking propensities. For the purpose of this study, entrepreneurs were defined as individuals who, within 3 months prior to the study, had ceased working for their employers in order to manage their own business ventures. Thirty-one individuals of each type were selected to participate in the study. Each was asked to complete a questionnaire which required the respondent to choose between a safe alternative and a more attractive but risky one. Test scores were designed to measure risk-taking propensity. (Lower scores are associated with greater conservatism in risk-taking situations.) Summary statistics for the test scores of the three groups are given in the table.

GROUP	SAMPLE SIZE	SAMPLE MEAN	STANDARD DEVIATION	GROUP TOTALS
Entrepreneurs	31	71.00	11.94	2,201
Newly hired managers	31	72.52	12.19	2,248
Promoted managers	31	66.97	10.84	2,076
	93			6,525

a. The following partial ANOVA table is calculated from the data. Complete the table.

SOURCE	df	SS	MS	F
Treatments		509.87		
Error		12,259.96		
Total				

b. Do the data provide sufficient evidence to indicate differences in the mean risk-taking propensities among the three groups? Test using $\alpha = .05$.

c. What assumptions must be satisfied in order for the test of part **b** to be valid?

d. Would you advise conducting tests to compare the individual pairs of means? Explain.

e. Would you classify this experiment as observational or designed? Explain.

15.22 The application of *management by objectives* (MBO), a method of performance appraisal, is the object of a study by Y. K. Shetty and H. M. Carlisle (1974) of Utah State University. The study dealt with the reactions of a university faculty to an MBO program. One hundred nine faculty members were asked to comment on whether they thought the MBO program was successful in improving their performance within their respective departments and the university. Each response was assigned a score from 1 (significant improvement) to 5 (significant decrease). The table shows the sample sizes, sample totals, mean scores, and sum of squares of deviations *within* each sample for samples of scores corresponding to the four academic ranks. Assume that the four samples in the table can be viewed as independent random samples of scores selected from among the four academic ranks.

a. Given that SST = 6.816 and SSE = 29.710, construct an ANOVA table for this experiment.

b. Do the data provide sufficient evidence to conclude there is a difference in mean scores among the four academic ranks? Test using $\alpha = .05$.

	INSTRUCTOR	ASSISTANT PROFESSOR	ASSOCIATE PROFESSOR	PROFESSOR
Sample size	15	41	29	24
Sample total	42.960	145.222	92.249	73.224
Sample mean	2.864	3.542	3.181	3.051
Within-sample sum of squared deviations	2.0859	14.0186	7.9247	5.6812

c. Use the Bonferroni technique for comparing the pairs of means, using an overall significance level of .05.

15.23 Most new products are test marketed in several locations, frequently using different advertising techniques (Klompmaker, Hughes, and Haley, 1976). Suppose the table represents the number of sales for a new product at each of three locations during each of the last 4 months.

LOCATION		
I	II	III
456	441	501
421	419	467
397	415	520
419	420	493

a. Treat this as a completely randomized design, and test to determine whether there is a difference among the mean sales at the three locations. Use Appendix D or a computer program to perform the calculations, and test at $\alpha = .05$.

b. Estimate the difference in the mean sales between locations I and III using a 90% confidence interval.

15.24 In Exercise 12.54, regression analysis was applied to the data in the table to investigate whether the mean debt-to-equity ratio varies among the insurance, publishing, electric utilities, and banking industries. In this exercise, we will use analysis of variance to perform the same investigation.

INSURANCE		PUBLISHING	
Firm	Debt-to-equity	Firm	Debt-to-equity
Chubb	.24	Deluxe Check	.03
Kemper	.09	New York Times	.32
St. Paul Cos.	.12	Times Mirror	.51
Lincoln National	.09	Dow Jones	.15
USF & G	.00	Gannett	.73
Aetna Life & Cas.	.07		

ELECTRIC UTILITIES		BANKING	
Firm	Debt-to-equity	Firm	Debt-to-equity
Pacific G & E	.84	U.S. Bancorp	.54
Houston Ind.	1.00	Sun Trust Banks	.29
Florida Progress	1.03	Mellon Bank	.82
Penn Power & Light	1.11	Michigan National	.13
North States Power	.83	Southeast Banking	.41
Ohio Edison	1.16		
Orange & Rockland	.70		

Source: "39th Annual Report on American Industry," *Forbes*, Jan. 12, 1987.

a. Identify the response, factor, factor type, treatments, and experimental units for this experiment.

b. Construct a dot diagram for the four sets of sample data. Indicate the location of each sample mean. Based on the information reflected in your dot diagram, do you believe that a difference exists among the industry means? Explain.

c. The SPSS ANOVA printout is reproduced here. Is there sufficient evidence to indicate that the mean debt-to-equity ratio varies among the four industries? Test using $\alpha = .01$.

SPSS Printout for Exercise 15.24

			Analysis of Variance			
Source	D.F.	Sum of Squares	Mean Squares	F Ratio	F Prob.
Between Groups	3	2.5187	.8396	20.3000	.000
Within Groups	19	.7858	.0414		
Total	22	3.3044			

Note: SPSS uses "Between Groups" instead of "Treatments," and "Within Groups" instead of "Error."

d. What assumptions must be satisfied in order for the test of part **c** to be valid?

e. Does the result of part **c** indicate that the individual pairs of means should be compared? If so, use $\alpha = .05$ and the Bonferroni technique to compare them.

15.25 A company that employs a large number of salespeople is interested in learning which of the salespeople sell the most: those strictly on commission, those with a fixed salary, or those with a reduced fixed salary plus a commission. The previous month's records for a sample of salespeople are inspected and the amount of sales (in dollars) is recorded for each, as shown in the table.

COMMISSIONED	FIXED SALARY	COMMISSION PLUS SALARY
$425	$420	$430
507	448	492
450	437	470
483	432	501
466	444	
492		

a. Construct a dot diagram for these data. Indicate the location of each sample mean.

b. The Minitab ANOVA printout for these data is shown here. Is there sufficient evidence to indicate that the mean sales differ among the three types of compensation?

Minitab Printout for Exercise 15.25

```
ANALYSIS OF VARIANCE
SOURCE     DF      SS        MS       F
FACTOR      2     4195      2098     3.17
ERROR      12     7945      662
TOTAL      14    12140
```

c. Use a 90% confidence interval to estimate the mean sales for salespeople who receive a commission plus salary.

15.26 How does flextime, which allows workers to set their individual work schedules, affect worker job satisfaction? Researchers recently conducted a study to compare a measure of job satisfaction for workers using three types of work scheduling: flextime, staggered starting hours, and fixed hours. Workers in each group worked according to their specified work scheduling system for 4 months. Although each worker filled out job satisfaction questionnaires both before and after the 4-month test period, we will examine only the post-test-period scores. The sample sizes, means, and standard deviations of the scores for the three groups are shown in the table.

	GROUP		
	Flextime	Staggered	Fixed
Sample size	27	59	24
Mean	35.22	31.05	28.71
Standard deviation	10.22	7.22	9.28

a. Assume that the data were collected according to a completely randomized design. Identify the response, the factor, the factor type, the treatments, and the experimental units.
b. Use the sample means to calculate the Sum of Squares for Treatments, SST.
c. Use the standard deviations to calculate the Sum of Squares for Error, SSE. Remember that SSE is the sum of squared differences between each measurement in the experiment and the corresponding treatment mean.
d. Construct an ANOVA table for this experiment.
e. Do the data provide sufficient evidence that the three groups differ with respect to their mean job satisfaction? Test using $\alpha = .05$.
f. Use the Bonferroni technique to compare the pairs of means corresponding to the three treatments. Use $\alpha = .05$ as the overall significance level for the comparisons.

15.27 In Exercise 9.30 we compared the mean bond price changes over a 12-month period for two underwriters. Similar data providing a comparison of five underwriting firms were extracted from the paper by D. Logue and R. Rogalski (1979) and are shown in the table. Suppose the data represent independent random samples from the five populations.

	UNDERWRITER				
	1	2	3	4	5
Sample size	27	20	23	11	15
Sample mean	$-.0491$	$-.0479$	$-.0307$	$-.0438$	$-.0051$
Sample variance	.009800	.006459	.002465	.001462	.002834

a. Given that SST = .023015 and SSE = .486047, construct an ANOVA table for this experiment.
b. Do the data provide sufficient evidence to indicate differences among the mean bond price changes over the 12-month period for the five underwriters? Use $\alpha = .05$.
c. Do the results of the test justify a comparison of the individual pairs of means? Explain.

15.28 Auditors may be called upon to perform compilations, reviews, or audits for their nonpublic clients. To do compilations and reviews of financial statements requires auditors an average of 25% and 50% fewer hours, respectively, than to perform an annual audit. Both auditors and clients fear that users of compilations and reviews—both of which are new forms of auditor association with a firm's financial reports—might not recognize the limited nature of these reports and might assume that the auditor was accepting the same degree of responsibility for them as for audited annual financial statements. To investigate these fears, Johnson, Pany, and White (1983) conducted an experiment designed to measure bankers' reactions to these three alternative forms of auditors' reports. Ninety-eight loan officers responded to a questionnaire in which they were asked to review a financial statement and background information for a commercial loan applicant. Thirty-one of the officers received a compilation, 25 received a review, 27 an audit, and 15 received a financial statement with no auditor association. One of the questions dealt with the loan officers' level of confidence about the financial statement's conformance to generally accepted accounting principles (GAAP). They were asked to indicate their level of confidence on a scale from 0 to 10 (no confidence to extreme confidence). Johnson, Pany, and White hypothesized that the mean level of confidence would not be the same for all four types of financial statements. Further, they hypothesized that the mean level of confidence associated with the audit would be greater than with the next highest form of auditor association, the review. Some of the data obtained from their experiment are summarized in the tables.

SOURCE	df	SS
Type of report	3	273
Error	94	494
Total	97	767

	TREATMENT MEANS
No auditor association	3.9
Compilation	5.5
Review	6.1
Audit	8.8

a. Do the data provide sufficient evidence to conclude that the mean level of confidence differs among the four forms of auditor association? Test using $\alpha = .05$.

b. Report the approximate p-value of your test.

c. Do the data provide sufficient evidence to conclude that the mean level of confidence associated with the audit is significantly higher than the mean for the review? Test using $\alpha = .05$.

d. Use the Bonferroni technique to compare the pairs of means using an overall significance level of $\alpha = .05$.

e. Relate your findings for parts **a**, **c**, and **d** to the expressed fear that users of financial reports might assume auditors were accepting the same degree of responsibility for each form of financial report.

S E C T I O N 15.3

THE RANDOMIZED BLOCK DESIGN

If the completely randomized design results in nonrejection of the null hypothesis that the treatment means differ because the sampling variability (as measured by MSE) is large, we may want to consider an experimental design that better controls the variability. In contrast to the selection of independent samples of experimental units specified by the completely randomized design, the **randomized block design** utilizes experimental units that are *matched sets*, assigning one from each set to each treatment. The matched sets of experimental units are called **blocks**. The concept of the randomized block design is that the sampling variability of the experimental units in each block will be reduced, in turn reducing the measure of error, MSE.

DEFINITION 15.8

The **randomized block design** consists of a two-step procedure:

1. Matched sets of experimental units, called **blocks**, are formed, each block consisting of p experimental units (where p is the number of treatments). Each of the b blocks should consist of experimental units that are as similar as possible.
2. One experimental unit from each block is randomly assigned to each treatment, resulting in a total of $n = bp$ responses.

For example, if we wish to compare SAT scores of female and male high school seniors, we could select independent random samples of five females and five males, and analyze the results of the completely randomized design as outlined in Section 15.2. Or, we could select matched pairs of females and males according to their scholastic records and analyze the SAT scores of the pairs. For example, we could select pairs of students with approximately the same GPAs from the same high school. Five such pairs (blocks) are depicted in Table 15.5. Note that this is just a **paired difference experiment**, first discussed in Section 9.4.

TABLE 15.5
Randomized Block Design: SAT Score Comparison

BLOCK	FEMALE SAT SCORE	MALE SAT SCORE	BLOCK MEAN
1 (School A, 2.75 GPAs)	540	530	535
2 (School B, 3.00 GPAs)	570	550	560
3 (School C, 3.25 GPAs)	590	580	585
4 (School D, 3.50 GPAs)	640	620	630
5 (School E, 3.75 GPAs)	690	690	690
TREATMENT MEAN	606	594	

As before, the variation between the treatment means is measured by squaring the distance between each treatment mean and the overall mean, multiplying each squared distance by the number of measurements for the treatment, and summing over treatments:

$$\text{SST} = \sum_{i=1}^{p} b(\bar{y}_{T_i} - \bar{y})^2$$
$$= 5(606 - 600)^2 + 5(594 - 600)^2 = 360$$

where $\bar{y}_{T_i}$ represents the sample mean for the ith treatment, b (the number of blocks) is the number of measurements for each treatment, and p is the number of treatments.

The blocks also account for some of the variation among the different responses. That is, just as SST measures the variation between the female and male means, we can calculate a measure of variation among the five block means representing different schools and scholastic abilities. Analogous to the computation of SST, we sum the squares of the differences between each block mean and the overall mean, multiplying each squared difference by the number of measurements for each block, and sum over blocks, to calculate the **Sum of Squares for Blocks** (SSB):

$$SSB = \sum_{i=1}^{b} p(\bar{y}_{B_i} - \bar{y})^2$$

$$= 2(535 - 600)^2 + 2(560 - 600)^2 + 2(585 - 600)^2 + 2(630 - 600)^2 + 2(690 - 600)^2$$

$$= 30,100$$

where $\bar{y}_{B_i}$ represents the sample mean for the ith block and p (the number of treatments) is the number of measurements in each block. As we expect, the variation in SAT scores attributable to Schools and Levels of scholastic achievement is apparently large.

As before, we want to compare the variability attributed to treatments with that which is attributed to sampling variability. In a randomized block design, the sampling variability is measured by subtracting that portion attributed to Treatments and Blocks from the Total Sum of Squares, SS(Total). The total variation is the sum of squared differences of each measurement from the overall mean:

$$SS(Total) = \sum_{i=1}^{n} (y_i - \bar{y})^2$$

$$= (540 - 600)^2 + (530 - 600)^2 + (570 - 600)^2 + (550 - 600)^2 + \cdots + (690 - 600)^2$$

$$= 30,600$$

Then the variation attributable to sampling error is found by subtraction:

$$SSE = SS(Total) - SST - SSB = 30,600 - 360 - 30,100$$

$$= 140$$

In summary, the Total Sum of Squares, 30,600, is divided into three components: 360 attributed to Treatments (gender), 30,100 attributed to Blocks (scholastic ability), and 140 attributed to sampling error.

The mean squares associated with each source of variability are obtained by dividing the sum of squares by the appropriate number of degrees of freedom. The partitioning of the Total Sum of Squares and of the total degrees of freedom for a randomized block experiment is summarized in Figure 15.12.

To determine whether we can reject the null hypothesis that the treatment means are equal in favor of the alternative that at least two of them differ, we calculate

$$MST = \frac{SST}{p - 1} = \frac{360}{2 - 1} = 360$$

$$MSE = \frac{SSE}{n - b - p + 1} = \frac{140}{10 - 5 - 2 + 1} = 35$$

FIGURE 15.12 Partitioning of the Total Sum of Squares for the Randomized Block Design

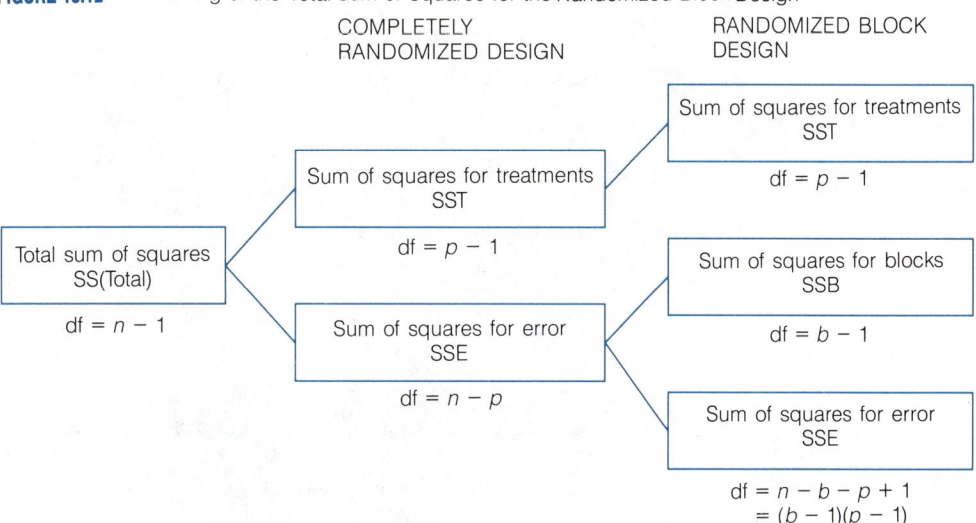

The F ratio that is used to test the hypothesis is

$$F = \frac{\text{MST}}{\text{MSE}} = \frac{360}{35} = 10.29$$

Comparing this ratio to the tabled F value corresponding to $\alpha = .05$, with $\nu_1 = (p - 1)$ degrees of freedom in the numerator and $\nu_2 = (n - b - p + 1) = 4$ degrees of freedom in the denominator, we find that

$$F = 10.29 > F_{.05} = 7.71$$

which indicates that we should reject the null hypothesis and conclude that the mean SAT scores differ for females and males.

If you review Section 9.4, you will find that the analysis of a paired difference experiment results in a one-sample t test on the difference between the treatment responses within each block. Applying the procedure to the differences between female and male scores in Table 15.5, we find

$$t = \frac{\bar{y}_d}{s_d / \sqrt{n_d}} = \frac{12}{\sqrt{70}/\sqrt{5}} = 3.207$$

At the .05 level of significance with $(n_d - 1) = 4$ degrees of freedom,

$$t = 3.21 > t_{.025} = 2.776$$

Since $t^2 = (3.207)^2 = 10.29$ and $t_{.025}^2 = (2.776)^2 = 7.71$, we find that the paired difference t test and the ANOVA F test are equivalent, with both the calculated test statistics and the rejection region related by the formula $F = t^2$. The difference between the tests is that the paired difference t test can be used to compare only two treatments in a randomized block design, while the F test can be applied to two or more treatments in a randomized block design. The F test is summarized in the box.

TEST TO COMPARE p TREATMENT MEANS: RANDOMIZED BLOCK DESIGN

H_0: $\mu_1 = \mu_2 = \cdots = \mu_p$

H_a: At least two treatment means differ

Test statistic: $F = \dfrac{\text{MST}}{\text{MSE}}$

Rejection region: $F > F_\alpha$ where F_α is based on $(p - 1)$ numerator degrees of freedom and $(n - b - p + 1)$ denominator degrees of freedom.

Assumptions: 1. The probability distributions of the observations corresponding to all the block–treatment combinations are normal.

2. The variances of all the probability distributions are equal.

Note that the assumptions concern the probability distributions associated with each block–treatment combination. The experimental unit selected for each combination is assumed to have been randomly selected from all possible experimental units for that combination, and the response is assumed to be normally distributed with the same variance for each of the block–treatment combinations. For example, the F test comparing female and male SAT score means requires the scores for each combination of gender and scholastic ability (e.g., females with 3.25 GPAs) to be normally distributed with the same variance as the other combinations employed in the experiment.

The calculation formulas for randomized block designs are given in Appendix D. We will rely on the computer programs available to analyze randomized block designs and to obtain the necessary elements for testing the null hypothesis that the treatment means are equal.

EXAMPLE 15.5

Refer to Examples 15.3 and 15.4. Suppose the USGA wants to compare the mean distances associated with the four brands of golf balls when struck by a driver, but wishes to employ human golfers rather than the robot, Iron Byron. Assume that ten balls of each brand are to be utilized in the experiment.

a. Explain how a completely randomized design could be employed.
b. Explain how a randomized block design could be employed.
c. Which design is likely to provide more information about the differences among the mean distances associated with the brand?

SOLUTION

a. Since the completely randomized design calls for independent samples, we can employ such a design by randomly selecting 40 golfers and then randomly assigning ten golfers to each of the four brands. Finally, each golfer will strike the ball of the assigned brand, and the distance will be recorded. The design is illustrated in Figure 15.13(a).

b. The randomized block design employs blocks of relatively homogeneous experimental units. For example, we could randomly select ten golfers and permit each golfer to hit four balls, one of each brand, in a random sequence. Then each golfer is a block, with each treatment (brand) assigned to each block (golfer). The design is summarized in Figure 15.13(b).

c. Because we expect much more variability among distances generated by "real" golfers than by Iron Byron, we would expect the randomized block design to control the variability more than the completely randomized design. That is, with 40 different golfers, we would expect more sampling variability among the measured distances within each brand than we would among the four distances generated by each of ten golfers hitting one ball of each brand.

FIGURE 15.13
Illustration of Completely Randomized Design and Randomized Block Design for Comparison of Four Golf Ball Brands

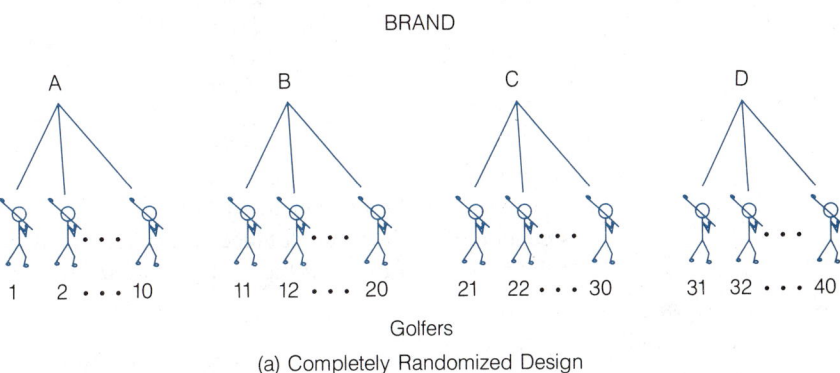

(a) Completely Randomized Design

(b) Randomized Block Design

EXAMPLE 15.6

Refer to Example 15.5. Suppose the randomized block design of part (b) is employed, utilizing a random sample of ten golfers, with each golfer using a driver to hit four balls, one of each brand, in a random sequence.

a. Set up a test of the null hypothesis that the mean distance associated with the four brands differ. Use $\alpha = .05$.

b. The data for the experiment are given in Table 15.6. Use a computer program to analyze the data, and conduct the test set up in part a.

TABLE 15.6

Distance Data for Randomized Block Design

GOLFER (Block)	BRAND A	BRAND B	BRAND C	BRAND D
1	202.4	203.2	223.7	203.6
2	242.0	248.7	259.8	240.7
3	220.4	227.3	240.0	207.4
4	230.0	243.1	247.7	226.9
5	191.6	211.4	218.7	200.1
6	247.7	253.0	268.1	244.0
7	214.8	214.8	233.9	195.8
8	245.4	243.6	257.8	227.9
9	224.0	231.5	238.2	215.7
10	252.2	255.2	265.4	245.2
Means	227.0	233.2	245.3	220.7

SOLUTION

a. We want to test whether the data in Table 15.6 provide sufficient evidence to conclude that the mean distances associated with the four brands differ. Denoting the population mean of the ith brand by μ_i, we test

H_0: $\mu_1 = \mu_2 = \mu_3 = \mu_4$

H_a: The mean distances differ for at least two of the brands

The test statistic compares the variation among the four treatment (brand) means to the sampling variability within each of the treatments.

Test statistic: $F = \dfrac{MST}{MSE}$

Rejection region: $F > F_\alpha = F_{.05}$, with $\nu_1 = (p - 1) = 3$ numerator degrees of freedom and $\nu_2 = (n - p - b + 1) = 27$ denominator degrees of freedom. From Table VIII of Appendix B, we find $F_{.05} = 2.96$. Thus, we will reject H_0 if $F > 2.96$.

The assumptions necessary to assure the validity of the test are as follows: (1) The probability distributions of the distances for each brand–golfer combination are normal. (2) The variances of the distance probability distributions for each brand–golfer combination are identical.

b. The Minitab program for ANOVA was utilized to analyze the data in Table 15.6, and the result is shown in Figure 15.14.

FIGURE 15.14
Minitab Printout for
Randomized Block
Design: Golf Ball Brand
Comparison

```
ANALYSIS OF VARIANCE ON DISTANCE

SOURCE      DF        SS        MS
GOLFER       9    12073.9    1341.5
BRAND        3     3298.7    1099.6
ERROR       27      546.6      20.2
TOTAL       39    15919.2
```

The values of MST and MSE (shaded on the printout) are 1,099.6 and 20.2, respectively. The F ratio is not given on the printout, but is equal to $F = \text{MST}/\text{MSE} = 1{,}099.6/20.2 = 54.4$, which exceeds the tabled value 2.96. We therefore reject the null hypothesis at the .05 level of significance, concluding that at least two of the brands differ with respect to mean distance travelled when struck by the driver. ∎

The results of an ANOVA can be summarized in a simple tabular format, similar to that utilized for the completely randomized design in Section 15.2. The general form of the table is shown in Table 15.7, and that for Example 15.6 is given in Table 15.8. Note that the randomized block design is characterized by three sources of variation—Treatments, Block, and Error—which add to the Total Sum of Squares. We hope that employing blocks of experimental units will reduce the error variability, and thereby make the test for comparing treatment means more powerful.

TABLE 15.7
ANOVA Summary Table
for a Randomized Block
Design

SOURCE	df	SS	MS	F
Treatment	$p - 1$	SST	MST	MST/MSE
Block	$b - 1$	SSB	MSB	
Error	$n - p - b + 1$	SSE	MSE	
Total	$n - 1$	SS(Total)		

TABLE 15.8
ANOVA Table for
Example 15.6

SOURCE	df	SS	MS	F
Treatments	3	3,298.7	1,099.6	54.4
Blocks	9	12,073.9	1,341.5	
Error	27	546.6	20.2	
Total	39	15,919.2		

When the F test results in the rejection of the null hypothesis that the treatment means are equal, we will usually want to compare the various pairs of treatment means to determine which specific pairs differ. We can employ the same Bonferroni procedure as in Section 15.2. The differences are that the degrees of freedom for error are $(n - p - b + 1)$ and that the number of measurements per treatment is equal to b, the number of blocks (because each treatment is observed once in each block). The Bonferroni procedure for a randomized block design is summarized in the box at the top of page 886.

> **BONFERRONI PROCEDURE FOR MULTIPLE COMPARISONS: RANDOMIZED BLOCK DESIGN**
>
> 1. Specify the overall level of significance α or, equivalently, the overall confidence level $100(1 - \alpha)\%$.
> 2. If there are c pairs of treatment means to be compared, specify the level of significance for each comparison at α/c.
> 3. Calculate the $100(1 - \alpha/c)\%$ confidence interval for each of the c pairs of means using the following formula:
>
> $$(\bar{y}_i - \bar{y}_j) \pm t_{\alpha/(2c)}\, s \sqrt{\left(\frac{1}{b} + \frac{1}{b}\right)}$$
>
> where $s = \sqrt{\text{MSE}}$ and $t_{\alpha/(2c)}$ is the tabulated value of t (from Table VI of Appendix B) that locates $\alpha/(2c)$ in the upper tail of the t distribution with $(n - p - b + 1)$ degrees of freedom (the number of degrees of freedom associated with error in the ANOVA).
> 4. Summarize the results of the multiple comparisons by ranking the treatment means, showing which pairs are significantly different.
>
> *Assumptions:*   Same as for the ANOVA.

EXAMPLE 15.7

Use the Bonferroni procedure to compare the mean distances of the four golf ball brands in Example 15.6. Use an overall significance level of $\alpha = .05$.

SOLUTION

The number of confidence intervals necessary to compare all pairs of brand means is $c = 4(3)/2 = 6$. The Bonferroni procedure therefore will use a level of significance equal to $\alpha/c = .05/6 = .00833$. The two-tailed confidence intervals will use a tabled t value with $(n - p - b + 1) = (40 - 4 - 10 + 1) = 27$ degrees of freedom, and an area of $\alpha/(2c) = .00417$ in each tail. The closest area in Table VI is .005, and $t_{.005}$ with 27 degrees of freedom is 2.771.*

The standard deviation s used in the comparisons is $s = \sqrt{\text{MSE}} = \sqrt{20.2} = 4.5$. Beginning with brands A and B, we calculate

$$(\bar{y}_A - \bar{y}_B) \pm t_{\alpha/(2c)}\, s \sqrt{\left(\frac{1}{b} + \frac{1}{b}\right)}$$

$$(227.0 - 233.2) \pm (2.771)(4.5)(.447)$$

$$-6.2 \pm 5.6 \quad \text{or} \quad (-11.8, -.6)$$

Thus, we conclude that the mean distance for brand B is between .6 and 11.8 yards greater than for brand A when struck with a driver. All the comparisons have the same confidence interval half-width of 5.6, and they are summarized as follows:

---

*Note that we have rounded up slightly, from .00417 to .005. Thus, we are not *guaranteed* an overall significance level of .05, but the conservative nature of the Bonferroni method virtually assures it.

$$
\begin{array}{llll}
A - C: & -18.3 \pm 5.6 & \text{or} & (-23.9, -12.7) \\
A - D: & 6.3 \pm 5.6 & \text{or} & (.7, 11.9) \\
B - C: & -12.1 \pm 5.6 & \text{or} & (-17.7, -6.5) \\
B - D: & 12.5 \pm 5.6 & \text{or} & (6.9, 18.1) \\
C - D: & 24.6 \pm 5.6 & \text{or} & (19.0, 30.2)
\end{array}
$$

Note that we are 95% confident that all the brand means differ, because none of the intervals contains zero. The listing of the brand means in Figure 15.15 has no vertical lines connecting them, because there are no nonsignificant differences at the .05 level.

FIGURE 15.15

Listing of Brand Means for Randomized Block Design

BRAND	MEAN
C	245.3
B	233.2
A	227.0
D	220.7

Note: All differences are statistically significant.

Unlike the completely randomized design, the randomized block design cannot, in general, be used to estimate individual treatment means. Whereas the completely randomized design employs a random sample for each treatment, the randomized block design does not necessarily employ a random sample of experimental units for each treatment. The experimental units within the blocks are assumed to be randomly selected, but the blocks themselves may not be randomly selected.

We can, however, test the hypothesis that the block means are significantly different. We simply compare the variability attributable to differences among the block means to that associated with sampling variability. The ratio of MSB to MSE is an F ratio similar to that formed in testing treatment means. The F statistic is compared to a tabled value for a specific value of α, with numerator degrees of freedom $(b - 1)$ and denominator degrees of freedom $(n - p - b + 1)$. The test is usually given on the same printout as the test for treatment means. Refer to the Minitab printout in Figure 15.14, and note that the test statistic for comparing the block means is

$$
F = \frac{\text{MSB}}{\text{MSE}} = \frac{\text{MS(Golfers)}}{\text{MS(Error)}}
$$

$$
= \frac{1,341.5}{20.2} = 66.4
$$

Since this calculated value exceeds 2.25, the tabled value of $F_{.05}$ with $\nu_1 = 10 - 1 = 9$ df and $\nu_2 = 27$ df, we conclude that the block means are different at the $\alpha = .05$ level of significance. The results of the test are summarized in the ANOVA Table 15.9 on page 888.

TABLE 15.9
ANOVA Table for
Randomized Block
Design: Test for Blocks
Included

SOURCE	df	SS	MS	F
Treatments	3	3,298.7	1,099.6	54.4
Blocks	9	12,073.9	1,341.5	66.4
Error	27	546.6	20.2	
Total	39	15,919.2		

In the golf ball example the test for block means confirms our suspicion that the golfers vary significantly and therefore that the use of the block design was a good decision. However, be careful not to conclude that the block design was a mistake

---

STEPS FOR CONDUCTING AN ANOVA FOR A RANDOMIZED BLOCK DESIGN

1. Be sure the design consists of blocks of homogeneous experimental units and that each treatment is randomly assigned to one experimental unit in each block.
2. Create an ANOVA summary table that specifies the variability attributable to Treatments, Blocks, and Error, and leads to the calculation of the F statistic to test the null hypothesis that the treatment means are equal in the population. Use either an ANOVA computer program or the calculation formulas in Appendix D to obtain the necessary numerical elements.
3. If the F test leads to the conclusion that the means differ, use the Bonferroni procedure to conduct multiple comparisons of as many of the pairs of means as you wish. Use the results to summarize the statistically significant differences among the treatment means. Remember that, in general, the randomized block design cannot be used to form confidence intervals for individual treatment means.
4. If the F test leads to the nonrejection of the null hypothesis that the treatment means are equal, several possibilities exist:
   a. The treatment means are equal—that is, the null hypothesis is true.
   b. The treatment means really differ, but other important factors affecting the response are not accounted for by the randomized block design. These factors inflate the sampling variability, as measured by MSE, resulting in smaller values of the F statistic. Either increase the sample size for each treatment or conduct an experiment that accounts for the other factors affecting the response (as in Section 15.4).
   Be careful not to automatically reach conclusion **a**, since the possibility of a Type II error must be considered if you accept H_0.
5. If desired, conduct the F test of the null hypothesis that the block means are equal. Rejection of this hypothesis lends statistical support to the utilization of the randomized block design.

if the F test for blocks does not result in rejection of the null hypothesis that the block means are the same. Remember that the possibility of a Type II error exists, and we are not controlling its probability as we are the probability α of a Type I error. If the experimenter believes that the experimental units are more homogeneous within blocks than between blocks, then he or she should use the randomized block design regardless of the results of a single test comparing the block means.

The procedure for conducting an analysis of variance for a randomized block design is summarized in the preceding box. Remember that the hallmark of this design is the utilization of blocks of homogeneous experimental units in which each treatment is represented.

| | | | | | | | | | | | | |

EXERCISES
15.29–15.41

LEARNING THE MECHANICS

15.29 A randomized block design yielded the following ANOVA table:

SOURCE	df	SS	MS	F
Treatments	4	501	125.25	9.109
Blocks	2	225	112.50	8.182
Error	8	110	13.75	
Total	14	836		

a. How many blocks and treatments were used in the experiment?
b. How many observations were collected in the experiment?
c. Specify the null and alternative hypotheses you would use to compare the treatment means.
d. What test statistic should be used to conduct the hypothesis test of part **c**?
e. Specify the rejection region for the test of parts **c** and **d**. Use $\alpha = .01$.
f. Conduct the test of parts **c–e**, and state the proper conclusion.
g. What assumptions are necessary to assure the validity of the test you conducted in part **f**?

15.30 An experiment was conducted using a randomized block design. The data from the experiment are displayed in the table.

TREATMENT	BLOCK		
	1	2	3
1	2	3	5
2	8	6	7
3	7	6	5

a. Fill in the missing entries in the analysis of variance table.

SOURCE	df	SS	MS	F
Treatments		21.5555		
Blocks				
Error				
Total		30.2222		

b. Specify the null and alternative hypotheses you would use to investigate whether a difference exists among the treatment means.

c. What test statistic should be used in conducting the test of part b?

d. Describe the Type I and Type II errors associated with the hypothesis test of part b.

e. Conduct the hypothesis test of part b using $\alpha = .05$.

15.31 A randomized block design was used to compare the mean responses for three treatments. Four blocks of three homogeneous experimental units were selected, and each treatment was randomly assigned to one experimental unit within each block. The data are shown in the accompanying table.

TREATMENT	BLOCK			
	1	2	3	4
A	3.4	5.5	7.9	1.3
B	4.4	5.8	9.6	2.8
C	2.2	3.4	6.9	.3

The Minitab ANOVA printout for this experiment is given below:

Minitab Printout for Exercise 15.31

ANALYSIS OF VARIANCE ON RESPONSE

SOURCE	DF	SS	MS
TREATMNT	2	12.032	6.016
BLOCK	3	71.749	23.916
ERROR	6	0.708	0.118
TOTAL	11	84.489	

a. Use the printout to fill in the entries in the following ANOVA table.

SOURCE	df	SS	MS	F
Treatments				
Blocks				
Error				
Total				

b. Do the data provide sufficient evidence to indicate that the treatment means differ? Use $\alpha = .05$.

c. Do the data provide sufficient evidence to indicate that blocking was effective in reducing the experimental error? Use $\alpha = .05$.

d. Use the Bonferroni technique to compare all pairs of treatment means. Use an overall significance level of $\alpha = .10$. How do the equal sample sizes for each treatment reduce the computations involved?

e. What assumptions are necessary to assure the validity of the inferences made in parts **b**, **c**, and **d**?

15.32 The analysis of variance for a randomized block design produced the ANOVA table shown here.

SOURCE	df	SS	MS	F
Treatments	3	28.2		
Blocks	5		13.80	
Error		34.1		
Total				

a. Complete the ANOVA table.

b. Do the data provide sufficient evidence to indicate a difference among the treatment means? Test using $\alpha = .01$.

c. Do the data provide sufficient evidence to indicate that blocking was a useful design strategy for this experiment? Explain.

d. If the sample means for treatments A and B are $\bar{y}_A = 9.7$ and $\bar{y}_B = 12.1$, respectively, find a 90% confidence interval for $(\mu_A - \mu_B)$. Interpret the interval.

15.33 Suppose an experiment utilizing a randomized block design has four treatments and nine blocks, for a total of $4 \times 9 = 36$ observations. Assume that the Total Sum of Squares for the response is SS(Total) = 500. For each of the following partitions of SS(Total), test the null hypothesis that the treatment means are identical, and the null hypothesis that the block means are identical. Use $\alpha = .05$ for each test.

a. The Sum of Squares for Treatments (SST) is 20% of SS(Total), and the Sum of Squares for Blocks (SSB) is 30% of SS(Total).

b. SST is 50% of SS(Total), and SSB is 20% of SS(Total).

c. SST is 20% of SS(Total), and SSB is 50% of SS(Total).

d. SST is 40% of SS(Total), and SSB is 40% of SS(Total).

e. SST is 20% of SS(Total), and SSB is 20% of SS(Total).

APPLYING THE CONCEPTS

15.34 The traditional retail store audit is one of the most widely used marketing research tools among consumer package-goods companies. It involves periodic audits of a sample of retail stores to monitor inventory and purchases of a particular product. V. K. Prasad, W. R. Casper, and R. J. Schieffer (1984) conducted a study to compare the market share estimates yielded by the traditional retail store audit with two alternative, less costly auditing procedures—weekend selldown audits and store purchases audits. Weekend selldown audits involve conducting opening inventory audits on Friday and closing inventory audits on Monday at a sample of retail stores. Weekend sales are measured as the difference between opening and closing inventory sales. Store purchase audits involve the monitoring of purchases by a sample of retail stores rather than by the stores' consumers. The market shares of six major brands of beer distributed in eastern cities were estimated using data generated by each of the three store audit methods. The estimates are presented in the table.

BRAND	TRADITIONAL STORE AUDIT	WEEKEND SELLDOWN AUDIT	STORE PURCHASES AUDIT
1	18.0	19.0	20.7
2	15.3	17.3	14.0
3	8.9	8.5	10.1
4	6.5	4.9	6.1
5	5.3	6.1	4.6
6	3.4	3.0	3.1
Means	9.57	9.80	9.76

a. The SAS printout for this experiment is shown here. Construct an ANOVA summary table using the printout.

SAS Printout for Exercise 15.34

General Linear Models Procedure

Dependent Variable: MKTSHARE

Source	DF	Sum of Squares	Mean Square	F Value	Pr > F
Model	7	605.888889	86.555556	66.33	0.0001
Error	10	13.048889	1.304889		
Corrected Total	17	618.937778			

R-Square	C.V.	Root MSE	MKTSHARE Mean
0.978917	11.762993	1.14232	9.71111111

Source	DF	Type I SS	Mean Square	F Value	Pr > F
AUDIT	2	0.19111	0.09556	0.07	0.9299
BRAND	5	605.69778	121.13956	92.84	0.0001

b. Is there sufficient evidence to indicate a difference in the mean estimated market shares produced by the three auditing methods? Test using $\alpha = .05$.

c. Use the Bonferroni technique to compare all pairs of treatment means. Use an overall significance level of $\alpha = .05$.

d. What assumptions must hold for your inferences of parts b and c to be valid?

15.35 A large clothing manufacturer conducted an experiment to study the effect on productivity of increases in its employees' hourly wages. Four treatments were used in the experiment:

Treatment 1:   No increase in hourly wage

Treatment 2:   Increase hourly wage by $.50

Treatment 3:   Increase hourly wage by $1.00

Treatment 4:   Increase hourly wage by $1.50

Twelve employees were selected and grouped into three blocks of size four according to the length of time they had been with the company. The four treatments were randomly assigned to the four employees in each block. The employees were observed for 3 weeks, and their productivity was measured as the average number of nondefective garments each produced per hour. The resulting productivity measures appear in the table.

	TREATMENT			
	1	2	3	4
Group 1 (less than 1 year)	2.4	3.0	3.1	3.2
Group 2 (1–5 years)	4.8	6.1	5.9	5.7
Group 3 (over 5 years)	5.1	7.0	7.2	7.3

a. What type of experimental design was used in this study? Why do you think such a design was employed?

b. The SPSS printout for this experiment is reproduced here. Is there evidence that the mean productivity levels differ among the four pay programs? Use $\alpha = .05$.

SPSS Printout for Exercise 15.35

```
            * * *   A N A L Y S I S   O F   V A R I A N C E   * * *

            NUMBER
      BY    TREATMNT
            GROUP


                              Sum of                 Mean                Signif
Source of Variation           Squares      DF        Square      F       of F

Main Effects                  33.362        5        6.672     45.236    .000
    TREATMNT                   3.740        3        1.247      8.452    .014
    GROUP                     29.622        2       14.811    100.412    .000

Explained                     33.362        5        6.672     45.236    .000

Residual                        .885        6         .148

Total                         34.247       11        3.113
```

Note: SPSS shows the combined Treatment and Block effects in both the "Main Effects" and "Explained" rows of the printout. Also, SPSS uses "Residual" instead of "Error."

c. What is the observed significance level for the test you conducted in part b?

d. Use the Bonferroni technique to compare all the pairs of treatment means. Use an overall significance level of $\alpha = .10$.

MONTH	STATION		
	1	2	3
May	28	31	53
June	25	22	61
July	37	30	56
August	20	26	48

15.36 A power plant that uses water from the surrounding bay for cooling its condensers is required by the Environmental Protection Agency (EPA) to determine whether discharging its heated water into the bay has a detrimental effect on the flora (plant life) in the water. The EPA requests that the power plant make its investigation at three strategically chosen locations, called *stations*. Stations 1 and 2 are located near the plant's discharge tubes, while station 3 is farther out in the bay. During one randomly selected day in each of 4 months, a diver descends to each of the stations, randomly samples of square meter area of the bottom, and counts the number of blades of the different types of grasses present. The results for one important grass type are shown in the table.

a. Use Appendix D to perform the computations necessary to create an ANOVA summary table for this experiment.

b. Is there sufficient evidence to indicate a difference among the mean numbers of blades found per square meter per month for the three stations? Use $\alpha = .05$.

c. Is there sufficient evidence to indicate a difference among the mean numbers of blades found per square meter for the 4 months? Use $\alpha = .05$.

d. Place a 90% confidence interval on the difference in means between stations 1 and 3.

STORE	ECONOMY BRAND	
	1	2
1	4.1	3.9
2	5.2	5.1
3	5.0	5.0
4	4.9	4.7
5	6.1	5.9

15.37 A food chain sells a particular item at all its stores. Each store carries three brands, two of which are economy brands. The management decides to discontinue selling one of the economy brands. It has decided to look at the *turn time* of each brand— i.e., the average time between successive purchases of the same brand. Five of the stores in the chain are selected, and an employee in each store reports the turn time (in minutes) for each brand. The results are reported in the table.

a. Is there a difference in the mean turn times for the two economy brands? Use $\alpha = .05$.

b. What is the purpose of the blocks in this experiment? Why is the mean square for blocks so large?

c. Recall that a randomized block design with $p = 2$ treatments is a paired difference experiment (Chapter 9). Analyze the data as a paired difference experiment using a t test to compare the treatment means. Test using $\alpha = .05$.

d. Compare the computed F and t values from parts **a** and **c**, and verify that $F = t^2$. Also verify that for the rejection region values of F and t, $F_\alpha = t^2_{\alpha/2}$ and, hence, the F test (of the randomized block analysis) and the t test (of the paired difference analysis) are equivalent for $p = 2$ treatments.

15.38 AT&T's long-distance phone charges may appear to be exorbitant when compared with some of its competitors, but this is because a comparison of charges between competing companies is often analogous to comparing apples and eggs. AT&T's charges for individuals are on a per-call basis. In contrast, its competitors often charge a monthly minimum long-distance fee, reduce the charges as the usage rises, or both. Shown in the table is a sampling of long-distance charges from Orlando, Florida, to 12 cities for three other companies offering long-distance service. The data were contained in an advertisement in the *Orlando Sentinel*, Mar. 19, 1984. A note in fine print below the advertisement states that the rates are based on "30 hours of usage" for each of the servicing companies.

The data in the table are pertinent for companies making phone calls to large cities. Therefore, assume that the cities receiving the calls were randomly selected from among all large cities in the United States.

FROM ORLANDO TO:	TIME	LENGTH OF CALL (Minutes)	COMPANY		
			1	2	3
New York	Day	2	$ .77	$ .79	$ .66
Chicago	Evening	3	.69	.71	.59
Los Angeles	Day	2	.87	.88	.66
Atlanta	Evening	1	.22	.23	.20
Boston	Day	3	1.15	1.19	.99
Phoenix	Day	5	1.92	1.98	1.65
West Palm Beach	Evening	2	.49	.42	.40
Miami	Day	3	1.12	1.05	.99
Denver	Day	10	3.85	3.96	3.30
Houston	Evening	1	.22	.23	.20
Tampa	Day	3	1.06	1.00	.99
Jacksonville	Day	3	1.06	1.00	.99
Means			1.12	1.12	.97

a. What type of design was used for the data collection?

b. The Minitab ANOVA printout for this experiment is shown here. Do the data provide sufficient evidence to indicate that the mean charges differ among the three companies? Use $\alpha = .05$.

Minitab Printout for Exercise 15.38

```
ANALYSIS OF VARIANCE ON CHARGE

SOURCE      DF       SS        MS
COMPANY      2    0.1820    0.0910
CITY        11   29.2598    2.6600
ERROR       22    0.2208    0.0100
TOTAL       35   29.6626
```

c. Company 3 placed the advertisement. Compare all the pairs of means using the Bonferroni technique with an overall significance level of $\alpha = .05$, and interpret the results.

15.39 A construction firm employs three cost estimators. Usually only one estimator works on each potential job, but it is advantageous to the company if the estimators are consistent enough that it does not matter which of the three estimators is assigned to a particular job. To check on the consistency of the estimators, several jobs are selected and all three estimators are asked to make estimates. The estimates (in thousands of dollars) for each job by each estimator are given in the table. The values of SST = .941, SSB = 10,239.969, and SSE = 13.919 are computed using the formulas in Appendix D.

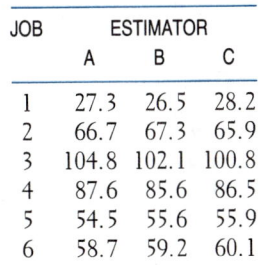

JOB	ESTIMATOR		
	A	B	C
1	27.3	26.5	28.2
2	66.7	67.3	65.9
3	104.8	102.1	100.8
4	87.6	85.6	86.5
5	54.5	55.6	55.9
6	58.7	59.2	60.1

a. Do the data provide sufficient evidence that the means for at least two of the estimators differ? Use $\alpha = .05$.

b. Find the approximate observed significance level for the test, and interpret it.

c. Present the complete ANOVA summary table for this experiment.

d. Does the analysis indicate that a comparison of the pairs of estimator means is worthwhile? Explain.

15.40 The table lists the number of strikes that occurred per year in five U.S. manufacturing industries over the period from 1976 to 1981. Only work stoppages that continued for at least 1 day and involved six or more workers were counted.

YEAR	FOOD AND KINDRED PRODUCTS	PRIMARY METAL INDUSTRY	ELECTRICAL EQUIPMENT AND SUPPLIES	FABRICATED METAL PRODUCTS	CHEMICALS AND ALLIED PRODUCTS
1976	227	197	204	309	129
1977	221	239	199	354	111
1978	171	187	190	360	113
1979	178	202	195	352	143
1980	155	175	140	280	89
1981	109	114	106	203	60

Source: *Statistical Abstract of the United States.*

a. Is this a designed or an observational experiment? Explain.

b. If the objective is to compare the mean number of strikes for the five industries, what type of ANOVA should be utilized? Explain.

c. The SAS ANOVA printout for these data is shown here. Is there evidence that the mean number of strikes differs among the five industries? Use $\alpha = .10$.

SAS Printout for Exercise 15.40

General Linear Models Procedure

Dependent Variable: STRIKES

Source	DF	Sum of Squares	Mean Square	F Value	Pr > F
Model	9	170505.667	18945.074	45.96	0.0001
Error	20	8243.533	412.177		
Corrected Total	29	178749.200			

R-Square	C.V.	Root MSE	STRIKES Mean
0.953882	10.662886	20.3021	190.400000

Source	DF	Type I SS	Mean Square	F Value	Pr > F
YEAR	5	40726.80	8145.36	19.76	0.0001
INDUSTRY	4	129778.87	32444.72	78.72	0.0001

d. Use the Bonferroni technique to compare the means for all pairs of industries, with an overall significance level of $\alpha = .10$. Interpret the results.

15.41 According to an advertisement in the *Gainesville Sun* (Mar. 18, 1984), shopping at local supermarket A can save you up to 21%. As proof of the statement, the advertising supermarket gives the results of a survey taken February 29, 1984, of the prices of 49 grocery items at supermarket A and at three of its competitors. Ten of these price comparisons are shown in the table. The data for a particular item represent a matched set of four observations, one for each of the supermarkets. As a result, the data in the table can be viewed as having been collected according to a randomized block design.

ITEM	SUPERMARKET			
	A	B	C	D
Hi C fruit drink	$ .59	$ .63	$ .79	$ .63
Cheerios cereal	1.10	1.18	1.39	1.18
Hunt's tomato paste	.31	.33	.43	.33
Del Monte green beans	.35	.40	.57	.38
Dole pineapple	.36	.39	.39	.39
Muellers spaghetti	.31	.33	.41	.33
Charmin	1.20	1.12	1.29	1.23
Duncan Hines cake mix	.77	.85	1.12	.77
Dial soap	.52	.60	.63	.55
Heinz ketchup	1.12	1.24	1.39	1.15

For these data, SST = .1858, SSB = 5.0607, and SSE = .0778.

a. Construct an analysis of variance table for the data.

b. Although you can see in the table that supermarket A's prices are as low as or lower than its competitors on nine of the ten items, test to determine if there is sufficient evidence to indicate that the mean prices of items differ among the four supermarkets. Test using $\alpha = .05$.

c. Use the Bonferroni technique to compare the mean prices for all pairs of supermarkets. Use an overall significance level of $\alpha = .05$.

d. Do the results of supermarket A's survey convince you that you can save money by shopping at supermarket A? Explain.

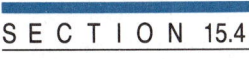

SECTION 15.4

FACTORIAL EXPERIMENTS

All of the experiments discussed in Sections 15.2 and 15.3 are **single-factor experiments**. The treatments were levels of a single factor, with the sampling of experimental units performed using either a completely randomized or randomized block design. However, most responses are affected by more than one factor, and we will therefore often wish to design experiments involving more than one factor.

Consider an experiment in which the effects of two factors on the response are being investigated. Assume that factor A is to be investigated at a levels, and factor B at b levels. Recalling that treatments are factor-level combinations, you can see that the experiment has, potentially, ab treatments that could be included in the experiment. A **complete factorial experiment** is one in which all possible ab treatments are utilized.

> **DEFINITION 15.9**
>
> A **complete factorial experiment** is one for which every factor-level combination is utilized. That is, the number of treatments in the experiment equals the total number of factor-level combinations.

For example, suppose the USGA wants to determine not only the relationship between distance and brand of golf ball, but also between distance and the club used to hit the ball. If they decide to use four brands and two clubs (say, driver and five-iron) in the experiment, then a complete factorial would call for utilizing all $4 \times 2 = 8$ brand–club combinations. This experiment is referred to more specifically as a **complete 4 × 2 factorial**. A layout for a two-factor factorial experiment (we are henceforth referring to a *complete factorial* when we use the term *factorial*) is given in Table 15.10. The factorial experiment is also referred to as a **two-way classification**, because it can be arranged in the row–column format exhibited in Table 15.10.

TABLE 15.10

Schematic Layout of Two-Factor Factorial Experiment

	LEVEL	FACTOR B AT b LEVELS				
		1	2	3	$\cdots$	b
	1	Trt. 1	Trt. 2	Trt. 3	$\cdots$	Trt. b
	2	Trt. $b + 1$	Trt. $b + 2$	Trt. $b + 3$	$\cdots$	Trt. $2b$
FACTOR A AT a LEVELS	3	Trt. $2b + 1$	Trt. $2b + 2$	Trt. $2b + 3$	$\cdots$	Trt. $3b$
	$\vdots$	$\vdots$	$\vdots$	$\vdots$	$\cdots$	$\vdots$
	a	Trt. $(a - 1)b + 1$	Trt. $(a - 1)b + 2$	Trt. $(a - 1)b + 3$	$\cdots$	Trt. ab

In order to complete the specification of the experimental design, the treatments must be assigned to experimental units. If the assignment of the ab treatments in the factorial experiment is random and independent, the design is completely randomized. For example, if the machine Iron Byron is used to hit 80 golf balls, 10 for each of the eight brand–club combinations, in a random sequence, the design would be completely randomized. On the other hand, if the assignment is made within homogeneous blocks of experimental units, then the design is a randomized block. For example, if ten golfers are employed to hit each of the eight golf balls, and each golfer hits all eight brand–club combinations in a random sequence, then the design is randomized block, with the golfers serving as blocks. In the remainder of this section, we confine our attention to factorial experiments utilizing completely randomized designs.

If we utilize a completely randomized design to conduct a factorial experiment with ab treatments, we can proceed with the analysis in exactly the same way we did in Section 15.2. That is, we calculate (or let the computer calculate) the measure of treatment mean variability (MST) and the measure of sampling variability (MSE), and use the F ratio of these two quantities to test the null hypothesis that the treatment means are equal. However, if this hypothesis is rejected so that we conclude some differences exist among the treatment means, important questions remain. Are both factors affecting the response, or only one? If both, do they affect the response independently, or do they interact to affect the response?

For example, suppose the distance data indicate that at least two of the eight treatment (brand–club combinations) means differ in the golf experiment. Does the Brand of ball (factor A) or the Club utilized (factor B) affect mean distance, or do both affect it? Several possibilities are shown in Figure 15.16. In Figure 15.16(a), the Brand means (only three are shown for the purpose of illustration) are the same, but the distances differ for the two levels of factor B (club). Thus, there is no effect of Brand on distance, but a Club main effect is present. In Figure 15.16(b), the Brand means differ, but the Club means are identical for each brand. Here a Brand main effect is present, but no effect of Club is present.

FIGURE 15.16 Illustration of Possible Treatment Effects: Factorial Experiment

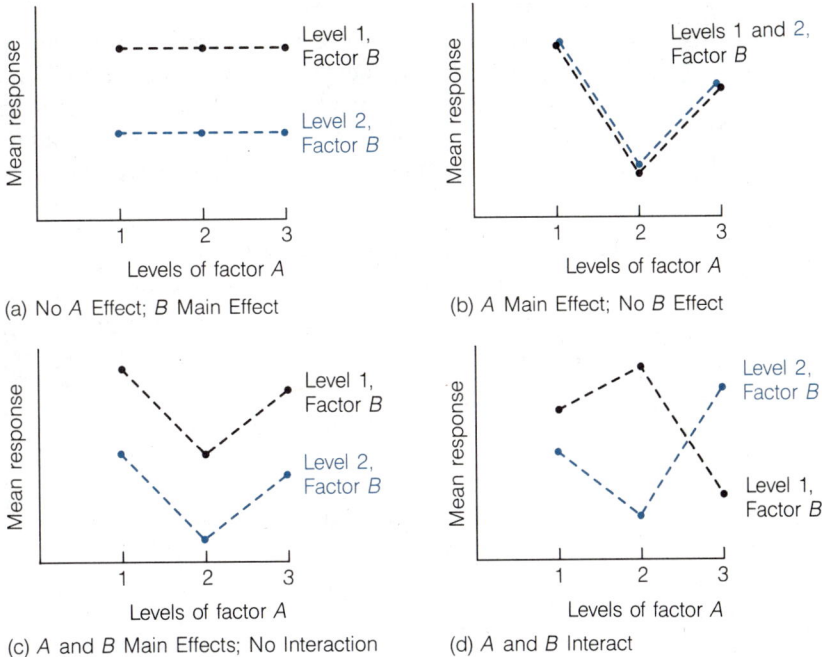

(a) No A Effect; B Main Effect

(b) A Main Effect; No B Effect

(c) A and B Main Effects; No Interaction

(d) A and B Interact

Figures 15.16(c) and 15.16(d) illustrate cases in which both factors affect the response. In Figure 15.16(c) the mean distance between clubs does not change for the three brands, so the effect of Brand on distance is independent of Club. That is, the two factors Brand and Club **do not interact**. In contrast, Figure 15.16(d) shows that the difference between mean distances between clubs varies with brand. Thus, the effect of Brand on distance depends on Club, and therefore the two factors **do interact**.

In order to determine the nature of the treatment effect, if any, on the response in a factorial experiment, we need to break the treatment variability into three components: interaction between factors A and B, main effect of factor A, and main effect of factor B. The interaction component is used to test whether the factors combine to affect the response, while the main effect components are used to determine whether the factors separately affect the response.

The partitioning of the Total Sum of Squares into its various components is illustrated in Figure 15.17. Notice that at stage 1 the components are identical to those in the one-factor completely randomized designs of Section 15.2; the Sums of Squares for Treatment and Error add to the Total Sum of Squares. The degrees of freedom for treatments is equal to $(ab - 1)$, 1 less than the number of treatments. The degrees of freedom for error is equal to $(n - ab)$, the total sample size minus the number of treatments. Only at stage 2 of the partitioning is the factorial experiment differentiated from those previously discussed. Here we divide the Treatment Sum of Squares into its three components: interaction and the two main effects. These components can then be utilized to test the nature of the differences, if any, among the treatment means.

There are a number of ways to proceed in the testing and estimation of factors in a factorial experiment. We present one approach in the accompanying box.

We assume the completely randomized design is **balanced**, meaning that the same number of observations are made for each treatment. That is, we assume that r experimental units are randomly and independently selected for each treatment.

FIGURE 15.17   Partitioning the Total Sum of Squares for a Two-Factor Factorial

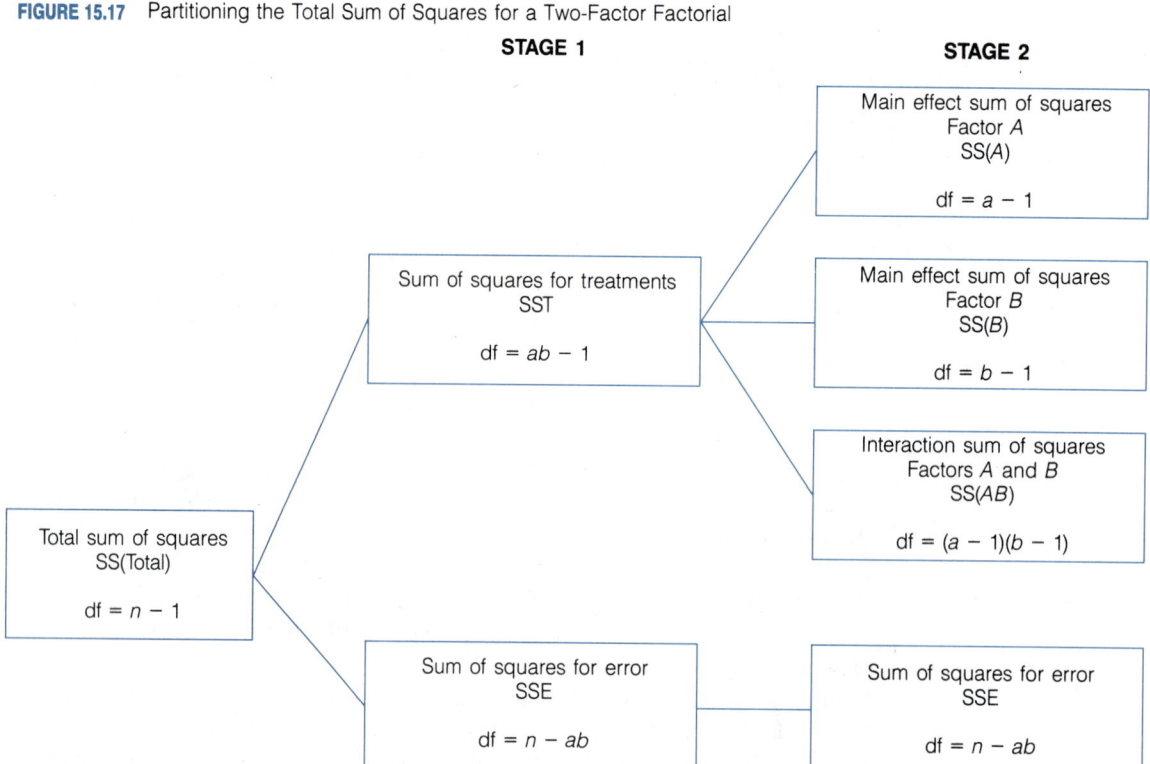

PROCEDURE FOR ANALYSIS OF TWO-FACTOR FACTORIAL EXPERIMENT

1. Partition the Total Sum of Squares into the Treatment and Error components (stage 1 of Figure 15.17). Use either a computer program or the calculation formulas in Appendix D to accomplish the partitioning.

2. Use the F ratio of Mean Square for Treatments to Mean Square for Error to test the null hypothesis that the treatment means are equal.*

   a. If the test results in nonrejection of the null hypothesis, consider refining the experiment by increasing the number of replications or introducing other factors. Also consider the possibility that the response is unrelated to the two factors.

   b. If the test results in rejection of the null hypothesis, proceed to step 3.

3. Partition the Treatment Sum of Squares into the Main Effect and Interaction Sum of Squares (stage 2 of Figure 15.17). Use either a computer program or the calculation formulas in Appendix D to accomplish the partitioning.

4. Test the null hypothesis that factors A and B do not interact to affect the response by computing the F ratio of the Mean Square for Interaction to the Mean Square for Error.

   a. If the test results in nonrejection of the null hypothesis, proceed to step 5.

   b. If the test results in rejection of the null hypothesis, conclude that the two factors interact to affect the mean response. Proceed to step 6a.

5. Conduct tests of two null hypotheses that the mean response is the same at each level of factor A and factor B. Compute two F ratios by comparing the Mean Square for each Factor Main Effect to the Mean Square for Error.

   a. If one or both tests result in rejection of the null hypothesis, conclude that the factor affects the mean response. Proceed to step 6b.

   b. If both tests result in nonrejection, an apparent contradiction has occurred. Although the treatment means apparently differ (step 2 test), the interaction (step 4) and main effect (step 5) tests have not supported that result. Further experimentation is advised.

6. a. If the test for interaction (step 4) is significant, use the Bonferroni multiple comparisons procedure to compare any or all pairs of the treatment means.

   b. If the test for one or both main effects (step 5) is significant, use the Bonferroni technique to compare the pairs of means corresponding to the levels of the significant factor(s).

---

*Some analysts prefer to proceed directly to test the interaction and main effect components, skipping the test of treatment means. We begin with this test to be consistent with our approach to regression analyses (global F test) and one-factor ANOVAs.

TESTS CONDUCTED IN ANALYSES OF FACTORIAL EXPERIMENTS: COMPLETELY RANDOMIZED DESIGN, r REPLICATES PER TREATMENT

TEST FOR TREATMENT MEANS

H_0: No difference among the ab treatment means

H_a: At least two treatment means differ

Test statistic: $F = \dfrac{\text{MST}}{\text{MSE}}$

Rejection region: $F \geq F_\alpha$, based on $(ab - 1)$ numerator and $(n - ab)$ denominator degrees of freedom [*Note:* $n = abr$]

TEST FOR FACTOR INTERACTION

H_0: Factors A and B do not interact to affect the response mean

H_a: Factors A and B do interact to affect the response mean

Test statistic: $F = \dfrac{\text{MS(AB)}}{\text{MSE}}$

Rejection region: $F \geq F_\alpha$, based on $(a - 1)(b - 1)$ numerator and $n - ab$ denominator degrees of freedom

TEST FOR MAIN EFFECT OF FACTOR *A*

H_0: No difference among the a mean levels of factor A

H_a: At least two factor A mean levels differ

Test statistic: $F = \dfrac{\text{MS(A)}}{\text{MSE}}$

Rejection region: $F \geq F_\alpha$, based on $(a - 1)$ numerator and $(n - ab)$ denominator degrees of freedom

TEST FOR MAIN EFFECT OF FACTOR *B*

H_0: No difference among the b mean levels of factor B

H_a: At least two factor B mean levels differ

Test statistic: $F = \dfrac{\text{MS(B)}}{\text{MSE}}$

Rejection region: $F \geq F_\alpha$, based on $(b - 1)$ numerator and $(n - ab)$ denominator degrees of freedom

ASSUMPTIONS FOR ALL TESTS

1. The response distribution for each factor-level combination (treatment) is normal.
2. The response variance is constant for all treatments.
3. Random and independent samples of experimental units are associated with each treatment.

The numerical value of r must exceed 1 in order to have any degrees of freedom with which to measure the sampling variability. [Note that if $r = 1$, then $n = ab$, and the degrees of freedom associated with Error (Figure 15.17) is df $= n - ab = 0$.] The value of r is often referred to as the number of **replicates** of the factorial experiment, since we assume that all ab treatments are repeated, or replicated, r times. Whatever approach is adopted in the analysis of a factorial experiment, several tests of hypotheses are usually conducted. The tests are summarized in the preceding box.

EXAMPLE 15.8

Suppose the USGA tests four different brands (A, B, C, D) of golf ball and two different clubs (driver, five-iron) in a completely randomized design. Each of the eight brand–club combinations (treatments) is randomly and independently assigned to four experimental units, each experimental unit consisting of a specific position in the sequence of hits by Iron Byron. The distance response is recorded for each of the 32 hits, and the results are shown in Table 15.11.

a. Use a computer program to partition the Total Sum of Squares, SS_{yy}, into the components necessary to analyze this 4×2 factorial experiment.
b. Follow the steps for analyzing a two-factor factorial experiment, and interpret the results of your analysis. Use $\alpha = .10$ for the tests you conduct.

TABLE 15.11
Distance Data for 4×2 Factorial Golf Experiment

		BRAND			
		A	B	C	D
CLUB	Driver	226.4	238.3	240.5	219.8
		232.6	231.7	246.9	228.7
		234.0	227.7	240.3	232.9
		220.7	237.2	244.7	237.6
	Five-iron	163.8	184.4	179.0	157.8
		179.4	180.6	168.0	161.8
		168.6	179.5	165.2	162.1
		173.4	186.2	156.5	160.3

SOLUTION

a. The SAS printout that partitions the Total Sum of Squares for this factorial experiment is given in Figure 15.18 (page 904). As described previously, the partitioning takes place in two stages. First, the Total Sum of Squares is partitioned into the Treatment and Error Sums of Squares at the top of the printout. Note that SST is 33,659.8 with 7 degrees of freedom, and SSE is 822.2 with 24 degrees of freedom, adding to 34,482.0 and 31 degrees of freedom. In the second stage of partitioning, the Treatment Sum of Squares is further divided into the Main Effect and Interaction Sums of Squares. At the bottom of the printout we see that SS(Club) is 32,093.1 with 1 degree of freedom, SS(Brand) is 800.7 with 3 degrees of freedom, and SS(Club $\times$ Brand) is 766.0 with 3 degrees of freedom, adding to 33,659.8 and 7 degrees of freedom.

FIGURE 15.18

SAS Printout for Factorial
Golf Experiment

Dependent Variable: DISTANCE

Source	DF	Sum of Squares	Mean Square	F Value	Pr > F
Model	7	33659.8087	4808.5441	140.35	0.0001
Error	24	822.2400	34.2600		
Corrected Total	31	34482.0487			

R-Square	C.V.	Root MSE	DISTANCE Mean
0.976155	2.8964608	5.85320	202.081250

Source	DF	Type I SS	Mean Square	F Value	Pr > F
CLUB	1	32093.11	32093.11	936.75	0.0001
BRAND	3	800.74	266.91	7.79	0.0008
CLUB*BRAND	3	765.96	255.32	7.45	0.0011

b. Once partitioning is accomplished, our first test is:

H_0:   The eight treatment means are equal

H_a: At least two of the eight means differ

Test statistic: $F = \dfrac{MST}{MSE} = 140.35$ (top of printout)

Rejection region: $F > F_{.10} = 1.98$, with $(ab - 1) = 7$ numerator degrees of freedom, and $(n - ab) = (32 - 8) = 24$ denominator degrees of freedom.

We reject this null hypothesis and conclude that at least two of the brand–club combinations differ in mean distance.

After accepting the hypothesis that the treatment means differ, and therefore that the factors Brand and/or Club somehow affect the mean distance, we want to determine how the factors affect the mean response. We begin with a test of interaction between Brand and Club:

H_0:   The factors Brand and Club do not interact to affect the mean response

H_a: Brand and Club interact to affect mean response

Test statistic: $F = \dfrac{MS(AB)}{MSE}$

$$= \dfrac{MS(\text{Brand} \times \text{Club})}{MSE} = \dfrac{255.32}{34.26}$$

$= 7.45$ (bottom of printout)

Rejection region: $F > F_{.10} = 2.33$, with $(a - 1)(b - 1) = 3$ numerator degrees of freedom and 24 denominator degrees of freedom.

Since the test statistic falls in the rejection region, we conclude that the factors Brand and Club interact to affect mean distance.

Because the factors interact, we need not test the main effects for Brand and Club. Instead, we compare the treatment means in an attempt to learn the nature of the interaction. Rather than compare all $8(7)/2 = 28$ pairs of treatment means, we test for differences only between pairs of brands within each club. That differences exist between clubs can be assumed. Therefore, only $4(3)/2 = 6$ pairs of means need to be compared for each club, or a total of 12 comparisons for the two clubs. If the Bonferroni methodology is used with an overall $\alpha = .10$, each of the comparisons will use $\alpha = .10/12 = .0083$. Then the appropriate t value for the two-sided confidence intervals is based on $\alpha/2 = .0083/2 = .0042 \approx .005$. From Table VI of Appendix B, with 24 degrees of freedom, we find $t_{.005} = 2.797$.

Rather than producing confidence intervals for all 12 comparisons, we note that the half-width of each interval will be constant:

$$\text{Half-width} = t_{.005}\, s\sqrt{(\tfrac{1}{4} + \tfrac{1}{4})}$$
$$= (2.797)(5.85)(.707)$$
$$= 11.6$$

where we find $s = 5.85$ labeled "Root MSE" on the printout, and we have $n_i = 4$ observations for each club–brand combination. We can declare all pairs of treatment means that differ by more than 11.6 yards statistically significantly different at the $\alpha = .10$ level of significance. The means are listed in descending order in Figure 15.19, and those not significantly different are connected by a vertical line.

FIGURE 15.19

Comparison of Treatment Means for Factorial Golf Experiment

CLUB	BRAND	MEAN
Driver	C	243.10
	B	233.72
	D	229.75
	A	228.42
Five-iron	B	182.68
	A	171.30
	C	167.18
	D	160.50

As shown in Figure 15.19, the picture is unclear with respect to Brand means. For the driver, the brand C mean significantly exceeds all except brand B. However, when hit with a five-iron, brand B's mean distance exceeds all except brand A's. Note the nontransitive nature of the multiple comparisons. For example, for the five-iron the brand B mean can be "the same" as the brand A mean, and the brand A mean "the same" as the brand C mean, and yet the brand B mean can significantly exceed the brand C mean. The reason lies in the definition of "the same": we must be careful not to conclude two means are equal simply because they are connected by a vertical line. The line indicates only that *the connected means are not significantly different.* You should conclude (at the overall α level of significance) only that means *not* connected are different, while withholding judgment on those that are connected. The picture of which means differ and by how much will become clearer as we increase the number of replicates of the factorial experiment.

The Club $\times$ Brand interaction can be seen in the plot of means in Figure 15.20. Note that the brand C mean drops considerably more than the others when going from driver to five-iron, while brand A gains relative to the others. Only brands B and D seem relatively consistent in their positions relative to the others, with brand B near the top and brand D near the bottom for both clubs.

FIGURE 15.20

Sample Mean Plot for Golf Factorial Experiment

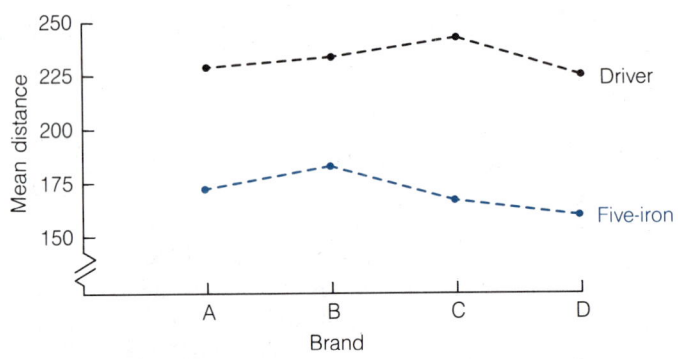

EXAMPLE 15.9

Refer to Example 15.8. Suppose the same factorial experiment is performed on four other brands (E, F, G, and H), and the results are as shown in Table 15.12. Repeat the factorial analysis and interpret the results.

TABLE 15.12

Distance Data for Second Factorial Golf Experiment

		BRAND			
		E	F	G	H
CLUB	Driver	238.6	261.4	264.7	235.4
		241.9	261.3	262.9	239.8
		236.6	254.0	253.5	236.2
		244.9	259.9	255.6	237.5
	Five-iron	165.2	179.2	189.0	171.4
		156.9	171.0	191.2	159.3
		172.2	178.0	191.3	156.6
		163.2	182.7	180.5	157.4

SOLUTION

The printout for the second factorial experiment is shown in Figure 15.21. The F ratio for Treatments is $F = 290.1$, which exceeds the tabled value of $F_{.10} = 1.98$ for 7 numerator and 24 denominator degrees of freedom. (Note that the same rejection regions will apply in this example as in Example 15.8, since the factors, treatments, and replicates are the same.) We conclude that at least two of the brand–club combinations are associated with different mean distances.

FIGURE 15.21

SAS Printout for Second
Factorial Golf Experiment

Dependent Variable: DISTANCE

Source	DF	Sum of Squares	Mean Square	F Value	Pr > F
Model	7	49959.3747	7137.0535	290.12	0.0001
Error	24	590.4075	24.6003		
Corrected Total	31	50549.7822			

R-Square	C.V.	Root MSE	DISTANCE Mean
0.988320	2.3515897	4.95987	210.915625

Source	DF	Type I SS	Mean Square	F Value	Pr > F
CLUB	1	46443.90	46443.90	1887.94	0.0001
BRAND	3	3410.32	1136.77	46.21	0.0001
CLUB*BRAND	3	105.16	35.05	1.42	0.2600

We next test for interaction between Brand and Club:

$$F = \frac{\text{MS(Brand} \times \text{Club})}{\text{MSE}} = 1.42$$

Since this F ratio does not exceed the tabled value of $F_{.10} = 2.33$ with 3 and 24 df, we cannot conclude at the .10 level of significance that the factors interact. In fact, note that the observed significance level (on the SAS printout) for the test of interaction is .26. Thus, at any level of significance lower than $\alpha = .26$, we could not conclude that the factors interact. We therefore proceed to test the main effects for Brand and Club.

We first test the Brand main effect:

H_0: No difference exists among the true Brand mean distances

H_a: At least two Brand mean distances differ

Test statistic: $F = \dfrac{\text{MS(Brand)}}{\text{MSE}} = \dfrac{1,136.77}{24.60} = 46.21$

Rejection region: $F > F_{.10} = 2.33$, with $(a - 1) = (4 - 1) = 3$ numerator degrees of freedom, and 24 denominator degrees of freedom.

Since 46.21 exceeds 2.33, we conclude that at least two of the Brand means differ. We will subsequently determine which brands mean differ using the Bonferroni technique. First, we also want to test the Club main effect:

H_0: No difference exists between the Club mean distances

H_a: The Club mean distances differ

Test statistic: $F = \dfrac{\text{MS(Club)}}{\text{MSE}} = \dfrac{46,443.9}{24.60} = 1,887.94$

Rejection region: $F > F_{.10} = 2.93$, with $(b - 1) = (2 - 1) = 1$ numerator degree of freedom, and 24 denominator degrees of freedom.

Since $F = 1,887.94$ exceeds 2.93, we conclude that the two clubs are associated with different mean distances. Since only two levels of Club were utilized in the experiment, this F test leads to the inference that the mean distance differs for the two clubs. It is no surprise (to golfers) that the mean distance for balls hit with the driver is significantly greater than that for those hit with the five-iron.

To determine which of the Brand's mean distances differ, we wish to compare the four Brand means. Since $4(3)/2 = 6$ pairs of means are to be compared, the Bonferroni multiple comparisons procedure requires that we specify $\alpha = .10/6 = .0167$ in order to have an overall significance level of (at least) .10. The two-tailed tabled t statistic has $\alpha/2 = .0167/2 = .0083 \approx .005$, with 24 degrees of freedom—that is, $t_{.005} - 2.797$. Each of the comparison intervals will have the same half-width:

$$\text{Half-width} = t_{.005}\, s\sqrt{(\tfrac{1}{8} + \tfrac{1}{8})}$$
$$= (2.797)(4.96)(.5)$$
$$= 6.9$$

where we find $s = 4.96$ labeled "Root MSE" on the printout, and we have $n_i = 8$ observations for each brand. The Brand means are shown in descending order in Figure 15.22, with a vertical line connecting the means that are not significantly different. Brands F and G are apparently associated with significantly greater mean distances than brands E and H, but we cannot distinguish between brands F and G or between brands E and H utilizing these data. Since the interaction between Brand and Club was not significant, we conclude that this difference among brands applies to both clubs. The sample means for all club–brand combinations are shown in Figure 15.23, and appear to support the conclusions of the tests and comparisons. Note that the Brand means maintain their relative positions for each Club: Brands F and G dominate brands E and H for both driver and five-iron.

FIGURE 15.22

Comparison of Mean Distances by Brand

BRAND	MEAN
G	223.59
F	218.44
E	202.44
H	199.20

FIGURE 15.23

Sample Mean Plot for Second Factorial Golf Experiment

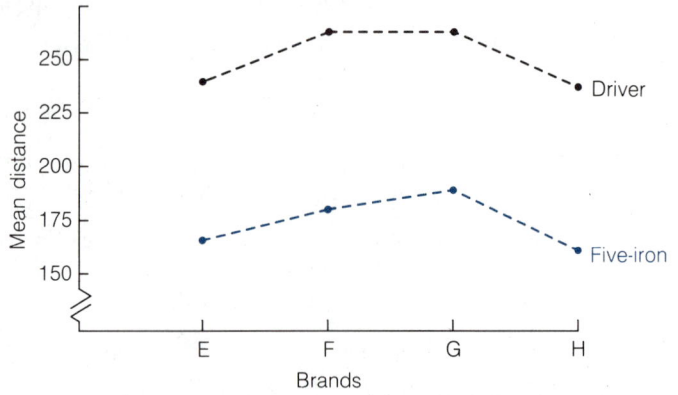

The analysis of factorial experiments can become complex if the number of factors is increased. Even the two-factor experiment becomes more difficult to analyze if some factor combinations have different numbers of observations than others. We have provided an introduction to these important experiments using two-factor factorials with equal numbers of observations for each treatment. Although similar principles apply to most factorial experiments, you should consult the references at the end of the chapter if you need to design and analyze more complex factorials.

EXERCISES
15.42–15.51

LEARNING THE MECHANICS

15.42 Suppose you conduct a 4×3 factorial experiment.
    a. How many factors are utilized in the experiment?
    b. Can you determine the factor type(s)—qualitative or quantitative—from the information given? Explain.
    c. Can you determine the number of levels utilized for each factor? Explain.
    d. Describe a treatment for this experiment, and determine the number of treatments employed.
    e. What problem is caused by utilizing a single replicate of this experiment? How is the problem solved?

15.43 The partially completed ANOVA table for a 3×4 factorial experiment with two replications is shown here.

SOURCE	df	SS	MS	F
A		.8		
B		5.3		
AB		9.6		
Error				
Total		17.0		

    a. Complete the analysis of variance table.
    b. Which sums of squares are combined to find the Sum of Squares for Treatment? Do the data provide sufficient evidence to indicate that the treatment means differ? Use $\alpha = .05$.
    c. Does the result of the test in part **b** warrant further testing? Explain.
    d. What is meant by factor interaction, and what is the practical implication if it exists?
    e. Test to determine whether these factors interact to affect the response mean. Use $\alpha = .05$, and interpret the results.
    f. Does the result of the interaction test warrant further testing? Explain.

15.44 The partially completed ANOVA table given here is for a two-factor factorial experiment.

SOURCE	df	SS	MS	F
A	3		.75	
B	1	.95		
AB			.30	
Error				
Total	23	6.5		

a. Give the number of levels for each factor.
b. How many observations were collected for each factor-level combination?
c. Complete the analysis of variance table.
d. Test to determine whether the treatment means differ. Use $\alpha = .10$.
e. Conduct the tests of factor interaction and main effects, each at the $\alpha = .10$ level of significance. Which of the tests are warranted as part of the factorial analysis? Explain.

15.45 The two-way table gives data for a 2×3 factorial experiment with two observations for each factor-level combination.

		FACTOR B		
	LEVEL	1	2	3
FACTOR A	1	3.1, 4.0	4.6, 4.2	6.4, 7.1
	2	5.9, 5.3	2.9, 2.2	3.3, 2.5

a. Identify the treatments for this experiment. Calculate and plot the treatment means, using the response variable as the y-axis, and the levels of factor B as the x-axis. Use the levels of factor A as plotting symbols. Do the treatment means appear to differ? Do the factors appear to interact?
b. The Minitab ANOVA printout for this experiment is reproduced here. Test to determine whether the treatment means differ at the $\alpha = .05$ level of significance. Does the test support your visual interpretation from part a?

Minitab Printout for Exercise 15.45

ANALYSIS OF VARIANCE ON RESPONSE

SOURCE	DF	SS	MS
A	1	4.441	4.441
B	2	4.127	2.063
INTERACTION	2	18.007	9.003
ERROR	6	1.475	0.246
TOTAL	11	28.049	

c. Does the result of the test in part b warrant a test for interaction between the two factors? If so, perform it using $\alpha = .05$.
d. Do the results of the previous tests warrant tests of the two factor main effects? If so, perform them using $\alpha = .05$.
e. Interpret the results of the tests. Do they support your visual interpretation from part a?
f. Use the Bonferroni technique to compare all pairs of treatment means. Use an overall significance level of $\alpha = .10$.

15.46 The accompanying two-way table gives data for a 2×2 factorial experiment with two observations per factor-level combination.

a. Identify the treatments for this experiment. Calculate and plot the treatment means, using the response variable as the y-axis, and the levels of factor B as the x-axis. Use the levels of factor A as plotting symbols. Do the treatment means appear to differ? Do the factors appear to interact?

		FACTOR B	
	LEVEL	1	2
FACTOR A	1	29.6 35.2	47.3 42.1
	2	12.9 17.6	28.4 22.7

b. Use the computational formulas in Appendix D to create an ANOVA table for this experiment.

c. Test to determine whether the treatment means differ at the $\alpha = .05$ level of significance. Does the test support your visual interpretation from part **a**?

d. Does the result of the test in part **b** warrant a test for interaction between the two factors? If so, perform it using $\alpha = .05$.

e. Do the results of the previous tests warrant tests of the two factor main effects? If so, perform them using $\alpha = .05$.

f. Interpret the results of the tests. Do they support your visual interpretation from part **a**?

g. Given the results of your tests, which pairs of means, if any, should be compared? Use the Bonferroni technique to compare them, with an overall significance level of $\alpha = .05$. Interpret the results.

15.47 Suppose a 3×3 factorial experiment is conducted with three replications. Assume that SS(Total) = 1,000. For each of the following scenarios, form an ANOVA table, conduct the appropriate tests, and interpret the results.

a. The Sum of Squares of factor A main effect [SS(A)] is 20% of SS(Total), the Sum of Squares for factor B main effect [SS(B)] is 10% of SS(Total), and the Sum of Squares for interaction [SS(AB)] is 10% of SS(Total).

b. SS(A) is 10%, SS(B) is 10%, and SS(AB) is 50% of SS(Total).

c. SS(A) is 40%, SS(B) is 10%, and SS(AB) is 20% of SS(Total).

d. SS(A) is 40%, SS(B) is 40%, and SS(AB) is 10% of SS(Total).

APPLYING THE CONCEPTS

15.48 Most short-run supermarket strategies such as price reductions, media advertising, and in-store promotions and displays are designed to increase unit sales of particular products temporarily. Factorial designs have been employed by Curhan (1974) and Wilkinson, Mason, and Paksoy (1982) to evaluate the effectiveness of such strategies. Two of the factors examined by Wilkinson et al. were Price level (regular, reduced price, cost to supermarket) and Display level (normal display space, normal display space plus end-of-aisle display, twice the normal display space). A complete factorial design based on these two factors involves nine treatments. Suppose each treatment was applied three times to a particular product at a particular supermarket. Each application lasted a full week and the dependent variable (response) of interest was unit sales for the week. To minimize treatment carryover effects, each treatment was preceded and followed by a week in which the product was priced at its regular price and was displayed in its normal manner. The table at the top of page 912 reports the data collected.

		PRICE		
		Regular	Reduced	Cost to supermarket
	Normal	989 1,025 1,030	1,211 1,215 1,182	1,577 1,559 1,598
DISPLAY	Normal plus	1,191 1,233 1,221	1,860 1,910 1,926	2,492 2,527 2,511
	Twice normal	1,226 1,202 1,180	1,516 1,501 1,498	1,801 1,833 1,852

a. The SAS ANOVA printout for this experiment is shown on page 913. Use the printout to complete the following diagram partitioning the Total Sum of Squares:

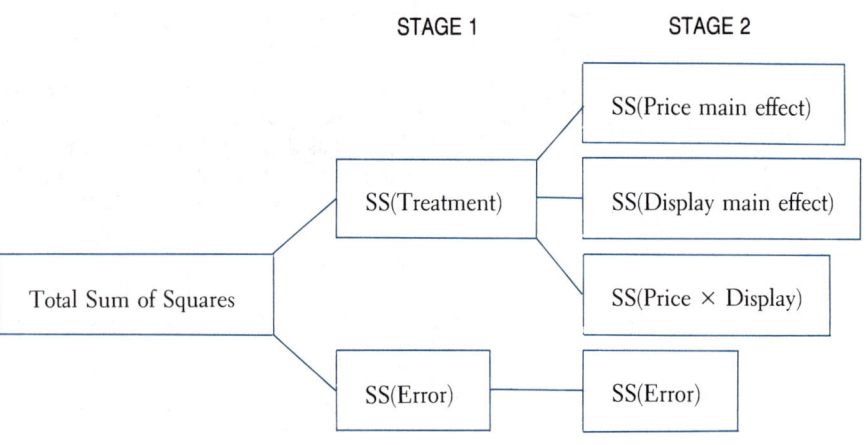

b. Do the data indicate that the mean sales differ among the nine treatments? Test using $\alpha = .10$.
c. Is the test of interaction between the factors Price and Display warranted as a result of the test in part b? If so, conduct the test using $\alpha = .10$.
d. Are the tests of the main effects for Price and Display warranted as a result of the previous tests? If so, conduct them using $\alpha = .10$.
e. Which pairs of treatment means should be compared as a result of the tests in parts b–d? Use the Bonferroni technique with an overall significance level of $\alpha = .10$ to compare them.
f. The treatment means are graphically portrayed in the accompanying SAS ANOVA printout. Use the graph to interpret the results of your tests.
g. What assumptions must hold in order for the inferential procedures you used to be appropriate?

SAS Printout for Exercise 15.48

Dependent Variable: SALES

Source	DF	Sum of Squares	Mean Square	F Value	Pr > F
Model	8	5291151.19	661393.90	1336.85	0.0001
Error	18	8905.33	494.74		
Corrected Total	26	5300056.52			

	R-Square	C.V.	Root MSE	SALES Mean
	0.998320	1.4344689	22.2428	1550.59259

Source	DF	Type I SS	Mean Square	F Value	Pr > F
DISPLAY	2	1691392.5	845696.3	1709.37	0.0001
PRICE	2	3089053.9	1544526.9	3121.89	0.0001
DISPLAY*PRICE	4	510704.8	127676.2	258.07	0.0001

Graph of Treatment Means for Exercise 15.48

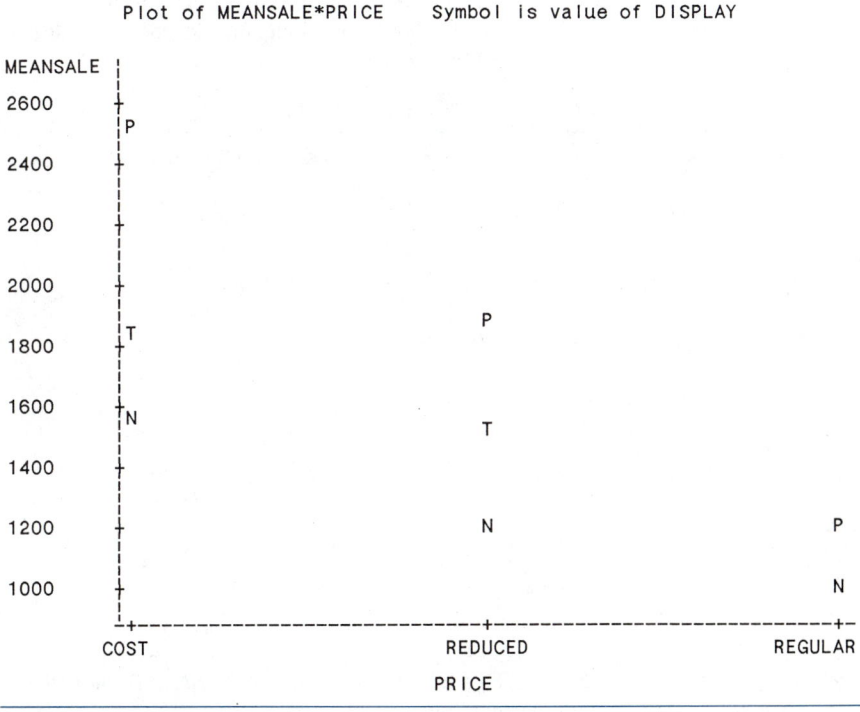

Plot of MEANSALE*PRICE   Symbol is value of DISPLAY

Note: The means for the Normal plus (P) and Twice normal (T) displays are nearly equal for Regular price. Both are represented by the symbol P at this level of price.

15.49 A beverage distributor wanted to determine the combination of advertising agency (two levels) and advertising medium (three levels) that would produce the largest increase in sales per advertising dollar. Each of the advertising agencies prepared copy or film, as required for each of the media—newspaper, radio, and television. Twelve small towns of roughly the same size were selected for the experiment and two each were assigned to receive an advertisement prepared and transmitted by each of the six agency–medium combinations. The dollar increases in sales per advertising dollar, based on a 1-month sales period, are shown in the table.

		ADVERTISING MEDIUM		
		Newspaper	Radio	Television
AGENCY	1	15.3 12.7	20.1 17.4	12.7 16.2
	2	18.9 22.4	24.3 28.8	12.5 9.4

a. Describe the experiment. Include the identification of the response variable, factors, factor types, factor levels, treatments, and experimental units. What is the experiment called? What type of design was employed?

b. The SPSS ANOVA printout for this experiment is reproduced here. Use it to perform a complete analysis of the experiment. Be sure to conduct relevant tests, and use the Bonferroni technique to compare relevant pairs of means. Use $\alpha = .10$ throughout.

SPSS Printout for Exercise 15.49

*** A N A L Y S I S   O F   V A R I A N C E ***

```
            SALES
     BY     AGENCY
            MEDIUM
```

Source of Variation	Sum of Squares	DF	Mean Square	F	Signif of F
Main Effects	238.299	3	79.433	13.934	.004
AGENCY	39.967	1	39.967	7.011	.038
MEDIUM	198.332	2	99.166	17.395	.003
2-way Interactions	77.345	2	38.672	6.784	.029
AGENCY MEDIUM	77.345	2	38.672	6.784	.029
Explained	315.644	5	63.129	11.074	.005
Residual	34.205	6	5.701		
Total	349.849	11	31.804		

Note: SPSS uses "Explained" instead of "Treatment" in the factorial analysis. Also, SPSS uses "Residual" instead of "Error."

c. Graph the treatment means, placing advertisement medium on the horizontal axis, and using agency as a plotting symbol. Do the results of your analysis in part **b** appear reasonable? Use the graph to interpret the results of your analysis.

15.50 How do women compare with men in their ability to perform laborious tasks that require strength? Some information on this question is provided in a study, by M. D. Phillips and R. L. Pepper (1982), of the firefighting ability of men and women. Phillips and Pepper conducted a 2×2 factorial experiment to investigate the effect of the factor Sex (male or female) and the factor Weight (light or heavy) on the length of time required for a person to perform a particular firefighting task. Eight persons were selected for each of the $2 \times 2 = 4$ Sex–Weight categories of the 2×2 factorial experiment and the length of time (in minutes) needed to complete the task was recorded for each of the 32 persons. The means and standard deviations of the four samples are shown in the table.

	LIGHT		HEAVY	
	Mean	Standard deviation	Mean	Standard deviation
FEMALE	18.30	6.81	14.50	2.93
MALE	13.00	5.04	12.25	5.70

a. Calculate the total of the $n = 8$ time measurements for each of the four categories of the 2×2 factorial experiment.

b. Calculate CM. (See Appendix D for computational formulas.)

c. Use the results of parts **a** and **b** to calculate the sums of squares for Sex, Weight, and for the Sex $\times$ Weight interaction.

d. Calculate each sample variance. Then calculate the sum of squares of deviations *within* each sample for each of the four samples.

e. Calculate SSE. [*Hint:*   SSE is the pooled sum of squares of the deviations calculated in part **d**.]

f. Now that you know SS(Sex), SS(Weight), SS(Sex $\times$ Weight), and SSE, find SS(Total).

g. Summarize the calculations in an analysis of variance table.

h. Conduct a complete analysis of these data. Use $\alpha = .05$ for any inferential techniques you employ. Interpret your conclusions graphically.

i. What assumptions are necessary to assure the validity of the inferential techniques you utilized? State them in terms of this experiment.

15.51 Refer to Exercise 15.50. Phillips and Pepper (1982) give data on another 2×2 factorial experiment utilizing 20 males and 20 females. The experiment involved the same treatments with ten persons assigned to each Sex–Weight category. The response measured for each person was the pulling force the person was able to exert on the starter cord of a P-250 fire pump. The means and standard deviations of the four samples (corresponding to the $2 \times 2 = 4$ categories of the experiment) are shown in the table.

	LIGHT		HEAVY	
	Mean	Standard deviation	Mean	Standard deviation
FEMALES	46.26	14.23	62.72	13.97
MALES	88.07	8.32	86.29	12.45

a. Use the procedures outlined in Exercise 15.50 to perform an analysis of variance for the experiment. Display your results in an analysis of variance table.

b. Conduct a complete analysis of these data. Use $\alpha = .05$ for any inferential techniques you employ. Interpret your conclusions graphically.

c. Note that the observed standard deviations for the four treatments vary from 8.32 to 14.23. Since the validity of the inferences requires that the treatment standard deviations be equal, how do you refute the criticism that the unequal standard deviations invalidate your analysis?

SECTION 15.5

USING REGRESSION ANALYSIS FOR ANOVA (OPTIONAL)

We have seen in previous sections that analysis of variance involves the statistical analysis of the relationship between a response and one or more factors. In Chapters 10–12 we performed very similar analyses using regression techniques, so it is not surprising that the two methodologies are closely related. In fact, each of the ANOVAs performed in this chapter can be formulated as a regression model, and the same analysis can be performed in a regression context. We will illustrate the ANOVA–regression correspondence in this section.

A one-factor experiment utilizing a completely randomized design provides the simplest example of this correspondence. Suppose the factor has p levels—that is, there are p treatments in the experiment. Assuming the factor is qualitative, the corresponding regression model is

$$y = \beta_0 + \beta_1 x_1 + \beta_2 x_2 + \cdots + \beta_{p-1} x_{p-1} + \varepsilon$$

where $x_1, x_2, \ldots, x_{p-1}$ are $(p-1)$ dummy variables describing the p levels of the factor. If the factor is quantitative, then the x's become powers of the variable, i.e., $x_i = x^i$. In either case, to test the null hypothesis of interest in the ANOVA, we test

$$H_0: \quad \mu_1 = \mu_2 = \cdots = \mu_p$$

using the F ratio between Mean Square for Treatments and Mean Square for Error. This is precisely the same test as the global model test in the corresponding regression example:

$$H_0: \quad \beta_1 = \beta_2 = \cdots = \beta_{p-1} = 0$$

which we test using the F ratio of the Mean Square for (Regression) Model to the Mean Square for Error. Both tests result in the same F ratio and have the same numerator and denominator degrees of freedom, $(p-1)$ and $(n-p)$, respectively.

To better understand the equivalence of these two procedures, note that if all the β's are equal to 0 in the regression model, the mean response is equal to β_0 *no matter which level of the qualitative factor is utilized as the base level.* That is, the mean response is equal for all p treatments, which is exactly the ANOVA null hypothesis. For example, in Section 15.2 we used ANOVA to test the null hypothesis that four brands of golf balls travel the same distance, on average, when hit by Iron Byron with a driver. We tested

$$H_0: \quad \mu_A = \mu_B = \mu_C = \mu_D$$

where μ_i is the mean distance for brand i. Equivalently, we could have defined the following regression model:

$$E(y) = \beta_0 + \beta_1 x_1 + \beta_2 x_2 + \beta_3 x_3$$

with

$$x_1 = \begin{cases} 1 & \text{if brand A} \\ 0 & \text{otherwise} \end{cases} \qquad x_2 = \begin{cases} 1 & \text{if brand B} \\ 0 & \text{otherwise} \end{cases} \qquad x_3 = \begin{cases} 1 & \text{if brand C} \\ 0 & \text{otherwise} \end{cases}$$

Then β_0 represents the mean distance for brand D, β_1 the difference between the mean distances of brand A and brand D, etc. The null hypothesis

$$H_0: \quad \beta_1 = \beta_2 = \beta_3 = 0$$

is equivalent to the ANOVA null hypothesis that the Brand means are equal.

EXAMPLE 15.10

Refer to Example 15.3, in which we used ANOVA for a completely randomized design to test the hypothesis that the mean distances associated with four brands of golf balls were the same. Ten experimental units were randomly and independently assigned to each brand, and the distance was recorded. The data are repeated in Table 15.13. Use a regression program to fit the equivalent regression model, and conduct the global F test for the model's usefulness in predicting the response y. Compare the test to that conducted in Example 15.3, and interpret the results. Use $\alpha = .10$.

TABLE 15.13
Distance Data for Golf Experiment: Completely Randomized Design

	BRAND A	BRAND B	BRAND C	BRAND D
	251.2	263.2	269.7	251.6
	245.1	262.9	263.2	248.6
	248.0	265.0	277.5	249.4
	251.1	254.5	267.4	242.0
	265.5	264.3	270.5	246.5
	250.0	257.0	265.5	251.3
	253.9	262.8	270.7	262.8
	244.6	264.4	272.9	249.0
	254.6	260.6	275.6	247.1
	248.8	255.9	266.5	245.9
Means	251.28	261.06	269.95	249.42

SOLUTION

The Minitab printout for the regression model using three dummy variables to describe Brand is given in Figure 15.24. To test the model's usefulness, we calculate the F ratio of the Regression (Model) Mean Square to the Residual (Error) Mean Square, both shown shaded in Figure 15.24. This is

$$F = \frac{\text{Regression Mean Square}}{\text{Residual Mean Square}} = \frac{903.07}{25.22} = 35.81$$

FIGURE 15.24

Minitab Printout for
Regression Model of
Example 15.10

```
THE REGRESSION EQUATION IS
DISTANCE = 249 + 1.86 X1 + 11.6 X2 + 20.5 X3

                               ST. DEV.      T-RATIO =
COLUMN        COEFFICIENT      OF COEF.      COEF/S.D.
                 249.420         1.588        157.06
X1                 1.860         2.246          0.83
X2                11.640         2.246          5.18
X3                20.530         2.246          9.14

S = 5.022

R-SQUARED = 74.9 PERCENT
R-SQUARED = 72.8 PERCENT, ADJUSTED FOR D.F.

ANALYSIS OF VARIANCE

   DUE TO      DF          SS       MS=SS/DF
   REGRESSION   3      2709.20       903.07
   RESIDUAL    36       907.86        25.22
   TOTAL       39      3617.06
```

The tabled value $F_{.10} \approx 2.25$, with 3 model degrees of freedom (number of independent variables in the model) in the numerator and 36 error degrees of freedom (n minus the number of β's) in the denominator. We therefore reject the null hypothesis that the model is not useful and conclude at the .10 level of significance that brand is related to mean distance.

The regression F value is identical to that obtained in Example 15.3 using ANOVA. The conclusion may sound somewhat different, but on closer inspection they are also identical. In the ANOVA we concluded that the mean distances differ for at least two of the four brands, while here we conclude that the model is useful for predicting distance. But for the model to be useful, the factor Brand must have some association with mean distance, since Brand is the only factor in the regression model. That is, the mean distance must vary with Brand, which implies that at least two of the brands have different mean distances.

It is instructive to interpret the least squares estimates of the β parameters in Figure 15.24. The value of $\hat{\beta}_0 = 249.42$ is the sample mean distance for brand D (Table 15.13), since brand D is the base level at which all three dummy variables are equal to 0. The value of $\hat{\beta}_1 = 1.86$ is the difference between the sample mean distances of brand A and brand D ($251.28 - 249.42 = 1.86$). Similarly, you can see that $\hat{\beta}_2$ and $\hat{\beta}_3$ are the sample mean differences between brand B and brand D and between brand C and brand D, respectively. Finally, note that the standard deviation of the regression model, $s = 5.022$, is the same as $\sqrt{\text{MSE}}$ in the ANOVA, and that both measure the variation in distance within a brand. In fact, we would expect to be able to predict the distance Iron Byron will hit a specified brand to within about ± 10 yards using this model. Or, to put it another way, we expect the actual distances for a specific brand to vary within a range of about 10 yards around the mean. Of course, this variability might be partially attributable to other factors that could be included in an expanded regression model. ∎

The assumptions necessary to assure the validity of an ANOVA and a regression analysis are also identical, although usually phrased differently. Using the golf experiment as an example, we assumed that the distance probability distributions are normal with equal variances for each brand when we conducted the ANOVA. The regression assumption that the error component is normally distributed with constant variance for all settings of the independent variables is seen to be equivalent when we recognize that the error component of this particular model describes the random variability of distance within a particular brand. Thus, we are assuming that the distance measurements within each brand are normally distributed with the same variance for each brand. Similarly, the ANOVA assumption that the samples are random and independent is matched by the regression assumption that the error components are independent with mean 0.

The following example demonstrates that the regression approach may also be used to model the randomized block design.

EXAMPLE 15.11

Refer to Example 15.6, in which a randomized block design was used to compare the mean distances of four brands of golf balls. Each of ten golfers (the blocks) hit balls of each brand using a driver, and the distance was recorded. The data are repeated in Table 15.14.

TABLE 15.14
Distance Data for
Randomized Block
Design

GOLFER (BLOCK)	BRAND A	BRAND B	BRAND C	BRAND D
1	202.4	203.2	223.7	203.6
2	242.0	248.7	259.8	240.7
3	220.4	227.3	240.0	207.4
4	230.0	243.1	247.7	226.9
5	191.6	211.4	218.7	200.1
6	247.7	253.0	268.1	244.0
7	214.8	214.8	233.9	195.8
8	245.4	243.6	257.8	227.9
9	224.0	231.5	238.2	215.7
10	252.2	255.2	265.4	245.2
Means	227.05	233.18	245.33	220.73

SOLUTION

a. The regression model must contain terms to account for both the Treatment and Block effects on the response. That is, the distance, y, must be modeled as a function of the Brand and Golfer utilized. Since Brand and Golfer are both qualitative variables, we use dummy variables to describe them:

$$\overbrace{y = \beta_0 + \beta_1 x_1 + \beta_2 x_2 + \beta_3 x_3}^{\text{Brands (Treatments)}}$$

$$\underbrace{+\ \beta_4 x_4 + \beta_5 x_5 + \cdots + \beta_{12} x_{12} + \varepsilon}_{\text{Golfers (Blocks)}}$$

where x_1 to x_3 are the three dummy variables representing the four brands (using brand D as the base or default level at which all three dummy variables are equal to 0), and x_4 to x_{12} are the nine dummy variables representing the ten golfers (using golfer 10 as the base or default level at which all nine dummy variables are equal to 0).

Note that no interaction between Brand and Golfer is included in the model. There are several reasons for this: (1) The presence of interaction between Treatments and Blocks would mean that the difference between Brand mean distances depends on the particular golfer. While this might in fact be true, the objective of the experiment is to estimate Brand mean differences independent of the particular golfer, and the utilization of interaction defeats that purpose.[*] (2) If interaction were included in the model, no degrees of freedom would remain with which to estimate the variance σ^2 of the random error component. To see this, note that the interaction would involve $3 \times 9 = 27$ combinations of dummy variables, bringing the total number of terms in the model to $3 + 9 + 27 = 39$. But this is equal to the total degrees of freedom $(n - 1)$ in the experiment. Thus, we assume that no interaction exists in a randomized block design in order to use those degrees of freedom for estimating sampling variability.

b. The SPSS printout for this regression model is shown in Figure 15.25. The Sum of Squares for the Model is shown shaded, and it includes the variability attributable to *both* Blocks and Treatments. We are interested in testing the null hypothesis

$$H_0: \quad \beta_1 = \beta_2 = \beta_3 = 0$$

which is equivalent to the ANOVA null hypothesis that the Brand mean distances are the same. In order to test this part of the regression model, we fit the reduced model

$$y = \beta_0 + \beta_4 x_4 + \beta_5 x_5 + \cdots + \beta_{12} x_{12} + \varepsilon$$

The SPSS least squares printout for this model is given in Figure 15.26.

The partial F test of Section 12.4 gives us the test of the null hypothesis that the Treatment means are equal:

$$F = \frac{[\text{SSE(Reduced model)} - \text{SSE(Complete model)}]/3}{\text{MSE(Complete model)}}$$

$$= \frac{(3{,}845.2775 - 546.6208)/3}{20.2452} = 54.31$$

The tabled F value is $F_{.05} = 2.96$, with 3 numerator degrees of freedom and 27 denominator degrees of freedom. Since $F = 54.31 > 2.96$, we reject H_0 at the .05 level of significance and conclude that the mean distances associated with the four brands differ. Both the calculated F ratio and the tabled F value are identical (within rounding differences) to those we obtained in the randomized block ANOVA of the same data in Example 15.6.

---

[*]We are treating Blocks as fixed rather than random effects. Technically, this implies that our inferences about treatments are conditioned on the particular Block levels utilized. While this results in some loss of generality, mixed fixed–random effect models are beyond the scope of this text.

FIGURE 15.25

SPSS Printout for
Randomized Block
Design: Complete Model
(Brands and Golfers)

```
Multiple R            .98268
R Square              .96566
Adjusted R Square     .95040
Standard Error        4.49947

Analysis of Variance
                     DF      Sum of Squares      Mean Square
Regression           12         15372.53900      1281.04492
Residual             27           546.62075        20.24521

F =      63.27644       Signif F =  .0000

------------------ Variables in the Equation ------------------

Variable             B          SE B        Beta        T     Sig T

X1              6.32000      2.01222      .13718     3.141    .0041
X2             12.45000      2.01222      .27023     6.187    .0000
X3             24.60000      2.01222      .53396    12.225    .0000
X4            -46.27500      3.18160     -.69589   -14.545    .0000
X5             -6.70000      3.18160     -.10075    -2.106    .0447
X6            -30.72500      3.18160     -.46204    -9.657    .0000
X7            -17.57500      3.18160     -.26429    -5.524    .0000
X8            -49.05000      3.18160     -.73762   -15.417    .0000
X9             -1.30000      3.18160     -.01955     -.409    .6861
X10           -39.67500      3.18160     -.59663   -12.470    .0000
X11           -10.82500      3.18160     -.16279    -3.402    .0021
X12           -27.15000      3.18160     -.40828    -8.533    .0000
(Constant)    243.65750      2.56509                94.990    .0000
```

FIGURE 15.26

SPSS Printout for
Randomized Block
Design: Reduced Model
(Golfers Only)

```
Multiple R            .87089
R Square              .75845
Adjusted R Square     .68598
Standard Error        11.32148

Analysis of Variance
                     DF      Sum of Squares      Mean Square
Regression            9         12073.88225      1341.54247
Residual             30          3845.27750       128.17592

F =      10.46642       Signif F =  .0000

------------------ Variables in the Equation ------------------

Variable             B          SE B        Beta        T     Sig T

X4            -46.27500      8.00550     -.69589    -5.780    .0000
X5             -6.70000      8.00550     -.10075     -.837    .4093
X6            -30.72500      8.00550     -.46204    -3.838    .0006
X7            -17.57500      8.00550     -.26429    -2.195    .0360
X8            -49.05000      8.00550     -.73762    -6.127    .0000
X9             -1.30000      8.00550     -.01955     -.162    .8721
X10           -39.67500      8.00550     -.59663    -4.956    .0000
X11           -10.82500      8.00550     -.16279    -1.352    .1864
X12           -27.15000      8.00550     -.40828    -3.391    .0020
(Constant)    254.50000      5.66074                44.959   0.0
```

Inferences about specific Brand mean differences can be made using the least squares estimates of the Brand β parameters. Note that $\hat{\beta}_1 = 6.32$ is the difference between the sample mean distances for brands A and D: $227.05 - 220.73 = 6.32$ (Table 15.14). Similarly, $\hat{\beta}_2$ and $\hat{\beta}_3$ equal the sample mean differences between brands B and D and brands C and D, respectively. ∎

The utilization of a regression model for a randomized block design is very similar to the method we will employ in the next example to describe a factorial experiment. The difference is that the interaction between Treatments and Blocks in a randomized block design is utilized to estimate sampling variability, while the interaction between factors in a factorial experiment is an integral part of the model.

EXAMPLE 15.12

Refer to Example 15.8, in which we employed a factorial experiment in a completely randomized design to evaluate the effects of Brand of golf ball and Club utilized on mean distance. Four brands and two clubs were utilized, and the 4×2 factorial was replicated four times. The data are repeated in Table 15.15.

TABLE 15.15
Data for 4×2 Factorial
Golf Experiment

			BRAND		
		A	B	C	D
CLUB	Driver	226.4	238.3	240.5	219.8
		232.6	231.7	246.9	228.7
		234.0	227.7	240.3	232.9
		220.7	237.2	244.7	237.6
	Five-iron	163.8	184.4	179.0	157.8
		179.4	180.6	168.0	161.8
		168.6	179.5	165.2	162.1
		173.4	186.2	156.5	160.3

a. Write a regression model to represent the factorial experiment.
b. Use a computer program to fit the regression model. Conduct an analysis similar to that employed in the ANOVA of Example 15.8.

SOLUTION

a. The two factors Brand and Club are qualitative, so we employ dummy variables to describe them. Also, in contrast to the randomized block experiment, we are very interested in knowing whether the factors interact—that is, whether the difference in the Brand mean distances depends on the Club. Therefore, the model is

$$y = \beta_0 + \overbrace{\beta_1 x_1 + \beta_2 x_2 + \beta_3 x_3}^{\text{Brand main effect}} + \overbrace{\beta_4 x_4}^{\text{Club main effect}}$$

$$+ \underbrace{\beta_5 x_1 x_4 + \beta_6 x_2 x_4 + \beta_7 x_3 x_4}_{\text{Brand} \times \text{Club interaction}} + \varepsilon$$

where x_1 to x_3 are dummy variables representing the three brands (using brand D as the base level at which all three dummy variables are equal to 0), and x_4 is the dummy variable representing the two clubs (using five-iron as the base level at which the dummy variable is equal to 0). Note that the interaction terms are just the cross-products of the Brand dummy variables with the Club dummy variable.

b. The SAS printout for the regression model is shown in Figure 15.27. The global F test for the model tests the null hypothesis

$$H_0: \quad \beta_1 = \beta_2 = \cdots = \beta_7 = 0$$

In general, this hypothesis implies that the model has no predictive value. In this example, it implies that the mean distance is unrelated to Brand and Club; that is, the mean distance is the same for all brand–club combinations. This should be the same test as the ANOVA test that the $4 \times 2 = 8$ treatment means are all equal. We see that the value $F = 140.35$ for the SAS printout (Figure 15.27) is the same value we calculated using the ANOVA $F = MST/MSE$ in Example 15.8. The SAS printout gives the observed significance level of the test statistic as .0001 based on 7 numerator and 24 denominator degrees of freedom, implying that the null hypothesis is rejected at any level of significance greater than .0001. Therefore, using $\alpha = .10$ (as in Example 15.8), we reach the same conclusion as in the ANOVA analysis: There is sufficient evidence that the mean distance varies among two or more brand–club combinations.

FIGURE 15.27

SAS Printout for Factorial Golf Experiment: Complete Model Including Interaction

Model: MODEL1
Dep Variable: DISTANCE

Analysis of Variance

Source	DF	Sum of Squares	Mean Square	F Value	Prob>F
Model	7	33659.80875	4808.54411	140.354	0.0001
Error	24	822.24000	34.26000		
C Total	31	34482.04875			

Root MSE	5.85320	R-Square	0.9762	
Dep Mean	202.08125	Adj R-Sq	0.9692	
C.V.	2.89646			

Parameter Estimates

Variable	DF	Parameter Estimate	Standard Error	T for H0: Parameter=0	Prob > |T|
INTERCEP	1	167.175000	2.92660213	57.123	0.0001
X1	1	4.125000	4.13884042	0.997	0.3289
X2	1	15.500000	4.13884042	3.745	0.0010
X3	1	-6.675000	4.13884042	-1.613	0.1199
X4	1	75.925000	4.13884042	18.345	0.0001
X1X4	1	-18.800000	5.85320425	-3.212	0.0037
X2X4	1	-24.875000	5.85320425	-4.250	0.0003
X3X4	1	-6.675000	5.85320425	-1.140	0.2654

The next test in a two-factor factorial experiment involves the interaction between the factors. In the regression context the null hypothesis that the factors Brand and Club do not interact is

$$H_0: \quad \beta_5 = \beta_6 = \beta_7 = 0$$

This test requires a comparison of the complete model in Figure 15.27 with the reduced model *without* the interaction terms. The SAS printout for the reduced model is shown in Figure 15.28. We then calculate

$$F = \frac{[\text{SSE(Reduced model)} - \text{SSE(Complete model)}]/3}{\text{MSE(Complete model)}}$$

$$= \frac{(1,588.201 - 822.240)/3}{34.260} = 7.45$$

FIGURE 15.28

SAS Printout for Factorial Golf Experiment: Reduced Model Without Interaction

Model: MODEL2
Dep Variable: DISTANCE

Analysis of Variance

Source	DF	Sum of Squares	Mean Square	F Value	Prob>F
Model	4	32893.84750	8223.46187	139.802	0.0001
Error	27	1588.20125	58.82227		
C Total	31	34482.04875			

Root MSE	7.66957	R-Square	0.9539	
Dep Mean	202.08125	Adj R-Sq	0.9471	
C.V.	3.79529			

Parameter Estimates

Variable	DF	Parameter Estimate	Standard Error	T for H0: Parameter=0	Prob > |T|
INTERCEP	1	173.468750	3.03166282	57.219	0.0001
X1	1	-5.275000	3.83478384	-1.376	0.1803
X2	1	3.062500	3.83478384	0.799	0.4315
X3	1	-10.012500	3.83478384	-2.611	0.0146
X4	1	63.337500	2.71160166	23.358	0.0001

This F ratio is identical to that obtained in the ANOVA in Exercise 15.8, and again results in rejection of the null hypothesis at the $\alpha = .10$ level of significance, since $F_{.10} = 2.33$ with 3 numerator and 24 denominator degrees of freedom. We therefore conclude that both Brand and Club affect the mean distance, and that the differences among the Brand mean distances depend on the Club (i.e., the factors interact). The presence of interaction makes the tests of the factor main effects unnecessary.

We can use the least squares estimates of the β parameters in the complete model to estimate particular treatment means and differences between particular pairs of means. However, it is probably easier to use the Bonferroni procedure or some other multiple comparisons procedure, as we did in Example 15.8.

Note that the variability of the random error component is measured by the standard deviation of the regression model, $s = 5.85$. This measure is identical to that we obtained in the ANOVA, $\sqrt{\text{MSE}} = \sqrt{34.260} = 5.85$. The implication is that the model can be used to predict the distance Iron Byron will hit the ball to within approximately $\pm 2s = \pm 11.7$ yards if we specify the brand and club. Furthermore, the model accounts for $R^2 = .976$, or 97.6%, of the total sample variation in distance. ∎

Most designed experiments can be analyzed using either ANOVA or regression techniques with identical results. The recommended method of analysis depends on a number of factors, including the ultimate objectives of the analysis (e.g., prediction of response or estimation of differences among treatment means), the type of computer software available, and, perhaps most important, the methodology with which the analyst is more familiar. Whichever analytical tools are employed, it is important to recognize that both ANOVA and regression relate the response mean to one or more factors, and that both provide an estimate of sampling variability with which to make inferences.

**EXERCISES
15.52–15.57**

LEARNING THE MECHANICS

[*Note:* Starred (*) exercises require the use of a computer.]

15.52 Suppose you conduct an experiment with one factor at five levels, utilizing a completely randomized design. Each level of the factor is randomly assigned three experimental units.

a. Write a regression model for this experiment. Interpret the β parameters of the model.

b. How many degrees of freedom will be available for estimating the standard deviation σ of the error component?

c. Write the null hypothesis that the treatment means are equal in terms of the β parameters of the model.

d. What is the rejection region for the test in part c using an $\alpha = .10$ level of significance?

15.53 Suppose the following regression model were proposed to describe a designed experiment:

$$y = \beta_0 + \beta_1 x_1 + \beta_2 x_2 + \beta_3 x_3 + \varepsilon$$

where

$$x_1 = \begin{cases} 1 & \text{if treatment A} \\ 0 & \text{otherwise} \end{cases} \qquad x_2 = \begin{cases} 1 & \text{if treatment B} \\ 0 & \text{otherwise} \end{cases} \qquad x_3 = \begin{cases} 1 & \text{if block 1} \\ 0 & \text{if block 2} \end{cases}$$

a. What type of design was employed?
b. Interpret each of the β parameters.
c. Describe how you would test the null hypothesis that the treatment means are equal. Include the hypotheses, test statistic, and rejection region. (Assume that each block–treatment combination is represented by a single response measurement.)

15.54 Suppose the following regression model were proposed to describe a designed experiment:

$$y = \beta_0 + \beta_1 x_1 + \beta_2 x_2 + \beta_3 x_3 + \beta_4 x_1 x_3 + \beta_5 x_2 x_3 + \varepsilon$$

where

$$x_1 = \begin{cases} 1 & \text{if factor A, level 1} \\ 0 & \text{otherwise} \end{cases} \qquad x_2 = \begin{cases} 1 & \text{if factor A, level 2} \\ 0 & \text{otherwise} \end{cases} \qquad x_3 = \begin{cases} 1 & \text{it factor B, level 1} \\ 0 & \text{if factor B, level 2} \end{cases}$$

a. What type of experiment does the model describe?
b. Describe how you would test the null hypothesis that the treatment means are equal. Include the hypotheses, test statistic, and rejection region. (Assume that three experimental units are randomly and independently assigned to each factor-level combination.)
c. Describe how you would test for interaction.
d. Why does the model in this exercise include interaction terms, while that in Exercise 15.53 does not?

APPLYING THE CONCEPTS

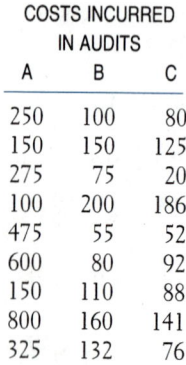

COSTS INCURRED
IN AUDITS

A	B	C
250	100	80
150	150	125
275	75	20
100	200	186
475	55	52
600	80	92
150	110	88
800	160	141
325	132	76
230	233	200

15.55 Refer to Exercise 15.20, in which a completely randomized design was utilized to compare the mean costs associated with the audits of firms in three size groups. The data are repeated in the accompanying table.
The model

$$E(y) = \beta_0 + \beta_1 x_1 + \beta_2 x_2$$

where

$$x_1 = \begin{cases} 1 & \text{if group A} \\ 0 & \text{otherwise} \end{cases} \qquad x_2 = \begin{cases} 1 & \text{if group B} \\ 0 & \text{otherwise} \end{cases}$$

was fit using the SAS regression program. The result is shown at the top of page 927.
a. Use the regression printout to conduct the test of the null hypothesis that the mean costs are the same for the three groups, and compare the result to that you obtained in Exercise 15.20.
b. Compare the observed significance levels of the tests.
c. Use the regression printout to partition the Total Sum of Squares into the Sums of Squares for Treatments and Error. Compare them with those in Exercise 15.20.
d. Interpret the value of R^2 and the least squares estimates of the β parameters.

SAS Printout for Exercise 15.55

Model: MODEL1
Dep Variable: COST

Analysis of Variance

Source	DF	Sum of Squares	Mean Square	F Value	Prob>F
Model	2	318861.66667	159430.83333	8.438	0.0014
Error	27	510163.00000	18894.92593		
C Total	29	829024.66667			

Root MSE	137.45882	R-Square	0.3846
Dep Mean	190.33333	Adj R-Sq	0.3390
C.V.	72.22004		

Parameter Estimates

Variable	DF	Parameter Estimate	Standard Error	T for H0: Parameter=0	Prob > |T|
INTERCEP	1	106.000000	43.46829411	2.439	0.0216
X1	1	229.500000	61.47345106	3.733	0.0009
X2	1	23.500000	61.47345106	0.382	0.7052

---

*15.56 Refer to Exercise 15.34, in which a randomized block design was employed to compare the mean market shares of six brands of beer estimated by three different auditing techniques. The data are repeated in the table.

BRAND	TRADITIONAL STORE AUDIT	WEEKEND SELLDOWN AUDIT	STORE PURCHASES AUDIT
1	18.0	19.0	20.7
2	15.3	17.3	14.0
3	8.9	8.5	10.1
4	6.5	4.9	6.1
5	5.3	6.1	4.6
6	3.4	3.0	3.1
Means	9.57	9.80	9.76

a. Write the appropriate regression model to analyze this randomized block design.
b. Use a computer program to fit the model. Interpret the least squares estimates of the β parameters.
c. What is the appropriate null hypothesis to test whether the auditing techniques differ in their estimates of mean market share? Test the hypothesis using $\alpha = .05$, and compare the result with that you obtained in Exercise 15.34.

15.57 Refer to Exercise 15.48, in which a 3 × 3 factorial experiment was utilized to examine the effects of the factors Price and Display on the unit sales of a product. The data are repeated in the accompanying table.

| | | PRICE | | |
		Regular	Reduced	Cost to supermarket
	Normal	989	1,211	1,577
		1,025	1,215	1,559
		1,030	1,182	1,598
DISPLAY	Normal plus	1,191	1,860	2,492
		1,233	1,910	2,527
		1,221	1,926	2,511
	Twice normal	1,226	1,516	1,801
		1,202	1,501	1,833
		1,180	1,498	1,852

The following regression model is proposed for the analysis of this experiment:

$$y = \beta_0 + \beta_1 x_1 + \beta_2 x_2 + \beta_3 x_3 + \beta_4 x_4 + \beta_5 x_1 x_3 + \beta_6 x_1 x_4 + \beta_7 x_2 x_3 + \beta_8 x_2 x_4 + \varepsilon$$

where

$$x_1 = \begin{cases} 1 & \text{if Price = Regular} \\ 0 & \text{otherwise} \end{cases} \qquad x_2 = \begin{cases} 1 & \text{if Price = Reduced} \\ 0 & \text{otherwise} \end{cases}$$

$$x_3 = \begin{cases} 1 & \text{if Display = Normal} \\ 0 & \text{otherwise} \end{cases} \qquad x_4 = \begin{cases} 1 & \text{if Display = Normal plus} \\ 0 & \text{otherwise} \end{cases}$$

The Minitab printout for this model is shown at the top of page 929.

a. Interpret the least squares estimates by using them to estimate the mean response for each of the 3 × 3 = 9 treatments.

b. Interpret the standard deviation and R^2 for the model.

c. Is there evidence that the mean unit sales differ for the nine treatments? Specify the null hypothesis in terms of the regression model and test it using $\alpha = .10$. Compare the result with that obtained in Exercise 15.48.

d. What is the appropriate null hypothesis to test for interaction between the factors Price and Display? What are the test statistic and rejection region for this test at $\alpha = .10$?

e. A second Minitab printout for the reduced model without interaction is also reproduced on page 929. Conduct the test you set up in part **d**, and compare the result with that obtained for the test of interaction in Exercise 15.48.

f. Do you recommend any further testing of the parameters in the model? Explain.

Minitab Printout for Exercise 15.57: Complete Model

```
THE REGRESSION EQUATION IS
SALES = 1829 - 626 X1 - 324 X2 - 251 X3 + 681 X4 + 62.7 X1X3 - 669 X1X4
        - 51.7 X2X3 - 288 X2X4
```

COLUMN	COEFFICIENT	ST. DEV. OF COEF.	T-RATIO = COEF/S.D.
	1828.67	12.84	142.40
X1	-626.00	18.16	-34.47
X2	-323.67	18.16	-17.82
X3	-250.67	18.16	-13.80
X4	681.33	18.16	37.52
X1X3	62.67	25.68	2.44
X1X4	-669.00	25.68	-26.05
X2X3	-51.67	25.68	-2.01
X2X4	-287.67	25.68	-11.20

```
S = 22.24
```

```
R-SQUARED = 99.8 PERCENT
R-SQUARED = 99.8 PERCENT, ADJUSTED FOR D.F.
```

ANALYSIS OF VARIANCE

DUE TO	DF	SS	MS=SS/DF
REGRESSION	8	5291151	661394
RESIDUAL	18	8905	495
TOTAL	26	5300057	

Minitab Printout for Exercise 15.57: Reduced Model

```
THE REGRESSION EQUATION IS
SALES = 1934 - 828 X1 - 437 X2 - 247 X3 + 362 X4
```

COLUMN	COEFFICIENT	ST. DEV. OF COEF.	T-RATIO = COEF/S.D.
	1933.74	66.13	29.24
X1	-828.11	72.45	-11.43
X2	-436.78	72.45	-6.03
X3	-247.00	72.45	-3.41
X4	362.44	72.45	5.00

```
S = 153.7
```

```
R-SQUARED = 90.2 PERCENT
R-SQUARED = 88.4 PERCENT, ADJUSTED FOR D.F.
```

ANALYSIS OF VARIANCE

DUE TO	DF	SS	MS=SS/DF
REGRESSION	4	4780446	1195112
RESIDUAL	22	519610	23619
TOTAL	26	5300057	

Designed experiments are those for which the analyst controls the method of sampling and the key variables that he or she believes will affect the results of the sampling effort. The **response, factors, levels, treatments, and experimental units** are elements of a designed experiment with which you should now be familiar. A **completely randomized design** is one for which treatment assignments are made to experimental units in a random and independent manner. A **randomized block design** is one for which treatment assignments are made within homogeneous groups, or **blocks,** of experimental units, with the assignments made randomly within the blocks.

Factorial experiments are those which involve more than one factor, and **complete factorials** utilize all combinations of the factors. F tests can be conducted to determine whether treatment means differ, and, if so, whether the factors interact or independently affect the response mean. The **Bonferroni procedure** can be used to compare pairs of treatment means.

Most designed experiments can also be analyzed in a regression context. Both ANOVA and regression models relate the response mean to one or more factors, and provide an estimate of the sampling variability with which inferences can be made.

SUPPLEMENTARY EXERCISES 15.58–15.81

[Note: Starred (*) exercises require the use of a computer.]

15.58 Mowen, Keith, Brown, and Jackson (1985) hypothesized that in evaluating salespeople, sales managers underutilize data on the salesperson's territory and instead rely on data associated with how much effort the salesperson expended. To investigate their hypothesis, they developed four scenarios based on the four factor-level combinations associated with the factors Salesperson's effort (high and low) and Sales territory difficulty (high and low). A random sample of 120 sales managers was selected, and 30 were randomly assigned to each of the four factor-level combinations. Each group was presented with the scenario corresponding to the appropriate factor-level combination, and then each sales manager was asked to evaluate the performance of a salesperson using a 7-point scale (1 = extremely high, 7 = extremely low). The authors summarized their results in the accompanying partial ANOVA table.

SOURCE	df	F
Sales territory difficulty	1	.39
Effort	1	53.27
Interaction	1	1.95

a. What type of experimental design was used in this study?

b. How many degrees of freedom are associated with SSE?

c. Do the data suggest that a sales manager's perceptions of a salesperson's performance are subject to an interaction effect between the difficulty of the salesperson's territory and the salesperson's level of effort? Test using $\alpha = .05$.

d. Do the data indicate that sales managers' performance ratings are influenced by sales territory data? Test using $\alpha = .05$.

e. Do the data indicate that the level of effort expended by a salesperson influences perceived performance? Test using $\alpha = .05$.

f. Do the results of parts **d** and **e** tend to confirm or to refute the hypothesis of Mowen et al.? Explain.

g. What assumptions must hold in order for the tests you conducted in parts **c**, **d**, and **e** to be valid?

15.59 The set of activities and decisions through which a firm moves from its initial awareness of an innovative industrial procedure to its final adoption or rejection of the innovation is referred to as the *industrial adoption process*. The process can be described as having five stages: (1) awareness, (2) interest (additional information requested), (3) evaluation (advantages and disadvantages compared), (4) trial (innovation is tested), and (5) adoption. As part of a study of the industrial adoption process, Ozanne and Churchill (1971) hypothesized that firms use a greater number of informational inputs to the process (e.g., visits by salespersons) in the later stages than in the earlier stages. In particular, they tested the hypotheses that a greater number of informational inputs are used in the interest stage than in the awareness stage, and that a greater number are used in the evaluation stage than in the interest stage. Ozanne and Churchill collected the information given in the table on the number of informational inputs used by a sample of 37 industrial firms that recently adopted a particular new automatic machine tool.

COMPANY	NUMBER OF INFORMATIONAL SOURCES USED			COMPANY	NUMBER OF INFORMATIONAL SOURCES USED		
	Awareness stage	Interest stage	Evaluation stage		Awareness stage	Interest stage	Evaluation stage
1	2	2	3	20	1	1	1
2	1	1	2	21	1	2	1
3	3	2	3	22	1	4	3
4	2	1	2	23	2	3	3
5	3	2	4	24	1	1	1
6	3	4	6	25	3	1	4
7	1	1	2	26	1	1	5
8	1	3	4	27	1	3	1
9	2	3	3	28	1	1	1
10	3	2	4	29	3	2	2
11	1	2	4	30	4	2	3
12	3	3	3	31	1	2	3
13	4	4	4	32	1	3	2
14	4	2	7	33	1	2	2
15	3	2	2	34	2	3	2
16	4	5	4	35	1	2	4
17	2	1	1	36	2	2	3
18	1	1	3	37	2	3	2
19	1	3	2				

a. What type of experimental design was employed? Identify the response, factor(s), factor type(s), treatments, and experimental units.

b. Given that SST = 15.730, SSE = 63.604, and SS(Total) = 165.297, construct a complete ANOVA summary table.

c. Do the data provide sufficient evidence to indicate that differences exist in the mean number of informational inputs of the three stages of the industrial adoption process studied by Ozanne and Churchill? Test using $\alpha = .05$.

d. Use the Bonferroni technique to compare the pairs of means corresponding to the three informational sources. Use an overall significance level of $\alpha = .05$.

e. What assumptions are necessary to assure the validity of the inferential procedures utilized in the analysis? What do you think of the normality assumption considering the nature of the data? [*Note:*   ANOVA procedures that do not require the normality assumption are presented in Chapter 16.]

15.60 Higher wholesale beef prices over the past few years have resulted in the sale of ground beef with higher fat content in an attempt to keep retail prices down. Four different supermarket chains were chosen, and four 1-pound packages of ground beef were randomly selected from each. The percentage of fat content was measured for each package, with the results shown in the table.

SUPERMARKET			
A	B	C	D
22	25	30	18
20	27	20	20
23	24	23	17
25	24	27	17

a. What type of experimental design does this represent?

b. Do the data provide sufficient evidence at the $\alpha = .05$ level of significance that the mean percentage of fat content differs for at least two of the supermarket chains? Use the calculation formulas in Appendix D.

c. Does the result of the test in part b warrant the comparison of the individual pairs of means? If so, use the Bonferroni technique to compare all pairs of means, using an overall level of significance $\alpha = .10$.

15.61 The table shows the partially completed analysis of variance for a two-factor factorial experiment.

SOURCE	df	SS	MS	F
A	3	2.6		
B	5	9.2		
AB			3.1	
Error		18.7		
Total	47			

a. Complete the analysis of variance table.

b. How many levels were utilized for each factor? How many treatments were employed? How many replications were performed?

c. Find the value of the Sum of Squares for Treatments. Test to determine whether the data provide evidence that the treatment means differ. Use $\alpha = .05$.

d. Is further testing of the nature of the factor effects warranted? If so, test to determine whether the factors interact. Use $\alpha = .05$. Interpret the result.

15.62 Advertising agencies are continually faced with the problem of creating attractive gimmicks to use in advertising campaigns. One agency recently decided to run an experiment to compare the preferences of financial analysts for three different brands of financial calculators, including its client's brand (brand A). It hoped to show that (1) financial analysts were not indifferent toward the three brands of calculators and (2) its client's brand was preferred to the industry leader, brand C. Three financial analysts were selected to perform an identical series of calculations on each of the three brands of calculators, A, B, and C. To avoid the possibility of fatigue, a suitable time period separated each set of calculations, and the calculators were used in random order by each analyst. A preference rating, based on a 0–100 scale, was recorded for each machine–analyst combination. These data are shown in the table.

ANALYST	BRAND		
	A	B	C
1	85	90	95
2	70	70	75
3	65	60	80

a. What type of design was employed?

b. Why is Analyst treated as a block rather than a factor in this experiment?

c. The Minitab printout for these data is reproduced here. Do the data provide sufficient evidence to indicate a difference among the mean preferences for the three brands? Use $\alpha = .10$.

Minitab Printout for Exercise 15.62

ANALYSIS OF VARIANCE ON RATING

SOURCE	DF	SS	MS
BRAND	2	200.0	100.0
ANALYST	2	816.7	408.3
ERROR	4	83.3	20.8
TOTAL	8	1100.0	

d. Use the Bonferroni technique with an overall 90% confidence level to compare both brands B and C to brand A. Should the advertising agency use the results of this experiment in its advertising campaign for brand A?

e. Why did the experiment call for each analyst to test all three brands of calculators, rather than assigning different analysts to each calculator?

*15.63 It has been hypothesized that treatment, after casting, of a plastic used in optic lenses will improve wear. Four different treatments are to be tested. To determine whether any differences in mean wear exist among treatments, 28 castings from a single formulation of the plastic were made, and seven castings were randomly assigned to each of the treatments. Wear was determined by measuring the increase in "haze" after 200 cycles of abrasion (better wear being indicated by small increases). The results are given in the table at the top of page 934.

a. What type of experiment was utilized? Identify the response, factors(s), factor type(s), treatments, and experimental units.

TREATMENT			
A	B	C	D
9.16	11.95	11.47	11.35
13.29	15.15	9.54	8.73
12.07	14.75	11.26	10.00
11.97	14.79	13.66	9.75
13.31	15.48	11.18	11.71
12.32	13.47	15.03	12.45
11.78	13.06	14.86	12.38

b. Use either a computer program or the computational formulas in Appendix D to analyze the data. Is there evidence of a difference in mean wear among the treatments? Use $\alpha = .05$.

c. What is the observed significance level of the test? Interpret it.

d. Use the Bonferroni technique to compare all the pairs of treatment means with an overall significance level of $\alpha = .10$.

e. Use a 90% confidence interval to estimate the mean wear for lenses receiving treatment A.

15.64 A large citrus products company is interested in purchasing several new orange juice extractors. To help them with their purchase decisions, six different manufacturers of juice extractors have agreed to let the company conduct an experiment to compare the yields of juices for the six different brands of extractors. Because of the possibility of a variation in the amount of juice per orange from one truckload of oranges to another, equal weights of oranges from a single truckload were assigned to each extractor, and this process was repeated for 15 loads. The amount of juice recorded for each extractor for each truckload produced the sums of squares shown in the table.

SOURCE	df	SS	MS	F
Extractors		84.71		
Truckloads		159.29		
Error		95.33		
Total		339.33		

a. Complete the ANOVA table.

b. Do the data provide sufficient evidence to indicate a difference among the mean amounts of juice extracted by the six extractors? Use $\alpha = .05$.

15.65 Explain the differences between a paired difference design (Chapter 9) and a randomized block design.

15.66 The data shown in the table are for a 4×3 factorial experiment with two replications.

a. Use a computer program or the calculation formulas in Appendix D to perform an analysis of variance for the data. Display the results in an ANOVA table.

b. Do the data indicate that the treatment means differ? Use $\alpha = .05$.

c. Is a test of factor interaction warranted? If so, perform it using $\alpha = .05$.

d. Is it necessary to perform tests of the main effects? If so, perform the tests using $\alpha = .05$.

	LEVEL OF B		
	1	2	3
LEVEL OF A 1	2	5	1
	4	6	3
2	5	2	10
	4	2	9
3	7	1	5
	10	0	3
4	8	12	7
	7	11	4

e. Use the Bonferroni technique to compare all pairs of treatment means using an $\alpha = .10$ level of significance.

15.67 Several companies are experimenting with the concept of paying production workers (generally paid by the hour) on a salary basis. It is believed that absenteeism and tardiness will increase under this plan, yet some companies feel that the working environment and overall productivity will improve. Fifty production workers under the salary plan are monitored at company A, and 50 under the hourly plan are monitored at company B. The number of work-hours missed due to tardiness or absenteeism over a 1-year period is recorded for each worker. The results are partially summarized in the table.

SOURCE	df	SS	MS	F
Companies		3,237.2		
Error		16,167.7		
Total	99			

a. Fill in the information missing from the table.
b. Is there evidence at $\alpha = .05$ that the mean number of hours missed differs for employees of the two companies?
c. Is there sufficient information given to form a confidence interval for the difference between the mean number of hours missed at the two companies?

15.68 From time to time, one branch office of a company must make shipments to a certain branch office in another state. There are three package delivery services between the two cities where the branch offices are located. Since the price structures for the three delivery services are quite similar, the company wants to compare the delivery times. The company plans to make several different types of shipments to its branch office. To compare the carriers, each shipment will be sent in triplicate, one with each carrier. The results listed in the table at the top of page 936 are the delivery times in hours.

a. What type of experiment was utilized? Is it a designed or observational experiment?
b. Given that SST = 8.8573, SSB = 3.9773, and SSE = .4227, construct a complete ANOVA summary table.
c. Is there evidence that the mean delivery times differ among the three carriers? Use $\alpha = .05$.

SHIPMENT	CARRIER		
	I	II	III
1	15.2	16.9	17.1
2	14.3	16.4	16.1
3	14.7	15.9	15.7
4	15.1	16.7	17.0
5	14.0	15.6	15.5

d. What assumptions are necessary to assure the validity of the test conducted in part **c**?

15.69 Sixteen workers were randomly selected to participate in an experiment to determine the effects of work scheduling and method of payment on attitude toward the job. Two types of scheduling were employed, the standard 8–5 workday and a modification whereby the worker was permitted to start the day at either 7 or 8 A.M. and to vary the starting time as desired; in addition, the worker was allowed to choose, on a daily basis, either a $\frac{1}{2}$-hour or 1-hour lunch period. The two methods of payment were a standard hourly rate and a reduced hourly rate with an added piece rate based on the worker's production. Four workers were randomly assigned to each of the four scheduling–payment combinations, and each completed an attitude test after 1 month on the job. The test scores are shown in the table.

		PAYMENT		
		Hourly rate		Hourly and piece rate
SCHEDULING	8-5	54, 68		89, 75
		55, 63		71, 83
	Worker-modified schedule	79, 65		83, 94
		62, 74		91, 86

a. What type of experiment was performed? Identify the response, factor(s), factor type(s), treatments, and experimental units.

b. The SAS printout for this experiment is shown on page 937. Is there evidence that the treatment means differ? Use $\alpha = .05$.

c. If the test in part **b** warrants further analysis, conduct the appropriate tests of interaction and main effects. Interpret your results.

d. What assumptions are necessary to assure the validity of the inferences? State the assumptions in terms of this experiment.

DIVISION		
1	2	3
1.1	1.4	.4
.9	1.6	.3
.8	1.0	.5

15.70 A corporation manages a very large number of stores that it classifies into three geographic divisions. Three stores are randomly selected from each division and a study is made to determine the mean inflation rate for the items in inventory. The inflation rate is recorded in the table as a percentage change in price over a year's time.

a. Is there sufficient evidence to indicate a difference in the mean inflation rates among the stores in different divisions?

b. In division 1, the numbers 1.1, .9, and .8 represent a random sample from what population?

SAS Printout for Exercise 15.69

General Linear Models Procedure

Dependent Variable: SCORE

Source	DF	Sum of Squares	Mean Square	F Value	Pr > F
Model	3	1806.00000	602.00000	12.29	0.0006
Error	12	588.00000	49.00000		
Corrected Total	15	2394.00000			

R-Square	C.V.	Root MSE	SCORE Mean
0.754386	9.3959732	7.00000	74.5000000

Source	DF	Type I SS	Mean Square	F Value	Pr > F
SCHEDULE	1	361.0000	361.0000	7.37	0.0188
PAYMENT	1	1444.0000	1444.0000	29.47	0.0002
SCHEDULE*PAYMENT	1	1.0000	1.0000	0.02	0.8888

15.71 One indicator of employee morale is the length of time employees stay with a company. A large corporation has three factories located in similar areas of the country. Although the corporation management attempts to maintain uniformity in management, working conditions, and employee relations at its various factories, it realizes that differences may exist among the various factories. To study this phenomenon, employee records are randomly selected at each of the three factories, and the length of employee service with the company is recorded. A summary of the data is given in the table. Is there evidence of a difference in mean length of service among the three factories? Use $\alpha = .05$.

Factory	1	2	3
Number in sample	15	21	17

SST = 421.74; SSE = 3,574.06

8-HOUR DAY	10-HOUR DAY	12-HOUR DAY
87	75	95
96	82	76
75	90	87
90	80	82
72	73	65
86		

15.72 England has experimented with different 40-hour work weeks to maximize production and minimize expenses. A factory tested a 5-day week (8 hours per day), a 4-day week (10 hours per day), and a $3\frac{1}{3}$-day week (12 hours per day), with the weekly production results shown in the table (in thousands of dollars worth of items produced).

a. What type of experiment was employed?

b. Use the SPSS printout on page 938 to test the null hypothesis that the mean productivity level is the same for the three lengths of workday. Use $\alpha = .10$.

c. Is further comparison of the pairs of means warranted? If so, use the Bonferroni technique to compare the pairs of mean productivity levels. Use an overall $\alpha = .10$.

SPSS Printout for Exercise 15.72

Variable   PRDCTION

By Variable   DAY

Analysis of Variance

Source	D.F.	Sum of Squares	Mean Squares	F Ratio	F Prob.
Between Groups	2	57.6042	28.8021	.3375	.719
Within Groups	13	1109.3333	85.3333		
Total	15	1166.9375			

15.73 In-town mileage tests were performed to compare three different brands of regular gasoline. Four different automobiles were used in the experiment, and each brand of gas was used in each car until the mileage was determined. The results in miles per gallon are shown in the table.

		AUTOMOBILE			
		1	2	3	4
	A	20.2	18.7	19.7	17.9
BRAND	B	19.7	19.0	20.3	19.0
	C	18.3	18.5	17.9	21.1

a. What type of experimental design was used in this study?
b. Use a computer program or the computational formulas in Appendix D to construct the ANOVA summary table for this experiment.
c. Is there evidence of a difference in the mean mileage rating among the three brands of gasoline? Use $\alpha = .05$.
d. Is there evidence of a difference in the mean mileage among the four models; i.e., is blocking important in this type of experiment? Use $\alpha = .05$.

15.74 Refer to Exercise 15.73. Use the Bonferroni technique to compare the pairs of Brand means. Use an overall significance level of $\alpha = .10$.

15.75 To be able to provide its clients with comparative information on two large suburban residential communities, a realtor wants to know the average home value in each community. Eight homes are selected at random within each community and are appraised by the realtor. The appraisals are given in the table (in thousands of dollars). Can you conclude that the average home value is different in the two communities? You have three ways of analyzing this problem.

COMMUNITY	
A	B
43.5	73.5
49.5	62.0
38.0	47.5
66.5	36.5
57.5	44.5
32.0	56.0
67.5	68.0
71.5	63.5

a. Use the two-sample t statistic (Section 9.2) to test $H_0: \mu_A = \mu_B$.
b. Consider the regression model

$$y = \beta_0 + \beta_1 x + \varepsilon$$

where

$$x = \begin{cases} 1 & \text{if community B} \\ 0 & \text{if community A} \end{cases} \qquad y = \text{Appraised price}$$

Since $\beta_1 = \mu_B - \mu_A$, testing H_0: $\beta_1 = 0$ is equivalent to testing H_0: $\mu_A = \mu_B$. Use the partial reproduction of the SAS printout shown here to test H_0: $\beta_1 = 0$. Use $\alpha = .05$.

SAS Regression Printout for Exercise 15.75

SOURCE	DF	SUM OF SQUARES	MEAN SQUARE	F VALUE	PR > F
MODEL	1	40.64062500	40.64062500	0.21	0.6501
ERROR	14	2648.71875000	189.19419643		ROOT MSE
CORRECTED TOTAL	15	2689.35937500		R-SQUARE	13.75478813
				0.015112	

PARAMETER	ESTIMATE	T FOR H0: PARAMETER = 0	PR > ¦T¦	STD ERROR OF ESTIMATE
INTERCEPT	53.25000000	10.95	0.0001	4.86305198
X	3.18750000	0.46	0.6501	6.87739406

c. The SAS ANOVA printout for this experiment is also given here. Is there evidence that the treatment means differ? Use $\alpha = .05$.

SAS ANOVA Printout for Exercise 15.75

General Linear Models Procedure

Dependent Variable: PRICE

Source	DF	Sum of Squares	Mean Square	F Value	Pr > F
Model	1	40.6406250	40.6406250	0.21	0.6501
Error	14	2648.718750	189.1941964		
Corrected Total	15	2689.359375			

R-Square	C.V.	Root MSE	PRICE Mean
0.015112	25.079956	13.7548	54.8437500

Source	DF	Type I SS	Mean Square	F Value	Pr > F
COMMUNTY	1	40.6406	40.6406	0.21	0.6501

d. Using the results of the three tests in parts **a–c**, verify that the tests are the equivalent (for this special case, $p = 2$) of the completely randomized design in terms of the test statistic value and rejection region.

15.76 Methods of displaying goods can affect their sales. The manager of a large produce market would like to try three different display types for a certain fruit. The manager chooses the locations for the three displays so that each display type will be equally accessible to the customers. The three displays will be set up for five 1-week periods. Between each of these periods will be a 2-week period when a standard display is used. For each of the 5 experimental weeks, the sales (in dollars) of fruit from each display is determined, with the results shown in the table at the top of page 940.

	DISPLAY		
	A	B	C
PERIOD 1	$525	$653	$408
2	450	500	380
3	400	510	335
4	602	699	475
5	306	422	315

a. What type of experimental design was used in this study? Identify the response, factor(s), factor type(s), treatments, and experimental units.

b. The SPSS printout for this experiment is shown here. Is there evidence of a difference among mean sales for the three types of displays? Use $\alpha = .05$.

SPSS Printout for Exercise 15.76

```
            * * *   A N A L Y S I S   O F   V A R I A N C E   * * *

            SALES
      BY    DISPLAY
            PERIOD
```

Source of Variation	Sum of Squares	DF	Mean Square	F	Signif of F
Main Effects	187191.467	6	31198.578	24.761	.000
DISPLAY	76436.133	2	38218.067	30.332	.000
PERIOD	110755.333	4	27688.833	21.976	.000
Explained	187191.467	6	31198.578	24.761	.000
Residual	10079.867	8	1259.983		
Total	197271.333	14	14090.810		

c. Do the data indicate that the use of weeks as blocks was necessary? Test at $\alpha = .05$.

d. Use the Bonferroni technique to compare the mean sales for all pairs of displays. Use an overall significance level of $\alpha = .05$.

15.77 Random samples of size 36, 36, and 15 were drawn from stocks listed on the New York Stock Exchange (NYSE), the American Stock Exchange (ASE), and from those traded over-the-counter (OTC), respectively. The closing prices of all 87 stocks on December 15, 1986, are listed in the table.

a. Is this experiment designed or observational? What type of experiment is it?

b. Given that the Total Sum of Squares is 30,935.513 and the Sum of Squares for Treatments is 2,759.905, complete the ANOVA summary table for the experiment.

c. Is there evidence that the mean closing price differed among the three markets? Use $\alpha = .10$.

d. Use the Bonferroni technique to compare the mean closing prices for all pairs of markets. Use an overall significance level of $\alpha = .10$.

NYSE			ASE			OTC		
$14\frac{1}{8}$	23	$15\frac{1}{2}$	$8\frac{1}{2}$	$8\frac{1}{2}$	$14\frac{3}{4}$	$27\frac{1}{4}$	$11\frac{3}{4}$	$8\frac{3}{4}$
$15\frac{7}{8}$	95	$43\frac{3}{4}$	$2\frac{5}{8}$	$3\frac{1}{4}$	$16\frac{3}{4}$	9	32	$11\frac{3}{4}$
$24\frac{1}{8}$	$5\frac{3}{8}$	$80\frac{1}{2}$	$11\frac{1}{4}$	$41\frac{7}{8}$	$\frac{3}{8}$	3	$9\frac{1}{8}$	17
28	$4\frac{1}{8}$	$11\frac{7}{8}$	$14\frac{1}{8}$	$12\frac{5}{8}$	86	18	9	$14\frac{1}{2}$
$77\frac{1}{2}$	$12\frac{3}{4}$	$4\frac{1}{2}$	$5\frac{5}{8}$	$9\frac{1}{2}$	$6\frac{1}{4}$	$8\frac{1}{2}$	$32\frac{1}{2}$	$21\frac{1}{4}$
$30\frac{3}{4}$	$35\frac{1}{4}$	$16\frac{1}{8}$	$4\frac{7}{8}$	$5\frac{5}{8}$	$26\frac{1}{2}$			
32	$24\frac{1}{4}$	25	$30\frac{3}{4}$	5	$\frac{1}{2}$			
$11\frac{5}{8}$	18	$10\frac{3}{4}$	$4\frac{3}{8}$	$24\frac{1}{8}$	$1\frac{1}{2}$			
$11\frac{1}{8}$	$21\frac{1}{8}$	$30\frac{3}{8}$	$6\frac{5}{8}$	$6\frac{1}{4}$	$7\frac{7}{8}$			
$20\frac{1}{8}$	$14\frac{3}{4}$	$83\frac{5}{8}$	$10\frac{1}{4}$	$21\frac{5}{8}$	$13\frac{1}{4}$			
$31\frac{7}{8}$	$19\frac{7}{8}$	$21\frac{1}{4}$	$1\frac{5}{8}$	$17\frac{7}{8}$	$12\frac{3}{4}$			
13	8	$6\frac{7}{8}$	$24\frac{1}{8}$	$16\frac{1}{4}$	$35\frac{1}{4}$			

Source: *Wall Street Journal*, Dec. 16, 1986.

15.78 Refer to Exercise 15.77.

a. Write a regression model to describe the experiment. Define all the variables you use.

b. The SAS regression printout for this experiment is shown here. Interpret the least squares estimates, and deduce the coding that was utilized for the dummy variables.

SAS Printout for Exercise 15.78

Model: MODEL1
Dep Variable: PRICE

Analysis of Variance

Source	DF	Sum of Squares	Mean Square	F Value	Prob>F
Model	2	2759.90461	1379.95230	4.114	0.0197
Error	84	28175.60868	335.42391		
C Total	86	30935.51329			

Root MSE	18.31458	R-Square	0.0892	
Dep Mean	19.47270	Adj R-Sq	0.0675	
C.V.	94.05260			

Parameter Estimates

Variable	DF	Parameter Estimate	Standard Error	T for H0: Parameter=0	Prob > |T|
INTERCEP	1	15.558333	4.72880473	3.290	0.0015
X1	1	10.601389	5.62840342	1.884	0.0631
X2	1	-1.141667	5.62840342	-0.203	0.8398

c. Use the regression printout to test the null hypothesis that the three markets had the same mean closing prices on December 15, 1986. Use $\alpha = .10$. What is the observed significance level of the test?

15.79 In Case Study 15.1, Dickson used an analysis of variance to evaluate a questionnaire that he had designed to measure the computer expertise of middle managers. The data he employed in the analysis are displayed in Table 15.4. Use the data to replicate the analysis performed by Dickson—including the confidence intervals. Compare the results with those presented in Figure 15.11.

15.80 In Example 12.6, we used a multiple regression analysis to analyze data for a 2 × 3 factorial experiment. The experiment measured worker productivity for each of two types of manufacturing plant (union and nonunion) and for each of three levels of bonus compensation. Three workers were randomly assigned to each of the six plant type–incentive factor level combinations. The objective of the experiment was to examine the relationship between worker productivity, y, and incentive level, x, for the two types of plants. Since the data in the table are the results of a 2 × 3 factorial experiment with the same number of observations per factor level combination, we can analyze the data using an analysis of variance.

TYPE OF PLANT	INCENTIVE								
	20¢/casting			30¢/casting			40¢/casting		
Union	1,435,	1,512,	1,491	1,583,	1,529,	1,610	1,601,	1,574,	1,636
Nonunion	1,575,	1,512,	1,488	1,635,	1,589,	1,661	1,645,	1,616,	1,689

a. Construct an analysis of variance table for the data. Use either a computer program or the formulas in Appendix D.
b. Do the data provide sufficient evidence to indicate an interaction between incentive level and type of plant? Test using $\alpha = .05$.
c. What are the practical implications of the test results of part **b**?
d. Compare your SSE with the one obtained in the multiple regression analysis, Figure 12.22. Why do they differ?
e. Compare the results of the analysis of Example 12.6 with those of your analysis of variance. Then consider the following: If one (or both) of the factors in a two-factor factorial experiment is a quantitative independent variable, a multiple regression analysis provides more practical information than an analysis of variance. Why? [*Note:* If both factors are qualitative variables, the two methods of analysis yield the same results.]

*15.81 Refer to Exercise 15.80.
a. Specify the regression model that could be utilized to yield results identical to those obtained in the ANOVA. [*Hint:* The model needs to have the same number of degrees of freedom for the main effects and the interaction as in the ANOVA summary table of Exercise 15.80, part **a.**]
b. Fit the model, and compare the value of SSE to that in Exercise 15.80.
c. Fit a reduced model and test to determine whether interaction terms contribute to the model. Test using $\alpha = .05$, and compare the result to part **b** of Exercise 15.80.

ON YOUR OWN . . .

Due to ever-increasing food costs, consumers are becoming more discerning in their choice of supermarkets. It usually is more convenient to shop at only one market, as opposed to buying different items at different markets. Thus, it would be useful to compare the mean food expenditure for a market basket of food items from store to store. Since there is a great deal of variability in the prices of products sold at any supermarket, we will consider an experiment that blocks on products.

Choose three (or more) supermarkets in your area that you want to compare; then choose approximately ten (or more) food products you typically purchase. For each food item, record the price each store charges in the following manner:

FOOD ITEM 1	FOOD ITEM 2 . . .	FOOD ITEM 10
Price store 1	Price store 1 . . .	Price store 1
Price store 2	Price store 2 . . .	Price store 2
Price store 3	Price store 3 . . .	Price store 3

Use the data you obtain to test

H_0: Mean expenditures at the stores are the same

H_a: Mean expenditures for at least two of the stores are different

Also, test to determine whether blocking on food items is advisable in this kind of experiment. Fully interpret the results of your analysis.

USING THE COMPUTER . . .

Do home values differ among the four census regions in the United States? Treat the 1,000 zip codes in the data set of Appendix C as if they were the population of all U.S. zip codes.

a. Use a completely randomized design with equal samples of size 50 to generate data to answer the above question. Construct a dot diagram and stem and leaf display for each of the four samples. What do your graphs suggest about the differences in mean home values among census regions?

b. Use ANOVA to answer the question of interest.

c. Use a multiple comparisons procedure to compare the four treatment means. Draw the appropriate conclusions.

d. What assumptions was it necessary to make in order to employ the inferential procedures of parts b and c? Which of those assumptions are suspect in this problem?

REFERENCES

Aplin-Brownlee, V. "Many workers facing terror of 'technophobia'." *Minneapolis Star and Tribune*, Jan. 10, 1984, 1C.

Brockhaus, R. H. "Risk-taking propensity of entrepreneurs." *Academy of Management Journal*, Vol. 23 (Sept. 1980), 509–520.

Curhan, R. C. "The effects of merchandising and temporary promotional activities on the sales of fresh fruit and vegetables in supermarkets." *Journal of Marketing Research*, Vol. 11 (Aug. 1974), 286–294.

Johnson, D. A., Pany, K., and White, R. "Audit reports and the loan decision: Actions and perceptions." *Auditing: A Journal of Practice and Theory*, Spring, 1983, 2, 38–51.

Klompmaker, J. E., Hughes, G. D., and Haley, R. I. "Test marketing in new products development." *Harvard Business Review*, May–June 1976, 128.

Logue, D. and Rogalski, R. *Harvard Business Review*, July–Aug. 1979.

Mendenhall, W. *Introduction to Linear Models and the Design and Analysis of Experiments*. Belmont, Calif.: Wadsworth, 1968. Chapter 8.

Mendenhall, W. and Sincich, T. *A Second Course in Business Statistics: Regression Analysis*, 2nd ed. San Francisco: Dellen, 1986.

Miller, R. B. and Wichern, D. W. *Intermediate Business Statistics: Analysis of Variance, Regression, and Time Series*, New York: Holt, Rinehart and Winston, 1977. Chapter 4.

Mowen, J. C., Keith, J. E., Brown, S. W., and Jackson, D. W., Jr. "Utilizing effort and task difficulty information in evaluating salespeople." *Journal of Marketing Research*, Vol. 22 (May 1985), 185–191.

Neter, J. and Wasserman, W. *Applied Linear Statistical Models*. Homewood, Ill.: Richard D. Irwin, 1974.

Ozanne, U. B. and Churchill, G. A. "Adoption research: Information sources in the industrial purchasing decision." In R. L. Day and T. E. Ness (eds.), *Marketing Models, Behavioral Science Applications*. Scranton, Pa.: Intext Educational Publishers, 1971. pp. 249–265.

Phillips, M. D. and Pepper, R. L. "Shipboard fire-fighting performance of females and males." *Human Factors*, 1982, 24(3).

Prasad, V. K., Casper, W. R., and Schieffer, R. J. "Alternatives to the traditional retail store audit: A field study." *Journal of Marketing*, Vol. 48, (Winter 1984), 54–61.

Scheffé, H. *The Analysis of Variance*. New York: Wiley, 1959.

Shetty, Y. K. and Carlisle, H. M. "Organizational correlates of a management by objectives program." *Academy of Management Journal*, 1974, 17(1).

Wilkinson, J. B., Mason, J. B., and Paksoy, C. H. "Assessing the impact of short-term supermarket strategy variables." *Journal of Marketing Research*, Vol. 19, Feb. 1982, 72–86.

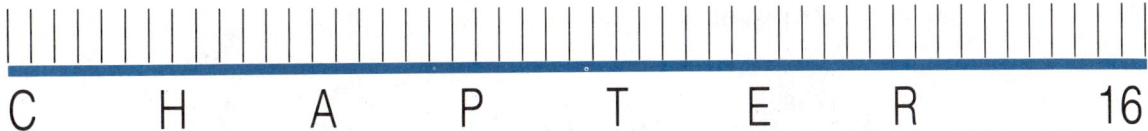

C H A P T E R 16

WHERE WE'VE BEEN...

Chapters 8, 9, and 15 presented techniques for making inferences about the mean of a single population and for comparing the means of two or more populations. Chapters 10–12 treated simple and multiple regression—the problem of relating the mean of a population of y values to a set of independent variables $x_1, x_2, \ldots, x_k$. Most of the techniques discussed in Chapters 8–12 and 15 are based on the assumption that the sampled populations have probability distributions that are approximately normal with equal variances. But how can you analyze data from populations that do not satisfy these assumptions? Or, how can you make comparisons between populations when you cannot assign specific numerical values to your observations?

WHERE WE'RE GOING...

In this chapter, we present techniques for comparing two or more populations that are based on an ordering of the sample measurements according to their relative magnitudes. These techniques, which require fewer or less stringent assumptions concerning the nature of the probability distributions of the populations, are called **nonparametric statistical methods**. We will present nonparametric statistical techniques for comparing two or more populations using two of the experimental designs described in Chapter 15—the completely randomized and the randomized block designs.

NONPARAMETRIC STATISTICS

CONTENTS

The t and F tests for comparing two or more populations (Chapters 9 and 15) are unsuitable for some types of business data. These data fall into two categories: The first are data sets that do not satisfy the assumptions upon which the t and F tests are based. For both tests, we assume that the random variables being measured have normal probability distributions with equal variances. Yet in practice, the observations from one population may exhibit much greater variability than those from another, or the probability distributions may be decidedly nonnormal. For example, the distribution might be very flat, peaked, or strongly skewed to the right or left. When any of the assumptions required for the t and F tests are seriously violated, the computed t and F statistics may not follow the standard t and F distributions. If this is true, the tabulated values of t and F (Tables VI–X of Appendix B) are not applicable, the correct value of α for the test is unknown, and the t and F tests are of dubious value.

The second type of data for which t and F tests are inappropriate are responses that are not susceptible to measurement but that can be *ranked in order of magnitude*. For example, if we want to compare the managerial ability of two executives based on subjective evaluations of trained observers, despite the fact that we cannot give an exact value to the managerial ability of a single executive, we may be able to decide that executive A has more ability than executive B. If executives A and B are evaluated by each of ten observers, we have the standard problem of comparing the probability distributions for two populations of ratings, one for executive A and one for B. But the t test of Chapter 9 would be inappropriate because the only data that can be recorded are preferences; i.e., each observer decides either that A is better than B or vice versa.

Consider another example of this type of data. Most firms that plan to market a new product nationally first test the product in a few cities or regions to determine its acceptability. For a food product this may entail taste tests in which consumers rank the new product in order of preference with respect to one or more currently popular brands. A consumer probably has a preference for each product, but the strength of the preference is difficult, if not impossible, to measure. Consequently, the best we can do is have each consumer examine the new product along with a few established products, and rank them according to preference: 1 for the most preferred, 2 for second, etc.

The **nonparametric** counterparts of the t and F tests compare the probability distributions of the sampled populations, rather than specific parameters of these populations (such as the means or variances). For example, nonparametric tests can be used to compare the probability distribution of the strengths of preferences for a new product to the probability distributions of the strengths of preferences for the currently popular brands. If it can be inferred that the distribution for the new product lies above (to the right of) the others (see Figure 16.1), the implication is that the new product tends to be more preferred than the currently popular products. Such an inference might lead to a decision to market the product nationally.

Many nonparametric methods use the **relative ranks** of the sample observations rather than their actual numerical values. These tests are particularly valuable when we are unable to obtain numerical measurements of some phenomena but are able

FIGURE 16.1
Probability Distributions of Strengths of Preference Measurements (New Product Is Preferred)

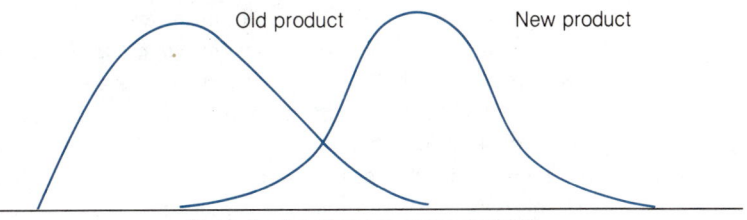

to rank them in comparison to each other. Statistics based on ranks of measurements are called **rank statistics**. In Sections 16.1 and 16.3, we present rank statistics for comparing two probability distributions using independent samples. In Sections 16.2 and 16.4, the matched pairs and randomized block designs are used to make nonparametric comparisons of populations. Finally, in Section 16.5 we present a nonparametric measure of correlation between two variables—**Spearman's rank correlation coefficient**. For a more complete discussion of tests based on rank statistics, see the references at the end of this chapter.

| | | | | | | | | | | | | |

SECTION 16.1

COMPARING TWO POPULATIONS: WILCOXON RANK SUM TEST FOR INDEPENDENT SAMPLES

Suppose two independent random samples are to be used to compare two populations, and the *t* test of Chapter 9 is inappropriate for making the comparison. Either we are unwilling to make assumptions about the form of the underlying probability distributions, or we are unable to obtain exact values of the sample measurements but can rank them in order of magnitude. For either of these situations, if the data can be ordered, the **Wilcoxon rank sum test** (developed by Frank Wilcoxon) can be used to test a hypothesis that the probability distributions associated with the two populations are equivalent.

For example, suppose six economists who work for the federal government and seven university economists are randomly selected, and each is asked to predict next year's percentage change in cost of living as compared with this year's figure. The objective of the study is to compare the government economists' predictions to those of the university economists. The data are shown in Table 16.1.

TABLE 16.1
Percentage Cost of Living Change, as Predicted by Government and University Economists

GOVERNMENT ECONOMIST		UNIVERSITY ECONOMIST	
Prediction	Rank	Prediction	Rank
3.1	4	4.4	6
4.8	7	5.8	9
2.3	2	3.9	5
5.6	8	8.7	11
0.0	1	6.3	10
2.9	3	10.5	12
		10.8	13

The two populations of predictions are those that would be obtained from *all* government and *all* university economists if they could all be questioned. To compare their probability distributions, we first *rank the sample observations as though they were all drawn from the same population.* That is, we pool the measurements from both samples and then rank the measurements from the smallest (a rank of 1) to the largest (a rank of 13). The ranks of the economists' predictions are indicated in Table 16.1.

The test statistic for the Wilcoxon test is based on the totals of the ranks for each of the two samples—that is, on the **rank sums**. If the two rank sums are nearly equal, the implication is that there is no evidence that the probability distributions from which the samples were drawn are different. On the other hand, if the two rank sums are very different, the implication is that the two samples may have come from different populations

In the economists' predictions example, we arbitrarily denote the rank sum for government economists by T_A and that for university economists by T_B. Then

$$T_A = 4 + 7 + 2 + 8 + 1 + 3 = 25$$
$$T_B = 6 + 9 + 5 + 11 + 10 + 12 + 13 = 66$$

The sum of T_A and T_B will always equal $n(n + 1)/2$, where $n = n_1 + n_2$. So, for this example, $n_1 = 6$, $n_2 = 7$, and

$$T_A + T_B = \frac{13(13 + 1)}{2} = 91$$

Since $T_A + T_B$ is fixed, a small value for T_A implies a large value for T_B (and vice versa) and a large difference between T_A and T_B. Therefore, the smaller the value of one of the rank sums, the greater is the evidence to indicate that the samples were selected from different populations.

Values that locate the rejection region for the rank sum associated with the smaller sample are given in Table XI of Appendix B. A partial reproduction of this table is shown in Table 16.2. The columns of the table represent n_1, the first sample size, and the rows represent n_2, the second sample size. *The T_L and T_U entries in the table are the boundaries of the lower and upper regions, respectively, for the rank*

TABLE 16.2 Reproduction of Part of Table XI of Appendix B

a. $\alpha = .025$ one-tailed; $\alpha = .05$ two-tailed

n_2 \ n_1	3		4		5		6		7		8		9		10	
	T_L	T_U	T_L	T_U	T_L	T_U	T_L	T_U	T_L	T_U	T_L	T_U	T_L	T_U	T_L	T_U
3	5	16	6	18	6	21	7	23	7	26	8	28	8	31	9	33
4	6	18	11	25	12	28	12	32	13	35	14	38	15	41	16	44
5	6	21	12	28	18	37	19	41	20	45	21	49	22	53	24	56
6	7	23	12	32	19	41	26	52	28	56	29	61	31	65	32	70
7	7	26	13	35	20	45	28	56	37	68	39	73	41	78	43	83
8	8	28	14	38	21	49	29	61	39	73	49	87	51	93	54	98
9	8	31	15	41	22	53	31	65	41	78	51	93	63	108	66	114
10	9	33	16	44	24	56	32	70	43	83	54	98	66	114	79	131

sum associated with the sample that has fewer measurements. If the sample sizes n_1 and n_2 are the same, either rank sum may be used as the test statistic. To illustrate, suppose $n_1 = 8$ and $n_2 = 10$. For a two-tailed test with $\alpha = .05$, we consult part **a** of the table and find that the null hypothesis will be rejected if the rank sum of sample 1 (the sample with fewer measurements), T, is less than or equal to $T_L = 54$ *or* greater than or equal to $T_U = 98$. The Wilcoxon rank sum test is summarized in the box.

WILCOXON RANK SUM TEST: INDEPENDENT SAMPLES*

ONE-TAILED TEST

H_0: Two sampled populations have identical probability distributions

H_a: The probability distribution for population A is shifted to the right of that for B

Test statistic: The rank sum, T, associated with the sample with fewer measurements (if sample sizes are equal, either rank sum can be used)

Rejection region: Assuming the smaller sample size is associated with distribution A (or, if sample sizes are equal, we use the rank sum T_A), we reject H_0 if $T_A \geq T_U$, where T_U is the upper value given by Table XI in Appendix B for the chosen *one-tailed* α value.

TWO-TAILED TEST

H_0: Two sampled populations have identical probability distributions

H_a: The probability distribution for population A is shifted to the left *or* to the right of that for B

Test statistic: The rank sum, T, associated with the sample with fewer measurements (if the sample sizes are equal, either rank sum can be used)

Rejection region: $T \leq T_L$ or $T \geq T_U$, where T_L is the lower value given by Table XI in Appendix B for the chosen *two-tailed* α value, and T_U is the upper value from Table XI.

[*Note:* If the one-sided alternative is that the probability distribution for A is shifted to the *left* of B (and T_A is the test statistic), we reject H_0 if $T_A \leq T_L$.]

Assumptions: 1. The two samples are random and independent.
            2. The two probability distributions from which the samples are drawn are continuous.

Ties: Assign tied measurements the average of the ranks they would receive if they were unequal but occurred in successive order. For example, if the third-ranked and fourth-ranked measurements are tied, assign each a rank of $(3 + 4)/2 = 3.5$.

Note that the assumptions necessary for the validity of the Wilcoxon rank sum test do not specify the shape or type of probability distribution. However, the distributions are assumed to be continuous so that the probability of tied measurements is 0 (see Chapter 6), and each measurement can be assigned a unique rank.

*Another statistic used for comparing two populations based on independent random samples is the **Mann–Whitney U statistic**. The U statistic is a simple function of the rank sums. It can be shown that the Wilcoxon rank sum test and the Mann–Whitney U test are equivalent.

In practice, however, rounding of continuous measurements will sometimes produce ties. As long as the number of ties is small relative to the sample sizes, the Wilcoxon test procedure will still have approximate significance level α. The test is not recommended to compare discrete distributions for which many ties are expected.

EXAMPLE 16.1

Test the hypothesis that the university economists' predictions of next year's percentage change in cost of living tend to be higher than the government economists'. Conduct the test using the data in Table 16.1 and $\alpha = .05$.

SOLUTION

H_0: The probability distributions corresponding to the government and university economists' predictions of inflation rate are identical

H_a: The probability distribution for the university economists' predictions lies above (to the right of) that for the government economists' predictions*

Test statistic: Since fewer government economists ($n_1 = 6$) than university economists ($n_2 = 7$) were sampled, the test statistic is T_A, the rank sum of the government economists' predictions.

Rejection region: Since the test is one-sided, we consult part **b** of Table XI for the rejection region corresponding to $\alpha = .05$. We will reject H_0 only for $T_A \leq T_L$, the lower value from Table XI, since we are specifically testing that the distribution of the government economists' predictions lies *below* the distribution of university economists' predictions, as shown in Figure 16.2. Thus, we will reject H_0 if $T_A \leq 30$.

FIGURE 16.2
Alternative Hypothesis and Rejection Region for Example 16.1

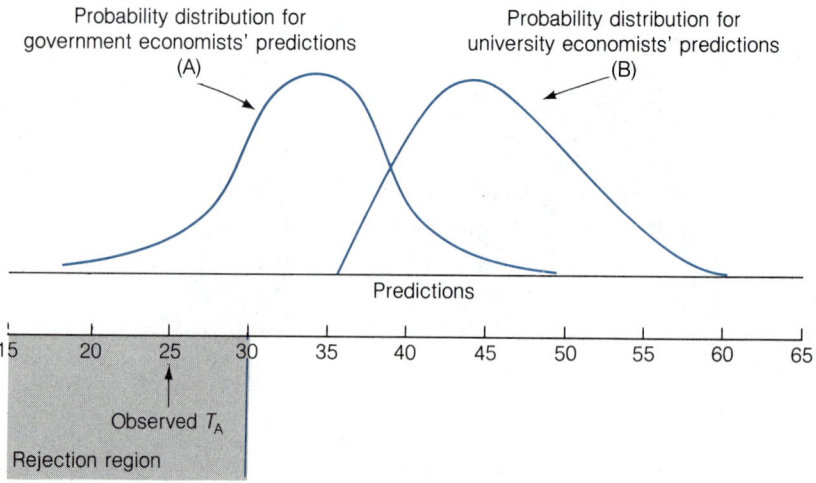

*The alternative hypotheses in this chapter will be stated in terms of a difference in the *location* of the distributions. However, since the shapes of the distributions may also differ under H_a, some of the figures (e.g., Figure 16.2) depicting the alternative hypothesis will show probability distributions with different shapes.

Since T_A, the rank sum of the government economists' predictions in Table 16.1, is 25, it is in the rejection region (see Figure 16.2). Therefore, we can conclude that the university economists' predictions tend, in general, to exceed the government economists' predictions. ∎

Table XI of Appendix B gives values of T_L and T_U for sample sizes n_1 and n_2 less than or equal to 10. When both sample sizes are 10 or larger, the sampling distribution of T_A can be approximated by a normal distribution with mean and variance

$$E(T_A) = \frac{n_1(n_1 + n_2 + 1)}{2} \quad \text{and} \quad \sigma_{T_A}^2 = \frac{n_1 n_2(n_1 + n_2 + 1)}{12}$$

Therefore, for $n_1 \geq 10$ and $n_2 \geq 10$, we can conduct the Wilcoxon rank sum test using the familiar z test of Chapters 8 and 9. The large-sample test is summarized in the box.

WILCOXON RANK SUM TEST: LARGE INDEPENDENT SAMPLES

ONE-TAILED TEST

H_0: Two sampled populations have identical probability distributions

H_a: The probability distribution for population A is shifted to the right of that for B

Test statistic:

$$z = \frac{T_A - \dfrac{n_1(n_1 + n_2 + 1)}{2}}{\sqrt{\dfrac{n_1 n_2(n_1 + n_2 + 1)}{12}}}$$

Rejection region: $z > z_\alpha$

Assumptions: $n_1 \geq 10$ and $n_2 \geq 10$

TWO-TAILED TEST

H_0: Two sampled populations have identical probability distributions

H_a: The probability distribution for population A is shifted to the left *or* to the right of that for B

Test statistic:

$$z = \frac{T_A - \dfrac{n_1(n_1 + n_2 + 1)}{2}}{\sqrt{\dfrac{n_1 n_2(n_1 + n_2 + 1)}{12}}}$$

Rejection region: $z < -z_{\alpha/2}$ or $z > z_{\alpha/2}$

EXERCISES 16.1–16.11   LEARNING THE MECHANICS

16.1 Specify the test statistic and the rejection region for the Wilcoxon rank sum test for independent samples in each of the following situations:
  a. H_0: Two probability distributions, A and B, are identical
   H_a: Probability distribution for population A is shifted to the right or left of the probability distribution for population B
   $n_A = 8$, $n_B = 6$, $\alpha = .10$

b. H_0: Two probability distributions, A and B, are identical

H_a: Probability distribution for population A is shifted to the right of the probability distribution for population B

$n_A = 5, \quad n_B = 6, \quad \alpha = .05$

c. H_0: Two probability distributions, A and B, are identical

H_a: Probability distribution for population A is shifted to the left of the probability distribution for population B

$n_A = 10, \quad n_B = 8, \quad \alpha = .025$

d. H_0: Two probability distributions, A and B, are identical

H_a: Probability distribution for population A is shifted to the right or left of the probability distribution for population B

$n_A = 20, \quad n_B = 20, \quad \alpha = .05$

16.2 Suppose you want to compare two treatments, A and B. In particular, you wish to determine whether the distribution for population B is shifted to the right of the distribution for population A. You plan to use the Wilcoxon rank sum test.

a. Specify the null and alternative hypotheses you would test.

b. Suppose you obtained the following independent random samples of observations on experimental units subjected to the two treatments:

A: 36, 39, 33, 29, 42, 33, 35, 28, 34

B: 35, 48, 52, 66

Conduct a test of the hypotheses described in part **a**. Test using $\alpha = .05$.

16.3 Explain the difference between the one- and two-tailed versions of the Wilcoxon rank sum test for independent random samples.

16.4 Random samples of sizes $n_1 = 20$ and $n_2 = 15$ were drawn from populations 1 and 2, respectively. The measurements obtained are listed in the table.

POPULATION 1				POPULATION 2		
9.0	15.6	25.6	31.1	10.1	11.1	13.5
21.1	26.9	24.6	20.0	12.0	18.2	10.3
24.8	16.5	26.0	25.1	9.2	7.0	14.2
17.2	30.1	18.7	26.1	15.8	13.6	13.2
18.9	25.4	22.0	23.3	8.8	12.5	21.5

a. Conduct a hypothesis test to determine whether the probability distribution for population 2 is shifted to the left of the probability distribution for population 1. Use $\alpha = .05$.

b. What is the approximate p-value of the test of part **a**?

APPLYING THE CONCEPTS

16.5 The property taxes levied by local governments are based on a government tax assessor's judgment of the value of the property in question. Frequently, property owners challenge the tax assessor's judgment. As a result, tax assessors must be able to demonstrate the equity or fairness of their judgments across neighborhoods, property classes, and other property groups. One measure that is used in the evaluation of fairness is the *assessment* *ratio.* The assessment ratio is calculated by dividing a property's assessed value by its market value (or a proxy for market value such as a recent sale price) (Freedman, 1985). Equity issues are evaluated by private real estate appraisers and government agencies using procedures such as the Wilcoxon rank sum test and the Kruskal–Wallis test (described in Section 16.3) to compare the distributions of the assessment ratios for different groups of properties (*Improving Real Property: A Reference Manual*, 1978). The table lists the assessment ratios for random samples of ten properties in neighborhood A and eight properties in neighborhood B.

NEIGHBORHOOD A		NEIGHBORHOOD B	
.850	.880	.911	.835
1.060	.895	.770	.800
.910	.844	.815	.793
.813	.965	.748	.796
.737	.875		

a. Use the Wilcoxon rank sum test to investigate the fairness of the assessments between the two neighborhoods. Use $\alpha = .05$ and interpret your findings in the context of the problem.

b. Under what circumstances could the two-sample t test of Chapter 9 be used to investigate the fairness issue of part a?

c. What assumptions are necessary to assure the validity of the test you conducted in part a?

U.S. PLANTS	JAPANESE PLANTS
7.11%	3.52%
6.06	2.02
8.00	4.91
6.87	3.22
4.77	1.92

16.6 Recall that the variance of a binomial sample proportion, $\hat{p}$, depends on the value of the population parameter, p. As a consequence, the variance of a sample percentage, $(100\hat{p})\%$, will also depend on p. The relevance of this fact is that if you conduct an unpaired t test (Section 9.2) to compare the means of two populations of percentages, you may be violating the assumption that $\sigma_1^2 = \sigma_2^2$, upon which the t test is based. If the disparity in the variances is large, you will obtain more reliable test results using the Wilcoxon rank sum test for independent samples. In Exercise 9.31, we used a Student's t test to compare the mean annual percentages of labor turnover between U.S. and Japanese manufacturers of air conditioners. The annual percentage turnover rates for five U.S. and five Japanese plants are shown in the table. Do the data provide sufficient evidence to indicate that the mean annual percentage turnover for American plants exceeds the corresponding mean for Japanese plants? Test using the Wilcoxon rank sum test with $\alpha = .05$. Do your test conclusions agree with those of the t test in Exercise 9.31?

DAMAGE PER ACCIDENT	
Before right-turn law	After right-turn law
$150	$ 145
500	390
250	680
301	560
242	899
435	1,250
100	290
402	963

16.7 A state highway department has decided to investigate the increased severity of automobile accidents occurring at a particular urban intersection since the adoption of the right-turn-on-red law. From police records, they chose a random sample of eight accidents that occurred at the intersection before the law was enacted and a random sample of eight accidents that occurred after the law was enacted. They used the total damage estimate for each accident as a measure of the accident's severity. The damage estimates are recorded in the table. Use the Wilcoxon rank sum test to determine whether the damages tended to increase after the enactment of the law. Test using $\alpha = .05$. Draw appropriate conclusions.

16.8 A major razor blade manufacturer advertises that its twin-blade disposable razor will "get you a lot more shaves" than any single-blade disposable razor on the market. A rival blade company that has been very successful in selling single-blade razors wishes to test this claim. Independent random samples of eight single-blade shavers and eight twin-blade shavers are taken, and the number of shaves that each gets before indicating a preference to change blades is recorded. The results are shown in the table.

TWIN BLADES		SINGLE BLADES	
8	15	10	13
17	10	6	14
9	6	3	5
11	12	7	7

a. Do the data support the twin-blade manufacturer's claim? Use $\alpha = .05$.

b. Do you think this experiment was designed in the best possible way? If not, what design might have been better?

c. What assumptions are necessary for the validity of the test you performed in part **a**? Do the assumptions seem reasonable for this application?

16.9 A realtor wants to determine whether a difference exists between home prices in two subdivisions. Six homes from subdivision A and eight homes from subdivision B are sampled, and the prices (in thousands of dollars) are recorded in the table.

SUBDIVISION			
A		B	
43	39	57	88
48	47	39	46
42		55	41
60		52	64

a. Use the two-sample t test to compare the population mean prices per house in the two subdivisions. What assumptions are necessary for the validity of this procedure? Do you think they are reasonable in this case?

b. Use the Wilcoxon rank sum test to determine whether there is a shift in the locations of the probability distributions of house prices in the two subdivisions.

16.10 Thirty-six stocks were randomly selected from those listed on the New York Stock Exchange (NYSE), and 36 stocks were randomly selected from those listed on the American Stock Exchange (ASE). The closing prices of all 72 stocks on December 15, 1986, are listed in the table, with the *Wall Street Journal* abbreviation for each firm's name.

a. Use the two-sided Wilcoxon rank sum test (the large-sample procedure) to determine whether the data provide sufficient evidence to indicate a shift in the locations of the distributions of closing prices for the two stock exchanges. Specify your null and alternative hypotheses and interpret the result of your test in the context of the problem. Use $\alpha = .05$.

NYSE				ASE			
Firm	Closing Price	Firm	Closing Price	Firm	Closing Price	Firm	Closing Price
Algln	$14\frac{1}{8}$	Fuqua	$24\frac{1}{4}$	Andal	$8\frac{1}{2}$	IntBknt	5
BCelts	$15\frac{7}{8}$	BwnSh	18	BSD	$2\frac{5}{8}$	BinkMf	$24\frac{1}{8}$
Glenfed	$24\frac{1}{8}$	BellInd	$21\frac{1}{8}$	Fstcrp	$11\frac{1}{4}$	AmIntl	$6\frac{1}{4}$
Newell	28	TranEx	$14\frac{3}{4}$	UnvPat	$14\frac{1}{8}$	FreqE	$21\frac{5}{8}$
CartWl	$77\frac{1}{2}$	CinMil	$19\frac{7}{8}$	Endvco	$5\frac{5}{8}$	Duplex	$17\frac{7}{8}$
HCA	$30\frac{3}{4}$	Radice	8	Telsci	$4\frac{7}{8}$	Ducom	$16\frac{1}{4}$
FstChic	32	Dynlct	$15\frac{1}{2}$	Stepan	$30\frac{3}{4}$	Devlcp	$14\frac{3}{4}$
Becor	$11\frac{5}{8}$	ThmBet	$43\frac{3}{4}$	Zentec Corp	$4\frac{3}{8}$	Baker	$16\frac{3}{4}$
HeclaM	$11\frac{1}{8}$	ACyan	$80\frac{1}{2}$	Nantck	$6\frac{5}{8}$	Damson	$\frac{3}{8}$
Tonka	$20\frac{1}{8}$	Hecks	$11\frac{7}{8}$	Pantast	$10\frac{1}{4}$	CitFst	86
Chmpln	$31\frac{7}{8}$	Berkey	$4\frac{1}{2}$	Pentron	$1\frac{5}{8}$	Unimar	$6\frac{1}{4}$
UtdMM	13	OhMatr	$16\frac{1}{8}$	PropCT	$24\frac{1}{8}$	MMed	$26\frac{1}{2}$
HazLab	23	Purolat	25	VST	$8\frac{1}{2}$	Digicon	$\frac{1}{2}$
Upjohn	95	Wendy	$10\frac{3}{4}$	AmBld	$3\frac{1}{4}$	MSR	$1\frac{1}{2}$
Navistr	$5\frac{3}{8}$	Amern	$30\frac{3}{8}$	Valspar	$41\frac{7}{8}$	Nichols	$7\frac{7}{8}$
Wrld Ar	$4\frac{1}{8}$	KimCl	$83\frac{5}{8}$	WangB	$12\frac{5}{8}$	Gull	$13\frac{1}{4}$
Conrac	$12\frac{3}{4}$	SonyCp	$21\frac{1}{4}$	Hlthch	$9\frac{1}{2}$	Frnkln	$12\frac{3}{4}$
OklaGE	$35\frac{1}{4}$	KeysCo	$6\frac{7}{8}$	Lionel	$5\frac{5}{8}$	GrtLKC	$35\frac{1}{4}$

Source: *Wall Street Journal*, Dec. 16, 1986, pp. 54–56.

b. In the context of the problem, specify the Type I and Type II errors associated with the test of part a.

c. Using the testing procedure of part a, what is the probability of committing a Type I error?

16.11 A *management information system* (MIS) is a computer-based information-processing system designed to support the operations, management, and decision functions of an organization. The development of an MIS involves three stages: definition of the system, physical design of the system, and implementation of the system (Davis and Olson, 1985). Steven Alter and Michael Ginzberg (1978) have shown that the successful implementation of an MIS is related to the quality of the entire development process. The implementation of an MIS could fail due to inadequate planning by and negotiating between the designers and the future users of the system prior to the construction of the system. Or, it could fail simply because members of the organization were improperly trained to use the system effectively.

Thirty firms that recently implemented an MIS were surveyed: sixteen were satisfied with the implementation results, fourteen were not. Each firm was asked to rate the quality of the planning and negotiation stages of the development process. Quality was to be rated on a scale of 0 to 100, with higher numbers indicating better quality. In particular, a score of 100 indicates that all the problems that occurred in the planning and negotiation stages appear to have been successfully resolved, while a score of 0 indicates that none of the problems appear to have been resolved. The results obtained are shown in the table.

FIRMS WITH SUCCESSFUL MIS			FIRMS WITH UNSUCCESSFUL MIS		
52	59	95	60	40	90
70	60	90	50	55	85
40	90	86	55	65	80
80	75	95	70	55	90
82	80	93	41	70	
65					

Source: Based on Alter and Ginzberg (1978).

a. Ginzberg used the Mann–Whitney U test (a procedure equivalent to the Wilcoxon rank sum test) on similar data to investigate differences in the quality of the development processes of successfully and unsuccessfully implemented MIS's. Use the large-sample Wilcoxon rank sum test to determine whether the distribution of quality scores for successfully implemented systems lies above the distribution of scores for unsuccessfully implemented systems. Test using $\alpha = .05$.

b. Under what circumstances could you use the two-sample t test of Chapter 9 to conduct the same test?

| | | | | | | | | | | | | |

SECTION 16.2

COMPARING TWO POPULATIONS: WILCOXON SIGNED RANK TEST FOR THE PAIRED DIFFERENCE EXPERIMENT

Nonparametric techniques can also be used to compare two probability distributions when a paired difference design is used. For example, for some paper products, softness of the paper is an important consideration in determining consumer acceptance. One method of determining softness is to have judges give a sample of the products a softness rating. Suppose each of ten judges is given a sample of two products that a company wants to compare. Each judge rates the softness of each product on a scale from 1 to 10, with higher ratings implying a softer product. The results are shown in Table 16.3.

Since this is a paired difference experiment, we analyze the differences between the measurements (see Section 9.4). However, the nonparametric approach requires that we calculate the ranks of the absolute values of the differences between the measurements—i.e., the ranks of the differences after removing any minus signs. Note that tied absolute differences are assigned the average of the ranks they would receive if they were unequal but successive measurements. After the absolute differences are ranked, the sum of the ranks of the positive differences, T_+, and the sum of the ranks of the negative differences, T_-, are computed.

TABLE 16.3
Paper Softness Ratings

JUDGE	PRODUCT A	B	DIFFERENCE (A − B)	ABSOLUTE VALUE OF DIFFERENCE	RANK OF ABSOLUTE VALUE
1	6	4	2	2	5
2	8	5	3	3	7.5
3	4	5	−1	1	2
4	9	8	1	1	2
5	4	1	3	3	7.5
6	7	9	−2	2	5
7	6	2	4	4	9
8	5	3	2	2	5
9	6	7	−1	1	2
10	8	2	6	6	10

$$T_+ = \text{Sum of positive ranks} = 46$$
$$T_- = \text{Sum of negative ranks} = 9$$

We are now prepared to test the nonparametric hypotheses:

H_0: The probability distributions of the ratings for products A and B are identical

H_a: The probability distribution for product A is shifted to the right or left of the probability distribution of the ratings for product B

Test statistic: $T = $ Smaller of the positive and negative rank sums T_+ and T_-

The smaller the value of T, the greater will be the evidence to indicate that the two probability distributions differ in location. The rejection region for T can be determined by consulting Table XII of Appendix B. A portion of that table is shown in Table 16.4 (page 958). This table gives a value, T_0, for each value of n, the number of matched pairs. The values of T_0 are tabulated for both a one- and a two-tailed test. For a two-tailed test with $\alpha = .05$, we will reject H_0 if $T \leq T_0$. The T_0 value that locates the rejection region for the judges' ratings in Table 16.3 is the value indicated for $n = 10$ pairs of observations. This value of T_0 is 8. Therefore, the rejection region for the test (see Figure 16.3) is

Rejection region: $T \leq 8$ for $\alpha = .05$

FIGURE 16.3
Rejection Region for
Paired Difference
Experiment

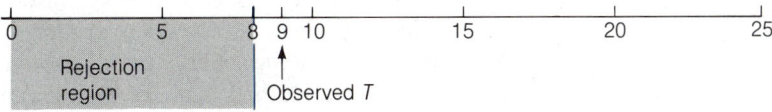

Since the smaller rank sum for the paper data, $T = T_- = 9$, does not fall within the rejection region, the experiment has not provided sufficient evidence to indicate that the two paper products differ with respect to their softness ratings at $\alpha = .05$.

TABLE 16.4

Reproduction of Part of Table XII of Appendix B

One-tailed	Two-tailed	$n = 5$	$n = 6$	$n = 7$	$n = 8$	$n = 9$	$n = 10$
$\alpha = .05$	$\alpha = .10$	1	2	4	6	8	11
$\alpha = .025$	$\alpha = .05$		1	2	4	6	8
$\alpha = .01$	$\alpha = .02$			0	2	3	5
$\alpha = .005$	$\alpha = .01$				0	2	3
		$n = 11$	$n = 12$	$n = 13$	$n = 14$	$n = 15$	$n = 16$
$\alpha = .05$	$\alpha = .10$	14	17	21	26	30	36
$\alpha = .025$	$\alpha = .05$	11	14	17	21	25	30
$\alpha = .01$	$\alpha = .02$	7	10	13	16	20	24
$\alpha = .005$	$\alpha = .01$	5	7	10	13	16	19
		$n = 17$	$n = 18$	$n = 19$	$n = 20$	$n = 21$	$n = 22$
$\alpha = .05$	$\alpha = .10$	41	47	54	60	68	75
$\alpha = .025$	$\alpha = .05$	35	40	46	52	59	66
$\alpha = .01$	$\alpha = .02$	28	33	38	43	49	56
$\alpha = .005$	$\alpha = .01$	23	28	32	37	43	49
		$n = 23$	$n = 24$	$n = 25$	$n = 26$	$n = 27$	$n = 28$
$\alpha = .05$	$\alpha = .10$	83	92	101	110	120	130
$\alpha = .025$	$\alpha = .05$	73	81	90	98	107	117
$\alpha = .01$	$\alpha = .02$	62	69	77	85	93	102
$\alpha = .005$	$\alpha = .01$	55	61	68	76	84	92

Note that, if a significance level of $\alpha = .10$ had been used, the rejection region would have been $T \leq 11$, and we would have rejected H_0. In other words, the samples do provide evidence that the probability distributions of the softness ratings differ at $\alpha = .10$.

The Wilcoxon signed rank test is summarized in the next box. Note that the difference measurements are assumed to have a continuous probability distribution so that the absolute differences will have unique ranks. Although tied (absolute) differences can be assigned ranks by averaging, the number of ties should be small relative to the number of observations to assure the validity of the test.

EXAMPLE 16.2

Suppose the U.S. Consumer Product Safety Commission (CPSC) wants to test the hypothesis that New York City electrical contractors are more likely to install unsafe electrical outlets in urban homes than in suburban homes. A pair of homes, one urban and one suburban and both serviced by the same electrical contractor, is chosen for each of ten randomly selected electrical contractors. A commission inspector assigns each of the 20 homes a safety rating between 1 and 10, with higher numbers implying safer electrical conditions. The results are shown in Table 16.5. Use the Wilcoxon signed rank test to determine whether the CPSC hypothesis is supported at the $\alpha = .05$ level.

WILCOXON SIGNED RANK TEST FOR A PAIRED DIFFERENCE EXPERIMENT

ONE-TAILED TEST

H_0: Two sampled populations have identical probability distributions

H_a: The probability distribution for population A is shifted to the right of that for population B

Test statistic: T_-, the negative rank sum (we assume the differences are computed by subtracting each paired B measurement from the corresponding A measurement)

Rejection region: $T_- \leq T_0$, where T_0 is found in Table XII in Appendix B for the one-tailed significance level α and the number of untied pairs, n.

TWO-TAILED TEST

H_0: Two sampled populations have identical probability distributions

H_a: The probability distribution for population A is shifted to the right *or* to the left of that for population B

Test statistic: T, the smaller of the positive and negative rank sums, T_+ and T_-

Rejection region: $T \leq T_0$, where T_0 is found in Table XII in Appendix B for the two-tailed significance level α and the number of untied pairs, n.

[*Note:* If the alternative hypothesis is that the probability distribution for A is shifted to the left of B, we use T_+ as the test statistic and reject H_0 if $T_+ \leq T_0$.]

Assumptions: 1. The sample of differences is randomly selected from the population of differences.
2. The probability distribution from which the sample of paired differences is drawn is continuous.

Ties: Assign tied absolute differences the average of the ranks they would receive if they were unequal but occurred in successive order. For example, if the third-ranked and fourth-ranked differences are tied, assign both a rank of $(3 + 4)/2 = 3.5$.

TABLE 16.5

Electrical Safety Ratings for Ten Pairs of New York City Homes

ELECTRICAL CONTRACTOR	URBAN HOME	SUBURBAN HOME	ELECTRICAL CONTRACTOR	URBAN HOME	SUBURBAN HOME
1	7	9	6	6	10
2	4	5	7	8	9
3	8	8	8	10	8
4	9	8	9	9	4
5	3	6	10	5	9

SOLUTION

The null and alternative hypotheses are

H_0: The probability distributions of home electrical ratings are identical for urban and suburban homes

H_a: The electrical ratings for suburban homes tend to exceed the electrical ratings for urban homes

These hypotheses can be tested for the data in Table 16.5 using the Wilcoxon signed rank test. Since a paired difference design was used (the homes were selected in urban–suburban pairs so that the electrical contractor was the same for both), we first calculate the difference between the ratings for each pair of homes, and then rank the absolute values of the differences (see Table 16.6). Note that one pair of ratings were the same (both 8), and the resulting 0 difference contributes to neither the positive nor the negative rank sum. Thus, we eliminate this pair from the calculation of the test statistic.

Test statistic: T_+, the positive rank sum

TABLE 16.6

Differences in Ratings and the Ranks of Their Absolute Values

RATING		DIFFERENCE	RANK OF ABSOLUTE DIFFERENCE
Urban	Suburban	(Urban − Suburban)	
7	9	−2	4.5
4	5	−1	2
8	8	0	(Eliminated)
9	8	1	2
3	6	−3	6
6	10	−4	7.5
8	9	−1	2
10	8	2	4.5
9	4	5	9
5	9	−4	7.5
		Positive rank sum = T_+ = 15.5	

In Table 16.6, we compute the urban minus suburban rating differences, and if the alternative hypothesis is true, we would expect most of these differences to be negative. Or, in other words, we would expect the *positive* rank sum T_+ to be small if the alternative hypothesis is true (see Figure 16.4).

FIGURE 16.4

The Alternative Hypothesis for Example 16.2; We Expect T_+ to Be Small

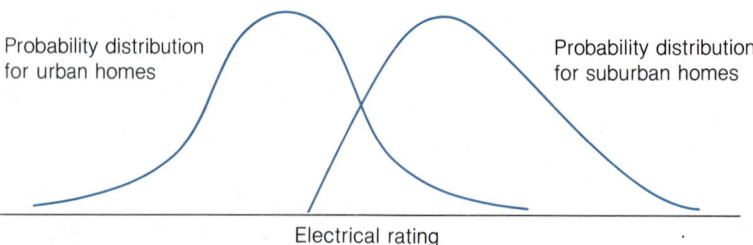

Probability distribution for urban homes   Probability distribution for suburban homes

Electrical rating

Rejection region: For $\alpha = .05$, from Table XII of Appendix B, we use $n = 9$ (remember, one pair of observations was eliminated) to find the rejection region for this one-tailed test:

$T_+ \le 8$

Since the computed value $T_+ = 15.5$ exceeds the critical value of 8, we conclude that this sample provides insufficient evidence at $\alpha = .05$ to support the alternative hypothesis. We *cannot* conclude on the basis of this sample information that suburban homes have safer electrical outlets than urban homes. ∎

As in the case for the rank sum test for independent samples, the sampling distribution of the signed rank statistic can be approximated by a normal distribution when the number n of paired observations is large (say $n \geq 25$). The large-sample z test is summarized in the box.

WILCOXON SIGNED RANK TEST FOR A PAIRED DIFFERENCE EXPERIMENT: LARGE SAMPLE

ONE-TAILED TEST

H_0: Two sampled populations have identical probability distributions

H_a: The probability distribution for population A is shifted to the right of that for population B

Test statistic:

$$z = \frac{T_+ - \dfrac{n(n+1)}{4}}{\sqrt{\dfrac{n(n+1)(2n+1)}{24}}}$$

Rejection region: $z > z_\alpha$

Assumptions: $n \geq 25$

TWO-TAILED TEST

H_0: Two sampled populations have identical probability distributions

H_a: The probability distribution for population A is shifted to the right *or* to the left of that for population B

Test statistic:

$$z = \frac{T_+ - \dfrac{n(n+1)}{4}}{\sqrt{\dfrac{n(n+1)(2n+1)}{24}}}$$

Rejection region: $z < -z_{\alpha/2}$ or $z > z_{\alpha/2}$

EXERCISES
16.12–16.24

LEARNING THE MECHANICS

16.12 Specify the test statistic and the rejection region for the Wilcoxon signed rank test for the paired difference design in each of the following situations:

a. H_0: Two probability distributions, A and B, are identical
H_a: Probability distribution for population A is shifted to the right or left of the probability distribution for population B
$n = 25$, $\alpha = .10$

b. H_0: Two probability distributions, A and B, are identical
H_a: Probability distribution for population A is shifted to the right of the probability distribution for population B
$n = 41$, $\alpha = .05$

c. H_0: Two probability distributions, A and B, are identical
H_a: Probability distribution for population A is shifted to the left of the probability distribution for population B
$n = 8$, $\alpha = .005$

16.13 Suppose you want to test a hypothesis that two treatments, A and B, are equivalent against the alternative hypothesis that the responses for A tend to be larger than those for B. You plan to use a paired difference experiment and to analyze the resulting data using the Wilcoxon signed rank test.

a. Specify the null and alternative hypotheses you would test.
b. Suppose the paired difference experiment yielded the data in the table. Conduct the test of part **a** using $\alpha = .025$.

PAIR OF EXPERIMENTAL UNITS	TREATMENT A	B	PAIR OF EXPERIMENTAL UNITS	TREATMENT A	B
1	56	42	6	76	75
2	62	45	7	74	63
3	98	87	8	29	30
4	45	31	9	63	59
5	82	71	10	80	82

16.14 Explain the difference between the one- and two-tailed versions of the Wilcoxon signed rank test for the paired difference experiment.

16.15 In order to conduct the Wilcoxon signed rank test, why do we need to assume that the probability distribution of differences is continuous?

16.16 A paired difference experiment with $n = 30$ pairs yielded $T_+ = 354$.
a. Specify the null and alternative hypotheses that should be used in conducting a hypothesis test to determine whether the probability distribution for population A is located to the right of that for population B.
b. Conduct the test of part **a** using $\alpha = .05$.
c. What is the approximate p-value of the test of part **b**?
d. What assumptions are necessary to assure the validity of the test you performed in part **b**?

APPLYING THE CONCEPTS

16.17 Which is the more effective means of dealing with complex group problem-solving tasks: face-to-face meetings or video teleconferencing? In an experiment similar to the one described in this exercise, Daniel K. Rosetti and Theodore J. Surynt of Stetson University concluded that video teleconferencing may be the more effective method. Ten groups of four people each were randomly assigned both to a specific communication setting (face-to-face or video teleconferencing) and to one of two specific complex problems. Upon completion of the problem-solving task, the same groups were placed in the alternative communication setting and asked to complete the second problem-solving task. The percentage of each problem task correctly completed was recorded for each group, with the results given in the accompanying table.

GROUP	FACE-TO-FACE	VIDEO TELECONFERENCING	GROUP	FACE-TO-FACE	VIDEO TELECONFERENCING
1	65%	75%	6	85%	90%
2	82	80	7	98	98
3	54	60	8	35	40
4	69	65	9	85	89
5	40	55	10	70	80

a. What type of experimental design was used in this study?
b. Specify the null and alternative hypotheses that should be used in determining whether the data provide sufficient evidence to conclude (as did Rosetti and Surynt) that the problem-solving performance of video teleconferencing groups is superior to that of groups that interact face-to-face.
c. Conduct the hypothesis test of part b. Use $\alpha = .05$. Interpret the results of your test in the context of the problem.
d. What is the p-value of the test in part c?

16.18 According to the American Bar Association, in 1982 there were 612,593 lawyers in the United States. About 70% of these lawyers were in private practice; about 15% worked in government as judges, prosecutors, legislators, etc.; and about 9% worked for businesses. Because of mushrooming government regulation, high outside legal fees, and complex litigation, the number of corporate lawyers has been growing at a rapid pace. The data shown in the table are the average salaries for lawyers with 8 years experience for a sample of ten U.S. cities.

CITY	CORPORATE LAWYERS	LAWYERS WITH LAW FIRMS
Atlanta	$45,500	$45,500
Chicago	43,000	48,000
Cincinnati	43,500	45,000
Dallas/Ft. Worth	49,500	46,500
Los Angeles	47,000	60,000
Milwaukee	37,500	50,000
Minneapolis/St. Paul	47,500	43,500
New York	43,500	54,000
Pittsburgh	42,000	44,000
San Francisco	47,500	59,500

Source: *The American Almanac of Jobs and Salaries,* 1984, pp. 365–374.

a. Use the Wilcoxon signed rank test to determine whether the data provide sufficient evidence to conclude that the salaries of corporate lawyers differ from those of lawyers working for law firms. Test using $\alpha = .05$.
b. Under what circumstances would it be appropriate to conduct the test in part a using the paired difference t test described in Chapter 9?

16.19 Traditionally, jobs in the United States have required employees to perform their work during a fixed 8-hour workday. A recent job-scheduling innovation that is helping managers to overcome the motivation and absenteeism problems associated with the fixed workday is a concept called *flexitime*. Flexitime is a flexible working hours program that permits employees to design their own 40-hour work week (Certo, 1980). The management of a large manufacturing firm is considering adopting a flexitime program for its hourly employees and has decided to base the decision on the success or failure of a pilot flexitime program. Ten employees were randomly selected and given a questionnaire designed to measure their attitude toward their job. These same people were then permitted to design and follow a flexitime workday. After 6 months, attitudes toward their jobs were again measured. The resulting attitude scores are displayed in the table at the top of page 964. The higher the score, the more favorable is the employee's attitude toward his or her work. Use a nonparametric test procedure to evaluate the success of the pilot flexitime program. Test using $\alpha = .05$.

EMPLOYEE	BEFORE FLEXITIME	AFTER FLEXITIME	EMPLOYEE	BEFORE FLEXITIME	AFTER FLEXITIME
1	54	68	6	82	88
2	25	42	7	94	90
3	80	80	8	72	81
4	76	91	9	33	39
5	63	70	10	90	93

WEEK	1986	1987
2	6	8
8	3	5
22	2	2
35	1	0
48	4	7
51	7	10

16.20 Refer to Exercise 16.7, in which a state highway department was interested in investigating the effects of the right-turn-on-red law at a particular urban intersection. The severity of the accidents that occur at the intersection was examined in Exercise 16.7. The highway department also wants to compare the probability distribution of the number of accidents that occurred per week before the law with the probability distribution of the number of accidents per week after the law. They randomly selected 6 weeks of the year and looked up the number of accidents that occurred in each of those weeks in the year preceding the adoption of the law (1986) and in the year following the adoption of the law (1987). The data are displayed in the table.

a. Use the Wilcoxon signed rank test for a paired difference experiment to investigate whether the probability distribution of the number of accidents per week after the law is located above (i.e., to the right of) the probability distribution of the number of accidents before the law.

b. Explain why a paired difference design was utilized in this study.

16.21 In Exercise 9.56, a paired difference test was used to compare the mean number of antitrust litigations per firm in the 1960's with the mean number per firm in the 1970's. The data are repeated in the table. Beckenstein, Gabel, and Roberts (1983) claim that, on average, companies faced more antitrust litigations in the 1970's than in the 1960's. Do the data support their claim? Test using the Wilcoxon signed rank test with $\alpha = .05$. Be sure to specify your null and alternative hypotheses.

FIRM	NUMBER OF LITIGATIONS 1960's	1970's	FIRM	NUMBER OF LITIGATIONS 1960's	1970's
1	10	10	6	7	6
2	8	12	7	6	11
3	9	8	8	9	12
4	7	16	9	8	11
5	8	14	10	7	12

LOCATION	A	B
1	879	1,085
2	445	325
3	692	848
4	1,565	1,421
5	2,326	2,778
6	857	992
7	1,250	1,303
8	773	1,215

16.22 A manufacturer of household appliances is considering one of two chains of department stores to be the sales merchandiser for its product in a particular region of the United States. Before choosing one chain, the manufacturer wants to make a comparison of the product exposure that might be expected for the two chains. Eight locations are selected where both chains have stores, and on a specific day, the number of shoppers entering each store is recorded. The data are shown in the table. Is there sufficient evidence to indicate that one of the chains tends to have more customers per day than the other? Test using $\alpha = .05$.

WEEK	MACHINE TYPE	
	A	B
1	14	12
2	17	13
3	10	14
4	15	12
5	14	9
6	9	11
7	12	11

16.23 A food vending company currently uses vending machines made by two different manufacturers. Before purchasing new machines, the company wants to compare the two types in terms of reliability. Records for 7 weeks are given in the table; the data indicate the number of breakdowns per week for each type of machine. The company has the same number of machines of each type. Is the probability distribution of the number of breakdowns for machine A shifted to the left or to the right of the probability distribution of the number of breakdowns for machine B? Use $\alpha = .05$.

16.24 Economic indexes provide measures of economic change. The table lists the producer commodity price indexes for January 1985 and January 1986, for six product categories. By comparing these two sets of indexes, you can obtain information regarding changes in the economy that occurred during 1985.

PRODUCT CATEGORY	JANUARY 1985	JANUARY 1986
Processed poultry	198.8	192.4
Concrete ingredients	331.0	339.0
Lumber	343.0	329.6
Gas fuels	1,073.0	1,034.3
Drugs and pharmaceuticals	247.4	265.9
Synthetic fibers	157.6	151.1

Source: *Standard & Poor's Statistical Service, Current Statistics,* Jan. 1987, pp. 12–13.

a. Conduct a paired difference *t* test to compare the mean values of these indexes for January 1985 and January 1986. Use $\alpha = .05$. What assumptions are necessary for the validity of this procedure? Why might these assumptions be in doubt?

b. Use the Wilcoxon signed rank test to determine whether the data provide evidence that the probability distribution of the economic indexes has changed. Use $\alpha = .05$. What assumptions are necessary to assure the validity of this test?

SECTION 16.3

KRUSKAL–WALLIS *H* TEST FOR A COMPLETELY RANDOMIZED DESIGN

Recall that a completely randomized design is one in which *independent* random samples are selected from each *p* populations (treatments) to be compared (Section 15.2). In Chapter 15, we used an analysis of variance and the *F* test to compare the means of the *p* populations (assuming the populations have normal probability distributions with equal variances). We now present a nonparametric technique that requires no assumptions concerning the population probability distributions to compare the *p* populations.

For example, suppose you want to compare the numbers of employees in companies representing each of three different business classifications: agriculture, manufacturing, and service. You sample ten companies from each type and record the number of employees in each sampled business (see Table 16.7 on page 966). You can see that the assumptions necessary for a parametric comparison of the means are doubtful for these data; the probability distributions are very likely to be skewed to the right, as indicated by the presence of some extremely large values. In addition, the variability in number of employees may not be constant for the different classifications. We therefore base our comparison on the rank sums for the classifica-

TABLE 16.7

Number of Employees in
Thirty Different Companies

AGRICULTURE	Rank	MANUFACTURING	Rank	SERVICE	Rank
10	5	244	25	17	9.5
350	27	93	19	249	26
4	2	3,532	30	38	15
26	13	17	9.5	5	3
15	8	526	29	101	20
106	21	133	22	1	1
18	11	14	7	12	6
23	12	192	23	233	24
62	17	443	28	31	14
8	4	69	18	39	16
$R_1 = 120$		$R_2 = 210.5$		$R_3 = 134.5$	

tions. The ranks are computed for each observation according to the relative magnitude of the measurements *when all p samples are combined* (see Table 16.7). Note that ties are handled in the usual manner, by assigning the average value of the ranks to each of the tied observations.

We test

H_0: All three populations have identical probability distributions

H_a: At least two of the three population probability distributions differ in location

If we denote the three sample rank sums by R_1, R_2, and R_3, the **test statistic** is given by

$$H = \frac{12}{n(n+1)} \sum_{j=1}^{p} \frac{R_j^2}{n_j} - 3(n+1)$$

where n_j is the number of measurements in the jth sample and n is the **total sample size** ($n = n_1 + n_2 + \cdots + n_p$). For the data in Table 16.7, we have $n_1 = n_2 = n_3 = 10$, and $n = 30$. The rank sums are $R_1 = 120$, $R_2 = 210.5$, and $R_3 = 134.5$. Thus,

$$H = \frac{12}{30(31)} \left[\frac{(120)^2}{10} + \frac{(210.5)^2}{10} + \frac{(134.5)^2}{10} \right] - 3(31)$$
$$= 99.097 - 93 = 6.097$$

The H statistic measures the extent to which the p samples differ with respect to their relative ranks. This is more easily seen by writing H in an alternative but equivalent form:

$$H = \frac{12}{n(n+1)} \sum_{j=1}^{p} n_j (\bar{R}_j - \bar{R})^2$$

where $\bar{R}_j$ is the mean rank corresponding to sample j, and $\bar{R}$ is the mean of all the ranks [i.e., $\bar{R} = (n+1)/2$]. Thus, the H statistic is 0 if all samples have the same mean rank, and becomes increasingly large as the distance between the sample mean ranks grows.

If the null hypothesis is true, the distribution of H in repeated sampling is approximately a χ^2 (**chi-square**) **distribution**. This approximation for the sampling distribution of H is adequate as long as each of the p sample sizes exceeds 5 (see the references for more detail). The χ^2 probability distribution is characterized by a single parameter, called the **degrees of freedom associated with the distribution**. Several χ^2 probability distributions with different degrees of freedom are shown in Figure 16.5. The degrees of freedom corresponding to the approximate sampling distribution of H will always be $(p - 1)$, one less than the number of probability distributions being compared. Because large values of H support the alternative hypothesis that at least two of the $(p - 1)$ population probability distributions differ in location, the rejection region for the test will be located in the upper tail of the χ^2 distribution, as shown in Figure 16.6.

FIGURE 16.5
Several χ^2 Probability Distributions

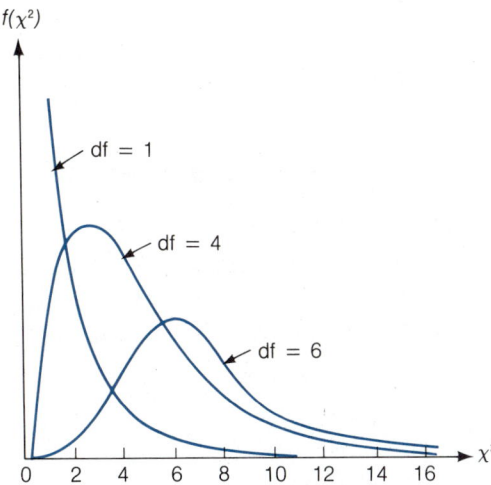

FIGURE 16.6
Rejection Region for the Comparison of Three Probability Distributions

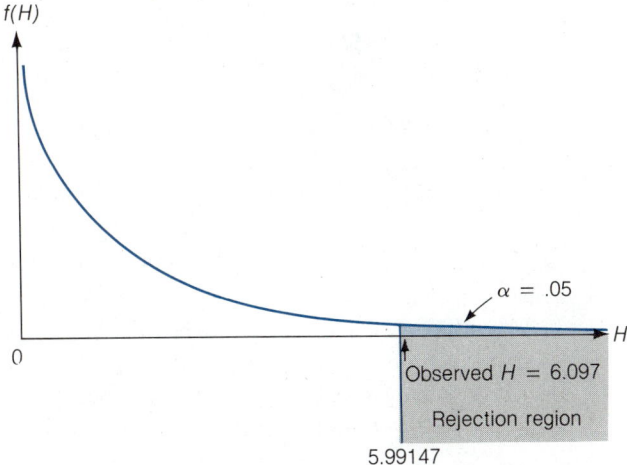

For the data of Table 16.7, the approximate distribution of the test statistic H is a χ^2 distribution with $(p - 1) = 2$ df. To determine how large H must be before we will reject the null hypothesis, we consult Table XIII in Appendix B; part of this table is shown in Table 16.8. Entries in the table give an upper-tail value of χ^2, call it χ_α^2, such that $P(\chi^2 > \chi_\alpha^2) = \alpha$. The columns of the table identify the value of α associated with the tabulated value of χ_α^2, and the rows correspond to the degrees of freedom. Thus, for $\alpha = .05$ and df $= 2$, we can reject the null hypothesis that the three probability distributions are the same if

$$H > \chi_{.05}^2 \quad \text{where } \chi_{.05}^2 = 5.99147$$

The rejection region is pictured in Figure 16.6. Since the calculated $H = 6.097$ exceeds the critical value of 5.99147, we conclude that at least two of the three probability distributions describing the number of employees for the three sampled business types differ in location.

The Kruskal–Wallis H test for comparing more than two probability distributions is summarized in the box. Note that we can use the Wilcoxon rank sum test to compare the separate pairs of populations if the Kruskal–Wallis H test supports the alternative hypothesis that at least two of the probability distributions differ.

KRUSKAL–WALLIS H TEST FOR COMPARING p PROBABILITY DISTRIBUTIONS

H_0: The p probability distributions are identical

H_a: At least two of the p probability distributions differ in location

Test statistic: $H = \dfrac{12}{n(n + 1)} \displaystyle\sum_{j=1}^{p} \dfrac{R_j^2}{n_j} - 3(n + 1)$

where

n_j = Number of measurements in sample j

R_j = Rank sum for sample j, where the rank of each measurement is computed according to its relative magnitude in the totality of data for the p samples

n = Total sample size = $n_1 + n_2 + \cdots + n_p$

Rejection region: $H > \chi_\alpha^2$ with $(p - 1)$ df

Assumptions: 1. The p samples are random and independent.
2. There are 5 or more measurements in each sample.
3. The p probability distributions from which the samples are drawn are continuous.

Ties: Assign tied measurements the average of the ranks they would receive if they were unequal but occurred in successive order. For example, if the third-ranked and fourth-ranked measurements are tied, assign both a rank of $(3 + 4)/2 = 3.5$. The number of ties should be small relative to the total number of observations.

TABLE 16.8

Reproduction of Part of
Table XIII of Appendix B

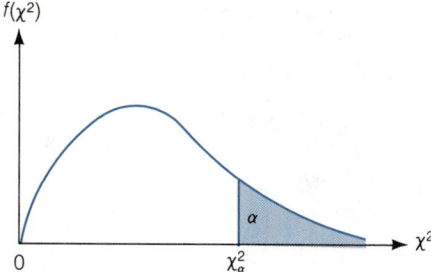

DEGREES OF FREEDOM	$\chi^2_{.100}$	$\chi^2_{.050}$	$\chi^2_{.025}$	$\chi^2_{.010}$	$\chi^2_{.005}$
1	2.70554	3.84146	5.02389	6.63490	7.87944
2	4.60517	5.99147	7.37776	9.21034	10.5966
3	6.25139	7.81473	9.34840	11.3449	12.8381
4	7.77944	9.48773	11.1433	13.2767	14.8602
5	9.23635	11.0705	12.8325	15.0863	16.7496
6	10.6446	12.5916	14.4494	16.8119	18.5476
7	12.0170	14.0671	16.0128	18.4753	20.2777
8	13.3616	15.5073	17.5346	20.0902	21.9550
9	14.6837	16.9190	19.0228	21.6660	23.5893
10	15.9871	18.3070	20.4831	23.2093	25.1882
11	17.2750	19.6751	21.9200	24.7250	26.7569

**EXERCISES
16.25–16.35**

LEARNING THE MECHANICS

16.25 Use Table XIII in Appendix B to find each of the following χ^2 values:
  a. $\chi^2_{.05}$, df = 20   **b.** $\chi^2_{.025}$, df = 15   **c.** $\chi^2_{.01}$, df = 36
  d. $\chi^2_{.10}$, df = 80   **e.** $\chi^2_{.05}$, df = 2   **f.** $\chi^2_{.005}$, df = 10

16.26 Use Table XIII in Appendix B to find each of the following probabilities:
  a. $P(\chi^2 \geq 3.07382)$ where df = 12
  b. $P(\chi^2 \leq 24.4331)$ where df = 40
  c. $P(\chi^2 \geq 14.6837)$ where df = 9
  d. $P(\chi^2 < 34.1696)$ where df = 20
  e. $P(\chi^2 < 6.26214)$ where df = 15
  f. $P(\chi^2 \leq .584375)$ where df = 3

16.27 Data were collected from three populations, A, B, and C, using a completely randomized design. The following describes the sample data:

$$n_A = n_B = n_C = 15$$
$$R_A = 235 \qquad R_B = 439 \qquad R_C = 361$$

  a. Specify the null and alternative hypotheses that should be used in conducting a test of hypothesis to determine whether the probability distributions of populations A, B, and C differ in location.
  b. Conduct the test of part **a**. Use $\alpha = .05$.
  c. What is the approximate *p*-value of the test of part **b**?

d. Calculate the mean rank for each sample and compute H according to the formula on page 966 that utilizes these means. Verify that this formula yields the same value of H that you obtained in part **b**.

16.28 Suppose you want to use the Kruskal–Wallis H test to compare the probability distributions of three populations. The following are independent random samples selected from the three populations:

I: 66 33   55 88 58 62 69 49
II: 22 31   16 25 30 33 40
III: 75 96 102 75 88 78

a. What type of experimental design was used?
b. Specify the null and alternative hypotheses you would test.
c. Specify the rejection region that would be used for your hypothesis test at $\alpha = .01$.
d. Conduct the test using $\alpha = .01$.

16.29 Under what circumstances does the χ^2 distribution provide an appropriate characterization of the sampling distribution of the Kruskal–Wallis H statistic?

APPLYING THE CONCEPTS

16.30 In choosing a mutual fund, an investor should compare his or her personal investment goals with those of the mutual fund. Among the hundreds of available mutual funds, three of the more prevalent advertised goals are income, growth, and maximum growth. *Income funds* seek to maximize current income; *growth funds* seek long-term capital appreciation, with current income a secondary goal; and *maximum growth funds* seek greater capital appreciation by taking larger risks. The table lists the total rate of return to investors for samples of seven mutual funds in each of these three categories.

INCOME		GROWTH		MAXIMUM GROWTH	
Mutual Fund	Rate of Return	Mutual Fund	Rate of Return	Mutual Fund	Rate of Return
Am. Nat. Income	8.3%	Babson Growth	20.2%	Fairfield	2.2%
Bull & Bear		Oppenheimer	1.9	44 Wall St.	
Equity–Income	19.0	Midamerica		Equity	16.9
Oppenheimer		Mutual	11.2	IDS Strategy–	
Equity–Income	15.6	Investment		Aggressive	
T. Rowe Price		Portfolios–		Equity	23.3
Equity–Income	26.8	Equity	4.5	Omega	12.1
Seligman Income	17.1	Franklin		Pilot	10.4
Safeco Income	20.1	Equity	19.3	Strong	
National Stock	14.9	Fidelity		Opportunity	59.9
		Contrafund	13.3	Lowry Market	
		Thomson		Timing	−9.3
		McKinnon			
		Growth	23.1		

Source: "Mutual fund scoreboard," *Business Week*, Feb. 23, 1987, pp. 70–103.

a. Do the data provide sufficient evidence to conclude that the rate-of-return distributions differ among the three types of mutual funds? Test using $\alpha = .05$.

b. What assumptions must hold for your test of part **a** to be valid?

c. Describe the Type I and Type II errors associated with the test of part **a** in the context of the problem.

d. Under what circumstances could the *F* test of Section 15.2 be employed to address the question of part **a**?

16.31 Random samples of seven lawyers employed by corporations were selected from each of three major cities. Their salaries were determined and are recorded in the table. You have been hired to determine whether differences exist among the salary distributions for corporate lawyers in the three cities.

ATLANTA	LOS ANGELES	WASHINGTON, D.C.
$45,500	$52,000	$41,500
47,900	72,000	40,100
43,100	41,000	39,000
42,000	54,000	56,500
49,000	33,000	37,000
52,000	42,000	49,000
39,000	50,000	43,500

Source: Based on *The American Almanac of Jobs and Salaries*, 1984, pp. 365–374.

a. Under what circumstances would it be appropriate to use the *F* test for a completely randomized design to perform the required analysis?

b. Which assumptions required by the *F* test are likely to be violated in this problem? Explain.

c. Use the Kruskal–Wallis *H* test to determine whether the salary distributions differ among the three cities. Specify your null and alternative hypotheses, and state your conclusions in the context of the problem. Use $\alpha = .05$. What assumptions are necessary to assure the validity of the nonparametric test?

16.32 CRS, a national car rental company, was interested in comparing the quality of service at its three largest airport locations. Six business travelers who frequently rent from a competitor were randomly selected at each airport and asked to use (free of charge) a CRS car the next time they rented a car at the airport. The travelers were asked to rate the service they received in four categories: (1) timeliness of service, (2) friendliness of employees, (3) cleanliness of the car, and (4) mechanical performance of the car. They were to use a rating scale of 1 to 10 for each category, with higher numbers indicating better service. An average score was computed for each traveler. These averages are reported in the table.

NEW YORK Kennedy	CHICAGO O'Hare	DALLAS Dallas–Fort Worth
7.50	7.25	9.25
6.25	6.75	8.75
7.50	5.75	9.00
4.75	8.00	9.25
6.25	5.75	8.75
7.00	6.00	8.75

URBAN	SUBURBAN	RURAL
4.3	5.9	5.1
5.2	6.7	4.8
6.2	7.6	3.9
5.6	4.9	6.2
3.8	5.2	4.2
5.8	6.8	4.3
4.7		

BRAND		
A	B	C
36	49	71
48	33	31
5	60	140
67	2	59
53	55	42

a. Use the Kruskal–Wallis H test to determine whether the level of service ratings differ among the three car rental outlets. Use $\alpha = .10$.

b. What experimental design was used by CRS?

16.33 An economist is interested in knowing whether property tax rates differ among three types of school districts—urban, suburban, and rural. A random sample of several districts of each type produced the data in the table (rate is in mills, where 1 mill = \$1/1,000). Do the data indicate a difference in the level of property taxes among the three types of school districts? Use $\alpha = .05$.

16.34 Three different brands of magnetron tubes (the key components in microwave ovens) were subjected to stressful testing, and the number of hours each operated without repair was recorded. Although these times do not represent typical lifetimes, they do indicate how well the tubes can withstand extreme stress.

a. Use the F test for a completely randomized design (Chapter 15) to test the hypothesis that the mean length of life under stress is the same for the three brands. Use $\alpha = .05$. What assumptions are necessary for the validity of this procedure? Is there any reason to doubt these assumptions?

b. Use the Kruskal–Wallis H test to determine whether evidence exists to conclude that at least two of the probability distributions of length of life under stress differ in location. Use $\alpha = .05$.

16.35 In Exercise 12.54 regression analysis was applied to investigate whether the mean debt-to-equity ratio varies among the insurance, publishing, electric utilities, and banking industries. In Exercise 15.24 an analysis of variance F test was used to perform the same investigation. The data are repeated in the table.

INSURANCE		PUBLISHING	
Firm	Debt-to-equity	Firm	Debt-to-equity
Chubb	.24	Deluxe Check	.03
Kemper	.09	New York Times	.32
St. Paul Cos.	.12	Times Mirror	.51
Lincoln National	.09	Dow Jones	.15
USF & G	.00	Gannett	.73
Aetna Life & Cas.	.07		

ELECTRIC UTILITIES		BANKING	
Firm	Debt-to-equity	Firm	Debt-to-equity
Pacific G & E	.84	U.S. Bancorp	.54
Houston Ind.	1.00	Sun Trust Banks	.29
Florida Progress	1.03	Mellon Bank	.82
Penn Power & Light	1.11	Michigan National	.13
North States Power	.83	Southeast Banking	.41
Ohio Edison	1.16		
Orange & Rockland	.70		

Source: "39th Annual Report on American Industry," *Forbes*, Jan. 12, 1987.

a. Compare the assumptions required by the analysis of variance F test and the Kruskal–Wallis H test. Which procedure can be more widely applied? Explain.

b. Use the Kruskal–Wallis H test to investigate whether debt-to-equity ratios differ among the four industries. Be sure to specify your null and alternative hypotheses and to state your conclusion in the context of the problem. Use $\alpha = .05$.

c. Assuming the Kruskal–Wallis H test indicates that differences exist among the four industries, which nonparametric procedure could be employed to compare the distribution of debt-to-equity ratios for the electric utility and banking industries?

SECTION 16.4

THE FRIEDMAN F_r TEST FOR A RANDOMIZED BLOCK DESIGN

In Section 15.3, we used relatively homogeneous blocks of experimental units to compare p population (treatment) means, based on a randomized block design. However, it was necessary to assume that the p populations had normal probability distributions and that their variances were equal. No such assumptions are required for the nonparametric counterpart, the **Friedman F_r test**, to compare the p probability distributions.

For example, suppose a marketing firm wants to compare the relative effectiveness of three different modes of advertising: direct-mail, newspaper ads, and magazine ads. For 15 clients, all three modes are used over a 1-year period, and the marketing firm records the year's percentage response to each type of advertising. That is, the firm divides the number of responses to a particular type of advertising by the total number of potential customers reached by the advertisements of that type. The results are shown in Table 16.9.

The 15 companies act as blocks in this experiment because we would expect the percentage responses to depend on the nature of the products of the company, its size, etc. Thus, we *rank the observations within each company (block)* and then *compute the rank sums for each of the three types of advertising (treatments)*.

TABLE 16.9

Percentage Response to Three Types of Advertising for 15 Different Companies

COMPANY	DIRECT-MAIL	RANK	NEWSPAPER	RANK	MAGAZINE	RANK
1	7.3	1	15.7	3	10.1	2
2	9.4	2	18.3	3	8.2	1
3	4.3	1	11.2	3	5.1	2
4	11.3	2	19.1	3	6.5	1
5	3.3	1	9.2	3	8.7	2
6	4.2	1	10.5	3	6.0	2
7	5.9	1	8.7	2	12.3	3
8	6.2	1	14.3	3	11.1	2
9	4.3	2	3.1	1	6.0	3
10	10.0	1	18.8	3	12.1	2
11	2.2	1	5.7	2	6.3	3
12	6.3	2	20.2	3	4.3	1
13	8.0	1	14.1	3	9.1	2
14	7.4	2	6.2	1	18.1	3
15	3.2	1	8.9	3	5.0	2
		$R_1 = 20$		$R_2 = 39$		$R_3 = 31$

The null and alternative hypotheses are

H_0:   The probability distributions of the response rates are identical for the three modes of advertising

H_a: At least two of the three probability distributions differ in location

The Friedman F_r **test statistic** is based on the rank sums:

$$F_r = \frac{12}{bp\,(p+1)} \sum_{j=1}^{p} R_j^2 - 3b(p+1)$$

where b is the number of blocks, p is the number of treatments, and R_j is the jth rank sum. For the data in Table 16.9,

$$F_r = \frac{12}{(15)(3)(4)} (R_1^2 + R_2^2 + R_3^2) - (3)(15)(4)$$

$$= \frac{12}{(15)(3)(4)} [(20)^2 + (39)^2 + (31)^2] - (3)(15)(4)$$

$$= 192.13 - 180 = 12.13$$

The Friedman F_r statistic measures the extent to which the p samples differ with respect to their relative ranks within the blocks. This is more easily seen by writing F_r in an alternative but equivalent form:

$$F_r = \frac{12}{bp(p+1)} \sum_{j=1}^{p} b(\overline{R}_j - \overline{R})^2$$

where $\overline{R}_j$ is the mean rank corresponding to treatment j, and $\overline{R}$ is the mean of all the ranks [i.e., $\overline{R} = (p+1)/2$]. Thus, the F_r statistic is 0 if all treatments have the same mean rank, and becomes increasingly large as the distance between the sample mean ranks grows.

As for the Kruskal–Wallis H statistic, the χ^2 distribution with $(p-1)$ degrees of freedom provides an approximation to the sampling distribution of F_r. We assume the approximation is adequate if either b (the number of blocks) or p (the number of treatments) exceeds 5. Then, for the advertising example, we use $\alpha = .10$ to form the rejection region:

Rejection region: $F_r > \chi^2_{.10}$ with $p - 1 = 2$ df

Consulting Table XIII in Appendix B, we find that $\chi^2_{.10}$ based on 2 df is 4.60517. Consequently, we will reject H_0 if $F_r > 4.60517$ (see Figure 16.7). Since the calculated $F_r = 12.13$ exceeds the critical value of 4.60517, we conclude that at least two of the response rate probability distributions for the three modes of advertising differ in location.

The Friedman F_r test for a randomized block design is summarized in the next box.

The Wilcoxon signed rank test for paired difference designs (Section 16.2) can be used to compare the pairs of treatments if the F_r statistic supports the alternative hypothesis that some of the probability distributions differ in location.

FIGURE 16.7
Rejection Region for the
Advertising Example

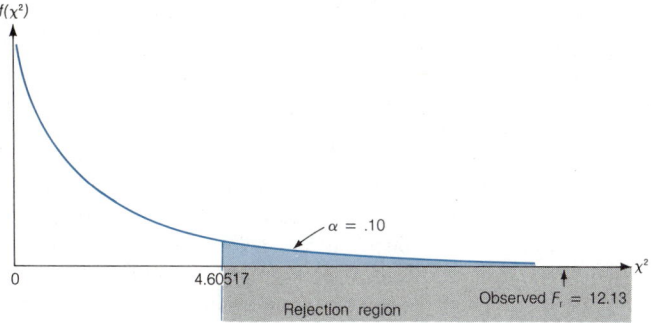

$f(\chi^2)$

$\alpha = .10$

0   4.60517 χ^2

Rejection region   Observed F_r = 12.13

FRIEDMAN F_r TEST FOR A RANDOMIZED BLOCK DESIGN

H_0: The probability distributions for the p treatments are identical

H_a: At least two of the probability distributions differ in location

Test statistic: $F_r = \dfrac{12}{bp(p + 1)} \sum\limits_{j=1}^{p} R_j^2 - 3b(p + 1)$

where

b = Number of blocks

p = Number of treatments

R_j = Rank sum of the jth treatment, where the rank of each measurement is computed relative to its position *within its own block*

Rejection region: $F_r > \chi_\alpha^2$ with $(p - 1)$ df

Assumptions: 1. The treatments are randomly assigned to experimental units within the blocks.
2. Either the number of blocks (b) or the number of treatments (p) should exceed 5 for the χ^2 approximation to be adequate.
3. The p probability distributions from which the samples within each block are drawn are continuous.

Ties: Assign tied measurements the average of the ranks they would receive if they were unequal but occurred in successive order. For example, if the third-ranked and fourth-ranked measurements are tied, assign both a rank of $(3 + 4)/2 = 3.5$. The number of ties should be small relative to the total number of observations.

CASE STUDY 16.1

CONSUMER RANKINGS OF PRODUCTS

Since consumers' images reflect to some extent actions taken by marketers in dealing with many marketing variables, it is frequently desirable to determine if these images have patterns.

McClure (1971) studied the pattern of consumers' images of three appliances. Five attributes—price, looks, need for repair, ease of use, and familiarity—were examined for each of two brands. For example, a consumer was asked to rank brand A's refrigerators, ranges, and automatic clothes washers in terms of the attribute "ease of use." McClure states:

> Conceptually, it was an investigation of the "halo effect" . . . which refers to the individual's supposed tendency to imbue his evaluations of specific characteristics of an appliance with the same direction of general feeling expressed about the brand. The halo effect would be considered operative to the extent that the individual's general image of Brand X influences his rating of an individual appliance of that brand when asked to evaluate it.

Responses of 282 female heads of households were obtained in a large midwestern city. For each of the ten responses (five attributes for two brands) a Friedman F_r test was conducted. The three types of appliances represented the treatments, and the 282 consumers represented the blocks. A significant value of the test statistic F_r would indicate consistent ranking of the three appliances by the 282 subjects for the attribute—i.e., that the probability distributions of ranks given the three appliances differ. A small value of F_r would lend credence to the hypothesis that the rankings are randomly performed—i.e., that they have approximately the same probability distributions. McClure found that the rank probability distributions differ for all attributes except "price" for brand A (at the $\alpha = .10$ level); brand A refrigerators consistently tend to obtain the highest ranking. However, brand B has no such clear pattern, with only the "familiarity" rankings showing significant differences. The subjects seemed to be most familiar with the brand B clothes washer but could not agree on the ranking of the other attributes.

McClure concluded his article:

> Many marketing researchers have not been aware of the Friedman two-way analysis of variance by ranks. However, it has potential for use in situations common to many consumer surveys in which sets of ordinal [rank] data are generated by each respondent.

EXERCISES
16.36–16.46

LEARNING THE MECHANICS

16.36 Use Table XIII in Appendix B to find each of the following χ^2 values:
  a. $\chi^2_{.05}$,   df $= 10$ b. $\chi^2_{.10}$,   df $= 15$
  c. $\chi^2_{.005}$,   df $= 50$ d. $\chi^2_{.01}$,   df $= 24$

16.37 Use Table XIII in Appendix B to find each of the following probabilities:
  a. $P(\chi^2 \geq 18.3070)$ where df $= 10$
  b. $P(\chi^2 < 25.9894)$ where df $= 18$
  c. $P(\chi^2 \geq 39.9968)$ where df $= 20$

16.38 Data were collected using a randomized block design with four treatments (A, B, C, and D) and $b = 6$. The following rank sums were obtained:

$R_A = 11$ $R_B = 21$ $R_C = 21$ $R_D = 7$

  a. How many blocks were used in the experimental design?
  b. Specify the null and alternative hypotheses that should be used in conducting a

hypothesis test to determine whether the probability distributions for at least two of the treatments differ in location.

c. Conduct the test of part **b**. Use $\alpha = .10$.

d. What is the approximate *p*-value of the test of part **c**?

e. Calculate the mean rank for each of the four treatments, and compute the value of the F_r test statistic according to the formula (page 974) that utilizes those means. Verify that the test statistic is the same as that you obtained in part **c**.

16.39 Suppose you have used a randomized block design to help you compare the effectiveness of three different treatments, A, B, and C. You obtained the data given in the table and plan to conduct a Friedman F_r test.

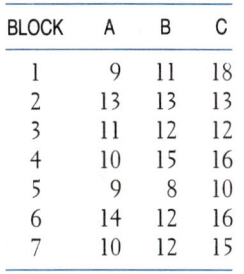

BLOCK	A	B	C
1	9	11	18
2	13	13	13
3	11	12	12
4	10	15	16
5	9	8	10
6	14	12	16
7	10	12	15

a. Specify the null and alternative hypotheses you would test.

b. Specify the rejection region for the test, using $\alpha = .10$.

c. Conduct the test using $\alpha = .10$.

APPLYING THE CONCEPTS

16.40 An *optical mark reader* (OMR) is a machine that is able to "read" pencil marks that have been entered on a scannable form. When connected to a computer, such systems are able to read and analyze data in one step. As a result, the keypunching of data into a machine-readable code can be eliminated. Eliminating this step reduces the possibility that the data will be contaminated by human error. OMRs are used by schools to grade exams and by survey research organizations to compile data from questionnaires (Oas, 1980). A manufacturer of OMRs believes its product can operate equally well in a variety of temperature and humidity environments. To determine whether operating data contradict this belief, the manufacturer asks a well-known industrial testing laboratory to test its product. Five recently produced OMRs were randomly selected and each was operated in five different environments. The number of forms each was able to process in an hour was recorded and used as a measure of the OMR's operating efficiency. These data appear in the table. Use the Friedman F_r test to determine whether evidence exists to indicate that the probability distributions for the number of forms processed per hour differ in location for at least two of the environments. Test using $\alpha = .10$.

MACHINE NUMBER	ENVIRONMENT				
	1	2	3	4	5
1	8,001	8,025	8,100	8,055	7,991
2	7,910	7,932	7,900	7,990	7,892
3	8,111	8,101	8,201	8,175	8,102
4	7,802	7,820	7,904	7,850	7,819
5	7,500	7,601	7,702	7,633	7,600

16.41 The table at the top of page 978 lists the values of five industries' inventories (in millions of dollars) as stated by the manufacturers. These figures represent book values of stocks on hand at the end of the period and include materials and supplies, goods in process, and finished goods. Suppose you wish to compare the locations of the probability distributions of the inventories among the five industries.

a. What can be learned about the five industries by conducting a Friedman F_r test? Identify the blocks and treatments of the experiment.

YEAR	FOOD AND KINDRED PRODUCTS	TOBACCO PRODUCTS	PAPER & ALLIED PRODUCTS	TEXTILE MILL PRODUCTS	CHEMICALS & ALLIED PRODUCTS
1979	$19,903	$3,434	$6,936	$6,131	$17,265
1980	21,737	3,649	7,812	6,623	19,621
1981	21,481	4,375	8,613	6,871	22,039
1982	20,769	4,411	8,559	6,192	19,907
1983	20,869	3,935	8,728	6,908	19,616
1984	21,500	3,558	9,691	7,017	21,872

Source: *Business Statistics: 1984*, Dept. of Commerce, Bureau of Economic Analysis, p. 16.

b. Conduct a Friedman F_r test using $\alpha = .05$. Specify the null and alternative hypotheses and state your conclusion in the context of the problem.

c. Find the approximate p-value for the test of part **b**, and interpret its value.

d. What assumptions must be made to assure the validity of the test?

e. Use the Wilcoxon signed rank test to compare all pairs of industries. Use $\alpha = .05$ for each comparison.

16.42 As part of the process of choosing a script for a television commercial, an advertising agency asks a panel of six executives to rate the scripts. The panel members are asked to use a rating scale of 1 to 10, with higher ratings indicating greater potential impact on the viewing audience. The ratings obtained are shown in the table.

a. Use the Friedman F_r test to determine whether evidence exists to indicate that the levels at which the executives rate the scripts differ. Test using $\alpha = .10$.

b. Use the appropriate Wilcoxon signed rank test to determine whether script B's probability distribution of ratings lies significantly above script C's. Test using $\alpha = .10$.

PANEL MEMBER	SCRIPT 1	SCRIPT 2	SCRIPT 3
1	7	8	4
2	8	7	5
3	6	8	7
4	5	6	6
5	8	9	7
6	8	8	6

16.43 Corrosion of different metals is a problem in many mechanical devices. Three sealers used to help retard the corrosion of metals were tested to see whether there were any differences among them. Samples of ten different metal compositions were treated with each of the three sealers, and the amount of corrosion was measured after exposure to the same environmental conditions for 1 month. The data are given in the table. Is there any evidence of a difference in the abilities of the sealers to prevent corrosion? Test using $\alpha = .05$.

METAL	SEALER 1	SEALER 2	SEALER 3
1	4.6	4.2	4.9
2	7.2	6.4	7.0
3	3.4	3.5	3.4
4	6.2	5.3	5.9
5	8.4	6.8	7.8
6	5.6	4.8	5.7
7	3.7	3.7	4.1
8	6.1	6.2	6.4
9	4.9	4.1	4.2
10	5.2	5.0	5.1

16.44 In recent years, domestic car manufacturers have devoted more attention to the small-car market. To compare the popularity of four domestic small cars within a city, a local trade organization obtained the information given in the table from four car dealers—one dealer for each of the four car makes. Is there evidence of differences in location among the probability distributions of the number of cars sold for each type? Use $\alpha = .10$.

MONTH	A	B	C	D
1	9	17	14	8
2	10	20	16	9
3	13	15	19	12
4	11	12	19	11
5	7	18	13	8

EAR	SPRAY		
	A	B	C
1	21	23	15
2	29	30	21
3	16	19	18
4	20	19	18
5	13	10	14
6	5	12	6
7	18	18	12
8	26	32	21
9	17	20	9
10	4	10	2

16.45 A serious drought-related problem for farmers is the spread of aflatoxin, a highly toxic substance caused by mold, which contaminates field corn. In higher levels of contamination, aflatoxin is potentially hazardous to animal and possibly human health. (Officials of the Food and Drug Administration have set a maximum limit of 20 parts per billion aflatoxin as safe for interstate marketing.) Three sprays, A, B, and C, have been developed to control aflatoxin in field corn. To determine whether differences exist among the sprays, ten ears of corn are randomly chosen from a contaminated corn field and each is divided into three pieces of equal size. The sprays are then randomly assigned to the pieces for each ear of corn, thus setting up a randomized block design. The table gives the amount (in parts per billion) of aflatoxin present in the corn samples after spraying.

a. Use the Friedman F_r test to determine whether there is evidence that the distributions of the levels of aflatoxin in corn differ for at least two of the three sprays. Test at $\alpha = .05$.

b. Do the results of the test warrant further comparison of the pairs of sprays? If so, conduct the comparison of the pairs of sprays using the Wilcoxon signed rank test and $\alpha = .01$ for each comparison. Can you conclude that one spray appears to best control the level of aflatoxin?

c. What assumptions are necessary to assure the validity of the procedures used in parts **a** and **b**?

BLOCK	TREATMENT			
	W	X	Y	Z
1	21	30	28	25
2	30	45	36	29
3	29	38	30	31
4	48	60	52	47
5	66	75	51	70
6	15	25	20	21

16.46 A randomized block design was used to collect the data in the table.

a. Do the data provide sufficient evidence to conclude that at least two of the probability distributions associated with the treatments differ in location? Test using $\alpha = .05$.

b. Describe the Type I and Type II errors associated with the test of part a.

c. Find the approximate p-value of the hypothesis test you conducted in part **a**.

S E C T I O N   16.5

SPEARMAN'S RANK CORRELATION COEFFICIENT

When economic conditions are favorable, many banks advertise special loan rates for new cars, appliances, and other items to attract customers. Suppose a bank wants to determine whether to aim its advertising at a broad spectrum of potential borrowers or to concentrate on a specific income group. It randomly samples ten noncommercial customers from recent files and ascertains the present income of each and the total amount each has borrowed over the past 3 years (excluding mortgages and business loans). The data are shown in Table 16.10.

TABLE 16.10
Income–Amount Borrowed Data

CUSTOMER	INCOME	RANK	TOTAL BORROWED	RANK
1	$14,800	5	$4,300	7
2	8,900	1	4,800	8
3	83,600	10	500	2
4	22,100	8	3,300	5
5	18,200	7	5,500	9
6	13,700	4	3,700	6
7	41,800	9	0	1
8	9,300	2	3,200	4
9	12,700	3	6,100	10
10	16,100	6	1,800	3

One method of determining whether a correlation exists between income and amount borrowed is to calculate the Pearson product moment correlation, r (Section 10.6). However, to make an inference about the population correlation ρ (Greek rho), we must assume that the two random variables, income and amount borrowed, are normally distributed. This assumption is usually inappropriate for incomes because they tend to have a relative frequency distribution that is heavily skewed to the right (Figure 16.8). Although the modal income (that with the largest relative frequency) may be relatively low, there are typically enough individuals with high incomes to make income distributions asymmetric.

FIGURE 16.8

Typical Relative Frequency Distribution of Incomes

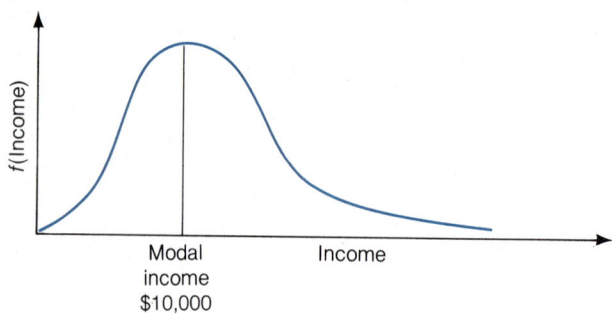

Thus, we turn to a nonparametric approach to correlation, which does not require underlying normal distributions. This nonparametric method, like those in the previous sections of this chapter, uses the ranks of the measurements to determine a measure of correlation. It is shown in the box.

SPEARMAN'S RANK CORRELATION COEFFICIENT

$$r_s = \frac{SS_{uv}}{\sqrt{SS_{uu}SS_{vv}}}$$

where

$$SS_{uv} = \sum (u_i - \bar{u})(v_i - \bar{v}) = \sum u_i v_i - \frac{\left(\sum u_i\right)\left(\sum v_i\right)}{n}$$

$$SS_{uu} = \sum (u_i - \bar{u})^2 = \sum u_i^2 - \frac{\left(\sum u_i\right)^2}{n}$$

$$SS_{vv} = \sum (v_i - \bar{v})^2 = \sum v_i^2 - \frac{\left(\sum v_i\right)^2}{n}$$

u_i = Rank of the ith measurement in sample 1

v_i = Rank of the ith measurement in sample 2

n = Number of pairs of measurements (number of measurements in each sample)

Note that the definition of Spearman's rank correlation coefficient is identical to the definition of Pearson's r (see page 514), except that Spearman's r_s uses ranks. You can use the shortcut formula shown in the next box for calculating r_s if there are no ties. The shortcut formula will also provide a satisfactory approximation to r_s when the number of ties is small relative to the number of pairs.

SHORTCUT FORMULA FOR r_s

$$r_s = 1 - \frac{6 \sum d_i^2}{n(n^2 - 1)}$$

where

$$d_i = u_i - v_i \quad \text{(difference in the ranks of the ith measurement for sample 1 and sample 2)}$$

EXAMPLE 16.3

Calculate Spearman's rank correlation coefficient, r_s, for the bank customer data given in Table 16.10.

SOLUTION

The ranks are reproduced in Table 16.11, along with the difference, d_i, for each pair of measurements. Then,

$$r_s = 1 - \frac{6 \sum d_i^2}{n(n^2 - 1)} = 1 - \frac{6(260)}{10(100 - 1)} = 1 - 1.576 = -.576$$

The sign of r_s indicates the nature of the relationship between the two variables. Positive values indicate a tendency for the variables to increase together, and negative values indicate a tendency for one variable to increase while the other decreases.

TABLE 16.11
Calculation of Spearman's Rank Correlation Coefficient

CUSTOMER	INCOME RANK u_i	TOTAL BORROWED RANK v_i	DIFFERENCE $d_i = u_i - v_i$	DIFFERENCE SQUARED d_i^2
1	5	7	−2	4
2	1	8	−7	49
3	10	2	8	64
4	8	5	3	9
5	7	9	−2	4
6	4	6	−2	4
7	9	1	8	64
8	2	4	−2	4
9	3	10	−7	49
10	6	3	3	9
			Total =	260

The strength of the relationship between the ranks is indicated by the numerical size of r_s. Since r_s is really a Pearson product moment correlation of the ranks, it must lie between -1 and $+1$. Recall that a correlation of 0 implies no linear relationship, while correlations of -1 and $+1$ imply perfect negative and positive relationships, respectively. In Example 16.3, we obtained $r_s = -.576$. Is this different enough from 0 to conclude that the variables are related in the population?

If we define ρ_s as the population Spearman rank correlation coefficient, this question can be answered by conducting the test

H_0: $\rho_s = 0$ (There is no population correlation between ranks)

H_a: $\rho_s \neq 0$ (There is a population correlation between ranks)

Test statistic: r_s, the sample Spearman rank correlation coefficient

To determine a rejection region, we consult Table XVI in Appendix B, which is partially reproduced in Table 16.12. Note that the left-hand column gives values of n, the number of pairs of measurements. The entries in the table are values for an upper-tail rejection region, since only positive values are given. Thus, for $n = 10$ and $\alpha = .05$, the value .564 is the boundary of the upper-tailed rejection region, so that $P(r_s > .564) = .05$ if in fact H_0 is true. That is, we would expect to see r_s exceed .564 only 5% of the time if in fact there is no relationship between the variables. The two-tailed rejection region is $r_s > .564$ or $r_s < -.564$, and the α value is therefore double the table value: $\alpha = 2(.05) = .10$. Thus, we have

Rejection region: $r_s < -.564$ or $r_s > .564$ for $\alpha = .10$

Since the calculated value from Example 16.3 is $r_s = -.576$, which is less than $-.564$, we reject H_0 at $\alpha = .10$ and conclude that the population rank correlation coefficient, ρ_s, differs from 0. In fact, it appears that bank customers at lower income

TABLE 16.12

Reproduction of Part of Table XVI of Appendix B

n	$\alpha = .05$	$\alpha = .025$	$\alpha = .01$	$\alpha = .005$
5	.900	—	—	—
6	.829	.886	.943	—
7	.714	.786	.893	—
8	.643	.738	.833	.881
9	.600	.683	.783	.833
10	.564	.648	.745	.794
11	.523	.623	.736	.818
12	.497	.591	.703	.780
13	.475	.566	.673	.745
14	.457	.545	.646	.716
15	.441	.525	.623	.689
16	.425	.507	.601	.666
17	.412	.490	.582	.645
18	.399	.476	.564	.625
19	.388	.462	.549	.608
20	.377	.450	.534	.591

levels tend to borrow more than those at higher income levels. Therefore, the bank would be wise to aim its loan advertising at those in the middle- and lower-income groups, unless it wants to attempt to entice those who rarely borrow to become customers, in which case higher-income groups should be the target.

A summary of Spearman's nonparametric test for correlation is shown in the box.

SPEARMAN'S NONPARAMETRIC TEST FOR RANK CORRELATION

ONE-TAILED TEST

H_0: $\rho_s = 0$

H_a: $\rho_s > 0$
  (or H_a: $\rho_s < 0$)

Test statistic: r_s, the sample rank correlation (formulas for calculating r_s are given on pages 980–981)

Rejection region:

  $r_s > r_{s,\alpha}$
  (or $r_s < -r_{s,\alpha}$ when H_a: $\rho_s < 0$)

where $r_{s,\alpha}$ is the value from Table XVI corresponding to the upper-tail area α and n pairs of observations

TWO-TAILED TEST

H_0: $\rho_s = 0$

H_a: $\rho_s \neq 0$

Test statistic: r_s, the sample rank correlation (formulas for calculating r_s are given on pages 980–981)

Rejection region:

  $r_s < -r_{s,\alpha/2}$ or $r_s > r_{s,\alpha/2}$

where $r_{s,\alpha/2}$ is the value from Table XVI corresponding to the upper-tail area $\alpha/2$ and n pairs of observations

Assumptions: 1. The sample of experimental units on which the two variables are measured is randomly selected.
         2. The probability distributions of the two variables are continuous.

Ties: Assign tied measurements the average of the ranks they would receive if they were unequal but occurred in successive order. For example, if the third-ranked and fourth-ranked measurements are tied, assign both a rank of $(3 + 4)/2 = 3.5$. The number of ties should be small relative to the total number of observations.

EXAMPLE 16.4

Manufacturers of perishable foods often use preservatives to retard spoilage. One concern is that too much preservative will change the flavor of the food. Suppose an experiment is conducted using samples of a food product with varying amounts of preservative added. Both length of time until the food shows signs of spoiling and a taste rating are recorded for each sample. The taste rating is the average rating for three tasters, each of whom rates each sample on a scale from 1 (good) to 5 (bad). Twelve sample measurements are shown in Table 16.13 (page 984). Use a nonparametric test to find out whether the spoilage times and taste ratings are negatively correlated. Use $\alpha = .05$.

TABLE 16.13

Data for Example 16.4

SAMPLE	DAYS UNTIL SPOILAGE	RANK	TASTE RATING	RANK
1	30	2	4.3	11
2	47	5	3.6	7.5
3	26	1	4.5	12
4	94	11	2.8	3
5	67	7	3.3	6
6	83	10	2.7	2
7	36	3	4.2	10
8	77	9	3.9	9
9	43	4	3.6	7.5
10	109	12	2.2	1
11	56	6	3.1	5
12	70	8	2.9	4

Note: Tied measurements are assigned the average of the ranks that would be given the measurements if they were different but consecutive.

SOLUTION

The test is one-tailed, with

H_0: $\rho_s = 0$

H_a: $\rho_s < 0$

Test statistic:* $r_s = 1 - \dfrac{6 \sum d_i^2}{n(n^2 - 1)}$

Rejection region: Reject H_0 if $r_s < -r_{s,.05}$, where from Table XVI, for $\alpha = .05$ and $n = 12$, $-r_{s,.05} = -.497$. [*Note:*   The value of α need not be doubled since the test is one-tailed.]

The first step in the computation of r_s is to sum the squares of the differences between ranks:

$$\sum d_i^2 = (2 - 11)^2 + (5 - 7.5)^2 + \cdots + (8 - 4)^2 = 536.5$$

Then

$$r_s = 1 - \frac{6(536.5)}{12(144 - 1)} = -.876$$

Since $-.876 < -.497$, we reject H_0 and conclude that the preservative does affect the taste of this food adversely. ∎

CASE STUDY 16.2

THE PROBLEM OF NONRESPONSE BIAS IN MAIL SURVEYS

Researchers who collect their sample data via mail questionnaires often run the risk that their respondents will not be a representative sample of the entire population. The reason for this, according to Rosenthal and Rosnow (1975), is the tendency for respondents to be (1) better educated, (2) of higher social-class status, (3) more intelligent, (4) in need of social approval, (5) more social, and (6) more interested in the research topic than nonrespondents.

*The shortcut formula is not exact when there are tied measurements, but it is a good approximation when the total number of ties is not large relative to n.

Researchers sometimes attempt to verify the representativeness of their sample by obtaining information about the demographic characteristics of the nonrespondents and comparing them to those of the sample of respondents. Finding similarities gives researchers confidence in the representativeness of their sample. Another approach is to contact a small sample of the nonrespondents and obtain responses to the questionnaire. These responses can be compared with those of the original respondents; the more similar the patterns of responses, the more confident the researcher can be that the original sample of returned questionnaires is representative of the population from which the sample came.

In a recent marketing research study by David W. Finn, Chih-Kang Wang, and Charles W. Lamb (1983), a random sample of 20 nonrespondents to their mail questionnaire were contacted by telephone and asked to respond to the following question on the questionnaire: "In general, what is your willingness to buy products made in each of the following countries?" They were asked to indicate their willingness by responding on a 5-point scale that ranged from "extremely willing" to "extremely unwilling." The mean willingness score was computed for each country and the countries were ranked accordingly. Similarly, a rank ordering was developed for the 273 respondents to their mail questionnaire. Both sets of rankings are displayed in Table 16.14.

TABLE 16.14

Ranking of Consumer Willingness to Buy Products Made in Indicated Countries

COUNTRY	RANK ORDER[a]	
	Respondent	Nonrespondent
United Kingdom	1	1
Japan	2	3
France	3	2
Taiwan	4	4
Brazil	5	5
India	6	6
Iran	7	7
Angola	8	8
USSR	9	9
Cuba	10	10

[a]Data were collected in Spring 1977.

Finn, Wang, and Lamb compared the rank orderings for the respondents and nonrespondents using Spearman's rank correlation coefficient. They obtained $r_s = .9879$ (p-value $< .01$) and concluded that "respondents and nonrespondents in this study did not differ attitudinally." Further, they noted that even if respondents and nonrespondents to a mail survey differ demographically, as was the case in their study, the Spearman rank correlation result indicates that such differences should not automatically be interpreted as signalling the existence of nonresponse bias. The sample of opinions obtained from the respondents may, in fact, be representative of the population of opinions even though respondents and nonrespondents differ demographically.

LEARNING THE MECHANICS

16.47 Use Table XVI of Appendix B to find each of the following probabilities:
   a. $P(r_s > .496)$ when $n = 23$
   b. $P(r_s > .496)$ when $n = 28$
   c. $P(r_s \leq .600)$ when $n = 9$
   d. $P(r_s < -.377$ or $r_s > .377)$ when $n = 20$

16.48 Compute Spearman's rank correlation coefficient for each of the following pairs of sample observations:

a. x	y	b. x	y	c. x	y	d. x	y
30	26	90	81	1	11	5	80
55	36	100	95	15	26	20	83
60	65	120	75	4	15	15	91
19	25	137	52	10	21	10	82
40	35	41	136			3	87

16.49 The sample data shown in the table were collected on variables x and y.

x	y
−1	−1
2	1
−5	−4
−1	1
3	3
0	1
4	2

   a. Specify the null and alternative hypotheses that should be used in conducting a hypothesis test to determine whether the variables x and y are correlated.
   b. Conduct the test of part **a** using $\alpha = .05$.
   c. What is the approximate p-value of the test of part **b**?
   d. What assumptions are necessary to assure the validity of the test of part **b**?

APPLYING THE CONCEPTS

16.50 The table reports the 1950 and 1985 sales (in millions of dollars) of a sample of 25 major U.S. corporations.

CORPORATION	1950 SALES	1985 SALES	CORPORATION	1950 SALES	1985 SALES
Abbott Laboratories	74	3,360	General Electric Corp.	2,233	28,285
Allis-Chalmers Corp.	344	961	Gillette Co.	99	2,400
American Cyanamid Co.	322	3,666	IBM Corp.	215	50,056
Armstrong World Industries	187	1,679	International Paper Co.	498	4,502
Boeing Co.	307	13,636	PepsiCo., Inc.	40	8,478
Borg-Warner Corp.	331	3,964	Philip Morris, Inc.	306	12,149
Bristol-Myers Co.	52	4,444	Phillips Petroleum Co.	533	15,676
Caterpillar Tractor Co.	337	6,769	R. J. Reynolds Industries, Inc.	758	13,533
Celanese Corp.	233	3,046	Sterling Drug, Inc.	139	1,848
Coca-Cola Co.	215	8,139	The Timken Co.	144	1,091
Corning Glass Works	117	1,691	Union Carbide Corp.	758	9,003
Eaton Corp.	148	3,675	Westinghouse Electric Corp.	1,020	10,700
Fruehauf Corp.	128	2,564			

   a. Calculate Spearman's rank correlation coefficient for the data, and interpret its value in the context of the problem.

b. Consider the hypotheses H_0: $\rho_s = 0$ and H_a: $\rho_s > 0$. Interpret these hypotheses in the context of the problem.

c. Test H_0: $\rho_s = 0$ against H_a: $\rho_s > 0$ using $\alpha = .05$. Interpret your conclusion in the context of the problem.

16.51 Public safety and millions of tax dollars are at stake in highway design decisions. Accordingly, Martin Helander (1978) argues that designers should consider both the physical and mental skills of drivers. He reports a study in which he investigated the correlation between the magnitude of brake pressure applied by drivers and various physiological variables such as heart rate and electrodermal response (EDR). The table lists 14 different traffic events to which 60 test drivers were exposed. The events are listed in rank order according to the average magnitude of brake pressure associated with each (rank 1 is the highest mean brake pressure and rank 14 is the lowest). The ranks of the events according to the drivers' average electrodermal response are also given (rank 1 is the highest response and rank 14 is the lowest). High electrodermal responses are associated with mental stress. Based on a one-sided hypothesis test employing Spearman's rank correlation coefficient, Helander concluded that events involving the use of the brake are perceived as stressful and, therefore, highway designs that minimize braking should be preferred.

	TRAFFIC EVENTS ORDERED IN RANK BY:	
TRAFFIC EVENT	Brake Pressure	EDR
1. Cyclist or pedestrian and oncoming car	1	1
2. Other car merges in front of own car	2	2
3. Multiple events	3	3
4. Leading car diverges	4	4
5. Cyclist or pedestrian	5	6
6. Own car passes other car with car following	6	5
7. Cyclist or pedestrian and car following	7	9
8. Car following and meeting other car	8	11
9. Meeting other car	9	8
10. Car following	10	10
11. Parked car	11	14
12. No event	12	12
13. Other car passes own car	13	7
14. Parked car and car following	14	13

Source: Helander (1978). Copyright 1978 by the American Psychological Association. Reprinted by permission.

a. Calculate Spearman's rank correlation coefficient and test its significance using $\alpha = .05$. Does your analysis confirm Helander's finding? Explain.

b. Describe the Type I and Type II errors associated with the hypothesis test.

c. What is the approximate p-value of the test of part **a**?

d. What assumptions are necessary to assure the validity of the test?

16.52 A *negotiable certificate of deposit* is a marketable receipt for funds deposited in a bank for a specified period of time at a specified rate of interest (Cook, 1983). The table lists the end-of-quarter interest rate for 3-month certificates of deposit during the period January 1979 through June 1986. The table also lists end-of-quarter values of Standard & Poor's 500 Stock Composite Average (an indicator of stock market activity) for the same time period.

YEAR	QUARTER	INTEREST RATE	S&P 500
1979	I	10.13%	101.59
	II	9.95	102.91
	III	11.89	109.32
	IV	13.43	107.94
1980	I	17.57	104.69
	II	8.49	114.24
	III	11.29	125.46
	IV	18.65	135.76
1981	I	14.43	136.00
	II	16.90	131.21
	III	16.84	116.18
	IV	12.49	122.55
1982	I	14.21	111.96
	II	14.46	109.61
	III	10.66	120.42
	IV	8.66	135.28
1983	I	8.69	152.96
	II	9.20	168.11
	III	9.39	166.07
	IV	9.69	164.93
1984	I	10.08	159.18
	II	11.34	153.18
	III	11.29	166.10
	IV	8.60	167.24
1985	I	9.02	180.66
	II	7.44	191.85
	III	7.93	182.08
	IV	7.80	211.28
1986	I	7.24	238.90
	II	6.73	250.84

Source: *Standard & Poor's Statistical Service, Current Statistics*, 1986, Standard & Poor's Corporation.

a. Compute Spearman's rank correlation coefficient to measure the strength of the relationship between the interest rate on certificates of deposit and the S&P 500.

b. Test the null hypothesis that the interest rate on certificates of deposit and the S&P 500 are not correlated against the alternative hypothesis that these variables are correlated. Use $\alpha = .10$.

c. Repeat parts **a** and **b** using monthly data instead of quarterly data. These can be obtained at your library in *Standard & Poor's Current Statistics*. Compare the results you obtained for monthly data with those you obtained using the quarterly data.

16.53 It has been conjectured that income is one of the primary determinants of an individual's satisfaction with his or her job. To investigate this theory, 15 employees of a particular firm are chosen at random and their gross salaries are noted. Each of the employees is then asked to complete a questionnaire designed to measure job satisfaction. The resulting scores (higher scores correspond to greater satisfaction) and gross incomes (in thousands of dollars) are given in the table.

EMPLOYEE	JOB SATISFACTION SCORE	INCOME	EMPLOYEE	JOB SATISFACTION SCORE	INCOME
1	92	29.9	9	45	16.0
2	51	18.7	10	72	25.0
3	88	32.0	11	53	17.2
4	65	15.0	12	43	9.7
5	80	26.0	13	87	20.1
6	31	9.0	14	30	15.5
7	38	11.3	15	74	16.5
8	75	22.1			

a. Compute Spearman's rank correlation coefficient for these data.
b. Is there evidence that job satisfaction and income are positively correlated? Use $\alpha = .05$.

16.54 Many large businesses send representatives to college campuses to conduct job interviews. To aid the interviewer, one company decides to study the correlation between the strength of an applicant's references (the company requires three references) and the performance of the applicant on the job. Eight recently hired employees are sampled, and independent evaluations of both references and job performance are made on a scale from 1 to 20. The scores are given in the table.

EMPLOYEE	REFERENCES	JOB PERFORMANCE
1	18	20
2	14	13
3	19	16
4	13	9
5	16	14
6	11	18
7	20	15
8	9	12

a. Compute Spearman's rank correlation coefficient for these data.
b. Is there evidence that strength of references and job performance are positively correlated? Use $\alpha = .05$.

16.55 The decision to build a new plant or to move an existing plant to a new location involves long-term commitment of both human and monetary resources. Accordingly, such decisions should be made only after carefully considering the relevant factors associated with numerous alternative plant sites. G. Michael Epping (1982) examined the relationship between the location factors deemed important by businesses that located in Arkansas and those that considered Arkansas but located elsewhere. A questionnaire that asked manufacturers to rate the importance of 13 general location factors on a 9-point scale was completed by 118 firms that had moved a plant to Arkansas in the period 1955–1977 and by 73 firms that had recently considered Arkansas but located elsewhere. Epping averaged the importance ratings and arrived at the rankings shown in the table. Calculate Spearman's rank correlation coefficient for these data and carefully interpret its value in the context of the problem.

FACTOR	MANUFACTURERS LOCATING IN ARKANSAS Rank	MANUFACTURERS NOT LOCATING IN ARKANSAS Rank
Labor	1	1
Taxes	2	2
Industrial site	3	4
Information sources and special inducements	4	5
Legislative laws and structure	5	3
Utilities and resources	6	7
Transportation facilities	7	8
Raw material supplies	8	10
Community	9	6
Industrial financing	10	9
Markets	11	12
Business services	12	11
Personal preferences	13	13

16.56 A large manufacturing firm wants to determine whether a relationship exists between the number of work-hours an employee misses per year and the employee's annual wages. A sample of 15 employees produced the data in the table, where annual wages are measured in thousands of dollars. Do these data provide evidence that the number of work-hours missed is related to annual wages? Use $\alpha = .05$.

EMPLOYEE	WORK-HOURS MISSED	ANNUAL WAGES	EMPLOYEE	WORK-HOURS MISSED	ANNUAL WAGES
1	49	12.8	9	191	7.8
2	36	14.5	10	6	15.8
3	127	8.3	11	63	10.8
4	91	10.2	12	79	9.7
5	72	10.0	13	43	12.1
6	34	11.5	14	57	21.2
7	155	8.8	15	82	10.9
8	11	17.2			

S U M M A R Y

We have presented several useful **nonparametric techniques** for comparing two or more populations. Nonparametric techniques are useful when the underlying assumptions for their parametric counterparts are not justified or when it is impossible to assign specific values to the observations. Nonparametric methods provide more general comparisons of populations than parametric methods because they compare the probability distributions of the populations rather than specific parameters.

Rank sums are the primary tools of nonparametric statistics. The **Wilcoxon rank sum statistic** can be used to compare two populations based on an independent sampling experiment, and the **Wilcoxon signed rank test** can be used for a **paired difference experiment**. The **Kruskal–Wallis H test** is applied when comparing p populations using a **completely randomized design**. The **Friedman F_r test** is used to compare p populations when a **randomized block design** is used.

The strength of nonparametric statistics lies in their general applicability. Few restrictive assumptions are required, and they may be used for observations that can be ranked but not measured exactly. Therefore, nonparametric methods provide useful alternatives to the parametric tests of Chapters 8, 9, and 15.

SUPPLEMENTARY EXERCISES 16.57–16.80

BEFORE	AFTER
12	4
5	2
10	7
9	3
14	8
6	

16.57 When is it inappropriate to use the t and F tests of Chapters 9 and 15 for comparing two or more population means?

16.58 A study was conducted to determine whether the installation of a traffic light was effective in reducing the number of accidents at a busy intersection. Samples of 6 months prior to installation and 5 months after installation of the light yielded the numbers of accidents per month listed in the table.

a. Is there sufficient evidence to conclude that the traffic light aided in reducing the number of accidents? Test using $\alpha = .025$.

b. Explain why this type of data might or might not be suitable for analysis using the t test of Chapter 9.

16.59 David J. Teece (1981) used the Wilcoxon signed rank test to examine the differential performance between organizations using an innovative decentralized and divisional organization structure known as the M-form, and their principal competitors. Teece identified the first firm in each of 20 industries that adopted the M-form structure. The principal competitor of each of these firms was identified. To qualify for inclusion in the study, the competing firm must also have adopted the M-form structure, but at a later date. A total of 14 pairs of firms qualified for the sample. The difference in performance within the pairs of firms was measured over two 3-year time periods: a "before" period in which only the M-form originator in each pair used the M-form, and an "after" period in which both firms had M-form structures. Differential performance in each time period was measured as the average difference in yearly return on stockholders' equity, where the differences were formed by subtracting the competitor's return on equity from the M-form originator's. The performance data appear in the table at the top of page 992.

INDUSTRY	AVERAGE DIFFERENCE IN YEARLY RETURN ON EQUITY	
	Before	After
Grocery	25.70%	10.78%
Chemicals	6.26	−.39
Textiles	5.84	1.94
Aluminum	4.16	.56
Meat packing	−2.43	−5.80
Packaged foods	.77	−1.18
Can manufacturing	3.79	2.49
Grain milling	3.86	4.81
Petroleum	3.29	2.38
Tires	−3.36	−2.46
Autos	13.48	14.38
Electrical equipment	−10.85	−10.40
Tobacco	1.50	1.90
Retail department stores	12.32	12.37

a. Teece concluded that ". . . the M-form innovation has been shown to display a statistically significant impact on firm performance." Do you agree? Test using $\alpha \doteq .05$.

b. What is the approximate p-value of the test of part a?

c. What assumptions are necessary to assure the validity of the test procedure you used in part a?

16.60 In Exercise 10.49, a calibration study undertaken by the Minnesota Department of Transportation to evaluate its newly installed weigh-in-motion scale was described. Pearson's product moment correlation coefficient was used to measure the strength of the relationship between the static weight of a truck and the truck's weight as measured by the weigh-in-motion equipment. The data are repeated in the table.

TRUCK NUMBER	STATIC WEIGHT OF TRUCK x (thousand pounds)	WEIGH-IN-MOTION READING PRIOR TO CALIBRATION ADJUSTMENT y_1 (thousand pounds)	WEIGH-IN-MOTION READING AFTER CALIBRATION ADJUSTMENT y_2 (thousand pounds)
1	27.9	26.0	27.8
2	29.1	29.9	29.1
3	38.0	39.5	37.8
4	27.0	25.1	27.1
5	30.3	31.6	30.6
6	34.5	36.2	34.3
7	27.8	25.1	26.9
8	29.6	31.0	29.6
9	33.1	35.6	33.0
10	35.5	40.2	35.0

Source: Adapted from data in Wright, Owen, and Pena (1983).

a. Calculate Spearman's rank correlation coefficient for x and y_1 and for x and y_2. Interpret, in the context of the problem, the values you obtain. Compare your results with those of Exercise 10.49, part c.

b. In the context of this problem, describe the circumstances that would result in Spearman's rank correlation coefficient being exactly 1. Being exactly 0.

16.61 An experiment was conducted to compare two print types, A and B, to determine whether type A is easier to read. Ten subjects were randomly divided into two groups of five. Each subject was given the same material to read, one group receiving the material printed with type A, the other group receiving print type B. The times necessary for each subject to read the material (in seconds) are shown below:

Type A: 95, 122, 101, 99, 108

Type B: 110, 102, 115, 112, 120

Do the data provide sufficient evidence to indicate that print type A is easier to read? Test using $\alpha = .05$.

16.62 The length of time required for a human to respond to a new pain killer was tested in the following manner: Seven randomly selected subjects were assigned to receive both aspirin and the new drug. The two treatments were spaced in time and assigned in random order. The length of time (in minutes) required for a subject to indicate that he or she could physically feel pain relief was recorded for both the aspirin and the drug. The data are shown in the table. Do the data provide sufficient evidence to indicate the new drug is more effective than aspirin in reducing pain? Test using $\alpha = .05$.

SUBJECT	1	2	3	4	5	6	7
Aspirin	15	20	12	20	17	14	17
Drug	7	14	13	11	10	16	11

16.63 Suppose a company wants to study how personality relates to leadership. Four supervisors with different types of personalities are selected. Several employees are then selected from the group supervised by each, and these employees are asked to rate the leader of their group on a scale from 1 to 20 (20 signifies highly favorable). The resulting data are shown in the table.

SUPERVISOR			
1	2	3	4
20	17	16	8
19	11	15	12
20	13	13	10
18	15	18	14
17	14	11	9
	16		10

a. What type of experimental design was employed? Identify the key elements of the experiment: response, factor(s), factor type(s), treatments, and experimental units.

b. Test to determine whether evidence exists that the probability distributions of ratings differ for at least two of the four supervisors. Use $\alpha = .05$.

c. What assumptions are necessary to assure the validity of the test?

d. Do the results of the test warrant further comparisons of the pairs of supervisors? If so, compare all pairs of probability distributions using $\alpha = .05$ for each comparison. Does one supervisor appear to be most popular?

16.64 A union wants to determine the preferences of its members before negotiating with management. Ten union members are randomly selected, and an extensive questionnaire is completed by each member. The responses to the various aspects of the questionnaire will enable the union to rank in order of importance the items to be negotiated. The rankings are shown in the table at the top of page 994.

PERSON	MORE PAY	JOB STABILITY	FRINGE BENEFITS	SHORTER HOURS
1	2	1	3	4
2	1	2	3	4
3	4	3	2	1
4	1	4	2	3
5	1	2	3	4
6	1	3	4	2
7	2.5	1	2.5	4
8	3	1	4	2
9	1.5	1.5	3	1
10	2	3	1	4

a. What type of experimental design was employed? Identify the key elements of the experiment: response, factor(s), factor type(s), treatments, and experimental units. [*Hint:* Be careful not to confuse the rankings of the response with the response itself.]

b. Test to determine whether evidence exists that the probability distributions of ratings differ for at least two of the four negotiable items. Use $\alpha = .05$.

c. What assumptions are necessary to assure the validity of the test?

d. Do the results of the test warrant further comparisons of the pairs of negotiable items? If so, compare all pairs of probability distributions using $\alpha = .01$ for each comparison. Does one item appear to be most important?

16.65 A national clothing store franchise operates two stores in one city—one urban and one suburban. To stock the stores with clothing suited to the customers' needs, a survey is conducted to determine the incomes of the customers. Ten customers in each store are offered significant discounts if they will reveal the annual income of their household. The results are listed in the table (in thousands of dollars). Is there evidence that the probability distributions of the customers' incomes differ in location for the two stores? Use $\alpha = .05$.

STORE 1		STORE 2	
18.8	29.5	12.3	10.3
27.9	16.3	19.2	15.6
12.2	22.1	6.3	9.8
85.3	15.7	24.5	8.6
13.1	24.0	11.0	19.3

16.66 An insurance company wants to determine whether a relationship exists between the number of claims filed by owners of family policies and the annual incomes of the families. A random sample of ten policies was selected for the study. The data are shown in the table. Do the data on claims and annual income provide sufficient evidence to conclude that a correlation exists between the number of claims per policy and the annual income of the policyholder? Use $\alpha = .10$.

FAMILY	CLAIMS 3-year period	ANNUAL INCOME Averaged over 3 years ($ thousand)
1	5	14.5
2	1	9.6
3	9	62.5
4	0	22.5
5	4	10.3
6	7	16.2
7	0	8.1
8	2	21.2
9	6	17.1
10	3	12.3

16.67 A state highway patrol was interested in knowing whether frequent patrolling of highways substantially reduced the number of speeders. Two similar interstate highways were selected for the study—one very heavily patrolled and the other only occasionally patrolled. After 1 month, random samples of 100 cars were chosen on each highway and the number of cars exceeding the speed limit was recorded. This process was repeated on 5 randomly selected days. The data are shown in the table.

DAY	HIGHWAY 1 Heavily patrolled	HIGHWAY 2 Occasionally patrolled
1	35	60
2	40	36
3	25	48
4	38	54
5	47	63

a. Use the paired t test with $\alpha = .10$ to compare the population mean number of speeders per 100 cars for the two highways. What assumptions are necessary for the validity of this procedure? Do you think the assumptions are reasonable in this situation?

b. Use a nonparametric procedure to determine whether the data provide evidence to indicate that heavy patrolling tends to reduce the number of speeders. Test using $\alpha = .10$.

16.68 A hotel had a problem with people reserving rooms for a weekend and then not honoring their reservations (no-shows). As a result, the hotel developed a new reservation and deposit plan that it hoped would reduce the number of no-shows. One year after the policy was initiated, management evaluated its effect in comparison with the old policy. Compare the records given in the table on the number of no-shows for the ten nonholiday weekends preceding the institution of the new policy and the ten nonholiday weekends preceding the evaluation time. Has the situation improved under the new policy? Test at $\alpha = .05$.

BEFORE	AFTER		
10	11	4	4
5	8	3	2
3	9	8	5
6	6	5	7
7	5	6	1

16.69 In recent years, many magazines have been forced to raise their prices because of increased postage, printing, and paper costs. Because magazines are now more expensive, some households may be subscribing to fewer magazines than they did 3 years ago. Ten households were selected at random, and the number of magazines subscribed to 3 years ago and now were determined. The results are listed in the table. Does this sample provide sufficient evidence to indicate that households tend to subscribe to fewer magazines now than they did 3 years ago? Use $\alpha = .05$.

HOUSEHOLD	3 YEARS AGO	NOW	HOUSEHOLD	3 YEARS AGO	NOW
1	8	4	6	6	5
2	3	5	7	4	3
3	6	4	8	2	2
4	3	3	9	9	6
5	10	5	10	8	2

16.70 A clothing manufacturer employs five inspectors who provide quality control of workmanship. Every item of clothing produced carries with it the number of the inspector who checked it. Thus, the company can evaluate an inspector by keeping records of the number of complaints received about products bearing his or her inspection number. The numbers of returns for 6 months are given in the table.

MONTH	INSPECTOR				
	1	2	3	4	5
1	8	10	7	6	9
2	5	7	4	12	12
3	5	8	6	10	6
4	9	6	8	10	13
5	4	13	3	7	15
6	4	8	2	6	9

a. What type of design was utilized for this experiment?

b. Use the appropriate nonparametric test to compare the treatments. Specify the hypotheses and interpret the results in terms of this experiment. Use $\alpha = .10$.

c. Does the result of your test warrant further comparison of the pairs of treatments? If so, compare the pairs using the appropriate nonparametric technique and $\alpha = .05$ for each comparison. Interpret the results in terms of this experiment.

d. What assumptions are necessary to assure the validity of the nonparametric procedures you employed? How do they differ from the assumptions that would have to be satisfied in order to use the corresponding parametric technique to analyze this experiment?

16.71 A savings and loan association is considering three locations in a large city as potential office sites. The company has hired a marketing firm to compare the incomes of people living in the area surrounding each site. The market researchers interview ten households chosen at random in each area to determine the type of job, length of employment, etc., of those in the households who work. This information will enable them to estimate the annual income of each household. The results in the table (estimated annual income in thousands of dollars) are obtained.

SITE 1		SITE 2		SITE 3	
14.3	16.2	19.3	22.2	14.5	18.3
15.5	23.5	25.5	83.5	9.3	23.3
12.1	14.7	30.2	27.9	17.2	16.7
8.3	18.0	52.1	21.2	13.2	20.0
20.5	15.1	28.6	24.0	12.6	15.2

a. What type of design was utilized for this experiment?

b. Use the appropriate nonparametric test to compare the treatments. Specify the hypotheses and interpret the results in terms of this experiment. Use $\alpha = .05$.

c. Does the result of your test warrant further comparison of the pairs of treatments? If so, compare the pairs using the appropriate nonparametric technique and $\alpha = .05$ for each comparison. Interpret the results in terms of this experiment.

d. What assumptions are necessary to assure the validity of the nonparametric procedures you employed? How do they compare with the assumptions that would have to be satisfied in order to use the appropriate parametric technique to analyze this experiment?

16.72 A manufacturer wants to determine whether the number of defectives produced by its employees tends to increase as the day progresses. Unknown to the employees, a complete inspection is made of every item that was produced on one day, and the hourly fraction defective is recorded. The resulting data are given in the table. Is there evidence that the fraction defective increases as the day progresses? Test at $\alpha = .05$.

HOUR	FRACTION DEFECTIVE
1	.02
2	.05
3	.03
4	.08
5	.06
6	.09
7	.11
8	.10

16.73 A businesswoman who is looking for a new investment considers a certain suburban community to be a good location for a new restaurant. She decides to survey some residents in the area to see what type of restaurant would be preferred. Ten people are chosen at random, and each is asked to estimate how many times in the past 6 months he or she has eaten in each of three types of restaurants—fast-food, family menu, and smorgasbord. The businesswoman then ranks the numbers for each person to obtain a preference ranking. The results are shown in the table at the top of page 998. Is there evidence of differences in the locations of the probability distributions of preferences for the three restaurant types? Use $\alpha = .05$.

| PERSON | PREFERENCE RANKING OF RESTAURANT TYPE | | |
	Fast-food	Family menu	Smorgasbord
1	1	2.5	2.5
2	2	1	3
3	3	2	1
4	3	1	2
5	3	1	2
6	2	3	1
7	1.5	1.5	3
8	1	2	3
9	3	1	2
10	3	2	1

16.74 Twelve samples of variously priced carpeting were selected and tested for wearability. The cost per square yard and the number of months of wear for each of the 12 samples of carpeting are listed in the table. Do the data provide sufficient evidence to indicate that wearability increases as the price increases? Test using $\alpha = .05$.

COST	MONTHS OF WEAR	COST	MONTHS OF WEAR
$ 6.95	32.5	$ 8.45	25.2
4.25	24.8	17.95	35.3
10.85	25.6	12.95	34.6
7.99	18.4	9.99	29.7
15.25	28.3	14.85	29.9
20.50	20.4	6.25	26.3

16.75 For many years, the Girl Scouts of America have sold cookies using various sales techniques. One troop experimented with several techniques, and reported the number of sales per scout listed in the table. Is there evidence that the probability distributions of number of sales differ in location for at least two of the four techniques? Use $\alpha = .10$.

DOOR-TO-DOOR	TELEPHONE	GROCERY STORE STAND	DEPARTMENT STORE STAND
47	63	113	25
93	19	50	36
58	29	68	21
37	24	37	27
62	33	39	18
		77	31

16.76 Refer to Exercise 16.75. Compare the locations of the probability distributions of the number of sales for the door-to-door and grocery store stand techniques. Use $\alpha = .05$.

16.77 Suppose the personnel director of a company interviewed six potential job applicants without knowing anything about their backgrounds and then rated them on a scale from 1 to 10. Independently, the director's supervisor made an evaluation of the background qualifications of each candidate on the same scale. The results are shown in the table. Is there evidence that candidates' qualification scores are related to their interview performance? Use $\alpha = .10$.

CANDIDATE	QUALIFICATIONS	INTERVIEW PERFORMANCE
1	10	8
2	8	9
3	9	10
4	4	5
5	5	3
6	6	6

16.78 A taste test conducted to compare three brands of beer utilized ten randomly selected beer drinkers. Each person was given three unmarked glasses of beer—one containing each brand—and was asked to rate each on a scale from 1 to 10 (a higher score indicates a better taste). Do the data given in the table indicate that one (or more) of the brands of beer is preferred to the others? Test using $\alpha = .05$. If the brands differ, try to determine which brand is preferred by using the appropriate nonparametric procedure.

PERSON	A	B	C	PERSON	A	B	C
1	5	7	3	6	10	9	8
2	8	8	5	7	6	8	7
3	6	7	7	8	5	5	4
4	9	6	7	9	6	8	5
5	9	8	5	10	7	6	4

16.79 Two car-rental companies have long waged an advertising war. An independent testing agency is hired to compare the number of rentals at one major airport. After 10 days, the agency has the data listed in the table. At this point, can either car-rental company claim to be number one at this airport? Use $\alpha = .05$.

DAY	RENTAL COMPANY A	RENTAL COMPANY B	DAY	RENTAL COMPANY A	RENTAL COMPANY B
1	29	22	6	16	20
2	26	29	7	35	30
3	19	30	8	43	45
4	28	25	9	29	38
5	27	26	10	32	40

16.80 David K. Campbell, James Gaertner, and Robert P. Vecchio (1983) investigated the perceptions of accounting professors with respect to the present and desired importance of various factors considered in promotion and tenure decisions at major universities. One hundred fifteen professors at universities with accredited doctoral programs responded to a mailed questionnaire. The questionnaire asked the professors to rate

(1) the *current* importance placed on 20 factors in the promotion and tenure decisions at their universities and (2) how they believe the factors *should* be weighted. Responses were obtained on a 5-point scale ranging from "no importance" to "extreme importance." The resulting ratings were averaged and converted to the rankings shown in the table. Calculate Spearman's rank correlation coefficient for the data and carefully interpret its value in the context of the problem.

FACTOR	CURRENT IMPORTANCE	IDEAL IMPORTANCE
I. Teaching (and related items):		
Teaching performance	6	1
Advising and counseling students	19	15
Students' complaints/praise	14	17
II. Research:		
Number of journal articles	1	6.5
Quality of journal articles	4	2
Refereed publications:		
a. Applied studies	5	4
b. Theoretical empirical studies	2	3
c. Educationally oriented	11	8
Papers at professional meetings	10	12
Journal editor or reviewer	9	10
Other (textbooks, etc.)	7.5	11
III. Service and professional interaction:		
Service to profession	15	9
Professional/academic awards	7.5	6.5
Community service	18	19
University service	16	16
Collegiality/cooperativeness	12	13
IV. Other		
Academic degrees attained	3	5
Professional certification	17	14
Consulting activities	20	20
Grantsmanship	13	18

ON YOUR OWN ...

In Chapters 15 and 16 we have discussed two methods of analyzing a randomized block design. When the populations have normal probability distributions and their variances are equal, we can use the analysis of variance described in Chapter 15. Otherwise, we can use the Friedman F_r test.

In the "On Your Own" section of Chapter 15, we asked you to conduct a randomized block design to compare supermarket prices, and to use an analysis of variance to interpret the data. Now use the Friedman F_r test to compare the supermarket prices.

How do the results of the two analyses compare? Explain the similarity (or lack of similarity) between the two results.

USING THE COMPUTER . . .

A large department store chain has decided to locate a new store in either Mississippi or Florida. As part of the information to be used in deciding between the two states, the firm would like confirmation of its belief that the Overall Purchasing Potential Index tends to be higher in Florida's zip codes than in Mississippi's. Assume that the Florida zip codes in the data set of Appendix C are a random sample of all of Florida's zip codes. Make a similar assumption for Mississippi. Use the Wilcoxon rank sum test to provide the desired information.

REFERENCES

Alter, S. and Ginzberg, M. "Managing uncertainty in MIS implementation." *Sloan Management Review*, Fall 1978, *20*, 23–31.

Beckenstein, A. R., Gabel, H. L. and Roberts, K. "An executive's guide to antitrust compliance." *Harvard Business Review*, Sept.–Oct. 1983, 94–102.

Campbell, D. K., Gaertner, J., and Vecchio, R. P. "Perceptions of promotion and tenure criteria: A survey of accounting educators." *Journal of Accounting Education*, Spring 1983, *1*, 83–92.

Certo, S. C. *Principles of Modern Management*. Dubuque, Iowa: Wm. C. Brown, 1980. Chapter 14.

Conover, W. J. *Practical Nonparametric Statistics*. New York: Wiley, 1971.

Cook, T. Q., ed. *Instruments of the Money Market*, 4th ed. Richmond, Va.: Federal Reserve Bank of Richmond, 1977.

Davis, G. B. and Olson, M. H. *Management Information Systems*, 2nd ed. New York: McGraw-Hill, 1985.

Epping, G. M. "Importance factors in plant location in 1980." *Growth and Change*, Apr. 1982, *13*, 47–51.

Finn, D. W., Wang, C.-K., and Lamb, C. W. "An examination of the effects of sample composition bias in a mail survey." *Journal of the Market Research Society*, Oct. 1983, *25*, 331–338.

Freedman, D. A. "The mean versus the median: A case study in 4-R Act litigation." *Journal of Business and Economic Statistics*, Jan. 1985, *3*, 1–13.

Gibbons, J. D. *Nonparametric Statistical Inference*. New York: McGraw-Hill, 1971.

Helander, M. "Applicability of drivers' electrodermal responses to the design of the traffic environment." *Journal of Applied Psychology*, 1978, *63*, 481–488.

Hollander, M. and Wolfe, D. A. *Nonparametric Statistical Methods*. New York: Wiley, 1973.

Improving Real Property Assessment: A Reference Manual. International Association of Assessing Officers, 1978.

Lehmann, E. L. *Nonparametrics: Statistical Methods Based on Ranks*. San Francisco: Holden-Day, 1975.

McClure, P. "Analyzing consumer image data using the Friedman two-way analysis of variance by ranks." *Journal of Marketing Research*, Aug. 1971, 8, 370–371.

Oas, J. A. "Processing, collection of primary data simplified by optimal mark reading." *Marketing News*, Dec. 12, 1980, 24.

Rosenthal, R. and Rosnow, R. L. *The Volunteer Subject*. New York: Wiley, 1975.

Rosetti, D. K. and Surynt, T. J. "Video teleconferencing and performance." *Journal of Business Communication*, Fall 1985, *22*, 25–31.

Siegel, S. *Nonparametric Statistics for the Behavioral Sciences*. New York: McGraw-Hill, 1956.

Teece, D. J. "Internal organization and economic performance: An empirical analysis of the profitability of principal firms." *The Journal of Industrial Economics*, Dec. 1981, *30*, 173–199.

Winkler, R. L. and Hays, W. L. *Statistics: Probability, Inference, and Decision*, 2nd ed. New York: Holt, Rinehart and Winston, 1975. Chapter 12.

Wright, J. L., Owen, F., and Pena, D. "Status of MN/DOT's weigh-in-motion program." St. Paul: Minnesota Department of Transportation, Jan. 1983.

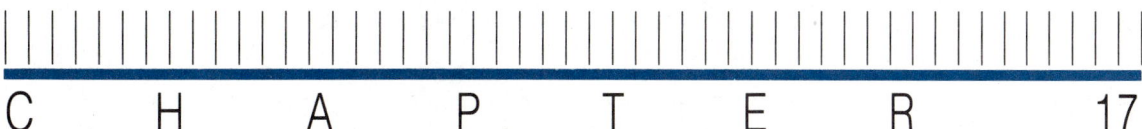

WHERE WE'VE BEEN . . .

The preceding chapters have presented statistical methods for analyzing many types of business data. Chapters 8–12 and 15 were appropriate for populations of data generated by quantitative random variables that were independent and had (at least approximately) normal probability distributions with a common variance. Nonparametric statistical procedures were presented in Chapter 16 to compare two or more populations when the assumptions of normality or common variance were likely to be violated or when the responses could be ranked only according to their relative magnitudes.

WHERE WE'RE GOING . . .

The methods of this chapter are appropriate for a type of data known as **count** or **classificatory data**. For example, a brokerage company might want to investigate the relationship between its customers' investment preferences (stocks, bonds, mutual funds, etc.) and its customers' occupations. To do this, the company would sample its customers and count the number in each preference–occupation category. Then, this data would be used to make inferences about the actual proportions of their population of customers in each category. Problems of this type, as well as others that involve count data, are the topic of Chapter 17.

THE CHI-SQUARE TEST AND THE ANALYSIS OF CONTINGENCY TABLES

CONTENTS

Many business experiments consist of enumerating the number of occurrences of some event. For example, we may count the number of defectives during a particular shift at a manufacturing plant, or the number of consumers who choose each of three brands of coffee, or the number of sales made by each of five automobile salespeople during the month of June.

In some instances, the objective of collecting the count data is to analyze the distribution of the counts in the various **classes** or **cells**. For example, we may want to estimate the proportion of smokers who prefer each of three different brands of cigarettes by counting the number in a sample of smokers who buy each brand. We will say that count data classified on a single scale have a **one-dimensional classification**. The analysis of one-dimensional count data is discussed in Section 17.1.

In many instances the objective of collecting the count data is to determine the relationship between two different methods of classifying the data. For example, we may be interested in knowing whether the size and model of the automobile purchased by new car buyers are related. Or the relationship between the shift and the number of defectives produced in a plant could be of interest. When count data are classified in a **two-dimensional** table, we call the result a **contingency table**. The analysis of general contingency tables is discussed in Section 17.2. In Section 17.3 we consider some special cases of contingency table analyses.

SECTION 17.1

ONE-DIMENSIONAL COUNT DATA: MULTINOMIAL DISTRIBUTION

Consumer preference surveys can be valuable aids in making marketing decisions. Suppose a large supermarket chain conducts a consumer preference survey by recording the brand of bread purchased by customers in its stores. Assume the chain carries three brands of bread—two major brands (A and B) and its own store brand. The brand preferences of a random sample of 150 buyers are observed, and the resulting count data appear in Table 17.1. Do these data indicate that a preference exists for any of the brands?

To answer this question, we have to know the underlying probability distribution of these count data. This distribution, called the **multinomial probability distribution**, is an extension of the binomial distribution (Section 5.4). The properties of the multinomial distribution are shown in the box.

PROPERTIES OF THE MULTINOMIAL PROBABILITY DISTRIBUTION

1. The experiment consists of n identical trials.
2. There are k possible outcomes to each trial.
3. The probabilities of the k outcomes, denoted by $p_1, p_2, \ldots, p_k$, remain the same from trial to trial, where $p_1 + p_2 + \cdots + p_k = 1$.
4. The trials are independent.
5. The random variables of interest are the counts $n_1, n_2, \ldots, n_k$ in each of the k cells.

TABLE 17.1

Consumer Preference Survey

A	B	STORE BRAND
61	53	36

You can see that the properties of the multinomial experiment closely resemble those of the binomial experiment and that, in fact, a binomial experiment is a multinomial experiment for the special case where $k = 2$.

In most practical applications involving a multinomial experiment, the true values of the k outcome probabilities, $p_1, p_2, \ldots, p_k$, will be unknown. The objective is therefore to make inferences about these probabilities.

Note that the consumer preference survey given in Table 17.1 satisfies the multinomial conditions. Suppose we want to test the null hypothesis that there is no preference for any of the three brands versus the alternative hypothesis that a preference exists for one or more of the brands. Then, letting

p_1 = Proportion of all customers who prefer brand name A

p_2 = Proportion of all customers who prefer brand name B

p_3 = Proportion of all customers who prefer the store brand

we want to test

H_0: $p_1 = p_2 = p_3 = \frac{1}{3}$ (No preference)

H_a: At least one of the proportions exceeds $\frac{1}{3}$ (A preference exists)

If the null hypothesis is true (i.e., if $p_1 = p_2 = p_3 = \frac{1}{3}$), then we would expect to see approximately $\frac{1}{3}$ of the customers in the sample purchase each brand. Or, more formally, the expected value (mean value) of the number of customers purchasing brand name A is given by

$$E(n_1) = np_1 = n\left(\tfrac{1}{3}\right) = 150\left(\tfrac{1}{3}\right) = 50$$

Similarly, $E(n_2) = E(n_3) = 50$ if no preference exists.

The following test statistic measures the degree of disagreement between the data and the null hypothesis:

$$X^2 = \frac{[n_1 - E(n_1)]^2}{E(n_1)} + \frac{[n_2 - E(n_2)]^2}{E(n_2)} + \frac{[n_3 - E(n_3)]^2}{E(n_3)}$$

$$= \frac{(n_1 - 50)^2}{50} + \frac{(n_2 - 50)^2}{50} + \frac{(n_3 - 50)^2}{50}$$

Note that the farther the observed numbers n_1, n_2, and n_3 are from their expected value (50), the larger X^2 will become. That is, large values of X^2 cast doubt on the null hypothesis and suggest that it is false.

We have to know the distribution of X^2 in repeated sampling before we can decide whether the data indicate that a preference exists. If in fact H_0 is true, X^2 can be shown to have approximately a χ^2 distribution with $(k - 1)$ degrees of freedom.* The χ^2 distribution was first introduced in Section 16.3, and the critical

---

*The derivation of the degrees of freedom for X^2 involves the number of linear restrictions imposed on the count data. We will simply give the degrees of freedom for each usage of X^2 and refer the interested reader to the references at the end of the chapter for more detail.

values are given in Table XIII of Appendix B. For the consumer preference survey in Table 17.1, with $\alpha = .05$ and $k - 1 = 3 - 1 = 2$ df, we will reject H_0 if

$$X^2 > \chi^2_{.05}$$

This value of χ^2 (found in Table XIII) is 5.99147 (see Figure 17.1). The computed value of the test statistic is

$$X^2 = \frac{(n_1 - 50)^2}{50} + \frac{(n_2 - 50)^2}{50} + \frac{(n_3 - 50)^2}{50}$$

$$= \frac{(61 - 50)^2}{50} + \frac{(53 - 50)^2}{50} + \frac{(36 - 50)^2}{50} = 6.52$$

FIGURE 17.1

Rejection Region for Consumer Preference Survey

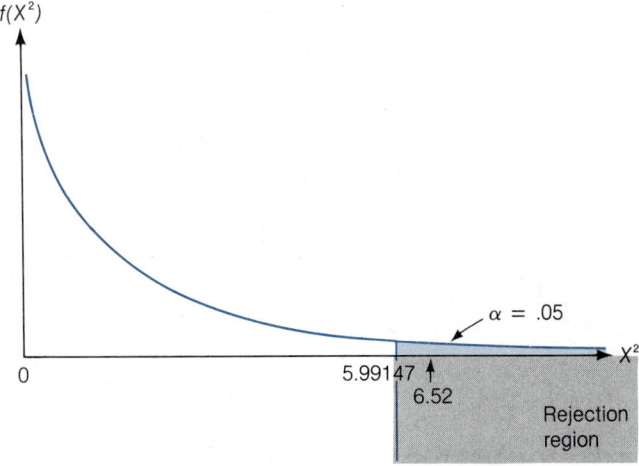

Since the computed $X^2 = 6.52$ exceeds the critical value of 5.99147, we conclude at $\alpha = .05$ that there is a customer preference for one or more of the brands of bread.

The general form for a test of hypothesis concerning multinomial probabilities is shown in the next box.

EXAMPLE 17.1

A large firm has established what is hopes is an objective system of deciding on annual pay increases for its employees. The system is based on a series of evaluation scores determined by the supervisors of each employee. Employees with scores above 80 receive a merit pay increase, those with scores between 50 and 80 receive the standard increase, while those below 50 receive no increase. The firm designed the plan with the objective that, on the average, 25% of its employees would receive merit increases, 65% would receive standard increases, and 10% would receive no increase.

After 1 year of operation using the new plan, the distribution of pay increases for the 600 company employees was as shown in Table 17.2. Test at the $\alpha = .01$

A TEST OF HYPOTHESIS ABOUT MULTINOMIAL PROBABILITIES

H_0: $p_1 = p_{1,0}, p_2 = p_{2,0}, \ldots, p_k = p_{k,0}$, where $p_{1,0}, p_{2,0}, \ldots, p_{k,0}$ represent the hypothesized values of the multinomial probabilities

H_a: At least one of the multinomial probabilities does not equal its hypothesized value

Test statistic: $X^2 = \sum_{i=1}^{k} \dfrac{[n_i - E(n_i)]^2}{E(n_i)}$

where $E(n_i) = np_{i,0}$, the expected number of outcomes of type i assuming H_0 is true. The total sample size is n.

Rejection region: $X^2 > \chi_\alpha^2$ where df $= k - 1$

Assumptions: The sample size n will be large enough so that, for every cell, the expected cell count, $E(n_i)$, will be equal to 5 or more.

---

level to determine whether these data indicate that the distribution of pay increases differs significantly from the proportions established by the firm.

TABLE 17.2
Distribution of Pay
Increases

NO INCREASE	STANDARD INCREASE	MERIT INCREASE
42	365	193

SOLUTION

Define

p_1 = Proportion of employees who receive no pay increase

p_2 = Proportion of employees who receive a standard increase

p_3 = Proportion of employees who receive a merit increase

Then the null hypothesis representing the firm's design is

H_0: $p_1 = .10, p_2 = .65, p_3 = .25$

and the alternative hypothesis is

H_a: At least two of the proportions differ from the firm's proposed plan

Test statistic: $X^2 = \sum \dfrac{[n_i - E(n_i)]^2}{E(n_i)}$

where

$E(n_1) = np_{1,0} = 600(.10) = 60$

$E(n_2) = np_{2,0} = 600(.65) = 390$

$E(n_3) = np_{3,0} = 600(.25) = 150$

Rejection region: For $\alpha = .01$ and df $= k - 1 = 2$, reject H_0 if $X^2 > \chi^2_{.01}$, where (from Table XIII of Appendix B) $\chi^2_{.01} = 9.21034$. We now calculate the test statistic:

$$X^2 = \frac{(42 - 60)^2}{60} + \frac{(365 - 390)^2}{390} + \frac{(193 - 150)^2}{150} = 19.33$$

Since this value exceeds the table value of χ^2 (9.21034), the data provide strong evidence ($\alpha = .01$) that the company's pay plan is not working as planned. ∎

By focusing on one particular outcome of a multinomial experiment, we can use the methods developed in Section 8.7 for a binomial proportion to establish a confidence interval for any one of the multinomial probabilities.* For example, if we want a 95% confidence interval for the proportion of the company's employees who will receive merit increases under the new system, we calculate

$$\hat{p}_3 \pm 1.96\sigma_{\hat{p}_3} \approx \hat{p}_3 \pm 1.96\sqrt{\frac{\hat{p}_3(1 - \hat{p}_3)}{n}} \quad \text{where } \hat{p}_3 = \frac{n_3}{n} = \frac{193}{600} = .32$$

$$= .32 \pm 1.96\sqrt{\frac{(.32)(1 - .32)}{600}} = .32 \pm .04$$

Thus, we estimate that between 28% and 36% of the firm's employees will qualify for merit increases under the new plan. It appears that the firm will have to raise the requirements for merit increases in order to achieve the stated goal of a 25% employee qualification rate.

CASE STUDY 17.1

INVESTIGATING RESPONSE BIAS IN A DIARY SURVEY

Marketing researchers sometimes collect data from consumers by asking them to keep diaries of their purchases or product usage. Such diary methods generally provide more accurate data than collection methods (such as the telephone interview) that require the consumer to recall from memory the details of his or her purchases and activities. However, diary methods are not problem-free. Frequently, response rates by consumers to requests for diary data are lower than for other types of market research surveys such as mailed questionnaires. In addition, recording biases (such as the nonrecording of events that occurred or the recording of inappropriate events) may be present in diary data (McKenzie, 1983).

John McKenzie (1983) recently studied the accuracy of telephone-call data collected by diary methods in Great Britain. As part of his study, he compared the demographic profile of a sample of 1,802 telephone users who responded to a request for diary data to the demographic profile of the population from which the sample of all those who were asked to keep diaries was selected. The population data were available from telephone company records. The particular population used in this study consisted of 29,507 households.

---

*Note that focusing on one outcome has the effect of combining the other $(k - 1)$ outcomes into a single group. Thus, we obtain, in effect, two outcomes—or a binomial experiment.

Table 17.3 shows the education profiles of the population and the sample. These data can be used to determine if the distribution of terminal education ages for the sample differs significantly from the distribution for the population. If a significant difference exists, then it can be inferred that the sample is not representative of the population and that the survey results have been affected by **response bias**.

TABLE 17.3

Terminal Education Age of Head of Household

	POPULATION		SAMPLE	
	Frequency	Relative frequency	Frequency	Relative frequency
Up to 15	13,137	$p_1 = .445$	791	.439
16–18	7,021	$p_2 = .238$	531	.295
19+	3,074	$p_3 = .104$	202	.112
Not known	6,275	$p_4 = .213$	278	.154

The sample data should lead to rejection of the following null hypothesis if the population and sample profiles differ:

H_0: $p_1 = .445, \ p_2 = .238, \ p_3 = .104, \ p_4 = .213$

H_a: At least one of the relative frequencies differs from its hypothesized value

The sample data yield

$$X^2 = \frac{(791 - 801.89)^2}{801.89} + \frac{(531 - 428.88)^2}{428.88} + \frac{(202 - 187.41)^2}{187.41} + \frac{(278 - 383.83)^2}{383.83} = 54.78$$

Since $X^2 = 54.78 > \chi^2_{.005} = 12.8381$ (df = 3), the null hypothesis is rejected and the existence of response bias is confirmed. McKenzie found similar discrepancies between the population and sample with respect to the age, sex, and social class of the head of household.

When response biases such as these can be identified or are suspected, they can generally be overcome by increasing the response rates of the survey. For a discussion of ways to increase response rates, see Sudman and Ferber (1971, 1974).

EXERCISES 17.1–17.16

LEARNING THE MECHANICS

17.1 Use Table XIII of Appendix B to find each of the following χ^2 values:
   a. $\chi^2_{.05}$ for df = 15   **b.** $\chi^2_{.990}$ for df = 100
   c. $\chi^2_{.10}$ for df = 12   **d.** $\chi^2_{.005}$ for df = 2

17.2 Find the following probabilities:
   a. $P(\chi^2 \leq .872085)$ for df = 6   **b.** $P(\chi^2 > 30.5779)$ for df = 15
   c. $P(\chi^2 \geq 82.3581)$ for df = 100   **d.** $P(\chi^2 < 13.7867)$ for df = 30

17.3 Find the rejection region for a one-dimensional χ^2 test of a null hypothesis concerning $p_1, p_2, \ldots, p_k$ if:
   a. $k = 3$ and $\alpha = .10$ **b.** $k = 5$ and $\alpha = .01$ **c.** $k = 4$ and $\alpha = .05$

17.4 What are the characteristics of a multinomial experiment? Compare the characteristics to those of a binomial experiment.

17.5 What conditions must n satisfy to make the χ^2 test valid?

CELL	n_i
1	48
2	69
3	83
4	61
5	39

17.6 A multinomial experiment with $k = 5$ cells and $n = 300$ produced the data shown in the table.
  a. Do these data provide sufficient evidence to contradict the null hypothesis that $p_1 = .15$, $p_2 = .25$, $p_3 = .30$, $p_4 = .20$, and $p_5 = .10$? Test using $\alpha = .05$.
  b. Find the approximate observed significance level for the test in part **a**.

17.7 A multinomial experiment with $k = 4$ cells and $n = 206$ produced the data shown in the table.

CELL	1	2	3	4
n_i	45	54	60	47

  a. Do these data provide sufficient evidence to conclude that the multinomial probabilities differ? Test using $\alpha = .05$.
  b. What are the Type I and Type II errors associated with the test of part **a**?

17.8 Refer to Exercise 17.7. Construct a 95% confidence interval for the multinomial probability associated with cell 3.

APPLYING THE CONCEPTS

U.S.	EUROPE	JAPAN
45	46	9

17.9 A 1985 Gallup survey portrays U.S. entrepreneurs as ". . . the mavericks, dreamers, and loners whose rough edges and uncompromising need to do it their own way set them in sharp contrast to senior executives in major American corporations" (Graham, 1985, p. lC). One of the many questions put to a sample of $n = 100$ entrepreneurs about their job characteristics, work habits, social activities, etc. concerned the origin of the car they personally drive most frequently. The responses given in the table were obtained.
  a. Do these data provide evidence of a difference in the preference of entrepreneurs for the cars of the United States, Europe, and Japan? Test using $\alpha = .05$.
  b. Do these data provide evidence of a difference in the preference of entrepreneurs for domestic versus foreign cars? Test using $\alpha = .05$.
  c. What assumptions must be made in order for your inferences of parts **a** and **b** to be valid?

17.10 Refer to Exercise 17.9. Use a 90% confidence interval to estimate the proportion of U.S. entrepreneurs who drive foreign cars.

17.11 Overweight trucks are responsible for much of the damage sustained by our local, state, and federal highway systems. Although illegal, overweight trucks proliferate. Truckers have learned to avoid weigh stations run by enforcement officers by taking back roads when weigh stations are open and/or by traveling during periods of the week when weigh stations are likely to be closed. A state highway planning agency recently monitored the movements of overweight trucks on a particular interstate highway using an unmanned, computerized scale that is built into the highway. Unknown to the truckers, the scale weighs their vehicles as they pass over it. For a particular week, each day's proportion of the week's total truck traffic (5-axle tractor truck semitrailers) was as shown in the table:

MONDAY	TUESDAY	WEDNESDAY	THURSDAY	FRIDAY	SATURDAY	SUNDAY
.191	.198	.187	.180	.155	.043	.046

Source: Dahlin and Owen (1984).

During the same week, the number of overweight trucks per day was as follows:

MONDAY	TUESDAY	WEDNESDAY	THURSDAY	FRIDAY	SATURDAY	SUNDAY
90	82	72	70	51	18	31

Source: Dahlin and Owen (1984).

a. The planning agency would like to know whether the number of overweight trucks per week is distributed over the 7 days of the week in direct proportion to the volume of truck traffic. Test using $\alpha = .05$.

b. Find the approximate p-value for the test of part **a**.

17.12 In 1960 there were 69,628,000 people in the U.S. civilian labor force, of whom 23,240,000 were women. The table describes the age distribution of those women.

AGE IN YEARS	RELATIVE FREQUENCY IN 1960
16–19	.089
20–24	.111
25–34	.178
35–44	.228
45–54	.227
55–64	.128
65 and over	.039

Source: *Statistical Abstract of the United States: 1986*, p. 392.

In 1984, there were 113,544,000 people in the U.S. civilian labor force, of whom 49,709,000 were women. A random sample of 500 working women in 1984 yielded the age distribution shown in the next table.

AGE IN YEARS	FREQUENCY IN 1984
16–19	38
20–24	75
25–34	143
35–44	110
45–54	72
55–64	50
65 and over	12

Source: Based on the relative frequencies for 1984 cited in *Statistical Abstract of the United States: 1986*, p. 392.

a. Do the sample data provide sufficient evidence to conclude that the 1984 age distribution of the female work force differs from the 1960 age distribution? Test using $\alpha = .05$.

b. In the context of the problem, specify the Type I and Type II errors associated with the hypothesis test of part a.

c. Find the approximate p-value for the test in part a.

d. Use 95% confidence intervals to estimate the proportion of women between the ages of 25 and 34 (inclusive) in the 1984 labor force and the proportion between 20 and 44 (inclusive).

17.13 A local manufacturing company utilizes a computerized sales invoice printing system. Each distinct bit of information on the invoice (e.g., sold-to address, ship-to address, sales tax, total sale) is referred to as a *field*. From a handwritten copy of each invoice, a keypunch operator transcribes the field data onto computer cards so that computer-printed invoices can be produced. The manager of data processing believes that the distribution of the number of errors per invoice appearing on printed invoices has changed dramatically since the physical arrangement of the fields on the invoices was changed 6 months ago. The table describes the distribution of the number of errors per invoice when the previous format was used:

ERRORS PER INVOICE	0	1	2	3	4	More than 4
PROPORTION OF FINISHED INVOICES	.90	.04	.03	.02	.005	.005

A random sample of 300 printed invoices was selected from those that were printed during the past week. Each invoice was examined for errors. The following data resulted:

ERRORS PER INVOICE	0	1	2	3	4	More than 4
NUMBER OF INVOICES	150	120	15	7	4	4

a. Do the data provide sufficient evidence to indicate that the proportions of printed invoices in the six error categories differ from the proportions using the previous format?

b. Find the approximate observed significance level for the test in part a.

17.14 A company that manufactures dice for gambling casinos in Nevada and New Jersey regularly inspects its product to be sure that only "fair" (i.e., balanced) dice are supplied to the casinos. One die was randomly chosen from a production lot and rolled 120 times. Counts of the numbers showing face up are recorded in the table. Do the data provide sufficient evidence to indicate that the die is unbalanced? Test using $\alpha = .10$.

NUMBERS FACE UP	1	2	3	4	5	6
FREQUENCY	28	27	20	18	15	12

17.15 Four inferential techniques, A, B, C, and D, are currently used by businesses to forecast demand for their product or service. To find out whether one technique is

A	B	C	D
48	68	45	39

preferred to any other, a random sample of 200 businesses were asked which technique they preferred. A summary of their responses is shown in the table. Is there sufficient evidence to indicate that there are differences in the proportions of businesses preferring each technique? Test using $\alpha = .05$.

17.16 Most companies target their advertising at specific income groups. To provide information to advertisers about its readers' incomes, a magazine decides to conduct a survey. A previous survey had indicated that 25% of the readers earned less than $15,000 per year, 60% earned from $15,000 to $25,000 per year, and 15% earned more than $25,000 per year. The income category breakdown for the 6,478 people who responded to the latest survey is shown in the table. Do these new survey results indicate that the proportions of the readership in the three categories have changed since the previous survey?

	INCOME CATEGORY		
	Less than $15,000	$15,000–25,000	More than $25,000
Number of respondents	1,653	3,946	879

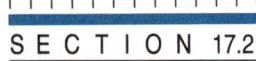

SECTION 17.2

CONTINGENCY TABLES

The energy shortage has made many consumers more aware of the size of the automobiles they purchase. Suppose an automobile manufacturer who is interested in determining the relationship between the size and manufacturer of newly purchased automobiles randomly samples 1,000 recent buyers of American-made cars. The manufacturer classifies each purchase with respect to the size and manufacturer of the purchased automobile. The data are shown in Table 17.4, which is an example of a **contingency table**. Contingency tables consist of **multinomial count data classified on two scales, or dimensions**.

TABLE 17.4
Contingency Table for
Automobile Size Example

	MANUFACTURER				TOTALS
	A	B	C	D	
Small	157	65	181	10	413
Intermediate	126	82	142	46	396
Large	58	45	60	28	191
TOTALS	341	192	383	84	1,000

Let the probabilities for the multinomial experiment in Table 17.4 be those shown in Table 17.5 (page 1014). Thus, p_{11} is the probability that a new-car buyer purchases a small car of manufacturer A. Note the probability totals, called **marginal probabilities**, for each row and column. The marginal probability p_1 is the probability that a small car is purchased, and the marginal probability p_A is the probability that a car of manufacturer A is purchased.

TABLE 17.5
Probabilities for
Contingency Table 17.4

	MANUFACTURER				TOTALS
	A	B	C	D	
Small	p_{11}	p_{12}	p_{13}	p_{14}	p_1
Intermediate	p_{21}	p_{22}	p_{23}	p_{24}	p_2
Large	p_{31}	p_{32}	p_{33}	p_{34}	p_3
TOTALS	p_A	p_B	p_C	p_D	1

Suppose we want to know whether the two classifications, manufacturer and size, are dependent. That is, if we know which size car a buyer will choose, does that information give us a clue about the manufacturer of the car the buyer will choose? In a probabilistic sense we know (Chapter 4) that independence of events A and B implies that $P(A \cap B) = P(A)P(B)$. Similarly, in the contingency table analysis, if the classifications are independent, the probability that an item is classified in any particular cell of the table is the product of the corresponding marginal probabilities. Thus, under the hypothesis of independence, in Table 17.5 we must have

$$p_{11} = p_1 p_A \qquad\qquad p_{12} = p_1 p_B$$

and so forth.

To test the hypothesis of independence, we use the same reasoning employed in the one-dimensional tests of Section 17.1. First, we calculate the expected (or mean) count in each cell assuming the null hypothesis of independence is true. We do this by noting that the expected count in the upper left-hand corner of the table, for example, is just the total number of multinomial trials, n, times the probability, p_{11}. Then

$$E(n_{11}) = np_{11}$$

and, if the classifications are independent,

$$E(n_{11}) = np_1 p_A$$

We can estimate p_1 and p_A by the sample proportions $\hat{p}_1 = n_1/n$ and $\hat{p}_A = n_A/n$. Thus, the estimate of the expected value $E(n_{11})$ is

$$\hat{E}(n_{11}) = n\left(\frac{n_1}{n}\right)\left(\frac{n_A}{n}\right) = \frac{n_1 n_A}{n}$$

Similarly,

$$\hat{E}(n_{12}) = \frac{n_1 n_B}{n}$$

$$\vdots$$

$$\hat{E}(n_{34}) = \frac{n_3 n_D}{n}$$

Using the data in Table 17.4, we find

$$\hat{E}(n_{11}) = \frac{n_1 n_A}{n} = \frac{(413)(341)}{1,000} = 140.833$$

$$\hat{E}(n_{12}) = \frac{n_1 n_B}{n} = \frac{(413)(192)}{1,000} = 79.296$$

$$\vdots$$

$$\hat{E}(n_{34}) = \frac{n_3 n_D}{n} = \frac{(191)(84)}{1,000} = 16.044$$

The observed data and the estimated expected values are shown in Table 17.6.

TABLE 17.6
Observed and Estimated
Expected (in Parentheses)
Counts

	MANUFACTURER				TOTALS
	A	B	C	D	
Small	157 (140.833)	65 (79.296)	181 (158.179)	10 (34.692)	413
Intermediate	126 (135.036)	82 (76.032)	142 (151.668)	46 (33.264)	396
Large	58 (65.131)	45 (36.672)	60 (73.153)	28 (16.044)	191
TOTALS	341	192	383	84	1,000

We now use the X^2 statistic to compare the observed and expected (estimated) counts in each cell of the contingency table:

$$X^2 = \frac{[n_{11} - \hat{E}(n_{11})]^2}{\hat{E}(n_{11})} + \frac{[n_{12} - \hat{E}(n_{12})]^2}{\hat{E}(n_{12})} + \cdots + \frac{[n_{34} - \hat{E}(n_{34})]^2}{\hat{E}(n_{34})}$$

$$= \sum_{i=1}^{3}\sum_{j=1}^{4} \frac{[n_{ij} - \hat{E}(n_{ij})]^2}{\hat{E}(n_{ij})}$$

Substituting the data of Table 17.6 into this expression yields

$$X^2 = \frac{(157 - 140.833)^2}{140.833} + \frac{(65 - 79.296)^2}{79.296} + \cdots + \frac{(28 - 16.044)^2}{16.044} = 45.81$$

Large values of X^2 imply that the observed and expected counts do not closely agree and therefore that the hypothesis of independence is false. To determine how large X^2 must be before it is too large to be attributed to chance, we make use of the fact that the sampling distribution of X^2 is approximately a χ^2 probability distribution when the classifications are independent. The number of degrees of freedom for the approximating χ^2 distribution will be $(r - 1)(c - 1)$, where r is the number of rows and c is the number of columns in the table.

For the size and manufacturer of automobiles example, the degrees of freedom for χ^2 is $(r - 1)(c - 1) = (3 - 1)(4 - 1) = 6$. Then, for $\alpha = .05$, we reject the hypothesis of independence if

$$X^2 > \chi^2_{.05} = 12.5916$$

Since the computed $X^2 = 45.81$ exceeds the value 12.5916, we conclude that the size and manufacturer of a car selected by a purchaser are dependent events.

The pattern of dependence can be seen more clearly by expressing the data as percentages. We first select one of the two classifications to be used as the base variable. In the automobile size preference example, suppose we select manufacturer as the classificatory variable to be the base. Next, we represent the responses for each level of the second categorical variable (size of automobile in our example) as a percentage of the subtotal for the base variable. For example, from Table 17.6 we convert the response for small car sales for manufacturer A (157) to a percentage of the total sales for manufacturer A (341). That is,

$$\left(\frac{157}{341}\right)100\% = 46\%$$

The conversion of all Table 17.6 entries is accomplished in the same way, and the values are shown in Table 17.7. The value shown at the right of each row is the row's total expressed as a percentage of the total number of responses in the entire table. Thus, the small car percentage is $\frac{413}{1,000}(100)\% = 41\%$ (rounded to the nearest percent).

TABLE 17.7

Percentages of Car Sizes by Manufacturer

	MANUFACTURER				ALL
	A	B	C	D	
Small	46	34	47	12	41
Intermediate	37	43	37	55	40
Large	17	23	16	33	19
TOTALS	100	100	100	100	100

If the size and manufacturer variables are independent, then the percentages in the cells of the table are expected to be approximately equal to the row percentages. Thus, we would expect the small car percentages for each of the four manufacturers to be approximately 41% if size and manufacturer were independent. The extent to which each manufacturer's percentage departs from this value determines the dependence of the two classifications, with greater variability of the row percentages meaning a greater degree of dependence. A plot of the percentages helps summarize the observed pattern. In Figure 17.2 we show the manufacturer (the base variable) on the horizontal axis and the size percentages on the vertical axis. The "expected" percentages under the assumption of independence are shown as horizontal lines, and each observed value is represented by a symbol indicating the size category.

FIGURE 17.2

Size as a Percentage of
Manufacturer Subtotals

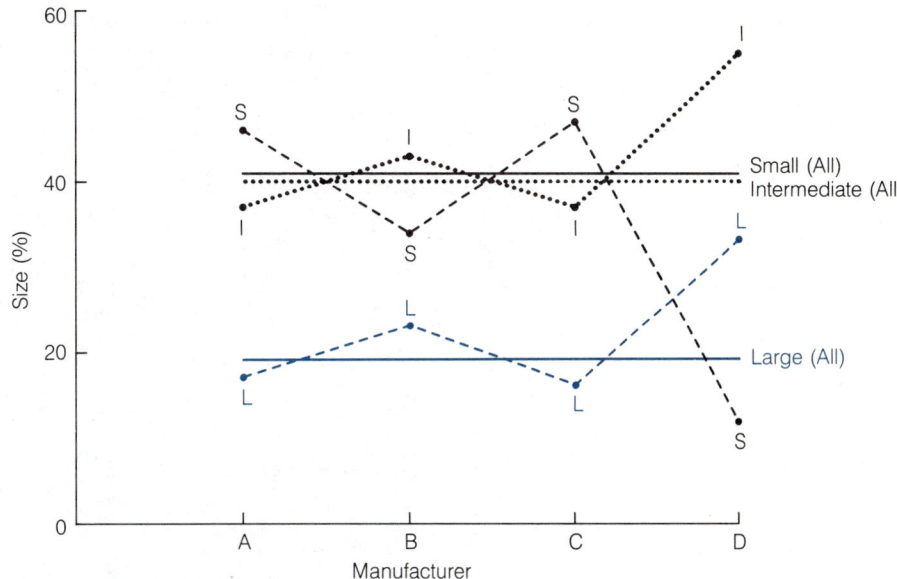

Figure 17.2 clearly indicates the reason that the test resulted in the conclusion that the two classifications in the contingency table are dependent. Note that the sales of manufacturers A, B, and C fall relatively close to the expected percentages under the assumption of independence. However, the sales of manufacturer D deviate significantly from the expected values, with much higher percentages for large and intermediate cars and a much smaller percentage for small cars than expected under independence. Also, manufacturer B deviates slightly from the expected pattern, with a greater percentage of intermediate than small car sales. Statistical measures of the degree of dependence and procedures for making comparisons of pairs of levels for classifications are available. They are beyond the scope of this text, but can be found in the references at the end of the chapter. We will, however, utilize descriptive summaries such as Figure 17.2 to examine the degree of dependence exhibited by the sample data.

The general form of a contingency table is shown in Table 17.8. Note that the

TABLE 17.8

General $r \times c$
Contingency Table

		COLUMN				ROW TOTALS
		1	2	$\cdots$	c	
ROW	1	n_{11}	n_{12}	$\cdots$	n_{1c}	r_1
	2	n_{21}	n_{22}	$\cdots$	n_{2c}	r_2
	$\vdots$	$\vdots$	$\vdots$		$\vdots$	$\vdots$
	r	n_{r1}	n_{r2}	$\cdots$	n_{rc}	r_r
COLUMN TOTALS		c_1	c_2	$\cdots$	c_c	n

observed count in the cell located in the ith row and the jth column is denoted by n_{ij}, the ith row total is r_i, the jth column total is c_j, and the total sample size is n. Using this notation, we give the general form of the contingency table test for independent classifications in the box.

GENERAL FORM OF A CONTINGENCY TABLE ANALYSIS: A TEST FOR INDEPENDENCE

H_0: The two classifications are independent

H_a: The two classifications are dependent

Test statistic: $X^2 = \sum_{i=1}^{r} \sum_{j=1}^{c} \dfrac{[n_{ij} - \hat{E}(n_{ij})]^2}{\hat{E}(n_{ij})}$

where

$$\hat{E}(n_{ij}) = \frac{r_i c_j}{n}$$

Rejection region: $X^2 > \chi_\alpha^2$, where χ_α^2 is based on $(r - 1)(c - 1)$ df

Assumption: The sample size, n, will be large enough so that, for every cell, the expected cell count, $E(n_{ij})$, will be equal to 5 or more.

EXAMPLE 17.2

A large brokerage firm wants to determine whether the service it provides to affluent customers differs from the service it provides to lower-income customers. A sample of 500 customers is selected, and each customer is asked to rate his or her broker. The results are shown in Table 17.9.

a. Test to determine whether there is evidence that broker rating and customer income are independent. Use $\alpha = .10$.

b. Plot the data and describe the patterns revealed. Is the result of the test supported by the plot?

TABLE 17.9
Observed and Estimated Expected (in Parentheses) Counts for Example 17.2

		CUSTOMER'S INCOME			TOTALS
		Under $20,000	$20,000–$50,000	Over $50,000	
	Outstanding	48 (53.856)	64 (66.402)	41 (32.742)	153
BROKER RATING	Average	98 (94.336)	120 (116.312)	50 (57.352)	268
	Poor	30 (27.808)	33 (34.286)	16 (16.906)	79
TOTALS		176	217	107	500

SOLUTION

a. The first step is to calculate estimated expected cell frequencies under the assumption that the classifications are independent. Thus,

$$\hat{E}(n_{11}) = \frac{r_1 c_1}{n} = \frac{(153)(176)}{500} = 53.856$$

$$\hat{E}(n_{12}) = \frac{r_1 c_2}{n} = \frac{(153)(217)}{500} = 66.402$$

and so forth. All the estimated expected counts are shown in Table 17.9.
We are now ready to conduct the test for independence:

H_0: The rating a customer gives his or her broker is independent of the customer's income

H_a: Broker rating and customer income are dependent

Test statistic: $X^2 = \sum\limits_{i=1}^{3}\sum\limits_{j=1}^{3} \frac{[n_{ij} - \hat{E}(n_{ij})]^2}{\hat{E}(n_{ij})}$

Rejection region: For $\alpha = .10$ and $(r - 1)(c - 1) = (2)(2) = 4$ df, reject H_0 if $X^2 > \chi^2_{.10}$, where $\chi^2_{.10} = 7.77944$.

The calculated value of X^2 is

$$X^2 = \frac{(48 - 53.856)^2}{53.856} + \frac{(64 - 66.402)^2}{66.402} + \cdots + \frac{(16 - 16.906)^2}{16.906}$$

$$= 4.28$$

Since $X^2 = 4.28$ does not exceed the critical value, 7.77944, there is insufficient evidence at $\alpha = .10$ to conclude that broker rating and customer income are dependent. This survey does not support the firm's alternative hypothesis that affluent customers get different broker service than lower-income customers.

b. The broker rating frequencies are expressed as percentages of income category frequencies in Table 17.10. The expected percentages under the assumption of independence are shown at the right of each row. The plot of the percentage data is shown in Figure 17.3 (page 1020), where horizontal lines represent the expected percentages assuming independence. Note that the response percentages deviate only slightly from those expected under the assumption of independence, supporting the result of the test in part a. That is, neither the descriptive plot

TABLE 17.10
Broker Ratings as Percentages of Income Class

| | | CUSTOMER'S INCOME | | | ALL |
		Under $20,000	$20,000–$50,000	Over $50,000	
BROKER RATING	Outstanding	27	29	38	31
	Average	56	55	47	54
	Poor	17	15	15	16
TOTALS		100	99[a]	100	101[a]

[a]Percentages do not add to 100 because of rounding.

nor the statistical test provide evidence that the rating given the broker services depends on (varies with) the customer's income.

FIGURE 17.3

Plot of Broker Rating–
Customer Income
Contingency Table

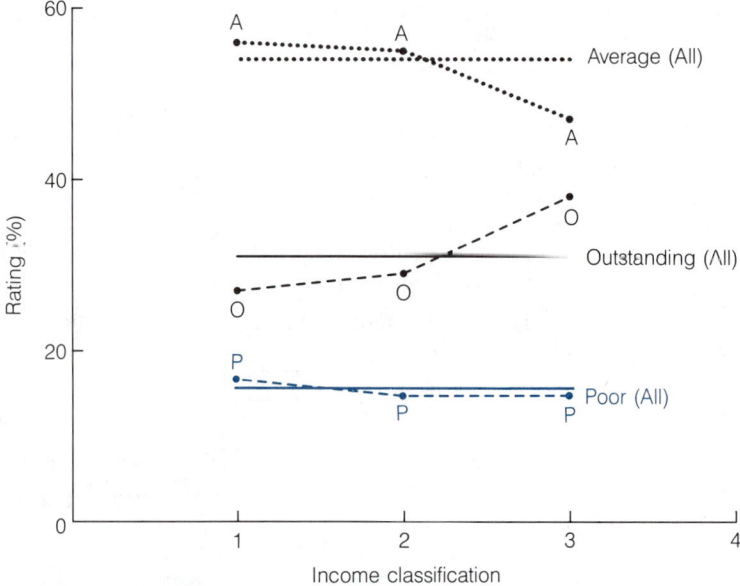

CASE STUDY 17.2

DECEIVED SURVEY
RESPONDENTS: ONCE
BITTEN, TWICE SHY

In their article "Deceived Respondents: Once Bitten, Twice Shy," Sheets et al. (1974) explore a situation sometimes encountered by marketing research personnel:

> For some time, people engaged in marketing and other field-based research have had to contend with the consequences of a fairly widely used ploy in the direct selling field: gaining a potential customer's attention and interest by requesting cooperation in some sort of false survey. Despite the efforts of the American Association for Public Opinion Research, the American Marketing Association, and other groups, and regardless of Federal Trade Commission orders, this gambit is still in use, although perhaps somewhat modified.

The authors hypothesized that a previous exposure to a false survey will increase the probability that a person will refuse to respond in a legitimate survey. They conducted an experiment in which 104 individuals were asked to cooperate in a marketing research study. The 54 people who agreed to participate were then given a low-key sales presentation for a fictitious encyclopedia. Between 2 and 4 days later, 49 of the original 54 participants (five were not available) and 70 completely new individuals (the control group) were interviewed. Each group was asked the same opening question. The results of this survey are presented in Table 17.11.

TABLE 17.11

Experimental and Control
Group Willingness to
Participate in True Market
Research

	EXPERIMENTAL	CONTROL	TOTALS
CONSENTED	12	36	48
REFUSED	37	34	71
TOTALS	49	70	119

A χ^2 test was used to analyze these count data. The χ^2 test statistic is found to be 8.691, significant at $\alpha = .005$. This indicates dependence of the refusal rate on previous exposure to false surveys. The interpretation given to these data by Sheets et al. (1974) is:

> The findings indicate support for the hypothesis: false market surveys have a deleterious effect upon respondent willingness to cooperate in subsequent market research studies. By inference, households that have been previously exposed to false research are half again as likely to refuse to cooperate in legitimate field research as those who have not. The implication for field researchers is either to stay away from areas that have had recent, heavy, direct, sales efforts or to plan for higher refusal rates in such areas.

**EXERCISES
17.17–17.32**

LEARNING THE MECHANICS

17.17 Find the rejection region for a test of independence of two classifications where the contingency table contains r rows and c columns and
a. $r = 5$, $c = 5$, $\alpha = .05$ **b.** $r = 3$, $c = 6$, $\alpha = .10$
c. $r = 2$, $c = 3$, $\alpha = .01$

17.18 Consider the accompanying 2×3 (i.e., $r = 2$ and $c = 3$) contingency table:
a. Specify the null and alternative hypotheses that should be used in testing the independence of the row and column classifications.
b. Specify the test statistic and the rejection region that should be used in conducting the hypothesis test of part **a**. Use $\alpha = .01$.
c. Assuming the row classification and the column classification are independent, find estimates for the expected cell counts.
d. Conduct the hypothesis test of part **a**. Interpret your result.

		COLUMN		
		1	2	3
ROW	1	10	32	53
	2	15	30	25

17.19 Refer to Exercise 17.18.
a. Convert the frequency responses to percentages by calculating the percentage of each column total falling in each row. Also convert the row totals to percentages of the total number of responses. Display the percentages in a table.
b. Create a graph with percentage on the vertical axis and the column number on the horizontal axis. Show the row total percentages as horizontal lines on the plot, and plot the cell percentages from part **a** using the row number as a plotting symbol.
c. What pattern do you expect to see if the rows and columns are independent? Does the plot support the results of the test of independence in Exercise 17.18?

		B		
		B_1	B_2	B_3
	A_1	39	75	42
A	A_2	63	51	70
	A_3	30	38	29

17.20 Test the null hypothesis of independence of the two classifications, A and B, of the 3×3 contingency table shown in the margin. Test using $\alpha = .05$.

17.21 Refer to Exercise 17.20. Convert the responses to percentages by calculating the percentage of each B class total falling into each A classification. Also, calculate the percentage of the total number of responses that constitute each of the A classification totals. Create a graph with percentage on the vertical axis and the B classification on the horizontal axis. Show the A classification total percentages as horizontal lines on the graph, and plot the individual cell percentages using the A class number as a plotting symbol. Does the graph support the result of the test of hypothesis in Exercise 17.20? Explain.

17.22 Test the null hypothesis of independence of the two classifications, A and B, of the 3×4 contingency table shown below. Test using $\alpha = .05$.

		B			
		B_1	B_2	B_3	B_4
	A_1	22	38	29	51
A	A_2	42	27	68	53
	A_3	26	85	102	68

17.23 A chi-square test was applied to each of the following contingency tables and the null hypothesis of independence was rejected. Construct a graph for each table that enables you to interpret the pattern of dependence in each case.

a.

		B	
		B_1	B_2
	A_1	50	10
A	A_2	10	40

b.

		D	
		D_1	D_2
	C_1	15	70
C	C_2	80	15

c.

		F		
		F_1	F_2	F_3
	E_1	10	30	15
E	E_2	30	10	15

17.24 In a contingency table test for independence, explain why the null hypothesis is independence and the alternative hypothesis is dependence (instead of vice versa).

APPLYING THE CONCEPTS

17.25 Are all employees equally prone to having accidents? Or, are certain employee groups— say, younger employees—more likely to have particular kinds of accidents? The effective implementation of hiring policies, training programs, and safety programs requires such knowledge. A study was recently conducted to address these questions for a particular manufacturing company. A portion of the research results is summarized in the accompanying contingency table.

		KIND OF ACCIDENT		
		Sprain	Burn	Cut
	Under 25	9	17	5
AGE	25 and over	61	13	12

Source: Derived from Parry, A. E. "Changing assumptions about loss frequency." *Professional Safety*, Oct. 1985, 39–43.

a. From the contingency table, the researcher concluded that there is a relationship between an employee's age and the kind of accident that the employee may have. Do you agree? Test using $\alpha = .05$.

b. According to the frequencies of the contingency table, which is the most frequent type of accident? Are younger or older employees more likely to have sprains? Burns? Justify your answers.

c. What assumptions must hold in order for your test of part **a** to be valid?

d. Plot the percentage of employees under 25 who are injured based on the total injuries for each kind of accident. Compare this to the percentage of the total number of employees under 25 who are injured based on the total of all kinds of accidents. What does the plot indicate about the pattern of dependence in the data?

17.26 Over the years, pollsters have found that the public's confidence in big business has been closely tied to the economic climate of the country. When businesses are growing and employment is increasing, public confidence is high. When the opposite occurs, public confidence is low. Harvey Kahalas (1981) explored the relationship between confidence in business and job satisfaction. He hypothesized that there is a relationship between level of confidence and job satisfaction, and that this is true for both union and nonunion workers. To test his hypothesis he used the sample data given in the tables (data were collected by the National Opinion Research Center).

I. Union members

		JOB SATISFACTION			
		Very satisfied	Moderately satisfied	Little dissatisfied	Very dissatisfied
CONFIDENCE IN MAJOR CORPORATIONS	A great deal	26	15	2	1
	Only some	95	73	16	5
	Hardly any	34	28	10	9

II. Nonunion Workers

		JOB SATISFACTION			
		Very satisfied	Moderately satisfied	Little dissatisfied	Very dissatisfied
CONFIDENCE IN MAJOR CORPORATIONS	A great deal	111	52	13	4
	Only some	246	142	37	18
	Hardly any	73	51	19	9

a. Kahalas concluded that his hypothesis was not supported by the data. Do you agree? Conduct the appropriate tests using $\alpha = .05$. Be sure to specify the null and alternative hypotheses of your tests.

b. Find and interpret the approximate p-values of the tests you conducted in part **a**.

17.27 In 1952, the National Broadcasting Company (NBC) conducted a study to determine the effects of television viewing on the purchase of products advertised on television. A sample of 2,452 women from the Davenport, Iowa, metropolitan area were interviewed, first in February and then again in May. Each time, the women were asked whether they watch a specific program and whether they purchase the product advertised during the program. Their responses are categorized in the table at the top of page 1024.

a. If a χ^2 test of independence were conducted for the table, what would be the null and alternative hypotheses?

b. Conduct the test referred to in part **a**. Test using $\alpha = .05$.

		BUYING, FEB./MAY				TOTALS
		Yes/Yes	Yes/No	No/Yes	No/No	
VIEWING FEB./MAY	Yes/Yes	460	173	191	351	1,175
	Yes/No	76	59	44	113	292
	No/Yes	86	27	53	80	246
	No/No	175	104	113	347	739
TOTALS		797	363	401	891	2,452

Source: *Journal of Marketing Research*, Feb. 1966, pp. 13–24.

c. What assumptions must you make so that the χ^2 test in part **b** will be valid?

17.28 In recent years, corporate boards of directors have been pressed to improve their monitoring of corporate economic performance and to become more careful overseers of the activities of management. In addition, boards are being asked to guide the long-term responsiveness of their respective organizations to the prevailing economic and social climate. These pressures have forced many boards of directors to become more articulate in defining the mission and strategies of their firms. To study the extent and nature of strategic planning being undertaken by boards of directors, Ahmed Tashakori and William Boulton (1983) questioned a sample of 119 chief executive officers of major U.S. corporations. One of the objectives of the study was to determine if a relationship exists between the composition of a board—where boards are classified as consisting of a majority of outside directors or a majority of inside directors—and its level of participation in the strategic planning process. To this end, the questionnaire data were used to classify the responding corporations according to the level of their board's participation in the strategic planning process:

Level 1: Board participates in formulation or implementation or evaluation of strategy

Level 2: Board participates in formulation and implementation, formulation and evaluation, or implementation and evaluation of strategy

Level 3: Board participates in formulation, implementation, and evaluation of strategy

The following results* were obtained:

LEVEL	1	2	3
NUMBER OF FIRMS	22	37	60

Of these 119 firms, 100 had boards where outside directors constitute a majority. Their levels of participation in strategic planning were as follows:

LEVEL	1	2	3
NUMBER OF FIRMS	20	27	53

*Reprinted by permission from the *Journal of Business Strategy*, Winter 1983, Vol. 3, No. 3. Copyright 1983, Warren, Gorham & Lamont, Inc., 210 South Street, Boston, Massachusetts. All rights reserved.

a. Tashakori and Boulton concluded that a relationship exists between a board's level of participation in the strategic planning process and the composition of the board. Do you agree? Construct the appropriate contingency table, and test using $\alpha = .10$.

b. In the context of the problem, specify the Type I and Type II errors associated with the test of part a.

c. Find the approximate p-value for the test in part a. Based on the p-value, would the null hypothesis have been rejected at $\alpha = .05$? Explain.

d. Construct a graph that helps to interpret the result of the test in part a.

17.29 An insurance company that sells hospitalization policies wants to know whether there is a relationship between the amount of hospitalization coverage a person has and the length of stay in the hospital. Records are selected at random at a large hospital by hospital personnel, and the information on length of stay and hospitalization coverage is given to the insurance company. The results are summarized in the table. Can you conclude that there is a relationship between length of stay and hospitalization coverage? Use $\alpha = .01$.

		LENGTH OF STAY (DAYS)			
		5 or under	6–10	11–15	Over 15
HOSPITALIZATION COVERAGE OF COSTS	Under 25%	26	30	6	5
	25–50%	21	30	11	7
	51–75%	25	25	45	9
	Over 75%	11	32	17	11

17.30 In late 1977, many farmers across the United States went on strike, protesting that the prices of farm products, chiefly grains, were less than the cost of production. Although the main strike goal was to receive 100% of parity prices for all farm products, a second controversial strike goal was to induce farmers to reduce production, thereby reducing surpluses and boosting prices. A sample survey of 100 farmers was conducted to determine whether a relationship exists between a farmer's decision to participate in the strike and the farmer's opinion concerning the necessity for a cutback in production. The results are shown in the table. Is there evidence of a relationship between a farmer's strike position and the stand on a cutback in production? Use $\alpha = .05$.

		ON STRIKE	
		Yes	No
50% CUTBACK IN PRODUCTION	Favor	21	7
	Undecided	37	2
	Opposed	22	11

17.31 Refer to Exercise 17.30. Convert the responses to percentages by calculating the percentage of farmers on strike based on each of the three production category totals. Also calculate the percentage of the total number surveyed who are on strike. Plot the percentages on the vertical axis and the production categories on the horizontal axis, showing the overall percentage on strike as a horizontal line on the plot. Show the percentage on strike for each production classification. Does the plot support the result of your test in Exercise 17.30?

17.32 One criterion used to evaluate employees in the assembly section of a large factory is the number of defective pieces per 1,000 parts produced. The quality control department wants to find out whether there is a relationship between years of experience and defect rate. Since the job is repetitious, after the initial training period, any improvement due to a learning effect might be offset by a decrease in the motivation of a worker. A defect rate is calculated for each worker for a yearly evaluation. The results for 100 workers are given in the table.

		YEARS OF EXPERIENCE (AFTER TRAINING PERIOD)		
		< 1[a]	1 < 5	5 < 10
	High	6	9	9
DEFECT RATE	Average	9	19	23
	Low	7	8	10

[a]The symbol "<" is read "less than," so "< 1" is read "less than 1" and "1 < 5" is read "1 to less than 5."

a. Is there evidence of a relationship between defect rate and years of experience? Use $\alpha = .05$.

b. Find the approximate observed significance level for the test in part **a**.

Suppose a national college placement firm wants to determine whether the job performance of college graduates is related to the region of the country in which the graduate attended college. The firm randomly selects 800 of last year's graduates who are currently employed, 200 from each of four regions: northeast (NE), southeast (SE), northwest (NW), southwest (SW). Then the employer of each graduate is contacted, and a rating of the employee's job performance is obtained. The results are shown in Table 17.12. The only difference between this contingency table and those in the previous section is that the row totals in Table 17.12 are all determined before the experiment is conducted, whereas in Section 17.2, the marginal totals were not known until after the experiment was run. Fortunately, this fact does not affect the analysis. Thus, to test dependence between job performance of college graduates and the region in which they attended college, we proceed as follows:

H_0: Job performance and region are independent

H_a: Job performance and region are dependent

Test statistic: $X^2 = \sum\limits_{i=1}^{4} \sum\limits_{j=1}^{3} \dfrac{[n_{ij} - \hat{E}(n_{ij})]^2}{\hat{E}(n_{ij})}$

Rejection region: For $\alpha = .05$ and $(r - 1)(c - 1) = 6$ df, we will reject H_0 if $X^2 > \chi^2_{.05} = 12.5916$

TABLE 17.12		JOB PERFORMANCE RATING			TOTALS
Results of Job Performance by Region		Unsatisfactory	Satisfactory	Outstanding	
REGION	NE	21 (16.75)	121 (134.75)	58 (48.5)	200
	NW	18 (16.75)	133 (134.75)	49 (48.5)	200
	SE	10 (16.75)	147 (134.75)	43 (48.5)	200
	SW	18 (16.75)	138 (134.75)	44 (48.5)	200
TOTALS		67	539	194	800

We calculate the estimated expected counts exactly as in Section 17.2:

$$\hat{E}(n_{11}) = \frac{r_1 c_1}{n} = \frac{(200)(67)}{800} = 16.75$$

$$\hat{E}(n_{12}) = \frac{r_1 c_2}{n} = \frac{(200)(539)}{800} = 134.75$$

and so forth. The estimated expected counts are shown in parentheses in Table 17.12. Then,

$$X^2 = \frac{(21 - 16.75)^2}{16.75} + \frac{(121 - 134.75)^2}{134.75} + \cdots + \frac{(44 - 48.5)^2}{48.5}$$

$$= 9.51$$

Since $X^2 = 9.51$ does not exceed the critical value of 12.5916, the placement firm cannot conclude at the $\alpha = .05$ level that job performance rating and region of college training are dependent. You can plot the percentages in each rating category across the four regions to see that the rating is independent of (relatively constant across) region.

SECTION 17.4

CAUTION

Because the X^2 statistic for testing hypotheses about multinomial probabilities is one of the most widely applied statistical tools, it is also one of the most abused statistical procedures. The user should always be certain that the experiment satisfies the properties of the multinomial experiment given in Section 17.1. Furthermore, the user should be certain that the sample is drawn from the correct population— that is, from the population about which the inference is to be made. If in Section 17.3 the placement firm had chosen 200 graduates from one college in each region, no valid inferences could be made about the entire region. We would obtain a comparison of four colleges, not four regions.

The use of the χ^2 probability distribution as an approximation to the sampling distribution for X^2 should be avoided when the expected counts are very small. The approximation can become very poor when these expected counts are small, and thus the actual value of α may be very different from the tabled value. As a rule of thumb, an expected cell count of at least 5 will mean that the χ^2 probability distribution can be used to determine an approximate critical value.

Finally, if the X^2 value does not exceed the established value of χ^2, *do not accept* the hypothesis of independence. You would be risking a Type II error (accepting H_0 if in fact it is false), and the probability, β, of committing such an error is unknown. The usual alternative hypothesis is that the classifications are dependent. Because there is literally an infinite number of ways two classifications can be dependent, it is difficult to calculate one or even several values of β to represent such a broad alternative hypothesis. Therefore, we avoid concluding that two classifications are independent, even when X^2 is small.

SUMMARY

The use of **count data** to test hypotheses about **multinomial probabilities** represents a very useful statistical technique. In a **one-dimensional table** we can use count data to test the hypothesis that the multinomial probabilities are equal to specified values. In a **two-dimensional contingency table**, we can test the independence of the two classifications. And these by no means exhaust the uses of the X^2 statistic. Many other applications can be found in the references at the end of this chapter.

Caution should be exercised to avoid misuse of the χ^2 procedure. The experiment must be multinomial,* and the expected counts should not be too small so that the χ^2 critical value may be used. Also, the X^2 statistic should not always be viewed as the final answer. If two classifications are found to be dependent, many measures of association exist for quantifying the nature and strength of their dependence (see the references).

SUPPLEMENTARY EXERCISES 17.33–17.63

17.33 Many investors believe that the stock market's directional change in January signals the market's direction for the remainder of the year. This so-called "January indicator" is frequently cited in the popular press. But is this indicator valid? If so, the well-known *random walk* and *efficient markets* theories (basically postulating that market movements are unpredictable) of stock-price behavior would be called into question. The accompanying table summarizes the relevant changes in the Dow Jones Industrial Average for the period December 31, 1927, through January 31, 1981. In a recent article, Joseph S. Martinich (1984) applied the chi-square test of independence to this data to investigate the January indicator.

		NEXT 11-MONTH CHANGE	
		Up	Down
JANUARY	Up	25	10
CHANGE	Down	9	9

*When the row (or column) totals are fixed, each row (or column) represents a separate multinomial experiment.

a. Examine the contingency table. Based solely on your visual inspection, do the data appear to confirm the validity of the January indicator? Explain.

b. Construct a plot of the percentage of years for which the 11-month movement is up based on the January change. Compare these two percentages to the percentage of times the market moved up during the last 11 months over all years in the sample. What do you think of the January indicator now?

c. If a chi-square test of independence is to be used to investigate the January indicator, what are the appropriate null and alternative hypotheses?

d. Conduct the test of part **c.** Use $\alpha = .05$. Interpret your results in the context of the problem.

e. Would you get the same result in part **d** if $\alpha = .10$ were used? Explain.

17.34 The classification of solder joints as acceptable or rejectable is a particularly difficult inspection task due to its subjective nature. Westinghouse Electric Company has experimented with different means of evaluating the performance of solder inspectors. One approach involves comparing an individual inspector's classifications with those of the group of experts that comprise Westinghouse's Work Standards Committee. In an experiment reported by Joseph J. Meagher and Joseph A. Scazzero (1985) of Westinghouse, 153 solder connections were evaluated by the committee and 111 were classified as acceptable. An inspector evaluated the same 153 connections and classified 124 as acceptable. Of the items rejected by the inspector, the committee agreed with 19.

a. Construct a contingency table that summarizes the classifications of the committee and the inspectors.

b. Based on a visual examination of the table you constructed in part **a**, does it appear that there is a relationship between the inspector's classifications and the committee's? Explain. (A plot of the percentage rejected by committee and inspector will aid your examination.)

c. Conduct a chi-square test of independence for these data. Use $\alpha = .05$. Carefully interpret the results of your test in the context of the problem.

17.35 Consumers have traditionally viewed products with warranties more favorably than products without warranties. In fact, several studies have demonstrated that, when given the choice between two similar products, one of which is warranted, consumers prefer the warranted product, even at a higher price. Thus, consumers generally perceive warranties as a kind of "value" added to the product. However, a substantial number of firms have been found to perceive their warranties primarily as legal disclaimers of responsibility and nothing more. As a result of the differences in perceptions by consumers and businesses, Congress passed the Magnuson–Moss Warranty Act, which took effect in 1977. Its purpose was to reform consumer product warranty practices. According to this act, all warranties must be designated as "full" or "limited" and must be clearly written in readily understood language. Further, it specified what was to be contained in warranties (McDaniel and Rao, 1982).

Recently, McDaniel and Rao (1982) undertook a study to investigate consumer satisfaction with warranty practices since the advent of the Magnuson–Moss Warranty Act. Using a mailed questionnaire, they sampled 237 midwestern consumers who had purchased a major appliance within the past 6 to 18 months. One of the questions they asked the consumers was, "Do most retailers and dealers make a conscientious effort to satisfy their customers' warranty claims?" One hundred fifty-six answered yes, 61 were uncertain, and 20 said no.

The population of consumers from which this sample was drawn had also been investigated 2 years prior to the Magnuson–Moss Warranty Act. At that time, 37.0% of the population answered yes to the same question, 53.3% were uncertain, and 9.7% said no.

a. As reflected in the answers to the above question, have consumer attitudes toward warranties changed since the pre-Magnuson–Moss Act study? Test using $\alpha = .05$.

b. Compare the pre- and post-Magnuson–Moss Warranty Act responses, and describe the changes that have occurred.

17.36 The U.S. Postal Service is investigating the effect of alternative mail-sorting procedures on the percentage of sorting errors. A sorting error occurs when a piece of mail is placed into an incorrect zip code category. The three alternatives to be evaluated are:

1. All manual (i.e., mail clerks sort all mail)

2. Mixed manual–automated (i.e., mail clerks sort handwritten addresses and optical scanning machines sort typewritten addresses)

3. All automated (i.e., optical scanning machines sort all mail)

To evaluate these three alternatives, the Postal Service randomly selects 1,500 pieces of mail from the main post office in Washington, D.C. Five hundred pieces of mail are randomly assigned to each of the three sorting procedures. The experiment produced these error rates for the three sorting procedures:

All manual:	19%
Mixed manual–automated:	17%
All automated:	24%

a. Based on these sample data, would you conclude that the error rates differ among the mail sorting procedures? Test using $\alpha = .05$.

b. Find the approximate observed significance level for the test.

17.37 When a buyer charges a purchase, the seller records the sale in his or her record books under a category called *accounts receivable*. Some retailers monitor the status of their accounts receivable by regularly classifying each as being in one of the following categories: current, 1–30 days late, 31–60 days late, over 60 days late, or uncollectable. Historical data indicate that the status of a particular retailer's accounts receivable can be described as follows:

Current	65%
1–30 days late:	15%
31–60 days late:	10%
Over 60 days late:	7%
Uncollectable:	3%

Six months after the interest rate charged to late accounts was increased, the status of the retailer's 200 accounts receivable was as follows:

Current:	78%
1–30 days late:	12%
31–60 days late:	5%
Over 60 days late:	2%
Uncollectable:	3%

a. Is there evidence to indicate that the increase in interest rates affected the timing of buyers' payments? Test using $\alpha = .10$.

b. Find the approximate observed significance level for the test.

	Yes	No
Men	63.16%	36.84%
Women	77.29%	22.71%

Source: Kaufman and Wolf (1982) and personal communication from Lois Kaufman.

17.38 In Case Study 9.2, we described part of the statistical analysis used by Kaufman and Wolf (1982) to examine the preferences and anxieties of men and women with respect to attending job interviews in hotel rooms. In this exercise, we supply you with additional data from their study and ask you to analyze it. A random sample of 302 students (95 men and 207 women) were asked whether they would be anxious or uncomfortable about interviewing in a hotel room. The responses are shown in the table. Do these data provide sufficient evidence to conclude that anxiety over hotel room interviewing is related to the interviewee's sex? Test using $\alpha = .05$.

17.39 Organizations that loan money to businesses are in effect gambling that the business will become (or remain) successful long enough so that the loan will not be defaulted. Loan applicants are carefully screened to weed out those firms with a high probability of defaulting. Characteristics of an applicant firm that might be examined by a loan officer include such things as firm's age and legal structure. The loan officer's evaluation will also consider the amount of the requested loan, the length of the loan, and the type of loan (e.g., for existing business, to buy existing business, or to start a new business).

Albert L. Page et al. (1977) conducted an empirical investigation of the past loan performance of the Greater Cleveland Growth Corporation, an affiliate of the Office of Minority Business Enterprise. Part of the study involved identifying demographic and firm characteristic variables that are related to the status of loans made by the Growth Corporation. Loans are classified as paid off, current, or defaulted. Loans that are either paid off or current are regarded as good loans.

Page et al. examined a sample of 64 loan histories and observed the frequencies for the six loan status and legal structure categories as shown in the table.

LOAN STATUS	FIRM'S LEGAL STRUCTURE	FREQUENCY
Defaulted	Sole proprietorship	14
Paid off or current	Sole proprietorship	13
Defaulted	Partnership	10
Paid off or current	Partnership	1
Defaulted	Corporation	12
Paid off or current	Corporation	14

Source: Reprinted from Page et al. (1977) by permission of the publisher. Copyright 1977 by Elsevier Science Publishing Company, Inc.

a. After applying a contingency table analysis to the data, Page et al. concluded that a relationship exists between the legal structure of an applicant firm and the success or failure of the loan. Test their conclusion using $\alpha = .05$.

b. Find the approximate observed significance level for the test in part **a**.

c. What assumptions must you make so that the test conclusions in part **a** will be valid?

d. Use a 95% confidence interval to estimate the proportion of loans made by the Growth Corporation that will be defaulted.

17.40 Along with the technological age comes the problem of workers being replaced by machines. A labor management organization wants to study the problem of workers displaced by automation in three industries. Case reports for 100 workers whose loss of job is directly attributable to technological advances are selected within each industry. For each worker selected, it is determined whether he or she was given another job within the same company, found a job with another company in the same industry, found a job in a new industry, or has been unemployed for longer than 6 months. The results are given in the table. Does the plight of automation-displaced workers depend on the industry? Use $\alpha = .01$.

		SAME COMPANY	NEW COMPANY, SAME INDUSTRY	NEW INDUSTRY	UNEMPLOYED
	A	62	11	20	7
INDUSTRY	B	45	8	38	9
	C	68	19	8	5

17.41 Refer to Exercise 17.40. Estimate the difference between the proportions of displaced workers who find work in another industry for industries A and C. Use a 95% confidence interval.

17.42 A computer used by a 24-hour banking service is supposed to randomly assign each transaction to one of five memory locations. A check at the end of a day's transactions gave the following counts to each of the five memory locations:

MEMORY LOCATION	1	2	3	4	5
NUMBER OF TRANSACTIONS	90	78	100	72	85

Is there evidence to indicate a difference in the proportions of transactions assigned to the five memory locations? Test using $\alpha = .025$.

17.43 Refer to Case Study 17.2.
  a. Verify that the value of the test statistic is 8.691.
  b. Conduct the χ^2 test described in the case study.
  c. Is the p-value of the test you conducted in part **b** less than or greater than .005? Explain.
  d. In the context of the problem, describe the Type I and Type II errors associated with the test of part **b**.

17.44 A restauranteur who owns restaurants in four cities is considering the possibility of building separate dining rooms for nonsmokers to accommodate customers who wish to dine in a smoke-free environment. Since this would involve significant expense, the restauranteur plans to survey the customers at each restaurant and ask them the following question: "Would you be more comfortable dining here if there was a separate dining room for nonsmokers only?" Suppose 75 people were randomly selected and surveyed at each restaurant with the results shown in the table. Is there sufficient evidence to indicate that customer preferences are different for the four restaurants (i.e., that customer preference and restaurant are dependent)? Use $\alpha = .10$.

		ANSWER TO QUESTION		
		Yes	No	It makes no difference
	1	38	32	5
RESTAURANT	2	42	26	7
	3	35	34	6
	4	37	30	8

17.45 It is commonly assumed that the more experience a job applicant has, the better that person will perform the necessary duties. Other factors, such as whether the person has a college degree or is male or female, also may be indicative of future performance. H. M. Greenberg and J. Greenberg (1980) argue that for sales jobs, the most important factor is the matching of the particular job requirement with an applicant's personal characteristics. This, they claim, will result in better retention of employees and produce higher levels of job performance. To validate this claim they studied two groups of recently hired sales personnel. In the first group, which numbered 1,980, all were job-matched; the 3,961 members of the second group were not. After 6 months they were evaluated, and the aggregate data are shown in the table, where 1 represents the highest level of performance and 4 represents the lowest. The tabulated values are the percentages of total sales personnel contained in the respective samples.

	PERFORMANCE					TOTALS
	1	2	3	4	Quit or fired	
Job-matched	9%	40%	32%	14%	5%	100%
Not job-matched	2%	17%	25%	31%	25%	100%

a. Use both the percentages given in the table and the sample sizes to construct a contingency table that shows the numbers of sales personnel falling in each category of the table.

b. Do the data provide sufficient evidence to indicate that the proportions of sales personnel falling in the performance categories depend on whether the people are job-matched? Test using $\alpha = .05$.

c. Do the data provide sufficient evidence to indicate the proportion of sales personnel receiving the highest rating (1) is larger if job-matched than if not? Test using $\alpha = .05$. [*Note:* This will require a one-sided test.]

17.46 An economist wanted to determine whether there is a relationship between a person's income and his or her political affiliation. The economist randomly sampled 265 registered voters and determined the income and political affiliation of each. A summary of the data is shown in the table. Do the data provide sufficient evidence to indicate a relationship between political affiliation and annual income? Test using $\alpha = .10$.

| | ANNUAL INCOME ($ THOUSAND) | | | |
	30 or over	20 < 30	10 < 20	Below 10
Republican	50	28	20	12
Democrat	14	35	35	41
Other	6	7	10	7

17.47 Despite a good winning percentage, a certain major league baseball team has not drawn as many fans as one would expect. In hopes of finding ways to increase attendance, management plans to interview fans who come to the games to find out why they come. One thing management might want to know is whether there are differences in support for the team among various age groups. Suppose the information in the table was collected during interviews with fans selected at random. Can you conclude that there is a relationship between age and number of games attended per year? Use $\alpha = .05$.

| | | NUMBER OF GAMES ATTENDED PER YEAR | | |
		1 or 2	3–5	Over 5
	Under 20	78	107	17
AGE	21–30	147	87	13
OF	31–40	129	86	19
FAN	41–55	55	103	40
	Over 55	23	74	22

17.48 If a company can identify times of day when accidents are most likely to occur, extra precautions can be instituted during those times. A random sampling of the accident report records over the last year at a plant gives the frequency of occurrence of accidents during the different hours of the workday. Can it be concluded from the data in the table that the proportions of accidents are different for the four time periods?

HOURS	1–2	3–4	5–6	7–8
NUMBER OF ACCIDENTS	31	28	45	47

17.49 *Product or service quality* is generally defined as fitness for use. This means the product or service meets the customer's needs. Generally speaking, fitness for use is based on five quality characteristics: technological (e.g., strength, hardness), psychological (taste,

SHIFT	NUMBER OF DEFECTIVES PRODUCED
First	25
Second	35
Third	80

beauty), time-oriented (reliability), contractual (guarantee provisions), and ethical (courtesy, honesty). The quality of a service may involve all these characteristics, while the quality of a manufactured product generally depends on technological and time-oriented characteristics (Schroeder, 1985). Following a barrage of customer complaints about the quality of its product, a manufacturer of gasoline filters for automobiles had its quality inspectors sample 600 filters—200 for each work shift—and check them for defects. The data in the table resulted.

a. Do the data indicate that the quality of the filters being produced may be related to the shift producing the filter? Test using $\alpha = .05$.

b. Estimate the proportion of defective filters produced by the first shift. Use a 95% confidence interval.

17.50 A national survey was conducted to determine the general public's view of the federal government's involvement in the regulation of private enterprise. Two hundred people from each of three income levels were asked if they thought the government was too involved, not involved enough, or involved just enough. A summary of their responses is shown in the table. Do the data provide sufficient information to indicate a relationship between income and view on government regulation of private enterprise? Test using $\alpha = .05$.

		INVOLVEMENT			TOTALS
		Too little	Just enough	Too much	
	Low	125	48	27	200
INCOME	Medium	103	58	39	200
	High	72	69	59	200
TOTALS		300	175	125	600

17.51 An appliance store is having a sale and wants to determine which modes of advertising are effective. A random sample of customers who learned about the sale indicated their source of information. A summary of the responses is given below:

Television:	53
Radio:	32
Newspaper:	36
Word of mouth:	48

Is there evidence that the proportions of customers who learned about the sale differ for the four modes of advertising? Use $\alpha = .05$.

17.52 Refer to Exercise 17.51. Estimate the proportion who learn about the sale by word of mouth. Use a 90% confidence interval.

17.53 Suppose an industrial security firm wants to conduct a study of criminal cases involving stolen company money in which employees have been found guilty. Among the data they record are the employee's salary (wages) and the amount of money stolen from the company for 400 recent cases. The results are given at the top of page 1036.

		AMOUNT STOLEN ($)			
		Under 5,000	5,000–9,999	10,000–19,999	20,000 or more
INCOME OF EMPLOYEE ($ THOUSAND)	Under 15	46	39	17	5
	15–25	78	79	61	19
	Over 25	5	14	25	12

a. Does this information provide evidence of a relationship between employee income and amount stolen? Use $\alpha = .05$.

b. Convert the responses to percentages by calculating the percentage in each income category based on the total number of cases in each amount stolen category. Also calculate the percentage of all 400 cases in each income category. Plot percentage on the vertical axis and amount stolen on the horizontal axis, showing the overall income category percentages as horizontal lines on the plot. Plot the income percentages for each cell using 1, 2, and 3 as the plotting symbols for the three categories. Does the plot support the result of the test in part **a**?

17.54 A local bank plans to offer a special service to its young customers. To determine their economic interests, a survey of 100 people under 30 years of age is conducted. Each person is asked to identify his or her top two financial priorities from the six choices shown in the table. Use the χ^2 test to determine whether the proportions of responses differ for the six pairs of priorities. Test at $\alpha = .10$.

FIRST PRIORITY	SECOND PRIORITY	NUMBER OF RESPONSES
Buy a car	Go on a trip	15
Car	Save money	14
Save	Car	22
Save	Trip	23
Trip	Car	10
Trip	Save	16

17.55 A corporation owns several convenience stores that are open 24 hours a day. It is interested in knowing whether there is a relationship between the time of day and the size of purchase. One of its stores is selected at random to be involved in a study. Store records are collected over a period of several weeks and then 300 purchases are randomly selected. Since the register also prints the time of the purchase, this random selection procedure yields both amount and time of purchase. The information is summarized in the table. Is there a relationship between time and size of purchase? Use $\alpha = .05$.

		SIZE OF PURCHASE		
		$2 or less	$2.01–$7	Over $7
TIME OF PURCHASE	8 A.M.–3:59 P.M.	65	38	14
	4 P.M.–11:59 P.M.	61	49	10
	12 midnight–7:59 A.M.	29	27	7

17.56 Refer to Exercise 17.55. Use a 90% confidence interval to estimate the difference between the proportions of customers who spend $2 or less for the periods 8 A.M.–3:59 P.M. and 12 midnight–7:59 A.M.

17.57 Five candidates have just entered the race for mayor of a large city. To determine whether any of the candidates has an early lead in popularity, 2,000 voters were polled and each was asked to indicate the candidate he or she preferred. A summary of their responses is shown in the table.

CANDIDATE	I	II	III	IV	V
VOTERS WHO PREFER CANDIDATE	385	493	628	235	259

   a. Do the data provide sufficient evidence to indicate a difference in preference for the five candidates? Test using $\alpha = .01$.
   b. Find the approximate observed significance level for the test in part **a**.

17.58 A city has three television stations. Each station has its own evening news program from 6:00 to 6:30 P.M. every weekday. An advertising firm wants to know whether there is an unequal breakdown of the evening news audience among the three stations. One hundred people are selected at random from those who watch the evening news on one of these three stations. Each is asked to specify which news program he or she watches. Do the results in the table provide sufficient evidence to indicate that the three stations do not have equal shares of the evening news audience? Use $\alpha = .05$.

STATION	1	2	3
NUMBER OF VIEWERS	35	43	22

17.59 Several life insurance firms have policies geared to college students. To get more information about this group, a major insurance firm interviewed college students to find out the type of life insurance they preferred, if any. The accompanying table was produced after surveying 1,600 students.

	Preferred a Term Policy	Preferred a Whole-life Policy	No Preference
Females	116	27	676
Males	215	33	533

   a. Is there evidence that the life insurance preference of students depends on their sex?
   b. Find the approximate observed significance level for the test in part **a**.
   c. Construct a plot of percentage responses that will help to interpret the result of the test in part **a**.

17.60 Refer to Exercise 17.59. Estimate the difference in the proportions of female and male college students who have no preference about life insurance.

17.61 A statistical analysis is to be done on a set of data consisting of 1,000 monthly salaries. The analysis requires the assumption that the sample was drawn from a normal distribution. A preliminary test, called the χ^2 **goodness-of-fit test**, can be used to help determine whether it is reasonable to assume that the sample is from a normal distribution. Suppose the mean and standard deviation of the 1,000 salaries are hypothesized to be $1,200 and $200, respectively. Using the standard normal table, we can approximate the probability of a salary being in the intervals listed in the table. The third column represents the expected number of the 1,000 salaries to be found in each interval if the sample was drawn from a normal distribution with $\mu = \$1,200$ and $\sigma = \$200$. Suppose the last column contains the actual observed frequencies in the sample. Large differences between the observed and expected frequencies cast doubt on the normality assumption.

INTERVAL	PROBABILITY	EXPECTED FREQUENCY	OBSERVED FREQUENCY
Less than $800	.023	23	26
$800 < $1,000	.136	136	146
$1,000 < $1,200	.341	341	361
$1,200 < $1,400	.341	341	311
$1,400 < $1,600	.136	136	143
$1,600 or above	.023	23	13

a. Compute the χ^2 statistic based on the observed and expected frequencies—just as you did in Section 17.1.
b. Find the tabulated χ^2 value when $\alpha = .05$ and there are 5 df (there are $k - 1 = 5$ df associated with this X^2 statistic).
c. Based on the X^2 statistic and the tabulated χ^2 value, is there evidence that the salary distribution is nonnormal?*
d. Find the approximate observed significance level for the test in part **c**.

17.62 Suppose a random variable is hypothesized to be normally distributed with mean 0 and standard deviation 1. A random sample of 200 observations on the variable yields frequencies in the listed intervals as shown in the table. Do the data provide sufficient evidence to contradict the hypothesis that x is normally distributed with $\mu = 0$ and $\sigma = 1$? Use the technique developed in Exercise 17.61.

INTERVAL	$x < -2$	$-2 \le x < -1$	$-1 \le x < 0$	$0 \le x < 1$	$1 \le x < 2$	$x \ge 2$
FREQUENCY	7	20	61	77	26	9

17.63 Refer to Exercise 9.99. Use contingency table analysis to test the hypothesis of interest to Dornoff and Tankersley. Use $\alpha = .01$. List the assumptions you made in conducting your hypothesis test, and comment on their appropriateness.

---

*If we want to test the null hypothesis that a population's relative frequency distribution is normal with unspecified mean and variance, we will need to estimate μ and σ in order to estimate the k cell probabilities. We lose 2 df corresponding to these estimates, so that the χ^2 rejection region will be based on $(k - 3)$ df.

ON YOUR OWN...

Market researchers rely on surveys to estimate the proportions of the consumer market that prefer various brands of a product. Choose a product with which you are familiar, and *guesstimate* the proportion of consumers you think favor the major brands of the product. (Choose a product for which there are at least three major brands sold in the same store.)

Now go to a store that carries these brands, and observe how many consumers purchase each brand. Be sure to observe long enough so that at least five (and preferably at least ten) purchases of each brand have been made. Also, quit sampling after a predetermined length of time or after a predetermined number of total purchases, rather than at some arbitrary time, which could bias your results.

Use the count data to test the null hypothesis that the true proportions of consumers who favor each brand equal your presampling guesstimates of the proportions. Would failure to reject this null hypothesis imply that your guesstimates are correct?

USING THE COMPUTER...

Are monthly homeowner costs related to the region of the country in which a homeowner lives? Use the zip code data of Appendix C to investigate this question.

a. Conduct a χ^2 test of independence to answer the above question. Treat the variable median monthly homeowner cost as if it had four classes: \$0–\$200.00, \$200.01–\$400.00, \$400.01–\$600.00, and \$600.01–\$800.00.

b. Construct a table similar to Table 17.10 and a graph similar to Figure 17.3 to help explain the results of the test you conducted in part **a**. Interpret your results.

REFERENCES

Conover, W. J. *Practical Nonparametric Statistics*. New York: Wiley, 1971.

Dahlin, C. and Owen, F. *An Analysis of Data Collected at the I-494 Weighing-in-motion Site*. St. Paul: Minnesota Department of Transportation, 1984.

Graham, E. "The entrepreneurial mystique," *Wall Street Journal*, May 20, 1985, p. 1C.

Greenberg, H. M. and Greenberg, J. "Job-matching for better sales performance." *Harvard Business Review*, Sept.–Oct. 1980.

Hollander, M. and Wolfe, D. A. *Nonparametric Statistical Methods*. New York: Wiley, 1973.

Kahalas, H. "The relationship between confidence in business and job satisfaction for union and nonunion members." *Baylor Business Studies*, Feb.–Apr. 1981, *127*, 45–53.

Kaufman, L. and Wolf, J. "Hotel room interviewing—Anxiety and suspicion." *Sloan Management Review*, Spring 1982, *23*, 57–64.

Martinich, J. S. "The January indicator: A nonrandom but unprofitable walk." *Mid-South Business Journal*, 1984, Vol. 4, No. 4.

McDaniel, S. W. and Rao, C. P. "Consumer attitudes toward and satisfaction with warranties and warranty performance—Before and after Magnuson–Moss." *Baylor Business Studies*. Nov.–Dec. 1982, *130*, 47–61.

McKenzie, J. "The accuracy of telephone call data collected by diary methods." *Journal of Marketing Research*. Nov. 1983, 20, 417–427.

Meagher, J. J. and Scazzero, J. A. "Measuring inspector variability." *1985 ASQC Quality Congress Transaction*, Baltimore, May 1985, 75–81.

Neter, J., Wasserman, W., and Whitmore, G. A. *Applied Statistics*, 2nd ed. Boston: Allyn & Bacon, 1982. Chapter 17.

Page, A. L., Trombetta, W. L., Werner, C., and Kulifay, M. "Identifying successful versus unsuccessful loans held by the minority small business clients of an OMBE affiliate." *Journal of Business Research*, June 1977, 5, 139–153.

Schroeder, R. G. *Operations Management*, 2nd ed. New York: McGraw-Hill, 1985. Chapter 21.

Sheets, T., Radlinski, A., Kohne, J., and Brunner, G. A. "Deceived respondents: Once bitten, twice shy." *Public Opinion Quarterly*, 1974, 18, 261–263.

Siegel, S. *Nonparametric Statistics for the Behavioral Sciences*. New York: McGraw-Hill, 1956. Chapter 9.

Sudman, S. and Ferber, R. "A comparison of alternative procedures for collecting consumer expenditure data for frequently purchased products." *Journal of Marketing Research*, May 1974, 11, 129–135.

Sudman, S. and Ferber, R. "Experiments in obtaining consumer expenditures by diary methods." *Journal of the American Statistical Association*, Dec. 1971, 66, 725–735.

Tashakori, A. and Boulton, W. "A look at the board's role in planning." *Journal of Business Strategy*, Winter 1983, 3, 64–70.

Winkler, R. L. and Hays, W. L. *Statistics: Probability, Inference and Decision*, 2nd ed. New York: Holt, Rinehart and Winston, 1975. Chapter 12.

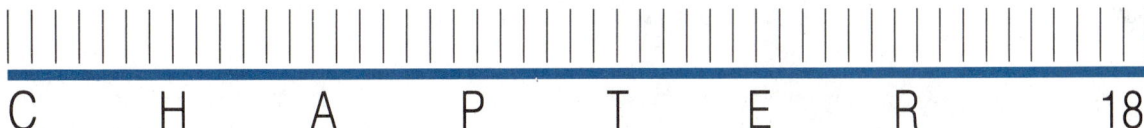

C H A P T E R 18

WHERE WE'VE BEEN . . .

In previous chapters we used a decision procedure to test hypotheses about population parameters. Using sample information, we decided to reject or accept the null hypothesis based on the calculated probabilities of making incorrect decisions—namely, the probability (α) of rejecting the null hypothesis if it was in fact true, and the probability (β) of accepting the null hypothesis if it was actually false. In this simplistic process, we assumed that a manager would be able to assess the gains or losses associated with each type of error and choose a test with acceptable values of α and β.

WHERE WE'RE GOING . . .

In Chapters 18 and 19, we present the basic concepts of a general theory for making decisions that explicitly accounts for the gains or losses associated with alternative decisions and the probabilities of the occurrence of these gains or losses. Chapter 18 is concerned with how to handle decision problems using only information that is currently available about the problem. In Chapter 19, we extend the analysis to include the case in which additional information can be obtained by sampling.

DECISION ANALYSIS USING PRIOR INFORMATION

CONTENTS

Suppose you have been given the responsibility of determining whether your firm should expand its sales region to include the southwestern part of the United States. Before making your decision, you would probably want answers to many questions. How large would the yearly demand for the product be? How many salespeople would be assigned to the new territory? How much and what types of advertising would be used? Are adequate warehousing facilities available? Who would the company's principal competitors be, and how would they react to new competition? Even if it were possible to obtain accurate answers (*perfect information*) to these and other pertinent questions, your decision problem would be extremely complex. Realistically, however, you cannot expect to receive perfect information. Thus, in making your decision, you will face the more complex problem of having to deal with answers about which you are uncertain.

How would you tackle such a decision problem? The most natural first step—and one that is used by most decision analysts—is to reduce the problem to a manageable size by considering only questions that bear significantly on the objective of your decision. In this case, your objective may be to increase corporate profit. It may turn out that the profitability of the decision to expand the sales region depends primarily on the extent of the demand for the product in the new territory during the first year after it has been introduced. However, since this demand is unknown, even with this simplified decision problem, you still must make your decision in the face of uncertainty. Given your uncertainty about the demand, how much information about demand in the new region would you want prior to making your decision, and how would you process such information? Answers to these questions are provided by the methodology referred to as **decision analysis**, which is the subject of this chapter and Chapter 19.

Decision analysis is a systematic approach to solving decision problems optimally under conditions of uncertainty. It *does not describe how or why* an individual makes a decision; rather, it *prescribes* a decision for the individual that is *consistent with his or her preferences and attitudes toward risk*. You might be asking yourself: "Why do I need to study decision-making as though it were a science, when I know most decisions (business or otherwise) are made on an intuitive level?" The answer is fourfold:

1. Yes, the vast majority of decisions made in business do not require, and are made without, formal analysis. But for that one crucial decision upon which "everything depends," it is very helpful to have a systematic, logical decision procedure to follow.
2. Most of us have had little experience intuitively processing the probabilistic and sample information that may confront us in a complex decision-making problem. Consequently, it is frequently more profitable to rely on the mechanically generated information of decision analysis to guide decision-making than on the less reliable information-processing capabilities of our intuition.
3. To use decision analysis, we are forced to consider carefully and logically all possible courses of action and the outcomes that could result from each. By so

doing, we may see a side of the problem not seen before, or we may even discover that we have been addressing the wrong problem. Thus, the information obtained from decision analysis may more than compensate for the effort expended in the analysis.

4. Another reason business and economics majors should study decision analysis is that many firms and government agencies use it on a regular basis. Consequently, you may very well be required to use decision analysis in your future employment.

One of the alternatives we face in making a decision is whether the decision should be made *now*—utilizing information we currently possess about the problem (we will refer to this as *prior information*)—or *postponed* until we have gathered additional information. In this chapter we study decision-making under uncertainty and assume that only prior information is available. In the next chapter, we will expand our study of decision-making to include situations in which additional information is available. We will discuss how to determine the value of additional information as well as when and how to use additional information in decision-making.

SECTION 18.1

THREE TYPES OF DECISION PROBLEMS

Although all decision problems involve the selection of a course of action from among two or more alternatives, we can classify them into one of three categories:

1. Decision-making under certainty
2. Decision-making under uncertainty
3. Decision-making under conflict

Decision-making under certainty entails the selection of a course of action when we *know* the result each alternative action will yield. If the number of alternatives being considered is small, such decisions may be easy to make. However, if the number of alternatives is large, the optimal decision may be difficult—if not impossible—to obtain. It may take too much time and/or be too costly to evaluate all the many alternatives individually and select the one with the most favorable results. Decision problems of this type are not addressed in this text, but here is a simple example of this type of problem:

A furniture company constructs and finishes tables and chairs. Each table produced by the company nets a profit of $100 and each chair a profit of $60. During 1 week the company has 305 work-hours available for assembly operations and 355 work-hours available for finishing. From past experience it is known that each chair requires 3 hours to be assembled and $1\frac{1}{2}$ hours of finishing, while each table requires 4 hours for assembly and 2 hours for finishing. How many tables and how many chairs should the company produce over the week in order to maximize profits?

Note that a unique solution to this problem exists that will maximize the firm's profit. Many problems involving decision-making under certainty, including this one, are solved using a technique known as **linear programming**.

Decision-making under uncertainty* entails the selection of a course of action when we do *not know* with certainty the results that each alternative action will yield. Furthermore, we assume that the outcome of whatever course of action we select is affected only by chance and not by an opponent or competitor. We discuss decision-making under uncertainty in detail in this chapter and Chapter 19. Our introductory example concerning the decision of whether to expand the sales region to the southwestern United States demonstrates decision-making under uncertainty.

Decision-making under conflict is similar to decision-making under uncertainty in that we do not know with certainty the result each available alternative course of action will yield. However, the reason for this uncertainty is different in the case of decision-making under conflict. In such cases, we are in effect "playing against" one or more opponents or competitors. The outcome of our chosen course of action depends on decisions made by our competitors. Decision problems of this type fall under the discipline known as **game theory**. Game theory is not discussed in this text; however, an example of this type of decision problem follows:

A local businessman is interested in purchasing real estate somewhere in the Dallas area for the purpose of building a fast-food restaurant. He has narrowed his alternatives to five suburban neighborhoods. He is certain his venture will be profitable as long as none of the major fast-food chains decides to locate near his restaurant. Thus, the businessman's decision problem involves choosing a parcel of land while knowing that the results of his decision depend on the expansion plans of his potential competitors in the fast-food industry. Furthermore, the location decision facing other fast-food chains interested in the Dallas area depends to some extent on the decision of the Dallas businessman. They, too, would prefer not to locate near another restaurant of the same type.

Since our objective throughout this text has been to make inferences when we have only partial (or *imperfect*) information, we concentrate on decision-making under uncertainty in the remaining sections.

| | | | | | | | | | | | | |

EXERCISES 18.1–18.4   APPLYING THE CONCEPTS

18.1 Compare and contrast decision-making under certainty, uncertainty, and conflict.

18.2 Describe two decision problems that you face every day. Categorize each as being a problem requiring a decision made under certainty, uncertainty, or conflict. Justify your categorization.

18.3 Categorize the following decision problems as decision-making under certainty, uncertainty, or conflict. Justify your categorization.
   a. The management of a bank is considering an application for a commercial loan. If they decide to make the loan but the customer defaults, the bank will lose the amount of the loan plus the lost profits. On the other hand, if the bank fails to grant the loan and the customer would have repaid it, the bank will lose the interest on the loan.

---

*Some texts distinguish between "decision-making under risk" and "decision-making under uncertainty" according to whether probabilities are available to describe the degree of uncertainty confronted by the decision-maker. We make no such distinction.

b. A manufacturer is currently facing a decision about the price of an electric lawn mower it makes. If the company sets the price too high, potential customers will purchase competitors' mowers. If, on the other hand, the price is set too low, the competitors will also drop their prices, thereby reducing everyone's profits.

c. A computer hardware company has two contracts for producing electronic components for the space program. Since the contracts are for a fixed number of components at a fixed price, the decision problem involves how to allocate fixed production resources to maximize profit.

18.4 Categorize the following decision problems as decision-making under certainty, uncertainty, or conflict. Justify your categorization.

a. A plant manager wants to replace an obsolete piece of machinery with a new model. There are two brands on the market from which to choose. Both brands are of equal quality, have the same guarantee, and produce the same number of items per hour. However, brand B is $500 cheaper than brand A.

b. A company is faced with the decision of whether or not to increase its production capacity by adding a new building to the existing facilities. If the company decides not to expand, they expect to make a profit of $550,000 for each of the next 3 years regardless of the state of the economy. If they build the addition and the economy continues to expand, the addition is expected to increase company profits to at least $650,000 a year for the next 3 years. If they build the addition and the economy remains stable or experiences a downward trend over the next 3 years, the company would incur a reduction in profits to $475,000 or less per year for the next 3 years.

S E C T I O N 18.2

DECISION-MAKING UNDER UNCERTAINTY: BASIC CONCEPTS

We will use the following example to introduce some basic concepts: A profit-motivated entrepreneur is committed to producing a concert that will feature a current rock star sometime during the latter part of June next year. The promoter has been unable to finalize plans, however, due to indecision over whether to take a chance on rain and hold the concert in Memorial Stadium (40,000 seats outdoors) or play it safe and hold the concert in the Civic Center (15,000 seats indoors). A sellout is expected at least a month in advance, whichever facility is chosen. If the stadium is chosen and the weather cooperates, the promoter will make a net profit of about $350,000 (ticket proceeds less costs, taxes, and other expenses). If it is raining at concert time, the rock star may choose not to perform (according to the contract) and the promoter will lose about $40,000 (stadium rental, commitment to the rock star, administrative costs, salaries of security personnel, advertising). However, if the Civic Center is chosen, the entrepreneur would make a net profit of about $150,000 regardless of the weather. Which option would you choose? We will use decision analysis to make our selection later in the chapter.

We can identify three specific elements of this decision problem: First, a choice must be made between two possible courses of action—rent the stadium or rent the Civic Center. We refer to these alternatives as **actions**. Second, it is uncertain which event will occur—rain or no rain. We refer to these events as **states of nature**. Third, depending on which action is chosen and which state of nature occurs on the evening of the concert, the decision-maker* will receive either a financial reward

---

*The term *decision-maker* is used to refer not only to an individual, but also to a corporation, a community, or in general, any entity faced with a decision problem.

or a penalty for the chosen action. The consequences of the decision problem are referred to as **outcomes**; these may be either positive or negative. For example, if the action chosen by the promoter is Rent the stadium and the state of nature that occurs is Rain, the outcome that results is −$40,000. That is, the action/state of nature combination Rent the stadium/Rain will *cost* the promoter $40,000. The combination Rent the stadium/No rain will yield a *profit* of $350,000. The reward (or penalty) corresponding to each action/state of nature combination is called the **outcome** or **payoff**.

We can conveniently summarize all three elements of a decision problem in a **payoff table** (Table 8.1). Each of the set of possible actions the decision-maker has chosen to consider is associated with a row of the *payoff table*. Each state of nature is associated with a column. The numbers in the table are the outcomes of the decision problem. For example, $350,000 is the outcome that would result from the implementation of the action associated with the top row of the table (Rent stadium) and the occurrence of the state of nature associated with the right-hand column of the table (No rain).

TABLE 18.1
Payoff Table for the Rock Concert Decision Problem

		STATE OF NATURE	
		Rain	No rain
ACTION	Rent stadium	−$40,000	$350,000
	Rent Civic Center	$150,000	$150,000

A decision problem can also be illustrated by a **decision tree**. A payoff table and a decision tree may display the same information, but as we will see in the next chapter, it is sometimes more convenient to use a decision tree. The decision tree in Figure 18.1 corresponds to Table 18.1. Conceptually, the promoter's movement through time toward the outcome of the decision problem is represented by movement from left to right through the decision tree. The ■ denotes a **decision fork** and signals that a decision must be made. At this position on the tree, the decision-maker must choose between the two actions, Rent stadium and Rent Civic Center. If Rent stadium is chosen, then from the decision fork we move along the upper branch of the tree. The ● denotes a **chance fork** and signals that the next branch of the tree the promoter will follow will be determined by the chance occurrence of a state of nature. If the decision-maker is positioned at the upper chance fork of Figure 18.1 and it rains on the day of the concert, then the upper branch of the chance fork (labeled Rain) will lead the promoter to the consequence (−$40,000) of the action/state combination Rent stadium/Rain.

Both the selection of actions to be considered in a decision problem and the choice of an action to implement are under the control of the decision-maker, but the state of nature is not. The decision-maker must choose a course of action *prior* to knowing which state of nature will occur and *without* being able to influence the random process generating the states of nature. *Note also that the states of nature*

FIGURE 18.1

Decision Tree for the Rock
Concert Decision Problem

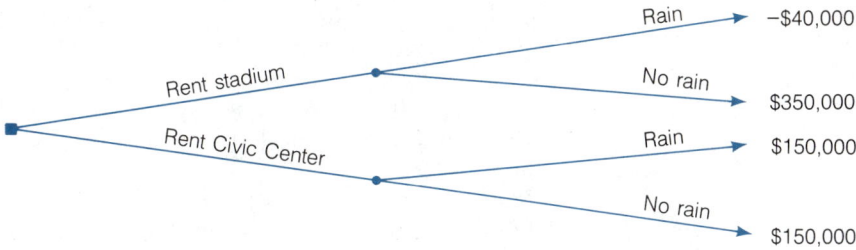

considered in any decision problem must be mutually exclusive and collectively exhaustive. That is, the state of nature that occurs must be clearly identifiable as one and only one of the listed states, and the list of states considered must include all possible states that can occur. The former constraint precludes any overlapping of, or vagueness in, state definitions; the latter precludes the possibility of any state occurrence not anticipated by the decision-maker. For practical purposes, the states of nature in the above example are mutually exclusive and collectively exhaustive. They are mutually exclusive because the weather on a June evening can be classified as either rainy or not rainy, but not both. They are collectively exhaustive because the list of possible weather patterns can be narrowed to include only rainy or not rainy.

All outcomes in a decision problem should be stated in terms of the same numerical quantity, and that quantity should be chosen to rank the outcomes relative to the decision-maker's overall objective. We refer to this measure of the outcomes of a decision problem as the **objective variable**. In the preceding example, the promoter's motive—or objective—for producing the rock concert was net profit. Accordingly, the decision outcomes are expressed in terms of net profit (dollars). If the objective had been to give as many people as possible an opportunity to see a live performance by the rock star, we would measure the outcomes of potential actions in terms of the number of people attending the concert.

We conclude by summarizing the concept of decision-making under uncertainty, the four elements common to this type of decision problem, and the methods for displaying these elements.

DECISION-MAKING UNDER UNCERTAINTY

If a decision-maker is faced with choosing one action from among two or more alternative actions, and each of these has possible outcomes that depend on the chance occurrence of one of a set of mutually exclusive and collectively exhaustive states of nature, the decision-maker is said to be faced with **decision-making under uncertainty**.

FOUR ELEMENTS COMMON TO DECISION PROBLEMS INVOLVING UNCERTAINTY

1. *Actions:* The set of two or more alternatives the decision-maker has chosen to consider. The decision-maker's problem is to choose one action from this set.
2. *States of nature:* The set of two or more mutually exclusive and collectively exhaustive chance events upon which the outcome of the decision-maker's chosen action depends.
3. *Outcomes:* The set of consequences resulting from all possible action/state of nature combinations.
4. *Objective variable:* The quantity used to measure and express the outcomes of a decision problem.

Table 18.2 and Figure 18.2 depict the general format for the payoff table and the decision tree, respectively. If the ith action is denoted by a_i and the jth state of nature by S_j, then the outcome resulting from the combination of the ith action with the jth state of nature is O_{ij}.

TABLE 18.2

General Form of a Payoff Table

		\multicolumn{4}{c}{STATE OF NATURE}
		S_1	S_2	...	S_m
	a_1	O_{11}	O_{12}	...	O_{1m}
	a_2	O_{21}	O_{22}	...	O_{2m}
ACTION	.	.	.		.
	a_n	O_{n1}	O_{n2}	...	O_{nm}

FIGURE 18.2

General Form of a Decision Tree

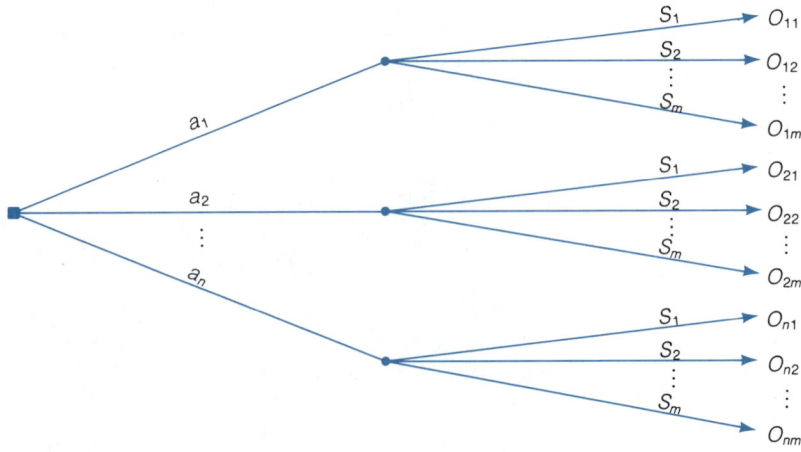

| | | | | | | | | | | | |

EXERCISES 18.5–18.10

APPLYING THE CONCEPTS

18.5 List and define the four primary elements of a decision problem under uncertainty.

18.6 The states of nature that are defined in a decision problem must be mutually exclusive and collectively exhaustive. In this context, what is meant by the phrase *mutually exclusive and collectively exhaustive*?

18.7 Describe a decision you make every day that is done in the face of uncertainty. Identify the actions and states of nature of your decision problem. Construct a decision tree to illustrate your decision problem.

18.8 For each of these decision problems, identify the actions, states of nature, outcomes, and objective variable:
  a. A winery is considering introducing a new low-cost dinner wine. The introduction of the wine will cost $3 million in promotional and fixed costs per year. Each bottle sold will contribute $.30 to profits. The management believes that sales could range from 5 to 25 million bottles per year.
  b. An administrator in the Environmental Protection Agency is trying to determine how much to fine companies that discharge a particular type of effluent into the waterways of the United States. If the fine is set too low, all companies will simply pay the fine because it will be less costly than installing pollution-control equipment. On the other hand, if the fine is set too high, small companies may be driven out of business because they cannot afford the capital expenditures necessary to meet the new EPA standards. Thus, the problem of the EPA administrator is to set the fine to minimize the total social cost—that is, the cost to society of (1) firms being driven out of business and (2) waterways being polluted.

18.9 Refer to part **b** of Exercise 18.4, in which a company was faced with the decision of whether to increase its production capacity. For this decision problem, identify the actions, states of nature, outcomes, and the objective variable.

18.10 A company that manufactures a well-known line of designer jeans is contemplating whether to increase its advertising budget by $1 million for next year. If the expanded advertising campaign is successful, the company expects sales to increase by $1.6 million next year. If the advertising campaign fails, the company expects sales to increase by only $400,000 next year. If the company decides not to increase its advertising expenditures, it expects sales to increase by $200,000 next year.
  a. Identify the actions, states of nature, outcomes, and the objective variable for this decision problem.
  b. Construct a decision tree that illustrates the jeans manufacturer's decision problem.

| | | | | | | | | | | | |

SECTION 18.3

TWO WAYS OF EXPRESSING OUTCOMES: PAYOFFS AND OPPORTUNITY LOSSES

The outcomes of the rock concert example discussed in the previous section were in terms of the net profit that would be realized by the promoter depending on the decision (action) concerning the location of the concert and on the weather (state of nature) on the day of the concert. That is, net profit was the objective variable. Recall that the objective variable can assume both positive and negative values. In general, we refer to outcomes that reflect the *actual* reward to the decision-maker in terms of the objective variable as **payoffs**.

Alternatively, outcomes can be expressed in terms of *opportunities* for higher profits that the decision-maker has *lost* as a result of the action selected. For example, in the rock concert example, if the weather is not rainy, a decision to hold the concert in the Civic Center will bring in a profit of $150,000 and a decision to use the stadium will bring in a profit of $350,000. If, in fact, the Civic Center was chosen, then by not choosing to rent the stadium, the promoter will have *lost the opportunity* to net an additional $200,000. We refer to this $200,000 as the **opportunity loss*** associated with the action/state combination Rent Civic Center/No rain. An opportunity loss may be determined in a similar fashion for each action/state combination of a decision problem, and an **opportunity loss table** may be constructed, as in Table 18.3. Notice that none of the opportunity losses of Table 18.3 is less than 0. A little thought should convince you that this is true in general for any opportunity loss. In Section 18.5, we will show that decision problems may be solved using outcomes expressed either as payoffs or as opportunity losses.

TABLE 18.3

Opportunity Loss Table for the Rock Concert Decision Problem

		STATE OF NATURE	
		Rain	No rain
ACTION	Rent stadium	$190,000	0
	Rent Civic Center	0	$200,000

DEFINITION 18.1

The **opportunity loss** is the difference between the payoff a decision-maker receives for a chosen action and the maximum that the decision-maker could have received for choosing the action yielding the highest payoff for the state of nature that occurred.

OPPORTUNITY LOSS DETERMINATION

Repeat the following procedure for each state of nature in a decision problem—i.e., each column of a payoff table:

1. Find the maximum payoff in a column. The opportunity loss associated with this payoff is 0.
2. The opportunity loss associated with any other payoff in this column is found by subtracting that payoff from the maximum payoff in the column.

EXAMPLE 18.1

A beer producer with breweries located in the western part of the United States and a distribution network that extends only as far east as the Mississippi River is considering expanding its sales region to include the northeastern part of the country.

*Sometimes called the **regret**.

To do so, the producer must build a new brewery in the Northeast in order to overcome refrigeration problems that would arise from having to transport its beer. The problem is to determine how large a brewery to construct. It has been decided that the size should be based on the projected gross profits (profit before taxes) for the fifth year of operation for each of the four sizes of breweries under consideration. The firm's marketing department recognizes that the company cannot possibly obtain more than a 15% market share during the fifth year of operation and has put together a payoff table for the firm's planning committee (Table 18.4). Construct the corresponding opportunity loss table.

TABLE 18.4

Payoff Table for the Brewer's Decision Problem

| | | STATE OF NATURE | | |
| | | Market share during the fifth year of operation | | |
	Brewery size	S_1: 0% < 5%	S_2: 5% < 10%	S_3: 10%–15%
ACTION	a_1: Small	$300,000	$350,000	$450,000
	a_2: Medium	$250,000	$700,000	$800,000
	a_3: Large	$200,000	$600,000	$1,000,000
	a_4: Very large	−$100,000	$100,000	$500,000

SOLUTION

For each column of the payoff table (Table 18.4) find the maximum payoff:

	COLUMN 1	COLUMN 2	COLUMN 3
Maximum payoff:	$300,000	$700,000	$1,000,000

The opportunity loss associated with each column maximum is 0. The opportunity associated with, for example, any other payoff in column 1 is found by subtracting that payoff from the column's maximum payoff, $300,000, as shown in Table 18.5. The resulting opportunity loss table is shown in Table 18.6 (page 1052).

TABLE 18.5 Calculation of Opportunity Losses for Example 18.1

| | Brewery size | STATE OF NATURE | | |
		S_1: 0% < 5%	S_2: 5% < 10%	S_3: 10%–15%
ACTION	a_1: Small	0	$700,000 − $350,000 = $350,000	$1,000,000 − $450,000 = $550,000
	a_2: Medium	$300,000 − $250,000 = $50,000	0	$1,000,000 − $800,000 = $200,000
	a_3: Large	$300,000 − $200,000 = $100,000	$700,000 − $600,000 = $100,000	0
	a_4: Very large	$300,000 − (−$100,000) = $400,000	$700,000 − $100,000 = $600,000	$1,000,000 − $500,000 = $500,000

TABLE 18.6
Opportunity Loss Table for
the Brewer's Decision
Problem

	Brewery size	STATE OF NATURE		
		S_1: 0% < 5%	S_2: 5% < 10%	S_3: 10%–15%
ACTION	a_1: Small	0	$350,000	$550,000
	a_2: Medium	$50,000	0	$200,000
	a_3: Large	$100,000	$100,000	0
	a_4: Very large	$400,000	$600,000	$500,000

After formulating and displaying a decision problem as just described, it may happen that an action has been included that should never be selected *no matter which state of nature occurs.* In Example 18.1, a_4 is such an action. Inspection of the payoff table (Table 18.4) or the opportunity loss table (Table 18.6) reveals that both actions a_2 and a_3 result in higher payoffs (and lower opportunity losses) than a_4 for each possible state of nature. Thus, a_4 is said to be **dominated** by both actions a_2 and a_3 and should never be chosen by the decision-maker. Accordingly, dominated actions should be dropped from consideration. Because dominated actions should not be admitted for consideration by the decision-maker, they are sometimes referred to as being **inadmissible actions.** We will see later that by eliminating inadmissible actions, we ease the computation burden necessary to solve a decision problem.

DEFINITION 18.2

Action a_i is said to **dominate** action a_j, thereby making a_j **inadmissible,** if each of the following is true:

1. For each state of nature, the payoff for action a_i is greater than or equal to the payoff for a_j.
2. For at least one state of nature, the payoff for action a_i is greater than the payoff for a_j.

The decision-maker should eliminate dominated actions (inadmissible actions) from all decision problems.

To use decision analysis to prescribe a decision in the face of uncertainty, we must characterize our uncertainty concerning the states of nature of the problem with a probability distribution. We discuss the determination of such distributions in the next section.

**EXERCISES
18.11–18.22**

LEARNING THE MECHANICS

18.11 Explain the difference between payoffs and opportunity losses.

18.12 Why should a dominated action be eliminated from a decision problem?

18.13 Eliminate any inadmissible actions from the accompanying payoff table and convert it to an opportunity loss table.

		STATE OF NATURE			
		S_1	S_2	S_3	S_4
ACTION	a_1	−60	−5	0	45
	a_2	−65	−10	40	75
	a_3	−70	−15	40	70
	a_4	−80	−40	−10	80

18.14 Identify any inadmissible actions in the decision tree shown here, and redraw the tree without the inadmissible actions. The objective variable is total sales (in thousands of dollars).

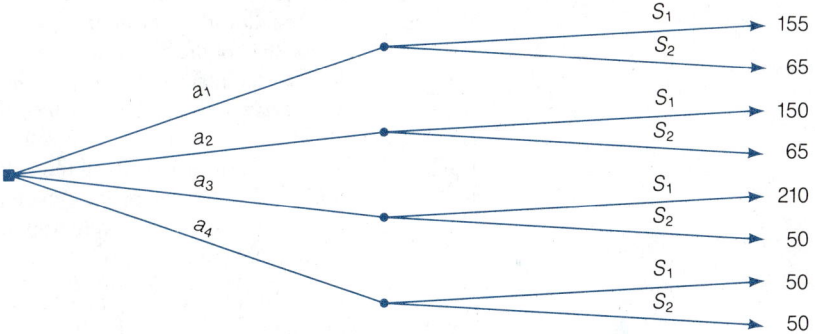

18.15 Shown in the margin is an opportunity loss table that has been derived from a payoff table.

    a. Which action(s) in the table is (are) inadmissible? Justify your answer.

    b. Why is it not possible to convert this opportunity loss table to the original payoff table?

	S_1	S_2	S_3
a_1	15	40	0
a_2	0	10	8
a_3	43	0	15
a_4	20	20	10

18.16 The outcomes displayed on the decision tree below are in terms of payoffs. Convert the outcomes to opportunity losses.

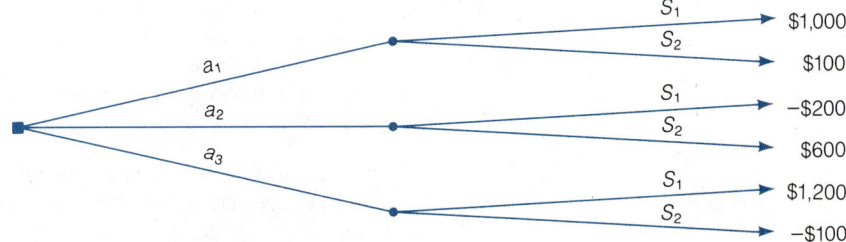

APPLYING THE CONCEPTS

18.17 A U.S. company that manufactures minicomputers is considering marketing its product in Europe. If it decides to do so, the company expects to attain a European market share of no more than 3% and no less than 1% over the next 3 years. The company's accounting department has determined that the payoff for this expansion should be an increase in total company profits for the next 3 years of $3 million if the product is able to attain a 3% market share, an increase of $1 million for a 2% market share, and a decrease of $1 million for a 1% market share. Total domestic profits for the next 3 years are expected to be $15 million regardless of whether or not the minicomputer is marketed in Europe.
a. Formulate the payoff table for this decision problem.
b. Convert the payoff table to an opportunity loss table.

18.18 A common problem in business is the management of perishable inventories—that is, items that lose the major portion of their economic value after a given date due to either obsolescence or spoilage. For example, if a corner newsstand orders too many papers and cannot sell them all, the excess papers have little value. Hence, these types of problems are called *newsboy problems*. Newsboy problems are common in retailing situations such as the following: A buyer for a large department store is trying to decide how many of a new style of dress to order. Because of rapid changes in fashion, she does not want to order too many dresses, but if she orders too few she will lose profits for her department. Each dress purchased will cost the store $30 and will sell for $50. Any dresses not sold at the end of the season will be sold at the annual half-price sale. The buyer believes the department will sell three, four, five, six, seven, or eight dozen dresses. Dresses must be purchased by the department store in lots of one dozen.
a. Formulate the payoff table for this decision problem.
b. Convert the payoff table to an opportunity loss table.

18.19 Refer to Exercise 18.10, in which a company that manufactures designer jeans must decide whether to increase its advertising budget for next year.
a. Formulate the payoff table for this decision problem.
b. Are either of the actions inadmissible? Explain.
c. Draw the decision tree that corresponds to your payoff table.

18.20 The purchasing agent for an automobile manufacturer is concluding a purchase agreement with a supplier of preformed body pieces. The pieces will be purchased in lots of 500, and the cost of a lot is $15,000. The body stamping has a frequent defect called a *burr*. If a piece is found to have a burr, it can be filed off by the automobile manufacturer at a cost of $10 per piece. Past experience with this supplier has shown that the proportion of defects tends to be 5%, 10%, or 50%. The supplier now offers a guarantee that it will assume the costs for all defective body moldings greater than 50 in a production lot of 500. This guarantee may be purchased for a cost of $1,000 before the lot begins production. The purchasing agent is interested in determining whether the guarantee should be purchased.
a. Formulate the payoff table for this problem.
b. Draw the decision tree that corresponds to your payoff table.

18.21 Computer products company A has sued computer products company B for patent infringements. The management of B is deciding whether they should settle the suit out of court. If they do so, the accounting and marketing departments calculate that company B would lose $50 million over the next 5 years in royalty payments. If they

go to court and win, they will have $1 million in court costs. However, if they lose the suit, they will have to pay company A $100 million in royalty payments.
a. Formulate the payoff table for this decision problem.
b. Draw the decision tree that corresponds to your payoff table.

18.22 A winery can introduce a new low-cost dinner wine for $3 million in fixed and promotional costs per year. Each bottle sold will contribute $0.30 to profits. The management believes that sales will be .5 million, 10 million, 15 million, 20 million, or 25 million bottles per year.
a. Formulate the payoff table for this problem.
b. Construct the opportunity loss table.

SECTION 18.4

CHARACTERIZING THE UNCERTAINTY IN DECISION PROBLEMS

In Chapter 4 we explained that probability distributions are used to characterize an individual's uncertainty about the outcomes of an experiment. In a decision problem, the observation of a state of nature associated with the problem may be regarded as an experiment and the various states of nature as experimental outcomes. Thus, it follows that a decision-maker can use a probability distribution to characterize the uncertainties associated with the states of nature that may occur in a decision problem. These probabilities measure the likelihood of the occurrence of the various states of nature and may be unknown. Consequently, we will assess these values using one of the following:

1. Information about the relative frequencies of the states
2. Judgmental (subjective) information about the states
3. A combination of relative frequency information and subjective information

In other words, we use any type of information that is available to assign probabilities to the states of nature.

For example, in the rock concert example of Section 18.2, the concert promoter could use historical weather data (relative frequency data) along with personal knowledge of the rock star's tendency to declare a day too rainy to perform (subjective information) to assess the probabilities of states for the decision problem. Or, if you wanted to decide whether to open your own business next year, you would be interested in knowing whether the economy would continue at a healthy pace for the next 3 years or whether it would fall into a recession. Since the observation of this experiment—observing the health of the economy over the next 3 years—can never be repeated, you must rely on your own experience or the advice of economic experts to assess the probabilities associated with these states of nature.

Caution: Great care must be taken in determining probabilities for the states of nature. Decision analyses based on poorly chosen probabilities may lead to inappropriate actions.

CASE STUDY 18.1

EVALUATING UNCERTAINTY IN RESEARCH AND DEVELOPMENT

Although all management functions must cope with uncertainty, R&D [Research and Development] is generally agreed to be the function involving the largest number and the widest range of uncertainties. Thus, the R&D manager faces huge problems not only in selecting the most promising avenues for R&D effort and expenditures but also in attempting to insure a steady flow of technically successful projects.

In the previous quote, Balthasar, Boschi, and Menke (1978) are describing how Sandoz, a Swiss pharmaceutical company, uses subjective probabilities as a basis for its research and development (R&D) planning and decisions. Twice a year a small group of experts is asked to assess the probability of technical success for each of Sandoz's R&D projects. The group of experts includes R&D line managers and other technical experts familiar with particular requirements for a project's success. When the program was begun, these probability assessments were obtained through interviews. Two basic methods of eliciting subjective probabilities were used—a direct method and an indirect method. A **direct method** is one that requires the expert to state the probability, or odds, of a project's success explicitly in numerical terms. An **indirect method** is one in which the expert's responses are not probabilities per se.

One indirect method used by Sandoz in the early stages of the program is the probability wheel. The wheel is a disk divided into two colored sections, one blue and one orange. The relative size of each section is adjustable. In the center of the disk is a pointer that can be spun and will stop in one of the two sections. The expert is asked which event is more likely: (1) the spinner will stop on the orange section, or (2) project A will succeed. If the answer is (1), the wheel is adjusted to decrease the relative size of the orange section. If the answer is (2), the wheel is adjusted to increase the relative size of the orange section. This procedure is repeated until the expert says the two events are equally likely. The relative size of the orange section is the expert's subjective probability of the success of project A.

Initially, all probabilities were elicited indirectly. As individual forecasters became more familiar with the process, probabilities were assigned directly to each project. When all the experts were sufficiently familiar with the technique, interviews were replaced by questionnaires. After individual probability assessments have been obtained from each expert, the entire group meets to discuss their assessments and arrive at one consensus probability for each R&D project. The advantage of the group assessment is that it brings together the opinions of individuals with different experience and information.

The consensus probabilities are then used in making decisions concerning R&D projects. There are many models and techniques for planning and controlling R&D that require input regarding the probability of a project's success. Obtaining subjective probability assessments allows management (as with Sandoz) to use these models as an aid in decision-making.

Sandoz is pleased with the results of using subjective probabilities in managing R&D. By comparing the experts' consensus success probabilities with the actual relative frequency of project successes over time, the researchers found that the probabilistic predictions are reliable estimates of future results. In line with their goal of "a steady flow of technically successful projects," Sandoz has found that decision-making based on subjective probabilities has reduced the variability in their expected success rate for projects in the early stage of development. The R&D managers at Sandoz conclude that explicit subjective probabilities are a useful input to assist in management planning and control in a highly uncertain environment.

SECTION 18.5

SOLVING THE DECISION PROBLEM USING THE EXPECTED PAYOFF CRITERION

Now that we have shown you how to structure a decision problem by constructing a payoff table or an opportunity loss table and how to characterize the uncertainty associated with the states of nature in a decision problem, we have to choose a rule for reaching a decision. Numerous rules have been proposed, but the one most commonly used in decision analyses uses a payoff table and chooses the action that produces the **maximum expected payoff**. This is called the **expected payoff criterion**. Equivalently, you could use an opportunity loss table and choose the action that produces the **minimum expected opportunity loss**. This is called the **expected opportunity loss criterion**. It can be shown that both criteria lead to the same solution.

To understand how the expected payoff criterion is used to reach a decision, recall the definition of the expected value of a discrete random variable (Chapter 5). If x is a discrete random variable with probability distribution $p(x)$, then the expected (or mean) value of x is

$$E(x) = \sum_{\text{All } x} xp(x)$$

As we will illustrate, the random variable, x, in a decision problem is the payoff, and the probabilities associated with x are the same as those that describe the likelihood of occurrence of states of nature.

For example, consider the brewery decision problem, Example 18.1 in Section 18.3. The payoff table is reproduced in Table 18.7. Note that we have eliminated the inadmissible action, a_4: Very large brewery.

TABLE 18.7
Payoff Table for the
Brewery Example

	Brewery size	STATE OF NATURE Market share during fifth year of operation		
		S_1: 0% < 5%	S_2: 5% < 10%	S_3: 10%–15%
	a_1: Small	$300,000	$350,000	$450,000
ACTION	a_2: Medium	$250,000	$700,000	$800,000
	a_3: Large	$200,000	$600,000	$1,000,000

Suppose the brewery had assessed the following probability distribution for the states of nature:

STATE	0% < 5%	5% < 10%	10%–15%
P(State will occur)	.4	.5	.1

Then if you choose action a_1 (see row 1 of Table 18.7), the payoffs can be $300,000, $350,000, or $450,000 with probabilities .4, .5, and .1, respectively. Thus, the probability distribution for the payoff, x, if you choose action a_1, is as shown in the next table.

PAYOFF x	$300,000	$350,000	$450,000
p(x)	.4	.5	.1

The expected payoff of action a_1, denoted by the symbol $EP(a_1)$, is

$$EP(a_1) = \sum xp(x)$$
$$= (\$300,000)(.4) + (\$350,000)(.5) + (\$450,000)(.1)$$
$$= \$340,000$$

This tells us that, if we were faced with this decision problem a very large number of times and chose action a_1 each time, the mean or expected payoff would be $340,000—assuming that the probabilities accurately reflect the likelihood of occurrence for the states of nature.

Similarly, we can write the probability distributions and expected payoffs for actions a_2 and a_3. For action a_2:

PAYOFF x	$250,000	$700,000	$800,000
p(x)	.4	.5	.1

$$EP(a_2) = \sum xp(x)$$
$$= (\$250,000)(.4) + (\$700,000)(.5) + (\$800,000)(.1)$$
$$= \$530,000$$

For action a_3:

PAYOFF x	$200,000	$600,000	$1,000,000
p(x)	.4	.5	.1

$$EP(a_3) = \sum xp(x)$$
$$= (\$200,000)(.4) + (\$600,000)(.5) + (\$1,000,000)(.1)$$
$$= \$480,000$$

Now examine the expected (mean) payoffs for each action, as summarized in Table 18.8. Which action would you choose? We think you would choose action a_2; i.e., you would recommend that the brewer construct a medium-sized brewery, because this strategy will produce the largest expected payoff—namely, $530,000.

TABLE 18.8
Expected Payoff Table for the Brewer's Decision Problem

ACTION a_i	EXPECTED PAYOFF FOR ACTION a_i $EP(a_i)$
a_1	$340,000
a_2	$530,000
a_3	$480,000

> **THE EXPECTED PAYOFF CRITERION**
>
> Choose the action that produces the largest expected payoff.

As noted at the beginning of this section, we will arrive at the same solution to the brewer's decision problem if we use an opportunity loss table, find the expected opportunity loss for each action, and then choose the action that produces the minimum expected opportunity loss. The solution is again action a_2.

> **THE EXPECTED OPPORTUNITY LOSS CRITERION**
>
> Choose the action that produces the smallest expected opportunity loss. (This will always lead to the same decision as the expected payoff criterion.)

EXAMPLE 18.2

A well-known cosmetics firm has been approached by a television producer to determine whether the firm would be interested in sponsoring a new television series next fall. The firm is faced with the choice of one of two actions. It can continue to sponsor a popular television show it has sponsored in the past, or it can shift to the producer's new prime-time show. The states of nature are the projected averages of the biweekly Nielsen ratings for the entire season. A Nielsen rating for a show is the percentage of homes in a sample taken by the A. C. Nielsen Co. that have watched the show for at least 6 minutes during the rating period. The objective variable is the firm's projected market share at the end of the television season next year. According to the expected payoff criterion, which show should the cosmetics firm sponsor? The firm's advertising agency and marketing department have come up with the representation of the firm's decision problem shown in Table 18.9.

TABLE 18.9   Cosmetics Firm's Decision Problem in Example 18.2

| | | STATE OF NATURE | | | |
| | | Projected average Nielsen rating for the new show (probabilities in parentheses) | | | |
		S_1: Below 12 (.1)	S_2: 12–17 (.2)	S_3: 18–29 (.4)	S_4: 30 or more (.3)
ACTION	a_1: Sponsor old show	.12	.12	.12	.12
	a_2: Sponsor new show	.06	.10	.14	.17

SOLUTION

Let x denote the payoff (market share) for a particular action and $p(x)$ its probability distribution. We now compute $E(x) = \sum_{\text{All } x} x p(x)$ for each action that produces the larger expected payoff. Thus, for action a_1:

$$EP(a_1) = .12(.1) + .12(.2) + .12(.4) + .12(.3) = .12$$

and for action a_2:

$$EP(a_2) = .06(.1) + .10(.2) + .14(.4) + .17(.3) = .133$$

Since the expected payoff for action a_2 is larger than that for a_1, the firm should sponsor the new show rather than the old show. Notice that in this example the objective variable was market share and not profit. The expected payoff and expected opportunity loss criteria can be used, as stated above, for any objective variable we want to maximize. For some situations, however, we may want to minimize the expected value of the objective variable (or, equivalently, maximize the expected opportunity loss). For example, if an objective variable such as time to process an order was used, we would prefer less time to more time. Then the expected payoff and expected opportunity loss criteria would have to be redefined, and we would seek the minimum expected payoff. ■

CASE STUDY 18.2

HURRICANES: TO SEED OR NOT TO SEED?

The seeding of hurricanes as a means of lessening their destructive power was suggested by R. H. Simpson in the early 1960's. Even though the results of experimental hurricane seeding were encouraging, government policy through the early 1970's prohibited the seeding of hurricanes that threatened coastal areas. Howard, Matheson, and North (1972) analyzed the seeding of hurricanes using decision analysis to determine (1) whether existing government policy should be modified and (2) whether further experiments on hurricane seeding should be carried out. We will discuss the second question in Case Study 19.1.

To answer the first question, Howard, Matheson, and North had to define the states of nature, assign probabilities to each of the possible states, and assess the consequences of each action/state combination. The actions specified were Seed a threatening hurricane and Do not seed a threatening hurricane. The states of nature were defined in terms of the percentage change in maximum wind speed over a 12-hour period. This measure was chosen because it is related to the primary cause of the destruction inflicted by most hurricanes, and it is this characteristic that seeding is expected to influence. Based on historical data, a probability distribution was assessed for changes in wind speed for a "representative hurricane" over a 12-hour period. Using Simpson's theoretical work on hurricane seeding and early experimental work on seeding, Howard, Matheson, and North also assessed a probability distribution for changes in wind speed of a seeded hurricane. These probability distributions are given in Table 18.10.

TABLE 18.10

Probability Distributions for States of Nature in Case Study 18.2

STATE Change in wind speed	STATE PROBABILITY Unseeded hurricanes	STATE PROBABILITY Seeded hurricanes
S_1: +25% or more	.054	.038
S_2: +10% to +25%	.206	.143
S_3: −10% to +10%	.480	.392
S_4: −25% to −10%	.206	.255
S_5: −25% or more	.054	.172

The consequences of the action/state combinations considered were property damage resulting from the hurricane and government liability in the case of seeded hurricanes. Using a least squares regression analysis (see Chapter 10) and past records of hurricane damage, the authors modeled the relationship between changes in wind speed and property damage (in millions of dollars) for a "representative hurricane." Results of their analysis appear in Table 18.11.

TABLE 18.11

Predicted Property Damage for Hurricane

CHANGE IN WIND SPEED	PREDICTED DAMAGE
+25% or more	335.8
+10% to +25%	191.1
−10% to +10%	100.0
−25% to −10%	46.7
−25% or more	16.3

Howard, Matheson, and North assumed that if a hurricane did not "improve" after seeding, the government would bear increased legal and social costs. They used a percentage of the property damage resulting from the states S_1, S_2, and S_3 to estimate these costs. The total cost associated with a particular state was found by adding these government responsibility costs to the predicted property damages. Their estimates were 50% for S_1, 30% for S_2, and 5% for S_3. Thus, the outcomes for each state under the decision to seed would be (including the $.25 million cost of seeding) as shown in Table 18.12. These quantities are expressed (in millions of dollars) in terms of negative gains, i.e., losses.

TABLE 18.12

Outcomes Associated with States of Nature in Case Study 18.2

STATE	OUTCOME
S_1: +25% or more	−503.95
S_2: +10% to +25%	−248.65
S_3: −10% to +10%	−105.25
S_4: −25% to −10%	−46.95
S_5: −25% or more	−16.55

The expected payoff criterion can now be applied as usual. Based on the given probabilities and outcomes, the expected payoffs are

$$EP(\text{No seeding}) = -\$116.0 \text{ million} \qquad EP(\text{Seeding}) = -\$110.78 \text{ million}$$

As a result of their analysis, Howard, Matheson, and North recommended that the government change its policy and allow seeding of hurricanes that threaten coastal areas. They noted that the decision to seed a particular hurricane, however, should be based on a decision analysis that uses all the meteorological and geographic factors relevant to the particular case.

LEARNING THE MECHANICS

18.23 Consider the payoff table shown here. Find the action that would be prescribed by the expected payoff criterion. Probabilities are in parentheses in the table.

		STATE OF NATURE			
		S_1 (.1)	S_2 (.3)	S_3 (.4)	S_4 (.2)
	a_1	-250	-400	0	300
ACTION	a_2	-300	-100	300	100
	a_3	-100	-40	-10	85

18.24 In the decision tree shown here, state probabilities are displayed in parentheses to the right of the state symbols, S_1 and S_2.

a. Identify any inadmissible actions, and explain why they are inadmissible.
b. The expected payoff for each action has been computed and appears to the left of the chance fork associated with each action in the decision tree. Verify that these expected payoffs are correct.
c. Identify the action that is prescribed by the expected payoff criterion.

18.25 Construct the decision tree for the decision problem described in Exercise 18.23. Include on your decision tree the state probabilities and the expected payoffs associated with each action. (For an example of such a tree, see Exercise 18.24.)

18.26 Consider the accompanying payoff table (probabilities are in parentheses).

		STATE OF NATURE		
		S_1 (.20)	S_2 (.30)	S_3 (.50)
	a_1	50	105	175
ACTION	a_2	-20	0	300

a. Use the expected payoff criterion to select the better action.
b. Convert the payoff table to an opportunity loss table and use the expected opportunity loss criterion to select the better action.

18.27 In the decision tree shown here, state probabilities are displayed in parentheses to the right of the state symbols, S_1 and S_2. The outcomes on the decision tree are expressed in terms of opportunity losses. According to the expected opportunity loss criterion, which action should be selected?

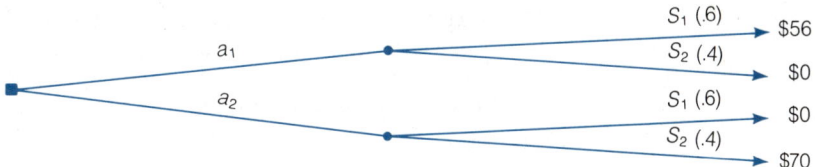

APPLYING THE CONCEPTS

18.28 *Materials requirements planning (MRP) systems* are computerized planning and control systems for manufacturing operations. They are used to manage raw materials and work-in-progress inventories. Since their introduction in the mid-1960's, MRP systems have made it possible to simultaneously reduce inventories and improve customer service (Schroeder, 1981). The operations manager for a company that produces a defense product wishes to obtain an MRP system. He has narrowed his choices to three different MRP vendors, X, Y, and Z, whose systems cost $5,000, $4,000, and $3,000, respectively. In addition, each vendor charges a fee for implementing the system (e.g., adapting the system to fit the purchaser's needs, installing the system, and debugging the system); the size of the fee depends primarily on the extent to which the system must be modified to meet the purchaser's needs. Systems X, Y, and Z are similar and would require basically the same modifications, but the implementation fees charged by the vendors differ significantly. These fees are described in the table. The manager, in consultation with the vendors, has determined that the probability of a major modification being required is .3, the probability of a moderate modification is .5, and the probability of a minor modification is .2. This uncertainty concerning the extent of the modification required exists because the operations manager has not finalized the specifications of the system he desires and the vendors will not specify the extent of the modification required until after the specifications have been made.

EXTENT OF MODIFICATION	IMPLEMENTATION FEE		
	X	Y	Z
Major	$3,000	$4,000	$6,000
Moderate	1,000	1,400	2,000
Minor	200	400	500

a. The objective variable for this decision problem is the initial cost plus the implementation fee for a system. Since we want to minimize (rather than maximize) the value of this objective variable, we need to convert our problem to one of maximization. This can be done by multiplying each value of system cost (intial cost plus implementation fee) by -1. The action that maximizes the value of this new objective variable will be the one that minimizes the system cost. Eliminate any inadmissible actions, and formulate the payoff table for this decision problem.

b. According to the expected payoff criterion, which MRP system should the manager purchase? Explain.

18.29 A company that manufactures hair dryers would like you to help it choose the travel case design for its new portable model. It has constructed the accompanying payoff table to characterize the decision problem. The objective variable is the contribution to the company's profits over the next year that results from marketing the new portable dryer. Use the expected payoff criterion to answer the questions.

		STATE OF NATURE	
		S_1: Consumers like design	S_2: Consumers dislike design
ACTION	a_1: Design A	$800,000	$-$200,000
	a_2: Design B	$660,000	$10,000

a. If $P(S_1) = P(S_2) = .5$, which design should the company choose?
b. If $P(S_1) = \frac{2}{3}$ and $P(S_2) = \frac{1}{3}$, which design should it choose?

MARKET SHARE	PROBABILITY
1%	.1
2%	.3
3%	.6

18.30 Reconsider Exercise 18.17, in which a minicomputer company is considering marketing its product in Europe. The company's marketing department has assessed the probability distribution in the table for the share of the European minicomputer market that the company will attain over the next 3 years. According to the expected payoff criterion, should the company enter the European market? Explain.

18.31 Reconsider Exercise 18.20, which concerned the purchase of body pieces by an automobile manufacturer. A lot of 500 pieces is purchased for $15,000, and the proportion of defects in each lot is .05, .10, or .50. The supplier offers a guarantee that it will assume the costs for all defective body moldings greater than 50 in a production lot of 500. This guarantee may be purchased for a cost of $1,000 before the lot begins production. The data in the table have been gathered on 100 past production lots of preformed body pieces. Given only this information, should the guarantee be purchased? Explain. [*Hint:* Use the sample of 100 lots to estimate the probabilities of the states of nature.]

PROPORTION OF DEFECTS	NUMBER OF TIMES OBSERVED
.05	55
.10	24
.50	21

SALES	PROBABILITY
3	.15
4	.25
5	.30
6	.15
7	.10
8	.05

18.32 Refer to the payoff table for the decision problem in Exercise 18.18. The dress buyer has assessed the probability distribution given in the table for the number of dresses (in dozens) she can sell. How many dozen dresses should the buyer order according to the expected payoff criterion?

18.33 Reconsider Exercises 18.10 and 18.19, in which a company that manufactures designer jeans is concerned about whether to increase its advertising budget for the year. The company has determined that, if the advertising budget is increased, the expanded advertising campaign will succeed with probability $\frac{2}{3}$ and will fail with probability $\frac{1}{3}$.

a. According to the expected payoff criterion, should the jeans manufacturer increase its advertising budget? Explain.

b. Convert the payoff table you developed in part **a** to an opportunity loss table.

c. According to the expected opportunity loss criterion, should the jeans manufacturer increase its advertising budget? Explain.

S E C T I O N   18.6

TWO NONPROBABILISTIC DECISION-MAKING CRITERIA: MAXIMAX AND MAXIMIN (OPTIONAL)

The expected payoff criterion is probabilistic because it requires that a probability distribution be assigned to the states of nature. There are several criteria that do not require the assignment of probabilities in order to reach a decision. Two common nonprobabilistic criteria are the **maximax** and **maximin decision rules**.

We will describe the maximax and maximin criteria by using the brewery decision problem first introduced in Example 18.1 (Section 18.3). The payoff table is reproduced in Table 18.13 (where the inadmissible action a_4 is omitted).

To use the **maximax rule**, we determine the maximum payoff associated with each action and choose the action that corresponds to the **maxi**mum of these **maxi**mum payoffs. Thus the name *maximax*. The easiest way to apply the maximax criterion is to find the maximum payoff in the entire table and choose the action corresponding to that payoff. Thus, the maximum payoff in Table 18.13 is $1,000,000, corresponding to action a_3: Large brewery. Therefore, the maximax criterion leads to the decision: Build a large brewery. Note that the maximax criterion ignores all information in the payoff table except the maximum value. That is, no state of nature except that associated with the maximum payoff is considered, and the size of the difference between the maximum payoff and the other payoffs is ignored. The maximax criterion is an optimistic one because it is based on the assumption (or hope) that the most favorable state of nature will occur.

TABLE 18.13
Payoff Table for the Brewer's Decision Problem

		STATE OF NATURE		
		Market share during fifth year of operation		
	Brewery size	S_1: 0% < 5%	S_2: 5% < 10%	S_3: 10%–15%
	a_1: Small	$300,000	$350,000	$450,000
ACTION	a_2: Medium	$250,000	$700,000	$800,000
	a_3: Large	$200,000	$600,000	$1,000,000

ACTION	MINIMUM PAYOFF
a_1	$300,000
a_2	$250,000
a_3	$200,000

To use the **maximin criterion**, we determine the minimum payoff associated with each action and select the action that corresponds to the **maxi**mum of these **mini**mum payoffs. In the brewery example, we find that the minimum payoffs are as indicated in the margin. Since the maximum of these minima is $300,000, corresponding to action a_1, the decision would be to build a small brewery if the maximin criterion was used. Note that the maximin criterion ignores all information except the minimum payoffs corresponding to each action. That is, no other states of nature except those corresponding to the minimum payoffs are considered. In the brewery example, all the minima are associated with the state of nature, less than 5%, so the other two states of nature are ignored by the maximin criterion.

Furthermore, the size of the difference between the minimum payoffs and the other payoffs is given no weight in determining which action to take. The maximin criterion is pessimistic because it is based on the assumption that the least favorable state of nature will occur.

Since much information in the payoff table is ignored by the maximax and maximin decision criteria, it is not surprising that they often lead to decisions that are intuitively unappealing and even nonsensical. In the brewery example, the maximin criterion leads to the decision to build a small brewery. But the slight loss that would be sustained if a medium or large brewery were built and less than 5% of the market were obtained appears to be more than compensated for by the gain in profit if the market share were at least 5% and a medium or large brewery were built. Thus, most decision analysts prefer the expected payoff criterion to these nonprobabilistic criteria. The use of the nonprobabilistic criteria is generally confined to decision problems for which the decision-maker is not willing to assign probabilities to the states of nature.

EXERCISES
18.34–18.41

LEARNING THE MECHANICS

18.34 Consider the following payoff table:

		STATE OF NATURE	
	S_1	S_2	S_3
a_1	75	105	60
a_2	70	80	60
a_3	−30	40	120
a_4	105	90	200

ACTION

a. Eliminate any inadmissible actions.
b. Find the action that would be prescribed by the maximax decision criterion.
c. Find the action that would be prescribed by the maximin decision criterion.

18.35 Consider the decison tree shown here.

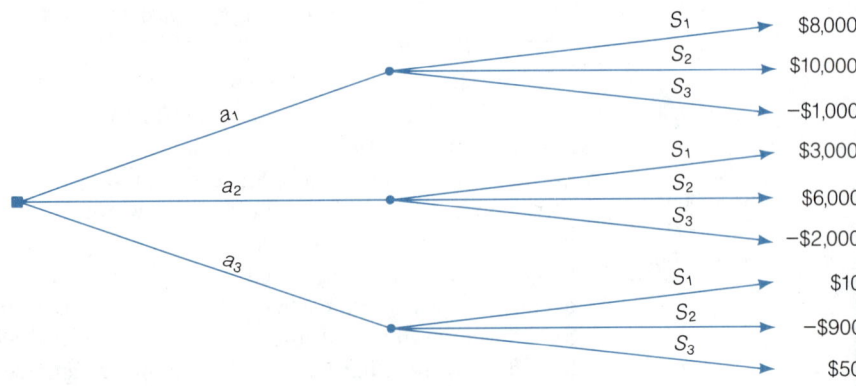

a. Find the action that would be prescribed by the maximax criterion.
b. Find the action that would be prescribed by the maximin criterion.
c. Explain why the action prescribed by the maximin criterion in part **b** is potentially unreasonable from a practical viewpoint.

18.36 Consider the decision tree shown here.

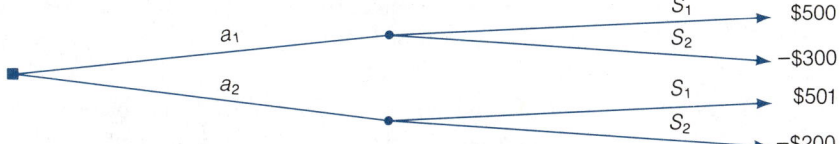

a. Find the action that would be prescribed by the maximax criterion.
b. Find the action that would be prescribed by the maximin criterion.

18.37 Consider the following payoff table:

| | | STATE OF NATURE | | | | | | |
		S_1	S_2	S_3	S_4	S_5	S_6	S_7
	a_1	50	35	10	0	−20	−80	−100
ACTION	a_2	100	30	20	10	−5	−40	−110
	a_3	105	0	−50	−100	−150	−300	−100

a. Find the action that would be prescribed by the maximax criterion.
b. Find the action that would be prescribed by the maximin criterion.

APPLYING THE CONCEPTS

18.38 Refer to Exercise 18.28, in which an operations manager is interested in purchasing an MRP system. Use the payoff table for the operations manager's decision problem to:
a. Find the action that would be prescribed by the maximax criterion.
b. Find the action that would be prescribed by the maximin criterion.

18.39 Refer to Exercise 18.17, in which a manufacturer of minicomputers is considering marketing its product in Europe. Use the payoff table for the company's marketing decision problem to:
a. Find the action that would be prescribed by the maximax criterion.
b. Find the action that would be prescribed by the maximin criterion.

18.40 Refer to Exercise 18.18, in which a buyer for a large department store is trying to decide how many of a new style of dress to order. Use the payoff table for the dress buyer's decision problem to:
a. Find the action that would be prescribed by the maximax criterion.
b. Find the action that would be prescribed by the maximin criterion.

18.41 Refer to Exercise 18.20, in which a purchasing agent for an automobile manufacturer is concluding a purchase agreement with a supplier of preformed body pieces. Use the payoff table for the purchasing agent's decision problem to:
a. Find the action that would be prescribed by the maximax criterion.
b. Find the action that would be prescribed by the maximin criterion.

SECTION 18.7

THE EXPECTED UTILITY CRITERION

The expected payoff (or opportunity loss) criterion sometimes fails to provide decisions that are consistent with the decision-maker's attitude toward risk. For example, suppose you were required to choose one of these two actions:

a_1: Deposit your $100,000 inheritance in the bank for 1 year at 7% interest.

a_2: Invest your $100,000 inheritance for 1 year with a .5 probability of having a total of $250,000 at the end of the year and a .5 probability of losing all your inheritance.

Which would you choose? Why? Most of us would probably choose action a_1, basing our decision on a "safety first" strategy. The usual argument goes, "Why pass up a chance for a sure $107,000 at the end of the year for a 50–50 chance to end up with nothing?" Notice that if you choose a_1, your decision is not in accord with the action prescribed by the expected payoff criterion, because

$$EP(a_1) = \$107,000(1) = \$107,000$$
$$EP(a_2) = \$250,000(.5) + 0(.5) = \$125,000$$

Thus, if we use the expected payoff criterion, we would select action a_2.

The expected payoff criterion can be adapted to reflect our attitude toward risk if we can express the outcomes of the decision problem in terms of an objective variable that better reflects the true relative values of outcomes. An objective variable that reflects the decision-maker's attitude toward risk is called a **utility function**, and the values assigned to the outcomes are referred to as **utility values**, or **utilities** (or sometimes **utiles**).

DEFINITION 18.3

A **utility function** is a rule that assigns numerical values to the potential outcomes of a decision problem in such a way that

1. The values rank the outcomes in accordance with the decision-maker's preferences.
2. The function itself describes the decision-maker's attitude toward risk.

We now illustrate how to assign utility values to the monetary outcomes of the inheritance example. The payoff table for this example is shown in Table 18.14. As noted previously, many of us would be reluctant to gamble on the outcome of the investment if the chance of success were only .5, even though the expected payoff is higher for action a_2 ($125,000) than for action a_1 ($107,000). *The key to assigning utility values to the action/state of nature combinations is to answer the following question: What would the probability of success for the investment have to be for you to value actions a_1 and a_2 equally?* The answer to this question is a probability, p, such that if the probability of the investment's success exceeds p, you would prefer to invest the money (a_2), but if the success probability is less than p, you would prefer to deposit the money in the bank and take the safe return. The probability p is called the **utility of the outcome** $107,000, and we write $U(107,000) = p$. The minimum payoff ($0) and the maximum payoff ($250,000) are assigned utility values of 0 and 1, respectively, so that $U(0) = 0$, and $U(250,000) = 1$.

TABLE 18.14

Payoff Table for the
Inheritance Example

		STATE OF NATURE Outcome of investment	
		S_1: Failure	S_2: Success
ACTION	a_1: Deposit $100,000 in bank	$107,000	$107,000
	a_2: Invest $100,000	0	$250,000

(TABLE 18.14 — Payoff Table for the Inheritance Example)

Suppose you decide that the probability of the investment's success would have to be .7 before the two actions are equally appealing. Then the utility value assigned to $107,000 is .7, and the payoff table with the utility values as outcomes is shown in Table 18.15.

TABLE 18.15

Payoff Table for the
Inheritance Example with
Utility Values as Payoffs

		STATE OF NATURE (PROBABILITIES IN PARENTHESES) Outcome of investment	
		S_1: Failure (.5)	S_2: Success (.5)
ACTION	a_1: Deposit $100,000	.7	.7
	a_2: Invest $100,000	0	1

The general rule for assigning utility values to monetary outcomes* is given in the box.

ASSIGNING UTILITY VALUES TO MONETARY OUTCOMES†

1. Identify the maximum and minimum payoffs in the decision table. Call them O_M and O_L, respectively.
2. Set $U(O_M) = 1$ and $U(O_L) = 0$, where $U(O)$ represents the utility value of outcome O.
3. To determine the utility value for any other outcome O_{ij} in the payoff table, determine the value of p that makes you have no preference between the following:
   a. Receiving O_{ij} with certainty
   b. Participating in a gamble in which you can win O_M with probability p or O_L with probability $(1 - p)$
   Then, $U(O_{ij}) = p$.

---

*Utility functions may be assessed for the outcomes corresponding to any objective variable, but because the objective variable in business decision-making is typically monetary (or can be converted to a monetary equivalent), we have restricted our discussion to utility functions for money.

†The choice of 0 and 1 as the minimum and maximum utility values is arbitrary. The same result is obtained if the utility values are all multiplied by the same constant or if a constant is added to each of the values. However, in either case, utility values lose their probabilistic interpretation. For more detail on the choice of scale, its implications for interpersonal and intrapersonal comparisons of utility values, and other methods for assessing utility functions, see the references at the end of this chapter.

We are now prepared to make a decision based on the **expected utility criterion**. We first find the expected utility for each action using the formula

$$EU(a_i) = \sum_{\substack{\text{All states} \\ \text{of nature}}} \left(\begin{array}{c} \text{Utility of the} \\ \text{action } a_i/\text{state of nature} \\ \text{combination} \end{array} \right) \left(\begin{array}{c} \text{Probability} \\ \text{of the} \\ \text{state of nature} \end{array} \right)$$

Thus, for the inheritance example,

$$EU(a_1) = .7(.5) + .7(.5) = .7$$
$$EU(a_2) = 0(.5) + 1(.5) = .5$$

We now choose the action with the higher expected utility, which is action a_1. You can see that this decision differs from that yielded by the expected payoff criterion and that it more accurately reflects a safety-first attitude toward risk.

EXPECTED UTILITY CRITERION

Choose the action that has the greatest expected utility, where the expected utility of action a_i is given by

$$EU(a_i) = \sum_{\substack{\text{All states} \\ \text{of nature}}} \left(\begin{array}{c} \text{Utility of the} \\ \text{action } a_i/\text{state of nature} \\ \text{combination} \end{array} \right) \left(\begin{array}{c} \text{Probability} \\ \text{of the} \\ \text{state of nature} \end{array} \right)$$

EXAMPLE 18.3

Recall the brewery size decision problem of Example 18.1. The payoff table is repeated in Table 18.16 (where the inadmissible action a_4 is left out). Suppose we assign the utility values shown in Table 18.17 to the monetary payoffs.

TABLE 18.16
Payoff Table for the Brewery Size Decision Problem

	Brewery size	STATE OF NATURE Market share during fifth year of operation		
		S_1: 0% < 5%	S_2: 5% < 10%	S_3: 10%–15%
	a_1: Small	$300,000	$350,000	$450,000
ACTION	a_2: Medium	$250,000	$700,000	$800,000
	a_3: Large	$200,000	$600,000	$1,000,000

TABLE 18.17
Utility Values for the Brewery Decision Problem

	Brewery size	STATE OF NATURE (PROBABILITIES IN PARENTHESES) Market share during fifth year of operation		
		S_1: 0% < 5% (.4)	S_2: 5% < 10% (.5)	S_3: 10%–15% (.1)
	a_1: Small	.35	.50	.65
ACTION	a_2: Medium	.20	.90	.95
	a_3: Large	0	.85	1.00

a. Interpret the utility value assigned to the $700,000 payoff, $U(700,000) = .9$.
b. Determine which action should be taken according to the expected utility criterion.

SOLUTION

a. Referring to the rules for assigning utility values, we see that the utility value of .9 represents the probability that makes us have no preference between receiving $700,000 with certainty and participating in a gamble in which we can gain $1,000,000 (the maximum payoff) with probability .9 or $200,000 (the minimum payoff) with probability $(1 - .9) = .1$. You can see that we have again adopted a conservative strategy, because the expected payoff for the gamble is $1,000,000(.9) + 200,000(.1) = \$920,000$, which exceeds the fixed payoff of $700,000. In other words, the assignment of $U(700,000) = .9$ reflects our desire to receive the fixed payoff unless the odds are very high that we will win the gamble. The rest of the utility values can be similarly interpreted, and they all reflect this conservative attitude toward risk.

b. The expected utilities are calculated as follows:

$$EU(a_1) = .35(.4) + .50(.5) + .65(.1) = .455$$

$$EU(a_2) = .20(.4) + .90(.5) + .95(.1) = .625$$

$$EU(a_3) = 0(.4) + .85(.5) + 1.0(.1) = .525$$

Thus, according to our assignment of utility values to the monetary outcomes, the expected utility criterion indicates that we should select action a_2 and build a medium-sized brewery. ■

You can see that the assignment of utility values is personal and subjective. If careful thought is given to this task, the result will be a utility function that reflects your preferences for outcomes and your attitude toward risk. Applying the expected utility criterion will then yield a decision consistent with your preferences and risk attitude.

EXERCISES
18.42–18.48

LEARNING THE MECHANICS

18.42 Suppose you are indifferent (i.e., have no preference) between receiving $100 with certainty and participating in a gamble in which you have a probability of .2 of receiving $1,000 and a probability of .8 of receiving nothing. You are also indifferent between receiving $500 with certainty and participating in a gamble in which you have a probability of .7 of receiving $1,000 and a probability of .3 of receiving nothing. Finally, you are also indifferent between receiving $800 with certainty and participating in a gamble in which you have a probability of .95 of receiving $1,000 and a probability of .05 of receiving nothing.
a. Let $U(\$1,000) = 1$ and $U(\$0) = 0$. Find $U(\$100)$, $U(\$500)$, and $U(\$800)$.
b. Use these utility values to plot your utility function for money over the range from $0 to $1,000.

18.43 Consider the utility function* for money,

$$U(x) = x^2 \qquad 0 \le x \le 100$$

and the payoff table shown here (probabilities in parentheses).

		STATE OF NATURE		
		S_1 (.25)	S_2 (.30)	S_3 (.45)
ACTION	a_1	$75	$50	$30
	a_2	$60	$80	0

a. Use the utility function to convert the outcomes in the payoff table from monetary values to utility values (utiles).

b. Which action would be prescribed by the expected utility criterion?

18.44 Consider the utility function for money,

$$U(x) = \frac{\ln(x + 51)}{5.525} \qquad -50 < x < 200$$

and the given payoff table (probabilities in parentheses).

		STATE OF NATURE		
		S_1 (.50)	S_2 (.40)	S_3 (.10)
	a_1	−$10	$50	$90
	a_2	0	$20	$20
ACTION	a_3	−$10	$60	$50
	a_4	−$10	−$50	$200

a. Eliminate any inadmissible actions.

b. Use the utility function to convert the outcomes in the payoff table from monetary values to utility values (utiles).

c. Which action is prescribed by the expected utility criterion?

d. Does this decision differ from the decision prescribed by the expected payoff criterion?

18.45 Consider the utility function for money,

$$U(x) = -1 + .01x \qquad 100 \le x \le 200$$

and the payoff table given here (probabilities are in parentheses).

		STATE OF NATURE			
		S_1 (.20)	S_2 (.15)	S_3 (.40)	S_4 (.25)
	a_1	$120	$180	$100	$100
ACTION	a_2	$105	$165	$190	$150
	a_3	$100	$120	$160	$200

*Notice that the minimum and maximum values of this utility function are not 0 and 1, respectively. Refer to the second footnote on page 1069.

a. Use the utility function to convert the outcomes in the payoff table from monetary values to utility values (utiles).
b. Which action is prescribed by the expected utility criterion?

APPLYING THE CONCEPTS

18.46 An investor is trying to decide whether to invest in a wildcat oil well. If the well is drilled and it is dry, she will lose $500,000. On the other hand, if the well is a gusher, she will make $1.5 million. It is also possible for the well to yield a lesser amount of oil than a gusher, in which case she will make $600,000.

a. A decision analyst asks her the following questions:

At what value of p would you be indifferent between the two situations: receive $600,000 with certainty, and receive $1.5 million with probability p or lose $500,000 with probability $(1 - p)$? She replies, at $p = .90$. What is the investor's utility value for $600,000?

At what value of p would you be indifferent between receiving $0 with certainty, and receiving $1.5 million with probability p or losing $500,000 with probability $(1 - p)$? She replies, at $p = .8$. Find her utility value for $0.

b. Graph the utility function for this investor for dollar values between $-\$500,000$ and $\$1,500,000$.

18.47 Refer to Exercise 18.46. The investor now must decide whether to keep the $500,000 or invest it in the oil well. She assesses the probabilities of the states to be as follows:

$$P(\text{Dry well}) = .5$$
$$P(\text{Moderate success}) = .3$$
$$P(\text{Gusher}) = .2$$

a. Set up the payoff table for this decision problem.
b. Use the expected payoff criterion to determine which action the investor should select.
c. Substitute the utility values for the monetary outcomes in the payoff table of part a and use the expected utility criterion to select the action.

18.48 Refer to Exercise 18.32, in which you selected an action for the dress purchaser using the expected payoff criterion. Now suppose the buyer's utility function for money is

$$U(x) = \frac{\sqrt{x + 2{,}000}}{110} \qquad -2{,}000 \le x \le 10{,}000$$

How many dresses should she order? Should she be using her own utility function to make this decision? Why or why not?

| | | | | | | | | | | | |

SECTION 18.8

CLASSIFYING DECISION-MAKERS BY THEIR UTILITY FUNCTIONS

The attitude of a decision-maker toward risk may be characterized by the type of utility function he or she uses. Recall the inheritance decision problem of Section 18.7, with the payoff table given in Table 18.14 and a table of utility values given in Table 18.15. These tables are repeated for convenience as Tables 18.18 and 18.19 on page 1074.

We know that these utility values reflect a conservative attitude toward risk, since the expected payoff of the investment must exceed $0(.3) + 250{,}000(.7) = \$175{,}000$

TABLE 18.18
Payoff Table for the
Inheritance Example

		STATE OF NATURE (PROBABILITIES IN PARENTHESES)	
		Investment fails (.5)	Investment succeeds (.5)
ACTION	a_1: Deposit $100,000 in bank	$107,000	$107,000
	a_2: Invest $100,000	0	$250,000

TABLE 18.19
Utility Value Table for the
Inheritance Example

		STATE OF NATURE (PROBABILITIES IN PARENTHESES)	
		Investment fails (.5)	Investment succeeds (.5)
ACTION	a_1: Deposit $100,000	.7	.7
	a_2: Invest $100,000	0	1

before the decision-maker prefers this gamble to the fixed $107,000 return. This attitude may be graphically portrayed by plotting the utility values on a vertical axis against the payoffs on a horizontal axis, as shown in Figure 18.3. The points are connected with a smooth curve. Utility functions with this shape are called **concave utility functions**. This shape is characteristic of a conservative attitude toward risk, and decision-makers whose utility functions are concave are called **risk-avoiders**.

FIGURE 18.3
Utility Function for the
Inheritance Example; a
Risk-Avoiding Attitude

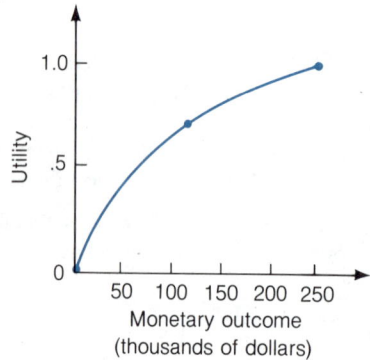

Now, suppose we decided to assign a utility value of .2 to the $107,000 payoff in Table 18.18. This value reflects a liberal, or gambling, attitude toward risk, since the interpretation is that we have no preference between a sure $107,000 and a .2 probability of receiving $250,000 (with a .8 probability of nothing). The expected payoff of the gamble is $0(.8) + \$250,000(.2) = \$50,000$, which is less than the fixed payoff if we deposit the money in the bank. The graph of this utility function is shown in Figure 18.4. Utility functions with this shape are called **convex utility functions**. This shape is characteristic of an aggressive attitude toward risk, and decision-makers who have convex utility functions are called **risk-takers**.

FIGURE 18.4
Utility Function for the
Inheritance Example; a
Risk-Taking Attitude

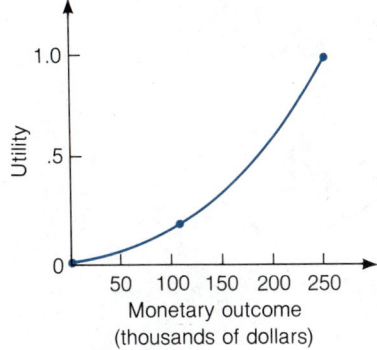

If the utility value of .428 is assigned to the $107,000 outcome, the implication is that we would have no preference between accepting $107,000 and gambling on receiving $250,000 with probability .428, a gamble that has an expected payoff of $0(.572) + 250,000(.428) = \$107,000$. The fact that both courses of action have the same expected payoff reflects a neutral attitude toward risk. The graph of the utility function in this case and, for **risk-neutral** decision-makers in general, is a straight line, as shown in Figure 18.5. Since the expected utility of action a_i is equal to the utility of the expected payoff for action a_i for straight-line utility functions, *the expected payoff criterion and the expected utility criterion are equivalent for risk-neutral decision-makers.*

FIGURE 18.5
Utility Function for the
Inheritance Example; a
Risk-Neutral Attitude

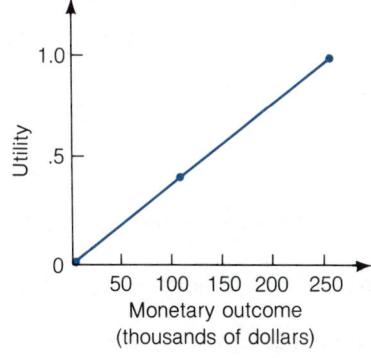

The three types of utility functions and the risk attitudes they characterize are summarized in the box.

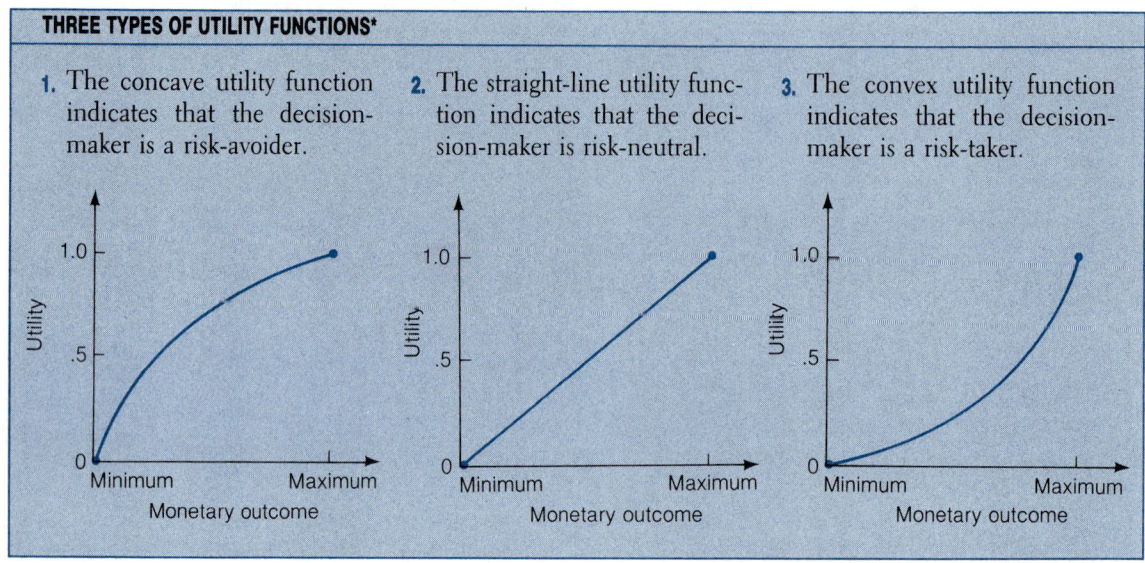

THREE TYPES OF UTILITY FUNCTIONS*

1. The concave utility function indicates that the decision-maker is a risk-avoider.

2. The straight-line utility function indicates that the decision-maker is risk-neutral.

3. The convex utility function indicates that the decision-maker is a risk-taker.

EXAMPLE 18.4

For the brewery decision problem (Examples 18.1 and 18.3), consider the two utility functions reflected in Table 18.20.

TABLE 18.20 Two Utility Functions for the Brewery Decision Problem

		SITUATION A			SITUATION B		
		STATE OF NATURE (PROBABILITIES IN PARENTHESES) Market share during fifth year of operation			STATE OF NATURE (PROBABILITIES IN PARENTHESES) Market share during fifth year of operation		
	Brewery size	S_1: 0% < 5% (.4)	S_2: 5% < 10% (.5)	S_3: 10%–15% (.1)	S_1: 0% < 5% (.4)	S_2: 5% < 10% (.5)	S_3: 10%–15% (.1)
	a_1: Small	.35	.50	.65	.04	.07	.11
ACTION	a_2: Medium	.20	.90	.95	.02	.30	.40
	a_3: Large	0	.85	1.00	0	.20	1.00

a. Graph the utility functions, and identify the attitude toward risk that each characterizes.

b. Determine the action that should be taken in each situation.

*Other types of utility functions may be obtained by combining these. For example, a decision-maker may be risk-neutral for relatively small monetary outcomes and a risk-avoider for relatively large monetary outcomes. Thus, the utility function might appear S-shaped.

SOLUTION

a. The utility functions are shown in Figure 18.6. Situation A is the same one we used and characterized as conservative in Example 18.3. You can see that the utility function in Figure 18.6(a) is *concave*, indicating a *risk-avoiding* attitude. However, the utility function corresponding to situation B is *convex*, which characterizes a *risk-taking* situation.

FIGURE 18.6 Utility Functions for the Brewery Decision Problem

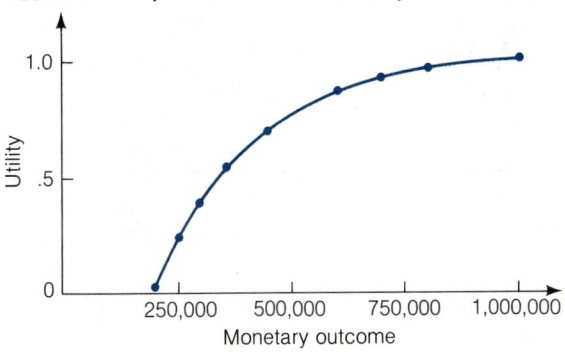

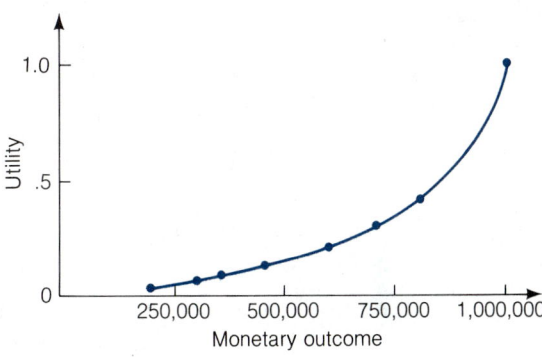

(a) Situation A   (b) Situation B

b. We showed in Example 18.3 that for situation A,

$$EU(a_1) = .455 \qquad EU(a_2) = .625 \qquad EU(a_3) = .525$$

so the expected utility criterion selects action a_2: Build a medium-sized brewery. For situation B, we find

$$EU(a_1) = .04(.4) + .07(.5) + .11(.1) = .062$$
$$EU(a_2) = .02(.4) + .30(.5) + .40(.1) = .198$$
$$EU(a_3) = 0(.4) + .20(.5) + 1.00(.1) = .200$$

Thus, the expected utility criterion selects action a_3: Build a large brewery. Note that the risk-taking utility function leads to the selection of a riskier action, a_3, than the risk-avoiding and risk-neutral functions, which both select a_2 (see Section 18.5 for the risk-neutral—that is, expected payoff—solution). ∎

CASE STUDY 18.3

AN AIRPORT: TO EXPAND OR NOT TO EXPAND?

The Mexican government received two conflicting recommendations on airport development for Mexico City. One study recommended expanding the present airport at Texcoco, and the other suggested transferring all operations as soon as possible to a new airport to be built at Zumpango. As a result of this disagreement, the Mexican Ministry of Public Works employed Keeney, Raiffa, and de Neufville to evaluate alternatives for airport development and recommend the most effective strategy.

The basic decision problem, as described by Keeney (1973), involved the determination of which types of aircraft (international, I; domestic, D; general, G; and military, M) should operate at each of the two locations (Texcoco, T, and Zumpango, Z) over the next 30 years. To simplify the problem, Keeney, Raiffa, and de Neufville assumed that changes in operations from one site to another could occur at only three times: 1975, 1985, and 1995. Each airport development strategy (action) indicated which types of aircraft would operate at each location for each time period. Some examples of possible strategies are given in Table 18.21. Strategies 1 and 2 are the recommendations of the previous studies. Strategy 3 is one possible intermediate strategy. Keeney, Raiffa, and de Neufville evaluated 100 such strategies.

TABLE 18.21

Possible Strategies for Airport Expansion

		1975		1985		1995	
		T	Z	T	Z	T	Z
STRATEGY	1	IDGM		IDGM		IDGM	
	2		IDGM		IDGM		IDGM
	3	DGM	I	MG	ID		IDGM

The outcome of each strategy was evaluated in terms of six attributes:

1. Cost
2. Capacity (number of aircraft operating)
3. Average access time to airport
4. Number of people killed or injured per aircraft accident
5. Number of people displaced by airport development
6. Number of people subjected to a high noise level

One step in the decision analysis was to evaluate utility functions for each of these attributes. The following description of the utility assessment for access time is representative of the other utility assessments. By questioning the clients, Keeney, Raiffa, and de Neufville determined that the bounds on average access time over all strategies were 12 minutes and 90 minutes. The best possible time, 12 minutes, was assigned a utility of 1; the worst possible time, 90 minutes, was assigned a utility of 0. The clients were then presented with a series of choices of the following type: 62 minutes access time for certain, or a lottery with p chance at 12 minutes and $(1 - p)$ chance at 90 minutes. The size of p was varied until the client indicated indifference between these two choices. In the case of 62 minutes, the client was indifferent for $p = .5$. Thus,

$$U(62 \text{ minutes}) = .5$$

FIGURE 18.7

Utility Function for Case Study 18.3

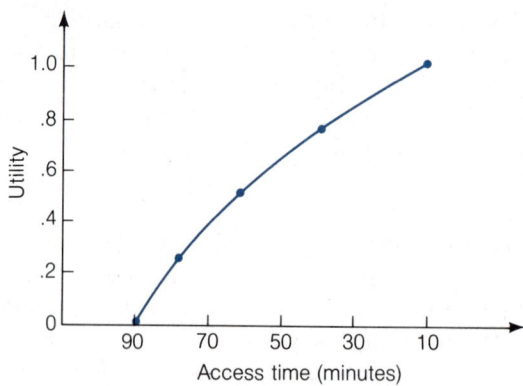

This process was repeated for several other access times, and the results were plotted as shown in Figure 18.7. You can see that the clients are risk-avoiders with respect to access time.

The results of the utility analysis for each attribute were combined to develop a set of utility functions. Probabilities for the various outcomes were also assessed. The utilities and probabilities were combined to determine an expected utility for each of the airport development strategies. The results of the decision analysis indicated that a strategy of gradual shift of operations from Texcoco to Zumpango over the next 30 years had the highest expected utility.

EXERCISES
18.49–18.54

LEARNING THE MECHANICS

18.49 Graph each of the following utility functions* for money and classify each as the utility function of a risk-avoiding, risk-taking, or risk-neutral decision-maker:

a. $U(x) = 2x^3, \quad 0 \le x \le 100$ **b.** $U(x) = \frac{1}{100}x, \quad 0 \le x \le 100$

c. $U(x) = \log(x + 100), \quad 10 \le x \le 100$ **d.** $U(x) = -1 + .01x, \quad 100 \le x \le 200$

18.50 Consider the following utility function for money:

$$U(x) = (x + 1{,}000)^{1/2} \qquad -1{,}000 \le x \le 1{,}000$$

a. Plot the function.

b. Is this the utility function of a risk-avoiding, risk-taking, or risk-neutral decision-maker?

c. Use this utility function and the expected utility criterion to identify the optimal action for the decision problem characterized by the payoff table shown here (probabilities in parentheses).

		STATE OF NATURE		
		S_1 (.2)	S_2 (.6)	S_3 (.2)
	a_1	−$500	$1,000	$250
ACTION	a_2	−750	900	810
	a_3	45	−100	650

18.51 Rework all parts of Exercise 18.50 using the utility function for money,

$$U(x) = (x + 1{,}000)^2 \qquad -1{,}000 \le x \le 1{,}000$$

Is the action prescribed by the expected utility criterion the same as the action prescribed in Exercise 18.50? If not, explain the reason for this disagreement.

18.52 Consider the utility function for money,

$$U(x) = .001x \qquad 0 \le x \le 1{,}000$$

and the payoff table given at the top of page 1080 (probabilities in parentheses).

*Notice that the minimum and maximum values of these utility functions are not 0 and 1, respectively. Refer to the second footnote on page 1069.

		STATE OF NATURE			
		S_1 (.10)	S_2 (.35)	S_3 (.35)	S_4 (.20)
	a_1	0	25	300	850
ACTION	a_2	1,000	500	200	80
	a_3	100	600	1,000	100

a. What does this utility function indicate about the decision-maker's attitude toward risk?

b. Which action is prescribed by the expected utility criterion? Should it differ from the action prescribed by the expected payoff criterion? Explain.

APPLYING THE CONCEPTS

18.53 Refer to Exercise 18.46, in which an investor is trying to decide whether to invest in a wildcat oil well.

a. Plot the utility function you constructed for the investor.

b. Categorize her as a risk-avoider, risk-taker, or risk-neutral.

18.54 Refer to Exercise 18.48, in which the dress purchaser used the utility function

$$U(x) = \frac{\sqrt{x + 2,000}}{110} \qquad -2,000 \le x \le 10,000$$

a. Plot the purchaser's utility function.

b. Is she a risk-taker, risk-avoider, or risk-neutral?

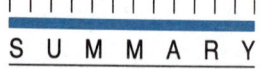

S U M M A R Y

Decision analysis can be used to prescribe a course of action when the decision-maker is uncertain about the outcome that will result from a chosen action. For the prescribed action to fully reflect the decision-maker's preferences regarding the outcomes *and* his or her attitudes toward risk, the **expected utility criterion** should be used. In order to use this criterion, it is necessary for the decision-maker to assess a **utility function**. It is through a utility function that the decision-maker is able to quantify his or her preferences regarding the potential outcomes of the decision problem, as well as his or her attitudes toward risk. In fact, a decision-maker can be classified as a **risk-avoider, risk-taker,** or **risk-neutral** by examining the shape of his or her utility function.

If a decision-maker is risk-neutral, then both the expected utility criterion and the **expected payoff criterion** prescribe the same action. The expected payoff criterion has the advantage of not requiring the assessment of a utility function. Thus, many decision-makers assume risk-neutrality and use the expected payoff criterion as an approximation to the expected utility criterion.

To obtain meaningful results from decision analysis, the decision-maker must take great care in developing the inputs to the analysis. The **actions, states of nature, outcomes,** and **objective variable** must be properly defined. Much careful, logical thought should go into the assessment of the utility function. The **probability** distribution of the states of nature should be assessed using as much information about the states of nature as it is possible to obtain. If conducted properly, decision analysis can provide valuable inputs to the decision-making process even if the decision it prescribes is not implemented. The detailed analysis and logical thinking it requires may uncover courses of action and/or states of nature not previously recognized. It may even be discovered that the wrong decision problem is being addressed.

SUPPLEMENTARY EXERCISES 18.55–18.69

[*Note: Starred (*) exercises refer to the optional section in this chapter.*]

18.55 A small company that produces auto parts for American-made cars expects to lose $1.5 million next year unless Congress passes a bill to limit the number of foreign-made cars that can be imported into the United States. If the bill passes, the company expects to make a profit of $2.5 million next year. The company's lobbyists in Washington believe that the probability that the bill will pass this year is .6. The auto parts company must decide whether to stay in business and risk a huge loss next year, or to lease its facility for a year to a Japanese firm for $500,000.
   a. Identify the objective variable, actions, and states of nature of the auto parts company's decision problem.
   b. Construct the payoff table for the decision problem.
   c. Convert the payoff table to an opportunity loss table.
   d. Verify that the expected payoff and expected opportunity loss criteria prescribe the same action.

18.56 Refer to Exercise 18.55. Assume that the auto parts manufacturer's utility function for money is as follows:

$$U(x) = .5 + .20x \qquad -2.5 \text{ million} \le x \le 2.5 \text{ million}$$

   a. Plot the utility function.
   b. Classify the utility function as being that of a risk-avoiding, risk-taking, or risk-neutral decision-maker.
   c. According to the expected utility criterion, which action should the auto manufacturer select?

18.57 Consider the following payoff table, with state probabilities shown in parentheses:

	S_1 (.5)	S_2 (.3)	S_3 (.2)
a_1	12	77	−27
a_2	24	2	12
a_3	−20	41	80
a_4	25	5	70
a_5	40	50	10

   a. Eliminate any inadmissible actions.
   b. Construct an opportunity loss table for this decision problem.

c. Calculate both the expected payoffs and the expected opportunity losses for the actions, and verify that both criteria select the same action.

d. Construct a decision tree for this problem that includes both state probabilities and expected payoffs. (See Exercise 18.24 for a description of where to locate these quantities on your tree.)

18.58 Indicate what type of information (subjective, relative frequency, or both) you would use to assign probabilities to the events described here. Explain the reasoning behind your choice. How would you obtain the necessary information to assign the probabilities?

a. Your firm is in the automobile insurance business, and you want to assess a probability distribution to characterize the total dollar value of damage to a certain model car when an 18- to 21-year-old driver is in an accident.

b. Your company is introducing a new product, and you are assigned the task of assessing a probability distribution for the market share the product will obtain.

c. A drug company is attempting to develop a new type of birth control pill. The research is still in the early stages, and you want to assess the probability that the research will be successful.

d. You are a banker and are considering giving a 30-year mortgage of $75,000 to a family with a yearly income of $25,000. You want to assess the probability that the family will default on the loan.

18.59 A bank is trying to decide whether to make a 1-year commercial loan of $75,000 to an automobile repair shop. Past experience has shown that one of three outcomes will result if the loan is made:

Outcome 1:   The customer will repay the loan plus the 10% interest with no complications.

Outcome 2:   The customer will have difficulty repaying the loan. The loan will eventually be repaid with the 10% interest, but a $1,000 collection cost will have been incurred by the bank.

Outcome 3:   The customer will declare bankruptcy, and the bank will recover only 60% of the amount loaned.

If the bank decides not to make the loan, the money will earn 8% for the year.

a. Construct the payoff table for this decision problem.

b. Convert the payoff table to an opportunity loss table.

c. Draw a decision tree for this problem.

18.60 Refer to Exercise 18.59. Suppose past records yield the frequency distribution for commercial loans shown in the table.

OUTCOME	FREQUENCY
1. Repaid	1,104
2. Repaid with difficulty	120
3. Defaulted	56

a. Use this information to assess the probabilities of the three outcomes.

b. Using the assessed probabilities and the expected payoff criterion, decide whether the bank should make the loan.

*18.61 Refer to Exercise 18.60. Suppose no probability estimates for the outcomes are available.
 a. Use the maximax criterion to decide whether to make the loan.
 b. Use the maximin criterion to make the decision.
 c. Critique these two nonprobabilistic criteria for making this decision.

18.62 Refer to Exercises 18.59 and 18.60. Suppose the bank uses the following utility function for money:

$$U(x) = \frac{\sqrt{x + 100,000}}{1,000} \qquad -100,000 \le x \le 900,000$$

 a. Graph this function over the given range of x, and classify the bank as risk-avoiding, risk-neutral, or risk-taking.
 b. Use this function to assign utility values to the monetary payoffs in the decision table of Exercise 18.59. Based on the expected utility criterion, should the loan be made?

18.63 A hosiery company must make a pricing decision regarding its new line of stockings. Two economists have been hired to develop forecasting equations that relate the quantity demanded to the price of the stockings. The forecasting equations the economists derive are

Economist 1: $q = 10 - 2p$
Economist 2: $q = 16 - 4p$

 where q is the quantity demanded in units of 100,000 and p is the price in dollars. Four prices are being considered: \$.99, \$1.98, \$2.75, and \$3.50. Assume that one of the economists' forecasting equations will be correct, but you do not know which one.
 a. Construct the payoff table for this problem.
 b. Construct the opportunity loss table for this problem.
 c. If the company believes the forecasting equations have an equal probability of being correct, which price should be charged according to the expected payoff criterion?

18.64 A medical doctor is involved in a \$1 million malpractice suit. He can either settle out of court for \$250,000 or go to court. If he goes to court and loses, he must pay the \$925,000 plus \$75,000 in court costs. If he wins in court, the plaintiffs pay the court costs.
 a. Construct a payoff table for this decision problem.
 b. Draw a decision tree for this problem.
 c. The doctor's lawyer estimates the probability of winning to be .2. Use the expected payoff criterion to decide whether the doctor should settle or go to court. Enter the lawyer's assessed probabilities and the expected payoffs associated with each action on your decision tree. (See Exercise 18.24 for a description of where to locate these quantities.)

*18.65 Refer to Exercise 18.64. Suppose no estimate is available for the probability of winning the suit.
 a. Use the maximax criterion to decide whether to settle or go to court.
 b. Use the maximin criterion to make the decision.

18.66 Refer to Exercise 18.64. Suppose the doctor's utility function for money is given by

$$U(x) = \frac{\sqrt{1,000,000 + x}}{1,400} \qquad -1,000,000 \le x \le 1,000,000$$

a. Convert the payoffs in Exercise 18.64 to utility values.
b. Use the expected utility criterion to decide whether the doctor should settle or go to court.

18.67 A common problem with management information systems (MIS) is that managers fail to use them, even when they are installed and are technically sound. Suppose a large computer firm is considering an MIS that will cost $1,000,000 to build and operate over a 5-year period. If the company's managers use the system, a savings of $750,000 per year will be realized.

a. Construct the payoff table for this decision problem.
b. Convert the payoff table to an opportunity loss table.
c. Suppose the probability that the MIS will be used is assessed to be only .05. Use the expected opportunity loss criterion to decide whether to install the MIS.

18.68 Graph each of the following utility functions and classify the decision-maker's attitude toward risk.

a. $U(x) = \dfrac{100 + .5x}{250}, \qquad -200 \le x \le 300$

b. $U(x) = \dfrac{\sqrt{x + 100}}{20}, \qquad -100 \le x \le 300$

c. $U(x) = \dfrac{x^3 - 2x^2 - x + 10}{800}, \qquad 0 \le x \le 10$

d. $U(x) = \dfrac{.1x^2 + 10x}{110}, \qquad 0 \le x \le 10$

	S_1	S_2	S_3
a_1	12	77	-10
a_2	-20	41	80
a_3	25	5	70
a_4	40	50	10

18.69 Consider the accompanying payoff table. The decision-maker is indifferent between the following pairs of events:

Receive $0 with certainty, and win $80 with probability .15 or lose $20 with probability .85

Receive $20 with certainty, and win $80 with probability .3 or lose $20 with probability .7

Receive $40 with certainty, and win $80 with probability .55 or lose $20 with probability .45

Receive $60 with certainty, and win $80 with probability .8 or lose $20 with probability .2

a. Plot the utility values versus monetary values between −$20 and $80 (assign these endpoints utility values of 0 and 1, respectively).
b. Draw a smooth curve through the points in part **a**, and use this curve to assign utility values to all the payoffs in the table.
c. If the state probabilities are $P(S_1) = .4$, $P(S_2) = .35$, and $P(S_3) = .25$, which action should be selected according to the expected utility criterion? How does this compare with the action selected by the expected payoff criterion?

ON YOUR OWN . . .

Suppose you are trying to decide whether to promote an outdoor event scheduled for this spring. Choose an event that can be held only in good weather (concert, baseball game, famous speaker, etc.) and for which your city has an available facility. Based on the expected attendance and a reasonable ticket price, determine the payoff if the event is held. Remember to subtract your costs for advertising, facility rental, etc. These costs represent the (negative) payoff in case you decide to promote the event and it must be cancelled. Construct the payoff table corresponding to the two-action/two-state decision problem.

Now, based on *your own* knowledge of the climate in the area at the proposed time of the concert, assign probabilities to the states of nature: Rain and No rain. Based on these probabilities, which action (Promote or Do not promote) is selected by the expected payoff criterion?

REFERENCES

Baird, B. F. *Introduction to Decision Analysis.* North Scituate, Mass.: Duxbury, 1978.

Balthasar, H. U., Boschi, R. A. A., and Menke, M. M. "Calling the shots in R&D." *Harvard Business Review*, May–June 1978, *56*, 151–160.

Brown, R. V., Kahr, A. S., and Peterson, C. *Decision Analysis for the Manager.* New York: Holt, Rinehart and Winston, 1974.

Bunn, D. *Applied Decision Analysis.* New York: McGraw-Hill, 1984.

Farquhar, P. H. "Utility assessment methods." *Management Science*, Nov. 1984, *30*(11), 1283–1300.

Howard, R. A., Matheson, J. E., and North, D. W. "The decision to seed hurricanes." *Science.* June 1972, *176*, 1191–1202.

Keeney, R. L. "A decision analysis with multiple objectives: The Mexico City airport." *Bell Journal of Economics and Management*, 1973, *4*, 101–117.

Keeney, R. L. and Raiffa, H. *Decisions with Multiple Objectives: Preferences and Value Tradeoffs.* New York: Wiley, 1976.

Luce, R. D. and Raiffa, H. *Games and Decisions.* New York: Wiley, 1957.

Raiffa, H. *Decision Analysis. Introductory Lectures on Choices Under Uncertainty.* Reading, Mass.: Addison-Wesley, 1968.

Raiffa, H. and Schlaifer, R. *Applied Statistical Decision Theory.* Cambridge, Mass.: MIT Press, 1961.

Schroeder, R. G. *Operations Management Decision Making in the Operations Function.* New York: McGraw-Hill, 1981. Chapter 14.

Winkler, R. L. *An Introduction to Bayesian Inference and Decision.* New York: Holt, Rinehart and Winston, 1972.

Winkler, R. L. and Hays, W. L. *Statistics: Probability, Inference, and Decision*, 2nd ed. New York: Holt, Rinehart and Winston, 1975. Chapters 2 and 9.

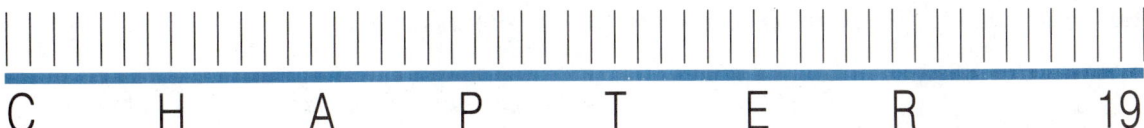

C H A P T E R 19

WHERE WE'VE BEEN...

In the previous chapter we discussed the use of the expected pay-off and the expected utility criteria for making decisions in the presence of uncertainty, and we noted that such decisions are precarious when we have limited prior information about the states of nature.

WHERE WE'RE GOING...

In this chapter we show how to incorporate sample information into the decision-making process. The sample information is used to revise (and hopefully improve) the prior probabilities of the states of nature. The expected monetary value of sample information is considered, and we present an example of a more complex decision problem involving two actions and an infinite number of states of nature.

CONTENTS

DECISION ANALYSIS USING PRIOR AND SAMPLE INFORMATION

In Chapter 18, we presented solutions to decision problems using only currently available information. In this chapter we incorporate **sample information** into the decision process. In Section 19.1, we discuss the use of sample information to revise state of nature probabilities, and then in Section 19.2 we solve a decision problem using these revised probabilities and the expected payoff (or expected opportunity loss) criterion. The maximum expected worth of the sample information is calculated in Section 19.3, and then we compute the actual expected worth of this information in Section 19.4. Finally, an example of a decision problem involving two actions and an infinite number of states of nature is discussed in Section 19.5.

Suppose you are trying to decide whether to purchase 100 shares of common stock in a particular company. You have decided that you want to make the purchase only if the probability is at least .75 that the stock's price will be higher a month from now. Based on your **prior information** about the company and stock market conditions, you assign a **prior probability** of only .6 to this event. Therefore, if you base your decision on your prior information, you will not buy the stock.

However, you may want to obtain **sample information** about the stock before making your decision. Your sample information might be an opinion from a reputable stock analyst whose predictions have the following reliability. Among the stocks that increase in price over a 1-month period, the analyst is able to predict the increase in 80% of the cases. But for stocks that will be stable or decrease in price, she predicts an increase in 40% of the cases. The analyst advises you that, in her opinion, the price of the stock in which you are interested will be higher 1 month from now. How should you revise your prior probability to incorporate this sample information?

Define the events

S₁: {Stock price will be higher 1 month from now}

S₂: {Stock price will be the same or lower 1 month from now}

I: {Analyst predicts the price will be higher 1 month from now}

We are interested in the conditional probability that the stock price will increase *given that* the analyst says it will increase—i.e., $P(S_1 \mid I)$. Recall that the definition of conditional probability stipulates

$$P(S_1 \mid I) = \frac{P(I \cap S_1)}{P(I)}$$

where $P(I \cap S_1)$ is the probability that the analyst says it will increase *and* it does in fact increase. We will find $P(S_1 \mid I)$ in two steps.

Step 1 Find $P(I \cap S_1)$. Recall that $P(I \cap S_1) = P(I \mid S_1) P(S_1)$. We know that $P(I \mid S_1)$—the probability that the analyst predicts a stock price will be higher in a month given that it will in fact be higher—is .8. Furthermore, $P(S_1)$ is the probability that the price will be higher a month from now, given no sample information; i.e., this is the prior probability of S_1, which we have assessed to be .6. Then

$$P(I \cap S_1) = P(I \mid S_1)P(S_1) = (.8)(.6) = .48$$

Step 2   Find $P(I)$. Note that $P(I)$ is the probability that the analyst predicts an increase in the stock's price. This will occur simultaneously with one of the two mutually exclusive, collectively exhaustive (i.e., no other states can occur) states of nature, S_1 and S_2. That is, either the stock will in fact be higher in price 1 month from now, or its price will be the same or lower. Thus,

$$P(I) = P(I \cap S_1) + P(I \cap S_2)$$

We have already found that $P(I \cap S_1) = .48$. In the same way,

$$P(I \cap S_2) = P(I \mid S_2)P(S_2)$$

We know that $P(I \mid S_2)$—the probability that the analyst predicts a stock will increase in price when in fact it will not—is .4. Also,

$$P(S_2) = 1 - P(S_1) = 1 - .6 = .4$$

Then

$$P(I \cap S_2) = P(I \mid S_2)P(S_2) = (.4)(.4) = .16$$

and

$$P(I) = P(I \cap S_1) + P(I \cap S_2) = .48 + .16 = .64$$

Finally, we combine steps 1 and 2 to find

$$P(S_1 \mid I) = \frac{P(I \cap S_1)}{P(I)} = \frac{.48}{.64} = .75$$

Thus, the probability that the stock will increase in price, given the sample information that the analyst predicts its rise, is .75. We refer to probabilities revised on the basis of sample information as **posterior probabilities**.* Since this posterior probability meets our previously established criterion tht the probability of increase be at least .75, we now decide to purchase the stock.

DEFINITION 19.1

A probability $P(S_i)$ of the state of nature S_i that does not incorporate sample information is called a **prior probability** of S_i.

A probability $P(S_i \mid I)$ of the state of nature S_i, given the sample information I, is called a **posterior probability** of S_i.

The process of revising prior probabilities to incorporate sample information is known as **Bayes' rule**. It is named for Thomas Bayes, an English Presbyterian minister and mathematician, who was one of the first to develop methods of calculating posterior probabilities. Bayes' rule for two states of nature and for k states of nature is given in the box at the top of page 1090. The states of nature are assumed to be mutually exclusive, collectively exhaustive events.

*We use the term *posterior* because the probabilities are determined *after* the sample information has been obtained, whereas *prior* probabilities are determined *before* sample information is obtained.

BAYES' RULE FOR CALCULATING POSTERIOR PROBABILITIES

Two states of nature: S_1 and S_2

$$P(S_1 \mid I) = \frac{P(I \cap S_1)}{P(I)} = \frac{P(I \mid S_1)P(S_1)}{P(I \mid S_1)P(S_1) + P(I \mid S_2)P(S_2)}$$

where I is the sample information.

k states of nature: $S_1, S_2, \ldots, S_k$

$$P(S_i \mid I) = \frac{P(I \cap S_i)}{P(I)} = \frac{P(I \mid S_i)P(S_i)}{\displaystyle\sum_{j=1}^{h} P(I \mid S_j)P(S_j)}$$

A **probability revision table** organizes the calculation of posterior probabilities. In Table 19.1 we show the probability revision table for the stock price example, and the general form of the table is given in Table 19.2.

TABLE 19.1
Probability Revision Table for the Stock Price Example

(1) State of nature	(2) Prior probability	(3) Conditional probability of sample information	(4) Probability of intersection of state and sample information (2) × (3)	(5) Posterior probability (4) ÷ Total of (4)
S_1	.6	.8	.48	.75
S_2	.4	.4	.16	.25
Total:	1.0		.64	1.00

TABLE 19.2
Probability Revision Table (General)

(1) State of nature	(2) Prior probability	(3) Conditional probability of sample information	(4) Probability of intersection of state and sample information (2) × (3)	(5) Posterior probability (4) ÷ Total of (4)
S_1	$P(S_1)$	$P(I \mid S_1)$	$P(I \cap S_1)$	$P(S_1 \mid I)$
S_2	$P(S_2)$	$P(I \mid S_2)$	$P(I \cap S_2)$	$P(S_2 \mid I)$
$\vdots$	$\vdots$	$\vdots$	$\vdots$	$\vdots$
S_k	$P(S_k)$	$P(I \mid S_k)$	$P(I \cap S_k)$	$P(S_k \mid I)$
Total:	1		$P(I)$	1

In Example 19.1, we show how to apply Bayes' rule to a decision problem with monetary outcomes.

EXAMPLE 19.1

A company has developed a home smoke detector that is considerably more reliable than those currently on the market, but it is also more expensive to produce. To market the detector competitively, the company will have to price it so low that the profit margin per unit will be quite small. Accordingly, the sales volume, and therefore the market share, will have to be high for the product to be worth marketing.

The company's accounting and marketing departments prepared Table 19.3 using the objective variable Net contribution to profit by the detector during its first 2 years on the market. The marketing department assessed the **prior state probabilities** (shown in parentheses in the table) using industry sales data for other detectors.

TABLE 19.3
Payoff Table for
Example 19.1

		STATE OF NATURE (PRIOR PROBABILITIES IN PARENTHESES) Market share after 2 years			
		S_1: .01 (.10)	S_2: .05 (.40)	S_3: .10 (.40)	S_4: .15 (.10)
ACTION	a_1: Market detector	$-\$1,500,000$	$-\$200,000$	$\$300,000$	$\$1,000,000$
	a_2: Do not market detector	0	0	0	0

Now suppose the company wants to incorporate **sample information** into the decision analysis. Twenty prospective detector purchasers are randomly sampled and asked their opinion of the new detector. Two of the 20 say they will purchase the detector if it is marketed. Use this sample information to *revise* the prior state probabilities by calculating the **posterior probabilities** associated with the states of nature.

SOLUTION

This decision problem has four states of nature, and the prior probabilities are known (see Table 19.3). The next step is to calculate the probability of the sample information given each state of nature (column 3 of the probability revision table). Note that the sample consists of a consumer preference survey and that we can regard its outcome as a binomial random variable (see Section 5.4). The number of trials is $n = 20$, and the probability, p, represents the true market share the company will obtain upon marketing the detector. Thus, each state of nature represents a different value of p. The binomial random variable, x, is the number of the sampled consumers who will purchase the new smoke detector, and the survey result is $x = 2$. We now calculate the conditional probabilities of this sample result (using Table II in Appendix B):

$$P(I \mid S_1) = P(x = 2 \mid p = .01) = \binom{20}{2}(.01)^2(.99)^{18} = .016$$

$$P(I \mid S_2) = P(x = 2 \mid p = .05) = \binom{20}{2}(.05)^2(.95)^{18} = .189$$

$$P(I \mid S_3) = P(x = 2 \mid p = .10) = \binom{20}{2}(.10)^2(.90)^{18} = .285$$

$$P(I \mid S_4) = P(x = 2 \mid p = .15) = \binom{20}{2}(.15)^2(.85)^{18} = .229$$

We are now prepared to use the probability revision table to calculate the posterior probabilities for the states of nature. The calculations are shown in Table 19.4. Note that the sample information revised the probabilities of states S_3 and S_4 *upward* and the probabilities of states S_1 and S_2 *downward*. The posterior probabilities indicate that the company can be more optimistic about the market share of the smoke detector than it was prior to obtaining the sample information.

TABLE 19.4
Probability Revision Table
for Example 19.1

	(1)	(2)	(3)	(4)	(5)
				Probability of	
			Conditional	intersection of	
			probability	state and sample	Posterior
	State of	Prior	of sample	information	probability
	nature	probability	Information	(2) × (3)	(4) ÷ Total of (4)
	S_1: $p = .01$	.10	.016	.0016	.0075
	S_2: $p = .05$	.40	.189	.0756	.3531
	S_3: $p = .10$	.40	.285	.1140	.5325
	S_4: $p = .15$	.10	.229	.0229	.1070
	Total:	1.00		.2141	1.0001[a]

[a]Rounding error.   ∎

The extent to which the prior probabilities are revised depends on the amount of information contained in the sample. For example, if 100 consumers had been surveyed in Example 19.1 and if the same *proportion* had responded favorably to the new smoke detector (i.e., $x = 10$), the four posterior state probabilities would be 0, .104, .826, and .069, respectively. Thus, the prior probability of state S_3: $p = .10$ is revised from .4 to .826. These probabilities are shown in Table 19.5 along with the posterior probabilities for a survey of $n = 20$ consumers (Example 19.1). As the sample size is increased, the weight given to the prior probabilities is diminished and more importance is attached to the sample information.

TABLE 19.5
Prior and Posterior
Probabilities for the
Smoke Detector Example:
Two Different Sample
Sizes

STATE	PRIOR PROBABILITY	POSTERIOR PROBABILITY	
		$n = 20$ ($x = 2$)	$n = 100$ ($x = 10$)
S_1: $p = .01$	.10	.008	.000
S_2: $p = .05$	.40	.353	.104
S_3: $p = .10$	.40	.532	.826
S_4: $p = .15$	.10	.107	.069

| | | | | | | | | | | | | |

EXERCISES 19.1–19.12   LEARNING THE MECHANICS

19.1 Give Bayes' rule for the case in which there are three states of nature, S_1, S_2, and S_3, and information I has been received by sampling.

19.2 Use the formula obtained in Exercise 19.1 to complete the following probability revision table:

(1) State of nature	(2) Prior probability	(3) Conditional probability of sample information	(4) (2) × (3)	(5) Posterior probability (4) ÷ Total of (4)
S_1	.25	.60		
S_2	.40	.80		
S_3	.35	.10		
Total:	1.00			

19.3 Complete the following probability revision table:

(1) State of nature	(2) Prior probability	(3) Conditional probability of sample information	(4) (2) × (3)	(5) Posterior probability (4) ÷ Total of (4)
S_1	.35	.90		
S_2	.15	.10		
S_3	.30	.65		
S_4	.20	.30		
Total:	1.00			

19.4 Two coins are placed in a hat. One coin has two heads, while the other has a head and a tail. One of the coins is drawn at random. It is flipped and a head is observed. What is the posterior probability that the coin contains two heads? If a tail is observed, what is the posterior probability that the coin contains two heads?

19.5 What role does sampling play in decision analysis?

19.6 Compare and contrast prior and posterior information.

19.7 Consider this prior distribution for x:

x	5	10	15	20	25
$p(x)$	.2	.4	.2	.1	.1

The following conditional probabilities are associated with the sample information, I:

$$P(I \mid x = 5) = .5 \qquad P(I \mid x = 10) = .8 \qquad P(I \mid x = 15) = .3$$
$$P(I \mid x = 20) = .2 \qquad P(I \mid x = 25) = .1$$

a. Find the posterior probability distribution for x.
b. Find the mean and variance of the posterior probability distribution for x.

APPLYING THE CONCEPTS

19.8 A systems analyst is concerned about the proportion, p, of records that are in error in the inventory control system she designed. She assessed the probability distribution shown in the table to characterize her beliefs regarding p.

PROPORTION OF ERRORS	PRIOR PROBABILITY
.09	.40
.10	.25
.11	.15
.12	.10
.13	.10

a. Find the mean and variance of the analyst's prior probability distribution for p.

b. A random sample of 25 records yielded three incorrect records. Find the analyst's posterior probability distribution for p.

c. Find the mean and variance of the analyst's posterior probability distribution for p.

d. The posterior probability distribution you computed in part **b** combines the analyst's beliefs about p *and* the information about p contained in the sample. Using your results from parts **a** and **c** and graphs of the prior and posterior probability distributions, describe how the posterior distribution differs from the prior distribution. In so doing, you are explaining what the analyst learned about p by sampling.

19.9 A press produces masks for use in the manufacture of color television tubes. If the press is correctly adjusted, it will produce masks with a scrap rate of 5%. If it is not adjusted correctly, it will produce scrap at a 50% rate. From past company records, the machine is known to be correctly adjusted 90% of the time. A quality control inspector randomly selects one mask from those recently produced by the press and discovers it is defective. What is the probability that the machine is incorrectly adjusted?

19.10 Refer to Exercise 19.9. The quality control inspector observes a second mask and finds that it is also defective. Using the posterior probabilities from Exercise 19.8 as prior probabilities, calculate the posterior probability that the machine is incorrectly adjusted. What assumptions are required to make this calculation?

19.11 Suppose the proportion of defectives produced by a certain production process is .01, .05, or .10. The prior probability that it is .05 is twice the probability that it is .10. The prior probability that it is .01 is equal to the probability that it is .10.

a. Determine the prior probabilities associated with the proportion of defectives produced by the production process.

b. A sample of five units drawn at random yields three defectives. Revise the prior distribution you obtained in part **a**.

19.12 [*Note: This exercise refers to optional Section 5.5.*] A replacement parts inventory manager for a computer manufacturer believes the demand for a certain component is distributed as a Poisson random variable with a mean monthly demand of two, three, or four units. Suppose experience suggests that a mean monthly demand for three units is twice as likely as a mean monthly demand of two units, and that a mean monthly demand of two units is as likely as a mean monthly demand of four units.

a. If the demand for this component last month was four units, find the posterior distribution for the mean monthly demand for next month.
b. If the demand in the following month was two units, find the posterior distribution for the mean monthly demand.
c. Comment on the ability of Bayes' rule to help a decision-maker "learn" about the shifting values of a parameter over time.

| | | | | | | | | | | | | |

SECTION 19.2

SOLVING DECISION PROBLEMS USING POSTERIOR PROBABILITIES

Recall that the expected payoff criterion (Section 18.5) selects the action with the highest expected payoff. The expected payoff for each action is computed using either the **prior probabilities** or the **posterior probabilities** of the states of nature, depending on whether sample information is available. For example, for the smoke detector decision problem (Example 19.1), the payoff table prior to obtaining sample information is given in Table 19.6.

The expected payoffs for the two actions are

$$EP(a_1) = \sum_{\text{All states}} (\text{Payoff})(\text{Prior probability of state})$$

$$= (-1,500,000)(.10) + (-200,000)(.40) + (300,000)(.40) + (1,000,000)(.10)$$

$$= -\$10,000$$

$$EP(a_2) = 0(.10) + 0(.40) + 0(.40) + 0(.10) = 0$$

Thus, the expected payoff criterion, using prior probabilities, selects action a_2 and indicates that the firm should not market the new smoke detector.

TABLE 19.6 Payoff Table for the Smoke Detector Decision Problem

		STATE OF NATURE (PRIOR PROBABILITIES IN PARENTHESES)			
		S_1: $p = .01$ (.10)	S_2: $p = .05$ (.40)	S_3: $p = .10$ (.40)	S_4: $p = .15$ (.10)
ACTION	a_1: Market detector	$-\$1,500,000$	$-\$200,000$	$\$300,000$	$\$1,000,000$
	a_2: Do not market detector	0	0	0	0

In Example 19.1 we revised these prior probabilities based on the sample information that 2 out of 20 randomly sampled consumers would purchase the new detector if it were marketed. The payoff table with the posterior state probabilities is shown in Table 19.7.

TABLE 19.7 Payoff Table for the Smoke Detector Decision Problem

		STATE OF NATURE (POSTERIOR PROBABILITIES IN PARENTHESES)			
		S_1: $p = .01$ (.008)	S_2: $p = .05$ (.353)	S_3: $p = .10$ (.532)	S_4: $p = .15$ (.107)
ACTION	a_1: Market detector	$-\$1,500,000$	$-\$200,000$	$\$300,000$	$\$1,000,000$
	a_2: Do not market detector	0	0	0	0

The expected payoffs for the actions using the posterior probabilities are

$$EP(a_1) = \sum_{\text{All states}} (\text{Payoff})(\text{Posterior probability of state})$$

$$= (-1,500,000)(.008) + (-200,000)(.353) + (300,000)(.532) + (1,000,000)(.107) = +\$184,000$$

$$EP(a_2) = 0$$

Thus, the expected payoff criterion using posterior probabilities selects action a_1, indicating that the company should market the new detector. The sample information has altered the selection of the expected payoff criterion, changing it from a_2: Do not market to a_1: Market. Since the **posterior decision analysis** incorporates both prior and sample information, while the **prior decision analysis** incorporates only prior information, *we prefer the posterior decision to the prior decision.*

The expected payoff criterion using posterior probabilities is summarized in the box.

EXPECTED PAYOFF CRITERION USING POSTERIOR PROBABILITIES

Choose the action with the maximum expected payoff, where the expected payoffs are calculated using the posterior state probabilities.

$$EP(a_i) = \sum_{\text{All states}} \begin{pmatrix} \text{Payoff for} \\ \text{action } a_i/\text{state} \\ \text{combination} \end{pmatrix} \begin{pmatrix} \text{Posterior} \\ \text{probability} \\ \text{of state} \end{pmatrix}$$

EXAMPLE 19.2

Suppose the company that is considering marketing the new smoke detector randomly samples 100 consumers and finds that 10 would purchase the new detector if it were marketed. The payoff table with the posterior probabilities is shown in Table 19.8. Which action should the company take if the expected payoff criterion is used?

TABLE 19.8 Payoff Table for the Smoke Detector Example

		STATE OF NATURE (POSTERIOR PROBABILITIES IN PARENTHESES)			
		S_1: $p = .01$ (.000)	S_2: $p = .05$ (.104)	S_3: $p = .10$ (.826)	S_4: $p = .15$ (.069)
ACTION	a_1: Market detector	$-\$1,500,000$	$-\$200,000$	$\$300,000$	$\$1,000,000$
	a_2: Do not market detector	0	0	0	0

SOLUTION

We calculate the expected payoffs using the posterior probabilities. Thus,

$$EP(a_1) = \sum_{\text{All states}} \begin{pmatrix} \text{Payoff for} \\ \text{action } a_1/\text{state} \\ \text{combination} \end{pmatrix} \begin{pmatrix} \text{Posterior} \\ \text{probability} \\ \text{of state} \end{pmatrix}$$

$$= (-1,500,000)(.000) + (-200,000)(.104) + (300,000)(.826) + (1,000,000)(.069) = \$296,000$$

$$EP(a_2) = 0(.000) + 0(.104) + 0(.826) + 0(.069) = 0$$

The expected payoff criterion selects action a_1. This selection incorporates both prior information and the information contained in the sample of 100 consumers.

■

The expected utility criterion presented in Section 18.7 may also be extended to incorporate sample information. Simply replace the prior probabilities by the posterior probabilities. This criterion will then prescribe an action that is consistent with the decision-maker's preferences for outcomes and risk attitude, as well as with the sample information. Thus, no matter which probabilistic criterion you use— expected payoff or expected utility—the incorporation of sample information requires only the substitution of posterior probabilities for prior probabilities.

**EXERCISES
19.13–19.17**

LEARNING THE MECHANICS

19.13 Use the posterior probabilities of Exercise 19.3 to select the action with the highest expected payoff in the following payoff table (outcomes are in thousands of dollars):

		STATE OF NATURE			
		S_1	S_2	S_3	S_4
	a_1	25	10	0	17
ACTION	a_2	30	10	2	0
	a_3	11	8	-16	21

19.14 Consider the payoff table given here (with prior probabilities in parentheses).

		STATE OF NATURE		
		S_1 (.6)	S_2 (.1)	S_3 (.3)
	a_1	10	20	-14
	a_2	-16	18	15
ACTION	a_3	21	12	9
	a_4	8	25	-20

a. Use the expected payoff criterion to choose an action.
b. Sampling information I has been obtained that suggests that S_1 is the true state of nature. The reliability of this information is reflected in the following probabilities:

$$P(I \mid S_1) = .80 \qquad P(I \mid S_2) = .30 \qquad P(I \mid S_3) = .10$$

Find the posterior state probabilities.
c. Compute the expected payoff for each action using the posterior probabilities.
d. Use your results of part **c** to determine which action is prescribed by the expected payoff criterion. Compare this action with the action selected in part **a** using prior probabilities.

19.15 Consider the accompanying payoff table.

		STATE OF NATURE		
		S_1 (.4)	S_2 (.3)	S_3 (.3)
	a_1	−160	200	100
ACTION	a_2	180	−172	90
	a_3	250	63	−80

a. Use the expected payoff criterion to choose an action.

b. Sampling information I has been obtained that suggests that S_2 is the true state of nature. The reliability of this information is reflected by the following probabilities:

$$P(I \mid S_1) = .20 \qquad P(I \mid S_2) = .60 \qquad P(I \mid S_3) = .20$$

Find the posterior state probabilities.

c. Given the sample information I described in part **b**, which action is implied by the expected payoff criterion?

APPLYING THE CONCEPTS

19.16 A large hospital is considering purchasing 100 new color television sets under one of two different purchase agreements. Under one agreement the TV sets would cost $460 each and all sets that are seriously defective would be replaced at no cost. Under the other agreement, the TV sets would cost $400 each and any seriously defective sets would have to be replaced by the hospital at $400 each. (Assume all replacement sets are nondefective.) The hospital's purchasing agent believes the probabilities shown in the table appropriately characterize the proportion of seriously defective sets in a shipment of 100 sets from the manufacturer.

PROPORTION DEFECTIVE	PROBABILITY
.00	.4
.05	.3
.10	.1
.15	.1
.20	.1

a. Construct the payoff table for this decision problem. [*Note:* Costs represent negative payoffs.]

b. Use the expected payoff criterion to determine which purchase agreement the hospital should choose.

c. Suppose the hospital was able to randomly sample one TV set from the incoming shipment of 100 sets before deciding which purchase agreement to choose. According to the expected payoff criterion, which agreement should they select if the sampled set was defective? Nondefective?

19.17 A small company that produces auto parts for American-made cars expects to lose $1.5 million next year unless Congress passes a bill to limit the number of foreign-made cars that can be imported into the United States. If the bill passes, the company expects to make a profit of $2.5 million next year. The company's lobbyists in Washington have reported that the probability of the bill passing this year is .6. The auto parts company must decide whether to stay in business and risk a huge loss next year or to lease its facility for a year to a Japanese firm for $500,000. The company has just received new information from "a source close to the White House" that indicates Congress will pass the bill. The company assesses the reliability of the source as follows:

P(Source says bill will pass | Bill will pass) = .8

P(Source says bill will pass | Bill will not pass) = .1

a. Construct the payoff table for the company's decision problem. Include the company's prior state probabilities.
b. Compute the posterior state probabilities.
c. Describe the effect of the sample information on the company's prior beliefs regarding the passage of the bill by comparing the company's prior and posterior state probabilities.
d. Given the sample information, which action is prescribed by the expected payoff criterion?

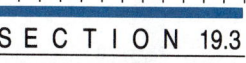

S E C T I O N   19.3

THE EXPECTED VALUE OF PERFECT INFORMATION

Now that we know how to incorporate sample information into a decision analysis, we will consider the cost of obtaining this information. How much should we be willing to pay for sample information? In this section we will find the maximum expected worth of the sample information—**the expected value of perfect information**. In the next section we will calculate a reasonable price to pay for the sample information.

The calculation of the maximum expected worth of sample information is made easier by the use of the opportunity loss table. Recall (Section 18.3) that the opportunity loss values are determined for each action/state combination by subtracting the payoff for that combination from the maximum payoff for that state of nature. For example, consider the smoke detector example introduced in the previous two sections. The payoff table is shown in Table 19.9.

TABLE 19.9   Payoff Table for the Smoke Detector Example

		STATE OF NATURE (PRIOR PROBABILITIES IN PARENTHESES)			
		S_1: $p = .01$ (.10)	S_2: $p = .05$ (.40)	S_3: $p = .10$ (.40)	S_4: $p = .15$ (.10)
ACTION	a_1: Market detector	−$1,500,000	−$200,000	$300,000	$1,000,000
	a_2: Do not market detector	0	0	0	0

To compute the opportunity losses, we subtract each entry in the payoff table from the maximum payoff in that column. Thus, for the state S_1, the maximum payoff is $0, and the opportunity loss (OL) for the (a_1, S_1) entry is

$$OL = 0 - (-1,500,000) = \$1,500,000$$

That is, if the company chooses to market the detector (a_1) and the state of nature S_1 eventuates, the loss will be $1,500,000 compared with the outcome of not marketing the detector (a_2). The entire opportunity loss table is given in Table 19.10.

TABLE 19.10 Opportunity Loss Table for the Smoke Detector Example

		STATE OF NATURE (PRIOR PROBABILITIES IN PARENTHESES)			
		S_1: $p = .01$ (.10)	S_2: $p = .05$ (.40)	S_3: $p = .10$ (.40)	S_4: $p = .15$ (.10)
ACTION	a_1: Market detector	−$1,500,000	$200,000	0	0
	a_2: Do not market detector	0	0	$300,000	$1,000,000

The expected payoff criterion and its equivalent, the expected opportunity loss (EOL) criterion, select action a_2, since using the prior probabilities,

$$EP(a_1) = -\$10,000 \qquad EOL(a_1) = \$230,000$$
$$EP(a_2) = 0 \qquad EOL(a_2) = \$220,000$$

Recall that the expected opportunity loss criterion selects the action that *minimizes* the expected opportunity loss.

We can determine the maximum expected value of sample information by determining the expected value of perfect information. To calculate the expected value of perfect information, we compute the expected opportunity loss, assuming the sample information will tell us *with certainty* which state will occur, and subtract it from the expected opportunity loss for the action selected by prior decision analysis. That is,

$$\left(\begin{array}{c} \text{Expected value of} \\ \text{perfect information} \\ \text{(EVPI)} \end{array} \right) = EOL \left(\begin{array}{c} \text{Action selected by} \\ \text{EOL criterion} \\ \text{using prior} \\ \text{probabilities} \end{array} \right) - EOL \left(\begin{array}{c} \text{Action dictated by} \\ \text{knowing state} \end{array} \right)$$

But the expected opportunity loss for the action dictated by knowing which state will occur will always be 0, since in this case we will *always* select the correct action and therefore suffer *no loss*. Thus, EVPI is just the expected opportunity loss for the action selected by the EOL criterion based on prior probabilities.

In the smoke detector example, the EOL criterion selected action a_2, with $EOL(a_2) = \$220,000$. Thus, the expected value of perfect information is

$$EVPI = EOL(a_2) = \$220,000$$

The EVPI indicates that the maximum expected gain from sample information is $220,000. This will be the expected gain *only if* the sample information is perfect; i.e., if it tells us which state will occur. Unfortunately, this will rarely, if ever, be the case. Since, realistically, the sample cannot be expected to yield perfect information, EVPI can be viewed as an upper limit on the amount we should be willing to pay for sample information.

EXPECTED VALUE OF PERFECT INFORMATION

$$EVPI = EOL(a')$$

where a' is the action chosen by the expected opportunity loss criterion prior to obtaining sample information.

EXAMPLE 19.3

TIME	
t (months)	p (Time)
$t \leq 12$	.2
$12 < t \leq 18$	.5
$18 < t \leq 24$	.3

A publisher of several different hobby-oriented magazines, airline magazines, and assorted newsletters is planning to computerize its warehouse. To do so will require extensive remodeling that, according to a consulting contractor, will take from 10 to 24 months to complete (depending on the availability of supplies and union labor). When asked to be more specific, the contractor gives the probability distribution shown in the table to characterize his uncertainty concerning how long the remodeling will take. While the remodeling is being completed, the publisher will have to rent temporary warehouse space. The owner of a vacant warehouse has presented four options concerning lease lengths and rent:

1. 24 months @ $20,000, with the option to sign renewable 6-month leases @ $7,000 for as long as necessary
2. 18 months @ $16,000, with the same option as above
3. 12 months @ $11,000, with the same option as above
4. 6 months @ $7,000, renewable as often as necessary (assuming renewals would always be for periods of 6 months)

a. If the publisher's objective is to maximize its expected cash inflow, which rental plan should be chosen?
b. If the publisher were to hire another consulting contractor to reevaluate the length of time the remodeling would take, what is the maximum expected worth of the consultant's information?

SOLUTION

a. The actions among which the publisher must choose are the four rental plans. The random variable upon which the outcome of the publisher's chosen action depends is the length of time it takes to complete the remodeling. The possible realizations of this variable as specified by the contractor are the states of nature in the publisher's decision problem. Thus, the skeleton payoff table is shown in Table 19.11 on page 1102.

TABLE 19.11

Skeleton of Payoff Table for Example 19.3

		STATE OF NATURE (PRIOR PROBABILITIES IN PARENTHESES) Time (t) to complete remodeling (in months)		
		$t \le 12$ (.2)	$12 < t \le 18$ (.5)	$18 < t \le 24$ (.3)
	Rental plan 1			
	Rental plan 2			
ACTION	Rental plan 3			
	Rental plan 4			

Since the publisher's objective is to maximize cash inflow, the objective variable is Cash inflow. The possible values the objective variable can assume are determined by the various rental plans. Notice that, since rental payments are cash outflows, all the possible values the objective variable can assume in this problem are negative. Thus, if plan 3 is chosen and remodeling is completed in 12 months or less, the firm's rent payments will amount to $11,000, a cash inflow of $-\$11,000$. Accordingly, $O_{31} = -\$11,000$. If, however, the remodeling takes from 12 to 18 months, the cash inflow will be $-\$18,000\,[-\$11,000 + (-\$7,000)]$, i.e., $O_{32} = -\$18,000$. The other outcomes can be determined similarly and are given in Table 19.12. Notice that plan 4 is an inadmissible action (why?) and can be eliminated from consideration.

TABLE 19.12

Payoff Table for Example 19.3

		STATE OF NATURE (PRIOR PROBABILITIES IN PARENTHESES)		
		$t \le 12$ (.2)	$12 < t \le 18$ (.5)	$18 < t \le 24$ (.3)
	Plan 1	$-\$20,000$	$-\$20,000$	$-\$20,000$
	Plan 2	$-\$16,000$	$-\$16,000$	$-\$23,000$
ACTION	Plan 3	$-\$11,000$	$-\$18,000$	$-\$25,000$
	Plan 4	$-\$14,000$	$-\$21,000$	$-\$28,000$

The decision problem has now been completely formulated and is ready to be solved. First, it is necessary to determine the expected payoff for each action.

$$EP(\text{Plan 1}) = -20,000(.2) + (-20,000)(.5) + (-20,000)(.3) = -\$20,000$$
$$EP(\text{Plan 2}) = -16,000(.2) + (-16,000)(.5) + (-23,000)(.3) = -\$18,100$$
$$EP(\text{Plan 3}) = -11,000(.2) + (-18,000)(.5) + (-25,000)(.3) = -\$18,700$$

Since the expected payoff criterion selects the action with the maximum expected payoff (which in this case means the action with the *lowest expected cash outflow*), plan 2 should be chosen.

b. To determine the maximum expected worth of additional information, we compute the expected value of perfect information (EVPI). First, we convert the payoff table to an opportunity loss table (Table 19.13). Even though the values assumed by the objective variable in this problem are negative, the same rules apply for determining opportunity losses. For example, the opportunity loss associated with the action/state combination (plan 1, $t \leq 12$) was found by subtracting the payoff associated with (plan 1, $t \leq 12$) from the maximum payoff under the state ($t \leq 12$), which is $-\$11,000$.

TABLE 19.13 Opportunity Loss Table

| | | STATE OF NATURE | | |
		$t \leq 12$ (.2)	$12 < t \leq 18$ (.5)	$18 < t \leq 24$ (.3)
	Plan 1	$-\$11,000 - (-\$20,000)$ $= \$9,000$	$-\$16,000 - (-\$20,000)$ $= \$4,000$	$-\$20,000 - (-\$20,000)$ $= 0$
ACTION	Plan 2	$-\$11,000 - (-\$16,000)$ $= \$5,000$	$-\$16,000 - (-\$16,000)$ $= 0$	$-\$20,000 - (-\$23,000)$ $= \$3,000$
	Plan 3	$-\$11,000 - (-\$11,000)$ $= 0$	$-\$16,000 - (-\$18,000)$ $= \$2,000$	$-\$20,000 - (-\$25,000)$ $= \$5,000$

The EVPI is the expected opportunity loss of the action that was chosen by the expected payoff (or EOL) criterion—i.e., EOL(Plan 2).

$$\text{EVPI} = \text{EOL(Plan 2)} = 5,000(.2) + 0(.5) + 3,000(.3) = \$1,900$$

Thus, even if the consulting contractor could tell the publishing company *with certainty* which state will occur, the expected value of this information is only $\$1,900$. In other words, the publishing company cannot expect to gain more than $\$1,900$ from the consultant's information and realistically can expect to gain less. ∎

In the next section we will discuss how to assign a more realistic value to the worth of sample information.

| | | | | | | | | | | | | |

EXERCISES 19.18–19.24

LEARNING THE MECHANICS

19.18 Consider the following payoff table (prior probabilities in parentheses):

| | | STATE OF NATURE | | |
		S_1 (.65)	S_2 (.20)	S_3 (.15)
	a_1	$\$50$	$-\$100$	$\$200$
ACTION	a_2	$-\$25$	$\$150$	$\$75$
	a_3	$\$90$	$-\$60$	$\$180$

a. Convert the payoff table to an opportunity loss table.

b. Find the expected value of perfect information and interpret your result.

19.19 Consider the decision tree shown here, with outcomes expressed as payoffs and prior probabilities in parentheses.

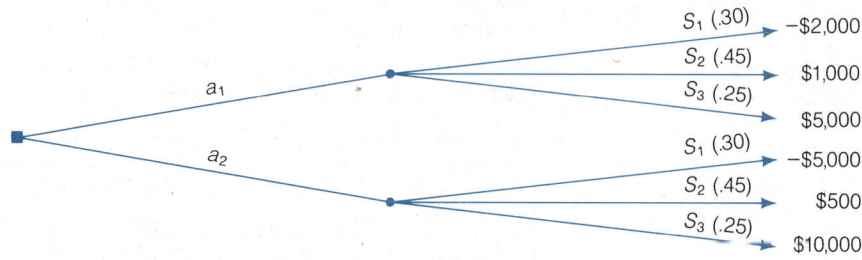

a. Convert the payoffs to opportunity losses.
b. Find the expected value of perfect information and interpret your result.
c. Suppose sample information I is obtained, and that the conditional probabilities associated with I are

$$P(I \mid S_1) = .10 \qquad P(I \mid S_2) = .95 \qquad P(I \mid S_3) = .15$$

Which state of nature does the sample information suggest is the true state of nature? Justify your response.
d. Using the sample information of part **c**, find the posterior probabilities for S_1, S_2, and S_3.
e. Using the results of part **d**, recalculate the expected value of perfect information. Explain why your result differs from the result you obtained in part **b**.

19.20 Consider the following payoff table (with prior probabilities in parentheses):

		STATE OF NATURE	
		S_1 (.15)	S_2 (.85)
ACTION	a_1	$6,000	−$3,000
	a_2	−$6,000	$20,000

a. Find the expected vaue of perfect information.
b. Sample information is obtained that suggests that S_1 is the true state of nature. The conditional probabilities of the sample information are

$$P(I_1 \mid S_1) = .7 \qquad P(I_1 \mid S_2) = .2$$

Using this sample information, calculate the revised EVPI. Does the EVPI increase or decrease? Why?
c. Suppose sample information had suggested that S_2 was the true state of nature. The conditional probabilities of the sample information are

$$P(I_2 \mid S_1) = .30 \qquad P(I_2 \mid S_2) = .80$$

Find the EVPI after receiving this sample result. Does the EVPI increase or decrease? Why?

d. Explain the difference in the changes in the EVPI for parts **b** and **c**. In general, under what conditions will the EVPI decrease after obtaining sample information? When will it increase?

APPLYING THE CONCEPTS

19.21 A petrochemical company has a distillation unit that occasionally goes out of control and produces batches of contaminated product. The cost of scrapping a contaminated batch of product is $25,000. If the contamination is only partial, then some of the finished product can be recovered by further processing at a cost of $15,000. Two actions are possible. The company can operate the unit under existing conditions, or it can perform maintenance on the distillation unit prior to each batch at a cost of $8,000 per batch. The maintenance procedure will ensure that each batch is acceptable. The probability that the distillation unit will be out of control is .05, and the probability that it will be partially out of control is .10.

a. Should management decide to install prebatch maintenance?

b. How much should the company be willing to pay for perfect knowledge about the state of the distillation unit for the next run?

19.22 Certified Public Accounting (CPA) firms are hired by private corporations to audit or certify their accounting records. In so doing, auditors check two main problem areas—possible fraud by company employees and acceptability of record-keeping practices. Upon completing its audit, the CPA firm either will certify that the client firm's records and financial statements are in order or will fail to certify them and will report on existing irregularities. If the auditor certifies the records when in fact errors and/or irregularities exist, the CPA firm may be sued by the client firm's stockholders for malpractice. If the CPA firm refuses to certify the records when in fact the records contain no errors or irregularities, it may be sued by the client.

Suppose you have audited a company's financial records, and you are faced with a certification decision. In your judgment, you have assessed the following costs for each: If you report incorrectly that the books are not certifiable when in fact they are, then you will lose the account and be sued by the client. You estimate the total cost for this error to be $2 million. If, on the other hand, you certify the accounts when irregularities exist, you will be sued for $10 million by the stockholders for malpractice.

a. Formulate the payoff table for this problem.

b. Describe how you would assess the state probabilities for this decision problem.

c. Assume you have assessed that the probability that the records are in order is .9. Which action would decision analysis prescribe?

d. What should the auditor be willing to pay for perfect information about the records?

19.23 Refer to Exercise 19.16, in which a hospital was considering purchasing 100 color television sets. On the basis of its prior information, how much should the hospital be willing to pay for perfect information about the number of seriously defective sets they would receive?

19.24 Refer to Exercise 19.17, in which an auto parts manufacturer is considering whether to stay in business or to lease its plant to another company. On the basis of its prior information, how much should the company be willing to pay for perfect information about whether Congress will pass the bill restricting the import of foreign cars?

THE EXPECTED VALUE OF SAMPLE INFORMATION: PREPOSTERIOR ANALYSIS (OPTIONAL)

We now want to make a realistic assessment of the expected worth of sample information. Of course, for the information to be of any use to us in deciding whether to obtain sample information, we must make the assessment *before* the sample is taken. Accordingly, the analysis is referred to as **preposterior analysis**.

For example, consider the smoke detector example of the previous sections. We repeat the payoff table with the prior state probabilities in Table 19.14. The expected payoff criterion using the prior probabilities yields $EP(a_1) = \$10,000$ and $EP(a_2) = 0$, so we would not market the detector (action a_2) based on the prior information.

TABLE 19.14 Payoff Table for the Smoke Detector Example

		STATE OF NATURE (PRIOR PROBABILITIES IN PARENTHESES)			
		S_1: $p = .01$ (.10)	S_2: $p = .05$ (.40)	S_3: $p = .10$ (.40)	S_4: $p = .15$ (.10)
ACTION	a_1: Market detector	$-\$1,500,000$	$-\$200,000$	$\$300,000$	$\$1,000,000$
	a_2: Do not market detector	0	0	0	0

TABLE 19.15
Probability Revision Table for Each Sample Outcome in the Smoke Detector Example

(1) State	(2) Prior probability	(3) Conditional probability of sample outcome given state	(4) Intersection of states and sample outcome (2) × (3)	(5) Posterior probability (4) ÷ Total of (4)
		$x = 0$		
S_1: $p = .01$	.10	.9801	.0980	.1146
S_2: $p = .05$	.40	.9025	.3610	.4221
S_3: $p = .10$	.40	.8100	.3240	.3789
S_4: $p = .15$	.10	.7225	.0722	.0844
Total:	1.00		.8552	1.0000
		$x = 1$		
S_1: $p = .01$	.10	.0198	.0020	.0145
S_2: $p = .05$	.40	.0950	.0380	.2764
S_3: $p = .10$	.40	.1800	.0720	.5236
S_4: $p = .15$	.10	.2550	.0255	.1855
Total:	1.00		.1375	1.0000
		$x = 2$		
S_1: $p = .01$	.10	.0001	.0000	.0000
S_2: $p = .05$	.40	.0025	.0010	.1389
S_3: $p = .10$	.40	.0100	.0040	.5556
S_4: $p = .15$	.10	.0225	.0022	.3056
Total:	1.00		.0072	1.0001[a]

[a]Rounding error.

Now suppose the company is considering taking a sample of two consumers to obtain their opinions about the detector. (We choose a small sample size to reduce the computational difficulty.) The value of x, the number of the sampled consumers who state they will purchase the detector, will be either 0, 1, or 2. *Preposterior analysis involves examining the decision problem for each possible sample outcome.* Thus, in Table 19.15 we use a probability revision table to derive the posterior probabilities for each of the three possible sample outcomes. The first two columns show the states and prior probabilities, respectively. Column 3 presents the conditional probability of observing the sample outcome given a particular state of nature. For the smoke detector example, these are binomial probabilities with $n = 2$ and a value of p corresponding to the market share of each state. Thus, for the first two entries in column 3 under the heading $x = 0$, we find

$$P(x = 0 \mid p = .01) = \binom{2}{x} p^x (1 - p)^{2-x} = \binom{2}{0}(.01)^0(.99)^2 = (.99)^2 = .9801$$

$$P(x = 0 \mid p = .05) = \binom{2}{0}(.05)^0(.95)^2 = (.95)^2 = .9025$$

and so on. In column 4 we find the probability of the intersection of the states and the sample outcome using the formula

$$P(x \cap S_i) = P(S_i)P(x \mid S_i)$$

which is the product of the respective entries in columns 2 and 3. Finally, we find the revised state probabilities—the posterior probabilities—in column 5 by applying Bayes' rule:

$$P(S_i \mid x) = \frac{P(x \cap S_i)}{\displaystyle\sum_{\text{All states}} P(x \cap S_j)} = \frac{P(S_i)P(x \mid S_i)}{\displaystyle\sum_{\text{All states}} P(S_j)P(x \mid S_j)}$$

Then each element in column 5 is equal to the corresponding element in column 4 divided by the column 4 total.

We now determine which action is dictated by each posterior distribution. That is, if none of the two sampled consumers will buy the detector (i.e., $x = 0$), which action should be selected? We use the expected payoff criterion with the posterior probabilities corresponding to $x = 0$:

$$EP(a_1 \mid x = 0) = \sum_{\text{All states}} (\text{Payoff})(\text{Posterior probability of state for } x = 0)$$

$$= (-1,500,000)(.1146) + (-200,000)(.4221) + (300,000)(.3789) + (1,000,000)(.0844) = -\$58,250$$

$$EP(a_2 \mid x = 0) = 0(.1146) + 0(.4221) + 0(.3789) + 0(.0844) = 0$$

Thus, the expected payoff criterion selects action a_2: Do not market the detector when $x = 0$ is the sample outcome.

Now suppose we observe $x = 1$ when the sample information is collected:

$$EP(a_1 \mid x = 1) = \sum_{\text{All states}} (\text{Payoff})(\text{Posterior probability of state for } x = 1)$$

$$= (-1,500,000)(.0145) + (-200,000)(.2764) + (300,000)(.5236) + (1,000,000)(.1855) = \$265,550$$

$$EP(a_2 \mid x = 1) = 0$$

So we choose a_1: Market the detector if $x = 1$.
    Finally,

$$EP(a_1 \mid x = 2) = (-1,500,000)(.0000) + (-200,000)(.1389) + (300,000)(.5556) + (1,000,000)(.3056) = \$444,500$$
$$EP(a_2 \mid x = 2) = 0$$

So the expected payoff criterion selections action a_1: Market the detector when $x = 2$.

    Table 19.16 summarizes our work to this point. Note that we included the probability of observing each of the sample outcomes in the last column of Table 19.16. These probabilities are the sums of the column 4 entries for each sample outcome in the probability revision table (Table 19.15). Thus,

$$P(x = 0) = \sum_{\text{All states}} P[(x = 0) \cap S_i]$$

$$= .0980 + .3610 + .3240 + .0722 = .8552$$

$P(x = 1)$ and $P(x = 2)$ can be computed in a similar manner. They form the **marginal** (or **predictive**) **probability distribution** for the random variable x. Note that the marginal probabilities are not dependent on the state, and they will sum to 1 if no rounding errors are present.

TABLE 19.16

Actions Selected and Expected Payoffs for the Smoke Detector Example

SAMPLE OUTCOME	ACTION SELECTED	EXPECTED PAYOFF	MARGINAL PROBABILITY OF SAMPLE OUTCOME
$x = 0$	a_2	0	.8552
$x = 1$	a_1	\$265,550	.1375
$x = 2$	a_1	\$444,500	.0072
			.9999[a] $\approx$ 1.0

[a]Rounding error.

    Now recall our ultimate objective. We are trying to determine how much is to be gained from sampling. Since we now know the expected payoff and the marginal probability of each sample outcome (Table 19.16), we can calculate the expected payoff of sampling (EPS):

$$EPS = \sum_{x} (EP \mid x)p(x) = \sum_{\substack{\text{All sample} \\ \text{outcomes}}} \begin{pmatrix} \text{Maximum expected} \\ \text{payoff for} \\ \text{sample outcome} \end{pmatrix} \begin{pmatrix} \text{Marginal} \\ \text{probability of} \\ \text{sample outcome} \end{pmatrix}$$

$$= (0)(.8552) + (265,550)(.1375) + (444,500)(.0072)$$
$$= \$39,713.53$$

or \$39,700, rounding to the nearest hundred dollars. Thus, the mean payoff is \$39,700 when $n = 2$ consumers are sampled.

    To determine the expected gain attributed to sampling, we compare the EPS to the expected payoff for no sampling (EPNS). But EPNS is the expected payoff of the action selected by the expected payoff criterion using prior probabilities. We showed in previous sections that action a_2 is chosen using prior decision analysis, so that

$$\text{EPNS} = EP(a_2) = 0$$

Finally, the expected value of sample information (EVSI) is the expected payoff of sampling less the expected payoff with no sampling; i.e.,

$$\text{EVSI} = \text{EPS} - \text{EPNS} = \$39{,}700 - \$0 = \$39{,}700$$

This figure represents the mean dollar amount the company will gain from sampling and therefore provides a figure it will be reluctant to exceed in the cost of obtaining the sample information.

The steps for computing the expected value of sample information are summarized in the box. The calculation of EVSI becomes very tedious when the number of possible sample outcomes is large. However, because the method is the same no matter how many possible outcomes exist, computers can be useful aids in determining EVSI.

COMPUTING THE EXPECTED VALUE OF SAMPLE INFORMATION (EVSI)

Step 1   Obtain the posterior state probabilities for each possible sample outcome using a probability revision table.

Step 2   Use the expected payoff criterion with the posterior probabilities to determine the action with the maximum expected payoff *for each sample outcome*.

Step 3   Find the marginal probability distribution for the sample outcomes by using

$$P(x) = \sum_{\text{All states}} P(x \cap S_i)$$

For each sample outcome, the marginal probability will be the sum of column 4 in the probability revision table (step 1).

Step 4   Find the expected payoff of sampling (EPS) by combining the results of steps 2 and 3:

$$\text{EPS} = \sum_{\substack{\text{All sample} \\ \text{outcomes}}} \begin{pmatrix} \text{Maximum} \\ \text{expected payoff} \\ \text{for sample outcome} \end{pmatrix} \begin{pmatrix} \text{Marginal} \\ \text{probability of} \\ \text{sample outcome} \end{pmatrix}$$

Step 5   Calculate the EVSI by

$$\text{EVSI} = \text{EPS} - \text{EPNS}$$

where EPNS is the expected payoff of no sampling, computed using the prior probabilities.

After the EVSI has been determined, it should be compared with the cost of sampling (CS) by computing the expected net gain for sampling (ENGS). For example, if the smoke detector company determines that a total cost of $500 will be incurred during the sampling of two consumers, the expected net gain of sampling is

$$\text{ENGS} = \text{EVSI} - \text{CS} = \$39,700 - \$500 = \$39,200$$

As long as the net gain is positive, the company can expect to gain by obtaining the sample information. We thus have the preposterior expected gain decision rule given in the box.

PREPOSTERIOR EXPECTED GAIN DECISION RULE

If the ENGS is greater than 0, the decision analyst should obtain the sample information before making a decision. The ENGS is calculated from the formula

$$\text{ENGS} = \text{EVSI} - \text{CS}$$

where CS is the cost of sampling.

Decision trees (Section 18.2) are often used to conduct a preposterior analysis or to summarize the results of a preposterior analysis. The decision tree for the smoke detector example is shown in Figure 19.1. The first decision fork (always reading from left to right) represents the decision of whether to sample. Along the upper branch (Do not sample) we place a decision fork representing the two possible actions of the decision problem. Beside this fork we write the expected payoff of no sampling (EPNS), which is 0 for this example. At the ends of the action branches, we place chance forks and branch into the four states of nature. Beside each chance fork we write the expected payoff for the associated action, where the expected payoffs are computed using the prior probabilities. The expected payoff is $-\$10,000$ for action a_1 and $\$0$ for action a_2 in our example. On the state branches, we write the prior probabilities, and at the ends of the branches are the outcomes corresponding to the action/state combinations.

The lower branch of the decision tree in Figure 19.1, the Sample branch, has an extra chance fork corresponding to the possible sample outcomes. The expected payoff of sampling (EPS), which is $\$39,700$ for this example, is written beside the sampling chance fork. The marginal probability for each sample outcome is written on the appropriate sample outcome branch, and the expected payoff for the optimal action corresponding to each sample outcome is recorded beside the decision fork at the end of each sample outcome branch. The remainder of the branching is into actions and then into states, just as for the Do not sample branch. The only difference is that posterior rather than prior probabilities are recorded on each state of nature branch. Note that two vertical lines are used to block the branches

FIGURE 19.1 Decision Tree for the Smoke Detector Example

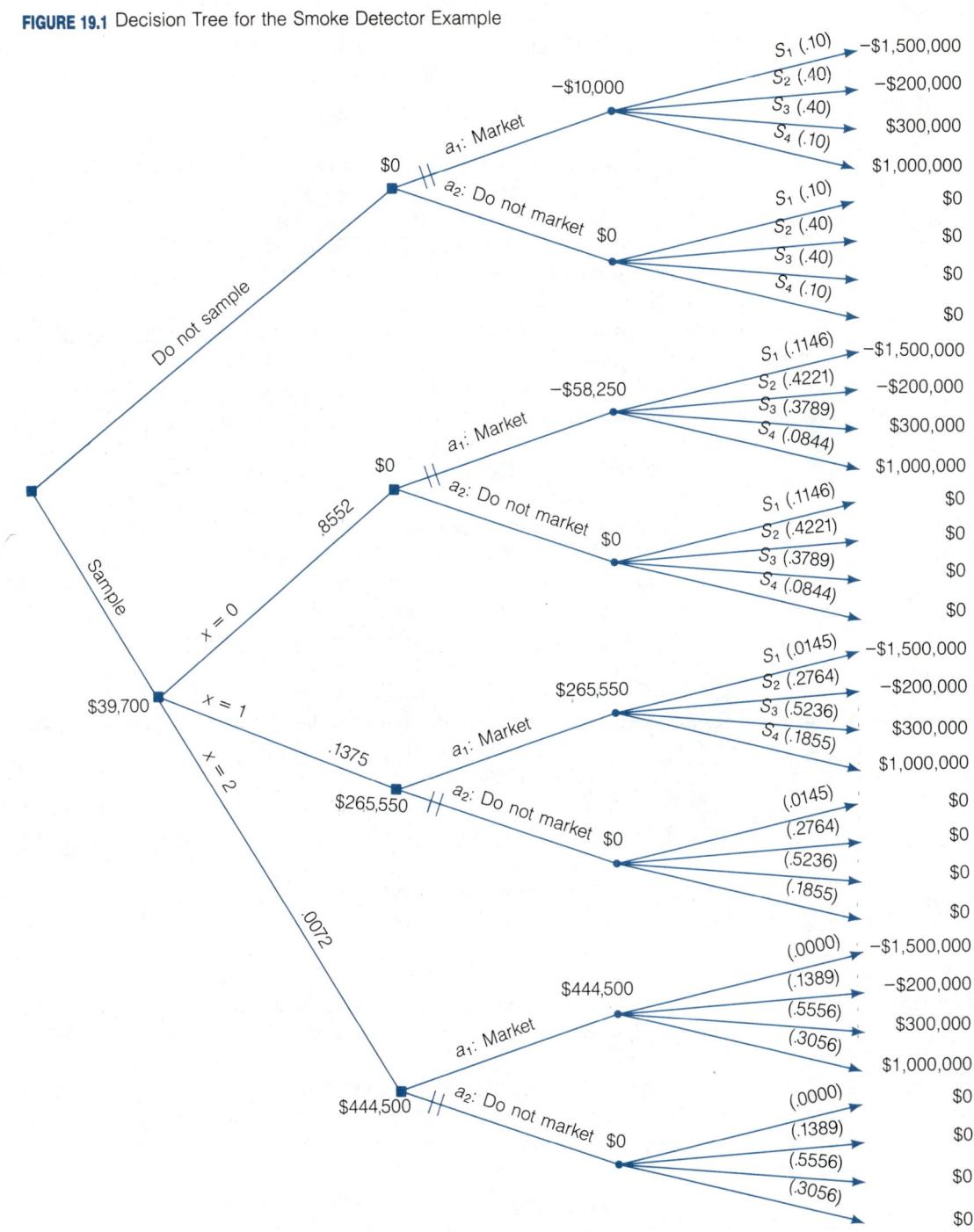

corresponding to actions with expected payoffs that are not maximum. This reminds us which action should be chosen at each decision fork.*

The steps to follow in drawing a decision tree are summarized in the box.

DRAWING A DECISION TREE TO SUMMARIZE A PREPOSTERIOR ANALYSIS

Step 1   Beginning at the left of the page, form two branches, the upper one corresponding to Do not sample and the lower to Sample. At the ends of the branches, record the expected payoff of no sampling (EPNS) and the expected payoff of sampling (EPS), respectively.

Step 2   At the first decision fork of the upper branch (Do not sample), draw a branch corresponding to each action of the decision problem. At the ends of each of these action branches, record the expected payoff corresponding to that action using the prior probabilities.

Step 3   At the chance fork of each action branch (still in the upper branch), create a branch for each state of nature. On each of the branches record the prior probability of each state. At the end of each branch, record the payoff corresponding to each action/state combination.

Step 4   At the sampling chance fork of the lower branch (Sample), draw a branch corresponding to each possible outcome of the sample. Place the marginal probability of each outcome on the corresponding branch. At the end of each sample branch, record the maximum expected payoff corresponding to the sample outcome.

Step 5   Repeat steps 2 and 3 for the lower branch, creating action and state of nature branches. The only change is that the posterior state probabilities (rather than the prior probabilities) corresponding to each sample outcome are recorded on the state of nature branches.

Step 6   Draw two vertical lines through any action branch with an expected payoff that is less than the expected payoff for another action branch originating from the same decision fork. These vertical lines are intended to block the branch, indicating that the action should not be taken.

EXAMPLE 19.4

In Section 18.2, we introduced the decision problem facing the promoter of a rock concert. The payoff table characterizing this decision problem is reproduced in Table 19.17. Suppose the promoter has the option of purchasing a long-range weather forecast for the night of the concert from a well-known meteorologist for $15,000. The meteorologist's track record in terms of the percentage of times in the past her predictions have or have not been accurate is shown in Table 19.18.

---

*The computations for EVSI are often performed directly from a decision tree such as Figure 19.1. Both EPNS and EPS can be determined by starting at the far right-hand side of the tree and taking expectations backward through the tree.

TABLE 19.17

Payoff Table for the Rock Concert Example

		STATE OF NATURE (PRIOR PROBABILITIES IN PARENTHESES)	
		Rain, S_1 $\left(\frac{1}{3}\right)$	No rain, S_2 $\left(\frac{2}{3}\right)$
ACTION	a_1: Rent stadium	−$40,000	$350,000
	a_2: Rent Civic Center	$150,000	$150,000

TABLE 19.18

Long-Range Prediction Record for Meteorologist

		ACTUAL WEATHER	
		Rain	No rain
METEOROLOGIST'S PREDICTION	Rain	85%	30%
	No rain	15%	70%

Eighty-five percent of the rainy days have been correctly predicted by the meteorologist and 70% of the days without rain have been correctly predicted by her. Should the promoter purchase this sample information (the meteorologist's opinion) or make a decision concerning which facility to rent utilizing just the prior information?

a. Calculate the EVSI.
b. Calculate the ENGS and make the decision about whether to obtain the meteorologist's prediction.
c. Summarize the results of this preposterior analysis using a decision tree.

SOLUTION

a. We will follow the five-step approach for finding the EVSI.

Step 1   The first step is to construct a probability revision table to obtain the posterior probability distribution corresponding to each sample outcome. The two possible sample outcomes in this decision problem are that the meteorologist will predict no rain and that she will predict rain. The probability revision tables corresponding to these outcomes are shown in Table 19.19 (page 1114). Thus, the posterior probability of rain is .586 if the meteorologist predicts rain, but it is only .097 if the meteorologist predicts no rain.

Step 2   We now use the expected payoff criterion and the posterior probabilities to determine the preferred action for each sample outcome.

$$EP(a_1 \mid \text{Predicts rain}) = (-40,000)(.586) + (350,000)(.414) = \$121,460$$
$$EP(a_2 \mid \text{Predicts rain}) = (150,000)(.586) + (150,000)(.414) = \$150,000$$
$$EP(a_1 \mid \text{Predicts no rain}) = (-40,000)(.097) + (350,000)(.903) = \$312,170$$
$$EP(a_2 \mid \text{Predicts no rain}) = (150,000)(.097) + (150,000)(.903) = \$150,000$$

Thus, if the meteorologist predicts rain, the expected payoff criterion selects action a_2: Rent Civic Center. But if the meteorologist predicts no rain, the criterion selects a_1: Rent stadium.

TABLE 19.19
Probability Revision Tables for the Rock Concert Example

(1)		(2)	(3)	(4)	(5)
			Conditional probability of sample outcome given state	Intersection of sample outcome and state	Posterior probability
State		Prior probability		$(2) \times (3)$	$(4) \div$ Total of (4)
			Predicts rain		
S_1:	Rain	.333	.85	.283	.586
S_2:	No rain	.667	.30	.200	.414
	Total:	1.000		.483	1.000
			Predicts no rain		
S_1:	Rain	.333	.15	.050	.097
S_2:	No rain	.667	.70	.466	.903
	Total:	1.000		.516	1.000

Step 3 We now find the marginal probabilities for each sample outcome by summing the probabilities of the combinations of sample outcome and state over all states. These are the column 4 sums in the probability revision table (Table 19.19). We find

$$P(\text{Predicts rain}) = .483$$
$$P(\text{Predicts no rain}) = .516$$

Step 4 We summarize the results of steps 2 and 3 in Table 19.20.

TABLE 19.20
Summary for the Rock Concert Example

SAMPLE OUTCOME	ACTION SELECTED	EXPECTED PAYOFF	MARGINAL PROBABILITY OF SAMPLE OUTCOME
Predicts rain	a_2	$150,000	.483
Predicts no rain	a_1	$312,170	.516

We now want to obtain the expected payoff of sampling:

$$\text{EPS} = \sum_{\substack{\text{All sample} \\ \text{outcomes}}} \left(\begin{array}{c} \text{Maximum expected payoff} \\ \text{for sample outcome} \end{array} \right) \left(\begin{array}{c} \text{Marginal probability of} \\ \text{sample outcome} \end{array} \right)$$

$$= (150,000)(.483) + (312,170)(.516)$$
$$= \$233,530 \text{ (rounding to the nearest dollar)}$$

Step 5 Finally, the EVSI is the difference between the EPS and the EPNS. Using the prior probabilities from Table 19.17, we find

$$EP(a_1) = (-40,000)(\tfrac{1}{3}) + (350,000)(\tfrac{2}{3}) = \$220,000$$
$$EP(a_2) = (150,000)(\tfrac{1}{3}) + (150,000)(\tfrac{2}{3}) = \$150,000$$

so that based on prior information we select action a_1, and the EPNS is $220,000.

Thus,

$$EVSI = EPS - EPNS$$
$$= \$233,530 - \$220,000 = \$13,530$$

The expected gain from sampling is $13,530.

b. We now want to decide whether the promoter should hire the meteorologist. The ENGS is

$$ENGS = EVSI - CS$$

The meteorologist will charge $15,000 for her prediction, so

$$ENGS = \$13,530 - \$15,000 = -\$1,470$$

Since the ENGS is negative, the promoter should not pay for the sample information.

c. The decision tree summarizing the results of the preposterior analysis is shown in Figure 19.2. Note that the upper branch represents the Do not sample decision. The action and state branches emanate from the Do not sample fork. The expected payoff of no sampling, the expected payoffs for each action, the prior probabilities, and finally, the payoffs are also shown. The fact that a_1 is the preferred action is indicated by the vertical lines through the a_2 branch.

FIGURE 19.2 Decision Tree for the Rock Concert Example

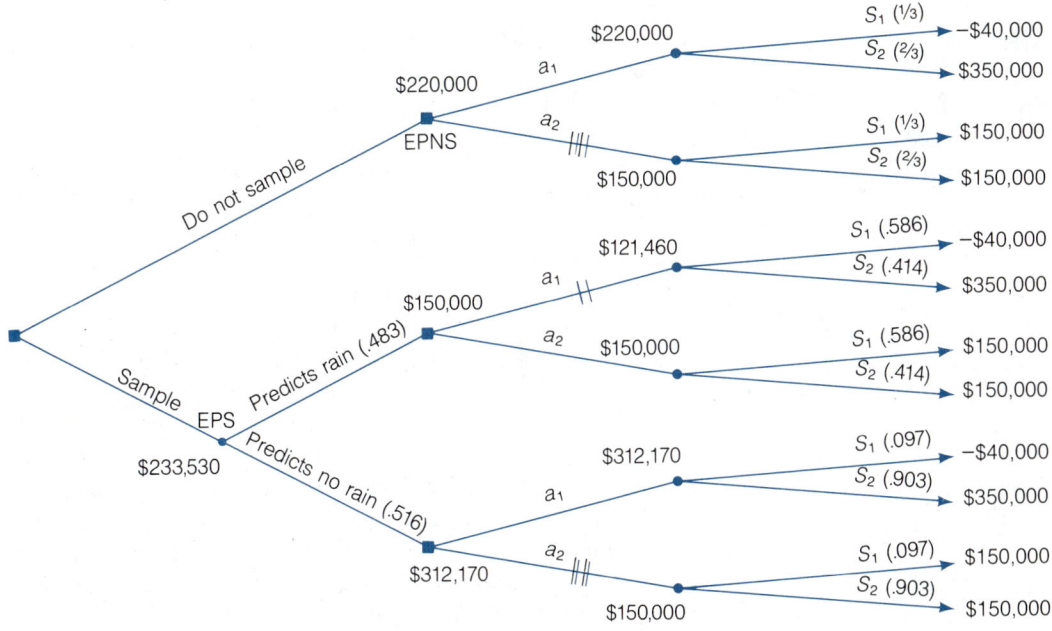

In the lower (Sample) branch, the first branching consists of the two possible sample outcomes, Predicts rain and Predicts no rain. The action and state branches emanate from the sample outcome branches. The expected payoff of sampling, the maximum expected payoff for each sample outcome, the expected payoff for each action, the posterior probabilities, and the actual payoffs are also recorded. We have again drawn vertical lines through action branches with expected payoffs that are exceeded by other actions emanating from the same decision fork.

∎

In the context of decision analysis, the term *sampling* means any procedure or process for gathering information. This includes statistical sampling, such as random sampling, as well as less technical methods, such as obtaining an expert opinion, as in the preceding example. If we are interested in random sampling, the optimal sample size can be determined by conducting a preposterior analysis for each potential sample size and choosing the sample size with the maximum ENGS. If this optimal sample size is 0 (or negative), the decision should be based on presently available information. Otherwise, the decision-maker should select a sample of the optimal size and then repeat the analysis to determine whether further sampling is expected to yield a positive net gain. Considerable effort is required to conduct a preposterior analysis when the sample size is moderate or large. Because the methodology remains the same for any sample size, computer programs have been written to conduct preposterior analyses.

CASE STUDY 19.1

AN EXAMPLE OF THE BENEFITS OF ADDITIONAL INFORMATION

In addition to studying whether hurricanes should be seeded (see Case Study 18.2) Howard, Matheson, and North (1972) addressed the question of whether more evidence on the effects of hurricane seeding should be gathered before the government policy decision on seeding is made. Would the expected loss from a representative hurricane be reduced by carrying out an additional seeding experiment?

First, the authors looked at the cost and possible outcomes of a seeding experiment. The cost to seed a hurricane and observe the results is $250,000. The possible outcomes are the same as those given in Case Study 18.2 for a seeded hurricane; the prior probabilities for these outcomes are also given in Case Study 18.2. Using the probabilities of each experimental outcome conditioned on each of the states of nature, the prior probabilities of the outcomes of seeding can be revised using Bayes' rule. The resulting posterior probabilities can then be used to evaluate the expected loss (EL) for seeding and not seeding. The decision to experiment or not can be made by comparing the expected loss of the "best" strategy *with* an experiment with the expected loss of the "best" strategy *without* an experiment. A condensed decision tree is shown in Figure 19.3. The tree indicates that the expected loss with the experiment is $2.83 million lower than without the experiment. Since the net gain from the experiment is greater than the cost ($.25 million), the experiment should be conducted. As a result of their analysis, Howard, Matheson, and North recommended that further experiments with hurricane seeding be conducted prior to making the government policy decision on seeding hurricanes.

FIGURE 19.3

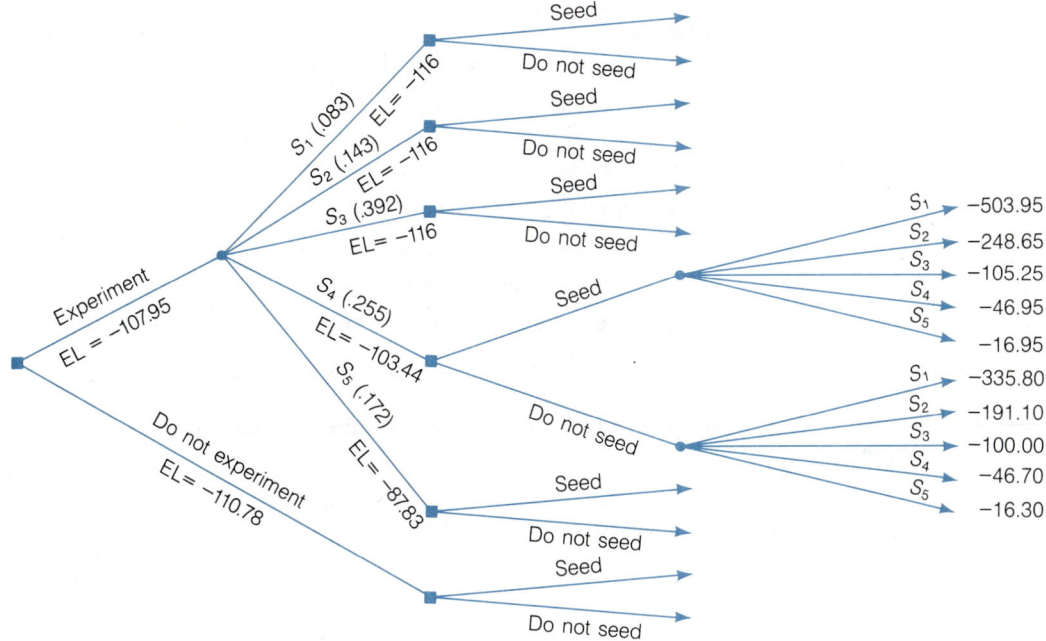

EXERCISES
19.25–19.32

LEARNING THE MECHANICS

19.25 Consider the following payoff table (prior probabilities are shown in parentheses):

		STATE OF NATURE	
		S_1 (.30)	S_2 (.70)
ACTION	a_1	$500	−$50
	a_2	−$100	$250

The decision-maker would like to decide whether to purchase sample information about the true state of nature prior to choosing an action. The sample information would cost $100. The reliability of the sample information is described by the following conditional probabilities:

P(Sample information indicates S_1 true | S_1 is true) = .8

P(Sample information indicates S_2 true | S_1 is true) = .2

P(Sample information indicates S_1 true | S_2 is true) = .1

P(Sample information indicates S_2 true | S_2 is true) = .9

a. Find the expected payoff of sampling (EPS) and the expected payoff from no sampling (EPNS).

b. Use the results of part **a** to find the expected value of sample information (EVSI).

c. Find the expected net gain of sampling (ENGS).

d. According to the ENGS, should the decision-maker purchase the sample information prior to making a decision? Explain.

19.26 Refer to Exercise 19.25. Suppose a second source of sample information was also available to the decision-maker. This information costs only $10, but it is much less reliable than the first source (as indicated by the following conditional probabilities):

P(Sample information indicates S_1 true | S_1 is true) = .6

P(Sample information indicates S_2 true | S_1 is true) = .4

P(Sample information indicates S_1 true | S_2 is true) = .4

P(Sample information indicates S_2 true | S_2 is true) = .6

a. Find the expected net gain of sampling (ENGS) for the second source of sample information.

b. From which source should the decision-maker purchase sample information? Explain.

19.27 Consider the following decision tree with outcomes expressed as payoffs and prior probabilities in parentheses:

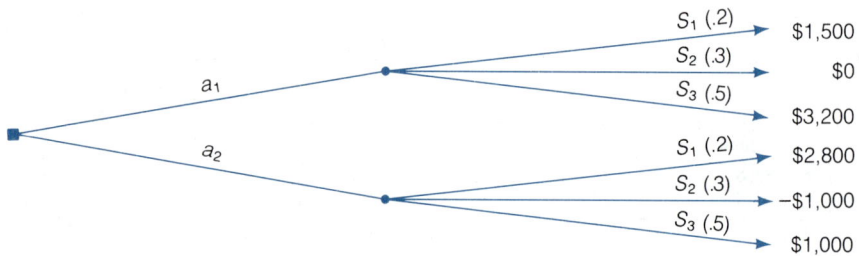

The decision-maker is considering purchasing sample information about the true state of nature prior to choosing an action. The sample information costs $250. The reliability of the sample information is described by the following conditional probabilities:

P(Sample information indicates S_1 true | S_1 true) = .7

P(Sample information indicates S_2 true | S_1 true) = .2

P(Sample information indicates S_3 true | S_1 true) = .1

P(Sample information indicates S_1 true | S_2 true) = .05

P(Sample information indicates S_2 true | S_2 true) = .9

P(Sample information indicates S_3 true | S_2 true) = .05

P(Sample information indicates S_1 true | S_3 true) = .1

P(Sample information indicates S_2 true | S_3 true) = .1

P(Sample information indicates S_3 true | S_3 true) = .8

a. What are the prior probabilities for S_1, S_2, and S_3?

b. Find the posterior probabilities for S_1, S_2, and S_3 when the sample information indicates S_1 is true.

c. Repeat part **b** for the case when the sample information indicates S_2 is true.
d. Repeat part **b** for the case when the sample information indicates S_3 is true.
e. Find the predictive (marginal) probabilities for the three sample results.
f. Expand the decision tree to include the decision-maker's sampling decision. Enter the payoffs on the tree along with all the probabilities you found in parts **a–e**. (For an example of such a tree, see Figure 19.2.)
g. Find EPNS and EPS, and enter them on your tree.
h. Find EVSI.
i. Use the preposterior expected gain decision rule to determine whether the decision-maker should purchase the sample information.

APPLYING THE CONCEPTS

19.28 The decision problem for a company that may market a new product is characterized in the following payoff table (prior probabilities in parentheses):

	PRODUCT'S STATUS AFTER 1 YEAR ON THE MARKET		
	Failure (.6)	Successful (.3)	Very successful (.1)
Market	−$200,000	$300,000	$600,000
Do not market	0	0	0

The company is considering whether to conduct a $30,000 market survey to gain information concerning the product's potential for success. One of four different conclusions would be yielded by the survey: "Product will be very successful," "Product will be successful," "Product will be a failure," or "Product's status after 1 year is uncertain." The reliability of the conclusions yielded by the survey can be described by the following conditional probabilities:

P(Survey concludes "very successful" | Product is very successful) = .6
P(Survey concludes "successful" | Product is very successful) = .2
P(Survey concludes "failure" | Product is very successful) = .1
P(Survey concludes "uncertain" | Product is very successful) = .1
P(Survey concludes "very successful" | Product is successful) = .2
P(Survey concludes "successful" | Product is successful) = .4
P(Survey concludes "failure" | Product is successful) = .2
P(Survey concludes "uncertain" | Product is successful) = .2
P(Survey concludes "very successful" | Product is a failure) = .1
P(Survey concludes "successful" | Product is a failure) = .1
P(Survey concludes "failure" | Product is a failure) = .5
P(Survey concludes "uncertain" | Product is a failure) = .3

a. Find the expected value of perfect information (EVPI), and interpret your result in the context of the problem.

b. Find the expected value of sample information (EVSI), and interpret your result in the context of the problem.

c. Find the expected net gain of sampling (ENGS), and use it to determine whether the company should undertake the proposed market survey.

19.29 Reconsider Exercises 18.18 and 18.32. The dress buyer expects to sell between three and eight dozen dresses with the prior probabilities shown in the table in the margin. A market research firm can be hired to forecast the demand for the new style of dress. Past records indicate that the research firm's conditional probabilities of forecasts, given the various states of nature (sales), are as shown in the following table:

SALES Dozen	PROBABILITY
3	.15
4	.25
5	.30
6	.15
7	.10
8	.05

		ACTUAL SALES (DOZENS)					
		3	4	5	6	7	8
FORECAST SALES (DOZENS)	3	.70	.10	.06	.04	0	0
	4	.20	.50	.10	.05	.05	0
	5	.10	.20	.50	.11	.05	.05
	6	0	.10	.20	.40	.20	.10
	7	0	.10	.04	.30	.60	.15
	8	0	0	.10	.10	.10	.70

Perform a preposterior analysis and indicate how much the buyer should be willing to spend for the sample information.

19.30 A computer company has the capacity to product seven computers per year. Its assessment of the prior probability distribution of next year's demand and the payoff distribution (in millions of dollars) constructed by the company's accounting department are shown in the accompanying tables.

DEMAND	PROBABILITY
0	.05
1	.15
2	.22
3	.22
4	.16
5	.10
6	.05
7	.02
>7	.03

		STATE OF NATURE Number of computers demanded								
		0	1	2	3	4	5	6	7	>7
ACTION	Produce 0	0	−1	−2	−3	−4	−5	−6	−7	−7
	Produce 1	−.2	1	0	−1	−2	−3	−4	−5	−5
	Produce 2	−.4	.8	2	1	0	−1	−2	−3	−3
	Produce 3	−.6	.6	1.8	3	2	1	0	−1	−1
	Produce 4	−.8	.4	1.6	2.8	4	3	2	1	1
	Produce 5	−1.0	.2	1.4	2.6	3.8	5	4	3	3
	Produce 6	−1.2	0	1.2	2.4	3.6	4.8	6	5	5
	Produce 7	−1.4	−.2	1.0	2.2	3.4	4.6	5.8	7	7

The computer company is considering hiring a market forecaster to predict the demand for computers in the coming year. The following table reflects his reliability:

		ACTUAL DEMAND							
	0	1	2	3	4	5	6	7	>7
FORECAST DEMAND 0	.70	.30	.15	0	0	0	0	0	0
1	.20	.50	.20	.10	.05	0	0	0	0
2	.10	.10	.40	.30	.20	.20	0	0	0
3	0	.05	.15	.40	.50	.40	.05	0	0
4	0	.05	.05	.15	.20	.30	.10	.05	0
5	0	0	.05	.05	.05	.10	.20	.05	0
6	0	0	0	0	0	0	.40	.10	0
7	0	0	0	0	0	0	.20	.50	.10
>7	0	0	0	0	0	0	.05	.30	.90

The market forecaster will provide the computer company with a demand forecast for $10,000. Should the computer company spend $10,000 for the forecast?

19.31 A shoe manufacturer is considering the possibility of introducing a new line of athletic shoes. The company's management estimates profit from the new shoes to be

$$\pi = 100p - 2.5$$

where π is profit in millions of dollars and p is the proportion of the market the new shoes will capture. Management's assessments of the probabilities of capturing different proportions of the market are given in the table. If the shoe company's management would like to purchase some marketing research, what should it be willing to spend for the information?

MARKET SHARE	PROBABILITY
0	.10
.01	.22
.02	.30
.03	.24
.04	.13
.05	.01

19.32 Refer to Exercise 19.16, in which a hospital was considering purchasing 100 color television sets. On the basis of its prior information, what is it worth to the hospital to be able to test one of the 100 television sets before selecting a purchase agreement?

| | | | | | | | | | | | |
S E C T I O N   19.5

AN EXAMPLE OF A TWO-ACTION, INFINITE-STATE DECISION PROBLEM (OPTIONAL)

The smoke detector example of the previous sections allowed us to introduce the concepts of posterior and preposterior decision analysis. However, the small number of different possible market shares used as states of nature made the example somewhat unrealistic. A better approach would be to permit the market share, p, of a product to take on any value between 0 and 1, inclusive. Then a company would be faced with a decision problem with two actions—Market the product and Do not market the product—and an infinite number of possible states of nature corresponding to the infinite number of values p could assume. In this section, we discuss an approach to solving a decision problem with two actions and an infinite number of states.

Clearly, a payoff table or a decision tree cannot be used to enumerate the potential outcomes associated with the two actions of this decision problem because the number of potential states of nature that could occur is infinite. Thus, it is necessary to describe the outcomes for each action as a function of the market share, p.

Suppose a company has estimated the total demand for a newly developed product to be 10,000 units and that its accountants have determined that (1) the company incurs a fixed production cost (setup cost) of \$50,000 whether it produces one unit or one million units, (2) it incurs a variable production cost of \$10 per unit, and (3) the price it would sell the product for would be \$35. Accordingly, the product's contribution to the profit, π, is

$$\pi = \text{Revenue} - \text{Expenditures}$$

where

$$
\begin{aligned}
\text{Revenue} &= (\text{Market share})(\text{Total market})(\text{Sale price}) \\
&= p(10,000)(\$35) \\
&= \$350,000p
\end{aligned}
$$

$$
\begin{aligned}
\text{Expenditures} &= \text{Fixed cost} + \text{Total variable cost} \\
&= \$50,000 + (\text{Market share})(\text{Total market})(\text{Variable cost}) \\
&= \$50,000 + (p)(10,000)(\$10) \\
&= \$50,000 + \$100,000p
\end{aligned}
$$

Thus,

$$
\begin{aligned}
\pi &= \$350,000p - (\$50,000 + \$100,000p) \\
&= -\$50,000 + \$250,000p
\end{aligned}
$$

Like the first row of the payoff table for the smoke detector example, this straight-line function describes the possible outcomes that could occur if the company chooses action a_1: Market the product. Thus, we will denote this function by π_{a_1}. If the company chooses action a_2: Do not market the product, the function describing the possible outcomes that could occur is simply $\pi_{a_2} = 0$ because there will be no profit or loss. Both these payoff functions are shown in Figure 19.4. Note that for p values greater than .20, action a_1 yields the higher contribution to profit, while action a_2 yields higher contributions when p is less than .20. The point at which the payoff functions intersect, $p = .20$, is referred to as the **breakeven value of p, p_{BE}.**

Although the best strategy is clear if the value of p is known, the market share will not be known in most realistic decision problems. We therefore need to specify a probability distribution for p to characterize our uncertainty concerning p. An important difference between the finite-state decision problems we discussed earlier and the infinite-state problem is that in the infinite-state problem we must utilize a continuous probability distribution rather than a discrete distribution to characterize the uncertainty about the potential states of nature. However, it can be shown

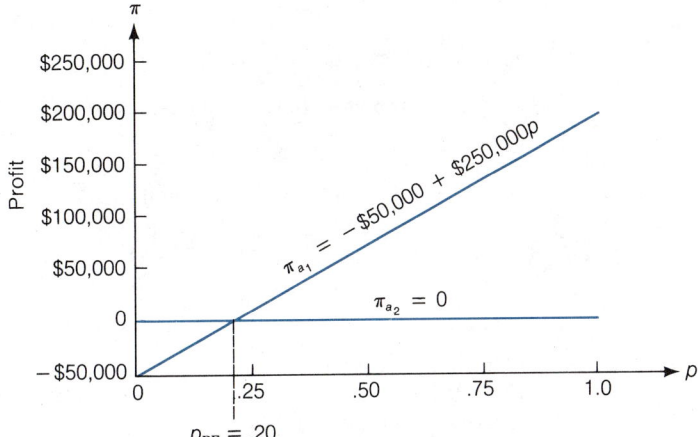

that in the case of two-action decision problems with linear payoff functions, the analysis requires only that we know the mean of the prior distribution of p. Thus, in the above example, the decision-maker need only assess $E(p)$.

If the expected value of p exceeds its breakeven value, then the expected payoff for action a_1 will exceed that for a_2. Conversely, if the expected value of p is less than the breakeven value, the expected payoff for action a_1 is less than that for a_2. Thus,

1. If $E(p) > p_{BE}$, then $EP(a_1) > EP(a_2)$ and we prefer action a_1 to action a_2.
2. If $E(p) < p_{BE}$, then $EP(a_1) < EP(a_2)$ and we prefer action a_2 to action a_1.
3. If $E(p) = p_{BE}$, then $EP(a_1) = EP(a_2)$ and both actions yield the same expected payoffs.

Suppose in the product marketing example the company assigns a prior distribution to p that is normal with a mean of .15 and a standard deviation of .05, as shown in Figure 19.5. Note that the mean is less than the breakeven value, $p_{BE} = .20$. Based on the prior information, the company should select action a_2: Do not market the product.

FIGURE 19.5
Prior Distribution for
Market Share, p

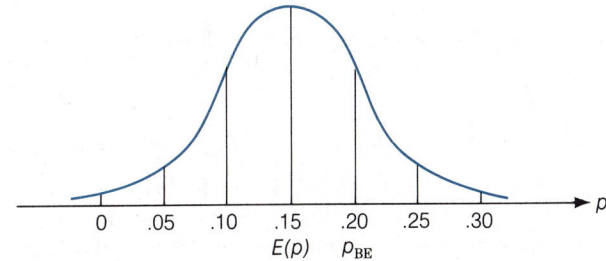

LEARNING THE MECHANICS

19.33 Consider the two profit functions for the actions A and B,

$$\pi_A = -8 + 2x \qquad \pi_B = -1 + 1x$$

and the prior distribution for x given in the table.

x	5	10	15	20	25
$p(x)$	.05	.10	.30	.40	.15

a. Graph the profit functions.
b. Find the breakeven point—the point where the decision-maker would be indifferent between actions A and B.
c. Find the expected value of x. Which action should the decision-maker take?

19.34 Refer to Exercise 19.33. Formulate the payoff table for the decision problem, and apply the expected payoff criterion. Does the action prescribed by the expected payoff criterion agree with your answer in part **c** of Exercise 19.33?

19.35 Consider the two profit functions for the actions A and B,

$$\pi_A = -1,000 + 5x \qquad \pi_B = .5x$$

and the prior distribution for x given in the table.

x	100	200	300	400	500
$p(x)$	.1	.2	.3	.2	.2

a. Find the breakeven value for x.
b. Find $E(x)$.
c. Should the decision-maker select action A or action B? Justify your answer.

19.36 Refer to Exercise 19.35. Suppose the prior distribution for x were normal with mean 320 and standard deviation 75. Which action should the decision-maker select? Justify your answer.

APPLYING THE CONCEPTS

19.37 A company is considering whether to replace a damaged bottling machine with a brand A machine or a brand B machine. Brand A costs $15,000 and brand B costs $10,000. The following cost functions describe the total cost (TC) of operating each brand for the next 3 years:

$$TC_A = 15,000 + 11x \qquad TC_B = 10,000 + 17x$$

where x is the number of hours the machine is used over the next 3 years. The company believes the probability distribution in the table characterizes the number of hours the new bottling machine would be used over the next 3 years.

x	2,000	2,500	3,000	3,500	4,000
$p(x)$	.1	.2	.3	.2	.2

a. Graph the total cost functions.
b. Find the breakeven point for this decision problem, and use it to construct a decision rule involving $E(x)$ for deciding which brand to purchase.
c. Compute $E(x)$.
d. According to the decision rule you constructed in part **b**, which brand should the company purchase?

19.38 The manager of a doughnut shop believes the shop's daily sales of glazed doughnuts (in dozens) can be characterized by a normal distribution with unknown mean and variance. Prior experience suggests that the probability of the sales being greater than 20 dozen is .50 and the probability of sales being greater than 30 dozen is .05.

a. Determine the mean and variance of the prior normal probability distribution of doughnut sales.
b. The manager is considering purchasing one of two automatic doughnut machines. The first machine can be purchased for $10,000 and costs 10¢ per dozen to operate. The second machine can be purchased for $15,000 and costs 5¢ per dozen to operate. If the shop is open 365 days per year, which machine should the manager order to minimize expected costs over the first year?

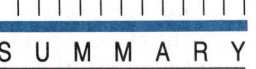

SUMMARY

In Chapter 18, we solved decision-making problems under uncertainty by assigning probabilities to the various states of nature and selecting the action that maximized the expected payoff or minimized the expected opportunity loss. Because the assessment of state probabilities is based on a limited amount of information, we may wish to collect additional information—a sample—and revise these probabilities. The state probabilities assessed before sampling are called **prior probabilities**. Their revised values, obtained using **sample information** and the procedures of this chapter, are called **posterior probabilities**. The decision analyst can compute the **expected value of sample information** *before* actually obtaining the sample; this is called a **preposterior analysis**. This quantity can then be compared to the **cost of sampling** to determine whether the sample information is worth its cost. If we decide to sample, we observe the sample outcome and use the appropriate posterior probabilities to make a decision.

We showed that the procedures for making decisions based on prior analysis (Chapter 18) or posterior analysis (Chapter 19) are identical once the probabilities of the states of nature have been assigned to the payoff table. In either case, use of the **expected utility criterion** will lead to decisions that are consistent with the decision-maker's preferences for outcomes and attitudes toward risk. Since the **expected payoff criterion** and the **expected utility criterion** are equivalent for risk-neutral decision-makers, the expected payoff criterion is frequently used as an approximation to the expected utility criterion.

Although most of the examples we considered are decision problems with a finite number of states of nature, the basic concepts of decision analysis remain unchanged when the number of states is infinite. Like most of the other topics we have discussed in this text, decision analysis embodies a variety of methods, criteria, and analytic techniques. Our objective was to present an introduction to this material. Consult the references at the end of this chapter for more detailed treatments.

SUPPLEMENTARY EXERCISES 19.39–19.50

[Note: Starred (*) exercises refer to the optional sections.]

19.39 Suppose you are sitting on a jury and hear the following evidence: A woman's body is found in a ditch in an urban creek following a violent argument with her boyfriend the previous evening. Investigation of the murder weapon shows a palm print that matches the boyfriend's print, but such evidence is not conclusive. A fingerprint expert asserts that such prints are possessed by 1 person in 1,000 (Finkelstein and Fairley, 1977).

a. What is the posterior probability of the boyfriend's guilt if the prior probability of guilt is .10? Repeat this procedure for each of the following prior probabilities: .2, .3, .4, .5, .6, .7, .8, and .9.

b. Graph the posterior probabilities you computed in part **a** versus their respective prior probabilities. Graphs of this type can be used to examine the sensitivity of a particular posterior probability to different prior probabilities. Comment on the importance of such sensitivity analyses for decision analysis.

19.40 The management of a bank must decide whether to install a commercial loan decision-support system (an on-line management information system) to aid its analysts in making commercial loan decisions. Experience suggests that each correct loan decision (accepting good loan applications and rejecting those that will eventually be defaulted) adds, on the average, approximately $25,000 per decision to the bank's profit. Further, it is estimated that the additional number, x, of correct loan decisions (per year) that could be attributed to the decision-support system has the probability distribution given in the table.

x	0	10	20	30	40	50	60	70	80	90
$p(x)$	.01	.04	.10	.15	.20	.15	.10	.10	.10	.05

a. If the decision-support system is estimated to have a useful life of 5 years, what would be the expected increase in profits that could be attributed to it?

b. The increase in profits will accrue only if the system is used by the analysts. Past experience with this type of system has shown that for various behavioral and political reasons the system was not used by analysts in 80% of the installations. Given this information and the fact that the system costs $1,500,000 to purchase, install, and maintain over a 5-year period, should the bank purchase the system?

*c. The bank is considering hiring a consulting firm to interview its loan analysts and then predict whether this particular group of analysts will use the decision-support system. The reliability of the firm's predictions is measured by the probabilities given in the accompanying table. The consulting firm charges $50,000 for its survey. Should the bank purchase the survey? Explain.

		ACTUAL OUTCOME	
		Used system	Did not use system
FORECAST	Will use system	.7	.1
	Will not use system	.3	.9

19.41 *Acceptance sampling* is commonly used by manufacturers to screen incoming lots of material for an excessive number of defective units. A sample is selected from each incoming lot of units, the number of defective units is counted, and the lot is either rejected or accepted depending on whether the number of defectives, x, exceeds a predetermined acceptance number (Montgomery, 1985). As an example of the decision analysis approach to acceptance sampling, assume the proportion of the number of defectives in an incoming lot is either 5% or 10%. The prior probability that the proportion of defectives will be 5% is .80. Suppose the cost of rejecting a lot with 10% defectives is $4,100. There are no costs for making a correct decision (i.e., accepting a lot with 5% defectives or rejecting a lot with 10% defectives).
   a. Formulate the payoff table for this problem.
   b. If no sample is drawn, should the lot be accepted or rejected?
   c. Calculate the EVPI for this problem. Should management consider a sampling inspection program?
   d. Assume a unit is drawn at random from the lot and is found to be defective. Should the lot be accepted or rejected?
   e. If the unit drawn in part **d** was not defective, should the lot be accepted or rejected?
   *f. Calculate the EVSI when the sample size is 1. Draw the corresponding decision tree.
   *g. A frequent measure of the goodness of a statistical procedure is its efficiency. In decision analysis, the *efficiency* of the sampling plan is

$$\text{Efficiency} = \left(\frac{\text{EVSI}}{\text{EVPI}}\right) 100\%$$

   What is the efficiency of a sampling plan if the sample size is 1?
   *h. It will cost management $10 in fixed costs plus $10 for every item inspected. Should management use a sample size equal to 1?

*19.42 Assume the same basic facts given in Exercise 19.41, but now consider a sample size of 3.
   a. Draw a decision tree for a preposterior analysis and calculate EVSI.
   b. What is the efficiency of this sampling plan?
   c. What is the ENGS?

*19.43 Repeat Exercise 19.41 with a sample size of 5. Then, using these results and the results of Exercises 19.41 and 19.42, draw the following graphs:
   a. Efficiency of sample versus sample size
   b. ENGS versus sample size

*19.44 A winery is considering the possibility of producing and marketing a new low-cost dinner wine. Introduction of the wine will cost $3 million in promotional and fixed costs per year, and each bottle sold will contribute 30¢ to profits. Management believes that sales would be below 5 million bottles with probability .10 and greater than 25 million bottles with probability .05. It also believes sales are approximately normally distributed. Given the winery's assessed probabilities, should it produce and market the new variety of dinner wine?

19.45 Given the payoff table shown here (state probabilities in parentheses), find the EVPI. Interpret this number.

		STATE OF NATURE		
		S_1 (.3)	S_2 (.5)	S_3 (.2)
ACTION	a_1	$5,000	$2,000	$6,000
	a_2	$7,000	$0	$8,000

19.46 Repeat Exercise 19.45 for the following payoff table:

		STATE OF NATURE			
		S_1 (.1)	S_2 (.3)	S_3 (.4)	S_4 (.2)
ACTION	a_1	$10,000	$3,000	$0	−$4,000
	a_2	$6,000	$3,000	$1,000	$2,000
	a_3	$3,000	$6,000	$1,000	$0
	a_4	$5,000	$3,000	$1,000	−$1,000

19.47 The chief forester of a large midwestern city must plan for the identification and removal next year of elm trees infected with Dutch elm disease. State law requires that all elms identified as being diseased must be removed and disposed of by October 30. For each diseased tree left standing after October 30, the city loses $300 in state funds that have been budgeted for its reforestation program. Since the forester's staff can cut only 15,000 trees per season, private contractors must be hired to cut the trees in excess of 15,000. Unfortunately, the private contractors must be hired at the beginning of the disease season (May) before the seriousness of the disease epidemic is known. There is a fixed cost of $2,000 per contract signed, and each contractor is paid $250 for each tree removed. Assume each contractor has the capacity to remove 6,000 trees per season. The forester has hired a consultant to help determine how many contractors to hire. The first action taken by the consultant is to assess a probability distribution that characterizes local disease experts' prior opinions regarding the number of elms that will be infected in the coming season. This distribution is given in the table.

NUMBER OF TREES	PROBABILITY
0– 5,000	.05
5,001–10,000	.20
10,001–20,000	.30
20,001–30,000	.20
30,001–40,000	.10
40,001–50,000	.08
50,001–60,000	.07

a. Using the midpoints of the intervals in the probability distribution as the states of nature, formulate the payoff table for this decision problem.

b. From your payoff table, construct a decision tree for the forester's decision problem.

c. Based on the prior probabilities, how many contractors should be hired to maximize the expected payoff?

d. How much should the forester be willing to pay for perfect information about the number of trees to be infected?

*19.48 Refer to Exercise 19.47. The consultant reports to the forester that a statistical model can be developed to predict the number of trees that will be infected. The conditional probabilities of the forecasts given the various states of nature are shown in the table. It will cost $5 million to develop the model. Conduct a preposterior analysis to determine whether the model should be developed.

		ACTUAL NUMBER OF TREES INFECTED						
		0–5,000	5,001–10,000	10,001–20,000	20,001–30,000	30,001–40,000	40,001–50,000	50,001–60,000
	0–5,000	.5	.3	.1	0	0	0	0
	5,001–10,000	.35	.4	.2	.1	0	0	0
	10,001–20,000	.15	.2	.5	.2	.1	0	0
FORECAST	20,001–30,000	0	.1	.2	.5	.2	0	0
	30,001–40,000	0	0	0	.2	.5	.1	.1
	40,001–50,000	0	0	0	0	.2	.5	.4
	50,001–60,000	0	0	0	0	0	.4	.5

19.49 Consider the following payoff table:

		STATE OF NATURE		
		S_1	S_2	S_3
	a_1	12	40	−5
ACTION	a_2	5	50	−10
	a_3	20	20	20

Given the prior probabilities

$$P(S_1) = .5 \qquad P(S_2) = .3 \qquad P(S_3) = .2$$

what is the EVPI for this problem?

19.50 Refer to Exercise 19.49.

a. Use the prior probabilities to determine which action is selected by the expected payoff criterion.

b. Suppose sample information has been purchased and the prior probabilities have been revised to yield the following posterior probabilities:

$$P(S_1) = .5 \qquad P(S_2) = .4 \qquad P(S_3) = .1$$

Use the posterior probabilities to determine which action is selected by the expected payoff criterion.

c. In which decision do you place more trust, the action selected using the prior probabilities in part **a**, or the action selected using the posterior probabilities in part **b**? Why?

ON YOUR OWN . . .

Refer to the "On Your Own" section of Chapter 18, in which you were to decide whether to promote a spring event that requires good weather. Ask a local meteorologist for a long-range forecast based on a careful study of all pertinent data. You need to know the meteorologist's prediction 60 days in advance of the scheduled date of the event. Before you can assess the worth of the meteorologist's prediction, you have to ask for two conditional probabilities: the probability that the meteorologist will predict rain 60 days in advance *given that* it actually will rain, and the probability that he or she will predict no rain *given that* it will not rain. Conduct a complete preposterior analysis using the meteorologist's conditional probabilities and your own prior probabilities from the Chapter 18 "On Your Own" section. Will the meteorologist's prediction affect the action selected by the expected payoff criterion? How much would you be willing to pay the meteorologist for the prediction? Explain.

REFERENCES

Baird, B. F. *Introduction to Decision Analysis*. North Scituate, Mass: Duxbury, 1978.

Brown, R. V., Kahr, A. S., and Peterson, C. *Decision Analysis for the Manager*. New York: Holt, Rinehart and Winston, 1974.

Bunn, D. *Applied Decision Analysis*. New York: McGraw-Hill, 1984.

Finkelstein, M. O. and Fairley, W. "A comment on 'Trial by mathematics.'" Reprinted in *Statistics and Public Policy*. W. Fairley and F. Mosteller (eds). Reading, Mass.: Addison-Wesley, 1977.

Howard, R. A., Matheson, J. E., and North, D. W. "The decision to seed hurricanes." *Science*, June 1972, *176*, 1191–1202.

Montgomery, D. C. *Introduction to Statistical Quality Control*. New York: Wiley, 1985. Chapter 10.

Raiffa, H. *Decision Analysis. Introductory Lectures on Choices Under Uncertainty*. Reading, Mass.: Addison-Wesley, 1968.

Raiffa, H. and Schlaifer, R. *Applied Statistical Decision Theory*. Cambridge, Mass.: MIT Press, 1961.

Winkler, R. L. *An Introduction to Bayesian Inference and Decision*. New York: Holt, Rinehart and Winston, 1972.

Winkler, R. L. and Hays, W. L. *Statistics: Probability, Inference, and Decision*, 2nd ed. New York: Holt, Rinehart and Winston, 1975. Chapters 8 and 9.

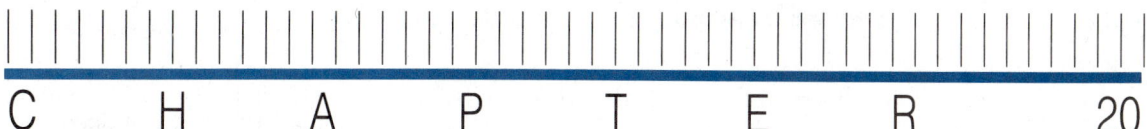

C H A P T E R 20

WHERE WE'VE BEEN ...

Although many methods are available for selecting a sample, the statistical methods described in the preceding chapters were based primarily on simple random sampling from populations of measurements that were large in relation to the sample size. Three exceptions to this method of data collection—the paired difference experiment and its generalization, the randomized block design (Chapters 9 and 15), and the factorial design (Chapter 15)—demonstrated the power of experimental design to increase the amount of information in sample data.

WHERE WE'RE GOING ...

The term *sample survey* is usually used in conjunction with sampling of people, households, businesses, etc. The Current Population Survey and the Gallup Poll are examples of such surveys. Special problems arise in survey sampling that may require more elaborate sampling designs than simple random sampling. This chapter introduces some of the problems encountered in survey sampling and the sampling designs and methods that have been developed to handle them.

SURVEY SAMPLING

CONTENTS

Almost all the statistical methods we have covered were based on simple random sampling (Section 4.6). Three exceptions to this method of data collection, the paired difference experiment, the randomized block design, and the factorial design, demonstrated that sampling designs other than simple random sampling can be used to increase the amount of information obtained in a sample. In this chapter, we present sampling designs and estimation procedures of a specific type, those used in **sample surveys**.

The term *sample survey* is usually used in conjunction with the sampling of collections of people, households, businesses, etc. A consumer preference poll is an example of a sample survey. Samplings conducted to estimate the general level of business inventories or to estimate the proportion of households that watched a particular television program are also examples of sample surveys. Sample survey designs may apply to either finite populations or infinite (conceptual) populations, which are sometimes called *processes* (refer to Case Study 3.2, page 81).

Most sample surveys are conducted to estimate one or more of three population parameters. For example, suppose we are interested in the market for seafood. One population parameter we might want to estimate is the mean amount, μ, of money spent monthly per household on seafood in a given market. A second population parameter of interest would be the total money, τ, spent on seafood per month in the market (i.e., the sum of the expenditures for all households in the market). Third, we might be interested in the proportion, p, of households that consume some seafood each month. Procedures for estimating μ and p for simple random samples were discussed in Chapter 8. We will discuss the estimation of the population total, τ, in this chapter. We present a summary of these estimation objectives in the box.

COMMON OBJECTIVES OF SAMPLE SURVEYS

1. Estimation of the population mean μ
2. Estimation of population total τ
3. Estimation of population proportion p

Sample surveys cost time and money, and sometimes they are almost impossible to conduct. For example, suppose we want to obtain an estimate of the proportion of households in the United States that plan to purchase new television sets next year, and we plan to base our estimate on the intentions of a random sample of 3,000 households. What are the problems associated with collecting these data? In order to use a random number table (Chapter 4) to select the sample, we would need a list of all the households in the United States. Obtaining such a list would be a monumental obstacle. After we obtain a list of households, we need to contact each of the 3,000 selected for the sample. Will all be at home when the surveyor reaches the household? And will all answer the surveyor's question? You can see that collecting a random sample is easier said than done. The large body of knowledge called **survey sampling** or **sample survey design** was developed to help solve some of the problems we have noted. It includes sample survey designs that will aid in reducing the cost and time involved in conducting a sample survey, and it includes the statistical estimation procedures associated with those designs. Since survey sampling is a course in itself (or several courses), we will present only a few of the most widely used sample survey designs and address only a few of the problems you might encounter. Further information on this important subject can be found in the references at the end of the chapter.

CASE STUDY 20.1

WHO DOES SAMPLE SURVEYS?*

We all know of the public opinion polls that are reported in the press and broadcast media. The Gallup Poll and the Harris Survey issue reports periodically, describing national public opinion on a wide range of current issues. State polls and metropolitan area polls, often supported by a local newspaper or television station, are reported regularly in many localities. The major broadcasting networks and national news magazines also conduct polls and report their findings.

But the great majority of surveys are not exposed to public view. The reason is that, unlike the public opinion polls, most surveys are directed to a specific administrative or commercial purpose. The wide variety of issues with which surveys deal is illustrated by the following listing of actual uses:

1. The U.S. Department of Agriculture conducted a survey to find out how poor people use food stamps.
2. Major television networks rely on surveys to tell them how many and what types of people are watching their programs.
3. Auto manufacturers use surveys to find out if people are satisfied with their cars.
4. The U.S. Bureau of the Census compiles a survey every month—the Current Population Survey (see Case Study 1.3 for more details)—to obtain information on employment and unemployment in the nation.
5. The National Center for Health Statistics sponsors a survey every year to determine how much money people are spending for different types of medical care.
6. Local housing authorities conduct surveys to ascertain satisfaction of people in public housing with their living accommodations.
7. The Illinois Board of Higher Education surveys the interest of Illinois residents in adult education.
8. Local, state, and national transportation authorities conduct surveys to acquire information on residents' commuting and travel habits.
9. Magazines and trade journals utilize surveys to find out what their subscribers are reading.
10. Surveys are used to ascertain the characteristics of people who use our national parks and other recreation facilities.
11. Sample surveys are used by marketing researchers to uncover new uses for products already on the market. Such information is helpful in redirecting existing advertising campaigns or creating new ones, as illustrated in the following examples (Cox, 1979):

*Most of this case study has been reproduced from the American Statistical Association's pamphlet, *What Is a Survey?* by R. Ferber, P. Sheatsley, A. Turner, and J. Waksberg (1980).

a. The producer of Ben-Gay, a topical analgesic, believed that consumers used Ben-Gay primarily for the relief of simple muscle aches. Data collected from a large sample of consumers revealed, however, that more than 50% of all consumers use Ben-Gay for arthritis relief (Davis, 1977).

b. More than 4,000 consumers are surveyed weekly by Lever Brothers. The sample information they obtained regarding Wisk detergent indicated that many consumers were using Wisk as a "pretreater" of shirt collars. This information spawned the familiar "ring-around-the-collar" advertising campaign (*Marketing News*, Feb. 10, 1978).

12. Bank auditors survey a bank's savings account customers to verify the accuracy of the bank's records.

SECTION 20.1

TERMINOLOGY

The terminology used in statistical survey sampling is slightly different from that used in other statistical applications. For example, the object upon which a measurement is taken is called an **element** instead of an **experimental unit**. The term **population** retains the same meaning: It is the collection of measurements about which we wish to make an inference.

DEFINITION 20.1

The object upon which a measurement is made is called an **element**.

Sometimes we may want to reduce the cost of sampling by taking measurements on collections of elements that are physically near one another or bear some other relationship that makes them more easily observed as a group. For example, if we plan to sample the opinions of all adults regarding some particular product, we might wish to randomly select households and then interview all the adults in the households. When nonoverlapping sets of elements are randomly selected and each element in the set is measured, the sets are called **sampling units**. For the product preference survey just described, a household would be a sampling unit and the adults in the household would be the elements. Note that each element in a sampling unit (a household) is measured and that the elements in one sampling unit do not occur in the elements of another.

In the preceding chapters, we randomly selected experimental units and made a single observation on each. Thus, the earlier chapters were restricted to the special case where each sampling unit contained only one element.

DEFINITION 20.2

A **sampling unit** is a collection of elements. The elements must satisfy the condition that those in any one sampling unit do not overlap with the elements in other sampling units.

In order to select a sample of sampling units from the total of those available, we must have a listing of them. Such a listing, which must include *all* the sampling units in the population of interest (to enable us to draw a sample that is representative of the population), is called a **frame**. Then, a **sample** is a subset of sampling units selected from a frame. The plan that specifies which sampling units will be included in a sample is called a **sampling design** or, for sample surveys, a **sample survey design**.

DEFINITION 20.3

A **frame** is a list of sampling units.

DEFINITION 20.4

A **sample** is a collection of sampling units selected from a frame.

We summarize the terminology in the box.

TERM	DEFINITION	EXAMPLE: PRODUCT PREFERENCE SURVEY
Element	Object on which a measurement is made	Individual consumer
Sampling unit	Collection of elements	Household
Frame	List of sampling units	List of all households in relevant population
Sample	Collection of sampling units selected from frame	Set of households from which product preferences are obtained

S E C T I O N   20.2

SAMPLE SURVEY DESIGNS

Several useful sampling designs are available for sample surveys. Two of the most common (in addition to random sampling), **stratified random sampling** and **cluster sampling**, are described in this section, and the appropriate procedures for estimating the population mean, μ, a proportion, p, and the total, τ, of all measurements in the population are presented in Sections 20.6 and 20.7. Two other sampling designs, **systematic sampling** and **randomized response sampling**, are discussed briefly in this section, but we refer you to the references at the end of the chapter for the associated estimation procedures.

Stratified random sampling is used when the sampling units associated with the population are physically separated into two or more groups of sampling units (called **strata**) where the within-strata response variation is less than the variation within the entire population. For example, if y is the rent paid for a two-bedroom apartment in a city, we might want to divide the city into regions (strata) where the rents within each stratum are relatively homogeneous. Then we would estimate a population mean, proportion, or total by selecting random samples from within each stratum and combining the strata estimates as explained in Section 20.6. Stratified random sampling often produces estimators with smaller standard errors than those achieved using simple random sampling. Furthermore, by sampling from each stratum we are more likely to obtain a sample representative of the entire population. In addition, the administrative and labor costs of selecting the strata samples are often less than those for simple random sampling.

It is often less costly to use **cluster sampling**, where we randomly select groups (or **clusters**) of sampling units rather than individual units. For example, suppose we wish to sample the opinions of voters in a city. It would cost approximately the same amount of money to have pollsters contact a random sample of 1,000 households as it would to contact 1,000 individual voters. Since each household could contain two or more voters, sampling 1,000 clusters (households) could produce the opinions of several thousand voters and do so at approximately the same cost as randomly sampling the opinions of 1,000 voters. The estimation procedures for cluster sampling are discussed in Section 20.7.

Sometimes it is difficult or too costly to select random samples. For example, it would be easier to obtain a sample of student opinions at a large university by selecting every hundredth name from the student directory, with the first name selected randomly from the first 100 names in the directory. Although **systematic samples** are usually easier to select than other types of samples, one difficulty is the possibility of a systematic sampling bias. For example, if every fifth item in an assembly line is selected for quality control inspection, and if five different machines are sequentially producing the items, all the items sampled may have been manufactured by the same machine. If we use systematic sampling we must be certain that no cycles (like every fifth item manufactured by the same machine) exist in the list of the sampling units.

Randomized response sampling is particularly useful when the questions of the pollsters are likely to elicit false answers. For example, suppose each person in a sample of wage earners is asked whether he or she cheated on his or her income tax return. A person who has not cheated most likely would give an honest answer to this question. A cheater might lie, thus biasing an estimate of the proportion of persons who cheat on their income tax return.

One method of coping with the false responses produced by sensitive questions is randomized response sampling. Each person is presented *two* questions; one question is the object of the survey and the other is an innocuous question to which the interviewee will give an honest answer. For example, each person might be asked these two questions:

1. Did you cheat on your income tax return?
2. Did you drink coffee this morning?

Then a procedure is used to randomly select which of the two questions the person is to answer. For example, the interviewee might be asked to flip a coin. If the coin shows a head, the interviewee answers the sensitive question, #1. If the coin shows a tail, the interviewee answers the innocuous question, #2. Since the interviewer never has the opportunity to see the coin, the interviewee can answer the question and feel assured that his or her guilt (if guilty) will not be exposed. Consequently, the random response procedure can elicit an honest response to a sensitive question.

Four of the most important sampling designs are discussed in this chapter. In each case, we will present the methodology for selecting the sample, calculating the estimates of population parameters, and measuring the standard error of the estimates. The size of the standard error will serve as a measure of the amount of information on a parameter that is provided by a specific sampling design.

CASE STUDY 20.2

METHODS OF DATA COLLECTION IN SAMPLE SURVEYS*

The survey designs described in this chapter prescribe methods for selecting elements from a frame. Once selected, the attribute of interest must be measured for each of the elements. That is, the data must be collected. Surveys that involve human populations can be classified by their method of data collection. Thus, there are **mail surveys, telephone surveys**, and **personal interview surveys**.

Mail surveys require the development of questionnaires that respondents complete on their own (i.e., self-enumeration). Mail surveys are seldom used to collect information from the general public because names and addresses are not often available and response rates tend to be low. However, this method may be effective with members of particular groups, such as subscribers to specialized magazines or members of a professional organization.

Telephone interviewing is an efficient method of collecting some types of data and is being increasingly used. Random samples of telephone numbers may be randomly or systematically selected from telephone directories, or a recent innovation called *random-digit dialing* may be employed. This approach involves using a random number generator to mechanically create the sample of phone numbers to be called. Random-digit dialing was developed to help overcome the sampling biases introduced into survey results by sampling from telephone directories (Glasser and Metzger, 1972).

Personal interviews are generally conducted in a respondent's home or office. They are much more expensive than either mail or telephone surveys, but may be necessary when complex information is being collected.

There are also newer methods of data collection by which information is recorded directly into computers. This includes the A. C. Nielsen Company's measurement of TV audiences using electronic devices—called *audimeters*—attached to a sample of TV sets. Nielsen places an audimeter in each of a sample of about 1,700 homes across the United States. The audimeter, usually located in a closet or in the basement, is wired to every TV set in the home and records when the sets are on or off and which channels are tuned in. The audimeter is connected via special telephone lines to Nielsen's computer. When you read in your newspaper that a particular TV show received, say, a "20 rating" for the week by A. C. Nielsen, it

*Portions of this case study have been reproduced from the American Statistical Association's pamphlet, *What Is a Survey?* by R. Ferber, P. Sheatsley, A. Turner, and J. Waksberg (1980).

means that 20% of the sample of Nielsen families tuned in to that show for at least 6 minutes (Chagall, 1978).

Some surveys combine various methods. Survey workers may use the telephone to screen eligible respondents (say, women of a particular age group) and then make appointments for a personal interview. The U. S. Bureau of the Census' monthly Current Population Survey (see Case Study 1.3) uses both telephone and personal interviews.

Because changes in attitude or behavior cannot be reliably ascertained from a single interview, some surveys use a *panel* of respondents who are interviewed two or more times. Such surveys are often used during election campaigns, or to chart a family's health or purchasing pattern over a period of time. The Nielsen families, for example, constitute a panel of respondents whose TV watching patterns are monitored over time. Panels are also used to trace changes in behavior over time, as with social experiments that study changes in the work behavior of low-income families in response to an income maintenance plan.

CASE STUDY 20.3

THE *LITERARY DIGEST* POLL: FDR VERSUS ALF LANDON

Regardless of the survey design and data collection method employed, great care must be exercised in implementing the survey. Poorly implemented surveys may yield disastrous results, as this case study illustrates.

In 1936, the *Literary Digest*, a popular magazine, mailed 10 million questionnaires to voters in the United States. The questionnaire asked which presidential candidate was preferred, the Democratic incumbent, F. D. Roosevelt, or the Republican governor of Kansas, Alfred Landon. The *Digest* had previously predicted the winner of the presidency in very election since 1916. Prior to receiving the responses to its questionnaire, the *Digest* boasted, "When the last figure has been totted and checked, if past experience is a criterion, the country will know to within a fraction of 1% the actual popular vote of forty million" (Aug. 22, 1936, p. 3). The *Digest* received 2.4 million responses—a sample size approximately 800 times larger than is currently used by the Gallup Poll. The sample results indicated Landon would win by a landslide: Landon 57% and FDR 43%. Unfortunately for Landon and the *Literary Digest*, the actual election results yielded a landslide for FDR: FDR 62% and Landon 38%.

What went wrong? How could such a large sample generate such misleading results? Part of the answer lies in the *Digest*'s choice of a sampling frame. The frame was constructed from sources such as telephone directories, club membership lists, magazine subscriber lists, and lists of car owners. Although use of such lists might not yield such misleading results today, the country was split politically along economic lines in 1936—Republicans were generally wealthier than Democrats. As a result, the majority of people listed in the *Digest*'s frame were probably Republicans. Accordingly, the sample was probably not representative of the population of voters in the United States; it was biased in favor of Republican voters.

Another reason for the lack of representativeness of the sample was the *Literary Digest*'s reliance on *voluntary response* to its poll. Respondents to mail questionnaires typically represent only that portion of the population with relatively intense interests in the subject matter of the questionnaire. The anti-Roosevelt voters—although a minority—apparently felt more strongly about the election than did the pro-Roosevelt majority. As a result, the *Literary Digest*'s poll was apparently affected by *nonresponse bias* (see Section 20.5).

If you have never heard of the *Literary Digest*, there is a reason: It is now defunct—thanks in part to the credibility lost as a result of its 1936 presidential poll (Bryson, 1976).

<table>
<tr><td>

| | | | | | | | | | | |

S E C T I O N 20.3

ESTIMATION IN SURVEY SAMPLING: BOUNDS ON THE ERROR OF ESTIMATION

</td></tr>
</table>

Estimation procedures developed for the various sample survey designs may differ from those presented in earlier chapters for two reasons. The standard errors of estimators presented in earlier chapters were based on the assumption that the number of sampling units, N, in the population is large relative to the sample size, n. This assumption may not hold in survey sampling and thus will necessitate a modification of the formulas given for the standard errors of the estimators.

The second difference is that the sampling distributions of estimators are often unknown. For this reason, it is difficult to construct exact confidence intervals for population parameters. The usual procedure (see Scheaffer, Mendenhall, and Ott, 1979) is to give an estimate along with an approximate upper limit on the error of estimation—i.e., on the difference that might occur between the estimate and the unknown value of the population parameter. This upper limit, which we call a **bound on the error of estimation**, is calculated using the Empirical Rule of Section 3.7. The logic is that, according to the Empirical Rule, most (approximately 95%) of the estimates produced by an unbiased estimator should lie within 2 standard errors of the estimated population parameter. Or, we could form an approximate large-sample confidence interval for a parameter using the logic of Section 8.1; that is, we will find the endpoints of the confidence interval by adding and subtracting 2 standard errors to the estimate. Consequently, we will present the formulas for estimators, the estimated bounds on the error of estimation, and approximate confidence intervals for each sample survey design using the procedure shown in the box.

GENERAL PROCEDURES FOR ESTIMATING POPULATION PARAMETERS BASED ON SAMPLE SURVEYS

1. Present a formula for calculating the estimate.
2. Give a bound on the error of estimation equal to 2 standard errors (or the sample estimate thereof) of the estimator.
3. Calculate an approximate 95% confidence interval for the parameter by forming the interval given by:

> Estimate ± (Bound on error)

that is,

> Estimate ± (2 estimated standard errors)

CASE STUDY 20.4

SAMPLING ERROR
VERSUS NONSAMPLING
ERROR*

In Chapter 7, we learned that the behavior of the sample mean, $\bar{x}$, in repeated sampling can be described by its sampling distribution. We described the difference between a particular value of the estimator, $\bar{x}$, and the true value of the population parameter, μ, as **estimation error**. This difference is also known as **sampling error**. It is not error in the sense that anyone or anything is at fault or deserves blame; it is simply due to the fact that $\bar{x}$ is computed from a subset of the population rather than from the entire population. The standard error of the sampling distribution of $\bar{x}$ is a measure of the magnitude of the sampling error (estimation error) that may be present in the results of a survey that has been conducted to estimate μ. Accordingly, the standard error of $\bar{x}$ is used to place a bound on the sampling error associated with $\bar{x}$. As we will see in Section 20.8, this bound can be tightened simply by increasing the sample size of the survey.

Unfortunately, the other types of errors that plague surveys known as **nonsampling errors**—are not so easily measured or controlled. Nonsampling errors are any phenomena other than sampling errors that cause a difference between an estimate and the true value of the population parameter. Nonsampling errors can be classified into two groups: **random errors** whose effects approximately cancel out if large samples are used and **biases** that tend to create errors in the same direction and thus do not cancel out over the entire sample.

Biases can arise from any aspect of the survey operation. Some of the main contributing causes are the following:

1. *Sampling operations.* There may be mistakes made in drawing the sample, or part of the population may be omitted from the sampling frame. (This was part of the problem with the *Literary Digest* poll, discussed in Case Study 20.3.)

2. *Noninterviews.* Information may be obtained for only part of the sample due to, for example, "not-at-homes" or nonresponse to mail questionnaires. This causes a problem because, typically, there are differences between the noninterviewed part of the sample and the part that is interviewed.

3. *Adequacy of respondent.* Sometimes respondents cannot be interviewed, and information is obtained about them from others; the proxy respondent is not always as knowledgeable about the facts.

4. *Understanding the concepts.* Some respondents may not understand what is wanted.

5. *Lack of knowledge.* Respondents in some cases do not know the information requested or do not try to obtain the correct information.

6. *Concealment of the truth.* Out of fear or suspicion of the survey, respondents may conceal the truth. In some instances, this concealment may reflect a respondent's desire to answer in a way that is socially acceptable, such as indicating that he or she is carrying out an energy conservation program when this is not actually so.

7. *Loaded questions.* The question may be worded to influence the respondents to answer in a specific (not necessarily correct) way.

*Portions of this case study have been reproduced from the American Statistical Association's pamphlet, *What Is a Survey?* by R. Ferber, P. Sheatsley, A. Turner, and J. Waksberg (1980).

8. *Processing errors.* These can include coding errors, data keying, computer programming errors, etc.
9. *Conceptual problems.* There may be differences between what is desired and what the survey actually covers. For example, the population or the time period may not be the one for which information is needed, but had to be used to meet a deadline.
10. *Interviewer errors.* Interviewers may misread the question or twist the answers in their own words and thereby introduce bias.

Although not every survey will be subject to all these biases, a good survey statistician would be aware of their possible existence and attempt to control as many as possible.

In the case of the U.S. Bureau of the Census' Current Population Survey (see Case Study 1.3), many safeguards have been built into the survey process to protect against biases due to interviewer errors (Taeuber, 1978):

1. The survey's 1,100 interviewers are continuously trained and retrained.
2. Each interviewer's works is reviewed each month.
3. Periodically, interviewers are accompanied by supervisory personnel.
4. Approximately twice each year, a sample of the addresses assigned to an interviewer is reinterviewed by a supervisor. The interviewers have no way of knowing when their work will be checked or which addresses will be reinterviewed.

These precautions not only protect against interviewer error but also provide a measure of quality of the Current Population Survey.

| | | | | | | | | | | | |

S E C T I O N   20.4

**ESTIMATION FOR
SIMPLE RANDOM
SAMPLING**

We discussed the estimation of a population mean, μ, and a proportion, p, based on simple random sampling in Chapter 8. The confidence intervals for these parameters were based on the assumption that the sample size, n, is sufficiently large and, although we did not state it, that the number, N, of sampling units in the population is large relative to the sample size, n.

In some sample surveys, the sample size, n, may represent 5% or perhaps 10% of the total number, N, of sampling units in the population. When the sample size is large relative to the number of measurements in the population, the standard errors of the estimators of μ and p (given in Chapter 8) should be multiplied by a **finite population correction factor**.

The form of the finite population correction factor depends on how the population variance σ^2 is defined. In order to simplify the formulas of the standard errors that are used in sample surveys, it is common to define σ^2 as division of the sum of squares of deviations by $N - 1$ rather than by N (analogous to the way we defined the sample variance). If we adopt this convention, the finite population correction factor becomes $\sqrt{(N - n)/N}$. Then the point estimators and the estimated bounds on the errors of estimation for μ and p are as shown in the boxes on page 1142.*

---

*For most sample surveys, the finite population correction factor is approximately equal to 1 and, if desired, can be safely ignored. However, if $n/N > .05$, the finite population correction factor should be included in the calculation of the standard error and the bound on the error of estimation.

ESTIMATION OF THE POPULATION MEAN, μ: SIMPLE RANDOM SAMPLING

Estimator of μ: $\bar{x} = \dfrac{\sum x_i}{n}$

Estimated bound on the error of estimation: $2\hat{\sigma}_{\bar{x}} = 2\dfrac{s}{\sqrt{n}} \sqrt{\dfrac{N - n}{N}}$

where

$s = \sqrt{\dfrac{\sum (x_i - \bar{x})^2}{n - 1}}$

N = Number of sampling units in the population

n = Number of sampling units in the sample

[*Note:*   In simple random sampling, each sampling unit contains only one element.]

Approximate 95% confidence interval: $\bar{x} \pm 2\hat{\sigma}_{\bar{x}}$

ESTIMATION OF THE POPULATION PROPORTION, p: SIMPLE RANDOM SAMPLING

Estimator of p: $\hat{p} = \dfrac{x}{n}$

where x is the number of sampling units that possess a specific attribute (in terms of the binomial distribution, x is the number of "successes").

Estimated bound on the error of estimation: $2\hat{\sigma}_{\hat{p}} = 2\sqrt{\dfrac{\hat{p}(1 - \hat{p})}{n}} \sqrt{\dfrac{N - n}{N}}$

where

N = Number of sampling units in the population

n = Number of sampling units in the sample

Approximate 95% confidence interval: $\hat{p} \pm 2\hat{\sigma}_{\hat{p}}$

The point estimator and the estimated bound on the error for estimating a population total, τ, were not presented in Chapter 8. Their formulas are shown in the next box.

EXAMPLE 20.1

A specialty manufacturer wants to purchase remnants of sheet aluminum foil. The foil, all of which is the same thickness, is stored on 7,462 rolls, each containing a varying amount of foil. To obtain an estimate of the total number of square feet of foil on all the rolls, the manufacturer randomly sampled 100 rolls and measured

---

ESTIMATION OF THE POPULATION TOTAL, τ: SIMPLE RANDOM SAMPLING

Estimator of τ: $\hat{\tau} = N\bar{x}$

where

N = Number of sampling units in the population

n = Number of sampling units in the sample

$\bar{x}$ = Sample mean

Estimated bound on the error of estimation: $2\hat{\sigma}_{\hat{\tau}} = 2\sqrt{N^2 \dfrac{s^2}{n}\left(\dfrac{N-n}{N}\right)}$

where s^2 is the sample variance:

$$s^2 = \frac{\sum (x_i - \bar{x})^2}{n - 1}$$

Approximate 95% confidence interval: $\hat{\tau} \pm 2\hat{\sigma}_{\hat{\tau}}$

---

the number of square feet on each roll. The sample mean was 47.4, and the sample variance was 153.1. Find an approximate 95% confidence interval for the total amount of foil on the 7,462 rolls.

SOLUTION

Each roll of foil is a sampling unit, and there are $N = 7,462$ units in the population and $n = 100$ in the sample. Further,

$\bar{x} = 47.4$ and $s^2 = 153.1$

Substituting these quantities into the formula for the confidence interval, we obtain (for $\hat{\tau} = N\bar{x}$):

$$\hat{\tau} \pm 2\sqrt{N^2 \frac{s^2}{n}\left(\frac{N-n}{N}\right)} = (7,462)(47.4) \pm 2\sqrt{(7,462)^2\,\frac{153.1}{100}\left(\frac{7,462 - 100}{7,462}\right)}$$

or, the approximate 95% confidence interval is

353,698.8 ± 18,341.8

Consequently, the manufacturer estimates the total amount of foil to be in the interval 335,357.0 square feet to 372,040.6 square feet. If the manufacturer wants to adopt a conservative approach, the bid for the foil will be based on the lower confidence limit, 335,357 square feet of foil. ∎

Examples of the estimation of a population mean, μ, and sample proportion, p, are not presented in this section because the examples would be identical to those presented in Chapter 8, except for the use of the finite population correction factor. We include exercises of this type at the end of this section.

| | | | | | | | | | | | |

EXERCISES 20.1–20.19   LEARNING THE MECHANICS

20.1 Distinguish among the following terms: element, sampling unit, and sample.

20.2 List three different methods of data collection used in surveys and describe an advantage associated with each.

20.3 What went wrong with the *Literary Digest's* 1936 election poll?

20.4 Distinguish between sampling error and nonsampling error, and give an example of each.

20.5 Calculate the percentage of the population sampled and the finite population correction factor for each of the following situations:
    a. $n = 1,000$, $N = 2,500$ **b.** $n = 1,000$, $N - 5,000$
    c. $n = 1,000$, $N = 10,000$ **d.** $n = 1,000$, $N = 100,000$

20.6 Suppose the standard deviation of the population is known to be $\sigma = 100$. Calculate the standard error of $\bar{x}$ for each of the situations described in Exercise 20.5.

20.7 Suppose $N = 5,000$, $n = 36$, and $s = 12$.
    a. Compare the size of the standard error of $\bar{x}$ computed with and without the finite population correction factor.
    b. Repeat part **a**, but this time assume $n = 500$.
    c. Theoretically, when sampling from a finite population, the finite population correction factor should always be used in computing the standard error of $\bar{x}$. However, when n is small relative to N, the finite population correction factor is close to 1 and can safely be ignored. Explain how parts **a** and **b** illustrate this point.

20.8 Suppose $N = 10,000$, $n = 1,000$, and $s = 20$.
    a Compute the standard error of $\bar{x}$ using the finite population correction factor.
    b. Repeat part **a** assuming $n = 5,000$.
    c. Repeat part **a** assuming $n = 10,000$.
    d. Compare parts **a**, **b**, and **c**, and describe what happens to the standard error of $\bar{x}$ as n is increased.
    e. The answer to part **c** is 0. This indicates that there is no sampling error in this case. Explain.

20.9 Suppose you want to estimate a population mean, μ, and $\bar{x} = 375$, $s = 11$, $N = 305$, and $n = 30$. Find an approximate 95% confidence interval for μ.

20.10 Suppose you want to estimate a population proportion, p, and $\hat{p} = .37$, $N = 4,000$, and $n = 900$. Find an approximate 95% confidence interval for p.

20.11 Suppose you want to estimate a population total, τ, and $\bar{x} = 39.4$, $s = 4.0$, $N = 3,500$, and $n = 100$. Find an approximate 95% confidence interval for τ.

20.12 A random sample of size $n = 30$ was drawn from a population of size $N = 2,000$. The following measurements were obtained:

21	33	19	29	22	38	58	29	52	36
18	35	42	36	41	35	36	33	38	29
38	39	54	42	42	37	30	53	37	29

a. Estimate τ and place a bound on the error of estimation.
b. Estimate μ and place a bound on the error of estimation.
c. Estimate p, the proportion of measurements in the population that are greater than 30. Place a bound on the error of estimation.

APPLYING THE CONCEPTS

20.13 CPA firms charge $30 to $50 per hour for the auditing services of their junior staff members. The rate for a senior staff member can be substantially higher. Because external audits can become quite expensive, many firms are creating or increasing the size of their existing internal audit departments. Internal auditors can lower the cost of an external audit by improving the company's accounting controls, performing financial examinations that support the outside auditors' activities, and providing general support for the outside auditors. As part of a study designed to determine the effect of internal audit activities on the cost of external audits, Professor Wanda A. Wallace (1984) questioned a sample of 32 large, diverse companies concerning expenditures for external audits. The mean external audit fee paid by the 32 companies in 1981 was $779,030; the standard deviation was $1,083,162.

a. Construct an approximate 95% confidence interval for the mean external audit fee in 1981 for the population of firms from which Professor Wallace drew her sample. Assume that $N = 1,500$.
b. Construct an approximate 95% confidence interval for the total amount spent by all firms in the population on external audits in 1981. Again, assume $N = 1,500$.
c. Carefully interpret your confidence interval of part b in the context of the problem.
d. What assumption must hold in order to assure the validity of the confidence intervals in parts a and b?

20.14 Organizations hire independent public accountants to perform audit examinations of their financial statements and to judge the fairness with which the financial statements characterize the financial position of the organization. The audit examination includes numerous reviews and tests that are designed to provide the auditor with evidence from which an opinion about the financial statements can be developed. In addition, this evidence provides the auditor with a basis for deciding whether the organization's financial statements have been prepared according to "generally accepted accounting principles." Since the early 1950's, auditors have relied to a great extent on sampling techniques, rather than 100% audits, to help them test and evaluate financial records. For example, sampling is frequently used to obtain an estimate of the total dollar value of an account—the account balance. The estimate can be used to check the account balance reported in the organization's financial statements. Such an examination of an account balance is known as a *substantive test* (Arkin, 1982). In order to evaluate the reasonableness of a firm's stated total value of its parts inventory, an auditor randomly samples 100 of the total of 5,000 parts in stock, prices each part, and reports the results shown in the table at the top of page 1146.

a. Find a point estimate of the total value of the parts inventory.
b. Estimate the bound on the error of estimation associated with your point estimate of part a. [*Hint:* $s = \$209.10$]
c. Construct an approximate 95% confidence interval for the total value of the parts inventory.

PART SERIAL NUMBER	PART PRICE	NUMBER IN SAMPLE	PART SERIAL NUMBER	PART PRICE	NUMBER IN SAMPLE
002	$108	3	271	$ 50	9
101	55	2	399	125	12
832	500	1	761	1,000	2
077	73	10	093	62	8
688	300	1	505	205	7
910	54	4	597	88	11
839	92	6	830	100	19
121	833	5			

d. The firm reported a total parts inventory value of $1,500,000. What does your confidence interval of part c suggest about the reasonableness of the firm's reported figure? Explain.

20.15 On Friday, February 3, 1984, the head of the Environmental Protection Agency (EPA), William Ruckelshaus, announced the banning of further use of the cancer-causing pesticide ethylene dibromide (EDB) as a fumigant for grain and flour-milling equipment. EDB is used to protect against infestation by microscopic roundworms called nematodes. In addition, Ruckelshaus announced maximum safe levels for EDB presence in raw grain, flour, cake mixes, cereals, bread, and other grain products now on supermarket shelves and in warehouses. Because the federal government does not have the authority to regulate the amount of chemicals in foods, these safe levels were intended as guidelines for state governments. Ruckelshaus estimated that, if state governments followed the EPA guidelines, approximately 7% of the existing corn products would have to be removed from supermarket and warehouse shelves. Following the announcement, state agriculture agencies began sampling the grain products sold in their respective states and testing for the presence of unsafe levels of EDB (Berg, Klauda, and Feyder, 1984). Of the 3,000 corn-related products sold in a particular state, tests indicated that 15 of a random sample of 175 had EDB residues above the safe level.

a. In the context of the problem, describe the population parameter, p, for which $\hat{p} = 15/175$ is a point estimate.

b. Estimate the bound on the error of estimation associated with $\hat{p}$ in part **a**. Interpret this bound in the context of the problem.

c. Construct an approximate 95% confidence interval for p.

d. Do the data provide sufficient evidence to indicate that more than 7% of the corn-related products in this state would have to be removed from shelves and warehouses? Test using $\alpha = .05$, and interpret your test results.

20.16 A sample survey is undertaken to determine the proportion of voters in a certain county who favor a proposal to create urban "enterprise job zones" that would seek to attract new business and job opportunities in declining areas of the county's cities. A random sample of 1,000 voters is selected from 50,840 eligible voters in the county. Of the 1,000 voters, 620 said they would favor the proposal. Use the techniques outlined in this section to find an approximate 95% confidence interval for the true proportion of the county's voters who favor the creation of urban enterprise job zones.

20.17 A small grocery chain, which stocks 410 items, conducted an audit to compare the dollar value of the inventory shown on its books with the actual value of the inventory on hand. Sixty items were randomly selected from the 410, each of the 60 items was inventoried, and the difference between the book and actual values of the inventory was recorded. The difference between the book and actual inventories for the 60 items had a mean equal to $330 and a standard deviation equal to $546.

a. Estimate the mean difference per item between the book and actual inventories using a 95% confidence interval.

b. Estimate the total difference between the book and actual inventories for the chain. Use a 95% confidence interval.

20.18 A wholesale shipment contains 800 boxes of light bulbs, with ten bulbs per box (a total of 8,000 bulbs). Before accepting the shipment, a retailer wants to estimate the total number of defective light bulbs in the shipment. The retailer randomly selects 50 boxes and determines the number of defectives in each box. If the number of defectives per box has a mean of .4 and a variance of 1.2, estimate the total number of defective bulbs in the shipment and place bounds on the error of estimation.

20.19 In an urban industrial community, 70,500 persons are classified as potential members of the work force. An economist who wishes to investigate the unemployment rate in the community interviews 6,150 potential members of the work force and finds that 572 are currently jobless. Estimate the current unemployment rate in the community, and place a bound on the error of estimation.

| | | | | | | | | | | | |

SECTION 20.5

SIMPLE RANDOM SAMPLING: NONRESPONSE

We have explained in Section 4.6 how to draw a simple random sample, but we did not comment on the physical problem of actually doing it. For example, we mentioned in the introduction to this chapter that it would be extremely difficult to select a random sample of 3,000 households from all the households in the United States. And, even if we had a frame, it would be costly to contact the selected households.

Two methods for reducing the cost of random sampling are to use a telephone survey or a mailed survey. This type of sampling eliminates transportation costs and reduces labor costs, but it introduces a serious difficulty, the problem of **nonresponse**. By this, we mean that sampling units contained in a sample do not produce sample observations. For example, an individual may not be at home when telephoned or may refuse to complete and mail back a questionnaire.

Nonresponse is a serious problem because it may lead to very biased results. There may be a high correlation between the type of response and whether or not a person responds. For example, most citizens in a community might have an opinion on a school bond issue, but the respondents in a mail survey might very well be those with vested interests in the outcome of the survey—say, parents with children of school age, or school teachers, or those whose taxes might be substantially affected. Others with no vested interests might have opinions on the issue but might not take the time to respond. For this example, the absence of the nonrespondents' data could lead to a larger estimate of the percentage in favor of the issue than was actually the case. In other words, the absence of the nonrespondents' data could lead to a biased estimate. This was one of the problems in the *Literary Digest* poll discussed in Case Study 20.3.

The problem of nonresponse identifies a very important sampling problem. If your sampling plan calls for a specific collection of sampling units, failure to acquire the responses from those units may violate your sampling plan and lead to biased estimates. If you intend to select a random sample and you cannot obtain the responses from some of the sampling units, then your sampling procedure is no longer random and the methodology based on it and the product of the methodology are suspect.

There are ways for coping with nonresponse. Most involve tracking down all or part of the nonrespondents and using the additional information to adjust for the missing nonrespondent data. For mailed surveys, however, it has been found that the inclusion of a monetary incentive with the questionnaire—even as little as 25¢—will substantially increase the response rate of the survey (Armstrong, 1975).

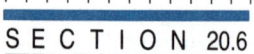

SECTION 20.6

STRATIFIED RANDOM SAMPLING

Suppose you were in the wholesale seafood business in a city that had three distinctly different market areas. To plan your purchasing, you wish to obtain an estimate of the mean monthly seafood consumption per household in the city.

If you base your estimate of the mean monthly consumption, μ, of seafood per household on the mean, $\bar{x}$, of a random sample of n households selected within the city, the standard error that measures the variation associated with your estimate is

$$\sigma_{\bar{x}} = \frac{\sigma}{\sqrt{n}}\sqrt{\frac{N-n}{N}}$$

One way to reduce $\sigma_{\bar{x}}$ and reduce the costs of collecting the sample is to select samples within the three markets. The seafood consumption per household is likely to be less in some neighborhoods than in others. Consequently, there will be a substantial amount of variability in the household consumption, x, within the city. In contrast, the variation in consumption within one of the relatively homogeneous (socially and economically) neighborhoods is likely to be less, as is also the variation in consumption within each of the other neighborhoods. This suggests an alternative to simple random sampling. We select a random sample from within each of the three relatively homogeneous marketing areas (called **strata**), estimate the mean consumption within each, and then combine these estimates to obtain an estimate of the mean monthly consumption per household for the whole city. This type of sampling plan, called **stratified random sampling**, has three advantages:

1. Stratified sampling provides additional information; that is, it gives estimates of the mean for *each* stratum as well as of the mean for the entire population.
2. Stratified sampling usually provides more accurate estimates of the population mean than does a simple random sample of the same size because the variability within the strata is usually less than the variability over the entire population.
3. The transportation and administrative costs of sampling within strata are usually less than the costs of sampling within the entire population. This is because the sampling units are frequently geographically closer when selected within the strata than when they are selected randomly from within the entire population.

To summarize, a stratified random sampling plan consists of partitioning the population into a group of k strata, each of which is more homogeneous than the population itself. This sampling plan usually results in more precise estimates (lower variability) at a lower cost. To implement a stratified sampling plan, select a random sample of n_1 sampling units from stratum 1, n_2 from stratum 2, . . . , and n_k from stratum k. Then, the total sample size selected from the population is $n = n_1 + n_2 + \cdots + n_k$. The notation and the formulas for parameter estimators are given in the following boxes.

NOTATION FOR STRATIFIED RANDOM SAMPLING

k = Number of strata

N_i = Number of sampling units in stratum i

N = Number of sampling units in the population
$= N_1 + N_2 + \cdots + N_k$

n_i = Number of sampling units selected from stratum i

n = Total number of sampling units in the sample
$= n_1 + n_2 + \cdots + n_k$

$\bar{x}_i$ = Mean of the sample for stratum i = $\dfrac{\sum\limits_{j=1}^{n_i} x_{ij}}{n_i}$

where x_{ij} is the jth measurement obtained from stratum i. Also,

s_i^2 = Sample variance for stratum i = $\dfrac{\sum\limits_{j=1}^{n_i} (x_{ij} - \bar{x}_i)^2}{n_i - 1}$

ESTIMATION OF THE POPULATION MEAN, μ: STRATIFIED RANDOM SAMPLING

Estimator of μ: $\bar{x}_{st} = \dfrac{1}{N}(N_1\bar{x}_1 + N_2\bar{x}_2 + \cdots + N_k\bar{x}_k)$

Estimated bound on the error of estimation: $2\hat{\sigma}_{\bar{x}_{st}} = 2\sqrt{\dfrac{1}{N^2}\sum\limits_{i=1}^{k} N_i^2 \left(\dfrac{N_i - n_i}{N_i}\right)\dfrac{s_i^2}{n_i}}$

Approximate 95% confidence interval: $\bar{x}_{st} \pm 2\hat{\sigma}_{\bar{x}_{st}}$

ESTIMATION OF THE POPULATION TOTAL, τ: STRATIFIED RANDOM SAMPLING

Estimator of τ: $\hat{\tau}_{st} = N\bar{x}_{st} = N_1\bar{x}_1 + N_2\bar{x}_2 + \cdots + N_k\bar{x}_k$

Estimated bound on the error of estimation: $2\hat{\sigma}_{\hat{\tau}_{st}} = 2\sqrt{\sum\limits_{i=1}^{k} N_i^2 \left(\dfrac{N_i - n_i}{N_i}\right)\dfrac{s_i^2}{n_i}}$

Approximate 95% confidence interval: $\hat{\tau} \pm 2\hat{\sigma}_{\hat{\tau}_{st}}$

ESTIMATION OF A POPULATION PROPORTION, p: STRATIFIED RANDOM SAMPLING

Estimator of p: $\hat{p}_{st} = \dfrac{1}{N}(N_1\hat{p}_1 + N_2\hat{p}_2 + \cdots + N_k\hat{p}_k)$

where $\hat{p}_i$ is the sample proportion for stratum i ($i = 1, 2, \ldots, k$). Also,

Estimated bound on the error of estimation: $2\hat{\sigma}_{\hat{p}_{st}} = 2\sqrt{\dfrac{1}{N^2}\sum_{i=1}^{k} N_i^2 \left(\dfrac{N_i - n_i}{N_i}\right)\dfrac{\hat{p}_i(1 - \hat{p}_i)}{n_i - 1}}$

Approximate 95% confidence interval: $\hat{p}_{st} \pm 2\hat{\sigma}_{\hat{p}_{st}}$

EXAMPLE 20.2

The seafood wholesaler described earlier selected random samples of $n_1 = n_2 = n_3 = 400$ households from within each of the three markets (strata) and obtained from each household an estimate of the dollar amount spent per month on seafood. The number of households in each market along with the sample means and variances are shown in the table.

NEIGHBORHOOD	N_i	$\bar{x}_i$	s_i^2
1	20,800	$5.31	16.83
2	6,400	$9.49	15.10
3	12,600	$6.75	23.78
	$N = 39,800$		

a. Estimate the total amount, τ, spent per month on seafood in the city.
b. Place bounds on the error of estimation.

SOLUTION

a. Substituting the values of N_i and $\bar{x}_i$ into the formula for $\hat{\tau}$, we obtain

$$\hat{\tau} = N\bar{x}_{st} = N_1\bar{x}_1 + N_2\bar{x}_2 + N_3\bar{x}_3$$
$$= (20,800)(5.31) + (6,400)(9.49) + (12,600)(6.75)$$
$$= \$256,234$$

b. The bound on the error of estimation is

$$2\hat{\sigma}_{\hat{\tau}} = 2\sqrt{\sum_{i=1}^{k} N_i^2 \left(\frac{N_i - n_i}{N_i}\right)\frac{s_i^2}{n_i}}$$

$$= 2\sqrt{(20,800)^2\left(\frac{20,800 - 400}{20,800}\right)\left(\frac{16.83}{400}\right) + (6,400)^2\left(\frac{6,400 - 400}{6,400}\right)\left(\frac{15.10}{400}\right) + (12,600)^2\left(\frac{12,600 - 400}{12,600}\right)\left(\frac{23.78}{400}\right)}$$

$$= \$10,666$$

Thus, we estimate the total monthly expenditure for seafood in the city (for the month sampled) to be $256,234, and an approximate 95% confidence interval is $256,234 \pm \$10,666$, or $245,568 to $266,900. ∎

EXAMPLE 20.3

Refer to Example 20.2, and estimate the mean monthly expenditure for seafood per household in neighborhood 2.

SOLUTION

Estimates of the mean expenditure per month per household for seafood for the three neighborhoods might play an important role in deciding how to allocate sales effort and in deciding where to locate retail markets. An estimate of the mean monthly expenditure per household for neighborhood 2 is

$$\bar{x}_2 = \$9.49$$

The estimated bound on the error of estimation is

$$2\hat{\sigma}_{\bar{x}_2} = 2\frac{s_2}{\sqrt{n_2}}\sqrt{\frac{N_2 - n_2}{N_2}} = 2\frac{\sqrt{15.10}}{\sqrt{400}}\sqrt{\frac{6,400 - 400}{6,400}} = \$.38$$

Thus, we estimate the mean monthly expenditure per household in neighborhood 2 for the sampled month to be $9.49. We are reasonably certain that the true mean monthly expenditure per household in neighborhood 2 is between $9.11 and $9.87.

∎

Examples 20.2 and 20.3 illustrate the methods for estimating parameters based on the stratified random sampling of $n_1, n_2, \ldots, n_k$ sampling units from the k strata. Without being specific, we know that the standard errors of the estimators will decrease as the total sample size, $n = n_1 + n_2 + \cdots + n_k$ increases, but we have not commented on the relative magnitudes of $n_1, n_2, \ldots, n_k$. As a general rule, we select larger samples from strata with greater variability. More precise determination of the sample size requires numerical estimates of the strata variances. Also, the cost of sampling for each stratum will usually play a role in determining strata sample sizes because the total cost of sampling must be kept within the budget for the project. An example of sample size determination is given in Section 20.8.

**EXERCISES
20.20–20.28**

LEARNING THE MECHANICS

20.20 A survey based on a stratified random sample produced the data shown in the table.

STRATUM	STRATUM SIZE	MEASUREMENTS
1	25,000	40, 70, 85, 63, 75, 82, 56, 49, 85, 98, 79, 90, 96, 88, 72, 66, 71, 79, 90, 79
2	10,000	26, 55, 42, 47, 58, 51, 62, 55, 45, 49, 65, 72, 33, 55, 61
3	5,000	10, 32, 30, 21, 40, 19, 23, 36, 30, 27

a. Find k, N_1, N_2, N_3, and N.
b. Find n_1, n_2, n_3, and n.
c. Find $\bar{x}_1$, $\bar{x}_2$, and $\bar{x}_3$.
d. Find s_1^2, s_2^2, and s_3^2.

e. Estimate the population mean and place a bound on the error of estimation.
f. Estimate the population total and place a bound on the error of estimation.
g. Estimate the proportion of measurements in the population that are over 50 and place a bound on the error of estimation.

20.21 A survey based on a stratified random sample produced the data shown in the table.

STRATUM	NUMBER OF SAMPLING UNITS IN STRATUM	MEASUREMENTS
1	4,000	10, 15, 5, 30, 25, 26, 38, 50, 10, 28
2	6,000	5, 33, 15, 45, 47, 36, 25, 40, 17, 31, 62, 28, 33, 45, 68
3	10,000	28, 75, 62, 43, 31, 48, 35, 26, 5, 81, 66, 18, 33, 38, 40, 45, 46, 18, 62, 40
4	15,000	45, 43, 15, 78, 92, 105, 38, 45, 49, 10, 36, 48, 17, 82, 76, 51, 39, 46, 40, 52, 88, 20, 40, 41, 50

a. Estimate the population mean, μ, and place a bound on the error of estimation.
b. Estimate the population total, τ, and place a bound on the error of estimation.
c. Estimate the proportion of the measurements in the population, p, that are between 35 and 55, inclusive, and place a bound on the error of estimation.

20.22 Refer to Exercise 20.21. Estimate the mean of stratum 4 and place a bound on the estimation error.

APPLYING THE CONCEPTS

20.23 How much does corporate America spend on employee training? This and many other training-related questions were addressed by *Training* magazine in its 1986 survey of

ORGANIZATION SIZE (Number of employees)	NUMBER OF ORGANIZATIONS IN STRATUM	NUMBER OF USABLE RESPONSES RECEIVED	MEAN OUTSIDE TRAINING EXPENDITURE PER ORGANIZATION (Thousands of $)	STANDARD DEVIATION OF OUTSIDE TRAINING EXPENDITURES PER ORGANIZATION (Thousands of $)[a]
50–99	114,464	87	11.7	2.0
100–499	91,754	444	26.6	4.3
500–999	11,011	357	42.3	7.1
1,000–2,499	7,340	553	89.6	10.8
2,500–9,999	3,670	575	142.5	15.3
10,000 or more	1,147	534	604.5	100.5
Total	229,386	2,550		

[a]Not reported by Gordon. The data in this column are fictitious.
Source: Gordon, J. "*Training* Magazine's Industry Report 1986," *Training*, Vol. 23, No. 10 (Oct. 1986), 26–42. Reprinted with permission from *Training*, The Magazine of Human Resources Development. Copyright 1986, Lakewood Publications, Inc., Minneapolis, Minnesota. All rights reserved.

U.S. organizations. A stratified sample of 15,210 companies was drawn from Dun and Bradstreet's directory of U.S. businesses. The strata and the number of usable responses received by *Training* magazine are described in the accompanying table. The table also presents data on the outside training expenditures (i.e., expenditures for seminars, conferences, audiovisual equipment, computer courseware, books, films, etc.) budgeted for 1986 by the survey respondents.

a. The use of these data to estimate the mean or the total budgeted outside training expenditure for the population of firms with 50 or more employees could yield a biased estimate. Explain.

b. Use an approximate 95% confidence interval to estimate the mean outside training expenditure for the population.

c. Use an approximate 95% confidence interval to estimate the total outside training expenditure for the population.

d. What assumptions would have to hold in order for your inferences of parts **b** and **c** to be valid?

e. Would the estimated bound on the error of estimation in part **b** change substantially if the finite population correction factor were ignored? Explain.

20.24 In 1980, the U.S. Department of Labor classified approximately 185,000 persons as health care administrators and estimated that another 105,000 positions would be created during the 1980's. Health care administrators include hospital administrators, managers of nursing homes, and managers of health maintenance organizations (Wright, 1982). In order to estimate the mean 1980 income of the head administrators of the 6,965 hospitals in the United States, a labor economist used stratified random sampling to select 30 administrators to be questioned about their incomes. The population was stratified according to the number of beds in each administrator's hospital. The results of the survey are shown here.

INCOME ($ THOUSAND)		
Under 100 beds ($N_1 = 3,210$)	100–299 beds ($N_2 = 2,015$)	300 beds and over ($N_3 = 1,740$)
32.0	39.2	69.2
39.1	55.4	65.0
35.6	51.6	58.9
36.2	48.0	49.3
38.7	37.5	70.5
	48.9	60.0
	46.1	54.8
	44.6	68.8
	45.2	57.3
	27.3	71.1
		68.1
		62.4
		45.0
		56.7
		59.5

Source: Salary data based on Wright, J. W., *The American Almanac of Jobs and Salaries.* New York: Avon Books, 1982. p. 608.

a. Find a point estimate for the mean 1980 income of hospital administrators.

b. Place bounds on the error of estimation associated with your point estimate in part a, and interpret the bounds in the context of the problem.

c. Find an approximate 95% confidence interval for the mean income of administrators of hospitals with 300 or more beds.

d. Examine the sample data and suggest a reason why the labor economist chose to allocate the sample size unevenly across the strata.

20.25 Since 1978, Internal Revenue Service (IRS) agents have been using sampling procedures to facilitate the auditing of tax returns of individuals and businesses. For example, sampling is used to estimate the total value of the error associated with a particular account balance reported on a tax return. Generally, agents use 95% confidence intervals to estimate such quantities. If substantial error is found, adjustments to the tax return will be suggested. For further details, see Brown (1982) and Hull and Everett (1982). In auditing the investment credit of a particular corporation, the IRS stratified the firm's population of 1,000 invoices containing the appropriate investment credit information into four strata according to the size of the expenditure involved: $0 to under $1,000, $1,000 to under $3,000, $3,000 to under $10,000, and $10,000 and over. Random samples of invoices of sizes $n_1 = 6$, $n_2 = 8$, $n_3 = 10$, and $n_4 = 15$ were drawn from each of the respective strata. Each sampled invoice was examined to determine whether it was properly treated by the firm in determining the firm's investment credit. The table describes the error associated with each sampled invoice as identified by the IRS. Positive errors reflect an overstatement by the firm of its investment credit and negative errors reflect an understatement.

INVESTMENT CREDIT ERRORS			
$0 to under $1,000 ($N_1 = 100$)	$1,000 to under $3,000 ($N_2 = 400$)	$3,000 to under $10,000 ($N_3 = 300$)	$10,000 and over ($N_4 = 200$)
$ 10	$ 0	$ 750	$ 0
0	0	0	0
0	100	0	5,000
−15	0	1,000	0
25	−50	0	0
20	0	0	0
	550	0	1,800
	0	1,500	0
		0	0
		2,000	0
			0
			0
			0
			0
			500

a. Find a point estimate for the total value of the error in the investment credit claimed by the firm.

b. Place bounds on the error of estimation associated with your point estimate of part a, and interpret the bounds in the context of the problem.

c. Find an approximate 95% confidence interval for the total value of the error.

d. The firm claimed an investment credit of $500,000. Based on your answers to parts a–c, approximately how much investment credit should the firm have claimed? Explain.

20.26 An economist wants to estimate the mean annual income of families in a mainly industrial community. Since one section of the city houses primarily factory workers, one mostly company executives, and the remaining area mostly farmers, the economist decides to use the three relatively homogeneous areas as strata. The economist selected random samples of 30 homes from within each of the three strata and gathered information on the annual income for each family. The number of households in each section of the city along with the sample means and variances are given in the table. Estimate the mean annual income for households in the community, and place a bound on the error of estimation.

CITY SECTION	N_i	$\bar{x}_i$ ($)	s_i^2
Factory workers	360	14,900	9,150,500
Executives	74	39,250	25,003,000
Farmers	95	23,800	16,801,100

20.27 The owners of a chain of department stores wish to estimate the proportion of customer accounts for which payments are 6 or more weeks overdue. A random sample of customers is taken in each of the chain's four stores, and the sample proportion of overdue accounts in each store is determined. These data and the total number of customer accounts are given in the table. Using the stores as strata, give an estimate of the true proportion of customer accounts that are 6 or more weeks overdue in this chain of department stores. Place a bound on the error of estimation.

STORE	N_i	n_i	$\hat{p}$, sample proportion of overdue accounts
1	1,572	100	.28
2	2,369	100	.31
3	3,007	120	.35
4	2,981	120	.10

20.28 Suppose you want to estimate the total amount of money spent on textbooks each quarter by students at your university. In order to reduce the variability in the data, you decide to consider student classes (freshman, sophomore, junior, senior) as strata. You randomly sample 50 students in each class and obtain an estimate of the total amount spent on textbooks during the quarter for each student. From the information given in the table, construct an approximate 95% confidence interval for the population total amount spent on textbooks per quarter by students at your university.

CLASS	N_i, total number of students	$\bar{x}_i$, average amount spent on textbooks	s_i^2, variance
Freshman	4,085	$75.20	62.50
Sophomore	3,520	62.00	86.50
Junior	5,525	45.15	31.40
Senior	5,070	42.85	39.70

CLUSTER SAMPLING

As explained in Section 20.2, **cluster sampling** involves the random selection of clusters of elements. Each sampling unit listed in the frame is a cluster of elements, and *all* the elements in a selected cluster are included in the sample.

Cluster sampling is often less costly than simple random sampling because it may be easier to construct a frame of clusters than a frame of the individual elements in a population. Second, it is frequently less costly when the elements within a cluster are geographically close to one another. This makes it easier and less costly for a pollster to obtain a response from each element. Sampling households rather than individual people is a good example of cluster sampling. If you are seeking the preferences, opinions, or buying habits of adult consumers in a city, it is easier to construct a frame of households than of individuals because all houses and apartment buildings would be listed at the local tax assessor's office. In contrast, no listing of the names of all adults in the city may be available. Also, it is less costly to interview two adults within one household than to travel and interview two persons selected at random from within the city.

The notation and the formulas for estimations based on cluster sampling are shown in the accompanying boxes.

NOTATION FOR CLUSTER SAMPLING

N = Number of clusters in the population

n = Number of clusters selected in a random sample

m_i = Number of elements in cluster i, $i = 1, 2, \ldots, n$

M = Number of elements in the population = $\sum\limits_{i=1}^{N} m_i$

$\overline{m}$ = Average cluster size for the sample = $\dfrac{\sum\limits_{i=1}^{n} m_i}{n}$

$\overline{M}$ = Average cluster size for the population = $\dfrac{M}{N}$

x_i = Total of all observations in cluster i

EXAMPLE 20.4

A heavy equipment manufacturer wished to estimate the mean cost of maintenance and repair for a new model of bulldozer sold last year. Although the manufacturer can locate the construction companies that have purchased the bulldozers, it is unlikely that individual maintenance records are kept for each and every machine. Consequently, it is easier to construct a frame of construction companies and to treat the collection of bulldozers within each company as a cluster. The total number of bulldozers sold last year was $M = 1,804$. Twenty construction companies were randomly selected from a set of $N = 279$ that purchased the new bulldozer last year. A listing of the number of bulldozers purchased by each construction company

ESTIMATION OF THE POPULATION MEAN, μ: CLUSTER SAMPLING

Estimator of μ: $\quad \bar{x} = \dfrac{\sum\limits_{i=1}^{n} x_i}{\sum\limits_{i=1}^{n} m_i}$

Estimated bound on the error of estimation: $\quad 2\hat{\sigma}_{\bar{x}} = 2\sqrt{\left(\dfrac{N-n}{Nn\overline{M}^2}\right) \dfrac{\sum(x_i - \bar{x}m_i)^2}{n-1}}$

where

$$\sum (x_i - \bar{x}m_i)^2 = \sum x_i^2 - 2\bar{x}\sum x_i m_i + \bar{x}^2 \sum m_i^2$$

If $\overline{M}$ is unknown, use $\overline{m}$ to approximate its value.

Approximate 95% confidence interval: $\quad \bar{x} \pm 2\hat{\sigma}_{\bar{x}}$

ESTIMATION OF THE POPULATION TOTAL, τ: CLUSTER SAMPLING

Estimator of τ: $\quad \hat{\tau} = M\bar{x} = M\left(\dfrac{\sum x_i}{\sum m_i}\right)$

Estimated bound on the error of estimation: $\quad 2\hat{\sigma}_{\hat{\tau}} = 2\sqrt{N^2\left(\dfrac{N-n}{Nn}\right) \dfrac{\sum(x_i - \bar{x}m_i)^2}{n-1}}$

where

$$\sum (x_i - \bar{x}m_i)^2 = \sum x_i^2 - 2\bar{x}\sum x_i m_i + \bar{x}^2 \sum m_i^2$$

Approximate 95% confidence interval: $\quad M\bar{x} \pm 2\hat{\sigma}_{\hat{\tau}}$

ESTIMATION OF A POPULATION PROPORTION, p: CLUSTER SAMPLING

Estimator of p: $\quad \hat{p} = \dfrac{\sum a_i}{\sum m_i}$

where a_i is the number of elements in cluster i that possess the characteristic of interest, $i = 1, 2, \ldots, n$.

Estimated bound on the error of estimation: $\quad 2\hat{\sigma}_{\hat{p}} = 2\sqrt{\left(\dfrac{N-n}{Nn\overline{M}^2}\right) \dfrac{\sum(a_i - \hat{p}m_i)^2}{n-1}}$

where

$$\sum (a_i - \hat{p}m_i)^2 = \sum a_i^2 - 2\hat{p}\sum a_i m_i + \hat{p}^2 \sum m_i^2$$

If $\overline{M}$ is unknown, use $\overline{m}$ to approximate its value.

Approximate 95% confidence interval: $\quad \hat{p} \pm 2\hat{\sigma}_{\hat{p}}$

along with the total cost of annual repair and maintenance (R&M) for the purchased bulldozers is shown in the table. Estimate the mean cost for the year per bulldozer for repair and maintenance.

COMPANY	NUMBER OF BULLDOZERS	R&M COST ($)	COMPANY	NUMBER OF BULLDOZERS	R&M COST ($)
1	3	1,270	11	2	494
2	10	5,860	12	6	1,980
3	6	4,310	13	10	7,740
4	10	7,940	14	3	1,144
5	2	500	15	15	12,130
6	3	968	16	3	1,770
7	4	1,490	17	4	1,052
8	3	2,710	18	4	2,617
9	2	390	19	12	3,985
10	5	1,785	20	6	2,463

SOLUTION

We are given

N = Number of construction companies (clusters) in the population = 279

n = Number of construction companies in the sample = 20

M = Total number of bulldozers in the population = 1,804

Then the average cluster size for the population is

$$\overline{M} = \frac{M}{N} = \frac{1,804}{279} = 6.47$$

We need to calculate

$$\sum m_i = 3 + 10 + 6 + \cdots + 6 = 113$$
$$\sum x_i = 1,270 + 5,860 + \cdots + 2,463 = 62,598$$
$$\sum m_i^2 = (3)^2 + (10)^2 + \cdots + (6)^2 = 907$$
$$\sum x_i^2 = (1,270)^2 + (5,860)^2 + \cdots + (2,463)^2 = 377,214,308$$
$$\sum x_i m_i = (1,270)(3) + (5,860)(60) + \cdots + (2,463)(6) = 553,603$$

Then, the estimated mean expenditure per bulldozer per year is

$$\bar{x} = \frac{\sum x_i}{\sum m_i} = \frac{62,598}{113} = 553.96$$

To calculate the estimated bound on the error of estimation, we first need to calculate

$$\sum (x_i - \bar{x}m_i)^2 = \sum x_i^2 - 2\bar{x}\sum x_i m_i + \bar{x}^2 \sum m_i^2$$
$$= 377,214,308 - 613,347,835.8 + 278,332,615.2$$
$$= 42,199,087.4$$

Then,

$$2\hat{\sigma}_{\bar{x}} = 2\sqrt{\left(\frac{N-n}{Nn\overline{M}^2}\right)\frac{\sum(x_i - \bar{x}m_i)^2}{n-1}}$$

$$= 2\sqrt{\frac{279-20}{(279)(20)(6.47)^2}\left(\frac{42,199,087.4}{20-1}\right)}$$

$$= 2\sqrt{2,462.7} = 99.25$$

Therefore, we estimate the mean annual expenditure per bulldozer to be $553.96, and an approximate 95% confidence interval for the true mean annual expenditure per bulldozer is 553.96 ± 99.25, or $454.71 to $653.21.

Thus, the salespeople for the heavy equipment manufacturer can conservatively advertise the mean annual cost of maintenance and repairs to be less than $653.21, the upper limit on the estimate of μ. ∎

EXERCISES
20.29–20.38

LEARNING THE MECHANICS

20.29 A population consists of 5,000 elements divided into 300 clusters. Four clusters were randomly selected and every element in each of the four clusters was measured. The resulting data are shown in the table.

CLUSTER	MEASUREMENTS
1	6, 10, 5, 4, 11, 9, 6, 11
2	21, 15, 9, 12, 18, 17, 17, 23, 29
3	30, 28, 36, 20, 31, 29, 30, 16, 20, 40, 31, 10
4	2, 40, 29, 18, 15, 11

a. Find M, N, n, and m_i $(i = 1, 2, \ldots, n)$.
b. Find $\bar{m}$ and $\overline{M}$.
c. Find x_i $(i = 1, 2, \ldots, n)$.
d. Find $\hat{\sigma}_{\bar{x}}$.
e. Estimate μ and place a bound on the error of estimation.
f. Find $\hat{\sigma}_{\hat{\tau}}$.
g. Estimate τ and place a bound on the error of estimation.

20.30 The table shows the results of a sample survey based on cluster sampling with $N = 400$, $n = 10$, and $M = 1,240$.

					CLUSTER					
	1	2	3	4	5	6	7	8	9	10
m_i	2	2	2	3	3	3	4	4	1	5
x_i	6.4	6.7	5.8	17.2	19.4	18.7	24.0	25.8	6.2	32.3

a. Find an approximate 95% confidence interval for the population mean, μ.
b. Find an approximate 95% confidence interval for the population total, τ.

20.31 A population of 2,000 elements was divided into 200 clusters. Six clusters were randomly selected, and every element in each of the six clusters was measured. The data that resulted are shown in the table.

CLUSTER	MEASUREMENTS
1	5, 18, 33, 22, 19, 18, 20, 8, 9, 25
2	2, 25, 16, 38, 11, 14, 17
3	22, 18, 33, 50, 35, 40
4	28, 24, 19, 31, 27, 26, 26, 24, 24, 26
5	13, 21, 19, 15, 26, 41, 12
6	41, 28, 30, 15, 31, 29, 29, 40, 21, 25, 18, 50

a. Estimate the population mean, μ, and place a bound on the error of estimation.
b. Estimate the population total, τ, and place a bound on the error of estimation.
c. Estimate the proportion of measurements in the population that are less than 30, and place a bound on the error of estimation.

APPLYING THE CONCEPTS

20.32 Over the last three decades the banking structure in the United States has undergone major changes due to the proliferation of bank holding companies (BHCs). By 1980, deposits of holding-company-affiliated banks accounted for more than two-thirds of the total deposits in commercial banks. Research has indicated, however, that the profitability of an individual bank does not improve after acquisition by a BHC. But what about the overall profitability of the BHC? Larry Frieder and Vincent Apilado (1983) addressed this question. Four BHCs were sampled and all affiliated banks in each BHC were examined. For each bank, it was determined whether the bank had a favorable (F) or an unfavorable (U) effect on the profitability of its BHC over the 3-year period after the bank was acquired. The resulting data are shown in the table.

BHC	IMPACT ON PROFITABILITY BY AFFILIATED BANKS
1	U, F, F, F, F, U, F, F, F, F, F, U, F, F
2	F, F, U, F, U, F, F, F, U, F, F, U, F, F, F, F, F, F, F, F
3	U, F, F, F, F, F, U, F, F, F, U, F, F, F, F, U, F, F, U, F
4	F, F, F, U, F, F, F, U, F, F, F, F, F, U, F, F, F, F, F, U, F, U, F, F, F, F, U, F, F, F, U, F, F, F, F

Source: Adapted from Frieder, L. A. and Apilado, V. P. "Bank holding company expansion: A refocus on its financial rationale." *The Journal of Financial Research*, Vol. 6, No. 1 (Spring 1983), 67–81.

a. Use an approximate 95% confidence interval to estimate the proportion of affiliated banks whose impact on the profitability of their BHC is favorable. Assume that the frame of BHCs from which Frieder and Apilado chose their sample contained 200 BHCs.
b. What assumption (other than the assumption made in part **a**) must hold in order for your inference of part **a** to be appropriate?

20.33 An important figure in the financial statements of a business is the value of its inventory. Inventory value may be critical in determining both the profit and net worth of the business. Businesses typically conduct physical inventories once a year. As a result, it is not practical to maintain a year-round staff of trained inventory-takers. Instead, for a day or two each year, workers from various areas in the firm are enlisted to conduct the inventory. Because of the job's boring nature and the workers' lack of knowledge about the items in the firm's inventory, errors in item identification, counts, and pricing frequently occur, particularly when workers are asked to conduct *100% physical inventories*. Since such errors nearly always occur, even 100% physical inventories yield only estimates of inventory value, not the actual value. Furthermore, since the extent of such errors is not known, it is not possible to determine the accuracy of the estimate. These problems may be alleviated to a great extent by estimating inventory value by statistical sampling rather than by a 100% physical inventory. Since fewer workers would be required to complete the inventory, it may be possible to avoid using unqualified workers. In addition, the smaller number of items to be processed would mean less opportunity for human error to affect the estimate. Further, by using statistical sampling, it is possible to obtain a bound on the difference between the actual inventory value and the estimated inventory value. A cluster sampling procedure that has been used in large factories involves dividing the factory floor into small zones—clusters— and assigning a number to each zone. Then all items (parts, tools, machinery, etc.) in each of a randomly selected set of areas are counted and valued (Arkin, 1982). The results of such a process are described in the table.

FACTORY FLOOR ZONE NUMBER	NUMBER OF ITEMS	TOTAL VALUE ($ thousand)
98	25	1.2
33	101	4.0
76	83	1.8
59	455	6.7
81	90	.5
3	22	.2
21	66	2.5
62	14	.1
42	299	5.0
17	387	10.2
6	46	1.1
70	33	.9

Note: The total number of zones is 100 and the total number of items on the factory floor is 12,968.

a. Estimate the total value of the inventory on the factory floor, and place a bound on the error of estimation. [*Note:* In practice, the number of items on the factory floor would probably not be known. We address this problem in Exercise 20.59.]

b. In the context of the problem, interpret the bound you obtained in part **a**.

c. A 100% physical inventory of the factory floor resulted in an estimate of $280,000 for the total inventory value. Does this seem reasonable given your result for part **a**? Explain.

20.34 Refer to Exercise 20.33. Estimate the mean dollar value per item on the factory floor, and place a bound on the error of estimation.

20.35 Managers are frequently faced with the need to change the behavior of the people within the organizations. One approach to effect such a change is to hire new people who bring new skills and interests to the organization or to retrain existing personnel. Another approach is to change the management system of the organization by implementing a program such as *management by objectives* (MBO). MBO is a technique "designed to create and maintain a routine of goal-setting at all levels of management, in order to facilitate motivation, evaluation, and decision-making" (Reitz, 1981, p. 148.) MBO is characterized by goal-setting that starts at the top level of the organization and moves downward. Each level sets its goals consistent with the level immediately above. An essential feature of MBO is frequent performance review (feedback). Managers and nonmanagers alike are evaluated with respect to the specific goals they set for themselves rather than against the performance of other managers or employees. The personnel director of a large corporation that recently implemented an MBO program has been charged with the task of estimating the proportion of the corporation's employees who believe MBO is an improvement over their previous management style. The corporation is physically located in 70 different plants and branch offices worldwide. Since the corporation's personnel records are decentralized, the construction of a frame for simple random sampling is not feasible given the deadline for completion of the study. Accordingly, it has been decided to use cluster sampling with each different location as a cluster. Ten locations were randomly selected, and all employees at each location were mailed a questionnaire asking for their opinions of the MBO program. A 100% response rate was obtained at each location. The results of the survey are shown in the table. Estimate the proportion of employees who prefer MBO to the previous management style, and place a bound on the error of estimation.

LOCATION	NUMBER OF EMPLOYEES	NUMBER OF EMPLOYEES PREFERRING MBO
Hartford, Connecticut	550	401
London, England	163	80
Dallas, Texas	780	580
San Jose, California	333	300
Toronto, Canada	495	395
Ann Arbor, Michigan	87	80
Osaka, Japan	212	210
New York, New York	1,001	683
Phoenix, Arizona	199	172
Dover, New Jersey	47	38

20.36 Before implementing a plan designed to reduce flight time and hence conserve fuel and energy, the Air Force needs an estimate of the total number of miles flown by a certain type of aircraft during a given month. Air Force records show that a total of 1,500 planes of this type in the fleet are harbored at 96 different airfields across the country.

The Air Force randomly selects six of the airfields and monitors the flight mileage of each plane at each airfield for 1 month. The data are shown in the table.

AIRFIELD	NUMBER OF AIRPLANES	MILES FLOWN (thousands)
1	20	36
2	10	25
3	18	16
4	18	24
5	10	15
6	16	20

a. Treat the collection of all aircraft of this type at each airfield as a cluster. Estimate the total number of miles flown per month, and place a bound on the error of estimation.

b. What are the advantages or disadvantages of this method of sampling compared to taking a simple random sample from the total of 1,500 airplanes in the fleet?

20.37 A city is divided into 180 small neighborhoods (clusters) for the purpose of estimating the average utility bill of households during the month of July. Ten neighborhoods are selected at random, and the July utility bills of every household in each of the ten neighborhoods are recorded. The data are summarized in the table. Obtain an estimate for the average July utility bill of all households in the city, and place a bound on the error of estimation.

NEIGHBORHOOD	NUMBER OF HOUSEHOLDS	TOTAL OF JULY UTILITY BILLS
1	20	$530
2	18	486
3	20	704
4	25	775
5	30	861
6	18	666
7	30	960
8	19	494
9	21	823
10	19	665

20.38 A real estate appraiser wants to estimate the proportion of adult males who own their homes in a section of a city. This information will then be used as part of an appraisal of the retail value of the homes. The area is divided into 200 clusters, each containing five city blocks. Eight clusters are selected at random, and information on the number of adult males who live in the area and who also own their homes is collected. Given the data in the table at the top of page 1164, estimate the proportion of adult males who own their homes in this city section with an approximate 95% confidence interval.

CLUSTER	NUMBER OF ADULT MALES	NUMBER WHO OWN HOMES
1	98	66
2	92	74
3	106	65
4	80	51
5	90	30
6	29	19
7	60	45
8	63	40

| | | | | | | | | | | | | |

SECTION 20.8

DETERMINING THE SAMPLE SIZE

To determine the sample size for sample surveys, you use essentially the same procedures as explained in Sections 8.2 and 8.8. Generally speaking, the bound on the error of estimation will be approximately inversely proportional to the number of sampling units. This relationship does not hold exactly because of the effect of the finite population correction factor and because, for stratified random sampling, you are really selecting k random samples, one corresponding to each stratum. Nevertheless, doubling the sample size, n, will decrease (even for stratified random sampling if you keep $n_1, n_2, \ldots, n_k$ in the same proportions), approximately, the bound on the error of estimation to $1/\sqrt{2}$ times its original value. If you quadruple the sample size n, you will cut the bound on the error of estimation in half.

To select the sample size to estimate a population parameter based on a specific sample survey design, first decide on the accuracy you desire in your estimate; that is, decide on the bound on the error of estimation that you are willing to tolerate. Set this number equal to the estimated bound on the error of estimation and solve the resulting equation for n. We will illustrate with an example.

EXAMPLE 20.5

Suppose the seafood wholesaler in Example 20.2 wanted to reduce the bound on the error of estimating the total, τ, of monthly seafood expenditure in the city to $5,000. That is, the wholesaler wants to estimate the total monthly expenditure to within $5,000 with approximate 95% confidence. If the wholesaler plans to use equal sample sizes, approximately how many households must be selected from within each stratum to estimate τ with a bound on the error of estimation of $5,000?

SOLUTION

We found in Example 20.2 that the bound on the error of estimation for $n_1 = n_2 = n_3 = 400$ was $10,666. Since $5,000 is slightly less than half of this value, we know (without calculating) that it will require approximately four times as many households to reduce the bound on the error of estimation to $1/\sqrt{4} = \frac{1}{2}$ its original size.

To solve the problem formally, let

$$2\hat{\sigma}_{\hat{\tau}} = 2\sqrt{\sum_{i=1}^{k} N_i^2 \left(\frac{N_i - n_i}{N_i}\right) \frac{s_i^2}{n_i}} = \$5,000$$

where (since we want equal sample sizes) we will let $n_1 = n_2 = n_3 = n_s$. We will also assume, for a first approximation to n_s, that $(N_i - n_i)/N_i \approx 1$. Substituting these values, along with $N_1 = 20{,}800$, $N_2 = 6{,}400$, $N_3 = 12{,}600$, and the sample variances from the earlier samples into the formula for the bound on the error of estimation yields

$$2\sqrt{\frac{(20{,}800)^2(16.83)}{n_s} + \frac{(6{,}400)^2(15.10)}{n_s} + \frac{(12{,}600)^2(23.78)}{n_s}} = 5{,}000$$

and solving for n_s yields $n_s = 1{,}868$.

This solution will be larger than the actual sample sizes required to achieve a $\$5{,}000$ bound on the error of estimation because the finite population correction factors will not equal 1. We now substitute $n_s = 1{,}868$ into the equation and re-solve it for n_s:

$$2\sqrt{(20{,}800)^2\left(\frac{20{,}800 - 1{,}868}{20{,}800}\right)\left(\frac{16.83}{n_s}\right) + (6{,}400)^2\left(\frac{6{,}400 - 1{,}868}{6{,}400}\right)\left(\frac{15.10}{n_s}\right) + (12{,}600)^2\left(\frac{12{,}600 - 1{,}868}{12{,}600}\right)\left(\frac{23.78}{n_s}\right)}$$
$$= 5{,}000$$

or $n_s = 1{,}645$.

This solution will be too small because we used $n_s = 1{,}868$ in the finite population correction factor. If we use this new value to calculate the finite population correction factors and re-solve the equation for n_s, we obtain $n_s = 1{,}672$. Thus, we would select approximately 1,672 households from each stratum in order to reduce the bound on the error of estimation to $\$5{,}000$. ∎

Example 20.5 illustrates that although the finite population correction factor does affect the sample size required to obtain a specified bound on the error of estimation, the effect is not large. The solution, assuming the finite population correction factor equals 1, was 1,868—a value not much larger than the solution obtained using the correction factor.

Often previous sample data will not be available from which to obtain estimates of the strata variances. These must then be more crudely estimated, possibly by using each stratum range divided by 4, or by calculating s_i^2 from small pilot samples drawn from each stratum. Although this will produce only crude estimates of the required sample sizes, the procedure is still worthwhile, since the general magnitude of the sample size can be determined. This, in turn, allows the researcher to determine an approximate relationship between sampling costs and the bound on the error of estimation.

**EXERCISES
20.39–20.44**

APPLYING THE CONCEPTS

20.39 Refer to Exercise 20.13. In estimating the mean external audit fee, suppose Professor Wallace wants the estimated bound on the error of estimation to be no more than $\$200{,}000$. How many additional firms should she sample?

20.40 Refer to Exercise 20.14. Suppose the auditor wants to estimate the total value of the parts inventory to within $100,000 of the true value with approximately 95% confidence. In order to obtain this result, how many of the firm's 5,000 inventory items should be sampled?

20.41 Refer to Exercise 20.24. Suppose the labor economist wants to estimate the mean 1980 income of the head administrators to within $1,000 with approximately 95% confidence.

a. If an equal number of administrators is to be sampled from each stratum, what is the total number of administrators that should be sampled?

b. If thirty administrators are to be sampled from each of the first two strata, "under 100 beds" and "100–299 beds," approximately how many should be sampled from the third stratum?

20.42 Refer to Exercise 20.28. Suppose you want to estimate the total amount of money spent on textbooks per quarter within $7,500 of the true value. If equal sample sizes are used, approximately how many students must be selected from each class in order to estimate the total with a bound on the error of estimation of $7,500 (with approximately 95% confidence)?

20.43 Refer to Exercise 20.26. The economist now desires to reduce the bound on the error of estimation to $600. Assuming equal sample sizes, how many homes from each city section should be sampled so that the economist estimates the mean annual income for households in the community to within $600 of the true value (with approximately 95% confidence)?

20.44 Refer to Exercise 20.16. It is desired to estimate the proportion of the county's voters who favor the creation of urban enterprise job zones to within .01 of the true value with approximate 95% confidence. In order to obtain this accuracy, how many of the county's 50,840 eligible voters should be sampled?

S U M M A R Y

This chapter introduced the important topic of **survey sampling**. We presented several sampling designs for reducing the cost of conducting a sample survey and also presented the associated methods of estimation.

The objective of a sample survey is usually to estimate one or more of three population parameters: a **population mean**, μ, a **population total**, τ, and (or) a **population proportion**, p. We introduced three sample survey designs for collecting a sample: simple random sampling, stratified random sampling, and cluster sampling.

Simple random sampling is conceptually easy to understand, but it is often difficult to conduct, and it may be more costly than other sample survey designs. This is because it may be difficult and costly to construct the frame, and the cost of collecting the sample may be relatively large due to geographic separation of the sampling units.

Stratified random sampling is used when the population can be subdivided into groups (strata) of sampling units that possess smaller variability within strata than between strata. Random samples are selected from within each stratum, and the information contained in these samples is then pooled to obtain an estimate of the desired population parameter. Stratified sampling has the advantage that it enables the sampler to obtain estimates of the individual stratum parameters. In addition, it may be less costly than simple random sampling because the strata frames are often easier to construct and the sampling units are often geographically closer to one another in comparison to simple random sampling within the entire population. Finally, stratified samples often result in more precise estimates because the variance within strata is less than the variance of the entire population.

Cluster sampling is a design in which you select a random sample of clusters of elements from the population and then include in the sample every element in the cluster. Thus, each cluster of elements is a sampling unit. Cluster sampling is advantageous because it is often easier and less costly to construct a frame of clusters than to construct a frame of individual elements. We obtain the responses from more elements at lower cost, but the responses within a cluster may be highly correlated (for example, the opinions of people in the same household may be similar). Consequently, we must sample an adequate number of clusters if we wish to obtain a good estimate of the desired population parameter.

There are many other sampling designs available to a sample surveyor; some are variations on stratified random and cluster sampling, and other designs are completely different. In addition, different types of estimators can be used with these designs. In this introduction to survey sampling, our intent was to present only the basic elements of a sample survey and several of the most important sample survey designs. More thorough presentations are given in textbooks devoted to this topic (see the references at the end of this chapter).

SUPPLEMENTARY EXERCISES 20.45–20.59

20.45 Describe how cluster sampling could be used to estimate the damage to a citrus farmer's orange crop.

20.46 How do the makers of Kleenex facial tissue determine how many tissues to include in a box or personal pack? In the case of the personal pack, John Koten (1984) reports that the marketing experts at Kimberly-Clark Corporation have ". . . little doubt that the company should put 60 tissues in each pack." By having hundreds of customers keep count of their Kleenex use in diaries, researchers determined that 60 is the average number of times a person uses a tissue during a cold. Suppose a 1987 study of a random sample of 250 people from a midwestern town of 5,000 people yielded the following summary data on the number of times they used a tissue when they had a cold:

$$\bar{x} = 57 \qquad s = 26$$

a. Estimate the mean number of times people in the midwestern town use a tissue during a cold. Place a bound on the estimation error.

b. Assume each person in the town had one cold during 1987. Use an approximate 95% confidence interval to estimate the total number of tissues used in the town in 1987 during colds. How might such information be used by Kimberly-Clark?

20.47 When a buyer charges a purchase, the seller records the sale in his or her accounting records under a category called *accounts receivable*. At a particular point in time, a large business may have many thousands of accounts receivable. A department store in Boston is interested in estimating (1) the mean age of its accounts receivable, (2) the total monetary value of its accounts receivable, and (3) the proportion of accounts receivable that are more than 90 days old. On September 10, the store's accounting department randomly selected 100 of its 15,887 accounts receivable and recorded the age, x, and the value, y, of each account. Define in the context of the problem (in words, not formulas) each of the following:

a. n and N
b. $\bar{x}$ and μ
c. $\hat{\tau}$ and τ
d. $\hat{p}$ and p
e. $\hat{\tau} \pm 2\hat{\sigma}_{\hat{\tau}}$

20.48 Refer to Exercise 20.47. In addition to the Boston location, the department store chain has stores in three other cities. It was decided to estimate the characteristics of the accounts receivable for the entire chain. Each store was treated as a stratum and the stratified sample design shown in the table was used to sample accounts receivable.

STRATUM	NUMBER OF ACCOUNTS RECEIVABLE ON SEPTEMBER 10	SAMPLE SIZE
1. Boston	15,887	75
2. New York	25,010	100
3. Atlanta	10,982	50
4. Dallas	18,931	80

The information collected from each account and the parameters to be estimated are described in Exercise 20.47. Define in the context of the problem (in words, not formulas) each of the following:

a. N
b. n_2 and n_4
c. $\bar{x}_3$ and $\bar{x}_{st}$ and μ
d. s_1 and σ_1
e. $\hat{p}_2$ and $\hat{p}_{st}$ and p

20.49 In recent years, corporations, labor unions and trade and medical associations have set up organizations known as *political action committees*—commonly called PACs—to raise money from their members for support of political candidates believed to represent their best interests. Approximately 3,500 PACs currently exist. Federal law limits the amount a PAC can contribute to a presidential candidate to $5,000 per election. However, a loophole in the law permits committees to spend as much as they want to elect or defeat a candidate as long as there is no cooperation or contact with the candidate or the candidate's authorized agents ("ABC's of How America Chooses a President," U.S. *News and World Report*, Feb. 20, 1984, pp. 39–46). In early 1984, in order to estimate the amount of money that would be spent in support of Ronald Reagan in the 1984 presidential election, a political economist hired by the Democratic party randomly sampled 30 PACs and asked them how much they expected to spend in support of Reagan. The following results were obtained (in thousands of dollars):

10	0	5	18	0	5	22	0	50	60
35	0	0	18	35	0	40	10	50	20
150	15	0	0	30	15	0	20	15	10

a. Estimate the total amount that PACs expected to spend in support of Reagan in 1984. Place a bound on the error of estimation.
b. Use an approximate 95% confidence interval to estimate the proportion of PACs that planned to support Reagan.
c. In addition to sampling error, what might cause your estimate for part **a** to be inaccurate?

20.50 A company has a long-standing tradition of not charging its customers interest for late payment of bills. However, economic realities have forced the company to reconsider its payment policy. To guide them in establishing a new policy, the company has decided to estimate the mean number of days between the date an invoice is mailed to a customer and the date payment is received from the customer. Because of the large variance in the number of days until payment and the wide range of invoice amounts, the company has decided to use stratified sampling to estimate this mean. The population of invoices for which full payment was received during the first 6 months of the year was sampled and the data shown in the table were obtained.

INVOICE AMOUNT	NUMBER OF INVOICES PER STRATUM	NUMBER OF DAYS UNTIL PAYMENT
Under $100	20,000	60, 50, 95, 30, 15, 92, 100, 65, 48, 49, 63, 82, 91, 110, 68, 30, 40, 35, 38, 50
$100 to under $500	15,000	40, 28, 33, 42, 61, 48, 31, 25, 33, 48, 51, 26, 15, 35, 63
$500 to under $1,000	5,000	38, 53, 61, 35, 36, 41, 41, 43, 20, 21
$1,000 and over	2,000	28, 15, 10, 30, 32, 27, 33, 35, 34, 30

a. Estimate the mean number of days until payment, and place a bound on the error of estimation.
b. In the context of the problem, interpret the bound you obtained in part **a**.
c. Construct an approximate 95% confidence interval for the mean number of days until payment of invoices whose amounts are under $100.
d. Use an approximate 95% confidence interval to estimate the proportion of invoices that are paid in full within 60 days. Within 90 days.

20.51 A large manufacturing company is considering a new health insurance plan for its employees. Before putting the plan into effect, the company wants to estimate the proportion of its employees who favor the new health insurance proposal. The company's sampling scheme is to randomly select nine of its 45 manufacturing plants scattered throughout the country and to use each plant as a cluster. The employees at each of the nine plants are interviewed, and their opinions regarding the new health insurance plan are recorded. Use the data in the table at the top of page 1170 to obtain an estimate of the true proportion of this company's employees who favor the new health insurance plan. Place a bound on the error of estimation.

PLANT	NUMBER OF EMPLOYEES	NUMBER FAVORING THE NEW PLAN
1	112	98
2	75	65
3	83	71
4	154	123
5	108	97
6	68	61
7	102	92
8	85	90
9	83	65

20.52 Refer to Exercise 20.33. If the actual total value of the inventory on the factory floor differs from the value derived from a 100% physical inventory, is the difference due to sampling error or nonsampling error? Explain.

20.53 When a poll reports, for example, that 61% of the public supports a program of national health insurance, it usually also reports the sampling error. For example, a poll might report that the estimate is accurate to within plus or minus 3%. An essay in *Time* magazine ("How Not to Read the Polls," Apr. 28, 1980, pp. 72–73) points out:

> Readers consistently misinterpret the meaning of this 'warning label.' . . . [The sampling error warning] says nothing about errors that might be caused by a sloppily worded question or a biased one or a single question that evokes complex feelings. Example: 'Are you satisfied with your job?' Most important of all, warning labels about sampling error say nothing about whether or not the public is conflict-ridden or has given a subject much thought. This is the most serious source of opinion poll misinterpretation.

Carefully explain the difference between sampling error and nonsampling error, both in general and in the context of the above quote.

20.54 Publishers of a weekly nationwide business magazine believe that a large proportion of their Florida subscribers invest in the stock market. They would like to be able to use this information to persuade brokerage firms in Florida to advertise in their magazine. The publishers send each of the 500,000 subscribers in Florida a questionnaire about their (the subscribers') stock market investments. A total of 10,000 of the questionnaires are returned, and of these, 9,296 subscribers responded that they do currently have stock market investments.

a. Use this information to estimate the proportion of Florida subscribers who invest in the stock market, and place a bound on the error of estimation.

b. As a brokerage firm in Florida, would you consider the resulting estimate in part **a** to be reliable? Explain.

20.55 Refer to Exercise 20.54. The publishers of the magazine have decided to alter their sampling scheme. Instead of sending out questionnaires, they will personally interview a random selection of their Florida subscribers. In order to estimate, with approximate 95% confidence, the true proportion investing in the stock market with a bound on the error of estimation of .05, how many of the magazine's 500,000 Florida subscribers should be included in the sample?

20.56 With the recent crackdown by many states on drunk drivers, the accuracy of devices used by state and local police to measure blood alcohol levels has been called into question. One particular device known as the *breathalyzer* has been under attack by lawyers in several states. As a result, state officials in Minnesota tested all 185 breathalyzers currently in use and found eight to yield inaccurate readings. Although only 4.2% of the devices failed to operate properly, lawyers argued that even 1% was too high and that a better instrument is needed (Homan, 1982). Anxious to demonstrate the reliability of its product, suppose the manufacturer of the breathalyzer subsidized the testing of the device in a random sample of ten other states, and all breathalyzers in use in those states were tested. The total number of breathalyzers in use in the United States is 6,000. The results obtained are shown in the table.

STATE	NUMBER OF DEVICES IN USE	NUMBER OF DEFECTIVE DEVICES
Nevada	50	2
New York	375	15
Virginia	150	4
Washington	105	0
South Carolina	60	1
Utah	100	5
Kansas	138	5
Wisconsin	155	6
Ohio	200	10
Georgia	105	2

a. What type of sampling design was used in this study?

b. Use an approximate 95% confidence interval to estimate the proportion of defective breathalyzers in use in the United States.

c. Does your confidence interval suggest that the Minnesota result was atypical? Explain.

d. Does your confidence interval indicate that the proportion defective could be as low as 1%? Explain.

20.57 A manufacturer of typewriters wants to estimate the average monthly repair cost of the typewriters sold by the company to certain businesses. Although a list of repair costs for each machine sold is not available, the company was able to obtain a list of businesses, how many typewriters they have purchased from the manufacturer, and the total amount spent on typewriter repairs during the past month. Of the total of 60 businesses that have dealt with the manufacturer, 12 were included in the list. The

BUSINESS	NUMBER OF TYPEWRITERS	TOTAL REPAIR COST FOR THE PAST MONTH	BUSINESS	NUMBER OF TYPEWRITERS	TOTAL REPAIR COST FOR THE PAST MONTH
1	8	$ 45	7	3	$16
2	9	36	8	8	49
3	5	23	9	5	29
4	12	55	10	10	62
5	10	47	11	9	51
6	14	101	12	7	27

data are given in the table. Using each business as a cluster, construct an approximate 95% confidence interval for the true average monthly repair cost of the typewriters.

20.58 In its next advertising campaign, a tobacco company will use an estimate of the average number of cigarettes smoked per day by its employees. The company expects a difference in amounts smoked by men and women, so it has decided to stratify on sex. From the company's 535 male employees, 50 are selected, and of the 366 female employees, 40 are sampled. An estimate of the number of cigarettes smoked per day by each is recorded. The accompanying table gives the respective means and variances of the samples. Estimate the average number of cigarettes smoked per day by the company employees, and place bounds on the error of estimation.

	N_i	n_i	$\bar{x}_i$	s_i^2
Men	535	50	8.5	16.8
Women	366	40	5.2	21.2

20.59 In Exercise 20.33, since the total number of items on the factory floor, M, was known to be 12,968, the point estimator $M\bar{x}$ could be used to estimate the total value, τ, of the inventory on the floor. In practice, however, it is unlikely that M would be known. As a result, a different point estimator must be used to estimate τ. It turns out that $N\bar{x}_t$ is also an unbiased estimator of τ, where $\bar{x}_t = \sum_{i=1}^{n} x_i/n$ is the average cluster total and N is the number of clusters in the population. The estimated bound on the error of estimation associated with $N\bar{x}_t$ is the same as the bound associated with $\hat{\tau}$ (see Section 20.7), except that the term $\bar{x}m_i$ is replaced by $\bar{x}_t$.

a. Assume M is unknown in Exericse 20.33, and estimate the total value of inventory on the factory floor using the approach described here. Place a bound on the error of estimation.

b. Compare your point estimate and bound with the results you obtained in part **a** of Exercise 20.33.

USING THE COMPUTER...

Treat the 1,000 zip codes in the data set of Appendix C as if they were the population of all U.S. zip codes. The four census regions geographically stratify this population.

a. Use stratified random sampling with $n_1 = 30$, $n_2 = 20$, $n_3 = 35$, and $n_4 = 15$ to estimate the mean number of households per zip code in the United States. Place a bound on the error of estimation. Since your statistical software package probably does not have a command to compute an estimator from a stratified sample, you will need to either (1) write a short program to compute the estimator and its bound, or (2) use the computer to find $\bar{x}_i$ and s_i^2 ($i = 1, 2, 3, 4$) and then calculate $\bar{x}_{st}$ and $\hat{\sigma}_{\bar{x}_{st}}$ by hand.

b. Use the sample data from part **a** to construct a 95% confidence interval for the mean number of households per zip code in the Northwest census region.

REFERENCES

Arkin, H. *Sampling Methods for the Auditor*. New York: McGraw-Hill, 1982. Chapters 1, 3, and 4.

Armstrong, J. S. "Monetary incentives in mail surveys." *Public Opinion Quarterly*, 1975, 39, 111–116.

Berg, A., Klauda, P., and Feyder, S. "EDB banned as grain pesticide." *Minneapolis Star and Tribune*, Feb. 4, 1984, 1A.

Brown, D. B. "Statistical sampling in the IRS examination of large cases." *The Tax Executive*, Apr. 1982, 34, 175–179.

Bryson, M. C. "The *Literary Digest* poll: Making of a statistical myth." *American Statistician*, Nov. 1976.

Chagall, D. "Can you believe the ratings?" *TV Guide*, June 24, 1978, 3.

Cochran, W. G. *Sampling Techniques*, 2nd ed. New York: Wiley, 1953.

Cox, E. P. III. *Marketing Research, Information for Decision-making*. New York: Harper & Row, 1979. p. 7.

Davis, L. A. "Grasp behavior before picking target market, says McC and McC exec." *Marketing News*, Nov. 18, 1977, 6.

Ferber, R., Sheatsley, P., Turner, A., and Waksberg, J. *What Is a Survey?* Washington, D.C.: American Statistical Association, 1980.

Freedman, D., Pisani, R., and Purves, R. *Statistics*, New York: Norton, 1978. Chapter 19.

Glasser, G. J. and Metzger, G. D. "Random-digit dialing as a method of telephone sampling." *Journal of Marketing Research*, Feb. 1972, 9, 59–64.

Greenberg, B. G., Kuebler, R. T., Abernathy, J. R., and Horvitz, D. G. "Application of randomized response technique in obtaining quantitative data." *Journal of the American Statistical Association*, 1971, 66.

Hansen, M. H., Hurwitz, W. N., and Madow, W. G. *Sampling Survey Methods and Theory*. Vol. 1. New York: Wiley, 1953.

Homan, S. "Breathalyzer battle is uncorked in district court." *Minnesota Daily*, Oct. 14, 1982, 1.

Huff, D. *How to Lie with Statistics*. New York: Norton, 1954. p. 20.

Hull, R. P. and Everett, J. O. "On the use of statistical sampling in tax audits." *The Tax Executive*, Oct. 1982, 35, 51–54.

Kish, L. *Survey Sampling*. New York: Wiley, 1965.

Koten, J. "Why do hot dogs come in packs of 10 and buns in 8's or 12's?" *Wall Street Journal*, Sept. 21, 1984, p. 1.

Reitz, H. J. *Behavior in Organizations*. Rev. ed., Homewood, Ill.: Richard D. Irwin, 1981. Chapters 14 and 18.

Scheaffer, R., Mendenhall, W., and Ott, R. L. *Elementary Survey Sampling*, 2nd ed. North Scituate, Mass.: Duxbury, 1979.

Taeuber, C. "Information for the nation from a sample survey." In Tanur, et al. (eds.) *Statistics: A Guide to the Unknown*, 2nd ed. San Francisco: Holden-Day, 1978.

Wallace, W. A. "Internal auditors can cut outside CPA costs." *Harvard Business Review*, Mar.–Apr. 1984, 16–20.

BASIC COUNTING RULES

Simple events associated with many experiments have identical characteristics. If you can develop a counting rule to count the number of simple events, it can be used to aid in the solution of many probability problems. For example, many experiments involve sampling n elements from a population of N. Then, as explained in Section 4.1, we can use the formula

$$\binom{N}{n} = \frac{N!}{n!(N - n)!}$$

to find the number of different samples of n elements that could be selected from the total of N elements. This gives the number of simple events for the experiment.

Here, we give you a few useful counting rules. You should learn the characteristics of the situation to which each rule applies. Then, when working a probability problem, carefully examine the experiment to see whether you can use one of the rules.

Learning how to decide whether a particular counting rule applies to an experiment takes patience and practice. If you want to develop this skill, try to use the rules to solve some of the exercises in Chapter 4. You will also find large numbers of exercises in the texts listed in the references at the end of Chapter 4. Proofs of the rules below can be found in the text by W. Feller listed in the references to Chapter 4.

1. **Multiplicative rule:** You have k sets of different elements, n_1 in the first set, n_2 in the second set, . . . , and n_k in the kth set. Suppose you want to form a sample of k elements **by taking one element from each of the k sets**. The number of different samples that can be formed is the product

$$n_1 \cdot n_2 \cdot n_3 \cdot \cdots \cdot n_k$$

EXAMPLE A.1

If a product can be shipped by four different airlines and each airline can ship via three different routes, how many ways can you ship the product?

SOLUTION

A method of shipment corresponds to a pairing of one airline and one route. Therefore, $k = 2$, the number of airlines is $n_1 = 4$, the number of routes is $n_2 = 3$, and the number of ways to ship the product is $n_1 \cdot n_2 = (4)(3) = 12$. ∎

How the multiplicative rule works can be seen by using a **decision tree**. The airline choice is shown by three branching lines in Figure A.1 (page 1176).

1175

FIGURE A.1
Decision Tree for
Example A.1

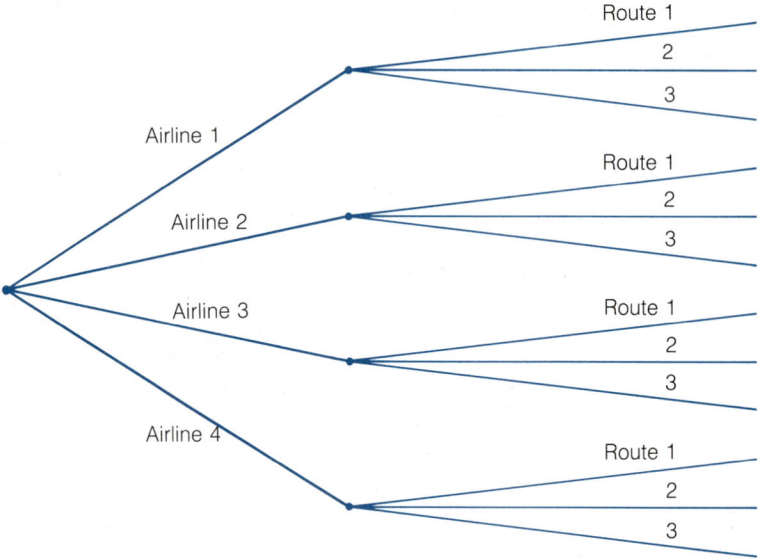

EXAMPLE A.2

You have twenty candidates for three different executive positions, E_1, E_2, and E_3. How many different ways could you fill the positions?

SOLUTION

For this example, there are $k = 3$ sets of elements:

Set 1:   The candidates available to fill position E_1

Set 2:   The candidates remaining (after filling E_1) that are available to fill E_2

Set 3:   The candidates remaining (after filling E_1 and E_2) that are available to fill E_3

The numbers of elements in the sets are $n_1 = 20$, $n_2 = 19$, $n_3 = 18$. Thus, the number of different ways to fill the three positions is $n_1 \cdot n_2 \cdot n_3 = (20)(19)(18) = 6{,}840$. ∎

2. **Partitions rule:**   You have a **single set** of N distinctly different elements, and you want to partition them into k sets, the first set containing n_1 elements, the second containing n_2 elements, . . . , and the kth containing n_k elements. The number of different partitions is

$$\frac{N!}{n_1! n_2! \cdots n_k!} \qquad \text{where} \quad n_1 + n_2 + n_3 + \cdots + n_k = N$$

EXAMPLE A.3

You have twelve construction workers and you want to assign three to job 1, four to job 2, and five to job 3. How many different ways could you make this assignment?

SOLUTION

For this example, $k = 3$ (corresponding to the $k = 3$ job sites), $N = 12$, and $n_1 = 3$, $n_2 = 4$, $n_3 = 5$. Then, the number of different ways to assign the workers to the job sites is

$$\frac{N!}{n_1!n_2!n_3!} = \frac{12!}{3!4!5!} = \frac{12 \cdot 11 \cdot 10 \cdot \cdots \cdot 3 \cdot 2 \cdot 1}{(3 \cdot 2 \cdot 1)(4 \cdot 3 \cdot 2 \cdot 1)(5 \cdot 4 \cdot 3 \cdot 2 \cdot 1)}$$

$$= 27,720$$ ■

3. **Combinations rule:**   The combinations rule given in Chapter 4 is a special case ($k = 2$) of the partitions rule. That is, sampling is equivalent to partitioning a set of N elements into $k = 2$ groups: elements that appear in the sample and those that do not. Let $n_1 = n$, the number of elements in the sample, and $n_2 = N - n$, the number of elements remaining. Then the number of different samples of n elements that can be selected from N is

$$\frac{N!}{n_1!n_2!} = \frac{N!}{n!(N - n)!} = \binom{N}{n}$$

This formula was given in Section 4.1.

EXAMPLE A.4

How many samples of four firemen can be selected from a group of ten?

SOLUTION

We have $N = 10$ and $n = 4$; then,

$$\binom{N}{n} = \binom{10}{4} = \frac{10!}{4!6!} = \frac{10 \cdot 9 \cdot 8 \cdot \cdots \cdot 3 \cdot 2 \cdot 1}{(4 \cdot 3 \cdot 2 \cdot 1)(6 \cdot 5 \cdot \cdots \cdot 2 \cdot 1)}$$

$$= 210$$ ■

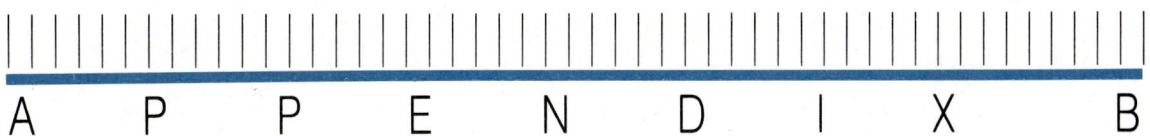

A P P E N D I X B

TABLES

CONTENTS

TABLE I Random Numbers

ROW	1	2	3	4	5	6	7	8	9	10	11	12	13	14
1	10480	15011	01536	02011	81647	91646	69179	14194	62590	36207	20969	99570	91291	90700
2	22368	46573	25595	85393	30995	89198	27982	53402	93965	34095	52666	19174	39615	99505
3	24130	48360	22527	97265	76393	64809	15179	24830	49340	32081	30680	19655	63348	58629
4	42167	93093	06243	61680	07856	16376	39440	53537	71341	57004	00849	74917	97758	16379
5	37570	39975	81837	16656	06121	91782	60468	81305	49684	60672	14110	06927	01263	54613
6	77921	06907	11008	42751	27756	53498	18602	70659	90655	15053	21916	81825	44394	42880
7	99562	72905	56420	69994	98872	31016	71194	18738	44013	48840	63213	21069	10634	12952
8	96301	91977	05463	07972	18876	20922	94595	56869	69014	60045	18425	84903	42508	32307
9	89579	14342	63661	10281	17453	18103	57740	84378	25331	12566	58678	44947	05585	56941
10	85475	36857	53342	53988	53060	59533	38867	62300	08158	17983	16439	11458	18593	64952
11	28918	69578	88231	33276	70997	79936	56865	05859	90106	31595	01547	85590	91610	78188
12	63553	40961	48235	03427	49626	69445	18663	72695	52180	20847	12234	90511	33703	90322
13	09429	93969	52636	92737	88974	33488	36320	17617	30015	08272	84115	27156	30613	74952
14	10365	61129	87529	85689	48237	52267	67689	93394	01511	26358	85104	20285	29975	89868
15	07119	97336	71048	08178	77233	13916	47564	81056	97735	85977	29372	74461	28551	90707
16	51085	12765	51821	51259	77452	16308	60756	92144	49442	53900	70960	63990	75601	40719
17	02368	21382	52404	60268	89368	19885	55322	44819	01188	65255	64835	44919	05944	55157
18	01011	54092	33362	94904	31273	04146	18594	29852	71585	85030	51132	01915	92747	64951
19	52162	53916	46369	58586	23216	14513	83149	98736	23495	64350	94738	17752	35156	35749
20	07056	97628	33787	09998	42698	06691	76988	13602	51851	46104	88916	19509	25625	58104
21	48663	91245	85828	14346	09172	30168	90229	04734	59193	22178	30421	61666	99904	32812
22	54164	58492	22421	74103	47070	25306	76468	26384	58151	06646	21524	15227	96909	44592
23	32639	32363	05597	24200	13363	38005	94342	28728	35806	06912	17012	64161	18296	22851
24	29334	27001	87637	87308	58731	00256	45834	15398	46557	41135	10367	07684	36188	18510
25	02488	33062	28834	07351	19731	92420	60952	61280	50001	67658	32586	86679	50720	94953

(continued)

TABLE I Continued

ROW \ COLUMN	1	2	3	4	5	6	7	8	9	10	11	12	13	14
26	81525	72295	04839	96423	24878	82651	66566	14778	76797	14780	13300	87074	79666	95725
27	29676	20591	68086	26432	46901	20849	89768	81536	86645	12659	92259	57102	80428	25280
28	00742	57392	39064	66432	84673	40027	32832	61362	98947	96067	64760	64584	96096	98253
29	05366	04213	25669	26422	44407	44048	37937	63904	45766	66134	75470	66520	34693	90449
30	91921	26418	64117	94305	26766	25940	39972	22209	71500	64568	91402	42416	07844	69618
31	00582	04711	87917	77341	42206	35126	74087	99547	81817	42607	43808	76655	62028	76630
32	00725	69884	62797	56170	86324	88072	76222	36086	84637	93161	76038	65855	77919	88006
33	69011	65795	95876	55293	18988	27354	26575	08625	40801	59920	29841	80150	12777	48501
34	25976	57948	29888	88604	67917	48708	18912	82271	65424	69774	33611	54262	85963	03547
35	09763	83473	73577	12908	30883	18317	28290	35797	05998	41688	34952	37888	38917	88050
36	91576	42595	27958	30134	04024	86385	29880	99730	55536	84855	29080	09250	79656	73211
37	17955	56349	90999	49127	20044	59931	06115	20542	18059	02008	73708	83517	36103	42791
38	46503	18584	18845	49618	02304	51038	20655	58727	28168	15475	56942	53389	20562	87338
39	92157	89634	94824	78171	84610	82834	09922	25417	44137	48413	25555	21246	35509	20468
40	14577	62765	35605	81263	39667	47358	56873	56307	61607	49518	89656	20103	77490	18062
41	98427	07523	33362	64270	01638	92477	66969	98420	04880	45585	46565	04102	46880	45709
42	34914	63976	88720	82765	34476	17032	87589	40836	32427	70002	70663	88863	77775	69348
43	70060	28277	39475	46473	23219	53416	94970	25832	69975	94884	19661	72828	00102	66794
44	53976	54914	06990	67245	68350	82948	11398	42878	80287	88267	47363	46634	06541	97809
45	76072	29515	40980	07391	58745	25774	22987	80059	39911	96189	41151	14222	60697	59583
46	90725	52210	83974	29992	65831	38857	50490	83765	55657	14361	31720	57375	56228	41546
47	64364	67412	33339	31926	14883	24413	59744	92351	97473	89286	35931	04110	23726	51900
48	08962	00358	31662	25388	61642	34072	81249	35648	56891	69552	48373	45578	78547	81788
49	95012	68379	93526	70765	10592	04542	76463	54328	02349	17247	28865	14777	62730	92277
50	15664	10493	20492	38391	91132	21999	59516	81652	27195	48223	46751	22923	32261	85653
51	16408	81899	04153	53381	79401	21438	83035	92350	36693	31238	59649	91754	72772	02338
52	18629	81953	05520	91962	04739	13092	97662	24822	94730	06496	35090	04822	86774	98289
53	73115	35101	47498	87637	99016	71060	88824	71013	18735	20286	23153	72924	35165	43040
54	57491	16703	23167	49323	45021	33132	12544	41035	80780	45393	44812	12515	98931	91202
55	30405	83946	23792	14422	15059	45799	22716	19792	09983	74353	68668	30429	70735	25499
56	16631	35006	85900	98275	32388	52390	16815	69298	82732	38480	73817	32523	41961	44437
57	96773	20206	42559	78985	05300	22164	24369	54224	35083	19687	11052	91491	60383	19746
58	38935	64202	14349	82674	66523	44133	00697	35552	35970	19124	63318	29686	03387	59846
59	31624	76384	17403	53363	44167	64486	64758	75366	76554	31601	12614	33072	60332	92325
60	78919	19474	23632	27889	47914	02584	37680	20801	72152	39339	34806	08930	85001	87820
61	03931	33309	57047	74211	63445	17361	62825	39908	05607	91284	68833	25570	38818	46920
62	74426	33278	43972	10119	89917	15665	52872	73823	73144	88662	88970	74492	51805	99378

63	09066	20795	00903	95452	92648	45454	09552	88815	16553	51125	79375	97596	16296	66092
64	42238	87025	12426	14267	20979	04508	64535	31355	86064	29472	47689	05974	52468	16834
65	16153	26504	08002	41744	81959	65642	74240	56302	00033	67107	77510	70625	28725	34191
66	21457	29820	40742	96783	29400	21840	15035	34537	33310	06116	95240	15957	16572	06004
67	21581	02050	57802	89728	17937	37621	47075	42080	97403	48626	68995	43805	33386	21597
68	55612	83197	78095	33732	05810	24813	86902	60397	16489	03264	88525	42786	05269	92532
69	44657	99324	66999	51281	84463	60563	79312	93454	68876	25471	93911	25650	12682	73572
70	91340	46949	84979	81973	37949	61023	43997	15263	80644	43942	89203	71795	99533	50501
71	91227	31935	21199	27022	84067	05462	35216	14486	29891	68607	41867	14951	91696	85065
72	50001	66321	38140	19924	72163	09538	12151	06878	91903	18749	34405	56087	82790	70925
73	65390	72958	05224	28609	81406	39147	25549	48542	42627	45233	57202	94617	23772	07896
74	27504	83944	96131	41575	10573	08619	64482	73923	36152	05184	94142	25299	84387	34925
75	37169	39117	94851	89632	00959	16487	65536	49071	39782	17095	02330	74301	00275	48280
76	11508	51111	70225	38351	19444	66499	71945	05422	13442	78675	84081	66938	93654	59894
77	37449	06694	30362	54690	04052	53115	62757	95348	78662	11163	81651	50245	34971	52924
78	46515	85922	70331	38329	57015	15765	97161	17869	45349	61796	66345	81073	49106	79860
79	30986	42416	81223	58353	21532	30502	32305	86482	05174	07901	54339	58861	74818	46942
80	63798	46583	64995	09785	44160	78128	83991	42865	92520	83531	80377	35909	81250	54238
81	82486	99254	84846	67632	43218	50076	21361	64816	51202	88124	41870	52689	51275	83556
82	21885	92431	32906	09060	64297	51674	64126	62570	26123	05155	59194	52799	28225	85762
83	60336	07408	98782	53458	13564	59089	26445	29789	85205	41001	12535	12133	14645	23541
84	43937	24010	46891	25560	86355	33941	25786	54990	71899	15475	95434	98227	21824	19585
85	97656	89303	63175	16275	07100	92063	21942	18611	47348	20203	18534	03862	78095	50136
86	03299	05418	01221	38982	55758	92237	26759	86367	21216	98442	08303	56613	91511	75928
87	79626	03574	06486	17668	07785	76020	79924	25651	83325	88428	85076	72811	22717	50585
88	85636	47539	68335	03129	65651	11977	02510	26113	99447	68645	34327	15152	55230	93448
89	18039	61337	14367	06177	12143	46609	32989	74014	64708	00533	35398	58408	13261	47908
90	08362	60627	15656	36478	65648	16764	53412	09013	07832	41574	17639	82163	60859	75567
91	79556	04142	29068	16268	15387	12856	66227	38358	22478	73373	88732	09443	82558	05250
92	92608	27072	82674	32534	17075	27698	98204	63863	11951	34648	88022	56148	34925	57031
93	23982	40055	25835	67006	12293	02753	14827	23235	35071	99704	37543	11601	35503	85171
94	09915	05908	96306	97901	28395	14186	00821	80703	70426	75647	76310	88717	37890	40129
95	59037	26695	33300	62247	69927	76123	50842	43834	86654	70959	79725	93872	28117	19233
96	42488	69882	78077	61657	34136	79180	97526	43092	04098	73571	80799	76536	71255	64239
97	46764	63003	86273	93017	31204	36692	40202	35275	57306	55543	53203	18098	47625	88684
98	03237	55417	45430	63282	90816	17349	88298	90183	36600	78406	06216	95787	42579	90730
99	86591	52667	81482	61582	14972	90053	89534	76036	49199	97548	43716	04379	46370	28672
100	38534	94964	01715	87288	65680	43772	39560	12918	86537	62738	19636	51132	25739	56947

Source: Abridged from W. H. Beyer (ed.). CRC Standard Mathematical Tables, 24th edition. (Cleveland: The Chemical Rubber Company), 1976. Reproduced by permission of the publisher.

TABLE II   Binomial Probabilities

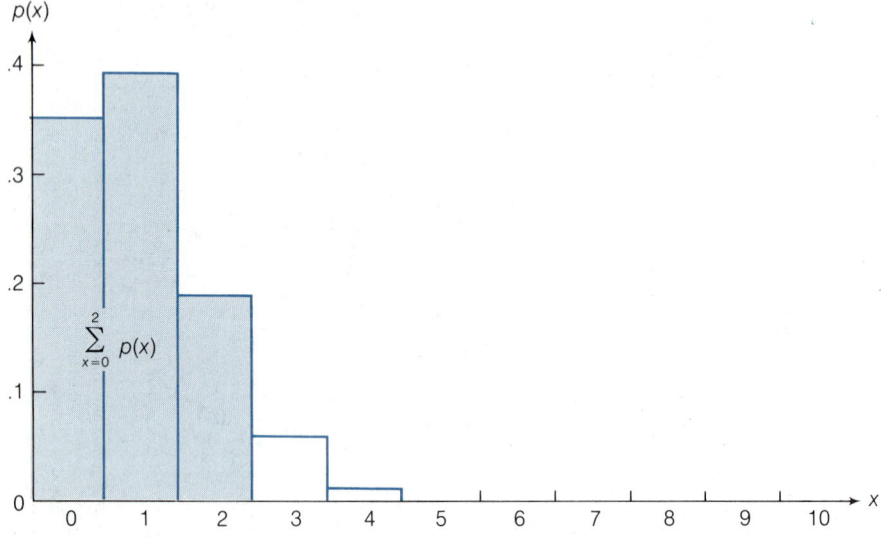

Tabulated values are $\sum\limits_{x=0}^{k} p(x)$. *(Computations are rounded at the third decimal place.)*

a. $n = 5$

k	0.01	0.05	0.10	0.20	0.30	0.40	0.50	0.60	0.70	0.80	0.90	0.95	0.99
0	.951	.774	.590	.328	.168	.078	.031	.010	.002	.000	.000	.000	.000
1	.999	.977	.919	.737	.528	.337	.188	.087	.031	.007	.000	.000	.000
2	1.000	.999	.991	.942	.837	.683	.500	.317	.163	.058	.009	.001	.000
3	1.000	1.000	1.000	.993	.969	.913	.812	.663	.472	.263	.081	.023	.001
4	1.000	1.000	1.000	1.000	.998	.990	.969	.922	.832	.672	.410	.226	.049

b. $n = 6$

k	0.01	0.05	0.10	0.20	0.30	0.40	0.50	0.60	0.70	0.80	0.90	0.95	0.99
0	.941	.735	.531	.262	.118	.047	.016	.004	.001	.000	.000	.000	.000
1	.999	.967	.886	.655	.420	.233	.109	.041	.011	.002	.000	.000	.000
2	1.000	.998	.984	.901	.744	.544	.344	.179	.070	.017	.001	.000	.000
3	1.000	1.000	.999	.983	.930	.821	.656	.456	.256	.099	.016	.002	.000
4	1.000	1.000	1.000	.998	.989	.959	.891	.767	.580	.345	.114	.033	.001
5	1.000	1.000	1.000	1.000	.999	.996	.984	.953	.882	.738	.469	.265	.059

c. $n = 7$

k \ p	0.01	0.05	0.10	0.20	0.30	0.40	0.50	0.60	0.70	0.80	0.90	0.95	0.99
0	.932	.698	.478	.210	.082	.028	.008	.002	.000	.000	.000	.000	.000
1	.998	.956	.850	.577	.329	.159	.063	.019	.004	.000	.000	.000	.000
2	1.000	.996	.974	.852	.647	.420	.227	.096	.029	.005	.000	.000	.000
3	1.000	1.000	.997	.967	.874	.710	.500	.290	.126	.033	.003	.000	.000
4	1.000	1.000	1.000	.995	.971	.904	.773	.580	.353	.148	.026	.004	.000
5	1.000	1.000	1.000	1.000	.996	.981	.937	.841	.671	.423	.150	.044	.002
6	1.000	1.000	1.000	1.000	1.000	.998	.992	.972	.918	.790	.522	.302	.068

d. $n = 8$

k \ p	0.01	0.05	0.10	0.20	0.30	0.40	0.50	0.60	0.70	0.80	0.90	0.95	0.99
0	.923	.663	.430	.168	.058	.017	.004	.001	.000	.000	.000	.000	.000
1	.997	.943	.813	.503	.255	.106	.035	.009	.001	.000	.000	.000	.000
2	1.000	.994	.962	.797	.552	.315	.145	.050	.011	.001	.000	.000	.000
3	1.000	1.000	.995	.944	.806	.594	.363	.174	.058	.010	.000	.000	.000
4	1.000	1.000	1.000	.990	.942	.826	.637	.406	.194	.056	.005	.000	.000
5	1.000	1.000	1.000	.999	.989	.950	.855	.685	.448	.203	.038	.006	.000
6	1.000	1.000	1.000	1.000	.999	.991	.965	.894	.745	.497	.187	.057	.003
7	1.000	1.000	1.000	1.000	1.000	.999	.996	.983	.942	.832	.570	.337	.077

e. $n = 6$

k \ p	0.01	0.05	0.10	0.20	0.30	0.40	0.50	0.60	0.70	0.80	0.90	0.95	0.99
0	.914	.630	.387	.134	.040	.010	.002	.000	.000	.000	.000	.000	.000
1	.997	.929	.775	.436	.196	.071	.020	.004	.000	.000	.000	.000	.000
2	1.000	.992	.947	.738	.463	.232	.090	.025	.004	.000	.000	.000	.000
3	1.000	.999	.992	.914	.730	.483	.254	.099	.025	.003	.000	.000	.000
4	1.000	1.000	.999	.980	.901	.733	.500	.267	.099	.020	.001	.000	.000
5	1.000	1.000	1.000	.997	.975	.901	.746	.517	.270	.086	.008	.001	.000
6	1.000	1.000	1.000	1.000	.996	.975	.910	.768	.537	.262	.053	.008	.000
7	1.000	1.000	1.000	1.000	1.000	.996	.980	.929	.804	.564	.225	.071	.003
8	1.000	1.000	1.000	1.000	1.000	1.000	.998	.990	.960	.866	.613	.370	.086

TABLE II   Continued

f. $n = 10$

p k	0.01	0.05	0.10	0.20	0.30	0.40	0.50	0.60	0.70	0.80	0.90	0.95	0.99
0	.904	.599	.349	.107	.028	.006	.001	.000	.000	.000	.000	.000	.000
1	.996	.914	.736	.376	.149	.046	.011	.002	.000	.000	.000	.000	.000
2	1.000	.988	.930	.678	.383	.167	.055	.012	.002	.000	.000	.000	.000
3	1.000	.999	.987	.879	.650	.382	.172	.055	.011	.001	.000	.000	.000
4	1.000	1.000	.998	.967	.850	.633	.377	.166	.047	.006	.000	.000	.000
5	1.000	1.000	1.000	.994	.953	.834	.623	.367	.150	.033	.002	.000	.000
6	1.000	1.000	1.000	.999	.989	.945	.828	.618	.350	.121	.013	.001	.000
7	1.000	1.000	1.000	1.000	.998	.988	.945	.833	.617	.322	.070	.012	.000
8	1.000	1.000	1.000	1.000	1.000	.998	.989	.954	.851	.624	.264	.086	.004
9	1.000	1.000	1.000	1.000	1.000	1.000	.999	.994	.972	.893	.651	.401	.096

g. $n = 15$

p k	0.01	0.05	0.10	0.20	0.30	0.40	0.50	0.60	0.70	0.80	0.90	0.95	0.99
0	.860	.463	.206	.035	.005	.000	.000	.000	.000	.000	.000	.000	.000
1	.990	.829	.549	.167	.035	.005	.000	.000	.000	.000	.000	.000	.000
2	1.000	.964	.816	.398	.127	.027	.004	.000	.000	.000	.000	.000	.000
3	1.000	.995	.944	.648	.297	.091	.018	.002	.000	.000	.000	.000	.000
4	1.000	.999	.987	.836	.515	.217	.059	.009	.001	.000	.000	.000	.000
5	1.000	1.000	.998	.939	.722	.403	.151	.034	.004	.000	.000	.000	.000
6	1.000	1.000	1.000	.982	.869	.610	.304	.095	.015	.001	.000	.000	.000
7	1.000	1.000	1.000	.996	.950	.787	.500	.213	.050	.004	.000	.000	.000
8	1.000	1.000	1.000	.999	.985	.905	.696	.390	.131	.018	.000	.000	.000
9	1.000	1.000	1.000	1.000	.996	.966	.849	.597	.278	.061	.002	.000	.000
10	1.000	1.000	1.000	1.000	.999	.991	.941	.783	.485	.164	.013	.001	.000
11	1.000	1.000	1.000	1.000	1.000	.998	.982	.909	.703	.352	.056	.005	.000
12	1.000	1.000	1.000	1.000	1.000	1.000	.996	.973	.873	.602	.184	.036	.000
13	1.000	1.000	1.000	1.000	1.000	1.000	1.000	.995	.965	.833	.451	.171	.010
14	1.000	1.000	1.000	1.000	1.000	1.000	1.000	1.000	.995	.965	.794	.537	.140

h. $n = 20$

k \ p	0.01	0.05	0.10	0.20	0.30	0.40	0.50	0.60	0.70	0.80	0.90	0.95	0.99
0	.818	.358	.122	.012	.001	.000	.000	.000	.000	.000	.000	.000	.000
1	.983	.736	.392	.069	.008	.001	.000	.000	.000	.000	.000	.000	.000
2	.999	.925	.677	.206	.035	.004	.000	.000	.000	.000	.000	.000	.000
3	1.000	.984	.867	.411	.107	.016	.001	.000	.000	.000	.000	.000	.000
4	1.000	.997	.957	.630	.238	.051	.006	.000	.000	.000	.000	.000	.000
5	1.000	1.000	.989	.804	.416	.126	.021	.002	.000	.000	.000	.000	.000
6	1.000	1.000	.998	.913	.608	.250	.058	.006	.000	.000	.000	.000	.000
7	1.000	1.000	1.000	.968	.772	.416	.132	.021	.001	.000	.000	.000	.000
8	1.000	1.000	1.000	.990	.887	.596	.252	.057	.005	.000	.000	.000	.000
9	1.000	1.000	1.000	.997	.952	.755	.412	.128	.017	.001	.000	.000	.000
10	1.000	1.000	1.000	.999	.983	.872	.588	.245	.048	.003	.000	.000	.000
11	1.000	1.000	1.000	1.000	.995	.943	.748	.404	.113	.010	.000	.000	.000
12	1.000	1.000	1.000	1.000	.999	.979	.868	.584	.228	.032	.000	.000	.000
13	1.000	1.000	1.000	1.000	1.000	.994	.942	.750	.392	.087	.002	.000	.000
14	1.000	1.000	1.000	1.000	1.000	.998	.979	.874	.584	.196	.011	.000	.000
15	1.000	1.000	1.000	1.000	1.000	1.000	.994	.949	.762	.370	.043	.003	.000
16	1.000	1.000	1.000	1.000	1.000	1.000	.999	.984	.893	.589	.133	.016	.000
17	1.000	1.000	1.000	1.000	1.000	1.000	1.000	.996	.965	.794	.323	.075	.001
18	1.000	1.000	1.000	1.000	1.000	1.000	1.000	.999	.992	.931	.608	.264	.017
19	1.000	1.000	1.000	1.000	1.000	1.000	1.000	1.000	.999	.988	.878	.642	.182

TABLE II Continued

i. $n = 25$

k \ p	0.01	0.05	0.10	0.20	0.30	0.40	0.50	0.60	0.70	0.80	0.90	0.95	0.99
0	.778	.277	.072	.004	.000	.000	.000	.000	.000	.000	.000	.000	.000
1	.974	.642	.271	.027	.002	.000	.000	.000	.000	.000	.000	.000	.000
2	.998	.873	.537	.098	.009	.000	.000	.000	.000	.000	.000	.000	.000
3	1.000	.966	.764	.234	.033	.002	.000	.000	.000	.000	.000	.000	.000
4	1.000	.993	.902	.421	.090	.009	.000	.000	.000	.000	.000	.000	.000
5	1.000	.999	.967	.617	.193	.029	.002	.000	.000	.000	.000	.000	.000
6	1.000	1.000	.991	.780	.341	.074	.007	.000	.000	.000	.000	.000	.000
7	1.000	1.000	.998	.891	.512	.154	.022	.001	.000	.000	.000	.000	.000
8	1.000	1.000	1.000	.953	.677	.274	.054	.004	.000	.000	.000	.000	.000
9	1.000	1.000	1.000	.983	.811	.425	.115	.013	.000	.000	.000	.000	.000
10	1.000	1.000	1.000	.994	.902	.586	.212	.034	.002	.000	.000	.000	.000
11	1.000	1.000	1.000	.998	.956	.732	.345	.078	.006	.000	.000	.000	.000
12	1.000	1.000	1.000	1.000	.983	.846	.500	.154	.017	.000	.000	.000	.000
13	1.000	1.000	1.000	1.000	.994	.922	.655	.268	.044	.002	.000	.000	.000
14	1.000	1.000	1.000	1.000	.998	.966	.788	.414	.098	.006	.000	.000	.000
15	1.000	1.000	1.000	1.000	1.000	.987	.885	.575	.189	.017	.000	.000	.000
16	1.000	1.000	1.000	1.000	1.000	.996	.946	.726	.323	.047	.000	.000	.000
17	1.000	1.000	1.000	1.000	1.000	.999	.978	.846	.488	.109	.002	.000	.000
18	1.000	1.000	1.000	1.000	1.000	1.000	.993	.926	.659	.220	.009	.000	.000
19	1.000	1.000	1.000	1.000	1.000	1.000	.998	.971	.807	.383	.033	.001	.000
20	1.000	1.000	1.000	1.000	1.000	1.000	1.000	.991	.910	.579	.098	.007	.000
21	1.000	1.000	1.000	1.000	1.000	1.000	1.000	.998	.967	.766	.236	.034	.000
22	1.000	1.000	1.000	1.000	1.000	1.000	1.000	1.000	.991	.902	.463	.127	.002
23	1.000	1.000	1.000	1.000	1.000	1.000	1.000	1.000	.998	.973	.729	.358	.026
24	1.000	1.000	1.000	1.000	1.000	1.000	1.000	1.000	1.000	.996	.928	.723	.222

TABLE III Poisson Probabilities

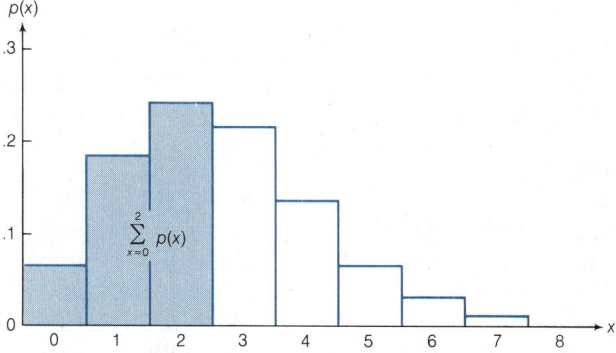

Tabulated values are $\sum_{x=0}^{k} p(x)$. (*Computations are rounded at the third decimal place.*)

λ \ x	0	1	2	3	4	5	6	7	8	9
.02	.980	1.000								
.04	.961	.999	1.000							
.06	.942	.998	1.000							
.08	.923	.997	1.000							
.10	.905	.995	1.000							
.15	.861	.990	.999	1.000						
.20	.819	.982	.999	1.000						
.25	.779	.974	.998	1.000						
.30	.741	.963	.996	1.000						
.35	.705	.951	.994	1.000						
.40	.670	.938	.992	.999	1.000					
.45	.638	.925	.989	.999	1.000					
.50	.607	.910	.986	.998	1.000					
.55	.577	.894	.982	.998	1.000					
.60	.549	.878	.977	.997	1.000					
.65	.522	.861	.972	.996	.999	1.000				
.70	.497	.844	.966	.994	.999	1.000				
.75	.472	.827	.959	.993	.999	1.000				
.80	.449	.809	.953	.991	.999	1.000				
.85	.427	.791	.945	.989	.998	1.000				
.90	.407	.772	.937	.987	.998	1.000				
.95	.387	.754	.929	.981	.997	1.000				
1.00	.368	.736	.920	.981	.996	.999	1.000			
1.1	.333	.699	.900	.974	.995	.999	1.000			
1.2	.301	.663	.879	.966	.992	.998	1.000			
1.3	.273	.627	.857	.957	.989	.998	1.000			
1.4	.247	.592	.833	.946	.986	.997	.999	1.000		
1.5	.223	.558	.809	.934	.981	.996	.999	1.000		

TABLE III Continued

λ \ x	0	1	2	3	4	5	6	7	8	9
1.6	.202	.525	.783	.921	.976	.994	.999	1.000		
1.7	.183	.493	.757	.907	.970	.992	.998	1.000		
1.8	.165	.463	.731	.891	.964	.990	.997	.999	1.000	
1.9	.150	.434	.704	.875	.956	.987	.997	.999	1.000	
2.0	.135	.406	.677	.857	.947	.983	.995	.999	1.000	
2.2	.111	.355	.623	.819	.928	.975	.993	.998	1.000	
2.4	.091	.308	.570	.779	.904	.964	.988	.997	.999	1.000
2.6	.074	.267	.518	.736	.877	.951	.983	.995	.999	1.000
2.8	.061	.231	.469	.692	.848	.935	.976	.992	.998	.999
3.0	.050	.199	.423	.647	.815	.916	.966	.988	.996	.999
3.2	.041	.171	.380	.603	.781	.895	.955	.983	.994	.998
3.4	.033	.147	.340	.558	.744	.871	.942	.977	.992	.997
3.6	.027	.126	.303	.515	.706	.844	.927	.969	.988	.996
3.8	.022	.107	.269	.473	.668	.816	.909	.960	.984	.994
4.0	.018	.092	.238	.433	.629	.785	.889	.949	.979	.992
4.2	.015	.078	.210	.395	.590	.753	.867	.936	.972	.989
4.4	.012	.066	.185	.359	.551	.720	.844	.921	.964	.985
4.6	.010	.056	.163	.326	.513	.686	.818	.905	.955	.980
4.8	.008	.048	.143	.294	.476	.651	.791	.887	.944	.975
5.0	.007	.040	.125	.265	.440	.616	.762	.867	.932	.968
5.2	.006	.034	.109	.238	.406	.581	.732	.845	.918	.960
5.4	.005	.029	.095	.213	.373	.546	.702	.822	.903	.951
5.6	.004	.024	.082	.191	.342	.512	.670	.797	.886	.941
5.8	.003	.021	.072	.170	.313	.478	.638	.771	.867	.929
6.0	.002	.017	.062	.151	.285	.446	.606	.744	.847	.916

λ	10	11	12	13	14	15	16
2.8	1.000						
3.0	1.000						
3.2	1.000						
3.4	.999	1.000					
3.6	.999	1.000					
3.8	.998	.999	1.000				
4.0	.997	.999	1.000				
4.2	.996	.999	1.000				
4.4	.994	.998	.999	1.000			
4.6	.992	.997	.999	1.000			
4.8	.990	.996	.999	1.000			
5.0	.986	.995	.998	.999	1.000		
5.2	.982	.993	.997	.999	1.000		
5.4	.977	.990	.996	.999	1.000		
5.6	.972	.988	.995	.998	.999	1.000	
5.8	.965	.984	.993	.997	.999	1.000	
6.0	.957	.980	.991	.996	.999	.999	1.000

TABLE III   Continued

x λ	0	1	2	3	4	5	6	7	8	9
6.2	.002	.015	.054	.134	.259	.414	.574	.716	.826	.902
6.4	.002	.012	.046	.119	.235	.384	.542	.687	.803	.886
6.6	.001	.010	.040	.105	.213	.355	.511	.658	.780	.869
6.8	.001	.009	.034	.093	.192	.327	.480	.628	.755	.850
7.0	.001	.007	.030	.082	.173	.301	.450	.599	.729	.830
7.2	.001	.006	.025	.072	.156	.276	.420	.569	.703	.810
7.4	.001	.005	.022	.063	.140	.253	.392	.539	.676	.788
7.6	.001	.004	.019	.055	.125	.231	.365	.510	.648	.765
7.8	.000	.004	.016	.048	.112	.210	.338	.481	.620	.741
8.0	.000	.003	.014	.042	.100	.191	.313	.453	.593	.717
8.5	.000	.002	.009	.030	.074	.150	.256	.386	.523	.653
9.0	.000	.001	.006	.021	.055	.116	.207	.324	.456	.587
9.5	.000	.001	.004	.015	.040	.089	.165	.269	.392	.522
10.0	.000	.000	.003	.010	.029	.067	.130	.220	.333	.458

	10	11	12	13	14	15	16	17	18	19
6.2	.949	.975	.989	.995	.998	.999	1.000			
6.4	.939	.969	.986	.994	.997	.999	1.000			
6.6	.927	.963	.982	.992	.997	.999	.999	1.000		
6.8	.915	.955	.978	.990	.996	.998	.999	1.000		
7.0	.901	.947	.973	.987	.994	.998	.999	1.000		
7.2	.887	.937	.967	.984	.993	.997	.999	.999	1.000	
7.4	.871	.926	.961	.980	.991	.996	.998	.999	1.000	
7.6	.854	.915	.954	.976	.989	.995	.998	.999	1.000	
7.8	.835	.902	.945	.971	.986	.993	.997	.999	1.000	
8.0	.816	.888	.936	.966	.983	.992	.996	.998	.999	1.000
8.5	.763	.849	.909	.949	.973	.986	.993	.997	.999	.999
9.0	.706	.803	.876	.926	.959	.978	.989	.995	.998	.999
9.5	.645	.752	.836	.898	.940	.967	.982	.991	.996	.998
10.0	.583	.697	.792	.864	.917	.951	.973	.986	.993	.997

	20	21	22
8.5	1.000		
9.0	1.000		
9.5	.999	1.000	
10.0	.998	.999	1.000

TABLE III Continued

λ \ x	0	1	2	3	4	5	6	7	8	9
10.5	.000	.000	.002	.007	.021	.050	.102	.179	.279	.397
11.0	.000	.000	.001	.005	.015	.038	.079	.143	.232	.341
11.5	.000	.000	.001	.003	.011	.028	.060	.114	.191	.289
12.0	.000	.000	.001	.002	.008	.020	.046	.090	.155	.242
12.5	.000	.000	.000	.002	.005	.015	.035	.070	.125	.201
13.0	.000	.000	.000	.001	.004	.011	.026	.054	.100	.166
13.5	.000	.000	.000	.001	.003	.008	.019	.041	.079	.135
14.0	.000	.000	.000	.000	.002	.006	.014	.032	.062	.109
14.5	.000	.000	.000	.000	.001	.004	.010	.024	.048	.088
15.0	.000	.000	.000	.000	.001	.003	.008	.018	.037	.070

λ \ x	10	11	12	13	14	15	16	17	18	19
10.5	.521	.639	.742	.825	.888	.932	.960	.978	.988	.994
11.0	.460	.579	.689	.781	.854	.907	.944	.968	.982	.991
11.5	.402	.520	.633	.733	.815	.878	.924	.954	.974	.986
12.0	.347	.462	.576	.682	.772	.844	.899	.937	.963	.979
12.5	.297	.406	.519	.628	.725	.806	.869	.916	.948	.969
13.0	.252	.353	.463	.573	.675	.764	.835	.890	.930	.957
13.5	.211	.304	.409	.518	.623	.718	.798	.861	.908	.942
14.0	.176	.260	.358	.464	.570	.669	.756	.827	.883	.923
14.5	.145	.220	.311	.413	.518	.619	.711	.790	.853	.901
15.0	.118	.185	.268	.363	.466	.568	.664	.749	.819	.875

λ \ x	20	21	22	23	24	25	26	27	28	29
10.5	.997	.999	.999	1.000						
11.0	.995	.998	.999	1.000						
11.5	.992	.996	.998	.999	1.000					
12.0	.988	.994	.997	.999	.999	1.000				
12.5	.983	.991	.995	.998	.999	.999	1.000			
13.0	.975	.986	.992	.996	.998	.999	1.000			
13.5	.965	.980	.989	.994	.997	.998	.999	1.000		
14.0	.952	.971	.983	.991	.995	.997	.999	.999	1.000	
14.5	.936	.960	.976	.986	.992	.996	.998	.999	.999	1.000
15.0	.917	.947	.967	.981	.989	.994	.997	.998	.999	1.000

TABLE III   Continued

λ \ x	4	5	6	7	8	9	10	11	12	13
16	.000	.001	.004	.010	.022	.043	.077	.127	.193	.275
17	.000	.001	.002	.005	.013	.026	.049	.085	.135	.201
18	.000	.000	.001	.003	.007	.015	.030	.055	.092	.143
19	.000	.000	.001	.002	.004	.009	.018	.035	.061	.098
20	.000	.000	.000	.001	.002	.005	.011	.021	.039	.066
21	.000	.000	.000	.000	.001	.003	.006	.013	.025	.043
22	.000	.000	.000	.000	.001	.002	.004	.008	.015	.028
23	.000	.000	.000	.000	.000	.001	.002	.004	.009	.017
24	.000	.000	.000	.000	.000	.000	.001	.003	.005	.011
25	.000	.000	.000	.000	.000	.000	.001	.001	.003	.006

λ \ x	14	15	16	17	18	19	20	21	22	23
16	.368	.467	.566	.659	.742	.812	.868	.911	.942	.963
17	.281	.371	.468	.564	.655	.736	.805	.861	.905	.937
18	.208	.287	.375	.469	.562	.651	.731	.799	.855	.899
19	.150	.215	.292	.378	.469	.561	.647	.725	.793	.849
20	.105	.157	.221	.297	.381	.470	.559	.644	.721	.787
21	.072	.111	.163	.227	.302	.384	.471	.558	.640	.716
22	.048	.077	.117	.169	.232	.306	.387	.472	.556	.637
23	.031	.052	.082	.123	.175	.238	.310	.389	.472	.555
24	.020	.034	.056	.087	.128	.180	.243	.314	.392	.473
25	.012	.022	.038	.060	.092	.134	.185	.247	.318	.394

λ \ x	24	25	26	27	28	29	30	31	32	33
16	.978	.987	.993	.996	.998	.999	.999	1.000		
17	.959	.975	.985	.991	.995	.997	.999	.999	1.000	
18	.932	.955	.972	.983	.990	.994	.997	.998	.999	1.000
19	.893	.927	.951	.969	.980	.988	.993	.996	.998	.999
20	.843	.888	.922	.948	.966	.978	.987	.992	.995	.997
21	.782	.838	.883	.917	.944	.963	.976	.985	.991	.994
22	.712	.777	.832	.877	.913	.940	.959	.973	.983	.989
23	.635	.708	.772	.827	.873	.908	.936	.956	.971	.981
24	.554	.632	.704	.768	.823	.868	.904	.932	.953	.969
25	.473	.553	.629	.700	.763	.818	.863	.900	.929	.950

λ \ x	34	35	36	37	38	39	40	41	42	43
19	.999	1.000								
20	.999	.999	1.000							
21	.997	.998	.999	.999	1.000					
22	.994	.996	.998	.999	.999	1.000				
23	.988	.993	.996	.997	.999	.999	1.000			
24	.979	.987	.992	.995	.997	.998	.999	.999	1.000	
25	.966	.978	.985	.991	.991	.997	.998	.999	.999	1.000

TABLE IV Normal Curve Areas

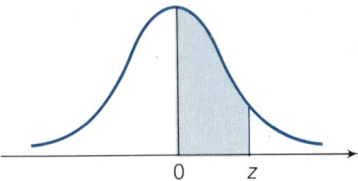

z	.00	.01	.02	.03	.04	.05	.06	.07	.08	.09
.0	.0000	.0040	.0080	.0120	.0160	.0199	.0239	.0279	.0319	.0359
.1	.0398	.0438	.0478	.0517	.0557	.0596	.0636	.0675	.0714	.0753
.2	.0793	.0832	.0871	.0910	.0948	.0987	.1026	.1064	.1103	.1141
.3	.1179	.1217	.1255	.1293	.1331	.1368	.1406	.1443	.1480	.1517
.4	.1554	.1591	.1628	.1664	.1700	.1736	.1772	.1808	.1844	.1879
.5	.1915	.1950	.1985	.2019	.2054	.2088	.2123	.2157	.2190	.2224
.6	.2257	.2291	.2324	.2357	.2389	.2422	.2454	.2486	.2517	.2549
.7	.2580	.2611	.2642	.2673	.2704	.2734	.2764	.2794	.2823	.2852
.8	.2881	.2910	.2939	.2967	.2995	.3023	.3051	.3078	.3106	.3133
.9	.3159	.3186	.3212	.3238	.3264	.3289	.3315	.3340	.3365	.3389
1.0	.3413	.3438	.3461	.3485	.3508	.3531	.3554	.3577	.3599	.3621
1.1	.3643	.3665	.3686	.3708	.3729	.3749	.3770	.3790	.3810	.3830
1.2	.3849	.3869	.3888	.3907	.3925	.3944	.3962	.3980	.3997	.4015
1.3	.4032	.4049	.4066	.4082	.4099	.4115	.4131	.4147	.4162	.4177
1.4	.4192	.4207	.4222	.4236	.4251	.4265	.4279	.4292	.4306	.4319
1.5	.4332	.4345	.4357	.4370	.4382	.4394	.4406	.4418	.4429	.4441
1.6	.4452	.4463	.4474	.4484	.4495	.4505	.4515	.4525	.4535	.4545
1.7	.4554	.4564	.4573	.4582	.4591	.4599	.4608	.4616	.4625	.4633
1.8	.4641	.4649	.4656	.4664	.4671	.4678	.4686	.4693	.4699	.4706
1.9	.4713	.4719	.4726	.4732	.4738	.4744	.4750	.4756	.4761	.4767
2.0	.4772	.4778	.4783	.4788	.4793	.4798	.4803	.4808	.4812	.4817
2.1	.4821	.4826	.4830	.4834	.4838	.4842	.4846	.4850	.4854	.4857
2.2	.4861	.4864	.4868	.4871	.4875	.4878	.4881	.4884	.4887	.4890
2.3	.4893	.4896	.4898	.4901	.4904	.4906	.4909	.4911	.4913	.4916
2.4	.4918	.4920	.4922	.4925	.4927	.4929	.4931	.4932	.4934	.4936
2.5	.4938	.4940	.4941	.4943	.4945	.4946	.4948	.4949	.4951	.4952
2.6	.4953	.4955	.4956	.4957	.4959	.4960	.4961	.4962	.4963	.4964
2.7	.4965	.4966	.4967	.4968	.4969	.4970	.4971	.4972	.4973	.4974
2.8	.4974	.4975	.4976	.4977	.4977	.4978	.4979	.4979	.4980	.4981
2.9	.4981	.4982	.4982	.4983	.4984	.4984	.4985	.4985	.4986	.4986
3.0	.4987	.4987	.4987	.4988	.4988	.4989	.4989	.4989	.4990	.4990

Source: Abridged from Table I of A. Hald, *Statistical Tables and Formulas* (New York: John Wiley & Sons, Inc.), 1952. Reproduced by permission of A. Hald and the publisher, John Wiley & Sons, Inc.

TABLE V   Exponentials

λ	$e^{-\lambda}$	λ	$e^{-\lambda}$	λ	$e^{-\lambda}$
.00	1.000000	2.35	.095369	4.70	.009095
.05	.951229	2.40	.090718	4.75	.008652
.10	.904837	2.45	.086294	4.80	.008230
.15	.860708	2.50	.082085	4.85	.007828
.20	.818731	2.55	.078082	4.90	.007447
.25	.778801	2.60	.074274	4.95	.007083
.30	.740818	2.65	.070651	5.00	.006738
.35	.704688	2.70	.067206	5.05	.006409
.40	.670320	2.75	.063928	5.10	.006097
.45	.637628	2.80	.060810	5.15	.005799
.50	.606531	2.85	.057844	5.20	.005517
.55	.576950	2.90	.055023	5.25	.005248
.60	.548812	2.95	.052340	5.30	.004992
.65	.522046	3.00	.049787	5.35	.004748
.70	.496585	3.05	.047359	5.40	.004517
.75	.472367	3.10	.045049	5.45	.004296
.80	.449329	3.15	.042852	5.50	.004087
.85	.427415	3.20	.040762	5.55	.003887
.90	.406570	3.25	.038774	5.60	.003698
.95	.386741	3.30	.036883	5.65	.003518
1.00	.367879	3.35	.035084	5.70	.003346
1.05	.349938	3.40	.033373	5.75	.003183
1.10	.332871	3.45	.031746	5.80	.003028
1.15	.316637	3.50	.030197	5.85	.002880
1.20	.301194	3.55	.028725	5.90	.002739
1.25	.286505	3.60	.027324	5.95	.002606
1.30	.272532	3.65	.025991	6.00	.002479
1.35	.259240	3.70	.024724	6.05	.002358
1.40	.246597	3.75	.023518	6.10	.002243
1.45	.234570	3.80	.022371	6.15	.002133
1.50	.223130	3.85	.021280	6.20	.002029
1.55	.212248	3.90	.020242	6.25	.001930
1.60	.201897	3.95	.019255	6.30	.001836
1.65	.192050	4.00	.018316	6.35	.001747
1.70	.182684	4.05	.017422	6.40	.001661
1.75	.173774	4.10	.016573	6.45	.001581
1.80	.165299	4.15	.015764	6.50	.001503
1.85	.157237	4.20	.014996	6.55	.001430
1.90	.149569	4.25	.014264	6.60	.001360
1.95	.142274	4.30	.013569	6.65	.001294
2.00	.135335	4.35	.012907	6.70	.001231
2.05	.128735	4.40	.012277	6.75	.001171
2.10	.122456	4.45	.011679	6.80	.001114
2.15	.116484	4.50	.011109	6.85	.001059
2.20	.110803	4.55	.010567	6.90	.001008
2.25	.105399	4.60	.010052	6.95	.000959
2.30	.100259	4.65	.009562	7.00	.000912

TABLE V Continued

λ	$e^{-\lambda}$	λ	$e^{-\lambda}$	λ	$e^{-\lambda}$
7.05	.000867	8.05	.000319	9.05	.000117
7.10	.000825	8.10	.000304	9.10	.000112
7.15	.000785	8.15	.000289	9.15	.000106
7.20	.000747	8.20	.000275	9.20	.000101
7.25	.000710	8.25	.000261	9.25	.000096
7.30	.000676	8.30	.000249	9.30	.000091
7.35	.000643	8.35	.000236	9.35	.000087
7.40	.000611	8.40	.000225	9.40	.000083
7.45	.000581	8.45	.000214	9.45	.000079
7.50	.000553	8.50	.000204	9.50	.000075
7.55	.000526	8.55	.000194	9.55	.000071
7.60	.000501	8.60	.000184	9.60	.000068
7.65	.000476	8.65	.000175	9.65	.000064
7.70	.000453	8.70	.000167	9.70	.000061
7.75	.000431	8.75	.000158	9.75	.000058
7.80	.000410	8.80	.000151	9.80	.000056
7.85	.000390	8.85	.000143	9.85	.000053
7.90	.000371	8.90	.000136	9.90	.000050
7.95	.000353	8.95	.000130	9.95	.000048
8.00	.000336	9.00	.000123	10.00	.000045

TABLE VI

Critical Values of t

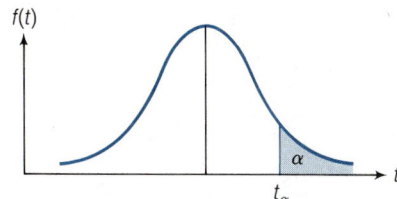

$f(t)$

α

t_α

t

ν	$t_{.100}$	$t_{.050}$	$t_{.025}$	$t_{.010}$	$t_{.005}$	$t_{.001}$	$t_{.0005}$
1	3.078	6.314	12.706	31.821	63.657	318.31	636.62
2	1.886	2.920	4.303	6.965	9.925	22.326	31.598
3	1.638	2.353	3.182	4.541	5.841	10.213	12.924
4	1.533	2.132	2.776	3.747	4.604	7.173	8.610
5	1.476	2.015	2.571	3.365	4.032	5.893	6.869
6	1.440	1.943	2.447	3.143	3.707	5.208	5.959
7	1.415	1.895	2.365	2.998	3.499	4.785	5.408
8	1.397	1.860	2.306	2.896	3.355	4.501	5.041
9	1.383	1.833	2.262	2.821	3.250	4.297	4.781
10	1.372	1.812	2.228	2.764	3.169	4.144	4.587
11	1.363	1.796	2.201	2.718	3.106	4.025	4.437
12	1.356	1.782	2.179	2.681	3.055	3.930	4.318
13	1.350	1.771	2.160	2.650	3.012	3.852	4.221
14	1.345	1.761	2.145	2.624	2.977	3.787	4.140
15	1.341	1.753	2.131	2.602	2.947	3.733	4.073
16	1.337	1.746	2.120	2.583	2.921	3.686	4.015
17	1.333	1.740	2.110	2.567	2.898	3.646	3.965
18	1.330	1.734	2.101	2.552	2.878	3.610	3.922
19	1.328	1.729	2.093	2.539	2.861	3.579	3.883
20	1.325	1.725	2.086	2.528	2.845	3.552	3.850
21	1.323	1.721	2.080	2.518	2.831	3.527	3.819
22	1.321	1.717	2.074	2.508	2.819	3.505	3.792
23	1.319	1.714	2.069	2.500	2.807	3.485	3.767
24	1.318	1.711	2.064	2.492	2.797	3.467	3.745
25	1.316	1.708	2.060	2.485	2.787	3.450	3.725
26	1.315	1.706	2.056	2.479	2.779	3.435	3.707
27	1.314	1.703	2.052	2.473	2.771	3.421	3.690
28	1.313	1.701	2.048	2.467	2.763	3.408	3.674
29	1.311	1.699	2.045	2.462	2.756	3.396	3.659
30	1.310	1.697	2.042	2.457	2.750	3.385	3.646
40	1.303	1.684	2.021	2.423	2.704	3.307	3.551
60	1.296	1.671	2.000	2.390	2.660	3.232	3.460
120	1.289	1.658	1.980	2.358	2.617	3.160	3.373
∞	1.282	1.645	1.960	2.326	2.576	3.090	3.291

Source: This table is reproduced with the kind permission of the Trustees of Biometrika from E. S. Pearson and H. O. Hartley (eds.), *The Biometrika Tables for Statisticians*, Vol. 1, 3rd ed., *Biometrika*, 1966.

TABLE VII   Percentage Points of the F Distribution, $\alpha = .10$

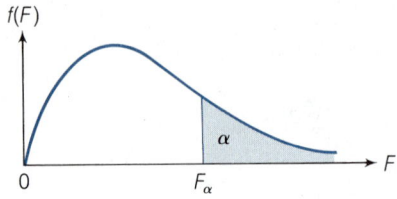

ν_1		NUMERATOR DEGREES OF FREEDOM							
ν_2	1	2	3	4	5	6	7	8	9
1	39.86	49.50	53.59	55.83	57.24	58.20	58.91	59.44	59.86
2	8.53	9.00	9.16	9.24	9.29	9.33	9.35	9.37	9.38
3	5.54	5.46	5.39	5.34	5.31	5.28	5.27	5.25	5.24
4	4.54	4.32	4.19	4.11	4.05	4.01	3.98	3.95	3.94
5	4.06	3.78	3.62	3.52	3.45	3.40	3.37	3.34	3.32
6	3.78	3.46	3.29	3.18	3.11	3.05	3.01	2.98	2.96
7	3.59	3.26	3.07	2.96	2.88	2.83	2.78	2.75	2.72
8	3.46	3.11	2.92	2.81	2.73	2.67	2.62	2.59	2.56
9	3.36	3.01	2.81	2.69	2.61	2.55	2.51	2.47	2.44
10	3.29	2.92	2.73	2.61	2.52	2.46	2.41	2.38	2.35
11	3.23	2.86	2.66	2.54	2.45	2.39	2.34	2.30	2.27
12	3.18	2.81	2.61	2.48	2.39	2.33	2.28	2.24	2.21
13	3.14	2.76	2.56	2.43	2.35	2.28	2.23	2.20	2.16
14	3.10	2.73	2.52	2.39	2.31	2.24	2.19	2.15	2.12
15	3.07	2.70	2.49	2.36	2.27	2.21	2.16	2.12	2.09
16	3.05	2.67	2.46	2.33	2.24	2.18	2.13	2.09	2.06
17	3.03	2.64	2.44	2.31	2.22	2.15	2.10	2.06	2.03
18	3.01	2.62	2.42	2.29	2.20	2.13	2.08	2.04	2.00
19	2.99	2.61	2.40	2.27	2.18	2.11	2.06	2.02	1.98
20	2.97	2.59	2.38	2.25	2.16	2.09	2.04	2.00	1.96
21	2.96	2.57	2.36	2.23	2.14	2.08	2.02	1.98	1.95
22	2.95	2.56	2.35	2.22	2.13	2.06	2.01	1.97	1.93
23	2.94	2.55	2.34	2.21	2.11	2.05	1.99	1.95	1.92
24	2.93	2.54	2.33	2.19	2.10	2.04	1.98	1.94	1.91
25	2.92	2.53	2.32	2.18	2.09	2.02	1.97	1.93	1.89
26	2.91	2.52	2.31	2.17	2.08	2.01	1.96	1.92	1.88
27	2.90	2.51	2.30	2.17	2.07	2.00	1.95	1.91	1.87
28	2.89	2.50	2.29	2.16	2.06	2.00	1.94	1.90	1.87
29	2.89	2.50	2.28	2.15	2.06	1.99	1.93	1.89	1.86
30	2.88	2.49	2.28	2.14	2.05	1.98	1.93	1.88	1.85
40	2.84	2.44	2.23	2.09	2.00	1.93	1.87	1.83	1.79
60	2.79	2.39	2.18	2.04	1.95	1.87	1.82	1.77	1.74
120	2.75	2.35	2.13	1.99	1.90	1.82	1.77	1.72	1.68
∞	2.71	2.30	2.08	1.94	1.85	1.77	1.72	1.67	1.63

DENOMINATOR DEGREES OF FREEDOM

Source: From M. Merrington and C. M. Thompson, "Tables of Percentage Points of the Inverted Beta (F)-Distribution," *Biometrika*, 1943, 33, 73–88. Reproduced by permission of the *Biometrika* Trustees.

| ν_1 | NUMERATOR DEGREES OF FREEDOM | | | | | | | | | |
ν_2	10	12	15	20	24	30	40	60	120	∞
1	60.19	60.71	61.22	61.74	62.00	62.26	62.53	62.79	63.06	63.33
2	9.39	9.41	9.42	9.44	9.45	9.46	9.47	9.47	9.48	9.49
3	5.23	5.22	5.20	5.18	5.18	5.17	5.16	5.15	5.14	5.13
4	3.92	3.90	3.87	3.84	3.83	3.82	3.80	3.79	3.78	3.76
5	3.30	3.27	3.24	3.21	3.19	3.17	3.16	3.14	3.12	3.10
6	2.94	2.90	2.87	2.84	2.82	2.80	2.78	2.76	2.74	2.72
7	2.70	2.67	2.63	2.59	2.58	2.56	2.54	2.51	2.49	2.47
8	2.54	2.50	2.46	2.42	2.40	2.38	2.36	2.34	2.32	2.29
9	2.42	2.38	2.34	2.30	2.28	2.25	2.23	2.21	2.18	2.16
10	2.32	2.28	2.24	2.20	2.18	2.16	2.13	2.11	2.08	2.06
11	2.25	2.21	2.17	2.12	2.10	2.08	2.05	2.03	2.00	1.97
12	2.19	2.15	2.10	2.06	2.04	2.01	1.99	1.96	1.93	1.90
13	2.14	2.10	2.05	2.01	1.98	1.96	1.93	1.90	1.88	1.85
14	2.10	2.05	2.01	1.96	1.94	1.91	1.89	1.86	1.83	1.80
15	2.06	2.02	1.97	1.92	1.90	1.87	1.85	1.82	1.79	1.76
16	2.03	1.99	1.94	1.89	1.87	1.84	1.81	1.78	1.75	1.72
17	2.00	1.96	1.91	1.86	1.84	1.81	1.78	1.75	1.72	1.69
18	1.98	1.93	1.89	1.84	1.81	1.78	1.75	1.72	1.69	1.66
19	1.96	1.91	1.86	1.81	1.79	1.76	1.73	1.70	1.67	1.63
20	1.94	1.89	1.84	1.79	1.77	1.74	1.71	1.68	1.64	1.61
21	1.92	1.87	1.83	1.78	1.75	1.72	1.69	1.66	1.62	1.59
22	1.90	1.86	1.81	1.76	1.73	1.70	1.67	1.64	1.60	1.57
23	1.89	1.84	1.80	1.74	1.72	1.69	1.66	1.62	1.59	1.55
24	1.88	1.83	1.78	1.73	1.70	1.67	1.64	1.61	1.57	1.53
25	1.87	1.82	1.77	1.72	1.69	1.66	1.63	1.59	1.56	1.52
26	1.86	1.81	1.76	1.71	1.68	1.65	1.61	1.58	1.54	1.50
27	1.85	1.80	1.75	1.70	1.67	1.64	1.60	1.57	1.53	1.49
28	1.84	1.79	1.74	1.69	1.66	1.63	1.59	1.56	1.52	1.48
29	1.83	1.78	1.73	1.68	1.65	1.62	1.58	1.55	1.51	1.47
30	1.82	1.77	1.72	1.67	1.64	1.61	1.57	1.54	1.50	1.46
40	1.76	1.71	1.66	1.61	1.57	1.54	1.51	1.47	1.42	1.38
60	1.71	1.66	1.60	1.54	1.51	1.48	1.44	1.40	1.35	1.29
120	1.65	1.60	1.55	1.48	1.45	1.41	1.37	1.32	1.26	1.19
∞	1.60	1.55	1.49	1.42	1.38	1.34	1.30	1.24	1.17	1.00

DENOMINATOR DEGREES OF FREEDOM

TABLE VIII   Percentage Points of the F Distribution, $\alpha = .05$

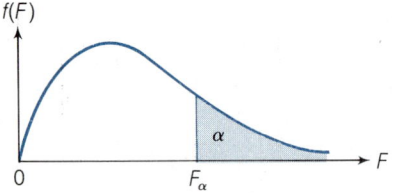

ν_2 $\backslash$ ν_1	NUMERATOR DEGREES OF FREEDOM								
	1	2	3	4	5	6	7	8	9
1	161.4	199.5	215.7	224.6	230.2	234.0	236.8	238.9	240.5
2	18.51	19.00	19.16	19.25	19.30	19.33	19.35	19.37	19.38
3	10.13	9.55	9.28	9.12	9.01	8.94	8.89	8.85	8.81
4	7.71	6.94	6.59	6.39	6.26	6.16	6.09	6.04	6.00
5	6.61	5.79	5.41	5.19	5.05	4.95	4.88	4.82	4.77
6	5.99	5.14	4.76	4.53	4.39	4.28	4.21	4.15	4.10
7	5.59	4.74	4.35	4.12	3.97	3.87	3.79	3.73	3.68
8	5.32	4.46	4.07	3.84	3.69	3.58	3.50	3.44	3.39
9	5.12	4.26	3.86	3.63	3.48	3.37	3.29	3.23	3.18
10	4.96	4.10	3.71	3.48	3.33	3.22	3.14	3.07	3.02
11	4.84	3.98	3.59	3.36	3.20	3.09	3.01	2.95	2.90
12	4.75	3.89	3.49	3.26	3.11	3.00	2.91	2.85	2.80
13	4.67	3.81	3.41	3.18	3.03	2.92	2.83	2.77	2.71
14	4.60	3.74	3.34	3.11	2.96	2.85	2.76	2.70	2.65
15	4.54	3.68	3.29	3.06	2.90	2.79	2.71	2.64	2.59
16	4.49	3.63	3.24	3.01	2.85	2.74	2.66	2.59	2.54
17	4.45	3.59	3.20	2.96	2.81	2.70	2.61	2.55	2.49
18	4.41	3.55	3.16	2.93	2.77	2.66	2.58	2.51	2.46
19	4.38	3.52	3.13	2.90	2.74	2.63	2.54	2.48	2.42
20	4.35	3.49	3.10	2.87	2.71	2.60	2.51	2.45	2.39
21	4.32	3.47	3.07	2.84	2.68	2.57	2.49	2.42	2.37
22	4.30	3.44	3.05	2.82	2.66	2.55	2.46	2.40	2.34
23	4.28	3.42	3.03	2.80	2.64	2.53	2.44	2.37	2.32
24	4.26	3.40	3.01	2.78	2.62	2.51	2.42	2.36	2.30
25	4.24	3.39	2.99	2.76	2.60	2.49	2.40	2.34	2.28
26	4.23	3.37	2.98	2.74	2.59	2.47	2.39	2.32	2.27
27	4.21	3.35	2.96	2.73	2.57	2.46	2.37	2.31	2.25
28	4.20	3.34	2.95	2.71	2.56	2.45	2.36	2.29	2.24
29	4.18	3.33	2.93	2.70	2.55	2.43	2.35	2.28	2.22
30	4.17	3.32	2.92	2.69	2.53	2.42	2.33	2.27	2.21
40	4.08	3.23	2.84	2.61	2.45	2.34	2.25	2.18	2.12
60	4.00	3.15	2.76	2.53	2.37	2.25	2.17	2.10	2.04
120	3.92	3.07	2.68	2.45	2.29	2.17	2.09	2.02	1.96
∞	3.84	3.00	2.60	2.37	2.21	2.10	2.01	1.94	1.88

DENOMINATOR DEGREES OF FREEDOM

Source: From M. Merrington and C. M. Thompson, "Tables of Percentage Points of the Inverted Beta (F)-Distribution," *Biometrika*, 1943, 33, 73–88. Reproduced by permission of the *Biometrika* Trustees.

ν_2 \ ν_1	NUMERATOR DEGREES OF FREEDOM									
	10	12	15	20	24	30	40	60	120	∞
1	241.9	243.9	245.9	248.0	249.1	250.1	251.1	252.2	253.3	254.3
2	19.40	19.41	19.43	19.45	19.45	19.46	19.47	19.48	19.49	19.50
3	8.79	8.74	8.70	8.66	8.64	8.62	8.59	8.57	8.55	8.53
4	5.96	5.91	5.86	5.80	5.77	5.75	5.72	5.69	5.66	5.63
5	4.74	4.68	4.62	4.56	4.53	4.50	4.46	4.43	4.40	4.36
6	4.06	4.00	3.94	3.87	3.84	3.81	3.77	3.74	3.70	3.67
7	3.64	3.57	3.51	3.44	3.41	3.38	3.34	3.30	3.27	3.23
8	3.35	3.28	3.22	3.15	3.12	3.08	3.04	3.01	2.97	2.93
9	3.14	3.07	3.01	2.94	2.90	2.86	2.83	2.79	2.75	2.71
10	2.98	2.91	2.85	2.77	2.74	2.70	2.66	2.62	2.58	2.54
11	2.85	2.79	2.72	2.65	2.61	2.57	2.53	2.49	2.45	2.40
12	2.75	2.69	2.62	2.54	2.51	2.47	2.43	2.38	2.34	2.30
13	2.67	2.60	2.53	2.46	2.42	2.38	2.34	2.30	2.25	2.21
14	2.60	2.53	2.46	2.39	2.35	2.31	2.27	2.22	2.18	2.13
15	2.54	2.48	2.40	2.33	2.29	2.25	2.20	2.16	2.11	2.07
16	2.49	2.42	2.35	2.28	2.24	2.19	2.15	2.11	2.06	2.01
17	2.45	2.38	2.31	2.23	2.19	2.15	2.10	2.06	2.01	1.96
18	2.41	2.34	2.27	2.19	2.15	2.11	2.06	2.02	1.97	1.92
19	2.38	2.31	2.23	2.16	2.11	2.07	2.03	1.98	1.93	1.88
20	2.35	2.28	2.20	2.12	2.08	2.04	1.99	1.95	1.90	1.84
21	2.32	2.25	2.18	2.10	2.05	2.01	1.96	1.92	1.87	1.81
22	2.30	2.23	2.15	2.07	2.03	1.98	1.94	1.89	1.84	1.78
23	2.27	2.20	2.13	2.05	2.01	1.96	1.91	1.86	1.81	1.76
24	2.25	2.18	2.11	2.03	1.98	1.94	1.89	1.84	1.79	1.73
25	2.24	2.16	2.09	2.01	1.96	1.92	1.87	1.82	1.77	1.71
26	2.22	2.15	2.07	1.99	1.95	1.90	1.85	1.80	1.75	1.69
27	2.20	2.13	2.06	1.97	1.93	1.88	1.84	1.79	1.73	1.67
28	2.19	2.12	2.04	1.96	1.91	1.87	1.82	1.77	1.71	1.65
29	2.18	2.10	2.03	1.94	1.90	1.85	1.81	1.75	1.70	1.64
30	2.16	2.09	2.01	1.93	1.89	1.84	1.79	1.74	1.68	1.62
40	2.08	2.00	1.92	1.84	1.79	1.74	1.69	1.64	1.58	1.51
60	1.99	1.92	1.84	1.75	1.70	1.65	1.59	1.53	1.47	1.39
120	1.91	1.83	1.75	1.66	1.61	1.55	1.50	1.43	1.35	1.25
∞	1.83	1.75	1.67	1.57	1.52	1.46	1.39	1.32	1.22	1.00

DENOMINATOR DEGREES OF FREEDOM

TABLE IX Percentage Points of the F Distribution, $\alpha = .025$

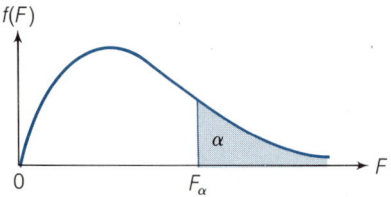

ν_1	NUMERATOR DEGREES OF FREEDOM								
ν_2	1	2	3	4	5	6	7	8	9
1	647.8	799.5	864.2	899.6	921.8	937.1	948.2	956.7	963.3
2	38.51	39.00	39.17	39.25	39.30	39.33	39.36	39.37	39.39
3	17.44	16.04	15.44	15.10	14.88	14.73	14.62	14.54	14.47
4	12.22	10.65	9.98	9.60	9.36	9.20	9.07	8.98	8.90
5	10.01	8.43	7.76	7.39	7.15	6.98	6.85	6.76	6.68
6	8.81	7.26	6.60	6.23	5.99	5.82	5.70	5.60	5.52
7	8.07	6.54	5.89	5.52	5.29	5.12	4.99	4.90	4.82
8	7.57	6.06	5.42	5.05	4.82	4.65	4.53	4.43	4.36
9	7.21	5.71	5.08	4.72	4.48	4.32	4.20	4.10	4.03
10	6.94	5.46	4.83	4.47	4.24	4.07	3.95	3.85	3.78
11	6.72	5.26	4.63	4.28	4.04	3.88	3.76	3.66	3.59
12	6.55	5.10	4.47	4.12	3.89	3.73	3.61	3.51	3.44
13	6.41	4.97	4.35	4.00	3.77	3.60	3.48	3.39	3.31
14	6.30	4.86	4.24	3.89	3.66	3.50	3.38	3.29	3.21
15	6.20	4.77	4.15	3.80	3.58	3.41	3.29	3.20	3.12
16	6.12	4.69	4.08	3.73	3.50	3.34	3.22	3.12	3.05
17	6.04	4.62	4.01	3.66	3.44	3.28	3.16	3.06	2.98
18	5.98	4.56	3.95	3.61	3.38	3.22	3.10	3.01	2.93
19	5.92	4.51	3.90	3.56	3.33	3.17	3.05	2.96	2.88
20	5.87	4.46	3.86	3.51	3.29	3.13	3.01	2.91	2.84
21	5.83	4.42	3.82	3.48	3.25	3.09	2.97	2.87	2.80
22	5.79	4.38	3.78	3.44	3.22	3.05	2.93	2.84	2.76
23	5.75	4.35	3.75	3.41	3.18	3.02	2.90	2.81	2.73
24	5.72	4.32	3.72	3.38	3.15	2.99	2.87	2.78	2.70
25	5.69	4.29	3.69	3.35	3.13	2.97	2.85	2.75	2.68
26	5.66	4.27	3.67	3.33	3.10	2.94	2.82	2.73	2.65
27	5.63	4.24	3.65	3.31	3.08	2.92	2.80	2.71	2.63
28	5.61	4.22	3.63	3.29	3.06	2.90	2.78	2.69	2.61
29	5.59	4.20	3.61	3.27	3.04	2.88	2.76	2.67	2.59
30	5.57	4.18	3.59	3.25	3.03	2.87	2.75	2.65	2.57
40	5.42	4.05	3.46	3.13	2.90	2.74	2.62	2.53	2.45
60	5.29	3.93	3.34	3.01	2.79	2.63	2.51	2.41	2.33
120	5.15	3.80	3.23	2.89	2.67	2.52	2.39	2.30	2.22
∞	5.02	3.69	3.12	2.79	2.57	2.41	2.29	2.19	2.11

DENOMINATOR DEGREES OF FREEDOM

Source: From M. Merrington and C. M. Thompson, "Tables of Percentage Points of the Inverted Beta (F)-Distribution," *Biometrika*, 1943, 33,73–88. Reproduced by permission of the *Biometrika* Trustees.

ν_1	NUMERATOR DEGREES OF FREEDOM									
ν_2	10	12	15	20	24	30	40	60	120	∞
1	968.6	976.7	984.9	993.1	997.2	1001	1006	1010	1014	1018
2	39.40	39.41	39.43	39.45	39.46	39.46	39.47	39.48	39.49	39.50
3	14.42	14.34	14.25	14.17	14.12	14.08	14.04	13.99	13.95	13.90
4	8.84	8.75	8.66	8.56	8.51	8.46	8.41	8.36	8.31	8.26
5	6.62	6.52	6.43	6.33	6.28	6.23	6.18	6.12	6.07	6.02
6	5.46	5.37	5.27	5.17	5.12	5.07	5.01	4.96	4.90	4.85
7	4.76	4.67	4.57	4.47	4.42	4.36	4.31	4.25	4.20	4.14
8	4.30	4.20	4.10	4.00	3.95	3.89	3.84	3.78	3.73	3.67
9	3.96	3.87	3.77	3.67	3.61	3.56	3.51	3.45	3.39	3.33
10	3.72	3.62	3.52	3.42	3.37	3.31	3.26	3.20	3.14	3.08
11	3.53	3.43	3.33	3.23	3.17	3.12	3.06	3.00	2.94	2.88
12	3.37	3.28	3.18	3.07	3.02	2.96	2.91	2.85	2.79	2.72
13	3.25	3.15	3.05	2.95	2.89	2.84	2.78	2.72	2.66	2.60
14	3.15	3.05	2.95	2.84	2.79	2.73	2.67	2.61	2.55	2.49
15	3.06	2.96	2.86	2.76	2.70	2.64	2.59	2.52	2.46	2.40
16	2.99	2.89	2.79	2.68	2.63	2.57	2.51	2.45	2.38	2.32
17	2.92	2.82	2.72	2.62	2.56	2.50	2.44	2.38	2.32	2.25
18	2.87	2.77	2.67	2.56	2.50	2.44	2.38	2.32	2.26	2.19
19	2.82	2.72	2.62	2.51	2.45	2.39	2.33	2.27	2.20	2.13
20	2.77	2.68	2.57	2.46	2.41	2.35	2.29	2.22	2.16	2.09
21	2.73	2.64	2.53	2.42	2.37	2.31	2.25	2.18	2.11	2.04
22	2.70	2.60	2.50	2.39	2.33	2.27	2.21	2.14	2.08	2.00
23	2.67	2.57	2.47	2.36	2.30	2.24	2.18	2.11	2.04	1.97
24	2.64	2.54	2.44	2.33	2.27	2.21	2.15	2.08	2.01	1.94
25	2.61	2.51	2.41	2.30	2.24	2.18	2.12	2.05	1.98	1.91
26	2.59	2.49	2.39	2.28	2.22	2.16	2.09	2.03	1.95	1.88
27	2.57	2.47	2.36	2.25	2.19	2.13	2.07	2.00	1.93	1.85
28	2.55	2.45	2.34	2.23	2.17	2.11	2.05	1.98	1.91	1.83
29	2.53	2.43	2.32	2.21	2.15	2.09	2.03	1.96	1.89	1.81
30	2.51	2.41	2.31	2.20	2.14	2.07	2.01	1.94	1.87	1.79
40	2.39	2.29	2.18	2.07	2.01	1.94	1.88	1.80	1.72	1.64
60	2.27	2.17	2.06	1.94	1.88	1.82	1.74	1.67	1.58	1.48
120	2.16	2.05	1.94	1.82	1.76	1.69	1.61	1.53	1.43	1.31
∞	2.05	1.94	1.83	1.71	1.64	1.57	1.48	1.39	1.27	1.00

DENOMINATOR DEGREES OF FREEDOM

TABLE X Percentage Points of the F Distribution, $\alpha = .01$

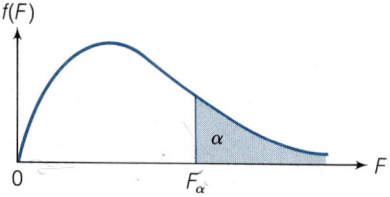

	ν_1	NUMERATOR DEGREES OF FREEDOM								
ν_2		1	2	3	4	5	6	7	8	9
1		4,052	4,999.5	5,403	5,625	5,764	5,859	5,928	5,982	6,022
2		98.50	99.00	99.17	99.25	99.30	99.33	99.36	99.37	99.39
3		34.12	30.82	29.46	28.71	28.24	27.91	27.67	27.49	27.35
4		21.20	18.00	16.69	15.98	15.52	15.21	14.98	14.80	14.66
5		16.26	13.27	12.06	11.39	10.97	10.67	10.46	10.29	10.16
6		13.75	10.92	9.78	9.15	8.75	8.47	8.26	8.10	7.98
7		12.25	9.55	8.45	7.85	7.46	7.19	6.99	6.84	6.72
8		11.26	8.65	7.59	7.01	6.63	6.37	6.18	6.03	5.91
9		10.56	8.02	6.99	6.42	6.06	5.80	5.61	5.47	5.35
10		10.04	7.56	6.55	5.99	5.64	5.39	5.20	5.06	4.94
11		9.65	7.21	6.22	5.67	5.32	5.07	4.89	4.74	4.63
12		9.33	6.93	5.95	5.41	5.06	4.82	4.64	4.50	4.39
13		9.07	6.70	5.74	5.21	4.86	4.62	4.44	4.30	4.19
14		8.86	6.51	5.56	5.04	4.69	4.46	4.28	4.14	4.03
15		8.68	6.36	5.42	4.89	4.56	4.32	4.14	4.00	3.89
16		8.53	6.23	5.29	4.77	4.44	4.20	4.03	3.89	3.78
17		8.40	6.11	5.18	4.67	4.34	4.10	3.93	3.79	3.68
18		8.29	6.01	5.09	4.58	4.25	4.01	3.84	3.71	3.60
19		8.18	5.93	5.01	4.50	4.17	3.94	3.77	3.63	3.52
20		8.10	5.85	4.94	4.43	4.10	3.87	3.70	3.56	3.46
21		8.02	5.78	4.87	4.37	4.04	3.81	3.64	3.51	3.40
22		7.95	5.72	4.82	4.31	3.99	3.76	3.59	3.45	3.35
23		7.88	5.66	4.76	4.26	3.94	3.71	3.54	3.41	3.30
24		7.82	5.61	4.72	4.22	3.90	3.67	3.50	3.36	3.26
25		7.77	5.57	4.68	4.18	3.85	3.63	3.46	3.32	3.22
26		7.72	5.53	4.64	4.14	3.82	3.59	3.42	3.29	3.18
27		7.68	5.49	4.60	4.11	3.78	3.56	3.39	3.26	3.15
28		7.64	5.45	4.57	4.07	3.75	3.53	3.36	3.23	3.12
29		7.60	5.42	4.54	4.04	3.73	3.50	3.33	3.20	3.09
30		7.56	5.39	4.51	4.02	3.70	3.47	3.30	3.17	3.07
40		7.31	5.18	4.31	3.83	3.51	3.29	3.12	2.99	2.89
60		7.08	4.98	4.13	3.65	3.34	3.12	2.95	2.82	2.72
120		6.85	4.79	3.95	3.48	3.17	2.96	2.79	2.66	2.56
∞		6.63	4.61	3.78	3.32	3.02	2.80	2.64	2.51	2.41

DENOMINATOR DEGREES OF FREEDOM

Source: From M. Merrington and C. M. Thompson, "Tables of Percentage Points of the Inverted Beta (F)-Distribution," *Biometrika*, 1943, 33, 73–88. Reproduced by permission of the *Biometrika* Trustees.

ν_2 \ ν_1	NUMERATOR DEGREES OF FREEDOM									
	10	12	15	20	24	30	40	60	120	∞
1	6,056	6,106	6,157	6,209	6,235	6,261	6,287	6,313	6,339	6,366
2	99.40	99.42	99.43	99.45	99.46	99.47	99.47	99.48	99.49	99.50
3	27.23	27.05	26.87	26.69	26.60	26.50	26.41	26.32	26.22	26.13
4	14.55	14.37	14.20	14.02	13.93	13.84	13.75	13.65	13.56	13.46
5	10.05	9.89	9.72	9.55	9.47	9.38	9.29	9.20	9.11	9.02
6	7.87	7.72	7.56	7.40	7.31	7.23	7.14	7.06	6.97	6.88
7	6.62	6.47	6.31	6.16	6.07	5.99	5.91	5.82	5.74	5.65
8	5.81	5.67	5.52	5.36	5.28	5.20	5.12	5.03	4.95	4.86
9	5.26	5.11	4.96	4.81	4.73	4.65	4.57	4.48	4.40	4.31
10	4.85	4.71	4.56	4.41	4.33	4.25	4.17	4.08	4.00	3.91
11	4.54	4.40	4.25	4.10	4.02	3.94	3.86	3.78	3.69	3.60
12	4.30	4.16	4.01	3.86	3.78	3.70	3.62	3.54	3.45	3.36
13	4.10	3.96	3.82	3.66	3.59	3.51	3.43	3.34	3.25	3.17
14	3.94	3.80	3.66	3.51	3.43	3.35	3.27	3.18	3.09	3.00
15	3.80	3.67	3.52	3.37	3.29	3.21	3.13	3.05	2.96	2.87
16	3.69	3.55	3.41	3.26	3.18	3.10	3.02	2.93	2.84	2.75
17	3.59	3.46	3.31	3.16	3.08	3.00	2.92	2.83	2.75	2.65
18	3.51	3.37	3.23	3.08	3.00	2.92	2.84	2.75	2.66	2.57
19	3.43	3.30	3.15	3.00	2.92	2.84	2.76	2.67	2.58	2.49
20	3.37	3.23	3.09	2.94	2.86	2.78	2.69	2.61	2.52	2.42
21	3.31	3.17	3.03	2.88	2.80	2.72	2.64	2.55	2.46	2.36
22	3.26	3.12	2.98	2.83	2.75	2.67	2.58	2.50	2.40	2.31
23	3.21	3.07	2.93	2.78	2.70	2.62	2.54	2.45	2.35	2.26
24	3.17	3.03	2.89	2.74	2.66	2.58	2.49	2.40	2.31	2.21
25	3.13	2.99	2.85	2.70	2.62	2.54	2.45	2.36	2.27	2.17
26	3.09	2.96	2.81	2.66	2.58	2.50	2.42	2.33	2.23	2.13
27	3.06	2.93	2.78	2.63	2.55	2.47	2.38	2.29	2.20	2.10
28	3.03	2.90	2.75	2.60	2.52	2.44	2.35	2.26	2.17	2.06
29	3.00	2.87	2.73	2.57	2.49	2.41	2.33	2.23	2.14	2.03
30	2.98	2.84	2.70	2.55	2.47	2.39	2.30	2.21	2.11	2.01
40	2.80	2.66	2.52	2.37	2.29	2.20	2.11	2.02	1.92	1.80
60	2.63	2.50	2.35	2.20	2.12	2.03	1.94	1.84	1.73	1.60
120	2.47	2.34	2.19	2.03	1.95	1.86	1.76	1.66	1.53	1.38
∞	2.32	2.18	2.04	1.88	1.79	1.70	1.59	1.47	1.32	1.00

DENOMINATOR DEGREES OF FREEDOM

TABLE XI   Critical Values of T_L and T_U for the Wilcoxon Rank Sum Test: Independent Samples

Test statistic is the rank sum associated with the smaller sample (if equal sample sizes, either rank sum can be used).

a. $\alpha = .025$ one-tailed; $\alpha = .05$ two-tailed

n_2 \ n_1	3		4		5		6		7		8		9		10	
	T_L	T_U	T_L	T_U	T_L	T_U	T_L	T_U	T_L	T_U	T_L	T_U	T_L	T_U	T_L	T_U
3	5	16	6	18	6	21	7	23	7	26	8	28	8	31	9	33
4	6	18	11	25	12	28	12	32	13	35	14	38	15	41	16	44
5	6	21	12	28	18	37	19	41	20	45	21	49	22	53	24	56
6	7	23	12	32	19	41	26	52	28	56	29	61	31	65	32	70
7	7	26	13	35	20	45	28	56	37	68	39	73	41	78	43	83
8	8	28	14	38	21	49	29	61	39	73	49	87	51	93	54	98
9	8	31	15	41	22	53	31	65	41	78	51	93	63	108	66	114
10	9	33	16	44	24	56	32	70	43	83	54	98	66	114	79	131

b. $\alpha = .05$ one-tailed; $\alpha = .10$ two-tailed

n_2 \ n_1	3		4		5		6		7		8		9		10	
	T_L	T_U	T_L	T_U	T_L	T_U	T_L	T_U	T_L	T_U	T_L	T_U	T_L	T_U	T_L	T_U
3	6	15	7	17	7	20	8	22	9	24	9	27	10	29	11	31
4	7	17	12	24	13	27	14	30	15	33	16	36	17	39	18	42
5	7	20	13	27	19	36	20	40	22	43	24	46	25	50	26	54
6	8	22	14	30	20	40	28	50	30	54	32	58	33	63	35	67
7	9	24	15	33	22	43	30	54	39	66	41	71	43	76	46	80
8	9	27	16	36	24	46	32	58	41	71	52	84	54	90	57	95
9	10	29	17	39	25	50	33	63	43	76	54	90	66	105	69	111
10	11	31	18	42	26	54	35	67	46	80	57	95	69	111	83	127

Source: From F. Wilcoxon and R. A. Wilcox, "Some Rapid Approximate Statistical Procedures," 1964, 20–23. Reproduced with the permission of American Cyanamid Company.

TABLE XII

Critical Values of T_0 in the Wilcoxon Paired Difference Signed Rank Test

ONE-TAILED	TWO-TAILED	$n = 5$	$n = 6$	$n = 7$	$n = 8$	$n = 9$	$n = 10$
$\alpha = .05$	$\alpha = .10$	1	2	4	6	8	11
$\alpha = .025$	$\alpha = .05$		1	2	4	6	8
$\alpha = .01$	$\alpha = .02$			0	2	3	5
$\alpha = .005$	$\alpha = .01$				0	2	3
		$n = 11$	$n = 12$	$n = 13$	$n = 14$	$n = 15$	$n = 16$
$\alpha = .05$	$\alpha = .10$	14	17	21	26	30	36
$\alpha = .025$	$\alpha = .05$	11	14	17	21	25	30
$\alpha = .01$	$\alpha = .02$	7	10	13	16	20	24
$\alpha = .005$	$\alpha = .01$	5	7	10	13	16	19
		$n = 17$	$n = 18$	$n = 19$	$n = 20$	$n = 21$	$n = 22$
$\alpha = .05$	$\alpha = .10$	41	47	54	60	68	75
$\alpha = .025$	$\alpha = .05$	35	40	46	52	59	66
$\alpha = .01$	$\alpha = .02$	28	33	38	43	49	56
$\alpha = .005$	$\alpha = .01$	23	28	32	37	43	49
		$n = 23$	$n = 24$	$n = 25$	$n = 26$	$n = 27$	$n = 28$
$\alpha = .05$	$\alpha = .10$	83	92	101	110	120	130
$\alpha = .025$	$\alpha = .05$	73	81	90	98	107	117
$\alpha = .01$	$\alpha = .02$	62	69	77	85	93	102
$\alpha = .005$	$\alpha = .01$	55	61	68	76	84	92
		$n = 29$	$n = 30$	$n = 31$	$n = 32$	$n = 33$	$n = 34$
$\alpha = .05$	$\alpha = .10$	141	152	163	175	188	201
$\alpha = .025$	$\alpha = .05$	127	137	148	159	171	183
$\alpha = .01$	$\alpha = .02$	111	120	130	141	151	162
$\alpha = .005$	$\alpha = .01$	100	109	118	128	138	149
		$n = 35$	$n = 36$	$n = 37$	$n = 38$	$n = 39$	
$\alpha = .05$	$\alpha = .10$	214	228	242	256	271	
$\alpha = .025$	$\alpha = .05$	195	208	222	235	250	
$\alpha = .01$	$\alpha = .02$	174	186	198	211	224	
$\alpha = .005$	$\alpha = .01$	160	171	183	195	208	
		$n = 40$	$n = 41$	$n = 42$	$n = 43$	$n = 44$	$n = 45$
$\alpha = .05$	$\alpha = .10$	287	303	319	336	353	371
$\alpha = .025$	$\alpha = .05$	264	279	295	311	327	344
$\alpha = .01$	$\alpha = .02$	238	252	267	281	297	313
$\alpha = .005$	$\alpha = .01$	221	234	248	262	277	292
		$n = 46$	$n = 47$	$n = 48$	$n = 49$	$n = 50$	
$\alpha = .05$	$\alpha = .10$	389	408	427	446	466	
$\alpha = .025$	$\alpha = .05$	361	379	397	415	434	
$\alpha = .01$	$\alpha = .02$	329	345	362	380	398	
$\alpha = .005$	$\alpha = .01$	307	323	339	356	373	

Source: From F. Wilcoxon and R. A. Wilcox, "Some Rapid Approximate Statistical Procedures," 1964, p. 28. Reproduced with the permission of American Cyanamid Company.

APPENDIX B TABLES

TABLE XIII

Critical Values of χ^2

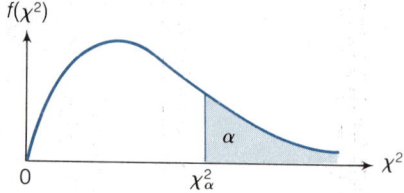

$f(\chi^2)$

α

χ_α^2

χ^2

0

DEGREES OF FREEDOM	$\chi_{.995}^2$	$\chi_{.990}^2$	$\chi_{.975}^2$	$\chi_{.950}^2$	$\chi_{.900}^2$
1	.0000393	.0001571	.0009821	.0039321	.0157908
2	.0100251	.0201007	.0506356	.102587	.210720
3	.0717212	.114832	.215795	.351846	.584375
4	.206990	.297110	.484419	.710721	1.063623
5	.411740	.554300	.831211	1.145476	1.61031
6	.675727	.872085	1.237347	1.63539	2.20413
7	.989265	1.239043	1.68987	2.16735	2.83311
8	1.344419	1.646482	2.17973	2.73264	3.48954
9	1.734926	2.087912	2.70039	3.32511	4.16816
10	2.15585	2.55821	3.24697	3.94030	4.86518
11	2.60321	3.05347	3.81575	4.57481	5.57779
12	3.07382	3.57056	4.40379	5.22603	6.30380
13	3.56503	4.10691	5.00874	5.89186	7.04150
14	4.07468	4.66043	5.62872	6.57063	7.78953
15	4.60094	5.22935	6.26214	7.26094	8.54675
16	5.14224	5.81221	6.90766	7.96164	9.31223
17	5.69724	6.40776	7.56418	8.67176	10.0852
18	6.26481	7.01491	8.23075	9.39046	10.8649
19	6.84398	7.63273	8.90655	10.1170	11.6509
20	7.43386	8.26040	9.59083	10.8508	12.4426
21	8.03366	8.89720	10.28293	11.5913	13.2396
22	8.64272	9.54249	10.9823	12.3380	14.0415
23	9.26042	10.19567	11.6885	13.0905	14.8479
24	9.88623	10.8564	12.4011	13.8484	15.6587
25	10.5197	11.5240	13.1197	14.6114	16.4734
26	11.1603	12.1981	13.8439	15.3791	17.2919
27	11.8076	12.8786	14.5733	16.1513	18.1138
28	12.4613	13.5648	15.3079	16.9279	18.9392
29	13.1211	14.2565	16.0471	17.7083	19.7677
30	13.7867	14.9535	16.7908	18.4926	20.5992
40	20.7065	22.1643	24.4331	26.5093	29.0505
50	27.9907	29.7067	32.3574	34.7642	37.6886
60	35.5346	37.4848	40.4817	43.1879	46.4589
70	43.2752	45.4418	48.7576	51.7393	55.3290
80	51.1720	53.5400	57.1532	60.3915	64.2778
90	59.1963	61.7541	65.6466	69.1260	73.2912
100	67.3276	70.0648	74.2219	77.9295	82.3581

Source: From C. M. Thompson, "Tables of the Percentage Points of the χ^2-Distribution," *Biometrika*, 1941, 32, 188–189. Reproduced by permission of the *Biometrika* Trustees.

DEGREES OF FREEDOM	$\chi^2_{.100}$	$\chi^2_{.050}$	$\chi^2_{.025}$	$\chi^2_{.010}$	$\chi^2_{.005}$
1	2.70554	3.84146	5.02389	6.63490	7.87944
2	4.60517	5.99147	7.37776	9.21034	10.5966
3	6.25139	7.81473	9.34840	11.3449	12.8381
4	7.77944	9.48773	11.1433	13.2767	14.8602
5	9.23635	11.0705	12.8325	15.0863	16.7496
6	10.6446	12.5916	14.4494	16.8119	18.5476
7	12.0170	14.0671	16.0128	18.4753	20.2777
8	13.3616	15.5073	17.5346	20.0902	21.9550
9	14.6837	16.9190	19.0228	21.6660	23.5893
10	15.9871	18.3070	20.4831	23.2093	25.1882
11	17.2750	19.6751	21.9200	24.7250	26.7569
12	18.5494	21.0261	23.3367	26.2170	28.2995
13	19.8119	22.3621	24.7356	27.6883	29.8194
14	21.0642	23.6848	26.1190	29.1413	31.3193
15	22.3072	24.9958	27.4884	30.5779	32.8013
16	23.5418	26.2962	28.8454	31.9999	34.2672
17	24.7690	27.5871	30.1910	33.4087	35.7185
18	25.9894	28.8693	31.5264	34.8053	37.1564
19	27.2036	30.1435	32.8523	36.1908	38.5822
20	28.4120	31.4104	34.1696	37.5662	39.9968
21	29.6151	32.6705	35.4789	38.9321	41.4010
22	30.8133	33.9244	36.7807	40.2894	42.7956
23	32.0069	35.1725	38.0757	41.6384	44.1813
24	33.1963	36.4151	39.3641	42.9798	45.5585
25	34.3816	37.6525	40.6465	44.3141	46.9278
26	35.5631	38.8852	41.9232	45.6417	48.2899
27	36.7412	40.1133	43.1944	46.9630	49.6449
28	37.9159	41.3372	44.4607	48.2782	50.9933
29	39.0875	42.5569	45.7222	49.5879	52.3356
30	40.2560	43.7729	46.9792	50.8922	53.6720
40	51.8050	55.7585	59.3417	63.6907	66.7659
50	63.1671	67.5048	71.4202	76.1539	79.4900
60	74.3970	79.0819	83.2976	88.3794	91.9517
70	85.5271	90.5312	95.0231	100.425	104.215
80	96.5782	101.879	106.629	112.329	116.321
90	107.565	113.145	118.136	124.116	128.299
100	118.498	124.342	129.561	135.807	140.169

TABLE XIV

Critical Values for the Durbin–Watson d Statistic, $\alpha = .05$

n	$k = 1$ d_L	$k = 1$ d_U	$k = 2$ d_L	$k = 2$ d_U	$k = 3$ d_L	$k = 3$ d_U	$k = 4$ d_L	$k = 4$ d_U	$k = 5$ d_L	$k = 5$ d_U
15	1.08	1.36	.95	1.54	.82	1.75	.69	1.97	.56	2.21
16	1.10	1.37	.98	1.54	.86	1.73	.74	1.93	.62	2.15
17	1.13	1.38	1.02	1.54	.90	1.71	.78	1.90	.67	2.10
18	1.16	1.39	1.05	1.53	.93	1.69	.82	1.87	.71	2.06
19	1.18	1.40	1.08	1.53	.97	1.68	.86	1.85	.75	2.02
20	1.20	1.41	1.10	1.54	1.00	1.68	.90	1.83	.79	1.99
21	1.22	1.42	1.13	1.54	1.03	1.67	.93	1.81	.83	1.96
22	1.24	1.43	1.15	1.54	1.05	1.66	.96	1.80	.86	1.94
23	1.26	1.44	1.17	1.54	1.08	1.66	.99	1.79	.90	1.92
24	1.27	1.45	1.19	1.55	1.10	1.66	1.01	1.78	.93	1.90
25	1.29	1.45	1.21	1.55	1.12	1.66	1.04	1.77	.95	1.89
26	1.30	1.46	1.22	1.55	1.14	1.65	1.06	1.76	.98	1.88
27	1.32	1.47	1.24	1.56	1.16	1.65	1.08	1.76	1.01	1.86
28	1.33	1.48	1.26	1.56	1.18	1.65	1.10	1.75	1.03	1.85
29	1.34	1.48	1.27	1.56	1.20	1.65	1.12	1.74	1.05	1.84
30	1.35	1.49	1.28	1.57	1.21	1.65	1.14	1.74	1.07	1.83
31	1.36	1.50	1.30	1.57	1.23	1.65	1.16	1.74	1.09	1.83
32	1.37	1.50	1.31	1.57	1.24	1.65	1.18	1.73	1.11	1.82
33	1.38	1.51	1.32	1.58	1.26	1.65	1.19	1.73	1.13	1.81
34	1.39	1.51	1.33	1.58	1.27	1.65	1.21	1.73	1.15	1.81
35	1.40	1.52	1.34	1.58	1.28	1.65	1.22	1.73	1.16	1.80
36	1.41	1.52	1.35	1.59	1.29	1.65	1.24	1.73	1.18	1.80
37	1.42	1.53	1.36	1.59	1.31	1.66	1.25	1.72	1.19	1.80
38	1.43	1.54	1.37	1.59	1.32	1.66	1.26	1.72	1.21	1.79
39	1.43	1.54	1.38	1.60	1.33	1.66	1.27	1.72	1.22	1.79
40	1.44	1.54	1.39	1.60	1.34	1.66	1.29	1.72	1.23	1.79
45	1.48	1.57	1.43	1.62	1.38	1.67	1.34	1.72	1.29	1.78
50	1.50	1.59	1.46	1.63	1.42	1.67	1.38	1.72	1.34	1.77
55	1.53	1.60	1.49	1.64	1.45	1.68	1.41	1.72	1.38	1.77
60	1.55	1.62	1.51	1.65	1.48	1.69	1.44	1.73	1.41	1.77
65	1.57	1.63	1.54	1.66	1.50	1.70	1.47	1.73	1.44	1.77
70	1.58	1.64	1.55	1.67	1.52	1.70	1.49	1.74	1.46	1.77
75	1.60	1.65	1.57	1.68	1.54	1.71	1.51	1.74	1.49	1.77
80	1.61	1.66	1.59	1.69	1.56	1.72	1.53	1.74	1.51	1.77
85	1.62	1.67	1.60	1.70	1.57	1.72	1.55	1.75	1.52	1.77
90	1.63	1.68	1.61	1.70	1.59	1.73	1.57	1.75	1.54	1.78
95	1.64	1.69	1.62	1.71	1.60	1.73	1.58	1.75	1.56	1.78
100	1.65	1.69	1.63	1.72	1.61	1.74	1.59	1.76	1.57	1.78

Source: From J. Durbin and G. S. Watson, "Testing for Serial Correlation in Least Squares Regression, II." *Biometrika*, 1951, 30, 159–178. Reproduced by permission of the *Biometrika* Trustees.

TABLE XV

Critical Values for the
Durbin–Watson d
Statistic, $\alpha = .01$

n	$k = 1$ d_L	$k = 1$ d_U	$k = 2$ d_L	$k = 2$ d_U	$k = 3$ d_L	$k = 3$ d_U	$k = 4$ d_L	$k = 4$ d_U	$k = 5$ d_L	$k = 5$ d_U
15	.81	1.07	.70	1.25	.59	1.46	.49	1.70	.39	1.96
16	.84	1.09	.74	1.25	.63	1.44	.53	1.66	.44	1.90
17	.87	1.10	.77	1.25	.67	1.43	.57	1.63	.48	1.85
18	.90	1.12	.80	1.26	.71	1.42	.61	1.60	.52	1.80
19	.93	1.13	.83	1.26	.74	1.41	.65	1.58	.56	1.77
20	.95	1.15	.86	1.27	.77	1.41	.68	1.57	.60	1.74
21	.97	1.16	.89	1.27	.80	1.41	.72	1.55	.63	1.71
22	1.00	1.17	.91	1.28	.83	1.40	.75	1.54	.66	1.69
23	1.02	1.19	.94	1.29	.86	1.40	.77	1.53	.70	1.67
24	1.04	1.20	.96	1.30	.88	1.41	.80	1.53	.72	1.66
25	1.05	1.21	.98	1.30	.90	1.41	.83	1.52	.75	1.65
26	1.07	1.22	1.00	1.31	.93	1.41	.85	1.52	.78	1.64
27	1.09	1.23	1.02	1.32	.95	1.41	.88	1.51	.81	1.63
28	1.10	1.24	1.04	1.32	.97	1.41	.90	1.51	.83	1.62
29	1.12	1.25	1.05	1.33	.99	1.42	.92	1.51	.85	1.61
30	1.13	1.26	1.07	1.34	1.01	1.42	.94	1.51	.88	1.61
31	1.15	1.27	1.08	1.34	1.02	1.42	.96	1.51	.90	1.60
32	1.16	1.28	1.10	1.35	1.04	1.43	.98	1.51	.92	1.60
33	1.17	1.29	1.11	1.36	1.05	1.43	1.00	1.51	.94	1.59
34	1.18	1.30	1.13	1.36	1.07	1.43	1.01	1.51	.95	1.59
35	1.19	1.31	1.14	1.37	1.08	1.44	1.03	1.51	.97	1.59
36	1.21	1.32	1.15	1.38	1.10	1.44	1.04	1.51	.99	1.59
37	1.22	1.32	1.16	1.38	1.11	1.45	1.06	1.51	1.00	1.59
38	1.23	1.33	1.18	1.39	1.12	1.45	1.07	1.52	1.02	1.58
39	1.24	1.34	1.19	1.39	1.14	1.45	1.09	1.52	1.03	1.58
40	1.25	1.34	1.20	1.40	1.15	1.46	1.10	1.52	1.05	1.58
45	1.29	1.38	1.24	1.42	1.20	1.48	1.16	1.53	1.11	1.58
50	1.32	1.40	1.28	1.45	1.24	1.49	1.20	1.54	1.16	1.59
55	1.36	1.43	1.32	1.47	1.28	1.51	1.25	1.55	1.21	1.59
60	1.38	1.45	1.35	1.48	1.32	1.52	1.28	1.56	1.25	1.60
65	1.41	1.47	1.38	1.50	1.35	1.53	1.31	1.57	1.28	1.61
70	1.43	1.49	1.40	1.52	1.37	1.55	1.34	1.58	1.31	1.61
75	1.45	1.50	1.42	1.53	1.39	1.56	1.37	1.59	1.34	1.62
80	1.47	1.52	1.44	1.54	1.42	1.57	1.39	1.60	1.36	1.62
85	1.48	1.53	1.46	1.55	1.43	1.58	1.41	1.60	1.39	1.63
90	1.50	1.54	1.47	1.56	1.45	1.59	1.43	1.61	1.41	1.64
95	1.51	1.55	1.49	1.57	1.47	1.60	1.45	1.62	1.42	1.64
100	1.52	1.56	1.50	1.58	1.48	1.60	1.46	1.63	1.44	1.65

Source: From J. Durbin and G. S. Watson, "Testing for Serial Correlation in Least Squares
Regression, II." *Biometrika*, 1951, 30, 159–178. Reproduced by permission of the *Biometrika*
Trustees.

TABLE XVI

Critical Values of
Spearman's Rank
Correlation Coefficient

The α values correspond to a one-tailed test of H_0: $\rho_S = 0$. The value should be doubled for two-tailed tests.

n	$\alpha = .05$	$\alpha = .025$	$\alpha = .01$	$\alpha = .005$
5	.900	—	—	—
6	.829	.886	.943	—
7	.714	.786	.893	—
8	.643	.738	.833	.881
9	.600	.683	.783	.833
10	.564	.648	.745	.794
11	.523	.623	.736	.818
12	.497	.591	.703	.780
13	.475	.566	.673	.745
14	.457	.545	.646	.716
15	.441	.525	.623	.689
16	.425	.507	.601	.666
17	.412	.490	.582	.645
18	.399	.476	.564	.625
19	.388	.462	.549	.608
20	.377	.450	.534	.591
21	.368	.438	.521	.576
22	.359	.428	.508	.562
23	.351	.418	.496	.549
24	.343	.409	.485	.537
25	.336	.400	.475	.526
26	.329	.392	.465	.515
27	.323	.385	.456	.505
28	.317	.377	.448	.496
29	.311	.370	.440	.487
30	.305	.364	.432	.478

Source: From E. G. Olds, "Distribution of Sums of Squares of Rank Differences for Small Samples," *Annals of Mathematical Statistics*, 1938, 9. Reproduced with the permission of the Editor, *Annals of Mathematical Statistics*.

DEMOGRAPHIC DATA SET

A demographic data set was assembled based on a systematic random sample of 1,000 United States zip codes. To obtain the sample, the more than 30,000 zip codes were sorted, and approximately every 30th was selected. The map in Figure C.1 shows the number of zip codes selected in each state. Note that each state is classified according to its census region: North Central, Northeast, South, and West.

Demographic data for each zip code area selected were supplied by CACI, an international demographic and market information firm, and are reproduced with its permission. CACI produces interim estimates of many of the demographic variables measured decennially by the Bureau of the Census. CACI also produces market information based on the Bureau of the Census Consumer Expenditure

FIGURE C.1   Number of Zip Code Areas Selected by State and Census Region

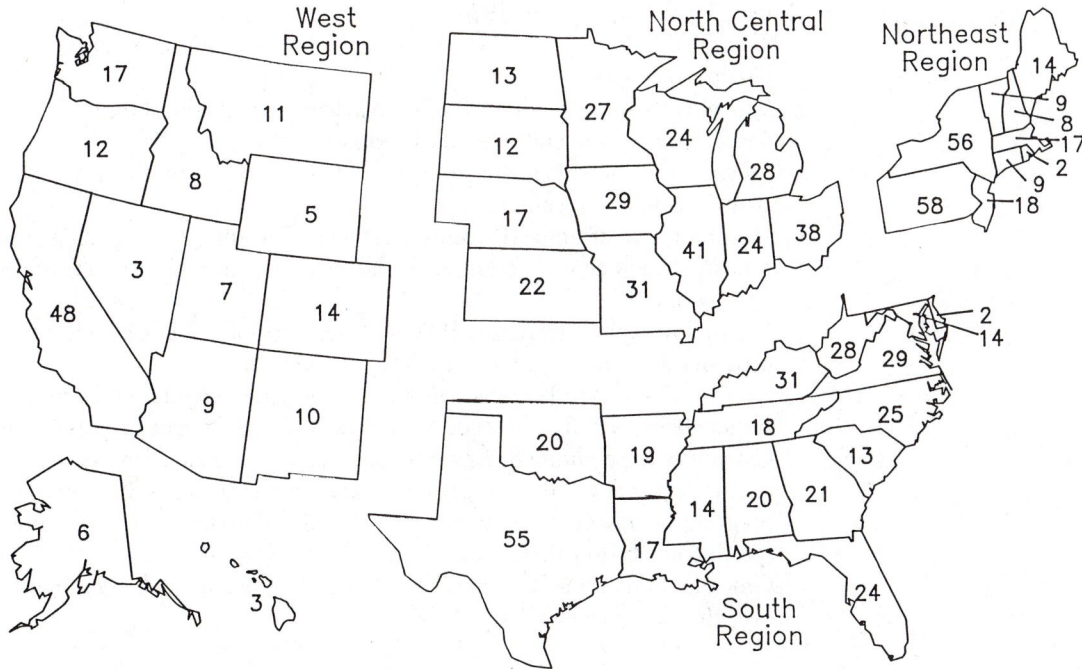

Survey (such as the four purchasing indexes given below for each zip code), which measure the relative propensity of a given market (zip code) to purchase the goods, a higher index value indicating a higher propensity to buy such goods as compared to other similar zip codes in the same census region. In general, the 1980 measurements are official U.S. Bureau of the Census estimates, while the 1986 measurements are CACI estimates.

Fifteen demographic measurements are presented for each zip code area. Portions of the data are referenced at the end of each chapter in "Using the Computer." The objectives are to enable the student to analyze real data in a relatively large sample using the computer, and to gain experience using statistical techniques and concepts on real data. Of course, neither the student nor the instructor need be bound by the suggestions in "Using the Computer"; the data are rich enough to support many more analyses than could be listed (or imagined) by the authors.

The following 15 measurements are reported for each zip code in the sample:

1. Population (1986):   Total population for the zip code, 1986.
2. Number of Households (1986):   Number of households for the zip code in 1986.
3. Age (Median, 1986):   Median age for the zip code in 1986.
4. Household Income (Median, 1986):   Median household income for the zip code in 1986.
5. Home Value (Median, 1980):   Median residential owner-occupied home value for the zip code in 1980.
6. Monthly Cost of Housing (Median, 1980):   Median monthly homeowner cost, including mortgage payments, real estate taxes, property insurance, utilities, and fuels, for the zip code in 1980.
7. Household Size (Average, 1980):   Average number of persons per household in the zip code, 1980.
8. Years of Education (Median, 1980):   Median years of education for adults (persons 25 years of age and over) in the zip code, 1980.
9. College Education (Percentage, 1980):   Percentage of adults in the zip code having a college education.
10. Women in Labor Force (Percentage, 1980):   Percentage of women age 16 and over in the zip code who are in the labor force (either having jobs or actively seeking one).
11. Unemployment (Percentage, 1980):   Percentage of the labor force in the zip code that is unemployed and actively seeking work.
12. Average Purchasing Potential (1986):   Purchasing potential index for all consumer items over the zip code area, based on the Bureau of Labor Statistics Consumer Expenditure Survey. The index compares a zip code to similar urban or rural zip codes in the same census region, with 100 set as the average value.
13. Sporting Goods Purchasing Potential (1986):   Purchasing potential index for sporting goods over the zip code area.
14. Groceries Purchasing Potential (1986):   Purchasing potential index for groceries over the zip code area.

15. Home Improvement Purchasing Potential (1986):   Purchasing potential index for home improvements over the zip code area.

Numerical codes have been assigned to each state and census region to facilitate computer utilization of these data. The codes were assigned alphabetically as shown in Tables C.1 and C.2.

TABLE C.1
Census Region Codes

CODE	CENSUS REGION
1	North Central region
2	Northeast region
3	South region
4	West region

TABLE C.2
State Codes

CODE	STATE		CODE	STATE	
01	AK:	Alaska	26	MT:	Montana
02	AL:	Alabama	27	NC:	North Carolina
03	AR:	Arkansas	28	ND:	North Dakota
04	AZ:	Arizona	29	NE:	Nebraska
05	CA:	California	30	NH:	New Hampshire
06	CO:	Colorado	31	NJ:	New Jersey
07	CT:	Connecticut	32	NM:	New Mexico
08	DE:	Delaware	33	NV:	Nevada
09	FL:	Florida	34	NY:	New York
10	GA:	Georgia	35	OH:	Ohio
11	HI:	Hawaii	36	OK:	Oklahoma
12	IA:	Iowa	37	OR:	Oregon
13	ID:	Idaho	38	PA:	Pennsylvania
14	IL:	Illinois	39	RI:	Rhode Island
15	IN:	Indiana	40	SC:	South Carolina
16	KS:	Kansas	41	SD:	South Dakota
17	KY:	Kentucky	42	TN:	Tennessee
18	LA:	Louisiana	43	TX:	Texas
19	MA:	Massachusetts	44	UT:	Utah
20	MD:	Maryland	45	VA:	Virginia
21	ME:	Maine	46	VT:	Vermont
22	MI:	Michigan	47	WA:	Washington
23	MN:	Minnesota	48	WI:	Wisconsin
24	MO:	Missouri	49	WV:	West Virginia
25	MS:	Mississippi	50	WY:	Wyoming

The demographic data base is contained in two ASCII files, which are both sorted by census region and state within census region. The file names and corresponding layouts are given in Tables C.3 and C.4 (page 1216). The data files are available on magnetic tape or diskette from the publisher.

TABLE C.3

ZIPCOD01.DAT

COLUMNS	DESCRIPTION
1–4	Observation number
7	Census region number
9–10	State number
12–16	Zip code
18–22	Population in '86
24–28	Number of households in '86
30–33	Median age in '86
35–39	Median household income in '86
41–46	Median home value in '80
48–50	Median monthly homeowner cost in '80
52–54	Median household size in '80
56–59	Median years of education in '80

TABLE C.4

ZIPCOD02.DAT

COLUMNS	DESCRIPTION
1–4	Observation number
7	Census region number
9–10	State number
12–16	Zip code
18–21	Percent college education in '80
23–26	Percent women in work force in '80
28–31	Percent unemployed in '80
33–37	Purchasing potential index '86: Overall average
39–43	Purchasing potential index '86: Sporting goods
45–49	Purchasing potential index '86: Grocery
51–55	Purchasing potential index '86: Home improvement

CALCULATION FORMULAS FOR ANALYSIS OF VARIANCE

CONTENTS

CM = Correction for mean

$$= \frac{(\text{Total of all observations})^2}{\text{Total number of observations}} = \frac{\left(\sum y_i\right)^2}{n}$$

$SS(\text{Total})$ = Total sum of squares

$$= (\text{Sum of squares of all observations}) - CM = \sum y_i^2 - CM$$

SST = Sum of squares for treatments

$$= \left(\begin{array}{c}\text{Sum of squares of treatment totals with} \\ \text{each square divided by the number of} \\ \text{observations for that treatment}\end{array}\right) - CM$$

$$= \frac{T_1^2}{n_1} + \frac{T_2^2}{n_2} + \cdots + \frac{T_p^2}{n_p} - CM$$

SSE = Sum of squares for error = $SS(\text{Total}) - SST$

MST = Mean square for treatments = $\dfrac{SST}{p - 1}$

MSE = Mean square for error = $\dfrac{SSE}{n - p}$

F = Test statistic = $\dfrac{MST}{MSE}$

where

n = Total number of observations

p = Number of treatments

T_i = Total for treatment i $(i = 1, 2, \ldots, p)$

CM = Correction for mean

$$= \frac{(\text{Total of all observations})^2}{\text{Total number of observations}} = \frac{\left(\sum y_i\right)^2}{n}$$

$SS(\text{Total})$ = Total sum of squares

$$= (\text{Sum of squares of all observations}) - CM = \sum y_i^2 - CM$$

SST = Sum of squares for treatments

$$= \left(\begin{array}{c}\text{Sum of squares of treatment totals with} \\ \text{each square divided by } b, \text{ the number of} \\ \text{observations for that treatment}\end{array}\right) - CM$$

$$= \frac{T_1^2}{b} + \frac{T_2^2}{b} + \cdots + \frac{T_p^2}{b} - CM$$

$\text{SSB} = \text{Sum of squares for blocks}$

$$= \left(\begin{array}{c} \text{Sum of squares of block totals with} \\ \text{each square divided by } k, \text{ the number} \\ \text{of observations in that block} \end{array} \right) - \text{CM}$$

$$= \frac{B_1^2}{p} + \frac{B_2^2}{p} + \cdots + \frac{B_p^2}{p} - \text{CM}$$

$\text{SSE} = \text{Sum of squares for error} = \text{SS(Total)} - \text{SST} - \text{SSB}$

$$\text{MST} = \text{Mean square for treatments} = \frac{\text{SST}}{p - 1}$$

$$\text{MSB} = \text{Mean square for blocks} = \frac{\text{SSB}}{b - 1}$$

$$\text{MSE} = \text{Mean square for error} = \frac{\text{SSE}}{n - p - b + 1}$$

$$F = \text{Test statistic} = \frac{\text{MST}}{\text{MSE}}$$

where

$n = \text{Total number of observations}$

$b = \text{Number of blocks}$

$p = \text{Number of treatments}$

$T_i = \text{Total for treatment } i \ (i = 1, 2, \ldots, p)$

$B_i = \text{Total for block } i \ (i = 1, 2, \ldots, b)$

S E C T I O N D.3

**FORMULAS FOR THE
CALCULATIONS FOR A
TWO-FACTOR
FACTORIAL
EXPERIMENT**

$\text{CM} = \text{Correction for the mean}$

$$= \frac{(\text{Total of all } n \text{ measurements})^2}{n} = \frac{\left(\sum_{i=1}^{n} y_i \right)^2}{n}$$

$\text{SS(Total)} = \text{Total sum of squares}$

$$= \text{Sum of squares of all } n \text{ measurements} - \text{CM} = \sum_{i=1}^{n} y_i^2 - \text{CM}$$

$\text{SS(A)} = \text{Sum of squares for main effects, factor A}$

$$= \left(\begin{array}{c} \text{Sum of squares of the totals } A_1, A_2, \ldots, A_a \\ \text{divided by the number of measurements} \\ \text{in a single total, namely } br \end{array} \right) - \text{CM}$$

$$= \frac{\sum_{i=1}^{a} A_i^2}{br} - \text{CM}$$

$SS(B)$ = Sum of squares for main effects, factor B

$$= \left(\begin{array}{c} \text{Sum of squares of the totals } B_1, B_2, \ldots, B_b \\ \text{divided by the number of measurements} \\ \text{in a single total, namely } ar \end{array} \right) - CM$$

$$= \frac{\displaystyle\sum_{i=1}^{b} B_i^2}{ar} - CM$$

$SS(AB)$ = Sum of squares for AB interaction

$$= \left(\begin{array}{c} \text{Sum of squares of the cell} \\ \text{totals } AB_{11}, AB_{12}, \ldots, AB_{ab} \\ \text{divided by the number of} \\ \text{measurements in a single} \\ \text{total, namely } r \end{array} \right) - SS(A) - SS(B) - CM$$

$$= \frac{\displaystyle\sum_{j=1}^{b}\sum_{i=1}^{a} AB_{ij}^2}{r} - SS(A) - SS(B) - CM$$

where

a = Number of levels of factor A

b = Number of levels of factor B

r = Number of replicates (observations per treatment)

A_i = Total for level i of factor A ($i = 1, 2, \ldots, a$)

B_i = Total of level i of factor B ($i = 1, 2, \ldots, b$)

AB_{ij} = Total for Treatment (i, j), i.e., for ith level of factor A and jth level of factor B

ANSWERS TO SELECTED EXERCISES

CHAPTER 2

2.1a. Quantitative   **b.** Qualitative   **c.** Quantitative   **d.** Quantitative   **e.** Qualitative   **f.** Quantitative
2.2a. Qualitative   **b.** Quantitative   **c.** Quantitative   **d.** Quantitative
2.3a. Qualitative   **b.** Quantitative   **c.** Quantitative   **d.** Qualitative   **e.** Quantitative
2.4a. Qualitative   **b.** Quantitative   **c.** Qualitative
2.5a. Quantitative   **b.** Quantitative   **c.** Quantitative   **d.** Qualitative   **2.8b.** Philip Morris; 211.8 billion
2.11c. Burnsville; St. Paul   **2.15c.** $105 million   **2.16b.** No. Total outlays are much larger than total receipts.
2.18a. $3.18 billion; $1.85 billion   **2.21c.** $.14   **2.22a.** 31.65%   **2.23c.** 6,026,000; 15,851,000   **2.25a.** 23
2.33b. .64; .64   **2.34a.** Frequency histogram   **b.** 14   **c.** 49   **2.38b.** .65   **2.39b.** .77
2.46a. Cumulative relative frequency   **b.** 5   **c.** No   **2.49a.** .9   **b.** .1   **2.50e.** .6; .34
2.51b. .66; .29   **c.** 63   **d.** 64   **2.54a.** Qualitative   **b.** Quantitative   **c.** Quantitative   **2.56d.** $3.84
2.62a. Frequency bar chart   **2.63b.** No   **2.65a.** Pie chart   **2.70c.** .31; .25

CHAPTER 3

3.1a. 34.35%   **b.** 30.6%   **c.** Skewed to the right   **3.2a.** 180; 162   **b.** Skewed to the right   **c.** No
3.3b. $-.22$; 0; 0 and $\frac{1}{8}$ **3.4** 26.303   **3.5a.** 135.24; 134.8; 3 modes: 131.2, 131.7, 139.0   **c.** 135.19
3.6a. McDonald's: $1,068,493.15; Burger King: $705,882.35   **b.** 41.67; 9.62; 1.37   **3.7** $.35
3.8a. $1,269; $943   **b.** $2,293; $431
3.11a. Skewed to the right; mean is 33.42; median is 29   **c.** Mean is 35.41; median is 29
3.12a. 2; 37.5; 6.124   **b.** 1.25; 1.835; 1.355
3.13a. 30; 200; 900   **b.** 9; 39; 81   **c.** 157; 10,069; 24,649   **d.** 9; 43; 81   **e.** 12; 102; 144
3.14a. 6; 5; 2.236   **b.** 1.5; 5.1; 2.258   **c.** 39.25; 1,302.25; 36.087   **d.** 1.8; 6.7; 2.588   **e.** 2; 15.6; 3.950
3.15a. 3.6; 5.3; 2.3   **b.** 4.33; 7.07; 2.66   **c.** 3.33; 4.27; 2.07   **d.** 2.25; 4.92; 2.22
3.16a. 6.2; 77.2; 8.786   **b.** 28.25; 2,304.92; 48.010
3.17a. 5.6; 17.3; 4.16   **b.** 13.75 feet; 152.25 square feet; 12.34 feet   **c.** -2.5; 4.3; 2.07
d. .33 ounce; .059 square ounce; .24 ounce   **3.22a.** 8.38%   **b.** 6.910; 2.629   **c.** Yes   **d.** No   **3.23** Increase; decrease
3.24a. 52.5; 299.29; 17.3   **b.** (millions of $); (millions of $)^2; (millions of $)   **c.** Decrease; decrease
3.25a. 316.2; 329.7   **b.** 21.4; 21.0   **c.** 6.99   **d.** Minneapolis/St. Paul
3.26a. A: 33.14; B: 34.57   **b.** A: 3.98; B: .98   **d.** A: .82; B: 2.12
3.27a. Germany: 101; Italy: 98; France: 80; U.K.: 181; Belgium: 32
b. Germany: 34.2; Italy: 30.8; France: 24.1; U.K.: 67.5; Belgium: 11.0
c. Belgium, France, Italy, Germany, U.K.; Belgium, France, Italy, Germany, U.K.   **d.** Yes; no
3.28a. Between 104.17 and 156.25   **3.29b.** 40.06; 4.74; 2.18   **d.** Yes   **e.** 68%; 95%; 100%   **f.** 75%; 94%; 100%
3.30a. 104   **b.** 12,321; 111   **d.** At most 100%; at most 75%; at most 11.1%   **e.** 15%; 5%; 0%
3.31a. .149; .027; .165   **b.** Chebyshev's theorem does not apply; 84%; 93.8%   **c.** 68%; 96%; 100%   **d.** 0
3.32a. No   **b.** At most 25%   **c.** 2%
3.33a. Chebyshev's theorem   **b.** $96,732; $45,322
c. At least 75%; at most 11.1%; 94.4% (34 measurements); 2.8% (1 measurement)   **d.** 4.26   **3.34** At most 13 weeks
3.35 .025   **3.36a.** Revenues: 88.44, 7,144.53; profits: -5.5556, 95.53; miles: 833.89, 1,193,296.6   **3.37** At least 88.9%
3.38 2.5%; 2.5%; 0%   **3.39a.** At least $\frac{8}{9}$ **b.** Approx. .16   **c.** Yes   **3.40** 11:30; 4:00   **3.41** Do not buy   **3.42** Median
3.43a. 1; sample   **b.** -1.5; sample   **c.** 1.765; population   **d.** $-.6$; population   **3.44a.** 3.8; 4.1   **b.** 23.5
3.45a. Brazil, -2.389; Egypt, 2.510; France, $-.064$; Italy, -1.547; Japan, -19.289; Mexico, -7.694; Panama, .411;
Soviet Union, 1.656; Sweden, $-.848$; Turkey, .463   **b.** Japan: -2.56; Soviet Union: .67
3.46a. 8.04; 3.68   **b.** U.S.: .42; Australia: .53; Japan: -1.45 **3.47** 1984: 86th percentile; 1981: 71st percentile
3.48a. .4   **b.** .4   **c.** No   **3.49a.** 68%; 100%   **b.** 31   **c.** First store   **3.50** 60; 10
3.51a. Greenspan: $-.20$; Wilson: 1.95   **3.52a.** At most 25%; cannot be determined; at most 11%   **b.** 1.5; 1.5
3.53a. 2.5%; 16%; 0%   **b.** 3.5   **c.** Over $190   **3.54** Upper quartile

3.55a. 39   **b.** 45; 31.5   **c.** 13.5   **d.** Skewed to the left   **e.** 50%; 75%   **3.58b.** 99   **3.61a.** 5.10; -2.14; 0; yes
3.65a. Honda   **b.** Honda   **c.** Ford   **d.** Honda; approx. 51.1
3.67a. 239; 29; 841   **b.** 634; 50; 2,500   **c.** 254; 2; 4   **d.** 10,004; 102; 10,404
3.68a. 17.7; 4.21   **b.** 33.5; 5.79   **c.** 50.67; 7.12   **d.** 2,467.67; 49.68   **3.69a.** 3.12   **b.** 9.02   **c.** 9.79
3.70a. 6; 27; 5.20   **b.** 6.25; 28.25; 5.32   **c.** 7; 37.67; 6.14   **d.** 3; 0; 0   **3.71a.** 12   **b.** 13   **c.** 15   **d.** 0
3.72a. 5.67; 1.066; 1.03   **b.** $-\$1.5$; 11.5 dollars squared; $3.39   **c.** .4125%; .088% squared; .30%
3.74a. 12.4; 19.42   **b.** 1.03; .08   **3.79a.** $1,552 million   **b.** $3,236 million   **3.81a.** At least 0%   **b.** At most 25%
3.82a. Approx. 68%   **b.** Approx. 2.5%   **3.85a.** $7,500   **b.** Over $120,000   **c.** Very unlikely   **3.86c.** 35.6; 33; 28
3.87 At least $\frac{8}{9}$ **3.88a.** .56; 3.30   **3.89** No information; $V = 0$ for all data sets
3.90a. -3.0 **b.** Approx. 0%   **c.** 90 or above
3.91a. Magazines   **b.** Magazines; newspapers   **c.** Magazines; newspapers   **e.** No
3.92a. At least .75   **b.** At least .89   **c.** At least .56   **d.** At most .13
3.93a. 26.6; 12.0   **b.** 26.3; 51.2   **3.94a.** 56.1; 21.9
3.95a. August, $R = 11$, while for June, $R = 10$; June, $s^2 = 17.48$, while for August, $s^2 = 11.9$; s^2 since R is affected by extreme values
b. $s^2 = 17.48$; no effect   **c.** $s^2 = 157.3$; s^2 is multiplied by the square of the constant

3.96a.

	U.S.	CANADA	U.K.	SWEDEN	FRANCE	W. GERMANY
Mean	41.08%	70.83%	60.42%	83.42%	66.33%	56.08%
Variance	330.08	529.24	634.26	511.17	1,051.88	907.17

b.

	MEAN	VARIANCE
Brewing	54.17	640.97
Cigarettes	91.00	142.00
Fabric	35.67	365.07
Paints	40.67	727.47
Petroleum	62.50	666.70
Shoes	20.33	71.87
Glass	85.83	210.97
Cement	67.67	805.07
Steel	60.67	352.67
Bearings	80.67	390.27
Refrigerators	77.50	202.70
Batteries	79.67	269.87

c. U.S., Sweden; antifriction bearings

d. Most competitive: shoes, fabric, paints; least competitive: cigarettes, glass bottles, antifriction bearings

CHAPTER 4

4.2a. Venn diagram   **b.** $P(A) = .3$; $P(B) = .2$ **c.** $P(A) = .25$; $P(B) = .3$ **4.3c.** $P(A) = \frac{1}{3}$; $P(B) = \frac{1}{2}$
4.4a. .4   **b.** .25   **c.** .6   **4.5a.** 120   **b.** 15   **c.** 56   **d.** 1   **e.** 1   **4.6b.** .152   **c.** .252
4.7 $P(A) = .16$; $P(B) = .64$ **4.8** $\frac{1}{20}$ **4.9** $P(A) = \frac{1}{6}$; $P(B) = \frac{5}{6}$ **4.11a.** $\frac{10}{35}$ **b.** $\frac{20}{35}$ **c.** $\frac{5}{35}$ **4.12c.** $\frac{35}{95}$
4.13a. .5   **b.** .225   **c.** .2   **4.14b.** $\frac{1}{18}$ **c.** $\frac{6}{18}$ **4.15a.** $\frac{1}{15}$, $\frac{1}{15}$, $\frac{1}{15}$ **b.** 0   **c.** 0; 1   **d.** 0; 1
4.16a. $\frac{1}{3}$ **b.** $\frac{1}{3}$ **c.** $\frac{1}{6}$ **d.** $\frac{1}{6}$ **4.17a.** 1 to 2   **b.** .5   **c.** .4
4.18a. .5   **b.** .19   **c.** .5   **d.** 1   **e.** .31   **f.** .69
4.19a. $A = \{(1, 4), (2, 3), (3, 2), (4,1)\}$; $B = \{(1, 3), (2, 3), (3, 3), (4, 3), (5, 3), (6, 3), (3, 1), (3, 2), (3, 4), (3, 5), (3, 6)\}$;
$A \cap B = \{(2, 3), (3, 2)\}$; $A \cup B = \{(1, 3), (2, 3), (3, 3), (4, 3), (5, 3), (6, 3), (3, 1), (3, 2), (3, 4), (3, 5), (3, 6), (1, 4), (4, 1)\}$
b. $\frac{4}{36}$; $\frac{11}{36}$ **c.** $\frac{2}{36}$; $\frac{13}{36}$ **4.20a.** $\frac{7}{8}$ **b.** $\frac{3}{8}$ **c.** $\frac{3}{8}$ **d.** 0
4.21a. $\frac{4,000}{8,000}$ **b.** $\frac{2,550}{8,000}$ **c.** $\frac{1,000}{8,000}$ **d.** $\frac{6,000}{8,000}$ **e.** $\frac{6,000}{8,000}$ **f.** $\frac{2,500}{8,000}$ **4.22a.** $A \cap F$ **b.** $B \cup C$; A^c
4.23a. $B \cap C$ **b.** A^c **c.** $C \cup B$ **d.** $A \cap C^c$
4.24a. Yes   **b.** $P(A) = .26$; $P(B) = .35$; $P(C) = .72$; $P(D) = .28$; $P(E) = .05$ **c.** .56; .05; .77   **d.** .74
4.25a. (1, R), (2, R), (3, R), (1, S), (2, S), (3, S), (1, E), (2, E), (3, E)   **b.** Sample space   **c.** .46   **d.** .05   **e.** .30
f. .61   **g.** .34   **4.26a.** .03   **b.** .62   **c.** .28   **d.** .32   **e.** .72   **4.27b.** .152; yes   **c.** .004   **d.** .514   **e.** .392

4.28a. $P \cap C \cap A$; 6 and 7 **b.** $\frac{1}{5}$ **c.** $A \cup C$; $\frac{4}{5}$ **d.** $P \cap C$; $\frac{3}{10}$
4.30a. $P(A) = .48$; $P(B) = .52$; $P(C) = .16$; $P(D) = .26$; $P(E) = .35$; $P(F) = .23$ **b.** 1 **c.** .05 **d.** .05 **e.** 0 **f.** 0
4.31a. .74 **b.** .59 **c.** .37 **d.** .11 **4.32b.** .30 **c.** .59 **d.** .41 **e.** .96
4.33a. .047 **b.** .287 **c.** .162 **d.** .688 **4.34a.** .65 **b.** 0 **c.** 0 **d.** .60 **e.** No
4.35a. $P(A) = .5, P(B) = .5; P(A \cap B) = 4$ **b.** $P(E_1|A) = \frac{.1}{.5} = .2; P(E_2|A) = \frac{.1}{.5} = .2; P(E_3|A) = \frac{.3}{.5} = .6$
c. .8 **d.** No **4.36a.** .37 **b.** .68 **c.** .15 **d.** .2206 **e.** 0 **f.** 0 **4.37** All pairs are dependent.
4.38a. $\frac{3}{8}$ **b.** $\frac{7}{8}$ **c.** $\frac{3}{8}$ **d.** $\frac{7}{8}$ **e.** 0 **f.** $\frac{4}{8}$ **4.39** $P(R \mid S) = 0$; $P(S \mid R) = 0$ **4.40** Yes
4.41a. $\frac{5}{8}$ **b.** $\frac{1}{2}$ **c.** $\frac{100}{125}$ **d.** $\frac{5}{8}$ **e.** Dependent **f.** No **4.42a.** .0118 **b.** .00314; .9906 **c.** .2067; .1003 **d.** .04
4.43a. .9883 **b.** .0117 **c.** .7917 **d.** .1183 **4.44** The events in a are mutually exclusive. **4.45** .005
4.46a. $\frac{83}{151}$ **b.** $\frac{68}{151}$ **c.** $\frac{41}{151}$ **d.** $\frac{6}{151}$ **e.** $\frac{6}{41}$ **4.47a.** .568 **b.** .24 **c.** .75 **d.** 1 **4.48** .6
4.49a. .18 **b.** .82 **c.** .31 **d.** .69 **e.** .26 **4.50a.** .2778 **b.** .58 **4.51** .1513
4.52a. .45 **b.** $\frac{1}{45}$ **4.55a.** $\frac{1}{12}$; $\frac{1}{6}$ **b.** 220 **c.** $\frac{1}{220}$, $\frac{1}{220}$
4.56a. {(0001, 0002), (0001, 0003), (0001, 0004), (0001, 0005), (0002, 0003), (0002, 0004), (0002, 0005), (0003, 0004), (0003, 0005), (0004, 0005)} **b.** $\frac{1}{10}$ **c.** $\frac{1}{10}$; $\frac{3}{10}$ **4.60a.** $\frac{1}{23,188,100}$, .48 **b.** $\frac{1}{6,227,993}$, .43 **c.** $\frac{1}{\binom{23,188,100}{2}}$ **4.62** Yes **4.63** .5

4.64 The events in c are mutually exclusive. **4.65a.** .8145 **b.** .0135 **d.** No **4.66** 462
4.67a. $\frac{150}{200}$ **b.** $\frac{23}{80}$ **c.** $\frac{90}{150}$ **d.** $\frac{12}{200}$ **4.68a.** No **c.** Yes **4.69** 252 **4.70a.** $\frac{124}{153}$, $\frac{111}{153}$ **b.** $\frac{101}{153}$, $\frac{19}{153}$ **c.** $\frac{33}{153}$, $\frac{120}{153}$
4.71 N: Noticed ad; N^c: Did not notice ad; L: Under 30; M: 30–50; H: Over 50;
a. $(L, N), (M, N), (H, N), (L, N^c), (M, N^c), (H, N^c)$ **b.** Sample space
c. $P(L, N) = .25$; $P(M, N) = .20$; $P(H, N) = .10$; $P(L, N^c) = .05$; $P(M, N^c) = .15$; $P(H, N^c) = .25$
4.73b. .95 **c.** .25 **d.** .50 **4.74a.** .26 **4.76a.** $\frac{6}{30}$ **b.** $\frac{6}{29}$ **c.** .634 **4.77a.** .7127 **b.** .2873
4.78a. .9639 **b.** .3078 **c.** .0361 **d.** 3 **4.80b.** .05 **4.81a.** $\frac{1}{15}$ **b.** $\frac{6}{15}$ **c.** $\frac{14}{15}$ **d.** $\frac{6}{14}$ **e.** $\frac{5}{14}$
4.82a. .429 **b.** .24 **c.** .221 **d.** .875 **4.84a.** $\frac{253}{650}$ **b.** $\frac{153}{650}$ **c.** $\frac{78}{650}$ **d.** $\frac{387}{650}$ **4.85a.** .236 **b.** .671
4.86a. .0000106 **b.** .07887 **c.** .0704 **4.87** .79 **4.88** No **4.91** False

CHAPTER 5

5.3a. Discrete **b.** Continuous **c.** Continuous **d.** Discrete
5.4a. Discrete **b.** Discrete **c.** Continuous **d.** Discrete **e.** Continuous

5.11a. HHH, THH, HTH, HHT, TTH, THT, HTT, TTT **b.**

x	0	1	2	3
p(x)	$\frac{1}{8}$	$\frac{3}{8}$	$\frac{3}{8}$	$\frac{1}{8}$

5.12a.

x	1	2	3	4	5	6
p(x)	$\frac{1}{6}$	$\frac{1}{6}$	$\frac{1}{6}$	$\frac{1}{6}$	$\frac{1}{6}$	$\frac{1}{6}$

5.13a. Invalid **b.** Valid **c.** Invalid **d.** Invalid **5.14a.** .8 **b.** .2 **c.** 1 **d.** .1 **e.** .5
5.15a. .10 **b.** 0 **c.** .25 **d.** .70 **e.** .90 **f.** .80 **5.16b.** .56; .32

5.17a.

x	0	1	2	3
p(x)	.000125	.007125	.135375	.857375

c. .99275 **5.18b.** .00006 **5.19**

x	0	1	2
p(x)	.2	.3	.5

5.21a. 4.25 **5.22a.** .5 **d.** No **5.23a.** 31; 169; 13 **c.** .95 **5.24a.** .3; .011; .105 **c.** .85 **d.** 1.0
5.25a. Expected total loss is $2,450 for both firms. **b.** Firm B **c.** Pure risk **5.26b.** $90,000 **5.27a.** .8 **b.** No
5.28a. 173.5; 147.54 **b.** 23.37; 21.73 **5.29a.** 4.978 **b.** 2.984 **c.** Approx. .949
5.30a. 3.26 **b.** 1.6324; 1.2777 **d.** .95 **5.31** $E(x) = \$13,000$; market new line **5.32** $1,000
5.33a. $7,200 **b.** $11,000 **c.** Too little

5.34a.

Total cost	$1,000	$2,000	$3,000
p(Total cost)	.25	.25	.50

b. .25 **c.** $2,250 **d.** $2,250

1224

5.35a. 10   **b.** 20   **c.** 1   **d.** 1   **e.** 6   **5.36a.** Discrete   **b.** Binomial   **d.** 2.4; 1.2

5.37a.

x	0	1	2	3	4	5
p(x)	.32768	.4096	.2048	.0512	.0064	.00032

5.38a. .1852   **b.** .3456   **c.** .04

5.39a.

x	0	1	2	3	4	5	6	7	8	9	10	11
p(x)	.000488	.005371	.026855	.080566	.161133	.225586	.225586	.161133	.080566	.026855	.005371	.000488

b. 5.5; 2.75   **d.** .93457   **5.40a.** .11328   **b.** .5   **c.** .005859
5.41a. .005   **b.** .973   **c.** .403   **d.** .966   **e.** .009   **f.** .207
5.42e. $p = .5$, symmetric; $p < .5$, skewed to the right; $p > .5$, skewed to the left   **5.43** .009
5.44a. .02; .08   **b.** .0922; .00384   **c.** .2866; .0544   **d.** .8129   **e.** Independence   **5.45** .0005
5.46a. .000672 or approx. 0   **5.47a.** Approx. 0   **b.** .151   **5.48a.** .013   **b.** .783
5.49b. 2.4; 1.47   **c.** Binomial with $p = .90$, $q = .10$, and $n = 24$; 21.6; 1.47
5.50 $\mu = .5$; $\sigma = .707$; no; guarantee is probably inaccurate
5.51a. 0   **b.** Approx. 0   **c.** Approx. 0   **d.** .098   **e.** .873   **f.** 1.0
5.52a. 0   **b.** Approx. 0   **c.** Approx. 0   **d.** .234   **e.** .966   **f.** 1.0   **5.53** $\mu = 48$; $\sigma = 6.5$; no; no
5.54a. Discrete   **b.** Poisson   **d.** $\mu = 2$; $\sigma = 1.414$ **e.** 2; 1.414   **5.55a.** .9197   **b.** .6767   **c.** .4232   **d.** Decreases
5.56a. .80885   **b.** .442175   **c.** .25102   **d.** .22313   **e.** .77687   **f.** .004456
5.57b. $\mu = 2$; $\sigma = 1.414$; 2 ± 2.828 **c.** .947345   **5.58b.** $\mu = 4$; $\sigma = 2$; 4 ± 4 **c.** .978656
5.59 Binomial: $p(0) = .277$, $p(1) = .365$, $p(3) = .231$; Poisson: $p(0) = .2865$, $p(1) = .35813$, $p(2) = .22383$
5.60a. $\mu = 8.5$; $\sigma = 2.915$ **b.** .1104   **c.** .5145   **d.** .3856   **5.61a.** $\sigma = 2$ **b.** No; $P(x > 10)$ very small
5.62a. .000177   **b.** $E(x) = 8.64$; $\sigma = 2.94$ **c.** Approx. 0   **5.63** .1929; .6602
5.64 .090; .91   **5.65** .2510; .5578; .0111   **5.66** .0803; yes   **5.67** .632   **5.68a.** .76493   **b.** .11615
5.71a. Discrete   **b.** Hypergeometric   **d.** $\mu = 2.8$; $\sigma = .748$ **e.** 2.8; .748
5.72a. .267857   **b.** .017857   **c.** .178571   **d.** 0

5.73a.

x	2	3	4	5
p(x)	$\frac{21}{252}$	$\frac{105}{252}$	$\frac{105}{252}$	$\frac{21}{252}$

b. 3.5; .5833   **c.** 3.5 ± 1.53 **d.** 1

5.74a.

x	2	3	4	5	6
p(x)	$\frac{15}{495}$	$\frac{120}{495}$	$\frac{225}{495}$	$\frac{120}{495}$	$\frac{15}{495}$

b. 4; .8528   **c.** 4 ± 1.71 **d.** .9394

5.75a. 0   **b.** .4545   **c.** .7273   **d.** .2727   **e.** .0303   **f.** 0   **5.76a.** Hypergeometric   **b.** Binomial   **5.77** .25

5.78a. 0   **b.** Weaken, claim not true   **5.79** .6; .333; .167; .0714;

5.82a.

x	0	1	2	3	4
p(x)	$\frac{70}{495}$	$\frac{224}{495}$	$\frac{168}{495}$	$\frac{32}{495}$	$\frac{1}{495}$

5.83a. 113.24; 4.19   **5.84a.** Discrete   **b.** Geometric   **d.** $\mu = 10$; $\sigma = 9.487$ **e.** 10; 9.487
5.85a. .16   **b.** .128   **c.** .488   **d.** 1   **e.** .1024   **f.** .64   **5.86b.** 1.429; .612   **c.** .91
5.87b. 3.333; 7.778   **c.** .8824   **5.88a.** .0625   **b.** .9375   **c.** $\mu = 2$; $\sigma = 1.414$ **d.** Small
5.89b. .02544; .01424   **d.** $\mu = 35$; $\sigma = 34.496$; 103.993   **5.90** .7536   **5.91a.** .995   **b.** Claim probably not valid
5.92 .729; 10   **5.93** .0819   **5.94a.** Poisson   **b.** Geometric   **c.** Hypergeometric   **d.** Binomial   **e.** Binomial
5.95a. .1536   **b.** .0768   **c.** .125   **5.96a.** .5357   **b.** .0667   **c.** .8   **5.97a.** .147   **b.** .0384   **c.** .16
5.98a. .2240   **b.** .1804   **c.** .0126   **5.99a.** .0183   **b.** .1954   **c.** .0733   **d.** .1563   **e.** .2381   **f.** .9084

5.100a.

x	1	2	3	4	5	6	7
p(x)	.4	.24	.144	.086	.052	.031	.019

b. 2.5; 1.936; -1.37 to 6.37   **c.** .953

5.101a. Discrete   **b.** Continuous   **c.** Continuous   **d.** Continuous
5.102 y has the largest variance; x has smallest variance

5.103 x	0	1	2
$p(x)$	$\frac{4}{9}$	$\frac{4}{9}$	$\frac{1}{9}$

5.104a. x	0	1	2	3	4	5
$p(x)$	.590	.329	.072	.009	.000	.000

b. .5; .45

5.105a. $E(x) = 497{,}000$ **b.** $E[(x - \mu)^2] = 6.6056 \times 10^{11}$ **5.106a.** True   **b.** .017   **5.107a.** $50.03   **5.108** .512
5.109a. A: 4.6; B: 3.7   **b.** A: $46,000; B: $55,500   **c.** A: $\sigma^2 = 1.34$, $\sigma = 1.16$; B: $\sigma^2 = 1.21$, $\sigma = 1.10$ **d.** A: .95; B: .95
5.110a. .2   **b.** .003   **c.** .002; probably not   **5.111** .4019; .1608   **5.112** $33,333.33   **5.113** .0996; .0738
5.114 .033; .985; claim probably invalid
5.115 $\frac{11}{12}$, $\frac{1}{30}$ **5.116** .303   **5.117a.** .001   **d.** Yes; $P(x = 20) = .012$ **5.118** .0357   **5.119** .051   **5.120** .0315; .1067
5.121a. .277   **b.** .723   **c.** 2.11; 2.34   **d.** No   **5.122a.** .34272   **b.** .0039   **5.123** .1715   **5.124** .265; .1755
5.125a. 5; 4   **b.** .617   **c.** .006   **5.126** .346; .683

CHAPTER 6

6.1a. .04, $20 \le x \le 45$ **b.** 32.5; 52.08   **c.** 1
6.2a. .6   **b.** .6   **c.** .4   **d.** 0   **e.** .2   **f.** .8   **g.** .36   **h.** .38   **i.** .6   **j.** .02
6.3a. $\frac{1}{3}$, $2 \le x \le 5$ **b.** 3.5; .75   **c.** .577   **6.4a.** 3.5   **b.** 2.6   **c.** 2.0   **d.** 4   **e.** 4.7   **6.5a.** 0   **b.** 1   **c.** 1
6.6 $c = 8.268$, $d = 11.732$; .289, $8.268 \le x \le 11.732$ **6.8b.** .5; .2; .2; 0
6.9b. .5; .083   **c.** .05; .95   **d.** Uniform distribution with $c = .90$ and $d = .95$ **6.10a.** 2, 1.33; .3125   **6.11** .333; 15 min.
6.12a. Continuous   **c.** 7, .289; $\mu \pm 2\sigma$ is 6.423 to 7.577   **6.13a.** .4772   **b.** .4987   **c.** .4332   **d.** .2881
6.14a. .7745   **b.** .9544   **c.** .6247   **d.** .9974
6.15a. .0919   **b.** .4895   **c.** .4772   **d.** .9179   **e.** .2417   **f.** .2857
6.16a. .0013   **b.** .0548   **c.** .05   **d.** .5   **e.** .1587   **f.** .05   **6.17a.** 0   **b.** .8413   **c.** .8413   **d.** .1587
6.18a. .6826   **b.** .95   **c.** .9   **d.** .9544   **6.19a.** .6826   **b.** .95   **c.** .9   **d.** .9544
6.20a. 1.645   **b.** 1.96   **c.** -1.96 **d.** 1.28   **e.** 1.28
6.21a. -1.88 **b.** 1.96   **c.** 1.645   **d.** 1.0   **e.** $-.42$ **f.** 1.47
6.22a. -2.5 **b.** 0   **c.** $-.625$ **d.** -3.75 **e.** 1.25   **f.** -1.25
6.23a. 1.25 below   **b.** 1.875 above   **c.** 0   **d.** 1.5 above
6.24a. .9544   **b.** .0918   **c.** .0228   **d.** .8607   **e.** .0927   **f.** .7049   **6.25a.** .0456; .0026   **b.** .6826; .9544
6.26a. 53   **b.** 55.88   **c.** 45.065   **d.** 53.3   **6.27** 182   **6.28a.** .2514   **b.** 90.375
6.29a. .1151   **b.** .6554   **c.** 10.50   **6.30** .0024   **6.31a.** 456   **b.** 3,174   **c.** 3,174
6.32a. XYZ   **b.** ABC: $105; XYZ: $107   **c.** ABC   **6.33a.** .0307   **b.** .0893   **6.34a.** .0301   **b.** .0301
6.35a. .2843   **b.** .0228   **6.36** 5.068
6.37a. $z_U = .675$; $z_L = -.675$ **b.** 2.7; -2.7 **c.** 4.725; -4.725 **d.** .007; approx. 0
6.39a. .367879   **b.** .082085   **c.** .000553   **d.** .223130   **6.40a.** .002479   **b.** .011109   **c.** .000123   **d.** .259240
6.41a. .999447   **b.** .999955   **c.** .981684   **d.** .632121   **6.42a.** .5; .25; .950213
6.43a. .0183156   **b.** .950213   **c.** .383401   **6.44a.** 20 minutes   **b.** .22313   **c.** .049787; .95021
6.45 .2231   **6.46** .095
6.47a. .367879; .212248   **b.** .049787   **c.** Less   **d.** 40.6
6.48a. $\mu = \sigma = 23.81$ **b.** 23.81   **c.** .716346   **d.** .8647   **e.** .9502
6.50a. $e^{-.5x}$ **b.** .13534   **c.** .7788   **d.** Yes   **e.** 820; 3,934   **f.** Approx. 37 days
6.53a. No   **b.** Yes   **c.** No   **d.** Yes   **e.** Yes   **f.** Yes   **6.54a.** Yes   **b.** $\mu = 10$, $\sigma^2 = 6$ **c.** .726   **d.** .7291
6.55a. .5; .5   **b.** .212; .2119   **c.** .831; .8301   **6.56a.** .5398   **b.** .731   **c.** .9345   **6.57a.** .488   **b.** .2334   **c.** 0
6.58 .8907   **6.59a.** .0559   **c.** No   **6.60a.** No   **b.** .6026   **c.** No; yes; yes   **6.61b.** Approx. 0   **c.** No   **d.** Yes
6.62 Approx. .0559; yes   **6.63** Approx. 0   **6.64a.** .0885   **b.** .7123
6.65a. $\frac{1}{80}$ **b.** 50; 23.1   **c.** 3.8 to 96.2   **d.** .625   **e.** 0   **f.** .875   **g.** .5775   **h.** .1875
6.66a. .95   **b.** .90   **c.** .9974   **d.** .9925   **e.** .0606   **f.** .9270   **6.67a.** .3446   **b.** .6179   **c.** .9989   **d.** .3612
6.68a. .6915   **b.** .1587   **c.** .1915   **d.** .3085   **e.** 0   **f.** Approx. 1.0   **6.69a.** $-.13$ **b.** .02   **c.** 1.04   **d.** $-.69$
6.70a. 47.68   **b.** 47.68   **c.** 30.13   **d.** 30.13   **e.** 41.53   **f.** 38.47
6.71a. .45119   **b.** .40657   **c.** 0   **d.** .87754   **e.** .27387   **f.** 0   **6.72a.** .9441   **b.** .9429   **c.** .9988   **d.** .0031
6.73a. .0918   **b.** 0   **c.** Lowered 4.87 decibels (to 95.13)   **6.74** .0122; .0062   **6.75** $9.6582   **6.76** .6065; .6321
6.77a. 288   **b.** .2483   **c.** Not possible   **6.79a.** 13.3 weeks   **b.** .8607   **c.** .59343   **6.80** .1056

6.81 Ranked high to low: **a.** bank 3, bank 1, bank 2   **b.** bank 3, bank 1, bank 2   **c.** bank 1, bank 3, bank 2
6.82 .6321   **6.83** Approx. .9817   **6.84a.** .1922   **b.** .4681   **c.** $(.4681)^3$ **6.85a.** .135   **b.** .5940
6.86a. .0548   **b.** .6006   **c.** .3446   **d.** 6,503.80   **6.87** 320, 16; $P(x \geq 400) \approx 0$
6.88a. .8264   **b.** Approx. 17 times   **c.** .6217   **d.** 27,907; 28,067

CHAPTER 7

7.1a. 20; .949   **b.** 100; 1.581   **c.** 25; .577   **d.** 400; .90   **7.2a.** 15; .6   **b.** Normal; no   **c.** .833   **d.** -1.67
7.3a. .0475   **b.** .9525   **c.** .9082   **d.** .905   **e.** .0475   **7.4a.** $\mu_{\bar{x}} = 50$; $\sigma_{\bar{x}} = .7071$ **b.** Normal; yes
7.5a. .5   **b.** .5   **c.** .9599   **d.** 1.0   **e.** 0   **f.** .4929   **g.** .6876   **h.** .0401   **7.6d.** 4.78; 1.642   **7.7b.** 4.5; 1.172
7.8b. 4.84; 1.313   **7.9b.** 4.68; .8270   **7.10a.** $\mu_{\bar{x}} = 2.5$; $\sigma_{\bar{x}} = .1904$ **b.** Normal; yes
7.14a. 6; .3536   **b.** .5222   **c.** .0793   **d.** Smaller standard deviation
7.15a. 6; (chip #1, chip #2), (chip #1, chip #3), (chip #1, chip #4), (chip #2, chip #3), (chip #2, chip #4), (chip #3, chip #4), where chip #1 is marked 1, chip #2 is marked 2, chip #3 is marked 2, and chip #4 is marked 3

b. $\frac{1}{6}$ **c.** 1.5, 1.5, 2, 2, 2.5, 2.5   **d.**

$\bar{x}$	1.5	2	2.5
$p(\bar{x})$	$\frac{2}{6}$	$\frac{2}{6}$	$\frac{2}{6}$

7.16a. Two 0's; one 0 and one 1; two 1's   **b.** 0; .5; 1   **c.**

$\bar{x}$	0	.5	1
$p(\bar{x})$	$\frac{1}{4}$	$\frac{2}{4}$	$\frac{1}{4}$

7.17a. Three 0's; three 1's; two 0's and one 1; one 0 and two 1's   **b.** 0; 1; $\frac{1}{3}, \frac{2}{3}$ **c.**

$\bar{x}$	0	$\frac{1}{3}$	$\frac{2}{3}$	1
$p(\bar{x})$	$\frac{1}{8}$	$\frac{3}{8}$	$\frac{3}{8}$	$\frac{1}{8}$

7.19 No   **7.20a.** 36.515   **b.** 120   **c.** 54
7.21a. .0031; claim probably not true   **b.** More likely; less likely   **c.** Less likely; more likely
7.22 Approximately normal with $\mu_{\bar{x}} = 10$ and $\sigma_{\bar{x}} = 2.05$ **b.** $P(\bar{x} \geq 13.49) \approx P(z \geq 1.70) = .0446$
7.23a. Increase   **b.** Risk reduced by half   **7.24a.** 95.25%   **b.** 100%   **7.25** .0456
7.26a. .35; .6538   **b.** Approx. normal with $\mu_{\bar{x}} = .35$ and $\sigma_{\bar{x}} = .03269$ **c.** .35; 03269   **d.** .0630
7.27a. .0013   **b.** No   **7.28** .4514   **7.29** .0082
7.30a. Approx. normal with unknown mean and $\sigma_{\bar{x}} = .0717$ **b.** Approx. normal with unknown mean and $\sigma_{\bar{y}} = .0536$ **c.** 501
7.31b. .5   **c.** .8904   **7.32a.** 1.342%   **b.** Approx. .0026   **c.** Approx. .2236

CHAPTER 8

8.1a. 1.645   **b.** 2.575   **c.** 1.96   **d.** 1.28   **8.2a.** 95%   **b.** 90%   **c.** 99%   **d.** 80%   **e.** 68%
8.3a. 7.8125 ± 1.8433 **8.4a.** 14.1 ± 5.70 **b.** $14.1 \pm .749$ **c.** Increase   **d.** Yes
8.5a. $14.1 \pm .901$ **b.** Increase; decrease   **8.10b.** $1.5\% \pm .587\%$ **8.11a.** .65%   **b.** $4.3 \pm .154$ **8.12a.** 31.5 ± 3.82
8.13 $.72 \pm .11$ **8.14** 480 ± 3.016 **8.15a.** 23.43 ± 1.41 **8.16a.** $3.23 \pm .041$ **8.17** 586   **8.18** 2,344
8.19a. .98, .784, .56, .392, .196   **8.20a.** 68   **b.** 31   **8.21a.** No; $n = 139$ **b.** Yes; $n = 98$
8.22a. 656   **b.** Wider   **c.** 38.3%   **8.23** 97   **8.24** 55   **8.25** 139   **8.26** 271   **8.32** No
8.33a. $z = 1.67$; reject H_0 **b.** Fail to reject H_0 **8.34a.** $z = -1.57$; reject H_0 **b.** Fail to reject H_0
8.35a. Reject H_0 if $z > 1.5$ **b.** .0668   **8.36a.** $z = 2.5$; reject H_0 **b.** $z = 2.5$; reject H_0
8.37b. Type I: .370; Type II: .240   **8.38a.** .0183   **b.** .0015   **c.** $z = 4.23$; yes   **8.39a.** $z = 7.02$; reject H_0
8.40a. H_0: $\mu = 35$ mpg; H_a: $\mu > 35$ mpg   **b.** $z = 1.8$; yes
8.41a. H_0: $\mu = \$8.00$; H_a: $\mu > \$8.00$ **c.** $z = .87$; do not reject H_0
8.42a. H_0: $\mu = 10$; H_a: $\mu < 10$ **c.** $z = -2.33$; reject H_0 **8.43** Power $= 1 - \beta$
8.45b. $\bar{x}_0 = 1,032.9$ **d.** $\beta = .7405$ **e.** .2595   **8.46a.** $\beta = .3613$ **b.** .6387
8.47a. $\mu_{\bar{x}} = 50$; $\sigma_{\bar{x}} = 2.5$ **b.** $\mu_{\bar{x}} = 45$; $\sigma_{\bar{x}} = 2.5$ **c.** $\beta = .2358$ **d.** .7642
8.48 .3156; smaller   **8.49a.** $\mu_{\bar{x}} = 10$; $\sigma_{\bar{x}} = .1$ **b.** $\mu_{\bar{x}} = 9.9$; $\sigma_{\bar{x}} = .1$ **c.** .83   **d.** .83

8.50

μ	49	47	45	43	41
a. β	.8106	.5319	.2358	.0642	.0102
d. Power	.1894	.4681	.7642	.9358	.9898

8.51a. .1963; Type II error　　**b.** .05; Type I error　　**c.** .8037　　**8.52** .1075

8.53a.

μ	35.5	36.0	36.5	37.0	37.5
Power	.1261	.2595	.4423	.6387	.8037

d. Approx. 0

8.54 Power becomes larger in each case; power curve shifts upward.
8.55a. No　　**b.** No　　**c.** Yes　　**d.** Yes　　**e.** No　　**f.** Yes
8.56a. No　　**b.** Yes　　**c.** Yes　　**d.** No　　**e.** No　　**8.58** .06　　**8.59** .0119　　**8.60** .0376　　**8.61** .1335　　**8.62** .0643
8.63 .0930　　**8.64a.** No　　**b.** .0985　　**c.** Small　　**8.65** 0　　**8.66a.** H_0: $\mu = \$4,627$; H_a: $\mu > \$4,627$　　**b.** .1075
8.67a. H_0: $\mu = \$80,380$; H_a: $\mu > \$80,380$　　**b.** .0071　　**8.68** .0359　　**8.71a.** 2.306　　**b.** 2.764　　**c.** -2.898　　**d.** -1.761
8.72a. 2.201　　**b.** 2.201　　**c.** -1.796　　**d.** 1.725　　**e.** 3.707　　**8.74a.** .10　　**b.** .05　　**c.** .05
8.75a. $t = -1.63$; do not reject H_0　　**b.** $t = -1.63$; do not reject H_0　　**8.76a.** $.05 < p\text{-value} < .10$　　**b.** $.10 < p\text{-value} < .20$
8.77a. 4.8 ± 2.040　　**b.** 2.132　　**8.78a.** $t = -1.40$; do not reject H_0　　**b.** $t = -1.40$; do not reject H_0
8.79a. 1.833 ± 2.141　　**b.** 2.015　　**8.80a.** 100 ± 3.72　　**b.** 100 ± 5.83; interval is wider
8.81a. Population must be normally distributed　　**b.** $\$67,648.58 \pm \$9,939.55$
8.82a. $t = -3.0$; yes　　**b.** $.01 < p\text{-value} < .02$　　**d.** $19.7 \pm .186$
8.83b. $\$1,901.88 \pm \$213.74$　　**c.** $\$427.48$; wider　　**8.84** $\$69,644.86 \pm \$29,330.03$　　**8.85** $t = -4.19$; yes
8.86 $t = 2.66$; yes　　**8.87** 43.5 ± 6.21　　**8.88a.** $t = 1.50$; no　　**b.** $.05 < p\text{-value} < .10$
8.89a. $t = 1.11$ (perceived minus chronological); no
8.90a. H_0: $\mu = .004$; H_a: $\mu > .004$; $t = 1.4$; do not reject H_0
b. $t = 3.3$; do not reject H_0　　**c.** plant 1: $p\text{-value} > .10$; plant 2: $.05 < p\text{-value} < .10$　　**8.92a.** Yes　　**b.** $.42 \pm .068$
8.93a. Yes　　**b.** $.76 \pm .059$　　**8.94a.** Yes　　**b.** No　　**c.** Yes　　**d.** No　　**8.95b.** $z = -1.53$; do not reject H_0　　**c.** .063
8.96a. $z = -2.33$　　**c.** Reject H_0　　**d.** .0099　　**8.97a.** $.3 \pm .083$　　**8.98a.** $z = -2.83$; do not reject H_0　　**b.** .9977
8.99a. $.4 \pm .047$　　**b.** 9.4%; 80% confidence interval would be narrower　　**8.100a.** Yes　　**b.** Yes; $z = 2.31$　　**c.** .0104
8.101b. Yes　　**c.** $.4 \pm .06$　　**d.** Narrower　　**8.102b.** $z = 1.72$; yes　　**c.** $.1209 \pm .0065$　　**8.103** .0427
8.104a. $z = 2.30$; yes　　**b.** .0107　　**8.105** $.08 \pm .0242$　　**8.106** $.06 \pm .02$　　**8.107a.** $z = -1.24$; no　　**b.** .1075
8.108b. Normal by the Central Limit Theorem　　**c.** .3174　　**8.109** p other than .5　　**8.110a.** 1,083　　**b.** 1,692
8.111a. 225　　**b.** 267　　**8.112** 34　　**8.113** 425　　**8.114** No　　**8.115** 32,270　　**8.116** 271　　**8.117** 322
8.118 No; need 457 tires　　**8.119a.** -1.725　　**b.** 3.250　　**c.** 1.860　　**d.** 2.898
8.120a. $.025 < p\text{-value} < .05$　　**b.** $.005 < p\text{-value} < .01$　　**c.** $.10 < p\text{-value} < .20$　　**d.** $.01 < p\text{-value} < .02$　　**8.121** Small
8.123a. No; $p\text{-value} = .9803$　　**b.** $.633 \pm .1024$　　**8.125a.** 12.2 ± 1.645　　**b.** 166　　**c.** $z = 1.3$; no; $p\text{-value} = .0968$
8.126a. 722　　**b.** 174　　**8.127a.** $z = 1.41$; no　　**c.** .0793
8.128b. .8212　　**c.** .1788　　**d.** $\beta = .3103$; power $= .6897$; power becomes larger
8.129a. No　　**b.** $\beta = .5910$; power $= .4090$　　**c.** Increase　　**8.130** $z = 1.07$; no　　**8.131** .1423
8.132 48 if $p = .02$; 601 if $p = .5$　　**8.133a.** $\$12,522 \pm \$784$　　**8.134a.** $z = -3.03$; yes　　**8.135** .0012
8.136 $\beta = .7422$; power $= .2578$　　**8.137** 260　　**8.138** $z = 3.79$; discount store's claim is supported　　**8.139** $4.5 \pm .122$

CHAPTER 9

9.1a. 150 ± 6　　**b.** 150 ± 8　　**c.** Mean 0; standard deviation 5　　**d.** 0 ± 10
9.2a. $\mu_{\bar{x}_1} = 12$; $\sigma_{\bar{x}_1} = .5$　　**b.** $\mu_{\bar{x}_2} = 10$; $\sigma_{\bar{x}_2} = .375$　　**c.** 2; .625　　**d.** Yes
9.3a. $-.8$　　**b.** .7881　　**c.** .2119　　**d** .9452
9.4a. 1.281　　**c.** Yes　　**d.** $\bar{x}_1 - \bar{x}_2 > 7.51$ or $\bar{x}_1 - \bar{x}_2 < 2.49$　　**e.** $z = 11.7$; reject H_0　　**9.5** 20 ± 2.51
9.6a. $\mu_{(\bar{x}_1 - \bar{x}_2)} = -5$; $\sigma_{(\bar{x}_1 - \bar{x}_2)} = 1.4$　　**b.** $z = -6.43$; reject H_0　　**c.** 0　　**9.7a.** Yes; $z = -2.76$　　**b.** .0029
9.8a. $z = -2.78$; yes　　**b.** .0027　　**9.9** -1.9 ± 1.12　　**9.10a.** Yes; $z = 12.71$; $p\text{-value} \approx 0$; reject H_0
9.12 $z = 1.96$; reject H_0; workers with belief in class systems are more satisfied with their jobs
9.13b. $z = 4.27$; yes　　**c.** Approx. 0　　**9.14** $1.9 \pm .87$; wider　　**9.15a.** $z = 2.77$; yes　　**9.16** $z = 2.17$; yes

9.17c. $z = 2.45$; yes   **d.** \$102,100 $\pm$ \$41,634.12   **9.19a.** No   **b.** No   **c.** No   **d.** Yes   **e.** No

9.21a. 95   **b.** 13.29   **c.** .181   **d.** 2,241.94   **9.22a.** .443   **b.** $t = -2.18$; reject H_0 **c.** $.025 < p\text{-value} < .05$

9.23a. $-.88 \pm .74$ **9.24a.** 15.36   **b.** $t = -5.35$; yes   **c.** Approx. 0   **9.25** -8 ± 3.077

9.26a. No; $z = .013$ **b.** 10.025 $\pm$ 4.817   **9.27a.** $t = 3.19$; yes   **b.** 5 ± 2.68 **9.28a.** $t = 2.10$; no   **c.** 4 ± 3.304

9.29a. $t = -1.96$; yes   **c.** Approx. .05   **d.** -7.4 ± 6.224 **9.30a.** $t = -8.81$; no   **b.** $-.0184 \pm .0446$

9.31a. $t = 4.46$; yes   **b.** $p\text{-value} < .005$ **9.32a.** $s_1^2 = 88.62$, $s_2^2 = 68.49$ **c.** $t = .61$; no

9.33a. $t = -.66$; no   **c.** 1.01 ± 2.796 **9.35a.** 4.10   **b.** ≈ 3.57 **c.** 8.80   **d.** 3.21

9.36a. .01   **b.** .95   **c.** .95   **d.** .01

9.39a. $F = 2.26$; no   **b.** $.10 < p\text{-value} < .20$ **9.40a.** $F = 4.29$; no   **b.** $.10 < p\text{-value} < .20$

9.41a. $F = 2.26$; no   **b.** $p\text{-value} \approx .10$ **9.42** $F = 2.63$; yes   **9.43** $F = 10$; yes

9.44a. $F = 5.03$; not appropriate   **d.** $.02 < p\text{-value} < .05$ **9.45** $F = 1.53$; no

9.46a. $F = 1.11$; do not reject H_0 **b.** Type II error   **9.47a.** $t > 1.833$ **b.** $t > 1.328$ **c.** $t > 2.776$ **d.** $t > 2.896$

9.48a. -7.6; 4.56   **b.** $74.8 - 82.4 = -7.6$ **c.** -3.73; yes   **d.** $.02 < p\text{-value} < .05$ **9.49** -7.6 ± 5.66

9.50a. $t < -1.333$ **b.** $t = -3.24$; reject H_0 **d.** -3.5 ± 1.88 **9.52a.** $t = .81$; do not reject H_0 **b.** $p\text{-value} > .20$

9.53c. $t = 2.327$; people do not receive asking price   **9.54a.** $t = -1.92$; yes   **9.55a.** $t = -.10$; no   **b.** $p\text{-value} > .10$

9.56a. $t = .078$; no   **b.** 3.3 ± 1.874 **9.57a.** $t = 7.68$; yes   **b.** $p\text{-value} < .01$ **d.** $.4167 \pm .1395$

9.58a. $t = .40$; no   **b.** $.417 \pm 2.2963$ **9.59a.** 662.5 ± 995.35 **9.60a.** $.239 \pm .0731$

9.64a. -2.33 **b.** -1.96 **c.** -1.645 **d.** -1.282

9.65a. .19; .323   **b.** $z < -1.645$ or $z > 1.645$ **c.** $z = -3.10$; reject H_0 **d.** $p\text{-value} < .002$ **9.66** $-.133 \pm .0978$

9.67a. $z < -1.96$ **b.** $z = -3.49$; reject H_0 **c.** Approx. 0   **9.68** $-.104 \pm .0379$ **9.69** $z = 1.10$; do not reject H_0

9.70a. Yes   **b.** $z = 4.81$; yes   **9.71** $-.535 \pm .086$

9.72a. $z = -5.08$; there is a difference   **b.** $-.215 \pm .019$ **9.73a.** $z = 4.48$; reject H_0 **b.** Approx. 0

9.74a. $.17 \pm .0353$ **9.75** $z = 7.79$; yes

9.76a. $.1732 \pm .0076$ (95% confidence interval)   **b.** $.0756 \pm .0064$ (95% confidence interval)   **c.** $.0976 \pm .0099$

9.77 $z = 17.97$; yes; company B   **9.78** Less likely   **9.80** $z = .79$; do not reject H_0 **9.81** $n_1 = n_2 = 41$ **9.82** 1,504

9.83 $n_1 = n_2 = 812$ **9.84** No; required sample size is 136   **9.85a.** $n_1 = n_2 = 911$ **b.** No

9.86 $n_1 = n_2 = 34$; so 24 additional observations required   **9.87** $n_1 = n_2 = 542$ **9.88** $n_1 = n_2 = 4,802$ **9.89** 34

9.90a. $z = 2.395$; R&D expenditures have increased   **9.91** 2.55 ± 2.409 **9.92** $z = -.86$; no

9.93 $F = 2.79$; do not reject H_0 **9.94a.** $z = 2.60$; yes   **b.** .0047   **9.96** $n_1 = n_2 = 542$

9.97a. $t = 1.06$; do not reject H_0 **b.** $p\text{-value} > .20$ **9.98** $F = 1.21$; do not reject H_0 **9.99a.** $.52 \pm .1128$ **9.100** 49

9.101a. $.69 \pm .039$ **b.** $-.01 \pm .058$ **9.102a.** $z = 3.77$; yes

9.103a. $t = -4.02$; yes   **b.** $p\text{-value} < .005$ **9.104** $z = 1.67$; yes   **9.105** $.4 \pm .616$ **9.106** $z = -4$; yes

9.107 Reject H_0 in parts **a, c, d**; do not reject H_0 in parts **b, e, f**   **9.108** 193   **9.109a.** $t = 2.27$; yes   **9.110** 9.9 ± 8.937

9.111 $.0308 \pm .0341$ **9.112a.** $-1,141.67 \pm 2,018.76$ **9.113** $.0047 \pm .0839$ **9.114a.** $t = -1.38$; no   **9.115** $F = 1.82$; no

CHAPTER 10

10.2 $y = 3.25 + .25x$ **10.3a.** $y = x$ **b.** $y = 2 - x$ **c.** $y = \frac{9}{4} + \frac{1}{8}x$ **d.** $y = \frac{13}{7} + \frac{5}{7}x$

10.5a. 2, 3   **b.** $-2, 3$ **c.** $2, -3$ **d.** $-1, 0$ **e.** 2, 0   **f.** 1.25, .5

10.7b. -26.2857 **c.** 33.7143   **d.** $-.780$ **e.** 3.429, 4.429   **f.** 7.102   **g.** $\hat{y} = 7.102 - .78x$

10.8a. SSE = 1.2201   **10.9d.** $y = 1 + x$ **e.** $\hat{y} = 1 + x$ **10.10c.** .020; .918   **10.11b.** $\hat{y} = 575.03 - .00219x$

10.12a. $\hat{y} = 34.143 + 3.2213x$ **c.** 59.91 cents per gallon   **10.13b.** 24.45; 2.38   **c.** 33.38

10.14b. $-.125$; 3.125   **d.** 15.5   **10.15a.** Positive   **c.** $\hat{y} = -55.56 + 522.367x$ **10.16a.** $\hat{y} = 133.09 - .265x$

10.17 .0507   **10.18a.** 57.5; 3.59375   **b.** 248.57143; 7.53247   **c.** 9.2882; .5160

10.19 1.2204, .2440, .4940, 5.1349, 1.0270, 1.0134   **10.20** c

10.21b. A: $\hat{y} = 6.62 - .0727x$; B: $\hat{y} = 9.31 - .1077x$ **c.** A: 19.056, 1.466, 1.211; B: 4.833, .372, .610

d. 1.53 ± 2.422; 1.77 ± 1.220 **e.** Brand B

10.22a. 53.281, 7.299   **b.** 82.46 cents, 7.299 cents   **c.** \$1.307, 7.299 cents

10.23a. Yes; $\hat{y} = 74.72 - 1.2886x$ **c.** 6,342.5, 352.4, 18.77; ± 37.54

10.24a. $\hat{y} = 18.89 + .01087x$ **b.** 23.24 years   **c.** 19.43; 3.886

10.25a. $\hat{y} = -.246 + 1.3152x$ **c.** 7.65, 11.59   **d.** 12.4365, 2.4873, 1.5771   **e.** ± 3.1542

10.26a. 31 ± 1.17, $31 \pm .94$ **b.** 64 ± 5.08, 64 ± 4.16 **c.** $-8.4 \pm .75$, $-8.4 \pm .62$

10.27b. $\hat{y} = .536 + .821x$ **d.** H_0: $\beta_1 = 0$; H_a: $\beta_1 \neq 0$ **e.** $t = 3.65$, 5   **f.** Reject H_0; model is useful

10.29 $t = -.33$, no   **10.30b.** $\hat{y} = 44.13 + .2366x$ **c.** 6,400.6, 376.5, 19.40   **e.** $t = 1.27$, no   **f.** $.2366 \pm .3935$

10.31a. $\hat{y} = 12.71 + 1.50x$ **b.** $t = 18.26$; yes

10.32a. Reject H_0 for each firm; model is useful   **b.** .14, .076   **c.** Conoco   **10.33** $t = -.96$; do not reject H_0, yes

10.34 $3.2213 \pm .4179$ **10.35** 5.638 ± 1.448 **10.36** $\hat{y} = 19.3 + 7.70x$, $t = 2.56$; yes ($\alpha = .05$)

10.39a. $\hat{y} = 40.926 - .1343x$ **b.** $t = -4.21$; yes   **c.** $-.1343 \pm .0578$

10.42a. .985, .971   **b.** $-.993$, .987   **c.** 0, 0   **d.** 0, 0   **10.43a.** .944   **b.** .877   **c.** .727   **10.44** .917, .841

10.45 .0109, .0001   **10.46** $r_1 = -.967$, $r_2 = -.110$; x_1 **10.47a.** .600, .360   **b.** $t = 2.60$, yes   **10.48** .392, $-.626$

10.49c. $r_1 = .965$; $r_2 = .996$ **d.** Yes   **10.50a.** $\hat{y} = -768.6 + 6.203x_1$ **b.** $t = 3.35$; yes   **c.** .584

10.51a. $\hat{y} = -235.1 + 1.273x_2$ **b.** $t = 18.30$; yes   **c.** .977   **d.** $\hat{y} = -235.1 + 1.273x_2$ **e.** x_2

10.52a. .815   **b.** $t = 3.98$; yes   **10.53a.** $\hat{y} = 44.17 - .0255x$ **c.** $t = -.03$; do not reject H_0 **d.** No   **e.** .0003

10.54b. $\hat{y} = 2 + x$ **c.** .8   **d.** $6 \pm .681$ **e.** 6 ± 1.927 **f.** Prediction

10.55c. 5.70 ± 1.457 **d.** $-.5 \pm .777$; -5.72 ± 1.296 **10.56a.** $-.5 \pm 3.039$ **d.** $t = 11.61$, reject H_0

10.57a. $\hat{y} = 1.4 + .8x$ **c.** 1   **d.** .1   **e.** $2.2 \pm .247$ **f.** $2.6 \pm .737$

10.58b. $\hat{y} = 4.861 - .3466x$ **c.** $t = -5.91$; yes   **d.** 1.74 ± 1.09 **e.** $2.61 \pm .38$ **10.59** 130.78 ± 8.467

10.60a. $\hat{y} = 5.325 + .5861x$ **c.** $t = 15.35$; model is useful   **d.** 28.183 ± 1.273; 28.183 ± 4.629

10.61a. A: $3.35 \pm .587$, B: $4.46 \pm .296$ **b.** A: 3.35 ± 2.224, B: 4.46 ± 1.120 **c.** A: $-.647 \pm 3.606$

10.63b. $\hat{y} = -.0817 + .1253x$ **c.** $t = 8.21$; yes   **d.** .9455, .894   **e.** $1.798 \pm .206$ **f.** $1.798 \pm .624$

10.64a. 575.03, $-.00219$ **b.** 293,208.46, 22,554.497, 150.18155   **c.** .0664

d. $t = -.96$; p-value $= .3538$; do not reject H_0 **e.** 498.5 ± 83.8

10.65a. $\hat{y} = 44.13 + .2366x$ **b.** 19.40375   **c.** .0865   **d.** $t = 1.27$, p-value $= .2216$, do not reject H_0

10.66a. $\hat{y} = 27.78 - .6x$ **c.** 7   **d.** .5385   **e.** $-.6 \pm .184$ **f.** $26.7 \pm .3479$ **g.** 26.7 ± 1.3453

10.67c. $\hat{y} = x$ **d.** SSE $= 0$

10.68b. $r = .7825$ **c.** $\hat{y} = -29.5 + 8.865x$ **d.** .612   **e.** $t = 6.03$; reject H_0, model is useful   **f.** Yes; no

g. 679.7 ± 369.3 **h.** 679.7 ± 80.1

10.69a. $-99,045$, 102.814   **b.** $t = 6.48$; yes   **c.** p-value $= .0001/2 = .00005$ **d.** .6776   **e.** $-631,806 \le y \le 2,078,740$

10.70a. 57.11 ± 34.811 **c.** 45   **10.71a.** $-.8145$, .6634   **b.** $t = -4.44$, yes

10.72a. $\hat{y} = 22.325 + 2.6497x$ **c.** $t = 6.28$, yes   **d.** .9421, .8875   **e.** 109.76 ± 2.155 **f.** 109.76 ± 6.088

10.73a. $\hat{y} = 46.399x$ **b.** $\hat{y} = 478.44 + 45.1525x$ **d.** $t = .91$, no   **10.74a.** .8429   **b.** $t = 2.71$; no ($\alpha = .05$)

10.75a. $\hat{y} = 16.283 + 2.0778x$ **c.** $t = 6.67$; yes   **d.** p-value $< .01$ **e.** 57.84 ± 6.146

10.76a. $\hat{y} = 18.224 + 1.1660x$ **c.** $t = 4.97$, yes   **d.** p-value $< .01$ **e.** .8044   **f.** 123.16 ± 19.09

10.77b. $\hat{y} = -15,124 + 76.175x$ **c.** .9185   **d.** $t = 15.74$, p-value $= .0001$, yes   **e.** 76.175 ± 10.035, yes   **f.** .0001

g. \$152,460; $146,713 \le y \le 158,206$

10.78 $\hat{y} = 918.82 + 62.725x$; $t = 7.18$, reject H_0, model is useful; 95% prediction interval: $7,191 \pm 5,080$ **10.79** $-.503$

10.80b. Inverse relationship

CHAPTER 11

11.1a. 506.346, -941.9, -429.06 **b.** $\hat{y} = 506.346 - 941.9x_1 - 429.06x_2$ **c.** 151,015.72376, 8,883.27787, 94.25114

d. $t = -3.424$, p-value $= .0032$, reject H_0 **e.** -429.06 ± 801.432

11.2a. $t = 1.45$, do not reject H_0 **b.** $t = 3.21$, reject H_0

11.3a. $t = 3.13$, reject H_0 **b.** $t = 3.13$; reject H_0 if $t > 1.717$; yes   **c.** $F = 9.82$ **d.** No

11.5a. $\hat{y} = 1.433 + .010x_1 + .379x_2$ **b.** $t = 3.15$, p-value $= .0046$, reject H_0

11.6a. 55.6214   **b.** $t = -2.8$, yes   **c.** $.005 < p$-value $< .01$

11.7a. $\hat{y} = 20.09 - .6705x + .0095x^2$ **c.** $t = 1.51$, p-value $= .1576$; no   **d.** $\hat{y} = 19.279 - .4449x$ **e.** $-.4449 \pm .0728$

11.8a. $t = -3.33$; yes   **11.9** $t = 2.11$, reject H_0 **11.10a.** $t = 3.04$, reject H_0 ($\alpha = .05$)   **b.** ± 5.0

11.11a. $\hat{y} = 93,074 + 4,152x_1 - 855x_2 + .92x_3 + 2,692x_4 + 15.5x_5$ **b.** 33,226

c. $t = 2.78$, p-value $= .0059$, reject H_0 **f.** $t = -2.86$, reject H_0 **g.** p-value $= .005$

11.12a. .8911   **b.** $F = 65.48$, reject H_0 **c.** p-value $< .0001$ **d.** $t = -6.803$; p-value $= .0001$, reject H_0

11.13b. $F = 48.53$, reject H_0 **11.14a.**

Source	df	SS	MS	F
Model	2	1.26	.6300	7.29
Error	17	1.47	.0865	
Total	19	2.73		

; $R^2 = .4615$ **b.** $F = 7.29$, reject H_0

11.16a. $F_1 = 1.62$, do not reject H_0; $F_2 = 2.15$, do not reject H_0 **b.** p-value $> .10$
11.17a. .98   **b.** $F = 190.75$, reject H_0 **c.** p-value $< .0001$ **11.18a.** $F = 31.98$, yes   **b.** $t = -2.27$, yes
11.19a. $\hat{y} = 131.924 + 2.726x_1 + .047x_2 - 2.587x_3$ **b.** $F = 17.87$, p-value $= .0001$, reject H_0
c. $t = .51$, p-value $= .6199$, do not reject H_0 **d.** .7701   **e.** \$19.41   **f.** ± 19.62 **11.20** $F = 1.06$, do not reject H_0
11.21b. $F = 58.56$, yes   **c.** $t = -50$, yes
11.22a. $\hat{y} = 295.327 - 480.838x_1 - 829.465x_2 + .0079x_3 + 2.3608x_4$ **b.** 46.76747
c. $F = 28.18$, p-value $= .0001$, reject H_0 **d.** $t = -3.197$, yes
11.23a. $\hat{y} = 2.41 + 1.43x_1 - .366x_2$ **b.** $F = 54.70$, reject H_0 **c.** .940   **d.** $\hat{y} = -.349 + 2.07x_1 + .0215x_2 - .0919x_1x_2$
e. .986   **g.** $t = -4.54$, yes   **h.** No
11.24a. $\hat{y} = 95,521 - 33x + .0201x^2$ **c.** 13,976   **d.** $F = 63.64$, reject H_0 **e.** $t = 2.38$; yes
11.25a. Mean   **11.26a.** $\hat{y} = 506 - 942x_1 - 429x_2$ **b.** 94.25   **c.** $F = 7.22$; yes   **d.** -429 ± 660.85 **e.** .459
11.27a. $\hat{y} = 506 - 942x_1 - 429x_2$ **b.** 94.25   **c.** $F = 7.22$; yes   **d.** -429 ± 660.85 **e.** .459   **11.28** Yes, yes
11.30a. $\hat{y}_1 = 1.43 + .00999x_1 + .379x_2$, $\hat{y}_2 = 1.16 + .0187x_1 + .406x_2$ **b.** .2769, .5581
c. $F_1 = 100.80$, reject H_0; $F_2 = 33.88$, reject H_0 **d.** $.379 \pm .0575$; $.406 \pm .1149$ **11.31a.** Yes
11.33a. 1.14301; 1.09089   **b.** .9003, .9162   **c.** $F_1 = 117.43$, reject H_0; $F_2 = 65.59$, reject H_0
d. $t = 1.507$, p-value $= .1576$, do not reject H_0, no   **11.34** Yes
11.35a. $\hat{y} = 90.1 - 1.836x_1 + .285x_2$ **b.** .916   **c.** $F = 64.91$, yes   **d.** $t = -5.01$, reject H_0 **e.** 10.7
11.36a. $\hat{y} = -13.062 + .742x_1 + 18.603x_2 + 13.410x_3$ **b.** .9098, $F = 188.33$, reject H_0
11.37a. $\hat{y} = .601 + .595x_1 - 3.725x_2 - 16.232x_3 + .235x_1x_2 + .308x_1x_3$ **b.** .928, $F = 139.42$, reject H_0
11.38a. $F = 165.84$, reject H_0 **c.** $t = 3.64$, yes; $t = -2.25$; no, $\beta_3 < 0$
11.39a. $r_{x_1x_5} = .878$ **b.** $\hat{y} = 52,291 - 665x_2 + 5.80x_3 + 4,172x_4 + 18.7x_5$, $s = 38,425$ **c.** $F = 176.83$, reject H_0
d. $521,190 \pm 38,413$ **11.41** $t = -2.06$, no
11.42a. $\hat{y} = .0562 + .273x_1 + .0006x_2$ **b.** $F = 164.74$, yes   **c.** $t = 4.34$, reject H_0 **d.** 14.25
11.43a. H_0: $\beta_2 = \beta_3 = 0$ **b.** $t = -3.33$, reject H_0 **11.44** $F = 15.32$, reject H_0
11.45a. $\hat{y} = 53.6 - .198x + .0003x^2$; $F = .61$, do not reject H_0 **b.** $t = .78$; no   **11.46a.** $F = 1.28$, no   **b.** No
11.47a. 13.68   **11.48a.** $\hat{y} = -1,187 + 1,333x - 45.6x^2$ **b.** .973   **c.** $F = 220.18$, yes   **d.** $t = -5.84$, reject H_0
11.49 $t = -1.10$, do not reject H_0 **11.50b.** $F = 52.21$, yes   **c.** $t = -3.33$; yes ($\alpha = .05$)   **d.** $t = 4$, yes ($\alpha = .05$)
11.52b. $t = 40.54$, yes   **c.** 17.79   **11.53b.** $F = 16.10$, yes   **c.** $t = 2.5$, yes   **d.** 945   **11.54b.** $t = 5$, yes   **c.** 825
11.56a. $F = 6.61$, yes   **b.** ± 1.0 **c.** No
11.57b. $\hat{y}_2 = 238 - .46x_1 - 19.5x_4 + 18x_5$ **c.** .649   **d.** H_0: $\beta_1 = \beta_2 = \beta_3 = 0$
e. $F = 8.65$, p-value $< .01$, reject H_0 **f.** $t = -2.10$; reject H_0 **g.** $-.46 \pm 3.130$ **h.** 192.54
11.58a. $\hat{y} = 1.68 + .444x_1 - .0793x_2$ **b.** $F = 150.65$, reject H_0 **c.** .977   **d.** $-.0793 \pm .0113$
e. 1.20 hours   **f.** 12 hours   **g.** $1.81 \pm .172$ **11.59b.** $s = .40228$, $R^2 = .681$ **c.** $F = 39.51$, p-value $= 0$, yes
11.61a. $s_1 = .40228$, $s_2 = .18714$, $\pm .804$, $\pm .374$ **b.** $F = 100.41$, p-value $= 0$, reject H_0
c. $t = 1.675$, p-value $= .1032$, do not reject H_0 **11.62** Second-order
11.63b. $\hat{y} = 344 - 15.8x_1$ **c.** $F = .13$, do not reject H_0, $R^2 = .008$ **d.** $\hat{y} = -855 + 190x_1 + 9.60x_2$
e. $F = 5.16$, reject H_0, $R^2 = .392$ **g.** 493.12

CHAPTER 12

12.1a. Quantitative   **b.** Qualitative   **c.** Quantitative   **d.** Quantitative   **e.** Qualitative
12.2a. Quantitative   **b.** Qualitative   **c.** Qualitative   **d.** Qualitative   **e.** Quantitative
12.3a. Quantitative   **b.** Quantitative   **c.** Qualitative   **d.** Qualitative   **e.** Qualitative   **f.** Qualitative   **g.** Quantitative
h. Qualitative   **12.5** y: quantitative; x_1: quantitative; x_2: quantitative; x_4: qualitative
12.6a. First-order, $\beta_0 = 10$, $\beta_1 = .5$ **b.** First-order, $\beta_0 = 80$, $\beta_1 = -.5$
12.7a. Second-order, $\beta_0 = 4$, $\beta_2 < 0$ **b.** Second-order, $\beta_0 = 8$, $\beta_2 > 0$
12.8a. First   **b.** Second   **c.** Third   **d.** Third   **e.** Second   **f.** First   **12.9** $E(y) = \beta_0 + \beta_1 x + \beta_2 x^2$
12.10a. $\hat{y} = 63.1 + 1.17x - .0374x^2$ **c.** H_0: $\beta_2 = 0$; H_a: $\beta_2 < 0$ **d.** $t = -3.03$, reject H_0
12.11a. $E(y) = \beta_0 + \beta_1 x + \beta_2 x^2$ **b.** $\beta_0 > 0$, $\beta_2 > 0$ **12.12a.** $E(y) = \beta_0 + \beta_1 x + \beta_2 x^2$ **b.** $\beta_0 > 0$, $\beta_2 < 0$
12.13 $E(y) = \beta_0 + \beta_1 x + \beta_2 x^2$ **12.14** $t = -6.60$, yes   **12.15a.** $E(y) = \beta_0 + \beta_1 x + \beta_2 x^2$
12.16b. First-order, first-order, second-order
12.17b. $E(y) = \beta_0 + \beta_1 x + \beta_2 x^2$ **c.** $\hat{y} = 4.08 + .084x - .00325x^2$ **d.** $F = 42.59$, reject H_0
12.18b. $\hat{y} = 79.75 + 6.48x - .81x^2$ **c.** $t = -4.08$, yes

12.19a. $\hat{y} = 11.9 - 1.13x$; $R^2 = .253$ **d.** $\hat{y} = 11.9 - .927x - .167x^2$; $R^2 = .256$ **f.** $t = -.27$, do not reject H_0

12.20a. $E(y) = \beta_0 + \beta_1 x_1 + \beta_2 x_1^2$ **b.** $F = 49.69$, yes   **c.** p-value $= .0001$ **d.** $t = -6.998$, yes   **e.** p-value $= .0001$

12.21a. $E(y) = \beta_0 + \beta_1 x_1 + \beta_2 x_2$ **b.** $E(y) = \beta_0 + \beta_1 x_1 + \beta_2 x_2 + \beta_3 x_1 x_2$

c. $E(y) = \beta_0 + \beta_1 x_1 + \beta_2 x_2 + \beta_3 x_1 x_2 + \beta_4 x_1^2 + \beta_5 x_2^2$ **12.22e.** 4   **12.23a.** Second-order   **f.** -4

12.25a. $\hat{y} = -2.55 + 3.82 x_1 + 2.63 x_2 - 1.29 x_1 x_2$ **e.** H_0: $\beta_3 = 0$, H_a: $\beta_3 \neq 0$ **f.** $t = -8.06$, reject H_0

12.26a. Both quantitative   **b.** $E(y) = \beta_0 + \beta_1 x_1 + \beta_2 x_2$

c. $E(y) = \beta_0 + \beta_1 x_1 + \beta_2 x_2 + \beta_3 x_1 x_2 + \beta_4 x_1^2 + \beta_5 x_2^2$ **d.** H_0: $\beta_3 = \beta_4 = \beta_5 = 0$

12.27a. Both quantitative   **b.** $E(y) = \beta_0 + \beta_1 x_1 + \beta_2 x_2$ **c.** $E(y) = \beta_0 + \beta_1 x_1 + \beta_2 x_2 + \beta_3 x_1 x_2$

d. $E(y) = \beta_0 + \beta_1 x_1 + \beta_2 x_2 + \beta_3 x_1 x_2 + \beta_4 x_1^2 + \beta_5 x_2^2$

12.29a. $\hat{y} = 149 + .472 x_1 - .0993 x_2 - .0005 x_1^2 + .000015 x_2^2$ **c.** $F = 31.66$, yes   **d.** Yes

12.30a. $\hat{y} = 1,333 - .151 x_1 - 2.63 x_2 + .052 x_1 x_2$ **b.** $F = 241.76$, reject H_0 **c.** $t = 7.57$, reject H_0

12.31a. At least one of the parameters β_3, β_4, or β_5 differs from 0.   **c.** 3, 24   **12.32** $F = 1.15$, do not reject H_0

12.34a. $\hat{y}_1 = 14.6 - .611 x_1 + .439 x_2 - .08 x_3 - .064 x_4$; $\hat{y}_2 = 14 - .642 x_1 + .396 x_2$ **b.** 152.66, 160.44   **c.** 5, 3

d. H_0: $\beta_3 = \beta_4 = 0$ **e.** $F = .38$, do not reject H_0 **f.** p-value $> .10$

12.35a. H_0: $\beta_1 = \beta_2 = \beta_3 = \beta_4 = \beta_5 = 0$ **b.** H_0: $\beta_3 = \beta_4 = \beta_5 = 0$ **c.** $F = 18.24$, reject H_0 **d.** $F = 8.46$, reject H_0

12.36a. $-215 - .910 x_1 + 78.7 x_2 + .143 x_1 x_2 - .0001 x_1^2 - 5.99 x_2^2$ **b.** $F = 298.8$; yes   **c.** 167,068

12.37 $F = 24.19$, yes   **12.38a.** $F = 10.52$, reject H_0 **b.** p-value $< .01$

12.39a. $\hat{y} = 1.119 - .0026 x_1 + .7 x_2 + .00018 x_1^2 - .03026 x_2^2 - .00149 x_1 x_2$ **b.** $F = 77.05$, yes

c. $F = 6.93$, yes   **d.** $t = -.65$, no

12.40a. $E(y) = \beta_0 + \beta_1 x_1 + \beta_2 x_2 + \beta_3 x_3 + \beta_4 x_1 x_2 + \beta_5 x_1 x_3 + \beta_6 x_2 x_3 + \beta_7 x_1^2 + \beta_8 x_2^2 + \beta_9 x_3^2$

b. $\hat{y} = 655.81 - 57.33 x_1 - 3.39 x_2 - 28.3 x_3 + .22 x_1 x_2 + 2.20 x_1 x_3 + .09 x_2 x_3 + .45 x_1^2 + .004 x_2^2 + .21 x_3^2$

c. $F = 10.25$, p-value $= .0006$, yes   **d.** $F = 2.25$; no

12.41a. $\hat{y} = 93,074 + 4,152 x_1 - 855 x_2 + .92 x_3 + 2,692 x_4 + 15.5 x_5$ **b.** $F = 190.75$, yes

c. $\hat{y} = 114,369 + 5,036 x_1 - 1,057 x_2 + 15 x_5$ **d.** $F = 303.71$, yes   **12.42** $E(y) = \beta_0 + \beta_1 x_1$

12.43 $E(y) = \beta_0 + \beta_1 x_1 + \beta_2 x_2$ **12.44a.** 10.2   **b.** 6.2   **c.** 22.2   **d.** 12.2   **e.** H_0: $\beta_1 = \beta_2 = \beta_3 = 0$

12.45a. $\hat{y} = 80 + 16.8 x_1 + 40.4 x_2$ **d.** $F = 24.72$, reject H_0 **12.46b.** $E(y) = \beta_0 + \beta_1 x_1$; base level: public colleges

12.47a. $E(y) = \beta_0 + \beta_1 x_1 + \beta_2 x_2$; base level: OTC   **b.** $\hat{y} = 15.6 + 10.6 x_1 - 1.14 x_2$ **c.** $F = 4.11$; yes

12.48 $E(y) = \beta_0 + \beta_1 x_1 + \beta_2 x_2 + \beta_3 x_3$; base level: 0

12.49a. $F = 4.80$, yes   **b.** $\$11,400$ **c.** $\$20,000$ **12.50** 3.80 ± 7.306

12.51a. Qualitative   **b.** $E(y) = \beta_0 + \beta_1 x_1 + \beta_2 x_2$; base level: B_1 **d.** $\beta_0 + \beta_2$ **12.52a.** 60   **b.** 510

12.53a. $E(y) = \beta_0 + \beta_1 x_1 + \beta_2 x_2 + \beta_3 x_3 + \beta_4 x_4$ **b.** $\hat{y} = 20.35 + 4.75 x_1 + 8.6 x_2 + 11.3 x_3 + 2.1 x_4$

c. H_0: $\beta_1 = \beta_2 = \beta_3 = \beta_4 = 0$ **d.** $F = 23.97$, p-value $= .0001$, reject H_0 **e.** 2.1 ± 2.846

12.54a. $E(y) = \beta_0 + \beta_1 x_1 + \beta_2 x_2 + \beta_3 x_3$; base level: insurance   **b.** $F = 20.30$, reject H_0 **c.** $t = 7.52$, yes

12.55a. $E(y) = \beta_0 + \beta_1 x_1$ **b.** $E(y) = \beta_0 + \beta_1 x_1 + \beta_2 x_2 + \beta_3 x_3$; base level: third level

c. $E(y) = \beta_0 + \beta_1 x_1 + \beta_2 x_2 + \beta_3 x_3 + \beta_4 x_1 x_2 + \beta_5 x_1 x_3$ **d.** $\beta_4 = \beta_5 = 0$ **e.** $\beta_2 = \beta_3 = \beta_4 = \beta_5 = 0$

12.56a. $E(y) = \beta_0 + \beta_1 x_1$, $E(y) = \beta_0 + \beta_1 x_1 + \beta_2 x_2$, $E(y) = \beta_0 + \beta_1 x_1 + \beta_3 x_3$ **b.** $\hat{y} = 44.80 + 2.17 x_1 + 9.41 x_2 + 15.63 x_3$

c. $\hat{y} = 44.80 + 2.17 x_1$, $\hat{y} = 44.80 + 2.17 x_1 + 9.41 x_2$, $\hat{y} = 44.80 + 2.17 x_1 + 15.63 x_3$ **d.** H_0: $\beta_2 = \beta_3 = 0$

e. $F = 3.33$, do not reject H_0

12.57a. x_1: quantitative; type: qualitative   **b.** $E(y) = \beta_0 + \beta_1 x_1 + \beta_2 x_2$, base level: brand B

c. $E(y) = \beta_0 + \beta_1 x_1 + \beta_2 x_2 + \beta_3 x_1 x_2$ **e.** H_0: $\beta_3 = 0$, H_a: $\beta_3 \neq 0$ **12.58a.** $F = 25.96$, yes   **b.** 10,329

12.59a. $E(y) = \beta_0 + \beta_1 x_1 + \beta_2 x_2 + \beta_3 x_3$; x_1: number of units; $x_2 = 1$ if E, 0 if not; $x_3 = 1$ if G, 0 if not

c. $\hat{y} = 36,388 + 15,617 x_1 + 152,487 x_2 + 49,441 x_3$;   E: $\hat{y} = 188,875 + 15,617 x_1$;

G: $\hat{y} = 85,829 + 15,617 x_1$; F: $\hat{y} = 36,388 + 15,617 x_1$ **e.** $F = 8.43$, yes

12.60a. $E(y) = \beta_0 + \beta_1 x_1 + \beta_2 x_2 + \beta_3 x_3$; base level: diet A   **b.** $E(y) = \beta_0 + \beta_1 x_1 + \beta_2 x_2 + \beta_3 x_3 + \beta_4 x_1 x_2 + \beta_5 x_1 x_3$

c. $\hat{y} = -10.8 + .101 x_1 + 3.48 x_2 - 4.36 x_3$; $\hat{y} = -8.27 + .0884 x_1 - 1.75 x_2 - 7.52 x_3 + .0246 x_1 x_2 + .0151 x_1 x_3$

d. $F = .22$; no   **e.** H_0: $\beta_2 = \beta_3 = \beta_4 = \beta_5 = 0$ **f.** $F = 4.99$, reject H_0

12.61a. $E(y) = \beta_0 + \beta_1 x_1$ **b.** $E(y) = \beta_0 + \beta_1 x_1 + \beta_2 x_2 + \beta_3 x_3$ **c.** $E(y) = \beta_0 + \beta_1 x_1 + \beta_2 x_2 + \beta_3 x_3 + \beta_4 x_1 x_2 + \beta_5 x_1 x_3$

12.62a. $E(y) = \beta_0 + \beta_1 x_1 + \beta_2 x_2 + \beta_3 x_3$ **b.** H_0: $\beta_2 = \beta_3 = 0$ **12.63** $F = 2.60$; do not reject H_0

12.64a. H_0: $\beta_2 = \beta_3 = 0$ **b.** $F = 6.99$, reject H_0

12.65b. $E(y) = \beta_0 + \beta_1 x_1 + \beta_2 x_2 + \beta_3 x_1 x_2$; base level; after OPEC   **c.** $\hat{y} = 475 - 110 x_1 + 409 x_2 - 502 x_1 x_2$

e. H_0: $\beta_2 = \beta_3 = 0$, $F = 54.32$, yes   **f.** $t = -5.96$, yes

12.66a. $E(y) = \beta_0 + \beta_1 x_1 + \beta_2 x_1^2$ **b.** Add $\beta_3 x_2 + \beta_4 x_3$ **c.** Add $\beta_5 x_1 x_2 + \beta_6 x_1 x_3 + \beta_7 x_1^2 x_2 + \beta_8 x_1^2 x_3$

12.67a. $\beta_5 = \beta_6 = \beta_7 = \beta_8 = 0$ **b.** $\beta_2 = \beta_5 = \beta_6 = \beta_7 = \beta_8 = 0$ **c.** $\beta_3 = \beta_4 = \beta_5 = \beta_6 = \beta_7 = \beta_8 = 0$

12.68 $E(y) = \beta_0 + \beta_1 x_1 + \beta_2 x_1^2 + \beta_3 x_2 + \beta_4 x_3 + \beta_5 x_4$

12.69a. $\hat{y} = 48.8 - 3.36x_1 + .0749x_1^2 - 2.36x_2 - 7.6x_3 + 3.71x_1x_2 + 2.66x_1x_3 - .0183x_1^2x_2 - .0372x_1^2x_3$
b. $\hat{y} = 48.8 - 3.36x_1 + .0749x_1^2$; $\hat{y} = 46.44 + .35x_1 + .0566x_1^2$; $\hat{y} = 41.2 - .70x_1 + .0377x_1^2$
d. $H_0: \beta_3 = \beta_4 = \beta_5 = \beta_6 = \beta_7 = \beta_8 = 0$ **e.** $F = 242.21$, reject H_0
12.70a. $F = 4.90$, reject H_0 **b.** $\beta_4 = \beta_5 = 0$ **c.** $F = 2.36$; no   **12.71** $F = 6.71$; yes   **12.72** $F = 149$; reject H_0
12.73a. $H_0: \beta_4 = \beta_5 = 0$ **b.** $\beta_3 = \beta_4 = \beta_5 = 0$ **12.74a.** $E(y) = \beta_0 + \beta_1x_1 + \beta_2x_1^2 + \beta_3x_2 + \beta_4x_1x_2 + \beta_5x_1^2x_2$
12.75 $F = 1.21$; do not reject H_0
12.76b. $E(y) = \beta_0 + \beta_1x_1 + \beta_2x_1^2 + \beta_3x_2 + \beta_4x_1x_2 + \beta_5x_1^2x_2$; base level: wholesale
c. $\hat{y} = 64.87 + 16.26x_1 - .854x_1^2 - 67.76x_2 + 14.9x_1x_2 - 1.45x_1^2x_2$
d. $R^2 = .878$ **e.** $H_0: \beta_3 = \beta_4 = \beta_5 = 0$ **f.** $F = 20.64$; reject H_0 **i.** 11 calls
12.77c. $E(y) = \beta_0 + \beta_1x_1 + \beta_2x_1^2 + \beta_3x_2 + \beta_4x_3 + \beta_5x_1x_2 + \beta_6x_1x_3 + \beta_7x_1^2x_2 + \beta_8x_1^2x_3$
d. $\hat{y} = .311 + .0035x_1 - .000002x_1^2 + .105x_2 + .365x_3 - .0016x_1x_2 - .0017x_1x_3 + .000001x_1^2x_2 + .000001x_1^2x_3$, base level: West
e. $F = 49.72$, yes   **f.** $F = 18.91$, yes   **g.** 1.84   **12.82b.** $\hat{y} = 148.055 - 7.57x_1$; $F = 196.615$, reject H_0
12.83a. $\log(\hat{y}) = 5.164 - .1138x_1$; $F = 799.634$, reject H_0
12.84b. $\log(\hat{y}) = 5.304 - .1145x_1 - .0217x_2$; $F = 462.333$, reject H_0 **c.** $t = -2.28$, reject H_0 **d.** $\pm.26$
12.85a. $\hat{y} = 57,094 + 20.4x$; $F = 343.42$, reject H_0 **c.** $\log(\hat{y}) = 11.7 + .000059x$; $F = 81.58$, reject H_0
12.86a. $E(y) = \beta_0 + \beta_1x_1 + \beta_2x_2 + \beta_3x_1x_2$, base level: bull market   **c.** $H_0: \beta_3 = 0$ **d.** $H_0: \beta_2 = \beta_3 = 0$
12.87a. $E(y) = \beta_0 + \beta_1x_1 + \beta_2x_2 + \beta_3x_3$; base level: low   **c.** 20
12.88b. $\hat{y} = 5.37 + .581x + .0001x^2$ **c.** $H_0: \beta_1 = \beta_2 = 0$; $F = 111.33$, reject H_0; $H_0: \beta_2 = 0$, $t = .03$, do not reject H_0
12.89a. Qualitative: x_7, x_8, x_9, x_{10} **b.** 5   **d.** $F = 11.96$, reject H_0 **e.** $H_0: \beta_7 = \beta_8 = \beta_9 = \beta_{10} = 0$
12.90a. $F = 222.173$, reject H_0 **b.** $H_0: \beta_2 = \beta_3 = \beta_4 = 0$; $F = 13.49$, reject H_0
e. $E(y) = \beta_0 + \beta_1x_1 + \beta_2x_2 + \beta_3x_3 + \beta_4x_4 + \beta_5x_1x_2 + \beta_6x_1x_3 + \beta_7x_1x_4$
f. $\hat{y} = .709 + .109x_1 - .252x_2 - .618x_3 - 1.20x_4 - .0156x_1x_2 + .0055x_1x_3 + .0049x_1x_4$ **g.** $F = 1.78$, no
12.91a. $H_0: \beta_2 = \beta_5 = 0$ **b.** $H_0: \beta_3 = \beta_4 = \beta_5 = 0$ **12.92** $F = 2.15$, do not reject H_0
12.93a. $\hat{y} = 10.2 + .683x_1 + 2.02x_2 + .5x_3$ **b.** $F = 62.78$, yes   **12.94a.** $E(y) = \beta_0 + \beta_1x + \beta_2x^2$ **b.** No
12.95a. $E(y) = \beta_0 + \beta_1x_1 + \beta_2x_1^2 + \beta_3x_2 + \beta_4x_3 + \beta_5x_1x_2 + \beta_6x_1x_3 + \beta_7x_1^2x_2 + \beta_8x_1^2x_3$
12.98a. 2   **b.** $E(y) = \beta_0 + \beta_1x_1 + \beta_2x_2$; base level: self-service
12.101a. $E(y) = \beta_0 + \beta_1x_1 + \beta_2x_2$; base level: air conditioning   **b.** $E(y) = \beta_0 + \beta_1x_1 + \beta_2x_2 + \beta_3x_1^2 + \beta_4x_1x_2 + \beta_5x_1^2x_2$
c. $H_0: \beta_3 = \beta_4 = \beta_5 = 0$ **12.103a.** $F = 2.49$; no   **b.** $F = 209$; reject H_0

CHAPTER 13

13.2a. Base year = 1975:   base year = 1980:

1975	1978	1979	1981	1983
100.00	113.72	148.68	188.96	136.02

1976	1977	1979	1980	1981
50.35	48.78	72.95	100.00	92.71

	1970	1971	1972	1973	1974	1975	1976	1977	1978	1979	1980	1981	1982	1983	1984
13.3	78.06	80.59	82.87	87.16	91.61	94.19	96.01	100.00	105.04	108.04	113.84	113.61	115.07	114.59	112.74
13.5	68.57	70.79	72.80	76.56	80.47	82.74	84.34	87.84	92.27	94.90	100.00	99.79	101.08	100.66	99.04

13.4 Quantity index

13.6a.	1973	1974	1975	1976	1977	1978	1979	1980	1981	1982	1983	1984	1985	1986
	100.00	101.75	82.71	98.50	113.28	67.67	82.46	113.28	54.14	77.69	90.73	83.71	137.09	91.48

b. Declined by 8.52%; increased by 23.81%

13.7a.

	1980	1981	1982	1983	1984	1985	1986
Jan.	100.0000	114.3243	120.8108	109.2792	108.1081	103.1531	107.2072
Feb.	106.8468	121.8918	118.7387	105.4054	107.4774	101.6216	100.8108
Mar.	110.8108	125.0450	114.2342	102.2522	107.5675	104.0540	88.5586
Apr.	111.8918	124.4144	109.0090	107.9279	109.0990	108.0180	80.6306
May	112.0720	123.4234	110.2702	111.9819	110.0000	110.1801	83.5135
Jun.	112.2522	122.7027	116.7567	113.6036	109.3693	111.0810	86.3063
Jul.	112.3423	121.8918	118.7387	114.5945	107.8378	111.0810	80.6306
Aug.	111.9819	121.4414	118.0180	114.3243	106.6666	110.0900	76.3964
Sep.	110.9009	122.3423	116.6666	113.2432	107.1171	108.9189	78.1081
Oct.	110.1801	121.8918	115.3153	111.6216	107.6576	107.9279	
Nov.	110.0900	121.7117	114.2342	110.2702	107.4774	108.1981	
Dec.	110.9009	121.4414	112.0720	109.4594	106.2162	108.3783	

d. Price index

13.8a.

b. Nonagricultural industries　　　**c.** Quantity indexes

YEAR	AGRICULTURAL	NONAGRICULTURAL
1974	100.0	100.0
1975	97.0	99.0
1976	94.8	102.6
1977	93.4	106.6
1978	96.4	111.3
1979	95.2	114.6
1980	95.7	115.2
1981	95.8	116.5
1982	96.8	115.4
1983	96.2	117.0
1984	94.5	122.1

13.10

Jan.	Feb.	Mar.	Apr.	May	Jun.	Jul.	Aug.	Sep.	Oct.	Nov.	Dec.
100.0	98.2	91.8	98.4	88.2	84.1	82.7	81.4	81.5	80.4	83.1	80.9

13.12a.

1961	1962	1963	1964	1965	1966	1967	1968	1969	1970	1971	1972
68.3	72.5	76.4	81.6	87.7	94.8	100.0	109.3	118.2	126.8	136.3	149.5

1973	1974	1975	1976	1977	1978	1979	1980	1981	1982	1983	1984
164.9	180.7	199.2	220.2	245.6	274.6	307.4	340.2	377.1	404.8	439.7	477.6

b.

1961	1962	1963	1964	1965	1966	1967	1968	1969	1970	1971	1972
37.8	40.1	42.3	45.2	48.5	52.5	55.3	60.5	65.4	70.2	75.4	82.7

1973	1974	1975	1976	1977	1978	1979	1980	1981	1982	1983	1984
91.3	100.0	110.2	121.9	136.0	152.0	170.1	188.3	208.7	224.0	243.4	264.4

13.14

1961	1962	1963	1964	1965	1966	1967	1968	1969	1970	1971	1972
71.5	74.8	77.8	82.6	88.2	95.5	100.0	108.5	116.8	125.9	132.8	143.9

1973	1974	1975	1976	1977	1978	1979	1980	1981	1982	1983	1984
159.9	178.6	195.2	212.4	233.2	258.7	292.7	326.7	359.4	379.0	404.9	435.0

	Jan.	Feb.	Mar.	Apr.	May	Jun.	Jul.	Aug.	Sep.	Oct.	Nov.	Dec.
13.16a. Price	100.0	101.3	105.8	108.6	103.4	103.4	103.7	101.9	96.7	93.7	99.6	94.1
Quantity	100.0	104.1	117.5	116.6	118.8	105.1	99.9	93.8	80.5	86.4	88.4	85.7
b.	100.0	103.3	117.1	114.9	116.9	104.5	98.8	93.4	81.1	85.4	89.0	85.8

13.18 7-point moving average

13.19

	1971	1972	1973	1974	1975	1976	1977	1978	1979	1980	1981	1982	1983	1984
$N = 3$	45.33	66.0	105.67	139.66	148.66	144.67	155.66	216.67	371.66	460.33	483.0	420.0	387.0	367.7
$N = 5$		78.8	103.8	120.6	138.4	157.6	187.2	277.6	344.6	390.2	436.2	446.8	387.8	

13.21

	1971	1972	1973	1974	1975	1976	1977	1978	1979	1980	1981	1982	1983
3-pt.	137.3	142.4	148.7	155.1	160.2	164.9	171.1	177.9	185.8	190.7	194.7	195.1	194.6
5-pt.		143.3	148.8	154.1	159.9	166.0	171.6	178.3	184.3	189.5	192.7	194.3	

13.22a.

Year, Qtr.	1980,1	1980,2	1980,3	1980,4	1981,1	1981,2	1981,3	1981,4	1982,1	1982,2	1982,3	1982,4
Uncentered			297.78	302.93	310.6	283.63	239.20	220.73	214.13	226.83	252.48	285.0	327.43
Centered				300.36	306.77	297.12	261.42	229.97	217.43	220.48	239.66	268.74	306.22

Year, Qtr.	1983,1	1983,2	1983,3	1983,4	1984,1	1984,2	1984,3	1984,4	1985,1	1985,2	1985,3
Uncentered	368.20	389.93	402.90	415.08	407.83	401.9	396.58	400.78	412.75		
Centered	347.82	379.07	396.42	408.99	415.18	408.59	399.24	398.68	406.77		

13.23

	1971	1972	1973	1974	1975	1976	1977	1978	1979	1980	1981	1982	1983	1984
$N = 3$	7.0	8.1	7.7	7.3	7.3	8.1	9.0	9.2	8.5	7.6	6.7	6.9	7.4	8.4
$N = 7$			7.4	7.9	8.1	8.3	8.1	8.1	8.0	7.9	7.8	7.8		

13.24b.

	N = 3	N = 5	N = 7
Jan. 1981			
Feb.	1,151		
Mar.	1,301	1,295	
Apr.	1,411	1,355	1,331
May	1,476	1,415	1,340
Jun.	1,435	1,407	1,432
Jul.	1,337	1,451	1,488
Aug.	1,460	1,479	1,461
Sep.	1,518	1,471	1,474
Oct.	1,583	1,496	1,492
Nov.	1,510	1,568	1,553
Dec.	1,487	1,584	1,683

	N = 3	N = 5	N = 7
Jan. 1982	1,649	1,681	1,698
Feb.	1,826	1,783	1,747
Mar.	1,941	1,860	1,901
Apr.	1,970	2,043	1,912
May	2,092	1,999	2,041
Jun.	2,008	2,069	2,135
Jul.	2,175	2,194	2,187
Aug.	2,193	2,298	2,219
Sep.	2,468	2,228	2,230
Oct.	2,355	2,305	2,230
Nov.	2,192	2,308	2,340
Dec.	2,192	2,286	2,402

	N = 3	N = 5	N = 7
Jan. 1983	2,290	2,371	2,397
Feb.	2,578	2,444	2,374
Mar.	2,583	2,499	2,479
Apr.	2,548	2,576	2,521
May	2,535	2,560	2,630
Jun.	2,482	2,627	2,629
Jul.	2,791	2,610	2,576
Aug.	2,638	2,653	2,692
Sep.	2,704	2,743	2,720
Oct.	2,727	2,777	
Nov.	2,761		
Dec.			

13.25

	1980	1981	1982	1983	1984	1985	1986
Jan.		129.3	132.5	120.6	120.5	116.0	107.8
Feb.		132.4	129.7	119.2	120.3	116.1	102.3
Mar.	120.2	135.2	127.2	119.2	120.4	117.0	97.6
Apr.	123.0	137.1	126.3	120.1	120.7	118.8	93.2
May	124.2	137.1	126.3	122.2	120.7	120.9	90.5
Jun.	124.4	136.3	127.2	124.9	120.5	122.2	89.9
Jul.	124.2	135.8	128.9	126.0	120.1	122.4	
Aug.	123.8	135.5	130.0	126.0	119.6	121.9	
Sep.	123.3	135.3	129.4	125.2	119.2	121.3	
Oct.	123.0	135.2	127.9	124.1	118.8	120.7	
Nov.	123.5	135.0	126.0	122.7	118.0	120.0	
Dec.	126.0	134.2	123.5	121.4	116.8	118.2	

13.26 $w = .2$

13.27a.

1977	1978	1979	1980	1981	1982	1983	1984	1985
	2.241	2.237	1.989	1.717	1.351	1.320	1.388	3.135

13.28a.

Year	Quarter		Year	Quarter	
1980	I	244.2	1983	I	284.1
	II	257.0		II	363.9
	III	299.2		III	405.7
	IV	330.2		IV	389.4
1981	I	284.8	1984	I	368.4
	II	292.6		II	430.4
	III	263.0		III	424.4
	IV	223.3		IV	386.9
1982	I	194.4	1985	I	356.5
	II	234.2		II	432.8
	III	259.2		III	449.6
	IV	272.7			

13.29

Year	w = .2	w = .8
1970	133.10	133.10
1971	133.96	136.54
1972	135.43	140.35
1973	138.06	146.95
1974	141.69	154.35
1975	145.47	159.35
1976	149.12	162.83
1977	153.39	168.97
1978	158.54	177.07
1979	163.67	182.77
1980	169.75	191.83
1981	174.54	193.33
1982	178.87	195.63
1983	182.18	195.43
1984	184.19	192.87

13.30a.

1971		1976		1981	
1971	100.31	1976	100.28	1981	133.08
	99.53		103.08		130.27
	98.70		104.59		122.22
	101.07		106.60		120.06
1972	105.36	1977	100.87	1982	117.98
	106.61		100.60		112.27
	109.37		97.75		118.93
	115.45		95.90		132.92
1973	112.70	1978	91.22	1983	146.95
	106.79		94.24		161.76
	107.94		100.05		164.78
	100.67		97.29		164.88
1974	95.99	1979	100.30	1984	160.89
	89.00		102.13		155.49
	71.18		107.16		162.92
	69.35		107.71		165.94
1975	79.16	1980	106.06	1985	176.25
	90.38		111.42		187.17
	85.82		122.42		183.61
	94.48		128.74		202.98

13.31

Year	
1970	36
1971	39
1972	49
1973	73
1974	117
1975	139
1976	132
1977	140
1978	167
1979	237
1980	425
1981	443
1982	409
1983	417
1984	389
1985	353

13.32

Year	Ag.	Nonag.
1974	3,515	83,279
1975	3,472	82,943
1976	3,416	83,934
1977	3,363	85,854
1978	3,372	88,577
1979	3,362	91,337
1980	3,363	93,177
1981	3,365	94,718
1982	3,379	95,281
1983	3,381	96,149
1984	3,357	98,363

13.33a.

Year	$w = .2$	$w = .8$	Year	$w = .2$	$w = .8$
1960	42.4	42.4	1973	84.7	112.1
1961	42.9	44.3	1974	91.3	116.7
1962	43.8	46.8	1975	98.9	126.9
1963	44.9	49.0	1976	110.2	149.5
1964	46.8	53.2	1977	124.0	173.3
1965	49.1	57.4	1978	138.8	193.1
1966	51.4	59.8	1979	154.9	214.1
1967	53.8	62.6	1980	171.3	232.1
1968	56.9	68.0	1981	189.3	255.6
1969	60.6	74.2	1982	204.9	265.0
1970	64.6	79.3	1983	222.3	286.5
1971	70.2	89.7	1984	241.8	312.9
1972	77.2	102.3			

13.34a.

Year	Cold	Hot	Galvanized
1971			
1972	10.62	8.09	10.36
1973	11.54	8.63	11.29
1974	13.30	9.54	12.59
1975	15.66	10.81	14.42
1976	18.19	12.37	16.32
1977	20.55	13.84	18.21
1978	23.02	15.46	20.30
1979	25.05	17.01	22.22
1980	27.85	18.55	24.36
1981	30.42	19.80	25.84
1982	33.64	21.06	27.35
1983	32.52	22.26	28.49
1984			

13.35a.

Year	Cold	Hot	Galvanized
1971	10.00	7.48	9.61
1972	10.39	7.94	10.25
1973	10.73	8.17	10.42
1974	11.76	8.64	11.40
1975	13.89	9.88	13.10
1976	16.03	11.04	14.59
1977	18.21	12.42	16.34
1978	20.66	13.97	18.41
1979	23.10	15.51	20.36
1980	24.80	16.99	22.12
1981	28.15	18.57	24.50
1982	30.70	19.68	25.63
1983	33.44	20.96	27.03
1984	30.79	22.35	28.66

13.36a.

Year	
1971	51.82
1972	57.48
1973	57.52
1974	65.55
1975	80.26
1976	88.81
1977	100.00
1978	113.06
1979	124.18
1980	131.68
1981	150.21
1982	154.55
1983	166.09
1984	157.23

b. Price index

13.37a.

	1984	1985
Jan.	100	80
Feb.	102	78
Mar.	103	80
Apr.	101	87
May	99	89
Jun.	97	91
Jul.	94	98
Aug.	93	99
Sep.	89	96
Oct.	87	101
Nov.	88	102
Dec.	84	102

13.38

1960	33	1973	89
1961	35	1974	91
1962	37	1975	100
1963	38	1976	120
1964	42	1977	139
1965	45	1978	153
1966	47	1979	170
1967	49	1980	183
1968	54	1981	202
1969	59	1982	207
1970	62	1983	226
1971	71	1984	247
1972	81		

13.39

Year	a.	b.
1974	100.00	43.59
1975	104.84	45.70
1976	120.46	52.51
1977	166.10	72.40
1978	199.08	87.12
1979	230.27	100.37
1980	229.42	100.00
1981	247.22	107.76
1982	264.29	115.20
1983	301.45	131.40
1984	361.74	157.68

c. Price indexes

d.

Year	
1974	48.40
1975	50.98
1976	60.34
1977	73.89
1978	90.55
1979	103.74
1980	100.00
1981	106.77
1982	112.57
1983	127.54
1984	153.83

13.40

Year	a. 3-pt.	b. $w = .3$
1974		43.59
1975	47.27	44.22
1976	56.87	46.71
1977	70.68	54.42
1978	86.63	64.23
1979	95.83	75.07
1980	102.71	82.55
1981	107.65	90.11
1982	118.12	97.64
1983	134.76	107.77
1984		122.74

13.41a.

Year	Laspeyres Index
1980	100.00
1981	108.69
1982	116.24
1983	134.55
1984	164.44

b. Revolving credit

13.42a.

Year	Paasche Index
1980	100.00
1981	108.50
1982	115.58
1983	133.00
1984	161.47

13.43

	a. Simple Composite	b. Net Business Formation	Industrial Production	New Housing Units
Year				
1970	100.00	100.00	100.00	100.00
1971	115.90	101.97	101.67	143.06
1972	128.54	108.93	111.04	164.32
1973	123.95	107.99	120.41	142.70
1974	105.20	102.63	119.94	93.33
1975	96.74	100.56	109.28	80.90
1976	112.27	108.65	121.06	107.21
1977	127.68	115.79	128.20	138.56
1978	131.80	120.49	133.67	140.81
1979	127.89	120.58	141.47	121.71
1980	113.59	115.04	136.36	90.09
1981	108.70	111.47	140.07	75.59
1982	102.71	106.39	128.57	74.05
1983	121.22	107.89	136.92	118.74
1984	127.86	110.06	151.48	121.98

13.44a.

Year	$w = .2$	$w = .8$
1970	106.4	106.4
1971	106.8	108.1
1972	108.6	114.3
1973	109.9	114.8
1974	109.8	110.3
1975	109.2	107.7
1976	110.5	114.0
1977	113.0	121.4
1978	116.1	126.8
1979	118.5	128.0
1980	119.3	123.5
1981	119.1	119.6
1982	118.0	114.5
1983	117.3	114.7
1984	117.3	116.6

b. $w = .2$

13.45a. $17,196.91; $14,276.85   **b.** $55,408.43

CHAPTER 14

14.2a.

1975	1978	1980	1983	1984	1985
1,160	1,714	1,429	1,362	1,512	1,646

14.3a.

Year	Forecast w = .3	Error	Forecast w = .7	Error
1982	182.52	13.68	192.70	3.50
1983	182.52	12.86	192.70	2.68
1984	182.52	9.71	192.70	−.47

b.

Year	Forecast w = .7, v = .3	Error	Forecast w = .3, v = .7	Error
1982	201.12	−4.92	203.41	−7.21
1983	206.71	−11.33	209.64	−14.26
1984	212.30	−20.07	215.87	−23.64

14.4 Forecasted value = 201.24

14.5

1985, Qtr.	a. Forecast	b. Error
I	165.94	14.72
II	165.94	25.91
III	165.94	16.14
IV	165.94	45.34

c.

1985, Qtr.	Forecast	Error
I	161.11	19.55
II	161.11	30.74
III	161.11	20.97
IV	161.11	50.17

d.

1985, Qtr.	Forecast w = .7, v = .5	Error	Forecast w = .3, v = .5	Error
I	168.28	12.38	173.78	6.88
II	170.70	21.15	174.68	17.17
III	173.13	8.95	175.58	6.50
IV	175.55	35.73	176.47	34.81

14.6

Model	MAD	RMSE
w = .7	25.5	28.3
w = .3	30.36	32.73
w = .7, v = .5	19.55	22.12
w = .3, v = .5	16.34	19.98

14.7

		a. w = .3	w = .7	b. w = .3, v = .5	w = .7, v = .5
1986	I	187.09	202.98	202.08	215.98
	II	187.09	202.98	208.98	226.82
	III	187.09	202.98	215.87	237.65
	IV	187.09	202.98	222.76	248.49

14.8

		Jan.	Feb.	Mar.	Apr.	May	Jun.	Jul.	Aug.	Sep.	Oct.	Nov.
a.	Forecast	331.3	331.3	331.3	331.3	331.3	331.3	331.3	331.3	331.3	331.3	331.3
b.	Error	−28.6	−32.1	−26.9	−6.0	−14.8	−14.5	−13.1	−.9	−8.4	−5.1	−5.6
c.	Forecast	331.3	317.0	308.1	306.3	315.8	316.1	316.5	317.3	323.9	323.4	324.8
d.	Error	−28.6	−17.8	−3.7	19.05	.72	.66	1.7	13.07	−.97	2.82	.91
e.	Forecast	317.9	310.9	303.9	296.9	289.8	282.8	275.8	268.8	261.8	254.7	247.7
	Error	−15.2	−11.7	.5	28.4	26.7	34.0	42.4	61.6	61.6	71.5	78.0
	Forecast (1-step ahead)	317.9	299.5	288.4	289.5	309.4	316.8	320.6	322.6	331.7	330.2	330.2
	Error	−15.2	−.3	16.0	35.8	7.1	0.0	−2.4	7.8	−8.8	−4.0	−4.5

14.9a.

Model	MAD	RMSE
$w = .5$	13.71	16.86
$w = v = .5$	42.81	50.31

b.

Model	MAD	RMSE
$w = .5$	7.70	11.91
$w = v = .5$	8.99	13.02

14.11a. $\hat{\beta}_0 = 6.609231$; $\hat{\beta}_1 = 1.704198$ **b.** $\hat{Y}_{1985} = 32.1722$; $\hat{Y}_{1986} = 33.8764$

c.

Year	Prediction Interval
1985	32.1722 ± 2.1018
1986	33.8764 ± 2.1571

14.12a. $Y = \beta_0 + \beta_1 t + \beta_2 x_1 + \beta_3 x_2 + \beta_4 x_3 + \varepsilon$

where $x_1 = \begin{cases} 1 & \text{if Qtr. 1} \\ 0 & \text{otherwise} \end{cases}$; $x_2 = \begin{cases} 1 & \text{if Qtr. 2} \\ 0 & \text{otherwise} \end{cases}$; $x_3 = \begin{cases} 1 & \text{if Qtr. 3} \\ 0 & \text{otherwise} \end{cases}$; $t = 1, \ldots, 40$

b. $\hat{Y} = 11.49 + .51t - 3.95x_1 - 2.09x_2 - 4.52x_3$; $F = 1,241.25$; model appears useful

c.

Qtr.	Forecast
1	28.45
2	30.82
3	28.90
4	33.93

14.13a. $\hat{Y}_t = -2.7138 + 11.3412t$

b.

Year	Forecast	Prediction Interval
1985	280.8212	280.8212 ± 62.08
1986	292.1624	292.1624 ± 62.63
1987	303.5036	303.5036 ± 63.21

14.14a. $F = 117.817$; model is useful

c.

		Forecast	95% Lower Limit	95% Upper Limit
1988	I	729.0	662.8	795.1
	II	706.0	639.8	772.1
	III	605.1	539.0	671.3
	IV	516.1	450.0	582.3

14.15a. $\hat{Y}_t = 82,336.9 + 2,279.1t$

Year	**b.** Forecast	**c.** 95% Lower Limit	95% Upper Limit
1985	118,802.82	116,890.82	120,714.82
1986	121,081.94	119,125.63	123,038.25

14.17a. Very strong negative autocorrelation   **b.** Very strong positive autocorrelation   **c.** Residuals are probably uncorrelated
14.18a. Cannot reject H_0 **b.** Cannot reject H_0 **c.** Reject H_0 **d.** Cannot reject H_0
14.19a. Yes   **b.** $d = .52$; reject H_0 ($\alpha = .05$)

14.24 and **14.25**

	14.24		Forecast	**14.25**	95%	95%	
	Year	Forecast	Error		Lower Limit	Upper Limit	
	1984	1,082.18	-129.39		1,876	1,288	Yes
	1985	1,047.06	-499.61		835	1,259	No
	1986	1,047.03	-848.92		834	1,260	No

14.26a. Strong, p-value $< 1\%$

14.27

Year	Variable	**a.** Forecast $w = .5$	**b.** Forecast $w = .5, v = .5$
1985	Agric.	3,352	3,364
1985	Nonagric.	99,174	101,694

14.28

Year	Forecast
1985	337.0
1986	356.7
1987	376.3

14.29

Year	a. Forecast	Expected Gain/Loss	b. Forecast	Expected Gain/Loss
1987	49.126	+3.501/share	50.802	+5.177/share
1988	49.126	+3.501/share	50.245	+4.62/share

14.30a. $\hat{Y}_t = 45.747 + .0492t$

Year	c. Forecast	d. Prediction Interval
1987	46.485	19.495, 73.475
1988	46.534	18.834, 74.230

14.32

Quarter	Forecast	Forecast Error
1	4,178.5	−29.3
2	4,249.9	−74.3
3	4,321.3	−87.0
4	4,392.7	−132.1

14.33a.

Quarter	Forecast	95% Lower Limit	95% Upper Limit
1	3,882.6	3,599.7	4,165.5
2	3,934.7	3,651.4	4,218.0
3	3,986.7	3,702.9	4,270.6
4	4,038.8	3,754.5	4,323.1

b. $F. = .011$

c.

Quarter	Forecast
1	3,887.1
2	3,935.9
3	3,987.2
4	4,034.2

14.34a. $d = .05$

14.35a. $\ln(\widehat{GNP}) = 6.93 + .0237t$ (simple)
$\ln(\widehat{GNP}) = 6.93 + .0237t + .00437x_1 + .00300x_2 + .00237x_3$

b.

Model	Quarter	Forecast	95% Lower Limit	95% Upper Limit
Simple	1	4,340.31	4,283.39	4,397.54
	2	4,444.40	4,384.80	4,504.72
	3	4,550.99	4,488.62	4,614.23
	4	4,660.14	4,594.89	4,726.31
Seasonal	1	4,367.78	4,288.96	4,447.06
	2	4,464.89	4,384.37	4,546.90
	3	4,570.83	4,488.17	4,655.01
	4	4,686.03	4,601.33	4,771.90

14.36

Model	MAD	RMSE
Holt–Winters	28.725	37.6
SLR	114.59	134.42
Seasonal	114.63	134.39
Simple log	53.844	70.303
Seasonal log	53.818	69.9628

14.37a. $\hat{Y}_t = 36.7 + .692t$

c.

1971	1972	1973	1974	1975
54.692	55.384	56.076	56.768	57.460

d.

Year	Primary	Secondary
1955	42.5	56.4
1975	57.5	Less than 42.5

; Projections are in agreement.

14.38a.

Model	Year	Forecast	95% Lower Limit	95% Upper Limit
Automobile	1985	170.5	150.93	190.07
	1986	181.4	161.00	201.80
Mobile home	1985	25.007	22.066	27.948
	1986	26.1415	23.08	29.21
Revolving credit	1985	101.97	87.86	116.08
	1986	110.38	95.68	125.08

b.

Model	Forecast 1985	1986
Automobile	187.5	209.2
Mobile home	26.3	27.5
Revolving credit	112.4	127.4

14.39b.

	Jan.	Feb.	Mar.
	124.383	124.383	124.383

14.40b. $B = s \cdot t_{n-2, \alpha/2} \sqrt{1 + \dfrac{1}{n} + \dfrac{(x_p - \bar{x})^2}{SS_{xx}}} \approx 10.6093(1.96)(1) \approx 20$ (confidence level = 95%)

c.

	Jan.	Feb.	Mar.
Forecast	145.4	146.2	147.1
Forecast interval	(122.6, 168.2)	(123.4, 169.1)	(124.1, 170.1)

e. $d = .410$, reject H_0

14.41

	Forecast	95% Lower Limit	95% Upper Limit
Jan.	126.04	112.46	139.63
Feb.	130.55	113.83	147.27
Mar.	133.97	115.86	152.08

CHAPTER 15

15.9a. 19.00   **b.** 99.00   **c.** 1.61   **d.** 3.87   **15.10a.** .99   **b.** .01   **c.** .05   **d.** .95   **15.11** Diagram b
15.12a. 9, 14; 9, 14   **b.** 75; 75   **c.** 20; 144   **d.** 95; 219; 78.9%, 34.2%   **e.** $F_1 = 37.5; F_2 = 5.21$
f. Reject H_0 in both cases
15.13a. $s_p^2 = 2; s_p^2 = 14.4$ **b.** $t_1^2 = (6.12)^2 = F_1; t_2^2 = (2.28)^2 = F_2$ **c.** $t_{.025}$ with 10 df is 2.228   **d.** Reject H_0

15.14

Source	df	SS	MS	F
Treatments	1	75	75	37.5
Error	10	20	2	
Total	11	95		

Source	df	SS	MS	F
Treatments	1	75	75	5.21
Error	10	149	14.4	
Total	11	219		

15.15a. $F = 1.5$, do not reject H_0 **b.** $F = 6$, reject H_0 **c.** $F = 24$, reject H_0 **d.** Increases

15.16a.

Source	df	SS	MS	F
Treatments	6	16.9	2.817	3.48
Error	35	28.3	.809	
Total	41	45.2		

b. 7   **c.** $F = 3.48$; yes   **d.** p-value $< .01$ **e.** $t = -.83$; no   **f.** $-.4 \pm .79$ **g.** $3.7 \pm .56$
15.17a. 4; 38   **b.** $F = 14.80$; reject H_0
15.18 $T_1 - T_2: -69.3 \pm 39.33; T_1 - T_3: -.9 \pm 40.87; T_1 - T_4: -88.6 \pm 43.08; T_2 - T_3: 68.4 \pm 36.89;$
$T_2 - T_4: -19.3 \pm 39.33; T_3 - T_4: -87.7 \pm 40.87$

15.19a

Source	df	SS	MS	F
Treatments	2	12.301	6.1505	2.931
Error	9	18.888	2.0987	
Total	11	31.189		

b. $F = 2.93$; do not reject H_0 **c.** $T_1 - T_2: -.62 \pm 3.158; T_1 - T_3: 1.98 \pm 3.438; T_2 - T_3: 2.6 \pm 3.596$
15.20b. $F = 8.44$, p-value $= .0014$, reject H_0 **c.** .0014
d. $A - B: 206 \pm 170.343; A - C: 229.5 \pm 170.343; B - C: 23.5 \pm 170.343$

15.21a.

Source	df	SS	MS	F
Treatments	2	509.87	254.935	1.87
Error	90	12,259.96	136.222	
Total	92	12,769.83		

b. $F = 1.87$, no   **d.** No

15.22a.

Source	df	SS	MS	F
Treatments	3	6.816	2.272	8.03
Error	105	29.710	.283	
Total	108	36.526		

b. $F = 8.03$, yes

c. $T_1 - T_2$: $-.678 \pm .424$; $T_1 - T_3$: $-.317 \pm .447$; $T_1 - T_4$: $-.187 \pm .462$; $T_2 - T_3$: $.361 \pm .341$; $T_2 - T_4$: $.491 \pm .361$; $T_3 - T_4$: $.130 \pm .388$

15.23a. $F = 16.95$; reject H_0 **b.** -72 ± 26.08
15.24c. $F = 20.3$; p-value $= 0$, reject H_0 **e.** $T_1 - T_2$: $-.246 \pm .352$; $T_1 - T_3$: $-.851 \pm .324$; $T_1 - T_4$: $-.336 \pm .352$;
$T_2 - T_3$: $-.605 \pm .341$; $T_2 - T_4$: $-.090 \pm .368$; $T_3 - T_4$: $.515 \pm .341$ **15.25b.** $F = 3.17$; no ($\alpha = .05$)   **c.** 473.25 ± 22.925

15.26b. 571.967 **c.** $7,719.83$ **d.**

Source	df	SS	MS	F
Treatments	2	571.97	285.985	3.96
Error	107	7,719.93	72.148	
Total	109	8,291.80		

e. $F = 3.96$, yes

f. $T_1 - T_2$: 4.17 ± 4.737; $T_1 - T_3$: 6.51 ± 5.719; $T_2 - T_3$: 2.34 ± 4.935

15.27a.

Source	df	SS	MS	F
Treatments	4	.023015	.005754	1.08
Error	91	.486047	.005341	
Total	95	.509062		

b. $F = 1.08$; no   **c.** No

15.28a. $F = 17.32$; yes   **b.** p-value $< .01$ **c.** $t = 4.24$; yes
d. $T_1 - T_2$: -1.6 ± 1.904; $T_1 - T_3$: -2.2 ± 1.977; $T_1 - T_4$: -4.9 ± 1.949; $T_2 - T_3$: $-.6 \pm 1.627$; $T_2 - T_4$: -3.3 ± 1.593;
$T_3 - T_4$: -2.7 ± 1.680
15.29a. 3, 5   **b.** 15   **c.** H_0: $\mu_1 = \mu_2 = \mu_3 = \mu_4 = \mu_5$ **d.** $F = 9.109$ **e.** $F > 7.01$ **f.** Reject H_0

15.30a.

Source	df	SS	MS	F
Treatments	2	21.5555	10.7778	5.54
Block	2	.8889	.4445	.23
Error	4	7.7778	1.9445	
Total	8	30.2222		

b. H_0: $\mu_1 = \mu_2 = \mu_3$ **c.** $F = \frac{\text{MST}}{\text{MSE}}$ **e.** $F = 5.54$, do not reject H_0

15.31a. $F_T = 50.98$, $F_B = 202.68$ **b.** $F = 50.98$, yes   **c.** $F = 202.68$, yes
d. $A - B$: $-1.125 \pm .763$; $A - C$: $1.325 \pm .763$; $B - C$: $2.45 \pm .763$

15.32a.

Source	df	SS	MS	F
Treatments	3	28.2	9.40	4.13
Block	5	69.0	13.80	6.07
Error	15	34.1	2.273	
Total	23	131.3		

c. $F = 6.07$, yes ($\alpha = .01$)   **d.** -2.4 ± 1.526

15.33a. $F_T = 3.20$, reject H_0; $F_B = 1.80$, do not reject H_0 **b.** $F_T = 13.33$, reject H_0; $F_B = 2.00$, do not reject H_0
c. $F_T = 5.33$, reject H_0; $F_B = 5.00$; reject H_0 **d.** $F_T = 16.00$, reject H_0; $F_B = 6.00$, reject H_0
e. $F_T = 2.67$, do not reject H_0; $F_B = 1.00$, do not reject H_0

15.34a.

Source	df	SS	MS	F
Treatments	2	.1911	.0956	.07
Block	5	605.6978	121.1396	92.84
Error	10	13.0489	1.3049	
Total	17	618.9378		

b. $F = .07$, p-value $= .9299$, no

c. $T_1 - T_2$: $-.23 \pm 2.09$; $T_1 - T_3$: $-.19 \pm 2.09$; $T_2 - T_3$: $.04 \pm 2.09$
15.35a. Randomized block design   **b.** $F = 8.452$, yes   **c.** p-value $= .014$
d. $T_1 - T_2$: -1.27 ± 1.164; $T_1 - T_3$: -1.3 ± 1.164; $T_1 - T_4$: -1.3 ± 1.164; $T_2 - T_3$: $-.03 \pm 1.164$; $T_2 - T_4$: $-.03 \pm 1.164$;
$T_3 - T_4$: 0 ± 1.164

15.36a.

Source	df	SS	MS	F
Treatments	2	1,962.2	981.1	39.40
Block	3	143.6	47.9	1.92
Error	6	149.2	24.9	
Total	11	2,254.9		

b. $F = 39.40$, yes   **c.** $F = 1.92$, no   **d.** -27 ± 6.86

15.37a. $F = 12.25$, yes   **c.** $t = 3.5$, reject H_0
15.38b. $F = 9.1$, yes   **c.** $T_1 - T_2$: $-.002 \pm .115$; $T_1 - T_3$: $.15 \pm .115$; $T_2 - T_3$: $.152 \pm .115$

15.39a. $F = .34$, no   **b.** p-value $> .10$ **c.**

Source	df	SS	MS	F
Treatment	2	.941	.4705	.34
Block	5	10,239.969	2,047.9938	1,471.37
Error	10	13.919	1.3919	
Total	17	10,254.829		

15.40c. $F = 78.72$, p-value $= .0001$, yes
d. $T_1 - T_2$: -8.83 ± 33.35; $T_1 - T_3$: 4.50 ± 33.35; $T_1 - T_4$: -132.83 ± 33.35; $T_1 - T_5$: 69.33 ± 33.35;
$T_2 - T_3$: 13.33 ± 33.35; $T_2 - T_4$: -124.00 ± 33.35; $T_2 - T_5$: 78.17 ± 33.35; $T_3 - T_4$: -137.33 ± 33.35;
$T_3 - T_5$: 64.83 ± 33.35; $T_4 - T_5$: 202.17 ± 33.35

15.41a.

Source	df	SS	MS	F
Treatments	3	.1858	.06193	21.49
Block	9	5.0607	.56230	195.14
Error	27	.0778	.00288	
Total	39	5.3243		

b. $F = 21.49$, reject H_0

c. $A - B$: $-.044 \pm .067$; $A - C$: $-.178 \pm .067$; $A - D$: $-.031 \pm .067$; $B - C$: $-.134 \pm .067$; $B - D$: $.013 \pm .067$;
$C - D$: $.147 \pm .067$

5.42a. 2   **c.** 4, 3

15.43a.

Source	df	SS	MS	F
A	2	.8	.4000	3.69
B	3	5.3	1.7667	16.31
AB	6	9.6	1.6000	14.77
Error	12	1.3	.1083	
Total	23	17.0		

b. $F = 13.18$, yes   **c.** Yes   **e.** $F = 14.77$, reject H_0

15.44a. A: 4, B: 2   **b.** 3   **c.**

Source	df	SS	MS	F
A	3	2.25	.75	5.00
B	1	.95	.95	6.33
AB	3	.90	.30	2.00
Error	16	2.40	.15	
Total	23	6.50		

d. $F = 3.90$, reject H_0 **e.** $F(AB) = 2.00$, do not reject H_0; $F(A) = 5.00$, reject H_0; $F(B) = 6.33$, reject H_0
15.45b. $F = 21.61$, reject H_0 **c.** $F = 36.60$, reject H_0 **d.** No
f. Half-width: 1.839; Pairs of treatment means that differ: (A_1B_1 and A_1B_3), (A_1B_2 and A_1B_3), (A_1B_1 and A_2B_1), (A_1B_2 and A_2B_2), (A_2B_1 and A_2B_2), (A_2B_1 and A_2B_3), (A_2B_2 and A_1B_3), (A_2B_3 and A_1B_3)

15.46b.

Source	df	SS	MS	F
A	1	658.845	658.845	46.65
B	1	255.380	255.380	18.08
AB	1	2.000	2.000	.14
Error	4	56.490	14.1225	
Total	7	972.715		

c. $F = 21.63$, reject H_0 **d.** $F = .14$, do not reject H_0

e. $F(A) = 46.65$, reject H_0; $F(B) = 18.08$, reject H_0

15.47a.

Source	df	SS	MS	F
A	2	200	100	3.00
B	2	100	50	1.50
AB	4	100	25	.75
Error	18	600	33.333	
Total	26	1,000		

$F(\text{Treatment}) = 1.5$, do not reject H_0 ($\alpha = .05$)

b.

Source	df	SS	MS	F
A	2	100	50	3.00
B	2	100	50	3.00
AB	4	500	125	7.50
Error	18	300	16.667	
Total	26	1,000		

$F(\text{Treatment}) = 5.25$, reject H_0; $F(AB) = 7.50$, reject H_0

c.

Source	df	SS	MS	F
A	2	400	200	12.00
B	2	100	50	3.00
AB	4	200	50	3.00
Error	18	300	16.667	
Total	26	1,000		

$F(\text{Treatment}) = 5.25$, reject H_0; $F(AB) = 3.00$, reject H_0

d.

Source	df	SS	MS	F
A	2	400	200	36.00
B	2	400	200	36.00
AB	4	100	25	4.50
Error	18	100	5.556	
Total	26	1,000		

$F(\text{Treatment}) = 20.25$, reject H_0; $F(AB) = 4.50$, reject H_0

15.48a. SS(Total) $= 5,300,056.52$; SS(Treatment) $= 5,291,151.19$; SS(Error) $= 8,905.33$; SS(Price) $= 3,089,053.9$;
SS(Display) $= 1,691,392.5$, SS(Display $\times$ Price) $= 510,704.8$ **b.** $F = 1,336.85$, reject H_0 **c.** $F = 258.07$, reject H_0
d. No   **e.** Pairs of treatment means: half-width $= 52.268$

15.49b. F(Treatment) = 11.074, reject H_0; F(Interaction) = 6.784, reject H_0; Pairs of treatment means: half-width = 8.85

15.50a.

	Light	Heavy
Female	146.40	116.00
Male	104.00	98.00

b. 6,739.605   **c.** SS(Sex) = 114.005, SS(Weight) = 41.405, SS(Sex × Weight) = 18.605

d.

	Variance	SS
Female, Light	46.3761	324.6327
Female, Heavy	8.5849	60.0943
Male, Light	25.4016	177.8112
Male, Heavy	32.4900	227.4300

e. SSE = 789.968   **f.** SS(Total) = 963.983

g.

Source	df	SS	MS	F
Sex	1	114.005	114.005	4.04
Weight	1	41.405	41.405	1.47
Sex × Weight	1	18.605	18.605	.66
Error	28	789.968	28.213	
Total	31	963.983		

h. F(Treatments) = 2.06, do not reject H_0

15.51a.

Source	df	SS	MS	F
Sex	1	10,686.361	10,686.361	68.74
Weight	1	538.756	538.756	3.47
Sex × Weight	1	831.744	831.744	5.35
Error	36	5,596.908	155.470	
Total	39	17,653.769		

b. F(Treatments) = 25.80, reject H_0; F(Sex × Weight) = 5.35, reject H_0

15.52a. $E(y) = \beta_0 + \beta_1 x_1 + \beta_2 x_2 + \beta_3 x_3 + \beta_4 x_4$ **b.** 10   **c.** H_0: $\beta_1 = \beta_2 = \beta_3 = \beta_4 = 0$ **d.** Reject H_0 if $F > 2.61$

15.53a. Randomized block   **c.** H_0: $\beta_1 = \beta_2 = 0$

15.54b. H_0: $\beta_1 = \beta_2 = \beta_3 = 0$, reject H_0 if $F > 3.49$ ($\alpha = .05$)   **c.** H_0: $\beta_4 = \beta_5 = 0$, reject H_0 if $F > 3.89$ ($\alpha = .05$)

15.55a. $F = 8.438$, reject H_0 **b.** p-value = .0014   **c.** SS(Treatment) = 318,861.667; SS(Error) = 510,163

15.56a. $E(y) = \beta_0 + \beta_1 x_1 + \beta_2 x_2 + \beta_3 x_3 + \beta_4 x_4 + \beta_5 x_5 + \beta_6 x_6 + \beta_7 x_7$

b. $\hat{y} = 3.22 - .2x_1 + .033x_2 + 16.1x_3 + 12.4x_4 + 6x_5 + 2.67x_6 + 2.17x_7$ **c.** H_0: $\beta_1 = \beta_2 = 0$; $F = .07$, do not reject H_0

15.57c. H_0: $\beta_1 = \beta_2 = \beta_3 = \beta_4 = \beta_5 = \beta_6 = \beta_7 = \beta_8 = 0$; $F = 1,336.15$, yes

d. H_0: $\beta_5 = \beta_6 = \beta_7 = \beta_8 = 0$; reject H_0 if $F > 2.29$ **e.** $F = 257.93$, reject H_0 **f.** No

15.58a. Completely randomized   **b.** 116   **c.** $F = 1.95$; no   **d.** $F = .39$, no   **e.** $F = 53.27$, yes

15.59a. Randomized block   **b.**

Source	df	SS	MS	F
Treatments	2	15.730	7.865	8.91
Block	36	85.963	2.388	2.70
Error	72	63.604	.883	
Total	110	165.297		

c. $F = 8.91$, yes

d. $T_1 - T_2$: $-.243 \pm .524$; $T_1 - T_3$: $-.892 \pm .524$; $T_2 - T_3$: $-.649 \pm .524$

15.60a. Completely randomized   **b.** $F = 6.30$, yes

c. $A - B$: -2.5 ± 5.68; $A - C$: -2.5 ± 5.68; $A - D$: 4.5 ± 5.68; $B - C$: 0 ± 5.68; $B - D$: 7 ± 5.68; $C - D$: 7 ± 5.68

15.61a.

Source	df	SS	MS	F
A	3	2.6	.8667	1.11
B	5	9.2	1.8400	2.36
AB	15	46.5	3.1000	3.98
Error	24	18.7	.7792	
Total	47	77.0		

b. A: 4; B: 6, 24, 2   **c.** 58.3, $F = 3.25$, reject H_0
d. Yes; $F = 3.98$, reject H_0

15.62a. Randomized block   **c.** $F = 4.81$, reject H_0 **d.** $A - B$: 0 ± 10.34; $A - C$: -10 ± 10.34; no significant difference
15.63a. Completely randomized   **b.** $F = 4.88$, yes   **c.** p-value $< .01$
d. $A - B$: -2.10 ± 2.369; $A - C$: $-.44 \pm 2.369$; $A - D$: 1.08 ± 2.369; $B - C$: 1.66 ± 2.369; $B - D$: 3.18 ± 2.369;
$C - D$: 1.52 ± 2.369 **e.** 11.99 ± 1.025

15.64a.

Source	df	SS	MS	F
Extractor	5	84.71	16.9420	12.44
Truckload	14	159.29	11.3779	8.35
Error	70	95.33	1.3619	
Total	89	339.33		

b. $F = 12.44$, yes

15.66a.

Source	df	SS	MS	F
A	3	74.33	24.78	16.52
B	2	4.08	2.04	1.36
AB	6	168.92	28.15	18.77
Error	12	18.00	1.50	
Total	23	265.333		

b. $F = 14.99$, yes   **c.** $F = 18.77$, reject H_0 **d.** No   **e.** Half-width: 4.813

15.67a.

Source	df	SS	MS	F
Company	1	3,237.2	3,237.2	19.62
Error	98	16,167.7	164.977	
Total	99	19,404.9		

b. $F = 19.62$, yes   **c.** No

15.68a. Randomized block   **b.**

Source	df	SS	MS	F
Carrier	2	8.8573	4.4287	83.82
Block	4	3.9773	.9943	18.82
Error	8	.4227	.0528	
Total	14	13.2573		

c. $F = 83.82$, yes

15.69b. $F = 12.29$, yes   **c.** F(Interaction) $= .02$, do not reject H_0; F(Schedule) $= 7.37$, reject H_0; F(Payment) $= 29.47$, reject H_0
15.70 $F = 15.59$, yes   **15.71** $F = 2.95$, no   **15.72a.** Completely randomized   **b.** $F = .3375$, do not reject H_0 **c.** No

15.73a. Randomized block   **b.**

Source	df	SS	MS	F
Brand	2	.63	.32	.20
Block	3	.86	.29	.18
Error	6	9.84	1.64	
Total	11	11.33		

c. $F = .20$, no   **d.** $F = .18$, no

15.74 $A - B$: $-.38 \pm 2.846$; $A - C$: $.175 \pm 2.846$; $B - C$: $.55 \pm 2.846$
15.75a. $t = -.46$, do not reject H_0 **b.** $t = .46$, do not reject H_0 **c.** $F = .21$, no **d.** $t^2 = (\pm.46)^2 = .21 = F$
15.76a. Randomized block **b.** $F = 30.332$, yes **c.** $F = 21.976$, yes
d. $A - B$: -100.2 ± 75.32; $B - C$: 74 ± 75.32; $D - C$: 174.2 ± 75.32

15.77a. Completely randomized **b.**

Source	df	SS	MS	F
Treatment	2	2,759.9	1,380.0	4.11
Error	84	28,175.6	335.4	
Total	86	30,935.5		

c. $F = 4.09$, yes

d. NYSE $-$ ASE: 11.74 ± 9.194; NYSE $-$ OTC: 10.60 ± 11.988; ASE $-$ OTC: -1.14 ± 11.988
15.78a. $E(y) = \beta_0 + \beta_1 x_1 + \beta_2 x_2$; base level: OTC **b.** $\hat{y} = 15.56 + 10.6x_1 - 1.14x_2$; base level: OTC
c. $F = 4.11$, $.01 < p\text{-value} < .05$, reject H_0

15.80a.

Source	df	SS	MS	F
Incentive	2	52,003.11	26,001.56	17.43
Plant	1	10,706.72	10,706.72	7.17
Interaction	2	69.78	34.89	.023
Error	12	17,902.00	1,491.83	
Total	17	80,681.61		

b. $F = .023$, no

15.81a. $E(y) = \beta_0 + \beta_1 x_1 + \beta_2 x_2 + \beta_3 x_3 + \beta_4 x_1 x_3 + \beta_5 x_2 x_3$
b. $\hat{y} = 1,650 - 125x_1 - 21.7x_2 - 46.3x_3 + .67x_1 x_3 - 8x_2 x_3$, SSE $= 17,902$
c. $\hat{y} = 1,651 - 125x_1 - 25.7x_2 - 48.8x_3$; $F = .023$, do not reject H_0

CHAPTER 16

16.1a. $T_B \leq 32$ or $T_B \geq 58$ **b.** $T_A \geq 40$ **c.** $T_B \geq 98$ **d.** $z < -1.96$ or $z > +1.96$ **16.2** $T_B = 42.5$; reject H_0
16.4a. $z = -4.2$; reject H_0 **b.** $p\text{-value} < .001$ **16.5a.** $T_B = 53$; reject H_0 **16.6** $T_{\text{U.S.}} = 39$; yes
16.7 $T_{\text{after}} = 86$; reject H_0 **16.8a.** $T_{\text{twin}} = 82$; no
16.9a. $t = -1.26$; do not reject H_0 ($\alpha = .05$) **b.** $T_A = 37.5$; do not reject H_0 ($\alpha = .05$)
16.10a. $z = 3.396$; reject H_0 **c.** $p\text{-value} > .001$ **16.11a.** $z = 1.77$; reject H_0
16.12a. $T \leq 101$ **b.** $T_- \leq 303$ **c.** $T_+ \leq 0$ **16.13b.** $T_- = 4.5$; reject H_0 **16.16b.** $z = 2.499$; reject H_0 **c.** $.0012$
16.17a. Randomized **c.** $T_+ = 3.5$; reject H_0 **d.** $.01 < p\text{-value} < .025$ **16.18a.** $T_+ = 7$; reject H_0 **16.19** $T_+ = 2$; reject H_0
16.20a. $T_+ = 1$; reject H_0 ($\alpha = .05$) **16.21** $T_+ = 3$; yes **16.22** $T_+ = 6$; no **16.23** $T_- = 8$; no
16.24a. $t = .740$; do not reject H_0 **b.** $T_- = 8$; do not reject H_0
16.25a. 31.4104 **b.** 27.4884 **c.** 58.5713 (by interpolation) **d.** 96.5782 **e.** 5.99147 **f.** 25.1882
16.26a. $.995$ **b.** $.025$ **c.** $.100$ **d.** $.975$ **e.** $.025$ **f.** $.100$
16.27b. $H = 8.190$; reject H_0 **c.** $.01 < p\text{-value} < .025$
16.28a. Completely randomized design **c.** Reject H_0 if $H > 9.21034$ **d.** $H = 15.9445$; reject H_0
16.30a. $H = .8089$; do not reject H_0 **16.31c.** $H = 1.2764$; do not reject H_0
16.32a. $H = 11.415$; reject H_0 **b.** Completely randomized design **16.33** $H = 5.85$; no
16.34a. $F = 1.33$; do not reject H_0 **b.** $H = 1.22$; do not reject H_0 **16.35b.** $H = 16.505$; reject H_0
16.36a. 18.3070 **b.** 22.3072 **c.** 79.4900 **d.** 42.9798 **16.37a.** $.050$ **b.** $.90$ **c.** $.005$
16.38a. 6 blocks **c.** $F_r = 15.2$; reject H_0 **d.** $<.005$ **16.39b.** $F_r > 4.60517$ **c.** $F_r = 6.9286$; reject H_0
16.40 $F_r = 12.8$; reject H_0 **16.41b.** $F_r = 22.9334$; reject H_0 **c.** $p\text{-value} < .005$
16.42a. $F_r = 4.75$; reject H_0 **b.** $T_- = 0$; reject H_0 **16.43** $F_r = 6.35$; yes **16.44** $F_r = 12.3$; yes
16.45a. $F_r = 7.85$; reject H_0 **16.46a.** $F_r = 12.2$; yes **c.** $.005 < p\text{-value} < .01$
16.47a. $.01$ **b.** $.005$ **c.** $.95$ **d.** $.10$ **16.48a.** $r_s = 1$ **b.** $r_s = -.9$ **c.** $r_s = 1$ **d.** $r_s = .2$
16.49b. $r_s = .898$; reject H_0 **c.** $p\text{-value} < .02$ **16.51a.** $r_s = .863736$ **c.** $p\text{-value} < .005$
16.52a. $r_s = -.667964$ **b.** Reject H_0 **16.53a.** $.861$ **b.** Yes **16.54a.** $.4524$ **b.** No **16.55** $r_s = .9341$
16.56 $r_s = -.8536$; yes **16.58a.** $T_{\text{after}} = 19$; yes **16.59a.** $T = 22$; reject H_0 **b.** $.025 < p\text{-value} < .05$

16.60a. x and y_1: .9848; x and y_2: .9879　　**16.61** $T_A = 21$; no　　**16.62** $T_- = 3$; yes　　**16.63b.** $H = 14.61$; yes
16.64b. $F_r = 6.21$; no　　**16.65** $T_1 = 135$; yes　　**16.66** $r_s = .4091$; no　　**16.67a.** $t = -2.96$; reject H_0　　**b.** $T_+ = 1$; yes
16.68 $T_{before} = 132.5$; yes　　**16.69** $T_- = 3.5$; yes　　**16.70b.** $F_r = 11.77$; yes　　**16.71b.** $H = 16.39$; yes　　**16.72** $r_s = .9286$; yes
16.73 $F_r = 1.55$; no　　**16.74** $r_s = .28$; no　　**16.75** $H = 12.79$; yes　　**16.76** $T_{door} = 28.5$; do not reject H_0
16.77 $r_s = .7714$; no　　**16.78a.** $F_r = 6.35$; yes　　**16.79** $T_- = 17.5$; no　　**16.80** $r_s = .858$

CHAPTER 17

17.1a. 24.9958　　**b.** 70.0648　　**c.** 18.5494　　**d.** 10.5966　　**17.2a.** .01　　**b.** .01　　**c.** .90　　**d.** .005
17.3a. $X^2 > 4.60517$　　**b.** $X^2 > 13.2767$　　**c.** $X^2 > 7.81473$　　**17.6a.** $X^2 = 3.941$; no　　**b.** p-value $> .10$
17.7a. $X^2 = 2.7379$; no　　**17.8** $.2293 \leq p \leq .3533$　　**17.9a.** $X^2 = 26.6627$; yes　　**b.** $X^2 = 1.0$; no　　**17.10** $.55 \pm .08184$
17.11a. $X^2 = 12.374$; do not reject H_0　　**b.** $.05 < p$-value $< .10$
17.12a. $X^2 = 61.826$; reject H_0; yes　　**c.** $p < .005$
d. 25–34 age group: $(.2464 \leq p \leq .3256)$; 20–44 age group: $(.6144 \leq p \leq .6976)$　　**17.13a.** $X^2 = 1{,}038$; yes　　**b.** Approx. 0
17.14 $X^2 = 10.3$; yes　　**17.15** $X^2 = 9.48$; yes　　**17.16** $X^2 = 10.44$; yes
17.17a. $X^2 > 26.2962$　　**b.** $X^2 > 15.9871$　　**c.** $X^2 > 9.21034$

17.18b. $X^2 > 9.21034$　　**c.**

14.39	35.69	44.90
10.60	26.30	33.09

d. $X^2 = 7.50$; do not reject H_0

17.20 $X^2 = 15.27$; reject H_0　　**17.22** $X^2 = 38.02$; reject H_0　　**17.25a.** $X^2 = 20.78$; yes　　**b.** Sprain; older; younger
17.26a. Union members: $X^2 = 13.36$, reject independence; nonunion workers: $X^2 = 9.16$, do not reject independence
b. Union members: $.025 < p$-value $< .05$; nonunion members: p-value $> .10$　　**17.27b.** $X^2 = 89.09$; reject H_0
17.28a. $X^2 = 4.97$; yes　　**c.** $.05 < p$-value $< .10$　　**17.29** $X^2 = 40.70$; yes　　**17.30** $X^2 = 9.50$; yes
17.32a. $X^2 = 1.35$; no　　**b.** p-value $> .10$　　**17.33a.** Yes　　**d.** $X^2 = 2.373$; do not reject H_0　　**e.** Yes
17.34c. $X^2 = 26.04$; reject H_0　　**17.35a.** $X^2 = 87.38$; yes　　**17.36a.** $X^2 = 8.13$; yes　　**b.** $.01 < p$-value $< .025$
17.37a. $X^2 = 18.54$; yes　　**b.** p-value $< .005$　　**17.38** $X^2 = 6.58$; yes
17.39a. $X^2 = 6.66$; reject H_0　　**b.** $.025 < p$-value $< .05$　　**d.** $.56 \pm .12$　　**17.40** $X^2 = 31.85$; yes　　**17.41** $.12 \pm .0947$
17.42 $X^2 = 5.51$; no　　**17.43b.** Reject H_0　　**c.** p-value $< .005$　　**17.44** $X^2 = 2.60$; no

17.45a.

	PERFORMANCE				
	1	2	3	4	Quit or fired
Job-Matched	178	792	634	277	99
Not Job-Matched	79	673	990	1,228	991

b. $X^2 = 895.79$; yes　　**c.** $z = 12.5$; yes

17.46 $X^2 = 42.54$; yes　　**17.47** $X^2 = 103.08$; yes　　**17.48** $X^2 = 7.384$; no $(\alpha = .05)$
17.49a. $X^2 = 47.979$; yes　　**b.** $.125 \pm .046$　　**17.50** $X^2 = 30.507$; yes　　**17.51** $X^2 = 6.93$; no　　**17.52** $.284 \pm .0571$
17.53a. $X^2 = 38.68$; yes　　**17.54** $X^2 = 7.4$; do not reject H_0　　**17.55** $X^2 = 3.13$; no　　**17.56** $.095 \pm .128$
17.57a. $X^2 = 269.91$; yes　　**b.** p-value $< .005$　　**17.58** $X^2 = 6.74$; yes　　**17.59a.** $X^2 = 46.25$; yes　　**b.** p-value $< .005$
17.60 $.143 \pm .042$　　**17.61a.** $X^2 = 9.647$　　**b.** 11.0705　　**c.** No $(\alpha = .05)$　　**d.** $.05 < p$-value $< .10$
17.62 $X^2 = 9.47$; no $(\alpha = .05)$　　**17.63** $X^2 = 67.15$; reject H_0

CHAPTER 18

18.3a. Uncertainty　　**b.** Conflict　　**c.** Certainty　　**18.4a.** Certainty　　**b.** Uncertainty
18.5 Actions, States of nature, Outcomes, Objective variable

18.13 a_3 is inadmissible;

	S_1	S_2	S_3	S_4
a_1	0	0	40	35
a_2	5	5	0	5
a_3	20	35	50	0

18.14 a_2 and a_4 are inadmissible

18.15a. a_4 inadmissible   **18.16** a_1, S_1: \$200; a_1, S_2: \$500; a_2, S_1: \$1,400; a_2, S_2: \$0; a_3, S_1: \$0; a_3, S_2: \$700

18.17a. Payoffs in millions of dollars:

	STATE OF NATURE		
	1%	2%	3%
Market	14	16	18
Do not market	15	15	15

b. Opportunity losses in millions of dollars:

	1%	2%	3%
Market	1	0	0
Do not market	0	1	3

18.18a.

		STATE OF NATURE					
		3 doz.	4 doz.	5 doz.	6 doz.	7 doz.	8 doz.
	3 doz.	720	720	720	720	720	720
	4 doz.	660	960	960	960	960	960
	5 doz.	600	900	1,200	1,200	1,200	1,200
ACTION	6 doz.	540	840	1,140	1,440	1,440	1,440
	7 doz.	480	780	1,080	1,380	1,680	1,680
	8 doz.	420	720	1,020	1,320	1,620	1,920

b.

		STATE OF NATURE					
		3 doz.	4 doz.	5 doz.	6 doz.	7 doz.	8 doz.
	3 doz.	0	240	480	720	960	1,200
	4 doz.	60	0	240	480	720	960
	5 doz.	120	60	0	240	480	720
ACTION	6 doz.	180	120	60	0	240	480
	7 doz.	240	180	120	60	0	240
	8 doz.	300	240	180	120	60	0

18.19a.

	STATE OF NATURE	
	Successful	Unsuccessful
Increase advertising	600,000	−600,000
Do not increase advertising	200,000	200,000

b. No

18.20a.

	STATE OF NATURE		
	5%	10%	50%
Purchase	−1,250	−1,500	−1,500
Do not purchase	−250	−500	−2,500

18.21a. Payoffs in millions of dollars:

	STATE OF NATURE	
	Win	Lose
Court	−1	−101
No court	−50	−50

18.22a. Payoffs in millions of dollars:

	STATE OF NATURE				
	.5	10	15	20	25
Introduce	-2.85	0	1.50	3	4.5
Do not introduce	0	0	0	0	0

b. Opportunity losses in millions of dollars:

	STATE OF NATURE				
	.5	10	15	20	25
Introduce	2.85	0	0	0	0
Do not introduce	0	0	1.5	3	4.5

18.23 a_2 **18.24a.** a_2 **c.** a_1 **18.26a.** a_2 **b.** a_2 **18.27** a_1

18.28a.

	STATE OF NATURE		
	Major	Moderate	Minor
Y	$-8,000$	$-5,400$	$-4,400$
Z	$-9,000$	$-5,000$	$-3,500$

b. Z   **18.29a.** Design B   **b.** Design A

18.30 Yes   **18.31** Do not purchase   **18.32** 6 doz.
18.33a. Both actions have the same expected payoff

b.

	Successful	Unsuccessful
Increase advertising	0	800,000
Do not increase advertising	400,000	0

c. Both actions have the same expected opportunity loss.

18.34a. a_2 and a_3 are inadmissible   **b.** a_4 **c.** a_4 **18.35a.** a_1 **b.** a_3 **18.36a.** a_2 **b.** a_2 **18.37a.** a_3 **b.** a_1
18.38a. Z   **b.** Y   **18.39a.** Market   **b.** Do not market   **18.40a.** 8 doz.   **b.** 3 doz.
18.41a. Do not purchase   **b.** Purchase

18.42a. $U(100) = .2; U(500) = .7; U(800) = .95$ **18.43a.**

	S_1	S_2	S_3
a_1	5,625	2,500	900
a_2	3,600	6,400	9

b. a_2

18.44a. None inadmissible   **b.**

	S_1	S_2	S_3
a_1	.672	.835	.896
a_2	.712	.772	.772
a_3	.672	.852	.835
a_4	.672	0	1.000

c. a_3 **d.** Yes

18.45a.

	S_1	S_2	S_3	S_4
a_1	.2	.8	0	0
a_2	.05	.65	.90	.50
a_3	0	.20	.60	1.00

b. a_2 **18.46a.** $U(600,000) = .9; U(0) = .8$

18.47a.

	Dry well	Moderate success	Gusher
Invest	−500,000	600,000	1,500,000
Do not invest	0	0	0

b. Invest   **c.** Do not invest

18.48 6 doz.   **18.49a.** Risk-taking   **b.** Risk-neutral   **c.** Risk-avoiding   **d.** Risk-neutral   **18.50b.** Risk-avoiding   **c.** a_1
18.51b. Risk-taking   **c.** a_2 **18.52a.** Risk-neutral   **b.** a_3; no   **18.53b.** Risk-avoider   **18.54b.** Risk-avoider

18.55b. Payoffs in millions of dollars:

	Bill will pass	Bill won't pass
Don't lease	2.5	−1.5
Lease	.5	.5

c. Opportunity losses in millions of dollars:

	Bill will pass	Bill won't pass
Don't lease	0	2
Lease	2	0

d. Don't lease

18.56b. Risk-neutral   **c.** Don't lease

18.57a. a_2 inadmissible   **b.**

	S_1	S_2	S_3
a_1	28	0	107
a_3	60	36	0
a_4	15	72	10
a_5	0	27	70

c.

	Expected Payoff	Expected Opportunity Loss
a_1	23.7	35.4
a_3	18.3	40.8
a_4	28.0	31.1
a_5	37.0	22.1

18.58a. Relative frequency   **b.** Subjective   **c.** Subjective   **d.** Both

18.59a.

	S_1	S_2	S_3
Loan	7,500	6,500	−30,000
Don't loan	6,000	6,000	6,000

b.

	S_1	S_2	S_3
Loan	0	0	36,000
Don't loan	1,500	500	0

18.60a. $P(S_1) = .8625$; $P(S_2) = .09375$; $P(S_3) = .04375$ **b.** Don't make loan   **18.61a.** Make loan   **b.** Don't make loan

18.62a. Risk-avoiding   **b.**

	S_1	S_2	S_3
Loan	.3279	.3263	.2646
Don't loan	.3256	.3256	.3256

; No

18.63a.

	1	2
$.99	7.94	11.92
$1.98	11.96	16.00
$2.75	12.38	13.75
$3.50	10.50	7.00

b.

	1	2
$.99	4.44	4.08
$1.98	.42	0
$2.75	0	2.25
$3.50	1.88	9.00

c. $1.98

18.64a. Payoffs in thousands of dollars:

	Win	Lose
Settle	−250	−250
Court	0	−1,000

c. Settle   **18.65a.** Go to court   **b.** Settle

18.66a.

	Win	Lose
Settle	.6186	.6186
Court	.7143	0

b. Settle

18.67a. Payoffs in millions of dollars:

	Use	Don't use
Install	2.75	−1
Don't install	0	0

b. Opportunity losses in millions of dollars:

	Use	Don't use
Install	0	1
Don't install	2.75	0

c. Don't install

18.68a. Risk-neutral   **b.** Risk-avoider   **c.** Risk-taker   **d.** Risk-taker

18.69b.

	S_1	S_2	S_3
a_1	.24	.95	.09
a_2	0	.55	1.00
a_3	.35	.19	.90
a_4	.55	.66	.22

c. a_4; same under expected payoff criterion

CHAPTER 19

19.2

(4)	(5)
.150	.297
.320	.634
.035	.069
.505	1.000

19.3

(4)	(5)
.315	.538
.015	.026
.195	.333
.060	.103
.585	1.000

19.4 $\frac{2}{3}$; 0

19.7a.

x	5	10	15	20	25
$p(x)$	.196	.627	.118	.039	.020

b. 10.3; 16.16

19.8a. .1025; .000179   **b.**

p	Posterior Probability
.09	.377
.10	.253
.11	.158
.12	.107
.13	.105
	1.000

c. .1031; .000181   **19.9** .5263   **19.10** .9174

19.11a.

p	Posterior Probability
.01	.25
.05	.50
.10	.25

b.

p	Posterior Probability
.01	.00096
.05	.21767
.10	.78137

19.12a.

Mean Demand	Posterior Probability
2	.145
3	.541
4	.314

b.

Mean Demand	Posterior Probability
2	.190
3	.586
4	.224

19.13 a_2 **19.14a.** a_3 **b.** $P(S_1) = .888$, $P(S_2) = .056$, $P(S_3) = .056$ **c.** 9.226; -12.376; 19.845; 7.392   **d.** a_3
19.15a. a_3 **b.** $P(S_1) = .25$, $P(S_2) = .5625$, $P(S_3) = .1875$ **c.** a_1

19.16a.

		STATE OF NATURE				
		.00	.05	.10	.15	.20
ACTION	Agreement 1	$-46,000$	$-46,000$	$-46,000$	$-46,000$	$-46,000$
	Agreement 2	$-40,000$	$-42,000$	$-44,000$	$-46,000$	$-48,000$

b. Agreement 2   **c.** Agreement 2 in both cases

19.17a. Payoffs in millions of dollars:

		STATE OF NATURE	
		(.6)	(.4)
		Bill passes	Bill doesn't pass
ACTION	Don't lease	2.5	-1.5
	Lease	.5	.5

b.

State of Nature	Posterior Probability
Bill passes	.923
Bill doesn't pass	.077

d. Don't lease

19.18a.

	S_1	S_2	S_3
a_1	$\$40$	$\$250$	$\$0$
a_2	$\$115$	$\$0$	$\$125$
a_3	$\$0$	$\$210$	$\$20$

b. $45.00

19.19a.

	S_1	S_2	S_3
a_1	0	0	5,000
a_2	3,000	500	0

b. $1,125   **c.** S_2 **d.** $P(S_1) = .061; P(S_2) = .864; P(S_3) = .075$ **e.** $375

19.20a. $1,800   **b.** $4,584   **c.** $744   **19.21a.** Do not install   **b.** $1,550

19.22a. Payoffs in millions of dollars:

		STATE OF NATURE	
		Error	No error
ACTION	Certify	-10	0
	Don't certify	0	-2

c. Certify   **d.** $1 million

19.23 $200   **19.24** $.8 million
19.25a. EPS = $267.96; EPNS = $145   **b.** EVSI = $122.96   **c.** ENGS = $22.96   **d.** Yes
19.26a. $13.97   **b.** First source
19.27a. Prior probabilities: $P(S_1) = .2; P(S_2) = .3; P(S_3) = .5$
b. Posterior probabilities: $P(S_1) = .6829; P(S_2) = .0732; P(S_3) = .2439$
c. Posterior probabilities: $P(S_1) = .1111; P(S_2) = .75; P(S_3) = .1389$
d. Posterior probabilities: $P(S_1) = .0460; P(S_2) = .0345; P(S_3) = .9195$
e. P(Sample information indicates S_1 true) = .205; P(Sample information indicates S_2 true) = .360; P(Sample information indicates S_3 true) = .435   **g.** EPNS = $1,900; EPS = $1,956.94   **h.** EVSI = $56.94   **i.** Should not purchase
19.28a. $120,000   **b.** $47,958   **c.** $17,958; purchase survey   **19.29** $49.17   **19.30** ENGS = $350,000; yes
19.31 Up to $.34 million   **19.32** $0   **19.33b.** $x = 7$ **c.** $E(x) = 17.5$; A

19.34a.

		STATE OF NATURE				
		5	10	15	20	25
ACTION	A	2	12	22	32	42
	B	4	9	14	19	24

b. Choose action A; yes

19.35a. $x = $222.22   **b.** $E(x) = $320   **c.** A   **19.36** A   **19.37b.** $x = 833.33$ **c.** 3,100   **d.** Brand A
19.38a. $\mu = 20; \sigma^2 = 36.95$ **b.** First machine

19.39a.

Prior Probability of Guilt	Posterior Probability of Guilt
.1	.9911
.2	.9960
.3	.9977
.4	.9985
.5	.9990
.6	.9993
.7	.9996
.8	.9998
.9	.9999

19.40a. $5,987,500   **b.** No   **c.** Purchase survey

19.41a.

		STATE OF NATURE	
		(.8) $p = .05$	(.2) $p = .1$
ACTION	Accept	0	$-4,100$
	Reject	$-1,000$	0

b. Rejected   **c.** 800   **d.** Rejected   **e.** Accepted

f. $22   **g.** 2.75%   **h.** Yes   **19.42a.** $89.22   **b.** 11.15%   **c.** $49.22   **19.44** Yes   **19.45** $1,000   **19.46** $1,100

19.47a. Payoffs in thousands of dollars:

		STATE OF NATURE						
		(.05) S_1	(.20) S_2	(.30) S_3	(.20) S_4	(.10) S_5	(.08) S_6	(.07) S_7
ACTION	a_1	0	0	0	$-3,000$	$-6,000$	$-9,000$	$-12,000$
	a_2	-2	-2	-2	$-2,702$	$-5,702$	$-8,702$	$-11,702$
	a_3	-4	-4	-4	$-2,504$	$-5,404$	$-8,404$	$-11,404$
	a_4	-6	-6	-6	$-2,506$	$-5,106$	$-8,106$	$-11,106$
	a_5	-8	-8	-8	$-2,508$	$-5,008$	$-7,808$	$-10,808$
	a_6	-10	-10	-10	$-2,510$	$-5,010$	$-7,510$	$-10,510$
	a_7	-12	-12	-12	$-2,512$	$-5,012$	$-7,512$	$-10,212$
	a_8	-14	-14	-14	$-2,514$	$-5,014$	$-7,514$	$-10,014$

c. 7   **d.** $10,620
19.48 Develop the model   **19.49** 9   **19.50a.** a_3 **b.** a_1 or a_2

CHAPTER 20

20.5a. 40%; .77   **b.** 20%; .89   **c.** 10%; .95   **d.** 1%; .99   **20.6a.** 77   **b.** 89   **c.** 95   **d.** 99
20.7a. 1.99; 2   **b.** .509; .537   **20.8a.** .6   **b.** .2   **c.** 0   **20.9** 375 $\pm$ 3.814   **20.10** .37 $\pm$ .028
20.11 137,900 $\pm$ 2,759.71   **20.12a.** 72,066.67 $\pm$ 7,115.52   **b.** 36.03 $\pm$ 3.56   **c.** .7 $\pm$ .166
20.13a. 779,030 $\pm$ 378,848.7165   **b.** 1,168,545,000.00 $\pm$ 568,273,074.8
20.14a. $782,300   **b.** $206,993.77   **c.** $782,300 $\pm$ $206,993.77   **d.** Unreasonable
20.15b. .041   **c.** .086 $\pm$ .041   **d.** $z = .81$; no   **20.16** .62 $\pm$ .0304
20.17a. $330 $\pm$ $130.25   **b.** $135,300 $\pm$ $53,403.90   **20.18** $\hat{\tau} = 320$; bound = 240   **20.19** $\hat{p} = .093$; bound = .0071
20.20a. $k = 3$; $N_1 = 25,000$, $N_2 = 10,000$, $N_3 = 5,000$, $N = 40,000$ **b.** $n_1 = 20$, $n_2 = 15$, $n_3 = 10$, $n = 45$
c. $\bar{x}_1 = 75.650$, $\bar{x}_2 = 51.7333$, $\bar{x}_3 = 26.80$ **d.** $s_1^2 = 232.3447$, $s_2^2 = 145.2095$, $s_3^2 = 77.5111$
e. $\hat{\mu} = 63.5646$; bound = 10.5187   **f.** $\hat{\tau} = 2,542,583.33$; bound = 5,801.7118   **g.** $\hat{p} = .7125$; bound = .5059
20.21a. $\hat{\mu} = 42.13$; bound = 5.21   **b.** $\hat{\tau} = 1,474,400$; bound = 182,206.5   **c.** $\hat{p} = .45$; bound = .119
20.22 $\hat{\mu}_4 = 49.84$; bound = 1.978   **20.23b.** $26.6784 $\pm$ $.27514   **c.** $6,119,651 $\pm$ $63,113.4136
20.24a. $44,844   **b.** $2,128   **c.** $44,840 $\pm$ $2,128
20.25a. $285,500.00   **b.** $200,377.33   **c.** $285,500 $\pm$ $200,377.33
20.26 $\bar{x}_{st} = $19,905$; bound = $779   **20.27** $\hat{p}_{st} = .2543$; bound = .0399   **20.28** $992,135.25 $\pm$ $17,993.29
20.29a. $M = 5,000$, $N = 300$, $n = 4$, $m_1 = 8$, $m_2 = 9$, $m_3 = 12$, $m_4 = 6$
b. $\bar{m} = 8.75$, $\bar{M} = 16.667$ **c.** $x_1 = 62$, $x_2 = 161$, $x_3 = 321$, $x_4 = 115$ **d.** $\hat{\sigma}_{\bar{x}} = 2.2410$ **e.** 18.8286 $\pm$ 4.4820
f. $\hat{\sigma}_{\hat{\tau}} = 11,205$ **g.** 94,143 $\pm$ 22,410   **20.30a.** 5.6034 $\pm$ .7225   **b.** 6,948.2 $\pm$ 895.9
20.31a. $\bar{x} = 24.17$; bound = 4.297   **b.** $\hat{\tau} = 48,346.15$; bound = 8,594.65   **c.** $\hat{p} = .75$; bound = .1433

20.32a. .7865 ± .02254   **20.33a.** $\hat{\tau}$ = \$273,600; bound = \$75,822   **20.34** $\bar{x}$ = \$21; bound = \$5.85

20.35 $\hat{p}$ = .76; bound = .0597   **20.36a.** $\hat{\tau}$ = 2,217; bound = 566.42   **20.37** $\bar{x}$ = \$31.65; bound = \$2.65

20.38 .631 ± .110   **20.39** 77   **20.40** 402   **20.41a.** 129   **b.** 202   **20.42** 274   **20.43** 48   **20.44** 8,001

20.46a. 57 ± 3.2055   **b.** 285,000 ± 16,027.48   **20.49a.** $\hat{\tau}$ = \$73,850; bound = \$37,845.89   **b.** .7 ± .167

20.50a. $\bar{x}_{st}$ = 48.55; bound = 6.31   **c.** 60.55 ± 11.996   **d.** 60 days: .5975 ≤ p ≤ .8549; 90 days: .7863 ≤ p ≤ .9757

20.51 $\hat{p}$ = .876; bound = .047   **20.54a.** $\hat{p}$ = .9296; bound = .005   **20.55** Approx. 400

20.56a. Cluster sampling   **b.** .035 ± .009   **20.57** \$5.41 ± \$0.64

20.58 $\bar{x}_{st}$ = 7.16; bound = .861   **20.59a.** $N\bar{x}_t$ = \$285,000; 167,896.79

Text designer: Janet Bollow
Cover designer: John Williams
Technical artist: Reese Thornton
Production manager: Susan Reiland
Typesetter: Typeset in 10/12 Electra
by Jonathan Peck Typographers, Ltd.

SYMBOL	DESCRIPTION	PAGE
σ (lowercase sigma)	Population standard deviation	96
σ^2	Population variance	96
$\sigma_{\hat{\beta}_1}$	Standard deviation of the sampling distribution of β_1	506
$\sigma_{\hat{p}}$	Standard deviation of the sampling distribution of $\hat{p}$	399
$\sigma_{(\hat{p}_1 - \hat{p}_2)}$	Standard deviation of the sampling distribution of $(\hat{p}_1 - \hat{p}_2)$	464
$\sigma_{\bar{x}}$	Standard deviation of the sampling distribution of $\bar{x}$	318
$\sigma_{(\bar{x}_1 - \bar{x}_2)}$	Standard deviation of the sampling distribution of $(\bar{x}_1 - \bar{x}_2)$	420
$\sigma_{\hat{y}}$	Standard deviation of the sampling distribution of $\hat{y}$	526
$\sigma_{(y - \hat{y})}$	Standard deviation of prediction error when $\hat{y}$ is used to predict a particular value of y	526
Σ (uppercase sigma)	Symbol for summation	78
t	Statistic used for small-sample tests of hypotheses	390
t_α	Value of t distribution with an area α to its right	387
T_A, T_B	Rank sums corresponding to the two samples in a Wilcoxon rank sum test	949
T_L, T_U	Upper and lower rejection region values in a Wilcoxon rank sum test	949
T_t	Secular trend of a time series	796
T_0	Rejection region value in a Wilcoxon signed rank test	959
T_+, T_-	Sum of the ranks of the positive and negative differences in a Wilcoxon signed rank test	959
τ (tau)	Population total	1132
$\hat{\tau}$	Estimator of population total τ	1143
$U(O_{ij})$	Utility value corresponding to action i and state of nature j outcome in a decision analysis	1069
$\bar{x}$	(1) Sample mean	79
	(2) Estimator of population mean μ computed from a cluster sample	1157
x_i	(1) ith sample measurement	78
	(2) Total of all observations in cluster i of a cluster sample	1156
$\bar{x}_i$	Sample mean for stratum i in a stratified random sample	1149
$\bar{x}_D$	Sample mean of differences in a paired difference experiment	452
$\bar{x}_{st}$	Estimator of population mean μ computed from a stratified random sample	1149
X^2	Test statistic for multinomial and contingency table tests	1005
Y_t	Value of a time series Y at time t	758
$\bar{y}$	Mean of all observations in an analysis of variance	858
$\bar{y}_i$	Sample mean for ith treatment in an analysis of variance	856
$\hat{y}$	Least squares prediction of y using a regression model	493
z	z-score	117
$z_{\alpha/2}$	Value of standard normal variable with an area $\alpha/2$ to its right	345